D1262095

Robert Dautray Jacques-Louis Lions

Mathematical Analysis and Numerical Methods for Science and Technology

Volume 1

Physical Origins and Classical Methods

With the Collaboration of
Philippe Bénilan, Michel Cessenat, André Gervat,
Alain Kavenoky, Hélène Lanchon

Translated from the French by Ian N. Sneddon

Springer-Verlag
Berlin Heidelberg New York
London Paris Tokyo Hong Kong

Robert Dautray
Ecole Polytechnique
F-92128 Palaiseau Cedex, France

Jacques-Louis Lions
Collège de France
11 place Marcelin Berthelot
F-75005 Paris, France

Title of the French original edition:
Analyse mathématique et calcul numérique pour les sciences et les techniques, Masson, S.A.

With 41 Figures

Mathematics Subject Classification (1980): 31-XX, 35-XX, 41-XX, 42-XX, 44-XX, 45-XX, 46-XX, 47-XX, 65-XX, 73-XX, 76-XX, 78-XX, 80-XX, 81-XX

ISBN 3-540-50207-6 Springer-Verlag Berlin Heidelberg New York
ISBN 0-387-50207-6 Springer-Verlag New York Berlin Heidelberg

Library of Congress Cataloging-in-Publication Data
Dautray, Robert. Mathematical analysis and numerical methods for science and technology.
Translation of: Analyse mathématique et calcul numérique pour les sciences et les techniques.
Includes bibliographical references.
Contents: v. 1. Physical Origins and Classical Methods — v. 2. Functional and variational methods/ with the collaboration of Michel Artola ... [et al]
1. Mathematical analysis. 2. Numerical analysis. I. Lions, Jacques Louis. I. Title.
QA300.D34313 1990 515 88-15089
ISBN 0-387-19045-7 (U.S.: v. 2)

Printed in Germany
Typesetting: Macmillan India Limited, Bangalore
Printing and binding: Konrad Triltsch GmbH, Würzburg
2141/3140-543210 – Printed on acid-free paper

Preface

In the first years of the 1970's Robert Dautray engaged in conversations with Jacques Yvon, High-Commissioner of Atomic energy, of the necessity of publishing mathematical works of the highest level to put at the disposal of the scientific community a synthesis of the modern methods of calculating physical phenomena.

It is necessary to get away from the habit of treating mathematical concepts as elegant abstract entities little used in practice. We must develop a technique, but without falling into an impoverishing utilitarianism. The competence of the Commissariat à l'Energie Atomique in this matter can provide a support of exceptional value for such an enterprise.

The work which I have the pleasure to present realises the synthesis of mathematical methods, seen from the angle of their applications, and of use in designing computer programs. It should be seen as complete as possible for the present moment, with the present degree of development of each of the subjects. It is this specific approach which creates the richness of this work, at the same time a considerable achievement and a harbinger of the future. The encounter to which it gives rise among the originators of mathematical thought, the users of these concepts and computer scientists will be fruitful for the solution of the great problems which remain to be treated, should they arise from the mathematical structure itself (for example from non-linearities) or from the architecture of computers, such as parallel computers.

This task has led to planning, spread over ten consecutive years of strenous work, by two exceptional men – the physicist Robert Dautray and the mathematician Jacques-Louis Lions. In addition, they have enlisted the assistance of younger research workers, so it is fair to include them in our thanks for a work which, deemed indispensible thoughout, does not seem to me to have been undertaken quite at this level anywhere else in the world.

Jean Teillac
High Commissioner of Atomic Energy

General Introduction

1. A very great number of the problems of mathematical physics can be "modelled" by partial differential equations. By a "model", we mean a set of equations (or inequations) which, together with boundary conditions (expressed on the boundary of the spatial domain where the phenomenon is studied) and, when the phenomenon is evolutionary, with initial conditions, allow us to define the state of the system. This is also called modelling by "distributed systems".
Naturally the description of the model (or of a model, since the same phenomenon can often be described, in conditions not always strictly equivalent, by different state variables) is an important – but not decisive – step.
Further, we must "study" the model, i.e. deduce qualitative or quantitative properties which
(a) recover, in simple conditions, observations (measurements) already made.
(b) give supplentary information about the system.
It has been observed for a long time that the majority of the phenomena of mathematical physics are *non-linear,* among the most celebrated cases being Boltzmann's equation in statistical mechanics, the Navier-Stokes in fluid mechanics (equations which moreover constitute an *approximation* to Boltzmann's equation) von Karman's equations governing the large displacements of flat plates, etc.
However, having the possibility of using in a systematic – and almost "commonplace" – way the procedures for calculating *approximate solutions* of the state of the system, precise results can generally only be obtained in *the linear cases.*
Certain physical problems can be modelled *directly* (i.e. without *approximations*) by linear equations: this is notably the case of the equation of transport of neutrons. Other phenomena can be deduced from "truly" non-linear systems by neglecting certain terms (which is valid in certain situations: "small" displacements, "slow" motions ...) produced by linearisation about a particular solution.
As, in addition, the *methods* brought into play for the solution of linear problems play an essential role in all the non-linear situations known to this day, it is indispensible to begin with the study of *linear distributed models,* or again with *boundary value problems for linear partial differential equations* (with brief incursions into the domain of *linear* integral equations, equations which we can deduce from linear partial differential equations, or can appear directly in this form).
It is the aim of this work to study linear distributed models, completely concentrating in particular on *physical examples* (from various sources), by *the general*

methods of linear analysis (stating very clearly the application of these methods to physically important situations).
We have tried to render the material accessible to a reader the level of whose knowledge is pretty nearly that of the Lebesgue integral; the (indispensible) theory of distributions is recalled in the Appendix to Volume 2.

2. The theory of partial differential equations constitutes today one of the important topics of scientific understanding.
The principal reasons for this state of affairs are, on the one hand the progress of *mathematical analysis* and, on the other hand, the arrival of *the technique of numerical calculus* which remained, for partial differential equations, almost totally inadequate until the 1950's. In effect, the arrival of computers, and their immense and unceasing progress, have allowed us – for the first time in history – to *calculate, beginning with the models,* quantities which, formerly we were able only to estimate very approximately and, perhaps over all, to calculate them *accurately* and rapidly, and hence the (fundamental) possibility for research workers and engineers to be able to use the numerical results for the modification or adaptation of scientific arguments, of experiments or of constructions in progress.
All that explains why, in very differing subjects, *modelling by partial differential equations,* followed by theoretical analysis, then numerical analysis, and then in its turn with comparison with experiment has become a *basic method of procedure.* Every aspect of technical and industrial activity is concerned; this procedure is indispensible in the preparation or experiments and of trials and their interpretation, technical studies, the development of manufacturing processes, maintenance, reliability, etc. ... Thus:
Modern equipment has to operate in high performance with certain materials. In the 1950's, the calculation of the strength of materials was carried out with high safety factors (for example, 5 or more) on the stress experienced by the material of a given piece at a given point. Today, when we calculate a stress with precision, the safety factor which we take is of the order of 1.4 or less (for example in aeronautical, nuclear, automobile engineering etc. ...) and then in the very best conditions for users.
Similarly, the reliability and security demanded by many modern techniques, from nuclear engineering to aeronautics, from aerospace to large public works (high speed rail transport, highway construction, generation and distribution of electricity, etc. ...) require the accumulation both of the safety factors and, as well require that each of the details is studied and is represented with great precision. No element is any longer "neglected", then and only then the faithful mathematical representation allows us to examine closely the least detail and underline the predictions.
Modelling by distributed systems has become similarly the basis of many disciplines in physics (plasmas, new materials, etc. ...) in the space and earth sciences (astrophysics, geophysics etc. ...), in chemistry and obviously in all branches of mechanics (a number of which have already been cited above).
Without wishing to draw up here an exhaustive list, we should add that, by the intervention, notably of dynamic programming, (non-linear) partial differen-

tial equations play an important role in the *management sciences* (stocks, energy, etc. ...).
Distributed models are similarly involved, and more and more, in the *life sciences.*

3. Plan of the Work. We give here a general sketch of the content of the chapters grouped by volumes. Each volume begins with a slightly more detailed account of its contents.
We begin by giving in Chapter I a list of mathematical models, important in applications to physics and to the mechanics of continuous media which can lead to linear problems.
The study of *stationary linear problems* begins with a review, in Chapter II, of the possibilities of making use of *classical methods.* We discover that their limits are quickly reached. We examine in Chapter III the possibilities of applying *functional transformations* (Fourier series, the Fourier, Mellin and Hankel transforms etc. ...). We touch there similarly on the limitations on their application. These limitations show the usefulness of working on sets (of distributions) very much more "extensive" than the sets of continuous functions considered in Chapter II: these are the spaces of general distributions introduced in Chapter III and the *Sobolev spaces* studied in Chapter IV.
The study of *differential operators* in the spaces of general distributions allows us to distinguish the properties of these operators (elliptic, parabolic and hyperbolic operators; local character of mathematical models using differential operators; characteristics etc. ...) which will serve us well throughout this work; this is the subject matter of Chapter V.
Throughout the whole of this work we shall have to handle *operators;* "operations" on these operators and their approximations are explained, in Chapter VI, in the mathematical situations used in the present work.
The mathematical techniques thus gathered together allow us to treat *variational methods,* which make up the subject of Chapter VII and whose potential for application extends to many non-linear problems.
Numerous *spectral* problems arise in applications (calculation of energy levels and states in quantum mechanics, critical conditions in neutronics, transmission in a wave guide or in an optical fibre etc. ...). The spectral theory which enables us to treat such problems is seen in Chapter VIII within the perspective of typical applications; it includes especially the study of the continuous spectrum, source of many difficulties. Examples of applications are give in Chapter IX.
A problem which is elliptic or hyperbolic according to the value of a parameter is treated in Chapter X: *Tricomi's problem* (in fluid mechanics it corresponds to the passage from subsonic flight to supersonic flight).
Mathematical models involving integrals permit the representation of actions at a distance (in physical space, electric potential; in time, the memory of a viscoelastic body; in the space of velocities, change of velocity as a result of collisions). *Integral equations* which come into play require the methods treated in Chapter IX.
Finally, the *numerical methods* to treat stationary problems form the subject matter of Chapters XII and XIII.

Linear evolution problems are treated first of all in the whole physical space in Chapter XIV.
The *diagonalisation* method, using the spectral theory of operators, which is the basis of several practical methods (giving rise to the decomposition into modes), is treated in Chapter XV. The method of the *Laplace transform* can be used to treat numerous evolution problems; it is considered in Chapter XVI.
The solution $u(t)$ of a large class of evolution problems can be written in the form $u(t) = G(t)\, u_0$, where u_0 is the initial value of u and u_0 is the initial value of u and $G(t)$ a family of operators forming a *semi-group*. The types of evolution of solutions can then be an examination of the various families of semi-groups $G(t)$. This provides a method, in certain ways more general than the preceding ones, to treat evolution problems. This is the subject of Chapter XVII.
Finally, the constructive methods of solving evolution problems (using constructions of solution in finite-dimensional spaces), the *variational methods,* are seen in Chapter XVIII [1]. The *Navier-Stokes problem* (in the linearised case) requires particular variational methods. These are described in Chapter XIX.
Chapter XX presents the *numerical methods* for linear evolution problems.
The problems involving a *transport equation* are not included in the categories treated in Chapters XIV to XVIII, since they take into account the very particular type of properties of the transport operator (transport of neutrons, transport of molecules and Boltzmann's equation, transport of charged particles and Vlasov's equation). A special chapter, Chapter XXI, is therefore devoted to these problems.
Later chapters study other aspects of certain of the problems studied in the present work (relations between problems of partial differential equations and probabilities, propagation of waves, etc. ...).

4. The writing of this work has been conceived with the object of making it accessible to an engineer or to an aspiring research worker taking only the information he needs to treat his problem; a restricted reading is therefore possible if the reader is guided by the index, the table of contents and the table of notations.

5. In producing this work the undersigned have benifitted from the collaboration of many colleagues: Michel Artola, Marc Authier, Claude Bardos, Philippe Bénilan, Michel Bernadou, Michel Cessenat, Jean-Michel Combes, André Gervat, Alain Kavenoky, Hélène Lanchon, Patrick Lascaux, Bertrand Mercier, Jean-Claude Nédélec, Olivier Pironneau, Jacques Planchard, Bruno Scheurer, Claude Wild, Claude Zuily.
Their contributions and the contributions on specific points due to several other colleagues will be acknowledged at the beginning of each volume.
The manuscript was read with particular care by Michel Cessenat whom we thank most warmly. Considering the size and the diversity of this work, the task he performed is considerable. In addition, Michel Cessenat proposed complementary or corrected texts, valuable contributions which we have often retained.

---

[1] These methods can be similarly extended to non-linear problems.

Our thanks go similarly to Jean-Marie Moreau for his important work in compiling the bibliographies for each volume, for reading the text and bringing it to the point of publication.
This work would not have seen the light of day without the support of the Atomic Energy Commission (C.E.A.): Jacques Yvon, then High Commissioner of Atomic Energy accepted our proposition immediately, as he could foresee its future development. He made its publication one of the scientific enterprises of the C.E.A. Our respective experiences had, in effect, as early as the end of the 1960's, confirmed our belief in the importance of the existence of a work of reference of this type. By the beginning of 1970's, we had elaborated our ideas into a plan, taking account of the needs of engineers, physicists and workers in mechanics etc. ... Jacques Yvon together with ourselves, wished to spread and put within their grasp the abundant recent work of mathematicians and numerical analysts.
In the initial period, at the time of preliminary drafts and launching the project, we benifitted from the initiative of Robert Lattes, who was then Scientific Adviser of the C.E.A.
We are grateful to Paul Bonnet, Inspector General of the C.E.A. for having inaugurated the C.E.A. collection with this work.
We have greatly valued, and are immensely grateful for, the initial help and encouragement of Jules Horowitz, Director at the C.E.A. who with his great experience in mathematical physics showed an immediate understanding of our aims.
Nothing would have been achieved in reaching the final result without the clear and active understanding of Jacques Chevallier, Director at the C.E.A.
We thank here also Michel Pecquer and Gérard Renon, Administrator General, as well as Jean Teillac, High Commissioner, of the C.E.A. whose constant and manifest approval, personally expressed, has been a source of permanent encouragement.

R. Dautray, J.-L. Lions

Practical Guide for the Reader

(1) Designation of the subdivisions of the text:
Number of chapters: in roman numerals;
Number of major divisions: the sign § followed by a numeral;
Number of sections: a numeral following the preceding;
Number of sub-sections: a numeral following the preceding;
etc. ...
For example: II, § 3.5.2, denotes Chapter II, § 3, section 5, subsection 2.
(2) *In the interior of each division* (§), the equations, definitions, theorems, propositions, corollaries, lemmas, remarks and examples are numbered separately in sequence beginning with the *numeral 1.*
(3) The *table of the notations used* appears at the end of each volume.

Introduction to Volume 1

Chapter I gives the *principal physical examples* studied in this work (these examples come from physics, from mechanics, from chemistry, etc. ...). A first (rudimentary) attempt at the classification of the problems is made.
In *all* the phenomena modelled by partial differential equations, and for reasons that are given in the text, a very important role is played by the *Laplacian operator*

$$\Delta = \frac{\partial^2}{\partial x_1^2} + \frac{\partial^2}{\partial x_2^2} + \frac{\partial^2}{\partial x_3^2}$$

in rectangular coordinates: This is why Chapter II is devoted to a direct study of the *principal questions linked with this operator,* "direct" *signifying here: without the use of techniques other than those of classical analysis.*
We give below the authors of various contributions to these two chapters.

Chapter I: H. Lanchon, M. Cessenat, A. Gervat, A. Kavenoky.
Chapter II: P. Bénilan, sole author of this Chapter.

We similarly wish to thank R. Balian, C. Bardos, A. Bossavit, C. Cohen-Tannoudji, G. Fournet, A. Kavenoky and E. Roubine for reading certain portions of the text and for their advice on modifying them.

The reader wishing to acquaint himself rapidly with the essential mathematical and numerical methods should be able to make use of this volume and the subsequent Vol. 2 by leaving for a later, deeper study §§ 5–8 of Chapter II of this volume and §§ 4, 5 of Chapter V of Vol. 2. These divisions are denoted by an asterisk * placed at their beginning, an asterisk which, moreover, appears in the table of contents.
We have placed the table of notations at the end of this volume.

R. Dautray, J.-L. Lions

Table of Contents

Chapter I. Physical Examples

Chapter II. The Laplace Operator

Chapter I. Physical Examples

Introduction

The present chapter comprises two parts:
Part A classifies models* according to physical themes. Thus there are examined, in sequence, linear (or linearized) boundary value problems arising from models in fluid mechanics, in elasticity and viscoelasticity, in electromagnetism, in neutronics and in quantum physics. (Certain very simple ideas of mechanics are recalled in an Appendix to help the reader to follow the explanations leading to the models.)
The language and notation used to treat each physical theme will be those established by use. It will not necessarily be consistent from one theme to the other. We shall be led occasionally, in describing problems, to make use of mathematical ideas which will be made precise later in the work.
Part B classifies models according to mathematical type. The classification of mathematical models is based on that of the partial differential equations encountered: diffusion equation (parabolic), wave equation (hyperbolic), transport equations, Schrödinger equations, stationary equations (recalling briefly each type appropriate to the principal physical examples).
Quite obviously, that would not be an exhaustive list: many other situations in physics, mechanics and chemistry etc. . . . , lead to partial differential equations – and, for the Part B, there are many other interesting equations which are not encountered in any of the types called to mind in Part A. Again, we add that Part B is content to *pose* problems. The *methods of solution* will be introduced in the subsequent chapters.

---

* *Physical and mathematical models.* To reply to the questions we pose in the study of a phenomenon, we construct a picture using physical concepts that we are able to render assimilable. We shall call this picture the *physical model* of the phenomenon. Several physical models can be set up to describe the same phenomenon; for example, a flow of gas in a nozzle can be "modelled" by considering it as assembly of N atoms, each of which is described by a point in the phase space $\mathbb{R}^{6N}$, or as a (continuous) fluid with the concepts of mass density, pressure, etc. The type of physical model chosen depends on the kind of questions concerning the phenomenon that the physicist poses.
Starting from the physical model, we can construct, with mathematical concepts – space $\mathbb{R}^3$ (position), scalar (pressure), vector (velocity), tensor (stress), positive real functions (phase densities etc.) – a mathematical description of the physical model. We shall call this description the *mathematical model.* It is not, in general, unique, and according to the questions we pose, we employ this or that mathematical description of our physical model. Thus the flow of a gas along a nozzle, for the physical model of a continuous, will be represented by the non-linear Navier–Stokes model IA (1.12) or else by the linear Stokes model IA (1.14) or again by the Euler model IA (1.18).
We shall use the word "modelling" to denote the construction of a model.

Part A. The Physical Models

§1. Classical Fluids and the Navier–Stokes System

1. Introduction. Mechanical Origin

The *general equations* which govern the motion of a *homogeneous, viscous, isotropic* fluid in a domain $\Omega \subset \mathbb{R}^3$ occupied by the fluid are[1]

$$\frac{d\rho}{dt} + \rho \operatorname{div} \mathbf{U} = 0 \quad \text{(conservation of mass)}^1, \tag{1.1}$$

$$\left.\begin{aligned} \rho \frac{du_i}{dt} &= \sigma_{ij,j} + f_i \\ \sigma_{ij} &= \sigma_{ji} \end{aligned}\right\} \quad \text{(conservation of momentum)}^{1,2}, \tag{1.2}$$

$$\sigma_{ij} = -p\delta_{ij} + \lambda . \varepsilon_{ll}(u) . \delta_{ij} + 2\mu . \varepsilon_{ij}(u) \quad \text{(constitutive equation)}, \tag{1.3}$$

where

$\mathbf{U}(\mathbf{x}, t)$ is the *velocity vector* represented by $\mathbf{u} = \{u_1, u_2, u_3\}$ in the reference frame considered[3],
$\mathbf{x} = \{x_1, x_2, x_3\}$ denote the Eulerian coordinates of the particle considered at the instant t[4],
$\rho(\mathbf{x}, t)$ is the *density*, $(\rho > 0)$
$\mathbf{f}(\mathbf{x}, t) = \{f_j(\mathbf{x}, t)\}$ is the density, per unit volume, of the external forces acting on the fluid (usually these are gravitational forces),
$p(\mathbf{x}, t)$ is the *pressure* of the fluid $(p > 0)$
$\varepsilon_{ij} = \frac{1}{2}(u_{i,j} + u_{j,i})$ are the components, in the chosen base, of the *rate of strain tensor* (sometimes called the *tensor of velocities of deformation*)[4],

---

[1] See Remark 1 below.

[2] Throughout the treatment of mechanics we adopt, on the one hand, the summation convention on repeated indices, and on the other hand, the simplified notation for the partial derivative $\frac{\partial \varphi}{\partial x_i} = \varphi_{,i}$. (See Appendix "Mechanics.")

[3] In §1, 2 and 3, concerning mechanics, we shall denote by $\mathbf{U}$ (or $\mathbf{V}$) an intrinsic vector (i.e. independent of the reference frame), and by $\mathbf{u}$ (or $\mathbf{v}$) the representation of this vector in a given frame, i.e. $\mathbf{u} = \{u_1, u_2, u_3\} \in \mathbb{R}^3$ (or $\mathbf{v} = \{v_1, v_2, v_3\} \in \mathbb{R}^3$).

[4] See Appendix "Mechanics."

σ_{ij} are the components, in the chosen base, of the *stress tensor*[5],
λ and μ are the *coefficients of viscosity*,

$\dfrac{\mathrm{d}}{\mathrm{d}t}$ denotes the *particle derivative* (also known as the *derivative following the motion*).

If the fluid is *incompressible*, i.e. if ρ is independent of both $\mathbf{x}$ and t, equation (1.1) is replaced by the *equation of state for an incompressible medium*

(1.4) $$\operatorname{div} \mathbf{U} = 0 .$$

Remark 1. The conservation laws written above (relations (1.1) and (1.2)) are the local expressions of the *balances concerning quantities* $\mathscr{Q}$ (here mass and momentum, later energy) valid on each domain $\mathscr{D}$ strictly interior to all the continuous medium whose motion we are following. These balances can be expressed as follows: "what we supply to $\mathscr{D}$ relatively to $\mathscr{Q}$, serves, on the one hand, to compensate for the loss across the boundary of $\mathscr{D}$, and, on the other hand, to control the variation of $\mathscr{Q}$ when we follow the motion of $\mathscr{D}$".
Mathematical arguments applied to the integral relation thus obtained then allows us to obtain three kinds of local information:
(1) partial differential equations and local relations of the type (1.1) and (1.2),
(2) boundary conditions,
(3) discontinuity relations.
The conservation laws are generally applicable to all "continuous" media (i.e. those which can reasonably be considered from a macroscopic point of view). The constitutive equations and equations of state, like those expressed by the relations (1.3) and (1.4) are, on the contrary, characteristic of the particular medium. Details of all these concepts will be found, for example, in Germain [2] or [3]. □

The boundary conditions resulting respectively from the laws of conservation of mass and of momentum are

(1.5) $$\mathbf{U} = \mathbf{0} \quad \text{on} \quad \partial\Omega$$

if $\partial\Omega$ is a fixed wall (otherwise it is the relative velocity which is zero),

(1.5 *bis*) $$\sigma_{ij} n_j = F_i \quad \text{on} \quad \partial\Omega$$

$\mathbf{n}$ being the unit vector of the external normal to $\partial\Omega$ and $\mathbf{F}$ the surface density of the forces applied by the wall $\partial\Omega$ on the fluid at the point considered.

---

[5] The stress tensor $\sigma_{ij}(\mathbf{x}, t)$ is the linear operator which associates with each unit vector $\mathbf{k}$, with origin $\mathbf{x}$, the stress vector $\mathbf{T}(\mathbf{x}, t, \mathbf{k})$. This represents the surface density of the contact forces exerted on a surface element passing through $\mathbf{x}$, normal to $\mathbf{k}$, by the elements of Ω situated in the region into which the vector $\mathbf{k}$ points; thus

$$T_i(\mathbf{x}, t, \mathbf{k}) = \sigma_{ij}(\mathbf{x}, t) k_j \quad \forall \mathbf{k} \quad \text{and} \quad \forall \mathbf{x} \in \Omega$$

and, in particular,

$$\sigma_{ij}(\mathbf{x}, t) = T_i(\mathbf{x}, t, \mathbf{e}_j)$$

$\mathbf{e}_j$ being one of the unit base vectors of the absolute cartesian reference frame adopted.

Remark 2. In fact, the law of conservation of mass imposes only

$$\mathbf{U}.\mathbf{n} = 0, \quad \text{on} \quad \partial\Omega ;$$

it is by taking account of the viscous friction on the wall which leads in (1.5) to the vanishing of the tangential velocity. □

The discontinuity conditions across a surface of contact (separating two immiscible fluids and in particular a free surface in contact with the atmosphere) are

$$\mathbf{U}^{(1)} = \mathbf{U}^{(2)}, \tag{1.6}$$

$$\sigma_{ij}^{(1)} N_j = \sigma_{ij}^{(2)} N_j . \tag{1.7}$$

N being normal to the surface of separation oriented, for example, from the medium 1 towards the medium 2.
These relations following also from the two quoted conservation laws, Remark 2 is valid again for (1.6), and the relation (1.7) expresses the continuity of the stress vector $\mathbf{T}$ across the surface of contact (see footnote 5).

Remark 3. The relations (1.6) and (1.7), with all those which have been written above, ought to allow us to determine the surface of contact which is in fact a supplementary unknown of the problem. In the case of two fluids at rest we find (taking account of the simpler form of (1.3) then obtained), that the relation (1.7) is replaced by the equality of the pressures on both sides of the surface of contact; experience shows however that such a relation cannot explain the phenomena of the "meniscus" and of the "angle of contact" appearing in a crucial way in capillary situations[6] (flow in very thin channels, drops, bubbles etc.). We have to take account here of "fluid–fluid" or "fluid–solid" *interfacial tensions* (*wetting* phenomena) arising from intermolecular forces due to the difference in the nature of the media involved. The jump in the pressures between the two fluids, which is not zero, is called the *capillary pressure*; it depends on the interfacial tension, on the *wetting angle*, and is inversely proportional to the diameter of the container.
In non-capillary situations, this capillary pressure is generally negligible; this justifies the use of (1.7). □

Remark 4. We can imagine (and they effectively exist) constitutive laws more complex than (1.3) for a fluid; they are in general of the form

$$\sigma_{ij} = K_0(\mathbf{x}, \varepsilon_I, \varepsilon_{II}, \varepsilon_{III})\delta_{ij} + K_1(\mathbf{x}, \varepsilon_I, \varepsilon_{II}, \varepsilon_{III})\varepsilon_{ij} + K_2(\mathbf{x}, \varepsilon_I, \varepsilon_{II}, \varepsilon_{III})\varepsilon_{il}\varepsilon_{lj}$$

where $\varepsilon_I, \varepsilon_{II}, \varepsilon_{III}$ are the three elementary invariants of the rate of strain tensor ε_{ij}

$$\varepsilon_I = \varepsilon_{jj}, \quad \varepsilon_{II} = \tfrac{1}{2}(\varepsilon_{jj}\varepsilon_{kk} - \varepsilon_{jk}\varepsilon_{jk}) \quad \text{and} \quad \varepsilon_{III} \det = \bar{\bar{\varepsilon}}^{7} .$$

If the fluid is homogeneous, the K_n do not depend explicitly on $\mathbf{x}$. In the case of "classical" or "newtonian" fluids considered in this chapter, a hypothesis of

---

[6] See for example: Brun *et al.* [1] or Landau–Lifschitz [3].

[7] We denote by $\bar{\bar{T}}$ or $\bar{\bar{\Sigma}}$ or $\bar{\bar{\varepsilon}}$ an intrinsic tensor, i.e. independent of any reference frame; see Appendix "Mechanics" §1.

linearity has been made between the stress tensor $\overline{\overline{\Sigma}}$ and the rate of strain tensor, which leads to (1.3). In all other cases, the fluid is said to be "*non newtonian*". □

2. Corresponding Mathematical Problem

The data of the problem are in general:

- The form of the domain Ω (for example the container of the fluid, or the space $\mathbb{R}^3$ less the obstacle in the aerodynamic case),
- the external forces $\mathbf{f}$, often neglected,
- the coefficients of viscosity λ, μ.

The fundamental unknown is the velocity vector $\mathbf{U}$, the secondary unknowns being ρ and p.

The Navier–Stokes equations. The incompressibility condition (1.4) and the homogeneity implying the invariance of ρ with the space variable, make (1.1) reduce to

$$\rho = \rho_0 \,, \quad \text{a given constant,} \tag{1.8}$$

and imply that the stress tensor takes the form

$$\sigma_{ij} = - p\delta_{ij} + 2\mu\varepsilon_{ij}(\mathbf{u}) \tag{1.9}$$

from which there follows the new form of (1.2):

$$\rho_0 \left(\frac{\partial u_i}{\partial t} + u_{i,j} u_j \right) + p_{,i} = f_i + \mu \Delta u_i \tag{1.10}$$

Dividing throughout by ρ_0 we can also write this equation in the vector form:

$$\frac{\partial \mathbf{U}}{\partial t} + \operatorname{grad}\left(\frac{1}{2} u^2 \right) + \operatorname{curl} \mathbf{U} \wedge \mathbf{U} + \operatorname{grad}(p/\rho_0) = \mathbf{f}/\rho_0 + \nu \Delta \mathbf{U} \tag{1.11}$$

where $\nu = \mu/\rho_0$ is called the *kinematical viscosity* of the fluid.

The system of non-linear equations (1.4) and (1.11) is known under the name *Navier–Stokes*; it consists of four partial differential equations (of which three are of second order) for the determination of the four unknowns u_1, u_2, u_3, p from the initial data on $\mathbf{U}$ and p and the boundary conditions (1.5) or (1.6), (1.7). Theoretically, that ought to be sufficient to solve the problem completely by using the constitutive equation (1.9) to determine the stress field[8].

In the case in which the fluid occupies a domain Ω, limited partially by a wall Γ_1 and partially by a free surface Γ_2 (itself an unknown of the problem since, in general, it depends on the solution), such that:

$$\Gamma_1 \cup \Gamma_2 = \Gamma, \text{ the boundary of } \Omega,$$

[8] In fact, because of the incompressibility relation (1.4) which can also be written $\varepsilon_{ii}(\mathbf{u}) = 0$ and constitutes an internal constraint, the pressure p at each instant can be determined only to within a constant. See Germain [2], Chap. VI.1.

we are led to solve the following mathematical problem:
Find the velocity field **U** *and the pressure gradient* grad p[9], *so as* Γ_{2T} *is the free part of the boundary, satisfying*

(1.12)

$$\begin{cases} \text{(i) } \operatorname{div} \mathbf{U} = \mathbf{0} \quad in\ \Omega_T \overset{\text{def}}{=} \Omega \times]0, T[^{10} \\ \text{(ii) } \dfrac{\partial U}{\partial t} + \operatorname{grad}\left(\dfrac{1}{2} u^2\right) + (\operatorname{curl} \mathbf{U}) \wedge \mathbf{U} + \operatorname{grad}(p/\rho_0) = \mathbf{f}/\rho_0 + \nu \Delta \mathbf{U} \quad in\ \Omega_T, \\ \text{(iii) } \mathbf{U}(\mathbf{x}, 0) = \mathbf{U}_0(\mathbf{x}), \\ \text{(iv) } \mathbf{U}(\mathbf{x}, t) = \mathbf{0} \quad on \quad \Gamma_{1T} \overset{\text{def}}{=} \Gamma_1 \times]0, T[\,, \\ \text{(v) } \mathbf{U}(\mathbf{x}, t) = \mathbf{U}^{(2)}(\mathbf{x}, t) \quad on \quad \Gamma_{2T} \overset{\text{def}}{=} \Gamma_2 \times]0, T[, \\ and \\ \text{(vi) } -p\mathbf{N} + \mu(\overline{\overline{\nabla \mathbf{U}}}.\mathbf{N} + \mathbf{N}.\overline{\overline{\nabla \mathbf{U}}}) = \mathbf{T}^{(2)}(\mathbf{N})^{11,12} \quad on\ \Gamma_{2T}, \end{cases}$$

$\mathbf{U}^{(2)}$ *and* $\mathbf{T}^{(2)}$ *being given, but* Γ_2 *and hence* **N** (*external normal to* Ω *on* Γ_2) *being unknowns.*

The boundary conditions on the free surface Γ_{2T} are supposed here to be known, the evolution of medium (2) being near to that of the fluid considered along the part Γ_2 of the boundary; we write on the one hand that because of the viscosity, the velocities of the two fluids are the same along Γ_2; on the other hand, the continuity of the *stress vector*, expressed in (1.7) is taken into account in the velocity field with the help of the constitutive relation (1.9):

(1.13) $$\sigma_{ij} N_j = -\ pN_i + 2\mu\varepsilon_{ij}(u) N_j = -\ pN_i + \mu(u_{i,j} N_j + u_{j,i} N_j)\,;$$

$\overline{\overline{\nabla \mathbf{U}}}$ is defined by

$$\mathrm{d}\mathbf{U} = \overline{\overline{\nabla \mathbf{U}}}\,.\,\mathrm{d}\mathbf{M}$$

and hence associates with the infinitesimal displacement d**M** of the point **M** the variation d**U** of the vector **U**; from this we have the alternative definition

$$\overline{\overline{\nabla \mathbf{U}}} = \overline{\overline{\operatorname{grad} \mathbf{U}}}\,,$$

or in components in a given reference frame

$$(\overline{\overline{\nabla \mathbf{U}}})_{ij} = u_{i,j} = \frac{\partial u_i}{\partial x_j}\,.$$

---

[9] The pressure p itself can be determined at each instant when it is given at a point or in a region (at infinity, for example).

[10] The stated problem is considered here only for a finite interval $(0, T)$.

[11] $\overline{\overline{\nabla \mathbf{U}}}\,.\mathbf{N}$ is the vector with i-th component $u_{i,j}N_j$, while the i-th component of $\mathbf{N}.\overline{\overline{\nabla \mathbf{U}}}$ is $N_j u_{j,i}$. Notice that $\overline{\overline{\nabla \mathbf{U}}}.\mathbf{N} \neq \mathbf{N}.\overline{\overline{\nabla \mathbf{U}}}$.

[12] A reader wishing to take capillary pressure into account (see Remark 3), should consult Joseph [1].

In addition

$$\mathbf{T}^{(2)}(\mathbf{N}) = \sigma_{ij}^{(2)} N_j \mathbf{e}_i$$

is the stress vector in the medium (2).

Remark 5. The condition (1.5 *bis*) valid on Γ_{1T} is not useful in solving the problem, but it serves *a posteriori* to determine the pressure and frictional forces exerted by the fluid on the wall. □

3. Linearisation. Stokes' Equations

A linearised version of the preceding problem (problem (1.12)) appears naturally enough when, on the one hand $\Gamma_1 = \Gamma$, and on the other, the motion is sufficiently slow for u_i and $u_{i,j}$ to be considered as small; we are then led to the following simplification of problem (1.12):

$$(1.14) \quad \begin{cases} \text{(i)} \ \operatorname{div} \mathbf{U} = 0 \quad \text{on} \quad \Omega_T , \\ \text{(ii)} \ \dfrac{\partial \mathbf{U}}{\partial t} - \nu \Delta \mathbf{U} = \dfrac{\mathbf{f}}{\rho_0} - \operatorname{grad} \dfrac{p}{\rho_0} \quad \text{on} \quad \Omega_T , \\ \text{(iii)} \ \mathbf{U}(\mathbf{x}, 0) = \mathbf{U}_0(\mathbf{x}) \quad \text{on} \quad \Omega , \\ \text{(iv)} \ \mathbf{U} = \mathbf{0} \quad \text{on} \quad \Gamma_T . \end{cases}$$

This system constitutes the *Stokes approximation* which is in fact particularly used in the *stationary* case ($\partial \mathbf{U}/\partial t = \mathbf{0}$); in this last case, the initial condition (iii) is removed.

Remark 6. The volume density of external forces $\mathbf{f}$ corresponding in general to the weight of the fluid, is often negligible or depending on a field of gradient $\mathscr{U}$ with the result that we can replace the right hand side of equation (1.14) (ii) by grad $\mathscr{K}$ with:

$$(1.15) \qquad \mathscr{K} = \frac{1}{\rho_0} [\mathscr{U} - p] ;$$

we note then that on applying the divergence operator to both sides of equation (1.14) (ii), and taking account of the incompressibility, we obtain

$$(1.16) \qquad \Delta \mathscr{K} = 0 \quad \text{on} \quad \Omega_T .$$

□

4. Case of a Perfect Fluid. Euler's Equations

A fluid is called *perfect*, when the effects of its viscosity can be supposed to be negligible; the constitutive equation (1.3) is then reduced to

$$(1.17) \qquad \sigma_{ij} = - p \delta_{ij} ;$$

the equations of motion can be written

$$\rho\left[\frac{\partial \mathbf{U}}{\partial t}+\frac{1}{2}\operatorname{grad} U^2+(\operatorname{curl}\mathbf{U})\wedge\mathbf{U}\right]+\operatorname{grad} p=\mathbf{f} \quad \text{in} \quad \Omega_T \tag{1.18}$$

and are called "Euler's equations".

(i) *In the homogeneous, incompressible case*, it is necessary to add as previously

$$\rho=\rho_0 \quad \text{and} \quad \operatorname{div}\mathbf{U}=0\,. \tag{1.19}$$

The five relations (1.18), (1.19) together with the boundary conditions (1.12) (iv), (v) (which we see have to be modified) and the initial conditions (1.12) (iii), enable us, in principle, to determine the unknowns u_i and, to within a constant, the unknown p.
(ii) *If the fluid is compressible*, the pair (1.19) is replaced by the single equation

$$\frac{\partial \rho}{\partial t}+\operatorname{div}(\rho\mathbf{U})=0\,. \tag{1.20}$$

We need additional information: for example, the fluid can be *barotropic*, i.e. there exists a relation:

$$p=g(\rho), \quad g \text{ given}\,; \tag{1.21}$$

this is a law of a thermodynamic nature, called an *equation of state*, and satisfying, in general, the following inequalities:

$$g>0, \quad C^2=\frac{\mathrm{d}g}{\mathrm{d}\rho}>0 \quad \text{and} \quad \frac{\mathrm{d}^2 g}{\mathrm{d}\rho^2}>0\,; \tag{1.22}$$

(the scalar quantity C has the dimensions of a speed and is in fact the speed of sound in the fluid under consideration).
This type of law includes, in particular, the two classic cases:

— $p=k\rho$ (k constant), of a perfect gas with constant specific heat in isothermal expansion;
— $p=k\rho^\gamma$ (k and γ constants), of a perfect gas with constant specific heat in adiabatic expansion.

The boundary conditions (1.5) (or (1.6), (1.7)) are more than is necessary; it is necessary to take account here of the fact that the order of the derivatives in the equations of motion has been reduced from two to one with the disappearance of the term $\Delta\mathbf{U}$. We are satisfied then with the conditions:

(1.23) $\mathbf{U}\,.\,\mathbf{n}=0$ if $\partial\Omega$ is wall (slipping condition)

(1.24) $\mathbf{U}\,.\,\mathbf{N}=\varphi$ if $\partial\Omega$ is a surface of contact, φ and ψ being supposed and

(1.25) $p=\psi$ given by the knowledge of the neighbouring medium.

These conditions are in fact the sole direct consequences of the conservation laws.

5. Case of Stationary Flows. Examples of Linear Problems

5.1. Definition and Properties of a Steady Flow

A flow is said to be *stationary* (or *steady*) if the velocity does not depend explicitly on the time; i.e. if the components u_i are functions of the Eulerian coordinates[13] x_i and not of t. We therefore have

$$\frac{\partial u_i}{\partial t}(\mathbf{x}, t) \equiv 0 ,$$

but the acceleration vector is not zero as $\mathbf{x} = \varphi(\mathbf{a}, t)$ and so

$$\gamma_i(\mathbf{x}) = \frac{\mathrm{d}u_i}{\mathrm{d}t}(\mathbf{x}) = u_{i,\,j}(\mathbf{x})\frac{\partial \varphi_j}{\partial t}(\mathbf{a}, t) = \mathbf{U}(\mathbf{x}) \,.\, \mathrm{grad}\, u_i(\mathbf{x}) \,.$$

In a stationary flow it is clear that the trajectories coincide with the flow lines; in effect[14], since the velocity field is the same at every instant t, we can write the relations (2.18) and (2.19) of the Appendix "Mechanics" in the unique form

$$\frac{\mathrm{d}x_1}{v_1(\mathbf{x})} = \frac{\mathrm{d}x_2}{v_2(\mathbf{x})} = \frac{\mathrm{d}x_3}{v_3(\mathbf{x})} \,. \tag{1.26}$$

In a stationary flow, the Navier–Stokes problem (1.12) is reduced to:

$$\text{(1.27)} \quad \begin{cases} \text{(i) } \mathrm{div}\, \mathbf{U} = 0 & \text{in } \Omega \subset \mathbb{R}^3 \\ \text{(ii) } \mathrm{grad}\, \dfrac{U^2}{2} + (\mathrm{curl}\, \mathbf{U}) \wedge \mathbf{U} + \mathrm{grad}\, \dfrac{p}{\rho_0} = \dfrac{\mathbf{f}}{\rho_0} + \nu \Delta \mathbf{U} & \text{in } \Omega \\ \text{(iii) } \mathbf{U}(\mathbf{x}) = \mathbf{0} & \text{on } \Gamma_1 \\ \left.\begin{array}{l} \text{(iv) } \mathbf{U}(\mathbf{x}) = \mathbf{U}^{(2)}(\mathbf{x}) \\ \quad \text{and} \\ \text{(v) } -pN + \mu(\overline{\overline{\nabla \mathbf{U}}} . \mathbf{N} + \mathbf{N} . \overline{\overline{\nabla \mathbf{U}}}) = \mathbf{T}^{(2)}(\mathbf{N}) \end{array}\right\} & \text{on } \Gamma_2 \,. \end{cases}$$

Remark 7. Above, we have defined stationary motions in a somewhat "abstract" way as how can we know *a priori* whether or not the velocity depends explicitly on the time before having solved the problem? In fact, we are led to enquire if there exists a stationary solution when the data themselves, $\mathbf{f}$, $\mathbf{U}^{(2)}$, $\mathbf{T}^{(2)}$ do not depend explicitly on the time. If such a solution, exists we see from the formulation (1.12) that p also will be stationary. We therefore arrive, in general, at a more precise definition of a stationary motion by supposing *a priori* that all the physical quantities in play: p, ρ, T (temperature) do not depend explicitly on the time. □

---

[13] See Appendix "Mechanics."

[14] See Appendix "Mechanics."

Taking account of the preceding remark, we can specify *the stationary motion of a barotropic perfect fluid* by the set of equations:

$$(1.28)\quad \begin{cases} \left.\begin{array}{l} \operatorname{div}(\rho \mathbf{U}) = 0 \\ \dfrac{1}{2}\operatorname{grad} U^2 + (\operatorname{curl}\mathbf{U}) \wedge \mathbf{U} + \dfrac{1}{\rho}\operatorname{grad} p = \dfrac{\mathbf{f}}{\rho} \end{array}\right\} \text{ in } \Omega \subset \mathbb{R}^3, \\ p = g(\rho), \\ \mathbf{U}.\mathbf{n} = 0 \quad \text{on } \Gamma_1 \text{ (wall)}, \\ \left.\begin{array}{l} \mathbf{U}.\mathbf{N} = \varphi \\ \quad \text{and} \\ p = \psi \end{array}\right\} \text{ on } \Gamma_2 \text{ (surface of contact); } \varphi \ \&\ \psi \text{ given}. \end{cases}$$

5.2. Stationary Irrotational Flow of an Incompressible Perfect Fluid

The nature of the data sometimes leads us to look for an irrotational velocity field, i.e. a $\mathbf{U}$ such that

$$(1.29)\qquad \operatorname{curl}\mathbf{U} = \mathbf{0}.$$

In this case we deduce the existence of a potential function Φ such that

$$(1.30)\qquad \mathbf{U} = \operatorname{grad}\Phi,$$

Φ depending only on $\mathbf{x}$ in the stationary case.
The incompressibility, on the one hand, and the boundary conditions (1.28) on the other hand, are then expressed by:

$$(1.31)\quad \begin{cases} \text{(i) } \rho = \rho_0, & \\ \text{(ii) } \Delta\Phi = 0 & \text{in } \Omega \subset \mathbb{R}^3, \\ \text{(iii) } \dfrac{\partial\Phi}{\partial n} = 0 & \text{on } \Gamma_1, \\ \left.\begin{array}{l} \text{(iv) } \dfrac{\partial\Phi}{\partial N} = \varphi \\ \text{(v) } p = \psi \end{array}\right\} & \text{on } \Gamma_2. \end{cases}$$

Finally Euler's equation leads to:

$$(1.32)\qquad \operatorname{grad}\left(\frac{U^2}{2} + \frac{p}{\rho_0}\right) = \frac{\mathbf{f}}{\rho_0} \quad \text{in } \Omega \subset \mathbb{R}^3,$$

with the result that if:

$$\frac{\mathbf{f}}{\rho_0} = \operatorname{grad}\mathscr{U} \quad \text{(see Remark 6)},$$

we have the relation

$$\frac{U^2}{2} + \frac{p}{\rho_0} - \mathscr{U} = \text{constant}^{15}, \quad \mathbf{x} \in \Omega \subset \mathbb{R}^3 . \tag{1.33}$$

The relations (1.31) and (1.33) enable us to determine **U**, then p (to the extent that p is known at a point or at infinity).

5.3. Flow of a Perfect Fluid Around a Solid Obstacle in Given Uniform Motion

We consider the flow generated by an undeformable solid obstacle moving with a given velocity in the midst of a fluid initially at rest. As previously, we seek an irrotational solution of an incompressible perfect fluid, but here:

$$\begin{cases} \Omega \text{ (the exterior of the obstacle } \mathcal{O}\text{) is unbounded ,} \\ \Gamma_1, \text{ part of the barrier partially limiting } \Omega, \text{ is in motion .} \end{cases}$$

In (1.31), the conditions on Γ_2 are then replaced by:

$$\operatorname{grad} \Phi = \mathbf{0} \text{ at infinity ,}$$

which expresses that far from the obstacle, the fluid remains undisturbed. The condition (1.31) (iii) on Γ_1 is replaced by:

$$\frac{\partial \Phi}{\partial n}(P) = (\mathbf{V}_A + \Omega \wedge \overrightarrow{AP}).\mathbf{n} \quad \text{on} \quad \partial\mathcal{O}$$

where P is the boundary point considered, A a fixed point of the obstacle with given velocity $\mathbf{V}_A$ and $\boldsymbol{\Omega}$ is the given instantaneous angular velocity of the obstacle. If $\mathbf{V}_A$ and Ω are constants, the fluid motion will be stationary and the potential function Φ of the fluid velocity will be the solution of:

$$\begin{cases} \text{(i) } \Delta\Phi = 0 \quad \text{in} \quad \Omega \subset \mathbb{R}^3 , \\ \text{(ii) } \dfrac{\partial \Phi}{\partial n}(P) = (\mathbf{V}_A + \Omega \wedge \overrightarrow{AP}).\mathbf{n} , \quad \forall P \in \mathcal{O} , \\ \text{(iii) } \operatorname{grad} \Phi = \mathbf{0} \text{ at infinity.} \end{cases} \tag{1.34}$$

We are concerned here with an *exterior Neumann problem.*
The quantities ρ and p are determined as before (to within a constant in the case of p).

5.4. Stationary Flow of an Incompressible Viscous Fluid in a Cylindrical Duct

Constant pressures p_1 and p_2 are imposed respectively at the two ends of the duct (see Fig. 1) and we suppose that the external forces are negligible in comparison with the pressure gradient (i.e. we suppose that $\mathbf{f} = \mathbf{0}$).

---

[15] This relation expresses Bernoulli's first theorem: see Germain [1].

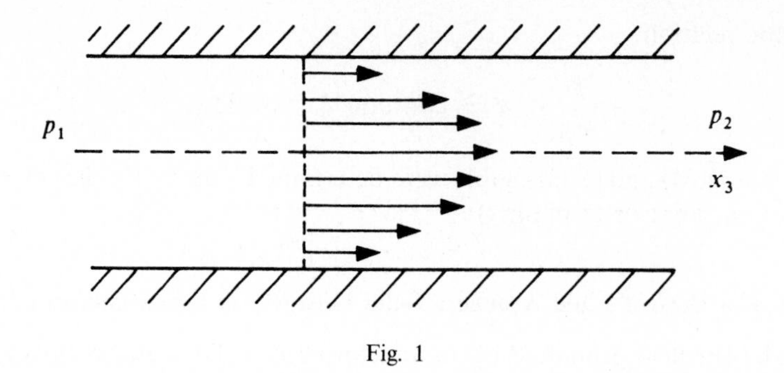

Fig. 1

In the case in which the length L of the duct can be considered to be "large", we model this duct by an infinite cylinder $\Omega \times \mathbb{R}_{x_3}$[16]. We then replace the effective data p_1, p_2 and L by the linear fall of pressure

$$-\pi = \frac{p_2 - p_1}{L}$$

(L the effective length of the tube).
We then look for a solution $\mathbf{U}$ of the Navier–Stokes problem (1.27) of the type

$$u_1 = 0; \quad u_2 = 0; \quad u_3(\mathbf{x}) \equiv u(x_1, x_2) \equiv u(\mathbf{x}) \quad \text{for} \quad \mathbf{x} \in \Omega \subset \mathbb{R}^2 ,$$

which reduce equation (1.27) (i) to

$$p_{,1} = 0 = p_{,2} \quad p_{,3} = \mu \Delta u ;$$

p thus depends only on x_3 and as u depends only on x_1, x_2, we have $p_{,3} = \mu \Delta u = \textit{constant}$ (which, sign apart, is no other than the linear fall of pressure). Finally, taking account of the wall condition (1.27)(iii), problem (1.27) reduces here to looking for $u(x_1, x_2)$ such that

$$(1.35) \qquad \begin{cases} \text{(i) } \Delta u = -\dfrac{\pi}{\mu} & \text{in} \quad \Omega \subset \mathbb{R}^2 , \\ \text{(ii) } u = 0 & \text{on} \quad \Gamma, \text{ boundary of } \Omega . \end{cases}$$

5.5. Plane Stationary Flow, Near to a Uniform Flow

We are interested here in the perturbation (supposed small) of a uniform flow in $\mathbb{R}^3$ by a cylindrical obstacle, supposed to be of infinite length, whose axis is perpendicular to the direction of the flow[17]. *The fluid is supposed to be perfect and*

---

[16] Here, Ω is no longer the domain occupied by the fluid but that on which we shall be led to work, i.e. the cross-section of the tube.

[17] In fact, in reality, it is often the obstacle (aircraft wing, for example) which moves with a uniform cruising velocity in a fluid; however, by a change of the reference frame, we are led to the situation described here in which the obstacle is considered as fixed. In addition, it is this model which we represent in wind-tunnels and which we then compare with the calculations.

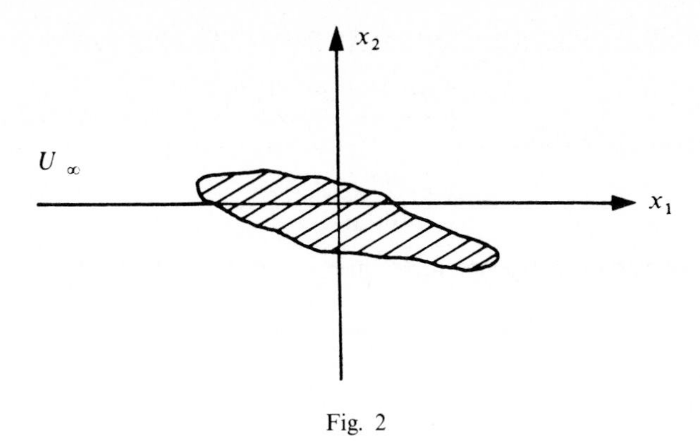

Fig. 2

barotropic compressible. We choose the cross section of the obstacle as the plane Ox_1x_2 (Fig. 2).

Thus, let $\mathbf{U}_\infty$ be the uniform velocity of the unperturbed flow (we choose Ox_1 parallel to $\mathbf{U}_\infty$ and denote by U_∞ the speed $|\mathbf{U}_\infty|$ of this flow); let ρ_∞, p_∞ and C_∞ denote respectively, the corresponding density, pressure and speed of sound (see (1.22)).

The quantity

$$M = \frac{U_\infty}{C_\infty} \tag{1.36}$$

is called the *Mach number* of the undisturbed flow.

We denote by

$$U_\infty + \eta u_1\,, \quad \eta u_2\,, \quad p_\infty + \eta p\,, \quad \rho_\infty + \eta\rho\,,$$

the quantities characterising the perturbed flow, η being an "infinitely" small parameter whose physical significance has to be made precise in each particular problem; here it is linked to the fact that U_∞ is in general large and that the object is small with respect to the "infinitely large" domain in which it is immersed. The linearisation of the equations (conservation of mass and Euler's equation) in the stationary case, then leads to the system:

$$(1.37)\quad \begin{cases} \text{(i) } U_\infty \dfrac{\partial\rho}{\partial x_1} + \rho_\infty\left(\dfrac{\partial u_1}{\partial x_1} + \dfrac{\partial u_2}{\partial x_2}\right) = 0 \quad (x_1, x_2) \in \Omega \subset \mathbb{R}^2\ {}^{18}\,, \\ \text{(ii) } U_\infty \dfrac{\partial u_1}{\partial x_1} + \dfrac{1}{\rho_\infty}\dfrac{\partial p}{\partial x_1} = 0\,, \\ \text{(iii) } U_\infty \dfrac{\partial u_2}{\partial x_1} + \dfrac{1}{\rho_\infty}\dfrac{\partial p}{\partial x_2} = 0\,, \end{cases}$$

[18] Here, Ω is the complement in $\mathbb{R}^2$ of the cross-section of the obstacle.

but the equation of state, $p = g(\rho)$, of the barotropic fluid considered implies

$$p_\infty + \eta p = g(\rho_\infty + \eta\rho),$$

and hence to the first order:

$$p = \left(\frac{\mathrm{d}g}{\mathrm{d}\rho}\right)_\infty \rho = C_\infty^2 \rho. \tag{1.38}$$

The natural conditions at infinity in $\mathbb{R}^2$ are:

$$U_\infty + \eta u_1 = U_\infty\,;\quad \eta u_2 = 0\,;\quad p_\infty + \eta p = p_\infty\,;\quad \rho_\infty + \eta\rho = \rho_\infty\,;$$

which imply:

$$u_1|_\infty = 0\,;\quad u_2|_\infty = 0\,;\quad p|_\infty = 0\,;\quad \rho|_\infty = 0\,.$$

Taking account of (1.37) (ii), we therefore have

$$p = -\rho_\infty U_\infty u_1\,. \tag{1.39}$$

From (1.37) (iii) we then deduce:

$$\operatorname{curl}\mathbf{u} = 0 \quad \text{with} \quad \mathbf{u}(\mathbf{x}) = (u_1, u_2) \quad \text{and} \quad \mathbf{x}\in\Omega\subset\mathbb{R}^2$$

and hence the possibility of introducing a potential $\varphi(\mathbf{x})$ such that

$$\mathbf{u} = \operatorname{grad}\varphi\,; \tag{1.40}$$

more precisely, we consider

$$\varphi(\mathbf{x}) = \int_{-\infty}^{x_1} u(\xi, x_2)\,\mathrm{d}\xi \quad (x_1, x_2)\in\Omega\,; \tag{1.40$'$}$$

then (1.37) (i) implies that

$$(1 - M^2)\frac{\partial^2\varphi}{\partial x_1^2} + \frac{\partial^2\varphi}{\partial x_2^2} = 0\,, \tag{1.41}$$

an equation which is elliptic, parabolic or hyperbolic[19] according as the speed at infinity is subsonic ($U_\infty < C_\infty$), trans-sonic ($U_\infty = C_\infty$) or supersonic ($U_\infty > C_\infty$)[20].

Since the fluid is perfect, the natural boundary condition on the obstacle is the *slipping condition* which can be written:

$$\mathbf{U}.\mathbf{n} = 0\,, \quad \text{on } \Gamma, \text{ the boundary of the obstacle.}$$

But is obvious physically that the perturbation of the flow at infinity will be small (and hence the linearisation legitimate) only if the tangent at each point of Γ makes an "infinitely" small angle with the Ox_1 axis and if the thickness of the profile is itself small[21]. More precisely, (cf. Fig. 3), we suppose that the profile is defined by:

$$x_2 = \eta F^+(x_1) \quad \text{and} \quad x_2 = \eta F^-(x_1) \quad \text{for} \quad a \leqslant x_1 \leqslant b\,,$$

---

[19] These concepts will be defined in Chap. V, §2.

[20] This problem will be studied in Chap. X (Tricomi's problem).

[21] In this case, the small parameter η introduced previously in a formal manner characterises the smallness of the angle and the width of the obstacle.

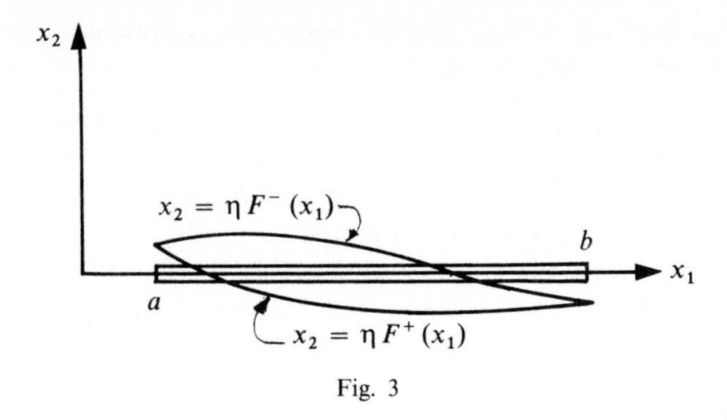

Fig. 3

with

$$F^+(x_1) > F^-(x_1) \quad \text{and} \quad F^+(a) = F^-(a)\,; \quad F^+(b) = F^-(b)\,.$$

The linearisation then leads (see Germain [1]) to the slipping condition on the two edges of the cut: $x_2 = \pm\, 0$, $a \leqslant x_1 \leqslant b$ account always being taken of the slope $\eta dF^{\pm}/\mathrm{d}x_1$ of the tangent to the profile with respect to the axis Ox_1. If we put

$$\delta^{\pm}(x_1) = \frac{\mathrm{d}F^{\pm}}{\mathrm{d}x_1}\,,$$

the linearised condition of slip is then reduced to

$$u_2(x_1, +\,0) = \delta^+(x_1)\,; \quad u_2(x_1, -\,0) = \delta^-(x_1)\,.$$

The perturbed flow is thus determined in terms of $\varphi(\mathbf{x})$ defined on $\Omega = \mathbb{R}^2 - \{c\}$ ($\{c\}$ denoting the above cut) to be the solution of

$$(1.42)\qquad \begin{cases} (1 - M^2)\dfrac{\partial^2\varphi}{\partial x_1^2} + \dfrac{\partial^2\varphi}{\partial x_2^2} = 0 \quad \underline{x} \in \Omega \subset \mathbb{R}^2\,, \\ \operatorname{grad}\varphi = 0, \text{ at infinity in } \mathbb{R}^2\,, \\ \dfrac{\partial\varphi}{\partial x_2}(x_1, +\,0) = \delta^+(x_1) \quad \dfrac{\partial\varphi}{\partial x_2}(x_1, -\,0) = \delta^-(x_1)\,; \end{cases}$$

from which we have

$$(1.43)\qquad \mathbf{u} = \operatorname{grad}\varphi\,; \quad p = -\,\rho_\infty U_\infty u_1\,; \quad \rho = \frac{p}{C_\infty^2}\,.$$

6. Non-Stationary Flows Leading to the Equations of Viscous Diffusion

The viscosity of the fluid exercises a damping action on the particles; in general, therefore, it is necessary to impose pressure gradients to maintain a stationary flow (see the example given in §1.5.4). We shall consider here flows under *constant pressure* which will allow us to exhibit the diffusion effects caused by the viscosity.

6.1. Flow of an Incompressible, Viscous, Homogeneous Fluid in a Very Long Cylindrical Duct

(We extrapolate the condition on the length by assuming the duct to be infinite, which means, in fact that we are studying the flow far from the ends.)
We suppose, in addition, that we can neglect the weight per unit volume with respect to the inertia term $\rho \mathrm{d}\mathbf{U}/\mathrm{d}t$, or that the axis of the cylinder is vertical. If we take Ox_3 to be the direction of the axis of the cylinder (of unit vector $\mathbf{k}_3$), and Ω to denote the cross-section, in the case in which the initial velocity is of the type

$$\mathbf{U}(\mathbf{x}, 0) = u_0(x_1, x_2)\mathbf{k}_3 \quad \mathbf{x} = \{x_1, x_2, x_3\} \in \Omega \times \mathbb{R} ,$$

it is legitimate to look for a solution of the type

$$\mathbf{U}(\mathbf{x}, t) = u(\mathbf{x}, t)\mathbf{k}_3$$

The condition of incompressibility (1.4), namely

$$\operatorname{div} \mathbf{U} = 0$$

then implies

$$\frac{\partial u}{\partial x_3} = 0 , \quad \rho = \rho_0$$

(ρ_0 constant, taking account of the homogeneity). This leads us to a two-dimensional problem on Ω and, since the pressure is supposed to be constant, the Navier–Stokes system (1.12) reduces to:

$$\text{(1.44)} \qquad \begin{cases} \dfrac{\partial u}{\partial t} - \nu \Delta u = \dfrac{f}{\rho_0} & \text{in } \Omega \times]0, \infty[, \\ u(\underset{\sim}{x}, 0) = u_0(\underset{\sim}{x}) & \text{in } \Omega , \\ u = 0 & \text{on } \Gamma \text{ (boundary of } \Omega) , \end{cases}$$

with: $\underset{\sim}{x} = \{x_1, x_2\}$; $\Delta u = \dfrac{\partial^2 u}{\partial x_1^2} + \dfrac{\partial^2 u}{\partial x_2^2}$ and $f = 0$ or $f = \rho_0 g$,

according as, following the above hypotheses, the weight of the fluid is neglected or not (g being the acceleration due to gravity).
We can eliminate the coefficient of Δu by a change of variable of the type: $x_i \to x_i \nu^{-\frac{1}{2}}$.

6.2. Flow of an Incompressible and Viscous, Homogeneous Fluid Between Two Infinite Parallel Plates

We suppose the system to be at rest, until at an instant, taken as origin of the time, when one of the plates is set in rectilinear motion parallel to the other. With the same hypotheses on the weight of the fluid as in the preceding example, we shall here look for a solution of the type

$$\mathbf{U}(\mathbf{x}, t) = u(x_3, t)\mathbf{k}_1$$

where $\mathbf{k}_1$ is the unit vector in the direction of the velocity of the moving plate (the other being kept fixed) and Ox_3 an axis perpendicular to the two plates, separated by a distance h, with the result that $x_3 \in]0, h[\subset \mathbb{R}$. The scalar function u is then a solution of the following simplified Navier–Stokes system:

$$(1.45) \quad \begin{cases} \left.\begin{aligned} &\frac{\partial u}{\partial t} - \nu \frac{\partial^2 u}{\partial x^2} = f \\ &u(x, 0) = 0 \end{aligned}\right\} \quad \text{on} \quad]0, h[\text{ (we have put } x_3 = x\text{)}, \\ u(0, t) = 0, \\ u(h, t) = v(t), \text{ velocity imposed on the moving plate.} \end{cases}$$

6.3. Flow with a Free Surface

In the application 6.2 we can suppose that it is the lower plate which is moving; if the other plate (at $x_3 = h$) is removed, leaving the corresponding fluid surface free, the boundary conditions have then to be modified as follows:

$$u(0, t) = v(t)$$

and, conforming with (1.6) and (1.7) we must have *on the free surface*:

$$\mathbf{U}^{(1)} = \mathbf{U}^{(2)}$$

and

$$(\sigma_{ij}^{(1)} - \sigma_{ij}^{(2)})N_j = 0,$$

say; on denoting by $\mathbf{U}$ the common velocity of the air (for example) and of the fluid considered to be in contact, and taking (1.13) into account, we have

$$-(p^{(1)} - p^{(2)})N_i + (\mu^{(1)} - \mu^{(2)})(u_{i,j} + u_{j,i})N_j = 0.$$

If we suppose, which is reasonable as long as the velocities are not too high (this depending essentially on the given $v(t)$), that the free surface remains defined by $x_3 = h$, then $\mathbf{N} = \mathbf{k}_3$ and the above vector relation is reduced to the following two scalar relations:

(i) for $i = 1$, $\partial u/\partial x_3(h, t) = 0$, we seek a solution of the type $\mathbf{U}(\mathbf{x}, t) = u(x_3, t)\mathbf{k}_1$
(ii) for $i = 3$, $p^{(1)} = p^{(2)}$.

The system (1.45) is then replaced by

$$(1.46) \quad \begin{cases} \left.\begin{aligned} &\frac{\partial u}{\partial t} - \nu \frac{\partial^2 u}{\partial x^2} = f \\ &u(x, 0) = 0 \end{aligned}\right\} \quad \text{for} \quad x = x_3 \in]0, h[\subset \mathbb{R}, \\ u(0, t) = v(t), \\ \dfrac{\partial u}{\partial N}(h, t) = 0, \end{cases}$$

which corresponds to a mixed boundary value problem.

7. Conduction of Heat. Linear Example in the Mechanics of Fluids[22]

7.1. General Introduction

Until now we have taken account of only two conservation laws: that of mass expressed by (1.1) and that of momentum expressed by (1.2). This supposes implicitly that the temperature of the fluid is known and constant (at least in first approximation). However, current experience shows that variations in the temperature involve dilatations, that is to say variations of volume and this generates, particularly in fluids, convective motions (lighter hot air has a tendency to rise and give way to colder air); similarly, the characteristic coefficients of a medium depend on the temperature (for example, hot oil is more fluid than cold oil). Conversely, internal or external friction in the course of motion generates variations in temperature. All of this shows that the *variations in pressure, in density, in displacement, in velocity and in temperature are coupled*; to take account of this fact, it is therefore necessary to introduce new relations.

The first family of relations valid in general for continuous media express *the fundamental principles of thermodynamics*[22]: the first principle is expressed by *the law of the conservation of energy*, and the second by the *fundamental inequality of Clausius and Duhem.*

On the contrary, the second family of relations will be characteristic of the particular medium studied, it contains, on the one hand, the expression of the energy as a function of the "thermodynamic" variables of the medium, and, on the other hand, the laws of dissipation of this medium.

7.2. The Conservation of Energy and the Fundamental Inequality of Clausius and Duhem

The conservation of energy is expressed locally by:

$$\rho \frac{\mathrm{d}e}{\mathrm{d}t} = \sigma_{ij} u_{i,j} - q_{i,i} + r \quad \text{in} \quad \Omega \tag{1.47}$$

where

$e(\mathbf{x}, t)$ denotes the specific *internal energy* (i.e. per unit mass) of the medium; $\mathbf{q}(\mathbf{x}, t)$ is the *heat flux* (with components q_i in the considered reference frame); the quantity $q(\mathbf{x}, t, \mathbf{n})$ defined by

$$q(\mathbf{x}, t, \mathbf{n}) = -\ \mathbf{q}(\mathbf{x}, t) . \mathbf{n} , \tag{1.48}$$

$\mathbf{n}$ a unit vector, represents the *rate at which heat is crossing a unit element of (fictitious) surface passing through* $\mathbf{x}$ *and perpendicular to* $\mathbf{n}$ (the passage being made in the sense and direction of the vector $\mathbf{q}$); here, it is a question of heat transmitted by conduction to the interior of Ω;

---

[22] For the concepts of thermodynamics used here, see, for example, Germain [2] or [3].

$r(\mathbf{x}, t)$ is a density per unit volume defining *a rate of heat supplied by external elements to the medium considered* (radiation, Joule effect, exothermic chemical reaction, ... etc.); this term, often called the "source" function is supposed to be given and is in fact null in a certain number of applications[23]; the terms

$$\frac{1}{\rho}\sigma_{ij}u_{i,j} = \frac{1}{\rho}\sigma_{ij}\varepsilon_{ij}(u) \quad \text{(because of the symmetry of } \sigma_{ij})^{24}$$

and

$$\frac{1}{\rho}(r - q_{i,i})$$

appearing on the right-hand side of (1.47), after division by ρ, are respectively the rate (per unit volume) of the *specific energy due to the internal forces* and of the *specific heat received.*

The boundary condition associated with this conservation law is

$$q = -q_i n_i = \varpi - F_i v_i \quad \text{on} \quad \partial\Omega, \tag{1.49}$$

where

ϖ is the rate of surface heat supplied by the exterior of Ω to the boundary point considered,

$\mathbf{F}$ (with components F_i) is the surface density of contact forces (pressure, friction...) applied by the exterior of Ω to the boundary point considered,

$\mathbf{V}$ (with components v_i) denotes the relative velocity of the medium with respect to the wall; $F_i v_i$ thus represents the energy dissipated by friction at the wall.

The Clausius–Duhem inequality expressing now the second principle of thermodynamics can be written locally

$$\rho\frac{\mathrm{d}s}{\mathrm{d}t} + \operatorname{div}\frac{\mathbf{q}}{T} - \frac{r}{T} = \rho\dot{\eta} \geqslant 0 \quad \text{in} \quad \Omega^{25}, \tag{1.50}$$

where $s(\mathbf{x}, t)$ is the *specific entropy*, $T(\mathbf{x}, t)$ is the *absolute temperature*; $\dot{\eta}$ is called the rate of *irreversible production of entropy per unit volume.*

Remark 8. Having introduced several new concepts into the laws previously discussed, such as e, q, r, s and T, it is important to analyse in what manner we can connect them with each other and with other unknowns already mentioned in this §1. This will be in part the role of the laws of behaviour.

The elementary concepts of thermodynamics indicate already that

$$\mathrm{d}e = T\mathrm{d}s + w, \tag{1.51}$$

$T\mathrm{d}s$ represents the *received reversible elementary heat* and w the *received reversible elementary work.*

---

[23] When r and $\mathbf{q}$ are null at the same time, the evolution of the medium is called *adiabatic*.

[24] See relation (1.2).

A combination of relations (1.47), (1.50) and (1.51) then shows that

$$\Phi \overset{\text{def}}{=} \rho T \dot{\eta} = \sigma_{ij}\varepsilon_{ij}(u) - \rho\dot{w} - \dot{\mathbf{q}} . \frac{\text{grad}\, T}{T} \geqslant 0^{25} ; \tag{1.52}$$

$$\Phi_1 \overset{\text{def}}{=} \sigma_{ij}\varepsilon_{ij}(u) - \rho\dot{w} \tag{1.53}$$

is called the *intrinsic dissipation*, $\dot{w}$ being the *received reversible power* and

$$\Phi_2 \overset{\text{def}}{=} - \frac{1}{T} \mathbf{q} . \text{grad}\, T \tag{1.54}$$

is the *thermal dissipation per unit volume*; $\Phi = \Phi_1 + \Phi_2$ is then the *total dissipation*. □

7.3. Law of Thermomechanical Behaviour: Case of Fluids

The behaviour of a medium is completely defined as soon as we know: on the one hand, the law of variation of e (or of another potential) as a function of a system of $m + 1$ thermodynamic variables characteristic of the medium – this leads to *m laws of state*; on the other hand, the nature of the dissipative terms, which leads to *supplementary laws.*

In the case of a fluid, the most appropriate thermodynamic variables are T and $\tau = 1/\rho$, $(m = 1)$; the elementary work is then defined by

$$w = - p\mathrm{d}\tau , \tag{1.55}$$

where p is the thermodynamic pressure; and the most natural thermodynamic potential is the *specific free energy*

$$\psi = e - Ts ; \tag{1.56}$$

so

$$\mathrm{d}\psi = - s\mathrm{d}T - p\mathrm{d}\tau ,$$

from which we derive the equation of state

$$p = - \frac{\partial \psi}{\partial \tau}(T, \tau) \tag{1.57}$$

when the expression for the potential ψ is given.

Now let us calculate the dissipation; taking account of (1.1), we have

$$\dot{w} = - p \frac{\mathrm{d}\tau}{\mathrm{d}t} = \frac{p}{\rho^2}\frac{\mathrm{d}\rho}{\mathrm{d}t} = - \frac{p}{\rho} \,\text{div}\, \mathbf{U} , \tag{1.58}$$

from which we deduce

$$\Phi_1 = (\sigma_{ij} + p\delta_{ij})\varepsilon_{ij}(u) .$$

[25] By convention, $\dot{\varphi} = \mathrm{d}\varphi/\mathrm{d}t$ for each symbol.

Thus, adopting a law of the type (1.3):

$$\sigma_{ij} = - p\delta_{ij} + \tau_{ij}\,, \tag{1.59}$$

where

$$\tau_{ij} = \lambda\varepsilon_{ll}(\mathbf{u})\delta_{ij} + 2\mu\varepsilon_{ij}(\mathbf{u}) \tag{1.60}$$

represents the *tensor of viscous stresses*, we see that

$$\Phi_1 = \tau_{ij}\varepsilon_{ij}(\mathbf{u}) \tag{1.61}$$

is a *purely viscous dissipation*.
To make precise the term for the thermal dissipation, we often adopt, as a first approximation, *Fourier's law of the conduction of heat* which in general can be written

$$q_i = - K_{ij}T_{,j}\,. \tag{1.62}$$

where the *thermal conduction tensor* $\bar{\bar{\mathbb{K}}}$ depends, in general, on the temperature. In fact, in the case of an *isotropic medium*, (a medium for which no direction of space plays a privileged role with respect to the other), which, in general, is the case for fluids

$$K_{ij} = k\delta_{ij} \quad \text{and so} \quad \mathbf{q} = - k \operatorname{grad} T\,. \tag{1.63}$$

In this case, the term for thermal diffusion can be written

$$\Phi_2 = \frac{k}{T}\,|\operatorname{grad} T|^2\,. \tag{1.64}$$

Remark 9. Without going into detail of the theory of the thermal properties of fluids, which is complex and leads in general to non-linear problems, we note that we have given here the essential elements to obtain a complete mathematical model. Knowledge of the potential $\psi(T, \tau)$ infers in effect that of p (by (1.57)) and

$$s = - \frac{\partial\psi}{\partial T}(T, \tau) \tag{1.65}$$

(though it is rather p and s which are known *a priori*); we deduce $e(T, \tau)$ from (1.56), with the result that, taking account of (1.58), we find the energy equation (1.47) becomes a partial differential equation in T depending, in general, strongly on the other unknowns.
The boundary conditions on T are then of the type:

$$T = T_d \tag{1.66}$$

on the part $\partial\Omega_T \subset \partial\Omega$ where the temperature T_d is imposed, and of the type

$$k\frac{\partial T}{\partial n} = \varpi - F_i v_i \quad \text{on} \quad \partial\Omega_\varpi \tag{1.67}$$

where $\partial\Omega_\varpi \subset \partial\Omega$ is the part on which the heat flux is imposed, account having been taken of (1.49) and of Fourier's law (1.63).

Finally, we note that it will often be necessary to take account of the term f_i, representing in equation (1.2) the external forces acting on the fluid, and of the variations in the density due to the thermal dilatation; they even generate by their existence a *motion* of the fluid called "*natural convection*"; it is convenient to add to the term representing the weight (when this is not neglected) a term to take account of this new driving force with the result that

$$f_i = \rho g_i[1 + \beta(T - T_e)] \tag{1.68}$$

where T_e is the equilibrium temperature which can be chosen as the initial temperature, $\mathbf{g}$ (with components g_i) is the acceleration due to gravity, and β is the coefficient of the thermal dilatation of the fluid defined by

$$\frac{\rho_e - \rho}{\rho} = \beta(T - T_e) \tag{1.69}$$

ρ_e being the density of the fluid in the equilibrium state corresponding to the temperature T_e.

7.4. Particular Case of an Incompressible, Homogeneous, Viscous Fluid, with Constant Specific Heat

An incompressible fluid is, from the thermodynamic point of view, a degenerate fluid in the sense that the pressure p in it is indeterminate (see footnote[8] above). Given k and μ, the coefficients of dissipation, supposed to be generally constant when the fluid is incompressible, it is sufficient to define the behaviour that we write

$$\sigma_{ij} = -\ p\delta_{ij} + 2\mu\varepsilon_{ij}(u) \tag{1.70}$$

$$q_i = -\ kT_{,i} \tag{1.71}$$
[26]

The relation

$$\operatorname{div} \mathbf{U} = u_{i,i} = 0 \tag{1.72}$$

takes the place of the equation of state since it implies, in particular, that

$$\tau = \frac{1}{\rho_0} = \text{constant} \quad (\text{see §1.2}). \tag{1.73}$$

It is enough therefore to know $e(T)$, or what comes to the same thing, the specific heat of the fluid. If this is constant and equal to c, we have

$$e = cT. \tag{1.74}$$

The system of equations (1.12) (supposing here that $\partial\Omega = \Gamma_1$) enables us, in general, to determine $\mathbf{U}$ and grad p. The thermal effects are here decoupled from the mechanical effects. In the absence of a source of heat ($r = 0$), the energy equation

---

[26] We note that the condition (1.52) for all $\mathbf{U}$ and T, together with (1.61) and (1.64), imply that k and μ are positive.

can, as a result of the preceding relations, be written:

$$\rho_0 c \frac{dT}{dt} = \mu[u_{i,j} + u_{j,i}]u_{i,j} + k\Delta T$$

thus

(1.75) $$\rho_0 c \frac{\partial T}{\partial t} - k\Delta T + \rho_0 c \mathbf{U}.\text{grad}\, T = \mu(u_{i,j} + u_{j,i})u_{i,j}\,;$$

this relation completed by initial conditions and boundary conditions of the type (1.66) and (1.67) permits the determination of T, account being taken of the previous determination of $\mathbf{U}$.

In the particular case of a perfect fluid, $\mu = 0$; if, in addition, we make the hypothesis of small perturbations (low velocities and small temperature gradients), then equation (1.75) reduces to the equation called *the heat equation* which is of same mathematical type as the equations of viscous diffusion encountered in §1.6, an equation which we thus write

(1.75)′ $$\rho_0 c \frac{\partial T}{\partial t} - k\Delta T = 0\,,$$

or else

(1.75)″ $$\rho_0 c \frac{\partial T}{\partial t} - k\Delta T = f,$$

if there exists a source of heat f per unit volume.

8. Example of Acoustic Propagation

We have already introduced in §1.4 the idea of a *barotropic perfect fluid*; it is suitable now to note that the relation

(1.76) $$p = g(\rho)$$

constitutes another example of the equation of state. The inequalities

(1.77) $$g > 0\,, \quad C^2 = \frac{dg}{d\rho} > 0 \quad \text{and} \quad \frac{d^2 g}{d\rho^2} > 0\,,$$

being imposed by conditions of thermodynamic stability[27], we verify that the relation (1.76), taken with Euler's equations (1.18), (1.20) and with the boundary conditions (1.23), (1.24), (1.25) enables us to determine $\mathbf{U}$, ρ and p.

We therefore now consider a *barotropic perfect fluid* [§1.4] occupying a domain Ω of the space and *immersing a body vibrating with small amplitude* (sound source) itself occupying the domain Ω_1. We study the propagation of these small perturbations in the fluid, and hence the propagation of the sound.

---

[27] See Germain [2].

We suppose that the forces of gravity $\mathbf{f}$ are negligible in comparison with the forces of inertia

$$\mathbf{f} \cong \mathbf{0}. \tag{1.78}$$

It is legitimate also to assume *irrotational flow*

$$\operatorname{curl} \mathbf{U} = \mathbf{0}\,; \tag{1.79}$$

in effect if the fluid remains at rest up to the instant $t = 0$ (as we suppose), there is no *a priori* reason why the small vibrations of the source, which produce small velocities normal to $\partial\Omega_1$, should produce a rotation; we know then (see Germain [1]) that the flow will remain irrotational.
The relation (1.79) implies the existence of a potential φ of the velocities such that

$$\mathbf{U} = \operatorname{grad} \varphi. \tag{1.80}$$

The equations of the conservation of mass and Euler's equations are then written

$$\frac{\partial \rho}{\partial t} + \rho_{,i}\varphi_{,i} + \rho \Delta\varphi = 0\,, \tag{1.81}$$

$$\frac{\partial}{\partial x_i}\left[\frac{\partial \varphi}{\partial t} + \frac{U^2}{2}\right] + \frac{1}{\rho}\frac{\partial p}{\partial x_i} = 0 \quad \text{with} \quad U^2 = |\operatorname{grad} \varphi|^2\,. \tag{1.82}$$

The vibrations having small amplitudes, we can make the classic hypotheses for small perturbations, namely

(i) the velocities u_i are "small" as well as their derivatives $u_{i,j}$ and $\partial u_i/\partial t$,
(ii) the pressure p and the density ρ vary little from their values p_0 and ρ_0 in the initial state of rest.

We can then linearise equations (1.81) and (1.82) which become

$$\frac{\partial \rho}{\partial t} + \rho \Delta\varphi = 0\,, \tag{1.83}$$

$$\frac{\partial}{\partial x_i}\left(\frac{\partial \varphi}{\partial t}\right) + \frac{1}{\rho}\frac{\mathrm{d}p}{\mathrm{d}\rho}\frac{\partial \rho}{\partial x_i} = 0\,; \tag{1.84}$$

as $\partial\varphi/\partial t$ is small as also is $\partial\rho/\partial x_i$, we can replace ρ and $\mathrm{d}p/\mathrm{d}\rho$ by the constants ρ_0 and C_0^2 (see relation (1.22)) from which we have

$$\frac{\partial \rho}{\partial t} + \rho_0 \Delta\varphi = 0\,, \tag{1.85}$$

$$\frac{\partial \varphi}{\partial t} + C_0^2 \frac{\rho}{\rho_0} = k(t)\,; \tag{1.86}$$

but φ can be modified by adding a function of t without changing the value of $\mathbf{U}$; in effect, if we put

$$\Phi = \varphi - \int_0^t k(s)\,\mathrm{d}s\,, \tag{1.87}$$

we always have

$$\mathbf{U} = \operatorname{grad} \Phi \quad \text{and} \quad \Delta\varphi = \Delta\Phi. \tag{1.88}$$

A combination of the relations (1.85), (1.86) and (1.87) shows that Φ is determined by:

the wave equation

$$\frac{1}{C_0^2}\frac{\partial^2 \Phi}{\partial t^2} - \Delta\Phi = 0\,; \tag{1.89}$$

the initial conditions, which can be expressed by

$$\Phi(\mathbf{x}, 0) = 0\,, \quad \frac{\partial \Phi}{\partial t}(\mathbf{x}, 0) = 0 \quad \text{(fluid at rest)}\,; \tag{1.90}$$

the boundary conditions

$$\left\{\begin{array}{l} \dfrac{\partial \Phi}{\partial n} = 0 \quad \text{on } \partial\Omega \text{ expressing the fact that the fluid does not} \\ \text{cross the boundary: } \mathbf{U}.\mathbf{n} = 0 = \operatorname{grad} \Phi.\mathbf{n} = \dfrac{\partial \Phi}{\partial n}\,, \\ \dfrac{\partial \Phi}{\partial n} = u_1 \quad \text{on } \partial\Omega_1 \text{ expressing the fact that the normal} \\ \text{component of the relative velocity of the fluid on the} \\ \text{boundary of the source is zero, } u_1 \text{ being the normal velocity} \\ \text{of the boundary } \partial\Omega_1 \text{ of the source, i.e.:} \end{array}\right. \tag{1.91}$$

$$(\mathbf{U} - \mathbf{U}_1).\mathbf{n} = \frac{\partial \Phi}{\partial n} - u_1 = 0. \tag{1.92}$$

9. Example with Boundary Conditions on Oblique Derivatives

We study the motion of water, supposed here to be a perfect, incompressible and homogeneous fluid, in a canal whose base is described by a given surface defined by

$$x_3 = -\, h(x_1, x_2)\,;$$

the free surface Γ_t is an unknown depending on the time, supposedly described by $x_3 = \eta(x_1, x_2, t)$. The domain occupied by the water is then

$$\Omega_t = \{\mathbf{x} \in \mathbb{R}^3 \quad \text{such that} \quad -h(x_1, x_2) < x_3 < \eta(x_1, x_2, t)\}\,.$$

The water is modelled as heavy; we denote by g the acceleration due to gravity and we take $\rho = 1$.

The unperturbed motion is the state of rest when the free surface satisfies $x_3 = 0$ and is in contact with an atmosphere at rest, supposed at constant pressure p_0 (which remains constant throughout subsequent time).

We suppose that the perturbed motion (due, for instance, to the atmosphere) is small and irrotational, with the result that we may, on the one hand neglect the terms in $|U|^2$ in equation (1.18), and on the other hand deduce (for the same reasons as before) the existence of a potential function φ such that

$$\mathbf{U} = \operatorname{grad} \varphi \,. \tag{1.93}$$

The *equations of mechanics of fluids* (1.18), (1.19) then give

$$\begin{cases} \text{(i) } \Delta\varphi = 0 \quad \text{(consequence of incompressibility)}\,, \\ \text{(ii) } \dfrac{\partial \varphi}{\partial t} + p = - g x_3 + p_0 \,. \end{cases} \tag{1.94}$$

The second relation obtained by spatial integration of Euler's equations serves only to determine the unknown pressure p.

It remains to state the *boundary conditions*:

(*a*) on the free surface Γ_t, we express, on the one hand, that the water does not "cross" this surface; on the other hand (in conformity with (1.25)) that $p = p_0$, from which, taking account of (1.93) and (1.94), we deduce

$$\begin{cases} \text{(i) } \dfrac{\partial \eta}{\partial t} + \dfrac{\partial \eta}{\partial x_1}\dfrac{\partial \varphi}{\partial x_1} + \dfrac{\partial \eta}{\partial x_2}\dfrac{\partial \varphi}{\partial x_2} - \dfrac{\partial \varphi}{\partial x_3} = 0 \quad \text{on } \Gamma_t\,, \quad t > 0 \\ \text{(ii) } \dfrac{\partial \varphi}{\partial t} + g\eta = 0 \qquad \text{on } \Gamma_t\,, \quad t > 0\,; \end{cases} \tag{1.95}$$

(*b*) on the base of the canal Γ_0, we have only that the velocity of the fluid is tangential to the base, which leads to

$$\frac{\partial \varphi}{\partial x_3} + \frac{\partial h}{\partial x_1}\frac{\partial \varphi}{\partial x_1} + \frac{\partial h}{\partial x_2}\frac{\partial \varphi}{\partial x_2} = 0 \quad \text{on} \quad \Gamma_0 \,. \tag{1.96}$$

The problem described in (1.93), (1.94), (1.95), (1.96) is non-linear, but the hypothesis of small perturbations will allow us to continue the linearisation process with the suppression of the term $|\mathbf{U}|^2$ in (1.18); the free surface being estimated to be close to the "equilibrium" position $x_3 = 0$, η and $\mathbf{U}$ are supposed to be sufficiently small for the products of these functions and of their derivatives to be neglected. The above problem then reduces to finding $\varphi(\mathbf{x}, t)$ defined to within a constant on

$$\Omega = \{\mathbf{x} \in \mathbb{R}^3, \quad \text{such that} \quad -h(x_1, x_2) < x_3 < 0\}$$

by:

$$\begin{cases} \text{(i) } \Delta\varphi = 0 & \text{in} \quad \Omega\,, \\ \text{(ii) } \dfrac{\partial^2 \varphi}{\partial t^2} + g\dfrac{\partial \varphi}{\partial x_3} = 0 & \text{for } x_3 = 0\,, \\ \text{(iii) } \dfrac{\partial \varphi}{\partial x_3} + \dfrac{\partial h}{\partial x_1}\dfrac{\partial \varphi}{\partial x_1} + \dfrac{\partial h}{\partial x_2}\dfrac{\partial \varphi}{\partial x_2} = 0 & \text{for } x_3 = - h(x_1, x_2)\,, \end{cases} \tag{1.97}$$

the relation (ii) being obtained from (1.95) by linearisation.

The velocity $\mathbf{U}$ is then defined by (1.93) and the pressure p by (1.94).

Remark 10. The search for solutions of (1.97) called "monochromatic" (or "harmonic") that is to say of the form

$$\varphi(\mathbf{x}, t) = \operatorname{Re}(e^{i\omega t}\Phi(\mathbf{x})) \tag{1.98}$$

leads to the *stationary problem, with oblique derivative*, consisting of determining Φ defined on Ω such that

$$\begin{cases} \text{(i) } \Delta\Phi = 0 & \text{in } \Omega \subset \mathbb{R}^3 , \\ \text{(ii) } \dfrac{\partial\Phi}{\partial x_3} - \dfrac{\omega^2}{g}\Phi = 0 & \text{on } x_3 = 0 , \\ \text{(iii) } \dfrac{\partial\Phi}{\partial x_3} + \dfrac{\partial h}{\partial x_1}\cdot\dfrac{\partial\Phi}{\partial x_2} + \dfrac{\partial h}{\partial x_2}\dfrac{\partial\Phi}{\partial x_2} = 0 & \text{on } x_3 = -h(x_1, x_1) . \end{cases} \tag{1.99}$$

□

Review

The mechanics of fluid arises in aerodynamics, for aircraft and other vehicles, in turbomachines, jet engines and the propulsion units of rockets, in chemical reactors, in heat exchangers, nuclear reactors, etc., and is most often used in the form of the non-linear Navier–Stokes equation (nevertheless, Euler's equation is used, for example, to study the flow round the blades of a turbomachine). The *linearised* models cited often serve in phenomena where the speeds are slow, and, for example: to study the flow of air round the fuselage and airfoils in motions with speeds slow compared with the speed of sound, to model the circulation of fluids in porous solids, or, again, to "approximate" models involving the equations of transport or of diffusion (in the case of magneto-hydrodynamics, for example)[28]. *Stationary* problems depending on the equations of Sect. 5 can be treated by the methods of Chap. II for certain among them[29] and more generally by the methods of Chap. VII. *Stationary* subsonic and supersonic flow in a nozzle which leads to Tricomi's problem will be studied in Chap. X. Certain *stationary* problems of this section can be treated with the help of integral equations and will be studied in Chap. XIA, §1 (floating dock), §4 (thin airfoil) and XIB §5 (Stokes' system).
The *linearised Navier–Stokes* equations, introduced in Sect. 3 are treated in Chap. XIX which concerns the stationary problem, called the Stokes problem or the evolution problem. The variational approach, the functional approach and

---

[28] We note, also, that in practical calculations in the aeronautical industry, according to the computer available, a compromise is made between the degree of approximation used for the equation of the fluid "air" and the level of complication with which we represent the geometry of the aircraft; use is made therefore of:

— the Navier–Stokes equation to treat the flow round a wing section;
— Euler's equation for a sub-ensemble of the aircraft;
— the linearised equation to treat the complete aircraft.

[29] The problem (1.31) for example.

regularity theorems are likewise examined in Chap. XIX. The numerical aspect is seen in Chap. XX.
Linearised *evolution* problems leading to those of *viscous diffusion* seen in Sect. 6, are examined by diagonalisation methods in Chap. XV, by the method of the Laplace transform in Chap. XVI, by that of semi-groups in Chap. XVII and by variational methods in Chap. XVIII. The numerical aspect of these problems is seen in Chap. XX.
The stationary problems in Sect. 7 of the *conduction of heat* can be treated by the "classical" methods of Chap. II, by the functional transformations (Fourier series for example) of Chap. III and by the variational methods of Chap. VII.
In the case of the *evolution* problem, such problems in the conduction of heat are treated in the chapters dealing with the method of diagonalisation (Chap. XV), of the Laplace transformation (Chap. XVI), of semi-groups (Chap. XVII) and in a much more general way by variational methods (Chap. XVIII); the numerical solution of these problems is considered in Chap. XX.

§2. Linear Elasticity

1. Introduction: Elasticity; Hyperelasticity

Elastic media are characterised by the fact that they have a *privileged* memory, that is to say, in some way, they remember only a single configuration which can be chosen as the *reference state.* In most cases, we suppose that in this, the medium is free from stress ($\sigma_{ij} = 0$) and that the temperature is uniform ($T = T_0$); we then say that the reference state is the natural state of the medium and we choose as the initial instant $t = 0$, the instant at which the elastic body considered passes from this reference state.
Let us suppose that subsequently the body is subjected to imposed forces or displacements; its state of stress at the instant t therefore depends only on its state of deformation at that instant t with respect to the natural state, and not to all the intermediate states.
If we denote by $\mathbf{a} = \{a_\alpha\}$ and $\mathbf{x} = \{x_i\}$ respectively the Lagrange and Euler coordinates of the same particle[30] with respect to an absolute reference frame, the motion of the continuous medium is defined by three relations:

$$x_i = \varphi_i(\mathbf{a}, t), \quad i = 1, 2, 3 .$$

The matrix F with components:

$$F_{i\alpha} = \frac{\partial x_i}{\partial a_\alpha} \equiv x_{i,\alpha} \quad i = 1, 2, 3 ; \quad \alpha = 1, 2, 3 \tag{2.1}$$

[30] See Appendix "Mechanics", §2.

expresses the *gradient of the transformation* and:

(2.2) $$\bar{\bar{\mathbb{L}}} = \frac{1}{2}[{}^t\bar{\bar{\mathbb{F}}}.\bar{\bar{\mathbb{F}}} - \bar{\bar{\mathbb{I}}}]^{31} \text{ with components } L_{\alpha\beta} = \frac{1}{2}(F_{i\alpha}F_{i\beta} - \delta_{\alpha\beta})$$

is the *Green–Lagrange deformation tensor*[32].

If we denote by $\bar{\bar{\Sigma}}$ the Cauchy stress tensor (or components σ_{ij}), i.e. that already introduced in §1.1, the law of behaviour of an elastic medium is expressed then by a relation of the type:

(2.3) $$\bar{\bar{\Sigma}}(\mathbf{x}, t) = \mathscr{G}[\mathbf{a}, \bar{\bar{\mathbb{L}}}(\mathbf{a}, t), T(\mathbf{a}, t)]$$

where T denotes the *temperature* and where the initial dependence on $\mathbf{a}$ expresses the possible *inhomogeneity* of the medium; the latter is said to be *isotropic* if the function $\mathscr{G}$ is invariant under every orthogonal transformation of the reference frame.

The usual elastic media are all *hyperelastic*, i.e. *they do not dissipate energy other than thermal energy*[33]. The medium is then characterised by knowledge of a *deformation energy* $W(\mathbf{a}, \bar{\bar{\mathbb{F}}}, T)$ such that:

(2.4) $$\bar{\bar{\Sigma}} = \frac{\rho}{\rho_0}\frac{\partial W}{\partial \bar{\bar{\mathbb{F}}}}.{}^t\bar{\bar{\mathbb{F}}} \quad \text{or, in components:} \quad \sigma_{ij} = \frac{\rho}{\rho_0}\frac{\partial W}{\partial F_{i\alpha}}F_{j\alpha}$$

where ρ_0 and ρ are the densities at instants 0 and t respectively.

2. Linear (not Necessarily Isotropic) Elasticity

The linear theory of elasticity is contained in the following framework.

(i) *The mechanical and thermal effects are decoupled*, the first being able to be studied independently of the second; in particular, this is possible when the evolution of the motion is either *isothermal* or *adiabatic*.

(ii) *The perturbations remain small*, with the result that we can proceed to a linearisation; there follow two important simplifications.

(*a*) On the one hand, the tensor $\bar{\bar{\mathbb{L}}}$ can be replaced by the *tensor of linearised deformations* $\bar{\bar{\varepsilon}}$, (or, simply, the *strain tensor*) with components

(2.5) $$\varepsilon_{ij}(u) = \frac{1}{2}(u_{i,j} + u_{j,i})^{34},$$

a linear functional of the *displacement field* defined by:

(2.6) $$u_i(\mathbf{a}, t) = x_i - a_\alpha\delta_{i\alpha}.$$

---

[31] We recall that, in general, $\bar{\bar{\mathbb{T}}}.\bar{\bar{\mathbb{S}}} = \bar{\bar{\mathbb{R}}}$ denotes the contracted product of the two second order tensors $\bar{\bar{\mathbb{T}}}$ and $\bar{\bar{\mathbb{S}}}$, i.e. in components $t_{ik}s_{kj} = r_{ij}$. We denote by ${}^t\bar{\bar{\mathbb{F}}}$ the transpose of the tensor $\bar{\bar{\mathbb{F}}}$ (i.e. the tensor with components ${}^tF_{ij} = F_{ji}$).

[32] A concrete interpretation of $\mathbb{L}$ will be found in Suquet [1].

[33] See §1.7 and §2.6.

[34] See Appendix "Mechanics", §2.

(b) On the other hand, the equations and the boundary conditions are taken, at each instant, to hold on the reference configuration Ω_0 which deviates only slightly from Ω_t; in particular, this implies that, in first approximation, we no longer distinguish between the Euler coordinates x_i and the Lagrange coordinates a_i.
(iii) *The law of elastic behaviour* is expressed by a *linear relation* between the stress tensor $\bar{\bar{\Sigma}}$ and the strain tensor $\bar{\bar{\varepsilon}}$; in components, let

(2.7) $$\sigma_{ij} = a_{ijkh}\varepsilon_{kh}(\mathbf{u}) .$$

The a_{ijkh}, called the *elastic constants*, are independent of the strain tensor $\bar{\bar{\varepsilon}}$, but dependent on $\mathbf{x}$ when the medium is not homogeneous; they have the following symmetry properties:

(2.8) $$a_{ijkh} = a_{jikh} = a_{ijhk} ,$$

resulting from the symmetry of the tensors $\bar{\bar{\Sigma}}$ (as in §1) and $\bar{\bar{\varepsilon}}$, and

(2.9) $$a_{ijkh} = a_{khij} ,$$

resulting from the existence of a strain energy[35].
There are therefore *a priori* twenty-one elastic constants (since we are in $\mathbb{R}^3$). The existence of a positive definite elastic energy implies then:

(2.10) $$a_{ijkh}X_{ij}X_{kh} > 0 \quad \forall X \text{ non zero} \in \mathbb{R}^9 \quad \text{with} \quad X_{ij} = X_{ji}.$$ [35]

We suppose also in the greater part of the applications that there exists a constant $\alpha > 0$ such that:

(2.11) $$a_{ijkh}X_{ij}X_{kh} \geqslant \alpha X_{ij}X_{ij} \quad \forall X \in \mathbb{R}^9 \quad \text{with} \quad X_{ij} = X_{ji} .$$

(coercivity inequality)[36].
In any case, (2.10) and (2.11) imply the invertibility of (2.7) and we have

(2.12) $$\varepsilon_{ij}(\mathbf{u}) = A_{ijkh}\sigma_{kh} ;$$

the A_{ijkh} called *coefficients of compliance* have the same properties as the a_{ijkh} (with the possibility of choosing the same α for coercivity).
As in the case of a viscous fluid (§1), we have to formulate the conservation of mass and of momentum; but here the linearisation leads to the following simplified equations:

(2.13) $$\frac{\rho - \rho_0}{\rho_0} = -\operatorname{div}\mathbf{u} \quad \text{on} \quad \Omega_0 \times \mathbb{R}^+ ,$$

(2.14) $$\rho_0 \frac{\partial^2 u_i}{\partial t^2} = \sigma_{ij,\,j} + f_i \quad \text{on} \quad \Omega_0 \times \mathbb{R}^+ ,$$

($u = \{u_i\}$ now denoting the displacement and not the velocity as in the Navier–Stokes equation), ρ and ρ_0 denoting the densities respectively in the

---

[35] See Appendix "Mechanics", §2.
[36] Coercivity will be studied later.

deformed and undeformed states; ρ_0 is in principle known, so the relation (2.13) serves only to determine ρ when $\mathbf{U}$ has been determined by the other relations; in fact, by the hypothesis of small perturbations, div $\mathbf{U}$ is small compared with unity with the result that $\rho \simeq \rho_0$. For this reason, the relation (2.13) is usually not included in the set for the solution of a problem in elasticity under the hypothesis of small perturbations.

It is important to insist again on the fact that linearisation enables us to apply these relations in the domain Ω_0 occupied by the *medium in its undeformed state* instead of having to apply them in Ω_t (the configuration at the instant t) which is, in fact, an unknown of the problem. In the same way, the *boundary conditions* can be applied to the boundary Γ_0 of Ω_0 (instead of Γ_t)[37].

If Γ_{F_k} denotes the part of Γ on which the component F_k of the *external surface forces* is given, then we must have

(2.15) $$\sigma_{kj} n_j = F_k \quad \text{on} \quad \Gamma_{F_k} \quad (\mathbf{n}, \text{ unit exterior normal to } \Gamma)\,.$$

If $\Gamma_{\bar{U}_k}$ denotes the part of Γ on which the *component* $\bar{U}_k$ *of the displacement* is prescribed, we obviously must write

(2.16) $$u_k = \bar{U}_k \quad \text{on} \quad \Gamma_{\bar{U}_k}.$$

The problem is said to be *regular* if

(2.17) $$\Gamma_{F_{\underline{k}}} \cap \Gamma_{\bar{U}_{\underline{k}}} = \varnothing \quad \Gamma_{F_{\underline{k}}} \cup \Gamma_{\bar{U}_{\underline{k}}} = \Gamma\,, \quad \forall k = 1, 2, 3^{38},$$

i.e. if *at each point of Γ, we have effectively three given complementary components of displacement or of surface force.*

The relations (2.5), (2.7), (2.13), (2.14), (2.15) and (2.16) finally have to be completed by *initial conditions* of the type

(2.18) $$\rho(\mathbf{x}, 0) \equiv \rho_0(\mathbf{x})\,,$$

(2.19) $$\mathbf{u}(\mathbf{x}, 0) = \mathbf{u}^0(\mathbf{x})\,; \quad \frac{\partial \mathbf{u}}{\partial t}(\mathbf{x}, 0) = \mathbf{u}^1(\mathbf{x})\,.$$

Remark 1. Replacing σ_{ij} in (2.14) by its expression (2.7), we obtain the following system of partial differential equations for the components u_i of $\mathbf{u}$:

(2.20) $$\rho_0 \frac{\partial^2 u_i}{\partial t^2} - \frac{\partial}{\partial x_j}\left(a_{ijkh} \frac{\partial u_k}{\partial x_h}\right) = f_i \quad i = 1, 2, 3 \quad x \in \Omega \subset \mathbb{R}^3 \quad t \in \mathbb{R}^+\,,$$

strongly coupled in $\mathbf{u}$[39], but independent of ρ and σ_{ij}; thus in the case in which the displacements are completely given on the whole boundary (i.e. since the problem is

---

[37] In what follows, we denote Ω_0 and Γ_0 more simply by Ω and Γ (remembering that they refer to a fixed domain and its boundary).

[38] We shall underline two repeated indices on each occasion when we wish to emphasise that there is no summation with respect to these two indices.

[39] We here give the name coupled differential system to one such that the derivatives $\partial u_i/\partial t$, $\partial^2 u_i/\partial t^2$ do not depend on u_i; it is said to be strongly coupled if these derivatives depend *a priori* on all the other components.

supposed regular $\Gamma_{F_k} = \emptyset, \forall k = 1, 2, 3$) the boundary condition:

$$u_j = \bar{U}_j \quad \forall j, \text{ on } \Gamma$$

is sufficient to determine directly the solution $\mathbf{u}$ from which we deduce ρ and σ_{ij} by (2.13), (2.5) and (2.7). □

3. Isotropic Linear Elasticity (or Classical Elasticity)

We speak of *isotropic behaviour* of a medium, when the latter presents no preferred directions; this can be expressed concretely by the fact that every principal reference frame for the strain tensor is likewise one for the stress tensor. Within the framework of linear elasticity, the symmetry of these two tensors allows us to apply the *theorem of isotropic functions*[40] with the result that the relation bet ween $\bar{\bar{\Sigma}}$ and $\bar{\bar{\varepsilon}}$ is necessarily of the form:

$$(2.21) \qquad \bar{\bar{\Sigma}} = g_0(\varepsilon_I, \varepsilon_{II}, \varepsilon_{III})\bar{\bar{\mathbb{I}}} + g_1(\varepsilon_I, \varepsilon_{II}, \varepsilon_{III})\bar{\bar{\varepsilon}} + g_2(\varepsilon_I, \varepsilon_{II}, \varepsilon_{III})\bar{\bar{\varepsilon}}^2$$

where $\varepsilon_I, \varepsilon_{II}, \varepsilon_{III}$ are the elementary invariants of $\bar{\bar{\varepsilon}}$, namely:

$$\varepsilon_I = \text{trace } \bar{\bar{\varepsilon}} = \varepsilon_{kk}\,; \quad \varepsilon_{II} = \frac{1}{2}(\varepsilon_{ii}\varepsilon_{jj} - \varepsilon_{ij}\varepsilon_{ij})\,; \quad \varepsilon_{III} = \det \bar{\bar{\varepsilon}}\,,$$

and the g_i are scalar-valued functions of these invariants.
The combination of (2.7), (2.8), (2.9) and (2.21) leads finally to:

$$(2.22) \qquad a_{ijkh} = \lambda\delta_{ij}\delta_{kh} + \mu(\delta_{ik}\delta_{jh} + \delta_{ih}\delta_{jk})$$[41]

which reduces to two the number of coefficients involved in isotropic elasticity: λ and μ are called *Lamé's coefficients*[42]. The relation (2.7) which, in fact is equivalent to six scalar relations, can then be written in the form of *Hooke's law*:

$$(2.23) \qquad \sigma_{ij} = \lambda\varepsilon_{kk}(\mathbf{u})\delta_{ij} + 2\mu\varepsilon_{ij}(\mathbf{u})$$

or, introducing the spherical parts $\bar{\bar{\Sigma}}^S$, $\bar{\bar{\varepsilon}}^S$ and the deviatoric parts $\bar{\bar{\Sigma}}^D$, $\bar{\bar{\varepsilon}}^D$ of tensors $\bar{\bar{\Sigma}}$ and $\bar{\bar{\varepsilon}}$ we have[43]:

$$(2.24) \qquad \sigma^s = (3\lambda + 2\mu)\varepsilon^s = 3K\varepsilon^s = -3K\frac{\rho - \rho_0}{\rho_0}$$

$$(2.25) \qquad \sigma^D_{ij} = 2\mu\varepsilon^D_{ij}\,.$$

We can interpret K and μ physically as respectively the *bulk modulus* (or the *compression modulus*) and the *rigidity modulus* at each point.

---

[40] For example, see Germain [2], p. 307.

[41] δ_{ij} is the Kronecker symbol defined by $\delta_{ij} = \begin{cases} 1 & \text{if } i = j \\ 0 & \text{if } i \neq j\,. \end{cases}$

[42] λ and μ can *a priori* depend on $\mathbf{x}$; if they do not they are called *Lamé's constants*.

[43] Every second order tensor $\bar{\bar{\mathbb{T}}}$ can be decomposed, in a unique manner, into its spherical part $\bar{\bar{\mathbb{T}}}^s = t^s\bar{\bar{\mathbb{I}}}$, where $t^s = \frac{1}{3}t_{kk}$ and $\mathbb{I}$ is the unit tensor, and its deviatoric part (or deviator) $\mathbb{T}^D = \mathbb{T} - \mathbb{T}^S$. (See Germain [2]). Notice that a deviator is a tensor with null trace.

With (2.24), the relations (2.23) are easily inverted to give

$$\varepsilon_{ij}(u) = \frac{1+\nu}{E}\sigma_{ij} - \frac{\nu}{E}\sigma_{kk}\,\delta_{ij}\,, \tag{2.26}$$

after we have introduced

$$E = \frac{3K\mu}{\lambda+\mu} \quad (\textit{Young's modulus}) \tag{2.27}$$

and

$$\nu = \frac{\lambda}{2(\lambda+\mu)} \quad (\textit{Poisson's ratio}) \tag{2.28}$$

The physical interpretation of simple problems of the type of loading by a hydrostatic pressure or by simple traction leads to the following inequalities:

$$K \geqslant 0\,,\quad E \geqslant 0\,,\quad \mu \geqslant 0 \quad \text{and} \quad 0 \leqslant \nu \leqslant \frac{1}{2} \quad \text{and so} \quad \lambda \geqslant 0\,. \tag{2.29}$$

Remark 2. *Homogeneous material.* In the case of a *homogeneous* material the various coefficients introduced above are constant. In this case, equation (2.20) can be written in vector form in $\Omega \times \mathbb{R}^+$ in one of two equivalent ways:

$$\rho_0 \frac{\partial^2 \mathbf{U}}{\partial t^2} - \mu\Delta\mathbf{U} - (\lambda+\mu)\,\mathrm{grad}\,\mathrm{div}\,\mathbf{U} = \mathbf{f}\,. \tag{2.30}$$

$$\rho_0 \frac{\partial^2 \mathbf{U}}{\partial t^2} - (\lambda+2\mu)\,\mathrm{grad}\,\mathrm{div}\,\mathbf{U} + \mu\,\mathrm{curl}(\mathrm{curl}\,\mathbf{U}) = \mathbf{f}\,. \tag{2.31}$$

□

4. Stationary Problems in Classical Elasticity

4.1. Definition of Stationary Problems in Solid Mechanics and Classification of Regular Problems

The mathematical term “stationary” does not have the same significance in the mechanics of solids as it has in fluid mechanics. In the fluids case the principal kinematic unknown is the velocity; as we have seen (§1.5), the word “stationary” does not signify the absence of motion (*i.e.* static), but *independence of the velocity with respect to the time.* In the case of solids, the principal kinematic unknown is, in general, the displacement, and the word “stationary” means “equilibrium”: we then have “*elastostatics*” (which is an approximation since a displacement occurs with a lapse of time) or *motion sufficiently slow* for each configuration to be able to be considered as in “equilibrium”; we then have “*quasi-statical elasticity.*” In both cases, the acceleration term disappears and the system (2.14) which expresses the conservation of momentum is reduced to:

$$\sigma_{ij,j} + f_i = 0\,,\quad x \in \Omega \subset \mathbb{R}^3 \quad i = 1, 2, 3 \tag{2.32}$$

which leads clearly to a “stationary” problem in the mathematical sense.

We distinguish three kinds of stationary problems in classical elasticity.
(*A*) *Problems of type I.* The data are $\mathbf{f}$ in Ω and $\mathbf{U}$ on Γ; it is then often indicated to take $\mathbf{U}$ as the principal unknown when the equations of the problem (2.30) are *Navier's equations*

$$(2.33) \qquad -\mu\Delta\mathbf{U} - (\lambda + \mu)\,\mathrm{grad\,div}\,\mathbf{U} = \mathbf{f} \quad \mathbf{x} \in \Omega \subset \mathbb{R}^3$$

or their equivalent (see (2.31)):

$$(2.33)' \qquad -(\lambda + 2\mu)\,\mathrm{grad\,div}\,\mathbf{U} + \mu\,\mathrm{curl}(\mathrm{curl}\,\mathbf{U}) = \mathbf{f} \quad \text{in} \quad \Omega$$

and the *boundary conditions*:

$$(2.34) \qquad \mathbf{U}(\mathbf{x}) = \bar{\mathbf{U}}(\mathbf{x}) \quad \mathbf{x} \in \Gamma\,,$$

$\bar{\mathbf{U}}$ being given on Γ. Starting from a solution of (2.33), (2.34) or (2.33)′, (2.34) we determine the stress field by applying successively (2.5) and (2.23).
(*B*) *Problems of type II.* Here the data are $\mathbf{f}$ in Ω and $\mathbf{F}$ on Γ. We then naturally take the stress field $\bar{\bar{\Sigma}}$ as the principal unknown; the equations of the problem are then *a priori*: *the equations of equilibrium*:

$$\left.\begin{array}{ll} (2.35) & \sigma_{ij,j} + f_i = 0 \\ (2.36) & \sigma_{ij} = \sigma_{ji} \end{array}\right\} \quad \text{in} \quad \Omega\,,$$

and *the boundary conditions*

$$(2.37) \qquad \sigma_{ij} n_j = F_i \quad \text{on} \quad \Gamma\,.$$

These relations are insufficient to determine the six unknowns σ_{ij}, they determine in fact the set of *statically admissible* stress fields; with each of these fields we can, by the relation (2.26) associate six scalar functions ε_{ij}; but we must then ensure the integrability of the system:

$$(2.38) \qquad u_{i,j} + u_{j,i} = 2\varepsilon_{ij}$$

capable of determining the displacement field $\mathbf{U}$, which can be expressed in terms of ε_{ij} by the (necessary and sufficient[44]) compatibility conditions

$$(2.39) \qquad \Delta\varepsilon_{ij} + \mathscr{E}_{,ij} - (\varepsilon_{ik,kj} + \varepsilon_{jk,ki}) = 0\,, \quad \text{with} \quad \mathscr{E} = \varepsilon_{kk}\,.$$

Taking account of (2.26) and (2.35), we can write these six compatibility relations in terms of the stress components in the form of *Beltrami's equations*

$$(2.40) \qquad \Delta\sigma_{ij} + \frac{1}{1+\nu}\Sigma_{,ij} + \frac{\nu}{1-\nu}\delta_{ij} f_{k,k} + f_{i,j} + f_{j,i} = 0\,,$$

$$\text{with} \quad \Sigma \equiv \sigma_{kk}, \quad \mathbf{x} \in \Omega \subset \mathbb{R}^3\,.$$

We verify this time that the number of equations (2.35), (2.40) (three plus six in the case $\Omega \subset \mathbb{R}^3$) is greater than the number of unknown σ_{ij} (six in the case of $\mathbb{R}^3$ when the symmetry is taken into account), but account must be taken of the fact that $\mathbf{f}$

[44] See, for example, Germain [1], Chap. V.

and $\mathbf{F}$ cannot be arbitrary: a necessary condition for the existence of a state of equilibrium is that the set of external forces applied to Ω must be self-equilibrating, which leads to their resultant wrench being zero[45] and this can be expressed by the two vector equations

$$(2.41)\qquad \begin{cases} \displaystyle\int_{\Omega} \mathbf{f}(M)\mathrm{d}M + \int_{\Gamma} \mathbf{F}(P)\mathrm{d}P = \mathbf{0}\,, \\ \displaystyle\int_{\Omega} \overline{OM} \wedge \mathbf{f}(M)\mathrm{d}M + \int_{\Gamma} \overline{OP} \wedge \mathbf{F}(P)\mathrm{d}P = \mathbf{0}\,. \end{cases}$$

Thus, for this type of problem, the data $\mathbf{f}$, $\mathbf{F}$ and Ω must *a priori satisfy* (2.41); the stress field $\underline{\underline{\Sigma}}$ is theoretically obtained from (2.36), (2.35), (2.40) and (2.37) and the displacement field $\mathbf{U}$ is obtained (to within a translation and a global rotation of Ω)[46] by integration of (2.38) with account being taken of (2.26).

(C) Problems of type III. In this case the data are: $\mathbf{f}$ in Ω, F_k on a part Γ_{F_k} of Γ and U_j on the complementary part Γ_{u_j} of Γ, the word complementary being taken in the sense of the conditions (2.17) of a regular problem; in this case $\underline{\underline{\Sigma}}$ and $\mathbf{U}$ cannot be determined independently, the one from the other.

4.2. Simplification Elements Used in the Solution of Certain Problems

The preceding developments show that, even in stationary cases the equations of elasticity present a complexity such that, in general, it is out of the question to find directly an explicit solution. We therefore rely on physical considerations and uniqueness results to look for particular solutions which will be established *a posteriori* to be the effective solution or a very acceptable approximate solution. In particular, we often try to reduce the problem to that of finding a single scalar function.

4.2.1. Saint–Venant's Principle. A first simplification device consists of allowing a certain flexibility in formulating the boundary conditions; for that, we base our argument on Saint–Venant's principle which can be stated as follows: "*If we replace a first distribution of given forces* $\mathbf{F}(P)$ *acting on a part* Γ_1 *of the boundary* Γ *by a second, these two distributions forming equal wrenches*[47] *(the other conditions on the complement of* Γ_1 *in* Γ *remaining unchanged), then in every region of* Ω *sufficiently far from* Γ_1*, the stress and displacement fields are practically unchanged.*" This principle (all the better verified if the dimensions of Γ_1 are small and the points of Ω at which we compare the two solutions are far from Γ_1) gives best results if applied to slender bodies like beams and if it is possible to choose the simplest possible distributions of surface forces $\mathbf{F}$ on the end faces of the boundary.

---

[45] See Appendix "Mechanics" §1.4.
[46] See Appendix "Mechanics", §2.
[47] See Appendix "Mechanics", §1.4.

4.2.2. Plane Strain Problems. The nature of the data (form of domain Ω, **f**, **F** and **U**) sometimes leads to the search for a *displacement field parallel to a plane*, for example with $U_3 = 0$, and for which U_1 and U_2 *only depend on* x_1 *and* x_2. We then have $\varepsilon_{13} = \varepsilon_{23} = \varepsilon_{33} = 0$, i.e. the strain tensor is plane. In this case

$$\sigma_{ij,3} = 0\,, \quad \sigma_{13} = \sigma_{23} = 0\,, \quad \sigma_{33} = \nu(\sigma_{11} + \sigma_{22})$$

If, in addition, **f** is derived from a potential $\mathscr{U}(x_1, x_2)$, $\mathbf{f} = \operatorname{grad} \mathscr{U}$, the first two equations of equilibrium (2.35) leads to the assumption of three scalar functions $\varphi(x_1, x_2)$, $\psi(x_1, x_2)$ and $\chi(x_1, x_2)$ such that:

$$(\sigma_{11} + \mathscr{U})\,dx_2 - \sigma_{12}\,dx_1 = d\varphi$$

$$-\sigma_{12}\,dx_2 + (\sigma_{22} + \mathscr{U})\,dx_1 = d\psi$$

with

$$\psi\,dx_1 + \varphi\,dx_2 = d\chi\,.$$

(The third of the equilibrium equations is satisfied).
Thus, the stress field is therefore given in terms of a single function χ (called the *Airy stress function*) defined to within an affine function by the relations

$$\text{(2.42)} \qquad \begin{cases} \sigma_{13} = \sigma_{23} = 0\,; & \sigma_{33} = \nu(\sigma_{11} + \sigma_{22}) \\ \sigma_{11} = \chi_{,22} - \mathscr{U}\,; & \sigma_{12} = -\chi_{,12}\,; \quad \sigma_{22} = \chi_{,11} - \mathscr{U}\,. \end{cases}$$

We verify then that the Beltrami equations (2.40) can only be satisfied if and only if:

$$\text{(2.43)} \qquad \Delta(\Delta\chi) = \frac{1 - 2\nu}{1 - \nu}\,\Delta\mathscr{U} \quad \text{on} \quad \Omega$$

(Ω denotes here the intersection of the (x_1, x_2) plane with the domain occupied by the material).
The boundary conditions relative to the stresses, in the framework of a problem of type III can be written finally:

$$\text{(2.44)} \qquad \begin{cases} \chi_{,22}n_1 - \chi_{,12}n_2 = F_1 + \mathscr{U}n_1 & \text{on} \quad \partial\Omega_{F_1} = \Gamma_{F_1} \cap \mathbb{R}^2_{(x_1, x_2)} \\ -\chi_{,12}n_1 + \chi_{,11}n_2 = F_2 + \mathscr{U}n_2 & \text{on} \quad \partial\Omega_{F_2} = \Gamma_{F_2} \cap \mathbb{R}^2_{(x_1, x_2)}\,. \end{cases}$$

4.2.3. Plane Stress Problems. As previously, physical reasons can lead to a search for a tensor of plane stresses as the solution of the given problem, i.e. a tensor such that

$$\text{(2.45)} \qquad \sigma_{13} = \sigma_{23} = \sigma_{33} = 0 \quad \text{and} \quad \sigma_{ij,3} = 0 \quad \text{for} \quad i, j = 1 \quad \text{or} \quad 2\,.$$

Exactly as before, the equilibrium equations lead us to introduce a scalar function $\chi(x_1, x_2)$ connected with σ_{ij} as in (2.42) apart from the fact that here $\sigma_{33} = 0$. We can easily show that Beltrami's equations can only be satisfied in the case in which

$$\Delta\mathscr{U} = 0 \quad (\operatorname{div} \mathbf{f} = 0)$$

and if and only if:

$$\text{(2.46)} \qquad \Delta\chi = 2\mathscr{U} + a_0 + a_1 x_1 + a_2 x_2\,,$$

χ remaining subject to the two conditions (2.44). We note that the displacement field is not, in general, a plane field.

4.2.4. Problems with Spherical or Cylindrical Symmetry. In the case where the data of the problem lets us foresee a symmetry of revolution of the solution (e.g., the equilibrium of a spherical chamber subject to uniform internal and external pressures), we look, in general for a displacement field of the form:

$$\mathbf{U}(\mathbf{x}) = u(r)\mathbf{e}_r$$

r being the first coordinate in the spherical (or cylindrical) system and $\mathbf{e}_r$ the corresponding unit vector. The principal unknown is the scalar function u and Navier's equations are considerably simplified.

5. Dynamical Problems in Classical Elasticity

5.1. General Dynamical Equations in Linear Elasticity

In the case in which the reference state of an elastic medium undergoes at a given instant (chosen as the initial instant) a small perturbation or when the data **f**, **F**, and **U** vary slightly (but eventually rapidly) with the time, it is necessary to take the acceleration term into account, but the hypotheses of linear elasticity are still in force. The displacement field is then given by (2.20) and it is necessary to take account of initial conditions.
If we suppose the medium to be occupying the whole space, we have to solve a system of the type:

$$(2.47)\quad \begin{cases} \rho_0 \dfrac{\partial^2 u_i}{\partial t^2} - \dfrac{\partial}{\partial x_j}\left(a_{ijkh} \dfrac{\partial u_k}{\partial x_h} \right) = f_i\,, \quad \mathbf{x} \in \mathbb{R}^3\,, \quad t \in \mathbb{R}\,, \\[2ex] \mathbf{U}(\mathbf{x}, 0) = \mathbf{U}^0(\mathbf{x})\,; \quad \dfrac{\partial \mathbf{U}}{\partial t}(\mathbf{x}, 0) = \mathbf{U}^1(\mathbf{x})\,, \quad \mathbf{x} \in \mathbb{R}^3\,. \end{cases}$$

The Cauchy problem so formulated is *strongly coupled*, i.e. each scalar relation involves all the components of the unknown **U**; we shall therefore look, in general, to simplify it in order to obtain more information *a priori* on the solution.
We shall suppose here that the medium is homogeneous and isotropic (§2.3) and write the system (2.47) in one of the two equivalent vector forms of Navier's equations (2.30) and (2.31):

$$(2.48)\quad \begin{cases} \text{(i)}\quad \rho_0 \dfrac{\partial^2 \mathbf{U}}{\partial t^2} - \mu \Delta \mathbf{U} - (\lambda + \mu)\operatorname{grad}(\operatorname{div}\mathbf{U}) = \mathbf{f}\,, \\[1ex] \text{or} \\[1ex] \text{(i)}'\quad \rho_0 \dfrac{\partial^2 \mathbf{U}}{\partial t^2} - (\lambda + 2\mu)\operatorname{grad}\operatorname{div}\mathbf{U} + \mu \operatorname{curl}(\operatorname{curl}\mathbf{U}) = \mathbf{f}\,, \\[1ex] \text{always with the initial conditions} \\[1ex] \text{(ii)}\quad \mathbf{U}(\mathbf{x}, 0) = \mathbf{U}^0(\mathbf{x}) \text{ and} \\[1ex] \text{(iii)}\quad \dfrac{\partial \mathbf{U}}{\partial t}(\mathbf{x}, 0) = \mathbf{U}^1(\mathbf{x})\,. \end{cases}$$

5.2. Dilatation Waves and Shear Waves

Here, **f** corresponds for most of the time to the gravity field, and is derived then from a potential $\mathscr{U}$

$$\mathbf{f} = \operatorname{grad} \mathscr{U} .$$

In fact **f** is often neglected in comparison with the other forces in play.
To obtain decoupled equations, we shall take as new unknowns[48]: the *dilatation* $\theta = \operatorname{div} \mathbf{U}$ and the "*rotation*" *vector* $\mathbf{w} = \frac{1}{2}\operatorname{curl} \mathbf{U}$. To obtain the decoupling, we apply successively the operators div and curl to the system (2.48) (i); taking account of the new hypotheses and initial conditions, we have:

$$\text{(2.49)} \quad \begin{cases} \rho_0 \dfrac{\partial^2 \theta}{\partial t^2} - (\lambda + 2\mu) \Delta \theta = \Delta \mathscr{U} , \\ \theta(\mathbf{x}, 0) = \theta^0(\mathbf{x}) \quad \text{with} \quad \theta^0 = \operatorname{div} \mathbf{U}^0 , \\ \dfrac{\partial \theta}{\partial t}(\mathbf{x}, 0) = \theta^1(\mathbf{x}) \quad \text{with} \quad \theta^1 = \operatorname{div} \mathbf{U}^1 , \end{cases}$$

and

$$\text{(2.50)} \quad \begin{cases} \rho_0 \dfrac{\partial^2 \mathbf{w}}{\partial t^2} - \mu \Delta \mathbf{w} = 0 \quad (\text{with } \operatorname{div} \mathbf{w} = 0) , \\ \mathbf{w}(\mathbf{x}, 0) = \mathbf{w}^0(\mathbf{x}) \quad \text{with} \quad \mathbf{w}^0 = \dfrac{1}{2} \operatorname{curl} U^0 , \quad \operatorname{div} \mathbf{w}^0 = 0 , \\ \dfrac{\partial \mathbf{w}}{\partial t}(\mathbf{x}, 0) = \mathbf{w}^1(\mathbf{x}) \quad \text{with} \quad \mathbf{w}^1 = \dfrac{1}{2} \operatorname{curl} U^1 , \quad \operatorname{div} \mathbf{w}^1 = 0 ; \end{cases}$$

θ and the components w_i of the rotation **w** are defined in a completely uncoupled fashion by the problems (2.49) and (2.50) which are both of the type

$$\text{(2.51)} \quad \begin{cases} \dfrac{1}{a^2} \dfrac{\partial^2 \psi}{\partial t^2} - \Delta \psi = g , \\ \psi(\mathbf{x}, 0) = \psi^0(\mathbf{x}) , \\ \dfrac{\partial \psi}{\partial t}(\mathbf{x}, 0) = \psi^1(\mathbf{x}) , \end{cases}$$

with:

$$\text{(2.52)} \quad \begin{cases} a = c_1 = \sqrt{\dfrac{\lambda + 2\mu}{\rho_0}} ; \quad \psi = \theta \text{ for the equation of the dilatation} \\ a = c_2 = \sqrt{\dfrac{\mu}{\rho_0}} ; \quad \psi = w_i \text{ for the equations of the rotation.} \end{cases}$$

---

[48] See Appendix "Mechanics", §2.

Finally, we can reduce (2.51) to the standard type of the wave equation by the change of variable $\xi_i = x_i/a$ and putting $u(\xi_i, t) = \psi(a\xi_i, t)$; we obtain

$$(2.53)\qquad \begin{cases} \dfrac{\partial^2 u}{\partial t^2} - \Delta u = f\,, \\ u(\xi, 0) = u^0(\xi)\,, \\ \dfrac{\partial u}{\partial t}(\xi, 0) = u^1(\xi)\,; \end{cases}$$

c_1 and c_2 are interpreted as the *speeds of propagation* respectively of the *dilatation* and of the *rotation.*

$\mathbf{w}$ and θ having been determined, we must next take up again consideration of the field $\mathbf{U}$. We can adopt a point of view suggested by the linearity of the problem (2.48) by considering $\mathbf{U}$ as the superposition of two fields $\mathbf{U}^{(1)}$ and $\mathbf{U}^{(2)}$: the first, $\mathbf{U}^{(1)}$ is irrotational ($\operatorname{curl}\mathbf{U}^{(1)} = \mathbf{0}$) which, from (2.48) (i)′ satisfies

$$(2.54)\qquad \rho_0 \frac{\partial^2 \mathbf{U}^{(1)}}{\partial t^2} - (\lambda + 2\mu)\Delta \mathbf{U}^{(1)} = \mathbf{f}$$

(always with the condition $\mathbf{f} = \operatorname{grad} \mathscr{U}$).

The second $\mathbf{U}^{(2)}$ is divergence free ($\operatorname{div}\mathbf{U}^{(2)} = 0$) which, from (2.48) (i), satisfies

$$(2.55)\qquad \rho_0 \frac{\partial^2 \mathbf{U}^{(2)}}{\partial t^2} - \mu \Delta \mathbf{U}^{(2)} = 0\,.$$

These equations are again decoupled; $\mathbf{U}^{(1)}$ represents a *pure dilatation* and $\mathbf{U}^{(2)}$ a *distortion.* We find again the speed c_1 and c_2 introduced previously.

5.3. Plane Waves (External Force Neglected)

If a perturbation is produced at a point of an isotropic elastic medium, this perturbation propagates outwards from this point in all directions and we can locally, at a great distance from the point at which the perturbation originated, consider these waves as plane; that means that, if we take as origin the point of the initial perturbation, and as direction Ox_1 at M that of the vector $\overrightarrow{OM}$, $\mathbf{U}$ can be considered to be independent of x_2 and x_3, from which by (2.48)(i) and (2.52) we have

$$(2.56)\qquad \begin{cases} \dfrac{1}{c_1^2}\dfrac{\partial^2 U_1}{\partial t^2} - \dfrac{\partial^2 U_1}{\partial x_1^2} = 0\,, \\ \dfrac{1}{c_2^2}\dfrac{\partial^2 U_2}{\partial t^2} - \dfrac{\partial^2 U_2}{\partial x_1^2} = 0\,, \\ \dfrac{1}{c_2^2}\dfrac{\partial^2 U_3}{\partial t^2} - \dfrac{\partial^2 U_3}{\partial x_1^2} = 0\,. \end{cases}$$

We see then that c_1 and c_2 can again be interpreted respectively as the *speeds of longitudinal and transverse waves.*

The system (2.56) to which must be added the initial conditions associated with the nature of the initial perturbation, is the typical model of the equation of one-dimensional waves.

5.4. Example with Time-Dependent Coefficients

It happens frequently that, the temperature field of an elastic solid being determined independently of the strain experienced by the material (which often provides a valuable approximation), the elastic coefficients and the density then become slowly varying functions of the time through the intermediary of the temperature; we are then to solve a problem of elasticity with time-dependent coefficients. This problem is static or dynamic, according to the nature of the external forces. In the most general (dynamical) case, the equations are then

$$
(2.57) \quad \begin{cases}
\rho(\mathbf{x}, t) \dfrac{\partial^2 u_i}{\partial t^2} - \dfrac{\partial}{\partial x_j}\left[a_{ijkh}(\mathbf{x}, t) \dfrac{\partial u_k}{\partial x_h} \right] = f_i(\mathbf{x}, t) \text{ in } \Omega , \\
u_j = U_j \quad \text{on} \quad \Gamma_{u_j} , \\
\sigma_{ij} = a_{ijkh}(\mathbf{x}, t) \dfrac{\partial u_k}{\partial x_h} \quad \text{in} \quad \Omega , \\
\sigma_{kj} n_j = F_k \quad \text{on} \quad \Gamma_{F_k} , \\
u_i(\mathbf{x}, 0) = u_i^0(\mathbf{x}) , \\
\dfrac{\partial u_i}{\partial t}(\mathbf{x}, 0) = u_i^1(\mathbf{x}) .
\end{cases}
$$

The quasi-static case is obtained, as usual, by neglecting the inertia terms.

5.5. Transverse Vibrations of an Elastic String

We consider a *perfectly elastic and flexible, homogeneous, heavy* string *held horizontally* between two fixed points A and B at a distance l_0 apart; at the instant $t = 0$, we draw the string lightly from its equilibrium position, then release it with a given small initial velocity. Supposing that this perturbation takes place in the *vertical plane* containing AB we study the subsequent motion of the string.
If x_1 and x_3 denote the coordinates at the instant t of the particle P of the string which was at $x_1 = a$ in the equilibrium position, we have:

$$x_i = \varphi_i(a, t) \quad i = \{1, 3\} . \quad a \in [0, l_0] \text{ is Lagrange's parameter}^{49} .$$

The *perfect flexibility* implies that the wrench of the forces exerted by the part PB of the string on the part PA is reduced to a simple tension $T(a, t)\boldsymbol{\tau}$, where $\boldsymbol{\tau}$ is the unit tangent vector at P. Under these conditions, *the fundamental principle of dynamics* is

[49] See Appendix "Mechanics", §2.1.

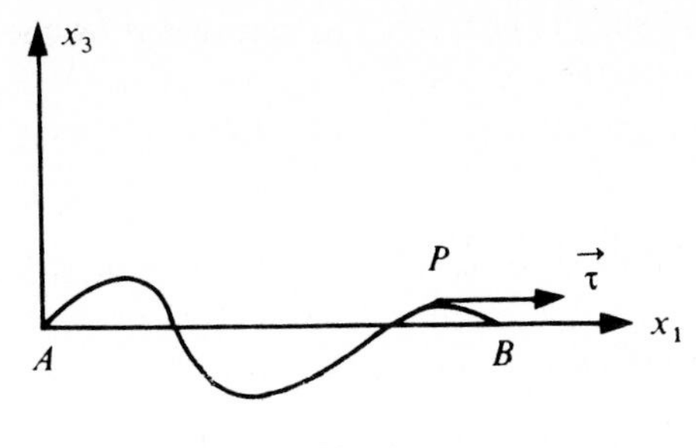

Fig. 4

expressed in Lagrange variables by the vector equation

(2.58) $$\frac{\partial}{\partial a}(T\boldsymbol{\tau}) - \rho_0 g \mathbf{k}_3 = \rho_0 \frac{\mathrm{d}^2 \overrightarrow{AP}}{\mathrm{d}t^2}$$ [50]

(g is the acceleration due to gravity, and ρ_0 is the *linear density* (mass per unit length) of the string before perturbation).
In components, we have

(2.59) $$\frac{\partial}{\partial a}\left[T\frac{\partial x_1}{\partial a}\Big/\sqrt{\left(\frac{\partial x_1}{\partial a}\right)^2 + \left(\frac{\partial x_3}{\partial a}\right)^2}\,\right] = \rho_0 \frac{\partial^2 x_1}{\partial t^2}$$

(2.60) $$\frac{\partial}{\partial a}\left[T\frac{\partial x_3}{\partial a}\Big/\sqrt{\left(\frac{\partial x_1}{\partial a}\right)^2 + \left(\frac{\partial x_3}{\partial a}\right)^2}\,\right] = \rho_0 \left[\frac{\partial^2 x_3}{\partial t^2} + g\right].$$

The *conservation of mass* is expressed by

(2.61) $$\frac{\rho_0}{\rho} = \sqrt{\left(\frac{\partial x_1}{\partial a}\right)^2 + \left(\frac{\partial x_3}{\partial a}\right)^2}.$$

$\rho(a, t)$ being the linear density at P [51].
The perfect elasticity implies that the tension at P is uniquely determined by the local variation of length

(2.62) $$e = \sqrt{\left(\frac{\partial x_1}{\partial a}\right)^2 + \left(\frac{\partial x_3}{\partial a}\right)^2} - 1\,;$$

hence, let

(2.63) $T(a, t) = \mathscr{T}[e(a, t)]$, where $\mathscr{T}$ is a given functional describing the mechanical behaviour of the string; we shall denote by T_0 the value $\mathscr{T}(0)$ of the tension in the original equilibrium position.

---

[50] This equation which is equivalent to (2.14) is introduced here as resting on the fundamental principle of mechanics, but it could have been introduced through the principle of virtual work: see in the Appendix "Mechanics", §3.5, the relations (3.33) which with the hypothesis of perfect flexibility ($\mathbf{M} \equiv \mathbf{0}$), reduces to the first vector equation in $\mathbf{T}(= T\boldsymbol{\tau})$.

[51] Equation equivalent to (2.13).

Finally the relations (2.59) to (2.63) must be completed by the *initial conditions*

$$(2.64)\quad \begin{cases} x_1(a,0) = a\,, & \dfrac{\partial x_1}{\partial t}(a,0) = 0\,, \\[2ex] x_3(a,0) = u_0(a)\,, & \dfrac{\partial x_3}{\partial t}(a,0) = u_1(a)\,, \end{cases}$$

and the *boundary conditions*

$$(2.65)\qquad x_1(0,t) = 0\,,\quad x_1(l_0,t) = l_0\,,\quad x_3(0,t) = x_3(l_0,t) = 0\,.$$

The problem thus formulated is difficult to solve so it is necessary to introduce the hypotheses of small perturbations justified by the nature of the initial conditions. Thus we suppose essentially:

(i) that the longitudinal displacement is negligible

$$x_1 = \varphi_1(a,t) \cong a \quad \frac{\partial x_1}{\partial t} \cong 0\,,$$

(ii) that the slope of the string with respect to the segment AB remains small; this is expressed by

$$\frac{\partial x_3}{\partial a} \cong \frac{\partial \varphi_3}{\partial a} \cong 0\,.$$

We can then put: $x_3 = u(x_1, t)$, u representing the transverse displacement of the particular abscissa $x_1 \simeq a$; as a consequence we have:

$$e = \frac{\partial x_1}{\partial a}\sqrt{1 + \left(\frac{\partial u}{\partial x_1}\right)^2} - 1 \cong 0$$

(small deformation). We show finally that the equations (2.59) and (2.60) can be replaced by the single equation

$$T_0\frac{\partial^2 u}{\partial x_1^2} - \rho_0\frac{\partial^2 u}{\partial t^2} = \rho_0 g\,;$$

(see Weinberger [1]).
It is enough then to put $C = \sqrt{(T_0/\rho_0)}$ to be led to the following problem (denoting x_1 by x):

$$(2.66)\quad \begin{cases} \text{(i)}\ \dfrac{\partial^2 u}{\partial t^2} - C^2\dfrac{\partial^2 u}{\partial x^2} = g\,, \\[2ex] \text{(ii)}\ u(x,0) = u_0(x)\,, \\[2ex] \text{(iii)}\ \dfrac{\partial u}{\partial t}(x,0) = u_1(x)\,, \\[2ex] \text{(iv)}\ u(0,t) = 0\,, \\[1ex] \text{(v)}\ u(l_0,t) = 0\,. \end{cases}$$

The function u describes the form of the string at each instant; and (2.61) serves finally only to determine $\rho(a, t)$ *a posteriori*. The mathematical model (2.66) is of the same simple type as (2.56).

5.6. Longitudinal Vibrations of an Elastic Rod

We consider a cylindrical, elastic rod with axis Ox, subjected to *tensions* or *imposed longitudinal contractions*, depending on the time (for example *displacement imposed at the two ends* of the rod through the intermediary of a press). In addition we suppose that these displacements are uniform over each end face with the result that each cross-section of the rod moves as a whole in the direction Ox; it is thus represented by an abscissa at each instant; a at the instant, $(0 \leqslant a \leqslant l_0)$ and $x = \varphi(a, t)$ at the subsequent instant $t > 0$. Here we neglect the transverse displacements and forces, especially the forces due to gravity. The only non-zero component of the displacement is therefore:

$$u(a, t) = \varphi(a, t) - a \tag{2.67}$$

and under these conditions, equation (2.58) above becomes

$$\frac{\partial T}{\partial a} = \rho_0 \frac{\partial^2 u}{\partial t^2} \, . \tag{2.68}$$

In effect $\boldsymbol{\tau} = \mathbf{k}$, unit vector along the Ox-axis.
If we adopt *Hooke's law of classical elasticity* (which assumes that the imposed displacement remains small)

$$T(a, t) = E(a) \frac{\partial u}{\partial a} \tag{2.69}$$

(E is the Young's modulus of the material and depends on a, as well as ρ if the rod is inhomogeneous).
Equation (2.68) can then be written

$$\frac{\partial}{\partial a}\left(E(a) \frac{\partial u}{\partial a} \right) - \rho(a) \frac{\partial^2 u}{\partial t^2} = 0 \, . \tag{2.70}$$

If we suppose that the rod is in equilibrium up to the instant $t = 0$ we have

$$u(a, 0) = 0 \, ; \quad \frac{\partial u}{\partial t}(a, 0) = 0 \, . \tag{2.71}$$

Assuming that the end $x = 0$ of the rod remains fixed, we have the boundary conditions

$$u(0, t) = 0 \, ; \quad u(l_0, t) = g(t) \tag{2.72}$$

This problem gives an example of the one-dimensional wave equation a little more general than (2.56) and (2.66).

In the case of a homogeneous bar both E and ρ are constant and we have the simple form as in the previous examples with

$$C = \sqrt{\frac{E}{\rho}}\,. \tag{2.73}$$

5.7. Vibrations of an Elastic Plate

5.7.1. Description of the Problem. We consider a thin elastic homogeneous plate, of constant thickness; by being clamped over part of the boundary it is maintained in a *horizontal direction*, over that part, the other part being free. This plate is heavy but *no force is applied to its surface*. The support to which it is rigidly attached *vibrates vertically*. We study the subsequent motion of the plate after the clamping. Let $\mathscr{V}$ be the domain occupied by the plate in its undeformed state. With respect to a suitably chosen orthonormal reference system, we can write

$$\mathscr{V} = \left\{\mathbf{x}\,;\ \mathbf{x} = (x_1, x_2, x_3) \in \mathbb{R}^3\,;\ (x_1, x_2) \in \Omega\,;\ -\frac{h}{2} < x_3 < \frac{h}{2}\right\}$$

where Ω is a bounded open set in $\mathbb{R}^2$, with relatively regular boundary Γ and where the thickness h *is supposed small* in comparison with the dimensions of Ω.

5.7.2. Equation of Motion in the Love–Kirchhoff Theory.[52] Being given that the imposed displacements (vibrations of the support) and the imposed forces (weight of the plate) are supposed to be of little importance, it is reasonable to base the posed problem on a linearised theory.

We, therefore, adopt Hooke's law of elastic behaviour, namely

$$\sigma_{ij} = \lambda\varepsilon_{ll}\delta_{ij} + 2\mu\varepsilon_{ij}\,, \tag{2.74}$$

where $\varepsilon_{ij}(u)$ is the linearised strain tensor introduced into the *Love–Kirchhoff* theory.

We assume, in addition, that the displacement $\mathbf{u}$ is such that

$$u_\alpha(x_1, x_2, 0, t) = 0 \quad \alpha = 1, 2^{53} \quad x_1, x_2 \in \Omega\,. \tag{2.75}$$

The principal unknown is then the *deflexion*[54]

$$u_3(\mathbf{x}, t) \equiv w(x_1, x_2, t) \quad \mathbf{x} \in \mathscr{V} \subset \mathbb{R}^3\,, \tag{2.76}$$

because $u_\alpha(\mathbf{x}, t) = -\,x_3 w_{,\alpha}(x_1, x_2, t)$; we have

$$\varepsilon_{ij} = 0 \quad \text{except} \quad \varepsilon_{\alpha\beta} = -\,x_3 w_{,\alpha\beta}{}^{53}\,, \tag{2.77}$$

and taking account of (2.74) and the relations linking the flexure and tension

---

[52] See Appendix "Mechanics", §3.6 and Duvaut–Lions [1].

[53] By convention, the Greek indices range only over the set $\{1, 2\}$.

[54] See Appendix "Mechanics", §3, relations (3.50).

tensors to the stress tensor[54] we obtain

$$(2.78)\quad \begin{cases} M_{\alpha\beta} = -\dfrac{h^3}{12}[\lambda . \Delta w . \delta_{\alpha\beta} + 2\mu w_{,\alpha\beta}] , \\ N_{\alpha\beta} = 0 , \\ Q_\alpha = 0 . \end{cases}$$

The equations of motion (3.61) and (3.63) and §3 of the Appendix "Mechanics" then reduces to

$$(2.79)\quad M_{\alpha\beta,\alpha\beta} - \rho g h = \rho h \frac{\partial^2 w}{\partial t^2} ; \quad (g, \text{ acceleration due to gravity})$$

(those in $N_{\alpha\beta}$ being satisfied identically) so that putting, as in (2.52)

$$(2.80)\quad a = \sqrt{\frac{\lambda + 2\mu}{\rho}} ,$$

$$(2.81)\quad \frac{\partial^2 w}{\partial t^2} + \frac{a^2 h^2}{12} \Delta(\Delta w) = -g .$$

5.7.3. Initial Conditions and Boundary Conditions. The vibrations commencing exactly at the (initial) instant $t = 0$ of the clamping we can write

$$(2.82)\quad w(x_1, x_2, 0) = \frac{\partial w}{\partial t}(x_1, x_2, 0) = 0$$

which implies that the plate has no previous deformation by its own weight.

(i) Let Γ_1 be the clamped part of Γ in the vibrating support; we can write the equality of the displacement and of the slope between the plate and the support to obtain

$$(2.83)\quad \begin{cases} w(x_1, x_2, t)|_{\Gamma_1} = \varphi(x_1, x_2, t) , \\ \left.\dfrac{\partial w}{\partial n}\right|_{\Gamma_1} = \psi(x_1, x_2, t) , \end{cases}$$

where $\varphi(x_1, x_2, t)$ represents the height from a datum line of the corresponding point of the support, and ψ the slope of the normal to Γ_1 on the support.

(ii) On Γ_2, the complement of Γ_1 in Γ, *which is free from force*, we have to take the imposed external forces $\bar{F}$ and $\bar{T}$ to be zero, which can be expressed (see Appendix "Mechanics", §3, (3.58), (3.59)) by

$$(2.84)\quad M_n = M_{\alpha\beta} n_\beta n_\alpha = 0 \quad \text{on} \quad \Gamma_2 ,$$

$$(2.85)\quad \frac{\partial M_\tau}{\partial \tau} + Q = \frac{\partial}{\partial \tau}[M_{\alpha\beta} n_\beta \varepsilon_{\alpha 3\gamma} n_\gamma] + M_{\alpha\beta,\beta} n_\alpha = 0 \quad \text{on} \quad \Gamma_2 .$$

5.7.4. Boundary Value Problem Associated with the Physical Situation Described. Putting

(2.86) $$\alpha^2 = \frac{a^2 h^2}{12}$$ [55]

we are finally led to solve, taking (2.78) into account, the following boundary value problem.

Find w defined on $\Omega \times]0, T[$ and satisfying:

(2.87)

$$\begin{cases} \text{(i)} \ \dfrac{\partial^2 w}{\partial t^2} + \alpha^2 \Delta(\Delta w) = -g & \text{on } \Omega \times]0, T[\,, \\[2ex] \text{(ii)} \ w(x_1, x_2, 0) = \dfrac{\partial w}{\partial t}(x_1, x_2, 0) = 0\,, & \text{on } \Omega\,, \\[2ex] \text{(iii)} \ w = \varphi \ \text{ and } \ \dfrac{\partial w}{\partial n} = \psi \ (\varphi \ \& \ \psi \text{ given}) & \text{on } \Gamma_1 \times]0, T[\,, \\[2ex] \text{(iv)} \ \Delta w + \dfrac{2\mu}{\lambda + 2\mu}\left[2 n_1 n_2 \dfrac{\partial^2 w}{\partial x_1 \partial x_2} - n_2^2 \dfrac{\partial^2 w}{\partial x_2^2} - n_1^2 \dfrac{\partial^2 w}{\partial x_1^2}\right] = 0 & \\ & \text{on } \Gamma_2 \times]0, T[\,, \\[2ex] \dfrac{\partial \Delta w}{\partial n} + \dfrac{2\mu}{\lambda + 2\mu}\dfrac{\partial}{\partial \tau}\left[n_1 n_2\left[\dfrac{\partial^2 w}{\partial x_2^2} - \dfrac{\partial^2 w}{\partial x_1^2}\right] + (n_2^2 - n_1^2)\dfrac{\partial^2 w}{\partial x_1 \partial x_2}\right] = 0 & \\ & \text{on } \Gamma_2 \times]0, T[\,. \end{cases}$$

5.7.5. Examples of Associated Stationary Problems. In the case in which the plate described previously is first deformed under the action of its own weight, it is reasonable to think that this deformation is realised in a "quasi-statical" fashion (see §2.4); we are led then to determine the corresponding deflexion w_0 from the solution of a problem of the type (2.87) from which the term $\dfrac{\partial^2 w}{\partial t^2}$ has disappeared as well as the initial conditions, namely

(2.88) $$-\Delta(\Delta w_0) = q \quad \text{with} \quad q = \frac{12 g}{a^2 h^2} > 0$$

with the statical clamping conditions on Γ_1, which comes down to replacing (2.87)

[55] Do not confuse this use of α with the previous use as an index.

(iii) by

$$\left\{\begin{aligned} & w_0(x_1, x_2)|_{\Gamma_1} = 0\,, \\ & \left.\frac{\partial w_0}{\partial n}\right|_{\Gamma_1} = 0\,, \end{aligned}\right. \tag{2.89}$$

(2.87) (iv) remains unchanged *a priori.*

We can also consider the case in which the plate is deformed by the action of a statical distribution of surface loads which, in the method of representing the plate by its middle surface plays a role equivalent to that of the load which previously we have integrated through the thickness; we are thus led to a similar problem.

6. Problems of Thermal Diffusion. Classical Thermoelasticity

6.1. General Laws of Thermoelasticity[56]

Until now we have been interested in the cases of elastic media undergoing changes in the course of which the dissipation of heat was zero; this hypothesis is justifiable in particular in all the cases in which we can neglect variations in temperature. *Thermoelastic behaviour* on the contrary, characterises a medium for which the *intrinsic dissipation is always identically zero* (a characteristic of elastic media) *but the thermal dissipation, in general, is different from zero.*

As we have noted in introducing elasticity, the natural thermodynamic variables are here $\mathbb{L}$, the Green–Lagrange stress tensor (2.2) and the temperature T, from which we have the practical utilisation of the free energy ψ depending on T and $L_{\alpha\beta}$ as a thermodynamic potential. Taking account of (1.51) and (1.56), the reversible power received in the course of a change is then:

$$\dot{w} = \frac{\partial \psi}{\partial L_{\alpha\beta}} \frac{\mathrm{d}L_{\alpha\beta}}{\mathrm{d}t}\,;^{57} \tag{2.90}$$

it is determined by all the possible changes of a particular medium when we know the six partial derivatives

$$P_{\alpha\beta} = \frac{\partial \psi}{\partial L_{\alpha\beta}}\,, \tag{2.91}$$

i.e. the *six equations of state* (in effect the components $L_{\alpha\beta}$ are symmetrical and the symmetry of the $P_{\alpha\beta}$ follows).

To obtain the complementary laws, we have to exploit, in particular the hypothesis of the nullity of the intrinsic dissipation, characteristic of elasticity; taking account

---

[56] See §2.1, §2.2 and §1.7.

[57] We have in effect $\mathrm{d}\psi = -s\,\mathrm{d}T + \dfrac{\partial \psi}{\partial L_{\alpha\beta}}\mathrm{d}L_{\alpha\beta}$.

of (1.53) and (2.90) we thus have:

(2.92) $$\Phi_1 = \sigma_{ij}\varepsilon_{ij}(\dot{\mathbf{u}}) - \rho \frac{\partial\psi}{\partial L_{\alpha\beta}} \frac{\mathrm{d}L_{\alpha\beta}}{\mathrm{d}t} = 0 \qquad \forall \mathbf{u}^{58} ;$$

but using (2.2), (2.1) and (2.6), we can show that

(2.93) $$\frac{\mathrm{d}L_{\alpha\beta}}{\mathrm{d}t} = F_{i\alpha}\varepsilon_{ij}(\dot{\mathbf{u}})F_{j\beta} \qquad F_{i\alpha}\frac{\partial\psi}{\partial L_{\alpha\beta}} = \frac{\partial\psi}{\partial F_{i\beta}} ;$$

thus (2.92) can be written

(2.94) $$\left(\sigma_{ij} - \rho \frac{\partial\psi}{\partial F_{i\alpha}} F_{j\alpha}\right)\varepsilon_{ij}(\dot{\mathbf{u}}) = 0 \qquad \forall \mathbf{u} ,$$

from which we have the expression

(2.95) $$\sigma_{ij} = \rho \frac{\partial\psi}{\partial F_{i\alpha}} F_{j\alpha} = \frac{\rho}{\rho_0} \frac{\partial W}{\partial F_{i\alpha}} F_{j\alpha} ,$$

introduced in (2.4) after having put

(2.96) $$W = \rho_0 \psi .$$

To the complementary laws (2.95), we have to add the *law of thermal dissipation* which is in general the *Fourier law*, already introduced in (1.62) namely

(2.97) $$q = - \bar{\bar{\mathbb{K}}}(T).\operatorname{grad} T^{59} .$$

6.2. Laws of State of Classical Thermoelasticity

We are concerned here, as in Sect. 2, with the hypothesis of small perturbations, which implies in particular that we can replace $\bar{\bar{\mathbb{L}}}$ by the tensor of linearised deformations with components

(2.98) $$\varepsilon_{ij}(u) = \frac{1}{2}(u_{i,j} + u_{j,i}) .$$

We also assume the existence of a natural state, free from constraints, with uniform temperature T_0 and with uniform specific entropy $s = 0$. We put $\theta = T - T_0$, this quantity also being supposed small. Recalling, finally, that under these assumptions the density can be considered as remaining constant to ρ_0, the free energy ψ is then a function of θ and ε_{ij} and

(2.99) $$\dot{w} = \frac{\partial\psi}{\partial\varepsilon_{ij}} \frac{\mathrm{d}\varepsilon_{ij}}{\mathrm{d}t} = \frac{\partial\psi}{\partial\varepsilon_{ij}} \varepsilon_{ij}(\dot{u}) ;$$

---

[58] We note that here **u** denotes the displacement field, whereas in (1.53) **u** denotes the velocity field.

[59] $\bar{\bar{\mathbb{K}}}$ is the heat conduction tensor of the medium.

thus, the vanishing of the intrinsic dissipation implies

(2.100) $$\sigma_{ij} = \rho_0 \frac{\partial \psi}{\partial \varepsilon_{ij}} = \frac{\partial}{\partial \varepsilon_{ij}}(\rho_0 \psi) = \frac{\partial W}{\partial \varepsilon_{ij}} .$$

In the *isothermal* case, we shall have (see (2.7)):

(2.101) $$\sigma_{ij} = a_{ijkh}\varepsilon_{kh}(u) \quad \text{and thus} \quad W(0, \varepsilon_{ij}) = \frac{1}{2} a_{ijkh}\varepsilon_{ij}(u)\varepsilon_{kh}(u) ,$$

always with the condition that $a_{ijkh} = a_{khij}$ whence the necessity of this hypothesis in (2.9) to preserve the existence of the strain energy W. The condition (2.10) having its origin in the principles of thermostatics expresses the need for the strain energy to be a positive definite quadratic form. In the non-isothermal case we write W again as a quadratic form in the variables ε_{ij} and θ and hence

(2.102) $$W(\theta, \varepsilon_{ij}) = \rho_0 \psi(\theta, \varepsilon_{ij}) = \frac{1}{2} a_{ijkh}\varepsilon_{ij}\varepsilon_{kh} + b_{kh}\varepsilon_{kh}\theta + b\theta^2 .$$

We then have

(2.103) $$\sigma_{ij} = \frac{\partial W}{\partial \varepsilon_{ij}} = a_{ijkh}\varepsilon_{kh} + b_{ij}\theta , \quad \text{and} \quad \rho_0 s = -\rho_0 \frac{\partial \psi}{\partial \theta} = -b_{kh}\varepsilon_{kh} - b\theta .$$

If, finally, we make the assumption of *isotropy* the theorem on isotropic functions[60] imposes the form (2.22) on a_{ijkh}, and, taking account of the necessary positivity on the entropy we see that this implies that

(2.104) $$b_{kh} = -3K\alpha\delta_{kh} \qquad b = -\beta ,$$

with the result that

(2.105) $$W(\theta, \varepsilon_{ij}) = \rho_0 \psi(\theta, \varepsilon_{ij}) = \frac{1}{2}(\lambda\varepsilon_{ii}\varepsilon_{jj} + 2\mu\varepsilon_{ij}\varepsilon_{ij}) - 3K\alpha\varepsilon_{ii}\theta - \beta\theta^2 ,$$

λ, μ, K are the coefficients already introduced into isothermal elasticity (relations (2.22) and (2.24)); α and β are two new positive coefficients whose interpretation will be given later.
The constitutive relations of classical thermoelasticity are thus

(2.106) $$\sigma_{ij} = \lambda\varepsilon_{kh}\delta_{ij} + 2\mu\varepsilon_{ij} - 3K\alpha\theta\delta_{ij} ;$$

the term $-3K\alpha\theta$ corresponding to the stresses of purely thermal origin. Inverting the relation (2.106) as in (2.26), we find

(2.107) $$\varepsilon_{ij} = \frac{1+\nu}{E}\sigma_{ij} - \frac{\nu}{E}\sigma_{kk}\delta_{ij} + \alpha\theta\delta_{ij}$$

with the result that α is readily interpreted as the *linear coefficient of dilatation*; if the body is free of stress, we obtain in effect

(2.108) $$\varepsilon_{ij} = \alpha\theta\delta_{ij} = \alpha(T - T_0)\delta_{ij} .$$

---

[60] See Germain [2], p. 307.

The coefficient of volume dilatation being 3α, the assumption of small perturbations implies here that α is independent of θ.
To interpret β, we note the specific heat at constant strain c_ε (quantity of reversible heat received per unit mass for a rise of one degree of temperature in the course of a transformation with constant strain) is given by

$$c_\varepsilon = \frac{\partial e}{\partial T}(T, \varepsilon_{ij}) = \frac{\partial}{\partial T}(\psi + sT) = T\frac{\partial s}{\partial T}(\theta, \varepsilon_{ij}) = \frac{\beta T}{\rho_0}, \tag{2.109}$$

account being taken of (2.103) and (2.104). Thus, since we are dealing with small perturbations, β can be considered to be given by the relation

$$\beta = \frac{\rho_0 c_\varepsilon}{T_0}. \tag{2.110}$$

6.3. General Equations of Classical Thermoelasticity

By substituting from (2.106) into the dynamical equations (2.14) we obtain first of all on the assumption that the medium is homogeneous:

$$\rho_0 \frac{\partial^2 \mathbf{U}}{\partial t^2} - \mu \Delta \mathbf{U} - (\lambda + \mu)\,\mathrm{grad}\,\mathrm{div}\,\mathbf{U} + 3K\alpha\,\mathrm{grad}\,\theta = \mathbf{f}. \tag{2.111}$$

We must next write the *energy equation*[61] (1.47) taking the relations (1.51) and (1.53) into account: these imply that

$$\rho\frac{\mathrm{d}e}{\mathrm{d}t} = \rho T\frac{\mathrm{d}s}{\mathrm{d}t} + \rho\dot{w} = \sigma_{ij}\varepsilon_{ij}(\dot{u}) - \Phi_1 \tag{2.112}$$

and from the fact that the intrinsic energy Φ_1 is zero we see then that the energy equation reduces to

$$\rho T\frac{\mathrm{d}s}{\mathrm{d}t} + \mathrm{div}\,q = r. \tag{2.113}$$

Finally, we must write *Fourier's law* (2.97) in the isotropic case, namely, since $K_{ij} = k\delta_{ij}$

$$\mathbf{q} = -k\,\mathrm{grad}\,\theta. \tag{2.114}$$

The assumptions of homogeneity and of small perturbations imply here that k is constant and that $\rho = \rho_0$.

[61] We have already seen in §2.2 that there was no reason to write the equation of conservation of mass. Moreover here, we have considered ρ to be the constant ρ_0. In addition, we recall that in §1, $\mathbf{U}$ represents the velocity field which here we denote by $\dot{\mathbf{U}}$.

On inserting into (2.113) the expressions (2.103), (2.104) and (2.110) which define s, as well as (2.114), after linearisation in $T\,(T \simeq T_0)$, we obtain the *heat equation*

(2.115) $$\rho_0 c_\varepsilon \frac{\partial \theta}{\partial t} - k \Delta \theta = 3 K \alpha T_0 \operatorname{div}\left(\frac{\partial \mathbf{U}}{\partial t}\right) + r\,.$$

The system (2.111), (2.115). to which we must add initial and boundary conditions, theoretically allows us to determine $\mathbf{U}$ and θ.
The boundary conditions and initial conditions on $\mathbf{U}$ are those already mentioned in elasticity §2.4.1 and §2.5.1. (In the case of problems of types II and III, we should look upon σ_{ij} as the principal unknowns.)
For the temperature θ, we allow here $\theta = 0$, as the initial condition, and take the boundary conditions to be of two types:

(i) if the temperature $\bar{T}$ is imposed on a part of $\partial\Omega$; we then have

(2.116) $$\theta = \bar{T} - T_0\,;$$

(ii) in the case in which the rate of surface heat (or the flux of heat across the boundary is prescribed), interpreting (1.49), we have in general a condition of the type

$$- q_i n_i = \varpi \quad \text{given}\,,$$

which, because of (2.114), becomes

(2.117) $$\frac{\partial \theta}{\partial n} = g \quad \text{given}\,.$$

Remark 3. In numerous cases, we can justify the neglect of the member on the right of (2.115); this equation is then reduced to the classical equation involving only θ; this function can therefore be determined independently of $\mathbf{U}$; putting it back into (2.111) we then have to deal with a simple problem in elasticity. □

Review

The use of linear elasticity to evaluate the resistance of materials is present in all the technical subjects. Certain problems of linear elasticity can be treated by the classical mathematical techniques of Chap. II. The solution of the more general stationary problems of elasticity can be obtained by variational methods (see Chap. VII, §2).
Problems which can be formulated as integral equations are treated in Chap. XIA, §4 and XIB, §4.
The numerical solution of these problems is discussed in Chap. XII which treats the theme of numerical methods.
Evolution problems of linear elasticity are treated by more and more powerful methods: the method of diagonalisation (Chap. XV, §4), of the Laplace transform

(Chap. XVI, §5), of semigroups (Chap. XVII, A, §6) and especially by variational methods (Chap. XVIII, §6). The numerical solution is looked at in §§3 and 5 of Chap. XX. The thermoelastic problem of Sect. 6 is studied by the variational methods of Chap. XVIII, §6.

§3. Linear Viscoelasticity

1. Introduction[62]

In the two general cases treated previously (viscous fluid in §1 and elasticity in §2), the stress tensor was in fact a function of the tensor chosen to describe the deformation and not a true functional; that is due to the very selective character of the memory of these media (infinitely short memory for fluids where the stress depends only on the strain velocity at the instant considered; excessive memory of a privileged state for elastic materials). The *viscoelastic materials*, on the contrary, are in general endowed with a "*continuous memory*" in this sense that the state of stress at the instant t depends on the *whole history of the strains* (or of the loads) experienced by the material at the previous instants.

As in the case of elasticity, *linear viscoelasticity* is formulated in the context of a thermal decoupling (isothermal changes) and of a *theory of small perturbations* in particular admitting only the intervention of the tensor of linearised strains and its derivatives with respect to the time; in addition, we assume that the law linking the history of the stresses to the history of the strains is *linear*[63].

A relatively large class of viscoelastic materials can be described by a constitutive law of "rate" type, that is to say:

$$\sigma_{ij} + \sum_{l=1}^{n_1} A_{ijkh}^{(l)} \frac{\partial^l \sigma_{kh}}{\partial t^l} = \sum_{l=0}^{n_2} a_{ijkh}^{(l)} \frac{\partial^l \varepsilon_{kh}}{\partial t^l} \quad (3.1)$$[63]

where the σ and ε are identically zero for $t < 0$ and where the coefficients $A^{(l)}$ and $a^{(l)}$ have to be determined experimentally, account being taken of certain necessary properties of the type (2.8) ... (2.11).

In the greater part of the applications, we are content with a law of the type:

$$\sigma_{ij} + A_{ijkh}\dot{\sigma}_{kh} = a_{ijkh}^{(0)}\varepsilon_{kh} + a_{ijkh}^{(1)}\dot{\varepsilon}_{kh} \quad (3.2)$$[64, 65]

This simplified model allows in effect to describe, relatively well, the behaviour of certain high polymers in *creep* (the evolution of the strains when beginning at the instant $t = 0$, we apply a constant load which we eventually remove at $t = t_1$) *and*

[62] See, for example: Gurtin–Sternberg [1], Duvault–Lions [1].

[63] See Freudenthal–Geiringer [1].

[64] Convention: $\dot{\varphi}$ denotes the derivatives of φ with respect to the time.

[65] We say that there is "viscosity" since the stress tensor depends on the rates of strain; hence the name "viscoelasticity".

in relaxation (the evolution of the stresses under imposed constant strain beginning at $t = 0$) of some high polymers.

2. Materials with Short Memory

When $A_{ijkh} \equiv 0$, the medium is said to have "*short memory*" since the state of stress at the instant t only depends on the strain at that instant and at the immediately preceding instants. The coefficients $a^{(0)}_{ijkh}$ and $a^{(1)}_{ijkh}$ play respectively the rôles of *coefficients of elasticity* and of *viscosity*; it is then natural and conforms to experience to suppose that there exists a positive constant α such that:

$$\left.\begin{array}{ll} (3.3) & a^{(0)}_{ijkh} X_{ij} X_{kh} \geqslant \alpha X_{ij} X_{ij} \\ \text{and} & \\ (3.4) & a^{(1)}_{ijkh} X_{ij} X_{kh} \geqslant \alpha X_{ij} X_{ij} \end{array}\right\} \quad \forall \mathbf{X} \in \mathbb{R}^9, \qquad X_{ij} = X_{ji} .$$

The justification of (3.3) is the same as in elasticity (§2.2); as for the inequality (3.4), it simply expresses that there exists a viscous dissipation of which the expression is:

$$a^{(1)}_{ijkh} \dot{\varepsilon}_{ij} \dot{\varepsilon}_{kh} ,$$

necessarily positive when (and only if) $\dot{\varepsilon}_{ij} \neq 0$ (the inequality (3.4) is in fact a little stronger).
Finally we have, as in elasticity and for similar reasons

$$a^{(l)}_{ijkh} = a^{(l)}_{jikh} = a^{(l)}_{khij} , \qquad l = 0, 1 . \tag{3.5}$$

Proceeding as for (2.20) in §2, we obtain the *equations of motion*:

$$\rho_0 \frac{\partial^2 u_i}{\partial t^2} - \frac{\partial}{\partial x_j}\left(a^{(0)}_{ijkh} \frac{\partial u_k}{\partial x_h} \right) - \frac{\partial}{\partial t} \frac{\partial}{\partial x_j}\left(a^{(1)}_{ijkh} \frac{\partial u_k}{\partial x_h} \right) = f_i \quad \text{in} \quad \Omega \times]0, T[. \tag{3.6}$$

The *boundary conditions* are again (with the same notation as in §2)

$$\sigma_{kj} n_j = F_k \quad \text{on} \quad \Gamma_{F_k} , \tag{3.7}$$

$$u_l = U_l \quad \text{on} \quad \Gamma_{U_l} , \tag{3.8}$$

and the initial conditions

$$u_i(\mathbf{x}, 0) = u_i^0(\mathbf{x}) ; \quad \frac{\partial u_i}{\partial t}(x, 0) = u_i^1(\mathbf{x}) . \tag{3.9}$$

3. Materials with Long Memory

The constitutive law (3.2) can also be written

$$\sigma_{ij}(t) = a_{ijkh} \varepsilon_{kh}(t) + \int_0^t b_{ijkh}(t - s) \varepsilon_{kh}(s) \, ds . \tag{3.10}$$

We see that the state of stresses at the instant t depends on the strain at the instant

t, but also on the strains at the instants previous to t, from which we have the qualification of "*long memory*".
The first term of the right hand side represents an *instantaneous elastic effect* and the coefficients a_{ijkh} satisfy:

$$(3.11)\qquad \begin{cases} a_{ijkh} \in L^{\infty}(\Omega)\,, \\ a_{ijkh} = a_{jikh} = a_{khij}\,, \\ a_{ijkh}X_{ij}X_{kh} \geqslant \alpha X_{ij}X_{ij}\,; \qquad \alpha > 0 \qquad \forall \mathbf{X} \in \mathbb{R}^9\,, \qquad X_{ij} = X_{ji}\,. \end{cases}$$

The coefficients b_{ijkh} take account of the memory effects of the material; in the usual practical examples, they are linear combinations of exponentials with negative exponents which expresses especially that the memory decreases very quickly with increasing time; these coefficients satisfy:

$$(3.12)\qquad \begin{cases} b_{ijkh}, \dfrac{\partial b_{ijkh}}{\partial t} \text{ and } \dfrac{\partial^2 b_{ijkh}}{\partial t^2} \in L^{\infty}(\Omega_T)\,; \quad \Omega_T = \Omega \times \,]0, T[\quad \Omega \subset \mathbb{R}^3\,, \\ b_{ijkh} = b_{jikh} = b_{khij}\,. \end{cases}$$

A viscoelastic material with long memory is said to be of *solid* type if:

$$(3.13)\qquad \int_0^{\infty} b_{ijkh}(t)\,\mathrm{d}t \overset{\text{def}}{=} b^{\infty}_{ijkh} \quad \text{is bounded}\,.$$

The coefficients

$$(3.14)\qquad a^{\infty}_{ijkh} \overset{\text{def}}{=} a_{ijkh} + b^{\infty}_{ijkh}$$

supposed to satisfy the properties (3.11) are called "*coefficients of retarded elasticity*", as we can show that under certain conditions, the solution $u(t)$ of a classical boundary value converges, when $t \to \infty$ to the solution of the elasticity problem corresponding to the coefficients a^{∞}_{ijkh}.
The equations of motion are here:

$$(3.15)\qquad \rho_0 \frac{\partial^2 u_i}{\partial t^2} - \frac{\partial}{\partial x_j}\left(a_{ijkh}\frac{\partial u_k}{\partial x_h}\right) - \int_0^t \frac{\partial}{\partial x_j}\left[b_{ijkh}(t-s)\frac{\partial u_k}{\partial x_h}\right]\mathrm{d}s = 0$$

with $\mathbf{x} \in \Omega \in \mathbb{R}^3$, $t \in \mathbb{R}$ and the boundary and initial conditions are the same as in (3.7), (3.8) and (3.9).

4. Particular Case of Isotropic Media[66]

When maintaining the preceding hypotheses of linearity, we confine ourselves to isotropic viscoelastic materials satisfying (3.2), in general we adopt one of the

[66] See Germain [2].

models described below following the particular characteristics of the medium studied.

4.1. Kelvin–Voigt Viscoelastic Medium or "Viscoelastic Solid"

This is a medium with *short memory* for which the law (3.2) (with $A_{ijkh} \equiv 0$ and the assumptions of isotropy), decomposed into a spherical part and a deviator (as in §2.3) allows us to write:

$$(3.16) \qquad \begin{cases} \sigma^s = 3K(\varepsilon^s + \theta_c \dot{\varepsilon}^s)\,, \\ \sigma^D_{ij} = 2\mu(\varepsilon^D_{ij} + \theta_g \dot{\varepsilon}^D_{ij})\,, \end{cases}$$

K, μ, θ_c, θ_g being positive coefficients eventually capable of depending on $\mathbf{x}$ but independent of t.
We can invert these relations by multiplying respectively by: $\exp(t/\theta_c)$ and $\exp(t/\theta_g)$. We then obtain

$$(3.17) \qquad \begin{cases} \varepsilon^s(t) = \displaystyle\int_0^t \frac{\sigma^s(\tau)}{3K\theta_c} \exp\left\{ -\frac{t-\tau}{\theta_c} \right\} d\tau\,, \\ \varepsilon^D_{ij}(t) = \displaystyle\int_0^t \frac{\sigma^D_{ij}(\tau)}{2\mu\theta_g} \exp\left\{ -\frac{t-\tau}{\theta_g} \right\} d\tau\,. \end{cases}$$

If we seek to describe a "creep" experiment, that is to say the one in which we impose a constant stress $\sigma_{ij}(0)$ beginning at the instant $t = 0$ we take

$$\sigma_{ij}(t) = Y(t)\sigma_{ij}(0)\,^{67}\,,$$

the strains are given by

$$(3.18) \qquad \begin{cases} \varepsilon^s(t) = \dfrac{\sigma^s(0)}{3K}\left[1 - \exp\left(-\dfrac{t}{\theta_g} \right) \right], \\ \varepsilon^D_{ij}(t) = \dfrac{\sigma^D_{ij}(0)}{2\mu}\left[1 - \exp\left(-\dfrac{t}{\theta_g} \right) \right], \end{cases}$$

and the coefficients K and μ can be interpreted as the coefficients of retarded elasticity corresponding to the behaviour when $t \to \infty$.
We notice that there is no instantaneous elasticity, that is to say no strain at $t = 0$. θ_c and θ_g are called respectively the *retardation time under compression and under shear.*
On the contrary, if we impose a constant strain $\varepsilon_{ij}(0)$ from the instant $t = 0$, (a "*relaxation*" experiment) we see from (3.16) that the response in the stress is purely elastic: the Kelvin–Voigt model does not exhibit the phenomenon of relaxation. It

[67] $Y(t)$ denotes the Heaviside function: $Y(t) = \begin{cases} 0 & \text{for } t < 0\,, \\ 1 & \text{for } t > 0\,. \end{cases}$

corresponds only to a first approximation to the behaviour in creep of a visco-elastic solid.

4.2. Maxwell's Viscoelastic Medium or "Viscoelastic Liquid"

This is a *particular case* of a material with long memory in the sense that we suppose that $a^{(0)}_{ijkh} \equiv 0$; the hypotheses of isotropy then lead to:

$$(3.19)\quad \begin{cases} \dot{\varepsilon}^s = \dfrac{1-2\nu}{E}\left[\dot{\sigma}^s + \dfrac{\sigma^s}{\tau_c}\right] & \text{with } \varepsilon^s(0) = \dfrac{1-2\nu}{E}\sigma^s(0)\,, \\[2ex] \dot{\varepsilon}^D_{ij} = \dfrac{1+\nu}{E}\left[\dot{\sigma}^D_{ij} + \dfrac{\sigma^D_{ij}}{\tau_g}\right] & \text{with } \varepsilon^D_{ij}(0) = \dfrac{1+\nu}{E}\sigma^D_{ij}(0)\,, \end{cases}$$

where, as previously, ν, E, τ_c, τ_g are positive coefficients which may depend on **x**, but are independent of t. (If τ_c and τ_g are infinitely large, we recover the laws of classical elasticity).

These relations can be inverted as in the preceding section and after some manipulation, we obtain:

$$(3.20)\quad \begin{cases} \sigma^s(t) = 3K\varepsilon^s(t) + \displaystyle\int_0^t \frac{\mathrm{d}}{\mathrm{d}\tau}\left\{3K\exp\left(-\frac{\tau}{\tau_c}\right)\right\}\varepsilon^s(t-\tau)\,\mathrm{d}\tau\,, \\[2ex] \sigma^D_{ij}(t) = 2\mu\varepsilon^D_{ij}(t) + \displaystyle\int_0^t \frac{\mathrm{d}}{\mathrm{d}\tau}\left\{2\mu\exp\left(-\frac{\tau}{\tau_g}\right)\right\}\varepsilon^D_{ij}(t-\tau)\,\mathrm{d}\tau\,, \end{cases}$$

if, as in elasticity, we have put:

$$(3.21)\quad \frac{1-2\nu}{E} = \frac{1}{3K}\,, \qquad \frac{1+\nu}{E} = \frac{1}{2\mu}\,.$$

A "*relaxation*" experiment with

$$\varepsilon_{ij}(t) = Y(t)\varepsilon_{ij}(0)$$

leads to the relations

$$(3.22)\quad \begin{cases} \sigma^s(t) = 3K\varepsilon^s(0)\exp\left(-\dfrac{t}{\tau_c}\right), \\[2ex] \sigma^D_{ij}(t) = 2\mu\varepsilon^D_{ij}\exp\left(-\dfrac{t}{\tau_g}\right), \end{cases}$$

and to the interpretation:

(i) the constants τ_c and τ_g as *relaxation times* corresponding respectively to the *cubical dilatation* and to the *sliding*,

(ii) the functions $K(\tau) = K\exp(-\tau/\tau_c)$ and $\mu(\tau) = \mu\exp(-\tau/\tau_g)$ as *relaxation* moduli corresponding to the *dilatation* and to the *sliding*.

Finally, the relaxation experiment shows that Maxwell's model exhibits a zero retarded elasticity (when $t \to \infty$).

In a "creep" experiment

$$(3.23)\qquad \begin{cases} \varepsilon^s(t) = \dfrac{1-2\nu}{E}\,\sigma^s(0)\left(1 + \dfrac{t}{\tau_c}\right), \\[2ex] \varepsilon^D_{ij}(t) = \dfrac{1+\nu}{E}\,\sigma^D_{ij}(0)\left(1 + \dfrac{t}{\tau_g}\right), \end{cases}$$

there are *two instantaneous compliances*: $(1 - 2\nu)/E$ in dilatation and $(1 + \nu)/E$ in sliding and a creep linear in t.

Maxwell's model is in fact a first approximation to the behaviour of a "viscoelastic liquid".

4.3. More General Model with Three Parameters

The preceding models can be illustrated by the following simple "rheological" schemes, analogic systems which let us simulate experimentally the constitutive laws. In Fig. 5 we show the schemes appropriate to the Maxwell and Kelvin–Voigt models; each of them contains a spring of stiffness G_e (equivalent to $3K$ or 2μ) and a dash-pot of viscosity η (equivalent to $3K\theta_c$ and $2\mu\theta_g$ in the Kelvin–Voigt model, or $3K\tau_c$ and $2\mu\tau_g$ in the Maxwell model). However, we have seen that these models only constitute first approximations to real viscoelastic media. We can improve these models by combining such simple elements. The model called "*the model with three parameters*" with long memory presents a good generalisation; it is based on the factual observation that the relaxation on the spherical parts is often negligible and as a result τ_c is infinite; we therefore write

$$(3.24)\qquad \sigma^s = 3K\varepsilon^s$$

and, on the contrary, we refine the description of the relaxation phenomenon in

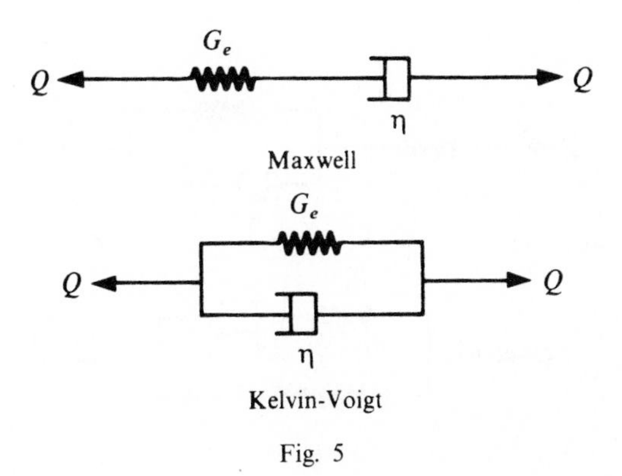

Fig. 5

sliding by putting *a priori*

$$\dot{\sigma}_{ij}^D + a\sigma_{ij}^D = b\dot{\varepsilon}_{ij}^D + c\varepsilon_{ij}^D, \tag{3.25}$$

which can be represented by either of the two schemes shown in Fig. 6. The physical interpretation of the coefficients a, b, c then lead to the two equivalent differential relations:

$$\dot{\sigma}_{ij}^D + \frac{\sigma_{ij}^D}{\tau_g} = 2\mu_0\dot{\varepsilon}_{ij}^D + 2\frac{\mu_\infty}{\tau_g}\varepsilon_{ij}^D, \tag{3.26}$$

$$\tau_g\dot{\sigma}_{ij}^D + \sigma_{ij}^D = 2\mu_\infty(\varepsilon_{ij}^D + \theta_g\dot{\varepsilon}_{ij}^D), \tag{3.27}$$

where:

μ_0 and μ_∞ are respectively the instantaneous and retarded relaxation moduli in sliding;

τ_g is the relaxation time in sliding;

$\theta_g = (\mu_0/\mu_\infty)\tau_g$ (hence the name of model with three parameters) is the retarded time in shear;

$$\mu(t) = \mu_\infty + (\mu_0 - \mu_\infty)\exp\left(-\frac{t}{\tau_g}\right) \tag{3.28}$$

is the relaxation modulus in sliding, and finally

$$J(t) = J_\infty + (J_0 - J_\infty)\exp\left(-\frac{t}{\theta_g}\right) \quad (\text{with } 2\mu_\infty J_\infty = 2\mu_0 J_0 = 1), \tag{3.29}$$

is the retardation function in shear.

With certain supplementary hypotheses and the preceding definitions, the constitutive laws (3.26), (3.27) can be written in the equivalent integral forms:

$$\sigma_{ij}^D(t) = 2\mu(0)\varepsilon_{ij}^D(t) + \int_0^t 2\frac{d\mu}{d\tau}(\tau)\varepsilon_{ij}^D(t-\tau)\,d\tau \tag{3.30}$$

$$\varepsilon_{ij}^D(t) = J(0)\sigma_{ij}^D(t) + \int_0^t \frac{dJ}{d\tau}(\tau)\sigma_{ij}^D(t-\tau)\,d\tau\,. \tag{3.31}$$

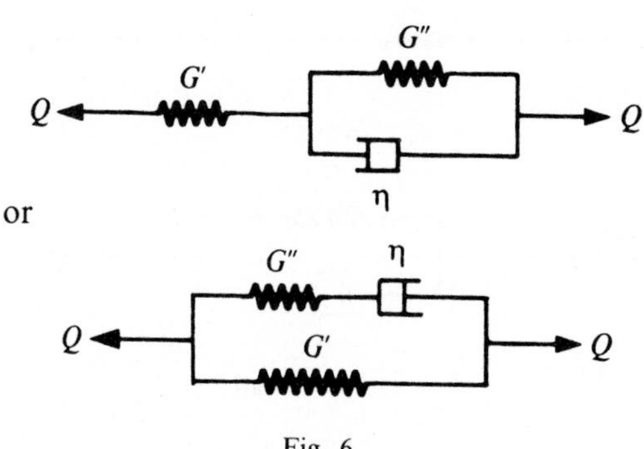

Fig. 6

This constitutive law is a particular case of the law (3.10) for materials with long memory and it generalises the Kelvin–Voigt and Maxwell media which we can recover by putting, respectively $\mu_\infty = 0$ (θ_g is then infinite) or $\tau_g = 0$ and μ_0 infinite.

5. Stationary Problems in Classical Viscoelasticity

As in classical elasticity, stationary problems are obtained in the case of quasi-statical evolution; the constitutive laws (3.16), (3.19), (3.26) or (3.27) involves rates of strain and rates of stress, but the classic method of solution, by the Laplace transform, enables us to reduce the solution to that of a family of problems in elasticity depending on a parameter p. The stationary problems encountered are therefore *a priori* the same as those in elasticity.

Review

Polymers, which play a considerable rôle in technology, often have complicated properties (and strongly different from those of metallic materials). The problems met with in their fabrication, their shaping and their use, going from fluids to solids, requires mathematical modelling of the type discussed in this §3.
Numerous materials treated by the food industries are viscoelastic; examples are gelatinous liquids and gels, doughs prepared from wheat flour and various cooked products (see H. G. Muller [1]).
The study of biomechanics involves many viscoelastic fluids (protoplasm, mucus, or phlegm, in the respiratory track, saliva, synovial fluid, etc. . . .) as well as viscoelastic "solids" (collagen, biological tissues, skin, etc.).
An evolution problem in $\mathbb{R}^3$ involving a constitutive law of Kelvin–Voigt type for a viscoelastic material with short memory, is treated in §3 of Chap. XIV by the Fourier transforms; the Laplace transform is used in §5 of Chap. XVI and the theory of semigroups in Chap. XVIIB. Variational methods are used in §6 and especially in §7 of Chap. XVIII for more general problems involving delay and memory. The numerical solution of such problems is discussed in §3 of Chap. XX.

§4. Electromagnetism and Maxwell's Equations

1. Fundamental Equations of Electromagnetism

1.1. Introduction

General references are: Roubine [1], Roubine–Bolomey [1], Fournet [1], C. Müller [1], Jackson [1], Jones [1], Landau–Lifschitz [2], Duvaut–Lions [1], Rocard [1].

Electromagnetic phenomena *in vacuo* are described with the help of two functions[68] **E** and **B**[69] *defined on the whole space* $\mathbb{R}_x^3 \times \mathbb{R}_t$,[70] with (vector) values in $\mathbb{R}^3$ – called respectively the *electric field* and the *magnetic induction*.
These functions **E** and **B** are linked with two functions (or distributions) ρ and **j** defined likewise on $\mathbb{R}_x^3 \times \mathbb{R}_t$, with $\rho(\mathbf{x}, t) \in \mathbb{R}$ and $\mathbf{j}(\mathbf{x}, t) \in \mathbb{R}^3$ – called respectively *charge density* and *current density* – by the equations, called *Maxwell's equations*:

$$(4.1) \quad \begin{cases} \text{(i)} \; -\dfrac{\partial \mathbf{E}}{\partial t} + \operatorname{curl} \mathbf{B} - \mathbf{j} = \mathbf{0} & \text{(the Maxwell–Ampère law)}, \\ \text{(ii)} \; \operatorname{div} \mathbf{E} - \rho = 0 & \text{(Gauss' electrical law)}, \\ \text{(iii)} \; \dfrac{\partial \mathbf{B}}{\partial t} + \operatorname{curl} \mathbf{E} = \mathbf{0} & \text{(the Maxwell–Faraday law)}, \\ \text{(iv)} \; \operatorname{div} \mathbf{B} = 0 & \text{(Gauss' magnetic law)}, \end{cases}$$

with the usual notation (for $\mathbf{E} = (E_1, E_2, E_3)$, $\mathbf{x} = (x_1, x_2, x_3)$)

$$\begin{cases} \operatorname{div} \mathbf{E} = \displaystyle\sum_{i=1}^{3} \frac{\partial E_i}{\partial x_i}, \\ \operatorname{curl} \mathbf{E} = \left(\dfrac{\partial E_3}{\partial x_2} - \dfrac{\partial E_2}{\partial x_3}, \dfrac{\partial E_1}{\partial x_3} - \dfrac{\partial E_3}{\partial x_1}, \dfrac{\partial E_2}{\partial x_1} - \dfrac{\partial E_1}{\partial x_2} \right)^{71}. \end{cases}$$

These equations can model as well as a microscopic study of electromagnetic phenomena (charge and current densities will then generally be concentrated at points – see Example 1 to 4 below), a macroscopic study (isotropic case) of these phenomena as we shall see in section 2.
We now make some remarks on the system (4.1).

Remark 1. *On the form of equations* (4.1).
(1) *The units employed.*
Here we have employed (and we shall use throughout this §4) the system of units called "*natural*" in which:

$$\begin{cases} \text{the speed of light } \textit{in vacuo } (c) = 1, \\ \text{permeability of the vacuum } (\mu_0) = 1^{72}. \end{cases}$$

In what follows in this §4, we shall suppose the absolute value of the charge of the electron, denoted by e, is equal to 1. We add that this charge is negative.

---

[68] Or possibly distributions on $\mathbb{R}_x^3 \times \mathbb{R}_t$ with values in $\mathbb{R}^3$.
[69] In some specialised works, use is also made of the notations **E** and **H** for these quantities.
[70] We suppose that we have chosen a direct orthonormal reference frame in the three-dimensional euclidean space E_3 representing the physical space ($\mathbb{R}_x^3$).
[71] We can similarly denote div **E** by $\nabla . \mathbf{E}$ and curl **E** by $\nabla \times \mathbf{E}$ where ∇ denotes the vector operator $\nabla = (\partial/\partial x_1, \partial/\partial x_2, \partial/\partial x_3)$.
[72] See Fournet [1], p. 5. *In vacuo*, we have, by definition $\mathbf{B} = \mu_0 \mathbf{H}$ (see below for the definition of **H** and (4.37)(ii)).

We shall indicate – and for practical calculations we shall use – in Chap. II other systems of units other than the "natural" system of units – what is presented **here** has the advantage of not involving constants in the writing of the equations (4.1). We recall that the legal system of practical units is the *système international* (S.I.) formerly called M.K.S.A.
(2) The equations (4.1) are often written in a different form from that presented here, making obvious the invariance of these equations with respect to a group of transformations: the Lorentz group.
We then make the time play a more symmetrical rôle with respect to the other variables; in addition $(\mathbf{E}, \mathbf{B})$ is considered as being made into an antisymmetric tensor – and $(\rho, \mathbf{j})$ a vector in $\mathbb{R}^4$.
We thus obtain a formulation of Maxwell's equations called the *tensor form* of these equations[73]. □

Remark 2
(1) The equations (4.1) are of hyperbolic type in **E** and **B** (for the notion of a hyperbolic equation see §2 of Chap. V).
(2) Relation between charge density and current density: By differentiating (4.1)(*ii*) with respect to t, and by taking the divergence of both sides of (4.1)(*i*) we obtain the relation called *the continuity relation for electricity* or the *equation of the conservation of electricity*:

$$\frac{\partial \rho}{\partial t} + \operatorname{div} \mathbf{j} = 0 . \tag{4.2}$$

Below (Remark 6), we shall similarly see that the equations (4.1) are not completely independent. □

We now define a quantity which plays a very important rôle in the various phenomena of electromagnetism: *the electromagnetic energy of the vacuum.*
First of all, we define the *electromagnetic energy density*, $\mathscr{E}_v(\mathbf{x}, t)$ in $(\mathbf{x}, t)$ for $(\mathbf{E}, \mathbf{B})$ by

$$\mathscr{E}_v(\mathbf{x}, t) = \frac{1}{2}(|E(\mathbf{x}, t)|^2 + |B(\mathbf{x}, t)|^2)^{74} , \qquad (\mathbf{x} \in \mathbb{R}^3, t \in \mathbb{R}) . \tag{4.3}$$

The *electromagnetic energy* for the "field" $(\mathbf{E}, \mathbf{B})$ is then defined by

$$W_v(t) = \int_{\mathbb{R}^3} \mathscr{E}_v(\mathbf{x}, t) \, d\mathbf{x} . \tag{4.4}$$

The suffix v in $\mathscr{E}_v(\mathbf{x}, t)$ and $W_v(t)$ (referring to the vacuum) can be omitted if there is

[73] See Landau–Lifschitz [2]. There exist formulations of Maxwell's equations making use of more recent mathematical formalisms. The history of these is given in Thirring [1] , p. 29. In this §4 of Chap. IA, we cite the formulation most often used in practical technology.

[74] With the notation $|\mathbf{E}|^2 = \sum_{j=1}^{3} E_j^2$.

no risk of confusion with the electromagnetic energy of continuous media which will be introduced later (Sect. 2).
Hence, the physical systems for which the electromagnetic energy of the "field" $(\mathbf{E}, \mathbf{B})$ remains finite during the course of time are those whose functions $(\mathbf{E}, \mathbf{B})$ satisfy (for all $t \in \mathbb{R}$):

$$W_v(t) = \frac{1}{2}\int (|E(x,t)|^2 + |B(x,t)|^2)\,dx < \infty . \tag{4.5}$$

Remark 3. *Local balance of the electromagnetic energy.*
Now, we define the vector $\mathbf{S}$ (in fact the vector field $(\mathbf{x}, t) \in \mathbb{R}^4 \to \mathbf{S}(\mathbf{x}, t) \in \mathbb{R}^3$)– called the *Poynting vector*–by

$$\mathbf{S} = \mathbf{E} \wedge \mathbf{B} \text{ [75]} . \tag{4.6}$$

This vector can be interpreted as the flux vector of the electromagnetic energy, account being taken of the units employed (Remark 1). By differentiating the expression (4.3) with respect to the time, and making use of the relations (4.1), (i) and (iii), we find

$$\frac{\partial \mathscr{E}_v}{\partial t} + \operatorname{div} \mathbf{S} = - \mathbf{E}.\mathbf{j} \text{ [76]} . \tag{4.7}$$

This relation expresses the (local) balance of the electromagnetic energy of the vacuum. □

Remark 4. *Maxwell's stress tensor.*
We also introduce "Maxwell's stress tensor T" (or the electromagnetic stress tensor):

$$T \overset{\text{def}}{=} (T_{ij}) = (E_i E_j + B_i B_j) - \frac{1}{2}\delta_{ij}(|E|^2 + |B|^2) \text{ [77]} .$$

We can verify the relation

$$\frac{\partial S}{\partial t} = \operatorname{Div} T - j \wedge B - \rho E \tag{4.7'}$$

with $\operatorname{Div} T$ denoting the vector with components $\sum_i \frac{\partial}{\partial x_i} T_{ij}$.
In particular the tensor T enables us to calculate the *resultant* $\mathbf{F}$ *of the forces* acting on the matter to the exterior of a domain $\Omega \subset \mathbb{R}^3$ as a result of the electromagnetic

---

[75] $\mathbf{E} \wedge \mathbf{B}$ denotes the vector with components $(E_2 B_3 - E_3 B_2, E_3 B_1 - E_1 B_3, E_1 B_2 - E_2 B_1)$– called the vector product of $\mathbf{E}$ and $\mathbf{B}$.

[76] With the notation $\mathbf{E}.\mathbf{j} = \sum_{k=1}^{3} E_k j_k$ (which will be used throughout this §4) for the scalar product of $\mathbf{E}$ and $\mathbf{j}$ in $\mathbb{R}^3$.

[77] With $\delta_{ij} = \begin{cases} +1 & \text{if } i = j, \\ 0 & \text{if } i \neq j. \end{cases}$

field in Ω[78], by its components

$$F_j \stackrel{\text{def}}{=} \int_\Gamma \sum_i T_{ij}.\nu_i \, d\Gamma = \int_\Omega \left(\sum_i \frac{\partial}{\partial x_i} T_{ij} \right) dx$$

where ν denotes the unit normal to $\Gamma = \partial\Omega$, exterior to Ω. Thus $f_j \stackrel{\text{def}}{=} \sum_i \frac{\partial}{\partial x_i} T_{ij}$ appears as the j-th component of the *force density* $\mathbf{f}$ i.e. $\mathbf{f} = \operatorname{Div} T$. □

Remark 5. It is usual to deduce Maxwell's equations from a Lagrangian $\mathscr{L}$ by application of the principle of least action (Itzykson–Zuber [1], Landau–Lifschitz [2], [5]) then using the invariance of $\mathscr{L}$ under certain transformation groups, such as the Lorentz group, to deduce – by application of Noether's theorem[79] – the various conservation equations (4.2), (4.7), (4.7)'.
This presentation proves to be particularly useful in the case of coupled systems, for relations with thermodynamics etc . . . (see Balian [1], Itzykson–Zuber [1]).
A geometrical formulation of this presentation is featured in the following references: Souriau [1], von Westenholz [1], Thirring [1]. □

Let us now give various types of electromagnetic problem.

1.2. Case in which the "Sources" are Supposed Known

We make the following assumptions:
(1) the charge and current densities ρ and $\mathbf{j}$ are known[80] and satisfy the relation (4.2);
(2) at the instant $t = 0$, the electric field $\mathbf{E}_0(\mathbf{x})$ and the magnetic induction $\mathbf{B}_0(\mathbf{x})$ are both known and satisfy the relations:

$$\text{(4.8)} \qquad \begin{cases} \operatorname{div} E_0 = \rho_0 \quad (\text{with } \rho_0(x) = \rho(x, 0); \quad x \in \mathbb{R}^3), \\ \operatorname{div} B_0 = 0 \,. \end{cases}$$

Under these assumptions, we now consider the *Cauchy problem for Maxwell's equations*:
Find the fields $\mathbf{E}$ *and* $\mathbf{B}$ *satisfying the equations* (4.1), *and such that*

$$\text{(4.9)} \qquad \begin{cases} \text{(i) } \mathbf{E}(\mathbf{x}, 0) = \mathbf{E}_0(\mathbf{x}) \quad \mathbf{x} \in \mathbb{R}^3 \,, \\ \text{(ii) } \mathbf{B}(\mathbf{x}, 0) = \mathbf{B}_0(\mathbf{x}) \,. \end{cases}$$

We must indicate the function spaces to which the different given functions must belong[81] to make precise the mathematical framework of the problem (4.9).
We shall content ourselves here with making a few remarks.

---

[78] Application: in particular this allows us to calculate the force exerted on the surface of a body called a "conductor" by an electromagnetic field (see Fournet [1], pp. 238 and 295).
[79] See Thirring [1], p. 47.
[80] Here we envisage a simple case. In cases useful to technology, it is equally often the distribution of charges and of the induced current that we seek to calculate.
[81] That will be done in later chapters.

Remark 6. "*Redundant*" *equations in the system* (4.1).
In the present case in which $\mathbf{j}$ and ρ are known (and linked by (4.2)), the system (4.1) comprises eight scalar equations for the six unknown scalars which are the components of $\mathbf{E}$ and $\mathbf{B}$.
We can expect from this that there are two "excess" equations. In effect, by applying the divergence to equations (4.1) (*i*) and (*iii*) and taking account of (4.2), we obtain

$$(4.10) \qquad \begin{cases} \dfrac{\partial}{\partial t}(\operatorname{div}\mathbf{E} - \rho) = 0 \\ \\ \dfrac{\partial}{\partial t}(\operatorname{div}\mathbf{B}) = 0 \,. \end{cases}$$

Thus if we suppose that at the initial instant $t = 0$, the relations (4.8) are satisfied, then we shall have in reality that for all $t \in \mathbb{R}$ (in $\mathbb{R}^3$):

$$\begin{cases} \operatorname{div}\mathbf{E} - \rho = 0 \\ \operatorname{div}\mathbf{B} = 0 \end{cases}$$

that is to say the relations (4.1)(*ii*) and (*iv*).
We see that, after all, the evolution of the electromagnetic field $\{\mathbf{E}, \mathbf{B}\}$ is given by the two *vector* equations (4.1)(*i*) and (*iii*) – hence to six *scalar* equations. □

Remark 7. *Relation to the wave equation.*
For the moment, let us suppose that $\rho = 0, \mathbf{j} = \mathbf{0}$.
Differentiating (4.1)(*i*) with respect to the time and using the formula

$$(4.11) \qquad \operatorname{curl}\operatorname{curl}\mathbf{E} = -\Delta\mathbf{E} + \operatorname{grad}\operatorname{div}\mathbf{E}\,,$$

we see that $\mathbf{E}$ satisfies the wave equation

$$(4.12) \qquad \frac{\partial^2\mathbf{E}}{\partial t^2} - \Delta\mathbf{E} = \mathbf{0}\,.$$

In the same way, by differentiating (4.1) (*iii*) with respect to the time, we show that $\mathbf{B}$ also satisfies the wave equation. □

Some Typical Examples[82]**:**

Example 1. *Free electromagnetic field in the absence of charge.*
This case corresponds to:

$$\mathbf{j} = \mathbf{0}\,, \qquad \rho = 0\,.$$

Example 2. *Electromagnetic field created by an immobile charged particle, placed at a point* $\mathcal{O} \in \mathbb{R}^3$ *(taken as the origin of coordinates).* In this case

$$\mathbf{j} = \mathbf{0}\,, \qquad \rho(\mathbf{x}) = e\delta(\mathbf{x})\,, \qquad \mathbf{x} \in \mathbb{R}^3\,,$$

with e denoting the (constant) electric charge.

---

[82] Examples 1, 2, 3 are very simple but allow us to approach more complicated and more realistic later.

Example 3. *Electromagnetic field created by a moving point charge (of electric charge e).*
We suppose that this motion is known, and given by a known (regular) function $t \to \mathbf{x}(t) \in \mathbb{R}^3$.
In this case

$$\rho(\mathbf{x}, t) = e\delta(\mathbf{x} - \mathbf{x}(t)), \qquad \mathbf{j}(\mathbf{x}, t) = e\mathbf{v}(t)\delta(\mathbf{x} - \mathbf{x}(t)),$$

with

$$\mathbf{v}(t) = \frac{\mathrm{d}}{\mathrm{d}t}\mathbf{x}(t)$$

denoting the velocity of the charged particle, which is therefore known.

Example 4. *General case of the electromagnetic field created by a system of N particles of electric charge e, whose motion is known and given*[83] *by a family of known (regular) functions $t \to \mathbf{x}_i(t) \in \mathbb{R}^3$ with $i = 1$ to N.*
Thus the velocity $\mathbf{v}_i(t) = \mathrm{d}\mathbf{x}_i(t)/\mathrm{d}t$ of each particle is known; the charge density ρ and the current density $\mathbf{j}$ are then given by

$$(4.13) \quad \begin{cases} \rho(\mathbf{x}, t) = \sum_{i=1}^{N} e\delta(\mathbf{x} - \mathbf{x}_i(t)) & (\mathbf{x}, t) \in \mathbb{R}^4, \\ \mathbf{j}(\mathbf{x}, t) = \sum_{i=1}^{N} e\mathbf{v}_i(t)\delta(\mathbf{x} - \mathbf{x}_i(t)). \end{cases}$$

In addition to the Cauchy problem (4.1), (4.9), we sometimes seek to determine the electric field **E** and the magnetic induction **B** (created by such given charges) which are electromagnetic waves of the kind called *outgoing* (in particular, see formula (4.103) in this §2 and Chap. II). □

We note, in particular, that in Examples 2, 3, 4 above, ρ and $\mathbf{j}$ being Dirac distributions concentrated on the trajectories (supposed known) of the particles, the solution $(\mathbf{E}(\mathbf{x}, t), \mathbf{B}(x, t))$ of the problem (4.1) (4.9) corresponding to these examples cannot have finite energy $W_v(t)$ (this is linked to the imperfection of the physical model adopted).

2. Macroscopic Equations: Electromagnetism in Continuous Media

2.1. Maxwell's Equations in Continuous Media

In the problems encountered in electromagnetism of continuous medium, the charge and current densities are not known, or are only partially known. We can

[83] Note that, in reality, the motion of the N charged particles cannot be *given* as the electromagnetic field created by each charge influences the motion of the other charges. Taking this into account we obtain a system of non-linear equations.

model the situation by decomposing $\mathbf{j}$ and ρ each into two parts $\tilde{\mathbf{j}}$ and $\tilde{\boldsymbol{\rho}}$ (current and charge densities in the matter on the atomic scale, supposed unknown) and $\mathbf{J}$ and $\boldsymbol{\rho}$ ("external" current and charge densities on the large scale, assumed – for the moment – to be known); $\tilde{\mathbf{j}}$ and $\tilde{\boldsymbol{\rho}}$ can be considered as being created in the matter by the electromagnetic field, and then creating in their turn an electromagnetic field. Taking account of the relations[84]

$$(4.14) \qquad j = \tilde{j} + J\,, \qquad \rho = \tilde{\rho} + \boldsymbol{\rho}\,,$$

we therefore have to rewrite the system (4.1) in the form:

$$(4.15) \qquad \begin{cases} \text{(i)} \ -\dfrac{\partial \mathbf{E}}{\partial t} + \operatorname{curl} \mathbf{B} - \tilde{j} = J \\ \text{(ii)} \ \operatorname{div} \mathbf{E} - \tilde{\rho} = \boldsymbol{\rho} \\ \text{(iii)} \ \dfrac{\partial \mathbf{B}}{\partial t} + \operatorname{curl} \mathbf{E} = 0 \\ \text{(iv)} \ \operatorname{div} \mathbf{B} = 0\,. \end{cases}$$

Introducing, now, two vector fields $(\mathbf{x}, t) \to \mathbf{P}(\mathbf{x}, t) \in \mathbb{R}^3$ and $(\mathbf{x}, t) \to \mathbf{M}(\mathbf{x}, t) \in \mathbb{R}^3$ linked to the densities $\tilde{\mathbf{j}}$ by the relations

$$(4.16) \qquad \begin{cases} \text{(i)} \ -\dfrac{\partial}{\partial t}(-\mathbf{P}) + \operatorname{curl} \mathbf{M} = \tilde{j} \\ \text{(ii)} \ \operatorname{div}(-\mathbf{P}) = \tilde{\rho}\,. \end{cases}$$

(We can, in some sense, consider $-\mathbf{P}$ and $\mathbf{M}$ respectively as the electric and magnetic fields generated by $\tilde{\mathbf{j}}$ and $\tilde{\rho}$ by comparison with the formulae (4.1)(i) and (ii)[85]). The vector field $\mathbf{P}$ is usually called the *polarisation* (*vector*), and $\mathbf{M}$ the *magnetisation* (*vector*).

We remark that with a choice like (4.16) we again have the relation

$$(4.17) \qquad \frac{\partial \tilde{\rho}}{\partial t} + \operatorname{div} \tilde{\mathbf{j}} = 0\,.$$

Let us put

$$(4.18) \qquad \begin{cases} \mathbf{D} \overset{\text{def}}{=} \mathbf{E} + \mathbf{P}\,, \\ \mathbf{H} \overset{\text{def}}{=} \mathbf{B} - \mathbf{M}\,. \end{cases}$$

The field $\mathbf{D}$ is called the *electric induction* (or the *electric displacement*), and it is usual to call $\mathbf{H}$ the *magnetic field*.

---

[84] We shall explain later, in Sect. 2.5, different possible interpretations of $\tilde{\rho}$ and $\tilde{j}$. (See also Fournet [1], p. 455.)

[85] However, we do not suppose that $\mathbf{P}$ and $\mathbf{M}$ satisfy (4.1)(iii) and (iv).

Substituting then the expressions (4.16) into (4.15) and taking account of (4.18), we obtain the new system (in $\mathbb{R}_x^3 \times \mathbb{R}_t$)

$$(4.19) \quad \begin{cases} \text{(i)} \ -\dfrac{\partial \mathbf{D}}{\partial t} + \operatorname{curl} \mathbf{H} = \mathbf{J}\,, \\ \text{(ii)} \ \operatorname{div} \mathbf{D} = \rho\,, \\ \text{(iii)} \ \dfrac{\partial \mathbf{B}}{\partial t} + \operatorname{curl} \mathbf{E} = 0\,, \\ \text{(iv)} \ \operatorname{div} \mathbf{B} = 0\,. \end{cases}$$

The equations (4.19) make up Maxwell's macroscopic equations – or alternatively *Maxwell's equations in continuous media.* In this system, the fields $\mathbf{E}(\mathbf{x}, t)$, $\mathbf{B}(\mathbf{x}, t)$, $\mathbf{D}(\mathbf{x}, t)$ and $\mathbf{H}(\mathbf{x}, t)$ are unknown. If ρ and $\mathbf{J}$ are *supposed known,* there are twelve scalar unknowns but only eight scalar equations (only six are independent, Remark 6 being again valid here). We add to these equations (4.19) relations – called *constitutive relations* – between **E** and **D** on the one hand and between **B** and **H** on the other hand, which we obtain from physical considerations describing the properties of the material considered. We shall later (in Sect. 2.3) indicate the most current constitutive relations. We finally obtain a system of twelve equations for the twelve scalar unknowns. □

Remark 8
(1) Starting from the system (4.19), we verify that we recover the system (4.1) by putting $\mathbf{P} = \mathbf{M} = \mathbf{0}$, hence $\mathbf{D} = \mathbf{E}$ and $\mathbf{H} = \mathbf{B}$.
(2) Starting from **E**, **B**, **D**, **H**, we can determine (via (4.18) and (4.16)) the current and charge densities $\tilde{\mathbf{j}}$ and ρ. □

We shall now deduce from the equations (4.19) consequences which are independent of the (possible) constitutive relations.

2.2. First Consequences of Maxwell's Equations of Continuous Media

2.2.1. Conservation of Electric Charge. From equations (4.19)(i) and (ii), we deduce, as in (4.2) the relation called the conservation of electric charge

$$(4.20) \qquad \frac{\partial \rho}{\partial t} + \operatorname{div} J = 0\,.$$

2.2.2. Formulation of the Other Conservation Laws. Now let Ω be a domain of $\mathbb{R}^3$ with fixed boundary Γ, supposed regular[86]. We integrate the equations (4.19) over the domain Ω and assume that the functions occurring in (4.19) are such that the resulting integrals make sense. We then make use of the relation

$$\int_\Omega \operatorname{curl} \mathbf{H} \, \mathrm{d}x = \int_\Gamma \nu \wedge \mathbf{H} \, \mathrm{d}\Gamma$$

[86] This notion will be made precise later in this work.

where ν is the unit vector of the normal to Γ, oriented towards the exterior of Ω, to obtain:

$$(4.21)\quad \begin{cases} \text{(i)} \quad -\dfrac{\mathrm{d}}{\mathrm{d}t}\displaystyle\int_{\Omega} \mathbf{D}\,\mathrm{d}x + \int_{\Gamma} (\nu \wedge \mathbf{H})\,\mathrm{d}\Gamma = \int_{\Omega} \mathbf{J}\,\mathrm{d}x\,, \\ \text{(ii)} \quad \dfrac{\mathrm{d}}{\mathrm{d}t}\displaystyle\int_{\Omega} \mathbf{B}\,\mathrm{d}x + \int_{\Gamma} (\nu \wedge \mathbf{E})\,\mathrm{d}\Gamma = 0\,, \\ \text{(iii)} \quad \displaystyle\int_{\Gamma} D\,.\,\nu\,\mathrm{d}\Gamma = \int_{\Omega} \rho\,\mathrm{d}x\,, \\ \text{(iv)} \quad \displaystyle\int_{\Gamma} \mathbf{B}\,.\,\nu\,\mathrm{d}\Gamma = 0 \end{cases}$$

and with (4.20):

$$(4.22)\quad \frac{\mathrm{d}}{\mathrm{d}t}\int_{\Omega} \rho\,\mathrm{d}x = -\int_{\Gamma} J\,.\,\nu\,\mathrm{d}\Gamma\,.$$

2.2.3. Transmission Conditions[87]**.** Let ρ_{Σ} and $\mathbf{J}_{\Sigma}$ denote the charge and current densities *concentrated* on a surface Σ, intersecting Ω and dividing it into two domains Ω_1 and Ω_2 (see Fig. 7). We apply (4.21)(iii) to each of the sub-domains Ω_1 and Ω_2:

$$(4.23)\quad \begin{cases} \displaystyle\int_{\partial\Omega_1} D\,.\,\nu\,\mathrm{d}(\partial\Omega_1) = \int_{\Omega_1} \rho\,\mathrm{d}x \\ \displaystyle\int_{\partial\Omega_2} D\,.\,\nu\,\mathrm{d}(\partial\Omega_2) = \int_{\Omega_2} \rho\,\mathrm{d}x\,. \end{cases}$$

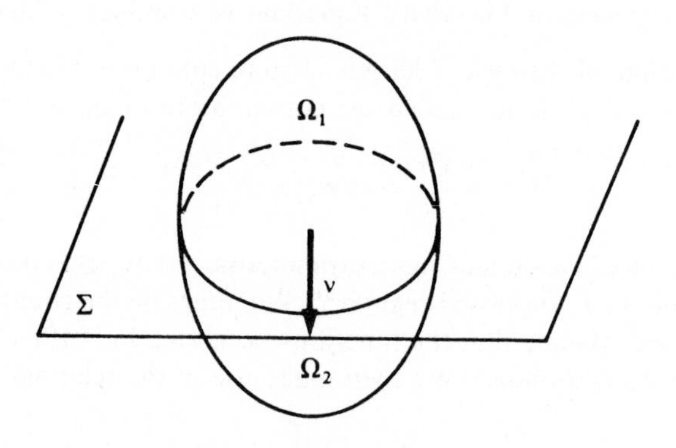

Fig. 7

[87] Also called *linking* conditions.

On noting that

$$(4.23)' \qquad \int_{\Omega} \rho \, \mathrm{d}x = \int_{\Omega_1} \rho \, \mathrm{d}x + \int_{\Omega_2} \rho \, \mathrm{d}x + \int_{\Sigma \cap \Omega} \rho_\Sigma \, \mathrm{d}\Sigma ,$$

and orienting the normal to Σ from Ω_1 to Ω_2, and denoting by $\mathbf{D}^{(1)}$ and $\mathbf{D}^{(2)}$ the values of $\mathbf{D}$ on both sides of Σ respectively in the media Ω_1 and Ω_2, we obtain, again with the help of (4.21)(iii):

$$(4.24) \qquad \int_{\Sigma \cap \Omega} \rho_\Sigma \, \mathrm{d}\Sigma = - \int_{\Sigma \cap \Omega} D^{(1)} . \nu \, \mathrm{d}\Sigma + \int_{\Sigma \cap \Omega} D^{(2)} . \nu \, \mathrm{d}\Sigma ;$$

from which we deduce on Σ (and for all $t \in \mathbb{R}$)

$$(4.25) \qquad \text{(i)} \quad (\mathbf{D}^{(2)} - \mathbf{D}^{(1)}) . \nu = \rho_\Sigma .$$

Operating in the same way with the other relations, we likewise find on Σ (and for all $t \in \mathbb{R}$):

$$(4.25) \qquad \begin{cases} \text{(ii)} \ (\mathbf{H}^{(2)} - \mathbf{H}^{(1)}) \wedge \boldsymbol{\nu} = - \mathbf{J}_\Sigma \\ \text{(iii)} \ (\mathbf{B}^{(2)} - \mathbf{B}^{(1)}) . \boldsymbol{\nu} = 0 \\ \text{(iv)} \ (\mathbf{E}^{(2)} - \mathbf{E}^{(1)}) \wedge \boldsymbol{\nu} = 0 . \end{cases}$$

The interest of the integral forms (4.21) and (4.22) is principally that they provide a direct physical interpretation. Thus for (4.22): *the variation per unit time of the total electric charge contained in a domain Ω is produced by the flux of charges across $\partial\Omega$.* The formulae (4.25) will give us the *transmission conditions* at the interface Σ between different continuous media.

2.2.4. Ampère's Theorem and Faraday's Law. Now let Σ be a regular orientable surface (a two-dimensional variety[88] in $\mathbb{R}^3$) with boundary $\partial\Sigma$, and whose unit normal $\boldsymbol{\nu}$ orients the boundary $\partial\Sigma$.
By an application of Stokes' formula[88]:

$$\int_\Sigma \operatorname{curl} \mathbf{H} . \boldsymbol{\nu} \, \mathrm{d}\Sigma = \int_{\partial\Sigma} \mathbf{H} \, \mathrm{d}(\partial\Sigma) ,$$

and by projection of (4.19)(i) and (iii) onto ν and integration over Σ, we obtain (for all $t \in \mathbb{R}$):

$$(4.26) \qquad \begin{cases} \text{(i)} \ \displaystyle\int_\Sigma \left(\frac{\partial \mathbf{D}}{\partial t} + \mathbf{J} \right) . \boldsymbol{\nu} \, \mathrm{d}\Sigma = \int_{\partial\Sigma} \mathbf{H} \, \mathrm{d}(\partial\Sigma) \\ \text{(ii)} \ \displaystyle\frac{\mathrm{d}}{\mathrm{d}t} \int_\Sigma \mathbf{B} . \nu \, \mathrm{d}\Sigma + \int_{\partial\Sigma} \mathbf{E} . \mathrm{d}(\partial\Sigma) = 0 . \end{cases}$$

---

[88] See Schwartz [2], Dieudonné [1] and Choquet-Bruhat *et al.* [1].

The relation (4.26)(i) is known as *Ampère's theorem*, while the relation (4.26)(ii) is called *Faraday's law*; it can be stated:

Faraday's Law: *The derivative with respect to the time of the flux of the magnetic induction* $\mathbf{B}$ *across a (fixed) surface* Σ *with boundary* $\partial\Sigma$ *is opposite to the circulation of the electric field along the contour* $\partial\Sigma$ [89].

Remark 9. The laws of Faraday and of Ampère (relations (4.26) with (4.22)), supposed satisfied for every regular surface Σ, imply Maxwell's equations (4.19) with (4.20).
In effect, let us choose for Σ the boundary (supposed regular) of an open set Ω: $\Sigma = \partial\Omega$ (hence $\partial\Sigma$ is *empty*); (4.26) implies, with (4.22):

$$\int_\Sigma \frac{\partial \mathbf{D}}{\partial t} \cdot \nu \, \mathrm{d}\Sigma = -\int_\Sigma \mathbf{J} \cdot \nu \, \mathrm{d}\Sigma = -\frac{\mathrm{d}}{\mathrm{d}t} \int_\Omega \rho \, \mathrm{d}x \tag{4.27}$$

or again:

$$\frac{\mathrm{d}}{\mathrm{d}t} \int_\Omega \operatorname{div} \mathbf{D} \, \mathrm{d}\Sigma = -\frac{\mathrm{d}}{\mathrm{d}t} \int_\Omega \rho \, \mathrm{d}x \,, \tag{4.28}$$

that is to say:

$$\frac{\partial}{\partial t} (\operatorname{div} \mathbf{D} - \rho) = 0 \,. \tag{4.29}$$

In the same way (4.26)(ii) implies:

$$\frac{\mathrm{d}}{\mathrm{d}t} \int_\Sigma \mathbf{B} \cdot \nu \, \mathrm{d}\Sigma = 0 \,, \tag{4.30}$$

and hence

$$\frac{\partial}{\partial t} (\operatorname{div} \mathbf{B}) = 0 \,. \tag{4.31}$$

Thus, if $\mathbf{D}$ and $\mathbf{B}$ are such that at time t_0, we have

$$\begin{cases} \operatorname{div} \mathbf{D}_{t_0} = \rho_{t_0} \,, \\ \operatorname{div} \mathbf{B}_{t_0} = 0 \,, \end{cases} \tag{4.32}$$

then (4.18) are satisfied at every instant t.
We easily see that the other relations (4.19) and (4.20) are also satisfied. □

2.2.5. Flux of Electromagnetic Energy of Continuous Media. The vector (density) of flux of electromagnetic energy in continuous media, again called the *Poynting*

---

[89] This statement is true whether $\partial\Sigma$ is fixed or moving. In the case in which the boundary $\partial\Sigma$ is moving, denote it by $\partial\Sigma(t)$ and by $\partial\Sigma(t + \mathrm{d}t)$ the boundary at the time $t + \mathrm{d}t$; let σ be an arbitrary point of $\partial\Sigma(t)$ and $u(\sigma) . \mathrm{d}t$ the vector (in the reference frame R_0) taking σ to a corresponding point, denoted by $\sigma + \mathrm{d}\sigma$, of $\partial\Sigma(t + \mathrm{d}t)$. The vector $\mathbf{E}$ featuring in (4.26)(ii) has to be evaluated in a moving frame with a translation $\mathbf{u}(\sigma)$ with respect to the reference frame R_0. (See the proof in Fournet [1], p. 311.)

vector is defined by:

$$\mathbf{S} \overset{\text{def}}{=} \mathbf{E} \wedge \mathbf{H}, \tag{4.33}$$

(that is to say: $\mathbf{S}(\mathbf{x}, t) = \mathbf{E}(\mathbf{x}, t) \wedge \mathbf{H}(\mathbf{x}, t), (\mathbf{x}, t) \in \mathbb{R}^4$) .

We shall compare the formula of the definition (4.33) (continuous medium) with the formula (4.6) (relatively to the vacuum).
Making use of the relation

$$\operatorname{div} \mathbf{S} = (\operatorname{curl} \mathbf{E}) . \mathbf{H} - \mathbf{E} . (\operatorname{curl} \mathbf{H}) , \tag{4.34}$$

we obtain with (4.19)

$$\operatorname{div} \mathbf{S} = - \left(\mathbf{H} . \frac{\partial \mathbf{B}}{\partial t} + \mathbf{E} . \frac{\partial \mathbf{D}}{\partial t} \right) - \mathbf{E} . \mathbf{J} . \tag{4.35}$$

This relation expresses the (local) balance of electromagnetic energy in continuous media (we shall compare it with the relation (4.7) corresponding to the vacuum). In particular, the expression $(\mathbf{H}\, \partial \mathbf{B}/\partial t + \mathbf{E}\, \partial \mathbf{D}/\partial t)$ appears as the "variation" (per unit time) of the *electromagnetic free energy*[90] (per unit volume) of the field $\{\mathbf{E}, \mathbf{B}, \mathbf{D}, \mathbf{H}\}$ in the medium considered.
We write generally (in a formal way) that the *variation of the density of electromagnetic free energy* $\delta \mathscr{E}$ is given by:

$$\delta \mathscr{E}(x, t) = (H(x, t) . \delta B(x, t) + E(x, t) . \delta D(x, t)) . \tag{4.36}$$

The electromagnetic energy of the field $\{\mathbf{E}, \mathbf{B}, \mathbf{D}, \mathbf{H}\}$ will be defined in Remark 11 in the case of what are called perfect media (see Sect. 2.3.1). We note that it would be possible to make here remarks similar to Remarks 4 and 5.

2.3. Constitutive Relations (or Laws of Behaviour)

We now propose to describe the relations known as "constitutive relations" introduced in Sect. 2.1.
These constitutive relations, we recall, are characteristic of the medium considered and are given by experiment; completing Maxwell's equations (4.19) they enable us to determine the evolution of all the unknowns **E**, **B**, **D**, **H**.
In many applications (having as their aim the elaboration of models which we can treat explicitly – in contrast to treatment by computer) we assume – and, in general, it is a sufficiently good approximation – that the constitutive relations between **E** and **D** on the one hand, and between **B** and **H** on the other, are *linear*. On the contrary, the electrical technologist very often has to take account of hysteresis phenomena (see Remark 13).

---

[90] It is necessary to distinguish the energy carried in electromagnetic form (electromagnetic free energy) from the energy (simply) which includes a heat term linked to electromagnetic phenomena, this term involving the entropy. See Fournet [1], p. 7. Throughout, to simplify the writing (and since in this §4 we do not treat questions of heat) we shall, in the sequel, use the expression "electromagnetic energy" instead of "electromagnetic free energy" when there will be no risk of a wrong interpretation.

2.3.1. Perfect Media. We call a medium a *perfect medium* if the following constitutive relations are satisfied:

$$(4.37) \qquad \begin{cases} \text{(i)}\ \mathbf{D} = \varepsilon \mathbf{E} \\ \text{(ii)}\ \mathbf{H} = \dfrac{1}{\mu}\mathbf{B} \end{cases}$$

with ε and μ positive constants characteristic of the medium considered called respectively the *permittivity* (or the *dielectric constant*) and the *permeability* (or the *magnetic permeability*).

Such a medium is isotropic and homogeneous.

Suppose here – for simplicity – that the whole space $\mathbb{R}^3$ is occupied by a "perfect medium".

In this simple case, **E**, **D**, **H** and **B** must satisfy the system of equations (4.19) and (4.37). Making use of (4.37) to eliminate **D** and **H** from the equations (4.19), we find that, in $\mathbb{R}_x^3 \times \mathbb{R}_t$, **E** and **B** satisfy the equations:

$$(4.38) \qquad \begin{cases} -\varepsilon\mu \dfrac{\partial \mathbf{E}}{\partial t} + \operatorname{curl} \mathbf{B} = \mu \mathbf{J}\,, \\ \operatorname{div} \mathbf{E} = \dfrac{\rho}{\varepsilon}\,, \\ \dfrac{\partial \mathbf{B}}{\partial t} + \operatorname{curl} \mathbf{E} = 0\,, \\ \operatorname{div} \mathbf{B} = 0\,. \end{cases}$$

Remark 10

(1) If $\varepsilon\mu = 1$ (see below for the physical meaning of this quantity), the system (4.38) is identical with the system (4.1) (with **j** and ρ replaced throughout by $\mu\mathbf{J}$ and ρ/ε respectively).

(2) If $\rho = 0$, $\mathbf{J} = \mathbf{0}$, we show, as for equation (4.12), that **E** and **B** satisfy the wave equation:

$$(4.39) \qquad \begin{cases} \text{(i)}\ \varepsilon\mu \dfrac{\partial^2 B}{\partial t^2} - \Delta B = 0\,, \\ \text{(ii)}\ \varepsilon\mu \dfrac{\partial^2 E}{\partial t^2} - \Delta E = 0\,, \end{cases} \quad \text{in} \quad \mathbb{R}_x^3 \times \mathbb{R}_t$$

with a speed of propagation of the waves given by $v = (\varepsilon\mu)^{-\frac{1}{2}}$[91]; this is the speed of light in the medium[92]. We customarily put $n = (\varepsilon\mu)^{\frac{1}{2}}$; we call n the *index of refraction* of the medium. □

---

[91] $v < 1$ therefore implies that we must have $\varepsilon\mu > 1$. We recall that throughout this §4 we have taken the velocity of light *in vacuo* as the unit of speed.

[92] In the case in which we have abandoned the system of natural units with $c = 1$, we have $v = c(\varepsilon\mu)^{-\frac{1}{2}}$, $n = (\varepsilon\mu)$ and so $v = c/n$.

Remark 11. Energy Balance. In the present case of a "perfect" medium (for which the constitutive relations (4.37) are therefore satisfied) the *density of the electromagnetic free energy* of the continuous medium is defined by:

$$\mathscr{E}(\mathbf{x}, t) \overset{\text{def}}{=} \frac{1}{2}(\mathbf{E}(\mathbf{x}, t) . \mathbf{D}(\mathbf{x}, t) + \mathbf{H}(\mathbf{x}, t) . \mathbf{B}(\mathbf{x}, t)) . \tag{4.40}$$

With the help of (4.37), the expression (4.34) of the (local) *balance of the electromagnetic free energy* of the continuous medium can be written:

$$\frac{\partial \mathscr{E}}{\partial t} + \operatorname{div} \mathbf{S} = - \mathbf{E} . \mathbf{J} . \tag{4.41}$$

Thus for an identically zero current density $\mathbf{J}$, the total electromagnetic free energy of a continuous medium:

$$W(t) = \int_{\mathbb{R}^3} \mathscr{E}(x, t) \mathrm{d}x \tag{4.42}$$

is conserved during the course of time.

2.3.2. Anisotropic Media. In homogeneous anisotropic media such as crystals, the constitutive relations (4.37) (ii) are of the form

$$D_i = \sum_{j=1}^{3} \varepsilon_{ij} . E_j , \qquad i = 1 \text{ to } 3, \quad \text{in} \quad \mathbb{R}^3 , \tag{4.43}$$

where ε_{ij} is a tensor characteristic of the medium and of its symmetries. In a similar manner, $\mathbf{H}$ and $\mathbf{B}$ are related by

$$H_i = \sum_{j=1}^{3} \mu'_{ij} B_j , \qquad i = 1 \text{ to } 3 , \quad \text{in} \quad \mathbb{R}^3 . \tag{4.43$'$}$$

The tensors (ε_{ij}) and (μ'_{ij}) are called respectively the *permittivity tensor* and the *inverse permeability tensor* of the medium considered. □

Remark 12. Relations of the type (4.43) imply that a variation of the electric field at the instant t cannot have an influence on $\mathbf{D}$ at instants t' preceding t. This expresses a "causality principle" between the cause ($\mathbf{E}$) and the effect ($\mathbf{D}$). While respecting this "principle", we are reduced, in certain practical cases, to replacing the formula (4.43) by a more general expression. □

Remark 13. We are sometimes led (especially in magneto-optical effects) to consider coefficients ε_{ij} depending on $\mathbf{H}$. In the same way, in materials such as iron, subjected to magnetic fields of sufficiently high intensity, the coefficients μ'_{ij} can depend on $\mathbf{H}$. We then have non-linear constitutive relations[93]. In addition these time-dependent relations can depend on the previous history of the phenomenon (hysteresis), see Durand [1]. □

---

[93] See Arecchi–Schulz–Dubois [1], Kittel [1], Marcuse [1], Zernike–Midwinter [1].

2.4. Conditions at the Interface Σ Between Two Different Perfect Continuous Media (or Transmission Conditions for Perfect Media). Boundary Conditions for the "Perfect Conductor"

2.4.1. Introduction. In the case of two different media occupying the domains Ω_1 and Ω_2 of $\mathbb{R}^3$, separated by a common, fixed, (regular) surface Σ, we simply apply the formula (4.25) to Σ to obtain the transmission conditions for the electromagnetic field on this surface. We rewrite these conditions (on denoting by $\boldsymbol{\nu}$ the normal to Σ directed from Ω_1 to Ω_2, and by ρ_Σ and $\mathbf{J}_\Sigma$ respectively the charge and current densities concentrated on the interface Σ) as follows:

$$
(4.44) \quad \begin{cases} \text{(i)} \ (\mathbf{D}^{(2)} - \mathbf{D}^{(1)}).\boldsymbol{\nu} = \rho_\Sigma , \\ \text{(ii)} \ (\mathbf{H}^{(2)} - \mathbf{H}^{(1)}) \wedge \boldsymbol{\nu} = -\mathbf{J}_\Sigma , \\ \text{(iii)} \ (\mathbf{B}^{(2)} - \mathbf{B}^{(1)}).\boldsymbol{\nu} = 0 , \\ \text{(iv)} \ (\mathbf{E}^{(2)} - \mathbf{E}^{(1)}) \wedge \boldsymbol{\nu} = \mathbf{0} . \end{cases} \quad \text{on } \Sigma \times \mathbb{R}_t
$$

Interpretation of the relations (4.44).
With sufficient conditions on continuity with respect to $\mathbf{x}$ on the vector functions $\mathbf{E}(\mathbf{x}, t)$, $\mathbf{D}(\mathbf{x}, t)$, $\mathbf{B}(\mathbf{x}, t)$ and $\mathbf{H}(\mathbf{x}, t)$ for all $t \in \mathbb{R}$ in each of the domains Ω_1 and Ω_2 the relations (4.44) imply: *the normal component of* $\mathbf{B}$ *and the tangential component of* $\mathbf{E}$ *are continuous across* Σ; on the contrary if $\rho_\Sigma \neq 0$, $\mathbf{J}_\Sigma \neq \mathbf{0}$, the normal component of $\mathbf{D}$ and the tangential component of $\mathbf{H}$ are discontinuous.
In the case in which the surface Σ would be moving (media in motion), we refer the reader to Jackson [1].
Let us now give some particular cases, useful in applications.

2.4.2. Transmission Conditions for Perfect Media. We consider the case of two media such that:
(*a*) equation (4.37) is satisfied in each of the media, but with constants (ε_1, μ_1) for medium 1 (hence in Ω_1) and (ε_2, μ_2) for medium 2 (hence in Ω_2) (hypothesis of perfect media) which are different;
(*b*) in addition, $\rho_\Sigma = 0$, $\mathbf{J}_\Sigma = \mathbf{0}$ (neither charge nor current is concentrated on the surface Σ).
Then (4.44) implies that the normal component of $\mathbf{D}$ and the tangential component of $\mathbf{H}$ are continuous across Σ.
On the contrary, if $\varepsilon_1 \neq \varepsilon_2$ and $\mu_1 \neq \mu_2$, the components of $\mathbf{E}$, $\mathbf{D}$, $\mathbf{B}$, $\mathbf{H}$, not appearing in (4.44) are always discontinuous across Σ: in effect, starting from (4.44), we have, on $\Sigma \times \mathbb{R}_t$:

$$
(4.45) \quad \begin{cases} \varepsilon_1(\mathbf{E}^{(1)}.\boldsymbol{\nu}) = \varepsilon_2(\mathbf{E}^{(2)}.\boldsymbol{\nu}) , \\ \dfrac{1}{\mu_1}(\mathbf{B}^{(1)} \wedge \boldsymbol{\nu}) = \dfrac{1}{\mu_2}(\mathbf{B}^{(2)} \wedge \boldsymbol{\nu}) , \\ \mu_1(\mathbf{H}^{(1)}.\boldsymbol{\nu}) = \mu_2(\mathbf{H}^{(2)}.\boldsymbol{\nu}) , \\ \dfrac{1}{\varepsilon_1}(\mathbf{D}^{(1)} \wedge \boldsymbol{\nu}) = \dfrac{1}{\varepsilon_2}(\mathbf{D}^{(2)} \wedge \boldsymbol{\nu}) . \end{cases}
$$

2.4.3. Boundary Conditions for a "Perfect Conductor". We now examine the case in which one of the media (e.g. the medium 2) is a "*perfect conductor*" (see the definition of this notion below in Remark 17). In such a medium, we shall suppose that the *electric field is null*[94]; we therefore have:

$$\mathbf{E}^{(2)} = \mathbf{D}^{(2)} = \mathbf{0} \quad \text{in} \quad \Omega_2 \times \mathbb{R}_t \, .$$

In the presentation which follows we shall consider the usual case in which, in addition, $\mathbf{B}^{(2)} = \mathbf{H}^{(2)} = \mathbf{0}$.
From the equations (4.19), we see that it is not possible to have either charge or current in the interior of such a medium – on the contrary, there can exist densities of charge ρ_Σ and of current $\mathbf{J}_\Sigma$ on the surface of this perfect conductor; the equations (4.44) can be written on $\Sigma \times \mathbb{R}_t$:

$$(4.46) \qquad \begin{cases} \text{(i) } \mathbf{D}^{(1)} . \, \nu = -\rho_\Sigma \, , \\ \text{(ii) } \mathbf{H}^{(1)} \wedge \nu' = \mathbf{J}_\Sigma \, , \\ \text{(iii) } \mathbf{B}^{(1)} . \, \nu = 0 \, , \\ \text{(iv) } \mathbf{E}^{(1)} \wedge \nu = \mathbf{0} \, . \end{cases}$$

But the quantities ρ_Σ and $\mathbf{J}_\Sigma$ are in general unknown: to impose the conditions (4.46) reduces then to imposing the two conditions (4.46) (iii) and (iv).
Let us now consider the case in which the two media occupy the whole space: $\Omega_1 \cup \bar{\Omega}_2 = \mathbb{R}^3_x$, Ω_2 always being occupied by a perfect conductor. We put $\Omega = \Omega_1$, $\Gamma = \Sigma = \partial\Omega$ (eliminating the suffix 1 for simplicity). Then, if the material occupying the domain Ω is perfect (and hence the constitutive equations (4.37) are satisfied in Ω), the study of Maxwell's equations (4.19) in the whole space $\mathbb{R}^3_x \times \mathbb{R}_t$ reduces to that of Maxwell's equations (4.19) in the space $\Omega \times \mathbb{R}_t$ with (4.37), and with the boundary conditions

$$(4.47) \qquad \begin{cases} \text{(i) } \mathbf{B} . \, \nu|_\Gamma = \mathbf{0} \, . \\ \text{(ii) } \mathbf{E} \wedge \nu|_\Gamma = \mathbf{0} \, . \end{cases}$$

on $\Gamma \times \mathbb{R}_t$. In particular, every Cauchy problem in $\mathbb{R}^3_x \times \mathbb{R}_t$ with null data in the complement of $\bar{\Omega}$ is equivalent to a Cauchy problem in $\Omega \times \mathbb{R}$ (deduced trivially from the problem in $\mathbb{R}^3_x \times \mathbb{R}_t$).

2.4.4. Inhomogeneous Boundary Conditions. In many applications in which we have two media – medium 1 occupying the domain $\Omega \subset \mathbb{R}^3$ and medium 2 occupying the complement $\mathbb{R}^3 \backslash \bar{\Omega}$ of $\bar{\Omega}$ – we suppose that a known electromagnetic field is imposed in the medium 2[95].

---

[94] An ideal perfect conductor will always conserve, at each point, the magnetic induction at the value it had at the moment it became perfect. A problem of this type was posed historically for superconductors which we require to be perfect conductors. Experiment has shown that they are not generally perfect conductors, the property cited at the beginning of this note not being satisfied (Meissner effect).

[95] The case in which the electromagnetic field in medium 2 is known only *at a great distance* from the surface of separation $\partial\Omega$ of the two media, leads to what is called a *transmission problem* in the whole space, with an inhomogeneous transmission condition on $\partial\Omega$ – in practice a very useful problem.

The determination of the electromagnetic field in Ω, for all $t \in \mathbb{R}$ when, in addition, we know the electromagnetic field in Ω at the initial instant $t = 0$, can be obtained by the solution of a Cauchy problem in Ω, inhomogeneous for the boundary conditions on Γ, these conditions being obtained by an application of (4.44).

Remark 14. "Redundancy" of Boundary Conditions. We have seen in Remark 6 that, if the relation $\operatorname{div} \mathbf{B}_0 = 0$ is satisfied at the *initial* instant $t = 0$, then we deduce from (4.19) (*iii*), the relation $\operatorname{div} \mathbf{B} = 0$, $\forall t$; in the same way, if the relation (4.44) (*iii*) at the surface Σ (or the relation (4.47) (*i*) on the boundary Γ of a perfect conductor) is satisfied at the *initial* instant, we can deduce (4.44) (*iv*) (resp. (4.47)(*ii*)) and from the relation $\operatorname{div} \mathbf{B} = 0$, that this relation (4.44)(*iii*) (resp. (4.47) (*ii*)) is satisfied at each instant t.

Proof in the case of boundary Γ of a perfect conductor. For each (vector) function $\boldsymbol{\varphi}$ regular from $\mathbb{R}^3$ into $\mathbb{R}^3$(i.e. $\boldsymbol{\varphi} \in (\mathscr{D}(\mathbb{R}^3))^3$)[96] then making use of (4.19) (*ii*), we can write:

$$\int_\Omega \left(\frac{\partial \mathbf{B}}{\partial t} + \operatorname{curl} \mathbf{E} \right) . \boldsymbol{\varphi} \, d\mathbf{x} = 0 \qquad \forall \boldsymbol{\varphi} \in (\mathscr{D}(\mathbb{R}^3))^3 .$$

Integrating by parts and taking account of the relation (4.47) (*ii*), we obtain

$$\int_\Omega \left(\frac{\partial \mathbf{B}}{\partial t} . \boldsymbol{\varphi} + \mathbf{E} . \operatorname{curl} \boldsymbol{\varphi} \right) d\mathbf{x} = 0 . \tag{4.48}$$

Taking $\boldsymbol{\varphi}$ of the form $\boldsymbol{\varphi} = \operatorname{grad} \psi$ with $\psi \in \mathscr{D}(\mathbb{R}^3)$, we find that this last relation can therefore be written:

$$\int_\Omega \frac{\partial \mathbf{B}}{\partial t} . \operatorname{grad} \psi \, d\mathbf{x} = 0 \qquad \forall \psi \in \mathscr{D}(\mathbb{R}^3) .$$

Integrating again by parts, we have:

$$\int_\Gamma \frac{\partial}{\partial t} (B . n) \psi \, d\Gamma - \int_\Gamma \frac{\partial}{\partial t} (\operatorname{div} B) \psi \, dx = 0 ,$$

and with the help of (4.10) we deduce immediately that

$$\frac{\partial}{\partial t} (\mathbf{B} . \mathbf{n}) = 0 \tag{4.49}$$

from which the stated result follows. □

It follows that – under the boundary conditions – we can in these *transmission problems* eliminate the unknown $\mathbf{B}$, and obtain a wave equation for $\mathbf{E}$ (with the boundary conditions (4.47) (*ii*) holding only on $\mathbf{E}$). □

---

[96] In fact, we shall use the restriction to Ω of this function.

2.5. On the Passage from Maxwell's "Microscopic" Equations to Maxwell's "Macroscopic" Equations

It is usual, in physics, to make the quantities **E** and **B** occurring in the equations (4.19) appear as "means" of the fields **E** and **B** occurring in (4.1) – but the very definition of these means can be difficult to make precise – (see Fournet [1], pp. 455–457).

In fact, in the constitutive relations used, we incorporate this notion of mean: on taking account of the laws (4.37) with ε and μ constants, we obtain therefore the equations (4.38) which for **J** and ρ null (or regular) and regular initial conditions will give solutions, themselves regular, and, as a result, the unknown regular charge and current densities $\tilde{\rho}$ and $\tilde{\mathbf{j}}$; whereas the "microscopic" densities of charge and current have, in the context of "classical"[97] modelling, to be a sum of Dirac distributions. From the point of view of the constitutive laws, we have thus obtained a regularisation of the "microscopic" problem.

In certain situations, intermediate between "microscopic situations" and "macroscopic situations", we do not involve "constitutive" relations; by contrast the polarisation vector **P** and the magnetisation vector **M** are taken to be characteristic of the structures being considered. The charge and current densities $\tilde{\rho}$ and $\tilde{\mathbf{j}}$ can then be derived by use of the formulae (4.16), but their calculation, especially that of $\tilde{\rho}$, leads to expressions which are not of the "Dirac distribution" type. We then say that $\tilde{\rho}$ corresponds to *fictitious charges*.

2.6. Case in which the Current Density J is Unknown. Ohm's Law

In many applications, the current density **J** (involved in the equations (4.19)) which arises from displacement of charges in the media considered is itself *unknown*[98]. By contrast, the charge density $\rho(\mathbf{x}, t)$ is a known function – it is then the sole datum occurring in the equations (4.19) which involve three supplementary unknowns (the three components of **J**). *New constitutive relations* characterising the relations (in the materials considered) between the current **J** and the electromagnetic field have to be introduced: in the majority of the applications, we assume that **J** *depends linearly* on the electric field **E**[99].

A local law of the type $\mathbf{J} = \sigma\mathbf{E}$ can be valid only if l, the free distances between the carriers of the electric charges are small, within the scale of the typical distances governing the spatial variations in **E**. In the contrary case, we have, for example, a relation of the type[100]

$$\mathbf{J}(\mathbf{x}) = \frac{3\sigma}{4\pi l}\int_{\mathbb{R}^3} \frac{[\mathbf{y}\,.\,\mathbf{E}(\mathbf{x}-\mathbf{y})]\mathbf{y}}{|\mathbf{y}|^4}\exp\left(-\frac{|\mathbf{y}|}{l}\right)d\mathbf{y} \tag{4.50}$$

[97] As opposed to the modelling in quantum physics.

[98] To model certain problems, we can split **J** into two parts, one known ("external") and the other unknown. We recall that we have to have the continuity relation (4.20) linking ρ and **J**.

[99] See Fournet [1] and Jackson [1].

[100] If **E** varies slowly, in the scale of l, we clearly recover the relation $\mathbf{J} \simeq \sigma\mathbf{E}$.

with σ and l positive constants; for this relation (which is nonlocal) see Fournet [1], p. 193. This expression allows us to take account of an abnormal skin which occurs at low temperatures (when l becomes large as a consequence of the diminution of thermal agitation) for high frequencies (the spatial scale of the variation of $\mathbf{E}$ becoming smaller and smaller in proportion as the frequency of the phenomena increases). In superconductors, a similar formula expresses the relation between the current density $\mathbf{J}$ and the vector potential $\mathbf{A}$ (in place of $\mathbf{E}$).
Finally, we obtain a problem with the same number of unknowns as of (independent) equations.

2.6.1. In *isotropic and homogeneous media*, without memory, we assume in the majority of applications the simple relation:

$$\mathbf{J} = \sigma \mathbf{E}, \qquad \sigma \geqslant 0\,, \tag{4.51}$$

with σ a positive constant characteristic of the medium considered, called the *constant of conductivity*. The relation (4.51) is called *Ohm's law*.
For a medium supposed (for example) to occupy the whole space $\mathbb{R}_x^3$ and in which we have the constitutive relations (4.37) and (4.51), the equations (4.38) give by the elimination of $\mathbf{J}$:

$$\left\{\begin{array}{ll} \text{(i)}\ -\varepsilon\mu \dfrac{\partial \mathbf{E}}{\partial t} - \mu\sigma\mathbf{E} + \operatorname{curl}\mathbf{E} = 0\,, & \\ \text{(ii)}\ \operatorname{div}\mathbf{E} = \dfrac{\rho}{\varepsilon}\,, & \text{in } \mathbb{R}_x^3 \times \mathbb{R}_t \\ \text{(iii)}\ \dfrac{\partial \mathbf{B}}{\partial t} + \operatorname{curl}\mathbf{E} = 0\,, & \\ \text{(iv)}\ \operatorname{div}\mathbf{B} = 0\,. & \end{array}\right. \tag{4.52}$$

Once this problem has been solved, $\mathbf{J}$ can be calculated with the help of (4.51).

Remark 15. The relation of the conservation of charge becomes with (4.51) and (4.52) (in $\mathbb{R}_x^3 \times \mathbb{R}_t$):

$$\frac{\partial \rho}{\partial t} + \frac{\sigma}{\varepsilon}\rho = 0, \quad \text{hence} \quad \rho = e^{-\frac{\sigma}{\varepsilon}t}\rho_0\,, \tag{4.53}$$

where $\rho_0 : \mathbf{x} \to \rho_0(\mathbf{x})$ is the (given) charge density at the initial instant $t = 0$. There is hence a diminution with time of the charge[101]. □

Remark 16. We can show that the relation (4.51), in contrast to the relations (4.37), leads to a *diminution in the energy with time* (*hence to a dissipation*). This expresses the phenomenon known in physics under the name of the *Joule effect*. In

[101] The "theoretical" time constant ε/σ would be of the order 10^{-19} sec. for copper (which does not make sense physically).

effect, by supposing $\rho = 0$, we can easily show that **E** and **B** satisfy the damped wave equations[102]

$$\text{(4.54)} \quad \begin{cases} \text{(i)}\ \varepsilon\mu \dfrac{\partial^2 \mathbf{E}}{\partial t^2} + \mu\sigma \dfrac{\partial \mathbf{E}}{\partial t} - \Delta \mathbf{E} = 0\,, \\[2ex] \text{(ii)}\ \varepsilon\mu \dfrac{\partial^2 \mathbf{B}}{\partial t^2} + \mu\sigma \dfrac{\partial \mathbf{B}}{\partial t} - \Delta \mathbf{B} = 0\,. \end{cases}$$

In addition, the energy balance (4.41) becomes here:

$$\text{(4.55)} \quad \frac{\partial \mathscr{E}}{\partial t} = -\operatorname{div} \mathbf{S} - \sigma \mathbf{E}^2\,,$$

from which by integration over $\mathbb{R}^3$ (when it is admissible):

$$\text{(4.56)} \quad \int_{\mathbb{R}^3} \frac{\partial \mathscr{E}}{\partial t}(\mathbf{x}, t)\mathrm{d}\mathbf{x} = \frac{\mathrm{d}}{\mathrm{d}t} \int_{\mathbb{R}^3} \mathscr{E}(\mathbf{x}, t)\mathrm{d}\mathbf{x} = \frac{\mathrm{d}}{\mathrm{d}t} W(t) = -\sigma \int_{\mathbb{R}^3} \mathbf{E}^2 \mathrm{d}x < 0\,.$$

Thus the macroscopic energy $W(t)$ is a function which decreases with the time. □

Remark 17. *Conductor, dielectric and magnetic medium.* We often call (see Fournet [1], p. 93) by the name *conductor*, every medium possessing electrical charges capable of being displaced within the "macroscopic scale", under the effect of an electric field.
An *ohmic conductor* is a medium in which Ohm's law (4.51) holds with σ a non-zero constant. It is assumed to be isotropic and homogeneous.
A "*perfect*" *conductor* is a "fictitious" medium such that $\sigma \to \infty$. In the interior of a perfect conductor, the field is zero (see Roubine [1], p. 25). Metals come near to the concept of a perfect conductor (since the constant σ then has a very high value).
A "*perfect*" *insulator* is a medium in which $\sigma = 0$.
A medium in which there are no "free electrons capable of carrying an electric current" is called a *dielectric* (see Fournet [1], p. 148). A *perfect dielectric* is a medium in which the relation (4.37) (i): $\mathbf{D} = \varepsilon\mathbf{E}$ is satisfied (with ε constant).
A *perfectly insulating, perfect dielectric* is such that in addition $\mathbf{J} = \mathbf{0}$: thus (4.51) is satisfied with $\sigma = 0$ (see Fournet [1], p. 196).
A *perfect magnetic medium* is a medium in which the relation (4.37) (ii) $\mathbf{H} = (1/\mu)\mathbf{B}$ is satisfied (with μ = a constant).
We use other classifications of media in electricity; we refer the reader to Fournet [1] on this subject. □

Remark 18. In the modelling of electromagnetic phenomena linked with metallic materials, we sometimes neglect the terms of second order in $\partial/\partial t$[103], with

[102] See Chap. IB, §1, Chap. XV, §4 and Chap. XVIIB.
[103] From (4.54), in the case of periodic phenomena of frequency ω, that corresponds to $\varepsilon\omega \ll \sigma$; this inequality is always satisfied for metallic conductors and for the frequencies encountered in practice in electromagnetic technology.

coefficient $\varepsilon\mu$, in equation (4.54), in comparison with the terms of the first order in $\partial/\partial t$ with coefficient $\mu\sigma$.
The equations (4.54) then become equations of diffusion type (examples of applications in Sect. 4.2). □

2.6.2. In *anisotropic media*, the linear constitutive relation between **J** and **E** can be written generally:

(4.57) $$J_i = \sum_{j=1 \text{ to } 3} \sigma_{ij} . E_j , \qquad i=1 \text{ to } 3 , \quad \text{in} \quad \mathbb{R}^3$$

with σ_{ij} (with i, j = 1 to 3) characteristic of the medium; (σ_{ij}) is called the *conductivity tensor*.

Remark 19. To model certain phenomena, we are sometimes led to assume relations of the form:

(4.58) $$J_i = \sum_j \sigma_{ij}(\mathbf{H}) . E_j$$

with coefficients σ_{ij} depending on the magnetic field **H** (as in the Hall effect (see Landau–Lifschitz [5]) which leads to non-linear equations).
We similarly call attention to the relation

(4.59) $$J \sim \sqrt{E}$$

($J = |\mathbf{J}|$, $E = |\mathbf{E}|$), used in the study of the behaviour of superconductors in strong electric fields. □

3. Potentials. Gauge Transformation (Case of the Entire Space $\mathbb{R}^3_x \times \mathbb{R}_t$)

In many problems concerning Maxwell's microscopic equations (*in vacuo* and in the whole space $\mathbb{R}^3_x \times \mathbb{R}_t$), hence in the context of equations (4.1), we are to make use, not of the functions **E** and **B**, but of the two functions:

(4.60) $$\begin{cases} (\mathbf{x}, t) \to \mathbf{A}(\mathbf{x}, t) \in \mathbb{R}^3 \quad \text{called "\textit{the vector potential}"} , \\ \text{and} \\ (\mathbf{x}, t) \to \varphi(\mathbf{x}, t) \in \mathbb{R} \quad \text{called "\textit{the scalar potential}"} , \end{cases}$$

which are related to **E** and **B** by

(4.61) $$\begin{cases} \mathbf{B} = \operatorname{curl} \mathbf{A} , \\ \mathbf{E} = - \operatorname{grad} \varphi - \dfrac{\partial \mathbf{A}}{\partial t} . \end{cases}$$

For such functions, the equations (4.1) (iii) and (4.1) are satisfied identically. On substituting the expressions (4.61) into (4.1) (i) and (ii), we obtain the inhomogeneous linear system:

$$(4.62)\qquad \begin{cases} \text{(i)}\ \dfrac{\partial^2 \mathbf{A}}{\partial t^2} - \Delta A + \operatorname{grad}\left(\operatorname{div}\mathbf{A} + \dfrac{\partial \varphi}{\partial t}\right) = \mathbf{j}\,, \\ \text{(ii)}\ -\Delta\varphi - \dfrac{\partial}{\partial t}(\operatorname{div}\mathbf{A}) = \rho\,. \end{cases}$$

We observe that the functions $\mathbf{A}$ and φ are not defined in a unique manner by (4.61) beginning from $\mathbf{E}$ and $\mathbf{B}$: if $\mathbf{A}$ and φ satisfy (4.61), then for each arbitrary function F of $\mathbf{x}$ and t, $\mathbf{A}'$ and φ' defined by:

$$(4.63)\qquad \begin{cases} \mathbf{A}' = \mathbf{A} + \operatorname{grad} F\,, \\ \varphi' = \varphi - \dfrac{\partial F}{\partial t}\,, \end{cases}$$

also satisfy (4.61).
The transformation $(\mathbf{A}, \varphi) \to (\mathbf{A}', \varphi')$ given by (4.63) is called a *gauge transformation.*

3.1. The Lorentz Condition

As a result of (4.63), we have (always in $\mathbb{R}^3_x \times \mathbb{R}_t$)

$$(4.64)\qquad \operatorname{div}\mathbf{A}' + \frac{\partial \varphi'}{\partial t} = \operatorname{div}\mathbf{A} + \frac{\partial \varphi}{\partial t} + \Delta\mathbf{F} - \frac{\partial^2 \mathbf{F}}{\partial t^2}\,.$$

Taking for F a solution of the equation

$$(4.65)\qquad \Delta F - \frac{\partial^2 F}{\partial t^2} = -\left(\operatorname{div} A + \frac{\partial \varphi}{\partial t}\right)$$

(where $\mathbf{A}$ and φ are supposed known), we see that it is possible to choose a pair $(\mathbf{A}_L, \varphi_L)$ such that

$$(4.66)\qquad \operatorname{div}\mathbf{A}_L + \frac{\partial \varphi_L}{\partial t} = 0\,.$$

This relation is called the *Lorentz condition.*
With this choice, equations (4.62) can be written:

$$(4.67)\qquad \begin{cases} \dfrac{\partial^2 \mathbf{A}_L}{\partial t^2} - \Delta\mathbf{A}_L = \mathbf{j}\,, \\ \dfrac{\partial^2 \varphi_L}{\partial t^2} - \Delta\varphi_L = \rho\,. \end{cases}$$

We note that (4.66) with (4.67) does not determine a unique pair $(\mathbf{A}_L, \varphi_L)$ when $\mathbf{j}$

and ρ are known: if now F is a function such that:

(4.68) $$\frac{\partial^2 F}{\partial t^2} - \Delta F = 0\,,$$

then $(\mathbf{A}', \varphi')$ given by (4.63), with $(\mathbf{A}, \varphi)$ replaced by $(\mathbf{A}_L, \varphi_L)$, also satisfies (4.66).

3.2. The Coulomb Condition

In place of the relation (4.66), we can impose:

(4.69) $$\operatorname{div} \mathbf{A} = 0\,.$$

Then (4.62) (ii) implies that φ has to satisfy *Poisson's equation* (see Chap. II).

(4.70) $$\Delta \varphi = -\rho\,,$$

in $\mathbb{R}^3_x$, but with φ and ρ possibly depending on the time; in general, we take for the solution φ:

(4.71) $$\varphi(\mathbf{x}', t) = \int_{\mathbb{R}^3} \frac{\rho(\mathbf{x}, t)}{|\mathbf{x}' - \mathbf{x}|}\,\mathrm{d}\mathbf{x} \qquad (\mathbf{x}' \in \mathbb{R}^3, t \in \mathbb{R})\,.$$

This expression has a sense by convolution if ρ is a distribution with compact support (see Chap. II). In Chap. II, we shall call such a distribution φ a *Newtonian potential*. In electromagnetism, φ given by (4.71) is called the *Coulomb potential* of the charges of density ρ, and the condition (4.69) is called the *Coulomb condition*. As to the vector potential $\mathbf{A}$, it satisfies the equation (4.62); hence:

(4.72) $$\frac{\partial^2 \mathbf{A}}{\partial t^2} - \Delta \mathbf{A} = \mathbf{j} - \operatorname{grad} \frac{\partial \varphi}{\partial t}\,.$$

When there is no source (ρ and $\mathbf{j}$ zero), this gauge is very practical since (4.70) is Laplace's equation

(4.73) $$\Delta \varphi = 0\,.$$

in $\mathbb{R}^3_x$. Let us take the solution $\varphi = 0$. The vector potential $\mathbf{A}$ then satisfies the relation

(4.74) $$\frac{\partial^2 \mathbf{A}}{\partial t^2} - \Delta \mathbf{A} = \mathbf{0}\,,$$

everywhere in $\mathbb{R}^3_x \times \mathbb{R}_t$ and defines $\mathbf{E}$ and $\mathbf{B}$ by (4.61), so

(4.75) $$\begin{cases} \mathbf{E} = -\dfrac{\partial \mathbf{A}}{\partial t}\,, \\ \mathbf{B} = \operatorname{curl} \mathbf{A}\,. \end{cases}$$

Remark 20. We shall not study here the correspondence between the scalar and vector potentials and the fields $\mathbf{E}$, $\mathbf{B}$ in the case where Ω is bounded (a principal difficulty of this study resides in the determination of the boundary conditions which $(\mathbf{A}, \varphi)$ must satisfy in terms of the boundary conditions which $(\mathbf{E}, \mathbf{B})$ satisfy). $\square$

3.3. The Debye Potential

Besides the vector potential $\mathbf{A}$ and the scalar potential φ which we have just presented and which are currently used in applications, we sometimes make use of a procedure due to Debye which, starting from solutions of the wave equation, enables us to construct solutions of Maxwell's equations (4.1).
Let $u = u(\mathbf{x}, t) \in \mathbb{R}$ be a solution of the wave equation:

$$\frac{\partial^2 u}{\partial t^2} - \Delta u = 0 \quad \text{in} \quad \mathbb{R}_x^3 \times \mathbb{R}_t . \tag{4.76}$$

We put: $u^1 = u, u^2 = \partial u/\partial t$, then

$$\begin{cases} f_M = (\mathbf{E}^1, \mathbf{B}^1) = (\operatorname{curl}\operatorname{curl}(\mathbf{x}u^1), \operatorname{curl}(\mathbf{x}u^2)) , \\ f_E = (\mathbf{E}^2, \mathbf{B}^2) = (-\operatorname{curl}(\mathbf{x}u^2), \operatorname{curl}\operatorname{curl}(\mathbf{x}u^1)) , \end{cases}$$

can be shown to be solutions of Maxwell's equations (4.1) with $\rho = 0, \mathbf{j} = \mathbf{0}$; this is a consequence of the identity

$$\operatorname{curl}\operatorname{curl}\operatorname{curl}(\mathbf{x}\varphi) = -\operatorname{curl}(\mathbf{x}\Delta\varphi) .$$

f_M (resp. f_E) is called the *transverse magnetic* (resp. *electric*) *wave* and $U = (u^1, u^2)$ is called the *Debye potential.*
The working spaces have still to be specified. For this we refer the reader to Schulenberger [1].

4. Some Evolution Problems

4.1. Evolution Problems with a (Volume) Contribution of Charges

We point out the following example of such a problem:

$$\begin{cases} \dfrac{\partial \mathbf{D}}{\partial t} + \mathbf{J} - \operatorname{curl}\left(\dfrac{1}{\mu}\mathbf{B}\right) = \mathbf{G}_1 , \\ \dfrac{\partial \mathbf{B}}{\partial t} + \operatorname{curl}\left(\dfrac{1}{\varepsilon}\mathbf{D}\right) = \mathbf{G}_2 , \end{cases}$$

with $\mathbf{J} = \sigma\mathbf{E}$ (Ohm's law), $\mathbf{G}_1$ and $\mathbf{G}_2$ with $\operatorname{div}\mathbf{G}_2 = 0$ for which we refer the reader to, for example, Duvaut–Lions [1].

4.2. "Decoupled" Cauchy Problems

By a "decoupled" problem is meant one in which interest is focussed on the electric field only, or on the magnetic field only.

In certain problems, we are interested solely in the electric (resp. magnetic) field, with troubling about the magnetic (resp. electric) field. The elimination of the "undesirable" field from Maxwell's equations can, in general, be effected without difficulty, and often without loss of information (this is a consequence, because of the boundary conditions, of Remark 14).
We give here some examples.

4.2.1. Diffusion of an Electric Wave in a Conductor[104]**.** A cylindrical conductor occupying the domain Ω is placed in an imposed external electric field $\mathbf{E}_1(t)$ parallel to the axis of the conductor and constant through the whole of space. We seek to find the electric field in the interior of the conductor. The physical model of this situation leads to the neglect of the terms involving $\partial^2/\partial t^2$ in the equations (4.54) of the electromagnetic field: the equations governing the evolution of the electric field then become equations of diffusion type:

$$(4.77)\quad (i)\quad \frac{\partial \mathbf{E}}{\partial t} - \frac{1}{\mu\sigma}\Delta \mathbf{E} = 0 \quad \text{in} \quad \Omega_+ = \Omega \times \mathbb{R}^+; \qquad \mathbf{E}(\mathbf{x}, t) \in \mathbb{R}^3 ,$$

with the supplementary condition

$$(4.77)\quad (ii)\qquad \operatorname{div} \mathbf{E} = 0 \quad \text{in} \quad \Omega_+ .$$

The modelling of the boundary conditions leads us to assume that in the case considered, the tangential component of the electric field over the boundary Γ of Ω is known, and fixed by the external electric field; we thus have:

$$(4.77)\quad (iii)\quad \boldsymbol{\nu} \wedge \mathbf{E}(\mathbf{x}, t) = \boldsymbol{\nu} \wedge \mathbf{E}_1(\mathbf{x}, t) \overset{\text{def}}{=} \mathbf{h}(\mathbf{x}, t) \quad \text{for} \quad \mathbf{x} \in \Gamma, t \in \mathbb{R}^+ ,$$

with $\boldsymbol{\nu}$ the exterior normal to Γ in $\mathbf{x}$, and $\mathbf{E}_1$ (or $\mathbf{h}$) given.
The Cauchy problem consists of determining $\mathbf{E}(\mathbf{x}, t)$ satisfying (4.77) (*i*), (*ii*) and (*iii*) with the initial condition

$$(4.77)\quad (iv)\quad \mathbf{E}(\mathbf{x}, 0) = \mathbf{E}_0(\mathbf{x}) , \quad \text{with} \quad \mathbf{E}_0 \quad \text{such that} \quad \operatorname{div} \mathbf{E}_0 = 0 .$$

In general we seek an electric field $\mathbf{E}(\mathbf{x}, t)$ with energy

$$W(t) = \frac{1}{2}\int_\Omega |\mathbf{E}(\mathbf{x}, t)|^2 d\mathbf{x} \quad \text{finite for all} \quad t \in \mathbb{R}^+ ,$$

the energy of the initial electric field $\mathbf{E}_0(\mathbf{x})$ being itself finite.

4.2.2. Magnetic Induction in a Plasma[105]**.** We seek to determine the magnetic induction $\mathbf{B}$ in a plasma (occupying the domain Ω) of a Tokomak.

---

[104] See Landau–Lifschitz [5], Jackson [1], Rocard [1].
[105] See Boujot [1] and Delcroix [1].

We recall Maxwell's equations (4.19) relating to continuous media, and the constitutive laws (4.37) and (4.51):

$$\begin{cases} \mathbf{H} = \dfrac{1}{\mu}\mathbf{B}\,, \\ \mathbf{J} = \sigma\mathbf{E}\,, \end{cases} \qquad \mathbf{H}(\mathbf{x}, t),\ \mathbf{B}(\mathbf{x}, t),\ \mathbf{J}(\mathbf{x}, t),\ \mathbf{E}(\mathbf{x}, t) \in \mathbb{R}^3$$

where μ and σ are the "constants" of magnetic permeability and of conductivity respectively (possibly eventually depending on the space variable $\mathbf{x}$) of the plasma considered. Neglecting the term in $\partial\mathbf{D}/\partial t$ of (4.19) (*i*), this equation (4.19) (*i*) can be written

$$\sigma\mathbf{E} = \operatorname{curl}\frac{1}{\mu}\mathbf{B} \quad \text{in} \quad \Omega_+ = \Omega \times \mathbb{R}^+ \,. \tag{4.78}$$

Equation (4.19) (*iii*) then gives, by the use of (4.78)

$$\text{(4.79) (i)} \qquad \frac{\partial\mathbf{B}}{\partial t} + \operatorname{curl}\left[\frac{1}{\sigma}\operatorname{curl}\left(\frac{1}{\mu}\mathbf{B}\right)\right] = \mathbf{0} \qquad \mathbf{B}(\mathbf{x}, t) \in \mathbb{R}^3,\ \mathbf{x} \in \Omega,\ t \in \mathbb{R}^+ \,.$$

The magnetic induction $\mathbf{B}$ similarly has to satisfy:

$$\text{(4.79) (ii)} \qquad \operatorname{div}\mathbf{B} = 0 \quad \text{in} \quad \Omega_+ \,.$$

(see (4.19) (*iv*)). We notice that in the case in which μ and σ are independent of $\mathbf{x}$ (4.79) (*i*) simplifies, with the help of (4.79) (*ii*), to

$$\frac{\partial\mathbf{B}}{\partial t} - \frac{1}{\sigma\mu}\Delta\mathbf{B} = 0 \quad \text{in} \quad \Omega_+ \,.$$

We thus recover (4.54) (*ii*), where the term $\varepsilon\mu\partial^2\mathbf{B}/\partial t^2$ has been neglected. The conditions at the boundary of the plasma leads to the conclusion that the tangential component of the electric field vanishes on the boundary, i.e. that

$$\text{(4.79) (iii)} \qquad \nu \wedge \operatorname{curl}\mathbf{B} = 0 \quad \text{on} \quad \Gamma_+ = \Gamma \times \mathbb{R}_+ \,.$$

Thus the posed (Cauchy) problem consists of determining $\mathbf{B}$ satisfying (4.79) (*i*), (*ii*) and (*iii*) for a known initial magnetic induction $\mathbf{B}_0$ such that

$$\text{(4.79) (iv)} \qquad \operatorname{div}\mathbf{B}_0 = 0 \quad \text{in} \quad \Omega.$$

We seek in general a magnetic field $\mathbf{B}(\mathbf{x}, t)$ with energy

$$W(t) = \frac{1}{2}\int_\Omega |\mathbf{B}(\mathbf{x}, t)|^2\,\mathrm{d}x \quad \text{finite for all} \quad t \in \mathbb{R}^+ \,,$$

the initial magnetic induction being itself of finite energy.

4.2.3. Propagation of the Electric Field in a Wave Guide[106]**.** We propose to determine the electric field in air (supposed to be homogeneous throughout this

[106] See Jackson [1], Landau–Lifschitz [5].

domain) situated in the interior of a metallic tube, this latter being supposed to be a "perfect conductor".

If ε is the permittivity of the air and μ its (magnetic) permeability, the equations giving the evolution of the electric field are given by (4.39), that is to say by:

$$(4.80)\quad (i)\quad \frac{\partial^2 \mathbf{E}}{\partial t^2} - \frac{1}{\varepsilon\mu}\Delta\mathbf{E} = 0 \qquad \mathbf{x}\in\Omega,\ \mathbf{E}(\mathbf{x},t)\in\mathbb{R}^3,\ t\in\mathbb{R}^+ (\text{or } t\in\mathbb{R}),$$

with

$$(4.80)\quad (ii)\qquad \operatorname{div}\mathbf{E} = 0 \quad \text{in} \quad \Omega\times\mathbb{R}_t^+ \quad \text{or} \quad \Omega\times\mathbb{R}_t\,.$$

The boundary condition (which translates the continuity of the tangential component of $\mathbf{E}$ on Γ, the surface of separation between the air and the perfect conductor) can be written, from (4.47):

$$(4.80)\quad (iii)\qquad \boldsymbol{\nu}\wedge\mathbf{E}(\mathbf{x},t) = 0 \quad \text{with} \quad \mathbf{x}\in\Gamma,\ t\in\mathbb{R}^+ \quad \text{or} \quad t\in\mathbb{R}\,.$$

We propose to determine the electric field $\mathbf{E}$ satisfying (4.80) (i), (ii), (iii) and the initial conditions

$$(4.80)\quad (\text{iv})\qquad \mathbf{E}(\mathbf{x},0) = \mathbf{E}^0(\mathbf{x})\,,\quad \frac{\partial\mathbf{E}}{\partial t}(\mathbf{x},0) = \mathbf{E}^1(\mathbf{x}) \quad \text{with} \quad \mathbf{x}\in\Omega\,.$$

In addition $\mathbf{E}^0$ and $\mathbf{E}^1$ satisfy:

$$(4.80)\quad (\text{v})\qquad \operatorname{div}\mathbf{E}^0 = \operatorname{div}\mathbf{E}^1 = 0 \quad \text{in} \quad \Omega\,.$$

Remark 21. There are many problems in electromagnetism other than the Cauchy problems to be treated: in particular, we might wish:

— to find solutions of (4.1) of the outgoing wave type
— to find "monochromatic" solutions of Maxwell's equations.

We shall return to these questions in Sect. 6.2. □

4.3. Coupled Problems

In certain physical problems, we are interested equally in physical quantities other than the electromagnetic field. For example, the electromagnetic field acts on particles and more generally (or more globally) on the surrounding matter whose evolution we have also to determine: the equations governing the evolution of these particles (or of this matter) involves the fields $\mathbf{E}$ and $\mathbf{H}$, and the movement of the matter has an influence on the electromagnetic field by means of the charge density ρ and the current density $\mathbf{j}$, which in general are expressed in a simple fashion with respect to the parameters of the motion of the particles (or of the matter).

Thus, the evolution of the electromagnetic field and of the matter is described by coupled equations.[107] We now indicate some simple examples of this situation.

[107] A method of "condensing" the model has been indicated in Sect. 2 using constitutive relations appropriate to continuous media.

4.3.1. Modelling of an "Electron Gas" (in a Plasma of Electrons and Ions).[108] Suppose that $-e$ and m are respectively the charge and mass of an electron. On an electron at $(\mathbf{x}, t)$, having velocity $\mathbf{v}$, the electromagnetic field exerts a force, called the *Lorentz force*, given by:

$$\mathbf{F} \equiv -e(\mathbf{E}(\mathbf{x}, t) + \mathbf{v} \wedge \mathbf{B}(\mathbf{x}, t)) .$$

Suppose that $u(\mathbf{x}, \mathbf{v}, t)$ denotes the density of a gas of electrons of velocity $\mathbf{v}$ at the point $\mathbf{x}$ and at the time t. The evolution of this density is given by *Vlasov's equation*:

$$\frac{\partial u}{\partial t} + \mathbf{v} . \nabla_x u + \frac{1}{m}\mathbf{F} . \nabla_{\mathbf{v}} u = 0^{109} .$$

The evolution problem bearing upon the unknowns $\mathbf{E}$, $\mathbf{B}$ and u is determined by Maxwell's equations (4.1) and Vlasov's equation, with ρ and $\mathbf{j}$ defined by:

$$\begin{cases} \rho(\mathbf{x}, t) = -e \displaystyle\int_{\mathbb{R}^3} u(\mathbf{x}, \mathbf{v}, t)\mathrm{d}v + en_0 \quad (n_0 \text{ constant})^{109} \\ \mathbf{j}(\mathbf{x}, t) = -e \displaystyle\int_{\mathbb{R}^3} \mathbf{v} . u(\mathbf{x}, \mathbf{v}, t)\mathrm{d}v . \end{cases}$$

We remark that the considered system of equations is non-linear as a result of the last term of Vlasov's equation; for certain applications, we shall linearise these equations.

4.3.2. Magnetohydrodynamics (MHD). The electromagnetic field is coupled with a conducting fluid in motion, the motion of the fluid being described by an equation of the type of those studied in fluid mechanics in §1, with the term for the applied force depending on the electromagnetic field. (See Delcroix [1], Chap. X.5, Krall–Trivelpiece [1], Chap. III and Landau–Lifschitz [5], Chap. VIII).

5. Static Electromagnetism

In the so-called static problems, we study solutions of Maxwell's equations which are independent of time. Thus the equations (4.19) in continuous media become:

$$(4.81) \qquad \begin{cases} \text{(i) curl } \mathbf{H} = \mathbf{J} , \\ \text{(ii) div } \mathbf{D} = \rho , \\ \text{(iii) curl } \mathbf{E} = 0 , \\ \text{(iv) div } \mathbf{B} = 0 . \end{cases}$$

---

[108] See Delcroix [1] and Krall–Trivelpiece [1].

[109] More generally, we shall denote by u_α the density of charged particles of type α and describe it by Vlasov equations; the u_α occur in ρ and $\mathbf{j}$.

5.1. Electrostatics[110]

If $\mathbf{J} = \mathbf{B} = \mathbf{H} = \mathbf{0}$, we are in the domain of electrostatics. We thus have the equations:

(4.82)
$$\begin{cases} \text{(i) } \operatorname{div} \mathbf{D} = \boldsymbol{\rho}\,, \\ \text{(ii) } \operatorname{curl} \mathbf{E} = \mathbf{0}\,. \end{cases}$$

We shall now make precise the regions considered:

5.1.1. Problem in the Whole of Space. If the space is occupied by several different media, for example of perfect dielectric type (see Remark 17), with properties independent of the time, thus corresponding to the physical law $\mathbf{D}_i = \varepsilon_i \mathbf{E}_i$ with ε_i constant in each of the media i, then we can treat the problem either globally in $\mathbb{R}^3_x$ by taking a relation of the type:

(4.83)
$$\mathbf{D} = \varepsilon \mathbf{E} \quad \text{in} \quad \mathbb{R}^3_x\,,$$

with $\varepsilon \equiv \varepsilon(\mathbf{x})$ a piecewise constant function[111], or by treating the problem as a transmission problem, the linking conditions at the interfaces being given (see (4.44) for the notation) on the boundary S_{12} between the media 1 and 2 (for example) by:

(4.84)
$$\begin{cases} \text{(i) } \mathbf{E}^{(2)} \wedge \boldsymbol{\nu} - \mathbf{E}^{(1)} \wedge \boldsymbol{\nu} = \mathbf{0}\,, \\ \text{(ii) } D^{(2)}{}_{\boldsymbol{\nu}} - D^{(1)}{}_{\boldsymbol{\nu}} = \rho_{S_{12}} \end{cases} \qquad (\text{if } \rho_{S_{12}} \text{ is given on } S_{12}{}^{112})\,.$$

In particular, (4.84) (i) implies the continuity of the tangential component of $\mathbf{E}$ on the surface separating the two media considered.

We can reduce the problem (4.82) to a *scalar* problem by noticing that (4.82) (ii) leads to seeking the existence of a function or distribution φ, called a "scalar potential" or "potential" (see Sect. 3: potentials) such that

(4.85)
$$\mathbf{E} = -\operatorname{grad} \varphi\,.$$

If the space is occupied by a single medium obeying $\mathbf{D} = \varepsilon \mathbf{E}$ with ε constant (independent of $\mathbf{x}$) then equation (4.82) (i) implies that φ must satisfy Poisson's equation in $\mathbb{R}^3$ (see Chap. II):

(4.86)
$$\Delta \varphi = -\frac{1}{\varepsilon} \rho\,.$$

We observe the problem of finding a solution $\mathbf{E}$ of (4.82) with (4.83) can be posed in different function spaces, ρ being possibly a (sum of) Dirac distribution(s) or concentrated on a given surface, or even distributed throughout the whole space and such that

$$Q = \int_{\mathbb{R}^3} \rho \,\mathrm{d}x < +\infty\,, \quad \text{or also} \quad \int_{\mathbb{R}^3} \rho^2 \,\mathrm{d}x < +\infty\,.$$

---

[110] See Felici [1].

[111] See Duvaut–Lions [1].

[112] The commonest situation in which this quantity is unknown also gives rise to a transmission problem (see Chap. IXA, §2).

5.1.2. Problem (4.82) with (4.83) in a Bounded Open Set Ω of $\mathbb{R}^3$ Bounded by a "Perfect Conductor". We suppose, for simplicity, that Ω is connected with complement $\Omega' = \mathbb{R}^3 \backslash \Omega$ connected and that Ω is filled with a perfect dielectric medium of permittivity $\varepsilon = 1$ (the vacuum for instance). The boundary condition is here given by (4.47)[113], so:

$$\mathbf{E} \wedge \boldsymbol{\nu} = \mathbf{0} \quad \text{on} \quad \Gamma \equiv \partial\Omega \tag{4.87}$$

with ν normal to Γ (directed towards the exterior of Ω); (4.82) (ii) again implies the existence of a real function φ with

$$E = \operatorname{grad} \varphi \quad \text{in} \quad \Omega . \tag{4.88}$$

Substituting this expression in (4.82)(i), we obtain again Poisson's equation in the open set Ω (with a charge density ρ supposed given)

$$\Delta\varphi = -\rho , \tag{4.89}$$

with the boundary condition:

$$\operatorname{grad} \varphi \wedge \boldsymbol{\nu} = 0 \quad \text{on} \quad \Gamma \equiv \partial\Omega , \tag{4.90}$$

due to (4.87). This implies that the tangential derivative of φ vanishes on the surface $\partial\Omega$, hence that the function φ must remain constant on the boundary $\partial\Omega$. We can always, by subtraction, take this constant to be zero (with the hypotheses made on $\Omega, \Gamma = \partial\Omega$ is connected). The problem (4.82) with (4.83) is thus reduced to a Dirichlet problem in φ which will be studied in Chap. II, §4: *Determine the solution* φ *of* (4.89) *vanishing on the boundary* Γ.

5.2. Magnetostatics[114]

If in (4.81), we have $\rho = 0$, $\mathbf{E} = \mathbf{D} = \mathbf{0}$, we are in the domain of *magnetostatics*. The equations (4.81) then reduce to:

$$\begin{cases} \operatorname{curl} \mathbf{H} = \mathbf{J} , \\ \operatorname{div} \mathbf{B} = 0 . \end{cases} \tag{4.91}$$

An example of the application of these equations is to the magnetic induction created by an electromagnet permeated by an imposed current. A problem which might be encountered is the determination of the distribution of the windings with rotational symmetry allowing us to obtain in a prescribed volume relative variations of the magnetic induction less than a given limit (for example, 10^{-6} for purposes of scientific instrumentation). (See Fournet [1]).

5.2.1. Problem (4.91) in the Whole Space $\mathbb{R}^3$. If the space is made up of several different perfect magnetic media[115] (see Remark 17 and p. 278 of Fournet [1]), each obeying a law $\mathbf{H}_i = \mu_i \mathbf{B}_i$ (μ_i being constant in each of the domains Ω_i where

---

[113] Or again by (4.84) (i), the relation (4.84)(ii) $\mathbf{D}.\boldsymbol{\nu} = \rho_\Gamma$ does not yield a relation here, since we suppose that the surface distribution is unknown.

[114] See Durand [1].

[115] With possibly concentrations of charges at their boundaries.

$\bigcup_i \bar{\Omega}_i = \mathbb{R}^3$) we can again treat the problem globally by assuming a physical law $\mathbf{H} = \mu\mathbf{B}$, with $\mu = \mu(\mathbf{x})$ a piecewise constant function. We can also treat the problem as a transmission problem, the linking conditions at the interfaces being given (see (4.44)) on the interface S_{12} between the media 1 and 2 (for example) by:

$$(4.92) \qquad \begin{cases} \text{(i) } \mathbf{B}^{(2)}.\nu = \mathbf{B}^{(1)}.\nu\,, \\ \text{(ii) } (\mathbf{H}^{(2)} - \mathbf{H}^{(1)}) \wedge \nu = -\mathbf{J}_{S_{12}} \end{cases} \qquad (\text{if } \mathbf{J}_{S_{12}} \text{ is given on } S_{12})\,.$$

In particular (4.29) (i) implies (with the condition of sufficient regularity on both sides of S_{12}) the continuity of the normal component of $\mathbf{B}$ at the interface.

5.2.2. Problem (4.9.1) in a Bounded Open Set Ω of $\mathbb{R}^3$ Whose Boundary is a Perfect Conductor. The boundary conditions are given by (4.46), which can be written as $\mathbf{B}.\nu|_\Gamma = 0$ if the surface current $\mathbf{J}_\Gamma$ is unknown. If the domain Ω is occupied by a perfect magnetic medium, then $\mathbf{H} = \mu\mathbf{B}$ (μ constant), and the problem to be solved is then that of finding $\mathbf{B}$ solution of

$$\begin{cases} \operatorname{curl} \mathbf{B} = \dfrac{\mathbf{J}}{\mu}\,, \qquad \operatorname{div} \mathbf{B} = 0 \quad \text{in} \quad \Omega \\ \mathbf{B}.\nu|_\Gamma = 0\,. \end{cases}$$

Remark 22. *Magnetostatics in an open set not simply connected*[116]. We suppose that $\Omega \subset \mathbb{R}^3$ is bounded and has the shape of a torus[117], that $\Sigma \subset \Omega$ is a surface such that the complement Ω_Σ of Σ in Ω is a simply connected open set (Σ is a "cut" of Ω). Then there exists a non-zero magnetic field $\mathbf{B} \in (L^2(\Omega))^3$ such that:

$$(4.93) \qquad \begin{cases} \text{(i) } \operatorname{curl} \mathbf{B} = 0 \quad \text{in} \quad \Omega\,, \\ \text{(ii) } \operatorname{div} \mathbf{B} = 0 \quad \text{in} \quad \Omega\,, \end{cases}$$

$$(4.94) \qquad \mathbf{B}.\nu = 0 \quad \text{on} \quad \Gamma = \partial\Omega\,.$$

In effect, there exists[118] a function φ, unique to within a non-zero additive constant, satisfying

$$(4.95) \qquad \begin{cases} \Delta\varphi = 0 \quad \text{in} \quad \Omega_\Sigma\,, \\ \varphi \text{ has a jump } [\varphi]_\Sigma = a \text{ (non-zero constant) across } \Sigma\,, \\ \text{the normal derivative of } \varphi \text{ is continuous across } \Sigma\,, \\ \text{the normal derivative of } \varphi \text{ is zero on the boundary } \Gamma \text{ of } \Omega\,. \end{cases}$$

If we then put $\mathbf{B} = \operatorname{grad} \varphi$ on Ω_Σ, we see that $\mathbf{B}$ clearly satisfies (4.93) (i) and (ii) and (4.94). The reader should consult Foias–Temam [1] and Chap. IX for the general

[116] A simply connected open set is such that every continuous closed curve can be deformed continuously until it has shrunk to a point. An example of a simply connected open set in $\mathbb{R}^3$ is the exterior of a sphere; an example of an open set in $\mathbb{R}^3$ which is not simply connected is the interior or exterior of a torus.

[117] We can also envisage the case in which Ω is the space between two coaxial cylinders, occupied by the vacuum or by a perfect magnetic medium.

[118] For a proof, see, for example, Foias–Temam [1] and Chap. IX.

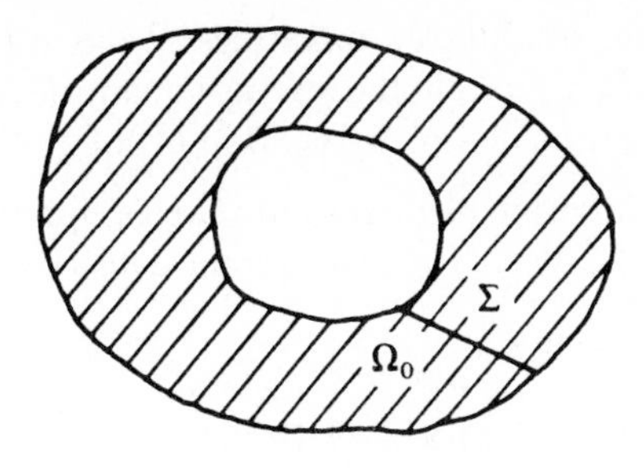

Fig. 8

case of an open set of more complex form (to which several "cuts" have to be made to render it simply connected). This remark is still valid if we replace $\Omega \subset \mathbb{R}^3$ by a domain $\Omega_0 \subset \mathbb{R}^2$ (for example, Ω_0 is the cross-section of a cylinder (see Fig. 8); this will be used in Sect. 6.2.2 to study the propagation of certain electromagnetic fields – called "TEM waves" – in wave guides.) □

6. Stationary Problems

6.1. Introduction and Inhomogeneous Stationary Problems

The expression "stationary problems" demands a precise definition. In this §4, we understand by that, the search for solutions of Maxwell's equations – either (4.1), or (4.19) with (4.37)[119], in the whole space or in an open set $\Omega \subset \mathbb{R}^3$ with (4.47) – which are of the form (see Remark 23 below):

$$\begin{cases} \mathbf{E}(\mathbf{x}, t) = \mathbf{E}_0(\mathbf{x})e^{i\omega t} \\ \mathbf{B}(\mathbf{x}, t) = \mathbf{B}_0(\mathbf{x})e^{i\omega t} \end{cases} \tag{4.96}$$

with ω a non-zero real constant, known or unknown according to the nature of the problem[120].

We say generally that an electromagnetic wave with such a solution is *monochromatic* – the electric field is then made up of functions which are periodic in time (but obviously not every periodic solution of Maxwell's equations is of this form). The constant ω is called the *pulsation* of the electromagnetic field, and $\omega/2\pi$ its *frequency*.

We shall immediately make two remarks.

Remark 23. Until now we have considered the electromagnetic field $\{\mathbf{E}, \mathbf{B}\}$ to be real (with values in $\mathbb{R}^3 \times \mathbb{R}^3$). There is no difficulty at mathematical level in

[119] We could also consider the case in which Ohm's law (4.51) is involved. We should then be led to seek solutions of Maxwell's equations of the form $\mathbf{E}(\mathbf{x}, t) = \mathbf{E}_0(\mathbf{x})e^{\alpha t}$ $\mathbf{B}(\mathbf{x}, t) = \mathbf{B}_0(\mathbf{x})e^{\alpha t}$, with $\alpha \in \mathbb{C}$ and $\operatorname{Re} \alpha < 0$.

[120] For $\omega = 0$, we recover the statical problems of Sect. 5.

representing it by means of complex quantities[121] (i.e. with values in $\mathbb{C}^3 \times \mathbb{C}^3$), which are based on (4.96). To obtain the physical quantities it is easy merely to take the real parts of the components of $\mathbf{E}$ and $\mathbf{B}$ in (4.96). □

Remark 24. Substituting from (4.96) into Maxwell's equations (4.1) (or (4.19) with (4.37)), we see the charge density ρ and current density $\mathbf{j}$ (resp. ρ and $\mathbf{J}$) can be represented by means of

$$(4.97) \qquad \begin{cases} \rho(\mathbf{x}, t) = \rho_0(\mathbf{x})e^{i\omega t} \\ \mathbf{j}(\mathbf{x}, t) = \mathbf{j}_0(\mathbf{x})e^{i\omega t} . \end{cases}$$

We shall therefore suppose in what follows that the charge and current densities are of the form (4.97). □

First of all we mention the following *asymptotic problem*: Being given charge and current densities ρ and $\mathbf{j}$ of the form (4.97) – hence vibrating with frequency $\omega/2\pi$ – will the solution $\{\mathbf{E}(\mathbf{x}, t), \mathbf{B}(\mathbf{x}, t)\}$ of the Cauchy problem formulated by (4.1) and (arbitrary) given initial conditions tend, as $t \to \infty$, to a solution of (4.1) of the form (4.96)? In other words will the electromagnetic field acquire, after a sufficiently long time, the frequency of the charge and current densities? This type of problem is sometimes called the *problem of the forced régime.*
We shall see that in the case where we consider the whole space $\mathbb{R}^3_x$, or the *exterior of a bounded obstacle* or again a *wave guide*, the reply to this question is positive in always giving a particular significance to the expression "to tend to"[122].
The quantities $\{\mathbf{E}_0, \mathbf{B}_0\}$ occurring in (4.96) will then be solutions of the *reduced* Maxwell equations (in the vacuum):

$$(4.98) \qquad \begin{cases} \text{(i)} \; - i\omega\mathbf{E}_0 + \operatorname{curl} \mathbf{B}_0 = \mathbf{j}_0 , \\ \text{(ii)} \; \operatorname{div} \mathbf{E}_0 = \rho_0{}^{123} , \\ \text{(iii)} \; i\omega\mathbf{B}_0 + \operatorname{curl} \mathbf{E}_0 = 0 , \\ \text{(iv)} \; \operatorname{div} \mathbf{B}_0 = 0 , \end{cases}$$

solutions of a particular type – called "outgoing waves" (the sense of this expression will be made precise later in this work).
The practical worker will be interested in the *asymptotic regime* and hence in the study of the system of equations (4.98) with suitable boundary conditions in domains $\Omega \subset \mathbb{R}^3$. In what follows in this Sect. 6 devoted to stationary problems, we shall treat of such problems in the frequently occurring one in which the charge density ρ_0 and current density $\mathbf{j}_0$ are identically zero in Ω. The passage to the case in which ρ_0 and $\mathbf{j}_0$ are non-zero raises no difficulties.

A stationary problem with inhomogeneous boundary conditions. We take the example in which Ω is the exterior of $\mathcal{O}$ a bounded (perfect) conductor, and in which we are interested in the reflexion of a known incident electromagnetic wave

[121] It should be noted that there is sometimes an interest in modelling the electromagnetic field in a complex form in another way, for example by setting $\mathbf{u} = \mathbf{E} + i\mathbf{B}$; see Chap. XVIIB and Itzykson–Zuber [1].

[122] It will be a question of weak convergence or of convergence in "the local energy norm". See Chap. XVIIB.

[123] With the condition: $i\omega\rho_0 + \operatorname{div}\mathbf{j}_0 = 0$ for (4.97) and (4.2).

("coming from infinity") $\{\mathbf{E}_I(\mathbf{x}, t), \mathbf{B}_I(\mathbf{x}, t)\}$ of the type (4.96), which we rewrite

$$(4.99) \qquad \begin{cases} \mathbf{E}_I(\mathbf{x}, t) = \mathbf{E}_1(\mathbf{x})e^{i\omega t} \\ \mathbf{B}_I(\mathbf{x}, t) = \mathbf{B}_1(\mathbf{x})e^{i\omega t} \end{cases}$$

(with $\{\mathbf{E}_1, \mathbf{B}_1\}$ satisfying the reduced Maxwell equations (4.98) with $\rho_0 = 0$ and $\mathbf{j}_0 = \mathbf{0}$ in Ω), a wave which is therefore incident upon the conducting body $\mathcal{O}$. The total electromagnetic field $\{\mathbf{E}(\mathbf{x}, t), \mathbf{B}(\mathbf{x}, t)\}$ is supposed to be of the form (4.96) and to satisfy the boundary conditions (4.47); we seek it in the form:

$$(4.100) \qquad \begin{cases} \mathbf{E}(\mathbf{x}, t) = \mathbf{E}_I(\mathbf{x}, t) + \mathbf{E}_R(\mathbf{x}, t)\,, \\ \mathbf{B}(\mathbf{x}, t) = \mathbf{B}_I(\mathbf{x}, t) + \mathbf{B}_R(\mathbf{x}, t)\,, \end{cases}$$

with $\{\mathbf{E}_R, \mathbf{B}_R\}$ the diffracted electromagnetic wave[124] which is also supposed to be of the form (4.96):

$$(4.101) \qquad \begin{cases} \mathbf{E}_R(\mathbf{x}, t) = \mathbf{E}_0(x)e^{i\omega t}\,, \\ \mathbf{B}_R(\mathbf{x}, t) = \mathbf{B}_0(\mathbf{x})e^{i\omega t}\,, \end{cases}$$

with $\{\mathbf{E}_0, \mathbf{B}_0\}$ a solution of the "reduced" Maxwell equations (4.98) in Ω (with $\rho_0 = 0, \mathbf{j}_0 = 0$) which will ensure that the sum of the incident field and of the "diffracted" field satisfies the reduced Maxwell equations (4.98). In addition, as the incident electromagnetic wave $\{\mathbf{E}_I, \mathbf{B}_I\}$ which is given, does not necessarily satisfy the boundary condition (4.47), $\{\mathbf{E}_0, \mathbf{B}_0\}$ has to satisfy the boundary conditions:

$$(4.102) \qquad \begin{cases} \mathbf{E}_0 \wedge \boldsymbol{\nu} = -\mathbf{E}_1 \wedge \boldsymbol{\nu}\,, \\ \mathbf{B}_0 \cdot \boldsymbol{\nu} = -\mathbf{B}_1 \cdot \boldsymbol{\nu}\,, \end{cases}$$

(with $\mathbf{E}_1$ and $\mathbf{B}_1$ known on Γ). These boundary conditions are said to be inhomogenous.

The conditions: $\{\mathbf{E}_0 . \mathbf{B}_0\}$ solution of (4.98) in Ω (with $\rho_0 = 0$, $\mathbf{j}_0 = 0$) and satisfying (4.102) on the boundary, are not sufficient to determine $\{\mathbf{E}_0, \mathbf{B}_0\}$ in a unique fashion; it is necessary to add a condition describing behaviour – which occurs in physics in a natural way – called "*Sommerfeld's (outgoing) radiation condition*" for ω[125]. This is written

$$(4.103) \qquad \begin{cases} \text{(i) } \operatorname{curl} \mathbf{E}_0 \wedge \dfrac{\mathbf{x}}{r} + i\omega\mathbf{E}_0 = o\left(\dfrac{1}{r}\right) \quad (\text{with } r = |\mathbf{x}| \to \infty)\,, \\ \text{(ii) } \operatorname{curl} \mathbf{B}_0 \wedge \dfrac{\mathbf{x}}{r} + i\omega\mathbf{B}_0 = o\left(\dfrac{1}{r}\right), \\ \text{(iii) } \mathbf{E}_0 = O\left(\dfrac{1}{r}\right), \mathbf{B}_0 = O\left(\dfrac{1}{r}\right)^{126}. \end{cases}$$

[124] The words "reflected" or "diffused" are sometimes used.

[125] This condition is also called "incoming" for $-\omega$ (see Remark 25); this corresponds to the usual convention in physics.

[126] Naturally, as in Remark 23, it would be convenient to take the real parts of the relations (4.103) for the physical quantities – giving the same result. For the meaning of the symbols o and O, see the table of notation $\left(\text{let us say simply that } o\left(\dfrac{1}{r}\right) \text{ denotes a quantity that tends to 0 faster than } \dfrac{1}{r} \text{ as } r \to \infty\right)$.

This condition expresses the fact that the *energy flux* across every sphere of radius R very large, in the sense of $|\mathbf{x}|$ increasing, is *positive* for the reflected electromagnetic wave satisfying the outgoing Sommerfeld condition (4.103). In effect, denoting by S_R the sphere with centre the origin and (very large) radius R, and integrating the energy flux vector (see (4.6) or (4.33)) over S_R we obtain for the flux of energy across S_R:

$$I_{S_R} \stackrel{\text{def}}{=} \int_{S_R} (\mathbf{E}_0 \wedge \mathbf{B}_0) . \boldsymbol{\nu} \, \mathrm{d}\gamma = \int_{S_R} (\mathbf{E}_0 \wedge \mathbf{B}_0) . \frac{\mathbf{x}}{r} \mathrm{d}\gamma$$

where $\mathrm{d}\gamma$ denotes the surface measure of S_R; but with (4.103) and (4.98) we have

$$\mathbf{B}_0 \wedge \frac{\mathbf{x}}{r} - \mathbf{E}_0 = o\left(\frac{1}{r}\right);$$

hence:

$$(\mathbf{E}_0 \wedge \mathbf{B}_0) . \frac{\mathbf{x}}{r} = \mathbf{E}_0 . \left(\mathbf{B}_0 \wedge \frac{\mathbf{x}}{r}\right) = |\mathbf{E}_0|^2 + o\left(\frac{1}{r}\right) \mathbf{E}_0 ;$$

from which, with (4.103) we have finally

$$I_{S_R} = \int_{S_R} |E_0|^2 \mathrm{d}\gamma + o\left(\frac{1}{R}\right) O\left(\frac{1}{R}\right) \times (4\pi R^2) > 0 . \qquad \square$$

The problem of determining $\{\mathbf{E}_0, \mathbf{B}_0\}$ satisfying all the preceding conditions reduces to determining $\mathbf{E}_0$ ($\mathbf{B}_0$ then being given by (4.98)(iii)), which is finally the solution of:

$$(4.104) \quad \begin{cases} \text{(i)} \ \Delta \mathbf{E}_0 + \omega^2 \mathbf{E}_0 = 0 & \text{in } \Omega , \\ \text{(ii)} \ \operatorname{div} \mathbf{E}_0 = 0 & \text{in } \Omega , \\ \text{(iii)} \ \mathbf{E}_0 \wedge \boldsymbol{\nu} = - \mathbf{E}_1 \wedge \boldsymbol{\nu} & \text{on } \Gamma, E_1 \text{ given} \\ \text{(iv)} \begin{cases} \operatorname{curl} \mathbf{E}_0 \wedge \dfrac{\mathbf{x}}{r} + i\omega \mathbf{E}_0 = o\left(\dfrac{1}{r}\right) & \textit{for } r = |\mathbf{x}| \to \infty , \\ \mathbf{E}_0 = O\left(\dfrac{1}{r}\right) . \end{cases} \end{cases}$$

Remark 25. There are other possible boundary conditions, with different physical interpretation: notably that called (Sommerfeld's) "*incoming radiation condition*" for ω[127]; for the electric field $\mathbf{E}_0$ this condition is expressed by:

$$(4.105) \qquad \operatorname{curl} \mathbf{E}_0 \wedge \frac{\mathbf{x}}{r} - i\omega \mathbf{E}_0 = o\left(\frac{1}{r}\right) \quad \text{as} \quad r \to \infty .$$

For such a condition, the energy flux I_{S_R} crossing a sphere S_R of very large radius R in the sense of increasing $|\mathbf{x}|$ should be *negative* corresponding clearly to the

[127] Or, again, outgoing for $- \omega$.

intuitive idea of an "incoming wave". This condition is used to model other types of problem than those considered above. □

The stationary problems with which we shall now be concerned are again *homogeneous* in having $\rho_0 = 0$, $\mathbf{j}_0 = \mathbf{0}$ but also have *homogeneous boundary conditions*. They lead to *spectral problems*.

6.2. Spectral Problems

6.2.1. Introduction. The spectral problems which we shall consider here will be essentially of two types, according to the nature of the open set Ω of $\mathbb{R}^3$ considered
(*a*) *Ω is bounded and regular*: this is the case of cavities. We are often interested in applications to open sets Ω of cylindrical form, bounded by plane sections.
(*b*) *Ω is an infinite cylinder in* $\mathbb{R}^3$: this is the case of wave guides, of power lines, and of optical fibres.
In these two cases, we shall suppose that on the boundary Γ of the open set Ω, the boundary conditions *imposed by the exterior of* Ω are those of the perfect conductor, hence giving the boundary conditions (4.47) which we rewrite below:

$$(4.106)\qquad \begin{cases} \text{(i) } \mathbf{E}_0 \wedge \boldsymbol{v}|_\Gamma = 0\,, \\ \text{(ii) } \mathbf{B}_0 \,.\, \boldsymbol{v}|_\Gamma = 0\,. \end{cases}$$

[128] □

For the sake of simplicity we shall here only treat the case in which Ω is occupied by the vacuum. The case in which Ω would be occupied by a perfect medium – hence such that (4.37) are satisfied with a permittivity ε and a permeability μ – can be deduced without any difficulty of a mathematical nature. Just as in the preceding monochromatic cases, Maxwell's equations (4.98) satisfied by $\{\mathbf{E}_0, \mathbf{B}_0\}$, in the domain Ω are

$$(4.107)\qquad \begin{cases} \text{(i) } \begin{cases} -\operatorname{curl} \mathbf{B}_0 = -i\omega \mathbf{E}_0\,, \\ \operatorname{curl} \mathbf{E}_0 = -i\omega \mathbf{B}_0\,, \end{cases} \\ \text{(ii) } \operatorname{div} \mathbf{E}_0 = \operatorname{div} \mathbf{B}_0 = 0\,, \end{cases}$$

and on putting:

$$(4.108)\qquad \begin{cases} \mathbf{u}_0 \stackrel{\text{def}}{=} \{\mathbf{E}_0, \mathbf{B}_0\}\,, \\ \mathscr{A}\mathbf{u}_0 \stackrel{\text{def}}{=} \{-\operatorname{curl} \mathbf{B}_0, \operatorname{curl} \mathbf{E}_0\}\,, \end{cases}$$

we see that Maxwell's equations in Ω become:

$$(4.109)\qquad \begin{cases} \text{(i) } \mathscr{A}\mathbf{u}_0 = -i\omega \mathbf{u}_0\,, \\ \text{(ii) } \operatorname{div} \mathbf{u}_0 = 0\,. \end{cases}$$

[128] In the problems encountered in technology we proceed in two stages. In the first, we suppose that the boundaries are realised by means of perfect conductors which allow us to impose (4.106). We deduce the electromagnetic field in the whole empty space, and in particular the tangential component of **B** on the boundary of the perfect conductor. In a second stage, the knowledge of this tangential component of **B** allows us to find the weak power transmitted to the real conductor, which, in particular, allows to calculate the attenuation of a wave guide (Fournet [1], p. 414).

We shall presently make precise the operator $\mathscr{A}$ and the chosen functional framework. It seems natural to look *a priori* for solutions $\mathbf{u}_0 = \{\mathbf{E}_0, \mathbf{B}_0\}$ of (4.109) (with (4.106)) of finite energy, so such that

$$(4.110) \qquad W = \frac{1}{2}\int_\Omega |\mathbf{u}_0(x)|^2 \mathrm{d}x \overset{\text{def}}{=} \frac{1}{2}\int_\Omega [|\mathbf{E}_0(x)|^2 + |\mathbf{B}_0(x)|^2]\mathrm{d}x < +\infty\,.$$

Let $\mathscr{H}_c$ be the complex Hilbert space of the couples $\{\mathbf{E}_0, \mathbf{H}_0\}$[129] satisfying (4.110) the scalar product in $\mathscr{H}_c$ being defined by

$$(4.111) \qquad (\mathbf{u}_0', \mathbf{u}_0)_{\mathscr{H}_c} = \int_\Omega (\mathbf{E}_0' : \bar{\mathbf{E}}_0 + \mathbf{B}_0' . \bar{\mathbf{B}}_0)\mathrm{d}x\,,$$

$$\mathbf{u}_0 = \{\mathbf{E}_0, \mathbf{B}_0\}, \quad \mathbf{u}_0' = \{\mathbf{E}_0', \mathbf{B}_0'\}\,.$$

We then define the operator $\mathscr{A}$ acting in the Hilbert space $\mathscr{H}_c$ by (4.108) and by its domain[130]:

$$(4.112) \qquad D(\mathscr{A}) \overset{\text{def}}{=} \{u_0 = \{\mathbf{E}_0, \mathbf{B}_0\} \in \mathscr{H}_c;$$

$$\mathscr{A}u_0 = \{-\operatorname{curl}\mathbf{B}_0, \operatorname{curl}\mathbf{E}_0\} \in \mathscr{H}_c, \mathbf{E}_0 \wedge v|_\Gamma = 0\}\,.$$

The study of this operator (see Chap. IXA) shows that the operator $i\mathscr{A}$ is self-adjoint[130] in $\mathscr{H}_c$.

Remark 26. We have not taken account, for the definition (4.112), of the conditions (4.107) (ii) and (4.106) (ii). We would be able to take account of this condition by considering the sub-space $\mathscr{H}_{1c}$ of $\mathscr{H}_c$ defined by:

$$\mathscr{H}_{1c} = \{\{\mathbf{E}_0, \mathbf{B}_0\} \in \mathscr{H}_c, \quad \operatorname{div}\mathbf{E}_0 = \operatorname{div}\mathbf{B}_0 = 0, \quad \mathbf{B}_0 . v|_\Gamma = 0\}$$[131].

In fact in the case studied here where $\omega \neq 0$, we deduce the relation (4.109) (ii) (or (4.107) (ii)) from the relation (4.109) (i) (or (4.107) (i)) (by application of the divergence operator to these formulas) since $\operatorname{div}\operatorname{curl}\mathbf{E}_0 = \operatorname{div}\operatorname{curl}\mathbf{B}_0 = \operatorname{div}\mathscr{A}\mathbf{u}_0 = 0$. In addition it is not necessary to impose the boundary condition (4.106) which is then automatically satisfied (see Remark 14). Thus, with this definition of the operator $\mathscr{A}$ by (4.112), the equation (4.109) (i) is equivalent, for $\omega \neq 0$, to (4.107) and (4.106) and the relations (4.109) (i), (4.107) (ii), and (4.106) (ii) are redundant. □

In this introduction to spectral problems we have not yet used the hypothesis: Ω bounded or not. But this hypothesis is fundamental for the results which we shall indicate presently and which will be proved later (see Chap. IX in the case in which Ω is bounded – see Sect. 6.2.2 of this §4 in the case where Ω is an infinite cylinder).

---

[129] We therefore take $\mathbf{E}_0(\mathbf{x}) \in \mathbb{C}^3$ and $\mathbf{B}_0(\mathbf{x}) \in \mathbb{C}^3$ for stationary problems (see Remark 23). For the other problems considered in the §4 (evolution problems, Sects. 1 to 4; statical problems, Sect. 5), the electric field and the magnetic induction have values in $\mathbb{R}^3$, and it is then more natural to consider the real Hilbert space $\mathscr{H}$ of the couples $\{\mathbf{E}_0, \mathbf{B}_0\}$ satisfying (4.110).

[130] See Chap. VI.

[131] See Chap. IXA. It should be noted that the words *eigenvalue* and *eigenvector* are used rather than *proper value* and *proper vector*.

(*a*) *The domain* Ω *is bounded and regular*. We show[131] that there exists a denumerable family of values $\omega_k \in \mathbb{R}$ (called *eigenvalues*) for which there exists a solution $\mathbf{u}_0 = \{\mathbf{E}_0, \mathbf{B}_0\}$ of (4.109) in $\mathscr{H}_c$ (hence of *finite energy*) called an *eigenvector* of $i\mathscr{A}$, or again a "*proper mode*". The set of the ω_k (with possibly adding O) constitutes the "spectrum" of the operator $i\mathscr{A}$.
Further:
(1) for each $\omega_k \neq 0$, there exists a finite number n_k of independent eigenvectors – we say that the *multiplicity* or the *degeneracy* of ω_k is finite. Let $(u_{0k,j})$, with $j = 1$ to n_k, be such a family of independent eigenvectors taken to be orthogonal and normalized to unity.
(2) The family of the $(u_{0k,j})$ for $j = 1$ to n_k, $k \in \mathbb{N}$, form an orthonormal base of the sub-space of $\mathscr{H}_c$ orthogonal to the "kernel" of $\mathscr{A}$ (denoted by ker $\mathscr{A}$) defined by:

$$\ker \mathscr{A} \overset{\text{def}}{=} \{\{\mathbf{E}_0, \mathbf{B}_0\} \in \mathscr{H}_c,\quad \operatorname{curl} \mathbf{E}_0 = \operatorname{curl} \mathbf{B}_0 = 0,\ \mathbf{E}_0 \wedge \boldsymbol{v}|_\Gamma = 0\}\,. \tag{4.113}$$

(3) The only points of accumulation (or limit points) of the ω_k are $\pm\infty$.
(*b*) *The domain* Ω *is an (infinite) cylinder*[132] *of* $\mathbb{R}^3$ *with axis* Ox_3.
We can show[133] that there exist no non-zero value $\omega \in \mathbb{R}$ or even $\omega \in \mathbb{C}$ with: $\mathbf{u}_0 = \{\mathbf{E}_0, \mathbf{B}_0\}$ a non-null solution of (4.109) in $\mathscr{H}_c$, so with finite energy.
On the contrary, we can find values $\omega \in \mathbb{R}^* = \mathbb{R}\backslash\{0\}$ for which there exists a solution $\mathbf{u}_0 = \{\mathbf{E}_0, \mathbf{B}_0\}$ of (4.109)[134] belonging not to $\mathscr{H}_c$ but to the space $D(\mathscr{A})_{\text{loc}}$ defined by:
$D(\mathscr{A})_{\text{loc}}$ is the set of couples $\{\mathbf{E}_0, \mathbf{B}_0\}$ with $\mathbf{E}_0(\mathbf{x}) \in \mathbb{C}^3$, $\mathbf{B}_0(\mathbf{x}) \in \mathbb{C}^3$ such that, for every compact set $K \subset \mathbb{R}^3$ we have the three conditions:

$$W_{0K} \overset{\text{def}}{=} \frac{1}{2}\int_{\Omega \cap K} [|\mathbf{E}_0|^2 + |\mathbf{B}_0|^2]\mathrm{d}\mathbf{x} < +\infty\,. \tag{4.114}$$

$$\int_{\Omega \cap K} [|\operatorname{curl} \mathbf{E}_0|^2 + |\operatorname{curl} \mathbf{B}_0|^2]\mathrm{d}\mathbf{x} < +\infty \tag{4.115}$$

with the boundary condition

$$\mathbf{E}_0 \wedge \boldsymbol{v} = 0 \quad \text{on} \quad \Gamma\,. \tag{4.116}$$

The condition (4.114) expresses that $\{\mathbf{E}_0, \mathbf{B}_0\}$ *has finite local energy right up to the boundary of the domain* Ω (the energy in every *bounded* domain possibly containing a part of the boundary of Ω is finite).
The condition (4.115) expresses that the operator $\mathscr{A}$ acts[134] on this space *with finite local energy*, which also allows us to give a meaning to the boundary condition (4.116) (see the idea of trace given in Chaps. IV, §4 and IXA).

---

[132] The results which we state are still valid when, instead of an (infinite) cylinder, we consider a connected open set Ω which is the complement of a bounded obstacle $\mathcal{O}$.
[133] This follows from the study made below in Sect. 6.2.2.
[134] Here $\mathscr{A}$ is considered to be a linear differential operator (see Chap. V) and not as an unbounded linear operator in $\mathscr{H}_c$ (see Chap. VI).

Every solution $\mathbf{u}_0 = \{\mathbf{E}_0, \mathbf{B}_0\}$ of (4.109) such that $\mathbf{u}_0 \notin \mathcal{H}_c$, but $\mathbf{u}_0 \in D(\mathcal{A})_{\text{loc}}$ satisfying in addition the following condition at infinity (in x_3):

(4.117) $\mathbf{u}_0$ is a tempered distribution[135] in x_3[136].

$\mathbf{u}_0$ is called a "*generalized eigenvector of* $\mathcal{A}$"[137] or again a "*proper mode*". We can show[138] that every $\omega \in \mathbb{C}^* = \mathbb{C}\backslash\{0\}$ for which a generalized eigenvector $\mathbf{u}_0$ exists is an element of the spectrum[137] $\sigma(i\mathcal{A})$ of the operator $i\mathcal{A}$. The study which follows enables us thus to obtain information on the spectrum of $i\mathcal{A}$ (see Remark 32), hence of the operator $\mathcal{A}$.

6.2.2. Spectral Problem for Wave Guides. We propose here to give rapid indications of the solution of spectral problems in the case in which Ω is an (infinite) cylinder in $\mathbb{R}^3$ (with axis Ox_3 and with cross section Ω_0) in the framework stated in Sect. 6.2 – that is to say Ω occupied by the vacuum and its boundary $\partial\Omega$ separating it from a perfect conductor.

Hence we propose to solve (4.109) in this particular case. As a result of the invariance of this problem by translation along Ox_3 (invariance of the operator $\mathcal{A}$ with its domain $D(\mathcal{A})$ and of the open set Ω by translation along Ox_3), we are led to seek the solutions $\{\mathbf{E}_0, \mathbf{B}_0\}$ of (4.109) of the form[139]

$$\text{(4.118)} \quad \begin{cases} \mathbf{E}_0(\mathbf{x}) \equiv \mathbf{E}_0(x_1, x_2, x_3) = e^{ikx_3}\tilde{\mathbf{E}}(x_1, x_2, k) & \tilde{\mathbf{E}} = (\tilde{E}_1, \tilde{E}_2, \tilde{E}_3)\,; \\ \mathbf{B}_0(\mathbf{x}) \equiv \mathbf{B}_0(x_1, x_2, x_3) = e^{ikx_3}\tilde{\mathbf{B}}(x_1, x_2, k) & \tilde{\mathbf{B}} = (\tilde{B}_1, \tilde{B}_2, \tilde{B}_3)\,, \end{cases}$$

with $\mathbf{x} \in \Omega$ and k *being a real unknown constant to be determined*, and with the conditions:

$$\text{(4.119)} \quad \tilde{\mathbf{E}}(x_1, x_2, k) \text{ and } \tilde{\mathbf{B}}(x_1, x_2, k) \in (L^2(\Omega_0))^3 \quad \Omega_0 \subset \mathbb{R}^2\,.$$

We should observe that the conditions (4.119) imply that the electromagnetic field $\{\mathbf{E}, \mathbf{B}\}$ given by (4.96) with (4.118) cannot have finite energy in Ω; on the contrary it will have finite energy per unit length – or again finite local energy (see (4.114) and (4.115)).

Remark 27. The equations (4.109) in the present case of wave guides model a situation in which we have a source of energy a long way off – said to be at infinity – called an (electromagnetic) generator of the wave guide. The pulsation ω of a solution of (4.109) is fixed by this generator; it is thus a "physical" datum, independent *a priori* of the shape and of the size of the wave guide. The generator

---

[135] See Appendix "Distributions" in Vol. 2.

[136] In the case where in place of an (infinite) cylinder, Ω is the complement of a bounded obstacle, the condition (4.117) must be replaced by:

$$\begin{cases} \mathbf{u}_0 \text{ is "tempered at infinity" (i.e. behaves at "infinity" like an element of } \mathcal{S}'\text{)} \\ \text{and} \\ \mathbf{u}_0 \text{ satisfies an outgoing (incoming) Sommerfeld radiation condition (4.103) ((4.105)).} \end{cases}$$

[137] See Chap. VIII.

[138] For that it is necessary that the conditions: "$\mathbf{u}_0 \in D(\mathcal{A})_{\text{loc}}$ and $\mathbf{u}_0$ satisfies (4.117)" are satisfied.

[139] Which enables us to carry out a Fourier transform in x_3.

provides a certain current $\mathbf{J}$ (*which is not given*) which we shall see is concentrated on the surface of the wave guide. □

We shall demonstrate that the solution of the problem (4.109), taking account of (4.118), leads to three types of solutions, independent, and orthogonal in $(L^2(\Omega_0))^3 \times (L^2(\Omega_0))^3$:
(1) solutions called "*transverse magnetic*" (or sometimes of "*electric type*" or again "*TM waves*") corresponding to a purely transverse magnetic field (hence the notation TM), i.e. such that $\tilde{\mathbf{B}}_3 = 0$ (and $\tilde{\mathbf{E}}_3 \neq 0$);
(2) solutions called "*transverse electric*" (or sometimes of "*magnetic type*" or again "*TE-waves*"), corresponding to a purely transverse electric field (hence the notation TE), i.e. such that $\tilde{\mathbf{E}}_3 = 0$ (and $\tilde{\mathbf{B}}_3 \neq 0$);
(3) solutions called "*transverse electromagnetic*" (or TEM-waves) such that $\tilde{\mathbf{E}}_3 = \tilde{\mathbf{B}}_3 = 0$.
For clarity of the exposition, we shall state several propositions. We shall thus show the necessity of studying these three types of waves – *TM*, *TE* and *TEM*. More precisely we shall prove the following result:

Proposition 1. *Let* $\mathbf{u}_0 = \{\mathbf{E}_0, \mathbf{B}_0\}$ *be a solution in* $D(\mathscr{A})_{\text{loc}}$[140] *of* (4.109) *of the form* (4.118) *for the constants* ω *and* k *both real. Then, with*

$$\alpha^2 = \omega^2 - k^2 \tag{4.120}$$

the component $\tilde{E}_3$ *of* $\mathbf{E}$ (*see* (4.118) *and* (4.119)) *satisfies Helmholtz's equation*

$$\begin{cases} \text{(i) } \Delta\tilde{E}_3 + \alpha^2\tilde{E}_3 = 0 & \textit{in} \quad \Omega_0\,(\textit{bounded}), \\ \text{(ii) } \tilde{E}_3 = 0 & \textit{on} \quad \Gamma_0 = \partial\Omega_0, \end{cases} \tag{4.121}$$
$$\textit{with} \quad \tilde{E}_3 \in L^2(\Omega_0);$$

the component $\tilde{B}_3$ *of* $\tilde{\mathbf{B}}$ *satisfies*

$$\begin{cases} \text{(i) } \Delta\tilde{B}_3 + \alpha^2\tilde{B}_3 = 0 & \textit{in} \quad \Omega_0, \\ \text{(ii) } \dfrac{\partial\tilde{B}_3}{\partial\nu} = 0 & \textit{on} \quad \Gamma_0 = \partial\Omega_0, \end{cases} \tag{4.122}$$
$$\textit{with} \quad \tilde{B}_3 \in L^2(\Omega_0).$$

Commentary[141]. Every solution $\tilde{E}_3 \neq 0$ with $\tilde{E}_3 \in L^2(\Omega_0)$ of the problem (4.121) is called an *eigenvector of the Laplacian* with the Dirichlet condition. Every solution $\tilde{B}_3 \neq 0$ with $\tilde{B}_3 \in L^2(\Omega_0)$ of the problem (4.122) is called an eigenvector of the Laplacian with the Neumann condition.
The value α^2 is called the *eigenvalue* of the operator $(-\Delta)$ with Dirichlet condition for (4.121) with Neumann condition for (4.122). The problem called "*spectral*" consists of determining the set of the eigenvalues α^2 and the eigenvectors $\tilde{E}_3$ of (4.121) (resp. $\tilde{B}_3$ of (4.122)); it will be studied in Chap. VIII.

---

[140] This contains the boundary condition (4.116).

[141] See Chap. VIII for the whole of this commentary.

We shall find in both cases (problems (4.121) and (4.122)) an infinite, denumerable set of eigenvalues α^2 positive without accumulation point at finite distance (and there exists a set of corresponding eigenvectors forming an orthonormal base of the Hilbert space $L^2(\Omega_0)$). The quantity α is determined by the geometry of the cross-section Ω_0 of the wave guide. We say also (see Roubine [1]) that α is a "mathematical" datum while ω is a "physical" datum, characteristic of the electromagnetic generator used (see Remark 27); starting with ω and α, we will calculate the real k, thanks to (4.120), by: $k^2 = \omega^2 - \alpha^2$.

Proof of Proposition 1. Applying the operator $\mathscr{A}$ to (4.109) – or the operator curl to (4.107)(i) – and using (4.109)(ii) – or (4.107)(ii) – we find that:

(4.123) $$\mathscr{A}^2\mathbf{u}_0 = + \Delta\mathbf{u}_0 = - \omega^2\mathbf{u}_0 .$$

It follows that each of the components $\tilde{E}_i$ and $\tilde{B}_j$ of $\mathbf{E}$ and $\mathbf{B}$ satisfies (4.120) and Helmholtz's equation

(4.124) $$\begin{cases} \Delta_2\tilde{E}_i + \alpha^2\tilde{E}_i = 0 , & i = 1 \text{ to } 3 , \\ \Delta_2\tilde{B}_j + \alpha^2\tilde{B}_j = 0 , & j = 1 \text{ to } 3 . \end{cases}$$

Now let us exploit the boundary conditions (4.106): (4.121)(ii) follows immediately from (4.106)(i). In addition, this condition (4.106)(i), together with (4.107) implies that

(4.125) $$\operatorname{curl}\mathbf{B}_0 \wedge \boldsymbol{v}|_\Gamma = \mathbf{0} .$$

Writing this out we have

(4.126) $$\begin{cases} \text{(i)} \ \dfrac{\partial}{\partial x_1}\tilde{B}_2 - \dfrac{\partial}{\partial x_2}\tilde{B}_1 = 0 & \text{on } \Gamma , \\ \text{(ii)} \ \left(v_1\dfrac{\partial}{\partial x_1}\tilde{B}_3 + v_2\dfrac{\partial}{\partial x_2}\tilde{B}_3\right) - ik(v_1\tilde{B}_1 + v_2\tilde{B}_2) = 0 & \text{on } \Gamma , \end{cases}$$

and, making use of the boundary condition (4.126)(ii), we find the condition (4.122)(ii). □

Consequences of Proposition 1. For there to exist a non-zero solution $\tilde{E}_3$ of (4.121) (resp. $\tilde{B}_3$ of (4.122)) it is necessary that we should have $\alpha^2 \geqslant 0$, (see Chap. VIII) and therefore

(4.127) $$|k| \leqslant |\omega| .$$

Thus for each fixed ω, there exists, at most, a finite number of possible values k satisfying (4.120).

Since there exists a smallest, strictly positive eigenvalue $(\alpha_0^D)^2$ in the case of the problem (4.121)[142], it is necessary that the data of the problem (4.109) are such that

(4.128) $$|\omega| \geqslant \alpha_0^D$$

[142] See Chaps. VIII and IXA.

if we are to have a non-zero solution of (4.121). The limiting pulsation α_0^D is called the *cut-off* pulsation of the wave guide.
In the case of the problem (4.122) the smallest eigenvalue is zero[142], and the corresponding eigenvector is $\tilde{B}_3$ = constant.
We shall now prove:

Corollary 1. *If* $\mathbf{u}_0 = \{\mathbf{E}_0, \mathbf{B}_0\}$ *is a solution in* $D(\mathscr{A})_{\text{loc}}$ *of* (4.109) *of the form* (4.118) *with* $\tilde{B}_3$ = *constant*[143], *then necessarily* $\tilde{B}_3 = 0$.

Proof. It follows immediately from the Proposition that such a solution must correspond to $\alpha = 0$ and $\tilde{E}_3 = 0$.
In addition, the formula (4.107)(i) implies:

$$-i\omega\tilde{B}_3 = \frac{\partial}{\partial x_1}\tilde{E}_2 - \frac{\partial}{\partial x_2}\tilde{E}_1\,,$$

from which, with the notation $|\Omega_0|$ = area of Ω_0, we deduce that

$$\int_{\Omega_0}\omega^2|\tilde{B}_3|^2\mathrm{d}x = \omega^2|\tilde{B}_3|^2|\Omega_0| = \int_{\Omega_0}\left|\frac{\partial}{\partial x_1}E_2 - \frac{\partial}{\partial x_2}E_1\right|^2\mathrm{d}x\,.$$

Integrating by parts, we obtain

$$\begin{aligned}\omega^2|\tilde{B}_3|^2|\Omega_0| = &\int_{\Omega_0}\Big[-E_2\frac{\partial}{\partial x_1}\left(\frac{\partial}{\partial x_1}\bar{E}_2 - \frac{\partial}{\partial x_2}\bar{E}_1\right)\\ &+ E_1\frac{\partial}{\partial x_2}\left(\frac{\partial}{\partial x_1}\bar{E}_2 - \frac{\partial}{\partial x_2}\bar{E}_1\right)\Big]\mathrm{d}x\\ &+\int_{\Gamma_0}(v_2E_1 - v_1E_2)\left(\frac{\partial}{\partial x_1}\bar{E}_2 - \frac{\partial}{\partial x_2}\bar{E}_1\right)\mathrm{d}\gamma\,.\end{aligned}$$

Now making use of the relations

$$\operatorname{div}\mathbf{E} = \frac{\partial E_1}{\partial x_1} + \frac{\partial E_2}{\partial x_2} = 0 \qquad \nu\wedge\mathbf{E}|_{\Gamma_0} = 0\,,$$

we obtain:

$$\omega^2|\tilde{B}_3|^2|\Omega_0| = -\int_{\Omega_0}(E_2\Delta_2\bar{E}_2 + E_1\Delta_2\bar{E}_1)\mathrm{d}x = 0 \quad \text{so that} \quad \tilde{B}_3 = 0\,. \quad \square$$

Corollary 1 implies that the first permissible value of α for (4.122) – i.e. leading to a solution $\mathbf{u}_0 = \{\mathbf{E}_0, \mathbf{B}_0\}$ of (4.109) with (4.118) for which $\tilde{B}_3 \neq 0$) – is the first non-zero eigenvalue $(\alpha_1^N)^2$ of the (Neumann) problem (4.122).
We therefore have:

Corollary 2. *For the existence of a solution* $\mathbf{u}_0$ *in* $D(\mathscr{A})_{\text{loc}}$ *of* (4.109) *with* (4.118) *such that* $\tilde{E}_3 \neq 0$ *(resp.* $\tilde{B}_3 \neq 0$*), it is necessary that the data of the problem* (4.109) *satisfy*

$$|\omega| \geqslant \alpha_0^D \quad (\text{resp. } |\omega| \geqslant \alpha_1^N)\,. \tag{4.129}$$

(α_0^D *is the cut-off pulsation*).

---

[143] That is to say, independent of x_1 and x_2.

Remark 28. *For the existence of a non-zero solution* $\mathbf{u}_0$ *of* (4.109) *with* (4.118) *such that* $\tilde{E}_3 = 0$ *and* $\tilde{B}_3 = 0$ *it is necessary that the numbers* k *and* ω *be linked by the relation* $|k| = |\omega|$.
In effect, if we put

(4.130) $$\chi \stackrel{\text{def}}{=} -\frac{k}{\omega}, \quad (\omega \neq 0),$$

we can then rewrite the relations (4.107) in the form

(4.131) $$\tilde{E}_1 = \chi \tilde{B}_2, \quad \tilde{E}_2 = -\chi \tilde{B}_1, \quad \tilde{B}_1 = -\chi \tilde{E}_2, \quad \tilde{B}_2 = \chi \tilde{E}_1,$$

and

(4.132) $$\frac{\partial}{\partial x_1}\tilde{B}_2 - \frac{\partial}{\partial x_2}\tilde{B}_1 = 0, \quad \frac{\partial}{\partial x_1}\tilde{E}_2 - \frac{\partial}{\partial x_2}\tilde{E}_1 = 0.$$

For the existence of a solution of (4.131), not identically zero, it is necessary to have $\chi^2 = 1$, and hence $|k| = |\omega|$. □

We shall now make precise the expressions for the various waves – TM, TE and TEM.

Proposition 2 (TM and TE waves). *Every non-zero solution* $\mathbf{u}_0$ *in* $D(\mathscr{A})_{\text{loc}}$ *of* (4.109) *with* (4.118) *such that:* (1) $\tilde{B}_3 = 0$ (*"TM-wave"*), *is given with the aid of its component* E_3 *non-zero solution of* (4.121) *by:*

(4.133) $$\tilde{\mathbf{E}}_T = \frac{ik}{\alpha^2}\operatorname{grad}_T E_3 \quad \text{with} \quad \tilde{\mathbf{E}}_T = (\tilde{E}_1, \tilde{E}_2),\ \operatorname{grad}_T E_3 = \left(\frac{\partial E_3}{\partial x_1}, \frac{\partial E_3}{\partial x_2}\right)$$

with k *defined from the given* ω *and from* α *by* $k^2 = \omega^2 - \alpha^2$ [144] (cf. (4.120)), $\alpha < |\omega|$

(4.134) $$\begin{cases} \tilde{B}_1 = -\theta \tilde{E}_2 \\ \tilde{B}_2 = \theta \tilde{E}_1 \end{cases} \quad \text{in } \Omega_0 .$$

where $\theta = -\omega/k$, $(k \neq 0,\ |\theta| > 1)$. *Thus the vector* $\tilde{\mathbf{B}}_T(\mathbf{x}) \stackrel{\text{def}}{=} \{\tilde{B}_1(\mathbf{x}), \tilde{B}_2(\mathbf{x})\}$ *is orthogonal in* $\mathbb{C}^2$ *to the vector* $\tilde{\mathbf{E}}_T(\mathbf{x}) = \{\tilde{E}_1(\mathbf{x}), \tilde{E}_2(\mathbf{x})\}$ *and*

$$|\tilde{\mathbf{B}}_T(\mathbf{x})| = |\theta|\,|\tilde{\mathbf{E}}_T(\mathbf{x})| > |\tilde{\mathbf{E}}_T(\mathbf{x})|, \qquad \forall \mathbf{x} \in \Omega_0 .$$

The TM-wave corresponding to $\mathbf{u}_0^{\pm}$ *is therefore given by*[145]

$$\mathbf{E}^{\pm}(\mathbf{x}, t) = [\exp i[\omega t \pm x_3(\omega^2 - \alpha^2)^{1/2}]].\{\tilde{\mathbf{E}}_T, \tilde{E}_3\}$$

$$\mathbf{B}^{\pm}(\mathbf{x}, t) = [\exp i[\omega t \pm x_3(\omega^2 - \alpha^2)^{1/2}]].\{\tilde{\mathbf{B}}_T, 0\};$$

(2) $\tilde{E}_3 = O$ (*"TE-wave"*), *is given in terms of its component* $\tilde{B}_3$, *non-zero (and non-constant) solution of* (4.122) *by:*

(4.135) $$\tilde{B}_T = \frac{ik}{\alpha^2}\operatorname{grad}_T \tilde{B}_3 \qquad \tilde{B}_T = (\tilde{B}_1, \tilde{B}_2),$$

---

[144] We are led to two independent solutions $\mathbf{u}_0^+$ and $\mathbf{u}_0^-$ of (4.109) and $\tilde{B}_3 = 0$; the solution. $\mathbf{u}_0^+$ (resp. $\mathbf{u}_0^-$) corresponding to $k > 0$ (resp. $k < 0$). We pass from one to the other by changing k by $-k$.
[145] Where $\{\tilde{\mathbf{E}}_T, \tilde{\mathbf{E}}_3\}$ denotes the vector with components $\tilde{\mathbf{E}}_T$ in the Ox_1x_2 plane and $\tilde{E}_3$ along Ox_3.

with k defined from the given ω and from α by $k^2 = \omega^2 - \alpha^2$ (cf. (4.120))[146] (so *that k can take the two values $\pm\sqrt{(\omega^2 - \alpha^2)}$ with $\alpha < |\omega|$.*

$$\tilde{E}_1 = \theta\tilde{B}_2, \quad \tilde{E}_2 = -\theta\tilde{B}_1\,, \tag{4.136}$$

where θ is defined as above.

Thus the vector $\tilde{\mathbf{E}}_T(\mathbf{x}) \overset{\text{def}}{=} \{\tilde{E}_1(\mathbf{x}), \tilde{E}_2(\mathbf{x})\}$ is orthogonal in $\mathbb{C}^2$ to the vector $\tilde{B}_T(\mathbf{x})$ and

$$|\tilde{\mathbf{E}}_T(\mathbf{x})| = |\theta||\tilde{\mathbf{B}}_T(\mathbf{x})| > |\tilde{\mathbf{B}}_T(\mathbf{x})| \qquad \forall \mathbf{x} \in \Omega_0\,.$$

The corresponding TE-wave is thus given by

$$\begin{aligned}\mathbf{E}^{\pm}(x, t) &= [\exp i[\omega t \pm x_3(\omega^2 - \alpha^2)^{1/2}]]\{\tilde{\mathbf{E}}_T, 0\}\\ \mathbf{B}^{\pm}(x, t) &= [\exp i[\omega t \pm x_3(\omega^2 - \alpha^2)^{1/2}]]\{\tilde{\mathbf{B}}_T, \tilde{B}_3\}\,.\end{aligned}$$

Proof. The proof of these results presents no particular difficulty. It is sufficient to write (4.109) – or (4.107), taking account of the simplifications due to (4.118) and to the hypothesis $\tilde{B}_3 = 0$ (resp. $\tilde{E}_3 = 0$). Thus we obtain in the case in which $\tilde{B}_3 = 0$, in Ω_0:

$$\begin{cases}\text{(i)}\ ik\tilde{B}_2 = -i\omega\tilde{E}_1\,,\\ \text{(ii)}\ -ik\tilde{B}_1 = -i\omega\tilde{E}_2\,,\\ \text{(iii)}\ -\dfrac{\partial}{\partial x_1}\tilde{B}_2 + \dfrac{\partial}{\partial x_2}\tilde{B}_1 = -i\omega\tilde{E}_3\,,\\ \text{(i)}'\ \dfrac{\partial}{\partial x_2}\tilde{E}_3 - ik\tilde{E}_2 = -i\omega\tilde{B}_1\,,\\ \text{(ii)}'\ ik\tilde{E}_1 - \dfrac{\partial}{\partial x_1}\tilde{E}_3 = -i\omega\tilde{B}_2\,,\\ \text{(iii)}'\ \dfrac{\partial}{\partial x_1}\tilde{E}_2 - \dfrac{\partial}{\partial x_2}\tilde{E}_1 = 0\,.\end{cases} \tag{4.137}$$

(4.134) follows immediately from (4.137) (i) and (ii); and (4.113) follows from (4.137)(i)′ and (ii)′. The equations (4.135) and (4.136) are obtained in a similar fashion. □

Proposition 3. *Let $\mathbf{u}_0 = \{\mathbf{E}_0, \mathbf{B}_0\}$ be a non-zero solution, in $D(\mathscr{A})_{\text{loc}}$ of* (4.109) *of the form* (4.118) *for the real constants ω and k; then one of the following four cases is necessarily realised*:

(1) *$|k| < |\omega|$, and $\alpha^2 = \omega^2 - k^2$ is an eigenvalue of the "Dirichlet" problem* (4.121), *but is not an eigenvalue of the "Neumann" problem* (4.122); *then $\mathbf{u}_0$ is a TM-wave* ($\tilde{B}_3 = 0$).

---

[146] This again leads to two independent solutions of (4.109) with (4.118) and $\tilde{E}_3 = 0$.

(2) $|k| < |\omega|$, *and* $\alpha^2 = \omega^2 - k^2$ *is an eigenvalue of the Neumann problem* (4.122), *but is not an eigenvalue of the Dirichlet problem* (4.121); *then* $\mathbf{u}_0$ *is a TE-wave* ($\tilde{E}_3 = 0$).
(3) $|k| < |\omega|$, *and* $\alpha^2 = \omega^2 - k^2$ *is an eigenvalue*[147] *simultaneously of the Dirichlet problem* (4.121) *and of the Neumann problem* (4.122); *then* $\mathbf{u}_0$ *can be written in a unique way*:

$$\mathbf{u}_0 = \mathbf{u}_0^{TM} + \mathbf{u}_0^{TE} \tag{4.138}$$

with: $\mathbf{u}_0^{TM} = \{\mathbf{E}_0^{TM}, \mathbf{B}_0^{TM}\}$ *such that* $\tilde{E}_3^{TM} = \tilde{E}_3$, $\tilde{B}_3^{TM} = 0$;

$\mathbf{u}_0^{TE} = \{\mathbf{E}_0^{TE}, \mathbf{B}_0^{TE}\}$ *such that* $\tilde{E}_3^{TM} = 0$, $\tilde{B}_3^{TE} = \tilde{B}_3$;

$\mathbf{u}_0^{TM}$ (resp. $\mathbf{u}_0^{TE}$) *is thus the TM- (resp. TE-) wave characterised by* $\tilde{E}_0^{TM} = \tilde{E}_3$ *solution of the problem* (4.121)(*resp.* $\tilde{B}_3^{TE} = \tilde{B}_3$, *solution of the problem* (4.122)), *the other components of* u_0^{TM} (*resp.* u_0^{TE}) *being given in terms of* $\tilde{E}_3^{TM}$ (*resp.* $\tilde{B}_3^{TE}$) *by the formulae* (4.133), (4.134) (*resp.* (4.135) *and* (4.136)).
(4) $|k| = |\omega|$. *Then* $\mathbf{u}_0$ *is a TEM-wave*[148].

Proof. Points (1) and (2) are immediate.
To prove (4.1.38), it is enough to observe that $\mathbf{u} - \mathbf{u}_0^{TM}$ is, by the same construction of $\mathbf{u}_0^{TM}$, a *TE*-wave and hence can be written in the form $\mathbf{u}_0^{TE}$ indicated.
Finally, the point (4) follows from Corollary 1. □

We have thus shown that every non-zero solution $\mathbf{u}_0$ in $D(\mathscr{A})_{\text{loc}}$ of (4.109) of the form (4.118) is – apart from the exceptional case (3) – a *TE* wave, a *TM*-wave or a TEM-wave.

Remark 29. Let us now calculate the scalar product (in $(L^2(\Omega_0))^3 \times (L^2(\Omega_0))^3$) of two solutions $\mathbf{u}_0 = \{\mathbf{E}_0, \mathbf{B}_0\}$ and $\mathbf{u}'_0 = \{\mathbf{E}'_0, \mathbf{B}'_0\}$ (belonging to $D(\mathscr{A})_{\text{loc}}$), of (4.109) of the form (4.118), (where x_3 is regarded as a parameter), corresponding, respectively to the constants (ω, k) and (ω', k'). Let us suppose, for example, that $\mathbf{u}'_0$ is a *TM*-wave.
This scalar product being given by

$$(\mathbf{u}'_0, \mathbf{u}_0) = \int_{\Omega_0} (\mathbf{E}'_0 . \bar{\mathbf{E}}_0 + \mathbf{B}'_0 . \bar{\mathbf{B}}_0) \mathrm{d}x\,, \tag{4.139}$$

we find on using the formulae (4.133) and (4.134) and on integrating by parts, that

$$(\mathbf{u}'_0, \mathbf{u}_0) = \mathrm{e}^{i(k'-k)x_3} \int_{\Omega_0} \left[\frac{ik'}{\alpha'^2} \left(-\frac{\partial \bar{\tilde{E}}_1}{\partial x_1} - \frac{\partial \bar{\tilde{E}}_2}{\partial x_2} \right) - \frac{i\omega'}{\alpha'^2} \left(\frac{\partial \bar{\tilde{B}}_1}{\partial x_2} - \frac{\partial \bar{\tilde{B}}_2}{\partial x_1} \right) + \bar{\tilde{E}}_3 \right] \tilde{E}'_3 \, \mathbf{dx}.$$

[147] When the cross-section Ω_T of the cylinder is rectangular, this condition is always satisfied.
[148] See, below, Proposition 4.

With $\operatorname{div} \mathbf{E}_0 = 0$ which can be written

$$\frac{\partial \tilde{E}_1}{\partial x_1} + \frac{\partial \tilde{E}_2}{\partial x_2} + ik\tilde{E}_3 = 0 \quad \text{and} \quad i\omega \tilde{E}_3 = \left(\frac{\partial \tilde{B}_2}{\partial x_1} - \frac{\partial \tilde{B}_1}{\partial x_2}\right),$$

we obtain

$$(u_0', u_0) = e^{i(k' - k)x_3} \int_\Omega \left(\frac{kk' + \omega\omega'}{\alpha'^2} + 1\right) \bar{\tilde{E}}_3 \tilde{E}_3' \, \mathrm{d}\mathbf{x} \,. \tag{4.140}$$

From this formula, we can draw the following consequences:
(1) The *TE* waves and the *TM* waves are orthogonal[149] in $[L^2(\Omega_0)]^3 \times [L^2(\Omega_0)]^3$.
(2) If $\mathbf{u}_0$ is *orthogonal* to all *TM* waves in $[L^2(\Omega_0)]^3 \times [L^2(\Omega_0)]^3$, then $\mathbf{u}_0$ is a *TE* or *TEM* wave. In effect, by (4.140), the relation $(\mathbf{u}_0, \mathbf{u}_0') = 0$ implies $\int \bar{\tilde{E}}_3 \tilde{E}_3' \mathrm{d}\mathbf{x} = 0$, if k' and ω are such that $kk' + \omega\omega' + \alpha'^2 \neq 0$. Now for each eigenvalue of (4.121), it is possible to find ω' and k' such that $\omega'^2 - k'^2 = \alpha'^2$ and $kk' + \omega\omega' + \alpha'^2 \neq 0$.
As the eigenvectors of the Laplacian (with the Dirichlet condition) form a complete base of the space $L^2(\Omega_0)$,[150] we deduce that $\tilde{E}_3 = 0$, so that $\mathbf{u}_0$ is a *TE* wave or a *TEM* wave.
(3) Thanks to (4.139) and (4.104), we obtain also the electromagnetic energy W_1 of a wave $\mathbf{u}_0 = \{\mathbf{E}_0, \mathbf{B}_0\} \in D(\mathscr{A})_{\mathrm{loc}}$ per unit length[151]. Here the calculation is carried out in the case in which $\mathbf{u}_0$ is a *TM*-wave:

$$\begin{aligned} W_1 &= \frac{1}{2}\int_{\Omega_0} (|\mathbf{E}_0|^2 + |\mathbf{B}_0|^2)\mathrm{d}x \\ &= \frac{1}{2}\left(\frac{k^2 + \omega^2}{\alpha^2} + 1\right)\int_{\Omega_0} |\tilde{E}_3|^2 \, \mathrm{d}x = \frac{\omega^2}{\alpha^2}\int_{\Omega_0} |\tilde{E}_3|^2 \, \mathrm{d}x \,. \end{aligned} \tag{4.141}$$

In addition by using the formulae (4.133) and (4.134), we have

$$\int_{\Omega_0} |\tilde{\mathbf{E}}_T|^2 \, \mathrm{d}\mathbf{x} = \frac{k^2}{\alpha^4}\int_{\Omega_0} |\operatorname{grad} \tilde{E}_3|^2 \, \mathrm{d}\mathbf{x} = \frac{k^2}{\alpha^2}\int_{\Omega_0} |\tilde{E}_3|^2 \, \mathrm{d}\mathbf{x}$$

from which we deduce

$$\begin{aligned} \int_{\Omega_0} |\mathbf{E}_0|^2 \, \mathrm{d}\mathbf{x} &= \int_{\Omega_0} (|\tilde{\mathbf{E}}_T|^2 + |\tilde{E}_3|^2)\, \mathrm{d}\mathbf{x} \\ &= \left(\frac{k^2}{\alpha^2} + 1\right)\int_{\Omega_0} |\tilde{E}_3|^2 \, \mathrm{d}\mathbf{x} = \frac{\omega^2}{\alpha^2}\int_{\Omega_0} |\tilde{E}_3|^2 \, \mathrm{d}\mathbf{x} \,, \end{aligned}$$

[149] The same is true for *TEM* and *TE* waves; the same for *TEM* and *TM* waves.
[150] That will be seen in Chap. VIII. §3.
[151] With the notation of (4.114), $W_1 = W_{0_K}$ with $K = \bar{\Omega}_0 \times [0, 1]$.

and in addition

$$(4.142)\qquad \int_{\Omega_0} |B_0|^2\,dx = \int_{\Omega_0} |B_T|^2\,dx = \theta^2 \int_{\Omega_0} |E_T|^2\,dx = \frac{\omega^2}{\alpha^2}\int_{\Omega_0} |\tilde{E}_3|^2\,dx = \int_{\Omega_0} |E_0|^2\,dx\,.$$

We have thus shown that *the electromagnetic energy*[152] W_1 *per unit length is equally divided between electrical energy and magnetic energy* – this is true for a *TM*-wave, or a *TE*-wave, or a *TEM*-wave. □

Remark 30. *Multiplicity (or degeneracy) of "eigen"-pulsations.* Let F_{TM} (resp. F_{TE}) denote the set of the $\omega \in \mathbb{R}$ for which there exists a non-zero solution $\mathbf{u}_0$ of the problem (4.109) with (4.118) for $\tilde{B}_3 = 0$ – *TM*-wave (resp. $\tilde{E}_3 = 0$ – *TE*-wave). This set is the complement in $\mathbb{R}$ of the interval $]-\alpha_0^D, +\alpha_0^D[$ (resp. $]-\alpha_1^D, +\alpha_1^D[$).

It is interesting to specify the number the N_ω^{TM} (resp. N_ω^{TE}) of independent *TM* waves (resp. *TE* waves), for given $\omega \in F_{TM}$ (resp. F_{TE}). This number N_ω^{TM} (resp. N_ω^{TE}) is also called the *TM multiplicity*, or the *TM degeneracy* (resp. *TE*) of ω. Now for ω fixed $\in F_{TM}$ (resp. F_{TE}) there exists a *finite* number N_ω^D (resp. N_ω^N)[153] of independent eigenvectors, $\tilde{E}_3$, solutions of the Dirichlet problem (4.121) (resp. of Neuman problem (4.122)) corresponding to the eigenvalues α^2 such that $\alpha^2 \leqslant \omega^2$, and *to each eigenvalue* α^2, *there correspond the two values of* k (see (4.120)) $k = \pm(\omega^2 - \alpha^2)^{1/2}$. We thus obtain

$$(4.143)\qquad \begin{cases} N_\omega^{TM} = 2N_\omega^D \\ N_\omega^{TE} = 2N_\omega^N \end{cases}$$

The independent *TM* (resp. *TE*) waves will be given, for each ω, in terms of the independent solutions $\tilde{E}_3$ of (4.121) (resp. $\tilde{B}_3$ of (4.122)) by the formulae (4.133), (4.134) (resp. (4.135), (4.136)).

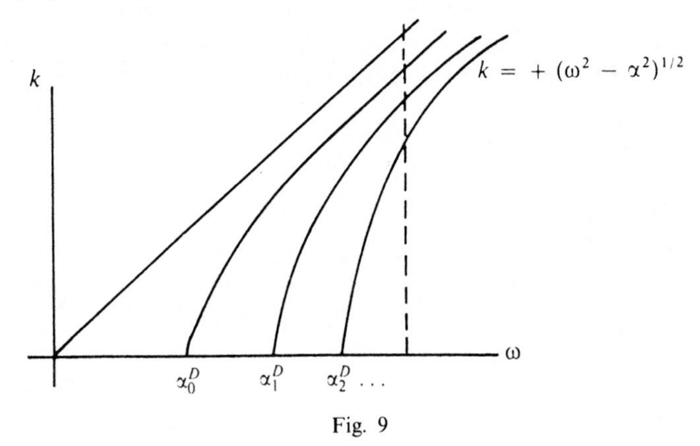

Fig. 9

[152] Linked to the empty domain.

[153] See Chap. VIII, §3.

In Fig. 9, Ω_0 is supposed given and hence also the set of discrete values α_0^D, α_1^D, etc. . . . *We can vary the given* ω *and represent the graph of the values* k *deduced from* (4.120) *as a function of* ω. It is this (or these) value(s) of k that we can adopt in (4.118). □

Finally we resume the study of *TEM* waves:

Proposition 4. *If the cross section* Ω_0 *of the cylinder* Ω *of* $\mathbb{R}^3$ *is simply connected, there does* **not** *exist a non-null TEM wave.*
If the cross section Ω_0 *of the cylinder* Ω *is multiply-connected, then there exist non-null TEM waves for all* ω *and* $k \in \mathbb{R}$ *such that* $k = \pm\omega$. *These waves are given by* $\tilde{\mathbf{E}}_T = \{\tilde{E}_1, \tilde{E}_2\}$, $\tilde{\mathbf{B}}_T = \{\tilde{B}_1, \tilde{B}_2\}$ *(and* $\tilde{E}_3 = \tilde{B}_3 = 0$*) with:*

$$\text{(4.144)}\qquad \begin{cases} \tilde{E}_1 = \varepsilon \tilde{B}_2 \quad \varepsilon = \pm 1\,, \\ \tilde{E}_2 = -\,\varepsilon \tilde{B}_1\,, \end{cases}$$

and $\tilde{\mathbf{B}}_T$ *satisfies the equations*

$$\text{(4.145)}\qquad \begin{cases} \text{(i) } \operatorname{curl}_2 \tilde{\mathbf{B}}_T = \mathbf{0} & \text{in } \ \Omega_0\,, \\ \text{(ii) } \operatorname{div}_2 \tilde{\mathbf{B}}_T = 0 & \text{in } \ \Omega_0\,, \\ \text{(iii) } \nu\,.\,\tilde{\mathbf{B}}_T = 0 & \text{on } \ \Gamma_0 = \partial\Omega_0\,. \end{cases}$$

Every TEM wave will finally be given by its electric field $\mathbf{E}(\mathbf{x}, t)$ *and its magnetic field* $\mathbf{B}(\mathbf{x}, t)$ *with*

$$\text{(4.146)}\qquad \begin{cases} \mathbf{E}(\mathbf{x}, t) = e^{i\omega(t \pm x_3)}\,.\,\tilde{\mathbf{E}}_T(x_1, x_2) & \mathbf{x} = \{x_1, x_2, x_3\} \in \Omega \\ \mathbf{B}(\mathbf{x}, t) = e^{i\omega(t \pm x_3)}\,.\,\tilde{\mathbf{B}}_T(x_1, x_2) & t \in \mathbb{R}\,. \end{cases}$$

Proof. The formulae (4.144) and (4.145) are (4.131) and (4.132) rewritten. In addition, (4.145) is identical with the formula (4.89) (see Remark 22), with here $\Omega_0 \subset \mathbb{R}^2$.
(a) If Ω_0 is simply connected (i.e. here for $\Omega_0 \subset \mathbb{R}^2$ without holes), the equation $\operatorname{curl}_2 \mathbf{B}_T = \mathbf{0}$ implies the existence of a function φ such that

$$\text{(4.147)}\qquad \mathbf{B}_T = \operatorname{grad} \varphi\,.$$

In terms of this function φ, probelm (4.145) becomes

$$\text{(4.148)}\qquad \begin{cases} \Delta\varphi = 0\,, \\ \left.\dfrac{\partial\varphi}{\partial\nu}\right|_{\partial\Omega} = 0\,. \end{cases}$$

We shall see, for example in Chap. II, that this Neumann problem admits only the solutions φ = constant, which implies that $\mathbf{E}_T = \mathbf{B}_T = \mathbf{0}$.
(b) If the section Ω_0 is multiply-connected (see, for example, Fig. 10), then – as a consequence of Remark 22 – there will exist non-null solutions of (4.145), hence of non-null *TEM* waves. □

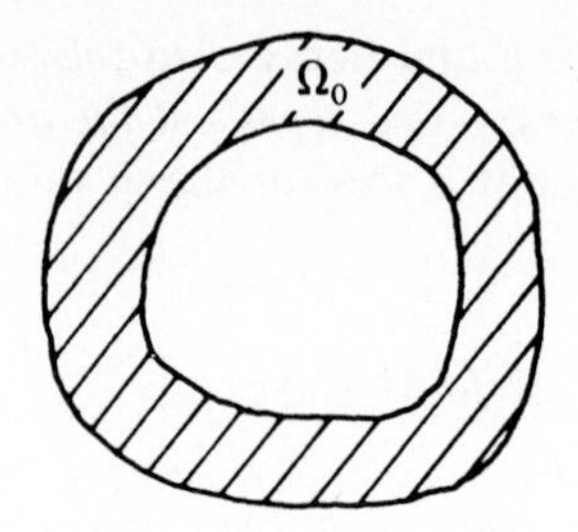

Fig. 10

Remark 31
(1) The number m of independent solutions $\mathbf{B}_T$ of (4.145) is equal to the number of cuts necessary to make the domain Ω_0 simply connected; it is also in the present case in which $\Omega_0 \subset \mathbb{R}^2$ the number of connected components less 1 of the boundary $\partial\Omega_0$. It follows that for each given $\omega \in \mathbb{R}$, the number of independent *TEM* waves will be $2m$. For example, in the case of Fig. 10, $m = 1$.
(2) It follows from (4.114) that the vectors $\tilde{\mathbf{E}}_T(x_1, x_2)$ and $\tilde{\mathbf{B}}_T(x_1, x_2)$ of a *TEM*-wave (hence $\mathbf{E}_0(\mathbf{x})$ and $B_0(\mathbf{x})$) are orthogonal in $\mathbb{R}^2$ (and in $\mathbb{R}^3$) and are such that

$$|\tilde{\mathbf{E}}_T(x_1, x_2)| = |\tilde{\mathbf{B}}_T(x_1, x_2)| \quad \text{for all} \quad \{x_1, x_2\} \in \Omega_0 . \tag{4.149}$$

In addition, we see from (4.145) that, $\tilde{\mathbf{E}}_T$ and $\tilde{\mathbf{B}}_T$ are independent of ω (and k). As a result of the formulae (4.146), the *fields* $\mathbf{E}(\mathbf{x}, t)$ *and* $\mathbf{B}(\mathbf{x}, t)$ *propagate as waves with speeds* ± 1 *along the* Ox_3 *axis.*[154] Finally the electric and magnetic energies per unit length are equal. □

Remark 32. *The spectrum*[155] *of the operator* $i\mathscr{A}$ *in* $\mathscr{H}_c$ *and the set of the given* ω *leads to the existence of a stationary solution of the problem* (4.109).
Since the set of the $\omega \in \mathbb{R}^*$ for which there exists a solution $\mathbf{u}_0 = \{\mathbf{E}_0, \mathbf{B}_0\}$ of (4.109) of the type (4.118) with (4.119) coincides with the spectrum[155] $\sigma\ (i\mathscr{A})$ (without the element 0) of the operator $i\mathscr{A}$ in $\mathscr{H}_c$ (see above), we deduce from the previous study the following results:
(1) If Ω_0 is simply connected, then (see (4.129) and Remarks 30),

$$\sigma(i\mathscr{A})\backslash\{0\} = F_{TM} \cup F_{TE} = \mathbb{R}\backslash] - \inf(\alpha_0^D, \alpha_1^N), + \inf(\alpha_0^D, \alpha_1^N)[\,^{156} . \tag{4.150}$$

(2) If Ω_0 is multiply connected, then

$$\sigma(i\mathscr{A}) = \mathbb{R} . \tag{4.151}$$

---

[154] Recall that in the chosen system of units, 1 is the speed of light
[155] For these concepts see Chap. VIII.
[156] In fact we can show (see Bandle [1], p. 128) that $\alpha_1^N \leqslant \alpha_0^D$, so $\alpha_1^N = \inf(\alpha_0^D, \alpha_1^N)$.

In addition, the multiplicity (or degeneracy) of the "eigenvalues"[157] N_ω for each $\omega \in \mathbb{R}$ – which is equal to the number of independent TE, TM, TEM waves – is given by (see (4.143), Remark 30 and Remark 31):

$$N_\omega = N_\omega^{TM} + N_\omega^{TE} + N_\omega^{TEM} = 2(N_\omega^D + N_\omega^N + m) . \qquad \square$$

Remark 33. *Cut-off frequencies and use of wave guides.* The different quantities $\alpha/2\pi$, with α^2 a (non-zero) eigenvalue of the Laplacian (see (4.121) and (4.122)), are called *cut-off frequencies.* The different possible uses of wave guides are
(1) "*Hollow tubes*"[158] (Ω_0 simply connected).
There is no frequency $f = \omega/(2\pi)$ below a certain threshold

$$f_m = \inf\left(\frac{\alpha_0^D}{2\pi}, \frac{\alpha_1^N}{2\pi}\right)^{156},$$

hence no propagation (without a very strong attenuation) of an electromagnetic wave of frequency $f < f_m$ (or again an electromagnetic generator "emitting" with a frequency $f < f_m$ cannot impose an electromagnetic wave of that frequency propagating in the tube).
For all given ω such that $|\omega| > 2\pi f_m$, we have seen that there exists a solution of (4.109) and thus a wave propagating in the cylinder. We have shown (Proposition 3) that every solution is a *TE* or *TM* wave (or the sum of a *TE*- and a *TM* wave) or a *TEM* wave.
(2) "*Coaxial*" (Ω_0 multiply connected) *operating at frequencies* $f < f_m$; the only electromagnetic waves which propagate are then of the "*TEM* waves" type. Such a wave guide used in this way is generally called a "line".
— "*Coaxial*" *operating at frequencies* f *above the threshold* f_n; this allows other than the *TEM* waves, the propagation of *TE* or *TM* waves.

Remark 34. "*Attenuation*" *of the waves.* We can likewise seek solutions $\mathbf{u}_0 = \{\mathbf{E}_0, \mathbf{B}_0\}$ of (4.109) of the form

$$\text{(4.152)} \qquad \begin{cases} E_0(x) = e^{\beta x_3}\tilde{E}(x_1, x_2, k) , \\ B_0(x) = e^{\beta x_3}\tilde{B}(x_1, x_2, k) , \end{cases} \quad \beta \in \mathbb{C}$$

with $\operatorname{Re}\beta < 0$, instead of (4.118).
Such waves are considered only for $x_3 > 0$; we thus model an *attenuation* of the wave, or again "*losses*"; for these concepts, we refer to Roubine [1].

Remark 35. *Energy flux, current intensity* (*voltage and impedance*). Finally, we call attention to the following quantities, which can be important in the use of a wave guide (see Roubine [1], p. 110).

---

[157] It is not strictly a question of speaking of eigenvalues so much as of elements of the spectrum of the operator $i\mathscr{A}$.
[158] We use the terminology of works already cited.

(1) *The currents on the various surfaces bounding the wave guide.*
Suppose that Γ^j is one of these surfaces, with cross-section Γ_0^j (so that we have $\Gamma^j \subset \Gamma \overset{\text{def}}{=} \partial\Omega$ and $\Gamma_0^j \subset \Gamma_0 \overset{\text{def}}{=} \partial\Omega_0$) with $j = 1, 2$, in the usual case of the "coaxials".
The current intensity (per unit length in the x_3-direction) crossing the section Γ_0^j will be given, as a result of the formula (4.46) (ii) by:

$$I_{\Gamma^j}(\mathbf{x}_3) = -\int_{\Gamma_0^j} (\mathbf{v} \wedge \mathbf{B}_0).\mathbf{e}_3 \, \mathrm{d}\Sigma_0 = -\, \mathrm{e}^{ikx_3} \int_{\Gamma_0^j} (\tilde{\mathbf{B}}_T.\tau) \, d\Sigma_0 {}^{159},$$

in which $\mathbf{e}_3$ denotes the unit vector along the x_3-axis, and τ the unit tangent vector to Γ_0^j (with suitable orientation).
In the case of a *TE* wave, we observe that this current intensity is zero: we have in effect with (4.135),

$$I_{\Gamma^j}(\mathbf{x}_3) = \frac{-ik}{\alpha^2} \mathrm{e}^{ikx_3} \int_{\Gamma_0^j} \frac{\partial \tilde{\mathbf{B}}_3}{\partial \tau} \, \mathrm{d}\Sigma_0 = 0^{160}.$$

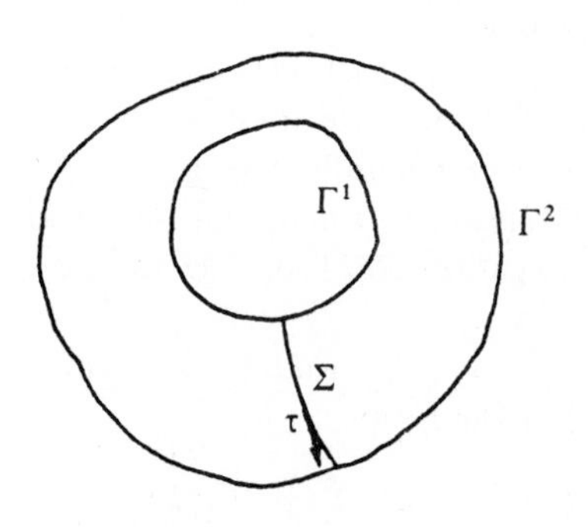

Fig. 11

(2) *The mean value (in the time) of the power $\mathscr{P}$ transmitted by the electromagnetic generator* (see Remark 27) in the wave guide – per unit length along the x_3-axis – is given by:

$$\mathscr{P} = \frac{1}{2} \mathrm{Re} \int_{\Omega_0} (E_0 \wedge \bar{B}_0).e_3 \, \mathrm{d}x .$$

[159] Recall that here $\mathbf{H}_0 = \mathbf{B}_0$ (in case of a vacuum in Ω); $\mathbf{e}_3$ denotes the unit vector in the x_3-direction.
[160] In which $\partial \tilde{B}_3/\partial\tau$ denotes the tangential derivative to Σ_0 of $\tilde{B}_3$.

We find (Fournet [1], p. 424, Landau–Lifschitz [5], p. 386) that:

$$\begin{cases} \mathscr{P}_{TM} = \dfrac{\omega k}{2\alpha^2} \displaystyle\int_{\Omega_0} |\tilde{\mathbf{E}}_3|^2 \, \mathrm{d}x & \text{for } TM\text{-waves}, \\[2ex] \mathscr{P}_{TE} = \dfrac{\omega k}{2\alpha^2} \displaystyle\int_{\Omega_0} |\tilde{\mathbf{B}}_3|^2 \, \mathrm{d}x & \text{for } TE\text{-waves}, \\[2ex] \mathscr{P}_{TEM} = \displaystyle\int_{\Omega_0} |\tilde{\mathbf{B}}_T|^2 \, \mathrm{d}x = \int_{\Omega_0} |\operatorname{grad} \varphi|^2 \, \mathrm{d}x & \text{for } TEM\text{-waves} \end{cases}$$

where φ is the function defined in Remark 22, such that $\mathbf{B}_T = \operatorname{grad} \varphi$ (we suppose here that $\partial\Omega_0$ has two connected components). We note that $\mathscr{P}_{TEM}$ is also equal to the energy W per unit length along the x_3 axis

$$W = \frac{1}{2} \int_{\Omega_0} [|\tilde{\mathbf{E}}_T|^2 + |\tilde{\mathbf{B}}_T|^2] \, \mathrm{d}\mathbf{x} = \int_{\Omega_0} |\tilde{\mathbf{B}}_T|^2 \, \mathrm{d}\mathbf{x} .$$

(3) In the case of "lines" (i.e. wave guides with TEM wave), we are led also to define the *voltage* $V(x_3)$ by

$$V(x_3) = \int_\Sigma \mathbf{E}_0(x_1, x_2, x_3) . \tau \, \mathrm{d}s = \mathrm{e}^{ikx_3} \int_\Sigma \tilde{\mathbf{E}}_T(x_1, x_2) . \tau \, \mathrm{d}s$$

(where Σ is a cut between Γ^1 and Γ^2 rendering Ω simply connected, and τ a unit vector tangent to Σ; see Fig. 11). We can then define the *characteristic impedance* Z_c of the line, as the ratio: $Z_c \overset{\text{def}}{=} V(x_3)/I_{\Gamma^j}(x_3)$.

We refer to Roubine [1] for developments of these fundamental ideas concerning "lines".

Review

In the very numerous applications of electromagnetism – technology of the production of electricity (alternators, transformers, circuit breakers, lines . . .) – use of strong currents (electric circuits and their various components, resistance, inductancies, capacities etc., motors, . . .) – high frequencies (generators, lines, antennae, receivers, circuits and their components, electron beams and their control) and their applications in radio, TV, radar, telecommunications – electronics (circuits and their components, generators, . . .) – automatic control – computing – telephone engineering – quantum electronics (including the optical cavity of lasers, their excited working) – instrumentation – linear and non-linear optics (fibre optics, the components of laser circuits, . . .) – electrochemistry, . . . , a certain number of physical and linear mathematical methods appeared:

— Propagation, scattering, reflexion, diffraction of waves, free or guided.
— Radiation (emission) of waves (antennae . . .).
— Resonant cavities.

—Variations of flux and induced currents.
—Foucault currents.
—Skin effect.
—Circuits of alternating or direct currents (induction, capacity, resistance)
—Distribution of the electromagnetic field (notably condensers, electrostatic lenses, calculation of a critical electrical field[161], rotating machines in static or moving situations).
—Approximation of Maxwell's equations by geometrical optics.

etc. . . .

Real situations are always more complex[162] than the mathematical models cited in this §4 as well as those which will appear later in this work. However, the structure of the phenomena given by these simple models is a useful guide for intuition. But we note above all that the numerical methods developed in this work allows us to treat more realistic cases, within the limit of capacity of presentday computers.
Problems of electromagnetism are presented in almost all the chapters of this work introducing mathematical techniques for the general application. We content ourselves here with underlining certain aspects. The problems of electrostatics are treated by reducing them to integral equations in Chap. II, Chap. IXA, §4 and IXB, §2 in a framework of Hilbert spaces. We have seen in Sect. 5 that problems involving an evolution in the time "sinusoidally" (monochromatic waves) lead in a great number of situations to a Helmholtz equation with various types of boundary conditions and of conditions at infinity (called "radiation conditions"). The *Helmholtz equation* is treated by classical methods in Chap. II. The corresponding operator $(-\Delta - \lambda^2)$ as a linear differential operator in general spaces of distributions is treated in Chap. V, and as an unbounded operator in the Hilbert space $L^2(\Omega)$ in Chap. IX from the spectral point of view. The variational methods of solution of these problems presented in Chap. VII, §2 lead to the numerical methods of Chap. XII.
The *wave equations* to which electromagnetism leads are studied with more and more powerful methods – diagonalisation in Chap. XV, §3 and §6, Laplace transform in Chap. XVI, §4 and §5, semi-groups in Chap. XVIIB, and especially the variational methods in Chap. XVIII, §5 and §6. These lead to the *numerical solutions* treated in Chap. XX, §3 and §5.
A more qualitative study of the waves arising in electromagnetism is given in the sections on wave equations[163]. Topics of interest are:–

—domain of influence of the initial conditions or of a source,
—finite speed of propagation of the singularities and the wave front,
—support of waves, propagation, Huyghens' principle,
—local decay of solutions; resonances,
—"eikonals", geometrical optics and theory of transport,

---

[161] For the corona effect, for example.
[162] Mainly because of the geometry of the devices and the constitutive relations of the materials, etc. . . .
[163] See Chap. XIV, §3; Chap. XV, §4.

— behaviour at infinity and scattering; reflexion, diffraction, transmission,
— vector character of Maxwell's equations and polarisation,
etc. . . .

§5. Neutronics. Equations of Transport and Diffusion

1. Problems of the Transport of Neutrons

1.1. The Integro-Differential Equation of Transport[164]

This equation describes the evolution of a population of neutrons in a domain X of $\mathbb{R}^3$ occupied by a medium interacting with the neutrons.
(1) A neutron is specified by

— its position $\mathbf{x} \in X \subset \mathbb{R}^3$;
— the direction $\omega \in S^2$ (unit sphere in $\mathbb{R}^3$) of its velocity $\mathbf{v}$[165]
— its kinetic energy $E = \frac{1}{2}mv^2$, (m the mass of the neutron, v its speed $|\mathbf{v}|$ with $E \in [\alpha, \beta]$. We put $\Omega_E = X \times S^2 \times [\alpha, \beta]$.

(2) The interaction between the neutrons and the atomic nuclei, and the production of neutrons by collision, are supposed to be described by the functions $\Sigma_t(\mathbf{x}, E)$, $\Sigma_s(\mathbf{x}, \omega' \to \omega, E' \to E)$, $\Sigma_f(\mathbf{x}, E')$ – called total, scattering and fission cross sections respectively – $\kappa(E)$ and $\nu(\mathbf{x}, E')$. Σ_t takes into account all categories of collisions of neutrons with kinetic energy E at $\mathbf{x}$; Σ_s takes into account the neutrons at $\mathbf{x}$ with kinetic energy E' and direction ω' which, as a result of scattering, take the kinetic energy E' and direction ω; Σ_f takes into account the neutrons with kinetic energy E' which induce fission at $\mathbf{x}$; $\kappa(E)$ is the spectrum of the neutrons emitted by fission, normalised to

$$\int \kappa(E)\,\mathrm{d}E = 1\,;$$

$\nu(\mathbf{x}, E')$ is the average number of neutrons emitted then from a fission at $\mathbf{x}$ due to a neutron of kinetic energy E'.
The functions Σ_t, Σ_s, Σ_f, κ, ν are supposed to be given, positive and bounded.
The *medium* is said to be *isotropic* if Σ_s depends on the directions ω and ω' through the scalar product $\omega' . \omega$ (the physical medium does not have a privileged direction in space). Throughout the rest of this account, we suppose that this is always the case. If, in addition, Σ_s is independent of $\omega' . \omega$, the *collision* is said to be *isotropic*.
(3) There exists a source of neutrons described by a given scalar function $S(\mathbf{x}, \omega, E, t)$.

---

[164] See Bussac–Reuss [1], Benoist [1], Duderstadt–Martin [1], Case *et al.* [1], Davison–Sykes [1], Weinberg–Wigner [1].
[165] Each neutron is supposed to move along a segment of the straight line between the collisions with the nuclei of atoms of the medium; the only changes of direction are due to these collisions.

(4) The population of neutrons is described by a scalar function $\bar{u}(\mathbf{x}, \omega, E, t)$, which is the *angular density of the number of neutrons at* $(\mathbf{x}, \omega, E) \in \Omega_E$ at the instant t. Similarly we define the *angular flux density*[166] $\varphi(\mathbf{x}, \omega, E, t)$ by

$$\varphi(\mathbf{x}, \omega, E, t) \stackrel{\text{def}}{=} |\mathbf{v}|\bar{u}(\mathbf{x}, \omega, E, t) .$$

Remark 1. We define the *total density* $\mathcal{N}(\mathbf{x}, E, t)$ and the *total flux* $\Phi(\mathbf{x}, E, t)$ of neutrons of energy E at a point x at the instant t in terms of $\bar{u}$ and φ by the formulae

$$\begin{cases} \text{(i) } \mathcal{N}(\mathbf{x}, E, t) = \displaystyle\int_{S^2} \bar{u}(\mathbf{x}, \omega, E, t)\,\mathrm{d}\omega , \\ \text{(ii) } (\mathbf{x}, E, t) = \displaystyle\int_{S^2} \varphi(\mathbf{x}, \omega, E, t)\,\mathrm{d}\omega . \end{cases} \qquad \square$$

The angular flux density $\varphi(\mathbf{x}, \omega, E, t)$ satisfies the *transport equation*

$$(5.1) \begin{cases} \dfrac{1}{|v|}\dfrac{\partial\varphi}{\partial t}(x, E, \omega, t) + \omega . \nabla_x\varphi(x, \omega, E, t) + \Sigma_t(x, E)\varphi(x, \omega, E, t) \\ \quad - \displaystyle\int_\alpha^\beta \mathrm{d}E' \int_{S^2} \Sigma_s(x, \omega' \to \omega, E' \to E)\varphi(x, \omega', E', t)\,\mathrm{d}\omega' \\ \quad - \dfrac{\kappa(E)}{4\pi}\displaystyle\int_\alpha^\beta \nu(x, E')\Sigma_f(x, E')\,\mathrm{d}E' \int_{S^2} \varphi(x, \omega', E', t)\,\mathrm{d}\omega' = S(x, \omega, E, t)^{167} . \end{cases}$$

In Chap. XXI we shall make use also of the variables $(\mathbf{x}, \mathbf{v})$, where the velocity $\mathbf{v}$ of the neutron belongs to a domain $V \subset \mathbb{R}^3$ and use the notation[168]

$$(5.2) \begin{cases} u(\mathbf{x}, \mathbf{v}, t) \stackrel{\text{def}}{=} \dfrac{m}{|v|}\bar{u}(\mathbf{x}, \omega, E, t) \text{ with } \omega = \mathbf{v}/|\mathbf{v}| \text{ and } E = \tfrac{1}{2}m|\mathbf{v}|^2 . \\ \Sigma(\mathbf{x}, \mathbf{v}) \stackrel{\text{def}}{=} |\mathbf{v}|\Sigma_t(\mathbf{x}, E)^{169} \\ f(\mathbf{x}, \mathbf{v}', \mathbf{v}) \stackrel{\text{def}}{=} m\dfrac{|\mathbf{v}'|}{|\mathbf{v}|}\left[\Sigma_s(\mathbf{x}, \omega' \to \omega, E' \to E) + \dfrac{\kappa(E)\nu(\mathbf{x}, E')\Sigma_f(\mathbf{x}, E')}{4\pi}\right]^{169} , \\ q(\mathbf{x}, \mathbf{v}, t) \stackrel{\text{def}}{=} \dfrac{m}{|\mathbf{v}|}S(\mathbf{x}, \omega, E, t)^{170} . \end{cases}$$

---

[166] The word *flux* employed here in neutronics denotes a *scalar* whose properties are different from the *flux vectors* encountered in §1 (flux of heat) and in §4 (electromagnetic flux). In this Chap. IA, we present each discipline with its own terminology and usual notation, even if they are not consistent among themselves.

[167] With the notation $\omega.\nabla_x\varphi = \sum_{i=1}^{3} \omega_i \partial\varphi/\partial x_i$. In the rest of this §5, ∇ always denotes the gradient taken with respect to the variable $\mathbf{x} \in X \subset \mathbb{R}^3$.

[168] To conform with the bulk of the publications in this subject.

[169] The medium being isotropic Σ depends only on $|\mathbf{v}|$ and f only on $\mathbf{v}'.\mathbf{v}$.

Taking account of (5.1) we see that the unknown function $u(\mathbf{x}, \mathbf{v}, t)$ must satisfy the *transport equation*

$$(5.3)\quad \begin{cases} \dfrac{\partial u}{\partial t}(\mathbf{x}, \mathbf{v}, t) + \mathbf{v} . \nabla u(\mathbf{x}, \mathbf{v}, t) + \Sigma(\mathbf{x}, \mathbf{v}) u(\mathbf{x}, \mathbf{v}, t) \\ \quad - \displaystyle\int_V f(\mathbf{x}, \mathbf{v}', \mathbf{v}) u(\mathbf{x}, \mathbf{v}', t) \mathrm{d}\mathbf{v}' = q(\mathbf{x}, \mathbf{v}, t)^{171} \\ \text{with } (\mathbf{x}, \mathbf{v}) \in \Omega \overset{\text{def}}{=} X \times V \quad \text{and} \quad t > 0 . \end{cases}$$

The unknown functions φ and u must be positive. In addition, the total number N of neutrons existing in the bounded domain X has to be finite at every instant t[172], namely

$$(5.4)\quad \begin{cases} N(t) = \displaystyle\int_\Omega u(\mathbf{x}, \mathbf{v}, t) \mathrm{d}\mathbf{x}\, \mathrm{d}\mathbf{v} = \int_{\Omega_E} \bar{u}(\mathbf{x}, \omega, E, t) \mathrm{d}\mathbf{x}\, \mathrm{d}\omega\, \mathrm{d}E \\ \quad = \displaystyle\int_{\Omega_E} \frac{1}{v(E)} \varphi(\mathbf{x}, \omega, E, t) \mathrm{d}\mathbf{x}\, \mathrm{d}\omega\, \mathrm{d}E < \infty^{173} . \end{cases}$$

The evolution problem for the transport equation, called the Cauchy problem, consists of determining the unknown function $\varphi(\mathbf{x}, \omega, E, t)$ satisfying (5.1) with the initial condition

$$(5.5)\quad \varphi(\mathbf{x}, \omega, E, 0) = \varphi_0(\mathbf{x}, \omega, E) ,$$

φ_0 being the angular flux at the instant $t = 0$, and with boundary conditions on the boundary ∂X of the domain X. Alternatively, we can take Cauchy problem to be: to determine $u(\mathbf{x}, \mathbf{v}, t)$ satisfying (5.3) with the initial condition

$$u(\mathbf{x}, \mathbf{v}, 0) = u_0(\mathbf{x}, \mathbf{v}) ,$$

$u_0(\mathbf{x}, \mathbf{v})$ being the density of neutrons at the instant $t = 0$, and with suitable boundary conditions.

Remark 2. *Multigroup transport equation.* In applications, we usually treat the influence of the parameter E on the equation (5.1) by dividing the interval of variation $[\alpha, \beta]$ of E into sub-intervals which are called *energy groups*:

$$[\alpha, \alpha_1][\alpha_1, \alpha_2] \ldots [\alpha_g, \alpha_{g+1}] \ldots [\alpha_{m-1}, \beta] .$$

---

[170] In the majority of applications S and hence q are positive; this we shall assume in this §5.

[171] In certain applications it is useful to put $f = \Sigma f_{(p)}$ with $f_{(p)}(\mathbf{x}, \mathbf{v}', \mathbf{v}) = \Sigma_{(p)}(\mathbf{x}, \mathbf{v}') g_{(p)}(\mathbf{v}', \mathbf{v}) c_{(p)}(\mathbf{x}, \mathbf{v}')$ where $c_{(p)}(\mathbf{x}, \mathbf{v}')$ is the average number of neutrons emitted by a collision at $\mathbf{x}$ by a neutron of velocity $\mathbf{v}'$, $g_{(p)}(\mathbf{v}', \mathbf{v})$ is the probability that a neutron with incident velocity $\mathbf{v}'$ emerges from a collision with velocity $\mathbf{v}$, and $\Sigma_{(p)}(\mathbf{x}, \mathbf{v}')$ is the effective macroscopic cross-section (corresponding to the process (p)) of a neutron with velocity v' at the point x with suitable normalisations. The processes already indicated are not the only ones which can appear in the applications [e.g. $(n, 2n)$, $(n, 3n)$, etc ...].

[172] We obviously assume that q is a source with a finite total number of neutrons at each instant t, namely $q(\cdot, \cdot, t) \in L^1(X \times V)$.

[173] With the notation $v(E) = |\mathbf{v}| = (2E/m)^{\frac{1}{2}}$.

The equation (5.1) is then replaced by a set of equations determining the *average flux in each energy group*. The effective cross-sections and the other data depending on E are replaced by the mean cross-sections for the energy group considered, called *multigroups*. To calculate these multigroup effective cross-sections is one of the essential steps in the calculation of the design of a reactor (see Bussac–Reuss [1]).
The same considerations apply equally to the equation (5.3). We then write the

$$\frac{\partial u^g}{\partial t} + v^g \nabla u^g + \Sigma^g u^g - \sum_{g'=1}^{m} f^{g' \to g} u^{g'} = q^g \quad \text{with} \quad g = 1, 2, \ldots, m\,. \tag{5.6}$$

1.2. Boundary Conditions

1.2.1. Neutronics in a Convex Domain $X \subset \mathbb{R}^3$, whose Exterior $\mathbb{R}^3 \backslash X$ is empty (or in a domain X whatsoever surround by a medium totally absorbing neutrons). We suppose here that every neutron arriving at a point on ∂X and coming from the interior of X disappears and that no neutron arrives at the exterior: the flux of neutrons entering into X is zero at every point of ∂X, that is to say

$$\begin{gathered} \varphi(\mathbf{x}, \omega, E, t) = 0 \quad \forall t \geqslant 0 \quad \text{and} \\ \text{p.p.}\ (\mathbf{x}, \omega, E) \in \partial X \times S^2 \times [\alpha, \beta] \text{ such that } \omega \,.\, \boldsymbol{\nu}(\mathbf{x}) < 0\,, \end{gathered} \tag{5.7}$$

(where $\boldsymbol{\nu}(\mathbf{x})$ denotes the normal at $\mathbf{x}$ to ∂X directed to the exterior of X).

1.2.2. Case of a Known Source Placed on the Boundary. Here the angular flux "entering" X through the boundary ∂X is known (and non zero)

$$\begin{gathered} \varphi(\mathbf{x}, \omega, E, t) = g_0(\mathbf{x}, \omega, E, t), \quad \forall t \geqslant 0 \quad \text{and} \\ \text{a.e.}\ (\mathbf{x}, \omega, E) \in \partial X \times S^2 \times [\alpha, \beta] \quad \text{such that} \quad \omega \,.\, \boldsymbol{\nu}(\mathbf{x}) < 0\,. \end{gathered} \tag{5.8}$$

1.2.3. Case of a Symmetric Boundary Condition. When the domain X and the data being considered, possess symmetry groups, the solution of the transport equation in which we are interested will have the same symmetry. We shall restrict ourselves to treating the "elementary motif"[174] and to apply boundary conditions deduced from those of the original problem (see Bussac–Reuss [1], fourth part).

1.3. The Integral Equation of Transport

We shall show in Chap. XXI that the transport Cauchy problem formed by the integro-differential equation (5.3) (with, for example, $X = \mathbb{R}^3$) with the initial condition

$$u(\mathbf{x}, \mathbf{v}, 0) = u_0(\mathbf{x}, \mathbf{v}) \quad \text{a.e.} \quad (\mathbf{x}, \mathbf{v}) \in \mathbb{R}^3 \times V\,, \tag{5.9}$$

[174] This "elementary motif" completed by the symmetry group of the problem, gives the total domain Ω.

is equivalent to the integral equation

$$
(5.10)\quad \begin{cases} u(\mathbf{x}, \mathbf{v}, t) = u_0(\mathbf{x} - \mathbf{v}t, \mathbf{v}) \exp\left(- \int_0^t \Sigma(\mathbf{x} - \mathbf{v}s, \mathbf{v}) \, \mathrm{d}s \right) \\ \quad + \int_0^t \left\{ Q(\mathbf{x} - \mathbf{v}(t - s), \mathbf{v}, s) \exp\left(- \int_0^{t-s} \Sigma(\mathbf{x} - \mathbf{v}\tau, \mathbf{v}) \, \mathrm{d}\tau \right) \right\} \mathrm{d}s \end{cases}
$$

with the notation

$$
(5.11)\quad Q(\mathbf{x}, \mathbf{v}, s) = \int_V f(\mathbf{x}, \mathbf{v}', \mathbf{v}) u(\mathbf{x}, \mathbf{v}', s) \, \mathrm{d}\mathbf{v}' + q(\mathbf{x}, \mathbf{v}, s) \,.
$$

In the case in which X differs from $\mathbb{R}^3$, the boundary conditions (5.7) lead to a formula of similar type (see Chap. XXI).
Equation (5.10), called the *integral equation of transport*, is very useful in numerous applications:

— it is the basis of the numerical methods of solution called methods of probability of collision (and derived methods) which are used extensively in the calculation codes for the design of the cells and of the assemblages of a nuclear reactor (see Hoffmann *et al.* [1]);
— the integral equation of transport allows us to study[175], in a simple way, certain properties of the transport operator T arising in (5.3) and made precise in (5.20), (5.21) below.

2. Problems of Neutron Diffusion

The capacities and speeds of existing computers do not yet permit the numerical solution of the transport equation for all the problems concerning the detailed modelling of a nuclear reactor. We shall show in Chap. XXI that for media which are only slightly heterogeneous and of size large in comparison with characteristic quantities[176] in the transport equation, we can obtain an approximate solution to the transport equation from the solution of an equation – called the *diffusion equation* – whose coefficients are calculated from those of the transport equation; for example, the total flux tends to the solution of a diffusion equation when $1/\Sigma_t$ and $1/\Sigma_s$ tend to zero.
The calculation of nuclear reactors is generally divided into parts: transport calculations allow us to obtain the homogenized characteristic quantities for each combustible assemblage in the reactor. This homogenized data is then used in the calculation of diffusion in one, two or three (spatial) to treat the whole reactor.

---

[175] That will be done in Chap. XXI.
[176] For example, the quantity $1/\Sigma_t$, which has the dimensions of a length.

2.1. The Diffusion Equation

We propose to describe the evolution of a population of neutrons in a domain $X \subset \mathbb{R}^3$ with the following changes with respect to Sect. 1:

— we restrict our attention to the calculation of the total neutron density $\mathcal{N}$ or of the total flux Φ (see Remark 1); we thus lose information concerning the direction of the velocity of the neutrons.
— the description of the neutron source also loses its angular dependence.

The interaction between the neutrons and the medium is then described by the given positive functions:–

$\Sigma_t(x, E)$: the total cross-section
$\Sigma_s^D(x, E' \to E)$: the differential cross-section of diffusion or scattering
$\Sigma_f(x, E')$: the cross-section of fission
$D(x, E)$: the diffusion coefficient
$\nu(x, E')$: mean number of neutrons emitted by fission.

The flux $\Phi(\mathbf{x}, E, t)$ satisfies the diffusion equation

$$(5.12)\quad \begin{cases} \dfrac{1}{|v|}\dfrac{\partial \Phi}{\partial t}(\mathbf{x}, E, t) - \operatorname{div}(D(\mathbf{x}, E)\nabla\Phi(\mathbf{x}, E, t)) + \Sigma_t(\mathbf{x}, E)\Phi(\mathbf{x}, E, t) \\ \qquad - \displaystyle\int_\alpha^\beta \Sigma_s^D(\mathbf{x}, E' \to E)\Phi(\mathbf{x}, E', t)\,\mathrm{d}E' \\ \qquad - \kappa(E)\displaystyle\int_\alpha^\beta \nu(\mathbf{x}, E')\Sigma_f(x, E')\Phi(\mathbf{x}, E', t)\,\mathrm{d}E' = S(\mathbf{x}, E, t) \\ \text{a.e. } \mathbf{x} \in X, \quad \text{a.e. } E \in [\alpha, \beta], \quad S \text{ given} . \end{cases}$$

The physical significance of Φ implies that it is positive and, since the total number of neutrons in a *bounded* medium has to be finite, [for a source S with a finite total number of neutrons at each instant t, namely $S \in L^1(X \times (\alpha, \beta))$], it is necessary that

$$\int_{X \times [\alpha, \beta]} \Phi\,\mathrm{d}\mathbf{x}\,\mathrm{d}E < \infty, \quad \forall t \geqslant 0 ,$$

if X is bounded.
The Cauchy problem for diffusion consists of determining the unknown function $\Phi(\mathbf{x}, E, t)$ knowing that it satisfies (5.12), the initial condition at time $t = 0$

$$(5.13)\qquad \Phi(\mathbf{x}, E, 0) = \Phi_0(\mathbf{x}, E) ,$$

and the boundary conditions on X whose precise nature will be discussed later.

Remark 3. When we obtain the diffusion equation (5.12) as the asymptotic form of the transport equation (see Chap. XXI), the diffusion coefficient D and the differential cross-section Σ_s^D are given by the theory as a function of the data arising in the transport equation; the effective cross-section Σ_t occurring in (5.12) is the same as that in (5.1) (in the usual case of an isotropic medium, where Σ_t does not depend on ω). If, also, the collision is isotropic, and hence that $\Sigma_s(\mathbf{x}, \omega' \to \omega, E' \to E)$ occurring in (5.1) depends neither on ω' nor on ω, the effective cross-section Σ_s^D will be identical with Σ_s, the transport value, to within a factor 4π. In the remainder of this §5 we shall suppose this to be the case and shall omit the D in Σ_s^D, denoting this quantity by Σ_s following usual practice. The Σ_f are identical in both types of equation. □

Remark 4. As for the transport equations, in applications the continuous parameter $E \in [\alpha, \beta]$ is very often eliminated from (5.12) by the *multigroup* method (see Remark 2). The diffusion equation (5.12) is replaced by m equations similar to (5.12), but from which the continuous parameter E has been eliminated, namely:

(5.14)

$$\begin{cases} \dfrac{1}{|\mathbf{v}^g|}\dfrac{\partial \Phi^g}{\partial t}(\mathbf{x}, t) - \operatorname{div}[D^g(\mathbf{x})\,\nabla\Phi^g(\mathbf{x}, t)] + \Sigma_t^g(\mathbf{x})\Phi^g(\mathbf{x}, t) \\ \quad - \displaystyle\sum_{g'=1}^{g'=m} \Sigma_s^{g'\to g}(\mathbf{x})\Phi^{g'}(\mathbf{x}, t) - \kappa^g \sum_{g'=1}^{m} \nu^{g'}(\mathbf{x})\Sigma_f^{g'}(\mathbf{x})\Phi^{g'}(\mathbf{x}, t) = S^g(\mathbf{x}, t) \\ \quad \text{a.e. } \mathbf{x} \in X, \quad \forall t \in \,]0, +\infty[\end{cases}$$

for $g = 1, 2, \ldots, m$ and the S^g given[177].
In practical calculations, the passage from the transport equation to the diffusion equation is accompanied by a reduction in the number of energy groups and frequently by a partial "homogenisation" of the media. It is therefore necessary to calculate effective cross-sections ensuring the best equivalence among these equations (see Amouyal *et al.* [1]). □

The theoretical study of the solutions of the diffusion equation (5.12) leads to the definition of problems for simplified models in which we suppose that all the neutrons have the same energy E (and hence the same $|\mathbf{v}|$); this is the *monokinetic* approximation which can be written:

$$(5.15) \quad \begin{cases} \dfrac{1}{|\mathbf{v}|}\dfrac{\partial \Phi}{\partial t}(\mathbf{x}, t) - \operatorname{div}[D(\mathbf{x})\,\nabla\Phi(\mathbf{x}, t)] + \Sigma_t(\mathbf{x})\Phi(\mathbf{x}, t) \\ \quad - \bar{\Sigma}_s(\mathbf{x})\Phi(\mathbf{x}, t) - \nu(\mathbf{x})\bar{\Sigma}_f(\mathbf{x})\Phi(\mathbf{x}, t) = S(\mathbf{x}, t)\,, \\ \quad \text{a.e. } \mathbf{x} \in X \subset \mathbb{R}^3\,, \quad \forall t \in \mathbb{R}_+ = \,]0, +\infty[\,. \end{cases}$$

Remark 5. We notice that in the case where D is a constant, the second term on the left-hand side of equation (5.15) becomes $-\ D\Delta\Phi$. We shall study the Laplacian Δ in subsequent chapters and particularly in Chap. II.
If, in addition, $\bar{\Sigma}_f$, $\bar{\Sigma}_s$ and $\bar{\Sigma}_t$ are constants the change

$$(5.16) \qquad w(\mathbf{x}, t) = \Phi(\mathbf{x}, t)\exp[(\Sigma_t - \bar{\Sigma}_s - \nu\bar{\Sigma}_f)|\mathbf{v}|t]$$

---

[177] The number of groups m should not be confused with the mass of the neutron also denoted by m.

of the dependent variable transforms equation (5.15) to

$$\frac{\partial w}{\partial t}(x, t) - |v| D \Delta w(x, t) = |v| S(x, t) \exp\{-|v|(\Sigma_t - \bar{\Sigma}_s - \nu\bar{\Sigma}_f)t\},$$

which we shall compare with the *heat equation* seen in §1 and §2 (see for example (2.115)) and which will be studied later in this work.
The exponential term in (5.16) takes account of absorptions and creations: workers in this field define the *effective macroscopic cross-section of absorption*, Σ_a, by the equation $\Sigma_a = \Sigma_t - \bar{\Sigma}_s$. □

2.2. Usual Boundary Conditions for the Diffusion Equation

(1) The study of the approximation of a problem in the transport of neutrons (see Sect. 1) by a problem in the diffusion of neutrons (see (5.12) or (5.14)) which will be made in Chap. XXI leads to the introduction of a function λ defined on the boundary ∂X of X, the domain studied, by:

$$\mathbf{x} \in \partial X \to \lambda(\mathbf{x}) \in \mathbb{R}^+ =]0, +\infty[$$

which is calculated from a knowledge of the data of the transport problem; we shall therefore assume λ to be known in the rest of his account. $\lambda(\mathbf{x})$ is called the *extrapolation length.*
The transport problem (5.1), (5.5) in a domain X with the boundary condition (5.7) on ∂X can be "approximated" by the diffusion problem made up of the equations (5.12) and (5.13)[178] in the domain X with the boundary condition

$$(5.17) \qquad \Phi(\mathbf{x}, E, t) + \lambda(\mathbf{x}) \frac{\partial \Phi}{\partial \nu}(\mathbf{x}, E, t) = 0 \quad \text{a.e.} \quad \mathbf{x} \in \partial X, \qquad \forall t > 0.$$

Problems involving this boundary condition are called *Robin problems*; they will be studied in Chap. VII, §2.
(2) In a good number of applications of the diffusion equation to neutronics, when the extrapolation length is small in comparison with the "size" of X[179], we make use of the extrapolation length $\lambda(\mathbf{x})$ to replace the boundary ∂X of the domain X, being studied, by a surface ∂X_e, called the *extrapolated surface*, exterior to the domain X, assumed convex and regular, the "distance" along the normal $\boldsymbol{\nu}(\mathbf{x})$, ($\mathbf{x} \in \partial X$), between ∂X and ∂X_e being taken to be $\lambda(\mathbf{x})$[180]. We then replace the domain X by the domain X_e interior to ∂X_e, in (5.12) and (5.13) while extrapolating to X_e the data D, Σ_t, Σ_s in X to the constant values equal to their "values on the boundary ∂X". Finally, we substitute for the boundary condition (5.17) the *simpler condition*

$$(5.18) \qquad \Phi(\mathbf{x}, E, t) = 0 \quad \text{a.e.} \quad \mathbf{x} \in \partial X, \quad \text{a.e.} \quad E \in [\alpha, \beta], \quad \forall t > 0,$$

[178] Φ_0 in (5.13) is deduced from φ_0 in (5.5) by the formula (ii) of the Remark 1.
[179] λ is of the order of a cm. and X can have a typical length of several metres.
[180] In a manner to be made precise, all operations are assumed to be regular.

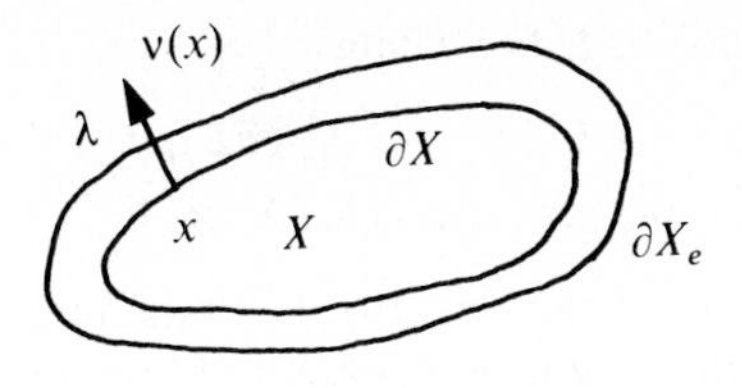

Fig. 12

called the *Dirichlet condition*. This approximation is valid if the approximation of the transport equation by the diffusion equation is itself valid (see Chap. XXI).
(3) When the considered domain X and the data possess symmetries in the space $\mathbb{R}^3$, we limit ourselves, as in the transport case, to treating the elementary motif by imposing compatible boundary value conditions (see Bouchard *et al.* [1]).

3. Stationary Problems

In neutronics, by the term stationary we usually mean the solutions of the transport equations (5.1) or (5.3) or of the diffusion equation (5.12) which are independent of the time. The phenomenon described by these functions φ, u or Φ, independent of the time, corresponds to the migration of neutrons in a permanent regime. In the theory of probability, these migrations are described by *stochastic processes*, which are *stationary* in the probabilistic sense of the word.

3.1. Stationary Problems for the Neutron Transport Equation

We shall consider the equation (5.3)[181] with q independent of t, with boundary conditions of the type (5.7) or (5.8), but g_0 no longer dependent on t. The stationary transport problem can then be stated: to find $u(\mathbf{x}, \mathbf{v})$ satisfying

$$(5.19) \qquad \begin{cases} \text{(i)} \ - Tu(\mathbf{x}, \mathbf{v}) = q(\mathbf{x}, \mathbf{v}) & \text{a.e.} \quad (\mathbf{x}, \mathbf{v}) \in X \times V\,, \\ \text{(ii)} \ u(\mathbf{x}, \mathbf{v}) = g(\mathbf{x}, \mathbf{v}) & \text{a.e.} \quad (\mathbf{x}, \mathbf{v}) \in \partial X \times V\,, \end{cases}$$

$(\mathbf{x}, \mathbf{v}) \in \partial X \times V$ satisfying $\mathbf{v} . \boldsymbol{\nu}(\mathbf{x}) < 0$ where $\boldsymbol{\nu}(\mathbf{x})$ is the external normal to ∂X at $\mathbf{x}$. Throughout the remainder of this §5, we shall take $g = 0$ which is an important case in applications. The operator T in (5.19) called the *transport operator* is such that

$$(5.20) \quad Tu(\mathbf{x}, \mathbf{v}) = - \mathbf{v} . \nabla u(\mathbf{x}, \mathbf{v}) - \Sigma(\mathbf{x}, \mathbf{v}) u(\mathbf{x}, \mathbf{v}) + \int_V f(\mathbf{x}, \mathbf{v}', \mathbf{v}) u(\mathbf{x}, \mathbf{v}') \mathrm{d}\mathbf{u}' \,.$$

It is defined in the space $L^1(X \times V)$ of integrable functions in $(\mathbf{x}, \mathbf{v})$ (we assume

[181] It would be possible also to consider (5.1).

always that X is bounded) and its *domain*[182] is

$$(5.21)\quad \begin{cases} D(T) = \{u; u \in L^1(X \times V), \quad Tu \in L^1(X \times V) \text{ and} \\ [u(\mathbf{x}, \mathbf{v}) = 0 \quad \text{a.e.} \quad (\mathbf{x}, \mathbf{v}) \in \partial X \times V \quad \text{and} \quad \mathbf{v}.\boldsymbol{\nu}(x) < 0]\} \,. \end{cases}$$

The study of the properties of the operator T contributes to our understanding of stationary transport, of the evolution problem of transport[183] and of the diffusion approximation[184].

Critical, Sub-critical and Super-Critical Nuclear Systems. We separate the function $f(\mathbf{x}, \mathbf{v}', \mathbf{v})$ occurring in (5.20) into two parts: the first $f_r(\mathbf{x}, \mathbf{v}', \mathbf{v})$ will describe the phenomena of slowing-down, of thermalisation and of diffusion (see Cadilhac *et al.* [1], Horowitz–Trétiakoff [1]) of the neutrons, and the second $f_f(\mathbf{x}, \mathbf{v}', \mathbf{v})$ the fission reactions which yield high-energy neutrons

$$(5.22)\qquad f(\mathbf{x}, \mathbf{v}', \mathbf{v}) = f_r(\mathbf{x}, \mathbf{v}', \mathbf{v}) + f_f(\mathbf{x}, \mathbf{v}', \mathbf{v}), \quad f_r, f_f \geqslant 0 \,.$$

The physical considerations justifying this partition similarly imply that

$$\int_V f_r(\mathbf{x}, \mathbf{v}', \mathbf{v}) d\mathbf{v} \leqslant \Sigma(\mathbf{x}, \mathbf{v}')^{185} \,.$$

The migration of a neutron born by fission, described by f_f is followed by slowing down and thermalisation which together are described by f_r; the disappearance of the neutrons at $(\mathbf{x}, \mathbf{v})$ is taken into account by Σ.
Similarly, we introduce the operator T_λ such that

$$(5.23)\quad \begin{cases} T_\lambda u(\mathbf{x}, \mathbf{v}) = -\mathbf{v}.\nabla u(\mathbf{x}, \mathbf{v}) - \Sigma(\mathbf{x}, \mathbf{v})\mathbf{u}(\mathbf{x}, \mathbf{v}) \\ \quad + \displaystyle\int_V f_r(\mathbf{x}, \mathbf{v}', \mathbf{v}) u(\mathbf{x}, \mathbf{v}') d\mathbf{v}' + \frac{1}{\lambda} \int_V f_f(\mathbf{x}, \mathbf{v}', \mathbf{v}) u(\mathbf{x}, \mathbf{v}') d\mathbf{v}', \quad (\lambda > 0) \,. \end{cases}$$

We have introduced previously coefficients $\boldsymbol{\nu}$ and c giving the number of neutrons created by collision. Bringing together these coefficients we shall see that the role played by λ is to correct the number of neutrons created by fission to act on the multiplication of the neutrons, arising from a fission.
Let us consider the eigenvalue problem[186]

$$(5.24)\quad \begin{cases} (i)\ T_\lambda u = 0 \quad \text{in} \quad \Omega = X \times V \\ (ii)\ \textit{the boundary condition } (5.19)\ (ii), \\ \quad \textit{where we recall we have taken } g = 0 \,. \end{cases}$$

---

[182] See Chap. VI for this concept and Chap. XXI for the mathematical study of the operator T.
[183] We shall see in Chap. XXI that the solution of (5.3) with suitable initial and boundary conditions and for a source $q \equiv 0$ can be written

$$u(\mathbf{x}, \mathbf{v}, t) = \exp[tT] u_0(\mathbf{x}, \mathbf{v}) \,,$$

where $\exp[tT]$ denotes the semi-group having T for infinitesimal generator.
[184] See Chap. XXI.
[185] See Bell–Glasstone [1].
[186] See Chap. XXI for the definition and study of this concept of operators of transport type.

We shall see in Chap. XXI that under wide assumptions concerning Σ, f_r, f_f and the domain X, there exists a value λ_0 of λ such that λ_0 is furthest to the "right" (on $\mathbb{C}$) of the values of λ such that the problem (5.24) admits a solution. In addition, there corresponds to λ_0 an "eigenvector" u_0 satisfying (5.24) which is positive[187]:

(5.25) $$u_0(\mathbf{x}, \mathbf{v}) \geqslant 0 \quad \text{a.e.} \quad (\mathbf{x}, \mathbf{v}) \in X \times V .$$

We call this (proper) value λ_0 the *effective multiplication factor* of the medium studied in the domain X and we denote it by K_{eff}. The following situations can occur:

(1) If $\lambda_0 = 1$, the system (5.24) is said to be *critical*. The equation (5.24) (*i*) is interpreted in the following manner

(5.26)
$$\mathbf{v} \,.\, \nabla u_0 \quad + \quad \Sigma u_0 \quad = \quad \int_V f_r u_0 \, d\mathbf{v}' \quad + \quad \int_V f_f u_0 \, d\mathbf{v}'$$

[Loss of neutrons at $(\mathbf{x}, \mathbf{v})$ by escape] + [Loss of neutrons at $(\mathbf{x}, \mathbf{v})$ by collision] = [Source of neutrons at $(\mathbf{x}, \mathbf{v})$ by slowing down and thermalisation] + [Source of neutrons at $(\mathbf{x}, \mathbf{v})$ by fission]

We shall see in Chap. XXI that in the particular case (essential for applications) in which
$$S = 0 \qquad \forall(\mathbf{x}, \mathbf{v}) \in X \times V$$
in (5.19) (*i*), the stationary problem[188] (5.19) with $g \equiv 0$, $\forall(\mathbf{x}, \mathbf{v}) \in \partial X \times V$, has a *unique* positive solution u_0 (*unique* to within a proportionality factor). If in addition, we consider the evolution problem formed by (5.3)[188] (where $q_- = 0$), the boundary condition (5.19) (*ii*) and the initial condition $u(\mathbf{x}, \mathbf{v}, 0) = \tilde{U}(\mathbf{x}, \mathbf{v})$, we shall see (Chap. XXI) that the solution $u(\mathbf{x}, \mathbf{v}, t)$ of this evolution problem tends exponentially as a function of t, when $t \to \infty$ to a function $u^\infty(\mathbf{x}, \mathbf{v})$ which is propor-

---

[187] More precisely, we shall see in Chap. XXI that under wide assumptions the operator T_λ admits a family of eigenvalues μ_λ^k (depending on λ), with $k = 1, 2, 3, \ldots$, of finite multiplicity and that the eigenvalue situated furthest to the right in the complex plane possesses the following properties: it is *real and simple*, a corresponding eigenvector $u(\mathbf{x}, \mathbf{v})$ is *positive* and possesses all the symmetries of the problem. We denote this eigenvalue by Λ_λ; it is a decreasing function of λ, which under wide hypotheses tends to infinity when λ tends to zero and is negative for large values of λ. One of the fundamental problems of neutronics is to determine the data of the problem (whence λ) for which we have
$$\Lambda_\lambda = 0 \quad \text{for} \quad \lambda = \lambda_0 ,$$
which is *characterised* by the existence of a positive function, not identically zero, possessing the symmetries of the problem and satisfying
$$T_{\lambda_0} u = 0$$
and
$$u(\mathbf{x}, \mathbf{v}) = 0 \quad \text{if} \quad \mathbf{x} \in \partial X \quad \text{and} \quad \mathbf{v} \,.\, \nu < 0 .$$

[188] Where we have put $f = f_r + f_f$.

tional to $u_0(\mathbf{x}, \mathbf{v})$; the factor of proportionality is deduced from the knowledge of $\tilde{U}(\mathbf{x}, \mathbf{v})$.
If $S \neq 0$ in the Problem (5.19), this problem does not, in general, have a solution.
(2) If $\lambda_0 < 1$, the system described by (5.19) is said to be *sub-critical*. The equation (5.24) (i) with λ_0 and u_0, implies that

$$(5.27) \qquad \begin{array}{ccc} \mathbf{v}\,.\,\nabla u_0 + \Sigma u_0 & > & \displaystyle\int_V f_r u_0 \,\mathrm{d}\mathbf{v}' + \int_V f_f u_0 \,\mathrm{d}\mathbf{v}' \\ \left[\begin{array}{c}\text{Loss of neutrons}\\ \text{by escape and}\\ \text{collision at } (\mathbf{x}, \mathbf{v})\end{array}\right] & > & \left[\begin{array}{c}\text{Source of neutrons by slow-}\\ \text{ing down, thermalisation and}\\ \text{fission at } (\mathbf{x}, \mathbf{v})\end{array}\right] \end{array}$$

We shall see in Chap. XXI that in this case, the problem (5.19)[188] has one and only one solution u_1 determined by the specification of S (with Σ and f given)[189].
Just as in the preceding case, we shall see (in Chap. XXI) that the evolution problem (5.3), with the boundary condition (5.19) (*ii*) and an initial condition $u(\mathbf{x}, \mathbf{v}) = \tilde{U}(\mathbf{x}, \mathbf{v})$, has a unique solution, which tends exponentially as a function of t, when $t \to +\infty$, to $u^\infty(\mathbf{x}, \mathbf{v})$, where $u^\infty(\mathbf{x}, \mathbf{v})$ is equal to the solution $u_1(\mathbf{x}, \mathbf{v})$ (of the problem (5.19)) which we have just cited.
(3) If $\lambda_0 > 1$, the system described by (5.19) is said to be *super-critical*. An interpretation analogous to (5.26) and (5.27) leads to the statement that, with u_0 as the density of the number of neutrons

$$(5.28) \qquad \left[\begin{array}{c}\text{loss of neutrons}\\ \text{by escape and collision}\\ \text{at } (\mathbf{x}, \mathbf{v})\end{array}\right] < \left[\begin{array}{c}\text{source of neutrons by}\\ \text{slowing-down and thermalisation}\\ \text{and fission at } (\mathbf{x}, \mathbf{v})\end{array}\right].$$

We shall see (in Chap. XXI) that the problem (5.19)[188] need not have a solution and that if it does have one, it is not unique; in addition, these solutions have no practical interest.
The evolution problem formulated by (5.3)[188], (5.19) (*ii*), and a suitable initial condition, has a solution which grows exponentially to $+\infty$ as a function of t.

Integral Formulation[190]. The calculations concerning transport of neutrons in heterogeneous combustible assemblages in nuclear reactors are almost always carried out for a stationary régime; they are often based on the integral form of the stationary transport equation which we write:

$$(5.29) \qquad u(\mathbf{x}, \mathbf{v}) = \int_0^\infty \mathrm{d}s\, Q_0(\mathbf{x} - s\mathbf{v}, \mathbf{v}) \exp - \left[\int_0^s \Sigma(\mathbf{x} - t\mathbf{v}, \mathbf{v})\,\mathrm{d}t\right]$$

where $$Q_0(\mathbf{x}, \mathbf{v}) = \int_V f(\mathbf{x}, \mathbf{v}' \to \mathbf{v})\, u(\mathbf{x}, \mathbf{v}')\,\mathrm{d}v' + q(\mathbf{x}, \mathbf{v})\,.$$

[189] For example, in the case where the condition S (or q) $\in L^1(X \times V)$ holds. Notice that in this sub-critical case, $S \neq 0$ is interesting for applications.
[190] See Bussac–Reuss [1], Benoist [2].

The multigroup formulation of this equation is

$$(5.30)\qquad u^g(\mathbf{x}, \omega) = \int_0^\infty \mathrm{d}s\, Q^g(\mathbf{x} - s\omega, \omega) \exp\left[- \int_0^s \Sigma_t^g(\mathbf{x} - t\omega)\, \mathrm{d}t \right]$$

with

$$Q^g(\mathbf{x}, \omega) = \sum_{g'=1}^{m} \int_{S^2} \Sigma_s^{g' \to g}(\mathbf{x}, \omega' \to \omega) u^{g'}(\mathbf{x}, \omega')\, \mathrm{d}\omega' + \frac{\kappa^g}{4\pi} \sum_{g'=1}^{m} \nu^{g'}(\mathbf{x}) \Sigma_f^{g'}(\mathbf{x}) \int_{S^2} u^{g'}(\mathbf{x}, \omega')\, \mathrm{d}\omega' + q^g(\mathbf{x}),$$

$g = 1, 2, \ldots, m$.
The equivalences between these formulations will be studied in Chap. XXI.

3.2. Stationary Problems for the Diffusion Equation

In a medium occupying the bounded domain X, characterised by the data D, Σ_t, Σ_s, with a volume source of neutrons $S(\mathbf{x}, E)$ independent of the time t, the stationary diffusion equation can be written

(5.31)
$$\begin{cases} - \operatorname{div}(D(x, E) \nabla \Phi(x, E)) + \Sigma_t(x, E) \Phi(x, E) \\ - \displaystyle\int_\alpha^\beta \Sigma_s(x, E' \to E) \Phi(x, E')\, \mathrm{d}E' - \kappa(E) \int_\alpha^\beta \nu(x, E') \Sigma_f(x, E') \Phi(x, E')\, \mathrm{d}E' \\ = S(x, E) \text{ a.e. } x \in X, \quad \text{a.e. } \quad E \in [\alpha, \beta], \end{cases}$$

with a boundary condition of the type (5.17) or (5.18). Let us adopt this latter case. Equation (5.31) can be written

$$(5.32)\qquad A\Phi(x, E) = S(x, E),$$

where the *diffusion operator* A is such that

$$(5.33)\ \text{(i)}\quad A\Phi = - \operatorname{div}(D \nabla \Phi) + \Sigma_t \Phi - \int \Sigma_s \Phi\, \mathrm{d}E' - \kappa \int \nu \Sigma_f \Phi\, \mathrm{d}E'$$

its domain $D(A)$ being for example

$$(5.33)\quad \begin{cases} \text{(ii)}\ D(A) = \{\Phi; \Phi \in L^1(X_e \times [\alpha, \beta]), A\Phi \in L^1(X_e \times [\alpha, \beta]) \text{ and} \\ \qquad [\Phi(x, E) = 0 \quad \text{a.e.} \quad x \in \partial X_e, \quad \text{a.e.} \quad E \in [\alpha, \beta]]\} \end{cases}$$

where X_e and ∂X_e are defined as in Sect. 2.2.
The neutron diffusion operator A plays a rôle comparable to that of T [see (5.20) and (5.21)]. Properties analogous to those of T, criticality, etc. . . . , can be defined. We put

$$(5.34)\qquad A_{\lambda'} \Phi \overset{\text{def}}{=} \operatorname{div}(D \nabla \Phi) - \Sigma_t \Phi + \int \Sigma_s \Phi\, \mathrm{d}E' + \frac{\kappa}{\lambda'} \int \nu \Sigma_f \Phi\, \mathrm{d}E',$$

and are led to study the eigenvalue problem

(5.35) $$A_{\lambda'}\Phi = 0\,.$$

The greatest eigenvalue λ'_0 is again called the coefficient of multiplication (for the diffusion equation) by workers in neutronics. Again, it permits us to study ideas of criticality of a nuclear system[191].

In the multigroup case, the stationary diffusion equation (5.31) can be written

(5.36) $$\begin{cases} -\operatorname{div}[D^g(\mathbf{x})\,\nabla\Phi^g(\mathbf{x})] + \Sigma_t^g(\mathbf{x})\,\Phi^g(\mathbf{x}) \\ \quad -\displaystyle\sum_{g'=1}^{g'=m} \Sigma_s^{g'\to g}(\mathbf{x})\,\Phi^{g'} - \kappa^g \sum_{g'=1}^{m} \nu^{g'}(\mathbf{x})\,\Sigma_f^{g'}(x)\,\Phi^{g'}(x) = S^g(\mathbf{x}), \\ g = 1 \text{ to } m\,, \end{cases}$$

and the definition corresponding to (5.34) is

(5.37) $$\begin{cases} A_{\lambda'}^g\Phi \overset{\text{def}}{=} -\operatorname{div}[D^g\,\nabla\Phi^g] + \Sigma_t^g\Phi^g - \displaystyle\sum_{g'=1}^{m} \Sigma_s^{g'\to g}\Phi^{g'} \\ \qquad\qquad -\dfrac{\kappa^g}{\lambda'}\displaystyle\sum_{g'=1}^{m} \nu^{g'}\Sigma_f^{g'}\Phi^{g'} \\ g = 1 \text{ to } m\,. \end{cases}$$

The equation corresponding to (5.35) can then be written

(5.38) $$A_{\lambda'}^g\Phi = 0 \qquad g = 1 \text{ to } m\,.$$

The monokinetic case seen in (5.15) yields

(5.39) $$\begin{aligned} &-\operatorname{div}[D(\mathbf{x})\,\nabla\Phi(\mathbf{x})] + \Sigma_t(\mathbf{x})\,\Phi(\mathbf{x}) - \bar{\Sigma}_s(\mathbf{x})\,\Phi(\mathbf{x}) \\ &-\nu(\mathbf{x})\,\bar{\Sigma}_f(\mathbf{x})\,\Phi(\mathbf{x}) = S(\mathbf{x})\,. \end{aligned}$$

Review

The mathematical models presented in this §5 are used for the calculation of nuclear reactors: this is done in several stages:

---

[191] More precisely: the operator $A_{\lambda'}$ is *a priori* simpler than the operator T_λ. It does not depend on the space V and in particular the boundary condition is defined $\mathbf{x} \in \partial X$ a.e. Its spectrum consists of a sequence of eigenvalues of finite multiplicity. Under sufficiently general conditions (see Chap. VIII, Appendix "Krein–Rutman" theorem), the operator A (like T_λ) has the following property: the eigenvalue furthest to the right in $\mathbb{C}$, $C_{\lambda'}$ say, is characterised by: $C_{\lambda'}$ is a simple real eigenvalue, whose eigenvector is of constant sign (taken to be positive) and possesses all the symmetry of the problem. In addition, $C_{\lambda'}$ yields an approximation to $A_{\lambda'}$; in particular, the coefficient λ'_0 which satisfies the relation $C_{\lambda'_0} = 0$, characterised by the existence of a positive function φ possessing the symmetries of the problem and satisfying (5.35) for $\lambda' = \lambda'_0$, is an approximation to the coefficient λ_0.

— the heterogeneous combustible assemblages are treated individually within transport theory by the "cell code"; this yields, in particular, the effective cross-sections for a small number of energy groups;

— the reactor is then treated globally by diffusion theory in one, two or three dimensions using these effective cross-sections. The principal calculations of diffusion concern the distribution of power in the reactor and the evolution of the combustible state in the course of operating (study of the combustible "cycle").

The evolution equations (called kinetic equations) are necessary for the determination of the control of the reactor and of its various transient régimes.

The transport equation presented in Sect. 1 involves an operator acting on functions of $2 \times n$ (6 in $\mathbb{R}^3$) variables in the phase space $X \times V$; this operator is neither self-adjoint, nor elliptic, nor normal in the space $L^2(X \times V)$: the general variational methods presented in Chap. VII to construct solutions do not apply to this operator. The properties of this operator T and the solution of problems involving a transport equation are treated [192] in the specialised Chap. XXI.

Analytical solutions of the transport problem are limited to problems in $\mathbb{R}$ (the monodimensional case) too simplified to be of practical use. We therefore have need of very powerful *numerical methods* to obtain results which can be used in practice. The deterministic numerical methods are very varied and form an area of research the essentials of which will be presented in the "numerical" Chap. XX. The simulation of transport processes by probabilistic methods[193] is also very commonly employed especially for very complex geometries (of the domains X) and in particular when we have to solve a "three-dimensional" problem (i.e. $X \subset \mathbb{R}^3$), there is no symmetry to allow the problem to be reduced to one in $\mathbb{R}^2$ or $\mathbb{R}$.

The diffusion equation, presented in Sect. 2 can be studied with the mathematical techniques of the subsequent chapters. We point out the study of the multigroup stationary equation by the variational methods of Chap. VII, §2.

Evolution problems related to the diffusion equation are studied by the diagonalisation method of Chap. XV, §3, the Laplace transform in Chap. XVI, §5, semi-groups in Chap. XVIIB and above all by *variational methods* in Chap. XVIII, §4.

The first methods of *numerical solution* of the diffusion equation have been based on the methods of finite differences; the development of the method of finite elements and of all the variational methods has led to a great acceleration in arriving at solutions and to making precise and available multi-dimensional calculations (in $\mathbb{R}^2$ and even $\mathbb{R}^3$) in routine cases. The method of finite elements is presented in Chap. XII as well as the calculation of useful proper values and proper vectors (Chap. XII, §6).

The numerical methods relating the evolution problems are presented in Chap. XXI for both problems of diffusion type and problems of transport type.

---

[192] However, see Milne's problem, treated by the Wiener–Hopf method, in Chap. XIA.

[193] For example, the Tripoli code (see Bauer *et al.* [1]).

§6. Quantum Physics

Introduction

Our aim being to present an introduction to the mathematical examples of quantum physics in the chapters which follow, we restrict ourselves to collecting below the very schematic broad lines of this discipline. For all explanations, justifications and details on the physics of the phenomena, we refer the reader to: Messiah [1], Cohen–Tannoudji *et al.* [1], Balian [1], Landau–Lifschitz [6], Gottfried [1], Schiff [1].
To facilitate the reading of this §6, in which are used mathematical concepts which will be developed only in the subsequent chapters, we give, at the end of §6, brief elements on the theory of operators in Hilbert spaces and on spectral theory.
The modelling of a quantum physical system puts in play particular mathematical ideas; that obliges us to introduce particular mathematical concepts for each aspect of the model being studied. Let us take the following example:

The Modelling of an Atom in Quantum Physics (1st Part A)

1. The Hydrogen Atom. The hydrogen atom is made up of a proton (of mass $m_p = 1.7 \times 10^{-27}$ Kg. and of charge $q = 1.6 \times 10^{-19}$ Coulomb) and an electron (of mass $m_e = 0.91 \times 10^{-30}$ Kg. and of charge $-q$). These two particles interact through an electrostatic energy (analogous to that described in §4), the electrostatic potential energy of this system of two particles, distant r apart, is $\mathscr{V}(r) = -(q^2/4\pi\varepsilon_0)(1/r) = -e^2/r$ (where we have put $e^2 = q^2/4\pi\varepsilon_0$). We can show (see Cohen–Tannoudji *et al.* [1]) that to study this system of two particles, it is sufficient to study, in a reference system with origin at the centre of mass of m_p and m_e, the motion of a "relative particle" of mass $\mu = m_e m_p/(m_e + m_p)$ subjected to the potential $\mathscr{V}(r)$[194]. Practically, taking into account of the large value m_p/m_e ($m_p/m_e \simeq 1{,}800$), this centre of mass coincides with the proton, and the "relative particle" can be considered to be the electron. In studying the relative system of the two particles, it is enough, therefore, to study the motion of a single particle.

2. Hydrogen-Like Ions. Physical models analogous to the preceding, consisting of an electron placed in a central electrostatic potential, are to be found in the study of "hydrogen-like ions", made up of an electron and a nucleus: the helium atom (He)[195], having lost an electron, the lithium atom (Li) having two electrons, the beryllium atom (Be) having lost three electrons, etc. . . . ; these atomic systems are not electrically neutral and their charge is respectively $+q$, $+2q$, $+3q$.

---

[194] If we have taken the two particles to be in the positions $\mathbf{x}_1$ and $\mathbf{x}_2$, we put $\mathbf{x} = \mathbf{x}_2 - \mathbf{x}_1$ and call $\mathscr{V}(\mathbf{x})$ the interaction potential between the two real particles.

[195] The helium atom has $Z = 2$ electrons, that of lithium has $Z = 3$ electrons, that of beryllium has $Z = 4$ electrons and the respective charges of their atomic nuclei are Z_q.

In a similar manner, to study the very excited states of the ionised matter, (plasma) for example an atom of silicon[196] having lost several (x) electrons, it is often useful to study phenomena involving an electron, the least bound ("the furthest out" from the nucleus) of those which remain. If this electron moves in a very large orbit, relatively to those of the other electrons, then the field across which it moves is close to the electrostatic field of a charge placed at the centre. The corrections with respect to this over-simple model can furthermore often be explicit to correct the results of the model obtained. These corrections can often be represented by a central field having its centre at the nucleus of the atom. It is practical then to model the silicon atom in the following manner: a central point brings together the nucleus of silicon (carrying the charge $Zq = 14q$) and $(Z - x - 1)$ electrons each with charge $(-q)$; the electron studied is then subject to the electrostatic field of the charge $(Z - x - 1)q$ and to the correction field considered above.
Let us now consider the situation of a "donor" atom of phosphorus (P)[197] situated in a crystal of silicon (S) – a situation often encountered in the study of semi-conductors. In the normal lattice of silicon, each silicon atom has four electrons – called valency electrons[198] – further out than the others, used to form four bonds with neighbouring silicon atoms in the lattice. If an atom like the phosphorus, which has at its disposal five valency electrons[199], is introduced into the crystal lattice of silicon in place of one of the silicon atoms, four of whose valency electrons will be taken into the bonds mentioned above. There will remain a disposable electron. This phosphorus atom, situated in the silicon lattice, can then be modelled[200], as far as its relations with the remaining (valency) electron are concerned, by a nucleus of charge q, in a hydrogen-like system whose centre is the nucleus of P. Many other situations lead to modelling an atom by a hydrogen-like atom and interest in this modelling is very general. □

It is now necessary to model the physical situation described above. This necessitates the use of "postulates" which are true for all quantum physical situations (and not only for hydrogen or hydrogen-like atoms) and which we shall present in Sect. 1 below. The physical ideas concerning a physical system (isolated or not), physical quantities, the measure of a physical quantity and the probability of finding such results, the results of experiments, the state of a physical system, the preparation of a physical system, etc. . . . are assumed to be known to the reader (see the references cited).

---

[196] The silicon atom has $Z = 14$ electrons.
[197] The phosphorus atom (P) has $Z = 15$.
[198] In the language of spectroscopy, these four valency electrons of silicon (Si) have the "configuration" $3s^2 3p^2$ (see Cohen–Tannoudji *et al.* [1]).
[199] The five valency electrons of the phosphorus atom P have the "configuration" $3s^2 3p^3$ (see Cohen–Tannoudji *et al.* [1]).
[200] In this modelling, the mass and the charge of the electron will have to be modified; the dielectric constant describes phenomenologically the effect of the silicon atoms contained in the volume explored by the valency electron and which are polarised by the phosphorus impurity. For the evaluation of the effective mass of the electron to take this into account, see Kittel [1], p. 219.

1. The Fundamental Principles of Modelling

Our modelling rules, based, of course, on physical considerations will be described below in the Schrödinger representation.

1.1. The Schrödinger Representation[201]

1.1.1. The Postulates[202]

Postulate 1. To every quantum physical system there corresponds a complex Hilbert space $\mathscr{H}$ with denumerable base.
In the sequel this space $\mathscr{H}$ will be defined precisely for each particular physical system.

Postulate 2. (i) To every physical quantity of a system (to which corresponds the Hilbert space $\mathscr{H}$) is associated a self-adjoint operator A in $\mathscr{H}$ [203].
(ii) The value of a physical quantity can only be one of the points of the spectrum of the operator with which it is associated.
In the treatment given here, we shall identify a physical quantity and the self-adjoint operator with which it is associated. The two together will be called an "*observable*".

Postulate 3. The state of a system at the time t is a functional F_t (called the mathematical expectation) denoted by $\langle . \rangle$, of the observables $A \in \mathscr{L}(\mathscr{H})$ of a system $\mathscr{H}$. F_t possesses the following properties:–
F_t is a continuous linear functional on the space $\mathscr{L}(\mathscr{H})$ of bounded operators on the Hilbert space $\mathscr{H}$ associated with the system, with values in $\mathbb{R}$.
$F_t(A)$ is positive or zero if the operator A is positive.
$F_t(I) = 1$ if I is the identity in $\mathscr{H}$.

---

[201] We likewise use the "Schrödinger point of view" or the "Schrödinger picture".

[202] Other systems of postulates than the customary approach which we present here are possible. Let us cite

(*i*) the *algebraic* approach in which the bounded observables A are taken as basic elements for the postulates, the collection of the A, called a Segal algebra satisfying certain postulates (Segal [1], Haag–Kastler [1]) whose physical sense is important. These postulates are completed to obtain a more operational mathematical structure and lead to *C^*-algebras*;

(*ii*) the approach by *quantum logic* in which the statements (called *propositions*) concerning the physical system studied are taken as the basic elements. Again, to approach closer to reality, it is necessary to add postulates, called of quantum systems (see Jauch [1], Mackey [1]). This theory yields easily to the setting up of the structure of a theory of *quantum probabilities* concerning the events which it describes, and of a space of quantum probabilities;

(*iii*) the approach by convexity in which the states are taken as the basic elements for the postulates and the set of states is provided with a convex structure (see Davies–Lewis [1]).

[203] For the simplicity of the present treatment, we shall assume that the operators A considered here are *bounded*, which we shall denote by $A \in \mathscr{L}(\mathscr{H})$. However, it is necessary later, for physical applications, to generalise to unbounded operators, since some fundamental physical quantities such as position, impulse, angular momentum, energy, etc. . . , are associated with unbounded self-adjoint operators.

The functional F_t can be put in the form

$$F_t(A) = \operatorname{tr}(\rho_t A) \tag{6.1}$$

where ρ_t is a self-adjoint, positive operator on $\mathscr{H}$ with trace unity[204].
Denoting by $\mathscr{L}^1(\mathscr{H})$ the space of trace operators[204] in $\mathscr{H}$, we note that $\rho_t \in \mathscr{L}^1(\mathscr{H})$. In the sequel, we shall call the functional F_t or the operator ρ_t (called the *density operator*) the *state* at the instant t.
Then let b be a Borel set in $\mathbb{R}$ ($b \in \mathscr{B}(\mathbb{R})$) and E^A the spectral measure[204] of the observable A. If we make an experiment on the system (at time t), the probability that it yields a value, situated in b, of the observable A is

$$\operatorname{prob}(t, A, b) \overset{\text{def}}{=} \operatorname{tr}(\rho_t E^A(b)) . \tag{6.2}$$

We verify that (6.2) clearly defines a law of probability compatible with the definition, given above of the mathematical expectation of the observable A.

Remark 1. The postulate 3 can be completed by the description of the state after observation relative to the observable A[205], with value in b; this state is then the operator[206]

$$\frac{E^A(b)\rho_t E^A(b)}{\operatorname{tr}[\rho_t E^A(b)]}$$

The state, denoted till now by F_t or $\langle\ \rangle$ or ρ_t, varies with t. Thus, the function

$$t \to \rho(t) \overset{\text{def}}{=} \rho_t \in \mathscr{L}^1(\mathscr{H})$$

is given by:

Postulate 4. When the evolution of the system is perfectly known[206], *and this system is isolated and invariant under translation in the time, the function $t \to \rho(t)$ which characterises its state at every t, is assumed to be continuous*[207] *and satisfies the relation*

$$\rho(t) = U(t - t_0)\rho(t_0)U^*(t - t_0), \qquad \forall t, \forall t_0 \in \mathbb{R} , \tag{6.3}$$

where $U(t)$ is a unitary operator[208] *on the Hilbert space $\mathscr{H}$ attached to the system; U is called the "evolution operator", an element of a unitary group $\{U(t)\}_{t\in\mathbb{R}}$ of class $\mathscr{C}^0$*[209]*. The infinitesimal generator B*[209] *of this evolution group is by definition, when it is multiplied by the constant $i\hbar$, the Hamiltonian H of the system. This operator H which is self-adjoint*[210], *is associated with the physical quantity called "energy".*

---

[204] See Chap. VIII.
[205] Which agrees with a preparation of state.
[206] See Balian [1].
[207] Vis-à-vis the norm due to the trace in $\mathscr{L}^1(\mathscr{H})$.
[208] See Chap. VI.
[209] See Chap. XVIIA.
[210] Because $U(t)$ is unitary. See Stone's theorem, Chap. XVIIA, §4.

The constant $\hbar$ is equal to $h/2\pi$ where h is Planck's constant. In the SI-system of units, $\hbar = 1{,}054 . 10^{-34}$ Joule Second. In the remainder of this §6, we shall suppose that we are working in the context of the hypotheses of this Postulate 4, unless the contrary is explicitly mentioned, and hence that the operator H does not depend on t.

Finally, we are able to watch the evolution of the states by

$$t \to \rho(t) \quad (\text{or } \rho_t)$$

or

$$t \to F_t \quad (\text{which we also denote by } F_{\rho(t)})$$

or

$$t \to \text{prob}(t, \ldots, b), \qquad b \in \mathscr{B}(\mathbb{R}) .$$

If a system is in the state $\rho(0)$ at time $t = 0$, the *probability* that the observable A has its value in $b \in \mathscr{B}(\mathbb{R})$ at time t is

$$\text{prob}(t, A, b) = \text{tr}[U(t)\rho(0)U^*(t)E^A(b)] \tag{6.4}$$

from (6.2) and (6.3).

Remark 2. We see throughout this §6 how to proceed to the choices imposed by the postulates in the case of the hydrogen atom (and hence also of the hydrogen-like atoms). However, in the more complex situations encountered in the practice of physics, it is often difficult to proceed to the mathematical choices indicated in the four postulates. If we study, for example, as in the introduction, the properties of a plasma of glass (SiO_2, with respectively $Z_{Si} = 14$ and $Z_O = 8$, electrons per atom) raised to a high temperature, what physical system to adopt to describe the atoms and their "clouds" of electrons? A set containing a high number of these atoms? The application of numerical techniques then leads today to an impasse[211]. Let us take the simplest path and restrict ourselves to the basic motif, namely the three atoms Si, O, O. How to model such a mixture? Let us simplify further and decide to model, as a first step, the Si atom in isolation, the other atoms and electrons featuring only through their interactions with this atom of Si. How to model these interactions? Let us simplify further the modelling and suppose that our atom of Si is "enclosed" in a bounded domain Ω of $\mathbb{R}^3$; can we be satisfied with imposing boundary conditions? If this approximation is made what boundary conditions do we impose? What space $\mathscr{H}$ do we choose? For the observables, we place at the outset in this above case of a Si atom enclosed in Ω, the elements of $\mathscr{H}$ which are functions of d variables where $d = 14 \times$ (number of variables per electron)[212]. Is it necessary that the *observable* operators act on elements of $\mathscr{H}$ with d variables? This is beyond the capacity of every computer. We therefore have to devise "variables" which are approximations of elements of the "true" observables[213] and which are, thanks to methods of separation of variables (themselves also obtained by an approximate method), relative to three space variables (this is the approxi-

[211] Let us note, however, the effectiveness of more and more powerful methods of statistical physics.

[212] E.g. three space variables if the electron is located in $\mathbb{R}^3$.

[213] "True", but relatively to the very approximate model of the atom "enclosed" in Ω.

mation of independent electrons moving under the action of an average potential, determined in a "self-consistent" fashion).
The same problems arise for the operators relating to the state (Postulate 3 and its sequels). In the simplest case that we shall see later (pure state), this operator is a projection P_φ on a vector φ of $\mathscr{H}$. But φ ought then to be a function of d variables; it is impossible to proceed in an exact manner with a practical numerical procedure. We must again simplify to reduce the actions on functions of d variables of the order of three variables (i.e. in $\mathbb{R}^3$).
We can continue, examining the *evolution* of the system (Postulate 4), this list of difficulties. The above example is not singular; the same situation exists in the treatment of a metallic, molecular condensed body or of an atomic nucleus. Everywhere, it is necessary to make very large approximations in numerical calculations, approximations justified by the physical understanding of the phenomena[214]; the mathematical objects that we construct to correspond (implicitly for most of the time) to the four postulates are successive approximations (as following the phenomenon studied, we use this or that approximation of the physical system, of $\mathscr{H}$, of the observables, of the states, of the operators of evolution). However, all these approximations use as basic elements, the mathematical objects which we have introduced into the postulates. We shall see in the remainder of this §6, elementary bricks necessary for the construction of more complicated models. □

The Liouville–von Neumann Equation. From the relation (6.3) and the definition of the infinitesimal generator B of the evolution group $U(t)$[215], a unitary group of class $\mathscr{C}^0$, we shall deduce the equation of which the function

$$t \to \rho(t) = U(t)\rho(0)U^*(t), \qquad \forall t \in \mathbb{R}, \qquad \rho(t), \rho(0) \in \mathscr{L}^1(\mathscr{H}),$$

is a solution.
In effect, by the definition of the infinitesimal generator B:

$$\frac{\mathrm{d}}{\mathrm{d}t} U(t)\Phi = BU(t)\Phi, \qquad \forall \Phi \in D(B), \tag{6.5}$$

$D(B)$, the domain of B, being the set of the $\Phi \in \mathscr{H}$ such that $\mathrm{d}/\mathrm{d}t(U(t)\Phi)$ exists. Put (following the postulate)

$$H \overset{\mathrm{def}}{=} i\hbar B. \tag{6.6}$$

We have seen that the operator H thus defined is the Hamiltonian operator of the system, which is self-adjoint, with domain $D(H) = D(B)$ and which is associated with the physical quantity called "energy".

---

[214] These approximations play an essential role in the treatment of the methods of quantum physics; they are called classical or semi-classical and are to be found in Messiah [1], Cohen–Tannoudji *et al.* [1], Landau–Lifschitz [6], Schiff [1].
[215] See Chap. XVII.

By deriving the relation (see (6.3))

$$\rho(t)\Phi = U(t)\rho(0)U^*(t)\Phi ,$$

we deduce by using (6.5) and (6.6) the formal equation[216]

$$(6.7) \qquad i\hbar\frac{\mathrm{d}}{\mathrm{d}t}\rho(t) = H\rho(t) - \rho(t)H \stackrel{\text{def}}{=} [H, \rho(t)] .$$

We shall study this equation, called the Liouville–von Neumann equation, in Chap. XVIIB.

1.1.2. Particular Case of States: the Pure States. An important particular case of state, with t given, is that in which the density operator $\rho(t)$ which it defines is an *orthogonal projection*[217] operator $P_{\Phi(t)}$ on the space

$$V_t = \{\psi(t)\,;\ \ \psi(t) = \lambda\Phi(t),\ \ \lambda \in \mathbb{C}\,,\ \ \Phi(t) \text{ a given element of } \mathscr{H}\}^{218}$$

generated by an element $\Phi(t)$ of the Hilbert space $\mathscr{H}$ associated with a system; we have therefore in this case

$$\rho(t) = P_{\Phi(t)} .$$

This state is determined in a unique manner at the instant t by $\Phi(t) \in \mathscr{H}$ being given.

We suppose here that $\Phi(t)$ and $\lambda\Phi(t)$ (for all $\lambda \in \mathbb{C}$, $\lambda \neq 0$) determine the same state at the instant t, and that if $\Phi_1(t)$ and $\Phi_2(t)$ are two elements of $\mathscr{H}$ which define the same state at the same instant t, they differ by a multiplicative constant $\lambda \in \mathbb{C}, \lambda \neq 0$[219]. It is convenient to choose in the "ray" V_t an element $\Phi(t)$ with unit norm. We then have:

$$(6.8) \qquad P_{\Phi(t)}\Phi_1 = (\Phi_1, \Phi(t))\Phi(t),\quad \forall\Phi_1 \in \mathscr{H}\,,\quad \Phi(t) \in \mathscr{H},\quad \|\Phi(t)\| = 1 .$$

We then introduce the notion of a *pure state*:

The state $t \to \rho(t)$ of a system is said to be **pure** *at the instant t if and only if $\rho(t)$ is an orthogonal projection operator $P_{\Phi(t)}$ (on the space generated by an element $\Phi(t)$ of the Hilbert space associated with the system*[220]*).*

We shall see later that a state evolving according to (6.3) which is pure at the instant t is pure at every instant. We shall call it simply a "pure state". A state which is not pure, is said to be a mixed state[221]. We shall call a pure state as well as the

[216] We employ the adjective "formal" to qualify such an operator whose domain is not made precise, or, say, a relation between operators whose domains are not made precise. In the rest of this §6, we shall omit this word.

[217] See Chap. VI.

[218] This space is called a "ray" of the Hilbert space $\mathscr{H}$; its dimension is 1. Note also the parameter λ can depend *a priori* on t.

[219] See Cohen–Tannoudji *et al.* [1], p. 219. In effect, $P_{\Phi(t)} = P_{\lambda\Phi(t)}$.

[220] A density operator ρ ($\operatorname{tr}\rho = 1, \rho \geqslant 0, \rho = \rho^*$) which is also an orthogonal projection operator ($\rho^2 = \rho$), necessarily project onto a space of dimension 1, also the state $t \to \rho(t)$ is pure at the instant t if and only if $\rho(t)$ is an orthogonal projection operator in $\mathscr{H}$.

[221] For the physical study of a mixed state, see Balian [1].

continuous function $t \to \rho(t) = P_{\Phi(t)}$ as the continuous function $t \to \Phi(t)$ with $\Phi(t) \in \mathscr{H}$[222].

In the absence of "superselection" (see later) every element of $\mathscr{H}$ determines a pure state at every instant t. If two pure states are defined at the instant t by Φ_1 and Φ_2, elements of $\mathscr{H}$, then $(\Phi_1 + \Phi_2)$, or more generally each complex combination $(\alpha\Phi_1 + \beta\Phi_2)$ (normalized) $(\alpha, \beta \in \mathbb{C})$ defines similarly a pure state at this instant t.

Mathematical Expectation of the Observables in the Case of Pure States. The probability $\text{prob}(t, A, b)$ of obtaining at time t the value of the observable A in $b \in \mathscr{B}(\mathbb{R})$ when the state of the system is the pure state $\Phi(t)$ (with $\|\Phi(t)\| = 1$) can be written, by using $\rho(t) = P_{\Phi(t)}$ and (6.2):

$$\text{prob}(t, A, b) = \text{tr}(P_{\Phi(t)} E^A(b)) = (\Phi(t), E^A(b)\Phi(t)) \tag{6.9}$$

The mathematical expectation at the time t of the observable A being in the pure state $\Phi(t)$ is given by the formula (6.1) (when A is bounded) with here $\rho(t) = P_{\Phi(t)}$, so

$$\langle A \rangle_{\rho_t} = \text{tr}(\rho_t A) = (\Phi(t), A\Phi(t))^{223} . \tag{6.10}$$

When the observable A is unbounded, the mathematical expectation of A being a pure state $\Phi(t)$ at the time t is again given by (6.10) if $\Phi(t) \in D(A)$ (with $D(A)$ denoting the domain of A).

Example 1. Let us take the example of a particle of mass m moving freely in the space $\mathbb{R}^3$. We shall see later that we associate with this system the Hilbert space $\mathscr{H} = L^2(\mathbb{R}^3)$ and that the self-adjoint operator associated with the energy of the system is:

$$H_0 = \frac{-\hbar^2}{2m}\Delta, \qquad D(H_0) = H^2(\mathbb{R}^3)^{224} .$$

Let us calculate the mean value of the energy at time t in a state $\Phi(t)$. If $\Phi(t) \in D(H_0)$, it has the value:

$$\langle H_0 \rangle_t = \left(\Phi(t), -\frac{\hbar^2}{2m}\Delta\Phi(t) \right) = -\frac{\hbar^2}{2m}\int_{\mathbb{R}^3} \Phi(t)\Delta\bar{\Phi}(t)\,\mathrm{dx} .$$

Evolution of Pure States. Let $\rho(0) = P_{\Phi(0)}$ be the state supposed pure at the instant $t = 0$ of a system. The state at the instant t will be, from the Postulate 4,

$$\rho(t) = U(t)\rho(0)U^*(t) = U(t)P_{\Phi(0)}U^*(t) .$$

---

[222] We can show in effect that the *continuity* of the function $t \in \mathbb{R} \to \rho(t) \in \mathscr{L}^1(\mathscr{H})$ with (6.3) implies the continuity of the function $t \in \mathbb{R} \to \Phi(t) \in \mathscr{H}$ with $\Phi(t) = U(t - t_0)\Phi(t_0)$ (see "Evolution of pure states" later).

[223] Where in the interest of precision we have used the notation $\langle\ \rangle_{\rho_t} \overset{\text{def}}{=} \langle\ \rangle_t$ in the state ρ_t.

[224] $H^2(\mathbb{R}^3)$ is defined in Chap. IV (Sobolev spaces); Δ is the Laplacian.

From the relation (6.8),

$$\rho(t)\Phi_1 = U(t)P_{\Phi(0)}U^*(t)\Phi_1 = U(t)(U^*(t)\Phi_1, \Phi(0))\Phi(0)$$
$$= (\Phi_1, U(t)\Phi(0))U(t)\Phi(0) \qquad \forall \Phi_1 \in \mathscr{H} ;$$

$\rho(t)$ is thus a *projection* operator, which "projects" on the space generated by the element $U(t)\Phi(0)$ of $\mathscr{H}$. The state $t \to \rho(t)$ is consequently a pure state. We conclude: a system, whose state at the instant $t = 0$ is the pure state $\Phi(0)$, is described at the instant t by the pure state $U(t)\Phi(0)$.
From the particular form of the Postulate 4, in the case of pure states:

(6.11) *The pure states $t \to \Phi(t)$ of a system satisfy the relation*

$$\Phi(t) = U(t)\Phi(0) , \qquad \forall t \in \mathbb{R} ,$$

where $U(t)$ is a unitary operator, element of a unitary group $U(t)_{t \in \mathbb{R}}$ of class $\mathscr{C}^0$ in the Hilbert space $\mathscr{H}$.

Schrödinger's Equation. The equation, of which the function $t \to \Phi(t)$ indicated by (6.11) is a solution, follows from the definition[225] of the infinitesimal generator B of the group $U(t)$ of class $\mathscr{C}^0$. This definition implies that for all $\Phi(0) \in D(B) \subset \mathscr{H}$:

$$\frac{\mathrm{d}}{\mathrm{d}t} U(t)\Phi(0) = BU(t)\Phi(0) , \tag{6.12}$$

with again $B = (1/i\hbar)H$ so that

$$i\hbar\frac{\mathrm{d}}{\mathrm{d}t}\Phi(t) = H\Phi(t) . \tag{6.13}$$

The equation (6.13) is called *Schrödinger's equation.*

Stationary Solutions of Schrödinger's Equation. The stationary solutions of Schrödinger's equation play an important role in physics: we now introduce this notion.
The state $t \to \rho(t)$ of a system is called a stationary state if and only if the density operator $\rho(t)$ is equal at every instant to the initial density operator $\rho(0)$:

$$\rho(t) = \rho(0) , \qquad \forall t \in \mathbb{R} .$$

We then have the following property:
A pure state $t \to \rho(t)$ is a stationary state if and only if it can be written

$$\Phi(t) = \mathrm{e}^{-\frac{iEt}{\hbar}}\Phi(0) \tag{6.14}$$

where E is an eigenvalue of the Hamiltonian operator H and $\Phi(0)$ an eigenvector associated with E.

---

[225] See Chap. XVIIA, §2.

Proof. A pure state $t \to \rho(t) = P_{\Phi(t)}$ is stationary if and only if $P_{\Phi(t)} = P_{\Phi(0)}$ i.e. if $\Phi(t)$ and $\Phi(0)$ determine the same state at every instant t. Two elements of $\mathscr{H}$, $\Phi(t)$ and $\Phi(0)$ define the same state only if they differ by a multiplicative constant $\lambda(t) \in \mathbb{C}$, $\lambda(t) \neq 0$; and since the operator $U(t)$ is unitary[226], a stationary pure state must satisfy:

$$(6.15) \quad \Phi(t) = \lambda(t)\Phi(0) \quad \text{with} \quad \lambda(t) \in \mathbb{C}, \quad |\lambda(t)| = 1\,, \quad \forall t \in \mathbb{R}\,.$$

Putting (6.15) into Schrödinger's equation (6.13) (if $\Phi \in D(H)$), we then verify that $\Phi(t)$ (and hence $\Phi(0)$) is an eigenvector[227] of the Hamiltonian operator H and that we may write

$$(6.16) \quad \Phi(t) = \mathrm{e}^{-\frac{iEt}{\hbar}}\Phi(0)\,,$$

where E is the proper value[228] corresponding to the proper vector $\Phi(0)$ of the Hamiltonian H. Conversely, if $\Phi(0)$ is the proper vector of the operator H corresponding to the proper value E, the vector Φ "evolves" according to (6.16) and therefore defines a stationary state. □

To find the stationary pure states of the system $\mathscr{H}$ with Hamiltonian H, is therefore to solve the problem:

$$(6.17) \quad \begin{cases} \text{find } \Phi \in D(H) \subset \mathscr{H} \quad \text{and} \quad E \in \mathbb{R} \quad \text{such that:} \\ H\Phi = E\Phi\,, \end{cases}$$

hence especially to study the spectrum of the operator H. That will be done in Chaps. VIII and IX.

Remark 3. Let us take up again an example like that of the Si atom "occupying the domain Ω" cited in Remark 2; the problem (6.17) results in solving an equation $H\Phi = E\Phi$ where Φ is a function of d variables (where $d = 14 \times$ number of variables per electron, supposing the silicon nucleus fixed). Today, there does not exist any method which allows us to solve this problem numerically. The methods used in practice consist of looking for approximate solutions, Φ_a, of the problem, constructed from functions of a single electron[229] from which we form antisymmetric tensor products and certain linear combinations of these products[230].
In the majority of practical cases, the operator H is bounded below and the beginning of the spectrum of H[231] is made up of a discrete sequence of proper values of finite multiplicity. The first proper value E_0 is called the "*ground level*" (or the "*fundamental level*"); the other proper values are called "*excited levels*".
An approximate calculation of E_0 is often made in practice by seeking the minimum of the expression $(\Phi, H\Phi)/(\Phi, \Phi)$ on a set of approximate functions Φ_a. □

---

[226] See Chap. VI.
[227] See Chap. VIII.
[228] Since the operator H is self-adjoint E is a real number. See Chap. VIII.
[229] To work on functions of a single space variable x.
[230] See Slater [1].
[231] In the centre of mass system.

Remark 4. In the study of an atom containing Z electrons (Z goes from 1 for hydrogen to 82 for lead and more for the actinides), we have seen that the Schrödinger equation to be solved relates to a wave function $\psi(\mathbf{x}_1, \ldots, \mathbf{x}_2)$ of $3Z$ variables. This situation arises in many other phenomena of quantum physics as, for example, in atomic nuclei; it is one of physical cases leading to a *problem of partial equations (PDE) in a space of very great* $(3Z)$ *dimension.*
In addition, PDE's in a space of high dimension occur often in econometrics. There are two reasons for that:
(1) *Economic* phenomena can be modelled by Wiener processes (these are earlier than Wiener and have been used for economic phenomena since the beginning of this century), i.e. by stochastic differential equations[232] in N dimensions, with N large (N can be the number of "products", or again the number of hydraulic barrages in the generation of hydraulic energy – see Colleter–Lederer [1] for the study of a hydro-electric system of about forty valleys with seasonal activity). We come out then with equations of diffusion type in N dimensions, with N large.
(2) Dynamic programming leads to partial differential equations where the number of space variables is similarly large. □

Remark 5. *Physical and non-physical quantities*[233].
(1) *Coherent subspaces of* $\mathscr{H}$. We now consider the situation at the fixed instant t, (for simplicity, we shall omit indication of the value of t in the expressions). To take account of the experimental facts, we are led to consider that certain elements of the Hilbert space associated with a physical system cannot define a physically realisable state. For example, if $\Phi^0 \in \mathscr{H}$ defines a state of a particle for which the electric charge of this particle is zero and if Φ^1 defines a state of this particle for which the electric charge is 1, the element:

$$\Phi = \alpha\Phi^0 + \beta\Phi^1 \qquad (\alpha, \beta \in \mathbb{C}) \tag{6.18}$$

does not define any experimentally observed physical state. To eliminate such undesirable linear combinations of states, we accept that the Hilbert space $\mathscr{H}$ is the closure of the direct sum of certain Hilbert sub-spaces $\mathscr{H}_\alpha$ ($\alpha = 1, 2, \ldots, n, \ldots$) called "coherent sub-spaces"

$$\mathscr{H} = \oplus_\alpha \mathscr{H}_\alpha ,$$

and that *the only elements of* $\mathscr{H}$ *susceptible of defining "physical" pure states are those which are "entirely contained" in a single one of the coherent spaces* $\mathscr{H}_\alpha$, *i.e. whose orthogonal projections on the* $\mathscr{H}_\alpha$ *differ from* 0 *on only a single coherent subspace.*
In the example of the charged particle, it is enough to suppose that Φ^0 and Φ^1 belong to two different coherent subspaces $\mathscr{H}_1$ and $\mathscr{H}_2$ to eliminate the linear combinations (6.18) from the set of physical states. *Self-adjoint operators which do not leave invariant the coherent sub-spaces* $\mathscr{H}_\alpha$ *cannot be associated with physical quantities.*

---

[232] See Cessenat *et al.* [1]
[233] See Cohen–Tannoudji *et al.* [1], Bogolubov *et al.* [1].

Such a selection of pure states and observables is called a *superselection*.
(2) *Domain of the observables.* The mean value of unbounded observables is not defined for all states. The mean value of an unbounded observable A in a pure state $\Phi(t)$, given by (6.10) when $\Phi(t) \in D(A)$, retains its meaning in more general cases, particularly if the operator A is defined making use of a sesquilinear form a[234], with domain $D(a)$; then (6.10) is replaced by

$$\langle A \rangle_{\rho_t} = a(\Phi(t), \Phi(t)) \quad \text{for} \quad \Phi(t) \in D(a) .$$

This is why, in certain formulations of quantum theories, we eliminate from the set of pure states, the elements of the Hilbert space which are not in the domains of the sesquilinear forms of the so-called "fundamental" observables (such as energy, impulse, momentum, . . .). We should then combine the two types of selection that we have just mentioned ($\Phi \in D(a)$ and $\Phi \in \mathscr{H}_\alpha$) and only retain in the coherent subspaces of the superselection the elements appertaining to the domains of the sesquilinear forms of the fundamental observables.
(3) *Gauge invariance of physical and non-physical quantities.* We have seen in §4 that *gauge* transformations can be applied to the electromagnetic potentials without changing the physical quantities (which are the electric and magnetic fields). Similar remarks can be made in classical mechanics and developed in relation to the Lagrangian and Hamiltonian formalisms.
In quantum mechanics, these gauge transformations are (unitary) transformations on the states, the operators (whence the Hamiltonian) and the evolution equation. It is necessary that the quantum theory be such the physical quantities remain invariant (constant) under these transformations; this furnishes a criterion to decide if operators can represent physical quantities[235]. □

Compatible Observables. After a measurement, the state of the system studied is modified. An observable A can then have probability 1 of having a definite value "a" in the spectrum (discrete, for example). We can then ask ourselves if there exist other observables B, C *etc.* having at the same time as A a well defined value? It should follow that an eigenvector (thus defining a pure state of the system studied) should be common to several observables. There exist in reality operators which do not commute[236] ($AB \neq BA$) and which have common eigenvectors; however, in general, this is not the case; we then say that A and B are not "simultaneously observable" or, alternatively, are incompatible. If A and B commute,[236] then such common eigenvectors exist[237]. □

Probabilities in Classical Mechanics and in Quantum Mechanics. We have just seen that for incompatible observables $A_1, A_2, A_3, \ldots$ it is not possible to speak of probabilities relative to the observables A_1 *and* A_2 *and* A_3, that is to say of *joint probabilities* of a set of random variables; in this case, the notions of conditional

---

[234] See Chap. VI.
[235] See Cohen–Tannoudji *et al.* [1] and Itzykson–Zuber [1].
[236] See Chaps. VI and VIII.
[237] In the case where A and B admit eigenvectors.

probabilities[238], of convergences, of the law of large numbers, of statistical inference, etc. . . . are not introduced as naturally as in the classical calculus of probabilities. There is thus a great difference between "*quantum probabilities*" Prob$(t, \ldots, b)$ relative to the mathematical expectations F_t cited in this §6 and the classical theory of probabilities. That arises from the fact that, in "quantum probability, the mathematical expectation (for example F_t) is defined as a *continuous linear functional on sets of operators* (which, in general, do not commute) whereas, in "classical probability", this notion is defined, for example, on the random variables

$$X_i : \Omega \to \mathbb{R}$$

which are mappings from a probability space $(\Omega, \mathscr{B}, \mu)$ into $\mathbb{R}$, that is to say on sets of functions (which commute).
The mathematical expectations $F_t(.)$ (or $\langle . \rangle$) and the probabilities prob$(t, ., b)$ are mathematical entities distinct from the expectations and probabilities treated in the classical developments of probabilities.
There exist several systems of axiom which permit the definition of theories of quantum probabilities compatible with the basic postulates of quantum physics. These systems can be constructed in such a manner that they admit as a special case the "classical" theory of probabilities which is used in statistical mechanics (see Gudder [1] and [2]).

1.2. The Heisenberg Representation

The relation (6.3) can be written

$$\text{Prob}(t, A, b) = \text{tr}(U(t)\rho(0)U^*(t)E^A(b)) = \text{tr}(\rho(0)(U^*(t)E^A(b)U(t))) .$$

For the interpretation of this expression, it is equivalent to consider that the system has passed from the state $\rho(0)$ at time $t = 0$, to the state $U(t)\rho(0)U^*(t)$ at the time t, the operator associated with the physical quantity being A whatever t is, or as well to consider that the system remains in the state $\rho(0)$ whatever t may be, but that the self-adjoint operator associated with the physical quantity is A at the time $t = 0$ and $U^*(t)AU(t)$ at the time t.
Also, without modifying the fundamental relation (6.3) of quantum physics, we can substitute for the postulates 1, 2, 3, 4 the following postulates:
Postulate 1 bis. Unchanged.
Postulate 2 bis.
(i) *With each physical quantity of a system to which there corresponds the Hilbert space $\mathscr{H}$ is associated a function $t \in \mathbb{R} \to A(t)$ with values self-adjoint operators $A(t)$ in $\mathscr{H}$.*
The self-adjoint operator $A(t)$ is the operator associated at *the instant t* with this physical quantity.
(ii) *The value of a physical quantity at the instant t can only be one of the points of the spectrum of the self-adjoint operator $A(t)$ with which it is associated.*

---

[238] And hence of independence.

In the rest of this text, we shall call observable, the function $t \to A(t)$ as well as the physical quantity with which A is associated. The operator $A(t)$ is "the observable at the instant t". Its "value" will be the value of the physical quantity which corresponds to it at the instant t.

Postulate 3 bis. Unchanged, just as is the introduction of the density operator ρ if only the functionals $F, \langle \ \rangle, \rho$ are independent of t, and the observable operator A is replaced by $A(t)$ and prob(t, A, b) by prob$(A(t), b)$, namely

$$\text{prob}(A(t), b) = \text{tr}(\rho E^{A(t)}(b)) . \tag{6.19}$$

Postulate 4 bis. When the system is isolated and its evolution is perfectly known and it is invariant under a translation in the time, the function $t \to A(t)$ satisfies the relation:

$$\begin{aligned} A(t) &= U^*(t - t_0)A(t_0)U(t - t_0) , \\ D(A(t)) &= U^*(t)(D(A(t_0))) \qquad \forall t, t_0 \in \mathbb{R} , \end{aligned} \tag{6.20}$$

where $U(t)$ is a unitary operator on $\mathscr{H}$, called the "evolution operator", an element of a group of class $\mathscr{C}^0$, $\{U(t)\}_{t \in \mathbb{R}}$, identical with the evolution group of the Schrödinger representation. The infinitesimal generator of this group $\{U(t)\}$ multiplied by $i\hbar$, is the Hamiltonian operator of the system, associated with the energy of the system.

If we associate with a physical quantity at the time $t = 0$, the operator $A(0) = A$ identical with the operator associated with this quantity in the Schrödinger representation, the relation between the function $t \in \mathbb{R} \to A(t)$ in the Heisenberg representation and the operator A in the Schrödinger representation, associated with the same physical quantity is then:

$$A(t) = U^*(t)AU(t) = \mathrm{e}^{+i\frac{Ht}{\hbar}} A \mathrm{e}^{-i\frac{Ht}{\hbar}} . \tag{6.21}$$

All that has been written in Sect. 1.1.2 concerning the pure states especially concerning physical or non-physical quantities, compatible observables, probabilities in quantum physics, remains valid in the Heisenberg representation when the time is not involved explicitly. On the other hand, the study of the evolution of a system in the course of time which leads in the Schrödinger representation to Schrödinger's equation for pure states and to the Liouville–von Neumann equation for an arbitrary state, leads in the Heisenberg representation to an equation – called the *Heisenberg equation* – for the observables, hence for the self-adjoint operators $A(t)$ satisfying (6.21).

This equation is written:

$$i\hbar \frac{\mathrm{d}A}{\mathrm{d}t}(t) = -(HA(t) - A(t)H) \stackrel{\text{def}}{=} -[H, A(t)]; \tag{6.22}$$

also, $A(t)$ is such that

$$A(0) = A ,$$

where H is the Hamiltonian operator of the system (the operators H, $A(t_0)$, $A(t)$ are self-adjoint).

The representations of Schrödinger and of Heisenberg are equivalent for the evaluation of physical quantities. According to the problem being studied we exercise a preference for one or the other of these representations. When we wish to underline the wave-function point of view, calculate transition probabilities or study imperfectly understood evolutions[239], the Schrödinger representation is used. The Heisenberg representation is basic to the study of quantities of the type $\langle A(t_1).A(t_2)\ldots\rangle$ and of very general systems, especially systems with an infinite number of particles. This technique is essential for the study of relativistic field theory[240].

2. Systems Consisting of One Particle

In the modelling of systems containing several particles, we are led very often, in numerical calculations, for the reasons stated in Remarks 3 and 4, to studies of systems involving a single particle. We shall therefore state precisely in this important particular case the ideas introduced in the postulates and especially of some observables in the Schrödinger representation (the relation (6.21) yields the expressions in the Heisenberg representation; see for example Messiah [1], Cohen–Tannoudji *et al.* [1]).

2.1. Particle with Zero Spin[241] (non-relativistic theory)

In non-relativistic quantum physics, we generally adopt as the Hilbert space $\mathscr{H}$ associated with the physical system consisting of a particle without spin, the space $L^2(\mathbb{R}^3)$, of (classes of) complex functions on $\mathbb{R}^3$, square integrable (relative to the Lebesgue measure) with the scalar product:

$$(\Phi_1, \Phi_2) \stackrel{\text{def}}{=} \int_{\mathbb{R}^3} \Phi_1(x)\bar{\Phi}_2(x)\,\mathrm{d}x\,,$$

and the associated norm $\|\Phi\| = (\Phi, \Phi)^{\frac{1}{2}}$.

We shall examine briefly the expression of the principal observables of a system with a spin-less particle, when the Hilbert space $\mathscr{H}$ is "realisable" by $L^2(\mathbb{R}^3)$.

Position Operator. The "classical" concept of *position* of a particle in the physical space $\mathbb{R}^3$ leads us to postulate the existence of three operators Q_j $(j = 1, 2, 3)$, which are self-adjoint in $\mathscr{H}$, whose spectra are the possible values of the coordinates x_j $(j = 1, 2, 3)$ of the particle in an orthogonal framework. We are led, in $\mathscr{H} = L^2(\mathbb{R}^3)$, to define these operators by the relations[242]:

$$(6.23)\quad \begin{cases} Q_j\Phi(x) = x_j\Phi(x)\,, & \forall\Phi \in D(Q_j)\,, \quad x_j \in \mathbb{R}\,, \quad j = 1, 2, 3 \\ D(Q_j) = \{\Phi;\, \Phi \in L^2(\mathbb{R}^3)\,, & x_j\Phi \in L^2(\mathbb{R}^3)\}\,; \end{cases}$$

---

[239] See Balian [1], for example if the Hamiltonian contains a random part.

[240] See Itzykson–Zuber [1].

[241] The spin of a particle will be introduced later.

[242] See Chap. VIII.

Q_j is called "the component operator on the x_j axis of the position vector". We shall verify[243] that Q_j is self-adjoint, and that its spectrum, purely continuous, is $\mathbb{R}$. The set of the three operators Q_j is called the "position operator" and denoted by Q.

Momentum Operator. The classical concept of momentum leads us to define the three operators P_j $(j = 1, 2, 3)$ in $\mathscr{H}$, called "the component operators on the x_j axes $(j = 1, 2, 3)$ of the momentum vector" by the relations[244]:

$$(6.24)\quad \begin{cases} P_j\Phi(x) = -i\hbar\dfrac{\partial\Phi}{\partial x_j}(x)\,, & \forall\Phi \in D(P_j)\,, \quad j = 1, 2, 3, \\ D(P_j) = \left\{\Phi;\, \Phi \in L^2(\mathbb{R}^3)\,, \quad \dfrac{\partial\Phi}{\partial x_j} \in L^2(\mathbb{R}^3)\right\}^{245} . \end{cases}$$

We shall verify that the operator P_j $(j = 1, 2, 3)$ is self-adjoint, and that its spectrum, purely continuous is $\mathbb{R}$[246]. The set of the three operators P_j is called the "momentum operator" and is denoted by P.
These operators are such that $(i/\hbar)P_j$, $(j = 1, 2, 3)$, is the infinitesimal generator of a unitary group of class $\mathscr{C}^0$ made up of "translation" operators $U_j(\lambda)$ defined by

$$(6.25)\quad U_j(\lambda)\Phi(\mathbf{x}) = \Phi(\mathbf{x} - \lambda\mathbf{e}_j)\,, \quad \Phi \in L^2(\mathbb{R}^3)\,, \quad \forall\lambda \in \mathbb{R}\,, \quad \forall\mathbf{x} \in \mathbb{R}^3\,, \quad \mathbf{e}_j \in \mathbb{R}^3\,,$$

where $\mathbf{e}_j$ is the unit vector in the direction of the x_j-axis. We therefore have in the usual notation[247]

$$(6.26)\qquad U_j(\lambda) = \mathrm{e}^{-\frac{i}{\hbar}P_j\lambda}\,.$$

We notice that the commutation relation between the operators Q_j and P_k which will be discussed in Remark 6 immediately below is

$$(6.27)\qquad P_kQ_j - Q_jP_k = -i\hbar\,\delta_{kj} \qquad k, j = 1, 2, 3$$

where δ_{kj} is the Kronecker symbol.

Remark 6. *Commutation relations.*
(1) *Physical meaning of the commutation relations.* The relation (6.26) can be taken as the relation defining the momentum P_j, valid whatever be the choice of the Hilbert space associated with the system if $U_j(\lambda)$ is the operator in $\mathscr{H}$ which "represents" (in the sense of group representations) the translation $(+\lambda\mathbf{e}_j)$ of the state of the system studied.
The operator $U_j(\lambda)$ is such that if $\Phi \in \mathscr{H}$ defines the pure state of the system at the instant t (for an observer using a reference frame $\mathcal{O}$ to evaluate the cartesian coordinates of the particles of the system), then $U_j(\lambda)\Phi$ defines the state obtained by the translation $+\lambda\mathbf{e}_j$ of the state of this system.

---

[243] See Chap. IX.
[244] See Chaps. VIII and IX.
[245] We observe that we have that $\bigcap_{j=1,2,3} D(P_j) = H^1(\mathbb{R}^3)$; see Chap. IX.
[246] See Chaps. VIII and IX.
[247] See Chap. XVIIA.

The relation (6.27) expresses the experimental fact that if the mean value of the observable position Q_j is $\langle Q_j \rangle = (\Phi, Q_j \Phi)$ for the first state, it is equal to $\langle Q_j \rangle + \lambda$ for the second state; in other words:

(6.28) $$(U_j(\lambda)\Phi, Q_j U_j(\lambda)\Phi) = (\Phi, Q_j \Phi) + \lambda, \qquad \forall \Phi \in D(Q_j) .$$

(2) *Weyl's relation.* More generally, the mean values of an observable $f(Q_j)$, a bounded function of Q_j [248] for the two states envisaged above, must satisfy

$$(U_j(\lambda)\Phi, f(Q_j) U_j(\lambda)\Phi) = (\Phi, f(Q_j + \lambda)\Phi) , \qquad \forall \Phi \in \mathscr{H} , \qquad \forall \lambda \in \mathbb{R} .$$

Using the expression (6.26) for $U_j(\lambda)$ and taking the particular case in which

$$f(Q_j) = e^{\frac{i}{\hbar} Q_j \mu} , \qquad \mu \in \mathbb{R}$$

this becomes

(6.29) $$e^{-\frac{i}{\hbar} Q_j \mu} e^{\frac{i}{\hbar} P_j \lambda} e^{\frac{i}{\hbar} Q_j \mu} e^{-\frac{i}{\hbar} P_j \lambda} e^{-\frac{i\mu\lambda}{\hbar}} = 1 \qquad j = 1, 2, 3 .$$

This form of the commutation relation between the operators P_j and Q_j is called *Weyl's relation* (see Prugovecki [1]). It requires only that the operators are bounded and hence has a precise meaning whereas (6.27) is only valid on domains of operators to be made precise. ☐

Angular Momentum Operator. The operators L_j ($j = 1, 2, 3$) in $\mathscr{H}$ such that

(6.30) $$\begin{cases} L_j \Phi(x) = -i\hbar \left(x_k \dfrac{\partial}{\partial x_l} - x_l \dfrac{\partial}{\partial x_k} \right) \Phi(x) , \text{ a.e. } x = (x_1, x_2, x_3) \in \mathbb{R}^3 , \\ j, k, l = \text{cyclic permutation of } 1, 2, 3 \\ D(L_j) = \{\Phi; \Phi \in L^2(\mathbb{R}^3) , \quad L_j \Phi \in L^2(\mathbb{R}^3)\} , \end{cases}$$

are called "component operators on the x_j-axis of the *angular momentum*". We shall see that these operators are self-adjoint and satisfy the commutation relations [249]

(6.31) $$L_j L_k - L_k L_j = i\hbar L_l \qquad (j, k, l = \text{cyclic permutation of } 1, 2, 3)$$

Just as the operator $(i/\hbar) P_j$ is the infinitesimal generator of a group representing translations, the operator $(i/\hbar) L_j$ is the infinitesimal generator of a group representing rotations round the x_j-axis. [250]
Putting

$$L^2 = L_1^2 + L_2^2 + L_3^2$$

we obtain [251] (formally)

(6.32) $$L^2 L_j - L_j L^2 = 0 \qquad j = 1, 2, 3 .$$

---

[248] For this idea, see the Appendix and Chaps. VIII and IX.
[249] On the intersection of the domains of these operators.
[250] See Chap. IX, and Normand [1], Miller [1] and van der Waerden [1].
[251] See the preceding references.

The observables which we have just introduced occur constantly in quantum physics. Thus, in the description of an isolated atom invariant under rotation, we can take among the eigenstates Φ, satisfying (6.17), those which are also eigenstates of the angular momentum operator (see, later, the introduction of spin).

Hamiltonian Operator. The Hamiltonian operator has been defined in Postulate 4 as the infinitesimal generator, multiplied by $i\hbar$, of the evolution group, and the observable associated with the energy of the system. The methods of passage from classical mechanics to quantum mechanics and the classical relation between the classical kinetic energy E and the momentum (p_1, p_2, p_3) of a *free particle* of mass m:

$$E = \sum_{j=1}^{j=3} \frac{p_j^2}{2m}, \tag{6.33}$$

lead us to adopt as the Hamiltonian operator of a free particle, the operator H_0 in $\mathscr{H}$ defined by:

$$\left\{\begin{aligned} &H_0 = \sum_{j=1}^{j=3} \frac{P_j^2}{2m}, \\ &\text{or, taking account of } P = -i\hbar\nabla, \\ &H_0 = -\frac{\hbar^2}{2m}\Delta. \end{aligned}\right. \tag{6.34}$$

We shall see[252] that $D(H_0) = H^2(\mathbb{R}^3)$, and that the operator H_0 thus defined is self-adjoint, positive and has $\mathbb{R}^+$ for its spectrum.
If the particle is *not free*, the action experienced by the particle is expressed by a complementary term $\mathscr{V}$ (called the *potential*), in the expression for the Hamiltonian operator of the particle considered. We put

$$H = -\frac{\hbar^2\Delta}{2m} + v; \tag{6.35}$$

in numerous applications, the operator $\mathscr{V}$ is multiplication by a real function. In other cases, this is a differential operator of the first order.

Examples of Hamiltonian Operators. Below, we present a list of Hamiltonian operators, useful in physics. Their statement brings into play the physical constants $\hbar$, c (speed of light), ε_0 (permittivity of the vacuum), μ_0 and the electric charge q. The legal system of units is the *système international* (SI), and in this, $\hbar = h/2\pi = 1.054 \times 10^{-34}$ Js, $c = 3 \times 10^8$ m s^{-1}, $\varepsilon_0 = 8.84 \times 10^{-12}$ F m^{-1} (farad per metre), $q = 1.6 \times 10^{-19}$ C (coulomb).
However, to simplify the formulae of the mathematical examples, we shall often choose a system of units – called "natural" – such that $\hbar = 1$, $c = 1$, $\varepsilon_0 = 1$, $\mu_0 = 1$.

Coulomb Potential. The Hamiltonian operator of a particle of mass $\frac{1}{2}$ and of electric charge $Q = -1$ in the presence of a point charge Z placed at $\mathbf{x} = \mathbf{0}$ in $\mathbb{R}^3$,

---

[252] See Chaps. VI, VIII and IX.

is written formally in the system of natural units:

$$H = -\Delta - \frac{Z}{4\pi|\mathbf{x}|}, \qquad \mathbf{x} \in \mathbb{R}^3 . \tag{6.36}$$

Application. Return to the Modelling of an Atom in Quantum Physics (2nd Part B)

The Hydrogen Atom. Let us develop a little further the case of the hydrogen atom and apply to it the fundamental principles seen above. The Hilbert space linked with this physical system in generally $L^2(\mathbb{R}^3)$[253]. The position of the electron is modelled by the operators Q_j ($j = 1, 2, 3$) (see (6.23)) while the momentum is modelled by the operators P_j ($j = 1, 2, 3$) (see (6.24)). The angular momentum is modelled by the operators L_j ($j = 1, 2, 3$) of (6.30). As to the energy operator, or Hamiltonian, it is modelled by the operator H of (6.36) with $Z = 1$. Adopting the SI-system of units, H can be written

$$H = -\frac{\hbar^2}{2\mu}\Delta - \frac{q^2}{4\pi\varepsilon_0|\mathbf{x}|}, \qquad \mathbf{x} \in \mathbb{R}^3 . \tag{6.36$'$}$$

Seeking the energy levels and the wave functions of the hydrogen atom in its stationary set we are led to solve (6.17) where H is given by (6.36). This is done especially in Cohen–Tannoudji *et al.* [1], Chap. VII. The results provide a point of departure from which to carry out simplified calculations of diverse phenomena (ionisation, interaction with electromagnetic radiation, etc.). □

Stark Effect. The Hamiltonian operator of a system consisting of a particle of mass $\frac{1}{2}$, of electric charge $Q = -1$ subject to the Coulomb potential $Z/|\mathbf{x}|$, and placed in addition in a uniform electric field $\mathscr{E}$, is defined formally by[254]:

$$H = -\Delta - \frac{Z}{4\pi|\mathbf{x}|} - \mathscr{E}.\mathbf{x}, \qquad \mathbf{x} \in \mathbb{R}^3, \qquad \mathscr{E} \in \mathbb{R}^3, \tag{6.37}$$

$\mathscr{E}.\mathbf{x}$ denoting the scalar product in $\mathbb{R}^3$.
The Stark effect shifts the spectroscopic rays of the atoms, modifying thus their interaction with the electromagnetic radiation. This field can be created by a source exterior to the material studied or just as well by neighbouring atoms of the atom that we are studying, the single particle studied being then an electron of this atom. In the modelling of the hydrogen-atom in the introduction (A), then in (B) mentioned above (Coulomb potential), and in the case of dense media, the addition of a Stark term simulating the action of particles exterior to the domain Ω considered in (A) (and in Remark 7 below) improves the modelling.

Electromagnetic Field (Constant in Time). The Hamiltonian operator of a system consisting of a particle of mass $\frac{1}{2}$, electric charge 1, placed in an electromagnetic field (see §4 of this Chap. IA) is defined formally by the expression:

$$H = (\mathbf{P} - \mathbf{A})^2 + v = \sum_{j=1}^{3}\left(\frac{1}{i}\frac{\partial}{\partial x_j} - A_j\right)^2 + v^{254} \tag{6.38}$$

---

[253] We recall that we have modelled the hydrogen again without introducing spin.
[254] In the system of natural units.

where we have been given:

$$\begin{cases} \text{the scalar potential } \mathscr{V}: \text{a real function on } \mathbb{R}^3, \\ \text{the vector potential } \mathbf{A} = (A_1, A_2, A_3) \text{ a function on } \mathbb{R}^3 \text{ with values in } \mathbb{R}^3. \end{cases}$$

Among the particular cases, other than the Coulomb potential and the Stark effect, already cited, we note the Hamiltonian operator of Zeeman effect obtained by taking the Coulomb potential (6.36) for $\mathscr{V}$, or a more realistic scalar potential if necessary, and for A the vector potential

$$\text{(6.39)} \qquad \mathbf{A}(\mathbf{x}) = -\frac{1}{2}\mathbf{x} \times \mathbf{B} \qquad \mathbf{A}(\mathbf{x}), \mathbf{x}, \mathbf{B} \in \mathbb{R}^3$$

where $\mathbf{B}$ is a uniform magnetic field, which is therefore a given vector, independent of $\mathbf{x}$.

Harmonic and Anharmonic Oscillators. The Hamiltonian operator of a particle of mass $\frac{1}{2}$ submitted to restoring force proportional to the displacement can be written formally:

$$\text{(6.40)} \qquad H = -\Delta + K\frac{|x|^2}{2}, \quad x \in \mathbb{R}^3, \quad K \in \mathbb{R}^+, \quad \mathscr{H} = L^2(\mathbb{R}^3)\text{[255]}.$$

Similarly we draw attention to the anharmonic oscillator

$$H = -\Delta + K\frac{|x|^2}{2} + \lambda|x|^4, \qquad x \in \mathbb{R}^3, \qquad K, \lambda \in \mathbb{R}^+.$$

This model is used in very many situations in the physics of molecules to describe their vibrations, in the physics of solids and in the quantum theory of fields.

Crystalline Potential. In the study of crystals considered as periodic media the theory called the "theory with one electron" (see Kittel [1]) leads to the Hamiltonian operator:

$$\text{(6.41)} \qquad H = -\Delta + \mathscr{V}\text{[256]}$$

in $\mathscr{H} = L^2(\mathbb{R}^3)$ where $\mathscr{V}$ is a periodic potential: we shall take for $\mathscr{V}$ a real function on $\mathbb{R}^3$ satisfying:

$$\text{(6.42)} \qquad \mathscr{V}(\mathbf{x} + \mathbf{R}) = \mathscr{V}(\mathbf{x}), \qquad \forall \mathbf{x} \in \mathbb{R}^3,$$

where $\mathbf{R}$ is one of the "vectors of the crystal lattice"[257]:

$$\text{(6.43)} \qquad \mathbf{R} = \sum_{i=1}^{i=3} n_i \mathbf{Q}_i, \qquad \mathbf{Q}_i \in \mathbb{R}^3, \qquad n_i \in \mathbb{Z}, \qquad i = 1, 2, 3,$$

the $\mathbf{Q}_i$ being the "base vectors" characteristic of the crystal lattice studied[257].

---

[255] In the system of natural units.

[256] Here we take the system of natural units with the mass of the electron $m_e = \frac{1}{2}$.

[257] See Kittel [1] and Brillouin [1].

For crystals of finite dimension, contained in a domain Ω of $\mathbb{R}^3$ we are led to define a Hamiltonian operator of type (6.41), but in $L^2(\Omega)$, with convenient boundary conditions. These potentials are used to describe crystalline solids.

Remark 7. The potentials cited are among the simplest of those ordinarily encountered in a precise modelling. In fact the physicist has, to describe phenomena, to construct much more complicated expressions; for example, for the modelling of the behaviour of matter at high pressure and temperature, there will be studied, in particular, a single atom "enclosed" in a domain Ω of $\mathbb{R}^3$ in order to represent the action of neighbouring atoms on that being studied; it will be necessary to imagine the boundary conditions on $\partial\Omega$, the boundary of Ω, for Schrödinger's equation in this domain Ω "occupied" by the atom so as to represent the interactions with the neighbouring atoms. These boundary conditions will not always be sufficient to represent the action of the external matter; in certain situations, for example, we must include in Ω, an electric field created by the atoms exterior to Ω acting on the atom.
The physicist must verify that the Hamiltonian H which he has constructed is a self-adjoint operator, as the postulates would impose if the system were isolated. Next, the physical problems which the physicist formulates, require for the most part a knowledge of the spectrum of Hamiltonian operator H (which yields the *energies* E_i *permitted to the system*) and the associated eigenwave functions Φ_i in order to facilitate the calculation of quantities such as the "matrix elements" (Φ_i, $A_k\Phi_j$), (the A_k being observables of the system studied) which permits the study of *the transition probabilities between states and of the cross-sections.* □

2.2. Particles with Non-Zero Spin (Non-Relativistic Theory)

Experience shows that for certain particles, the choice of the operators L_j defined by (6.30) and of the space $\mathscr{H} = L^2(\mathbb{R}^3)$ is not the right one: in particular the spectrum of L_j ($j = 1, 2, 3$) is different from the set of "experimental" values of the component along the x_j-axis of the angular momentum.

Choice of the space $\mathscr{H}$. The agreement with the experimental facts is restored if we choose $\mathscr{H} = L^2(\mathbb{R}^3, \mathbb{C}^n)$ where n is a positive integer which depends on the type of the particle, $L^2(\mathbb{R}^3, \mathbb{C}^n)$ being the space of classes on $\mathbb{R}^3$ with values in $\mathbb{C}^n$:

$$\mathbf{x} \to \Phi(\mathbf{x}) = (\Phi_1(\mathbf{x}), \Phi_2(\mathbf{x}), \ldots, \Phi_n(\mathbf{x}))\,, \qquad \mathbf{x} \in \mathbb{R}^3\,,$$

such that Φ_l ($l = 1, 2, 3$) belong to $L^2(\mathbb{R}^3)$.
Provided with the scalar product

$$(\Psi, \Phi) \overset{\text{def}}{=} \int (\Psi(x), \Phi(x))_{\mathbb{C}^n}\, \mathrm{d}x = \int \sum_{l=1}^{l=n} (\Psi_l(x)\bar{\Phi}_l(x))\, \mathrm{d}x\,,$$

$$\Psi_l, \Phi_l \in L^2(\mathbb{R}^3)\,; \qquad \Psi, \Phi \in L^2(\mathbb{R}^3, \mathbb{C}^n)\,,$$

and the associated norm:

$$\|\Phi\| = (\Phi, \Phi)^{1/2}\,,$$

the space $L^2(\mathbb{R}^3, \mathbb{C}^n)$ is a Hilbert space.

Remark 8. It is sometimes more convenient to use instead of $L^2(\mathbb{R}^3, \mathbb{C}^n)$ other spaces which are identifiable with it. We observe that an element $\Phi = (\Phi_1, \Phi_2, \ldots, \Phi_n)$ of $L^2(\mathbb{R}^3, \mathbb{C}^n)$ can be written in the form of a column

(6.44) $$\Phi = \begin{pmatrix} \Phi_1 \\ \Phi_2 \\ \cdots \\ \Phi_n \end{pmatrix}, \qquad \Phi_l \in L^2(\mathbb{R}^3), \quad l = 1, 2, \ldots, n$$

(Φ_l will be called the component of index l of the element $\Phi \in L^2(\mathbb{R}^3, \mathbb{C}^n)$, and a linear operator A in $L^2(\mathbb{R}^3, \mathbb{C}^n)$ can be put in the form of an $n \times n$ matrix: $\{A_{lm}\}$, ($l, m = 1, 2, \ldots n$) where A_{lm} is a linear operator in $L^2(\mathbb{R}^3)$. The component with index l of the element $A\Phi \in L^2(\mathbb{R}^3, \mathbb{C}^n)$ is written

(6.45) $$(A\Phi)_l = \sum_{m=1}^{m=n} A_{lm}\Phi_m \qquad l = 1, 2, \ldots, n\,.$$

The space $L^2(\mathbb{R}^3, \mathbb{C}^n)$ is *identifiable with* $L^2(\mathbb{R}^3) \otimes \mathbb{C}^n$. We can also identify $L^2(\mathbb{R}^3, \mathbb{C}^n)$ with the space $L^2(\mathbb{R}^3 \times \sigma_n)$ with $\sigma_n = \{1, 2, \ldots, n\}$ equipped with the scalar product

$$(\Psi, \Phi)_{L^2(\mathbb{R}^3 \times \sigma_n)} = \int_{\mathbb{R}^3} \sum_{j=1}^{j=n} \Psi(x, j)\bar{\Phi}(x, j)\,\mathrm{d}x\,, \qquad (x, j) \in \mathbb{R}^3 \times \sigma_n$$

is a Hilbert space.

In effect, to each element $\Phi = (\Phi_1, \Phi_2, \ldots, \Phi_n) \in L^2(\mathbb{R}^3, \mathbb{C}^n)$ we can establish bijectively a function Φ on $\mathbb{R}^3 \times \sigma_n$ defined by

$$(x, j) \overset{\Phi}{\rightarrow} \Phi(x, j) \overset{\text{def}}{=} \Phi_j(x)\,, \qquad (x, j) \in \mathbb{R}^3 \times \sigma_n\,. \qquad \square$$

Choice of the angular momentum operator. Experience leads us to postulate that the operators of the angular momentum J_j are given by

(6.46) $$J_j = L_j \otimes I_{\mathbb{C}^n} + I_{L^2(\mathbb{R}^3)} \otimes S_j\,; \qquad j = 1, 2, 3$$[258]

acting in $\mathscr{H} = L^2(\mathbb{R}^3) \otimes \mathbb{C}^n$. In the expression (6.46), $I_{\mathbb{C}^n}$ and $I_{L^2(\mathbb{R}^3)}$ are respectively the identity operators in $\mathbb{C}^n$ and $L^2(\mathbb{R}^3)$, L_j is the operator defined by (6.30) and called "the component of the orbital angular momentum" and S_j is an operator in $\mathbb{C}^n$, called the component of the *spin* angular momentum, such that the three operators S_j ($j = 1, 2, 3$) satisfy

$$S_jS_k - S_kS_j = ihS_l \quad (j, k, l = \text{cyclic permutation of } 1, 2, 3)$$

The S_j are self-adjoint operators in $\mathbb{C}^n$ whose spectrum is the discrete set

(6.47) $$\sigma(S_j) = \{-s, (-s+1), \ldots, (s-1), s\}$$

---

[258] We often write this in the simplified form $J_j = L_j + S_j$ ($j = 1, 2, 3$) or again

$$\mathbf{J} = \mathbf{L} + \mathbf{S}.$$

s being a positive whole number or one-half of positive odd integer ($s = m + \frac{1}{2}$, $m = 0, 1, 2, \ldots$) such that $n = 2s + 1$. The number s is called the *spin of the particle*. A particle "without spin" is one for which $s = 0$. We then have $n = 1$, and the Hilbert space associated with it is $\mathscr{H} = L^2(\mathbb{R}^3) \otimes \mathbb{C}^1 = L^2(\mathbb{R}^3)$. With a particle of spin $s = \frac{1}{2}$, for which we have $n = 2$, we associate the space $L^2(\mathbb{R}^3) \otimes \mathbb{C}^2$ (in non-relativistic theory). The particles with half-integer spin are called *fermions*: (examples: electrons, neutrinos, protons[259] . . .). The operators of spin in this case can be written

$$(6.48) \quad S_1 = \frac{\hbar}{2}\begin{pmatrix} 0 & 1 \\ 1 & 0 \end{pmatrix}, \quad S_2 = \frac{\hbar}{2}\begin{pmatrix} 0 & -i \\ +i & 0 \end{pmatrix}, \quad S_3 = \frac{\hbar}{2}\begin{pmatrix} 1 & 0 \\ 0 & -1 \end{pmatrix},$$

The particles with integer (or zero) spin are called *bosons* (example: photons, . . .)[259].

Application. The Modelling of an Atom in Quantum Physics (3rd Part C)

The Hydrogen Atom. The model of the hydrogen atom described in (**A**) and (**B**) permits us to check experimentally a great number of the properties of this atom (especially absorption and emission spectra). However, the great precision of experimental measurements necessitates, for certain phenomena (fine structure of spectral lines, "anomalous" Zeeman effect, etc.) the introduction of a spin $s = \frac{1}{2}$ of the electron with the spin operators (6.48). This electron possesses then a total angular momentum J given by $J = L + S$ (see (6.46), with (6.48)).

2.3. Particle in Relativistic Physics

Particles whose speeds are close to that of light are no longer described by a wave function solution of Schrödinger's equation but by a function (or distribution) on $\mathbb{R}^4$ called generally a field, a solution of the Klein–Gordon equation[260] or of the Dirac equation[260] according to the type of particle considered.
The *Klein–Gordon equation* is used in quantum field theory for scalar fields[261]. Again with the convention $\hbar = 1$, $c = 1$ and mass m, it can be written:

$$(6.49) \qquad \left(\frac{\partial^2}{\partial t^2} - \Delta + m^2\right)\Phi(x, t) = 0$$

For a free particle of spin $\frac{1}{2}$, the evolution of the "wave vector" Φ is given (with

[259] See the references cited at the beginning of this §6.

[260] These equations of evolution of the wave functions which aim to unify quantum mechanics and relativity lead to difficulties. However, the interest in these evolution equations is very great in the quantum theory of fields, where they are considered as equations of evolution, not of wave functions, but of "field operators" for particles; more precisely, the solutions sought are vector distributions (see Chap. XVI, §2) defined on $\mathscr{S}(\mathbb{R}^4)$ with operator values (for all these ideas see Itzykson–Zuber [1], the Dirac equation, p. 45.)

[261] Case $s = 0$ (zero spin), see Itzykson–Zuber [1], Chap. 3.1.2.

always $\hbar = 1, c = 1$), the *Dirac equation*

$$(6.50) \qquad i\frac{\partial \Phi}{\partial t} = \left(\frac{1}{i}\alpha . \nabla + \beta m\right)\Phi \overset{\text{def}}{=} H\Phi$$

where $\Phi(.,t)$ is an element of $L^2(\mathbb{R}^3) \otimes \mathbb{C}^4$, the matrices $\alpha = \{\alpha^1, \alpha^2, \alpha^3\}$ and β are given Hermitian matrices[262] (hence H is self-adjoint) and $\alpha . \nabla$ is by definition

$$\alpha , \nabla \overset{\text{def}}{=} \sum_{j=1}^{3} \alpha^j \frac{\partial}{\partial x_j}$$

With the notation

$$(6.51) \qquad \begin{cases} \gamma^0 = \beta\,, & \gamma^j = \beta\alpha^j \quad (j = 1,2,3) \\ \partial_\mu = \dfrac{\partial}{\partial x^\mu}\,, & x^0 \overset{\text{def}}{=} t \quad x^j \overset{\text{def}}{=} x_j \quad j = 1,2,3 \end{cases}$$

and the convention of summation of indices, Dirac's equation (6.50) becomes

$$(6.52) \qquad (i\gamma^\mu \partial_\mu - m)\Phi = 0\,,$$

and putting $\not{k} = k_\mu \gamma^\mu$ for each letter k this further reduces to

$$(6.53) \qquad (i\not{\partial} - m)\Phi = 0\,.$$

When the particle of spin $\frac{1}{2}$, of mass m, and of chare Q (with generally $Q = q$ or $Q = -q$) interacts with an electromagnetic field of vector potential

$$\{A_\mu\} = (A_0, A_1, A_2, A_3) \quad \text{with} \quad A_0 = \mathscr{V} \quad \text{and} \quad (A_1, A_2, A_3) = A$$

(see §4, relations (4.61)), Dirac's equation (6.53) becomes

$$(6.54) \qquad (i\not{\partial} - Q\not{A} - m)\Phi = 0^{263}\,.$$

---

[262] With the property that $((1/i)\alpha . \nabla + \beta m)^2 = (-\Delta + m^2)I_4$, (where I_4 is the unit matrix in $\mathbb{C}^4$, which gives the anticommutative relations for the matrices α^j and β:

$$[\alpha^k, \alpha^j]_a \overset{\text{def}}{=} \alpha^k\alpha^j + \alpha^j\alpha^k = 2\delta_{kj}\,,$$

$$[\alpha^j, \beta]_a \overset{\text{def}}{=} \alpha^j\beta + \beta\alpha^j = 0 \quad \text{with} \quad (\alpha^j)^2 = 1 \quad \text{and} \quad \beta^2 = 1$$

The reader should consult the reference Itzykson–Zuber [1] for supplementary remarks. We point out on the representation below:

$$\alpha^j = \begin{pmatrix} 0 & \sigma^j \\ \sigma^j & 0 \end{pmatrix}, \quad \beta = \begin{pmatrix} I & 0 \\ 0 & -I \end{pmatrix}, \quad \gamma^0 = \begin{pmatrix} I & 0 \\ 0 & -I \end{pmatrix}, \quad \gamma^j = \begin{pmatrix} 0 & \sigma^j \\ -\sigma^j & 0 \end{pmatrix}$$

where I is the 2×2 unit matrix and σ^j the Pauli matrices

$$\sigma^1 = \begin{pmatrix} 0 & 1 \\ 1 & 0 \end{pmatrix}, \quad \sigma^2 = \begin{pmatrix} 0 & -i \\ i & 0 \end{pmatrix}, \quad \sigma^3 = \begin{pmatrix} 1 & 0 \\ 0 & -1 \end{pmatrix}$$

They are linked to the matrices S_j by $S_j = \frac{1}{2}\hbar\sigma^j$.

[263] We have kept here the expression Q for the electric charge so that the physical meaning of these equations (6.53), (6.54), (6.55) may be more apparent.

This equation can also be written

$$(6.55) \qquad i\frac{\partial \Phi}{\partial t} = \left[\alpha\left(\frac{1}{i}\nabla - QA\right) + \beta m + Q\mathscr{V}\right]\Phi$$

namely

$$i\frac{\partial \Phi}{\partial t} = (H_0 + H_{\text{interaction}})\Phi$$

with

$$\begin{cases} H_0 = \alpha P + \beta m\,, \qquad P = \text{momentum} = \dfrac{1}{i}\nabla \\ H_{\text{interaction}} = -\,Q\alpha A + |Q\mathscr{V}| \end{cases}$$

which is near to the non-relativistic case.

Application. Modelling of an Atom in Quantum Physics

The Hydrogen Atom. The Dirac equation for the hydrogen atom is obtained from (6.55) by taking $\mathbf{A} \equiv \mathbf{0}$ and for A_0 the Coulomb electrostatic already used for the Schrödinger equation for the hydrogen atom. This modelling allows us to take account of the fine structure of the spectrum of the hydrogen atom with very great precision. It is remarkable that the exact solution of Dirac's equation for the hydrogen atom is possible (see Itzykson–Zuber [1]) and that it conforms to the experimental results concerning hydrogen-like atoms. Similarly, perturbation methods allow us to study with the advantage of giving a physical meaning to the differences of the model of Dirac's equation with respect to the modelling of Schrödinger's equation given in Application *A*, *B*, *C*. This improved model *D* similarly leads to the treatment of the Zeeman effect of the hyperfine structure of the ground level, the influence of the spin of the electron on the Zeeman effect of the resonance line of hydrogen, the importance for the Stark effect of a degeneracy between levels of different parity, etc. (see Cohen–Tannoudji *et al.* [1]). If we study the hydrogen-like ion with a high value of Z, for example, lead (Pb) the electrons most strongly linked have a very high energy, and hence a speed which is not negligible in comparison with that of light. We must therefore use Dirac's equation to describe one of these electrons linked to the hydrogen-like lead ion.

3. Systems of Several Particles

3.1. System of N Particles – Pauli's Principle

The Spaces Used. If the N particles ($N \in \mathbb{N}$) are *all distinct* and if $\mathbf{h}_i$ ($i = 1, 2, \ldots, N$) is the Hilbert space associated with the particle i, we adopt as the Hilbert space associated with the system of the N particles, the Hilbert tensor product:

$$(6.56) \qquad \mathbf{h}_1 \hat{\otimes} \mathbf{h}_2 \hat{\otimes} \ldots \hat{\otimes} \mathrm{h}_N \overset{\text{def}}{=} \hat{\otimes}_{i=1}^{i=N} \mathbf{h}_i$$

If the N particles are *all identical*, the indiscernability of the particles leads us to adopt as the Hilbert space, not the tensor product

$$\mathbf{h}^{\hat{\otimes} N} \overset{\text{def}}{=} \mathbf{h} \hat{\otimes} \mathbf{h} \ldots \hat{\otimes} \mathbf{h} \overset{\text{def}}{=} \hat{\otimes}^N \mathbf{h},$$

of N spaces $\mathbf{h}_i$ all identical to a same space $\mathbf{h}$ but to a certain subspace of $\mathbf{h}^{\hat{\otimes} N}$ which we shall now specify. First of all let us recall some definitions.
Let ψ be an element of $\mathbf{h}^{\hat{\otimes} N}$ of the form

$$(6.57) \qquad \psi_{\alpha_1, \alpha_2, \ldots \alpha_N} = \Phi_{\alpha_1}(1) \otimes \Phi_{\alpha_2}(2) \otimes \ldots \otimes \Phi_{\alpha_N}(N)$$
$$\text{with} \quad \Phi_{\alpha_i} \in \mathbf{h} \; (i = 1 \ldots N)^{264}$$

and P^π a "permutation" operator such that (for example)

$$(6.58) \quad P^\pi(\Phi_{\alpha_1}(1) \otimes \Phi_{\alpha_2}(2) \otimes \ldots \otimes \Phi_{\alpha_N}(N)) = \Phi_{\alpha_1}(2) \otimes \Phi_{\alpha_2}(1) \otimes \ldots \otimes \Phi_{\alpha_N}(m)$$

where $(2, 1, \ldots, m)$ is an example of a permutation of $(1, 2, \ldots, N)$ denoted by π.
We consider the elements of the form

$$(6.59) \qquad \psi_+ \overset{\text{def}}{=} k_+ \sum_\pi P^\pi(\Phi_{\alpha_1}(1) \otimes \Phi_{\alpha_2}(2) \otimes \ldots \otimes \Phi_{\alpha_N}(N)), \quad \Phi_{\alpha_i} \in \mathbf{h}$$

$$(6.60) \qquad \psi_- \overset{\text{def}}{=} k_- \sum_\pi P^\pi(-1)^{p_\pi}(\Phi_{\alpha_1}(1) \otimes \Phi_{\alpha_2}(2) \otimes \ldots \otimes \Phi_{\alpha_N}(N));$$

the coefficients k_+ and k_- permitting normalisation, the summation $\sum_\pi$ extends over all the permutations π and $(-1)^{p_\pi}$ is the "parity" of the permutation P^π [265].

Definition[266]. *The elements ψ_+ and $\psi_- \in \mathbf{h}^{\hat{\otimes} N}$ defined by the relations (6.59) and (6.60) are called respectively "symmetric elements" and "antisymmetric elements" of $\mathbf{h}^{\hat{\otimes} N}$. The Hilbert space generated in $\mathbf{h}^{\hat{\otimes} N}$ by the symmetric (resp. antisymmetric) elements is called a symmetric (resp. antisymmetric) Hilbert sub-space and is denoted by $\mathbf{h}^{\otimes N}_+$ (resp. $\mathbf{h}^{\otimes N}_-$).*

---

[264] Where we have written $1, 2, \ldots, i, \ldots, N$ for $x_1, x_2, \ldots, x_i, \ldots, x_N$ where x_i denotes the set of the coordinates of a particle. We imply that $h = L^2(\mathbb{R}^3_x)$ and (6.57), (6.58) signify that we have (a.e. $x_j \in \mathbb{R}^3$, $j = 1$ to N)

$$\Psi_{\alpha_1, \ldots, \alpha_N}(x_1, \ldots, x_N) \overset{\text{def}}{=} \Phi_{\alpha_1}(x_1) \ldots \Phi_{\alpha_N}(x_N)$$

$$P^\pi \Psi_{\alpha_1, \ldots, \alpha_N}(x_1, \ldots, x_N) = \Psi_{\alpha_1, \ldots, \alpha_N}(x_2, x_1, \ldots, x_m)$$

which we can also write for the permutation π taken as example

$$P^\pi \Psi_{\alpha_2, \alpha_1, \ldots, \alpha_m} = \Psi_{\alpha_1, \alpha_2, \ldots, \alpha_N}$$

[265] The parity (or signature) of the permutation P^π is $(-1)^\nu$ where $\nu = p_\pi$ is the number of transpositions of which the permutation π is the product (see Miller [1]).

[266] Instead of defining the *bosons* and the *fermions* by the parity of their intrinsic spin (and hence the values of s) as we have done after the relation (6.17), we should be able to define the bosons and fermions by the fact that their wave function is symmetric or antisymmetric. A theorem connecting *spin* and *statistics* should infer the "parity" of the spin; this last step corresponds better to the physical nature of certain phenomena.

We can now complete Postulate 1 by specifying the Hilbert space associated with systems of identical particles.

Postulate 1 (continuation)[266]. *With the systems of N identical particles with integer spin (particles called bosons) we associate the Hilbert space* $\mathbf{h}^{\hat{\otimes} N}_{+}$, *symmetric subspace of the tensor product* $\mathbf{h}^{\hat{\otimes} N} = \hat{\otimes}^N \mathbf{h}$ *where* $\mathbf{h}$ *is the Hilbert space associated with a single particle. With the systems of N identical particles with half-integer spin (particles called fermions) we associate the Hilbert space* $\mathbf{h}^{\hat{\otimes} N}_{-}$, *antisymmetric subspace of the tensor product* $\mathbf{h}^{\hat{\otimes} N} = \hat{\otimes}^N \mathbf{h}$ *where* $\mathbf{h}$ *is the Hilbert space associated with a single particle.*

The Observables. When the N particles of a physical system are identical, the indiscernability of the particles leads us to impose on the self-adjoint operators, capable of representing observables, that they must commute with all the permutation operators P^π (defined by the relation (6.58))

$$AP^\pi = P^\pi A\,, \qquad \forall \pi\,.$$

These operators A then leave invariant the subspaces $\mathbf{h}^{\hat{\otimes} N}_{-}$ and $\mathbf{h}^{\hat{\otimes} N}_{+}$, and we adopt as observables their restrictions to these subspaces.

Example of an Observable: The Hamiltonian Operator of a System of N Particles. For simplicity, we shall suppose that the particles are discernable and of zero spin. The Hilbert space associated with a system of N distinct particles of zero spin which we adopt here is

$$\mathbf{h}^{\hat{\otimes} N} = L^2(\mathbb{R}^3) \hat{\otimes} L^2(\mathbb{R}^3) \ldots \hat{\otimes} L^2(\mathbb{R}^3) = \hat{\otimes}^N L^2(\mathbb{R}^3)\,, \tag{6.61}$$

a space which can be identified with $L^2[(\mathbb{R}^3)^N]$.

We denote by $\mathbf{x} = (x_1, \ldots, x_\alpha, \ldots, x_N)$ an element of $(\mathbb{R}^3)^N$ with $x_\alpha \in \mathbb{R}^3$, $(\alpha = 1, \ldots, N)$ and by Δ_{x_α} the Laplacian operator "relative to the variable x_α", that is to say

$$\Delta_{x_\alpha}\psi(x_1, \ldots, x_\alpha, \ldots, x_N) = \sum_{i=1}^{i=3} \frac{\partial^2}{\partial x_{\alpha i}^2}\psi(x_1, \ldots, x_\alpha, \ldots, x_N) \tag{6.62}$$

where $x_{\alpha i}$ $(i = 1, 2, 3)$ are the "coordinates" of the point $x_\alpha \in \mathbb{R}^3$.

The Hamiltonian operator associated with the energy of a system of N *free* particles without spin, of mass m_α $(\alpha = 1, \ldots, N)$ is then

$$H_0 = -\frac{1}{2}\sum_{\alpha=1}^{\alpha=N} \frac{\Delta_{x_\alpha}}{m_\alpha}\,, \qquad D(H_0) = H^2[(\mathbb{R}^3)^N]^{267}\,. \tag{6.63}$$

If the particles interact, we shall suppose that the action undergone by the particles can be expressed by a supplementary term $\mathscr{V}$, called the "potential", in the expression for the Hamiltonian operator which can then be written:

$$H = -\frac{1}{2}\sum_{\alpha=1}^{\alpha=N} \frac{\Delta_{x_\alpha}}{m_\alpha} + \mathscr{V}\,. \tag{6.64}$$

---

[267] Or the restriction of H_0 to $\mathbf{h}^{\hat{\otimes} N}_{+}$ or to $\mathbf{h}^{\hat{\otimes} N}_{-}$, if the particles are identical.

The potential $\mathscr{V}$ will be, for example, of the form

(6.65) $$\mathscr{V}(x_1, \ldots, x_\alpha, \ldots, x_N) = \sum_{\alpha=1}^{\alpha=N} \mathscr{V}_\alpha(x_\alpha) .$$

or again

(6.66) $$\mathscr{V}(x_1, \ldots, x_\alpha, \ldots, x_N) = \sum_{\beta=1}^{N} \sum_{\alpha<\beta} \mathscr{V}_{\alpha\beta}(x_\alpha - x_\beta) .$$

This is the case of the "Coulomb" potential

(6.67) $$\mathscr{V}(x_1, \ldots, x_N) = \frac{1}{4\pi} \sum_{\alpha<\beta} \sum_{\beta=1}^{\beta=N} \frac{1}{|x_\alpha - x_\beta|}$$

(in the system of natural units); in this expression we have supposed that the electric charges Q are equal to 1. We shall see that the operator:

(6.68) $$H = H_0 + \mathscr{V} , \qquad D(H) = D(H_0) ,$$

where $\mathscr{V}$ is the Coulomb potential (6.67) and where H_0 is given by (6.63), is self-adjoint in $L^2(\mathbb{R}^{3N})$.

Application. The Modelling of an Atom in Quantum Physics (5th Part E)

The Atom Possessing Z Electrons. We have examined in application A, B, C, D the hydrogen-like atoms. When that approximation is not valid for modelling an atom (other than hydrogen), we can make use of the models with several electrons of this section. For example, the "isolated" atom of silicon, with $Z = 14$, and its $Z = 14$ electrons can be modelled by the nucleus of silicon of charge Zq placed at the origin O, the Z electrons interacting among each other and with the central nucleus of silicon through the Hamiltonian

$$H = H_0 + \mathscr{V} - \frac{1}{4\pi} \sum_{\alpha=1}^{Z} \frac{Zq^2}{|x_\alpha|} \quad \text{(with } \mathscr{V} \text{ given by (6.67) and with } N = Z)$$

with the conventions of the system of natural units (then $q^2 \simeq 1/137$). With the conventions of the system of SI units (Système International), the Hamiltonian H can be written:

$$H_{\text{Si}} = -\frac{\hbar^2}{2m_e} \sum_{\alpha=1}^{Z} \Delta_{x_\alpha} + \frac{q^2}{4\pi\varepsilon_0} \sum_{\beta=1}^{Z} \sum_{\alpha<\beta} \frac{1}{|x_\alpha - x_\beta|} - \frac{Zq^2}{4\pi\varepsilon_0} \sum_{\alpha=1}^{Z} \frac{1}{|x_\alpha|} \qquad \square$$

3.2. System with an Indeterminate Number of Identical Particles: Fock's Representation

The number of particles of a system is a physical quantity which is associated with a self-adjoint operator $[N]$ in $\mathscr{H}$, the Hilbert space of the system. The spectrum of $[N]$ having to be the set of possible values of the number of particles, must be discrete and part of $\mathbb{N}$, the set of positive integers or zero.

Let us denote by $\mathbf{h}(N)$ the space generated by the eigenvectors of $[N]$ corresponding to each eigenvalue N. In the so-called "Fock" representation, we demand of the Hilbert space $\mathscr{H}$ that it be such that $\mathbf{h}(N)$ is identical with the space $\mathbf{h}^{\otimes N}_{+}$ or $\mathbf{h}^{\otimes N}_{-}$ (considered in the preceding section) according as the particles are bosons or fermions. In numerous cases, we choose as a representation of *the Hilbert space associated with a system consisting of an indeterminate number of identical particles, the infinite direct sum of the Hilbert spaces* $\mathbf{h}^{\hat{\otimes} N}_{\pm}$ *associated with the systems of N particles*:

$$(6.69) \qquad \mathscr{H}_{\pm} \overset{\text{def}}{=} \mathbf{h}^0 \oplus \mathbf{h} \oplus \mathbf{h}^{\hat{\otimes} 2}_{\pm} \oplus \ldots \oplus \mathbf{h}^{\hat{\otimes} N}_{\pm} + \ldots \overset{\text{def}}{=} \bigoplus_{N \in \mathbb{N}} \mathbf{h}^{\hat{\otimes} N}_{\pm},$$

such a space is called a Fock representation.

The space $\mathbf{h}^0 = \mathbb{C}$, corresponding to the zero proper value of the operator $[N]$, represents the state called "the vacuum state" ($N = 0$).

Thus it is possible to generate the Hilbert space $\mathscr{H}$ ($= \mathscr{H}_+$ or $\mathscr{H}_-$) associated with a system of an indeterminate of identical particles starting from the space $\mathbf{h}$ associated with a single particle by means of the formula (6.69); this construction is the Fock representation; naturally the procedure is not the same for bosons and fermions.

The Fock representation has a direct physical interpretation. It is very convenient and adapted to the physical nature of the system to be described.

Application. Quantum Statistical Physics

(1) The use of systems with an indeterminate number of particles leads especially for equilibrium, to the definition of the principal quantities arising in thermodynamics and to the rediscovery of the laws of thermodynamics.

(2) In a like manner the vibrations of a crystal lattice which are called phonons can be 0, 1, 2, . . . phonons in number. It is the same with the photons at equilibrium in a domain Ω, i.e. with the electromagnetic radiation in equilibrium in Ω.

The phonons and the photons are bosons. Helium-three atoms (^{3}He) forming a liquid at low temperature ($T < 3°$K) give an example of particles which are fermions. The free electrons in the interior of a solid furnish another example of a system of fermions. In first approximation, these systems are quantum gases of indiscernable particles without interaction. These systems can be modelled by using (6.69).

Questions linked to the statistics of these particles interest the physicist, such as the state density, the energy of such a system in equilibrium, etc., questions which belong to quantum statistical physics[268]. □

Remark 9. In the physical systems presented here, the symmetries have played an essential part.

— The invariance under translation in the time stated in Postulate 4 implies the existence of an evolution group $U(t)$ with generator $B = iH$ and a conservation law – that of energy.

---

[268] See Balian [1], D. Kastler [1], Ruelle [1].

This evolution group similarly conserves the total probability $\int_{\mathbb{R}^3} |\Phi|^2 \, dx$.

— The invariance under translation in the space variables of a system of free particles implies the law of the conservation of momentum P in the course of the evolution (as P and H commute).
— The invariance under rotation in the space of a system of free particles implies the law of conservation of angular momentum L in the course of the evolution (as L and H commute). □

Review

The search for the energies E and the states Φ of a quantum system leads to the study of the stationary Schrödinger equation and hence to that of the spectrum of the *Hamiltonian operator* H. We shall study in Chap. IX the properties, especially spectral properties, of solutions appropriate to particular cases of this operator.
The evolution problem of the Schrödinger equation will be treated by the Fourier transform (Chap. XIV, §4), by diagonalisation (Chap. XV, §5), by the Laplace transform (Chap. XVI, §5), by semi-groups (Chap. XVIIB) and by variational methods (Chap. XVIII, §7).
The Liouville–von Neumann equation which governs, more generally, quantum systems, is studied in Chap. XVIIB.
Stone's theorem which links the self-adjoint Hamiltonian operator H and the unitary operator $U(t) = \exp(-itH)$ which governs the evolution of the physical system considered is studied in Chap. XVIIA, §4.
Unitary operators which therefore play a fundamental role in evolution, but also in the change of representation of the problems of quantum physics are seen in Chap. VI.
Operators *with trace*, used for representing states are defined and studied in Chap. VI and their spectra in Chap. VIII.

Appendix. Concise Elements Concerning Some Mathematical Ideas Used in this §6

The ideas introduced below will be recalled and developed in Chaps. VI and VIII.

1. Hilbert Space. A complex Hilbert space is a vector space on the field of complex numbers, given a scalar product, and complete within the topology of the norm associated with the scalar product.
We shall denote by $(.\,,.)$ or again $(.\,,.)_{\mathscr{H}}$ the scalar product in a Hilbert space $\mathscr{H}$; and by $\|.\|$ the associated norm, such $\|u\| = (u, u)^{\frac{1}{2}}$, $\forall\, u \in \mathscr{H}$. "Complete within the topology of the norm" means that every Cauchy sequence $\{u_i\}_i \subset \mathscr{H}$ converges to an element $u \in \mathscr{H}$:

$$\{u_i\}_i \subset \mathscr{H} \text{ such that } \lim_{i,j\to\infty} \|u_i - u_j\| = 0 \Rightarrow \exists\, u \in \mathscr{H} \text{ such that } \lim_{i\to\infty} \|u_i - u\| = 0.$$

The Hilbert spaces which we shall encounter, admit a denumerable orthonormal base, i.e. a sequence $(\Phi_i)_{i \in \mathbb{N}}$ such that:

(i) $\Phi_j \in \mathscr{H}$

(ii) $(\Phi_i, \Phi_j) = \delta_{ij} = \begin{cases} 1 & \text{if } i = j \\ 0 & \text{if } i \neq j \end{cases}$

(iii) for all $\Phi \in \mathscr{H}$, we have $\Phi = \Sigma(\Phi, \Phi_i)\Phi_i$.

If there exists one such base, there exists an infinity. These spaces are then said to be with denumerable bases (or to be *separable* – see Chaps. VI and VIII).

Example. The space $L^2(\mathbb{R})$ of the (classes of) measurable complex-valued functions f such that

$$\int_{\mathbb{R}} |f(x)|^2 \, dx < \infty$$

is a complex Hilbert space for the scalar product $(.\,,.)$ defined by:

$$(f, g) = \int_{\mathbb{R}} f(x)\bar{g}(x) \, dx .$$

This space is separable, as, for example, the sequence of the functions $x \to H_n(x)$, called the *Hermite functions*, defined by

$$H_n(x) = \frac{(-1)^n}{(\sqrt{\pi}2^n n!)^{\frac{1}{2}}} e^{\frac{x^2}{2}} \frac{d^n e^{-x^2}}{dx^n}, \qquad \forall x \in \mathbb{R},$$

is an orthonormal base in $L^2(\mathbb{R})$, as will be seen in Chap. VIII, §2. □

2. Bounded Operators. A *bounded* linear operator A in $\mathscr{H}$, is a linear mapping of $\mathscr{H}$ into $\mathscr{H}$ such that there exists $M > 0$ and

$$\|Av\| \leqslant M\|v\| , \qquad \forall v \in \mathscr{H} .$$

The set of bounded (linear) operators in $\mathscr{H}$ is denoted by $\mathscr{L}(\mathscr{H})$; the operator A is said to be *positive* if

$$(Av, v) \geqslant 0 , \quad \forall v \in \mathscr{H} .$$

Let A be a positive operator and $\{\Phi_i\}_{i \in \mathbb{N}}$ an orthonormal base of $\mathscr{H}$; we say that the quantity (finite or infinite)

$$\operatorname{tr}(A) = \sum_{i \in \mathbb{N}} (A\Phi_i, \Phi_i)$$

is the *trace* of A.

We can show that $\operatorname{tr}(A)$ is independent of the choice of the base. This idea of trace can be generalized to the case of operators which are bounded but not necessarily positive. We notice that if $\mathscr{H}$ has finite dimension n, the operator A is represented on the base (Φ_i) by an $n \times n$ matrix. The trace of A then coincides with the trace of this matrix. Operators with finite trace (or quite simply "trace operators") will be studied in Chap. VIII, §2.

By the *adjoint* of a bounded operator A in $\mathscr{H}$, we mean the operator denoted by[269] A^* such that

$$(u, Av) = (A^*u, v)\,, \qquad \forall u, v \in \mathscr{H}\,.$$

The bounded operator U in $\mathscr{H}$ is said to be *unitary* if

$$(u, v) = (Uu, Uv)\,, \qquad \forall u, v \in \mathscr{H} \quad \text{and} \quad U\mathscr{H} = \mathscr{H}$$

We verify that the operator U of $\mathscr{H}$ into $\mathscr{H}$ is unitary if and only if

$$UU^* = U^*U = I$$

where I is the identity operator in $\mathscr{H}$.

3. Unbounded Operators. An *unbounded* (linear) operator A in $\mathscr{H}$ is not defined on the whole of $\mathscr{H}$, but only on a sub-space $D(A)$ of $\mathscr{H}$, called the "domain of A". We shall encounter different types in Chap. VI.

The most useful operators in quantum mechanics are the *self-adjoint* operators. An unbounded operator A in $\mathscr{H}$ with domain $D(A)$, is said to be *self-adjoint* if it satisfies the following two conditions:

(i) $(Au, v) = (u, Av)\,, \qquad \forall u, v \in D(A)\,,$
(ii) its domain $D(A)$ is equal to the set of the $z \in \mathscr{H}$ such that there exist $y \in \mathscr{H}$ satisfying $(z, Av) = (y, v)$, $\forall v \in D(A)$.

Example. In the Hilbert space $L^2(\mathbb{R})$, the unbounded linear operator A, defined in terms of the differential operator $-\,id/dx$ by

$$\begin{cases} Au = -\,iu' \; (u' = \text{derivative of } u \text{ in the sense of distributions}) \\ D(A) = \{u \in L^2(\mathbb{R})\,, u' \in L^2(\mathbb{R})\} \overset{\text{def}}{=} \text{Sobolev space } H^1(\mathbb{R}) \end{cases}$$

is a self-adjoint operator. It is the same for the operator B, defined by

$$\begin{cases} Bu = -\,u'' \, (\text{in the sense of distributions}) \\ D(B) = \{u \in L^2(\mathbb{R})\,, u'' \in L^2(\mathbb{R})\} \overset{\text{def}}{=} \text{Sobolev space } H^2(\mathbb{R})\,. \end{cases}$$ □

4. Spectrum of an Operator. We define the *resolvant set* $\rho(A)$ of a self-adjoint operator in $\mathscr{H}$ to be the set of all $\lambda \in \mathbb{C}$ such that

$$R(\lambda, A) \overset{\text{def}}{=} (\lambda I - A)^{-1} \in \mathscr{L}(\mathscr{H})\,.$$

The *spectrum* of A, denoted by $\sigma(A)$, is the complement in $\mathbb{C}$ of the resolvant set. For self-adjoint operators A, we have that $\sigma(A) \subset \mathbb{R}$. The *point* spectrum $\sigma_p(A)$ is the set of all *eigenvalues* of A. In other words:

$$\lambda \in \sigma_p(A)\,, \qquad \text{if there exists } v \in D(A)\,, v \neq 0 \quad \text{such that} \quad Av = \lambda v\,.$$

---

[269] In physics, we often meet the notation A^+ for this adjoint operator.

We have $\sigma_p(A) \subset \sigma(A)$, but the inclusion is strict in general. All possible cases can occur: the spectrum can consist of the whole real axis, or only of a denumerable set of isolated eigenvalues, or indeed of a denumerable set and a continuous part (we see appearing here significant differences with the finite-dimensional case).

5. Spectral Family and Spectral Measure. If $\mathscr{H}$ is of dimension n, a self-adjoint operator A is represented by a hermitian matrix. There exist therefore orthogonal eigen sub-spaces V_k associated with the eigenvalues (λ_k), $1 \leqslant k \leqslant n$ of A such that

$$Au_k = \lambda_k u_k \quad \text{if} \quad u_k \in V_k \,.$$

We recall that the dimension of the eigen sub-space V_k is equal to the multiplicity of the eigenvalue λ_k, and that $\mathscr{H}$ is the direct sum of the eigen subspaces V_k:

$$\mathscr{H} = \bigoplus_{k=1}^{k=n} V_k$$

In particular if $P_k \colon \mathscr{H} \to \mathscr{H}$ denotes the orthogonal projection on the eigen sub-space V_k, we have the following two properties:

$$I = \sum_{1 \leqslant k \leqslant n} P_k \,, \qquad A = \sum_{1 \leqslant k \leqslant n} \lambda_k P_k$$

(where I denotes the identity operator on $\mathscr{H}$). With a view to generalizing these ideas to infinite-dimensional spaces, we introduce the family of sub-spaces $(M(\lambda))_{\lambda \in \mathbb{R}}$ where $M(\lambda)$ is defined as the direct sum of all the eigen sub-spaces V_k associated with the eigenvalues $\lambda_k \leqslant \lambda$

$$M(\lambda) = \bigoplus_{\lambda_k \leqslant \lambda} V_k \,.$$

We observe that

$$M(\lambda) \subset M(\mu) \,, \quad \text{for} \quad \lambda < \mu \,,$$

and that $M(-\infty) = \{0\}$ and $M(+\infty) = \mathscr{H}$.
Let $E(\lambda)$ be the orthogonal projection on $M(\lambda)$, then the family of orthogonal projections $(E(\lambda))_{\lambda \in \mathbb{R}}$ has the following properties:

$$\text{(6.76)} \qquad \begin{cases} \text{(i)}\ E(\lambda) . E(\mu) = E(\lambda) \quad \text{for} \quad \lambda < \mu \\ \text{(ii)}\ E(-\infty) = 0 \quad \text{and} \quad E(+\infty) = I \\ \text{(iii)}\ E(\lambda + 0) = E(\lambda) \end{cases}$$

where $E(\lambda \pm 0)$ are the two orthogonal projections defined by

$$E(\lambda \pm 0)x = \lim_{\varepsilon \to 0+} E(\lambda \pm \varepsilon)x \,.$$

Every family of orthogonal projections satisfying (6.76) is called a *spectral family*. Being given a spectral family $(E(\lambda))_{\lambda \in \mathbb{R}}$, we can associate with it a measure called a *spectral measure*, denoted again by E and defined in the following fashion:

$$\text{if} \quad b =]\lambda, \mu], \qquad \text{we put} \quad E(b) = E(\mu) - E(\lambda) \,,$$

$$\text{if} \quad b = \{\lambda\}, \qquad \text{we put} \quad E(b) = E(\lambda) - E(\lambda - 0) \,.$$

We use later the additivity property:

$$E(b_1 \cup b_2) = E(b_1) + E(b_2)$$

applicable when b_1 and b_2 are disjoint to extend the definition of E to all the intervals of $\mathbb{R}$ closed, open or other, and more generally to every Borel set. We can take the integral with respect to this measure and define just as the integral of a continuous real function on $\mathbb{R}$, the integral

$$T = \int_{\mathbb{R}} f(\lambda)\,\mathrm{d}E(\lambda)$$

whose values are operators in $\mathscr{H}$, and which is a self-adjoint operator for

$$D(T) = \left\{ u \in L^2(\mathbb{R}),\ \int_{\mathbb{R}} |f(\lambda)|^2\,\mathrm{d}(E(\lambda)u, u) < +\infty \right\}.$$

By definition, we have

$$\int_{\mathbb{R}} \mathrm{d}E(\lambda) = E(\mathbb{R}) = E(+\infty) - E(-\infty) = I\,.$$

In the finite-dimensional case, this integral reduces to a finite sum (the spectral measure is in this case a sum of Dirac measures) and we have

$$T = \sum_{1 \leqslant k \leqslant n} f(\lambda_k) P_k\,,$$

with the result that in the case where $f(\lambda) = \lambda$, we have $T = A$.
This formula (again with $f(\lambda) = \lambda$) can be generalized to the infinite-dimensional case in the following fashion:

Theorem. *Let $\mathscr{H}$ be a complex Hilbert space and A a self-adjoint operator in $\mathscr{H}$, then there exists a spectral family $E^A(\lambda)$ such that*

$$A = \int_{\mathbb{R}} \lambda\,\mathrm{d}E^A(\lambda)\,.$$

(See Chap. VIII, §3).

Appendix "Mechanics". Elements Concerning the Problems of Mechanics

§1. Indicial Calculus. Elementary Techniques of the Tensor Calculus[1]

The object of this §1 is to familiarise the reader with the very useful techniques of indicial calculus whose underlying justification is introduced in general within the context of the tensor calculus: first of all, there are two conventions to lighten the written notation:

(i) The automatic summation over each repeated index after the dimension of the space considered has been clearly defined: examples are:–

(a) $A_{ijkh}\sigma_{kh}$ denotes in fact $\sum_{k=1}^{n}\sum_{h=1}^{n} A_{ijkh}\sigma_{kh}$.

(b) If δ_{ij} is the Kronecker symbol ($= 0$ for $i \neq j$ and 1 if $i = j$), then

$$\delta_{ii} = 2 \quad \text{in} \quad \mathbb{R}^2, \qquad \delta_{ii} = n \quad \text{in} \quad \mathbb{R}^n .$$

(ii) Notation for partial derivatives: e.g.

$$\varphi_{,ij} = \frac{\partial^2 \varphi}{\partial x_i \partial x_j} \quad \text{and} \quad \varphi_{,ii} = \Delta\varphi .$$

1. Orientation Tensor or Fundamental Alternating Tensor in $\mathbb{R}^3$

1.1. Introduction by the Vector Product

Defining successively the tensors t_{ij} and ε_{ijk} by:

$$t_{ij}(\mathbf{v})x_j = (\mathbf{x} \times \mathbf{v})_i \quad \text{and} \quad t_{ij}(\mathbf{v}) = \varepsilon_{ijk}v_k, \quad \forall \mathbf{x} \quad \text{and} \quad \mathbf{v} \in \mathbb{R}^3$$

and proceeding by identification, we find that

$$(1.1) \qquad \varepsilon_{ijk} = \begin{cases} 1 & \text{if } (ijk) \text{ is an even permutation of } (123)\,, \\ -1 & \text{if } (ijk) \text{ is an odd permutation of } (123) \\ 0 & \text{if the same index is repeated} \end{cases}$$

---

[1] These notions valid, as they are, uniquely in the case of cartesian reference frames can be found with more detail in Germain [1] or [2].

and this definition remains the same in every direct orthonormal change of base; $\bar{\bar{\varepsilon}}^{(3)}$ is a pseudo tensor called, by an abuse of language, *the orientation tensor* or *the fundamental alternating tensor*; the index 3 denotes here that we are dealing with a tensor of order 3.

1.2. Important Relations

From the preceding definition, we derive successively the following ten identities ($\mathbf{e}_i$ denoting the unit vectors of the orthonormal base chosen in $\mathbb{R}^3$).

$$\mathbf{x} \times \mathbf{v} = \varepsilon_{ijk} x_j v_k \mathbf{e}_i \tag{1.2}$$

$$(\mathbf{u}, \mathbf{v}, \mathbf{w}) = (\mathbf{u} \,.\, \mathbf{v} \times \mathbf{w}) = \varepsilon_{ijk} u_i v_j w_k \tag{1.3}$$

$$\varepsilon_{lmn} \det A = \varepsilon_{ijk} a_{il} a_{jm} a_{kl} \tag{1.4}$$

$$\operatorname{curl} \mathbf{v} = \varepsilon_{ijk} \frac{\partial}{\partial \mathbf{x}_j} v_k \mathbf{e}_i = \varepsilon_{ijk} v_{k,j} \mathbf{e}_i \tag{1.5}$$

$$\varepsilon_{ijk} \varepsilon_{lmn} = \det \begin{pmatrix} \delta_{il} & \delta_{im} & \delta_{in} \\ \delta_{jl} & \delta_{jm} & \delta_{jn} \\ \delta_{kl} & \delta_{km} & \delta_{kn} \end{pmatrix} \qquad \delta_{ij} = \begin{cases} 0 & \text{if } \ i \neq j \\ 1 & \text{if } \ i = j \end{cases} \tag{1.6}$$

$$\varepsilon_{ijk} \varepsilon_{imn} = \delta_{jm} \delta_{kn} - \delta_{jn} \delta_{km} \tag{1.7}$$

$$\varepsilon_{ijk} \varepsilon_{ijn} = 2\delta_{kn} \tag{1.8}$$

$$\varepsilon_{ijk} \varepsilon_{ijk} = 6 \tag{1.9}$$

$$\det A = \frac{1}{6} \varepsilon_{ijk} \varepsilon_{lmn} a_{il} a_{jm} a_{kn} \tag{1.10}$$

$$(A^{-1})_{ij} = \frac{1}{2 \det A} \varepsilon_{jlm} \varepsilon_{ipq} a_{lp} a_{mq} \,. \tag{1.11}$$

These identities allow us to write in a condensed manner mathematical symbols which are usually very cumbersome when we wish to write them in terms of a given base (e.g. curl **v** and det A). The simplifications and subtleties of calculations appear much more neatly in such a condensed form. For example, we often have to make use of the following property:

1.3. Necessary and Sufficient Condition for a Tensor to be Symmetric

A necessary and sufficient condition for a tensor $\bar{\bar{\mathbb{T}}}$[2] *to be symmetric is that its components* t_{ij} *in every orthonormal base satisfy*

$$\varepsilon_{ijk} t_{jk} = 0 \,.$$

---

[2] $\bar{\bar{\mathbb{T}}}$ is the intrinsic symbol for a second order tensor independently of every base; in the different bases, it is represented by the matrices similar to the corresponding $T = \{t_{ij}\}$, that is to say that if T is the matrix representing $\bar{\bar{\mathbb{T}}}$ in an orthonormal reference frame (R), then after a change of base $(R) \to (R')$ characterised by Q, $\bar{\bar{\mathbb{T}}}$ will be represented by $T' = Q^{-1} T Q$.
We can put an "index" $\bar{\bar{\mathbb{T}}}^{(p)}$ when we are dealing with a tensor of order $p \neq 2$.

In effect,

$$t_{ij} = t_{ji} \quad \text{implies that} \quad \varepsilon_{ijk} t_{ij} = \varepsilon_{ijk} t_{ji} = -\varepsilon_{ijk} t_{ij}$$

which implies that $\varepsilon_{ijk} t_{ij} = 0$. Also

$$\varepsilon_{ijk} t_{jk} = 0 \quad \text{implies that} \quad \varepsilon_{ilm} \varepsilon_{ijk} t_{jk} = 0 = (\delta_{lj}\delta_{mk} - \delta_{lk}\delta_{mj})\, t_{jk} = t_{lm} - t_{ml} ,$$

so that $t_{lm} = t_{ml}$.

1.4. Rapid Writing of Some Identities

Making use of the relations (1.2), . . . , (1.11), we can write:

$$\operatorname{grad} \varphi = \varphi_{,i} \mathbf{e}_i ; \quad \operatorname{div} \mathbf{v} = v_{i,i} \quad \text{and} \quad \Delta\varphi = \varphi_{,ii}$$

We recover immediately the expressions for the classical symbols:

div(φ**v**), curl(φ**v**), div(**v** $\times$ **w**), curl(grad φ), div(grad φ), curl(curl **v**), curl(**v** $\times$ **w**).

1.5. Inversion of Certain Algebraic Relations

The identities (1.6) . . . (1.9) are very practical for certain calculations, for example:
(i) to pass from (1.4) to (1.10), we have multiplied (1.4) by ε_{lmm} and summing over the three repeated indices.
(ii) to obtain (1.11), we start from:

$$B = A^{-1} , \quad \text{namely} \quad b_{ij} a_{jk} = \delta_{ik} ,$$

so from (1.8):

$$2b_{ij} a_{jk} = \varepsilon_{ipq} \varepsilon_{kpq} \left(= \varepsilon_{ipq} \varepsilon_{kpq} \frac{\det A}{\det A} \right) ,$$

then, we have used (1.4) to express $\varepsilon_{kpq} \det A$ making a_{jk} appear in the right hand side and thus determining b_{ij} by identification.

2. Possibilities of Decompositions of a Second Order Tensor

2.1. Canonical Decomposition into Symmetrical and Antisymmetrical Parts

Every second order tensor $\bar{\bar{\mathbb{T}}}$ may be decomposed in a unique fashion into a sum of a symmetric tensor and of an antisymmetric tensor:

$$\bar{\bar{\mathbb{T}}} = \bar{\bar{\mathbb{T}}}_s + \bar{\bar{\mathbb{T}}}_a$$

with

$$\bar{\bar{\mathbb{T}}}_s = \tfrac{1}{2}[\bar{\bar{\mathbb{T}}} + {}^t\bar{\bar{\mathbb{T}}}] \quad \text{namely} \quad t_{(ij)} = \tfrac{1}{2}(t_{ij} + t_{ji})$$

and

$$\bar{\bar{\mathbb{T}}}_a = \tfrac{1}{2}[\bar{\bar{\mathbb{T}}} - {}^t\bar{\bar{\mathbb{T}}}] \quad \text{namely} \quad t_{[ij]} = \tfrac{1}{2}(t_{ij} - t_{ji})$$

2.2. Decomposition of a Tensor into a Spherical Part and a Deviator

A spherical tensor $\bar{\bar{\mathbb{S}}}$ is a tensor proportional to the unit tensor $\bar{\bar{\mathbb{I}}}$. A deviator $\bar{\bar{\mathbb{D}}}$ is a tensor whose trace is zero: $D_{kk} = 0$. An arbitrary tensor is decomposable in a unique fashion into the sum of a spherical part and a deviator:

$$\bar{\bar{\mathbb{T}}} = \bar{\bar{\mathbb{T}}}^s + \bar{\bar{\mathbb{T}}}^D$$

with, by denoting by T_t the quantity t_{kk} (the first invariant of $\bar{\bar{\mathbb{T}}}$) and by S the quantity $\frac{1}{3}t_{kk}$,

$$\bar{\bar{\mathbb{T}}}^s = \tfrac{1}{3}T_I\bar{\bar{\mathbb{I}}}\,, \quad \text{namely} \quad t^s_{ij} = \tfrac{1}{3}t_{kk}\delta_{ij} = S\delta_{ij}$$

and

$$\bar{\bar{\mathbb{T}}}^D = \bar{\bar{\mathbb{T}}} - \tfrac{1}{3}T_I\bar{\bar{\mathbb{I}}} \quad \text{namely} \quad t^D_{ij} = t_{ij} - S\delta_{ij}\,.$$

3. Generalized Divergence Theorem

This is a generalization of Ostrogradski's theorem (see Bass [1]).

Theorem 1. *Suppose that Ω is a bounded domain of $\mathbb{R}^p$ with a regular boundary, and that* **n** *denotes the unit vector normal to $\partial\Omega$ (oriented to the exterior of Ω). Finally, let $t_{ij\ldots q}$ denote a component of an arbitrary tensor field, continuously differentiable on Ω and continuous on $\bar{\Omega}$; then*:

$$\int_\Omega t_{ij\ldots q,r}\,\mathrm{d}v = \int_{\partial\Omega} t_{ij\ldots q}n_r\,\mathrm{d}\sigma\,. \tag{1.12}$$

Examples (we use here the notation and relations introduced at the beginning of this Sect. 1):

(i) $$\int_\Omega \operatorname{grad}\varphi\,\mathrm{d}v = \int_{\partial\Omega}\varphi\mathbf{n}\,\mathrm{d}\sigma\,,$$

(ii) $$\int_\Omega \operatorname{curl}\mathbf{u}\,\mathrm{d}v = \mathbf{e}_i\int_\Omega \varepsilon_{ijk}u_{k,j}\,\mathrm{d}v = \mathbf{e}_i\int_{\partial\Omega}\varepsilon_{ijk}u_k n_j\,\mathrm{d}\sigma = \int_{\partial\Omega}\mathbf{n}\times\mathbf{u}\,\mathrm{d}\sigma\,,$$

(iii) $$\begin{aligned}\int_\Omega \varphi\Delta\psi\,\mathrm{d}v &= \int_\Omega[(\varphi\psi_{,i})_{,i} - (\varphi_{,i}\psi)_{,i}]\mathrm{d}v\\ &= \int_{\partial\Omega}\varphi\psi_{,i}n_i\,\mathrm{d}\sigma - \int_\Omega \operatorname{grad}\varphi\,.\operatorname{grad}\psi\,\mathrm{d}v\\ &= \int_{\partial\Omega}\varphi\frac{\partial\psi}{\partial n}\mathrm{d}\sigma - \int_\Omega \operatorname{grad}\varphi\,.\operatorname{grad}\psi\,\mathrm{d}v\,.\end{aligned}$$

(iv) Let Σ be a bounded domain in $\mathbb{R}^2$, $\boldsymbol{\tau}$ ($= \mathbf{e}_3 \times \mathbf{n}$) the tangent vector to $\partial\Sigma$, then:

$$\begin{aligned}\int_\Sigma \mathbf{e}_3\,.\operatorname{curl}\mathbf{u}\,\mathrm{d}\sigma &= \int_\Sigma \delta_{i3}\varepsilon_{ijk}u_{k,j}\,\mathrm{d}\sigma = \int_{\partial\Sigma}\delta_{i3}\varepsilon_{ijk}u_k n_j\,\mathrm{d}s\\ &= \int_{\partial\Sigma}(\mathbf{e}_3, \mathbf{n}, \mathbf{u})\,\mathrm{d}s = \int_{\partial\Sigma}(\mathbf{e}_3\times\mathbf{n})\,.\mathbf{u}\,\mathrm{d}s = \int_{\partial\Sigma}(\mathbf{u}\,.\boldsymbol{\tau})\,\mathrm{d}s\,.\end{aligned}$$

4. Ideas About Wrenches

4.1. Preliminary Definitions

A *field of vectors* $\mathscr{H}$ defined on the affine Euclidean space $\mathscr{E}^n$ (the set of points of $\mathbb{R}^n$) is said to be *affine* if there exists a linear operator $\mathscr{L}$ on E^n, the Euclidean vector space associated with $\mathscr{E}^n$, such that

$$(1.13) \qquad \forall A \text{ and } B \in \mathscr{E}^n, \quad \mathscr{H}(B) - \mathscr{H}(A) = \mathscr{L}(\overrightarrow{AB}),$$

it is clear that this operator is unique.
By an *antisymmetric field*, we mean an affine field such that the associated operator $\mathscr{L}$ is antisymmetric; that is to say that $\mathscr{L}$ is such that

$$(1.14) \qquad \forall \mathbf{u} \text{ and } \mathbf{v} \in E^n \quad \mathbf{u}.\mathscr{L}(\mathbf{v}) = -\mathbf{v}.\mathscr{L}(\mathbf{u}).$$

If $\bar{\bar{\mathbb{L}}}$ is the tensor identified with the operator $\mathscr{L}$, we have then for all orthonormal bases:

$$(1.15) \qquad L_{ij} = -L_{ji}.$$

With every antisymmetric field $\mathscr{H}$ defined on the oriented[3] space $\mathscr{E}^3$ there corresponds one and only one vector $\mathbf{R}$ such that

$$(1.16) \qquad \forall A, B \in \mathscr{E}^n \quad \mathscr{H}(B) - \mathscr{H}(A) = \mathbf{R} \times \overrightarrow{AB};$$

in effect, $\mathscr{H}$ is then defined in terms of $\bar{\bar{\mathbb{L}}}$ by:

$$\bar{\bar{\mathbb{L}}}\mathbf{u} = \mathbf{R} \times \mathbf{u}, \quad \forall \mathbf{u} \in E^n,$$

namely, in components

$$L_{ij}u_j = \varepsilon_{ikj}r_k u_j$$

from which we deduce that

$$(1.17) \qquad L_{ij} = -\varepsilon_{ijk}r_k, \quad \text{and} \quad r_l = -\tfrac{1}{2}\varepsilon_{lij}L_{ij}.$$

4.2. Definition and Basic Properties of Wrenches on the Oriented Space $\mathscr{E}^3$

The set consisting of an antisymmetric field $\mathscr{M}$ on the oriented space $\mathscr{E}^3$ and its vector $\mathbf{R}$ is called a *wrench* $\mathscr{T}$ *on the oriented space* $\mathscr{E}^3$. We write

$$\mathscr{T} = \{\mathscr{M}, \mathbf{R}\};$$

$\mathbf{R}$ is called the "*general resultant*" and $\mathscr{M}$ the "*field of moments*" of the wrench $\mathscr{T}$. A wrench $\mathscr{T}$ is determined in a unique manner if its general resultant $\mathbf{R}$ and the value at an arbitrary point (say O) of its field of moments are given; in effect, from (1.16) and the above-mentioned definition, the field of moments $\mathscr{M}$ of $\mathscr{T}$ is then defined by

$$\mathscr{M}(P) = \mathscr{M}(O) + \mathbf{R} \times \overrightarrow{OP}, \qquad \forall P \in \mathscr{E}^3.$$

---

[3] The space $\mathscr{E}^n$ (resp. E^n) is said to be *oriented* if it admits only direct changes of bases.

The vectors $\mathbf{R}$ and $\mathscr{M}(P)$ (sometimes denoted by $\mathscr{M}_p$) are called "*elements of reduction*" of the wrench $\mathscr{T}$ at P.
A classical example given by mechanics is that of the wrench of the external forces $\mathscr{F}_e$ applied to a material system $\mathscr{T}$, defined on the one hand by a volume density $\mathbf{f}(M)$ and on the other hand by a surface density $\mathbf{F}(P)$, $\forall M \in \mathscr{S}$ and $\forall P \in \partial\mathscr{S}$. The elements of reduction of such a wrench are:
its general resultant:

$$\mathscr{R}[\mathscr{F}_e \to \mathscr{S}] = \int_{\mathscr{S}} \mathbf{f}(M)\,\mathrm{d}M + \int_{\partial\mathscr{S}} \mathbf{F}(P)\,\mathrm{d}P\,,$$

and its resultant moment at O:

$$\mathscr{M}_o[\mathscr{F}_e \to \mathscr{S}] = \int_{\mathscr{S}} \overrightarrow{OM} \times \mathbf{f}(M)\,\mathrm{d}M + \int_{\partial\mathscr{S}} \overrightarrow{OP} \times \mathbf{F}(P)\,\mathrm{d}P\,.$$

The set of wrenches with the operations "addition"

$$\mathscr{T}_1 + \mathscr{T}_2 = \{\mathscr{M}_1 + \mathscr{M}_2, \mathbf{R}_1 + \mathbf{R}_2\} \tag{1.19}$$

and "multiplication by a scalar":

$$\alpha\mathscr{T} = \{\alpha\mathscr{M}, \alpha\mathbf{R}\} \tag{1.20}$$

is a *vector space of dimension 6* (in effect, we have just seen, two vectors suffice to determine a wrench).
Finally, we can define the product of two wrenches $\mathscr{T}_1$ and $\mathscr{T}_2$ by the bilinear form[4]

$$\Phi(\mathscr{T}_1, \mathscr{T}_2) = \mathbf{R}_1 . \mathscr{M}_2(O) + \mathbf{R}_2 . \mathscr{M}_1(O)\,, \tag{1.21}$$

sometimes denoted by $\{\mathscr{T}_1\}\{\mathscr{T}_2\}$; the verification that this form does not depend on the choice of the point O is immediate.

§2. Notation, Language and Conventions in Mechanics

1. Lagrangian and Eulerian Coordinates

The *Lagrangian coordinates* of a particle in $\mathbb{R}^n$ is a set of n parameters which characterise the position of this particle at the initial instant $t = 0$; most of the time, we are concerned with the coordinates, in a fixed reference frame, of the point of $\mathbb{R}^n$ with which the chosen particle coincides at this instant $t = 0$. It will be convenient to denote them by $\mathbf{a} = \{a_\alpha\}$ with a Greek index α.
Similarly, the *Eulerian coordinates* of a particle are the members of a set of n parameters characterising its position at the instant t considered. In general it is a

[4] This bilinear form, which is symmetrical, is not positive definitive, in effect: $\Phi(\mathscr{T}, \mathscr{T}) = 2\mathbf{R} . \mathscr{M}(P)$ can be zero or strictly negative as well as positive. Hence Φ does not define a scalar product on the space of the wrenches.

question of the coordinates x_i (Latin index i), in the same reference system as previously, of the point of space coinciding with this particular particle at the instant t. We put $\mathbf{x} = \{x_i\}$ with a Latin index.
When we follow the same particle in a motion determined between the instants 0 and T, there exist scalar mappings φ_i and ψ_α such that (in $\mathbb{R}^3$ for example):

$$x_i = \varphi_i(\mathbf{a}, t) ; \qquad a_\alpha = \psi_\alpha(\mathbf{x}, t) , \qquad \forall t \in [0, T] \tag{2.1}$$

for $i = 1, 2, 3$; $\alpha = 1, 2, 3$ with

$$\varphi_i(\mathbf{a}, 0) = \delta_{i\alpha} a_\alpha . \tag{2.2}$$

The conventional notation:

$$x_{i,\alpha} = \frac{\partial \varphi_i}{\partial a_\alpha} \quad \text{and} \quad a_{\alpha,j} = \frac{\partial \psi}{\partial x_j} \tag{2.3}$$

is then clear.

2. Notions of Displacement and of Strain[5]

It is clear that with the ideas and conventions introduced above, the quantity:

$$u_i(\mathbf{a}, t) = x_i - \delta_{i\alpha} a_\alpha \tag{2.4}$$

denotes the i-th component of the *displacement* between the instants 0 and t, of the particle which found itself at $\mathbf{a}$ at the instant 0 and at $\mathbf{x}$ at the instant t.
When we seek to define the notion of *strain* of a continuous medium we easily show that it can be characterised by the tensor field $\bar{\bar{\mathbb{L}}}$, called the *Green–Lagrange tensor*, with components:

$$L_{\alpha\beta} = \tfrac{1}{2}[x_{i,\alpha} x_{i,\beta} - \delta_{\alpha\beta}] \tag{2.5}$$

or, again

$$L_{\alpha\beta} = \tfrac{1}{2}[u_{\alpha,\beta} + u_{\beta,\alpha} + u_{i,\alpha} u_{i,\beta}] ; \tag{2.6}$$

hence, this tensor (which is in fact a function of the Lagrangian variables) is expressed as a non-linear function of the Lagrangian derivatives of the displacement.
In numerous cases in the mechanics of solids, the displacements $\mathbf{u}$ vary very slowly when we pass from one point M_0 to a neighbouring point M; we can then neglect the quadratic terms involving the derivatives of $\mathbf{u}$ in the expression for $L_{\alpha\beta}$; we then obtain the *linearised strain tensor* which, when we consider it as a differential operator on $\mathbf{u}$, is defined in terms of components by:

$$\varepsilon_{\alpha\beta}(\mathbf{u}) = \tfrac{1}{2}[u_{\alpha,\beta} + u_{\beta,\alpha}] . \tag{2.7}$$

If, in addition, the displacements are small, we identify, in first approximation, the

---

[5] For more details, see Germain [2].

Eulerian coordinates and the Lagrangian coordinates; the relation (2.7) can again be written:

$$\varepsilon_{ij}(\mathbf{u}) = \tfrac{1}{2}(u_{i,j} + u_{j,i}) \,. \tag{2.8}$$

Remark 1. We notice that $\varepsilon_{ij}(\mathbf{u})$ represents the symmetric part (see §1.2) of the tensor $\overline{\overline{\text{grad}}}\ \mathbf{u}$ with components $u_{i,j}$ which we call the *displacement gradient tensor* and which we can decompose as follows:

$$u_{i,j} = \varepsilon_{ij}(\mathbf{u}) + \omega_{ij}(\mathbf{u})$$

with

$$\omega_{ij}(\mathbf{u}) = \tfrac{1}{2}(u_{i,j} - u_{j,i}) \,; \tag{2.9}$$

we show that

$$\varepsilon_{ij}(\mathbf{u}) \equiv 0$$

if and only if $\mathbf{u}$ is a *field of moments*:

$$\mathbf{u}(M', t) = \mathbf{u}(M, t) + \Omega(t) \times \overrightarrow{MM'}$$

which corresponds to the field of *displacements of a rigid body* rotating with angular velocity

$$\Omega(t) = \tfrac{1}{2}\,\text{curl}\,\mathbf{u}$$

with components $\Omega_i = \frac{1}{2}\varepsilon_{ijk}\omega_{kj}$. The tensor $\bar{\bar{\varepsilon}}(\mathbf{u})$ thus characterises the *pure deformation* in theory of small perturbations while tensor $\bar{\bar{\omega}}(\mathbf{u})$ or the vector $\Omega(t)$, which are fixed if t is fixed, each characterise a *pure rotation*.
An important consequence of the above is that if the six components $\varepsilon_{ij}(\mathbf{u})$ are given, they can determine the corresponding field $\mathbf{u}$ only to within a global rigid body displacement (that is to say within a field of moments). □

3. Notions of Velocity and of Rate of Strain[6]

The velocity of a particle is given in components by:

$$v_i(\mathbf{x}, t) = \frac{dx_i}{dt} = \frac{\partial \varphi_i}{\partial t}(\mathbf{a}, t) \qquad \mathbf{a} = \psi(\mathbf{x}, t) \,,$$

namely by

$$v_i(\mathbf{x}, t) = \frac{\partial u_i}{\partial t}(\mathbf{a}, t) \,, \qquad \mathbf{a} = \psi(\mathbf{x}, t)$$

and we can characterise in a natural fashion the *velocity of strain* directly by the tensor with components:

$$\varepsilon_{ij}(\mathbf{v}) = \tfrac{1}{2}(v_{i,j} + v_{j,i}) \tag{2.10}$$

called the *rate of strain tensor*.

---

[6] For more details, see Germain [2].

We note that the introduction of this tensor does not require any linearisation. As previously, the case

$$\varepsilon_{ij}(\mathbf{v}) \equiv 0$$

corresponds to a velocity field $\mathbf{v}$ of a rigid body for which the rate of rotation is characterised by $\omega_{ij}(\mathbf{v})$ or $\Omega_i(\mathbf{v}) = \frac{1}{2}\varepsilon_{ijk}\omega_{kj}(\mathbf{v})$.

4. Notions of Particle Derivative, of Acceleration and of Dilatation[7]

When a function f is expressed as a function of the Eulerian coordinates of a particle, we can take an interest in its variation with the course of time for a fixed particle. If we put:

$$f(\mathbf{x}, t) = f[\varphi(\mathbf{a}, t), t] = g(t)\,,$$

We call the *particle derivative* of f, its total derivative in t for $\mathbf{a}$ fixed; thus

$$\frac{\mathrm{d}f}{\mathrm{d}t} = \frac{\partial f}{\partial t} + \frac{\partial f}{\partial x_i}\cdot\frac{\mathrm{d}x_i}{\mathrm{d}t} = \frac{\partial f}{\partial t} + \mathbf{v}\,.\,\mathrm{grad}\,f\,. \tag{2.11}$$

In the case of a vector quantity $\mathbf{X}(\mathbf{x}, t)$, that leads to:

$$\frac{\mathrm{d}X_i}{\mathrm{d}t} = \frac{\partial X_i}{\partial t} + X_{i,j}v_j \tag{2.12}$$

namely

$$\frac{\mathrm{d}\mathbf{X}}{\mathrm{d}t} = \frac{\partial \mathbf{X}}{\partial t} + \overline{\overline{\mathrm{grad}}}\,\mathbf{X}\,.\,\mathbf{v} \tag{2.13}$$

and in the particular case in which the vector quantity coincides with the velocity field of the medium considered, we obtain the acceleration vector:

$$\gamma = \frac{\mathrm{d}\mathbf{v}}{\mathrm{d}t} = \frac{\partial \mathbf{v}}{\partial t} + \mathrm{grad}\left(\frac{1}{2}\mathbf{v}^2\right) + (\mathrm{curl}\,\mathbf{v}) \times \mathbf{v}\,. \tag{2.14}$$

We can also be interested in the variation with the passage of time of an integral taken over a domain $\mathscr{D}$ which evolves. Let us give as an example the result in the case of a volume domain $\mathscr{D}$ which evolves with the same velocity field as the medium considered: for

$$K(t) = \int_{\mathscr{D}} c(\mathbf{x}, t)\,\mathrm{d}v$$

then

$$\frac{\mathrm{d}K}{\mathrm{d}t} = \int_{\mathscr{D}}\left\{\frac{\partial c}{\partial t} + \mathrm{div}(c\mathbf{v})\right\}\mathrm{d}v = \int_{\mathscr{D}}\frac{\partial c}{\partial t}\mathrm{d}v + \int_{\partial\mathscr{D}} c\mathbf{v}\,.\,\mathbf{n}\,\mathrm{d}\sigma\,. \tag{2.15}$$

---

[7] See Germain [2].

If $K(t)$ is no other than $\mathscr{V}(\mathscr{D})$, the volume of the domain $\mathscr{D}$, we then have $c = 1$:

$$\frac{\mathrm{d}}{\mathrm{d}t}\mathscr{V}(\mathscr{D}) = \int_{\mathscr{D}} \operatorname{div} \mathbf{v} \, \mathrm{d}v \,, \tag{2.16}$$

from which we have the designations:

rate of volume dilatation for the quantity $\varepsilon_{jj}(\mathbf{v}) = \operatorname{div} \mathbf{v}$;

volume dilatation for the quantity $\varepsilon_{jj}(\mathbf{u}) = \operatorname{div} \mathbf{u}$;

incompressible medium for the case where $\operatorname{div} \mathbf{v} \equiv 0$.

5. Notions of Trajectory and of Stream Line

The trajectory of a particle M in a given motion is the locus of the positions M_t of this particle in the course of time. If we characterise a particle by its position $\mathbf{a}$ at the initial instant, its trajectory is defined parametrically by the relations

$$x_i = \varphi_i(\mathbf{a}, t) \,, \qquad (i = 1, 2, 3) \,.$$

already introduced in (2.1).
If the motion is characterised by its velocity field $\mathbf{v}(\mathbf{x}, t)$, then the trajectory of each particle is again defined parametrically by the solution $\mathbf{x}(t)$ of the system

$$\begin{cases} \text{(i)} \ \dfrac{\mathrm{d}\mathbf{x}}{\mathrm{d}t} = \mathbf{v}(\mathbf{x}, t) \\ \text{(ii)} \ \mathbf{x}(0) = \mathbf{a} \,. \end{cases} \tag{2.17}$$

The *streamlines at a fixed instant t* are the lines of the velocity vector field, that is to say, the lines which at each of their points, are tangent to the velocity vector of the particle situated at this point at the instant considered. These lines are therefore defined by the differential system:

$$\frac{\mathrm{d}x_1}{v_1(\mathbf{x}, t)} = \frac{\mathrm{d}x_2}{v_2(\mathbf{x}, t)} = \frac{\mathrm{d}x_3}{v_3(\mathbf{x}, t)} \tag{2.18}$$

in which t is a fixed parameter.
A particular streamline is characterised by (2.18) and the data of one of its points.

Remark 2. If we write the equations (2.17) (i) in the form:

$$\frac{\mathrm{d}x_1}{v_1(\mathbf{x}, t)} = \frac{\mathrm{d}x_2}{v_2(\mathbf{x}, t)} = \frac{\mathrm{d}x_3}{v_3(\mathbf{x}, t)} = \mathrm{d}t \,, \tag{2.19}$$

we should have the tendency to identify streamlines with trajectories, but in fact, despite their resemblance, the systems (2.18) and (2.19) differ fundamentally because of the fact that in the first case t is a fixed parameter, while in the second, t is a variable. The fact that the streamlines and the trajectories are coincident is characteristic of a stationary flow[8]. □

§3. Ideas Concerning the Principle of Virtual Power[9]

1. Introduction: Schematization of Forces

Let us quote here an extract from Germain [4]:
"In mechanics, there are two ways of schematizing, through mathematical concepts, forces acting at an instant t on a material system.
"The *first* consists of *representing a force by a vector*. We generalize here by representing the forces acting on a continuous system by a field associated with a measure: field of volume, surface, or line forces ... When the forces are thus schematized, it is clear that the fundamental principle of dynamics is completely indicated as the basis of the solution of problems.
"The *second way*[10] is that of virtual power (or virtual work) and is based on observations of the following type:

— When we wish to see if a suitcase is heavy, we lift it.
— To estimate the tension in a (stationary) transmission belt, we try to draw it aside from its equilibrium position.
— It is in driving a vehicle that we become aware of the frictional forces both external (running on rubber tires) and internal (is the hand-brake engaged?). In this second viewpoint, we detect the external and internal forces of a system, not directly, but by the effect produced when this system is subjected to a displacement, a velocity, or a deformation: we determine them through the power which they develop in an arbitrary motion. The essential underlying mathematical idea is that of "duality". This second view, inspired thus by the most common experience, has the advantage of calculating forces,in general ill-defined in the first point of view, as those of linkage; it is also very flexible, as, according as we shall choose a set of virtual motions more or less "great" we shall have a description of forces more or less small".

2. Preliminary Definitions

The term "*virtual motion*" which characterises the method described above and which will be introduced below, expresses the idea "*of arbitrariness*" with respect to the *real motion* to which the body will effectively be subjected in the context of a

[8] See §1.5.
[9] See Germain [2], [3], [4].
[10] Due, it would seem, to d'Alembert for what is the basic idea.

particular problem; but also the notion of "*virtually possible*" takes account of the simplifying hypotheses made in the context of different theories. Whence, thanks to this "arbitrariness", we have flexibility of this method which allows great choice but also gives the possibility of more or less limiting this arbitrariness.
We shall retain terms "*velocities*" and hence "*virtual power*" rather than "displacements" and "*virtual work*" as on the one hand the former are more general and imply the latter. In effect a motion always takes place in time, even in "quasistatic" cases; on the other hand, the quantity

$$\varepsilon_{ij}(\mathbf{u}) = \tfrac{1}{2}(u_{i,j} + u_{j,i})$$

which occurs in a fundamental way in the description of a continuous medium (see §2.2 and §2.3 of this appendix), corresponds to a rigorous idea (rate of deformations) when $\mathbf{u}$ denotes a velocity field and only to an approximate concept (linearized deformation) in the case in which $\mathbf{u}$ denotes a displacement.

2.1. Virtual Motion of a System S

Let S be a system in motion in a reference frame $\mathscr{R}$ and S^t its configuration at an instant t; *we define virtual motion of this system at the instant t in the frame* $\mathscr{R}$ *at each time that a vector field* $\hat{\mathbf{u}}$ *defined on* S^t *is given.* $\hat{\mathbf{u}}(M)$ *is called the "virtual velocity of* M"; *we suppose it to be piecewise-continuous on* S.
If $\hat{\mathbf{u}}^*$ *defines a virtual motion of* S *at the instant t in another reference frame* $\mathscr{R}^*$, *we say that* $\mathbf{u}$ *and* $\mathbf{u}^*$ *define the some virtual motion of* S *if* $\forall M \in S$, *we have*:

$$\hat{\mathbf{u}}(M) = \mathbf{u}_e(M) + \hat{\mathbf{u}}^*(M) \tag{3.1}$$

$\mathbf{u}_e(M)$ *being the velocity of transportation of* M (*velocity in* $\mathscr{R}$ *of the point bound to* $\mathscr{R}^*$, *coinciding with* M *at the instant* t).
The set of fields $\hat{\mathbf{u}}$ thus defined on S^t is a normed vector space $\hat{\mathscr{V}}$: $(\lambda\hat{\mathbf{u}}^{(1)} + \mu\hat{\mathbf{u}}^{(2)})$ being the field of vectors $(\lambda\hat{\mathbf{u}}^{(1)}(M) + \mu\hat{\mathbf{u}}^{(2)}(M))$ with, for example,

$$\|\hat{\mathbf{u}}\|_\infty = \sup_{M \in S^t} |\hat{\mathbf{u}}(M)| .$$

2.2. Virtual Power of the Action of a System Σ on a System S

An action of a system Σ *on a system* S *will be defined at the instant t for the space* $\hat{\mathscr{V}}$ *of virtual motions of* S *at the instant t, at each time that we are given a continuous linear form* $\mathscr{L}$ *defined on* $\hat{\mathscr{V}}$ (for the chosen topology). *The number*

$$\mathscr{P} = \mathscr{L}(\hat{\mathbf{u}}) \tag{3.2}$$

is then called the virtual power of the action of the system Σ *on the system* S *in the motion* $\hat{\mathbf{u}}$.

2.3. Idea of the Method and Examples of Simple Virtual Motions

The steps of the method of virtual power are very simple; they consist of constructing theories, more or less sharp, of the mechanics of continuous media and giving each a space $\mathscr{V}$ of virtual motions which fixes in some way the degree of

sharpness of the theory, the equivalent schematization of the forces being deduced by "duality". The equations of statics and dynamics are then obtained by the application of the principle of virtual power stated in Sect. 3 of this §3.
(i) If S is reduced to a point M, the space $\hat{\mathscr{V}}$ is the space of vectors $\mathbf{u}_M$ in $\mathbb{R}^3$; a linear form on $\hat{\mathscr{V}}$ is then

$$\hat{\mathscr{P}} = \mathscr{L}(\mathbf{u}_M) = \mathbf{F}_M \cdot \mathbf{u}_M$$

and fixes the vector $\mathbf{F}_M$ which represents the associated forces exerted on M.
(ii) If S is a *rigid body*, the velocity field of these points satisfies, at each instant t:

$$u_i(\mathbf{x}) = u_i(\mathbf{0}) + \Omega_{ij} x_j\,, \qquad \mathbf{0} \in S \tag{3.3}$$

where (Ω_{ij}) is the (constant, antisymmetric) matrix, image of the rate of rotation tensor[11]. A field u_i satisfying the identity (3.3) for all $\mathbf{x} \in S$ is the field of moments of a *wrench*[12], with general resultant:

$$\omega = -\frac{1}{2}\varepsilon_{kij}\Omega_{ij}\mathbf{e}_k \tag{3.4}$$

and of moment $\mathbf{u}(\mathbf{0})$ at $\mathbf{0}$. Such a field is called a *distributor* and is denoted by $\hat{\mathscr{C}}$. With t fixed, we can choose as the space of virtual motions, the vector space $\hat{\mathscr{V}}^R$ (of dimension 6) of the distributors $\hat{\mathscr{C}}$. These virtual motions *preserve the rigidity of* S; they are called "*virtual motions rigidifying* S" when S is deformable. We can show that a linear form defined on $\hat{\mathscr{V}}^R$ can be written

$$\hat{\mathscr{P}} = \mathscr{L}(\hat{\mathscr{C}}) = \{\mathscr{T}\}\{\hat{\mathscr{C}}\} \quad \text{for} \quad \hat{\mathscr{C}} \in \hat{\mathscr{V}}^R \tag{3.5}$$

where $\mathscr{T}$ is a wrench field defined at M by its (constant) general resultant $\mathbf{T}$ and its moment:

$$\mathbf{m}(M) = \mathbf{m}(0) + \mathbf{T} \wedge \overrightarrow{OM} \tag{3.6}$$

with the result that

$$\hat{\mathscr{P}} = \mathbf{T} \cdot \hat{\mathbf{u}}(M) + \mathbf{m}(M) \cdot \hat{\omega}\,. \tag{3.7}$$

The quantities $\mathbf{T}$ and $\mathbf{m}$ are then respectively interpreted as the resultant and the field of moments of the field of wrenches of the forces acting on the rigid body S.
(iii) In *classical analytical* mechanics, that is to say of *systems formed by a finite number of rigid solids*, the virtual motions depend again on a finite number of parameters $\hat{\varpi}_p$; $p = 1, 2, \ldots, q$. The forces are then determined by the coefficients Q_j of the linear form:

$$\hat{\mathscr{P}} = \sum_{p=1}^{q} Q_p \hat{\varpi}_p \tag{3.8}$$

defined on the vector space $\hat{\mathscr{V}}$ defined by the $\hat{\varpi}_p$. The Q_p are the components of the "generalized force" acting on the system.

---

[11] See §2.3.
[12] See §1.4.

2.4. Virtual Motion of a Deformable Continuous Medium

In this case, there are no longer particular relations between the velocities of two points of the medium (we shall only have regularity hypotheses), with the result that $\hat{\mathscr{V}}$ will necessarily be of infinite dimension. *When the forces exerted by Σ on S will be defined by a field* $\mathbf{f}(M)$ determining the density of forces with respect to $d\omega$ the measure (volume, surface or line) associated with S, it will be legitimate to take as the expression for $\mathscr{P}$:

$$\hat{\mathscr{P}} = \int_S \mathbf{f}(M).\hat{\mathbf{u}}(M)\,d\omega \,, \tag{3.9}$$

which we shall do, unless mention is made to the contrary. In the same way we shall define the virtual power of the accelerations

$$\hat{\mathscr{A}} = \int_S \gamma(M).\hat{\mathbf{u}}(M)\,\rho\,d\omega \tag{3.10}$$

where $\gamma(M)$ is the effective acceleration of the point M and ρ is the mass density associated with the measure $d\omega$.
However, when the forces are not defined directly, (cohesive forces between the particles of S, called internal forces in S, external forces at a distance, contact forces) it will be necessary to involve the variation, from one point to another, of the virtual velocities, in the definition of $\hat{\mathscr{P}}$. Thus, following the degree of subtlety sought by the theory, we shall incorporate the first derivatives $\hat{\mathbf{u}}_{i,j}$ (theory of the first gradient) or the derivatives up to order p (theory of the p-th gradient). In the first case, we choose for $\hat{\mathscr{V}}$ a sub-set of $H^1(S)$[13] and, to define $\hat{\mathscr{P}}$, we shall choose a linear form of the type:

$$\hat{\mathscr{P}} = \int_S (F_i(M)\hat{\mathbf{u}}_i(M) + F_{ij}\hat{\mathbf{u}}_{i,j})\,d\omega \,; \tag{3.11}$$

in the case of a p-gradient theory, it is $H^p(S)$ which will be involved with an adapted linear form.

3. Fundamental Statements

The application of the method rests on the following two statements.

Statement I. *Axiom of the virtual power of internal forces.*
The virtual power of the internal forces in a system S is zero in every virtual motion freezing S at the instant considered.

By the term "*the internal forces in S*" we mean the cohesive forces between the different parts of S; for a part $\mathscr{D}$ of S, which, in the limit, can be reduced to a single

---

[13] $H^1(S)$, $H^p(S)$ are Sobolev spaces defined in Chap. IV.

particle, these are the forces exerted on $\mathscr{D}$ by the parts of S exterior to $\mathscr{D}$[14]. In particular, the statement expresses that the virtual power of the internal forces in S during a given virtual motion, has a value independent of the reference frame in which the motion is observed; in effect, in the relation (3.1), the field of velocities of transportation $\mathbf{u}_e(M)$ is rigidifying.

Statement II. *Principle of Virtual Power.*
There exists at least one reference system[15] *(called absolute or Galilean) in which at each instant t, for each system S, and for each virtual motion of $\hat{\mathscr{V}}$, the virtual power of the acceleration quantities is equal to the sum of the virtual powers of all the forces applied to the system S, both internal and external, namely:*

$$\hat{\mathscr{A}} = \hat{\mathscr{P}}_{(i)} + \hat{\mathscr{P}}_{(e)}, \quad \forall \hat{\mathbf{u}} \in \hat{\mathscr{V}} . \tag{3.12}$$

To obtain a schematization of the forces, it is therefore necessary:
(i) to choose $\hat{\mathscr{V}}$ and the linear forms $\hat{\mathscr{P}}_{(i)}$ and $\hat{\mathscr{P}}_{(e)}$ in the spirit described in §3.2 with the result that Statement I is satisfied;
(ii) to apply Statement II, that is to say, write (3.12).
The conditions thus found will be the fundamental equations of motion for the medium studied; they depend on the quantities which appear as the "generalized coefficients" in the linear forms $\hat{\mathscr{P}}_{(i)}$ and $\hat{\mathscr{P}}_{(e)}$. These quantities make up, for the restricted schematization, a mathematical representation of internal forces and of external forces.

Remark 1. (1) We show easily that the statements I and II implies the fundamental principle of dynamics: in effect, the relation (3.12), with Axiom I taken into account, has as a particular case:
which leads immediately to the equality of the wrench of the acceleration quantities and of the external forces.
(2) It is necessary to underline the analogy between the formulation of the principle of virtual power and the weak-variational formulation of a problem; the reader is referred especially to Chap. VII, §2, to Chap. XV, §4 and to Chap. XVIII, §6. In this case the space of test functions is none other than a space of virtual motions, "generalized" and enlarged to a distribution space, which, in mechanics, is, in general, taken to be a regular space. The relations depending on the "generalized coefficients" on an application of the principle of virtual power, are equivalent to "more or less strong" local formulations of variational methods. □

4. Theory of the First Gradient

— We shall suppose here that S is three-dimensional and we shall apply (3.12) to every domain $\mathscr{D}$ interior to S without insisting here on regularity conditions on $\hat{\mathscr{V}}$.

[14] To go deeply into this critical notion, see for example Germain [2].
[15] The words "reference system" implies not only the choice of a reference frame in space, but also that of a time scale; this occurs essentially at the level of the definition (3.10) of $\hat{\mathscr{A}}$.

— We shall choose linear forms on $\hat{\mathscr{V}}$ of the type (3.11) to define $\hat{\mathscr{P}}_{(i)}$ and $\hat{\mathscr{P}}_{(e)}$, which will lead us to a slight generalization of the classical theory of the mechanics of continuous media.

4.1. Virtual Power of Internal Forces

Let us therefore take

$$\hat{\mathscr{P}}_{(i)}(\hat{\mathbf{u}}) = \int_{\mathscr{D}} (K_i(\mathbf{x})\hat{u}_i(\mathbf{x}) + K_{ij}(\mathbf{x})\hat{u}_{i,j}(\mathbf{x}))\,\mathrm{d}x\,; \qquad \forall \hat{\mathbf{u}} \in \mathscr{V} \tag{3.13}$$

In order to derive the consequences of Statement I, we must be able, in the expression (3.13) to make the characteristic properties of a rigidifying field appear naturally; from which we have the decomposition of the matrix $\hat{u}_{i,j}$ into:

— *symmetric part*

$$\hat{D}_{ij} = \frac{1}{2}[\hat{u}_{i,j} + \hat{u}_{j,i}]$$

which is zero for a rigidifying field[16];
— *antisymmetric part*:

$$\hat{\Omega}_{ij} = \frac{1}{2}[\hat{u}_{i,j} - \hat{u}_{j,i}]$$

which, when $\hat{D}_{ij} = 0$ is linked to the instantaneous angular velocity $\hat{\omega}$ by

$$\hat{\Omega}_{ij} = -\varepsilon_{ijk}\hat{\omega}_k{}^{17}$$

We have therefore (suppressing $\mathbf{x}$ for simplicity of the notation):

$$\hat{\mathscr{P}}_{(i)}(\hat{\mathbf{u}}) = \int_{\mathscr{D}} (K_i\hat{u}_i + K^{(a)}_{ij}\hat{\Omega}_{ij} + K^{(s)}_{ij}\hat{D}_{ij})\,\mathrm{d}x\,, \qquad \forall \hat{\mathbf{u}} \in \hat{\mathscr{V}}, \tag{3.14}$$

$K^{(a)}_{ij}$ and $K^{(s)}_{ij}$ denoting respectively the antisymmetric and symmetric parts of K_{ij} with the result that by putting

$$m_k = -\varepsilon_{ijk}K^{(a)}_{ij} \quad \text{that is to say} \quad K^{(a)}_{ij} = -\frac{1}{2}\varepsilon_{ijl}m_l\,, \tag{3.15}$$

we have

$$\hat{\mathscr{P}}_{(i)}(\hat{\mathbf{u}}) = \int_{\mathscr{D}} (\mathbf{K}.\hat{\mathbf{u}} + \mathbf{m}.\hat{\omega})\,\mathrm{d}x\,, \qquad \forall \hat{\mathscr{C}} \equiv \hat{\mathscr{V}}^R. \tag{3.16}$$

It is then immediate, by a judicious choice of $\hat{\mathbf{u}} \in \hat{\mathscr{V}}^R$, to show that Axiom I is satisfied only when $\mathbf{K}$ and $\mathbf{m}$ (hence in particular K^a_{ij}) are identically zero; thus:

$$\hat{\mathscr{P}}_{(i)}(\hat{\mathbf{u}}) = \int_{\mathscr{D}} K^{(s)}_{ij}\hat{D}_{ij}\mathrm{d}x, \qquad \forall \hat{\mathbf{u}} \in \hat{\mathscr{V}}. \tag{3.17}$$

---

[16] Refer on the one hand to §2.3 and on the other to Sect. 2.3 of this §3.

[17] We recall that with every distributor $\mathscr{C}$ is associated its moment $\mathbf{u}$ and its resultant, the instantaneous rotation vector.

4.2. Virtual Power of External Forces

The external forces acting on $\mathscr{D}$ are of two types:
(i) *Forces at a distance* acting on $\mathscr{D}$ by systems external to S which we seek to identify under the form of volume densities with component f_i and F_{ij} of such a kind that the corresponding virtual power can be expressed by:

$$\hat{\mathscr{P}}_{(d)}(\hat{\mathbf{u}}) = \int_{\mathscr{D}} (f_i \hat{u}_i + F_{ij} \hat{u}_{i,j}) \mathrm{d}\mathbf{x} , \qquad \forall \hat{\mathbf{u}} \in \hat{\mathscr{V}} ; \tag{3.18}$$

namely

$$\hat{\mathscr{P}}_{(d)}(\hat{\mathbf{u}}) = \int_{\mathscr{D}} (f_i - F_{ij,j}) \hat{u}_i \mathrm{d}\mathbf{x} + \int_{\partial\mathscr{D}} F_{ij} \hat{u}_i n_j \mathrm{d}\sigma \tag{3.19}$$

($\mathbf{n}$ denoting the exterior unit normal to $\partial\mathscr{D}$ and $\mathrm{d}\sigma$ the element of surface area);
(ii) the *forces* acting on $\mathscr{D}$ by the parts of S exterior to $\mathscr{D}$ and which we supposed to be exclusively transmitted by *contact*; we seek therefore to identify them in the form of surface densities with components T_i and T_{ij}[18] of such a nature that the corresponding virtual power can be written:

$$\hat{\mathscr{P}}_{(c)}(\hat{\mathbf{u}}) = \int_{\partial\mathscr{D}} (T_i \hat{u}_i + T_{ij} \hat{u}_{i,j}) \mathrm{d}\sigma . \tag{3.20}$$

4.3. Application of the Principle of Virtual Power

Statement II is written here

$$\hat{\mathscr{A}}(\hat{\mathbf{u}}) = \hat{\mathscr{P}}_{(i)}(\hat{\mathbf{u}}) + \hat{\mathscr{P}}_{(d)}(\hat{\mathbf{u}}) + \hat{\mathscr{P}}_{(c)}(\hat{\mathbf{u}}) , \qquad \forall \hat{\mathbf{u}} \in \hat{\mathscr{V}} ,$$

hence, with the usual integrations by parts and putting

$$\tau_{ij} = - K_{ij}^{(s)} - F_{ij} , \tag{3.21}$$

we obtain

$$\left\{ \begin{aligned} \int_{\mathscr{D}} \rho \gamma_i \hat{u}_i \mathrm{d}x &= \int_{\mathscr{D}} (f_i + \tau_{ij,j}) \hat{u}_i \mathrm{d}x + \int_{\partial\mathscr{D}} (T_i - \tau_{ij} n_j) \hat{u}_i \mathrm{d}\sigma \\ &\quad + \int_{\partial\mathscr{D}} T_{ij} \hat{u}_{i,j} \mathrm{d}\sigma, \qquad \forall \hat{\mathbf{u}} \in \hat{\mathscr{V}} \end{aligned} \right. \tag{3.22}$$

This relation being satisfied for every arbitrary domain $\mathscr{D}$, we obtain finally the following local relations:

$$\left\{ \begin{array}{lll} f_i + \tau_{ij,j} = \rho \gamma_i & \text{in} & S \\ T_i = \tau_{ij} n_j & \text{on} & \partial\mathscr{D} \text{ and, by passage to the limit, on } \partial S \\ T_{ij} = 0 & \text{on} & \partial\mathscr{D} \text{ and, by passage to the limit, on } \partial S . \end{array} \right. \tag{3.23}$$

[18] It is supposed that T_i and T_{ij} depend not only on the point $\mathbf{x} \in \partial\mathscr{D}$ but also on the orientation $\mathbf{n}$ of the exterior normal to $\partial\mathscr{D}$ at this point: $T_i(\mathbf{x}, \mathbf{n})$, $T_{ij}(\mathbf{x}, \mathbf{n})$.

The tensor with components τ_{ij} is called the *stress tensor* and the vector with components T_i the *stress vector* at every point of S for the direction $\mathbf{n}$. It is easy to see that these definitions are completely consistent with those of §1 of this Chap. IA.

Remark 2. We recover the classical theory by supposing that F_{ij} is identically zero; then in this case τ_{ij} is symmetric and we denote it by σ_{ij} with

$$\sigma_{ij} = - K_{ij}^{(s)} .$$

This is the stress tensor of the classical theory, called the "*intrinsic stress tensor*", in this sense that in all cases it characterises the internal forces. We deduce the expression for the virtual power of the internal forces:

$$\hat{\mathscr{P}}_{(i)} = - \int_{\mathscr{D}} \sigma_{ij} \hat{D}_{ij} \mathrm{d}x . \tag{3.24}$$

In the general theory of the first gradient, we interpret the tensor with components F_{ij} as the sum of a field of "volume couples" represented by its antisymmetric part and of a field of "volume double forces" represented by its symmetric part. □

4.4. Commentaries[19]

(i) It is possible by modifying the hypotheses and especially those relative to the space $\mathscr{V}$ to establish complemetary results similar to those obtained above, for example in the case of discontinuities.
(ii) We have just described one theory of continuous media more refined than the classical theory which shows *a posteriori* that there are multiple schematizations possible for the same physical situation.
(iii) The method of virtual power is particularly elegant and effective when it is a question of finding laws for particular media such as curvilinear media (wires, rods, bars, beams), plates and shells.

5. Application to the Formulation of Curvilinear Media

Let S be a system (wire, rod, bar, beam) capable of being pictured as an arc of the curve $\widehat{AB}$. We denote by P the current point with curvilinear abscissa s (growing from a to b as P describes AB). (cf. Fig. 13). In order to reduce to an arc of a curve, the schematization classically adopted consists of retaining for the space $\mathscr{V}$ only *virtual motions rigidifying each cross-section* Σ. Such a motion is determined on the section with abscissa s by two vector fields, Ω and $\mathbf{V}_s$, defined on $\widehat{AB}$ (s being the abscissa of the point P representing the cross-section Σ). We denote by $\hat{\mathscr{C}}$ the distributor function of s, whose elements of reduction at P (assumed continuous) are $\boldsymbol{\Omega}$ and $\mathbf{V}_s$.

---

[19] A part of the developments raised and a good bibliography will be found in the article already cited: Germain [3].

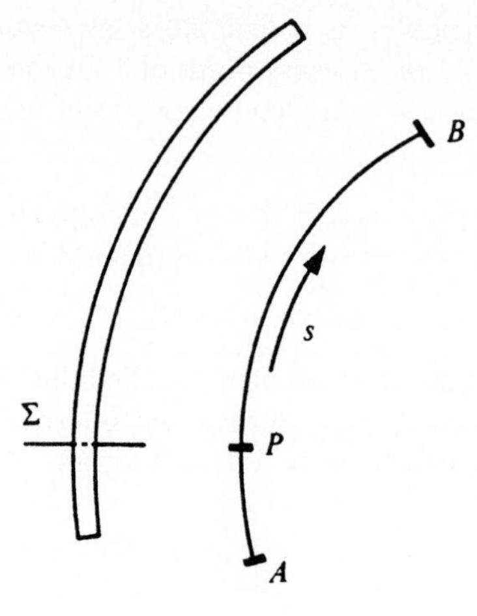

Fig. 13

5.1. Virtual Power of the Internal Forces

A theory of the first gradient leads us to choose for every arc $s_1 < s < s_2$:

$$\hat{\mathscr{P}}_{(i)} = -\int_{s_1}^{s_2}\left[\{\mathscr{S}\}\{\hat{\mathscr{C}}\} + \{\mathscr{T}\}\left\{\frac{\mathrm{d}\hat{\mathscr{C}}}{\mathrm{d}s}\right\}\right]\mathrm{d}s \tag{3.25}$$

(the choice of the negative sign will be justified *a posteriori*).
The Statement I implies that $\hat{\mathscr{P}}_{(i)}$ has to be invariant whatever $\hat{\mathscr{C}} = \hat{\mathscr{C}}_0$ may be constant, which implies that $\mathscr{S} = 0$, from which we have:

$$\hat{\mathscr{P}}_{(i)} = -\int_{s_1}^{s_2}\{\mathscr{T}\}\left\{\frac{\mathrm{d}\hat{\mathscr{C}}}{\mathrm{d}s}\right\}\mathrm{d}s\,. \qquad \forall\hat{\mathscr{C}}\in\hat{\mathscr{V}}\,. \tag{3.26}$$

Thus, at each point P, the *internal forces* are defined by a wrench $\mathscr{T}$ called the *wrench of the stresses* whose elements of reduction at P are denoted by $\mathbf{T}$ and $\mathbf{M}_s$. If $\boldsymbol{\tau}(s)$ is the unit tangent vector at P to $\widehat{AB}$, then

$\mathbf{T}.\boldsymbol{\tau} = \mathbf{T}_\tau$ is the normal traction, or compression, (at Σ),
$\mathbf{T} - \mathbf{T}_\tau\boldsymbol{\tau}$ is the shearing force,
$\mathbf{M}_s.\boldsymbol{\tau} = M_{s\tau}$ is the torsional moment,
and
$M_s - M_{s\tau}\tau$ is the bending moment.
An integration by parts of (3.26) lead to:

$$\hat{\mathscr{P}}_{(i)} = \int_{s_1}^{s_2}\left\{\frac{\mathrm{d}\mathscr{T}}{\mathrm{d}s}\right\}\cdot\{\hat{\mathscr{C}}\}\,\mathrm{d}s - \{\mathscr{T}(s_2)\}\{\hat{\mathscr{C}}(s_2)\} + \{\mathscr{T}(s_1)\}\{\hat{\mathscr{C}}(s_1)\}\,. \tag{3.27}$$

5.2. Virtual Power of External Forces

As previously we distinguish among the *external forces* acting on the arc $s_1 < s < s_2$, the line forces $\mathscr{F}$ exerted by the systems external to S and the contact forces $\mathscr{T}_1$ and $\mathscr{T}_2$ exerted on the sections Σ_{s_1} and Σ_{s_2} by the parts of S exterior to this arc. We suppose therefore that:

$$\hat{\mathscr{P}}_{(e)} = \hat{\mathscr{P}}_{(d)} + \hat{\mathscr{P}}_{(c)}$$

with

$$\hat{\mathscr{P}}_{(d)} = \int_{s_1}^{s_2} \{\mathscr{F}\}\{\hat{\mathscr{C}}\}\, ds \tag{3.28}$$

and

$$\hat{\mathscr{P}}_{(c)} = \{\mathscr{T}_1\}\{\hat{\mathscr{C}}(s_1)\} + \{\mathscr{T}_2\}\{\hat{\mathscr{C}}(s_2)\}\,.$$

5.3. Application of the Principle of Virtual Power in the Statical Case

The relation

$$\hat{\mathscr{P}}_{(i)} + \hat{\mathscr{P}}_{(d)} + \hat{\mathscr{P}}_{(c)} = 0\,, \qquad \forall \hat{\mathscr{C}} \text{ and } \forall s_1, s_2$$

then leads to the local relation

$$\frac{d\mathscr{T}}{ds} + \mathscr{F} = 0 \quad \text{on } \hat{AB} \tag{3.29}$$

and

$$\mathscr{T}_1 = -\mathscr{T}(s_1); \qquad \mathscr{T}_2 = +\mathscr{T}(s_2)\,. \tag{3.30}$$

If we denote by $\mathbf{f}$ and $\mathbf{m}_s$ the elements of reduction of $\mathscr{F}$ at P, we can translate (3.29) into two vector equations with the condition to take, before the derivation, the elements of reduction of $\mathscr{T}$ at a fixed point (A for example) and no longer at P which depends on s; those elements of reduction of $\mathscr{F}$ at A are:

$$\mathbf{T} \text{ and } \mathbf{M}_a = \mathbf{M}_s + \overrightarrow{AP} \times \mathbf{T}$$

and those of $\dfrac{d\mathscr{T}}{ds}$:

$$\frac{d\mathbf{T}}{ds} \text{ and } \frac{d\mathbf{M}_a}{ds} = \frac{d\mathbf{M}_s}{ds} + \boldsymbol{\tau} \times \mathbf{T} + \overrightarrow{AP} \times \frac{d\mathbf{T}}{ds}\,,$$

so that we may re-write (3.29) as:

$$\begin{cases} \text{(i)} & \dfrac{d\mathbf{T}}{ds} + \mathbf{f} = 0 \quad \text{on } [a, b] \\ \text{(ii)} & \dfrac{d\mathbf{M}_s}{ds} + \boldsymbol{\tau} \times \mathbf{T} + \mathbf{m}_s = 0 \quad \text{on } [a, b] \end{cases} \tag{3.31}$$

In the equation of moments, we have taken account of (3.31) (i) and of:

$$\mathbf{m}_a = \mathbf{m}_s + \overrightarrow{AP} \times \mathbf{f}$$

Finally, letting s_1 tend to a and s_2 to b in (3.30), after re-writing in vector form, we obtain the natural boundary conditions:

$$\mathbf{T}(a) = -\mathbf{F}_A\,; \quad \mathbf{T}(b) = \mathbf{F}_B; \quad \mathbf{M}_a(a) = -\mathbf{M}_A \text{ and } \mathbf{M}_b(b) = \mathbf{M}_B \tag{3.32}$$

where $\mathbf{F}_A$, $\mathbf{F}_B$, $\mathbf{M}_A$ and $\mathbf{M}_B$ are respectively the forces and couples exerted at A and B.

5.4. Examples of End Conditions

If the end B is *free* $\mathbf{F}_B = \mathbf{0}$ and $\mathbf{M}_B = \mathbf{0}$.
If the end B is *simply supported*: $\mathbf{M}_B = \mathbf{0}$, $\mathbf{F}_B$ is indeterminate, but the displacement $\mathbf{u}$ is then zero at B.
Finally, if the beam is clamped at A, i.e. is *encastré*: $\mathbf{F}_A$ and $\mathbf{M}_A$ are indeterminate *a priori* but the displacement $\mathbf{u}$ and the rotation $\boldsymbol{\Omega}$ are both zero at A. (Cf. Fig. 14 where in (1) the beam is simply supported at A and B, and in (2) the beam is built in at A and free at B).

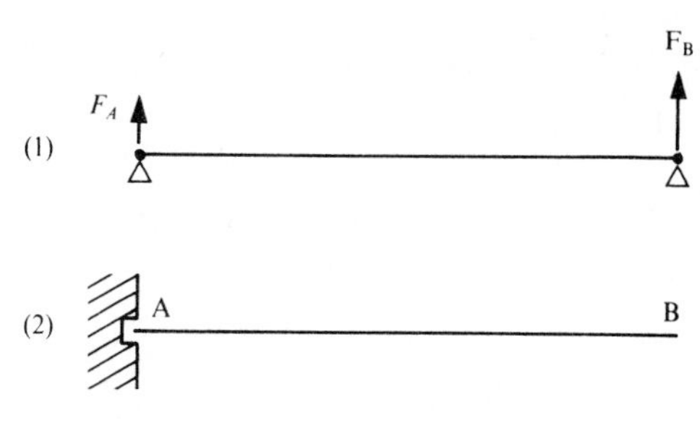

Fig. 14

5.5. Application of the Principle of Virtual Power in the Dynamical Case

In the *dynamical* case, Statement II is written:

$$\hat{\mathscr{P}}_{(i)} + \hat{\mathscr{P}}_{(d)} + \hat{\mathscr{P}}_{(s)} = \hat{\mathscr{A}}$$

with $\hat{\mathscr{A}}$, virtual power of the acceleration quantities, defined by:

$$\hat{\mathscr{A}} = \int_{s_1}^{s_2} \gamma(s) \, . \, \hat{\mathbf{V}}_s \rho \, \mathrm{d}s \, .$$

The relations (3.31) are then replaced by

$$\text{(3.33)} \qquad \rho\gamma = \frac{\mathrm{d}\mathbf{T}}{\mathrm{d}s} + \mathbf{f} \qquad \frac{\mathrm{d}\mathbf{M}_s}{\mathrm{d}s} + \boldsymbol{\tau} \times \mathbf{T} + \mathbf{m}_s = 0 \, ;$$

the data $\mathscr{F}$, $\mathbf{F}_A$, $\mathbf{M}_A$, $\mathbf{F}_B$ and $\mathbf{M}_B$ can depend on the time, but the relations (3.32) remain unchanged.

5.6. Remark 3

(i) In general, $\mathbf{f}$, the line density of forces, and $\mathbf{m}_s$, the line density of couples, are given, but often $\mathbf{m}_s = \mathbf{0}$.
(ii) In the case of a concentrated force $(\mathbf{F}_0, \mathbf{M}_0)$ at a point $\mathbf{P}_0$, we can apply integration by parts in (3.27) only on $[s_1, s_0\,[$ and $]\,s_0, s_2]$ and we obtain, at P_0 the

discontinuity relation:

(3.34) $[\![\boldsymbol{\tau}]\!]_{P_0} + \{\mathscr{F}_0\} = 0$ namely $[\![\mathbf{T}]\!]_{P_0} + \mathbf{F}_0 = \mathbf{0}\,,\ [\![\mathbf{M}_{S_0}]\!]_{P_0} + \mathbf{M}_0 = \mathbf{0}\,,$

$[\![\varphi]\!]_{P_0}$ denoting the jump of the quantity φ at P_0. □

6. Application to the Formulation of the Theory of Thin Plates[20]

A thin plate is a continuous system $\bar{S}$ which can be schematized geometrically by the part S of a plane which we shall choose as the plane $x_3 = 0$[21]: this is possible whenever the thickness of the plate is "small" compared with its other dimensions. We shall consider successively two theories each defined exactly by the choice made for the space of the virtual motions.

6.1. First Theory, Called "Natural"

We accept here that, as a result of the thinness of the plate, the *field of virtual velocities* can be chosen in the form:

(3.35) $$\hat{u}_\alpha(\mathbf{x}) = \hat{v}_\alpha(x_1, x_2) + x_3 \hat{l}_\alpha(x_1, x_2) \quad \text{and} \quad \hat{u}_3(\mathbf{x}) = \hat{w}(x_1, x_2)$$

(the Greek indices taking the values 1 and 2 only). A virtual motion is then determined by five functions of x_1 and x_2, namely: $\hat{v}_\alpha$, $\hat{l}_\alpha$ and $\hat{w}$.
The field of *associated rates of deformation*[22] is defined by:

(3.36) $$\begin{cases} \hat{D}_{\alpha\beta} = \hat{d}_{\alpha\beta} + x_3 \hat{K}_{\alpha\beta}\,;\, 2\hat{D}_{\alpha 3} = 2\hat{D}_{3\alpha} = \hat{l}_\alpha + \hat{w}_{,\alpha} = \hat{b}_\alpha;\, \hat{D}_{33} = 0 \\ \text{where} \\ \hat{d}_{\alpha\beta} = \dfrac{1}{2}[\hat{v}_{\alpha,\beta} + \hat{v}_{\beta,\alpha}] \quad \text{and} \quad \hat{K}_{\alpha\beta} = \dfrac{1}{2}[\hat{l}_{\alpha,\beta} + \hat{l}_{\beta,\alpha}]\,, \end{cases}$$

(3.37) $$\begin{cases} \text{with the result that for a motion } \hat{u} \text{ rigidifying } \bar{S} \\ \hat{d}_{\alpha\beta} = 0\,, \quad \hat{K}_{\alpha\beta} = 0 \quad \text{and} \quad \hat{b}_\alpha = 0\,. \end{cases}$$

(*a*) *Virtual power of internal forces.*
Taking account of (3.37) and the axiom of the virtual power of the internal forces we are led to put $\forall \mathscr{D} \subset S$:

(3.38) $$\hat{\mathscr{P}}_{(i)} = -\int_{\mathscr{D}} (N_{\alpha\beta}\hat{d}_{\alpha\beta} + M_{\alpha\beta}\hat{K}_{\alpha\beta} + Q_\alpha \hat{b}_\alpha)\,dx_1\,dx_2\,.$$

The symmetric tensors $N_{\alpha\beta}$, $M_{\alpha\beta}$, and Q_α then define the *intrinsic stresses*:
$N_{\alpha\beta}$: the tensor of the *tensions* (in the plane of S),
$M_{\alpha\beta}$: the tensor of the *bending moments*,
Q_α: the vector of the *shearing forces*,

---

[20] The reader should consult the works by Ciarlet [1], Destuynder [1] (Bernadou and Boisserie [1] on the theory of shell to obtain models leading to asymptotic developments.
[21] We denote by $\mathbf{k}_1, \mathbf{k}_2, \mathbf{k}_3$, the unit vectors corresponding to the chosen orthonormal reference frame.
[22] With the field of virtual velocities $\hat{\mathbf{v}}$.

(*b*) *Virtual power of external forces*

— For *forces at a distance* exerted on $\bar{\mathscr{D}}$ by systems external to $\bar{S}$, we put:

$$\hat{\mathscr{P}}_{(d)} = \int_{\mathscr{D}} (P_\alpha \hat{v}_\alpha + m_\alpha \hat{l}_\alpha + \bar{\omega}\hat{w}) \mathrm{d}x_1 \mathrm{d}x_2 , \tag{3.39}$$

(P_1, P_2) defining a field of *surface forces* acting in the plane of the plate, (m_1, m_2) a *distribution of external couples*, which are often considered to be zero, $\bar{\omega}$ a *surface density of normal forces*.

For the *contact forces* exerted on ∂D by the complement of $\bar{\mathscr{D}}$ in $\bar{S}$ we put:

$$\hat{\mathscr{P}}_{(c)} = \int_{\partial\mathscr{D}} (T_\alpha \hat{v}_\alpha + C_\alpha \hat{l}_\alpha + T\hat{w}) \mathrm{d}s , \tag{3.40}$$

(T_1, T_2) defining the *line forces* acting on ∂D in the plane of the plate. Taking account of the fact that the motion $\hat{\mathbf{u}}$ is rigidifying for every straight segment normal to the plane $x_3 = 0$, and that the resultant of the corresponding distributor is $\hat{\omega} = \mathbf{k}_3 \times \hat{\mathbf{l}}$, the *vector* $\mathbf{k}_3 \times \mathbf{C}$ *can be interpreted as a line density of couples* acting along $\partial\mathscr{D}$; the tangential component of this vector determines the bending couple and its normal component the torsional couple.

Finally T is a line density of shear forces (normal to the plate) acting along the length of ∂D.

The application of the principle of virtual power and several integrations by parts of type (1.12) carried out on (3.38) (and remembering that because of the symmetry of $N_{\alpha\beta}$ and $M_{\alpha\beta}$, $N_{\alpha\beta}\hat{d}_{\alpha\beta} \equiv N_{\alpha\beta}\hat{v}_{\alpha,\beta}$ and $M_{\alpha\beta}\hat{K}_{\alpha\beta} \equiv M_{\alpha\beta}\hat{l}_{\alpha,\beta}$) lead to the relations:

$$N_{\alpha\beta,\beta} + P_\alpha = 0 \tag{3.41}$$

$$M_{\alpha\beta,\beta} - Q_\alpha + m_\alpha = 0 \tag{3.42}$$

$$Q_{\alpha,\alpha} + \bar{\omega} = 0 , \tag{3.43}$$

valid at every point of $\bar{S}$, and

$$T_\alpha = N_{\alpha\beta} \nu_\beta \tag{3.44}$$

$$C_\alpha = M_{\alpha\beta} \nu_\beta \tag{3.45}$$

$$T = Q_\alpha \nu_\alpha , \tag{3.46}$$

valid along every line of the type $\partial\mathscr{D}$, $\boldsymbol{\nu}$ denoting the unit vector of the external normal to $\partial\mathscr{D}$.

For this theory the boundary conditions on ∂S are then written:

$$N_{\alpha\beta} \nu_\beta = \bar{T}_\alpha \tag{3.47}$$

$$M_{\alpha\beta} \nu_\beta = \bar{C}_\alpha \tag{3.48}$$

$$Q_\alpha \nu_\alpha = \bar{T} , \tag{3.49}$$

$\bar{T}_\alpha$, $\bar{C}_\alpha$ and $\bar{T}$ being the external forces applied to the boundary ∂S.

Remark 4. If we introduce the volume stresses σ_{ij} defined in the three-dimensional domain $\bar{S}$ itself defined by

$$(x_1, x_2) \in S \quad \text{and} \quad -h'(x_1, x_2) \leqslant x_3 \leqslant h''(x_1, x_2)$$

we have

$$\hat{\mathscr{P}}_{(i)} = -\int_{\bar{\mathscr{D}}} \sigma_{ij} \bar{D}_{ij} \, dx ,$$

which permits the interpretation $N_{\alpha\beta}$, $M_{\alpha\beta}$, Q_α as functions of σ_{ij} through the identification

(3.50)
$$N_{\alpha\beta} = \int_{-h'}^{h''} \sigma_{\alpha\beta} \, dx_3; \quad M_{\alpha\beta} = \int_{-h'}^{h''} x_3 \sigma_{\alpha\beta} \, dx_3 \quad \text{and}$$

$$Q_\alpha = \int_{-h'}^{h''} \sigma_{\alpha 3} \, dx_3 . \qquad \square$$

6.2. Second Theory, Called the Love–Kirchhoff Theory

This is a theory less refined than the previous one from which it can be deduced by taking:

(3.51)
$$\hat{b}_\alpha = 0 ,$$

i.e. by supposing $\hat{D}_{\alpha 3} = 0$. This expresses the hypothesis that each segment of matter originally normal to the plane $x_3 = 0$ remains normal to the deformed plane.

The essential consequence of this assumption for the application of the principle of virtual power is that $\hat{l}_\alpha = -\hat{w}_{,\alpha}$ and hence that $\hat{l}_\alpha$ and $\hat{w}_\alpha$ are no longer independent; we have in particular:

$$\hat{K}_{\alpha\beta} = -\hat{w}_{,\alpha\beta} .$$

In writing down the principle of virtual power, we eliminate $\hat{l}_\alpha$ from the integrals taken over $\mathscr{D}$ and $\partial\mathscr{D}$ and hence have the natural disappearance of the terms Q_α, m_α and C_α. On the other hand we have the appearance of the term $\partial\hat{w}/\partial n$ (rate of virtual variation of the slope of the deformed plate along $\partial\mathscr{D}$) in the integrals taken over $\partial\mathscr{D}$, which implies the appearance of a term Γ (bending couple).

We can decompose the integrations by parts on each regular piece of $\partial\mathscr{D}$, from which we have the appearance of terms of possible discontinuities in the passage to the angular points P_i of ∂S.

We obtain finally:

the *equations of equilibrium* which reduce to:

(3.52)
$$N_{\alpha\beta,\beta} + P_\alpha = 0$$

(3.53)
$$M_{\alpha\beta,\alpha\beta} + \bar{\omega} = 0$$

at each point of $\bar{S}$;

conditions along every arc $\partial\mathscr{D}$ with continuous tangent whose unit vector is $\boldsymbol{\tau} = \mathbf{k} \times \boldsymbol{\nu}$:

$$T_\alpha = N_{\alpha\beta}\nu_\beta \tag{3.54}$$

$$\Gamma = M_n \quad \text{where} \quad M_n = M_{\alpha\beta}\nu_\beta\nu_\alpha \tag{3.55}$$

$$T = \frac{\partial M_\tau}{\partial \tau} + Q \quad \text{where} \quad M_\tau = M_{\alpha\beta}\nu_\beta\tau_\alpha \quad \text{and} \quad Q = M_{\alpha\beta,\beta}\nu_\alpha\,; \tag{3.56}$$

finally conditions on the frontiers along ∂S are:

$$N_{\alpha\beta}\nu_\beta = \bar{T}_\alpha \tag{3.57}$$

$$\mathbf{M}_n = \bar{\Gamma} \tag{3.58}$$

$$\frac{\partial \mathbf{M}_\tau}{\partial \tau} + \mathbf{Q} = \bar{\mathbf{T}}\,, \tag{3.59}$$

with at the angular points P_i of ∂S, if they exist:

$$[\![\mathbf{M}_\tau]\!]_i = \bar{\mathbf{R}}_i\,, \tag{3.60}$$

where $\bar{\mathbf{R}}_i$ is the external point force applied at P_i, where $[\![\varphi]\!]_i$ denotes the discontinuity of the quantity φ at P_i along ∂S traversed in the direction of $\boldsymbol{\tau}$.

Remark 5

(i) The coupling of the type (3.42), (3.43) between $M_{\alpha\beta}$ and Q_α has disappeared which allows us to treat separately two simpler independent problems:

— A *tension problem* or the *membrane problem* governed by the relations (3.52) and (3.57). By definition, a membrane is a surface which can only support forces of the type P_α and $\bar{T}_\alpha$.

— A *pure bending problem* governed by the relations (3.53), (3.58), (3.59) and possibly (3.60).

(ii) The decoupling is possible here because it is a question of a statical problem. In the dynamical case in effect (and for the two theories described above) it is necessary to take account of the virtual power of the acceleration quantities defined by:

$$\hat{\mathscr{A}} = \int_{\mathscr{D}} \gamma(\mathbf{x})\hat{\mathbf{u}}(\mathbf{x})\rho(\mathbf{x})\,\mathrm{d}\mathbf{x}\,.$$

Expanding γ and ρ in the neighbourhood of $x_3 = 0$ and neglecting, after integration in x_3, terms of order x_3^2, we obtain

$$\hat{\mathscr{A}} = \int_{\mathscr{D}} \bar{\rho}h(\bar{\gamma}_\alpha\hat{v}_\alpha + \bar{\gamma}_3\hat{w})\,\mathrm{d}x_1\,\mathrm{d}x_2$$

where $h(x_1, x_2)$ is the thickness of the plate at the point (x_1, x_2) and where $\bar{\rho}(x_1, x_2) = \rho(x_1, x_2, 0)$; $\bar{\gamma}(x_1, x_2)' = \gamma(x_1, x_2, 0)$.
The application of the principle of virtual power:

$$\hat{\mathscr{P}}_{(i)} + \hat{\mathscr{P}}_{(d)} + \hat{\mathscr{P}}_{(c)} = \hat{\mathscr{A}}$$

leads then to the following modifications:
the relations (3.41) and (3.52) become:

$$N_{\alpha\beta,\beta} + P_\alpha = \bar{\rho} h \bar{\gamma}_\alpha , \tag{3.61}$$

and the relations (3.43) and (3.53) become respectively:

$$Q_{\alpha,\alpha} + \bar{\omega} = \bar{\rho} h \bar{\gamma}_3 \tag{3.62}$$

and

$$M_{\alpha\beta,\alpha\beta} + \bar{\omega} = \bar{\rho} h \bar{\gamma}_3 . \tag{3.63}$$

There is now coupling with the displacement field by means of $\hat{\gamma} = \partial^2 \hat{\mathbf{u}} / \partial t^2$; it is therefore necessary to take account of the constitutive relation to determine the stresses. □

6.3. Examples of Boundary Conditions

The external forces of the type **P**, **m** and ω are in general given; it is not always the same for the line forces applied to the edge of the plate (i.e. along ∂S). The current cases are the following:

— *The plate can be clamped* along an arc of ∂S: this means that the displacement and the slope of the deformed plate are zero on the arc, by contrast $\bar{\mathbf{T}}$ and $\bar{\Gamma}$ (the normal reaction and the couple at the clamping) in the Love–Kirchhoff theory, for example, are unknown.

— *The plate can be simply supported* along an arc of ∂S; in this case, the displacement and the bending couple $\bar{\Gamma}$ are both zero; by contrast $\bar{\mathbf{T}}$, the reaction of the support is unknown.

— *The plate can be free* along an arc of ∂S: in this case $\mathbf{T}$ and Γ are zero.

Linear and Non-Linear Problems in §1 to §6 of this Chapter IA

To throw light on their origins, we recapitulate below the situations in §1 to §6 which led to *linear* problems or else to *non-linear* problems.

The equations in §1 are in general non-linear. However, the study of motions with low speeds can be accomplished by "linearizing" the equations of flow (Stokes' equations). It is obvious that the complete study, and especially that at high speeds, of the problems of fluid mechanics, ought to be carried out within a non-linear context, but this lies out with the scope of the present book.

Small displacements of a solid can be treated by the mathematical models of linearized elasticity seen in §2. The study of large deformations forces us to treat non-linear elasticity and beyond that, plasticity. These considerations hold also for viscoelasticity and viscoplasticity (§3).

We have seen in §4 that Maxwell's equations in the vacuum are linear. In continuous media, the general equations of Maxwell and the simple constitutive relations which we have presented lead to linear problems which we shall look at in later chapters. However, in technical practice and for example, if we consider the

actual constitutive relation between the field and the magnetic induction, the induction can, at the instant t,
(1) depend equally on the values of the field at instants $s < t$ (linearly or not); this is a model of hysteresis.
(2) depend non-linearly on the field (magnetic behaviour of steel).
These phenomena occur in the saturation of magnetic circuits in transformers and alternators as well as heating by induction (see Bossavit [3] and [4]).
The transport of neutrons is represented by the transport equation which is linear in the most frequently occurring conditions, the interaction between neutron and neutron being taken to be zero and the nuclei with which the neutrons "collide" assumed to remain stationary. When we take the fluctuations (especially the thermal agitation) of these nuclei into account, the equations obtained remain linear[23]. However, when the ensemble made up of the atoms and their nuclei is in motion (in astrophysics, for example), the phenomenon is no longer linear (see Wilson *et al.* [1]).
The transport of the molecules in a gas, where the interactions in pairs (or more) of the moving molecules play an essential part, is described by the non-linear equation of Boltzmann[24]. In the case of weak interactions (dilute gases) linearization can be valid in the representation of certain phenomena. That will be seen in Chap. XXI.
The description of a plasma, as a gas of ionized particles and of electrons in an electromagnetic field, is similarly essentially non-linear (equations of MHD, Vlasov's equations, etc.). There again, certain phenomena of weak amplitude can be described by linearizing the equations of MHD and Vlasov's equations (see Krall–Trivelpiece [1], Van Kampen–Felderhof [1]). The linearized equations will be studied in Chap. XXI.
The theories of transport, referred to above, allow us to calculate coefficients of diffusion[24] of neutrons, of matter, of heat, of electricity, etc., diffusions which then can be modelled by equations, called precisely diffusion equations (which are less sharp than the transport equation or Boltzmann's equation). In the case of neutrons seen above (with, however, the non-linear exceptions already referred to) these diffusion equations are linear. In the case of the conduction of heat, we have seen in §1 that the thermal conductivity depends in general on temperature T. The diffusion satisfied in this case is therefore non-linear. However for small variations of the temperature, we can linearize it to obtain again a good description of the phenomena (see Chap. XV, §3).
Schrödinger's equation for an isolated physical system is, by its physical origin, essentially linear. That comes from the axioms of quantum mechanics explained in §6. It will therefore be studied in this work. By contrast, certain approximations of this equation diminishes the number of variables passing from a problem in $\mathbb{R}^{3Z}$ to

[23] See Cadilhac–Horowitz–Soulé–Trétiakoff [1] and Cadilhac–Soulé–Trétiakoff [1].
[24] For the study of the domain of validity of the transport equations, and of the calculation with these equations of the "coefficients of transport" (coefficients of diffusion, conductivity, viscosity, etc.), see Balescu [1].

one in $\mathbb{R}^3$, such as the equation of Hartree or of Hartree–Fock, but losing the linearity of the original problem. (See Landau–Lifschitz [6]).
The preceding mathematical models which give rise to linear problems which are dependent on the time (evolution problems) or not (stationary problems) are examined for a first time in Part B of this Chap. I (§1 and §2) which follows. They are further treated in the following chapters.

Coupled Problems

Numerous phenomena bring into play mathematical models involving simultaneously equations from several mechanical or physical disciplines (hence from several different sections of this Chap. IA), for example

— the circulation of a fluid in a hollow elastic pipe, or a boiler, or a heat exchanger, is described by equations derived in §1 and §2, the coupling being due especially to the effect of the deformations on the flow patterns (see Jeanpierre *et al.* [1]):
— the flow of an ionized gas in an electromagnetic field is described by the equations of §1 (if we consider as a fluid) and those of §4 (see the equations of M.H.D. in Krall–Trivelpiece [1]):
— the heating of an electric cable, of the magnetic circuits of transformers and of alternaters, heating by induction involve the heat equations of §2 and the electromagnetic equations of §4, the coupling being due principally to the variation with temperature of the constitutive relations (see above)[25].

We call such problems *coupled problems.*

Essentially Non-Linear Problems

Besides the disciplines, (mechanics, physics, engineering etc.), reviewed in §1 to §6 of this Chap. IA, other disciplines lead to important mathematical models. We have not presented them in this Chap. IA because they lead to mathematical problems which are essentially non-linear. We cite the following four examples (and this list is far from being exhaustive):

— *reaction–diffusion equations* which arise especially in *chemistry*. The equation which describes the reactions contains a term $g(u)$ representing the chemical activation as a function of a state variable u (for example the temperature) which is essentially non-linear. This non-linearity can be the cause of the creation of a propagation front (flame or detonation or combustion). The existence of convection terms can also contribute to the non-linearity. For work in this field the reader should consult: Fife [1], Strieder–Aris [1], Conley–Smoller [1].
— the phenomena of *biomechanics* form the object of mathematical modelling which often leads to non-linear problems. The possible linearization of the

[25] In addition to the hereditary effects (hysteresis) and the non-linearities, already cited, due to realistic constitutive relations, and to the influence of the temperature on the electromagnetic properties, the variation with temperature of the specific heat introduces new non-linearities. (see Bossavit [1], [2], [3]).

deformations of muscle, for example, is so unreliable that it is uncertain how to define a reference state starting from which we should evaluate the deviations to be linearized (see Fung [1], Holmes [1], Chadwick–Cole [1], Mow–Lai [1]).

— the densities (n and p) of the moving "carriers" (electrons and holes) in *semiconductors* are governed by non-linear models which describe their transport in the material (see Nag [1]). the non-linear terms arising primarily from the products of the density u by the electric field $\nabla\psi$ (ψ being the "electrostatic" potential). For these works either non-linear or only quasi-linear, we refer to Mock [1], [2], Van Roosbroeck [1], Vapaille [1].

— *quantum electronics*, especially the theory of the *laser*, bring into play the non-linear conditions which contribute to fixing a minimum threshold of the electromagnetic field and of the population of excited levels of the atoms from the oscillation of the laser arises. A non-linear terms is especially due to the interaction of the electromagnetic field propagating in the medium and its populations of electrons (see Yariv [1], Arecchi–Schultz–Dubois [1], Marcuse [1] and Martin [1]).

*

* *

Part B. First Examination of the Mathematical Models

The models springing from mechanics or physics which we have brought together in Chap. IA involve one or several partial differential equations and certain constraints.
(1) The partial differential equation (or system of equations) was written

$$P(D)u = f,$$

where u was the unknown, f was given on $\Omega_x \times \mathbb{R}_t$, and $P(D)$ a differential operator; in the sequel, we shall only be occupied with the case where $P(D)$ is linear. It will be necessary to give a mathematical meaning to this equation, and for example to take it in the sense of distributions in $\mathscr{D}'(\Omega_x \times \mathbb{R}_t)$, or $\mathscr{E}'(\Omega_x \times \mathbb{R}_t)$, or if $\Omega_x = \mathbb{R}^n_x$, $\mathscr{S}'(\mathbb{R}^{n+1})$, etc., or indeed in a more restricted space as we shall see later. For that it will be necessary that the coefficients of $P(D)$ are not irregular. The irregular case will be treated in *ad hoc* functional spaces (of the type of Sobolev spaces).
(2) We have seen that physical considerations imposed that certain quantities were finite. For example, in an electromagnetic problem, the energy (see Chap. IA, (4.110)) had to be finite, and that led to the *Hilbert space* $\mathscr{H}_c$ with scalar product (4.111). Or again the number of neutrons had to be finite in a bounded domain X (namely (5.4)) which led us to use the *Banach space* $L^1(X \times V)$.
The boundary conditions on all or part of the boundary of the domain studied, the conditions at infinity and the initial conditions at $t = 0$ should similarly come into play.
We shall example point (1) in §1 which follows and the point (2) in §2.

§1. The Principal Types of Linear Partial Differential Equations Seen in Chapter IA

We shall see in Chap. V that the types of linear differential operators $P(D)$ distinguishing among the linear problems of Chap. IA are the following:

(a) parabolic operators which we have encountered particularly in the equations of diffusion;
(b) hyperbolic operators which we have encountered in the wave equations;
(c) the Schrödinger operators which we have encountered in the equations of the same name:

(d) elliptic operators, among which we have encountered, in particular, the Laplacian operator.

The corresponding equations $P(D)u = f$ will similarly be called parabolic, hyperbolic, Schrödinger, elliptic.

1. Equation of Diffusion Type

We have encountered in Chap. IA numerous problems involving linear differential equations of the form:

$$\frac{\partial u}{\partial t} - \Delta_x u = f \quad \text{with } f \text{ given in} \quad \Omega_{x,t} = \Omega_x \times \mathbb{R}_{t+}{}^{1} \tag{1.1}$$

or more generally

$$\frac{\partial u}{\partial t} - \sum_{i,j=1}^{n} \frac{\partial}{\partial x_i} a_{ij} \frac{\partial}{\partial x_j} u = f(x,t)^2 \qquad (x,t) \in \Omega \times \mathbb{R}_+ . \tag{1.2}$$

where the a_{ij} are real functions of $x \in \Omega^3$ satisfying the following conditions:

$$(1.3) \begin{cases} \text{(i) } a_{ij}(x) = a_{ji}(x) \quad \text{for} \quad i,j = 1, \dots, n, x \in \Omega , \\ \text{(ii) the } a_{ij} \text{ are "regular" functions of } x \text{ in } \Omega\text{; more precisely we shall} \\ \quad \text{suppose: } a_{ij} \in \mathscr{C}^\infty(\Omega) \cap \mathscr{C}^1(\bar{\Omega}) . \end{cases}$$

This will allow us to give a meaning (that of the distributions $\mathscr{D}'(\Omega_{x,t})$) to the equation (1.2) for all $u \in \mathscr{D}'(\Omega_{x,t})$, (and also – see §2 below – of giving a meaning to the boundary conditions). These regularity hypotheses do not permit us to cover all the "physically interesting" cases and especially the case of materials (characterised by the a_{ij}) of different kind in different parts of the space domain $\Omega \subset \mathbb{R}^n_x$ considered.

We then have to consider functions a_{ij} which are, *for example*, different constants in two regions separated by an interface Σ, the functions a_{ij} being then *discontinuous*, but regular separately in each of the domains, which allows us to give a meaning to the *so-called transmission conditions* across the surface Σ.

For greater simplicity in this outline, we no longer shall consider such transmission problems in this §1.

---

[1] In the sequel we denote by Ω_x or Ω an open set in $\mathbb{R}^n$ and by $\mathbb{R}_{t+}$ or $\mathbb{R}_+$ the interval $]0, \infty[$.

[2] Or again

$$\frac{\partial u}{\partial t} - \sum_{i,j=1}^{n} \frac{\partial}{\partial x_i} a_{ij} \frac{\partial}{\partial x_j} u + \sum_{i=1}^{n} b_i \frac{\partial}{\partial x_i} u + a_{00} u = f(x,t)$$

with coefficients b_i and a_{00} satisfying the regularity conditions

$$b_i \in \mathscr{C}^\infty(\Omega) \cap \mathscr{C}^1(\bar{\Omega}), \quad a_{00} \in \mathscr{C}^\infty(\Omega) \cap \mathscr{C}^0(\bar{\Omega}) .$$

(See (1.3)).

[3] We can even suppose that these functions depend on the time – which will not be done here so as not to make the exposition too ponderous.

Remark 1. We notice that if we are constrained to consider only unknowns u in a certain functional space – for example the space $\mathscr{C}(\mathbb{R}^+, L^2(\Omega))$ of continuous functions of $t \in \mathbb{R}^+$ with values in the space $L^2(\Omega)_-$, then the condition

$$a_{ij} \in \mathscr{C}^1(\bar{\Omega}) \quad \text{for} \quad i,j = 1 \text{ to } n\,,$$

is sufficient to give a meaning (that of the distributions $\mathscr{D}'(\Omega_{x,t})$) to the equation (1.2) for all $u \in \mathscr{C}(\mathbb{R}^+, L^2(\Omega))$ – (see, for example, Richtmyer [1]). We shall confine ourselves here to the "regular" case (1.3). □

Finally, we shall suppose that the a_{ij} are such that

$$\sum_{i,j=1}^{n} a_{ij}(x)\xi_i\xi_j \geqslant \alpha\left(\sum_{i=1}^{n} |\xi_i|^2\right) \quad \text{for all} \quad x \in \bar{\Omega}\,, \tag{1.4}$$

$$\text{and} \quad \xi = (\xi_1, \ldots, \xi_n) \in \mathbb{R}^n\,,$$

where α is a strictly positive constant (independent of x). In the simple case:

$$a_{ij}(x) = D(x)\delta_{ij} \tag{1.3)'$$

(with $D(x)$ called the diffusion coefficient in certain applications), this condition (1.4) is equivalent to: there exists a constant $\alpha > 0$ such that:

$$D(x) \geqslant \alpha > 0\,.$$

We observe that, in particular, the condition (1.4) prohibits the coefficients a_{ij} all being zero on the boundary $\partial\Omega$ of Ω.
The physical interpretation of the condition (1.4), is often linked to the thermodynamics of the systems being considered (see Germain [2]).

1.1. Principal Examples of Equations of Diffusion Type

Let us recapitulate the principal examples of Chap. IA in which an equation of the type (1.2) features:
Conduction of heat. We have seen in §1, Sect. 7, that the temperature in a medium at rest – and when the variations of temperature are not too great – obeys the equation IA – (1.75) (or (1.75)′ or again (1.75)″), and that these equations are clearly of the type (1.2).
Diffusion of neutrons. The flux of neutrons – in the monokinetic approximation of diffusion (see IA, §5.2) satisfies the equation IA (5.15), which is again (except for the terms $v\bar{\Sigma}_f\Phi$, $\Sigma_t\Phi$ and $\bar{\Sigma}_s\Phi$) of the type (1.2), the relations (1.3), (1.4) being satisfied.
Mechanics of viscous fluids. We have seen in Sect. 6 of §1, Chap. IA, that the study of the flow of a homogeneous, incompressible, viscous fluid leads again to equations of the type (1.2): see the formulae IA (1.44) for the flow of such a fluid in a very long cylindrical pipe, and IA (1.45) or IA (1.46) for the case of such a fluid between two infinite parallel plates (or with one surface free). We have seen in IA, §1 that the relations (1.3) and (1.4) are satisfied.
Electromagnetism. In the framework of electromagnetism, for certain phenomena where Ohm's law IA (4.51) plays a fundamental role, and especially in the case of the diffusion of an electric wave in a conductor (see Sect. 4.2 of Chap. IA, §4), and

for the study of the magnetic induction in a plasma, we have obtained the equations IA (4.77) (i) and IA (4.79) (i) which again are of the type (1.2), with however a vector-valued unknown $\mathbf{u}(\mathbf{x}, t)$ (represented by the electric field $\mathbf{E}(\mathbf{x}, t)$ or the magnetic induction $\mathbf{B}(\mathbf{x}, t)$). The relations (1.3) and (1.4) are verified (see Fournet [1]).

1.2. Review

Equations of diffusion type are obtained in modelling very varied physical phenomena; the particular case of neutronics is a model case. This equation can be obtained, more particularly by the following methods:
(1) to strike a balance of the number of particles per unit volume, the term $D\Delta u\,d\mathbf{x}$ (to take the simple case) representing the flight of particles out of a domain of volume $d\mathbf{x}$, when the law of the neutron current is $J = D\,\text{grad}\,u$ (Fick's law) (see Bussac–Reuss [1]);
or else
(2) to describe the propagation of neutrons by the transport equation (see IA, §5) and study the behaviour of this equation when the free path $1/\Sigma$ tends to zero (see Chap. XXI);
or else
(3) to represent the paths of neutrons by stochastic processes of the type of Brownian motion with continuous absorption of the particles.
The above three types of models can be used similarly to derive the equation of the diffusion of heat. In these models, it is the molecules, the phonons, the electrons, for example, that are the agents of transmission. In this case, the reader will be able to establish for himself the model based on the balances noted (1); the transport model can be studied in the reference Ferziger–Kaper [1] for example; as for the use of random processes, it is studied in Tye [1] for example. It can be shown[4] that the probabilistic model cited in (3) above enables us to characterise the diffusion equations in a precise manner and hence to give a mathematical meaning to the word "diffusion".
The equations (1.1) and (1.2) which we have considered, and which we shall rewrite:–

(1.5) $$P(D)u = f \quad \text{in} \quad \Omega \times \mathbb{R}_+$$

where

(1.6) $$P(D) = \frac{\partial}{\partial t} - \sum_{i,j} \frac{\partial}{\partial x_i} a_{ij} \frac{\partial}{\partial x_j},$$

with coefficients a_{ij} satisfying (1.3) and (1.4) are said to be of parabolic type. The linear differential operator (*l.d.o.*) $P(D)$, is similarly said to be parabolic (the notion of a parabolic l.d.o. is in fact wider than (1.6) including, for example, the operators occurring in viscoelasticity, to characterise materials with short memory; see IA, §3, equation (3.6)).

[4] See Dautray [1].

The parabolic operators will be studied in Chaps. V. It is in Chap. XIV to XVIII that there will be treated the properties of the diffusion equations: "infinite" speed of propagation, dissipation, regularisation, irreversibility, decay, etc. (these concepts will be defined later as they arise).

2. Equation of the Type of Wave Equations

We have encountered in Chap. IA many problems involving linear differential equations of the form:

$$\frac{\partial^2 u}{\partial t^2} - \Delta u = f \quad \text{with } f \text{ given, in } \Omega_x \times \mathbb{R}_+ \text{ with } \Omega_x \text{ an open set of } \mathbb{R}^n \tag{1.7}$$

or more generally

$$\frac{\partial^2 u}{\partial t^2} - \sum_{i,j=1}^{n} \frac{\partial}{\partial x_i} a_{ij} \frac{\partial}{\partial x_j} u = f(x, t), \quad (x, t) \in \Omega_x \times \mathbb{R} \tag{1.8}$$

with f given, and the a_{ij} being real functions of $x \in \Omega$ satisfying (1.3) and (1.4).

2.1. Principal Examples of Wave Equations

We recall the principal examples of Chap. IA in which a wave equation appears. *Elasticity. Elastic waves and vibrations.* We have seen in Chap. IA, §2, Sect. 5, that the volume dilatation θ and the vorticity ω which are related to the displacement field $\mathbf{u}$ through the equations

$$\theta = \operatorname{div} \mathbf{u} \quad \text{and} \quad \omega = \tfrac{1}{2} \operatorname{curl} \mathbf{u},$$

in a homogeneous, isotropic elastic medium characterised by the Lamé coefficients λ and μ (assumed constant) and occupying the whole space $\mathbb{R}^3$, satisfy the wave equations IA (2.49) and (2.50) (this latter equation then being a vector equation). The displacement fields $\mathbf{u}^{(1)}$ and $\mathbf{u}^{(2)}$ (respectively of pure dilatation and of distortion) again obey the "vector" wave equations IA (2.54) and (2.55).
We have again obtained the (scalar) wave equation in the study of the vibrations of a heavy, homogeneous, perfectly elastic and flexible string (see IA, §2, Sect. 5.5, equation (2.66) (i)) and of a cylindrical elastic rod subjected to imposed longitudinal tensions or contractions, depending on the time (see IA, §2, Sect. 5.6, equation (2.68) or (2.70)).
In Chap. IA, §2 we have also encountered various extensions of equation (1.8), especially the equations of elasticity IA (2.57) which are of "vector" type (the unknown takes values in $\mathbb{R}^3$ instead of $\mathbb{R}$) and with coefficients depending on the time.
Electromagnetism. The wave equation, essentially in vector form, has been met with repeatedly in §4 of Chap. IA; we cite especially:

- for the propagation of electromagnetic waves *in vacuo*, (in the absence of charge and electric current), see equation IA (4.12), or in a perfect medium, characterised by constants of permittivity ε and (magnetic) permeability μ, see equation IA (4.39), or for the propagation of the electric field in a wave guide, see Chap. IA, Sect. 4.2.3, equation (4.80) (i),

— for the propagation of the vector potential $\mathbf{A}_L$ with the Lorentz condition, see IA (4.66), and of the scalar potential, see equation IA (4.67), or again of the vector potential $\mathbf{A}$ with the Coulomb condition IA (4.69), see the equations IA (4.74).

— for the Debye potential, see Chap. IA, §4, Sect. 3.3.

We have seen that these equations of wave type are often obtained (fairly) directly from systems of first order partial differential equations. It is therefore not surprising – and it is therefore what we shall verify in the rest of this work – that the properties of the phenomena which arise either from wave equations or from certain systems of first order differential equations, are similar.
We recall briefly the principal examples of such systems of first order differential equations.

2.2. Systems of Differential Equations of the First Order

Electromagnetism. Maxwell's equations (IA, §4), taken relatively either *in vacuo* (see equation IA (4.1)) or in continuous media (see equation IA (4.19)) with constitutive relations, for example those of a perfect medium IA (4.37), form such a system.
Neutronics. This is also the case for the transport equation IA (5.1) or IA (5.3) (yet the integral terms which figure in these equations may add an important perturbation to the properties of the linear differential operator $\mathbf{u} \to \mathbf{v} . \nabla\mathbf{u}$, called the *convection operator*)[5].
Quantum physics. Similarly, we cite Dirac's equations (see IA, §6) which constitute a first order system.

2.3. Review

The equations of wave type and the differential systems of the first order systems which we cited above are particular cases of equations of the type said to be hyperbolic. Hyperbolic operators will be defined and studied in Chap. V; the reader is referred there for the ideas that follow.
The wave equations encountered were of the form

$$P(D)u = f, \tag{1.9}$$

where the (hyperbolic) operator $P(D)$ was written:

$$P(D) = \frac{\partial^2}{\partial t^2} - \sum_{i,j=1}^{n} \frac{\partial}{\partial x_i} a_{ij} \frac{\partial}{\partial x_j} . \tag{1.10}$$

The first order differential systems were similarly of the form (1.9) with

$$P(D) = \frac{\partial}{\partial t} - \sum_{i=1}^{n} A_i \frac{\partial}{\partial x_i} - A_0 , \tag{1.11}$$

[5] We shall see in Chap. XXI that the transport equation shares *hyperbolic* properties with the wave equation.

with A_i ($i = 1$ to n) and A_0 denoting $n \times n$ matrices with complex coefficients (assumed here for simplicity to be independent of $\mathbf{x}$ and t) satisfying one of the following conditions:

(i) the matrix $A.\xi = \sum_{i=1}^{n} \xi_i A_i$, $\quad \forall \xi = (\xi_1, \ldots, \xi_n) \in \mathbb{R}^n$ is Hermitian;
(ii) the proper values of the matrix $A.\xi$ are real and distinct for all $\xi \in \mathbb{R}^n$.
The properties of these wave equations will be studied in Chaps. XIV to XVIII. In addition, a future book on wave equations will examine in more detail certain aspects of hyperbolic problems; propagation with finite velocity, wave front and compact support of the solutions; conservation; propagation of the singularities; "scattering"[6] by an obstacle; characteristics and bicharacteristics; reversibility, extinction, etc. These concepts will be defined and studied later.

3. Schrödinger Equation

These equations, encountered in §6 of Chap. IA, are written in the form:

$$i \frac{\partial u}{\partial t} - Hu = 0^7 \tag{1.12}$$

where H is a linear differential operator called the Hamiltonian. We have, in particular, encountered the following situation:

$$H = -\frac{1}{2m}\Delta + \mathscr{V}(\mathbf{x}), \quad \mathbf{x} \in \mathbb{R}^3, \tag{1.13}$$

where $\mathscr{V}(\mathrm{x})$ is the operator of multiplication by a real-valued function again denoted by $\mathscr{V}(\mathbf{x})$ or else a differential operator of order 1.
The case of a system of several particles led us to introduce

$$H = \sum_{i=1}^{N} -\frac{1}{2m_i}\Delta_{x_i} + \mathscr{V}(\mathbf{x}_1, \ldots, \mathbf{x}_N), \quad (\mathbf{x}_1, \ldots, \mathbf{x}_N \in \mathbb{R}^3)^8 \tag{1.13$'$}$$

$\mathscr{V}(\mathbf{x})$ being again a real function or a differential operator of order 1.
In many applications it will be essential to examine if the operator H is self-adjoint (for that it will be necessary to introduce the necessary definitions and theorems, which will be done in Chap. VI[9]). In this way the evolution of u governed by the equation (1.12) will be characterised by a unitary group conserving the quantity

$$\int |u(\mathbf{x}, t)|^2 \mathrm{d}\mathbf{x} = \|u(t)\|^2 \tag{1.14}$$

---

[6] The corresponding French word is "diffusion".
[7] Certain phenomena are modelled by taking $\mathbf{x} \in \Omega \subset \mathbb{R}^3$ or $(\mathbf{x}_1, \ldots, \mathbf{x}_N) \in \Omega \subset \mathbb{R}^{3N}$.
[8] In the system of natural units (in which $\hbar = 1$).
[9] The case where the operator H depends explicitly on the time is only looked at in Chap. XVIII (with the aid of variational methods).

(whose square root can be taken as norm at fixed t) which is the probability of the presence of the particle(s) in the whole space and that the energy of the system studied is

(1.14)′
$$W = \int \bar{u} H u \, \mathrm{d}\mathbf{x} \,.$$

The study of the behaviour, for very long times, of systems governed by Schrödinger's equation, which we shall call the study of "scattering" is essential for applications. Certain analogies, although limited, with the "scattering" by an object of hyperbolic systems which we shall see in the book to come on waves, should be noted.

If certain properties of reversible evolution, of conservation, of scattering, connect Schrödinger equations with hyperbolic equations, we note nevertheless similarities with diffusion equations: support of solutions in the whole space ($\mathbb{R}^3$ for example), "infinite speed" of propagation and absence of wave front etc. That will appear in Chap. XIV to XVIII.

4. The Equation $Au = f$ in which A is a Linear Operator not Depending on the Time and f is Given (Stationary Equations)

Introduction. In Chap. IA we have encountered problems involving the equation

(1.15)
$$-\Delta u(\mathbf{x}) = f(\mathbf{x}), \quad \mathbf{x} \in \Omega \in \mathbb{R}^n \ (f \text{ given})$$

(called *Poisson's equation*) with boundary conditions which are similarly independent of the time. More generally, we have an operator A such that

(1.16)
$$Au \overset{\text{def}}{=} -\sum_{i,j=1}^{n} \frac{\partial}{\partial x_i} a_{ij} \frac{\partial u}{\partial x_j} + \sum_{i=1}^{n} a_i \frac{\partial u}{\partial x_i} + a_0 u = f \quad \text{in} \quad \Omega \subset \mathbb{R}^3$$
$$(a_{ij}, a_i, a_0, f \text{ given})$$

where the hypotheses (1.3) and (1.4) were satisfied or even:

(1.17)
$$Bu = f$$

where B is a differential operator of given order m.

4.1. Examples of the Situations (1.15), (1.16), (1.17)

Electromagnetism. In electrostatics (see IA, §4), we have studied the potential $\Phi(x)$ in the interior of a perfect medium with a distribution of charges $\rho(x)$ supposed known. The equation satisfied by the unknown Φ, which is IA (4.86) in $\mathbb{R}^3$ (or IA (4.92) in a domain Ω of $\mathbb{R}^3$ with appropriate boundary conditions), is of the type (1.15).

Fluid mechanics. The search for the velocity potential in a stationary irrotational flow of an incompressible perfect fluid (see IA, §1) leads to the problem IA (1.31) involving equation IA (1.31) (ii) which is indeed of the type (1.15) – with $f = 0$. The flow of such a fluid round an obstacle (see Chap. IA, §1, Sect. 5.3) is an example of this situation.

The study of the stationary flow of an incompressible viscous fluid in a cylindrical pipe (see IA, §1, Sect. 5.4) leads to the problem IA (1.35) with the equation IA (1.35) (i) which is again of the form (1.15).

Finally, the study of thermal "equilibrium" leads to the search for solutions of the equation IA (1.75)″ when $\partial u/\partial t \equiv 0$, and this equation is of the type (1.15).

Linear elasticity. The problems of statical (or quasi-statical) elasticity seen in Chap. IA, §2, Sect. 4, involve the equations IA (2.33) or IA (2.33)′ or IA (2.40) in $\Omega \subset \mathbb{R}^3$ which are of the type (1.16) or (1.17).

In problems of plane deformations (IA, §2.4.2.2) and plane stresses (IA, §2.4.2.3) we have met with equations IA (2.43) and IA (2.46) which are of the type

$$\Delta(\Delta u) = f, \tag{1.18}$$

and hence enter into the category of the equations (1.17).

Neutronics. The stationary diffusion equation (see IA, §5) is of the type (1.16). However it contains an independent parameter E and terms of integral operators; these disappear in the multigroup equation IA (5.36) and in the monokinetic equation (obtained by putting $\partial \Phi/\partial t \equiv 0$ in IA (5.15)).

4.2. Review

Among the operators occurring on the left-hand side of the equations (1.15), (1.16) and (1.17) we retain only those which are *elliptic*. This concept will be defined and studied in Chap. V. We shall see that Δ is the fundamental prototype of elliptic operator; it will be studied in Chap. II.

The particular properties of elliptic equations: regularity of the solution u, "non-localized" support of u, extrema of the solution etc., will appear in Chap. II and in Chap. V, and the solution by variational methods will be carried out in Chap. VII.

Remark 2. In Chap. IA we have encountered problems of type (1.16) in which the operator A is not always elliptic. For example, we may cite:

- *Tricomi's problem* IA (1.41) involves an equation *changing its type* (elliptic–hyperbolic) according to the part of the domain. We shall study it in Chap. X.
- *The stationary transport equation* IA (5.19) (i) and IA (5.20) is never elliptic. That is due to the presence of the convection operator $\mathbf{v} . \nabla \mathbf{u}(\mathbf{x}, \mathbf{v})$ in the transport operator T (see IA (5.20)); we shall consider the properties of the convection operator in Chap. XXI. □

In all the examples cited in §§1 to 6 of Chap. IA, the problems (1.15), (1.16), (1.17) have been described in a variable fashion following the disciplines in which they arose (statical, quasi-statical, stationary, permanent, etc.). From now on in this work we shall describe all such problems as *stationary*.

4.3. Stationary Problems of Spectral Type

Among the problems whose expression did not depend on the time, which we have encountered in Chap. IA, are equations of the form

(1.19) $$-\Delta u - \lambda u = 0 \quad \text{in} \quad \Omega \subset \mathbb{R}^n$$

(with boundary conditions) where λ was an unknown parameter, or more generally

(1.20) $$Au \overset{\text{def}}{=} -\sum_{i,j=1}^{n} \frac{\partial}{\partial x_i} a_{ij} \frac{\partial u}{\partial x_j} + a_0 u + \lambda u = f^{10} \quad \text{in} \quad \Omega \subset \mathbb{R}^n$$

(and boundary conditions). In (1.20), the given coefficients a_{ij} again satisfy (1.3) and (1.4), the given a_0 is supposed regular (in fact we shall need to have $a_0 \in \mathscr{C}^\infty(\Omega) \cap \mathscr{C}^0(\bar{\Omega})$) and λ is an unknown number (real or complex according to the problem) to be determined. More generally, we have encountered the equation

(1.21) $$Bu + \lambda u = f$$

(and boundary conditions) where B is a self-adjoint operator[11] in a Hilbert space. First of all, let us recall some examples from Chap. IA.

Electromagnetism. In looking for solutions of the form IA (4.99) of Maxwell's equations (see IA, §4), we have been led, for example in the case of wave guides, to equations of Helmholtz type, namely IA (4.121) (i) and IA (4.122) (i) which are both of the type (1.19).

Neutronics. The problem of criticality for the equation of monokinetic diffusion IA (5.39) of IA, §5 can be reduced to a problem of the type (1.20)[12] in certain applications. In other applications, on the contrary, the problem to which IA (5.39) is led is again of the form (1.20)[12], but with λ known and hence fixed; what is then made to vary is the shape of the domain Ω.

In the case of the multigroup form (m groups) the equations of diffusion of neutrons IA (5.36), this problem which can be put formally in the form (1.20) involves an operator A which is not self adjoint[13]. We shall return to this point in Chap. XVIIB.

Quantum physics. In Chap. IA, §6, we have defined the stationary states and have shown that the stationary solutions of Schrödinger's equation IA (6.13) had the form IA (6.14). We have seen that they could be found by solving IA (6.17) which is indeed of the type (1.20) (with $f = 0$, and the role of the parameter λ being played by E). In these problems, the unknown parameters λ (hence E) are the energy levels of the system studied. □

We shall see in the chapters which follow that the search for the λ's in $\mathbb{R}$ (or $\mathbb{C}$) and for the solutions u in a suitable functional space X (a question to which we return in §2 which follows) involves the boundary conditions in a crucial way. This search for

[10] We have made no mention of the term in $\partial/\partial x_i$ since it can be eliminated by a change of the unknown function.

[11] See Chap. VI.

[12] With $f \equiv 0$.

[13] When the problem is formulated in a Hilbert space.

the λ's and the corresponding u's, called a *spectral problem*, will be treated in Chap. VIII after the necessary elements have been elaborated in Chap. VI.

§2. Global Constraints Imposed on the Solutions of a Problem: Inclusion in a Function Space; Boundary Conditions; Initial Conditions

1. Introduction. Function Spaces

We have seen in the problems of Chap. IA, that, in addition to one or more partial differential equations posed in the domain $\Omega \subset \mathbb{R}^n$ being considered, these problems can include conditions on the boundary $\partial\Omega$, initial conditions atthe time $t = 0$ and sometimes global constraints concerning the unknown. This is the situation that we now examine in this §2. First of all we examine these last constraints. In electromagnetic problems, we have generally to impose on the unknowns **E** and **B** of the system IA (4.1) the condition of being such that the total energy W_v (or W) given by IA (4.4) is finite. It is this that we have used, for example, in §4.6.2. in using a Hilbert space $\mathbb{X} = \mathscr{H}_c$ satisfying the condition IA (4.110).
In problems on the transport of neutrons seen in IA, §5.1, we have imposed the condition that the number of neutrons in a bounded domain X of $\mathbb{R}^3$ is finite at every instant, as expressed by IA (5.4); this implies that the unknown in problems of evolution (and of stationary problems) must take its values in $\mathbb{X} = L^1(X \times V)$.
Similarly we should note that the densities of number of neutrons must be positive and hence work in the set of positive functions of $L^1(X \times V)$ (i.e. in the cone $L^1_+(X \times V)$; that will not be useful for evolution problems and in Chap. XXI on transport, we shall see that it is enough to use $L^1(X \times V)$, the transport operator T being such that a function of $L^1(X \times V)$ positive at the instant $t = 0$ will remain positive for every instant $t > 0$.
The same condition, "the total number of neutrons must be finite," applied to the diffusion problems of Chap. IA, §5, leads to the demand that the unknown functions of the time, representing a density of neutrons or a flux, take their values in $L^1(\Omega)$.
For the applications of quantum physics seen in Chap. IA, §6 we have required that the probability of presence at each instant t,

$$\int \bar{\Phi}(t) \,.\, \Phi(t)\,\mathrm{d}x \,,$$

over the whole space ($\mathbb{R}^3$ for example for a particle or $\mathbb{R}^{3N}$ for N particles), is equal to 1, which has led us, for example, to take as the Hilbert space associated with the system studied $X = L^2(\mathbb{R}^3)$ or $X = L^2(\mathbb{R}^{3N})$, (see Sect. 2 of Chap. IA, §6) or, more generally a space of square integrable functions with vector values (for particles with non-zero spin).
Similar considerations can be developed for the thermal problems looked at in IA, §1, and the elasticity problems considered in IA, §2, etc.

2. Initial Conditions and Evolution Problems

In the problems arising from mechanics and physics which we have presented in Chap. IA, it was natural to suppose that the unknown function u was a continuous function of t (with values in the space $\mathbb{X}$ mentioned above in (1)). The continuity at the instant $t = 0$ permits us, in particular, to give a meaning to the initial condition

$$u(t = 0) = u_0 \quad \text{with} \quad u_0 \in \mathbb{X} . \tag{2.1}$$

In order that the solution $u(x, t)$ of an evolution problem which we write

$$\begin{cases} \text{(i)} \; P_{x,t}(D)\, u(\mathbf{x}, t) = f(\mathbf{x}, t), \quad (\mathbf{x}, t) \in \Omega_{x,t} \overset{\text{def}}{=} \Omega \times \mathbb{R}_{t+} , \\ \quad \text{with} \\ \text{(ii)} \; \text{conditions on the boundary } \partial\Omega \\ \quad \text{and} \\ \text{(iii)} \; \text{initial conditions} , \end{cases} \tag{2.2}$$

be a continuous function of the time (with values in $\mathbb{X}$) it is necessary that the data: the "source" f, the initial conditions u_0, the boundary conditions on $\partial\Omega$ and the coefficients of the linear differential operator $P_{x,t}(D)$ satisfy certain relations which we shall explain in Chaps. XIV to XVIII. In the great majority of the cases we treat the space variables x and time variable t play very different mathematical roles; in addition, the problems (2.2) which we encounter in Chapter IA reduce to a great extent to the following two cases:

$$\begin{cases} \text{(i)} \; \dfrac{\partial u}{\partial t}(\mathbf{x}, t) + P_x(D)\, u(\mathbf{x}, t) = f(\mathbf{x}, t), \qquad \forall (\mathbf{x}, t) \in \Omega \times \mathbb{R}_+ , \\ \quad \text{with} \\ \text{(ii)} \; \text{boundary conditions on the boundary } \partial\Omega \text{ of } \Omega \text{ and for all } t \in \mathbb{R}_+ , \\ \quad \text{and} \\ \text{(iii)} \; \text{the initial condition (called the } \textit{Cauchy condition}\text{)} \\ \quad u(\mathbf{x}, 0) = u_0(\mathbf{x}) , \qquad \forall \mathbf{x} \in \Omega , \end{cases} \tag{2.3}$$

and

$$\begin{cases} \text{(i)} \; \dfrac{\partial^2 u}{\partial t^2}(\mathbf{x}, t) + P_x(D).u(\mathbf{x}, t) = f(\mathbf{x}, t), \qquad \forall (\mathbf{x}, t) \in \Omega \times \mathbb{R}_+ \\ \quad \text{with} \\ \text{(ii)} \; \text{boundary conditions on } \partial\Omega, \text{ for all } t \in \mathbb{R}_+ \\ \quad \text{and} \\ \text{(iii)} \; \text{the initial conditions (called the } \textit{Cauchy data}\text{)} \\ \quad u(\mathbf{x}, 0) = u^0(\mathbf{x}) \\ \quad \dfrac{\partial u}{\partial t}(\mathbf{x}, 0) = u^1(\mathbf{x}) , \qquad \forall \mathbf{x} \in \Omega . \end{cases} \tag{2.4}$$

These problems (2.3) and (2.4) are called *Cauchy problems* or *mixed problems in the sense of Hadamard.*

The problem (2.4) can be put in a form similar to (2.3) by writing

$$(2.5)\qquad \begin{cases} U(\mathbf{x}, t) = \left\{ u(\mathbf{x}, t), \dfrac{\partial u}{\partial t}(\mathbf{x}, t) \right\} \\ U_0(\mathbf{x}) = \{u^0(\mathbf{x}), u^1(\mathbf{x})\} \\ F(\mathbf{x}, t) = \{0, f(\mathbf{x}, t)\} \\ \mathscr{P}_x(D) = \begin{pmatrix} 0 & I \\ P_x(D) & 0 \end{pmatrix}. \end{cases}$$

With this notation, the problem (2.4) can be written

$$(2.6)\qquad \frac{\partial U}{\partial t}(\mathbf{x}, t) + \mathscr{P}_x(D) U(\mathbf{x}, t) = F(\mathbf{x}, t),$$

which is similar to (2.3).

Remark 1. Being given a partial differential equation (for example (2.3)(i) or (2.4)(i)) with boundary conditions (for example (2.3)(ii) or (2.4)(ii)) we are often led[14] to ask if there exist solutions of (2.3)(i) and (ii) (or of (2.4)(i) and (ii)) of the form

$$(2.7)\qquad u(\mathbf{x}, t) = u_0(\mathbf{x}) e^{i\omega t}$$

where ω is a real number[15], known or unknown according to the types of problems. The existence of such solutions can impose constraints on the system studied: for example $f(x, t)$ has similarly to be of the form (2.7) (or zero), namely

$$(2.8)\qquad f = f_0(\mathbf{x}) e^{i\omega t}.$$

Account being taken of (2.7), the problem (2.13)(i) and (ii) is written:

$$(2.9)\qquad \begin{cases} \text{(i) } i\omega u_0 + P_x(D) u_0 = f_0 \text{ in } \Omega \subset \mathbb{R}^n, \quad \text{with} \\ \text{(ii) boundary conditions for } u_0 \text{ on } \partial\Omega; \end{cases}$$

and the problem (2.4)(i) and (ii) is written

$$(2.10)\qquad \begin{cases} \text{(i) } -\omega^2 u_0 + P_x(D) u_0 = f_0 \text{ on } \Omega, \quad \text{with} \\ \text{(ii) boundary conditions for } u_0 \text{ on } \partial\Omega, \end{cases}$$

a problem which we have called *of spectral type* in §1.4.3. We shal not return to the other examples of the change of unknown function (2.7) already mentioned in §1. □

[14] See, for example, §4.6 of Chap. IA.

[15] ω can be complex in some applications.

Remark 2. We shall encounter in Chap. XV, §4 and in Chap. XVIIIB a more general equation than (2.4)(i), namely

$$\frac{\partial^2 u}{\partial t^2}(\mathbf{x}, t) + 2\beta \frac{\partial u}{\partial t}(\mathbf{x}, t) + P_x(D).u(\mathbf{x}, t) = f(\mathbf{x}, t) \qquad (2.11)$$

$$\forall (\mathbf{x}, t) \in \Omega \times \mathbb{R}_+ ,$$

where β is a given positive constant. In the case where $P_x(D) = -\Delta$, we shall call equation (2.11) the equation of *damped waves*. We have encountered it already in IA, §4, in IA(4.54).

In the case where $n=1$ and

$$P_x(D) = -\frac{\partial^2}{\partial x^2} + \text{constant}$$

it is called the *telegraphists' equation*[16] and is often written in the form

$$\frac{\partial^2 u}{\partial t^2} + (\alpha + \beta) - c^2 \frac{\partial u^2}{\partial x^2} - \alpha\beta u = f, \quad x \in (0, L) \subset \mathbb{R}, \quad t \in \mathbb{R}_+ . \qquad (2.12)$$

The change of the unknown function

$$u = e^{-\beta t} v$$

reduces equation (2.11) to

$$\frac{\partial^2 v}{\partial t^2} + (P_x(D) - \beta^2) v = e^{\beta t} f(x, t) , \qquad \forall (x, t) \in \Omega \times \mathbb{R}_+ \qquad (2.13)$$

which is of the type (2.4)(i). □

3. Boundary Conditions

In the mathematical models met with in Chap. IA, we have found the existence of boundary conditions for the unknown function u. First of all, let us consider the case of the *stationary* partial differential equations mentioned in IB, §1.4.

The problems encountered in Chap. IA were written in the forms

$$(2.14) \quad \begin{cases} \text{(i) } P_x(D)u = f, \quad u \in \mathbb{X}, \quad \mathbf{x} \in \Omega \subset \mathbb{R}^n, \quad \text{with } f \text{ given, with:} \\ \text{(ii) boundary conditions for } u \text{ on } \partial\Omega, \text{ the boundary of } \Omega , \end{cases}$$

or else

$$(2.15) \quad \begin{cases} \text{(i) } P_x(D)u - \lambda u = f, \quad u| \in \mathbb{X}, \quad \mathbf{x} \in \Omega \subset \mathbb{R}^n, \quad \text{with } f \text{ given with :} \\ \text{(ii) boundary conditions for } u \text{ on } \partial\Omega \text{ where } \lambda \in \mathbb{R}^{17} \text{ is to} \\ \qquad \text{be determined (spectral problem) .} \end{cases}$$

[16] This equation is used in the study of transmission lines (see Roubine [1]).

[17] Or else $\lambda \in \mathbb{C}$ in certain problems.

Because of the important part played in Chap. IA by the elliptic linear differential operators $P_x(D)$ of the 2nd. order, we shall begin with them.

3.1. Problems (2.14) or (2.15) Associated with a Linear Differential Operator $P_x(D)$ of the Second Order

Throughout this section 3.1 we shall be concerned with the case already studied since the beginning of this Chap. IB in which $P_x(D)$ is of the form

$$(2.16)\quad \begin{cases} P_x(D) = -\sum_{i,j=1}^{n} \dfrac{\partial}{\partial x_i} a_{ij}(\mathbf{x}) \dfrac{\partial}{\partial x_j} + a_{00}(\mathbf{x}) \\ \text{with the coefficients } a_{ij} \text{ satisfying (1.3) and (1.4) and also} \\ a_{00} \in \mathscr{C}^{\infty}(\Omega) \cap \mathscr{C}(\bar{\Omega}) \,. \end{cases}$$

Among the boundary conditions (2.14)(ii) or (2.15)(ii) met with in Chap. IA, let us mention the

Dirichlet boundary conditions. The boundary condition

$$(2.17)\qquad u = g \quad \text{on the boundary} \quad \partial\Omega \text{ of } \Omega$$ [18]

is called an *inhomogeneous Dirichlet condition* and can equally be denoted by

$$u|_{\partial\Omega} = g \quad \text{or} \quad u|_{\Gamma} = g$$

g being given on $\partial\Omega$. The case $g \equiv 0$ on $\partial\Omega$ is called the *homogeneous Dirichlet condition.* These boundary conditions have been used in each of §§1 to 6 of Chap. IA. In particular let us recall the following cases:

— in §1, *fluid mechanics.* The condition (2.17) where u is the velocity of the fluid and, for example, $g = 0$ in the case of flow against a fixed wall has been encountered in numerous examples of Chap. IA, §1. To illustrate the case (2.14) with (2.16), we cite the condition IA (1.35)(ii) for flow in a cylindrical pipe. Similarly we quote the stationary Stokes problem IA (1.14) (with $\partial U/\partial t = 0$) with the boundary condition IA (1.14)(iv), which is of a type similar to (2.14).

The conduction of heat seen in Section 7 of Chap. IA, §1, gives rise to the stationary problem obtained by making $\partial T/\partial t = 0$ in IA (1.75)′ (or IA(1.75)″) with the boundary condition IA(1.66)[19] which is of type (2.17) when the temperature is imposed over the whole of $\partial\Omega$.

— in §2; *linear elasticity*, the displacements $\mathbf{U}(P)$ are given on $\partial\Omega$ by IA (2.34) in the problem of type IA (2.33).

[18] To simplify the writing we also write Γ for this boundary.

[19] This condition supposes that Ω is surrounded by an infinite *reservoir* of heat with an infinite conductivity. This image can be used in other physical domains.

— in §4; *electromagnetism.* We have seen that in electrostatics (Section 5.1) when we study the potential Φ in a domain Ω (which is a vacuum or else is occupied by a perfect dielectric) bounded by a perfect conductor, we have to solve the problem IA (4.92), with the potential Φ constant on the boundary $\partial\Omega$, which reduces to the problem (2.17) where $g = 0$ if the boundary $\partial\Omega$ is simply-connected. The frequently occurrying case in applications to electrostatics where $\partial\Omega$ is not simply-connected will be treated in Chap. II.

Similarly, we note that the wave-guide problem seen in IA, §4.6.2 led us to the problem IA (4.121) which contains a Dirichlet condition IA (4.121)(ii) (TM-wave).

— in §5; *neutronics.* The diffusion problem, with equation IA (5.31), in a domain Ω placed in a vacuum can be treated with the boundary condition IA (5.18) which is indeed of the type (2.17) (with $g = 0$).

— in §6; *quantum physics.* The simple examples given in IA, §6 lean on the axioms cited and were treated in the whole $\mathbb{R}^3$ space (or $\mathbb{R}^{3N}$ in the case of N particles). However, to model certain physical situations, we have pointed out that we are often led to treat problems involving a stationary Schrödinger equation IA (6.17) in which we imposed on the unknown wave function Φ the condition $\Phi = 0$ on $\partial\Omega$, the boundary of a domain Ω, which leads to the condition (2.17) (where $g = 0$) associated with the problem (2.15) (where $f = 0$). □

To apply the condition (2.17) in a problem of the type (2.14) or (2.15) like those we have mentioned in the examples, the value of the unknown u on the boundary $\partial\Omega$ must have a meaning which permits (2.17) to be written down. We shall see in Chap. IV, §4 that this is not always possible. For each problem (2.14) or (2.15) it will be necessary to verify the compatibility among

(1) the inclusion of the unknown u in the function space $\mathbb{X}$,
(2) the differential operator $P(D)$,
(3) the given function f,
(4) the boundary conditions (2.17) on the boundary $\partial\Omega$ of Ω.

To do this we shall need theorems[20] on *the trace of a distribution* (or function) *u on the boundary $\partial\Omega$* (of a domain Ω), u belonging to a function space $\mathbb{X}$, as well as $P_x(D)u$; these will permit us to make precise the boundary conditions on $\partial\Omega$ for u, which it is permissible to fix.

Neumann boundary conditions. We have similarly encountered in Chap. IA, the boundary condition

$$\frac{\partial u}{\partial \nu} = g \quad \text{on the boundary } \partial\Omega \text{ of } \Omega\,, \quad g \text{ given on } \partial\Omega\,, \tag{2.18}$$

(where $\partial/\partial\nu$ is the normal derivative with respect to $\partial\Omega$) called an *inhomogeneous Neumann boundary condition.* In the case $g(x) = 0\,,\quad \forall x \in \partial\Omega\,,$ this condition is called the *homogeneous Neumann boundary condition.* In particular we have encountered (2.18) associated with a problem (2.14) in which $P_x(D) = -\Delta$. More

[20] See Chap. IV, §4.

generally, when $P_x(D)$ was of the form (2.16) we have met with a boundary condition of the form

$$\sum_{i,j=1}^{n} a_{ij}\frac{\partial u}{\partial x_j}\nu_i = g \quad \text{on} \quad \partial\Omega\,, \quad g \text{ given on } \partial\Omega\,, \tag{2.19}$$

which in the rest of this work, we shall equally denote by

$$\left.\frac{\partial u}{\partial \nu_P}\right|_{\partial\Omega} = g \tag{2.19$'$}$$

(where ν_i $(i = 1, \ldots, n)$ denote the components of ν the unit normal vector to $\partial\Omega$ at x, exterior to Ω). The condition (2.19) will still be called more generally a *Neumann boundary condition.*
Now, let us cite some particular cases.

§1. *Fluid Mechanics.* The irrotational flow of an incompressible perfect fluid described in Sect. 5.2 of Chap. IA, §1, by the equations IA (1.31) for the potential Φ of the velocity $\mathbf{U}$ contains the boundary conditions IA (1.31)(iii) which are of the type (2.18), with $g = 0$, if Γ_1 [see IA (1.31)] is the whole of $\partial\Omega$.
We similarly have a condition of the type (2.18) in the problem IA (1.34) of the Sect. 5.3, but with $g \neq 0$.
In the case of the conduction of heat seen in Sect. 7 of IA, §1, the condition IA (1.67) is of the type (2.18). In the case of perfect "insulation", the right hand side of IA (1.67) is zero and this condition IA (1.67) is equivalent to (2.18) with $g = 0$.
§4. *Electromagnetism.* In electrostatics, the search for the potential Φ in the interior of a bounded domain Ω (occupied by a perfect dielectric or else a vacuum), whose boundary $\partial\Omega$ carries a given electric charge density $\rho(\mathbf{x})$, the domain exterior to $\bar{\Omega}$ being a perfect conductor, leads to a condition on Φ of type (2.18), as a consequence of the relations IA (4.47)(ii) and IA (4.91).
Similarly we note that the wave-guide problem treated in IA, §4.6.2, leads us to the problem IA (4.122) for the TE-waves which contains the Neumann condition IA (4.122)(ii).
§5. *Neutronics.* When the data Ω. D, Σ, etc., of the diffusion problem IA (5.31) (or IA (5.36) in the multigroup case) admit a plane of symmetry π, it follows as a consequence of the indications given in IA, §5.2.2, that problem IA (5.31) (or IA (5.36) in the multigroup case) can be treated in the half of Ω situated on one side of π and that the boundary conditions on π can be taken as in (2.18) (with $g = 0$). □

The Neumann conditions (2.19) (and the particular case (2.18)) require the existence of a normal derivative of u on the boundary; for that, the boundary must be sufficiently regular[21]. In addition we must give a meaning to this normal derivative which permits us to write (2.19). Just as we have already seen in connection with the Dirichlet conditions, we shall study that in Chap. IV, §4, with theorems giving the

[21] These notions will be made precise in Chap. IV.

trace on the boundary $\partial\Omega$ of the mathematical entities which interest us. This will lead us to study the compatability of the functional space $\mathbb{X}$ occurring in (2.14), of the operator $P_x(D)$ (given by (2.16)), of f, and of the boundary condition (2.19). This study will be resumed for general domains Ω, in the case $\mathbb{X} = L^2(\Omega)$, using the variational methods of Chap. VII.

Remark 3. *Compatability between the given f and g of a boundary value problem.* In the particular case $P_x(D) = -\Delta$, the problem (2.14) with ('2.18) is written

$$(2.20) \qquad \begin{cases} \text{(i)} \ -\Delta u = f \quad \text{in} \quad \Omega \subset \mathbb{R}^n, \\ \text{(ii)} \ \dfrac{\partial u}{\partial v} = g \quad \text{on} \quad \partial\Omega . \end{cases}$$

Suppose, in addition, that Ω is bounded and connected and that f and g satisfy:

$$(2.21) \qquad f \in \mathbb{X} = L^2(\Omega), \quad g \in L^2(\Gamma)^{22} .$$

We can show that it is necessary and sufficient for the existence of a solution u of (2.20) (with (2.21)), that the given functions f and g satisfy

$$(2.22) \qquad \int_\Omega f(x)\mathrm{d}x + \int_{\partial\Omega} g(x)\mathrm{d}\gamma = 0$$

(where $\mathrm{d}\gamma$ is the element of surface of $\partial\Omega$). We shall verify the necessity of (2.22) by applying Green's formula (see Chap. II). In particular, if $f = 0$, the given function g must satisfy

$$(2.22)' \qquad \int_{\partial\Omega} g\,\mathrm{d}\gamma = 0 .$$

This relation is known under the name of *Gauss' theorem*. It will be seen in Chap. II. □

Mixed boundary conditions. In going through the examples of Chap. IA, cited for Dirichlet conditions and for Neumann conditions we have seen that many of these contain the possibility of a Dirichlet condition on a part $\partial\Omega_1$ of the boundary of Ω and of a Neumann condition on its complement in $\partial\Omega$, denoted by $\partial\Omega_2$ (for example, the conduction of heat in IA, §1.7 with IA (1.66) on $\partial\Omega_1$ and IA (1.67) on $\partial\Omega_2$). We write these *boundary conditions*, called *mixed*

$$(2.22)'' \qquad \begin{cases} \text{(i)} \ u|_{\partial\Omega_1} = g_1, g_1 \quad \text{being given on} \quad \partial\Omega_1 \\ \text{(ii)} \ \left.\dfrac{\partial u}{\partial v}\right|_{\partial\Omega_2} = g_2{}^{23}, g_2 \quad \text{being given on} \quad \partial\Omega_2 \end{cases}$$

with

$$\partial\Omega_1 \cup \partial\Omega_2 = \partial\Omega .$$

---

[22] We can generalize with the aid of the Sobolev spaces of Chap. IV to $g \in H^{-\frac{1}{2}}(\Omega)$, for example.
[23] Or more generally (2.19).

The conditions (2.22)″ are significant only if $\partial\Omega_1$ and $\partial\Omega_2$ are not too "small." The mathematical meaning of the expression "small" will be made precise with the aid of notions of *capacities* (see Chap. II, §5); however, in many applications, the condition "$\partial\Omega_1$ and $\partial\Omega_2$ have non-empty interiors in $\partial\Omega$" is sufficient.

Remark 4. As in the case of the boundary conditions (2.17) or (2.18), we must make the mathematical meaning of (2.22)″ precise: this can be done again, for $L^2(\Omega)$, with the variational methods of Chap. VII.
For example, the problem

$$\Delta u = f \quad \text{in} \quad \Omega \tag{2.23}$$

with the boundary conditions (2.22)″, will always admit a "weak" or generalized solutions. Even if the given functions f, g_1, g_2 are very regular, there will in general be discontinuities in the first derivatives of the solution u along the interface between $\partial\Omega_1$ and $\partial\Omega_2$[24]. □

Robin's boundary condition. We have encountered in Chap. IA, §5.2, the boundary condition IA (5.17) for the diffusion equation. More generally, we cite the following boundary condition, called *Robin's condition*

$$\alpha(\mathbf{x})\,u(\mathbf{x}) + \beta(\mathbf{x})\frac{\partial u}{\partial \nu}(\mathbf{x}) = 0 \text{ [25]} \tag{2.24}$$

$\alpha(x)$ and $\beta(x)$ being two regular functions, given on $\partial\Omega$, not simultaneously zero.

Boundary conditions of oblique derivatives. The relation

$$\text{(2.25) (i)} \qquad \frac{\partial u}{\partial S}(\mathbf{x}) \overset{\text{def}}{=} \sum_{i=1}^{n} \frac{\partial u}{\partial x_i}(\mathbf{x}).S_i(\mathbf{x}) = g(\mathbf{x})\,, \quad \forall \mathbf{x} \in \partial\Omega\,,$$

where g is given on $\partial\Omega$ and $\mathbf{S}(\mathbf{x}) = (S_1(\mathbf{x}), \ldots, S_i(\mathbf{x}) \ldots, S_n(\mathbf{x}))$, $\forall \mathbf{x} \in \partial\Omega$, is a given vector field on $\partial\Omega$ with values in $\mathbb{R}^n$ such that

$$\text{(2.25) (ii)} \qquad \mathbf{S}(\mathbf{x}).\boldsymbol{\nu}(\mathbf{x}) \neq 0\,, \quad \forall \mathbf{x} \in \partial\Omega \text{ [26]}$$

associated with a problem (2.14) or (2.15) is called the boundary condition of *oblique derivatives.*
We note that the Neumann condition (2.19), corresponding to the operator $P_x(D)$ given by (2.16) is thus an oblique derivative condition, the vector field $\mathbf{S}(\mathbf{x})$ on $\partial\Omega$ being defined by

$$S_i(\mathbf{x}) = \sum_{j=1}^{n} a_{ji} v_j\,, \qquad i = 1 \text{ to } n\,. \tag{2.26}$$

The relation (2.25)(ii) clearly holds in this case since, taking account of (1.4) we find

[24] See, for example, Lions–Magenes [1].

[25] If $\alpha = 0$ (resp. $\beta = 0$), we recover the Neumann (resp. Dirichlet) condition.

[26] $\nu(x)$ is always the exterior normal to $\partial\Omega$ at x. When (2.25) is established, the vector field $S(x)$ is said to be *transverse* to $\partial\Omega$.

that

$$S(x).v(x) = \sum_{i,j=1}^{n} a_{ji}v_j v_i \geqslant \alpha\left(\sum_{i=1}^{n} |v_i|^2\right) = \alpha|v|^2 \neq 0 .$$

We shall find examples of boundary conditions of oblique derivative type in Chap. IA, §1.9 and in Bouligand *et al* [1], Fichera [1], Lienard [1], Lions [1].

Periodic boundary conditions: non-local boundary conditions. In the examples of neutronics, in Chap. IA, §5, we have cited the case of a problem whose data admitted a symmetry group: let us suppose, for example, that $\Omega = \mathbb{R}^3$, and that the neutronic system studied is a "network", repetition of a basic cell Ω_0 defined by

$$\Omega_0 = \{x = \{x_1, x_2, x_3\}\,;\;\; x_1 \in (0, l_1)\,,\;\; x_2 \in (0, l_2)\,,\;\; x_3 \in \mathbb{R}\}\,.$$

We suppose that the domain studied Ω $(= \mathbb{R}^3)$ is formed by all the domains deduced from Ω_0 by the translations (nl_1, ml_2), n and m being arbitrary integers, positive, negative or zero. Let us suppose that we study the multigroup diffusion equation (5.36) in $\mathbb{R}^2$, the data being such that, for example

$$D^g(x + nl_1 + ml_2) = D^g(x) \quad \forall \mathbf{x} \in \mathbb{R}^2\,, \quad n, m, \text{ integers}$$

and the same for the other data.

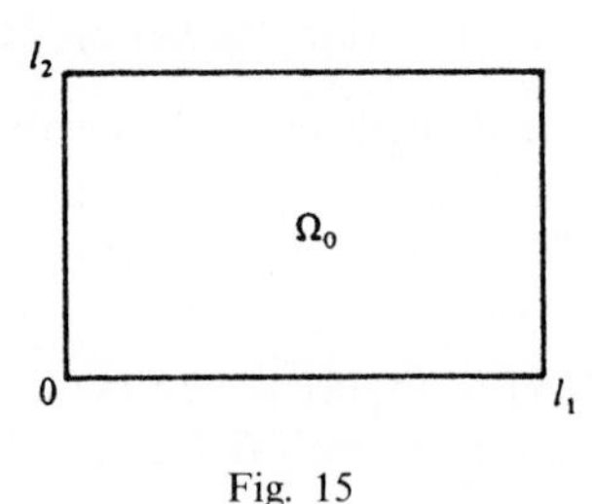

Fig. 15

The boundary conditions to which we are led by the physical considerations (see Bussac–Reuss [1]) are, for the unknown which we shall denote by U^g in the cell Ω_0:

$$\text{(2.27) (i)} \quad \begin{cases} U^g(0, x_2) = U^g(l_1\,, x_2)\,, & x_2 \in (0, l_2)\,, \\ U^g(x_1\,, 0) = U^g(x_1\,, l_2)\,, & x_1 \in (0, l_1)\,, \end{cases}$$

and

$$\text{(2.27) (ii)} \quad \begin{cases} \dfrac{\partial U^g}{\partial x_1}(0, x_2) = \dfrac{\partial U^g}{\partial x_1}(l_1\,, x_2) & x_2 \in (0, l_2) \\ \dfrac{\partial U^g}{\partial x_2}(x_1\,, 0) = \dfrac{\partial U^g}{\partial x_2}(x_1\,, l_2) & x_1 \in (0, l_1) \\ & g \in (1, 2, \ldots, m)\,. \end{cases}$$

We notice that these conditions are not local. Other non-local boundary conditions are likewise used in applications, especially:

$$\int_{\partial\Omega} u(\mathbf{x})\,\mathrm{d}\gamma = 0^{27} \quad \text{and} \quad \frac{\partial u}{\partial v} = \text{constant, not given on } \partial\Omega \tag{2.28}$$

or

$$\int_{\partial\Omega} \frac{\partial u}{\partial v}(\mathbf{x})\,\mathrm{d}\gamma = 0 \quad \text{and} \quad u = \text{constant,} \quad \text{not given on } \partial\Omega\,. \tag{2.29}$$

Similarly we point out the boundary condition

$$\frac{\partial u}{\partial v} + Ku = 0\,.$$

where K is an integral operator (to be compared with the capacity operator which will be studied in Chap. II, §5 which is used in electrotechnology (see Bossavit [1])).

Boundary conditions involving (tangential) derivatives of order greater than one. In certain applications, we are led to impose the boundary condition

$$\frac{\partial u}{\partial v} - \Delta_{\partial\Omega} u = 0\,, \qquad \forall \mathbf{x} \in \partial\Omega^{28} \tag{2.30}$$

where $\Delta_{\partial\Omega}$ is the Laplacian on $\partial\Omega$ (see Chap. V, §1), the boundary $\partial\Omega$ being a twice continuously differentiable variety. This condition is associated, for example, with (2.14) in which $P_x(D) = -\Delta$.

Boundary conditions for a vector-valued unknown. Let us end this list by noting that (2.27) could have been written in vector form for the vector

$$\mathbf{U} = (U^1, \ldots, U^g, \ldots, U^m)\,.$$

The vector form is similarly used in electromagnetism, in §4, where we have used it in particular in IA (4.77)(iii), (4.79)(iii), (4.80)(iii).

Conditions at infinity in a problem (2.14) *or* (2.15) *with* (2.16). We have examined a certain number of boundary conditions arising in applications for the problems (2.14) or (2.15), when the operator $P_x(D)$ was given by (2.16). When the domain Ω in which we wish to solve the problem (2.14) or (2.15) is unbounded physical considerations likewise lead us to impose constraints on the behavior at infinity of the solutions u.

In particular, this is the case for problems concerning radiation from *antennas* (see §4 and Roubine–Bolomey [1] and Roubine *et al.* [1]), where we arrive at problems which we shall study in Chapter II and which we state as follows: to find u

---

[27] $\mathrm{d}\gamma$ is the "surface" element of $\partial\Omega$.

[28] See Nussenzweig [1].

satisfying

$$(2.31)\quad \begin{cases} \text{(i) } \Delta u + k^2 u = 0 \quad \text{in} \quad \Omega \text{ (with } \mathbb{R}^n \backslash \bar{\Omega} = \Omega' \text{, bounded)},\\ \qquad \text{with } k \text{ given},\\ \text{(ii) } u = 0 \quad \text{on} \quad \partial\Omega,\\ \text{(iii) at infinity in } \Omega, u \text{ satisfies an } \textit{outgoing Sommerfeld radiation}\\ \qquad \textit{condition}, \text{ or else, an } \textit{incoming Sommerfeld radiation condition}^{29}. \end{cases}$$

We shall return to these problems (and their notation) in Chaps. II and XI.
We shall see that (2.31) possesses a unique solution u.
The electrostatic problems of Chap. IA, §4.5, in the case of an open set Ω, complement of a bounded domain Ω' will be seen in Chap. II. They lead to the problem: to find u satisfying:

$$(2.32)\quad \begin{cases} \Delta u = 0 \quad \text{in} \quad \Omega \subset \mathbb{R}^n,\\ \quad u = g \quad \text{on} \quad \partial\Omega \ (g \text{ given})\\ \text{with the condition at infinity :}\\ u(\mathbf{x}) \to 0 \quad \text{when} \quad |\mathbf{x}| \to \infty \quad \text{if} \quad n \geqslant 3,\\ u(\mathbf{x}) \ \text{ bounded when } \ |\mathbf{x}| \to \infty \quad \text{if} \quad n = 2. \end{cases}$$

Case of elliptic operators, degenerate on the boundary. We shall encounter in Chap. VIII, §2, in the context of Sturm–Liouville equations (and operators), operators such as the Legendre operator[30]

$$P_x(D) = -\frac{\mathrm{d}}{\mathrm{d}x}(1 - x^2)\frac{\mathrm{d}}{\mathrm{d}x}, \qquad x \in]-1, +1[\subset \mathbb{R},$$

which are of (elliptic) type "analogous" to the operator $P_x(D)$ featuring in (2.16), arising in problems (2.14) or (2.15). In such a problem we cannot impose conditions at $x = -1$ or $x = +1$ on the unknown function u.
More generally, when the operator (2.16) is such that

$$(2.33)\qquad a_{ij} = D(\mathbf{x})\,\delta_{ij}$$

where $D(x)$ is a positive function of x, regular in Ω, tending to 0 when $\mathbf{x} \in \Omega \to \mathbf{x}_0 \in \partial\Omega$, we are able, or not, to impose in (2.14) or (2.15) conditions on the boundary, according to the rapidity with which $D(\mathbf{x}) \to 0$ when $x \to \partial\Omega$. The operator $P_x(D)$ is then said to be *degenerate on the boundary*.

3.2. Problems (2.14) or (2.15) Associated with an Operator $P_x(D)$ of the 4th Order

We have seen in Chap. IA, §2, problems of plane deformations in the linear theory of elastic plates. To restrict ourselves here to the stationary case, we take equation

---

[29] See Chap. II.

[30] The Legendre operator arises naturally from the Laplacian in $\mathbb{R}^3$ when it is expressed in polar coordinates (see e.g. Trèves [1]).

IA (2.88). It is of the form (2.14) with

$$P_x(D) = \Delta\Delta \overset{\text{def}}{=} \Delta^2 .$$

In the case of $\partial\Omega$ (or $\partial\Omega_1$, a part of $\partial\Omega$) being *clamped*, we have two boundary conditions IA (2.83) of the type

$$(2.34) \qquad u = g_1 \quad \text{and} \quad \frac{\partial u}{\partial v} = g_2 \quad \text{on} \quad \partial\Omega \quad (\text{or else on } \partial\Omega_1)$$

where g_1 and g_2 are given

In the case of a boundary free of force on $\partial\Omega_2$ (complement of $\partial\Omega_1$ in $\partial\Omega$), we have the two conditions IA (2.84) which involve Δu and $\partial\Delta u/\partial v$ on $\partial\Omega_2$ (see IA (2.87)(iv)). We shall study such problems in the remainder of this work; the numerical aspect will be developed in Chap. XII.

3.3. Problems Associated with First Order Equations or with Systems of First Order Equations

The stationary problem of the *transport* of neutrons involves the transport equation IA (5.19)(i) and the boundary condition IA (5.19)(ii) (where, again, we take $g = 0$). We notice that this boundary condition does not involve a condition on the boundary for $\mathbf{v}$, which, from the mathematical point of view, is linked to the fact that equation (5.19)(i) contains no term involving differentiation with respect to the variable $\mathbf{v}$. We shall study these problems in Chap. XXI.

We note that many problems of *electromagnetism* which we have seen in this Chap. IB, §2, can be considered as boundary value problems associated with Maxwell's equations (which form a system of first order equations). We shall return to this topic in the future book devoted to wave equations.

4. Transmission Conditions

In the problems of *linear elasticity* of IA, §2, it happens frequently in applications that the domain Ω is occupied by several materials, whose elastic properties are

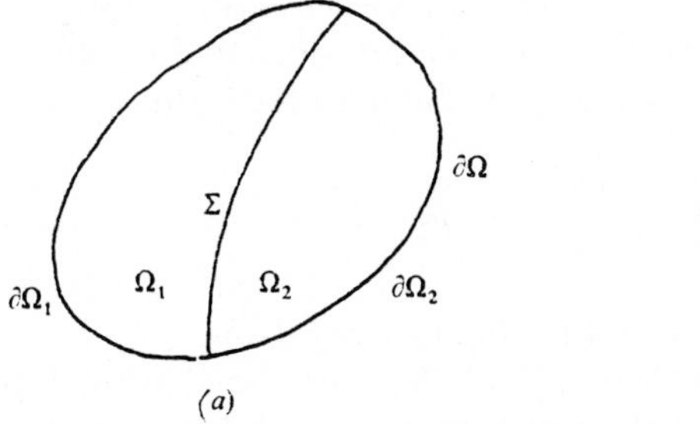

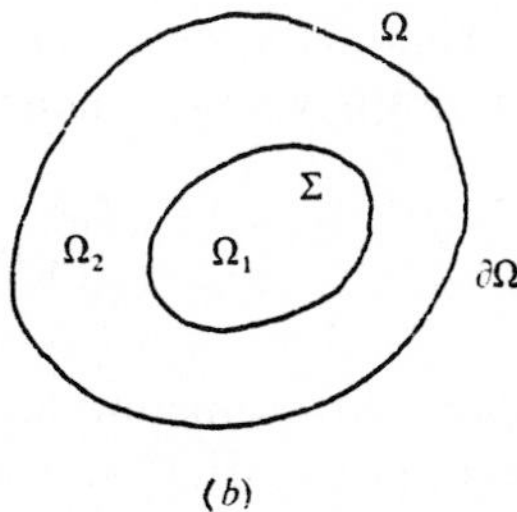

Fig. 16

different, joined together over the whole of a surface Σ as shown in the two cases (*a*) and (*b*) of Fig. 16, appropriate to the simple case $\Omega = \Omega_1 \cup \Omega_2 \cup \Sigma$.
This situation arises similarly in *electromagnetism* (IA, §4) and in *neutronics* (IA, §5) with the appropriate transpositions to electromagnetic or neutronic properties.
Let us examine this latter case and more particularly the *stationary problem of diffusion* of neutrons; for simplicity, let us take the monokinetic diffusion equation IA (5.39). Let us adopt, for example, the case of Fig. 16*a*, with the indices "1" and "2" respectively on each side of Σ. The physical considerations (see Bussac–Reuss [1]) impose on Σ the conditions:

$$(2.35)\qquad \begin{cases} \text{(i)}\ \varphi^1 = \varphi^2 \\ \text{(ii)}\ D^1(\mathbf{x})\dfrac{\partial \varphi^1}{\partial v} = D^2(\mathbf{x})\dfrac{\partial \varphi^2}{\partial v} \end{cases}$$

which expresses the continuity of the "flux" and of the 'current" of neutrons.
The problem to be solved is therefore to find φ^1 and φ^2, satisfying IA (5.39) in Ω_1 for φ^1, IA (5.39) in Ω_2 for φ^2, the boundary conditions IA (5.18) (for example) for φ^1 on $\partial\Omega_1$ and for φ^2 on $\partial\Omega_2$, and the transmission condition (2.35). We shall treat this type of problem later in this work. □

More generally, for the problems (2.14) or (2.15), with the operator $P_x(D)$ given by (2.16), the transmission conditions most frequently used in applications express the "continuity" of the unknown and of that of a flux across the transversal, namely

$$(2.36)\qquad \begin{cases} \text{(i)}\ u^1 = u^2 \quad \text{on} \quad \Sigma \\ \text{(ii)}\ \displaystyle\sum_{j=1}^{n} a_{ij}^1 \frac{\partial u^1}{\partial x_j} v_i = \sum_{j=1}^{n} a_{ij}^2 \frac{\partial u^2}{\partial x_j} v_i\,, \qquad i = 1 \text{ to } n\,, \quad \text{on } \Sigma \end{cases}$$

We note that (2.36)(i) can be considered as a junction of the functions (see Dirichlet's conditions) and that, taking account of (2.19)′, we may write (2.36)(ii) as

$$(2.37)\qquad \frac{\partial u^1}{\partial v_{P1}} = \frac{\partial u^2}{\partial v_{P2}} \quad \text{on} \quad \Sigma$$

which joins these derivatives (see Neumann conditions). Also, by using the vector fields $S^1(\mathbf{x})$ and $S^2(\mathbf{x})$ on Γ defined by (2.26) in terms of a_{ij}^1 and a_{ij}^2, and using the notation (2.25)(i), we can equally well write (2.36)(ii) in the form

$$(2.38)\qquad \frac{\partial u^1}{\partial S^1}(\mathbf{x}) = \frac{\partial u^2}{\partial S^2}(\mathbf{x}) \quad \text{on} \quad \Sigma\,,$$

which expresses the junction between the oblique derivatives.
We shall see in Chap. II, §8 and in Chap. VII, §2, that variational methods permit us to treat the problem (2.14) with (2.16), a discontinuity of the data (in Ω) on Σ and the transmission condition (2.36). □

Remark 5. We have formulated the boundary conditions and the transmission conditions for the stationary problems (2.14) and (2.15). The considerations so developed can be extended to the case of evolution problems. □

5. Problems Involving Time Derivatives of the Unknown Function u on the Boundary

5.1. The Time Derivatives of u Arising at the Same Time in the Equation and on the Boundary

Example 1. *Wave equation with a boundary condition involving derivatives with respect to the time.*
In an unbounded open set Ω of $\mathbb{R}^n$ with bounded regular boundary $\partial\Omega$ we seek u a solution of

$$(2.39)\quad \begin{cases} \text{(i)}\ \dfrac{\partial^2 u}{\partial t^2} - \Delta u = 0 \quad \text{in} \quad \Omega \times \mathbb{R}_+ , \\[2mm] \text{(ii)}\ \alpha \dfrac{\partial^2 u}{\partial t^2} - \dfrac{\partial u}{\partial \nu} + \beta u = -f , \qquad (\mathbf{x}, t) \in \partial\Omega \times \mathbb{R}_+ , \\[2mm] \text{(iii)}\ u(\mathbf{x}, 0) = u^0(\mathbf{x}) , \qquad \dfrac{\partial u}{\partial t}(\mathbf{x}, 0) = u^1(\mathbf{x}) , \qquad \mathbf{x} \in \Omega , \end{cases}$$

with α and β positive constants, f a given function on $\partial\Omega \times \mathbb{R}_+$, and ν being the normal directed towards the interior of Ω.

Application 1. *Vibrating string and harmonic oscillation*[31]. Suppose that a very long elastic string taken to be the half-line $]0, \infty[$, has its end $x = 0$ fixed to a support performing harmonic oscillations. Our problem is to find the transverse displacement $u(x, t)$ of each material point of the string. We see in Nuseenzveig [1] that u is the solution of:

$$(2.40)\quad \begin{cases} \text{(i)}\ \dfrac{\partial^2 u}{\partial t^2} - \dfrac{\partial^2 u}{\partial x^2} = 0 \quad \text{in} \quad \mathbb{R}_+ \times \mathbb{R}_+ \\[2mm] \text{(ii)}\ \dfrac{\partial^2 u}{\partial t^2} - 2\gamma \dfrac{\partial u}{\partial x} \to \omega^2 u = 0 \quad \text{for} \quad x = 0 , \quad t \in \mathbb{R}_+ \\[2mm] \quad (\gamma \text{ and } \omega \text{ given positive constants}) \\[2mm] \text{(iii)}\ u(x, 0) = u^0(x) , \qquad \dfrac{\partial u}{\partial t}(x, 0) = u^1(x) , \qquad x \in \mathbb{R} . \end{cases}$$

This problem will be solved in Chap. XVIII, §6.

[31] See Nussenzveig [1].

Application 2[32]**.** *Elastic waves produced by the displacements of a spherical source S_2 (in $\mathbb{R}^3$) in the exterior $\mathbb{R}^3 \backslash B$ of a unit ball B in $\mathbb{R}^3$.* This example again leads to the problem (2.39) as we shall see in Chap. XVIII, §6, where it will be solved.

5.2. Case in which the Time Derivative of u Appears Only on the Boundary

We have seen in fluid mechanics (see equation IA (1.97)) one of the examples in which the derivative of u with respect to the time does not occur in the equation in Ω, but only in that on the boundary $\partial\Omega$.
Let us give some other examples of this type:
to find a solution of the equation

$$\Delta u = 0 \quad \text{in} \quad \Omega \times \mathbb{R}_+ , \tag{2.41}$$

satisfying the conditions

$$\begin{cases} \text{(i)} \ \dfrac{\partial^2 u}{\partial t^2} + \dfrac{\partial u}{\partial v} = f \quad \text{on} \quad \Gamma \times \mathbb{R}^+ \\ \text{(ii)} \ u(\mathbf{x}, 0) = u^0(\mathbf{x}) , \quad \dfrac{\partial u}{\partial t}(\mathbf{x}, 0) = u^1(\mathbf{x}) \end{cases} \tag{2.42}$$

or, again,

$$\begin{cases} \text{(i)} \ \dfrac{\partial u}{\partial t} + \dfrac{\partial u}{\partial v} = f \qquad \text{on} \quad \Gamma \times \mathbb{R}^+ , \\ \text{(ii)} \ u(\mathbf{x}, 0) = u_0(\mathbf{x}) \qquad \text{on} \quad \Gamma . \end{cases} \tag{2.43}$$

This type of problem will be seen in Chap. XVIII, §6.

6. Problems of Time Delay

Example 1. We prescribe a scalar function $t \to \omega(t)$ (the delay) and denote by ψ the function

$$\psi(t) = t - \omega(t) , \qquad t \in]0, T[.$$

In addition, let

$$B(\mathbf{x}, t, \nabla_x) = \sum_{i=1}^{n} b_i(\mathbf{x}, t) \frac{\partial}{\partial x_i} + b_0(\mathbf{x}, t) I \qquad (I = \text{identity}).$$

We define for $u(\mathbf{x}, t)$, $(\mathbf{x}, t) \in \Omega_T \overset{\text{def}}{=} \Omega \times]0, T[$,

$$Mu(\mathbf{x}, t) = B\left(\mathbf{x}, t, \frac{\partial}{\partial \mathbf{x}}\right) u(\mathbf{x}, \psi(t)) ,$$

---

[32] See Rodean [1].

and we seek $u(x, t)$ satisfying

$$(2.44)\quad \begin{cases} \text{(i)}\ \dfrac{\partial u}{\partial t} - \Delta u + Mu = f \\ \text{(ii) a boundary condition on } \Gamma \\ \text{(iii)} \begin{cases} (a)\ u(\mathbf{x}, 0) = u_0(\mathbf{x}) \quad \mathbf{x} \in \Omega \\ (b)\ u(\mathbf{x}, t) = \tilde{u}(\mathbf{x}, t) \quad \text{in} \quad \Omega \quad \text{and} \quad t < 0\,,\ \tilde{u} \text{ given}\,. \end{cases} \end{cases}$$

— the condition (iii)(*a*) is called a Cauchy "point" condition,
— the condition (iii)(*b*) is called a "thick" Cauchy condition – indispensable because of the delay term which involves the past,
— such a "thick" problem can lead to a mixed problem with a given Cauchy "point" condition (see Chap. XVIII, §7).

Example 2. We suppose that on the boundary $\partial\Omega$ of the open set $\Omega \subset \mathbb{R}^n$, different conditions are imposed on the parts $\partial\Omega_1$ and $\partial\Omega_2$ (with $\partial\Omega = \partial\Omega_1 \cup \partial\Omega_2$); we suppose also that $\partial\Omega_1$ has a non-empty interior in $\partial\Omega$. We are given a family of operators $T_0(t)$, tangent to $\partial\Omega_2$.
We seek $u(x, t)$ satisfying:

$$(2.45)\quad \begin{cases} \text{(i)}\ \dfrac{\partial u}{\partial t} - \Delta u = f \quad \text{in} \quad \Omega_T = \Omega \times \left]0, T\right[\\ \text{(ii)} \begin{cases} u = 0 \quad \text{in} \quad \Omega \quad \text{and} \quad t < 0 \quad \text{(thick condition)} \\ u(\mathbf{x}, 0) = u_0(\mathbf{x}) \quad \text{in} \quad \Omega \end{cases} \\ \text{(iii)}\ u(\mathbf{x}, t) = 0 \quad \text{on} \quad \partial\Omega_1 \times \left]0, T\right[\\ \text{(iv)}\ \dfrac{\partial u}{\partial \nu} + T_0(t)u(\psi(t)) = 0 \quad \text{on} \quad \partial\Omega_2 \times \left]0, T\right[\,. \end{cases}$$

Here the delay arises in the boundary conditions. We could multiply the number of examples of this type encountered in phenomena of hysteresis type.

Example 3. We seek $u(x, t)$ satisfying

$$(2.46)\quad \begin{cases} \text{(i)}\ \dfrac{\partial u}{\partial t} - \Delta u + \displaystyle\int_{-\infty}^{t} K\left(\mathbf{x}, \sigma, t, \frac{\partial}{\partial \mathbf{x}}\right) u(\mathbf{x}, \sigma)\,d\sigma = f \\ \text{(ii) boundary condition on } \partial\Omega \\ \text{(iii)} \begin{cases} (a)\ u(\mathbf{x}, 0) = u_0(\mathbf{x}) \quad \text{in} \quad \Omega \\ (b)\ u(\mathbf{x}, t) = \tilde{u}(\mathbf{x}, t) \quad \text{in} \quad \Omega\,, \quad t < 0\,; \quad \tilde{u} \quad \text{given}\,. \end{cases} \end{cases}$$

— we have therefore a diffusion problem "with memory",
— a common case is

$$K(\mathbf{x}, \sigma, t, \nabla_x) = -\gamma(t - \sigma)\Delta$$

where γ is a twice continuously differentiable scalar function (see Chap. XVIII, §7).

In general $K\left(x, \sigma, t, \dfrac{\partial}{\partial x}\right)$ is a differential operator (kernel) of order less than 2. This can be a convolution, if

$$K\left(x, \sigma, t, \frac{\partial}{\partial x}\right) = \gamma(t - \sigma) . \qquad \square$$

Review of Chapter IB

The set of mathematical models obtained in Part A has given way to mathematical problems which have been classified in Part B. Several classifications appear:

Classification 1. *Classification into linear stationary problems and linear evolution problems.*

The first class brings together the mathematical problems in whose statement the time t does not appear. In the framework of Part IA which makes use of the language of physics and mechanics, we have introduced, by turns, the words stationary, statical, quasi-statical etc., these concepts not always having the same meaning in different disciplines. Nevertheless, it must do well for us, in the rest of this work, to adopt a unique nomenclature; to this end, all the mathematical problems of this nature will be called *stationary.*

Among these latter problems, we have distinguished two types of problems:

— problems containing the equation

$$Au = f \tag{1}$$

and boundary conditions, where A is a known linear operator, f a known function or distribution, u being the unknown function or distribution. These problems will be studied in Chap. II and in Chap. VII for A a linear differential operator and in Chap. XI for A a linear integral operator.

— problems containing the equation

$$Au - \lambda u = f \tag{2}$$

(and boundary conditions) where in addition, λ is an unknown complex constant. These problems will be capable of being treated by the techniques of Chap. VIII for A a linear differential operator and by those of Chap. XI for A a linear integral operator. The numerical methods relevant to all of these problems, will be treated in Chaps. XII and XIII.

The second class groups together the mathematical problems depending on time which we have called evolution problems. In particular, we have encountered problems containing equations of the type:

$$a\frac{\partial^2 u}{\partial t^2} + b\frac{\partial u}{\partial t} - Au = 0 , \tag{3}$$

with boundary conditions and initial conditions.

In the majority of cases envisaged in this Chap. I, the operator A was either a differential operator or an integro-differential operator. These evolution problems will be treated in Chap. XIV to Chap. XX (inclusive).

Classification 2. *Type of equations and types of boundary conditions.*
Starting from §1 of this Chap. IB it has appeared useful to make a rapid classification of the mathematical problems to be treated.
In the first instance (Chap. IB, §1), we have, in particular, distinguished the principal types of linear equations encountered in Chap. IA, by characterizing the operator A featuring in the left-hand side of each of the equations (1), (2), (3) above. Anticipating the study of the classification of linear differential operators carried out in Chap. V (where elliptic, parabolic, hyperbolic and Schrödinger operators are defined in general spaces of distributions) we have especially distinguished among the linear problems encountered in Part A:

— the heat equation or the diffusion equation (parabolic operators),
— the wave equation (hyperbolic operators),
— the Schrödinger equation,
— Laplace's equation (elliptic operators).

Other linear operators are studied in later chapters, particularly the linear integral operators; they will first be encountered in Chap. VI and the integral equations to which they give rise will be studied in Chap. XI.
In the second instance (§2), we have analysed other elements of the problems studied. A prime characteristic, often met with in Chap. IA, is that in the phenomena described by the equations (1) or (2) or (3) *a quantity* such as

$$\int_\Omega u\mathrm{d}x\,, \int_\Omega |u|^2\,\mathrm{d}x\,, \quad \int_\Omega |\nabla u|^2\,\mathrm{d}x\,, \quad \text{etc.},$$

must be finite, by the physical nature of the problem (the total energy must be finite, for example). This constraint on the unknown u of the problem leads to look for it (and hence to study the problem) in function spaces X, which are smaller than the mere consideration of the equations (1) or (2) or (3) alone will give. These spaces will, in particular, be presented and studied in Chap. IV.
A second characteristic of the mathematical problems encountered in Part A is the existence of *boundary conditions* on $\partial\Omega$ the boundary of the domain $\Omega(\Omega \subset \mathbb{R}^n_x)$ on which the problem is studied (Dirichlet conditions, Neumann conditions, mixed boundary conditions, conditions at infinity etc.). This new constraint can be taken into account in the definition of the operator A appropriate to equations (1), (2), (3) by restricting its domain of action (denoted by $D(A)$ with $D(A) \subset X$) to functions which satisfy it. This concept will be defined and studied in Chap. VI and in Chap. VIII.
In the first instance, we shall study in a "classical" framework stationary problems like (1), A being the Laplacian operator, so as, in particular, to survey the possibilities and the limitations of the "classical" methods: this is the object of Chap. II.

Chapter II. The Laplace Operator

Introduction

This chapter studies, by methods based on *classical function spaces* (continuously differentiable functions), the theory of harmonic functions and of other classical concepts attached to the Laplacian operator Δ: Newtonian potential, capacities, the Dirichlet problem; the radiation conditions – associated with the operator $\Delta + k^2$ – are similarly introduced.
The Laplacian operator admits a very rich structure, and is the source of numerous problems and numerous generalizations; certain of these are indicated here in sections (marked by $*$) which can be passed over, at least on a first reading.
"Functional" aspects, based on the calculus of variations and on the introduction of the Sobolev spaces are not used here (or scarcely touched on implicitly); they will be taken again, in a context going far beyond the Laplacian operator, in the following chapters.

§1. The Laplace Operator

1. Poisson's Equation

We give the name *the Laplace operator* or *the Laplacian* of dimension n, to the linear differential operator with constant coefficients in $\mathbb{R}^n$

$$\Delta = \frac{\partial^2}{\partial x_1^2} + \cdots + \frac{\partial^2}{\partial x_n^2} .$$

From an elementary point of view, being given a point $x = (x_1, \ldots, x_n) \in \mathbb{R}^n$ and a function u defined in the neighbourhood of x.

$$\Delta u(x) = \frac{\partial^2 u}{\partial x_1^2}(x) + \cdots + \frac{\partial^2 u}{\partial x_n^2}(x)$$

is defined under the condition of the existence of the partial derivatives $\partial^2 u(x)/\partial x_1^2, \ldots, \partial^2 u(x)/\partial x_n^2$; that is to say on the hypothesis of the existence of the *gradient of u*

$$\operatorname{grad} u = \left(\frac{\partial u}{\partial x_1}, \ldots, \frac{\partial u}{\partial x_n}\right),$$

in the neighbourhood of x, we have

$$\Delta u(x) = \operatorname{div}(\operatorname{grad} u)(x)$$

where for a field $p = (p_1, \ldots, p_n)$ defined in the neighbourhood of x, the *divergence of* p at x is defined by

$$\operatorname{div} p(x) = \frac{\partial p_1}{\partial x_1}(x) + \cdots + \frac{\partial p_n}{\partial x_n}(x),$$

with the assumption that the partial derivatives $\frac{\partial p_1}{\partial x_1}(x), \ldots, \frac{\partial p_n}{\partial x_n}(x)$, exist.

We notice immediately the fundamental property of *the invariance of the Laplacian under a Euclidean transformation.* A Euclidean transformation is an isometry of the *Euclidean distance in* $\mathbb{R}^n$

$$d(x, y) = |x - y| = ((x_1 - y_1)^2 + \cdots + (x_n - y_n)^2)^{1/2},$$

that is to say, a mapping T of a part D of $\mathbb{R}^n$ into $\mathbb{R}^n$ satisfying

$$|T(x) - T(y)| = |x - y| \qquad \forall x, y \in D .$$

It can be shown that a Euclidean transformation defined on a connected open set in $\mathbb{R}^n$ is composed of a translation and of an orthogonal transformation (see Dieudonné [2]).

We then have the following property, which is, moreover, characteristic of the Laplacian (see Chap. V, §2).

Proposition 1. *Let $x \in \mathbb{R}^n$ and let T be a Euclidean transformation defined in the neighbourhood of x and u a twice differentiable function defined in the neighbourhood of x. Then $Tu \overset{\text{def}}{=} u \circ T^{-1}$, defined in the neighbourhood of $y = Tx$ is twice differentiable in y and*

$$\Delta(Tu)(T(x)) = \Delta u(x) . \tag{1.1}$$

This follows immediately from the fact that the Euclidean transformation T of an open set Ω is of class $\mathscr{C}^\infty$ satisfying:

$T'(x)$ is an orthogonal transformation, $T''(x) = 0$ for all $x \in \Omega$, and the formula for the differentiation of composition of functions applied to the Laplacian gives

Lemma 1. *For $y \in \mathbb{R}^n$, let $h = (h^1, \ldots, h^n)$ be defined in the neighbourhood of y and be twice differentiable at y; also let u be defined in the neighbourhood of $x = h(y)$ and be twice differentiable at x. Then $u \circ h$ is twice differentiable at y, and*

$$\Delta\,(u \circ h)(y) = \sum_{k,l} \operatorname{grad} h^k(y) . \operatorname{grad} h^l(y) \frac{\partial^2 u}{\partial x_l \partial x_k}(x) + \sum_k \Delta h^k(y) \frac{\partial u}{\partial x_k}(x) . \tag{1.2}$$

In this formula, in the classical manner the dot between $\operatorname{grad} h^k(y)$ and $\operatorname{grad} h^l(y)$ denotes the scalar product of these two vectors: for two vectors $p = (p_1, \ldots, p_n)$ and $q = (q_1, \ldots, q_n)$ the scalar product is defined by the equation

$$p . q = p_1 q_1 + \cdots + p_n q_n .$$

We now consider Ω an open set in $\mathbb{R}^n$. For every positive integer $m = 0, 1, \ldots$, the Laplacian Δ maps $\mathscr{C}^{m+2}(\Omega)$ in $\mathscr{C}^m(\Omega)$, in the sense in which for $u \in \mathscr{C}^{m+2}(\Omega)$, the function

$$\Delta u: x \in \Omega \to \Delta u(x)$$

is of class $\mathscr{C}^m$. Thus Δ maps $\mathscr{C}^\infty(\Omega)$ into itself.
Being given a function f, on a prescribed set Ω, we give the name *Poisson's equation* to the partial differential equation

$$\Delta u = f \quad \text{on} \quad \Omega\,.$$

From an elementary point, a solution is a function u defined on Ω and possessing partial derivatives $\dfrac{\partial^2 u}{\partial x_k^2}$ on Ω and satisfying

$$\Delta u(x) = f(x)\,, \quad \forall x \in \Omega$$

In fact, to make use of the invariance under a Euclidean transformation, we shall need to suppose that u is twice differentiable on Ω. This elementary point of view is sometimes interesting: meanwhile consideration of the case $n = 1$, where Poisson's equation reduces to the ordinary differential equation $u'' = f$, shows clearly that this framework is not satisfying: the class of twice differentiable functions possesses very few properties. For that reason, we state the

Definition 1. Being given a function f defined on Ω, we say that u is a *classical solution* of Poisson's equation

$$\Delta u = f \quad \text{on} \quad \Omega$$

if $u \in \mathscr{C}^2(\Omega)$ satisfies

$$\Delta u(x) = f(x) \qquad \forall x \in \Omega\,.$$

The above definition imposes, for the existence of a solution of class $\mathscr{C}^2(\Omega)$, the continuity of the given f. Even in the case $n = 1$, the use of non-continuous data is interesting in describing the differential equation $u'' = f$. For example to remain in the mathematical domain, the *convex* functions on an open set of Ω of $\mathbb{R}$ can be characterised as the solutions of the differential equation

$$u'' = f \quad \text{on} \quad \Omega$$

with f a positive Radon measure on Ω. The notion of a solution must then be taken in the sense of distributions.
Let us recall that a distribution on Ω is a continuous linear form on the space $\mathscr{D}(\Omega)$ of $\mathscr{C}^\infty$-functions with compact support in Ω. Following the elementary theory of distributions[1], we formulate the

Definition 2. Being given a distribution f on Ω, we call a *distribution solution* of Poisson's equation

$$\Delta u = f \quad \text{on} \quad \Omega\,,$$

[1] See Appendix "Distributions" at the end of vol. 2, for the concept of a distribution.

each distribution u on Ω satisfying

$$\langle u, \Delta\zeta \rangle = \langle f, \zeta \rangle \qquad \forall \zeta \in \mathscr{D}(\Omega) .$$

Remark 1. We recall that a function $f \in \mathscr{C}^0(\Omega)$ defines a distribution

$$T_f \colon \zeta \in \mathscr{D}(\Omega) \to \int \zeta(x) f(x) \, \mathrm{d}x \tag{1.3}$$

and that the mapping $f \to T_f$ is injective with the result that we can identify f with the distribution that it defines. Being given $u \in \mathscr{C}^2(\Omega)$, we have

$$\langle u, \Delta\zeta \rangle = \int u(x) \Delta\zeta(x) \, \mathrm{d}x = \int \Delta u(x) \zeta(x) \, \mathrm{d}x = \langle \Delta u, \zeta \rangle$$

with the result that for $f \in \mathscr{C}^0(\Omega)$, *$u$ is a classical solution iff $u \in \mathscr{C}^2(\Omega)$ and is a distribution solution.* In the case $n = 1$ for a given $f \in \mathscr{C}^0(\Omega)$ every distribution solution of the differential equation $u'' = f$ is a classical solution. This situation is no longer true in dimension $n \geqslant 2$: being given an open set $\Omega \subset \mathbb{R}^n$ with $n \geqslant 2$, there exist functions $f \in \mathscr{C}^0(\Omega)$ such that there do **not** exist classical solutions of the corresponding Poisson's equation (see §3, Remark 5). □

From Proposition 1, being given T a Euclidean transformation of Ω, if u is a classical solution of $\Delta u = f$ on Ω, then v defined by $v = Tu = u \circ T^{-1}$ is a classical solution of $\Delta v = Tf$ on $T(\Omega)$. This extends to the body of distribution solutions through the use of the notion of an *image of a distribution*: let us consider a diffeomorphism h of Ω into $\mathbb{R}^n$; for $f \in \mathscr{C}^0(\Omega)$, we have for all $\zeta \in \mathscr{D}(h(\Omega))$,

$$\langle hf, \zeta \rangle = \int f(h^{-1}(y)) \zeta(y) \, \mathrm{d}y = \int f(x) \zeta(h(x)) \, |\mathrm{Det}\, h'(x)| \, \mathrm{d}x$$

Supposing that h is of class $\mathscr{C}^\infty$, $\zeta \circ h \, |\mathrm{Det}\, h'| \in \mathscr{D}(\Omega)$ and hence[2]

$$\langle hf, \zeta \rangle = \langle f, \zeta \circ h \, |\mathrm{Det}\, h'| \rangle . \tag{1.4}$$

For a distribution f on Ω, this formula defines a distribution on $h(\Omega)$ and serves as the definition of *the distribution hf, image of f by h.*
In the case of a Euclidean transformation T, $|\det T'| = 1$ and

$$\langle Tf, \zeta \rangle = \langle f, \zeta \circ T \rangle .$$

Since $\Delta(\zeta \circ T) = (\Delta\zeta) \circ T$, we have from Proposition 1:

Proposition 2. *Let T be a Euclidean transformation of Ω, f a distribution on Ω and u a distribution solution of $\Delta u = f$ on Ω. Then $v = Tu$ is a distribution solution of $\Delta v = Tf$ on $T(\Omega)$.*

Remark 2. For greater clarity we have considered general distributions. In fact, in practice, we shall use only the following four types of distributions:
(1) the continuous functions: they are defined everywhere; two continuous functions equal in the sense of distributions, are equal everywhere;

---

[2] In this Chap. II h' denotes the derivative of h.

(2) the locally integrable functions (in the sense of Lebesgue): they are defined almost everywhere (except on a set negligible in the sense of Lebesgue); they define a distribution by the same formula (1.3); two locally integrable functions equal in the sense of distributions (that is to say defining the same distribution) are equal almost everywhere;
(3) the Radon measures or distributions of order 0; an important example is the Dirac measure (or mass) at a point $x \in \Omega$:

$$\delta_x \colon \zeta \in \mathscr{D}(\Omega) \to \zeta(x) ;$$

(4) distributions of order 1, that is to say of the form

$$\zeta \in \mathscr{D}(\Omega) \to \langle f_0, \zeta \rangle + \left\langle f_1, \frac{\partial \zeta}{\partial x_1} \right\rangle + \cdots + \left\langle f_n, \frac{\partial \zeta}{\partial x_n} \right\rangle$$

where $f_0, f_1, \ldots, f_n$ are Radon measures on Ω. □

2. Examples in Mechanics and Electrostatics

Characteristic of the property of invariance under a Euclidean transformation, the Laplacian occurs in very many physical problems. Here we make precise the situations in mechanics and in physics in which Poisson's equation in $\mathbb{R}^3$ occurs directly.
In *mechanics*, a certain quantity of matter distributed in the space $\mathbb{R}^3$ with a (mass) *density* $\rho(x)$, gives rise at each point $x \in \mathbb{R}^3$ a *gravitational potential* $v_g(x)$ which satisfies

$$\Delta v_g = k_g \rho$$

where k_g is a constant depending on the chosen system of units. The *gravitational force* $f(x)$ being exerted on a mass m placed at $x \in \mathbb{R}^3$ is given by

$$f = -m \operatorname{grad} v_g .$$

The work W done by the force of gravity when we deplace the mass m along a Jordan curve $\gamma\colon [0, 1] \to \mathbb{R}^3$ is given by

$$W = \int_\gamma f(x)\,\mathrm{d}x = \int_0^1 f(\gamma(t))\gamma'(t)\,\mathrm{d}t = -\int_0^1 m \operatorname{grad} v_g(\gamma(t))\gamma'(t)\,\mathrm{d}t$$

$$= -m \int_0^1 \frac{\mathrm{d}}{\mathrm{d}t} v_g(\gamma(t))\,\mathrm{d}t = m(v_g(\gamma(0)) - v_g(\gamma(1))) .$$

In other words, the difference $v_g(x_1) - v_g(x_2)$ of the gravitational potential is the work done by the force of gravity when a unit mass is displaced along an *arbitrary* curve with an initial point x_1 and end-point x_2.
Let us now state precisely the *dimensions* with respect to the fundamental quantities: length (L), mass (M), time (T):
the mass density ρ has the dimension ML^{-3}, the force of gravity f has the dimension MLT^{-2} with the result that grad v_g must have the dimension LT^{-2} and the gravitational potential has the dimension L^2T^{-2}.

The constant k_g has the dimension $L^3M^{-1}T^{-2}$.
In fact (for reasons which will appear soon), we shall put

$$k_g = 4\pi G$$

where G, which has the same dimension $L^3M^{-1}T^{-2}$, is called the *universal gravitational constant*.
In the *c.g.s. system of units* where the fundamental units are the centimetre (cm) for L, the gram (g) for M and the second (s) for T the constant G is given by $G = 6.673 \times 10^{-3}\,\text{cm}^3\,\text{g}^{-1}\,\text{s}^{-2}$. In the *S.I. system* (*Système International*)[3], in which the fundamental units are the metre (m) for L, the kilogram (kg) for M, and the second (s) for T, we have

$$G = 6.673 \times 10^{-5}\,\text{m}^3\,\text{kg}^{-1}\,\text{s}^{-2}\,.$$

The unit of energy or of work is then the Joule ($1\text{J} = 1\,\text{m}^2\,\text{kg}\,\text{s}^{-2}$) and the gravitational potential is expressed in $\text{J}\,\text{kg}^{-1}$.
In *electrostatics*, a certain quantity of electric charge distributed in $\mathbb{R}^3$ with a density (of electric charge) $\rho(x)$ gives rise at each point an *electrostatic* (or *Coulomb*) *potential* v_c, solution of Poisson's equation

$$\Delta v_c = -k_c\rho\,.$$

The *electrostatic force* $f(x)$ exerted on an (electric) charge q placed at x is given by

$$f = -q\,\text{grad}\,v_c\,.$$

To the fundamental quantities of mechanics L, M, T must now be added another quantity which can be either the charge (Q) or the intensity (I), these two quantities being related by $Q = IT$. The charge density ρ has the dimension QL^{-3} and the electrostatic force MLT^{-3} with the result that:
the electrostatic potential has the dimension $ML^2T^{-2}Q^{-1} = ML^2T^{-3}I^{-1}$,
the constant k_c has the dimension $ML^3T^{-2}Q^{-2} = ML^3T^{-4}I^{-2}$.
Poisson's equation for the electrostatic potential is valid in a homogeneous, perfect dielectric[4] (see Chap. IA, §4) that we suppose here to occupy the whole space, the constant k_c depends on the permittivity of the medium[4]; we define the *permittivity constant* by $\varepsilon = 1/k_c$ (we consider also $\varepsilon' = 4\pi\varepsilon$) which has the dimension $M^{-1}L^{-3}T^2Q^2 = M^{-1}L^{-3}T^4I^2$. We have $k_c = 1/\varepsilon = 4\pi/\varepsilon'$.
We denote by ε_0 (resp. ε'_0), the *permittivity constant of the vacuum*. In the *system of units u.e.s.c.g.s.*, we use the units of the c.g.s. system and *we fix the permittivity constant of the vacuum*

$$\varepsilon_0 = \frac{1}{4\pi} \quad (\text{resp. } \varepsilon'_0 = 1)\,.$$

That defines the units of charge ($Q = M^{\frac{1}{2}}L^{\frac{3}{2}}T^{-1}$) and of intensity ($I = M^{\frac{1}{2}}L^{\frac{3}{2}}T^{-2}$).

---

[3] This is the official system in France.
[4] See Chap. IA, §4.

We then have for the vacuum

$$k_{c,o} = 4\pi .$$

In the *S.I. system of units*, we use in addition to the mechanical units, the ampère (A) as the unit of intensity. The unit of electric charge is the coulomb (C), 1C = 1 ampère second; the unit of electrostatic potential is the volt (V); 1V = 1 joule/second.
These correspond numerically[5] to

$$\varepsilon_0 = 8{,}8542 \times 10^{-12}\,\mathrm{m}^{-3}\,\mathrm{kg}^{-1}\,\mathrm{s}^4\,\mathrm{A}^2\,,^6$$
$$k_{c,o} = 1{,}1294 \times 10^{11}\,\mathrm{m}^3\,\mathrm{kg}\,\mathrm{s}^{-4}\,\mathrm{A}^{-2}\,.$$

3. Green's Formulae: The Classical Framework

We do not propose to give a full account of differential geometry here, but only to specify clearly the background against which we shall use the formulas of Green, basic tool for the study of the Laplacian and, more generally, for the study of boundary values.
We shall consider a bounded open set Ω with boundary Γ. The point of departure is *Ostrogradski's formula* for a field $p = (p_1, \ldots, p_n)$

$$\int_\Omega \operatorname{div} p\,\mathrm{d}x = \int_\Gamma p\,.\,n\,d\gamma \tag{1.5}$$

whose significance we shall make precise below.
Considering functions u, v and applying Ostrogradski's formula to the field $p = v\,\mathrm{grad}\,u$ and making use of the identity

$$\operatorname{div}(v\,\mathrm{grad}\,u) = \mathrm{grad}\,u\,.\,\mathrm{grad}\,v + v\,\Delta u\,, \tag{1.6}$$

we obtain

$$\int_\Omega v\,\Delta u\,\mathrm{d}x = \int_\Gamma v\,\frac{\partial u}{\partial n}\,d\gamma - \int_\Omega \mathrm{grad}\,u\,.\,\mathrm{grad}\,v\,\mathrm{d}x\,, \tag{1.7}$$

where we have put

$$\frac{\partial u}{\partial n} = \mathrm{grad}\;u\,.\,n\,. \tag{1.8}$$

We refer to (1.7) as *Green's formula for integration by parts*. In effect in the case $n = 1$, $\Omega =]a, b[$, the formula (1.7) can be written

$$\int_a^b u''(x)v(x)\,\mathrm{d}x = u'(b)v(b) - u'(a)v(a) - \int_a^b u'(x)v'(x)\,\mathrm{d}x\,.$$

---

[5] By using $\varepsilon_0\mu_0 c^2 = 1$, $\mu_0 = 4\pi\,.\,10^{-7}$ and the value of c, the speed of light (see Fournet [1]).
[6] Or again Fm^{-1} where F, the farad, is the unit of capacity. See §5.

Finally, interchanging u and v in the formula (1.7) and subtracting the two formulae, we obtain *Green's formula*

$$\int_\Omega (u\,\Delta v - v\,\Delta u)\,\mathrm{d}x = \int_\Gamma \left(u\frac{\partial v}{\partial n} - v\frac{\partial u}{\partial n}\right)\mathrm{d}\gamma\,. \tag{1.9}$$

3a. Elementary Theory

In the classical theory, which we are now considering, we suppose that Ω is a *regular open set* in the sense in which its boundary Γ is a hypersurface of class $\mathscr{C}^m$ with $m \geqslant 1$, Ω being locally on one side of this surface. In other words, for all $a \in \Gamma$, we can find an orthonormal reference frame R, a neighbourhood (open set) U of a in $\mathbb{R}^n$, an open set $\mathcal{O}$ of $\mathbb{R}^{n-1}$ and a function α of class $\mathscr{C}^m$ on $\mathcal{O}$, such that denoting by $(x', x_n') = (x_1', \ldots, x_{n-1}', x_n')$ the coordinates in the reference frame R, we would have

$$\Gamma \cap U = \{(x', \alpha(x'));\quad x' \in \mathcal{O}\}$$
$$\Omega \cap U = \{x \in U;\, x_n' < \alpha(x')\}\,.$$

The data of (R, U) defines $(\mathcal{O}, \alpha)$ perfectly; we can always suppose

$$U = \{(x', \alpha(x') + t);\quad x' \in \mathcal{O},\quad |t| < \delta\}\,.$$

We shall say that (R, U) defines a *normal local parametrisation* of Γ (in the neighbourhood of a) and we shall employ, without recalling the definitions, the notations $\mathcal{O}$, α, δ, (x', x_n').

For all $x \in \Gamma$, $n(x)$ denotes *the unit normal vector to Γ, exterior to Ω*. In a normal local parametrisation, for $x = (x', \alpha(x'))$, the vector $n(x)$ is given in the base of the reference frame R by

$$n(x) = \frac{(-\operatorname{grad}\alpha(x'), 1)}{(1 + |\operatorname{grad}\alpha(x')|^2)^{1/2}} \tag{1.10}$$

where here the gradient and its norm are taken in $\mathbb{R}^{n-1}$.

In particular the *normal field*: $x \in \Gamma \to n(x)$ is of class $\mathscr{C}^{m-1}$.

We denote by $\mathrm{d}\gamma$ the *surface measure* or the *element of area* on Γ. It is sufficient to define it locally; in a normal local parametrisation, we have for every function ζ continuous on Γ with compact support in U.

$$\int_\Gamma \zeta\,\mathrm{d}\gamma = \int_{\mathcal{O}} \zeta(x', \alpha(x'))(1 + |\operatorname{grad}\alpha(x')|^2)^{1/2}\,\mathrm{d}x'\,. \tag{1.11}$$

We then prove classically Ostrogradski's formula (1.5) for a field p of class $\mathscr{C}^1$ in the neighbourhood of $\bar{\Omega}$: first of all we localise, by using a partition of unity, reducing in a normal local parametrisation (R, U) to the case $\operatorname{supp} p \subset U$ (the case $\operatorname{supp} p \subset \Omega$ is trivial); in the reference frame R we denote $p(x', \alpha(x') + t) = (p'(x', t),$

$p'_n(x', t)$); using (1.10) and (1.11) we find that the formula (1.5) can be written

$$\int_{\mathcal{O}} \mathrm{d}x' \int_{-\delta}^{0} (\operatorname{div} p)(x', \alpha(x') + t)\,\mathrm{d}t = \int_{\mathcal{O}} (p'_n(x', 0) - p'(x', 0) \,.\, \operatorname{grad} \alpha(x'))\,\mathrm{d}x' \,.$$

We verify it immediately by using $\operatorname{supp}(p', p'_n) \subset \mathcal{O} \times]-\delta, 0]$ and

$$(\operatorname{div} p)(x', \alpha(x') + t) = (\operatorname{div}_{x'} p'(x', t) - \frac{\partial p'}{\partial t}(x', t) \,.\, \operatorname{grad} \alpha(x') + \frac{\partial p'_n}{\partial t}(x', t)) \,.$$

We deduce the formulae of Green (1.7) (resp. (1.9)) for functions u of class $\mathscr{C}^1$ (resp. $\mathscr{C}^2$) in the neighbourhood of $\bar{\Omega}$.

This formulation, which we shall call "elementary" of the formulae of Ostrogradski and Green is often not sufficient for applications: the regularity of the neighbourhood of the boundary, the regularity even of the functions on Ω or of the open set Ω are constraints which we should like to avoid. In §1.3c we give a statement within the classical framework: Ω regular and functions regular in Ω. We shall extend these considerations to the case of non-regular open sets Ω and to non-regular functions in Ω, in §6.

3b. Normal Differentiation

We propose to define $p.n$ and $\partial u/\partial n$ on Γ for fields p and functions u defined only on Ω. We remain always with the classical framework: *we suppose that Ω is regular, p a continuous field on Ω and u a function of class $\mathscr{C}^1$ on Ω and we define $p.n$ and $\partial u/\partial n$ as continuous functions on Γ*. Being given $a \in \Gamma$ and a *unit vector e leaving Ω* at a, that is to say such that $\cos(e, n(a)) > 0$, we can consider

$$(1.12) \qquad \lim_{\substack{x \to a,\, x \in \Gamma \\ t \to 0,\, t > 0}} p(x - te).n(x)$$

and prove the

Lemma 2. *If the limit (1.12) exists for at least one unit vector e leaving Ω at a, then it exists for all such vectors. More precisely, being given a compact set K of Γ such that for all $a \in K$ the limit (1.12) exists for at least one unit vector e leaving Ω at a, then for all $\varepsilon > 0$, the limit (1.12) exists* **uniformly** *for $a \in K$ and e a unit vector satisfying* $\cos(e, n(a)) \geqslant \varepsilon$.

This lemma permits us to *define $p.n(a)$ for $a \in \Gamma$ by*

$$(1.13) \qquad p.n(a) = \lim_{\substack{x \to a,\, x \in \Gamma \\ t \to 0,\, t > 0}} p(x - tn(a)).n(x)$$

if this limit exists and to prove

Proposition 3. *The function $p.n$ is continuous on the set of the points $a \in \Gamma$ for which it is defined (by (1.13)). More precisely, being given $\tilde{n}$ continuous on $\bar{\Omega}$ extending the normal field n, $p.n$ is the trace on Γ of the extension by continuity to $\bar{\Omega}$ of the function $p.\tilde{n}$ (continuous on Ω).*

We define $\partial u/\partial n$ by applying Proposition 3 to $p = \operatorname{grad} u$: *at a point $a \in \Gamma$, $\partial u/\partial n$ is defined by*

$$\frac{\partial u}{\partial n}(a) = \lim_{\substack{x \to a,\, x \in \Gamma \\ t \to 0,\, t > 0}} \operatorname{grad} u(x - tn(a)) . n(x) \tag{1.14}$$

or again from Proposition 3, being given $\tilde{n}$ continuous on $\bar{\Omega}$ extending the normal field n, *$\partial u/\partial n$ is the trace on Γ of the extension by continuity to $\bar{\Omega}$ of the function* $\tilde{n}.\operatorname{grad} u$ (continuous on Ω), this trace being independent of the choice of the extension. By construction, *if $\partial u/\partial n$ is defined on Γ, it is continuous on Γ.* We can connect this definition with the elementary notion of the *directional derivative* and show that if $\partial u/\partial n$ is defined on Γ, then

(1) u extends by continuity to $\bar{\Omega}$.

(2) $\displaystyle\frac{\partial u}{\partial n}(a) = \lim_{\substack{t \to 0 \\ t > 0}} \frac{u(a - tn(a)) - u(a)}{t}$, the limit being uniform for $a \in \Gamma$.

We shall use *the notation $\mathscr{C}^1_n(\bar{\Omega})$ to denote the class of functions $u \in \mathscr{C}^1(\Omega)$ with $\partial u/\partial n$ defined on Γ*; hence for $u \in \mathscr{C}^1_n(\bar{\Omega})$, u extends by continuity to $\bar{\Omega}$ ($u \in \mathscr{C}^0(\bar{\Omega})$) and for each continuous extension $\tilde{n}$ of the normal field to $\bar{\Omega}$, $\tilde{n}.\operatorname{grad} u$ extends by continuity to $\bar{\Omega}$ ($\tilde{n}.\operatorname{grad} u \in \mathscr{C}^0(\bar{\Omega})$).

3c. Statement of the Classical Formulae of Green

With these definitions of $p.n$ and $\partial u/\partial n$ which give a meaning to the integrals over Γ in the formulae of Ostrogradski and Green, we can extend the elementary domain (p of class $\mathscr{C}^1$ in the neighbourhood of $\bar{\Omega}$, u of class $\mathscr{C}^2$ in the neighbourhood of $\bar{\Omega}$) of application of these formulae, yet with the condition of giving a meaning to the integrals on Ω. Within the classical framework, we state:

Proposition 4 (Formulae of Ostrogradski and Green in the classical context). *Let Ω be a regular bounded open set and*

(1) *p a field of class $\mathscr{C}^1$ on Ω with $p.n$ defined on Γ and* $\operatorname{div} p$ *integrable on Ω. Then*

$$\int_\Omega \operatorname{div} p \, dx = \int_\Gamma p.n \, d\gamma \,;$$

(2) *$u \in \mathscr{C}^2(\Omega) \cap \mathscr{C}^1_n(\bar{\Omega})$ and $v \in \mathscr{C}^1(\Omega) \cap \mathscr{C}^0(\bar{\Omega})$ with $v\Delta u$ and* $\operatorname{grad} u.\operatorname{grad} v$ *integrable on Ω. Then*

$$\int_\Omega v \,\Delta u \, dx = \int_\Gamma v \frac{\partial u}{\partial n} d\gamma - \int_\Omega \operatorname{grad} u.\operatorname{grad} v \, dx \,;$$

(3) $u, v \in \mathscr{C}^2(\Omega) \cap \mathscr{C}^1_n(\bar{\Omega})$ with *$(v\Delta u - u\Delta v)$ integrable on Ω. Then*

$$\int_\Omega (v\,\Delta u - u\,\Delta v)\, dx = \int_\Gamma \left(v\frac{\partial u}{\partial n} - u\frac{\partial v}{\partial n}\right) d\gamma \,.$$

Let us show how to deduce Proposition 4 from the elementary formulation of Green's formulae (regular data in the neighbourhood of $\bar{\Omega}$) with the help of the following geometrical lemma:

Lemma 3. *Let Ω be a given regular open set (of class $\mathscr{C}^m$). There exists a function Φ of class $\mathscr{C}^m$ on an open neighbourhood U of Γ, regular on U (that is to say* $\operatorname{grad}\Phi(x) \neq 0$ *for all $x \in U$) such that*

$$\Gamma = \{x \in U;\quad \Phi(x) = 0\}$$

$$\Omega \cap U = \{x \in U;\quad \Phi(x) < 0\}\,.$$

Proof of Proposition 4. Using the function Φ of Lemma 3, let us put for $\delta > 0$

$$\Omega_\delta = (\Omega \backslash U) \cup \{x \in U;\quad \Phi(x) < -\delta\}\,.$$

For δ sufficiently small, this is a regular, relatively compact open set in Ω, with boundary $\Gamma_\delta = \{x \in U;\quad \Phi(x) = -\delta\}$. Let us put

$$\tilde{n}(x) = \frac{\operatorname{grad}\Phi(x)}{|\operatorname{grad}\Phi(x)|}\,.$$

The field $\tilde{n}$ is of class $\mathscr{C}^{m-1}$ on U and extends the normal field n of Γ, thus, also, the normal field n_δ of Γ_δ.
For a field p of class $\mathscr{C}^1$ on Ω, we can apply Ostrogradski's elementary formula on Ω_δ since Ω is a neighbourhood of Ω_δ

$$\int_{\Omega_\delta} \operatorname{div} p\,\mathrm{d}x = \int_{\Gamma_\delta} p\,.\,\tilde{n}\,\mathrm{d}\gamma_\delta \tag{1.15}$$

where $\mathrm{d}\gamma_\delta$ is the surface measure of Γ_δ.
In the limit when $\delta \to 0$, we obtain the Ostrogradski formula for p on Ω, since $\operatorname{div} p$ is integrable on Ω and $p\,.\,\tilde{n}$ is continuous on $\bar{\Omega} \cap U$[7].
The case of Green's formulae is treated in an identical manner. □

Remark 3. Recalling the above proof, being given p of class $\mathscr{C}^1$ on Ω, we have (1.15) for all $\delta > 0$. Suppose that $p\,.\,n$ is defined on Γ, the second integral and hence also the first integral converges when $\delta \to 0$. That shows that the integral

$$\int_\Omega \operatorname{div} p\,\mathrm{d}x$$

is convergent. Also, making use of the Fatou–Lebesgue theorem, *we are assured of the integrability of* $\operatorname{div} p$ *on Ω when* $\operatorname{div} p$ *is bounded below or bounded above by a function integrable on Ω.* □

In particular, recalling the reasoning for Green's formula of integration by parts with $v = u$, on applying to $p = u \operatorname{grad} u$, we obtain the

---

[7] For every function ζ continuous on $U \cap \bar{\Omega}$, $\displaystyle\int_\Gamma \zeta\,\mathrm{d}\gamma = \lim_{\delta \to 0} \int_{\Gamma_\delta} \zeta\,\mathrm{d}\gamma_\delta$. This is shown directly by local parametrisations.

Proposition 5. *Let $u \in \mathscr{C}^2(\Omega) \cap \mathscr{C}^1_n(\bar{\Omega})$. Suppose that $u\Delta u$ is bounded below by a function integrable on Ω. Then $|\operatorname{grad} u|^2$ and $u\Delta u$ are integrable Ω and*

$$\int_\Omega u\Delta u \, \mathrm{d}x = \int_\Gamma u \frac{\partial u}{\partial n} \mathrm{d}\gamma - \int_\Omega |\operatorname{grad} u|^2 \, \mathrm{d}x \, .$$

4. The Laplacian in Polar Coordinates

The Laplace operator being invariant under a Euclidean transformation, is, in particular, invariant under rotations in $\mathbb{R}^n$. For that reason, it will often be interesting to use the representation of $\mathbb{R}^n$ in polar coordinates.

Let us begin with the *case of the plane* $\mathbb{R}^2$. The mapping

$$h : (r, \theta) \to (r\cos\theta; r\sin\theta) \, .$$

is analytic on $\mathbb{R}^2\backslash\{0\} \times \mathbb{R}$. It is only locally bijective: for $\theta_0 \in \mathbb{R}$, h ia a diffeomorphism of $]0, \infty[\times]\theta_0 - \pi, \theta_0 + \pi[$ onto $\mathbb{R}^2\backslash D^-_{\theta_0}$ with $D^-_{\theta_0} = \{h(\lambda, \theta_0); \lambda \leqslant 0\}$; we shall call the inverse diffeomorphism, *the representation of proper polar coordinates of* $\mathbb{R}^2\backslash D^-_{\theta_0}$. However, it will be interesting to consider $r < 0$: for example h is a diffeomorphism of $\mathbb{R}^* \times]\theta_0, \theta_0 + \pi[$ on $\mathbb{R}^2\backslash D_{\theta_0}$ with $D_{\theta_0} = \{h(\lambda, \theta_0); \lambda \in \mathbb{R}\}$.

Now, let us take $(r_0, \theta_0) \in \mathbb{R}^2$ with $r_0 \neq 0$ and let us consider u defined in the neighbourhood of $x_0 = h(r_0, \theta_0)$ and twice differentiable at x_0. An elementary calculation shows that:

$$\Delta u(x_0) = \frac{\partial^2(u\circ h)}{\partial r^2}(r_0, \theta_0) + \frac{1}{r_0}\frac{\partial(u\circ h)}{\partial r}(r_0, \theta_0) + \frac{1}{r_0^2}\frac{\partial^2(u\circ h)}{\partial\theta^2}(r_0, \theta_0)$$

which, with abuse of notation, can be written

$$\Delta u = \frac{\partial^2 u}{\partial r^2} + \frac{1}{r}\frac{\partial u}{\partial r} + \frac{1}{r^2}\frac{\partial^2 u}{\partial\theta^2} \, . \tag{1.16}$$

This equation is called the *formula for the Laplacian in polar coordinates*. We note also that we have

$$\frac{\partial^2 u}{\partial r^2} + \frac{1}{r}\frac{\partial u}{\partial r} = \frac{1}{r}\frac{\partial}{\partial r}\left(r\frac{\partial u}{\partial r}\right) \tag{1.17}$$

This formula is still true if u is a distribution on an open set Ω of the plane not containing the origin: in the neighbourhood of a point of $h^{-1}(\Omega)$, h is a diffeomorphism with the result that we define the image distributions $u\circ h = h^{-1}u$ and $\Delta u\circ h = h^{-1}\Delta u$; we then have

$$(\Delta u)\circ h = \left(\frac{\partial^2}{\partial r^2} + \frac{1}{r}\frac{\partial}{\partial r} + \frac{1}{r^2}\frac{\partial^2}{\partial\theta^2}\right) \quad (u\circ h)$$

the derivatives and the products by $1/r$ $(r \neq 0)$ being taken in the sense of distributions. The validity of this formula can be assured by general arguments of continuity and density in $\mathscr{D}'$. In order to persuade a reader with slight acquaintance with the general theory of distributions, we give a direct proof. First of all

$$\begin{vmatrix} \dfrac{\partial}{\partial r} r\cos\theta & \dfrac{\partial}{\partial \theta} r\cos\theta \\ \dfrac{\partial}{\partial r} r\sin\theta & \dfrac{\partial}{\partial \theta} r\sin\theta \end{vmatrix} = r$$

with the result that by applying (1.4), and using $u = h(h^{-1}u)$, we have

$$\langle u, \zeta \rangle = \langle h^{-1}u, r\zeta \circ h \rangle \qquad \forall\, \zeta \in \mathscr{D}(\Omega)\,.$$

In particular,

$$\langle \Delta u, \zeta \rangle = \langle u, \Delta\zeta \rangle = \langle h^{-1}u, r(\Delta\zeta)\circ h \rangle\,.$$

Applying the formula for the Laplacian in plane polar coordinates to ζ and using (1.17) we obtain

$$\langle \Delta u, \zeta \rangle = \left\langle h^{-1}u, \frac{\partial}{\partial r}\left(r \frac{\partial \zeta\circ h}{\partial r}\right) + \frac{1}{r}\frac{\partial^2 \zeta\circ h}{\partial \theta^2} \right\rangle$$

or, by the definition of the differentiation of distributions,

$$\begin{aligned} \langle \Delta u, \zeta \rangle &= \left\langle \frac{\partial}{\partial r}\left(r \frac{\partial h^{-1}u}{\partial r}\right) + \frac{1}{r}\frac{\partial^2 h^{-1}u}{\partial \theta^2}, \zeta\circ h \right\rangle \\ &= \left\langle \frac{1}{r}\frac{\partial}{\partial r}\left(r \frac{\partial h^{-1}u}{\partial r}\right) + \frac{1}{r^2}\frac{\partial^2 h^{-1}u}{\partial \theta^2}, r\zeta\circ h \right\rangle \\ &= \left\langle h\left(\frac{1}{r}\frac{\partial}{\partial r}\left(r\frac{\partial}{\partial r}\right) + \frac{1}{r^2}\frac{\partial^2}{\partial\theta^2}\right)(h^{-1}u)\,, \zeta \right\rangle \end{aligned}$$

which, account being taken of (1.17), proves the formula. □

We now propose to generalize this formula in $\mathbb{R}^n$. We denote by Σ *the unit sphere in* $\mathbb{R}^n$, defined by

$$\Sigma = \{\sigma = (\sigma_1, \ldots, \sigma_n) \in \mathbb{R}^n\,; \qquad |\sigma| = (\sigma_1^2 + \cdots + \sigma_n^2)^{1/2} = 1\}\,.$$

This is an analytic hypersurface in $\mathbb{R}^n$, the boundary of the unit ball. The map $(r, \sigma) \to r\sigma$ is an analytical diffeomorphism of $]0, \infty[\, \times \Sigma$ onto $\mathbb{R}^n\backslash\{0\}$. We call the inverse diffeomorphism $x \in \mathbb{R}^n\backslash\{0\} \to (|x|, x/|x|)$ the *representation in polar coordinates in* $\mathbb{R}^n$. We note the ambiguity in the case $n = 2$: in the representation in polar coordinates (without other specifications), the argument $\sigma = x/|x|$ is a point on the unit circle in $\mathbb{R}^2$ (which we can identify with the group of rotations of the plane, with the group of complex numbers with modulus 1, or again with the additive group $\mathbb{R}/2\pi\mathbb{Z}$); in the representation in suitable polar coordinates, we choose for θ_0 a suitable parametrisation of the unit circle with the point with polar angle $\theta_0 + \pi$ removed.

We note also that in the case $n = 1$, $\Sigma = \{-1, +1\}$ and the representation in polar coordinates is

$$x \in \mathbb{R}^* \to (|x|, \operatorname{sign} x) \quad \text{where} \quad \operatorname{sign} x = \begin{cases} +1 & x > 0 \\ -1 & x < 0 \end{cases} = \frac{x}{|x|} .$$

Because of this case $n = 1$, we shall sometimes employ for $x \in \mathbb{R}^n \backslash \{0\}$, the notation $\operatorname{sign} x = x/|x|$.
We denote by $\mathrm{d}\sigma$ *the surface measure of* Σ. For every function u integrable on $\mathbb{R}^n$, we have the formula

$$(1.18) \qquad \int_{\mathbb{R}^n} u \,\mathrm{d}x = \int_0^{\infty} r^{n-1} \,\mathrm{d}r \int_{\Sigma} u(r\sigma) \,\mathrm{d}\sigma .$$

In the case $n = 1$ this formula can be written

$$\int_{\mathbb{R}} u \,\mathrm{d}x = \int_0^{\infty} (u(r) - u(-r)) \,\mathrm{d}r ,$$

and in the case $n = 2$,

$$(1.19) \qquad \int_{\mathbb{R}^2} u \,\mathrm{d}x = \int_0^{\infty} r \,\mathrm{d}r \int_0^{2\pi} u(r\cos\theta, r\sin\theta) \,\mathrm{d}\theta .$$

We note that being given a unit vector e, the map $(r, \sigma) \to r\sigma$ is an analytic diffeomorphism of $\mathbb{R}^* \times \Sigma^+(e)$ on $\mathbb{R}^n \backslash \Pi(e)$ where

$$\Sigma^+(e) = \{\sigma \in \Sigma\,;\, \sigma . e > 0\} \quad \text{and} \quad \Pi(e) = \{x \in \mathbb{R}^n\,;\, x . e = 0\} .$$

In this representation, the formula (1.18) can be written

$$(1.20) \qquad \int_{\mathbb{R}^n} u \,\mathrm{d}x = \int_{-\infty}^{\infty} r^{n-1} \,\mathrm{d}r \int_{\Sigma^+(e)} u(r\sigma) \,\mathrm{d}\sigma$$

which gives for $n = 2$

$$(1.21) \qquad \int_{\mathbb{R}^2} u \,\mathrm{d}x = \int_{-\infty}^{\infty} r \,\mathrm{d}r \int_0^{\pi} u(r\cos\theta, r\sin\theta) \,\mathrm{d}\theta .$$

We say that a function $u(x)$ is *radial* if it depends only on $r = |x|$. It is naturally defined on an annulus $I \times \Sigma$ where I is a part $[0, \infty[$; it is clear that a function defined on $I \times \Sigma$ is radial iff it is invariant under rotation. Since Δ is invariant under rotation, *if u is a radial function, Δu is also radial.* We have the following *formula for the Laplacian of a radial function*:

$$(1.22) \qquad \Delta u(r) = \frac{\partial^2 u}{\partial r^2}(r) + \frac{n-1}{r}\frac{\partial u}{\partial r}(r) = \frac{1}{r^{n-1}}\frac{\partial}{\partial r}\left(r^{n-1}\frac{\partial u}{\partial r}\right)(r) .$$

In the case $n = 3$, we have also

$$(1.23) \qquad \Delta u(r) = \frac{1}{r}\frac{\partial^2(ru)}{\partial r^2}(r) .$$

We can easily verify these formulae by elementary calculations.

Being given a function u defined on $I \times \Sigma$, it is often interesting to consider *the radialised function of u*,

$$\tilde{u}(x) = \int_\Sigma u(|x|\sigma)\,\mathrm{d}\sigma\,. \tag{1.24}$$

We then prove

Proposition 6. *The operation of radialisation commutes with the Laplacian, that is to say, with the definition* (1.24),

$$\Delta\tilde{u}(x) = \widetilde{\Delta u}(x) = \int_\Sigma (\Delta u)(|x|\sigma)\,\mathrm{d}\sigma\,. \tag{1.25}$$

Proof. We put $v(r) = \int_\Sigma u(r\sigma)\,\mathrm{d}\sigma$ with the result that $\tilde{u}(x) = v(|x|)$ and from (1.22),

$$\Delta\tilde{u}(x) = \frac{1}{r^{n-1}}\frac{\mathrm{d}}{\mathrm{d}r}\left(r^{n-1}\frac{\mathrm{d}v}{\mathrm{d}r}\right)(|x|)\,.$$

We can always suppose that u defined on $\mathbb{R}^n$ can be replaced by $\zeta(|x|)u(x)$ where $\zeta \in \mathscr{D}(I)$, $\zeta = 1$ on a neighbourhood of the norm of the point at which we evaluate $\Delta\tilde{u}$. We then have

$$\frac{\mathrm{d}v}{\mathrm{d}r}(r) = \int_\Sigma \operatorname{grad} u(r\sigma)\,.\,\sigma\,\mathrm{d}\sigma = \frac{1}{r^{n-1}}\int_{\partial B(0,r)} \frac{\partial u}{\partial n}\,\mathrm{d}\gamma\,,$$

and hence by Green's formula

$$\frac{\mathrm{d}v}{\mathrm{d}r}(r) = \frac{1}{r^{n-1}}\int_{B(0,r)} \Delta u\,\mathrm{d}x\,.$$

Using (1.18)

$$r^{n-1}\frac{\mathrm{d}v}{\mathrm{d}r}(r) = \int_0^r s^{n-1}\,\mathrm{d}s \int_\Sigma (\Delta u)(s\sigma)\,\mathrm{d}\sigma\,;$$

thus

$$\frac{\mathrm{d}}{\mathrm{d}r}\left(r^{n-1}\frac{\mathrm{d}v}{\mathrm{d}r}(r)\right) = r^{n-1}\int_\Sigma (\Delta u)(r\sigma)\,\mathrm{d}\sigma\,. \qquad \square$$

We say that a function $u(x)$ is *spherical* if it depends only on $\sigma = x/|x|$. It is naturally defined on a "cone" $\mathbb{R}^+ \times \mathcal{O}$ where $\mathcal{O}$ is an open set of Σ; it is clear that a function defined on $\mathbb{R}^+ \times \mathcal{O}$ is spherical iff it is invariant under the homotheties H_λ: $x \to \lambda x$ ($\lambda > 0$). Now for a function u defined on $\mathbb{R}^+ \times \mathcal{O}$, $\Delta(H_\lambda u) = (1/\lambda^2)H_\lambda(\Delta u)$. In particular, applying this result with $\lambda = |x|$, we obtain for a spherical function

$$\Delta u(x) = \frac{1}{|x|^2}(\Delta u)\left(\frac{x}{|x|}\right). \tag{1.26}$$

We call the *Laplacian on the sphere*, or *the Laplace–Beltrami operator* the trace Δ_σ of the Laplacian on the sphere: in other words, for each function u of class $\mathscr{C}^2$ on an open set $\mathcal{O}$ of Σ, $\Delta_\sigma u(\sigma) = \Delta_x u(x/|x|)_{x=\sigma}$.
For a function u of class $\mathscr{C}^2$ on an open set Ω of $\mathbb{R}^n$, we then have the *formula for the Laplacian in polar coordinates in* $\mathbb{R}^n$

$$(1.27)\qquad (\Delta u)(r\sigma) = \frac{1}{r^{n-1}}\frac{\partial}{\partial r}\left(r^{n-1}\frac{\partial}{\partial r}\right)u(r\sigma) + \frac{1}{r^2}\Delta_\sigma u(r\sigma)\,.$$

The simplest method of verifying this formula is to make use of the density in $\mathscr{C}^2(I \times \mathcal{O})$ of functions with separable variables $u(x) = v(|x|)w(x/|x|)$; the verification of (1.27) for such functions follows immediately from (1.22) and (1.26). We can also adapt the proof of Proposition 6. This formula is obviously still true for distributions.
In the case $n = 2$, the formula (1.27) coincides with the formula (1.16). That merely expresses that the Laplace–Beltrami operator on the unit circle of $\mathbb{R}^2$ is given by

$$\Delta_\sigma u(\cos\theta, \sin\theta) = \frac{\partial^2}{\partial\theta^2}u(\cos\theta, \sin\theta)\,.$$

In the case $n = 3$, we can refer locally to points on the unit sphere by the *Euler angles* (θ, φ) defined for $\sigma = (x, y, z)$ by

$$x = \sin\varphi\cos\theta\,,\qquad y = \sin\varphi\sin\theta\,,\qquad z = \cos\varphi\,.$$

The map $(\theta, \varphi) \to \sigma(\theta, \varphi)$ is an analytic diffeomorphism from $]\theta_0 - \pi, \theta_0 + \pi[\times]0, \pi[$ onto the sphere Σ deprived of the semi-circle $\{(-\sin\varphi\cos\theta_0, -\sin\varphi\sin\theta_0, \cos\varphi); 0 \leqslant \varphi \leqslant \pi\}$.
In this representation, the Laplace–Beltrami operator is given by the equation

$$(1.28)\quad \Delta_\sigma u(\sigma(\theta, \varphi)) = \left(\frac{1}{\sin^2\varphi}\frac{\partial^2}{\partial\theta^2} + \frac{1}{\sin\varphi}\frac{\partial}{\partial\varphi}\left(\sin\varphi\frac{\partial}{\partial\varphi}\right)\right)u(\sigma(\theta, \varphi))$$

called *the formula for the Laplacian in spherical coordinates.*
This formula can be generalized for the sphere in $\mathbb{R}^n$ with n arbitrary (see Dieudonné [1], Vol. III). Another classical representation of the sphere is the representation in *stereographic coordinates.* The stereographic projection is the mapping

$$x' = (x_1, \ldots, x_{n-1}) \in \mathbb{R}^{n-1} \to \sigma(x') = \frac{(2x', |x'|^2 - 1)}{1 + |x'|^2}\,.$$

This is a diffeomorphism from $\mathbb{R}^{n-1}$ onto the sphere Σ with the "north pole" $(0, \ldots, 0, 1)$ removed; it is in fact the trace on the hyperplane $\mathbb{R}^{n-1} \times \{0\}$ of the polar inversion, with pole the north pole and of radius $\sqrt{2}$. We shall return to the use of inversion for the study of the Laplacian in §2.
Finally we draw attention to the representation of $\mathbb{R}^n = \mathbb{R}^m \times \mathbb{R}^{n-m}$ $(2 \leqslant m < n)$ in *cylindrical coordinates* which consists of representing $x = (x', x'')$, $x' \in \mathbb{R}^m$, $x'' \in \mathbb{R}^{n-m}$ by the polar coordinates of x', $r = |x'|$, $\sigma = x'/|x'|$ and the

point $x'' \in \mathbb{R}^{n-m}$. In this representation

$$(1.29) \qquad (\Delta u)(r\sigma, x'') = \left(\frac{1}{r^{m-1}}\frac{\partial}{\partial r}\left(r^{m-1}\frac{\partial}{\partial r}\right) + \frac{1}{r^2}\Delta_\sigma + \Delta_{x''}\right)u(r\sigma, x'),$$

where Δ_σ is the Laplace–Beltrami operator of the unit sphere in $\mathbb{R}^m$.
In particular, in the case $n = 3$, the cylindrical coordinates of (x, y, z) are (r, θ, z) where (r, θ) are the polar coordinates in the (x, y) plane. We have

(1.30)

$$\Delta u(r\cos\theta, r\sin\theta, z) = \left(\frac{1}{r}\frac{\partial}{\partial r}\left(r\frac{\partial}{\partial r}\right) + \frac{1}{r^2}\frac{\partial^2}{\partial\theta^2} + \frac{\partial^2}{\partial z^2}\right)u(r\cos\theta, r\sin\theta, z),$$

—the *formula for the Laplacian in cylindrical coordinates* in the space $\mathbb{R}^3$.
This account of the passage to polar coordinates expressed in the classical framework remains valid within the framework of distributions.

§2. Harmonic Functions

1. Definitions. Examples. Elementary Solutions

Definition 1. Let Ω be an open set of $\mathbb{R}^n$. We define a *harmonic function on* Ω to be a function $u \in \mathscr{C}^2(\Omega)$ satisfying

$$\Delta u(x) = 0 \qquad \forall x \in \Omega .$$

In other words, a harmonic function on Ω is a classical solution of the equation

$$\Delta u = 0 \quad \text{on} \quad \Omega$$

called *Laplace's equation on* Ω.

Remark 1. We shall prove later (see Corollary 7) that every harmonic function on Ω is analytic on Ω. We shall prove also (see Proposition 1 of §3 and also Corollaries 7 and 10 of this §2) that every distribution solution of Laplace's equation is a harmonic function (and hence analytic on Ω). □

Case $n = 1$. The harmonic functions on an interval of $\mathbb{R}$ are the affine functions. This is immediate since $\mathrm{d}^2u/\mathrm{d}x^2 = 0$.
Case $n = 2$. Identifying the plane $\mathbb{R}^2$ with the field $\mathbb{C}$ of the complex numbers, we have the classical result linking a holomorphic function to a harmonic function. We denote here classically by (x, y) a point of $\mathbb{R}^2$ which we identify, topologically (but not algebraically) with the complex number $z = x + iy$ of $\mathbb{C}$.

Proposition 1.
(*a*) *Let Ω be an open set of $\mathbb{C}$ and f a holomorphic function of the complex variable z. Then the function $u(x, y) = \mathrm{Re} f(x + iy)$ is harmonic on Ω.*

(*b*) *Let Ω be a simply connected open set of $\mathbb{R}^2$ and u a real-valued harmonic function on Ω. Then there exists a holomorphic function f of the complex variable $z = x + iy$, $((x, y) \in \Omega)$ such that $u(x, y) = \operatorname{Re} f(x + iy)$ for all $(x, y) \in \Omega$.*

Proof. Being given $f(x + iy) = u(x, y) + iv(x, y)$ with u and v, real and differentiable in (x, y), the function f is holomorphic (that is to say differentiable with respect to the complex variable) in $z = x + iy$ iff the real functions u and v satisfy the Cauchy–Riemann conditions

$$\frac{\partial u}{\partial x} = \frac{\partial v}{\partial y}, \quad \frac{\partial u}{\partial y} = -\frac{\partial v}{\partial x} \quad \text{at} \quad (x, y).$$

(*a*) We make use of the fact that a holomorphic function on Ω is of class $\mathscr{C}^2$ (and even analytic) on Ω (with respect to the complex variable). We deduce immediately that u and v are of class $\mathscr{C}^2$ (and even $\mathscr{C}^\infty$) on Ω (with respect to the real variables). Differentiating the Cauchy–Riemann conditions we obtain

$$\Delta u = \frac{\partial^2 u}{\partial x^2} + \frac{\partial^2 u}{\partial y^2} = \frac{\partial^2 v}{\partial x\, \partial y} - \frac{\partial^2 v}{\partial y\, \partial x} = 0.$$

(*b*) We have to find a function v (at least of class $\mathscr{C}^1$), satisfying (with u given) the Cauchy–Riemann conditions, that is to say

$$\operatorname{grad} v = \left(-\frac{\partial u}{\partial y}, \frac{\partial u}{\partial x}\right).$$

Since Ω is simply-connected and $-\dfrac{\partial u}{\partial y}$ and $\dfrac{\partial u}{\partial x}$ are of class $\mathscr{C}^1$, that is equivalent to

$$\frac{\partial}{\partial y}\left(-\frac{\partial u}{\partial y}\right) = \frac{\partial}{\partial x}\left(\frac{\partial u}{\partial x}\right),$$

that is to say, to $\Delta u = 0$. ☐

Remark 2. Let us consider u harmonic on an open set Ω of $\mathbb{R}^2$. If Ω is simply connected, there exists $v \in \mathscr{C}^2(\Omega)$ such that $f(x + iy) = u(x, y) + iv(x, y)$ is holomorphic. The function v is defined to within an additive constant and is also harmonic ($v = \operatorname{Re}(-if)$): we sometimes call it *the harmonic conjugate function of u*. If Ω is not simply connected, the conjugate harmonic function is defined locally and is multiple-valued on Ω[8].

Now, every open set is locally simply connected, hence *every function, harmonic on an open set of $\mathbb{R}^2$ is locally the real part of a holomorphic function.* Making use, then, of the fact that a holomorphic function is analytic, we obtain a *simple proof of the analyticity of a harmonic function on an open set of the plane.*

---

[8] More generally, let Ω be an open set of $\mathbb{R}^n$ and p a field of class $\mathscr{C}^1(\Omega)$ satisfying

$$\operatorname{curl} p = \mathbf{0}, \quad \operatorname{div} p = 0.$$

Locally, p is the gradient of a harmonic function (defined to within a constant). If Ω is simply connected, p is the gradient of a function harmonic on Ω (defined to within a constant), but if Ω is not simply connected, this function can be multiple-valued.

In a general manner *the theory of holomorphic functions will be a powerful tool for the study of the Laplacian in* $\mathbb{R}^2$. We note that the reason that *the case* $n = 2$ *is so special*, is due to the decomposition of the Laplacian in $\mathbb{R}^2$:

$$\Delta = \frac{\partial^2}{\partial x^2} + \frac{\partial^2}{\partial y^2} = \left(\frac{\partial}{\partial x} + i\frac{\partial}{\partial y}\right)\left(\frac{\partial}{\partial x} - i\frac{\partial}{\partial y}\right).$$

The linear differential operator (with complex coefficients)

$$\bar{\partial} = \frac{1}{2}\left(\frac{\partial}{\partial x} + i\frac{\partial}{\partial y}\right)$$

is called the *Cauchy–Riemann operator* (the Cauchy–Riemann conditions can be written "in summary", $\bar{\partial}(u + iv) = 0$).
We can show easily that *for* $n \geqslant 3$, *it is impossible to write* Δ *as the product of two linear differential operators of the first order in* $\mathbb{C}$ (see Chap. V).
We now give some immediate properties of *the set of functions, harmonic in an open set* Ω *of* $\mathbb{R}^n$ *which we shall denote by* $\mathscr{H}(\Omega)$.

Proposition 2
(*a*) $\mathscr{H}(\Omega)$ *is a vector space: every liner combination of harmonic functions is again harmonic;*
(*b*) $\mathscr{H}(\Omega)$ *is stable under differentiation: the derivatives* $\partial u/\partial x_i$ *of a harmonic function* u *are harmonic functions* (while awaiting the proof that a harmonic function is $\mathscr{C}^\infty$ we have to suppose that $u \in \mathscr{C}^3(\Omega)$).
(*c*) $\mathscr{H}(\Omega)$ *is closed for the topology of uniform topology on every compact set* (*but also for the induced topology from that of distributions on* Ω): *every limit of harmonic functions is harmonic* (until we prove that a harmonic distribution is a harmonic function, we have to take the limit in $\mathscr{C}^2(\Omega)$, i.e. the uniform limit in every set of Ω, of u and its first and second derivatives). We shall return to this property later (see Harnack's theorems, Proposition 13 from which come the Corollaries 8 and 9).
(*d*) *If* $u, v \in \mathscr{H}(\Omega)$, *then* $uv \in \mathscr{H}(\Omega)$ *iff* $\operatorname{grad} u \,.\, \operatorname{grad} v = 0$. This result immediately follows from

$$\Delta(uv) = u\Delta v + 2\operatorname{grad} u \,.\, \operatorname{grad} v + v\Delta u\,. \tag{2.1}$$

We now give some (fundamental) examples of harmonic functions:

Example 1. *The affine functions are harmonic on* $\mathbb{R}^n$ (and hence on every open set of $\mathbb{R}^n$). In the case $n = 1$, these are the only harmonic functions (on an interval of $\mathbb{R}$). In the case $n \geqslant 2$, there exist *numerous harmonic polynomials*; we note that a polynomial is harmonic on a non-empty open set of $\mathbb{R}^n$ iff it is so on the whole of $\mathbb{R}^n$. *A quadratic form* $\sum_{i,j} a_{ij}x_i x_j$ *is harmonic iff its trace* $\sum a_{ii} = 0$. From Proposition 1, a polynomial in two variables $\sum_{k,l} a_{kl}x^k y^l$ is harmonic iff it is the real part of a polynomial of the complex variable $x + iy$, $\sum_k c_k(x + iy)^k$. We shall return to the study of harmonic polynomials in §7. Let us note

Proposition 3. *The harmonic polynomials are the only tempered distribution solutions of Laplace's equation on $\mathbb{R}^n$.*

Proof. We shall make use of the Fourier transform for tempered distributions (see Vol. 2, Appendix "Distributions"). Let us consider a tempered distribution solution u of

$$\Delta u = 0 .$$

Its Fourier transform $\hat{u}$ is a solution of the algebraic equation

$$y^2 \hat{u}(y) = 0 .$$

We then deduce that $\hat{u}$ is with support contained in $\{0\}$ and hence is a distribution of finite order $m \in \mathbb{N}^*$ and a linear combination of the Dirac measure at the origin δ and its derivatives (see Vol. 2, Appendix "Distributions"):

$$\hat{u} = \sum_{|\alpha| \leqslant m} a_\alpha \frac{\partial^\alpha \delta}{\partial y^\alpha}\,^{9} .$$

Using the inverse Fourier transform, we obtain for u:

$$u = \sum_{|\alpha| \leqslant m} a_\alpha x^{\alpha}\,^{10} . \qquad \square$$

We note that this proposition shows in particular that a tempered distribution solution of Laplace's equation is a (harmonic) analytic function. We note also the corollary, which we can obtain in the case $n = 2$ by means of *Liouville's theorem*:

Corollary 1. *A harmonic function bounded on the whole of $\mathbb{R}^n$ is constant.*

Proof. A (locally integrable) function, bounded on $\mathbb{R}^n$, is a tempered distribution (see Vol. 2, Appendix "Distributions"). From Proposition 2, a harmonic function bounded on $\mathbb{R}^n$ is hence a polynomial; being bounded, it is constant. $\square$

We shall later extend this result by another method to the case of harmonic functions on $\mathbb{R}^n$ bounded above or below (see Proposition 7).

Example 2. Let us now determine the *radial harmonic functions.* Suppose that we are given an interval I of $]0, \infty[$ and that we consider a radial function

$$u(x) = v(|x|)$$

defined on the ring $\Omega = \{x \in \mathbb{R}^n; |x| \in I\}$. Making use of the formula of the Laplacian of a radial function (see (1.22)), u is harmonic on Ω iff v is of class $\mathscr{C}^2$ on I

---

[9] We adopt here the multi-indicial notation

$$\alpha = (\alpha_1, \ldots, \alpha_n) \in N^n, \quad |\alpha| = \alpha_1 + \ldots + \alpha_n, \quad \frac{\partial^\alpha}{\partial y^\alpha} = \frac{\partial^{|\alpha|}}{\partial y_1^{\alpha_1} \ldots \partial y_n^{\alpha_n}}, \quad x^\alpha = x_1^{\alpha_1} \ldots x_n^{\alpha_n} .$$

[10] The coefficients a_α are obviously not arbitrary, the polynomials $\Sigma a_\alpha x^\alpha$ having to be harmonic. In the case $n = 1$, we have necessarily $m \leqslant 1$, but $n \geqslant 2$, there exist harmonic polynomials of arbitrary degree.

and

$$\frac{1}{r^{n-1}}\frac{\mathrm{d}}{\mathrm{d}r}\left(r^{n-1}\frac{\mathrm{d}v}{\mathrm{d}r}\right) = 0 \qquad r \in I\,;$$

this is equivalent to

$$\frac{\mathrm{d}v}{\mathrm{d}r} = \frac{c}{r^{n-1}} \tag{2.2}$$

on I with c an arbitrary constant. We distinguish

Case $n = 2$. The integration of (2.2) gives

$$v(r) = c \operatorname{Log} r + c_0\,^{11}\,.$$

Case $n \geqslant 3$. The integration of (2.2) gives

$$v(r) = -\frac{c}{(n-2)r^{n-2}} + c_0\,.$$

We can therefore state

Proposition 4. *Every radial function, harmonic on an annulus in $\mathbb{R}^n$ is of the form*

$$c \operatorname{Log}|x| + c_0 \quad \text{if} \quad n = 2$$

$$\frac{c}{|x|^{n-2}} + c_0 \quad \text{if} \quad n \geqslant 3\,,$$

where c_0, c are constants.

We note that this proposition proves that a radial harmonic function is analytic; also, *if a radial function, harmonic on $\{0 < |x| < r_0\}$ is bounded, then it is constant.* We remark also the preceding calculation remains valid for a radial distribution; this proves that *a radial distribution solution of Laplace's equation on $\{|x| \in I\}$, where I is an interval of $]0, \infty[$ is a (harmonic) analytic function.*

Now the function $\operatorname{Log}|x|$ in the case $n = 2$, and the function $|x|^{2-n}$ in the case $n \geqslant 3$, is integrable in the neighbourhood of the origin; it therefore defines a distribution u on $\mathbb{R}^n$.

Let us now calculate the distribution Δu defined by:

$$\langle \Delta u, \zeta \rangle = \langle u, \Delta\zeta \rangle = \int u\,\Delta\zeta\,\mathrm{d}x = \lim_{\varepsilon \to 0} \int_{|x| > \varepsilon} u\,\Delta\zeta\,\mathrm{d}x\,.$$

To calculate this limit, let us begin with the case $n = 2$. Changing to plane polar coordinates, we have

$$\int_{|x| > \varepsilon} u\,\Delta\zeta\,\mathrm{d}x = \int_\varepsilon^\infty r \operatorname{Log} r\,\mathrm{d}r \int_0^{2\pi} \Delta\zeta(r\cos\theta, r\sin\theta)\,\mathrm{d}\theta\,.$$

---

[11] The notation Log denotes the Naperian logarithm.

Using the formula for the Laplacian in polar coordinates we see that

$$\int_0^{2\pi} \Delta\zeta(r\cos\theta, r\sin\theta)\,\mathrm{d}\theta = \int_0^{2\pi} \left(\frac{1}{r}\frac{\partial}{\partial r}\left(r\frac{\partial\zeta}{\partial r}\right) + \frac{1}{r^2}\frac{\partial^2\zeta}{\partial\theta^2}\right)\mathrm{d}\theta$$
$$= \frac{1}{r}\frac{\mathrm{d}}{\mathrm{d}r}\left(r\frac{\mathrm{d}}{\mathrm{d}r}\int_0^{2\pi} \zeta(r\cos\theta, r\sin\theta)\,\mathrm{d}\theta\right)$$

since $\int_0^{2\pi} \frac{\partial^2\zeta}{\partial\theta^2}\,\mathrm{d}\theta = 0$, the function $\zeta(r\cos\theta, r\sin\theta)$ being periodic

Integrating by parts in the integral obtained by inserting the latter result into the integral of $u\Delta\zeta$ we obtain

$$\int_{|x|>\varepsilon} u\,\Delta\zeta\,\mathrm{d}x = \int_\varepsilon^\infty \operatorname{Log} r\frac{\mathrm{d}}{\mathrm{d}r}\left(r\frac{\mathrm{d}}{\mathrm{d}r}\int_0^{2\pi} \zeta(r\cos\theta, r\sin\theta)\,\mathrm{d}\theta\right)\mathrm{d}r$$
$$= -\,\varepsilon\operatorname{Log}\varepsilon\frac{\mathrm{d}}{\mathrm{d}r}\int_0^{2\pi} \zeta(r\cos\theta, r\sin\theta)\,\mathrm{d}\theta\Big|_{r=\varepsilon}$$
$$+ \int_0^{2\pi} \zeta(\varepsilon\cos\theta, \varepsilon\sin\theta)\,\mathrm{d}\theta\,.$$

When $\varepsilon \to 0$, the first term tends to 0 and the second to $2\pi\zeta(0)$. Therefore

$$\langle\Delta u, \zeta\rangle = 2\pi\zeta(0)\,,$$

that is to say, *finally*

$$\Delta u = 2\pi\delta\,,$$

where δ is the Dirac measure at the origin.

We note that we should be able to simplify this calculation by remarking that Δu is a radial distribution; hence it is enough to determine $\langle\Delta u, \zeta\rangle$ for a radial function $\zeta = \zeta(|x|)$. We shall carry out the calculation in the case $n \geqslant 3$. Here

$$\int_{|x|>\varepsilon} u\,\Delta\zeta\,\mathrm{d}x = \int_\varepsilon^\infty \frac{1}{r^{n-2}} \times r^{n-1}\,\mathrm{d}r \int_\Sigma \frac{1}{r^{n-1}}\frac{\mathrm{d}}{\mathrm{d}r}\left(r^{n-1}\frac{\mathrm{d}\zeta}{\mathrm{d}r}\right)\mathrm{d}\sigma$$
$$= \sigma_n\int_\varepsilon^\infty \frac{1}{r^{n-2}}\frac{\mathrm{d}}{\mathrm{d}r}\left(r^{n-1}\frac{\mathrm{d}\zeta}{\mathrm{d}r}\right)\mathrm{d}r\,,$$

where $\sigma_n = \int_\Sigma \mathrm{d}\sigma$ is the total surface area of the unit sphere in $\mathbb{R}^n$. Integrating by parts, we have

$$\int_{|x|>\varepsilon} u\,\Delta\zeta\,\mathrm{d}x = \sigma_n\left(-\,\varepsilon\frac{\mathrm{d}\zeta}{\mathrm{d}r}(\varepsilon) - (n-2)\zeta(\varepsilon)\right)$$

and hence in the limit as $\varepsilon \to 0$,

$$\Delta u = -\,(n-2)\sigma_n\delta\,.$$

This calculation leads us to denote

$$(2.3)\qquad \begin{cases} E_2 = \dfrac{1}{2\pi}\,\mathrm{Log}\,|x| \\[2ex] E_n = \dfrac{1}{k_n|x|^{n-2}} \quad \text{with} \quad k_n = -(n-2)\sigma_n\,. \end{cases}$$

With this notation, the (locally integrable) function E_n is a distribution solution of the Poisson equation

$$\Delta E_n = \delta \quad \text{on} \quad \mathbb{R}^n\,.$$

We pose the

Definition 2. An *elementary* (or *fundamental*) *solution* of the Laplacian in $\mathbb{R}^n$ is any distribution E on $\mathbb{R}^n$ which satisfies the Poisson equation

$$\Delta E = \delta \quad \text{on} \quad \mathbb{R}^n\,.$$

With this definition, *E_n is an elementary solution of the Laplacian*; since the restriction to $\mathbb{R}^n\backslash\{0\}$ of an elementary solution is a solution of Laplace's equation, taking account of Proposition 4 (see also the remark which follows), *E_n is even (to within an additive constant) the only elementary radial solution.* The notion of an elementary solution plays a fundamental role in theory of linear differential operators (see Chap. V). The elementary solution E_n will play a constant role throughout the whole of the theory of the Laplacian (see also its physical interpretation in §2.6). As we have noted above, *this elementary solution E_n is analytic on* $\mathbb{R}^n\backslash\{0\}$; using general theorems on linear differential operators, we deduce (see Chap. V, §2.2.4) that *every distribution on an open set Ω satisfying Laplace's equation on Ω (and hence, in particular, every harmonic function on Ω) is a (harmonic) analytic function on Ω.* We shall give another direct proof of this result later.
Finally, we have so far considered only the case $n \geqslant 2$. In the *case $n = 1$* it is immediate that

$$E_1 = \frac{1}{2}|x|$$

is an elementary solution of d^2/dx^2; to within a constant, it is the only even elementary solution.

Example 3. Every derivative of a harmonic function is again harmonic. In particular the function $\partial E_n/\partial x_n$ is harmonic on $\mathbb{R}^n\backslash\{0\}$; a simple calculation shows that for all $n \geqslant 2$

$$\frac{\partial E_n}{\partial x_n} = \frac{2x_n}{\sigma_n|x|^n}\,.$$

This cylindrical function (it depends only on $r = (x_1^2 + \ldots + x_{n-1}^2)^{\frac{1}{2}}$ and x_n) will be particularly useful in the study of the Laplacian for problems with cylindrical symmetry, or more generally for problems in which one of the directions in space (usually taken to be the x_n direction) is privileged. The Laplacian then behaves like an evolution operator

$$\Delta_x = \frac{\partial^2}{\partial t^2} + \Delta_x . \qquad \square$$

2. Gauss' Theorem. Formulae of the Mean. Maximum Principle

Let us consider a regular bounded open set and two functions $u, v \in \mathscr{H}(\Omega) \cap \mathscr{C}_n^1(\bar{\Omega})$, that is to say in the adopted nomenclature, harmonic on Ω with normal derivatives defined on Γ (in the classical sense of §1.3*b*). The application of Green's formula (Proposition 4 of §1.3*c*) gives

$$\int_\Gamma \left(v \frac{\partial u}{\partial n} - u \frac{\partial v}{\partial n} \right) \mathrm{d}\gamma = 0 , \tag{2.4}$$

which we shall call "*Green's formula for harmonic functions*".
Particularizing with $v = 1$, we obtain *Gauss' theorem* (stated in the classical framework).

Proposition 5. *Let Ω be a regular bounded set and $u \in \mathscr{H}(\Omega) \cap \mathscr{C}_n^1(\bar{\Omega})$. Then*

$$\int_\Gamma \frac{\partial u}{\partial n} \mathrm{d}\gamma = 0 .$$

We shall show later (see Proposition 14) that this property when it holds for *every open set* with a regular boundary, characterises the harmonic functions. We shall return also to §2.6 of this chapter for the important physical interpretation of this theorem.
Suppose that now we are given an open ball $B = B(x_0, r_0)$ of centre $x_0 \in \mathbb{R}^n$ and radius $r_0 > 0$:

$$B(x_0, r_0) = \{x \in \mathbb{R}^n ; |x - x_0| < r_0\} .$$

We denote its boundary by ∂B; we have

$$\partial B(x_0, r_0) = x_0 + r_0 \Sigma$$

where Σ always denotes the unit sphere in $\mathbb{R}^n$. We keep the notation $\mathrm{d}\gamma$ for the surface measure on ∂B; we have

$$\int_{\partial B} u \,\mathrm{d}\gamma = r_0^{n-1} \int_\Sigma u(x_0 + r_0 \sigma) \mathrm{d}\sigma , \tag{2.5}$$

where $\mathrm{d}\sigma$ always denotes the surface measure on Σ.
Now let $u \in \mathscr{H}(B)$. From Gauss' theorem (Proposition 5) we have, for all $0 < r < r_0$:

$$\int_{\partial B_r} \frac{\partial u}{\partial n} \mathrm{d}\gamma = 0 , \tag{2.6}$$

where $B_r = B(x_0, r)$.

Applying now Green's formula for harmonic functions to u and $v(x) = E_n(x - x_0)$, where E_n is the elementary solution defined by (2.3), on the annulus $\Omega = \{x; \varepsilon < |x - x_0| < r\}$ with $0 < \varepsilon < r < r_0$. The boundary Γ of Ω is made up of ∂B_r and ∂B_ε; the restriction to ∂B_r (resp. ∂B_ε) of the normal field to Γ exterior to Ω is the normal to ∂B_r exterior to B_r (resp. the opposite of the normal field to ∂B_ε exterior to B_ε). Since v is constant on ∂B_r and ∂B_ε we have from (2.6)

$$\int_\Gamma v \frac{\partial u}{\partial n} \mathrm{d}\gamma = \int_{\partial B_r} v \frac{\partial u}{\partial n} \mathrm{d}\gamma - \int_{\partial B_\varepsilon} v \frac{\partial u}{\partial n} \mathrm{d}\gamma = 0 .$$

Now

$$\operatorname{grad} E_n(x) = \frac{1}{\sigma_n |x|^{n-1}} \frac{x}{|x|},$$

this being true as well for $n = 2$ ($\sigma_2 = 2\pi$) as for $n \geqslant 3$. Hence

$$\int_\Gamma u \frac{\partial v}{\partial n} \mathrm{d}\gamma = \frac{1}{\sigma_n r^{n-1}} \int_{\partial B_r} u \, \mathrm{d}\gamma - \frac{1}{\sigma_n \varepsilon^{n-1}} \int_{\partial B_\varepsilon} u \, \mathrm{d}\gamma .$$

Using (2.5), we obtain in the limit when $r \to r_0$ and $\varepsilon \to 0$, the *formula of the mean* (stated in the classical framework):

Proposition 6. *Let $B = B(x_0, r_0)$ and $u \in \mathscr{H}(B) \cap \mathscr{C}^0(\bar{B})$. Then*

$$u(x_0) = \frac{1}{\sigma_n r_0^{n-1}} \int_{\partial B} u \, \mathrm{d}\gamma . \tag{2.7}$$

Remark 3. To be able to apply this formula, it is clear that it is not necessary to suppose that u is continuous on $\bar{B}$. Denoting for $\lambda \in]0, 1[$

$$u_\lambda(x) = u(\lambda x_0 + (1 - \lambda)x),$$

the function u_λ is harmonic in the neighbourhood of $\bar{B}$. Applying the formula of the mean to u_λ, since $u_\lambda(x_0) = u(x_0)$, it suffices to be able to pass to the limit when $\lambda \to 0$. That is true in particular from Lebesgue's theorem, if

$$\begin{cases} u(x) = \lim\limits_{\lambda \to 0} u(\lambda x_0 + (1 - \lambda)x) \quad \text{exists} & \mathrm{d}\gamma\text{-a.e. } x \in \partial B \\ |u(\lambda x_0 + (1 - \lambda)x)| \leqslant u_0(x) & \mathrm{d}\gamma\text{-a.e. } x \in \partial B \quad \text{with} \quad u_0 \in L^1(\partial B) \end{cases}$$

We shall return later to the extension of integral formulae in a more general context (see §6). □

We now give a variant of the formula (2.7): applying it for $0 < r < r_0$, we have for $u \in \mathscr{H}(B)$,

$$u(x_0) = \frac{1}{\sigma_n r^{n-1}} \int_{\partial B_r} u \, \mathrm{d}\gamma = \frac{1}{\sigma_n} \int_\Sigma u(x_0 + r\sigma) \mathrm{d}\sigma .$$

Hence if we suppose u integrable on B, we have

$$\int_B u \, \mathrm{d}x = \int_0^{r_0} r^{n-1} \mathrm{d}r \int_\Sigma u(x_0 + r\sigma) \mathrm{d}\sigma = \sigma_n u(x_0) \frac{r_0^n}{n} .$$

We can then state:

Corollary 2. *Let $B = B(x_0, r_0)$ and $u \in \mathscr{H}(B) \cap L^1(B)$. Then*

$$u(x_0) = \frac{n}{\sigma_n r_0^n} \int_B u \, dx . \tag{2.8}$$

We note that σ_n/n is the volume of the unit ball.
The formula of the mean is a powerful tool. A first example of its use is the generalisation of Corollary 1:

Proposition 7. *A harmonic function on $\mathbb{R}^n$, bounded above or below, is constant.*

Proof. Let us suppose, for example, that u is bounded below; by the addition of a constant, we can always suppose that $u \geqslant 0$.
Being given $x_1, x_2 \in \mathbb{R}^n$ and $r_1 \geqslant r_2 + |x_1 - x_2| \geqslant r_2 > 0$, we have

$$B_2 = B(x_2, r_2) \subset B(x_1, r_1) = B_1$$

Applying (2.8) we obtain

$$u(x_2) = \frac{n}{\sigma_n r_2^n} \int_{B_2} u \, dx \leqslant \frac{n}{\sigma_n r_2^n} \int_{B_1} u \, dx = \left(\frac{r_1}{r_2}\right)^n u(x_1) .$$

Taking $r_1 = r_2 + |x_1 - x_2|$ and letting $r_2 \to \infty$, we find that

$$u(x_2) \leqslant u(x_1) .$$

Interchanging the roles of x_1 and x_2, we have $u(x_1) \leqslant u(x_2)$ and hence $u(x_1) = u(x_2)$. □

Let us now state the *principle of the maximum*

Proposition 8. *Let Ω be a connected set and $u \in \mathscr{H}(\Omega)$. If there exists $x_0 \in \Omega$ such that*

$$u(x) \leqslant u(x_0) (\textit{resp.}\, u(x) \geqslant u(x_0)) \quad \textit{for all} \quad x \in \Omega ,$$

then u is constant on Ω.

In other words: a non-constant harmonic function on a connected open set can have neither a maximum nor a minimum on that open set.

Proof. First of all let us consider $x_0 \in \Omega$ and $r_0 > 0$ such that

$$\bar{B}(x_0, r_0) \subset \Omega \quad \text{and} \quad u(x) \leqslant u(x_0) \quad \text{for all} \quad x \in B(x_0, r_0)$$

and show that then u is constant on $B(x_0, r_0)$. We have in effect from (2.8), since $\sigma_n r_0^n/n$ is the volume of the ball $B = B(x_0, r_0)$

$$\int_B (u(x_0) - u(x)) \, dx = 0 .$$

If $u(x_0) - u(x) \geqslant 0$ for $x \in B$, then necessarily

$$u(x_0) - u(x) = 0 \qquad \forall x \in B .$$

Now let us consider

$$\Omega_0 = \{x_0 \in \Omega \, ; \quad u(x) \leqslant u(x_0) , \qquad \forall x \in \Omega\}$$

From the hypothesis $\Omega_0 \neq \emptyset$; from what has gone before, Ω_0 is open. Finally Ω_0 is closed in Ω, since

$$\Omega_0 = \left\{ x_0 \in \Omega \, ; \quad u(x_0) = \max_{\Omega} u \right\}.$$

Hence, since Ω is connected, $\Omega_0 = \Omega$. □

Remark 4. Note that in the above proof we have used only that u should satisfy the formula of the mean (2.8); we could just as well have used the formula of the mean (2.7); in fact we shall show later that these formulae characterise the harmonic functions (see Proposition 15).
Note also that, making use of the analyticity of a harmonic function, we see that if u is constant in the neighbourhood of a point, it is constant everywhere (on a connected open set); that shows that a non-constant harmonic function on a connected open set can have neither a *local* maximum nor a *local* minimum. □

Let us give some useful corollaries of this maximum principle.

Corollary 3. *Let Ω be a* **bounded** *open set and $u \in \mathscr{H}(\Omega) \cap \mathscr{C}^0(\bar{\Omega})$. Then*

$$\min_{\Gamma} u \leqslant u(x) \leqslant \max_{\Gamma} u \quad \textit{for all} \quad x \in \Omega \,.$$

If, in addition, Ω is connected and u is non-constant, then

$$\min_{\Gamma} u < u(x) < \max_{\Gamma} u \quad \textit{for all} \quad x \in \Omega \,.$$

Proof. The function u being continuous on the compact set $\bar{\Omega}$ attains its maximum at a point $x_0 \in \bar{\Omega}$. If $u(x_0)$ were strictly superior to $\max_{\Gamma} u$, x_0 would belong to Ω and u would not be constant on the connected component of x_0 in Ω; this contradicts the principle of the maximum.
The second assertion is an immediate corollary of the first assertion and of the principle of the maximum. □

This corollary is in general false for an unbounded open set: for $\Omega = \{x \in \mathbb{R}^n; |x| > 1\}$, the function $u(x) = E_2(x)$ (case $n = 2$) or $u(x) = 1/k_n - E_n(x)$ is harmonic and positive on Ω but is zero on $\partial\Omega = \Sigma$. It is clear that to generalise the statement of Corollary 3 to the case of an unbounded open set, we must introduce infinity as belonging to the boundary. We can state:

Corollary 4. *Let Ω be an unbounded open set and $u \in \mathscr{H}(\Omega) \cap \mathscr{C}^0(\bar{\Omega})$. Being given $M_-, M_+ \in \mathbb{R}$, $(M_+ > M_-)$ let us suppose that*

(i) $M_- \leqslant u(x) \leqslant M_+$ *for all* $x \in \Gamma$,

(ii) $M_- \leqslant \liminf_{|x| \to \infty} u(x) \leqslant \limsup_{|x| \to \infty} u(x) \leqslant M_+$.

Then

$$M_- \leqslant u(x) \leqslant M_+ \quad \textit{for all} \quad x \in \Omega \,.$$

Proof. Being given $\varepsilon > 0$, we deduce from hypothesis (i), that there exists $R > 0$ such that

$$x \in \Omega , \quad |x| \geqslant R \Rightarrow M_- - \varepsilon \leqslant u(x) \leqslant M_+ + \varepsilon .$$

We apply the preceding corollary to the restriction of u to the bounded open set $\Omega_R = \Omega \cap B(0, R)$; since the boundary $\partial\Omega_R \subset \Gamma \cup (\Omega \cap \partial B(0, R))$, we have

$$M_- - \varepsilon \leqslant \min_{\partial\Omega_R} u \leqslant \max_{\partial\Omega_R} u \leqslant M_+ + \varepsilon ,$$

so

$$M_- - \varepsilon \leqslant u(x) \leqslant M_+ + \varepsilon \quad \text{for all} \quad x \in \Omega_R .$$

We have also

$$M_- - \varepsilon \leqslant u(x) \leqslant M_+ + \varepsilon \quad \text{for all} \quad x \in \Omega \setminus \Omega_R = \{x \in \Omega ; |x| > R\} .$$

Therefore

$$M_- - \varepsilon \leqslant u(x) \leqslant M_+ + \varepsilon \quad \text{for all} \quad x \in \Omega .$$

In the limit when $\varepsilon \to 0$, we obtain the result. □

Finally we note the following useful results:

Corollary 5. *Let Ω be an open set and $u_1, u_2 \in \mathscr{H}(\Omega) \cap \mathscr{C}^0(\bar{\Omega})$. If Ω is unbounded, we suppose in addition that*

$$\lim_{|x| \to \infty} u_1(x) = \lim_{|x| \to \infty} u_2(x) = 0 .$$

Then

$$\begin{cases} u_1 \leqslant u_2 \quad \text{on} \quad \Gamma \Rightarrow u_1 \leqslant u_2 \quad \text{on} \quad \Omega \\ u_1 = u_2 \quad \text{on} \quad \Gamma \Rightarrow u_1 = u_2 . \end{cases}$$

Proof. It is sufficient to apply the preceding corollaries to $u = u_1 - u_2$.

Corollary 6. *Let Ω be an open set and (u_k) a sequence of functions harmonic on Ω and continuous on $\bar{\Omega}$. If Ω is unbounded, we suppose in addition that*

$$\lim_{|x| \to \infty} u_k(x) = 0 \quad \textit{for all} \quad k .$$

Then if $(u_k(x))$ converges uniformly for $x \in \Gamma$, the sequence (u_k) converges uniformly on Ω.

Proof. From Corollaries 3 and 4, we have for all k and l

$$\sup_{\Omega} |u_k - u_l| \leqslant \sup_{\Gamma} |u_k - u_l|$$

(if Ω is unbounded, $\lim_{|x| \to \infty} u_k - u_l = 0)$[12] . □

---

[12] We recall that a sequence of functions (f_k) defined on an (arbitrary) set converges uniformly if $\lim_{k, l \to \infty} \sup |f_k - f_l| = 0$.

3. Poisson's Integral Formula; Regularity of Harmonic Functions; Harnack's Inequality

Let us begin by proving the

Lemma 1. *Let* $B = B(x_0, r_0)$, $x \in B \setminus \{x_0\}$ *and* $u \in \mathscr{H}(B) \cap \mathscr{C}^0(\bar{B})$. *Then*

$$u(x) = \int_{\partial B} u(t) \frac{\partial}{\partial n}\left(E_n(t - x) - \left(\frac{r_0}{|x - x_0|}\right)^{n-2} E_n(t - y)\right) \mathrm{d}\gamma(t) \tag{2.9}$$

where

$$y = x_0 + \frac{r_0^2}{|x - x_0|^2}(x - x_0) . \tag{2.10}$$

Proof. We can always suppose that $u \in \mathscr{C}_n^1(\bar{B})$: it is then enough to apply the result to $B_r = B(x_0, r)$ for $0 < r < r_0$ and pass to the limit with $r \to r_0$.
We note first of all that $y \notin B$, with the result that the function $v(t) = E_n(t - y)$ is in $\mathscr{H}(B) \cap \mathscr{C}_n^1(\bar{B})$. Applying Green's formula for harmonic functions (2.4), we have

$$0 = \int_{\partial B} \left(u(t) \frac{\partial E_n}{\partial n}(t - y) - E_n(t - y) \frac{\partial u}{\partial n}(t) \right) \mathrm{d}\gamma(t) . \tag{2.11}$$

Now let us consider $0 < \varepsilon < r_0 - |x - x_0|$ and the open set

$$\Omega = \{t \in B ; |t - x| > \varepsilon\} ;$$

the function $v(t) = E_n(t - x)$ is in $\mathscr{H}(\Omega) \cap \mathscr{C}_n^1(\bar{\Omega})$. The boundary $\partial\Omega$ is made up of ∂B and $\partial B(x, \varepsilon)$. Using the fact that v is constant on $\partial B(x, \varepsilon)$ and Gauss' theorem for u, we have

$$\int_{\partial\Omega} v(t) \frac{\partial u}{\partial n}(t) \mathrm{d}\gamma(t) = \int_{\partial B} v(t) \frac{\partial u}{\partial n}(t) \mathrm{d}\gamma(t) .$$

Making use of the value of grad E_n, we have (see the proof of the formula of the mean):

$$\int_{\partial\Omega} u(t) \frac{\partial v}{\partial n}(t) \mathrm{d}\gamma(t) = \int_{\partial B} u(t) \frac{\partial v}{\partial n}(t) \mathrm{d}\gamma(t) - \frac{1}{\sigma_n} \int_{\Sigma} u(x + \varepsilon\sigma) \mathrm{d}\sigma \text{ [13]}.$$

Applying Green's formula for the harmonic functions u and v on Ω, and passing to the limit as $\varepsilon \to 0$, we obtain:

$$u(x) = \int_{\partial B} \left(u(t) \frac{\partial E_n}{\partial n}(t - x) - E_n(t - x) \frac{\partial u}{\partial n}(t) \right) \mathrm{d}\gamma(t) .$$

---

[13] We recall (see §1.4) that Σ is the unit sphere of $\mathbb{R}^n$ and $\mathrm{d}\sigma$ the surface measure of Σ.

Combining with (2.11) we deduce that for all $\alpha \in \mathbb{R}$,

$$\begin{cases} u(x) = \displaystyle\int_{\partial B} u(t)\left(\frac{\partial E_n}{\partial n}(t-x) - \alpha\frac{\partial E_n}{\partial n}(t-y)\right)\mathrm{d}\gamma(t) \\ \qquad - \displaystyle\int_{\partial B} (E_n(t-x) - \alpha E_n(t-y))\frac{\partial u}{\partial n}(t)\,\mathrm{d}\gamma(t)\,. \end{cases}$$

We can show that for

$$\alpha = \left(\frac{r_0}{|x - x_0|}\right)^{n-2},$$

the second integral is zero, which will prove the lemma. This comes from the identity

$$\frac{|t-x|}{|t-y|} = \left|\frac{x - x_0}{r_0}\right|, \qquad \forall t \in \partial B \tag{2.12}$$

which we can verify by making use of (2.10).
In the case $n \geqslant 3$, we therefore have

$$E_n(t-x) = \alpha E_n(t-y)$$

and the result is proved.
In the case $n = 2$, we have $\alpha = 1$ and

$$E_2(t-x) = E_2(t-y) + C \qquad C = \frac{1}{2\pi}\operatorname{Log}\left|\frac{x-x_0}{r_0}\right|,$$

with the result that the second integral reduces to

$$C\int_{\partial B}\frac{\partial u}{\partial n}(t)\,\mathrm{d}\gamma(t)$$

which is zero from Gauss' theorem. □

Making explicit the right hand side of (2.9), we prove now *Poisson's integral formula* (for a ball):

Proposition 9. *Let $B = B(x_0, r_0)$ and $u \in \mathscr{H}(B) \cap \mathscr{C}^0(\bar{B})$. Then, for all $x \in B$,*

$$u(x) = \frac{1}{r_0\sigma_n}\int_{\partial B}\frac{r_0^2 - |x - x_0|^2}{|t-x|^n}u(t)\,\mathrm{d}\gamma(t)\,. \tag{2.13}$$

Remark 5. We can apply Remark 3 on the non-necessity of the condition $u \in \mathscr{C}^0(\bar{B})$ (as for the formula of the mean). With respect to the formula of the mean, Poisson's integral formula is concerned with expressing the value of the harmonic function u at each point of B as a function of its values on ∂B; it is allied to Cauchy's integral formula for holomorphic functions, but has the inconvenience as far as this latter result is concerned of applying only for a ball. In the meantime we shall also obtain Poisson integral formulae for the exterior of a ball and for a half-space (see

Propositions 21 and 22), without forgetting the case $n = 2$ where we are able, at least in theory, to obtain integral formulae for any simply connected set whatsoever (see the end of §2.5). □

Proof. We resume the notation of the lemma and its proof. We have

$$\frac{\partial E_n}{\partial n}(t - x) = \operatorname{grad} E_n(t - x).n(t) = \frac{(t - x).n(t)}{\sigma_n |t - x|^n}.$$

Also, taking account of (2.10), (2.12) and $(t - x_0).n(t) = r_0$,

$$\frac{\partial E_n}{\partial n}(t - y) = \frac{(t - y).n(t)}{\sigma_r |t - y|^n} = \frac{1}{\sigma_n}\left(\frac{|x - x_0|}{r_0 |t - x|}\right)^n \left[r_0 - \frac{r_0^2}{|x - x_0|^2}(x - x_0).n(t)\right].$$

Hence, taking account of the value of α

$$\frac{\partial E_n}{\partial n}(t - x) - \alpha \frac{\partial E_n}{\partial n}(t - y) = \frac{1}{\sigma_n |t - x|^n}\left[(t - x_0).n(t) - \frac{|x - x_0|^2}{r_0}\right]$$

$$= \frac{r_0^2 - |x - x_0|^2}{r_0 \sigma_n |t - x|^n}.$$

Taking this back into (2.9), we obtain (2.13). □

First of all, let us apply Poisson's formula to give a direct proof of the analyticity of harmonic functions. We give a statement a little more refined than will be useful in the sequel:

Proposition 10. *Let $B = B(x_0, r_0)$ be a ball and (u_k) a sequence of functions, harmonic on B and continuous on $\bar{B}$. We suppose that (u_k) converges uniformly on ∂B when $k \to \infty$. Then (u_k) converges uniformly on $\bar{B}$, its derivatives converge uniformly on every compact set of B and its limit is harmonic and analytic.*

We note the corollary:

Corollary 7. *Every continuous function solution in the sense of distributions of Laplace's equation on an open set Ω of $\mathbb{R}^n$, is harmonic analytic on that open set.*

Proof of Proposition 10. The uniform convergence of (u_k) follows from the maximum principle; from Corollary 4, $\max_{\bar{B}} |u_k - u_l| \leqslant \max_{\partial B} |u_k - u_l|$. To simplify the notation, we suppose that $x_0 = 0$, $r_0 = 1$, which takes away nothing from the generality of the proof. Poisson's formula for the functions u_k can be written

$$u_k(x) = \frac{1}{\sigma_n} \int_\Sigma \frac{1 - |x|^2}{(1 - T(x, \sigma))^{n/2}} u_k(\sigma) \, d\sigma \tag{2.14}$$

with

$$T(x, \sigma) = 1 - |\sigma - x|^2.$$

For $(x, \sigma) \in \bar{B}(0, 1 - \varepsilon) \times \Sigma$, we have

$$|\sigma - x| \geqslant \varepsilon \quad \text{and hence} \quad |T(x, \sigma)| \leqslant 1 - \varepsilon^2.$$

Since T is a polynomial (of second degree) in (x, σ), developing $(1 - T^2)^{-n/2}$ in a power series on $\{|T| \leqslant 1 - \varepsilon^2\}$, we obtain a development in power series of

$$k(x, \sigma) = (1 - |x|^2)(1 - T(x, \sigma))^{-n/2}$$

converging normally on $\bar{B}(0, 1 - \varepsilon) \times \Sigma$. It follows from (2.14) by integration of this development, that u_k can be developed in power series which are convergent normally on $B(0, 1 - \varepsilon)$ and that these power series converge normally when $k \to \infty$; in particular all the derivatives of u_k converge uniformly on $B(0, 1 - \varepsilon)$ when $k \to \infty$.
This clearly proves the proposition. □

Proof of Corollary 7. Let us consider $u \in \mathscr{C}(\Omega)$ with

$$\int u \Delta \zeta \, dx = 0 \quad \text{for all} \quad \zeta \in \mathscr{D}(\Omega) .$$

It is enough to show that the function u is harmonic on every bounded open set Ω' with $\bar{\Omega}' \subset \Omega$. We fix such an open set Ω' and consider $0 < r_0 < \operatorname{dist}(\Omega', \partial\Omega)$. We *regularise u by convolution*: that is to say we prescribe a function $\rho \in \mathscr{D}(\mathbb{R}^n)$ with $\rho \geqslant 0$ and such that

$$\int \rho \, dx = 1 \quad \text{and} \quad \operatorname{supp} \rho \subset B(0, r_0) .$$

For all $k \geqslant 1$, we put $\rho_k(x) = k^n \rho(kx)$ with the result that $\rho_k \in \mathscr{D}(\mathbb{R}^n)$, and we have:

$$\rho_k \geqslant 0 , \qquad \int \rho_k \, dx = 1 \quad \text{and} \quad \operatorname{supp} \rho_k \subset B\left(0, \frac{r_0}{k}\right) .$$

We then put $u_k = u * \rho_k$, that is to say

$$u_k(x) = \int u(y) \rho_k(x - y) \, dy = \int u(x - y) \rho_k(y) \, dy$$

which is well defined for all $x \in \Omega'$. We have $u_k \in \mathscr{C}^2(\Omega')$ and from the hypothesis

$$\Delta u_k(x) = \int u(y) \Delta \rho_k(x - y) \, dy = 0 \quad \text{for all} \quad x \in \Omega' ,$$

that is to say

$$u_k \in \mathscr{H}(\Omega') .$$

We then obtain the result as a corollary of Proposition 10, since $u_k \to u$ uniformly on Ω'. This classical fact is immediately obvious by

$$|u_k(x) - u(x)| = \left| \int (u(x - y) - u(x)) \rho_k(y) \, dy \right| \leqslant \max_{|y| \leqslant \frac{r_0}{k}} |u(x - y) - u(x)| .$$

□

Let us now apply Poisson's formula to prove *Harnack's inequality* (*for a ball*).

Proposition 11. *Let* $B = B(x_0, r_0)$, $u \in \mathscr{H}(B)$ *with* $u \geqslant 0$, $x_1, x_2 \in B$. *Denoting* $r_1 = |x_1|$, $r_2 = |x_2|$, *we have*

$$u(x_1) \leqslant \frac{r_0 + r_1}{r_0 - r_2}\left(\frac{r_0 + r_2}{r_0 - r_1}\right)^{n-1} u(x_2)\,. \tag{2.15}$$

Proof. We can always suppose $u \in \mathscr{C}(\bar{B})$, it is sufficient in effect to apply the result to the ball $B(x_0, r)$ with $\max(r_1, r_2) < r < r_0$ and pass to the limit with $r \to r_0$. We thus have Poisson's formula

$$u(x_i) = \frac{1}{r_0 \sigma_n} \int_{\partial B} \frac{r_0^2 - r_i^2}{|t - x_i|^n} u(t)\,\mathrm{d}\gamma(t)\,, \qquad i = 1, 2\,.$$

Now, for $t \in \partial B$,

$$r_0 - r_i \leqslant |t - x_i| \leqslant r_0 + r_i\,,$$

with the result that

$$\frac{r_0 - r_i}{(r_0 + r_i)^{n-1}} = \frac{r_0^2 - r_i^2}{(r_0 + r_i)^n} \leqslant \frac{r_0^2 - r_i^2}{|t - x_i|^n} \leqslant \frac{r_0^2 - r_i^2}{(r_0 - r_i)^n} = \frac{r_0 + r_i}{(r_0 - r_i)^{n-1}}\,.$$

We deduce that

$$\frac{r_0^2 - r_1^2}{|t - x_1|^n} \leqslant \frac{r_0 + r_1}{r_0 - r_2}\left(\frac{r_0 + r_2}{r_0 - r_1}\right)^{n-1} \frac{r_0^2 - r_2^2}{|t - x_2|^n}\,.$$

Multiplying by $u(t) \geqslant 0$ and integrating, we obtain the result. $\square$

Considering $B = B(x_0, r_0)$, $u \in \mathscr{H}(B)$ with $u \geqslant 0$ and $0 < r < r_0$, we have by application of (2.15),

$$u(x_1) \leqslant \left(\frac{r_0 + r}{r_0 - r}\right)^n u(x_2) \quad \text{for all} \quad x_1, x_2 \in \bar{B}(x_0, r) \tag{2.16}$$

which can just as well be written

$$\max_{\bar{B}(x_0, r)} u \leqslant \left(\frac{r_0 + r}{r_0 - r}\right)^n \min_{\bar{B}(x_0, r)} u\,.$$

By an argument of connectedness, we obtain as a generalisation the *Harnack inequality for a connected open set.*

Proposition 12. *Let* Ω *be a connected open set and* K *a compact subset of* Ω. *There exists a constant* C *such that for all* $u \in \mathscr{H}(\Omega)$ *with* $u \geqslant 0$.

$$\max_K u \leqslant C \min_K u\,. \tag{2.17}$$

Proof. Using the connectedness of Ω and the compactness of K, we can find a finite sequence of balls $B(x_1, 2r_1), \ldots, B(x_N, 2r_N)$ contained in Ω such that the balls $\bar{B}(x_1, r_1), \ldots, \bar{B}(x_N, r_N)$ cover K and are connected two by two: $\bar{B}(x_i, r_j) \cap \bar{B}(x_{j+1}, r_{j+1})$ contains a point x_j' for $j = 1, \ldots, N - 1$.

We show that the proposition is satisfied with $C = 3^{nN}$. That reduces to showing that being given $u \in \mathscr{H}(\Omega)$ with $u \geqslant 0$, for all $x, y \in \bigcup_j \bar{B}(x_j, r_j)$, then $u(x) \leqslant 3^{nN} u(y)$.
The worst situation will arise when $x \in \bar{B}(x_1, r_1)$, $y \in \bar{B}(x_N, r_N)$; applying (2.16) we have

$$u(x) \leqslant 3^n u(x'_1), u(x'_1) \leqslant 3^n u(x'_2), \ldots, u(x'_{N-1}) \leqslant 3^n u(y) . \qquad \square$$

We note that Harnack's inequality implies the principle of the maximum: if $u \in \mathscr{H}(\Omega)$ and $u(x_0) = \min u$ for an $x_0 \in \Omega$, changing u to $u - u(x_0)$, we can suppose $u(x_0) = 0$; then Proposition 12 shows that $\max_K u \leqslant 0$ and hence $u = 0$ on every compact set containing x_0; obviously then $u = u(x_0)$ on Ω. Harnack's inequality is a much more powerful tool than the principle of the maximum, as will be shown in the proofs of the following *compactness theorems* called *Harnack's theorems.*
Given an open set Ω and a sequence (u_k) of harmonic functions on Ω, it follows from Proposition 10 that *if (u_k) converges uniformly on every compact set of Ω, then all the derivatives of u_k also converge uniformly on every compact set of Ω and the limit is harmonic on Ω*. We then say that u_k *converges in* $\mathscr{H}(\Omega)$.

Proposition 13. *Let Ω be a connected open set and (u_k) a sequence of harmonic functions on Ω. Suppose that:*
(i) there exists a function u_0 harmonic on Ω bounding the functions u_k below ($u_0(x) \leqslant u_k(x)$ for all $x \in \Omega$ and all k),
(ii) there exists $x_0 \in \Omega$ such that $(u_k(x_0))$ is bounded above.
Then there exists a sub-sequence convergent in $\mathscr{H}(\Omega)$.

Proof. Replacing u_k by $u_k - u_0$ we can suppose $u_k \geqslant 0$. Since convergence in $\mathscr{H}(\Omega)$ is uniform convergence on every compact set, from Ascoli's theorem[14], it is enough to prove that, for all $x \in \Omega$,
(*a*) $(u_k(x))$ is bounded,
(*b*) the functions u_k are equicontinuous in x, i.e. that the limit $\lim_{y \to x} u_k(y) = u_k(x)$ is uniform with respect to k.
From (*ii*) and Harnack's inequality, the sequence (u_k) is bounded on every compact K of Ω: in effect we can always suppose $x_0 \in K$; from Proposition 12 we then have

$$\max_K u_k \leqslant C \min_K u_k \leqslant C u_k(x_0) \leqslant C' .$$

Let us hence prove (*b*); to simplify the notation, we take $x = x_0$. For $r > 0$ with $\bar{B}(x_0, r) < \Omega$ we denote

$$m_k(r) = \min_{\bar{B}(x_0, r)} u_k, \qquad M_k(r) = \max_{\bar{B}(x_0, r)} u_k .$$

[14] For the topological ideas used here, we refer to Dieudonné [1] or Schwartz [2]. See also Lemma 4 of §4.4.

Applying (2.16) to the functions $u_k - m_k(r)$ and $M_k(r) - u_k$ harmonic and positive on $B(x_0, r)$, we obtain the inequalities:

$$M_k\left(\frac{r}{2}\right) - m_k(r) \leqslant 3^n\left(m_k\left(\frac{r}{2}\right) - m_k(r)\right)$$

$$M_k(r) - m_k\left(\frac{r}{2}\right) \leqslant 3^n\left(M_k(r) - M_k\left(\frac{r}{2}\right)\right);$$

from which, on adding, and putting $\omega_k(r) = M_k(r) - m_k(r)$, *the oscillation of* u_k *on* $\bar{B}(x_0, r)$, we have

$$\omega_k(r) + \omega_k\left(\frac{r}{2}\right) \leqslant 3^n\left(\omega_k(r) - \omega_k\left(\frac{r}{2}\right)\right);$$

which is equivalent to

$$\omega_k\left(\frac{r}{2}\right) \leqslant \theta\omega_k(r) \quad \text{with} \quad \theta = \frac{3^n - 1}{3^n + 1} < 1 .$$

By recurrence for all $j = 1, 2, \ldots$

$$\omega_k\left(\frac{r}{2^j}\right) \leqslant \theta^j\omega_k(r) \leqslant \theta^j M_k(r) . \tag{2.18}$$

Since $M_k(r)$ is bounded, and $0 < \theta < 1$, we deduce that $\omega_k\left(\frac{r}{2^j}\right) \to 0$ as $j \to \infty$ uniformly with respect to k, which proves (b). □

We note the corollaries:

Corollary 8. *Let Ω be a bounded open set and (u_k) a sequence of functions harmonic on Ω and continuous on $\bar{\Omega}$, satisfying*

$$|u_k(x)| \leqslant M \quad \textit{for all} \quad x \in \Gamma \quad \textit{and} \quad \textit{all } k .$$

Then there exists a sub-sequence of (u_k) converging in $\mathscr{H}(\Omega)$.

Proof. From the maximum principle we have

$$|u_k(x)| \leqslant M \quad \text{for all} \quad x \in \Omega \quad \text{and all} \quad k$$

and hence in particular the conditions (i) and (ii) of Proposition 13. In the case Ω is connected, the corollary is proved. In the case of an arbitrary Ω, it suffices to apply Proposition 13 and use a "diagonal procedure"[15] since Ω has at most a denumerable number of connected components. □

---

[15] The *diagonal procedure* is the following topological lemma: given a double sequence $(a_{k,l})_{k,l}$ (of a topological space) satisfying:
for all fixed l and for every sub-sequence $(a_{k,l})_k$ we can extract a convergent sequence;
then there exists an increasing sequence $(k_j)_j$ such that for all l, the sequence $(a_{k_j,l})_j$ is convergent.

Corollary 9. *Let Ω be a connected open set and $(u_k)_k$ a monotonic sequence (increasing or decreasing) of functions harmonic on Ω. Suppose that there exists $x_0 \in \Omega$ such that $(u_k(x_0))$ is bounded. Then the sequence (u_k) converges in $\mathscr{H}(\Omega)$.*

Proof. Considering for example (u_k) increasing, we have the conditions needed for the application of Proposition 13. But the sequence being monotonic, if a sub-sequence converges, the whole sequence converges.
We note that the monotonicity of the sequence permits a direct proof: for a compact set K of Ω (containing x_0), applying (2.17) to $u_k - u_l$ ((u_k) increasing $k \geqslant l$)

$$\max_K (u_k - u_l) \leqslant C \min_K (u_k - u_l) \leqslant C(u_k(x_0) - u_l(x_0)) .$$

Since the sequence $(u_k(x_0))$ is increasing and bounded above, it converges; hence (u_k) converges uniformly on K. □

4. Characterisation of Harmonic Functions. Elimination of Singularities

Given an open set Ω of $\mathbb{R}^n$, we have seen that a function $u \in \mathscr{H}(\Omega)$ satisfies:
(a) Gauss' theorem: for every regular bounded open set ω, with $\bar{\omega} \subset \Omega$,

$$\int_{\partial\omega} \frac{\partial u}{\partial n} \mathrm{d}\gamma = 0 ;$$

(b) the formula of the mean on spheres:

$$\text{for each point } x \text{ and } r > 0 \text{ with } \bar{B}(x, r) \subset \Omega, \quad u(x) = \frac{1}{\sigma_n r^{n-1}} \int_{\partial B(x, r)} u \, \mathrm{d}\gamma ;$$

(c) the formula of the mean on balls:

$$\text{for each point } x \text{ and } r > 0 \text{ with } \bar{B}(x, r) \subset \Omega , \qquad u(x) = \frac{n}{\sigma_n r^n} \int_{B(x, r)} u \, \mathrm{d}x .$$

We propose to show that *these propositions characterise the harmonic functions*. Let us begin with Gauss' theorem, which we note has a (classical) meaning when $u \in \mathscr{C}^1(\Omega)$. We prove:

Proposition 14. *Let $u \in \mathscr{C}^1(\Omega)$ possess the Gauss property: for every regular bounded open set ω with $\bar{\omega} \subset \Omega$, $\displaystyle\int_{\partial\omega} \frac{\partial u}{\partial n} \mathrm{d}\gamma = 0$. Then u is harmonic on Ω.*

Proof. We note that if $u \in \mathscr{C}^2(\Omega)$, the result follows immediately from the formula of Green (Ostrogradski): for $x_0 \in \Omega$ and $r > 0$ with $\bar{B}(x_0, r) \subset \Omega$, we have

$$\int_{B(x_0, r)} \Delta u \, \mathrm{d}x = \int_{\partial B(x_0, r)} \frac{\partial u}{\partial n} \mathrm{d}\gamma = 0 .$$

Hence there exists $x_r \in B(x_0, r)$ such that $\Delta u(x_r) = 0$. In the limit when $r \to 0$, $\Delta u(x_0) = 0$.

In the general case considered, u is only of class $\mathscr{C}^1$, we shall regularize u by convolution (see the proof of Corollary 7) using a radial function ρ. Hence we take

$$\rho \in \mathscr{D}(\mathbb{R}^+), \quad \rho \geqslant 0, \quad \int_0^\infty r^{n-1}\rho(r)\,\mathrm{d}r = \frac{1}{\sigma_n}, \quad \operatorname{supp}\rho \subset]0, 1],$$

$$\rho = 1 \text{ in the neighbourhood of } 0;$$

for all $k \geqslant 1$, we put $\rho_k(x) = k^n\rho(k|x|)$ which satisfies

$$\rho_k \in \mathscr{D}(\mathbb{R}^n), \quad \rho_k \geqslant 0, \quad \int \rho_k\,\mathrm{d}x = 1, \quad \operatorname{supp}\rho_k \subset \bar{B}\left(0, \frac{1}{k}\right).$$

We put $u_k = u * \rho_k$ which is defined and is of class $\mathscr{C}^\infty$ on

$$\Omega_k = \left\{x \in \Omega;\ \operatorname{dist}(x, \partial\Omega) > \frac{1}{k}\right\}.$$

We have for $x \in \Omega_k$

$$\begin{aligned}\Delta u_k(x) &= \int u(y)\Delta\rho_k(x - y)\,\mathrm{d}y = \int \operatorname{grad} u(y).\operatorname{grad}\rho_k(x - y)\,\mathrm{d}y \\ &= k^{n+1}\int \operatorname{grad} u(y).\frac{\mathrm{d}\rho}{\mathrm{d}r}(k|x - y|)\frac{x - y}{|x - y|}\,\mathrm{d}y \\ &= k^{n+1}\int_0^{1/k}\frac{\mathrm{d}\rho}{\mathrm{d}r}(kr)r^{n-1}\,\mathrm{d}r\int_\Sigma \operatorname{grad} u(x - r\sigma).\sigma\,\mathrm{d}\sigma.\end{aligned}$$

But

$$\int_\Sigma \operatorname{grad} u(x - r\sigma).\sigma\,\mathrm{d}\sigma = \frac{1}{r^{n-1}}\int_{\partial B(x,r)}\frac{\partial u}{\partial n}\,\mathrm{d}\gamma = 0.$$

Hence

$$\Delta u_k(x) = 0.$$

In other words, u_k is harmonic on Ω_k. From Proposition 10, since $u_k \to u$ uniformly on each compact set of Ω, $u \in \mathscr{H}(\Omega)$. □

Let us now consider the case of the formulae of the mean. We have seen in the proof of Corollary 2 that the formula of the mean on spheres implies the formula of the mean on balls. We shall therefore consider only this latter case. Note that the mean on a ball B of a function u can be defined when $u \in L^1(B)$. Hence, let us consider a function u, *locally integrable on* Ω and for all $x \in \Omega$ and $0 < r < \operatorname{dist}(x, \partial\Omega)$, define the function M by the equation

$$M(x, r) = \frac{n}{\sigma_n r^n}\int_{B(x,r)} u(y)\,\mathrm{d}y,$$

the mean of u integrable on $B(x, r)(\bar{B}(x, r) \subset \Omega)$. We can also write

$$M(x, r) = \frac{n}{\sigma_n}\int_B u(x + ry)\,\mathrm{d}y \tag{2.19}$$

where $B = B(0, 1)$. We see therefore[16] that the function

$$(x, r) \to M(x, r)$$

is continuous on the open set $U = \{(x, r);\ x \in \Omega,\ 0 < r < \operatorname{dis}(x, \partial\Omega)\}$ of $\Omega \times]0, \infty[$.

In the case $u \in \mathscr{C}^2(\Omega)$, we have $M \in \mathscr{C}^2(U)$ and by differentiation under the integral sign and the use of Green's formula, we obtain

$$\Delta_x M(x, r) = \frac{n}{\sigma_n} \int_B \Delta u(x + ry)\,\mathrm{d}y = \frac{n}{\sigma_n} \int_\Sigma \operatorname{grad} u(x + r\sigma).\sigma\,\mathrm{d}\sigma\,.$$

In the same way we have

$$\frac{\partial M}{\partial r}(x, r) = \frac{n}{\sigma_n} \int_B \operatorname{grad} u(x + ry).y\,\mathrm{d}y \tag{2.20}$$

$$= \frac{n}{\sigma_n r^{n+1}} \int_0^r \tau^n \,\mathrm{d}\tau \int_\Sigma \operatorname{grad} u(x + \tau\sigma).\sigma\,\mathrm{d}\sigma\,.$$

Therefore M satisfies the partial differential equations

$$\Delta_x M = \frac{1}{r^n} \frac{\partial}{\partial r}\left(r^{n+1} \frac{\partial M}{\partial r}\right) \quad \text{on} \quad U. \tag{2.21}$$

For u only in $L^1_{\mathrm{loc}}(\Omega)$, M always satisfies the partial differential equation (2.21) provided that it is taken in the sense of distributions in U. This follows from a general argument of the density of $\mathscr{C}^2(\Omega)$ in $L^1_{\mathrm{loc}}(\Omega)$, of the continuity $u \to M$ of $L^1_{\mathrm{loc}}(\Omega)$ into $\mathscr{C}(U)$ and of the continuity of the derivatives in $\mathscr{D}'(U)$; the reader will also be able to verify it by duality.

Now let us suppose

$$M(x, r) = u(x) \quad a.e. \quad (x, r) \in U\,. \tag{2.22}$$

In the case $u \in \mathscr{C}^2(\Omega)$, we deduce immediately from (2.21) that $\Delta_x M(x, r) = 0$ for all $(x, r) \in U$ and hence u is harmonic. In the general case $u \in L^1_{\mathrm{loc}}(\Omega)$ this is again true by using the following reasoning. From (2.22) $\partial M/\partial r = 0$ in $\mathscr{D}'(U)$ and hence from (2.21), $\Delta_x(M) = 0$ in $\mathscr{D}'(U)$. Consider a bounded open set ω with $\bar{\omega} \subset \Omega$, $U \supset \omega \times]0, r_0[$ with $r_0 = \operatorname{dist}(\omega, \partial\Omega)$; since M is continuous on $\omega \times]0, r_0[$, using the definition of $\Delta_x M$ in $\mathscr{D}'(\omega \times]0, r_0[)$ and Corollary 7, we see immediately that for all $r \in]0, r_0[$, the function $x \in \omega \to M(x, r)$ is harmonic.

We have thus proved

Proposition 15. *Let $u \in L^1_{\mathrm{loc}}(\Omega)$. Suppose that*

(2.23)

$$u(x) = \frac{n}{\sigma_n r^n} \int_{B(x, r)} u(y)\,\mathrm{d}y \quad a.e.\ (x, r) \quad with \quad x \in \Omega,\quad 0 < r < \operatorname{dist}(x, \partial\Omega)\,.$$

---

[16] We make use of the continuity of the translation in L^1.

Then there exists a function $\tilde{u}$, harmonic on Ω such that

$$u(x) = \tilde{u}(x) \quad a.e. \quad x \in \Omega .$$

Remark 6. We have used only very little of the information we have on the function M. If $u \in \mathscr{C}^0(\Omega)$, it follows from (2.19) that $M \in \mathscr{C}^0(U \cup \Omega \times \{0\})$ and

$$M(x,0) = u(x) \quad \text{for all} \quad x \in \Omega .$$

Similarly if $u \in \mathscr{C}^1(\Omega)$, then $M \in \mathscr{C}^1(U \cup \Omega \times \{0\})$ and

$$\frac{\partial M}{\partial r}(x,0) = \frac{n}{\sigma_n}\int_B \operatorname{grad} u(x).y\,\mathrm{d}y .$$

In other words, M is a solution of the "Cauchy problem"

$$(2.24)\quad \begin{cases} \dfrac{1}{r_n}\dfrac{\partial}{\partial r}\left(r^{n+1}\dfrac{\partial M}{\partial r}\right) = \Delta_x M & \text{on} \quad U \\[2ex] M(.,0) = u_0 = u\,, \quad \dfrac{\partial M}{\partial r}(.,0) = u_1 = \dfrac{n}{\sigma_n}\displaystyle\int_B \operatorname{grad} u.y\,\mathrm{d}y & \text{on} \quad \Omega . \end{cases}$$

Let us consider the case $\Omega = \mathbb{R}^n$, $U = \mathbb{R}^n \times \mathbb{R}^+$. We can show (see Delsarte–Lions [1]) that a solution of (2.24) cannot come back to two points $r_1 \neq r_2$ on its initial value u_0 without being constant, except for a discrete set of "spectral" values r_1 and r_2. In other words, if for $r_1 \neq r_2$, not belonging to a discrete set

$$u(x) = M(x, r_1) = M(x, r_2) \quad \text{for all} \quad x ,$$

then u is harmonic. □

We note the Corollary:

Corollary 10. *A locally integrable function on Ω, solution in the sense of distributions of Laplace's equation, is equal, almost everywhere to a harmonic function.*

Proof. Regularizing by convolution (see proof of Corollary 7) for every bounded open set Ω' with $\bar{\Omega}' \subset \Omega$, there exists a sequence of harmonic functions u_n on Ω' such that $u_n \to u$ in $L^1(\Omega')$. Passing to the limit in the formulae of the mean applied to the harmonic functions u_n, we see that u satisfies (2.23) on Ω'. The corollary then follows from Proposition 15. □

We shall use this corollary to "*eliminate the singularities*". The problem is the following: given an open set Ω and a closed subset F such that $\Omega' = \Omega \backslash F$ is dense in Ω (e.g. a variety of dimension $m < n$); given $u \in \mathscr{H}(\Omega')$ we seek to extend by continuity u to a harmonic function on Ω. If it is possible, we shall then say that we have eliminated the set of singularities F.
To specify the problem, we consider $\Omega = B(0, r_0)$, $F = \{0\}$ and a *radial function* u harmonic on the punctured ball $\Omega \backslash \{0\}$. From Proposition 4,

$$u(x) = cE_n(x) + c_0 \quad \text{on} \quad \Omega \backslash \{0\} ,$$

and u extends to a harmonic function on Ω iff $c = 0$. We see that a sufficient (and necessary) condition to eliminate the singularity at 0 is

$$\text{if } n \geqslant 3\,, \qquad \lim_{|x| \to 0} |x|^{n-2} u(x) = 0\,;$$

$$\text{if } n = 2\,, \qquad u(x) \text{ bounded when } x \to 0\,,$$

$$\text{if } n = 1\,, \qquad \frac{du}{dx}(x) \text{ converges when } x \to 0\,.$$

We now consider the *general case of an isolated singularity*: we are given an open set Ω of $\mathbb{R}^n$, $x_0 \in \Omega$ and $u \in \mathscr{H}(\Omega \setminus \{x_0\})$. The *case* $n = 1$ is immediate: u is linear to the left and to the right of x_0; u and du/dx possesses limits to the left and to the right; u extends to a harmonic function iff these limits are equal.
Let us now consider the *case* $n \geqslant 2$ and prove the

Proposition 16. *Let Ω be an open set in $\mathbb{R}^n$ ($n \geqslant 2$), $x_0 \in \Omega$ and $u \in \mathscr{H}(\Omega \setminus \{x_0\})$. Suppose that the function $|x - x_0|^{n-2} u(x)$ is bounded as $x \to x_0$. Then there exist $c, c_0 \in \mathbb{R}$ such that the function $\tilde{u}$ defined on Ω by*

$$\tilde{u}(x) = \begin{cases} u(x) - \dfrac{c}{|x - x_0|^{n-2}} & \text{if } x \in \Omega \setminus \{x_0\} \\[2ex] c_0 & \text{if } x = x_0 \end{cases}$$

is harmonic on Ω.

In particular, $c = \lim_{x \to x_0} |x - x_0|^{n-2} u(x)$ exists and if $n = 2$ or, when $n \geqslant 3$, if $c = 0$ the function u extends by continuity to a function harmonic on Ω.

Proof. For $0 < r < r_0 = \operatorname{dist}(x_0, \partial\Omega)$, we define

$$v(r) = \frac{1}{\sigma_n} \int_\Sigma u(x_0 + r\sigma)\, d\sigma\,.$$

We know (see Proposition 6 of §1) that the radial function $v(|x|)$ is harmonic on $B(0, r_0) \setminus \{0\}$. Hence (see Proposition 4 of this §2)

$$v(|x|) = c_1 E_n(x) + c_0\,.$$

We distinguish then:
in the case $n \geqslant 3$, we take $c = c_1/k_n$,
in the case $n = 2$, since $u(x)$ is bounded when $x \to x_0$, it is the same for $v(r)$ as $r \to 0$, with the result that $c_1 = 0$. We take $c = 0$.

Replacing $u(x)$ by $u(x) - \dfrac{c}{|x - x_0|^{n-2}} - c_0$, we can therefore suppose $v(r) = 0$ and we have to prove that u, extended by 0 for $x = x_0$, is harmonic on Ω. In fact u, defined almost everywhere on Ω is locally integrable on Ω since $|x - x_0|^{2-n}$ is locally integrable on $\mathbb{R}^n$. From Corollary 10, it is sufficient to show that u is a distribution solution of Laplace's equation on Ω; in effect there will then exist $\tilde{u}$, harmonic on Ω such that $u(x) = \tilde{u}(x)$ for almost all $x \in \Omega$ and we necessarily have

$$\begin{cases} \tilde{u}(x) = u(x) \quad \text{for all} \quad x \in \Omega \setminus \{x_0\} \\ \tilde{u}(x_0) = \dfrac{1}{\sigma_n} \displaystyle\int_\Sigma \tilde{u}(x_0 + r\sigma) \mathrm{d}\sigma = 0 \,. \end{cases}$$

We therefore suppose $v(r) = 0$ and show that given $\zeta \in \mathscr{D}(\Omega)$

$$\int u \Delta \zeta \, \mathrm{d}x = 0 \,.$$

Taking $\rho \in \mathscr{D}(\mathbb{R})$ with $0 \leqslant \rho < 1$, $\operatorname{supp} \rho \subset]-\infty, 1]$ and $\rho = 1$ in the neighbourhood of 0. For $k \geqslant 1$ we put

$$\zeta_k(x) = \zeta(x) \rho(k|x - x_0|) \,.$$

We have $\zeta_k \in \mathscr{D}(\Omega)$ and $\zeta_k = \zeta$ in the neighbourhood of x_0; since u is harmonic on $\Omega \setminus \{x_0\}$

$$\int u \Delta \zeta \, \mathrm{d}x = \int u \Delta \zeta_k \, \mathrm{d}x \quad \text{for all} \quad k \geqslant 1 \,.$$

We have

$$\Delta \zeta_k(x) = \rho(k|x - x_0|) \Delta \zeta(x) + 2k \frac{\mathrm{d}\rho}{\mathrm{d}r}(k|x - x_0|) \frac{x - x_0}{|x - x_0|} \cdot \operatorname{grad} \zeta(x)$$
$$+ k^2 \zeta(x) \frac{1}{r^{n-1}} \frac{\mathrm{d}}{\mathrm{d}r} \left(r^{n-1} \frac{\mathrm{d}}{\mathrm{d}r} \right) \rho(k|x - x_0|)$$

and

$$\operatorname{supp} \zeta_k \subset B\left(x_0, \frac{1}{k} \right).$$

Using the change of variables $x \to k(x - x_0)$ we have

$$\int u \Delta \zeta \, \mathrm{d}x = \int_{B(0,1)} u\left(x_0 + \frac{y}{k} \right) \left\{ \rho(|y|) \Delta \zeta \left(x_0 + \frac{y}{k} \right) \right.$$
$$+ 2k \frac{\mathrm{d}\rho}{\mathrm{d}r}(|y|) \frac{y}{|y|} \cdot \operatorname{grad} \zeta \left(x_0 + \frac{y}{k} \right)$$
$$\left. + k^2 \zeta \left(x_0 + \frac{y}{k} \right) \frac{1}{r^{n-1}} \frac{\mathrm{d}}{\mathrm{d}r} \left(r^{n-1} \frac{\mathrm{d}}{\mathrm{d}r} \right) \rho(|y|) \right\} \frac{\mathrm{d}y}{k^n} \,.$$

From the hypothesis $|u(x_0 + y/k)| \leqslant Ck^{n-2}$ for $y \in B(0, 1)$, it follows that the first two terms in the integral tend to 0 when $k \to \infty$; hence

$$\int u \Delta \zeta \, \mathrm{d}x = \lim_{k \to \infty} \int_{B(0,1)} u\left(x_0 + \frac{y}{k} \right) \zeta \left(x_0 + \frac{y}{k} \right) \frac{1}{r^{n-1}} \frac{\mathrm{d}}{\mathrm{d}r}$$
$$\times \left(r^{n-1} \frac{\mathrm{d}}{\mathrm{d}r} \right) \rho(|y|) \frac{\mathrm{d}y}{k^{n-2}}$$

which, on changing to polar coordinates, we may write

$$\int u \Delta \zeta \,\mathrm{d}x = \lim_{k \to \infty} \int_0^1 \frac{\mathrm{d}}{\mathrm{d}r}\left(r^{n-1} \frac{\mathrm{d}\rho}{\mathrm{d}r}(r)\right) \mathrm{d}r \int_\Sigma u\left(x_0 + \frac{r}{k}\sigma\right) \zeta\left(x_0 + \frac{r}{k}\sigma\right) \frac{\mathrm{d}\sigma}{k^{n-2}} .$$

Using $v(r) = 0$, we have

$$\left| \int_\Sigma u\left(x_0 + \frac{r}{k}\sigma\right) \zeta\left(x_0 + \frac{r}{k}\sigma\right) \frac{\mathrm{d}\sigma}{k^{n-2}} \right| = \left| \int_\Sigma u\left(x_0 + \frac{r}{k}\sigma\right) \right.$$

$$\left. \times \left(\zeta\left(x_0 + \frac{r}{k}\sigma\right) - \zeta(x_0)\right) \frac{\mathrm{d}\sigma}{k^{n-2}} \right| \leqslant \frac{\varepsilon_k}{r^{n-2}} ,$$

with

$$\varepsilon_k = \sigma_n \sup_{B(x_0, \frac{1}{k})} |x - x_0|^{n-2} |u(x)| \sup_{B(x_0, \frac{1}{k})} |\zeta(x) - \zeta(x_0)| .$$

From the hypothesis, $\lim \varepsilon_k = 0$.
Since

$$\int_0^1 \left| \frac{\mathrm{d}}{\mathrm{d}r}\left(r^{n-1} \frac{\mathrm{d}\rho}{\mathrm{d}r}(r)\right) \right| \frac{\mathrm{d}r}{r^{n-2}} < \infty ,$$

we deduce that

$$\int u \Delta \zeta \,\mathrm{d}x = 0 . \qquad \square$$

These techniques can be extended to the case of a variety in Ω. We shall first of all consider *a hypersurface* V *of* Ω. We suppose that V *is regular and separates* $\Omega \setminus V$ *into two open sets* Ω_+ *and* Ω_- *such that*

$$V = \partial\Omega_+ \cap \Omega = \partial\Omega_- \cap \Omega , \qquad \Omega \setminus V = \Omega_+ \cup \Omega_- .$$

Given a function u defined on $\Omega \setminus V$ we can consider its restrictions u_+ and u_- to Ω_+ and Ω_- respectively. If $u \in \mathscr{C}^0(\Omega \setminus V)$, we can consider the traces u_+ and u_- on V if they exist: for $x_0 \in V$

$$u_+(x_0) = \lim_{\substack{x \to x_0 \\ x \in \Omega_+}} u(x) , \qquad u_-(x_0) = \lim_{\substack{x \to x_0 \\ x \in \Omega_-}} u(x) .$$

The function u *can be extended by continuity to* Ω *iff* u_+ *and* u_- *exist and are equal on* V. Similarly, if $u \in \mathscr{C}^1(\Omega \setminus V)$, we can consider the normal derivatives of u_+ (resp. u_-) to V exterior to Ω_+ (resp. Ω_-): if they exist, *we denote by* $\partial u/\partial n_+$ *and* $\partial u/\partial n_-$ *these normal derivatives*. If $u \in \mathscr{C}^1(\Omega)$, it is clear that $\partial u/\partial n_+ = -\partial u/\partial n_-$ on V; by contrast, contrarily in the case of the traces of u_+ and u_- being zero, the existence of $\partial u/\partial n_+$, $\partial u/\partial n_-$ satisfying $\partial u/\partial n_+ = -\partial u/\partial n_-$ does not imply $u \in \mathscr{C}^1(\Omega)$: in effect that says nothing about the existence of tangential derivatives on V. Finally we recall (see §1.3*b*) that the existence of normal derivatives $\partial u/\partial n_+$ (resp. $\partial u/\partial n_-$) implies the existence of traces u_+ (resp. u_-). We now prove the

Proposition 17. *Suppose that* V *is a regular hypersurface in* Ω *and that* $u \in \mathscr{H}(\Omega \setminus V)$. *Suppose also that the normal derivatives* $\partial u/\partial n_+$ *and* $\partial u/\partial n_-$ *exist on*

V and that

$$u_+ = u_- \quad \text{and} \quad \frac{\partial u}{\partial n_+} = -\frac{\partial u}{\partial n_-} \quad \text{on} \quad V .$$

Then u extends by continuity to a function harmonic on Ω.

Proof. By the same hypothesis, u extends by continuity to Ω. From Corollary 7, it is sufficient to show that u is a solution in the sense of distributions of Laplace's equation on Ω, that is to say, for all $\zeta \in \mathscr{D}(\Omega)$

$$\int u \Delta \zeta \, dx = \int_{\Omega_+} u_+ \Delta \zeta \, dx + \int_{\Omega_-} u_- \Delta \zeta \, dx = 0 .$$

We apply Green's formula (§1, Proposition 4) on Ω_+; taking account of $\zeta \in \mathscr{D}(\Omega_+)$, u_+ harmonic on Ω_+, and that $\partial u_+/\partial n = \partial u/\partial n_+$ exists on $V = \partial\Omega_+ \cap \Omega$, we obtain

$$\int_{\Omega_+} u_+ \Delta \zeta \, dx = \int_V \left(u_+ \frac{\partial \zeta}{\partial n_+} - \zeta \frac{\partial u}{\partial n_+} \right) d\gamma .$$

Similarly

$$\int_{\Omega_-} u_- \Delta \zeta \, dx = \int_V \left(u_- \frac{\partial \zeta}{\partial n_-} - \zeta \frac{\partial u}{\partial n_-} \right) d\gamma .$$

Adding and using the hypotheses on u and $\partial\zeta/\partial n_+ = -\partial\zeta/\partial n_-$ we obtain $\int u \Delta \zeta \, dx = 0$. □

We note the corollary of this proposition concerning *the uniqueness of the Cauchy problem*:

Corollary 11. *Let Ω be a regular connected open with boundary Γ, $z_0 \in \Gamma$ and $u \in \mathscr{H}(\Omega)$. We suppose that $\partial u(z)/\partial n$ exists and $\partial u(z)/\partial n = u(z) = 0$ for all z in the neighbourhood of z_0 in Γ. Then $u \equiv 0$ on Ω.*

Proof. We fix $r > 0$ sufficiently small with the result that

$$\frac{\partial u}{\partial n}(z) \ \text{exists and} \ \frac{\partial u}{\partial n}(z) = u(z) = 0 \quad \text{for all} \quad z \in B(z_0, r) \cap \Gamma = V .$$

Since Ω is regular, V is a regular hypersurface of $B(z_0, r)$ which separates $B(z_0, r) \setminus V$ into two open sets Ω_+^0 and Ω_-^0 such that

$$V = \partial\Omega_+^0 \cap B(z_0, r) = \partial\Omega_-^0 \cap B(z_0, r) ;$$

we can suppose, for r sufficiently small, that $\Omega_+^0 = \Omega \cap B(z_0, r)$. We define u_0 on $B(z_0, r) \setminus V$ by

$$u_0 = u \quad \text{on} \quad \Omega_+^0 , \qquad u_0 = 0 \quad \text{on} \quad \Omega_-^0 .$$

From the hypothesis and Proposition 17, u_0 extends by continuity to a harmonic function $\tilde{u}_0$ on $B(z_0, r)$. From Corollary 7, $\tilde{u}_0$ is analytic on $B(z_0, r)$; being zero on

Ω_-^0, it is identically zero on $B(z_0, r)$. Hence u, which is analytic on Ω, being zero on $\Omega \cap B(z_0, r)$, is identically zero on Ω. □

Finally, we shall consider the following result for a variety V of dimension $m \leqslant n - 2$:

Proposition 18. *Let V be a closed regular variety in Ω of dimension $m \leqslant n - 2$ and $u \in \mathscr{H}(\Omega \setminus V)$. We suppose that*

$$\lim_{x \to x_0} d(x, V)^{n-m-2} u(x) = 0 \quad \text{for all} \quad x_0 \in V$$ [17]

Then u extends by continuity to a function harmonic on Ω.

Proof. The result being local, we can suppose that, possibly by changing the reference frame, that there exists an open set $\mathcal{O}$ of $\mathbb{R}^m$ and a regular map α of $\mathcal{O}$ into $\mathbb{R}^{n-m}$ such that

$$V = \{(x', \alpha(x'))\,;\ x' \in \mathcal{O}\}$$

$$\Omega = \{(x', \alpha(x')) + t\,;\ x' \in \mathcal{O}\,,\ t \in \mathbb{R}^{n-m},\ |t| < \delta_0\}\,.$$

Given K a compact sub-set of $\mathcal{O}$, we see, using the regularity of α, that there exists $0 < \delta < \delta_0$ and $c > 0$ such that

$$d(x, V) \geqslant c|t| \quad \text{for} \quad x = (x', \alpha(x') + t)\,, \quad x' \in K, |t| \leqslant \delta\,.$$

In particular the function $d(x, V)^{2-n+m}$ is locally integrable on Ω, since $|t|^{2-n+m}$ is locally integrable on $\mathbb{R}^{n-m}$. Using the hypothesis the function $d(x, V)^{n-m-2} u(x)$ can be extended by continuity to Ω and hence *a fortiori* is locally bounded; we deduce that u defined almost everywhere on Ω (V is negligible for the Lebesgue measure), is locally integrable on Ω. It suffices therefore, from Corollary 10, to prove that, for all $\zeta \in \mathscr{D}(\Omega)$,

$$\int u \Delta \zeta \,\mathrm{d}x = 0\,.$$

We follow the proof of Proposition 16:
We take $\rho \in \mathscr{D}(\mathbb{R}^{n-m})$ with $0 \leqslant \rho \leqslant 1$, $\operatorname{supp} \rho \subset B(0, 1)$ and $\rho = 1$ in the neighbourhood of 0, and we define

$$\zeta_k(x) = \zeta(x) \rho(kt) \quad \text{for} \quad x = (x', \alpha(x') + t)\,.$$

We have

$$\operatorname{supp} \zeta_k \subset \left\{(x', \alpha(x') + t); x' \in K, |t| \leqslant \frac{1}{k}\right\}$$

where K is the projection of the support of ζ onto $\mathbb{R}^m$.
Calculating $\Delta \zeta_k$, we see that

$$|\Delta \zeta_k| \leqslant C k^2$$

where C depends only on ζ, ρ and α but not on k.

[17] Where $d(x, V)$ is the distance of the point x from the variety V.

Then

$$\left|\int u \Delta \zeta \, \mathrm{d}x\right| = \left|\int u \Delta \zeta_k \, \mathrm{d}x\right| \leqslant \frac{C\varepsilon_k}{c^{n-m-2}} \int_K \mathrm{d}x' \int_{B(0,\frac{1}{k})} \frac{k^2 \, \mathrm{d}t}{|t|^{n-m-2}}$$

with

$$\varepsilon_k = \sup_{\substack{x = (x', \alpha(x') + t) \\ x' \in K, |t| \leqslant \frac{1}{k}}} d(x, V)^{n-m-2} u(x) .$$

The latter integral is independent of k; from the hypothesis, $\varepsilon_k \to 0$ when $k \to \infty$. Hence $\int u \Delta \zeta \, \mathrm{d}x = 0$. □

5. Kelvin's Transformation; Application to Harmonic Functions in an Unbounded Set; Conformal Transformation

Let us recall that an *inversion with pole* $x_0 \in \mathbb{R}^n$ *and radius* $r_0 > 0$ is the transformation, denoted by $I(x_0, r_0)$ of $\mathbb{R}^n \setminus \{0\}$ onto itself defined by

$$I(x_0, r_0): x \to y = x_0 + \frac{r_0^2}{|x - x_0|^2}(x - x_0) .$$

Inversion is involutive, that is to say that: $I(x_0, r)^{-1} = I(x_0, r)$; point by point, it conserves the sphere $\partial B(x_0, r_0)$, conserves globally the hyperplanes passing through x_0, and transforms the hyperplane passing through $x \neq x_0$ and orthogonal

$$\partial B\left(\frac{I(x_0, r_0)x + x_0}{2}, \frac{|I(x_0, r_0)x - x_0|}{2}\right)$$

punctured at the pole x_0. The inversion $I(x_0, r_0)$ is an (analytic) diffeomorphism of $\mathbb{R}^n \setminus \{0\}$

$$I(x_0, r_0)(x) = x_0 + r_0^2 \operatorname{grad} \operatorname{Log} |x - x_0| .$$

A simple calculation shows that the Jacobian matrix of $I(x_0, r_0)$ at the point x is

$$I(x_0, r_0)'(x) = \left(\frac{\partial y_j}{\partial x_i}(x)\right) = \frac{r_0^2}{|x - x_0|^2} T(x - x_0) , \tag{2.25}$$

with

$$T(x) = \left(\delta_{ij} - \frac{2x_i x_j}{|x|^2}\right)$$

which is an orthogonal matrix.
A simple calculation shows also that

$$\Delta I(x_0, r_0)(x) = \frac{-2(n-2)r_0^2}{|x - x_0|^4}(x - x_0) .$$

Using Lemma 1 of §1.1, we see that given a function u defined in the neighbourhood of $y = I(x_0, r_0)(x)$ and twice-differentiable in y, we can show that $w(x) = u(I(x_0 . r_0)(x))$ is twice differentiable in x and

$$(2.26) \qquad \Delta w(x) = \frac{r_0^4}{|x - x_0|^4} \Delta u(y) - \frac{2(n-2)r_0^2}{|x - x_0|^4} \operatorname{grad} u(y).(x - x_0) .$$

This relation leads to

Definition 3. *The Kelvin transformation of pole x_0 and radius $r_0 > 0$ is the name given to the map denoted $H(x_0, r_0)$ which maps any function u defined on a part Ω of $\mathbb{R}^n \setminus \{0\}$ to a $v = H(x_0, r_0)u$ defined on $I(x_0, r_0)^{-1}(\Omega) = I(x_0, r_0)(\Omega)$ by*

$$v(x) = \frac{r_0^{n-2} u(I(x_0, r_0)x)}{|x - x_0|^{n-2}} .$$

It is immediate that *Kelvin's transformation is involutive*: if $v = H(x_0, r_0)u$, then $u = H(x_0, r_0)v$. Thus *Kelvin's transformation leaves invariant the functions defined on the sphere $\partial B(x_0, r_0)$.*
We now prove the

Proposition 9. *For a function u defined in the neighbourhood of $y \neq x_0$ and twice differentiable at y, the Kelvin transformation $H(x_0, r_0)u$ is twice differentiable at $x = I(x_0, r_0)y$ and*

$$(2.27) \qquad \Delta H(x_0, r_0)u(x) = \frac{r_0^{n+2}}{|x - x_0|^{n+2}} \Delta u(I(x_0, r_0)x) .$$

Proof. Since the function $r_0^{n-2}|x - x_0|^{2-n}$ is harmonic on $\mathbb{R}^n \setminus \{x_0\}$, we have with the above notation

$$\Delta H(x_0, r_0)u(x) = \frac{r_0^{n-2}}{|x - x_0|^{n-2}} \Delta w(x) + 2 \operatorname{grad} \frac{r_0^{n-2}}{|x - x_0|^{n-2}} \cdot \operatorname{grad} w(x) .$$

From (2.25)

$$\operatorname{grad} w(x) = \frac{r_0^2}{|x - x_0|^2} T(x - x_0).\operatorname{grad} u(y) ,$$

and hence since $T(x - x_0)$ is orthogonal,

$$\operatorname{grad} \frac{r_0^{n-2}}{|x - x_0|^{n-2}} \cdot \operatorname{grad} w(x) = + \frac{(n-2)r_0^n}{|x - x_0|^{n+2}} \operatorname{grad} u(y).(x - x_0) ,$$

which, as a consequence of (2.26), prove (2.27). □

We deduce immediately *the invariance of harmonic functions under Kelvin's transformation*:

Corollary 12. *Kelvin's transformation $H(x_0, r_0)u$ of a function u, harmonic on an open set Ω of $\mathbb{R}^n \setminus \{x_0\}$, is harmonic on $I(x_0, r_0)(\Omega)$.*

Remark 7. For a distribution u on an open set Ω of $\mathbb{R}^n \setminus \{x_0\}$, we can define its Kelvin transformation $H(x_0, r_0)u$: $w = u \circ I(x_0, r_0)$ is defined as the distribution image $I(x_0, r_0)^{-1} u$ and the function $|x - x_0|^{2-n}$ being analytic on $\mathbb{R}^n \setminus \{x_0\}$, we can define the product

$$v = \frac{r_0^{n-2}}{|x - x_0|^{n-2}} w .$$

The formula (2.21) is obviously still true for distributions. □

Inversion maps the point at infinity into its pole x_0 so Kelvin's transformation is a very useful tool for reducing problems in unbounded open sets to problems in bounded open sets. We give here some applications of this method.

First of all we shall apply Proposition 14 *to the singularity at infinity of the exterior problem,* that is to say, of the problem on an open set Ω of $\mathbb{R}^n$ such that $\mathbb{R}^n \setminus \bar{\Omega}$ is a non-empty bounded set: such an open set is then unbounded; it contains also all $x \in \mathbb{R}^n$ with $|x|$ sufficiently large; also its boundary Γ is bounded (and hence compact).

We now prove the

Proposition 20. *Let Ω be an open set of $\mathbb{R}^n$ with $n \geqslant 2$ and $\mathbb{R}^n \setminus \bar{\Omega}$ a non-empty bounded set and $u \in \mathscr{H}(\Omega)$. Suppose $u(x)$ is bounded as $|x| \to \infty$. Then there exist c and $c_0 \in \mathbb{R}$ such that*

$$u(x) = c + \frac{c_0}{|x|^{n-2}} + o\left(\frac{1}{|x|^{n-2}}\right) \quad \textit{when} \quad |x| \to \infty\ ^{18} .$$

Proof. Let us fix $x_0 \in \mathbb{R}^n \setminus \bar{\Omega}$ and denote $\Omega' = I(x_0, 1)\Omega \cup \{x_0\}$: this is an open neighbourhood of x_0. From Corollary 12, the Kelvin transformation, $v = H(x_0, 1)u$ is harmonic on $\Omega' \setminus \{x_0\}$. Since $u(x)$ is bounded when $|x| \to \infty$, the function $|y - x_0|^{n-2} v(y) = u(I(x_0, 1)y)$ is bounded when $y \to x_0$. Using Proposition 16, there exist $c, c_0 \in \mathbb{R}$ such that

$$v(y) = \frac{c}{|y - x_0|^{n-2}} + c_0 + o(1) \quad \text{when} \quad y \to x_0 .$$

Returning to $u = H(x_0, 1)v$, we have

$$u(x) = c + \frac{c_0}{|x - x_0|^{n-2}} + o\left(\frac{1}{|x - x_0|^{n-2}}\right) \quad \text{when} \quad |x| \to \infty ,$$

from which the result follows, since

$$\frac{1}{|x - x_0|^{n-2}} = \frac{1}{|x|^{n-2}} + o\left(\frac{1}{|x|^{n-2}}\right) \quad \text{when} \quad |x| \to \infty .$$ □

We apply the same method to obtain *Poisson's integral formula for the exterior of a ball:*

Proposition 21. *Let $\Omega = \mathbb{R}^n \setminus \bar{B}$ be the exterior of a ball $B = B(x_0, r_0)$ of $\mathbb{R}^n$ and $u \in \mathscr{H}(\Omega) \cap \mathscr{C}^0(\bar{\Omega})$. Let us suppose u to be bounded and write $c = \lim_{|x| \to \infty} u(x)$*

[18] We recall that given two functions f and g defined in the neighbourhood of a point x_0 we denote by $f(x) = o(g(x))$ when $x \to x_0$ the situation in which $\lim_{\substack{x \to x_0 \\ g(x) \neq 0}} \{f(x)/g(x)\} = 0$. The limit $|x| \to \infty$ corresponds to $x_0 = \infty$.

(which exists by Proposition 20). Then

(2.28) $$u(x) = \left(1 - \frac{r_0^{n-2}}{|x - x_0|^{n-2}}\right)c + \frac{1}{r_0 \sigma_n}\int_{\partial B} \frac{|x - x_0|^2 - r_0^2}{|t - x|^n} u(t)\,\mathrm{d}\gamma(t)$$

for all $x \in \Omega$.

Proof. Let us use the inversion $I(x_0, r_0)$: given $t \in \partial B$, $x \in \Omega$ and $y = I(x_0, r_0)x$, we have $y \in B$ and (see (2.12))

$$|t - x| = \frac{r_0}{|y - x_0|}|t - y|\,.$$

Hence

$$\frac{|x - x_0|^2 - r_0^2}{|t - x|^n} = \frac{r_0^2 - |y - x_0|^2}{|t - y|^n} \times \frac{|y - x_0|^{n-2}}{r_0^{n-2}}\,.$$

In particular, applying Poisson's integral formula for the constant function $v(y) = c$ (see Proposition 9), we have:

$$\frac{1}{r_0 \sigma_n}\int_{\partial B} \frac{|x - x_0|^2 - r_0^2}{|t - x|^n} c\,\mathrm{d}\gamma(t) = \frac{|y - x_0|^{n-2}}{r_0^{n-2}} \frac{1}{r_0 \sigma_n}\int_{\partial B} \frac{r_0^2 - |y - x_0|^2}{|t - y|^n} c\,\mathrm{d}\gamma(t)$$

$$= c\frac{|y - x_0|^{n-2}}{r_0^{n-2}} = c\frac{r_0^{n-2}}{|x - x_0|^{n-2}}\,.$$

Replacing $u(x)$ by $u(x) - c$, we can therefore always suppose that $c = 0$. The Kelvin transformation $v = H(x_0, r_0)u$ is harmonic on the punctured ball $B \setminus \{x_0\} = I(x_0, r_0)(\Omega)$; but

$$\lim_{y \to x_0} |y - x_0|^{n-2} v(y) = \lim_{|x| \to \infty} r_0^{n-2} u(x) = 0\,.$$

From Proposition 16, v extends to a function, harmonic on Ω, which we again denote by v. On the other hand $v \in \mathscr{C}^0(\bar{B})$, since $\bar{B} \setminus \{x_0\} = I(x_0, r_0)(\bar{\Omega})$. We thus have the Poisson integral formula for v:

$$v(y) = \frac{1}{r_0 \sigma_n}\int_{\partial B} \frac{r_0^2 - |y - x_0|^2}{|t - y|^n} v(t)\,\mathrm{d}\gamma(t)$$

$$= \frac{|x - x_0|^{n-2}}{r_0^{n-2}} \frac{1}{r_0 \sigma_n}\int_{\partial B} \frac{|x - x_0|^2 - r_0^2}{|t - x|^n} u(t)\,\mathrm{d}\gamma(t)\,.$$

Returning to u, we obtain (2.28). □

Finally, using Kelvin's transformation in the proof of *Poisson's integral formula for a half space*, we have:

Proposition 22. *Let* $\Omega = \{x \in \mathbb{R}^n; (x - x_0).e > 0\}$ *with* $x_0 \in \mathbb{R}^n$ *and* e *a unit vector, be an open half-space, and let* $u \in \mathscr{H}(\Omega) \cap \mathscr{C}^0(\bar{\Omega})$ *be bounded on* $\bar{\Omega}$. *Then*

(2.29) $$u(x) = \frac{2(x - x_0).e}{\sigma_n}\int_{\Gamma} \frac{u(t)\,\mathrm{d}\gamma(t)}{|t - x + x_0|^n} \quad \text{for all} \quad x \in \Omega\,.$$

Note that, by a change of the reference frame, we can always take

$$\Omega = \{x = (x', x_n);\ x' \in \mathbb{R}^{n-1},\ x_n > 0\},$$

that is to say, $x_0 = 0$ and e is the n-th of the base vectors of the reference frame. The formula (2.29) can then be written

$$u(x', x_n) = \frac{2x_n}{\sigma_n}\int_{\mathbb{R}^{n-1}} \frac{u(t)\,dt}{(|t - x'|^2 + x_n^2)^{n/2}}. \tag{2.30}$$

It is clear that the integral has a meaning since u is bounded and $t \to (|t - x'|^2 + x_n^2)^{-n/2}$ is integrable on $\mathbb{R}^{n-1}$ for all $(x', x_n) \in \Omega$.
Finally we notice that by the change of variable $t \to (t - x')/x_n$ we can write the formula (2.30) as

$$u(x', x_n) = \frac{2}{\sigma_n}\int_{\mathbb{R}^{n-1}} \frac{u(x' + tx_n)\,dt}{(1 + |t|^2)^{n/2}}. \tag{2.31}$$

Proof of Proposition 22. We can suppose $\Omega = \{(x', x_n);\ x' \in \mathbb{R}^{n-1},\ x_n > 0\}$. From formula (2.31), it is sufficient to prove (2.30) (or (2.31)) for $x' = 0$, $x_n = 1$. We consider the inversion $I(y_0, 1)$ with $y_0 = (0, -1)$: it transforms Ω to the ball $B(\frac{1}{2}y_0, \frac{1}{2})$, the plane $\Gamma = \{(x', 0);\ x' \in \mathbb{R}^{n-1}\}$ to the punctured sphere $\partial B \backslash \{y_0\}$ and the point $(0, 1)$ to $\frac{1}{2}y_0$. The Kelvin transformation $v = H(y_0, 1)u$ is harmonic on B and continuous on $\bar{B}\backslash\{y_0\}$. We prove that although v is not continuous at $y_0 \in \partial B$, we have all the same the formula of the mean (see Proposition 6 and Remark 3)

$$v\left(\frac{y_0}{2}\right) = \frac{2^{n-1}}{\sigma_n}\int_{\partial B} v(s)\,d\gamma(s). \tag{2.32}$$

Admitting this formula for the moment and returning to u, we shall have

$$u(0, 1) = \frac{v(y_0/2)}{2^{n-2}} = \frac{2}{\sigma_n}\int_{\partial B} \frac{u(I(y_0, 1)s)}{|s - y_0|^{n-2}}\,d\gamma(s).$$

But carrying out the change of variables $s \to t = I(y_0, 1)s$, making use of (2.25) and $|s - y_0| = \dfrac{1}{(1 + |t|^2)^{\frac{1}{2}}}$, we shall have

$$u(0, 1) = \frac{2}{\sigma_n}\int_{\mathbb{R}^{n-1}} u(t)(1 + |t|^2)^{\frac{n-2}{2}} \frac{dt}{(1 + |t|^2)^{n-1}} = \frac{2}{\sigma_n}\int_{\mathbb{R}^{n-1}} \frac{u(t)\,dt}{(1 + |t|^2)^{n/2}}.$$

This calculation proves the proposition with the reservation that (2.32) is satisfied; it shows also that the integral in (2.32) is well defined.
To justify (2.32) it is sufficient (see Remark 3) to dominate the function $v_\lambda(s) = v(\frac{1}{2}\lambda y_0 + (1 - \lambda)s)$ by a function integrable on ∂B. Now, since u is bounded

$$|v_\lambda(s)| \leqslant \frac{C}{\left|\dfrac{y_0}{2} + (1 - \lambda)\left(\dfrac{y_0}{2} - s\right)\right|^{n-2}} \quad \text{with} \quad C = \sup_\Omega u$$

On the other hand

$$\left|\frac{y_0}{2} + (1-\lambda)\left(\frac{y_0}{2} - s\right)\right| \geqslant \left|\frac{y_0}{2}\right| = \frac{1}{2} \quad \text{for} \quad \left(\frac{y_0}{2} - s\right)\frac{y_0}{2} \geqslant 0$$

$$\left|\frac{y_0}{2} + (1-\lambda)\left(\frac{y_0}{2} - s\right)\right| \geqslant |y_0 - s| \quad \text{for} \quad \left(\frac{y_0}{2} - s\right)\frac{y_0}{2} \leqslant 0\,;$$

hence

$$|v_\lambda(s)| \leqslant C\left(2^{n-2} + \frac{1}{|y_0 - s|^{n-2}}\right).$$

From the above calculation, $|y_0 - s|^{2-n}$, which is the Kelvin transformation of the constant function 1, is integrable on ∂B. We have thus dominated $v_\lambda(s)$. □

Remark 8. Using the homogeneity, we have established Poisson's integral formula in a half-space, by applying Kelvin's transformation to the formula of the mean. Returning to the ball, we thus obtain another proof of Poisson's integral formula in a ball. □

Let us now consider *the particular case* $n = 2$. Kelvin's transformation is then simply the inversion map $I(x_0, r_0)$

$$H(x_0, r_0)u = u \circ I(x_0, r_0) = I(x_0, r_0)u$$

(do not forget that $I(x_0, r_0)^{-1} = I(x_0, r_0)$). In particular, the harmonic functions are invariant under inversion. In fact more generally, we have

Proposition 23. *Plane harmonic functions are invariant under conformal transformations. More precisely, given an open set Ω of $\mathbb{R}^2$ and a conformal transformation H defined on Ω, the function u defined on Ω is harmonic iff its image $Hu = u \circ H^{-1}$ is harmonic on $H(\Omega)$.*

We recall that a *conformal transformation* of an open set Ω (of $\mathbb{R}^n$ in general) is a diffeomorphism H of Ω into $\mathbb{R}^n$ preserving the angles; given two (regular) hypersurfaces Σ_1 and Σ_2 of Ω meeting at an angle θ[19] at the point x, their images $H(\Sigma_1)$ and $H(\Sigma_2)$ meet at $H(x)$ at the same angle θ. A linear conformal transformation is a (linear) similitude, product of a homothetic transformation by an orthogonal transformation.

It is clear that a diffeomorphism H of Ω into $\mathbb{R}^n$ is a conformal transformation iff its derivative $H'(x)$ is at every point $x \in \Omega$, a (linear) conformal transformation, that is to say, if

$$H'(x) = c(x)T(x) \quad \text{for all} \quad x \in \Omega \tag{2.33}$$

with $c(x) > 0$ and $T(x)$ an orthogonal transformation.

[19] The angle between two hypersurfaces Σ_1 and Σ_2 meeting at x is the angle between the normals to Σ_1 and Σ_2 respectively at x: the angle between two straight lines D_1 and D_2 being the number $\theta \in [0, \frac{1}{2}\pi]$ such that $\cos\theta = |e_1 . e_2|$ for e_1 and e_2 unit vectors of D_1 and D_2 respectively.

We return now to the case $n = 2$. If we identify topologically

$$\mathbb{R}^2 = \{(x, y)\} \quad \text{with} \quad \mathbb{C} = \{z = x + iy\},$$

a transformation of an open set $\Omega \in \mathbb{R}^2$ into $\mathbb{R}^2$ is identified with a (complex) function of a complex variable. If we put

$$H(x + iy) = U(x, y) + iV(x, y),$$

the condition (2.33) is written

$$\left(\frac{\partial U}{\partial x}\right)^2 + \left(\frac{\partial U}{\partial y}\right)^2 = \left(\frac{\partial V}{\partial x}\right)^2 + \left(\frac{\partial V}{\partial y}\right)^2 = c^2, \quad \frac{\partial U}{\partial x}\frac{\partial V}{\partial x} + \frac{\partial U}{\partial y}\frac{\partial V}{\partial y} = 0,$$

that is to say

$$\frac{\partial U}{\partial x} = \varepsilon \frac{\partial V}{\partial y}, \quad \frac{\partial U}{\partial y} = -\varepsilon \frac{\partial V}{\partial x} \quad \text{with} \quad \varepsilon = \pm 1.$$

We recognize there the Cauchy–Riemann equations: H or its conjugate $\bar{H} = U - iV$ is (locally) holomorphic. In other words: *a conformal transformation of the plane* $\mathbb{R}^2$ *is (to within a symmetry) a holomorphic transformation of* $\mathbb{C}$[20].
We can now give the

Proof of Proposition 23. Let us consider u harmonic on Ω and H a conformal transformation of Ω. The problem being local, we can suppose (see Proposition 1) that $u = \operatorname{Re} f$ where f is holomorphic on Ω.
Also the Laplacian being invariant by symmetry, we can suppose that H is a holomorphic transformation. Then $Hu = \operatorname{Re}(f \circ H^{-1})$ is the real part of a holomorphic function; hence Hu is harmonic. □

In a similar way to the holomorphic functions, the conformal transformations are a very useful tool for the study of the Laplacian in the plane. Principally by the use of the *theorem of conformal representation*: every simply connected open set Ω of the plane, distinct from the whole plane, is an image of the unit disk by a conformal transformation H. This theorem and Proposition 23, gives in theory an "integral formula" for an arbitrary simply connected open set Ω, a formula giving a conformal representation H mapping the unit disk onto Ω. We can meanwhile refer the reader to dictionaries of conformal transformations for particular open sets (see Kober [1], Nehari [1], Laventriev–Shabat [1]).
Let us give an example of the use of Proposition 23.

Example 4. Let us consider the holomorphic transformation $H: z \to \dfrac{2z}{1 + z^2}$.
We verify easily that it maps the semi-disk

$$\Omega = \{z; \operatorname{Im} z > 0, |z| < 1\}$$

onto the half-plane $\{z': \operatorname{Im} z' > 0\}$, the punctured semi-circle

$$\{z; \operatorname{Im} z \geqslant 0, |z| = 1\} \backslash \{i\}$$

---

[20] The equivalence is true locally or more generally in a simply connected open set.

onto the part $\{x; |x| \geqslant 1\}$ of the real axis and conserves (globally) the interval $\{x; -1 < x < 1\}$ of the real axis. Given $u \in \mathscr{H}(\Omega) \cap \mathscr{C}^0(\bar{\Omega})$ the image by H, $v = Hu$ is harmonic on the half-space $H(\Omega)$, continuous and bounded on $\overline{H(\Omega)}$. We can then apply to it Poisson's integral formula for a half-space (Proposition 22).

$$v(x' + iy') = \frac{y'}{\pi} \int_{-\infty}^{\infty} \frac{v(t)\,\mathrm{d}t}{(t - x')^2 + y'^2} \quad \text{for all} \quad x' \in \mathbb{R},\ y' > 0\,.$$

Returning to u we shall have

$$u(z) = \frac{\operatorname{Im} H(z)}{\pi} \int_{-\infty}^{\infty} \frac{u(H^{-1}(t))\,\mathrm{d}t}{|t - H(z)|^2}\,,$$

which gives, after an elementary calculation, *Poisson's integral formula for* the *half-circle* using the polar coordinates

$$x = r\cos\theta\,, \quad y = r\sin\theta\,, \quad 0 < r < 1\,, \quad 0 < \theta < \pi$$

that is

$$u(x, y) = \frac{(1 - r^2)y}{\pi} \left\{ \int_{-1}^{+1} \frac{(1 - \xi^2)u(\xi, 0)\,\mathrm{d}\xi}{R[r^2(1 + \xi^2)^2 - 2\xi x(1 + \xi^2)(1 + r^2) + \xi^2 R]} + \int_0^{\pi} \frac{2u(\cos\varphi, \sin\varphi)\sin\varphi\,\mathrm{d}\varphi}{R[R - 4x(1 + r^2)\cos\varphi + 4r^2\cos^2\varphi]} \right\}$$

with $$R = |1 + (x + iy)^2|^2 = 1 + 2r^2\cos 2\theta + r^4\,.$$ □

6. Some Physical Interpretations (in Mechanics and Electrostatics)

6a. Elementary solutions and the laws of Newton and Coulomb

We go back to the examples in mechanics and electrostatics of § 1.2.
In mechanics, a point mass m_0 placed at a point $x_0 \in \mathbb{R}^3$ is modelled by a Dirac distribution $m_0\delta_{x_0}$. From this we have said in § 1.2, it has given rise to a gravitational potential v_g in $\mathbb{R}^3$, solution of Poisson's equation

(2.34) $$\Delta v_g = k_g m_0 \delta_{x_0} \quad \text{in} \quad \mathscr{D}'(\mathbb{R}^3)\,.$$

A solution of (2.34) is

(2.35) $$v_g(x) = k_g m_0 E_3(x - x_0) = -\frac{k_g m_0}{4\pi|x - x_0|}\,.$$

We note from Proposition 4 that v_g, given by (2.35) is, to within an additive constant, the only radial solution of (2.34); we shall see[21] also that v_g is the only solution of (2.34) tending to zero at infinity.

---

[21] See Proposition 3 of §3; it follows from Corollaries 12 and 1 that v_g given by (2.35) is the only locally integrable solution of (2.34) tending to zero at infinity.

With this gravitational potential v_g, the force exerted on a point mass m placed at a point $x \in \mathbb{R}^3$ is then given by

$$f(x) = -m \operatorname{grad} v_g(x) = -\frac{k_g m m_0}{4\pi} \frac{x - x_0}{|x - x_0|^3}.$$

We thus obtain *Newton's law*: a point mass m_0 placed at point $x_0 \in \mathbb{R}^3$ exerts on a point mass m placed at $x \in \mathbb{R}^3$ an *attractive force* of intensity inversely proportional to the square of the distance $|x - x_0|$.

We note that *conversely* we can "recover" the gravitational potential v_g and Poisson's equation starting from Newton's law. More precisely, let us now take as our starting point the most directly physically comprehensible affirmation of Newton's law: a point mass m_0 placed $x_0 \in \mathbb{R}^3$ exerts on a point mass m placed at $x \in \mathbb{R}^3 \backslash \{x_0\}$ an attractive force (carried by the straight line joining x_0 to x) proportional to the masses m_0 and m, that is to say of the form

$$f(x) = -m_0 m h(x) \frac{x - x_0}{|x - x_0|}, \qquad h(x) > 0.$$

The work done by the force $f(x)$ when we displace the mass m from the point x_1 to the point x_2 is independent of the path taken iff $h(x)$ depends only on the distance $|x - x_0|$; the field $E(x) = m_0 h(x)(x - x_0)/|x - x_0|$ is then the gradient of a potential $v(x) = V(|x - x_0|)$. To express the of Newton's law, that is to say that f is inversely proportional to the square of the distance, leads to the result that the expression for $V(r)$ is of the form $-km_0/(4\pi r) + c$, that is to say from Proposition 4, that v is harmonic on $\mathbb{R}^3 \backslash \{0\}$, or again that v is a solution of Poisson's equation

$$\Delta v = k m_0 \delta_{x_0}.$$

We can develop similar considerations in *electrostatics* with the following changes: (i) a point charge q_0 placed at a point $x_0 \in \mathbb{R}^3$ creates an electrostatic potential v_c in $\mathbb{R}^3$ satisfying

$$\Delta v_c = -k_c q_0 \delta_{x_0} \quad \text{in} \quad \mathscr{D}'(\mathbb{R}^3); \tag{2.36}$$

(ii) We adopt as solution of (2.36),

$$v_c(x) = -k_c q_0 E_3(x - x_0) = \frac{k_c q_0}{4\pi |x - x_0|}$$

and the electrostatic force exerted on a point charge q placed at $x \in \mathbb{R}^3$ is given by

$$\mathbf{f}(x) = -q \operatorname{grad} v_c(x) = \frac{k_c q q_0}{4\pi} \frac{\mathbf{x} - \mathbf{x}_0}{|\mathbf{x} - \mathbf{x}_0|^3} = \frac{q q_0}{\varepsilon_0'} \frac{\mathbf{x} - \mathbf{x}_0}{|\mathbf{x} - \mathbf{x}_0|^3}\,[22];$$

(iii) thus we obtain Coulomb's law which is of a similar form to Newton's law, with the essential difference that the electrostatic force is *repulsive* if q and q_0 are of the same sign, *attractive* if these charges are of opposite signs.

---

[22] $\varepsilon_0' = 4\pi/k_c$ is the permittivity constant of the vacuum (see §1.2).

6b. Differential Equations (of Poisson or Laplace) and Integral Formulae (Gauss' Law)

We recall here (see Chap. IA §4.5) that Poisson's equation for an electrostatic field

$$\Delta v_c = -k_c\rho \qquad \text{(in an open set } \Omega\text{)}$$

is a consequence of Maxwell's equations

$$\begin{cases} \operatorname{div} \mathbf{E} = k_c\rho \\ \operatorname{curl} \mathbf{E} = 0 \end{cases}$$

with the assumption that the electric field is derivable globally from a potential:

$$\mathbf{E} = -\operatorname{grad} v_c\,.$$

This implies that the work W done by the electrostatic force when the charge q is displaced from a point $\mathbf{x}_0 \in \Omega$ to a point $\mathbf{x}_1 \in \Omega$ does not depend on the path followed (see § 1.2)[23].

In addition the integral form of the equation $\operatorname{div} \mathbf{E} = k_c\rho$, that is to say

$$(2.37) \qquad \int_{\partial\omega} \mathbf{E}\,.\,\mathbf{n}\,\mathrm{d}\gamma = \int_{\omega} k_c\rho\,\mathrm{d}\mathbf{x}$$

for every open set ω (with $\bar{\omega} \subset \Omega$) denotes that the flux of the electric field across every "regular closed" surface $\partial\omega$ is equal to the total charge contained in the domain ω "interior" to the surface $\partial\omega$ (with multiplicative constant k_c). In the special case $\rho = 0$ in Ω, and for $\mathbf{E} = -\operatorname{grad} v_c$, Proposition 14 implies that there is an equivalence, for $v_c \in \mathscr{C}'(\Omega)$ between the integral form (2.37), i.e.

$$\int_{\partial\omega} \frac{\partial v_c}{\partial n}\,\mathrm{d}\gamma = 0$$

(expressing that no charge can be found in the domain ω, following Gauss' law), and Laplace's equation:

$$\Delta v_c = 0\,.$$

The condition $v_c \in \mathscr{C}^1(\Omega)$, equivalent to $\mathbf{E} \in \mathscr{C}^0(\Omega)^3$, expresses that the motion of a point charge in the vacuum does not have a discontinuity in the acceleration. In fact, we know that then the potential, and hence also the field, will be analytic in Ω (Corollary 7), which implies that the motion of a point charge *in vacuo* is then analytic (it is the same for the gravitational field, and the motion of a point mass).

6c. Principle of the Maximum (Minimum) and Equilibrium

We have seen in Proposition 8 that a harmonic potential u in a connected open set Ω cannot achieve a minimum in Ω (we also say, in a colourful way that we cannot have potential wells in Ω). It is however possible to find points x_0 in the domain Ω such that $\operatorname{grad} u(x_0)$, that is to say points at which the force is zero. Such points

---

[23] Note that if the considered field is not derived globally from a potential, this result is false (see particularly in hydrodynamics and magnetostatics).

are equilibrium points: a charge or a point mass finding itself at x_0 at the instant $t = 0$ without initial velocity, will remain at this point for all $t > 0$. In a general way (see Arnold [1], p. 104) such equilibrium points are unstable. Let us illustrate this through the following example.

For $\Omega \subset \mathbb{R}^2$ with $0 \in \Omega$, the harmonic potential

$$u = \frac{k}{2}(x_1^2 - x_2^2) \tag{2.38}$$

is such that $\operatorname{grad} u(0) = 0$. The equations of motion of a point mass for the potential (2.38) are

$$\begin{cases} \dfrac{d^2x_1}{dt^2} = -kx_1 \\ \dfrac{d^2x_2}{dt^2} = +kx_2 . \end{cases} \tag{2.39}$$

For $k > 0$, we put $\omega = \sqrt{k}$; supposing that the initial velocities are zero, the motion is given by the equations

$$\begin{cases} x_1 = x_1^0 \cos \omega t \\ x_2 = x_2^0 \operatorname{ch} \omega t , \end{cases}$$

which shows that the point mass moves off indefinitely when $x_2^0 \neq 0$. (Cf. Fig. 1.)

More generally, we can show in a similar fashion that for every harmonic potential u, a point x_0 such that $\operatorname{grad} u(x_0) = 0$ and the Hessian $\partial^2 u(x_0)/\partial x_i \partial x_j \neq 0$ is an unstable equilibrium point.

Finally, we note that these considerations can be applied to potentials other than harmonic potentials; in particular, we can obtain the same result for superharmonic potential (see §4.1 and §4.6).

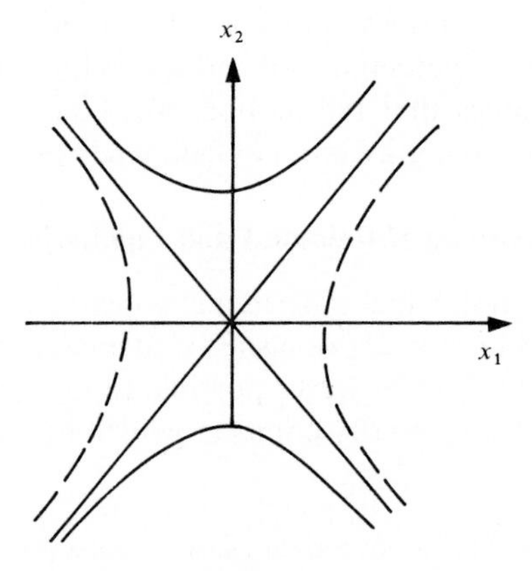

Fig. 1

§3. Newtonian Potentials

1. Generalities on Newtonian Potentials of a Distribution with Compact Support

We recall (see §2.1) that we have defined $E_n(x) = E_n(|x|)$

$$E_n(r) = \begin{cases} \dfrac{1}{2} r & \text{if } n = 1 \\ \dfrac{1}{2\pi} \operatorname{Log} r & \text{if } n = 2 \\ \dfrac{1}{k_n r^{n-2}} & \text{if } n \geqslant 3 \quad \text{with} \quad k_n = -(n-2)\sigma_n \end{cases}$$

which is analytic harmonic on $\mathbb{R}^n \setminus \{0\}$, locally integrable on $\mathbb{R}^n$ and an elementary solution of the Laplacian:

$$\Delta E_n = \delta \qquad \text{(in the sense of distributions).}$$

It is even (see Proposition 4 of § 2), to within an additive constant, the only radial elementary solution. In other words, E_n is the only radial elementary solution satisfying

$$\text{(3.1)} \qquad \begin{cases} \text{if } n = 1, & E_n(0) = 0 \\ \text{if } n = 2, & E_n(x) = 0 \quad \text{for} \quad |x| = 1 \\ \text{if } n \geqslant 3, & \lim\limits_{|x| \to \infty} E_n(x) = 0. \end{cases}$$

From the invariance under translation, we see that for all $y \in \mathbb{R}^n$, the function $u(x) = E_n(x - y)$ is a solution of Poisson's equation

$$\Delta u = \delta_y$$

where δ_y is the Dirac mass at y.

By linearity, for $y_1, \ldots, y_n \in \mathbb{R}^n$, $\alpha_1, \ldots, \alpha_n \in \mathbb{R}$, the function $u(x) = \Sigma \alpha_i E_n(x - y_i)$ is a solution of the Poisson equation

$$\Delta u = \sum \alpha_i \delta_{y_i}.$$

The distribution (measure) $f = \Sigma \alpha_i E_n(x - y_i)$ corresponds to a distribution of point masses at $y_1, \ldots, y_N$ of values $\alpha_1, \ldots, \alpha_N$. More generally, given $f \in \mathscr{E}'$[24], a distribution on $\mathbb{R}^n$, with compact support, we can define the *convolution distribution*

[24] Following usual practice $\mathscr{E}'$ denotes the space of distributions f on $\mathbb{R}^n$ with compact support, i.e. $f = 0$ on $\{x; |x| > R_0\}$ for R_0 sufficiently large.

of E_n by f[25]

(3.2) $$\langle E_n * f, \zeta \rangle = \langle f, E_n * \zeta \rangle .$$

This clearly has meaning since the convolution $E_n * \zeta$ of the locally integrable function E_n by a function ζ of class $\mathscr{C}^\infty$ with compact support is a function of class $\mathscr{C}^\infty$:

$$E_n * \zeta(x) = \int E_n(x - y)\zeta(y)\,dy = \int E_n(y)\zeta(x - y)\,dy .$$

Definition 1. Given $f \in \mathscr{E}'$, a distribution on $\mathbb{R}^n$ with compact support, the distribution $E_n * f$ is called the *Newtonian potential of f.*

We note some immediate properties:

Remarks 1

(*a*) E_n is the Newtonian potential of δ, the Dirac mass at the origin. More generally, $u(x) = \Sigma\alpha_i E_n(x - y_i)$ is the Newtonian potential of $f = \Sigma\alpha_i\delta_{y_i}$.

(*b*) The Newtonian potential is *linear*: if u_1, u_2 are the Newtonian potentials of $f_1, f_2 \in \mathscr{E}'_n$, then $\lambda_1 u_1 + \lambda_2 u_2$ is the Newtonian potential of $\lambda_1 f_1 + \lambda_2 f_2$ for all scalars λ_1 and λ_2.

(*c*) The derivative $\partial u/\partial x_i$ of the Newtonian potential u of $f \in \mathscr{E}'$ is the Newtonian potential of $\partial f/\partial x_i$. We note also that $\dfrac{\partial u}{\partial x_i} = \dfrac{\partial E_n}{\partial x_i} * f$, the convolution of $f \in \mathscr{E}'$ with the locally integrable function

$$\frac{\partial E_n}{\partial x_i}(x) = \frac{x_i}{\sigma_n |x|^n} .$$

These two properties are deducible from the general theorems concerning the differentiation of convolutions[25].

(*d*) The Newtonian potential u of $f \in \mathscr{E}'$ is a solution of Poisson's equation

(3.3) $$\Delta u = f \quad \text{on} \quad \mathbb{R}^n .$$

In effect by definition, $\forall \zeta \in \mathscr{E}$

$$\langle \Delta u, \zeta \rangle = \langle u, \Delta\zeta \rangle = \langle f, E_n * \Delta\zeta \rangle$$

$$E_n * \Delta\zeta(x) = \langle E_n, \Delta\zeta_x \rangle = \zeta_x(0) = \zeta(x) \quad \text{with} \quad \zeta_x(y) = \zeta(x - y) .$$

We shall see below how to characterise this solution of Poisson's equation (Proposition 3).

(*e*) *A distribution* $u \in \mathscr{E}'$ *is the Newtonian potential of its Laplacian* $f = \Delta u (\in \mathscr{E}')$. This is a particular case of Proposition 3 below, but it is seen directly from the identities concerning convolutions[26]:

$$u = u * \delta = u * \Delta E_n = \Delta u * E_n .$$

---

[25] We consider here the convolution of the *even function* E_n by a distribution: it is a simple particular case of the convolution of distributions (see Vol. 2, Appendix "Distributions").

[26] The formula for differentiating the convolution of distributions is deduced easily from the definition beginning with the formula for functions (see Vol. 2, Appendix "Distributions").

(f) The point (e) above shows that *every distribution is locally a Newtonian potential.* More precisely,

Lemma 1. *Let Ω be an open set of $\mathbb{R}^n$, $f \in \mathscr{D}'(\Omega)$ and u a solution (in the sense of distributions) of Poisson's equation $\Delta u = f$ on Ω.*
Then for every bounded open set Ω_1 with $\bar{\Omega}_1 \subset \Omega$ there exists $f_1 \in \mathscr{E}'$ such that

$$(3.4) \qquad f_1 = f \text{ on } \Omega\,, \quad u = \text{the Newtonian potential of } f_1 \text{ on } \Omega_1\,.$$

In effect, considering $\rho \in \mathscr{D}(\Omega)$ with $\rho = 1$ on Ω_1, the distribution ρu is the Newtonian potential of $f_1 = \Delta(\rho u)$ – see (e) above – and $f_1 = f$ on Ω_1.
The remark (f) permits us to reduce *the local study of any solutions whatsoever of Poisson's equation on an arbitrary open set Ω, to the local study of Newtonian potentials*[27]. □

In particular, we prove

Proposition 1. *Let Ω be an open set of $\mathbb{R}^n$ f a distribution on Ω and u a solution (in the sense of distributions) of Poisson's equation*

$$\Delta u = f \quad on \quad \Omega\,.$$

Then (1) *if $f = 0$ on Ω, u is harmonic on Ω;*
(2) *if $f \in \mathscr{C}^\infty(\Omega)$, $u \in \mathscr{C}^\infty(\Omega)$*[28].

Proof. Suppose that we are given a bounded open set Ω_1 with $\bar{\Omega}_1 \subset \Omega$. From Remark 1 ($f$), there exists $f_1 \in \mathscr{E}'$ such that

$$f_1 = f\,, \qquad u = E_n * f_1 \quad \text{on} \quad \Omega_1\,.$$

Then we select an open set Ω_2 with $\bar{\Omega}_2 \subset \Omega_1$ and $\rho \in \mathscr{D}(\Omega_1)$ with $\rho = 1$ on Ω_2. From the Remark 1(b), $u = u_1 + u_2$ on Ω_1 where u_1 and u_2 are the Newtonian potentials of ρf_1 and $(1 - \rho) f_2$ respectively.
For every $x \in \Omega_2$, the function $\varphi_x(y) = (1 - \rho(y)) E_n(x - y)$ is of class $\mathscr{C}^\infty$ on $\mathbb{R}^n$: we can therefore consider $\langle f_1, \varphi_x \rangle$. In fact, since the map $(x, y) \to \varphi_x(y)$ is of class $\mathscr{C}^\infty$ on $\Omega_2 \times \mathbb{R}^n$, the map $x \to \langle f_1, \varphi_x \rangle$ is of class $\mathscr{C}^\infty$ on Ω_2; we verify easily that this is the restriction of u_2 to Ω_2. In other words, u_2 is $\mathscr{C}^\infty$ on Ω_2. Since $\Delta u_2 = 0$ on Ω_2, the function u_2 is also harmonic on Ω_2 and hence analytic on Ω_2 (see Corollary 7 of § 2).
Let us suppose that $f = 0$ (resp. of class $\mathscr{C}^\infty$) on Ω. Then $\rho f_1 = 0$ on $\mathbb{R}^n$ (resp. $\rho f_1 \in \mathscr{D}(\mathbb{R}^n)$); we then deduce that $u_1 = 0$ (resp. of class $\mathscr{C}^\infty$) on $\mathbb{R}^n$ and hence $u = u_1 + u_2$ on Ω_1 is harmonic (resp. of class $\mathscr{C}^\infty$) on Ω_2; this being true for every open set Ω_2 with $\bar{\Omega}_2 \subset \Omega_1$ where Ω_1 is a bounded open set with $\bar{\Omega}_1 \subset \Omega$. □

---

[27] All the statements made in this Remark 1, can be applied without change to any elementary solution whatsoever of any linear differential operator (see Chap. V).

[28] This proposition is a particular case of general results concerning linear differential operators; the point (2) will serve as the definition of the notion of a *hypoelliptic* operator (see Chap. V, §2). We shall find also that the same property (2) is true on replacing "$\mathscr{C}^\infty$" by "analytic": if u is an elementary solution of Poisson's equation $\Delta u = f$ on Ω and f is analytic on Ω, then u is analytic on Ω (see Chap. V, §4).

It follows in particular from Proposition 1 that, given $f \in \mathscr{E}'$, the Newtonian potential u of f is harmonic analytic on the exterior of $B(0, R_0)$ for R_0 sufficiently large. We give the behaviour at infinity of u:

Proposition 2. *Let $f \in \mathscr{E}'$ and let u be the Newtonian potential of f. Then for every multi-index $\alpha \in \mathbb{N}$*[29]

$$\frac{\partial^\alpha u}{\partial x^\alpha}(x) = \langle f, 1 \rangle \frac{\partial^\alpha E_n}{\partial x^\alpha}(x) + O\left(\frac{1}{|x|^{n+|\alpha|-1}}\right)^{30} \quad \textit{when} \quad |x| \to \infty . \tag{3.5}$$

In particular

$$\textit{if} \quad n \geqslant 3, \quad \lim_{|x| \to \infty} u(x) = 0$$

$$\textit{if} \quad n \geqslant 2, \quad \lim_{|x| \to \infty} \operatorname{grad} u(x) = 0 .$$

We shall make use of the following lemma:

Lemma 2. *For every multi-index $\alpha \in \mathbb{N}^n$ with $m = |\alpha| \geqslant 1$, there exists a constant C (depending only on n and α) such that*

$$\left| \frac{\partial^\alpha E_n}{\partial x^\alpha}(x) \right| \leqslant \frac{C}{|x|^{n+m-2}}, \quad \textit{for all} \quad x \in \mathbb{R}^n \backslash \{0\} . \tag{3.6}$$

Proof of Lemma 2. The result is trivial for $n = 1$:

$$\left| \frac{\mathrm{d}E_1}{\mathrm{d}x}(x) \right| = \frac{1}{2}, \qquad \left| \frac{\mathrm{d}^2 E_1}{\mathrm{d}x^2}(x) \right| = 0$$

for all $x \neq 0$. Let us suppose hence that $n \geqslant 2$. We have then

$$\operatorname{grad} E_n(x) = \frac{x}{\sigma_n |x|^n} .$$

By recurrence on $m = |\alpha|$, we see that

$$\frac{\partial^\alpha E_n}{\partial x^\alpha}(x) = \frac{P_\alpha(x)}{|x|^{n+2(m-1)}},$$

where P_α is a polynomial of degree m.
Hence the lemma is proved. □

Proof of Proposition 2. Let us consider R_0 such that $\operatorname{supp} f \subset B(0, R_0)$. Given $x \in \mathbb{R}^n \backslash \bar{B}(0, R_0)$, the function $y \to E_n(x - y)$ is of class $\mathscr{C}^\infty$ in the neighbourhood of $\operatorname{supp} f$. Hence

$$u(x) = \langle f(y), E_n(x - y) \rangle$$

---

[29] For the multi-index notation see footnote [9] above.

[30] Given two functions $f(x)$ and $g(x)$, we say that $f = 0(g)$ when $x \to x_0$ if there exists a constant M such that $|f(x)| \leqslant M|g(x)|$ for all x sufficiently near to x_0. Here we apply the idea to the case $x_0 = \infty$.

and

$$(3.7)\qquad \frac{\partial^\alpha u}{\partial x^\alpha}(x) - \langle f, 1\rangle \frac{\partial^\alpha E_n}{\partial x^\alpha}(x) = \left\langle f(y), \quad \frac{\partial^\alpha E_n}{\partial x^\alpha}(x-y) - \frac{\partial^\alpha E_n}{\partial x^\alpha}(x) \right\rangle .$$

The distribution f is of finite order (see Vol. 2, Appendix, "Distributions") i.e. there exist measures (f_β), $\beta \in \mathbb{N}^n$, $|\beta| \leqslant m$ such that

$$f = \sum \frac{\partial^\beta f_\beta}{\partial x^\beta} \qquad \text{(in the sense of distributions)}^{31}$$

Obviously we can suppose that $\operatorname{supp} f_3 \subset \bar{B}(0, R_0)$.
Denoting by $R_\alpha(x)$ the two members of the equality (3.7) we have

$$\begin{aligned} R_\alpha(x) = &\int \left(\frac{\partial^\alpha E_n}{\partial x^\alpha}(x-y) - \frac{\partial^\alpha E_n}{\partial x^\alpha}(x) \right) \mathrm{d} f_0(y) \\ &+ \sum_{1 \leqslant |\beta| \leqslant m} \int \frac{\partial^{\alpha+\beta} E_n}{\partial x^{\alpha+\beta}}(x-y) \, \mathrm{d} f_\beta(y) . \end{aligned}$$

From Lemma 2, for $|y| \leqslant R_0 < R_1 \leqslant |x|$ we have

$$\left| \frac{\partial^{\alpha+\beta} E_n}{\partial x^{\alpha+\beta}}(x-y) \right| \leqslant \frac{C}{|x-y|^{n+|\alpha+\beta|-2}} \leqslant \frac{C_1}{|x|^{n+|\alpha|-1}} \quad \text{for} \quad 1 \leqslant |\beta| \leqslant m$$

$$\begin{aligned} \left| \frac{\partial^\alpha E_n}{\partial x^\alpha}(x-y) - \frac{\partial^\alpha E_n}{\partial x^\alpha}(x) \right| &\leqslant \sup_{0 \leqslant \lambda \leqslant 1} \left| \operatorname{grad} \frac{\partial^\alpha E_n}{\partial x^\alpha}(x - \lambda y) . y \right| \\ &\leqslant \sup_{0 \leqslant \lambda \leqslant 1} \frac{C|y|}{|x - \lambda y|^{n+|\alpha|-1}} \leqslant \frac{C_1}{|x|^{n+|\alpha|-1}} , \end{aligned}$$

where C_1 depends only on $|\alpha|$, m, R_0, R_1. We deduce

$$|R_\alpha(x)| \leqslant \frac{C_1 \sum M_\beta}{|x|^{n+|\alpha|-1}} \quad \text{for all} \quad x \notin B(0, R_1) ,$$

where

$$M_\beta = \sup_{|\zeta| \leqslant 1} \left| \int \zeta(x) \, \mathrm{d} f_\beta(x) \right|$$

is the total variation of the measure f_β. □

We then have characterisation:

Proposition 3. *Let $f \in \mathscr{E}'$ and u be a distribution on $\mathbb{R}^n$ with $n \geqslant 2$. Then u is the Neutonian potential of f iff u is a solution of*

$$\Delta u = f \quad \text{on} \quad \mathbb{R}^n$$

---

[31] In the examples that we shall consider, f will be a function (sect. 2), a measure (sect. 3), or a distribution of order 1 (sect. 4).

and satisfies the conditions at infinity:

$$\text{if } n \geqslant 3\,, \quad \lim_{|x|\to\infty} u(x) = 0\,,$$

$$\text{if } n = 2\,, \quad \lim_{|x|\to\infty} \left(u(x) - \frac{1}{2\pi} \langle f, 1\rangle \operatorname{Log}|x| \right) = 0\,.$$

Proof. The necessary condition is a consequence of Remark 1(d) and of Proposition 5. To show that the condition is sufficient, replacing u by $u - \bar{u}$ where $\bar{u}$ is the Newtonian potential of f, we can always suppose $f = 0$ and we then have to show that $u = 0$. In other words, taking Proposition 1 into account, we have a function u, harmonic on $\mathbb{R}^n$ which tends to zero at infinity. Using Corollary 1 (or Corollary 5) of §2 we deduce that u is identically zero.
We note the Corollary resulting from invariance under an orthogonal transformation of the characterisation of Proposition 3:

Corollary 1. *The Newtonian potential of a radial distribution with compact support (that is to say invariant under orthogonal transformation) is radial.*

Remark 2. We have excluded from the characterisation of Proposition 3, the case $n = 1$, which can (and ought to) be treated directly. The Newtonian potential u of a distribution f on $\mathbb{R}$ with support contained in $]-R_0, +R_0[$ is defined for $|x| \geqslant R_0$ by

$$u(x) = \left\langle f(y)\,, \frac{1}{2}|x-y| \right\rangle = \frac{1}{2}\langle f, 1\rangle\,|x| - \frac{1}{2}\langle f, y\rangle \frac{x}{|x|}\,.$$

To characterise the solution u of Poisson's equation $\dfrac{d^2 u}{dx^2} = f$ which is the Newtonian potential, it is necessary to have two independent boundary conditions since the solutions of Laplace's equation on $\mathbb{R}$ are affine functions depending on two arbitrary constants.
For example, we could choose the boundary conditions

$$\lim_{x\to+\infty} \frac{u(x)}{x} = \frac{1}{2}\langle f, 1\rangle$$

$$\lim_{x\to+\infty} u(x) + u(-x) = 0\,. \qquad \square$$

Remark 3. *In the case* $n = 2$, the Newtonian potential of a distribution $f \in \mathscr{E}'$ does not tend to zero at infinity. More precisely, if u is the Newtonian potential of $f \in \mathscr{E}'$, the following statements are equivalent (only in the case $n = 2$)

(i) $\lim_{|x|\to\infty} u(x) = 0$;

(ii) $u(x)$ is bounded as $|x| \to \infty$;

(iii) $\langle f, 1\rangle = 0$.

This follows immediately from Proposition 2. $\square$

We note also the following proposition which characterises the Newtonian potential to within an additive constant.

Proposition 4. *Let $f \in \mathscr{E}'$ and let u be a distribution on $\mathbb{R}^n$ with $n \geqslant 2$. Then, to within an additive constant, u is the Newtonian potential of f iff u is a solution of Poisson's equation*

$$\Delta u = f$$

and satisfies at infinity one or other of the following conditions:

$$\lim_{|x| \to \infty} \frac{u(x)}{|x|} = 0 \tag{3.8}$$

or

$$\lim_{|x| \to \infty} \operatorname{grad} u(x) = 0 \,. \tag{3.9}$$

Proof. Taking account of Proposition 2, and reasoning as in the proof of Proposition 3, we see that it is sufficient to show that a function, harmonic on $\mathbb{R}^n$ satisfying one or other of the conditions (3.8) or (3.9) is constant. In the case of the condition (3.9), since the derivatives $\partial u / \partial x_i$ are also harmonic, we deduce that they are zero: hence u is constant. In the case of the condition (3.8), we can remark that it implies that u is a tempered distribution and hence, from Proposition 3 of § 2, a polynomial; but only the constant polynomials satisfy (3.8). □

In Remark 1 (f), we have seen that every distribution is locally a Newtonian potential. Using Green's formula, we have in fact a global representation on a bounded open set Ω of a distribution on $\bar{\Omega}$: we state here the result within the classical framework.

Proposition 5. *Let Ω be a regular bounded open set and let $u \in \mathscr{C}^2(\Omega) \cap \mathscr{C}^1_n(\bar{\Omega})$ with $\Delta u \in L^1(\Omega)$. Then*

$$u = u_0 + u_1 + u_2 \quad on \quad \Omega$$

where u_0, u_1, u_2 are the Newtonian potentials of the distributions f_0, f_1, f_2 on $\mathbb{R}^n$ defined by

$$\langle f_0, \zeta \rangle = \int_\Omega \zeta \, \Delta u \, \mathrm{d}x$$

$$\langle f_1, \zeta \rangle = \int_\Gamma \zeta \left(- \frac{\partial u}{\partial n} \right) \mathrm{d}\gamma$$

$$\langle f_2, \zeta \rangle = \int_\Gamma \frac{\partial \zeta}{\partial n} u \, \mathrm{d}\gamma \,.$$

We note that $f_0, f_1, f_2 \in \mathscr{E}'$:
f_0 is an integrable function on $\mathbb{R}^n$ with support contained in $\bar{\Omega}$,
f_1 is a measure on $\mathbb{R}^n$ with support contained in Γ,
f_2 is a distribution of order 1 on $\mathbb{R}^n$ with support contained in Γ.

We say that u_1 is the *simple layer* (resp. *double layer*) *potential* defined by the function $-\partial u/\partial n$ (resp. u) continuous on Γ. In a general way, we have the

Definition 2. Given a regular bounded open set Ω of $\mathbb{R}^n$ and $\varphi \in \mathscr{C}^0(\Gamma)$, we give the name *simple layer* (resp. *double layer*) *potential of* φ to the Newtonian potential of the distribution f, with compact support, defined by

$$\langle f, \zeta \rangle = \int_\Gamma \zeta(t)\varphi(t)\,\mathrm{d}\gamma(t) \quad \left(\textit{resp.} \quad \int_\Gamma \frac{\partial \zeta}{\partial n}(t)\varphi(t)\,\mathrm{d}\gamma(t)\right).$$

We shall study these potentials, as well as their physical interpretation in the following sections.

Proof of Proposition 5. We fix $x_0 \in \Omega$ and denote for $0 < \varepsilon < \operatorname{dist}(x_0, \Gamma)$

$$\Omega_\varepsilon = \Omega \backslash \bar{B}(x_0, \varepsilon).$$

Applying Green's formula (see Proposition 4 of § 1) to u and $v(x) = E_n(x - x_0)$ on the open set Ω_ε, and noting that v is harmonic in the neighbourhood of $\bar{\Omega}_\varepsilon$, we have

$$0 = \int_{\Omega_\varepsilon} u\,\Delta v\,\mathrm{d}x = \int_{\Omega_\varepsilon} v\,\Delta u\,\mathrm{d}x + \int_{\partial\Omega_\varepsilon} v\left(-\frac{\partial u}{\partial n}\right)\mathrm{d}\gamma + \int_{\partial\Omega_\varepsilon} \frac{\partial v}{\partial n}u\,\mathrm{d}\gamma.$$

Let us determine the limit of each of these integrals when $\varepsilon \to 0$.

$$(1) \qquad \int_{\Omega_\varepsilon} v\,\Delta u\,\mathrm{d}x \to \int_\Omega E_n(x - x_0)\,\Delta u(x)\,\mathrm{d}x = u_0(x_0).$$

This follows from Lebesgue's theorem, since the function $E_n(x - x_0)\,\Delta u(x)$ is integrable on Ω

$$(2) \qquad \int_{\partial\Omega_\varepsilon} v\left(-\frac{\partial u}{\partial n}\right)\mathrm{d}\gamma \to \int_\Gamma E_n(x - x_0)\left(-\frac{\partial u}{\partial n}(x)\right)\mathrm{d}\gamma(x) = u_1(x_0).$$

In effect $\partial\Omega_\varepsilon$ is made up of Γ and $\partial B(x_0, \varepsilon)$. On $\partial B(x_0, \varepsilon)$ we have $v = v_\varepsilon$ where

$$v_\varepsilon = \begin{cases} \dfrac{1}{2}\varepsilon & \text{if } n = 1 \\[2ex] \dfrac{1}{2}\operatorname{Log}\varepsilon & \text{if } n = 2 \\[2ex] \dfrac{1}{k\varepsilon^{n-2}} & \text{if } n \geqslant 3. \end{cases}$$

On the other hand, since u is of class $\mathscr{C}^1$ in the neighbourhood of x_0, $|\partial u/\partial n| \leqslant C$ on $\partial B(x_0, \varepsilon)$ where C is independent of $\varepsilon \to 0$.
Therefore

$$\left|\int_{\partial B(x_0,\varepsilon)} v\left(-\frac{\partial u}{\partial n}\right)\mathrm{d}\gamma\right| \leqslant C v_\varepsilon \sigma_n \varepsilon^{n-1}.$$

In all cases $v_\varepsilon \varepsilon^{n-1} \to 0$ when $\varepsilon \to 0$, from which we have the limit.

$$(3) \qquad \int_{\partial\Omega_\varepsilon} \frac{\partial v}{\partial n} u \, \mathrm{d}\gamma \to \int_\Gamma \frac{\partial E_n}{\partial n}(x - x_0) u(x) \, \mathrm{d}\gamma(x) - u(x_0) = u_2(x_0) - u(x_0) .$$

This reduces to showing that

$$\int_{\partial B(x_0, \varepsilon)} \frac{\partial v}{\partial n} u \, \mathrm{d}\gamma \to - u(x_0) .$$

Now $\frac{\partial v}{\partial n} = - \frac{1}{\sigma_n \varepsilon^{n-1}}$ on $\partial B(x_0, \varepsilon)$ (we note also that the normal derivative is exterior to Ω_ε and hence interior to $B(x_0, \varepsilon)$).
Therefore we have

$$\int_{\partial B(x_0, \varepsilon)} \frac{\partial v}{\partial n} u \, \mathrm{d}\gamma = - \int_\Sigma u(x_0 + \varepsilon\sigma) \, \mathrm{d}\sigma \to - u(x_0) . \qquad \square$$

Remark 4. The Newtonian potentials u_1 and u_2 are harmonic on Ω since f_1 and f_2 are with support in Γ. Hence u and u_0 have the same regularity on Ω.
Given $u \in \mathscr{C}^2(\Omega) \cap \mathscr{C}^1_n(\bar{\Omega})$ with $\Delta u \in L^1(\Omega)$, the proposition states in particular that there exists u_0 the Newtonian potential of an integrable function f_0 with compact support, u_1 the simple layer potential of a function $\varphi_1 \in \mathscr{C}^0(\Omega)$ and u_2 the double layer potential of a function $\varphi_2 \in \mathscr{C}^0(\Gamma)$ such that

$$u = u_0 + u_1 + u_2 \quad \text{on} \quad \Omega .$$

We note though that this decomposition is not unique: $f_0 = \Delta u$ on Ω is determined on Ω only if we impose on it the condition of having compact support on $\bar{\Omega}$; even in this case φ_1 and φ_2 and hence u_1 and u_2 are not determined uniquely since the sum of a single layer potential and a double layer potential can be zero on Ω without each of the terms being zero. $\square$

2. Study of Local Regularity of Solutions of Poisson's Equation

We study the local regularity of solutions of Poisson's equation

$$\Delta u = f \quad \text{on} \quad \Omega$$

as a function of the regularity of the distribution f given on the open set Ω of $\mathbb{R}^n$. In the case $n = 1$, the problem is simple, since Poisson's equation reduces to the differential equation

$$\frac{\mathrm{d}^2 u}{\mathrm{d}x^2} = f .$$

If f is of class $\mathscr{C}^m$, then u is of class $\mathscr{C}^{m+2}$; if f is a bounded (measurable function), then u is of class $\mathscr{C}^1$ with Lipschitzian[32] derivative $\mathrm{d}u/\mathrm{d}x$; if f is an integrable function, then u is of class $\mathscr{C}^1$ with absolutely continuous derivative[33] $\mathrm{d}u/\mathrm{d}x$; if f is a bounded (Radon) measure, then u is Lipschitzian[33] with a derivative $\mathrm{d}u/\mathrm{d}x$ of bounded variation[33]; if f is a distribution of order 1, then u is of bounded variation[33]; etc.

The situation is entirely different in the case $n \geqslant 2$. Taking account of Proposition 5, we shall content ourselves to studying the regularity of solutions of Poisson's equation when f *is a distribution of at least order* 1. Let us note the

Proposition 6. *Let Ω be an open set of $\mathbb{R}^n$ $(n > 2)$, f a distribution at least of order* 1 *on Ω and u a distribution solution of Poisson's equation $\Delta u = f$ on Ω. Then u is a (locally integrable) function on Ω. More precisely*:
(1) *in the case of a general distribution f of order* 1, *$u \in L^p_{\mathrm{loc}}(\Omega)$ for all $p < n/(n-1)$*;
(2) *if f is a measure on Ω, then $u \in L^p_{\mathrm{loc}}(\Omega)$ for all $p < n/(n-2)$; the derivatives $\partial u/\partial x_i$ are (locally integrable) functions on Ω and $\partial u/\partial x_i \in L^p_{\mathrm{loc}}(\Omega)$ $\forall p < n(n-1)$.*
(3) *if f is a function belonging to $L^q_{\mathrm{loc}}(\Omega)$ with $q > \frac{1}{2}n$* (resp. $q > n$), *then u is continuous* (resp. *of class $\mathscr{C}^1$*) *on Ω*.

This proposition will follow, with the help of the Remark 1(f) from the following lemma on the *convolution of a measure and a function.*

Lemma 3[34]. *Let f be a measure on $\mathbb{R}^n$ with compact support and g a Borel function belonging to $L^p_{\mathrm{loc}}(\mathbb{R}^n)$ with $1 \leqslant p < \infty$. Then*
(1) *for almost all $x \in \mathbb{R}^n$, the function $y \to g(x-y)$ is integrable with respect to the measure f.*
(2) *the convolution $g * f$ is a function belonging to $L^p_{\mathrm{loc}}(\mathbb{R}^n)$ and defined almost everywhere on $\mathbb{R}^n$ by*

$$g * f(x) = \int g(x-y)\,\mathrm{d}f(y)\,;$$

---

[32] A function v on an interval I is:
(i) *Lipschitzian* if there exists C such that $|v(x) - v(y)| \leqslant C|x - y|$ for all $x, y \in I$;
(ii) *absolutely continuous* if $\forall \varepsilon > 0$, there exists $\delta > 0$ such that for every sequence of intervals $]a_k, b_k[\; \subset I$, two by two disjoint $\Sigma(b_k - a_k) < \delta \Rightarrow \Sigma|v(b_k) - v(a_k)| < \varepsilon$;
(iii) *of bounded variation* if there exists C that for every sequence $a_1 < a_2 < \ldots < a_N$ of elements of I, $\Sigma|v(a_k) - v(a_{k-1})| \leqslant C$.
From Lebesgue's differentiation theorem, every function of bounded variation is a.e. differentiable and its derivative (defined a.e.) is integrable. A function is absolutely continuous iff it is of bounded variation and the integral of its derivative; a function is Lipschitzian iff it is locally absolutely continuous and has an (essentially) bounded derivative.
Finally a distribution on I is defined by a Lipschitzian (*resp.* absolutely continuous, *resp.* of bounded variation) function iff its derivative (in the sense of distributions) is defined by bounded measurable function (*resp.* an integrable function, *resp.* a bounded measure) (see Schwartz [2], Rudin [1]).

[33] For these definitions see the preceding footnote.

[34] We grant a privileged place to the notation of distributions: a measure is a distribution of order 0; a measure f is a function if it is defined by a (locally integrable) function. However we make use also of the theory of integration with respect to a measure: a distribution f of order 0 defines a (unique) regular Borel measure which we shall always denote by f; we know then to say what it means that a Borel function is integrable with respect to f, what is its own integral, etc. (see Marle [1] or Rudin [1]).

(3) *if f is a function belonging to $L^q(\mathbb{R}^n)$ where q is the conjugate of p ($p^{-1} + q^{-1} = 1$), then $g * f$ is continuous on $\mathbb{R}^n$.*

Proof of Lemma 3. The points (1) and (2) result from Fubini's theorem. Taking $r_0 > 0$ such that $\bar{B}(0, r_0)$ contains supp f, denoting by μ the variation of f, we have $f = \varepsilon\mu$ where ε is a Borel function taking the values $+1$ or -1 (Lebesgue's decomposition theorem or the Radon–Nikodym theorem). Given $r > 0$, we have

$$\int_{B(0,r)} \mathrm{d}x \int |g(x - y)\varepsilon(y)| \,\mathrm{d}\mu(y) = \int_{\bar{B}(0,r_0)} \mathrm{d}\mu(y) \int_{B(y,r)} |g(x)| \,\mathrm{d}x$$

$$\leqslant \left(\int \mathrm{d}\mu\right)\left(\int_{B(0,r_0+r)r} |g(x)| \mathrm{d}x\right) < \infty .$$

Hence (by Fubini's theorem), for almost all $x \in B(0, r)$, the function $y \to g(x - y)\varepsilon(y)$ is integrable with respect to μ and the function

$$x \to \int g(x - y)\varepsilon(y)\,\mathrm{d}\mu(y) \quad \text{is integrable on} \quad B(0, r) .$$

In other words, we have the point (1) and the function defined a.e. on $\mathbb{R}^n$ by

$$u(x) = \int g(x - y)\,\mathrm{d}f(y)$$

is locally integrable on $\mathbb{R}^n$. To obtain the proof of (2), it is enough to note that

$$\langle g * f, \zeta\rangle = \int \mathrm{d}f(y) \int g(x - y)\zeta(x)\,\mathrm{d}x = \int u(x)\zeta(x)\,\mathrm{d}x$$

and for supp $\zeta \subset \bar{B}(0, r)$, by Hölder's inequality[35]

$$|\langle g * f, \zeta\rangle| \leqslant \left(\int \mathrm{d}\mu\right)\left(\int_{B(0,r_0+r)} |g(x)|^p \,\mathrm{d}x\right)^{\frac{1}{p}} \|\zeta\|_{L^q} .$$

Finally to prove the point (3), we can make use of

$$|g * f(x_1) - g * f(x_2)| \leqslant \|f\|_{L^q}\left(\int_{B(0,r_0+r)} |g(x + (x_1 - x_2)) - g(x)|^p \,\mathrm{d}x\right)^{\frac{1}{p}}$$

$$\text{a.e. } x_1, x_2 \in B(0, r)$$

and the continuity of the translation in $L^p_{\mathrm{loc}}(\mathbb{R}^n)$ $(1 \leqslant p < \infty)$. We can also use the fact that the map $g \to g * f$ from $L^p_{\mathrm{loc}}(\mathbb{R}^n)$ into $L^\infty_{\mathrm{loc}}(\mathbb{R}^n)$ is continuous; for g continuous $g * f$ is continuous; since the continuous functions are dense in

[35] *Hölder's inequality.* A function v, locally integrable on Ω, belongs to $L^p(\Omega)$ iff there exists C such that $|\int v\zeta\,\mathrm{d}x| \leqslant C\|\zeta\|_{L^q}$; the smallest constant C is $C = \|v\|_{L^p}$; $p^{-1} + q^{-1} = 1$.

$L^p_{loc}(\mathbb{R}^n)$, for all $g \in L^p_{loc}(\mathbb{R}^n)$, $g * f$ will be the uniform limit, on every compact set, of continuous functions and hence will be continuous.

Proof of Proposition 6. Given a bounded open set Ω_1 with $\bar{\Omega}_1 \subset \Omega$ and $\rho \in \mathscr{D}(\Omega)$ with $\rho = 1$ on Ω_1, we put $f_0 = \rho f$ and $f_1 = \Delta(\rho f) - f_0$. From Remark 1. (f), $u = u_0 + u_1$ on Ω_1 where u_0, u_1 are the Newtonian potentials of f_0, f_1. Since $f_1 = 0$ on Ω_1, u_1 is harmonic on Ω_1 so u and u_0 have the same regularity on Ω_1; also f_0 on $\mathbb{R}^n$ the same regularity as f on Ω_1. In other words *we can always suppose* $\Omega = \mathbb{R}^n$, f a distribution with compact support and u the Newtonian potential of f; this we do. If f is a distribution of order 1,

$$f = f_0 + \frac{\partial f_1}{\partial x_1} + \cdots + \frac{\partial f_n}{\partial x_n}$$

where $f_0, f_1, \ldots, f_n$ are measures with compact support,

$$u = u_0 + u_1 + \cdots + u_n$$

where $u_0, u_1, \ldots, u_n$ are the Newtonian potentials of f_0, $\partial f_1/\partial x_1, \ldots, \partial f_n/\partial x_n$ respectively.
Applying Lemma 3 with $f = f_0$ and $g = E_n$ which belongs to $L^p_{loc}(\mathbb{R}^n)$ for all $p < n/(n-2)$, we deduce that u_0 is a function belonging to $L^p_{loc}(\mathbb{R}^n)$ for all $p < n/(n-2)$. If, in addition, $f_0 \in L^q(\mathbb{R}^n)$ with $q > n/2$ which is the conjugate of $n/(n-2)$, then u_0 is continuous on $\mathbb{R}^n$.
Now, for $i = 1, \ldots, n$, $u_i = E_n * \dfrac{\partial f_i}{\partial x_i} = \dfrac{\partial E_n}{\partial x_i} * f_i$. Applying Lemma 3 with $f = f_i$ and $g = \dfrac{\partial E_n}{\partial x_i}$ which belongs to $L^p_{loc}(\mathbb{R}^n)$ for all $p < n/n - 1$, we deduce that $u_i \in L^p_{loc}(\mathbb{R}^n)$ for all $p < n/(n-1)$.
To complete the proof, note that $\dfrac{\partial u_0}{\partial x_i} = \dfrac{\partial E_n}{\partial x_i} * f_0$, with the result that $\dfrac{\partial u_0}{\partial x_i} \in L^p_{loc}(\mathbb{R}^n)$ for all $p < n/(n-1)$ and if in addition $f \in L^q(\mathbb{R}^n)$ with $q > n$ which is the conjugate of $n/(n-1)$, then $\partial u/\partial x_i$ is continuous on $\mathbb{R}^n$. □

It is clear that we can obtain by this method other results concerning the regularity of the solution as a function of the given f. In particular, by considering only the Lebesgue spaces L^p, we lose information since the elementary solution has more properties than those associated with belonging to $L^p_{loc}(\Omega)$ for all $p < n/(n-2)$. We can make this precise by making use of more sophisticated spaces: the spaces of Marcinkiewicz or more generally the spaces of Lorentz (see Vol. 4, Appendix "Singular Integrals" and Butzer–Berens [1]). We shall be content here to make these properties more precise in the case of *a radial solution*. We shall also be content to restrict ourselves to the case in which f is a function[36].
We shall prove the

[36] The case of a distribution (of order 0 or 1) will form the study of §3.

Proposition 7. *Let f be an integrable function on the ball $B(0, r_0)$ and u a* **radial** *solution of Poisson's equation*

$$\Delta u = f \quad \text{on} \quad B(0, r_0).$$

Then

(1) *u is of class $\mathscr{C}^1$ on $B(0, r_0) \backslash \{0\}$,*

$$\lim_{x \to 0} \frac{u(x)}{E_n(x)} = 0 \quad \text{and} \quad \lim_{x \to 0} |x|^{n-1} \operatorname{grad} u(x) = 0;$$

(2) *if $E_n(x) f(x)$ (resp. $f(x)/|x|^{n-1}$) is integrable on $B(0, r_0)$, then u is continuous (resp. of class $\mathscr{C}^1$) on $B(0, r_0)$;*
(3) *the second derivatives $\partial^2 u / \partial x_i \, \partial x_j$ are locally integrable functions on $B(0, r_0) \backslash \{0\}$. If $f(x) \operatorname{Log} |x|$ is integrable, then $\partial^2 u / \partial x_i \, \partial x_j$ are locally integrable on $B(0, r_0)$.*

If $f \in L^p(B(0, r_0))$ with $1 \leqslant p \leqslant \infty$, then $\dfrac{\partial^2 u}{\partial x_i \partial x_j} \in L^p_{\text{loc}}(B(0, r_0))$.

(4) *If f is of class $\mathscr{C}^m$ on $B(0, r_0)$, then u is of class $\mathscr{C}^{m+2}$ on $B(0, r_0)$.*

We shall use the

Lemma 4. *Let f be a* **radial integrable function** *on $\mathbb{R}^n$ with compact support. The Newtonian potential u of f is of class $\mathscr{C}'$ on $\mathbb{R}^n \backslash \{0\}$, and we have*

$$u(x) = E_n(x) \int_{B(0, |x|)} f(y) \, dy + \int_{\mathbb{R}^n \backslash B(0, |x|)} E_n(y) f(y) \, dy \tag{3.10}$$

$$\operatorname{grad} u(x) = \frac{x}{\sigma_n |x|^n} \int_{B(0, |x|)} f(y) \, dy. \tag{3.11}$$

The second derivatives are locally integrable on $\mathbb{R}^n \backslash \{0\}$ and

$$\frac{\partial^2 u}{\partial x_i \partial x_j}(x) = \frac{x_i x_j}{|x|^2} f(x) + \left(\frac{\delta_{ij}}{n} - \frac{x_i x_j}{|x|^2} \right) \frac{n}{\sigma_n |x|^n} \int_{B(0, |x|)} f(y) \, dy \tag{3.12}$$

where δ_{ij} is the Kronecker symbol

$$\delta_{ij} = \begin{cases} 1 & \text{if} \quad i = j \\ 0 & \text{if} \quad i \neq j. \end{cases}$$

Proof of Lemma 4. From Proposition 3, it will be enough to verify that the function u defined by (3.10) is clearly a solution of Poisson's equation satisfying the condition at infinity. It is more satisfying to "discover" these formulae. First of all, we note from Corollary 1 that the Newtonian potential u has to be radial; the fact that this is a function (locally integrable on $\mathbb{R}^n$) follows from Proposition 6. We therefore have

$$f(x) = g(|x|), \qquad u(x) = v(|x|)$$

where g, v are (measurable) functions defined a.e. on $]0, \infty[$. The function g has support contained in $[0, r_0]$ and for f to be integrable g satisfies

$$\int_0^\infty r^{n-1} g(r)\, dr < \infty .$$

The function $r^{n-1} v(r)$ is locally integrable on $]0, \infty[$ and the function v satisfies the differential equation

$$\frac{1}{r^{n-1}} \frac{d}{dr}\left(r^{n-1} \frac{dv}{dr} \right) = g \quad \text{on} \quad]0, \infty[. \tag{3.13}$$

This expresses that u satisfies Poisson's equation (see the Laplacian of a radial function in §1.4).
We see then that $r^{n-1} v'(r)$ is absolutely continuous on $]0, \infty[$; hence u is of class $\mathscr{C}^1$ on $\mathbb{R}^n \backslash \{0\}$ and the second derivatives $\partial^2 u / \partial x_i \, \partial x_j$ are locally integrable on $\mathbb{R}^n \backslash \{0\}$.
Now from Proposition 2, when $r \to \infty$,

$$\begin{aligned} \frac{dv}{dr}(r) &= \operatorname{grad} u(r\sigma).\sigma = \langle f, 1 \rangle \operatorname{grad} E_n(r\sigma).\sigma + O\left(\frac{1}{r^n}\right) \\ &= \frac{1}{r^{n-1}} \int_0^\infty t^{n-1} g(t)\, dt + O\left(\frac{1}{r^n}\right). \end{aligned}$$

We integrate (3.13) to obtain

$$\frac{dv}{dr}(r) = \frac{1}{r^{n-1}} \int_0^r t^{n-1} g(t)\, dt \quad \text{for all} \quad r > 0 .$$

This proves (3.11) and by differentiation (3.12). To obtain (3.10), it is enough to integrate (3.11), using the condition at infinity given by Proposition 2,

$$u(x) = E_n(x) \int f(y)\, dy + O\left(\frac{1}{|x|^{n-1}}\right) \quad \text{when} \quad |x| \to \infty . \qquad \square$$

Proof of Proposition 7. Taking up the reasoning of the beginning of the proof of Proposition 6, we can always reduce it to the case where f is integrable with compact support and u is the Newtonian potential of f; if ρ is radial ρu will be radial and so also f.
The point (1) is then included in Lemma 4 or results from the formulae (3.10) and (3.11) by the use of Lebesgue's theorem. To prove that $\lim\limits_{x \to 0} \dfrac{u(x)}{E_n(x)} = 0$, we observe that

$$\left| \frac{E_n(y)}{E_n(x)} f(y) \right| \leqslant |f(y)| \quad \text{for} \quad 0 < |x| \leqslant |y| \leqslant 1 .$$

Similarly the point (2) follows from (3.10) and (3.11), the use of Lebesgue's theorem and the inequalities

$$|E_n(x)f(y)| \leqslant |E_n(y)f(y)| \quad \text{and} \quad \left|\frac{x}{|x|^n}f(y)\right| \leqslant \frac{|f(y)|}{|y|^{n-1}}$$

$$\text{for} \quad 0 < |y| \leqslant |x| \leqslant 1 .$$

The first part of the point (3) is stated in the lemma and we have from (3.12)

$$\left|\frac{\partial^2 u}{\partial x_i \partial x_j}(x)\right| \leqslant |f(x)| + \left(1 - \frac{\delta_{ij}}{n}\right)\frac{n}{r^n}\int_0^r t^{n-1}g(t)\,dt$$

with $r = |x|$ and $g(r) = |f(x)|$.
We have trivially

$$\left\|\frac{\partial^2 u}{\partial x_i \partial x_j}\right\|_{L^\infty} \leqslant \left(2 - \frac{\delta_{ij}}{n}\right)\|f\|_{L^\infty} .$$

Also

$$\begin{aligned}\left\|\frac{\partial^2 u}{\partial x_i \partial x_j}\right\|_{L^1(B(0,1))} &\leqslant \|f\|_{L^1} + (n - \delta_{ij})\sigma_n\int_0^1 \frac{dr}{r}\int_0^r t^{n-1}g(t)\,dt \\ &= \|f\|_{L^1} + (n - \delta_{ij})\int_{B(0,1)} |f(y)|\,\mathrm{Log}\frac{1}{|y|}\,dy .\end{aligned}$$

Finally for $1 < p < \infty$,

$$\left\|\frac{\partial^2 u}{\partial x_i \partial x_j}\right\|_{L^p(B(0,R))} \leqslant \|f\|_{L^p} + \sigma_n(n - \delta_{ij})\left(\int_0^R \frac{dr}{r}\left(\int_0^r t^{n-1}g(t)\,dt\right)^p\right)^{1/p}$$

But from Hölder's inequality, with q the conjugate of p

$$\int_0^r t^{n-1}g(t)\,dt \leqslant \left(\int_0^r t^{n-1}\,dt\right)^{1/q}\left(\int_0^r t^{n-1}g(t)^p\,dt\right)^{1/p} \leqslant \left(\frac{r^n}{n}\right)^{1/q}\frac{\|f\|_{L^p}}{\sigma_n^{1/p}},$$

from which $\displaystyle\int_0^R \frac{dr}{r}\left(\int_0^r t^{n-1}g(t)\,dt\right)^p \leqslant n^{-p/q}\sigma_n^{-1}\|f\|_{L^p}^p\frac{q}{np}R^{\frac{np}{q}}$,

and

$$\left\|\frac{\partial^2 u}{\partial x_i \partial x_j}\right\|_{L^p(B(0,R))} \leqslant \|f\|_{L^p}\left(1 + \left(1 - \frac{\delta_{ij}}{n}\right)\frac{(\sigma_n R^n)^{1-\frac{1}{p}}}{(p-1)^{1/p}}\right). \tag{3.14}$$

This completes the proof of point (3). For the point (4), it is clear by (3.12) that u is of class $\mathscr{C}^{m+2}$ on $\mathbb{R}^n\setminus\{0\}$. Now using Taylor's formula

$$f(x) = \sum_{0 \leqslant k \leqslant \frac{m}{2}} a_k|x|^{2k} + o(|x|^m)^{37},$$

[37] If $f(x) = g(|x|) \in \mathscr{C}^m(\mathbb{R}^n)$, g is of class $\mathscr{C}^m$ on $]0, \infty[$ and the odd derivatives of g are zero at O. In the expansion $a_k = \dfrac{1}{(2k)!}\cdot\dfrac{d^{2k}}{dx^{2k}}g(O)$.

putting this back into (3.12), we obtain

$$\frac{\partial^2 u}{\partial x_i \partial x_j}(x) = \frac{a_0}{n}\delta_{ij} + \sum_{1 \leqslant k \leqslant \frac{m}{2}} \frac{a_k}{2k+n}|x|^{2(k-1)}(2k x_i x_j + \delta_{ij}|x|^2) + o(|x|^m),$$

from which the result follows by induction over m[38]. □

Let us return now to *the general case* and study the *local regularity of the second order derivatives.* We can generalize the point (3) of Proposition 7 for $1 < p < \infty$ and state:

Proposition 8. *Let Ω be an open set in $\mathbb{R}^n$, $1 < p < \infty$, $f \in L^p_{\text{loc}}(\Omega)$ and u a solution of Poisson's equation*

$$\Delta u = f \quad \text{on} \quad \Omega .$$

Then its second derivatives $\partial^2 u/\partial x_i\, \partial x_j$ are functions belonging to $L^p_{\text{loc}}(\Omega)$.

In the case of the statement of the Proposition: $1 < p < \infty$ arbitrary, this proposition reduces to the *Calderon–Zygmund theorem*: *there exists a constant C_p such that*

$$\left\| \frac{\partial^2 \zeta}{\partial x_i \partial x_j} \right\|_{L^p} \leqslant C_p \|\Delta \zeta\|_{L^p} \quad \textit{for all} \quad \zeta \in \mathscr{D}(\mathbb{R}^n)$$

(see Morrey [1] for a "quick" proof of this result: See also Volume 4. Appendix "singular integrals"). We give here a proof, in the case $p = 2$, by using the Fourier transform.

Proof of Proposition 8 in the case $p = 2$. As we have seen in the proof of Proposition 6, we can always suppose that $\Omega = \mathbb{R}^n$, $f \in L^2(\mathbb{R}^n)$ with compact support and u, the Newtonian potential of f. It is clear that u is a tempered distribution (if $n \geqslant 3$, $u(x)$ tends to zero at infinity: if $n = 2$ or 1, $u(x)$ grows at the most like $\operatorname{Log}|x|$ or $|x|$ at infinity. We can therefore make use of the Fourier transform (see Vol. 2, Appendix, "Distributions"). We have

$$\left| \mathscr{F}\left(\frac{\partial^2 u}{\partial x_i \partial x_j} \right) \right| = |-\, y_i y_j \mathscr{F}(u)| \leqslant |-\,|y|^2 \mathscr{F}(u)| = |\mathscr{F}(\Delta u)| = |\mathscr{F}(f)| .$$

Making use of Plancherel's theorem we have $\dfrac{\partial^2 u}{\partial x_i \partial x_j} \in L^2(\mathbb{R}^n)$ and

$$\left\| \frac{\partial^2 u}{\partial x_i \partial x_j} \right\|_{L^2} = \left\| \mathscr{F}\left(\frac{\partial^2 u}{\partial x_i \partial x_j} \right) \right\|_{L^2} \leqslant \|\mathscr{F}(f)\|_{L^2} = \|f\|_{L^2} .$$

Remark 5. The conclusion of Proposition 8 is not true for $p = 1$: this is clear even in the radial case from the formula (3.12). It is no longer true either in the general case for $p = \infty$; also *in general for a function f continuous on Ω, the solutions*

[38] A function $v \in \mathscr{C}^{m-1}(\Omega)$ and of class $\mathscr{C}^m$ on $\Omega \backslash \{0\}$ is of class $\mathscr{C}^m(\Omega)$ iff v admits an expansion limited to the order m at x_0.

of the corresponding Poisson's equation do not belong to the class $\mathscr{C}^2(\Omega)$; we note as we have seen in Proposition 7 that in the radial case, f continuous implies that the solutions of Poisson's equation are of class $\mathscr{C}^2$. We can show the impossibility of these properties in general, in the non-radial case by a functional method; using *reductio ad absurdum*, let us suppose the property to be true; for every bounded (resp. continuous) function f on $\mathbb{R}^n$ with compact support the functions $\partial^2(E_n * f)/\partial x_i \partial x_j$ would be locally bounded (resp. continuous) on $\mathbb{R}^n$; from the closed graph theorem[39]: for all $R > 0$, there would exist a constant C such that

$$\left\| \frac{\partial^2}{\partial x_i \partial x_j}(E_n * \zeta) \right\|_{L^\infty(B(0,R))} \leqslant C \|\zeta\|_{L^\infty}$$

for every regular function ζ with support in $\bar{B}(0, R)$. Now being given f integrable with support in $\bar{B}(0, R)$, we consider its Newtonian potential u:

$$\left\langle \frac{\partial^2 u}{\partial x_i \partial x_j}, \zeta \right\rangle = \left\langle f, \frac{\partial^2}{\partial x_i \partial x_j}(E_n * \zeta) \right\rangle \leqslant C \|f\|_{L^1} \|\zeta\|_{L^\infty}$$

which contradicts the fact that in general $\partial^2 u/\partial x_i \partial x_j$ is not integrable. □

To obtain the regularity of the second derivatives we must therefore use a stronger hypothesis than continuity: the class of *Hölder functions*[40] is well adapted for this purpose as is shown in the

Proposition 9. *Let Ω be an open set in $\mathbb{R}^n$, $0 < \alpha < 1$, $m \geqslant 1$. For every function $f \in \mathscr{C}^{m,\alpha}(\Omega)$[40], the solutions of Poisson's equation*

$$\Delta u = f \quad on \quad \Omega$$

are of class $\mathscr{C}^{m+2,\alpha}$ on Ω.

We prove, first of all, the

Lemma 5. *Let f be a Hölder function on $\mathbb{R}^n$ with compact support and u the Newtonian potential of f. Then u is of class $\mathscr{C}^2$ on $\mathbb{R}^n$. For every regular open set Ω*

---

[39] See Chap. VI. Here the map $f \to \dfrac{\partial^2}{\partial x_i \partial x_j}(E_n * f)$ is continuous from $\mathscr{E}'$ into $\mathscr{D}'$: if it maps the space X of bounded (resp. continuous) functions with support in $\bar{B}(0, R)$ into the space Y of bounded (resp. continuous) functions on $\bar{B}(0, R)$, it is continuous from X into Y.

[40] We recall that a function f defined on a part K of $\mathbb{R}^n$ is called a *Hölder function of order* α $(0 < \alpha < 1)$ *on* K if there exists a constant C such that

$$|f(x) - f(y)| \leqslant |x - y|^\alpha \qquad \forall x, y \in K$$

A function $f \in \mathscr{C}^{m,\alpha}(\Omega)$ if f is of class $\mathscr{C}^m$ and its derivatives of order m are Hölder functions of order α on every compact subset of Ω.

containing supp f *and* $x \in \Omega$, *we have*[41]

$$(3.15)\quad \begin{cases} \dfrac{\partial^2 u}{\partial x_i \partial x_j}(x) = \displaystyle\int_\Omega \frac{\partial^2 E_n}{\partial x_i \partial x_j}(x - y)(f(y) - f(x))\,\mathrm{d}y \\ \qquad - f(x) \displaystyle\int_{\partial\Omega} \frac{\partial E_n}{\partial x_i}(x - y) n_j(y)\,\mathrm{d}\gamma(y)\,. \end{cases}$$

We note that these two integrals clearly have a meaning: it is clear for the second since $x \notin \partial\Omega$; for the first, supposing f to be a Hölder function of order α, we have

$$(3.16)\quad \left| \frac{\partial^2 E_n}{\partial x_i \partial x_j}(x - y)(f(y) - f(x)) \right| \leqslant \frac{n}{\sigma_n} \frac{C}{|x - y|^{n-\alpha}}$$

which is integrable on Ω. In addition the function on the right is continuous at $x \in \Omega$.

Proof of Lemma 5. Given $\zeta \in \mathscr{D}(\Omega)$ we have

$$\left\langle \frac{\partial^2 u}{\partial x_i \partial x_j}, \zeta \right\rangle = - \left\langle \frac{\partial E_n}{\partial x_i} * f, \frac{\partial \zeta}{\partial x_j} \right\rangle = - \iint \frac{\partial E_n}{\partial x_i}(x - y) f(y) \frac{\partial \zeta}{\partial x_j}(x)\,\mathrm{d}x\,\mathrm{d}y$$

$$= \lim_{\varepsilon \to 0} \int f(y) v_\varepsilon(y)\,\mathrm{d}y$$

where

$$v_\varepsilon(y) = \int_{\Omega \setminus B(y,\varepsilon)} - \frac{\partial E_n}{\partial x_i}(x - y) \frac{\partial \zeta}{\partial x_j}(x)\,\mathrm{d}x\,.$$

By use of Ostrogradski's formula

$$v_\varepsilon(y) = \int_{\Omega \setminus B(y,\varepsilon)} \left\{ \zeta(x) \frac{\partial^2 E_n}{\partial x_i \partial x_j}(x - y) - \frac{\partial}{\partial x_j}\left(\zeta(x) \frac{\partial E_n}{\partial x_i}(x - y)\right) \right\} \mathrm{d}x$$

$$= \int_{\Omega \setminus B(y,\varepsilon)} \zeta(x) \frac{\partial^2 E_n}{\partial x_i \partial x_j}(x - y)\,\mathrm{d}x$$

$$- \int_{\partial(\Omega \setminus B(y,\varepsilon))} \zeta(x) \frac{\partial E_n}{\partial x_i}(x - y) n_j(x)\,\mathrm{d}\gamma(x)\,.$$

Therefore

$$\int f(y) v_\varepsilon(y)\,\mathrm{d}y = I_1^\varepsilon - I_2^\varepsilon + I_3^\varepsilon - I_4^\varepsilon\,,$$

[41] This formula is still true for a.e. $x \in \Omega$ if $f \in L^p(\mathbb{R}^n)$ with support in Ω (see Vol. 4, Appendix "singular integrals").

with

$$I_1^\varepsilon = \int_\Omega \mathrm{d}y \int_{\Omega\setminus B(y,\varepsilon)} \zeta(x) \frac{\partial^2 E_n}{\partial x_i \partial x_j}(x-y)(f(y)-f(x))\,\mathrm{d}x$$

$$I_2^\varepsilon = \int_\Omega \mathrm{d}y \int_{\partial(\Omega\setminus B(y,\varepsilon))} \zeta(x) \frac{\partial E_n}{\partial x_i}(x-y)(f(y)-f(x))\,n_j(x)\,\mathrm{d}\gamma(x)$$

$$I_3^\varepsilon = \int_\Omega \mathrm{d}y \int_{\Omega\setminus B(y,\varepsilon)} \zeta(x) f(x) \frac{\partial^2 E_n}{\partial x_i \partial x_j}(x-y)\,\mathrm{d}x$$

$$I_4^\varepsilon = \int_\Omega \mathrm{d}y \int_{\partial(\Omega\setminus B(y,\varepsilon))} \zeta(x) f(x) \frac{\partial E_n}{\partial x_i}(x-y) n_j(x)\,\mathrm{d}\gamma(x)\,.$$

We study the limit of each integral as $\varepsilon \to 0$:

(1)
$$I_1^\varepsilon \to \int_\Omega \zeta(x)\,\mathrm{d}x \int_\Omega \frac{\partial^2 E_n}{\partial x_i \partial x_j}(x-y)(f(y)-f(x))\,\mathrm{d}y\,;$$

this follows from Lebesgue's theorem and the estimate (3.16);

(2)
$$I_2^\varepsilon \to 0\,,$$

in effect using the fact that f is a Hölder function of order α

$$|I_2^\varepsilon| = \left| \int_\Omega \mathrm{d}y \int_{\partial B(y,\varepsilon)} \zeta(x)(f(y)-f(x)) \frac{(x_i-y_i)(x_j-y_j)}{\sigma_n \varepsilon^{n+1}}\,\mathrm{d}\gamma(x) \right|$$
$$\leqslant \int_\Omega \mathrm{d}y \int_\Sigma |\zeta(x+\varepsilon\sigma)| \frac{C\varepsilon^\alpha}{\sigma_n}\,\mathrm{d}\sigma\,;$$

(3)
$$I_3^\varepsilon = \int_\Omega \zeta(x) f(x)\,\mathrm{d}x \int_{\Omega\setminus B(x,\varepsilon)} \frac{\partial^2 E_n}{\partial x_i \partial x_j}(x-y)\,\mathrm{d}y$$
$$= -\int_\Omega \zeta(x) f(x)\,\mathrm{d}x \int_{\partial(\Omega\setminus B(x,\varepsilon))} \frac{\partial E_n}{\partial x_i}(x-y) n_j(y)\,\mathrm{d}\gamma(y)$$
$$= -\int_\Omega \zeta(x) f(x)\,\mathrm{d}x \left(\int_{\partial\Omega} \frac{\partial E_n}{\partial x_i}(x-y) n_j(y)\,\mathrm{d}\gamma(y) + c_{ij} \right),$$

where

$$c_{ij} = \int_{\partial B(x,\varepsilon)} \frac{\partial E_n}{\partial x_i}(x-y) \frac{x_j-y_j}{|x_j-y_j|}\,\mathrm{d}\gamma(y) = \frac{1}{\sigma_n}\int_\Sigma y_i y_j\,\mathrm{d}\sigma(y)$$

is independent of ε.

(4)
$$I_4^\varepsilon = \int_\Omega \mathrm{d}y \int_{\partial B(y,\varepsilon)} \zeta(x) f(x) \frac{\partial E_n}{\partial x_i}(x-y) \frac{y_j-x_j}{|y_j-x_j|}\,\mathrm{d}\gamma(x)$$
$$= -\int_\Omega \frac{\mathrm{d}y}{\sigma_n} \int_\Sigma \zeta(y+\varepsilon x) f(y+\varepsilon x) x_i x_j\,\mathrm{d}\sigma(x)\,.$$

Hence

$$I_4^{\varepsilon} \to - c_{ij} \int_{\Omega} \zeta(y) f(y) \, dy \, .$$

Grouping together these limits, we obtain (3.15). □

Remark 6. The formula (3.15) and the proof of Lemma 5 shows that a sufficient condition for the solutions of Poisson's equation

$$\Delta u = f \quad \text{on} \quad \Omega$$

to be continuous at x_0 is

$$|f(x) - f(x_0)| \leqslant \varepsilon(|x - x_0|) \quad \text{for all } x \text{ in the neighbourhood of } x_0{}^{32},$$

where $\varepsilon\colon \mathbb{R}^+ \to \mathbb{R}^+$ is (continuous) increasing and satisfies

$$\int_0 \varepsilon(r) \frac{dr}{r} < \infty \, .$$

We can verify that this condition[42] is necessary in the sense: given $\varepsilon\colon \mathbb{R}^+ \to \mathbb{R}^+$ increasing and continuous with

$$\int_0 \varepsilon(r) \frac{dr}{r} = \infty \, ,$$

there exists a continuous function f with compact support satisfying

$$|f(x)| \leqslant \varepsilon(|x|) \qquad \forall x \, ,$$

and such that the Newtonian potential of f is not twice differentiable at 0. □

Proof of Proposition 9. First of all it is sufficient to demonstrate the proposition for $m = 0$, since the derivatives of u are solutions of the Poisson equation corresponding to the derivatives of f. Also (see the proof of Proposition 6), we can suppose f has compact support in Ω and that u is the Newtonian potential of f. Let us consider two points $x_1, x_2 \in \Omega$ and put $\delta = |x_1 - x_0|$, $x_0 = \frac{1}{2}(x_1 + x_2)$. We can always suppose by translation that $x_0 = 0$, and can always take $\Omega = B(0, R)$ with $R \geqslant \delta$.
Applying the formula (3.15) we then have

$$\frac{\partial^2 u}{\partial x_i \partial x_j}(x_1) - \frac{\partial^2 u}{\partial x_i \partial x_j}(x_2) = I_1 - I_2 + (f(x_1) - f(x_2)) I_3 + I_4 + f(x_1)(I(x_2) - I(x_1)) + (f(x_2) - f(x_1)) I(x_2)$$

where

$$I_k = \int_{B(0,\,\delta)} \frac{\partial^2 E_n}{\partial x_i \partial x_j}(x_k - y)(f(y) - f(x_k)) \, dy \qquad k = 1, 2$$

---

[42] A function satisfying this condition is said to be continuous in the sense of Dini, or to satisfy Dini's condition at x_0.

$$I_3 = \int_{B(0,R)\setminus B(0,\delta)} \frac{\partial^2 E_n}{\partial x_i \partial x_j}(x_1 - y)\,dy$$

$$I_4 = \int_{B(0,R)\setminus B(0,\delta)} \left(\frac{\partial^2 E_n}{\partial x_i \partial x_j}(x_1 - y) - \frac{\partial^2 E_n}{\partial x_i \partial x_j}(x_2 - y)\right)(f(y) - f(x_2))\,dy$$

$$I(x) = \int_{\partial B(0,R)} \frac{\partial E_n}{\partial x_i}(x - y) n_j(y)\,d\gamma(y)\,.$$

We then have the estimates

$$(1)\quad |I_k| \leqslant \int_{B(x_k, \frac{3\delta}{2})} \frac{n}{\sigma_n} \frac{C}{|x_k - y|^{n-\alpha}}\,dy = \frac{Cn}{\alpha}\left(\frac{3\delta}{2}\right)^\alpha, \quad k = 1, 2\,,$$

$$(2)\quad |I_3| = \left|\int_{\partial(B(0,R) - B(0,\delta))} \frac{\partial E_n}{\partial x_i}(x_1 - y) n_j(y)\,d\gamma(y)\right|$$

$$\leqslant \frac{1}{\sigma_n}\int_{\partial(B(0,R))} \frac{d\gamma(y)}{|x_1 - y|^{n-1}} + \frac{1}{\sigma_n}\int_{\partial B(0,\delta)} \frac{d\gamma(y)}{|x_1 - y|^{n-1}} \leqslant 2^{n+1}\,,$$

$$(3)\quad |I_4| \leqslant \int_0^1 dt \int_{B(0,R)\setminus B(0,\delta)} \left|\operatorname{grad} \frac{\partial^2 E_n}{\partial x_i \partial x_j}(x_t - y)(x_1 - x_2)\right| C|y - x_2|^\alpha\,dy$$

with $x_t = tx_1 + (1 - t)x_2$.
Calculating the third derivatives of E_n, we find

$$|I_4| \leqslant C\frac{n(n+5)}{\sigma_n}\delta \int_0^1 dt \int_{B(0,R)\setminus B(0,\delta)} \frac{|y - x_2|^\alpha}{|y - x_t|^{n+1}}\,dy$$

$$\leqslant Cn(n+5)\delta \int_\delta^R \frac{\left(r + \frac{\delta}{2}\right)^\alpha}{\left(r - \frac{\delta}{2}\right)^{n+1}} r^{n-1}\,dr \leqslant C_1\delta^\alpha$$

with

$$C_1 = Cn(n+5)\int_1^\infty \frac{\left(r + \frac{1}{2}\right)^\alpha}{\left(r - \frac{1}{2}\right)^{n+1}} r^{n-1}\,dr$$

which depends only on C, n and α.

$$(4)\quad |I(x_2) - I(x_1)| \leqslant \int_0^1 dt \int_{\partial B(0,R)} \left|\operatorname{grad} \frac{\partial E_n}{\partial x_i}(x_t - y).(x_1 - x_2)\right| d\gamma(y)$$

$$\leqslant n\frac{\delta}{R}\left(\frac{R}{R - (\delta/2)}\right)^n \leqslant \frac{n2^n}{R}\delta\,;$$

$$(5)\quad |I(x_2)| \leqslant \left(\frac{R}{R - (\delta/2)}\right)^{n-1} \leqslant 2^{n-1}\,.$$

Regrouping these estimates, we obtain

$$\left|\frac{\partial^2 u}{\partial x_i \partial x_j}(x_1) - \frac{\partial^2 u}{\partial x_i \partial x_j}(x_2)\right| \leqslant C_2 \delta^\alpha + \frac{n 2^n}{R} |f(x_1)| \delta$$

with
$$C_2 = \frac{2Cn}{\alpha}\left(\frac{3}{2}\right)^\alpha + 2^{n+1}C + C_1 + 2^{n-1}C$$

depending only on C, n, α. In the limit when $R \to \infty$

$$\left|\frac{\partial^2 u}{\partial x_i \partial x_j}(x_1) - \frac{\partial^2 u}{\partial x_i \partial x_j}(x_2)\right| \leqslant C_2 |x_1 - x_2|^\alpha$$

for all $x_1, x_2 \in \mathbb{R}^n$.

3. Regularity of Simple and Double Layer Potentials

We consider a regular bounded open set Ω with boundary Γ. Given $\varphi \in \mathscr{C}^0(\Gamma)$, we have defined (see Definition 2) the simple layer potential of φ as the Newtonian potential u_1 of the measure $\varphi \mathrm{d}\gamma$ on Γ: this is a harmonic function on $\mathbb{R}^n$ defined by

$$u_1(x) = \int_\Gamma E_n(t - x)\varphi(t)\,\mathrm{d}\gamma(t)^{43}\,. \tag{3.17}$$

We have also defined the double layer potential of φ as the Newtonian potential u_2 of the distribution $-\operatorname{div}(\varphi n \mathrm{d}\gamma)$ with support on Γ: it is a harmonic function on $\mathbb{R}^n \backslash \Gamma$ defined by

$$u_2(x) = \int_\Gamma \frac{\partial}{\partial n} E_n(t - x)\varphi(t)\,\mathrm{d}\gamma(t) = \frac{1}{\sigma_n}\int_\Gamma \sum \frac{(t_i - x_i)n_i(t)}{|t - x|^n}\varphi(t)\,\mathrm{d}\gamma(t) \tag{3.18}$$

where $n(t)$ is the unit vector normal to Γ at t and exterior to Ω.
We propose to study *the regularity of u_1 and u_2 in the neighbourhood of Γ as a function of that of φ*. It is clear that the datum is the compact regular hypersurface Γ, rather than Ω: this hypersurface separates $\mathbb{R}^n$ into two open sets: the regular bounded set Ω with boundary Γ which by an abuse of language we shall call *the interior of Γ* and denote by Ω_i, the open set $\mathbb{R}^n \backslash \bar{\Omega}$ which we shall call, again by abuse of language[44], *the exterior of Γ* and denote by Ω_e[45]. Given a function u defined on $\mathbb{R}^n$, we shall denote by u^i (resp. u^e) *the restriction of u to Ω_i* (resp. Ω_e).
As usual the case $n = 1$ is immediate. A regular bounded open set of $\mathbb{R}$ is a finite union of open intervals; it is clear by additivity of the Newtonian potentials, that it is enough to study the case of an open interval $\Omega =]a, b[$; we then have

$$\Gamma = \{a, b\}, \quad \mathrm{d}\gamma = \delta_a + \delta_b, \quad n(a) = -1, \quad n(b) = +1\,.$$

---

[43] The n, dimension of the space $\mathbb{R}^n$ and the n normal vector are obviously unconnected.
[44] Topologically, the hypersurface Γ has an empty interior and its exterior is $\mathbb{R}^n \backslash \Gamma$.
[45] Ω_i and Ω_e are well defined by Γ: two regular bounded open sets which have the same boundary are equal.

For a function $\varphi = (\varphi(a), \varphi(b))$, we have therefore for $x \in \mathbb{R}\backslash\{a, b\}$

$$u_1(x) = \frac{1}{2}(\varphi(a)|x - a| + \varphi(b)|x - b|)$$

$$\frac{du_1}{dx}(x) = \frac{1}{2}(\varphi(a)\,\text{sign}\,(x - a) + \varphi(b)\,\text{sign}\,(x - b))$$

$$u_2(x) = \frac{1}{2}(\varphi(a)\,\text{sign}\,(x - a) - \varphi(b)\,\text{sign}\,(x - b))\,.$$

We see therefore that, *in the case* $n = 1$, u_1 is continuous on $\mathbb{R}$, du_1/dx and u_2 have a limit on the left and a limit on the right at every point z of Γ with a jump

$$\frac{du_1}{dx}(z+) - \frac{du_1}{dx}(z-) = \varphi(z)$$

$$u_2(z+) - u_2(z-) = -n(z)\varphi(z)\,.$$

In addition u_1^i, u_2^i (resp. u_1^e, u_2^e) are $\mathscr{C}^\infty$ on $\bar{\Omega}_i$ (resp. $\bar{\Omega}_e$).
The radial case is equally simple and gives rise to the same phenomena. A radial regular bounded open set in $\mathbb{R}^n$ is the union of a finite number (possibly zero) of *annuli* $B(0, r_1)\backslash B(0, r_0)$, $(0 < r_0 < r_1)$, and, if it contains 0, of a ball $B(0, r)$. We note that

$$\partial(B(0, r_1)\backslash B(0, r_0)) = \partial B(0, r_1) \cup \partial B(0, r_0)\,,$$

the exterior normal to the ring being exterior (resp. interior) on $\partial B(0, r_1)$ (resp. $\partial B(0, r_0)$) to the ball $B(0, r_1)$ (resp. $B(0, r_0)$). Because of the additivity of the Newtonian potentials, it suffices to study the case of a ball $\Omega = B(0, r)$; we suppose that φ is radial, i.e. constant on $\partial B(0, r)$.
We then prove the

Lemma 6. *The simple layer potential u_1 defined by the* **constant** *function φ on $\partial B(0, r)$ is continuous on $\mathbb{R}^n$, constant on $\bar{B}(0, r)$, and given by*

$$u_1(x) = r^{n-1}\sigma_n \varphi E_n(x) \quad \textit{for all} \quad x \in \mathbb{R}^n\backslash B(0, r)\,.$$

Proof. We know (see Corollary 1) that u_1 is radial. Being harmonic on $B(0, r)$ it is constant on $B(0, r)$ (see Proposition 4 of §2) so

$$u_1(x) = u_1(0) = \int_{\partial B(0, r)} \varphi E_n(t)\, d\gamma(t) \quad \text{for all} \quad x \in B(0, r)\,.$$

Being harmonic on $\mathbb{R}^n\backslash\bar{B}(0, r)$, it is of the form (see the same proposition)

$$u_1(x) = aE_n(x) + b \quad \text{for all} \quad x \in \mathbb{R}^n\backslash\bar{B}(0, r)\,,$$

where a and b are determined by the condition at infinity of Proposition 2

$$u_1(x) = \left(\left(\int_{\partial B(0, r)} \varphi\, d\gamma(t)\right) E_n(x) + O\left(\frac{1}{|x|^{n-1}}\right)\right) \quad \text{when} \quad |x| \to \infty\,.$$

Hence

$$a = \int_{\partial B(0,r)} \varphi \, d\gamma(t) = r^{n-1}\sigma_n \varphi, \quad b = 0 .$$

It is clear then that u_1 is continuous on $\mathbb{R}^n$. □

For the double layer potential, we have in a more general way the

Lemma 7. *The double layer potential u_2 defined by a* **constant** *function φ on the boundary Γ of a regular bounded open set Ω of $\mathbb{R}^n$ is given on $\mathbb{R}^n \backslash \Gamma$ by*

$$u_2 = \begin{cases} \varphi & \text{if } x \in \Omega_i \\ 0 & \text{if } x \in \Omega_e \end{cases}$$

Proof. Applying Proposition 5 to the constant function $u \equiv \varphi$ on Ω, we have $u_2 = u = \varphi$ on $\Omega_i = \Omega$. Now if $x \in \Omega_e$ the function $t \to E_n(t - x)$ is harmonic on Ω; by Gauss' theorem (see Proposition 5 of §2).

$$u_2(x) = \varphi \int_\Gamma \frac{\partial E_n}{\partial n}(t - x) \, d\gamma(t) = 0 .$$ □

We see therefore that again *in the radial case* (and more generally for the double layer potential of a function φ constant on each connected component of Γ), u_1 is continuous on $\mathbb{R}^n$, u_1^i, u_2^i (resp. u_1^e, u_2^e) are $\mathscr{C}^\infty$ on $\bar{\Omega}_i$ (resp. $\bar{\Omega}_e$), u_2 and $\partial u_1/\partial n$ have a jump at every point z of Γ, given by

$$u_2^i(z) - u_2^e(z) = \varphi(z)$$

$$\frac{\partial u_1^i}{\partial n}(z) - \frac{\partial u_1^e}{\partial n}(z) = - \varphi(z) .$$

We propose to generalize these properties. We begin with the *continuity of the simple layer potential*:

Proposition 10. *Let Γ be the boundary of a regular bounded open set and $\varphi \in \mathscr{C}^0(\Gamma)$. Then the simple layer potential u_1 of φ is continuous on $\mathbb{R}^n$.*

Proof. We have only to prove the continuity at a point $z \in \Gamma$. Let us consider a normal parametric representation $((R, U, \mathcal{O}, \alpha)$[46] in the neighbourhood of $z = (x_0', \alpha(x_0'))$; we fix $r_0 > 0$ sufficiently small that

$$\bar{B}(x_0', r_0) \subset \mathcal{O}, \bar{U}_{r_0} = \{(x', \alpha(x') + \tau); x' \in \bar{B}(x_0', r_0), |\tau| \leqslant r_0\} \subset U .$$

For $x \in \mathbb{R}^n \backslash \Gamma$, let us write $u_1(x) = I_1(x) + I_2(x)$

with

$$I_1(x) = \int_{\Gamma \backslash \bar{U}_{r_0}} E_n(t - x)\varphi(t) \, d\gamma(t), \qquad I_2(x) = \int_{\Gamma \cap \bar{U}_{r_0}} E_n(t - x)\varphi(t) \, d\gamma(t) .$$

[46] See §1.3.a. We recall that R is a reference frame, U a neighbourhood of z, $\mathcal{O}$ an open set of $\mathbb{R}^{n-1}$, α a function of class $\mathscr{C}^m$ on $\mathcal{O}$ such that $(x', x_n') \in \mathbb{R}^{n-1} \times \mathbb{R}$ being the coordinates in the reference frame R:

$$U (\text{resp. } U \cap \Gamma, U \cap \Omega) = \{(x', \alpha(x') + \tau); x' \in \mathcal{O}, |\tau| < \delta \, (\text{resp. } \tau = 0, -\delta < \tau < 0)\} .$$

Since $E_n(t - x)$ is bounded $(t, x) \in (\Gamma \setminus \bar{U}_{r_0}) \times \bar{U}_{r_0/2}$, we have, from Lebesgue's theorem,

$$I_1(x) \to \int_{\Gamma \setminus \bar{U}_{r_0}} E_n(t - z)\varphi(t)\,d\gamma(t) \quad \text{when} \quad x \to z\,.$$

We have therefore only to consider $I_2(x)$. For $x = (x', \alpha(x') + \tau)$, we have (see formula (1.11) of §1.3a)

$$I_2(x) = \int_{B(x'_0, r'_0)} E_n(t' - x', \alpha(t') - \alpha(x') - \tau)\,\varphi(t', \alpha(t'))(1 + |\operatorname{grad} \alpha(t')|^2)^{\frac{1}{2}}\,dt'\,.$$

Now the term under the integral sign is majorized by

$$Cg(t' - x')$$

where

$$C = \sup_{B(x'_0, r_0)} |\varphi(t', \alpha(t'))|\,(1 + |\operatorname{grad} \alpha(t')|^2)^{\frac{1}{2}}$$

and

$$g(t') = \begin{cases} \dfrac{1}{(n - 2)\sigma_n |t'|^{n-2}} & \text{if} \quad n \geqslant 3 \\ \dfrac{1}{2\pi} \operatorname{Log} \dfrac{1}{|t'|} & \text{if} \quad n = 2\text{[47]}. \end{cases}$$

Since g is locally integrable on $\mathbb{R}^{n-1}$, the function

$$t \in \Gamma \cap \bar{U}_{r_0} \to E_n(t - z)\,\varphi(t)$$

is integrable with respect to $d\gamma$ and

$$I_2(x) \to \int_{\Gamma \cap \bar{U}_{r_0}} E_n(t - z)\varphi(t)\,d\gamma(t) \quad \text{when} \quad x \to z\,. \qquad \square$$

Remark 7. In fact the function $g \in L^p_{\text{loc}}(\mathbb{R}^{n-1})$ for all $p < (n - 1)/(n - 2)$; it follows (see Lemma 3) that the result of the proposition remains true for the simple layer potential defined by a function $\varphi \in L^q(\Gamma)$ with $q > n - 1$. We study later the simple layer potentials defined by functions which are integrable on Γ or even by measures on Γ (see §6). $\square$

We now study the "*continuity*" *of double layer potentials.* We shall have to suppose that Γ is a little more than of class $\mathscr{C}^1$; we introduce the definition:

Definition 3. We shall say that a regular open set has *boundary Γ of class $\mathscr{C}^{1+\varepsilon}$* if in the neighbourhood of every point $z \in \Gamma$, there exists a normal parametric representation $(R, U, \mathcal{O}, \alpha)$ and an increasing continuous function ε: $\mathbb{R}^+ \to \mathbb{R}^+$ with $\displaystyle\int_0 \varepsilon(r)\frac{dr}{r} < \infty$ such that

$$\text{(3.19)} \qquad |\operatorname{grad} \alpha(x') - \operatorname{grad} \alpha(\bar{x}')| \leqslant \varepsilon(|x' - \bar{x}'|) \quad \text{for all} \quad x', \bar{x}' \in \mathcal{O}\,.$$

[47] We can always suppose that $|t' - x'| < 1$ with r_0 sufficiently small and x near to z.

We note that this condition is obviously satisfied if Γ is of class $\mathscr{C}^2$ or more generally of class $\mathscr{C}^{1,\theta}$ with $0 < \theta < 1$[48].

Proposition 11. *Let Γ be the boundary of class $\mathscr{C}^{1+\varepsilon}$ of a regular bounded open set and let $\varphi \in \mathscr{C}^0(\Gamma)$. Then:*

(1) *$\forall z \in \Gamma$, the function $t \to \frac{\partial E_n}{\partial n}(t - z)\varphi(t)$ is integrable on Γ (with respect to $\mathrm{d}\gamma$);*

(2) *the function w defined on Γ by*

$$w(z) = \int_\Gamma \frac{\partial E_n}{\partial n}(t - z)\varphi(t)\,\mathrm{d}\gamma(t)$$

is continuous on Γ.

(3) the internal double layer potential u_2^i (resp. external u_2^e) defined by φ extends by continuity to $\bar{\Omega}_i$ (resp. $\bar{\Omega}_e$) by putting on Γ

$$u_2^i(z) = w(z) + \frac{1}{2}\varphi(z)\left(\text{resp. } u_2^e(z) = w(z) - \frac{1}{2}\varphi(z)\right).$$

Proof. We fix $z \in \Gamma$ and consider a normal parametric representation $(R, U, \mathcal{O}, \alpha)$ in the neighbourhood of z satisfying (3.19).

With the notation of the proof of Proposition 10, for $x \in \mathbb{R}^n \backslash \Gamma$, $u_2(x) = I_1(x) + I_2(x)$ with

$$I_1(x) = \int_{\Gamma \backslash \bar{U}_{r_0}} \frac{\partial E_n}{\partial n}(t - x)\varphi(t)\,\mathrm{d}\gamma(t)$$

$$I_2(x) = \int_{\Gamma \cap \bar{U}_{r_0}} \frac{\partial E_n}{\partial n}(t - x)\varphi(t)\,\mathrm{d}\gamma(t)\,.$$

Also

$$I_1(x) \to \int_{\Gamma \backslash \bar{U}_{r_0}} \frac{\partial E_n}{\partial n}(t - z)\varphi(t)\,\mathrm{d}\gamma(t) \quad \text{when} \quad x \to z\,.$$

The existence of the integral does not pose a problem.

Now, at $x = (x', \alpha(x') + \tau)$, we have[49]

$$\begin{aligned} I_2(x) &= \frac{1}{\sigma_n}\int_{B(x_0', r_0)} \frac{\alpha(t') - \alpha(x') - \tau - \operatorname{grad}\alpha(t').(t' - x')}{(|t' - x'|^2 + |\alpha(t') - \alpha(x') - \tau|^2)^{n/2}}\varphi(t', \alpha(t'))\,\mathrm{d}t' \\ &= I_3(x', \tau) - I_4(x', \tau) \end{aligned}$$

---

[48] Γ is said to be of class $\mathscr{C}^{m,\theta}$ if in normal parametric representations $(R, U, \mathcal{O}, \alpha)$, the function α is of class $\mathscr{C}^{m,\theta}$ on $\mathcal{O}$.

[49] We make use of $\frac{\partial E_n}{\partial n}(t - x) = \frac{(t - x).n(t)}{\sigma_n |t - x|^n}$ and for $t = (t', \alpha(t'))$

$$n(t) = \frac{(-\operatorname{grad}\alpha(t'), 1)}{(1 + |\operatorname{grad}\alpha(t')|^2)^{\frac{1}{2}}} \quad \text{(see formula (1.10) of §1.3.a.)}$$

with

$$I_3(x', \tau) = \frac{1}{\sigma_n} \int_{B(x'_0, r_0)} \frac{\alpha(t') - \alpha(x') - \operatorname{grad} \alpha(t').(t' - x')}{(|t' - x'|^2 + |\alpha(t') - \alpha(x') - \tau|^2)^{n/2}} \varphi(t', \alpha(t'))\, dt'$$

and

$$I_4(x', \tau) = \frac{\tau}{\sigma_n} \int_{B(x'_0, r_0)} \frac{\varphi(t', \alpha(t'))\, dt'}{(|t' - x'|^2 + |\alpha(t') - \alpha(x') - \tau|^2)^{n/2}} .$$

Making use of the hypothesis (3.19), the term under the integral sign in I_3 is majorized by $Cg(t' - x')$ with

$$C = \sup_{B(x'_0, r_0)} |\varphi(t', \alpha(t')| \quad \text{and} \quad g(t') = \frac{\varepsilon(|t'|)}{|t'|^{n-1}} .$$

By the hypothesis on ε given in (3.19), the function g is locally integrable on $\mathbb{R}^{n-1}$. We deduce the points (1) and (2) of the proposition; and also

$$I_3(x', \tau) \to \int_{\Gamma \cap \bar{U}_{r_0}} \frac{\partial E_n}{\partial n}(t - z)\varphi(t)\, d\gamma(t) \quad \text{when} \quad x = (x', \tau) \to z .$$

To complete the proof, it is sufficient therefore to show that

$$I_4(x', \pm \tau) \to \pm \frac{\varphi(x'_0, \alpha(x'_0))}{2} \quad \text{when} \quad x' \to x'_0, \tau \to 0, \tau > 0 .$$

With the aim of showing that this limit is uniform with respect to x'_0, it is enough to consider a normal limit, since $w(z)$ and $\varphi(z)$ are continuous at z[50]. To simplify the calculations, we can also suppose that the reference frame is chosen with centre z and with $n(z)$ as the last of the base vectors, with the result that

(3.20) $$x'_0 = 0, \quad \alpha(0) = 0 \quad \text{and} \quad \operatorname{grad} \alpha(0) = 0$$

and we have to show that

$$I_\pm(\tau) = \frac{\tau}{\sigma_n} \int_{B(0, r_0)} \frac{\varphi(t', \alpha(t'))\, dt'}{(|t'|^2 + (\alpha(t') \mp \tau)^2)^{n/2}} \to \frac{\varphi(0)}{2} \quad \text{when} \quad \tau \to 0, \quad \tau > 0 .$$

Changing the variables we have

$$I_\pm(\tau) = \frac{1}{\sigma_n} \int_{B(0, \sigma_0/\tau)} \frac{\varphi(\tau t', \alpha(\tau t'))\, dt'}{\left(|t'|^2 + \left(1 \mp \frac{\alpha(\tau t')}{\tau}\right)^2\right)^{n/2}} .$$

But using (3.19) and (3.20), $\dfrac{|\alpha(\tau t')|}{\tau} \leqslant |t'|\varepsilon(\tau|t'|)$, with the result that, for τ sufficiently small, we obtain:

$$\left(|t'|^2 + \left(1 \mp \frac{\alpha(\tau t')}{\tau}\right)^2\right)^{-n/2} \leqslant \begin{cases} 2^{-n} & \text{on } B(0, 1) \\ |t'|^{-n} & \text{on } \mathbb{R}^n - \bar{B}(0, 1) . \end{cases}$$

---

[50] If a sequence (x_m) of $\mathbb{R}^n \backslash \Gamma$ converges to $z \in \Gamma$, there exists a sequence (z_m) of Γ converging and $\lambda_m \to 0$, such that $x_m = z_m + \lambda_m n(z_m)$.

In the limit,

$$I_{\pm}(\tau) \to \frac{1}{\sigma_n} \int_{\mathbb{R}^n} \frac{\varphi(0)\,dt'}{(1 + |t'|^2)^{n/2}} = \frac{\varphi(0)}{2}\,. \qquad \square$$

In the same way we study the "*continuity*" *of the normal derivative of a simple layer potential.*

Proposition 12. *Let Γ be the boundary of class $\mathscr{C}^{1+\varepsilon}$ of a regular bounded open set, and let $\varphi \in \mathscr{C}^0(\Gamma)$. Then*

(1) *for all $z \in \Gamma$, the function $t \to \operatorname{grad} E_n(z - t).n(z)\varphi(t)$ is integrable on Γ;*
(2) *the function h defined on Γ by*

$$h(z) = \int_\Gamma \operatorname{grad} E_n(z - t).n(z)\varphi(t)\,d\gamma(t) = \frac{1}{\sigma_n}\int_\Gamma \frac{(z - t).n(z)}{|z - t|^n}\varphi(t)\,d\gamma(t)$$

is continuous on Γ;
(3) *the normal derivative $\dfrac{\partial u_1^i}{\partial n}$ $\left(\text{resp. } \dfrac{\partial u_1^e}{\partial n}\right)$ of the interior (resp. exterior) simple layer potential defined by φ, is defined on Γ*[51]*, and given by*

$$\frac{\partial u_1^i}{\partial n}(z) = h(z) - \frac{1}{2}\varphi(z) \quad \left(\text{resp. } \frac{\partial u_1^e}{\partial n}(z) = h(z) + \frac{1}{2}\varphi(z)\right).$$

Remark 8. It is necessary to distinguish clearly between the function h defined in this proposition and the function $w(z)$ defined in Proposition 11. We note that

$$w(z) + h(z) = \int_\Gamma \operatorname{grad} E_n(t - z).(n(t) - n(z))\varphi(t)\,d\gamma(t)\,. \tag{3.21}$$
$\square$

Proof of Proposition 12. The points (1) and (2) are obtained as corollaries of the points (1) and (2) of Proposition 11 and of the formula (3.21) of Remark 8: in effect making use of the hypothesis (3.19), we have

$$|n(t) - n(z)| \leqslant \varepsilon(|t - z|)$$

with $\varepsilon\colon \mathbb{R}^+ \to \mathbb{R}^+$ continuous, increasing and satisfying $\displaystyle\int_0 \varepsilon(r)\frac{dr}{r} < \infty$. Making use of arguments from the proofs of the preceding propositions, we shall easily show that the term under the integral sign in (3.21) is integrable and $w + h$ thus defined is continuous.

Now considering $x \in \mathbb{R}^n \backslash \Gamma$, we have by differentiation under the integral sign

$$\operatorname{grad} u_1(x) = \int_\Gamma \operatorname{grad} E_n(x - t)\,\varphi(t)\,d\gamma(t)\,.$$

---

[51] In the sense of §1.3.b: $u_1^i \in \mathscr{C}_n^1(\bar{\Omega}_i)$, $u_1^e \in \mathscr{C}_n^1(\bar{\Omega}_e)$. We note however that n is directed along the external normal to $\Omega = \Omega_i$ and hence that $\dfrac{\partial u_1^e}{\partial n}(z)$ is here in the opposite direction to that of the external normal derivative to Ω_e.

Using the same argument as in the proof of Proposition 11, we see that it is sufficient to demonstrate that

$$\lim_{\substack{\tau \to 0 \\ \tau > 0}} \operatorname{grad} u_1(z \pm \tau n(z)).n(z) = h(z) \pm \frac{1}{2}\varphi(z) \quad \text{uniformly for} \quad z \in \Gamma .$$

We have

$$\operatorname{grad} u_1(z \pm \tau n(z)).n(z) = I_1(z, \tau) + I_2(z, \tau)$$

with

$$I_1(z, \tau) = \frac{1}{\sigma_n}\int_\Gamma \frac{(z - t).n(z)}{|z - t \pm \tau n(z)|^n}\varphi(t)\,\mathrm{d}\gamma(t)$$

$$I_2(z, \tau) = \frac{1}{\sigma_n}\int_\Gamma \frac{\tau}{|z - t \pm \tau n(z)|^n}\varphi(t)\,\mathrm{d}\gamma(t) .$$

It is clear that the problem is for $|z - t|$ small; reasoning geometrically, we see that $z - t$ is then tangent to Γ with the result that

$$|z - t \pm \tau n(z)|^2 \sim |z - t|^2 + \tau^2 .$$

We deduce that

$$I_1(z, \tau) \to h(z) \quad \text{when} \quad \tau \to 0 ,$$

and using the proof of Proposition that

$$I_2(z, \tau) \to \frac{1}{2}\varphi(z) \quad \text{when} \quad \tau \to 0 . \qquad \square$$

We propose finally to study the *regularity of the derivatives of interior and exterior simple layer potentials right up to the boundary* Γ. In the same way as for the local regularity considered in §3.2, the continuity of φ is not sufficient to ensure that u_1^i (resp. u_1^e) of class $\mathscr{C}^1$ on $\bar{\Omega}_i$ (resp. $\bar{\Omega}_e$); also $\varphi \in \mathscr{C}^1(\Gamma)$ is not sufficient to ensure that u_2^i (resp. u_2^e) is of class $\mathscr{C}^1$ on $\bar{\Omega}_i$ (resp. $\bar{\Omega}_e$). By combining the geometrical techniques used in the proofs of Propositions 10, 11 and 12 and the "singular integrals" techniques used in the proof of Proposition 9 we can demonstrate that: *if Γ is of class $\mathscr{C}^{m+1,\alpha}$ and φ of class $\mathscr{C}^{m,\alpha}$ with $m \in \mathbb{N}$ and $0 < \alpha < 1$, then the interior (resp. exterior) simple layer potential defined by φ is of class $\mathscr{C}^{m+1,\alpha}$ on $\bar{\Omega}_i$ (resp. $\bar{\Omega}_e$), and the interior (resp. exterior) double layer potential defined by φ is of class $\mathscr{C}^{m,\alpha}$ on $\bar{\Omega}_i$ (resp. $\bar{\Omega}_e$).*

We shall content ourselves by proving, by way of an example, the

Proposition 13. *Let Γ be the boundary of class $\mathscr{C}^{1+\varepsilon}$ of a regular bounded open set and φ defined on Γ satisfying*

$$\text{(3.22)} \quad \begin{cases} \textit{there exists } \varepsilon\colon \mathbb{R}^+ \to \mathbb{R}^+ \textit{ increasing and continuous with} \\ \displaystyle\int_0 \varepsilon(r)\frac{\mathrm{d}r}{r} < \infty \textit{ such that } |\varphi(z) - \varphi(t)| \leqslant \varepsilon(|z - t|), \\ \textit{for all} \quad t, z \in \Gamma\text{[52]} . \end{cases}$$

[52] That obviously implies that φ is continuous on Γ; it is clear also that this condition is satisfied if φ is a Hölder function on Γ.

Then the interior (resp. exterior) single layer potential u_1^i (resp. u_1^e) defined by φ is of class $\mathscr{C}^1$ on $\bar{\Omega}_i$ (resp. $\bar{\Omega}_e$).

Remark 9. Since $u_1^e = u_1^i$ (Proposition 10), the *tangential derivatives* of u_1^i *and* u_1^e coincide on Γ[53]. □

Proof. We know already that the normal derivatives $\partial u_1^i/\partial n$ and $\partial u_1^e/\partial n$ exist (Proposition 11); it is sufficient therefore to consider the tangential derivatives; since they have to be equal on Γ (Remark 8) we therefore have to show that, for all $z \in \Gamma$

$$\operatorname{grad}_{\text{tang}} u_1(x) = \operatorname{grad} u_1(x) - (\operatorname{grad} u_1(x) . n(z))n(z)$$

converges when $x \to z$, $x \in \mathbb{R}^n \backslash \Gamma$.
This comes to the same thing as showing that

$$\operatorname{grad}_{\text{tang}} u_1(z + \tau n(z)) \quad \text{converges when} \quad \tau \to 0, \quad \tau \neq 0$$

with the condition the convergence is uniform with respect to z.
To simplify the notation, we can consider a normal parametric representation $(R, U, \mathcal{O}, \alpha)$ centred at z ($z = (0, 0)$) and whose last base vector is $n(z)$. For $x = z + \tau n(z) = (0, \tau)$ and $t = (t', \alpha(t'))$

$$\operatorname{grad} E_n(x - t) = \frac{x - t}{\sigma_n |x - t|^n} = \frac{1}{\sigma_n |x - t|^n}(- t', \tau - \alpha(t')) .$$

We can always suppose that supp $\varphi \subset U \cap \Gamma$. We thus have

$$\operatorname{grad}_{\text{tang}} u_1(x) = \left(- \int_{\mathcal{O}} K(\tau, t')\psi(t')\,\mathrm{d}t', 0 \right),$$

with

$$K(\tau, t') = \frac{t'}{\sigma_n(|t'|^2 + (\tau - \alpha(t'))^2)^{n/2}}$$

$$\psi(t') = \varphi(t', \alpha(t'))(1 + |\operatorname{grad} \alpha(t')|^2)^{1/2} .$$

Using the hypotheses on Γ and φ, we see that ψ satisfies (3.22) on $\mathcal{O}$; also ψ is with compact support in $\mathcal{O}$.
The kernel K is singular in the sense in which

$$|K(\tau, t')| \sim \frac{1}{\sigma_n |t'|^{n-1}} \quad \text{when} \quad (\tau, t') \to 0 ,$$

with the result that we cannot pass to the limit directly be Lebesgue's theorem ($1/|t'|^{n-1}$ is not locally integrable). But

$$K(\tau, t') = \operatorname{grad}_{t'} E_n(t', \tau - \alpha(t')) + R(\tau, t')$$

---

[53] In the case of a double layer potential of a function $\varphi \in \mathscr{C}^{2+\varepsilon}(\Gamma)$, u_2^e and u_2^i are of class $\mathscr{C}^1$ on $\bar{\Omega}_e$ and $\bar{\Omega}_i$ respectively. As we have seen, $u_2^i - u_2^e = \varphi$ on Γ, with the result that the tangential derivatives of u_2^e and u_2^i differ on Γ by the tangential derivatives of φ; on the contrary, we can show that the normal derivatives $\partial u_2^i/\partial n$ and $\partial u_2^e/\partial n$ coincide on Γ. (See Gunther [1] and Courant–Hilbert [1]).

with

$$R(\tau, t') = \frac{(\tau - \alpha(t'))\,\mathrm{grad}\,\alpha(t')}{\sigma_n(|t'|^2 + (\tau - \alpha(t'))^2)^{n/2}}\,.$$

This kernel R is not singular and even

$$R(\tau, .) \to 0 \quad \text{in} \quad L^1_{\mathrm{loc}}(\mathcal{O}) \quad \text{when} \quad \tau \to 0\,.$$

In effect for a compact set $\mathscr{K}$ of $\mathcal{O}$ we have

$$\int_{\mathscr{K}} |R(\tau, t')|\,\mathrm{d}t' = \frac{1}{\sigma_n}\int_{\mathscr{K}/|\tau|} \frac{\left|1 - \dfrac{\alpha(\tau t')}{\tau}\right| |\mathrm{grad}\,\alpha(\tau t')|}{\left(|t'|^2 + \left(1 - \dfrac{\alpha(\tau t')}{\tau}\right)^2\right)^{n/2}}\,\mathrm{d}t'\,.$$

But $|\alpha(\tau t')| \leqslant \tau\varepsilon(\tau)$ for $|t'| \leqslant 1$, α and $\mathrm{grad}\,\alpha$ are bounded on $\mathscr{K}$, $\mathrm{grad}\,\alpha(\tau t') \to 0$ when $\tau \to 0$, for all t'; hence by Lebesgue's theorem

$$\int_{\mathscr{K}} |R(\tau, t')|\,\mathrm{d}t' \to 0 \quad \text{when} \quad \tau \to 0\,.$$

It follows from this calculation that $\mathrm{grad}_{\mathrm{tang}}\, u_1(x)$ converges with $\int_{\mathcal{O}} K_1(\tau, t')\psi(t')\,\mathrm{d}t'$, and that

$$\lim_{x \to z} \mathrm{grad}_{\mathrm{tang}}\, u_1(x) = \left(-\lim_{\tau \to 0}\int_{\mathcal{O}} K_1(\tau, t')\psi(t')\,\mathrm{d}t', 0\right)$$

where

$$K_1(\tau, t') = \mathrm{grad}_{t'}\, E_n(t', \tau - \alpha(t'))\,.$$

Let us take $\rho \in \mathscr{D}(\mathcal{O})$ with $\rho = 1$ in the neighbourhood of $\mathrm{supp}\,\psi$. We have

$$\int_{\mathcal{O}} K_1(\tau, t')\psi(t')\,\mathrm{d}t' = \int_{\mathcal{O}} K_1(\tau, t')\rho(t')\psi(t')\,\mathrm{d}t' = I(\tau) - \mathscr{I}(\tau)$$

with

$$I(\tau) = \int_{\mathcal{O}} \mathrm{grad}_{t'}(E_n(t', \tau - \alpha(t'))\rho(t'))(\psi(t') - \psi(0))\,\mathrm{d}t'$$

$$\mathscr{I}(\tau) = \int_{\mathcal{O}} E_n(t', \tau - \alpha(t'))\,\mathrm{grad}\,\rho(t')\psi(t')\,\mathrm{d}t'\,.$$

In addition to the formula for the differentiation of a product, we have used

$$\int_{\mathcal{O}} \mathrm{grad}_{t'}(E_n(t', \tau - \alpha(t'))\rho(t'))\,\mathrm{d}t' = 0\,.$$

From Lebesgue's theorem, these two integrals converge when $\tau \to 0$. In effect from the hypothesis (3.22)

$$|\psi(t') - \psi(0)| \leqslant \varepsilon(|t'|) \quad \text{with} \quad \int_0 \varepsilon(r)\frac{\mathrm{d}r}{r} < \infty$$

and it follows immediately that

$$|\text{grad}_{t'}(E_n(t', \tau - \alpha(t'))\rho(t'))| \leqslant C\left(1 + \frac{1}{|t'|^{n-1}}\right)$$

where C depends only on n and ρ. Since $\varepsilon(|t'|)\left(1 + \frac{1}{|t'|^{n-1}}\right)$ is locally integrable in $\mathbb{R}^{n-1}$, we have the result for $I(\tau)$. The result for $\mathscr{I}(\tau)$ is immediate since

$$|E_n(t', \tau - \alpha(t'))| \leqslant |E_n(t', -\alpha(t'))|$$

which is locally integrable on $\mathbb{R}^{n-1}$. □

The regularity results for simple layer potentials allows to obtain the regularity on the surface of discontinuity of the Newtonian potential of a function f having discontinuities of the first kind. Applying Propositions 12 and 13 we obtain the

Proposition 14. *Let Ω be a regular bounded open set with boundary of class $\mathscr{C}^{1+\varepsilon}$, $f \in \mathscr{C}(\bar{\Omega})$ and u the Newtonian potential of $f\chi_\Omega$*[54]*. We suppose that (in the sense of distributions in Ω) the derivatives satisfy $\partial f/\partial x_l \in L^p(\Omega)$ with $p > n$. Then the restriction u^i (resp. u^e) of u to Ω (resp. $\mathbb{R}^n \backslash \bar{\Omega}$) is of class $\mathscr{C}^2$ on $\bar{\Omega}$ (resp. $\mathbb{R}^n \backslash \Omega$). In addition, for $l = 1, \ldots, n$*

$$\sum_k \frac{\partial^2 u^i}{\partial x_k \partial x_l} n_k = \sum_k \frac{\partial^2 u^e}{\partial x_k \partial x_l} n_k + f n_l \quad \text{on} \quad \Gamma$$

where n_k denotes the k-th component of the (unit) normal vector exterior to Ω.

Proof. We have

$$\Delta\left(\frac{\partial u}{\partial x_l}\right) = \frac{\partial}{\partial x_l}(\Delta u) = \frac{\partial}{\partial x_l}(f\chi_\Omega) = \frac{\partial f}{\partial x_l}\chi_\Omega - f n_l \,\mathrm{d}\gamma \quad \text{in} \quad \mathscr{D}'(\mathbb{R}^n)$$

since from Ostrogradski's formula (see §1.3)

$$\left\langle \frac{\partial}{\partial x_l}\chi_\Omega, \zeta \right\rangle = -\left\langle \chi_\Omega, \frac{\partial \zeta}{\partial x_l} \right\rangle = -\int_\Omega \frac{\partial \zeta}{\partial x_l}\mathrm{d}x = -\int_\Gamma \zeta n_l \,\mathrm{d}\gamma \,.$$

On the other hand, since (see Proposition 2)

$$\lim_{|x| \to \infty} \frac{\partial u}{\partial x_l}(x) = 0 \,,$$

from Proposition 3, $\partial u/\partial x_l$ is the Newtonian potential of the distribution with compact support $(\partial f/\partial x_l)\chi_\Omega - f n_l \,\mathrm{d}\gamma$.
Since $(\partial f/\partial x_l)\chi_\Omega \in L^p(\mathbb{R}^n)$ with $p > n$, from Proposition (6.3), its Newtonian potential v_l is of class $\mathscr{C}^1$ on $\mathbb{R}^n$.

---

[54] Throughout this chapter we denote by χ_Ω the characteristic function of the set Ω: $\chi_\Omega(x) = 1$ if $x \in \Omega$, $\chi_\Omega(x) = 0$ otherwise. Hence the Newtonian potential of $f\chi_\Omega$ is $u(x) = \int_\Omega E_n(x-y)f(y)\,\mathrm{d}y$, for all $x \in \mathbb{R}^n$. We recall (see Proposition 6 point (3)) that u is of class $\mathscr{C}^1$ on $\mathbb{R}^n$.

From Propositions 12 and 13, the restriction w_l^i (resp. w_l^e) to Ω (resp. $\mathbb{R}^n \setminus \bar{\Omega}$) of the simple layer Newtonian potential of the restriction of f to Γ is in $\mathscr{C}^1(\bar{\Omega})$ (resp. $\mathscr{C}^1(\mathbb{R}^n \setminus \Omega)$) and

$$\frac{\partial w_l^i}{\partial n} = \frac{\partial w_l^e}{\partial n} - f \quad \text{on} \quad \Gamma .$$

Using $\dfrac{\partial u^i}{\partial x_l} = v_l - w_l^i$, $\dfrac{\partial u^e}{\partial x_l} = v_l - w_l^e$, we deduce the proposition. □

The above reasoning and the results on simple layer potentials stated above allows us to prove that *given a regular bounded open set Ω with boundary of class $\mathscr{C}^\infty$ and $f \in \mathscr{C}^\infty(\bar{\Omega})$, the restrictions to Ω and $\mathbb{R}^n \setminus \bar{\Omega}$ of the Newtonian potential of $f\chi_\Omega$ are of class $\mathscr{C}^\infty$ on $\bar{\Omega}$ and $\mathbb{R}^n \setminus \Omega$ respectively*. In effect, from what we have seen above, $\partial u/\partial x_l$ is the Newtonian potential of $(\partial f/\partial x_l)\chi_\Omega - fn_l \, d\gamma$; since the interior and exterior Newtonian potentials of $fn_l \, d\gamma$ are of class $\mathscr{C}^\infty$ on $\bar{\Omega}$ and $\mathbb{R}^n \setminus \Omega$ respectively (see the statement preceding Proposition 13), the above affirmation is deduced by induction

$$\begin{aligned}
&\forall f \in \mathscr{C}^\infty(\bar{\Omega}), \quad (E_n * f\chi_\Omega)^i \in \mathscr{C}^m(\bar{\Omega}) \\
&\quad \Rightarrow \forall f \in \mathscr{C}^\infty(\bar{\Omega}), \quad \forall l = 1, \ldots, n, \quad \left(E_n * \frac{\partial f}{\partial x_l}\chi_\Omega\right)^i \in \mathscr{C}^m(\bar{\Omega}) \\
&\quad \Rightarrow \forall f \in \mathscr{C}^\infty(\bar{\Omega}), \quad \forall l = 1, \ldots, n, \quad \frac{\partial}{\partial x_l}(E_n * f\chi_\Omega)^i \in \mathscr{C}^m(\bar{\Omega}) \\
&\quad \Rightarrow \forall f \in \mathscr{C}^\infty(\bar{\Omega}), (E_n * f\chi_\Omega)^i \in \mathscr{C}^{m+1}(\bar{\Omega}) .
\end{aligned}$$

4. Newtonian Potential of a Distribution Without Compact Support

We have defined the Newtonian potential of a distribution f with compact support as the distribution $u = E * f$, that is to say

$$\langle u, \zeta \rangle = \langle f, E_n * \zeta \rangle ;$$

this has clearly a meaning: for $\zeta \in \mathscr{D}(\mathbb{R}^n)$, we have $E_n * \zeta \in \mathscr{E}(\mathbb{R}^n)$ which allows the definition of $\langle f, E_n * \zeta \rangle$ for all $f \in \mathscr{E}'(\mathbb{R}^n)$ that is to say for every distribution f with compact support. But the function $E_n * \zeta$ is not just any function of $\mathscr{E}(\mathbb{R}^n)$: it has a well defined behaviour at infinity (see Proposition 2). This will enable us to define $\langle f, E_n * \zeta \rangle$ for distributions f without compact support, but having a certain behavior at infinity.
We pose the

Definition 4. We say that a *distribution f on $\mathbb{R}^n$ is regular at infinity* (with respect to the Laplacian) if there exists a distribution u on $\mathbb{R}^n$ such that

$$(3.23) \qquad \left\{\begin{array}{l}
\text{for all } \rho \in \mathscr{D}(\mathbb{R}^n) \text{ with } \rho = 1 \text{ in the neighbourhood of } 0 \\
\langle u, \zeta \rangle = \lim\limits_{\varepsilon \to 0} \langle f, \rho_\varepsilon(E_n * \zeta) \rangle \quad \text{for all} \quad \zeta \in \mathscr{D}(\mathbb{R}^n) \\
\text{where} \quad \rho_\varepsilon(x) = \rho(\varepsilon x) .
\end{array}\right.$$

The distribution u (which is obviously unique) is then called *the Newtonian potential of the distribution f regular at infinity.*

In other words, f is regular at infinity if the Newtonian potentials of the distributions ρf, with compact support, converge in $\mathscr{D}'(\mathbb{R}^n)$ when $\rho \to 1$, $\rho \in \mathscr{D}'(\mathbb{R}^n)$; the meaning of this convergence is specified by (3.23). The limit is the Newtonian potential of f.

This notion clearly generalizes that of the Newtonian potential of a distribution with compact support: if f is a distribution with compact support, for $\rho \in \mathscr{D}(\mathbb{R}^n)$ with $\rho = 1$ in the neighbourhood of O, for ε sufficiently small, $\rho_\varepsilon = 1$ in the neighbourhood of supp f and hence $\rho_\varepsilon f = f$. However *this notion does not cover all the cases of distributions for which we can consider a "natural" potential.* For example, let us consider for $1 \leqslant m \leqslant n-1$ the distribution $f = 1_{\mathbb{R}^m} \otimes \delta_{\mathbb{R}^{n-m}}$, that is to say

$$\langle f, \zeta \rangle = \int_{\mathbb{R}^m} \zeta(x', 0)\,\mathrm{d}x' \tag{3.24}$$

where $\mathbb{R}^n = \mathbb{R}^m \times \mathbb{R}^{n-m} = \{(x', x''); x' \in \mathbb{R}^m, x'' \in \mathbb{R}^{n-m}\}$. It will follow from Proposition 15 below that $1_{\mathbb{R}^m} \otimes \delta_{\mathbb{R}^{n-m}}$ *is regular iff* $m < n - 2$ (or $m = 0$ obviously); its Newtonian potential will then be the function

$$u(x', x'') = E_{n-m}(x'') \,. \tag{3.25}$$

It is clear that for all $0 \leqslant m \leqslant n-1$, the function u can be considered as the "Newtonian potential" $1_{\mathbb{R}^m} \otimes \delta_{\mathbb{R}^{n-m}}$. The problem encountered in the cases $m = n-2$ and $m = n-1$, which come in fact from *the particular properties of the Laplacian in $\mathbb{R}^2$ and $\mathbb{R}$*, will be considered later (see §3.4.b).

4a. Positive Functions and Measures, Regular at Infinity

We begin by studying the (locally integrable) functions and the positive (Radon) measures, regular at infinity on $\mathbb{R}^n$ and their Newtonian potentials. We have the characterisation:

Proposition 15. *Let f be a positive Radon measure on $\mathbb{R}^n$;*
(1) *f is regular at infinity iff*

$$\begin{cases} \text{case } n \geqslant 3, & \displaystyle\int \frac{\mathrm{d}f(x)}{(1+|x|)^{n-2}} < \infty \\ \text{case } n = 2, & \displaystyle\int \operatorname{Log}(1+|x|)\,\mathrm{d}f(x) < \infty \\ \text{case } n = 1, & \displaystyle\int |x|\,\mathrm{d}f(x) < \infty\,; \end{cases} \tag{3.26}$$

(2) *supposing this condition satisfied,*

(a) *for almost all $x \in \mathbb{R}^n$, the functions $y \to E_n(x-y)$ and $y \to \dfrac{\partial E_n}{\partial x_i}(x-y)$ are integrable with respect to the measure f.*

(b) *the Newtonian potential u of f and its derivatives* $\partial u/\partial x_i$ *are* (*locally integrable*) *functions on* $\mathbb{R}^n$ *given by*

$$u(x) = \int E_n(x - y)\,\mathrm{d}f(y) \qquad \text{a.e.} \quad x \in \mathbb{R}^n$$

$$\frac{\partial u}{\partial x_i}(x) = \int \frac{\partial E_n}{\partial x_i}(x - y)\,\mathrm{d}f(y) \qquad \text{a.e.} \quad x \in \mathbb{R}^n\,;$$

(c) *for every bounded Borel function* ρ *with compact support and with* $\rho = 1$ *in a neighbourhood of* 0,

$$\lim_{\varepsilon \to 0} \int_{B(0,R)} \left(|u(x) - u_\varepsilon(x)| + \sum \left| \frac{\partial u}{\partial x_i}(x) - \frac{\partial u_\varepsilon}{\partial x_i}(x) \right| \right) \mathrm{d}x = 0$$

for all $R > 0$, *where* u_ε *is the Newtonian potential of the measure with compact support* $\rho_\varepsilon f$ *with* $\rho_\varepsilon(x) = \rho(\varepsilon x)$.

We note the

Corollary 2. *Let f be a Radon measure* (*not necessarily positive*) *on* $\mathbb{R}^n$ *satisfying*:

$$(3.27) \qquad \begin{cases} \text{case } n \geqslant 3, \text{ the measure } (1 + |x|)^{-n+2} f \text{ is bounded on } \mathbb{R}^n; \\ \text{case } n = 2, \text{ the measure } \operatorname{Log}(1 + |x|) f \text{ is bounded on } \mathbb{R}^2; \\ \text{case } n = 1, \text{ the measure } xf \text{ is bounded on } \mathbb{R} \end{cases}$$

Then f is regular at infinity.

That follows immediately from the proposition, from the fact that the set of distributions, regular at infinity, is a vector space, and from the decomposition of a Radon measure into positive and negative parts (f^+ and f^- parts respectively):

$$\text{for } \zeta \geqslant 0, \quad \langle f^+, \zeta \rangle = \sup\{\langle f, \rho \rangle; \rho \leqslant \zeta\}$$

$$f^- = (-f)^+, \quad f = f^+ - f^-, \quad |f| = f^+ + f^- .$$

A measure g is bounded iff $\int \mathrm{d}|g|(x) < \infty$[55].

We note also that the Newtonian potential of a distribution f regular at infinity being a linear function of f, under the hypothesis of Corollary 2, we have the conclusion (2) of Proposition 15.

Remark 10. The essential difference between the cases $n \geqslant 3$ and $n = 1, 2$ appears clearly in the statements above. In particular:
(1) *in the case* $n \geqslant 3$, *every bounded Radon measure on* $\mathbb{R}^n$ *is regular at infinity.* There exist unbounded measures which are also regular at infinity: *every function* $f \in L^q(\mathbb{R}^n)$ *with* $q < n/2$ *is regular at infinity*; in effect the function $1/(1 + |x|)^{n-2}$ belongs to $L^p(\mathbb{R}^n)$ for all $p > n/(n - 2)$.

---

[55] See, for example, Bourbaki [2], Schwartz [2] or Marle [1] for the study of Radon measures. If f is a (locally integrable) function, these ideas are classical: $f^+ = \sup(0, f)$.

(2) *in the case* $n = 1, 2$, *every positive measure, regular at infinity, is necessarily bounded.* A bounded measure is not however necessarily regular at infinity. □

Before proving Proposition 15, let us give some examples:

Example 1. Let us consider a function f on $\mathbb{R}^n$ and suppose that

$$|f(x)| = O(g(|x|)) \quad \text{when} \quad |x| \to \infty$$

where $g(r)$ is a positive (measurable) function defined for $r > r_0$. Then the regularity at infinity of f is assured by

$$\text{(3.28)} \qquad \begin{cases} \text{if } n \neq 2, & \displaystyle\int_{r_0}^{\infty} r g(r) \, dr < \infty \\ \text{if } n = 2, & \displaystyle\int_{r_0}^{\infty} r \operatorname{Log}(1 + r) g(r) \, dr < \infty \, . \end{cases}$$

If when $|x| \to \infty$, $f \geqslant 0$ and $g(|x|) = O(f(x))$, then the condition (3.28) is necessary to ensure the regularity at infinity of f. In particular, we note that a function f on $\mathbb{R}^n$ satisfying

$$|f(x)| \leqslant \frac{C}{|x|^{2+\varepsilon}} \quad \text{for} \quad |x| > r_0 \quad \text{with} \quad \varepsilon > 0$$

is regular at infinity. □

Example 2. For $1 \leqslant m \leqslant n - 1$, the measure $1_{\mathbb{R}^m} \otimes \delta_{\mathbb{R}^{n-m}}$ is regular at infinity iff $m < n - 2$. It is in effect a positive measure and condition (3.26) reduces to $m < n - 2$.
The Newtonian potential u is then given by

$$u(x', x'') = \int_{\mathbb{R}^m} E_n(x' - y', x'') \, dy' = \frac{1}{k_n} \int_{\mathbb{R}^m} \frac{dy'}{(|x' - y'|^2 + |x''|^2)^{\frac{n}{2} - 1}}$$

$$= \frac{1}{k_n |x''|^{n-m-2}} \int_{\mathbb{R}^m} \frac{dy'}{(1 + |y'|^2)^{\frac{n}{2} - 1}} = E_{n-m}(x'') \, .$$

Example 3. Let us consider the case where we are given a *regular* open set Ω of $\mathbb{R}^n$, but *whose boundary* Γ *is unbounded*[56]. For $\varphi \in \mathscr{C}^0(\Gamma)$, the measure $f = \varphi d\gamma$ is regular at infinity under the hypothesis expressing (3.27):

$$\text{(3.29)} \qquad \begin{cases} \text{case } n \geqslant 3, & \displaystyle\int_{\Gamma} \frac{|\varphi(x)| \, d\gamma(x)}{(1 + |x|)^{n-2}} < \infty \\ \text{case } n = 2, & \displaystyle\int_{\Gamma} \operatorname{Log}(1 + |x|) |\varphi(x)| \, d\gamma(x) < \infty \, . \end{cases}$$

---

[56] For example a half-space or a cylinder.

If $\varphi \geqslant 0$, this is a necessary condition for $\varphi \mathrm{d}\gamma$ to be regular at infinity. The Newtonian potential of f is the simple layer potential u_1, harmonic on $\mathbb{R}^n \backslash \Gamma$, defined by

$$u_1(x) = \int_\Gamma E_n(x - t)\varphi(t)\,\mathrm{d}\gamma(t) \quad \text{for all} \quad x \in \mathbb{R}^n \backslash \Gamma .$$

Let us suppose that Ω is "*uniformly regular*", that is to say that the function $1/(1 + |x|)^k$ is integrable with respect to $\mathrm{d}\gamma$ if $k > n - 1$. In this case, the measure $\varphi \mathrm{d}\gamma$ is, in particular, regular at infinity if

$$|\varphi(x)| \leqslant \frac{C}{|x|^{1+\varepsilon}} \quad \text{for} \quad |x| > r_0, \quad x \in \Gamma \quad \text{with} \quad \varepsilon > 0 .$$

We shall now prove Proposition 15; we give in a lemma the principal argument of its proof:

Lemma 8. *Let f be a positive measure on $\mathbb{R}^n$, g a (Borel) function, locally integrable on $\mathbb{R}^n$ and ω: $\mathbb{R}^+ \to \mathbb{R}^+$, continuous and satisfying*

$$\omega(|x - y|) \leqslant C(|x|)\omega(|y|) \quad \textit{for all} \quad x, y \in \mathbb{R}^n , \tag{3.30}$$

with C: $\mathbb{R}^+ \to \mathbb{R}^+$ continuous.

(1) *Let us suppose*

$$|g(x)| = O(\omega(|x|)) \quad \textit{when} \quad |x| \to \infty$$

$$\int \omega(|x|)\,\mathrm{d}f(x) < \infty .$$

Then

(*a*) *for almost all $x \in \mathbb{R}^n$, the function $y \to g(x - y)$ is integrable with respect to f;*

(*b*) *the function v: $x \to \int g(x - y)\,\mathrm{d}f(y)$ is locally integrable on $\mathbb{R}^n$;*

(*c*) *for every bounded Borel function ρ with compact support and with $\rho = 1$ in the neighbourhood of O*

$$\lim_{\varepsilon \to 0} \int_{B(0,R)} |g * (\rho_\varepsilon f) - v|\,\mathrm{d}x = 0 \quad \textit{for all} \quad R > 0$$

where, as before: $\rho_\varepsilon(x) = \rho(\varepsilon x)$.

(2) *Let us suppose that when $|x| \to \infty$*

$$g(x) \geqslant 0 \quad \text{and} \quad \omega(|x|) = O(g(x)) .$$

Then $\int \omega(|x|)\,\mathrm{d}f(x) < \infty$ is a necessary condition for the existence of a distribution v on $\mathbb{R}^n$ satisfying

$$\begin{cases} \textit{for all } \rho \in \mathscr{D}(\mathbb{R}^n) \textit{ with } \rho = 1 \textit{ in the neighbourhood of } O \\ \qquad \langle \mathbf{v}, \zeta \rangle = \lim\limits_{\varepsilon \to 0} \langle g * \rho_\varepsilon f, \zeta \rangle \\ \textit{for all } \zeta \in \mathscr{D}(\mathbb{R}^n) . \end{cases}$$

Proof of Proposition 15. We note first of all that the functions

$$\omega_1(r) = 1 + r, \quad \omega_L(r) = 1 + \mathrm{Log}(1 + r), \quad \omega_{-k}(r) = \frac{1}{(1 + r)^k} \quad (k \geqslant 0)$$

satisfy the condition (3.30). In effect we have

$$(3.31) \quad \begin{cases} \omega_1(|x - y|) \leqslant 1 + |x| + |y| \leqslant \omega_1(|x|)\omega_1(|y|) \\ \omega_L(|x - y|) \leqslant 1 + \mathrm{Log}(1 + |x| + |y|) \leqslant \omega_L(|x|)\omega_L(|y|) \\ \omega_{-k}(|x - y|) \leqslant \omega_k(|x|)\omega_{-k}(|y|) \end{cases}$$

the last inequality resulting from

$$1 + |y| \leqslant 1 + |x - y| + |x| \leqslant (1 + |x|)(1 + |x - y|)\,.$$

Applying the lemma with

$$\begin{cases} \text{case } n \geqslant 3, & g = -E_n, & \omega = \omega_{2-n} \\ \text{case } n = 2, & g = E_2\;, & \omega = \omega_L \\ \text{case } n = 1, & g = E_1\;, & \omega = \omega_1\,, \end{cases}$$

we obtain the point (1) and the part of the point (2) relevant to the Newtonian potential u. Applying the lemma with $g = \partial E_n/\partial x_i$, $\omega = \omega_{1-n}$ we obtain the part of the point (2) relative to the derivatives $\partial u/\partial x_i$ of the Newtonian potential. □

Proof of Lemma 8. First of all we prove (1). The hypothesis $g(x) = O(\omega(|x|))$ can be written

$$(3.32) \qquad |g(x)| \leqslant C_0\omega(|x|) \quad \text{for} \quad |x| \geqslant r_0\,.$$

We fix $R > 0$ and ρ_0 continuous with compact support with $0 \leqslant \rho_0 \leqslant 1$ and $\rho_0 = 1$ on $B(0, R + r_0)$. The measure $\rho_0 f$ being of compact support we can apply Lemma 3 to it; this gives the results (a) and (b) on replacing f by $\rho_0 f$. Now $(1 - \rho_0)f$ has support in $\mathbb{R}^n \setminus \bar{B}(0, R + r_0)$; hence, using (3.32) and (3.30), for $x \in B(0, R)$ and $y \in \mathrm{supp}(1 - \rho_0)f$, we have

$$|g(x - y)| \leqslant C_0\omega(|x - y|) \leqslant C(R)\omega(|y|)$$

with

$$C(R) = C_0 \sup_{r \leqslant R} C(r)\,,$$

we deduce that

$$\int_{B(0,R)} \mathrm{d}x \int |g(x - y)|(1 - \rho_0(y))\,\mathrm{d}f(y) < \infty$$

and hence by Fubini's theorem, the points (a) and (b) on replacing f by $(1 - \rho_0)f$. Combining the two results we complete the proof of points (a) and (b); the point (c) is then proved easily by the use of Lebesgue's theorem. To prove (2), we observe that the hypothesis $\omega(|x|) = O(g|x|)$ can be written

$$g(x) \geqslant c\omega(|x|) \quad \text{for} \quad |x| \geqslant r_0 \quad \text{with} \quad c > 0\,.$$

We fix $\rho, \zeta \in \mathscr{D}(\mathbb{R}^n)$ with $0 \leqslant \rho \leqslant 1, \zeta \geqslant 0, \rho = 1$ on $B(0, r_0 + 1)$, $\operatorname{supp}\zeta \subset \bar{B}(0, 1)$. We have

$$\langle g * (\rho_\varepsilon f), \zeta \rangle = I_0 + I_\varepsilon$$

with

$$I_0 = \langle g * \rho f, \zeta \rangle$$

$$I_\varepsilon = \int \zeta(x)\,\mathrm{d}x \int g(x - y)(\rho(\varepsilon y) - \rho(y))\,\mathrm{d}f(y)\,.$$

For $\varepsilon > 0$ sufficiently small, $\rho_\varepsilon \geqslant \rho$ and $\rho_\varepsilon = \rho$ on $B(0, r_0 + 1)$. Then, since $|x - y| \geqslant r_0$ for $x \in \operatorname{supp}\zeta$ and $y \in \operatorname{supp}(\rho_\varepsilon - \rho)$

$$I_\varepsilon \geqslant \int \zeta(x)\,\mathrm{d}x \int c\omega(|x - y|)(\rho(\varepsilon y) - \rho(y))\,\mathrm{d}f(y)\,.$$

From (3.30),

$$\omega(|y|) = \omega(|x - (x - y)|) \leqslant C(|x|)\omega(|x - y|)\,,$$

hence

$$I_\varepsilon \geqslant c_1 \int \zeta(x)\,\mathrm{d}x \int \omega(|y|)(\rho(\varepsilon y) - \rho(y))\,\mathrm{d}f(y)\,,$$

with

$$c_1 = \frac{c}{\sup\limits_{r \leqslant 1} C(r)}\,.$$

This proves (2). □

We now propose to generalize the characterisations of Propositions 3 and 4. It is clear that the *Newtonian potential of a distribution f on $\mathbb{R}^n$ regular at infinity is a solution of Poisson's equation*

$$\Delta u = f \quad \text{on} \quad \mathbb{R}^n\,.$$

In effect, given $\zeta \in \mathscr{D}(\mathbb{R}^n)$ we choose $\rho \in \mathscr{D}(\mathbb{R}^n)$ such that $\rho = 1$ in the neighbourhood of 0; for ε sufficiently small, $\rho_\varepsilon = 1$ in the neighbourhood of $\operatorname{supp}\zeta$; now the Newtonian potential u_ε of $\rho_\varepsilon f$ satisfies the Poisson equation

$$\Delta u_\varepsilon = \rho_\varepsilon f \quad \text{on} \quad \mathbb{R}^n\,,$$

and hence

$$\Delta u_\varepsilon = f \quad \text{in the neighbourhood of } \operatorname{supp}\zeta$$

We deduce that if u is the Newtonian potential of f,

$$\langle \Delta u, \zeta \rangle = \langle u, \Delta\zeta \rangle = \lim_{\varepsilon \to 0} \langle u_\varepsilon, \Delta\zeta \rangle = \langle f, \zeta \rangle\,.$$

This being true for all ζ, u is clearly a solution of Poisson's equation. We wish to

characterise u by a condition at infinity: the situation is not however as simple as in the case of a distribution with compact support.
Let us prove the following result in the *radial case*:

Lemma 9. *Let f be a positive* **radial** *measure on $\mathbb{R}^n$, regular at infinity and u its Newtonian potential. Then when $|x| \to \infty$*[57]

case $n \geqslant 3$, $u(x) \to 0$

case $n = 2$, $u(x) - E_2(x) \int \mathrm{d}f(y) \to 0$

case $n \geqslant 2$, $|\operatorname{grad} u(x)| = O\left(\frac{1}{|x|}\right)$.

Proof. We have

$$u(x) = E_n(x) \int_{B(0, |x|)} \mathrm{d}f(y) + \int_{\mathbb{R}^n \setminus B(0,|x|)} E_n(y) \, \mathrm{d}f(y)$$

$$\operatorname{grad} u(x) = \frac{x}{\sigma_n |x|^n} \int_{B(0, |x|)} \mathrm{d}f(y) \, .$$

We see easily that the formulae established in Lemma 4 for a function with compact support are still valid for a positive measure regular at infinity.
In the case $n \geqslant 3$, we have for $|x| \geqslant R > 0$

$$|u(x)| \leqslant |E_n(x)| \int_{B(0, R)} \mathrm{d}f(y) + \int_{\mathbb{R}^n \setminus B(0, R)} |E_n(y)| \, \mathrm{d}f(y) \, .$$

Since

$$\int_{\mathbb{R}^n \setminus B(0, R)} |E_n(y)| \, \mathrm{d}f(y) \leqslant \frac{1}{|k_n|} \left(1 + \frac{1}{R} \right)^{n-2} \int_{\mathbb{R}^n \setminus B(0, R)} \frac{\mathrm{d}f(y)}{(1 + |y|)^{n-2}} \, ,$$

this integral tends to zero when $R \to \infty$. For $\varepsilon > 0$, we can therefore choose $R > 0$ such that

$$\int_{\mathbb{R}^n \setminus B(0, R)} |E_n(y)| \, \mathrm{d}f(y) < \frac{\varepsilon}{2} \, ;$$

R being fixed, we can find r_0 such that

$$|x| > r_0 \Rightarrow |E_n(x)| \int_{B(0, R)} \mathrm{d}f(y) < \frac{\varepsilon}{2} \, ,$$

and hence $|u(x)| < \varepsilon$. This shows that $u(x) \to 0$ when $|x| \to \infty$.

---

[57] For a function v defined a.e. on $\mathbb{R}^n$, we say that $v(x) \to 0$ when $|x| \to \infty$ if there exists r_0 and a function $\varepsilon(r)$ defined for all $r > r_0$ such that

$$\lim_{r \to \infty} \varepsilon(r) = 0, \quad |v(x)| \leqslant \varepsilon(|x|), \quad \text{a.e.} \quad x \text{ with } |x| > r_0 \, .$$

In the case $n = 2$, f is a bounded measure; we have for $|x| \geqslant 1$

$$0 \leqslant u(x) - E_2(x)\int \mathrm{d}f(y) = \frac{1}{2\pi}\int_{\mathbb{R}^2\setminus B(0,|x|)} \operatorname{Log}\frac{|y|}{|x|}\,\mathrm{d}f(y)$$

$$\leqslant \frac{1}{2\pi}\int_{\mathbb{R}^2\setminus B(0,|x|)} \operatorname{Log}(1 + |y|)\,\mathrm{d}f(y)\,.$$

Finally in the case $n \geqslant 2$, for $|x| > 1$

$$|\operatorname{grad} u(x)| = \frac{1}{\sigma_n |x|^{n-1}}\int_{B(0,|x|)} \mathrm{d}f(y) \leqslant \frac{2^{n-2}}{\sigma_n |x|(1 + |x|)^{n-2}}\int_{B(0,|x|)} \mathrm{d}f(y)$$

$$\leqslant \frac{2^{n-2}}{\sigma_n |x|}\int \frac{\mathrm{d}f(y)}{(1 + |y|)^{n-2}}\,. \qquad \square$$

Remark 11. *The results of the lemma still remain true for a measure f on $\mathbb{R}^n$, dominated at infinity by a radial positive measure g, regular at infinity*, that is to say that for an $r_0 > 0$, we should have

$$|\langle f, \zeta\rangle| \leqslant \int \zeta(x)\,\mathrm{d}g(x) \quad \text{for} \quad \zeta \geqslant 0 \quad \text{with} \quad \operatorname{supp}\zeta \subset \mathbb{R}^n\setminus \bar{B}(0, r_0)\,.$$

In the case of functions, that means (see Example 1 above)

$$|f(x)| \leqslant g(|x|) \quad \text{for} \quad |x| > r_0\,.$$

In effect, since the result is true for a distribution with compact support (see Proposition 2) we can always suppose that

$$\operatorname{supp} f \subset \mathbb{R}^n\setminus \bar{B}(0, r_0)\,.$$

It is clear also that f is regular at infinity and that its Newtonian potential is:

$$u(x) = \int E_n(x - y)\,\mathrm{d}f(y)\,.$$

In the case $n \geqslant 3$, we have immediately

$$|u(x)| \leqslant \int |E_n(x - y)|\,\mathrm{d}g(y) = -v(x)$$

where v is the Newtonian potential of g.
Therefore

$$u(x) \to 0 \quad \text{when} \quad |x| \to \infty\,.$$

We shall similarly prove the other results.
On the other hand, *the results of the lemma are false in general* if f is not radial: the example of $f = 1_{\mathbb{R}^m} \otimes \delta_{\mathbb{R}^{n-m}}$ shows this clearly; it is a positive measure, regular at infinity if $m < n - 2$, whose Newtonian potential

$$u(x', x'') = E_{n-m}(x'')$$

does not tend to zero at infinity (except obviously in the directions of x''); the regularity of the measure at finite distance has nothing to do with this property: we can easily construct a function $f \in \mathscr{C}^\infty(\mathbb{R}^n)$, regular at infinity and whose Newtonian potential, in the case $n \geqslant 3$, does not tend to zero at infinity: it will be sufficient that its support is contained in a band.

$$\{(x', x_n); x' \in \mathbb{R}^{n-1}, |x_n| \leqslant \delta\} . \qquad \square$$

However we can get round this difficulty by *replacing the topological limits by limits in mean.* We pose the

Definition 5. We say that a (locally integrable) function f on $\mathbb{R}^n$ *tends to zero at infinity in mean* if

$$\lim_{k \to \infty} \int_K |u(kx)| \, dx = 0 \qquad \text{for every compact set } K \text{ of } \mathbb{R}^n \backslash \{0\} .$$

It is clear that if $u(x) \to 0$ when $|x| \to \infty$ (in the topological sense), then it tends to zero in mean. But the converse is not true: a function $u \in L^p(\mathbb{R}^n)$ with $1 \leqslant p < \infty$ tends to zero at infinity in mean (whereas in general it does not tend to zero in the topological sense); in effect by Hölder's inequality:

$$\int_K |u(kx)| \, dx = k^{-n} \int_{k^{-1}K} |u(x)| \, dx \leqslant k^{-n} \left(\int_{k^{-1}K} dx \right)^{1 - \frac{1}{p}} \|u\|_{L^p}$$

$$= k^{-\frac{n}{p}} \left(\int_K dx \right)^{1 - \frac{1}{p}} \|u\|_{L^p} .$$

We now give the characterisation:

Proposition 16. *Let f be a positive measure on $\mathbb{R}^n$, regular at infinity and let u be a locally integrable function on $\mathbb{R}^n$;*
(1) Case $n \geqslant 3$; u is the Newtonian potential of f iff u is a solution of Poisson's equation

$$\Delta u = f \quad \text{on} \quad \mathbb{R}^n$$

and tends to zero at infinity in mean;
(2) Case $n = 2$; u is the Newtonian potential of f iff u is a solution of Poisson's equation

$$\Delta u = f \quad \text{on} \quad \mathbb{R}^2$$

and the function $u - E_2 \int df(y)$ tends to zero at infinity in mean.

(3) Case $n \geqslant 2$; u is, to within an additive constant, the Newtonian potential of f iff u is a solution of Poisson's equation

$$\Delta u = f \quad \text{on} \quad \mathbb{R}^n$$

and the function $u/(1 + |x|)$ tends to zero at infinity in mean.

Proof. First, we prove the necessary conditions. We have already seen that the Newtonian potential was a solution of Poisson's equation; we have thus only to

show that in the case $n \geq 3$ (resp. $n = 2$) u (resp. $u - E_2 \int df(y)$) tend to zero at infinity in mean.
In the case $n \geq 3$, the fundamental solution E_n, also u, is negative. The function u will tend to zero at infinity in mean iff it is the same for its radialized function

$$\tilde{u}(x) = \int_\Sigma u(|x|\sigma)\,d\sigma\,.$$

Now $\tilde{u}$ is the Newtonian potential of the radialized function $\tilde{f}$ of f: by commutation of the Laplacian and of the radialization (see Proposition 6 of §1), that follows from the characterization of Newtonian potentials in the case of a distribution with compact support (see Proposition 3); since u is the limit of Newtonian potentials of $\rho_\varepsilon f$, choosing ρ radial, that is true for an arbitrary distribution regular at infinity. From Lemma 9, $\tilde{u}(x) \to 0$ when $|x| \to \infty$ and hence a *fortiori* $\tilde{u}$ tends to zero at infinity in mean.
In the case $n = 2$,

$$u(x) - E_2(x)\int df(y) = \frac{1}{2\pi}\int \operatorname{Log}\frac{|x-y|}{|x|}\,df(y) \leq \frac{1}{2\pi}\int \operatorname{Log}\left(1 + \frac{|y|}{|x|}\right)df(y)\,.$$

From Lebesgue's theorem, since $\int \operatorname{Log}(1 + |y|)\,df(y) < \infty$, we have

$$\int \operatorname{Log}\left(1 + \frac{|y|}{|x|}\right)df(y) \to 0 \quad \text{when} \quad |x| \to \infty\,.$$

Hence $u - E_2 \int df(y)$ is majorised by a function tending to zero at infinity. It follows that $u - E_2 \int df(y)$ tends to zero at infinity in mean iff its radialized $\tilde{u} - E_2 \int df(y)$ tends to zero at infinity in mean. We take up again the above reasoning.
Finally we show that the conditions are sufficient. By linearity, it is enough to show that a harmonic function u on $\mathbb{R}^n$ tending to zero at infinity in mean (resp. such that $u/(1 + |x|)$ tends to zero at infinity in mean) is zero (resp. constant). Admitting for the moment that u is a tempered distribution; then from Proposition 3 of §2, u is a (harmonic) polynomial; denoting its degree by m and its principle part by u_m, we have

$$\int_K |u(kx)|\,dx \sim k^m \int_K |u_m(x)|\,dx$$

$$\left(\text{resp.} \int_K \frac{|u(kx)|}{1 + |kx|}\,dx \sim k^{m-1}\int_K \frac{|u_m(x)|}{|x|}\,dx\right) \quad \text{when} \quad k \to \infty\,.$$

Hence $m = 0$ and $u_m = 0$, that is to say $u = 0$ (resp. $m = 1$ and $u_m = 0$ or $m = 0$, that is to say $u =$ constant).
It remains therefore to show that a (locally integrable) function v tending to zero at

infinity in mean is a tempered distribution: that follows from the estimate

$$\int_{B(0,R)} |v(x)|\,dx \leqslant C(1 + R^n)\,.$$

We can establish this estimate by noting that

$$\int_{\{k \leqslant |x| \leqslant 2k\}} |v(x)|\,dx = k^n \int_{\{1 \leqslant |x| \leqslant 2\}} |v(kx)|\,dx \leqslant C_0 k^n\,.$$

Putting $C_1 = \int_{B(0,1)} |v(x)|\,dx$, we then have for $R > 1$, choosing $k \geqslant 1$ such that $2^{k-1} < R \leqslant 2^k$,

$$\int_{B(0,R)} |v(x)|\,dx = \int_{B(0,1)} |v(x)|\,dx + \sum_{j=1}^{k} \int_{\{2^{j-1} \leqslant |x| \leqslant 2^j\}} |v(x)|\,dx$$

$$\leqslant C_1 + C_0 \sum_{j=1}^{k} 2^{n(j-1)} \leqslant C_1 + C_0 R^n\,. \qquad \square$$

The limit at infinity in mean, if it gives a simple characterization of the Newtonian potentials of a positive measure, regular at infinity, is not a very precise instrument. We can in fact *evaluate the behaviour of the Newtonian potential at infinity* in a much more precise fashion.

We indicate two methods here: a method of *weights* and a *truncation* method. We shall restrict ourselves to the case $n \geqslant 3$.

Proposition 17. *Let f be a positive measure, regular at infinity on $\mathbb{R}^n$ with $n \geqslant 3$ and u its Newtonian potential. Then we have the estimates*[58]:

(1)

$$\int \frac{|u(x)|\,dx}{(1 + |x|^2)^{\frac{n}{2}+1}} \leqslant \frac{1}{n(n-2)} \int \frac{df(x)}{(1 + |x|^2)^{\frac{n}{2}-1}}\,, \tag{3.33}$$

(2) *for* $0 < \alpha < n - 2$,

$$\int \frac{|u(x)|\,dx}{(1 + |x|^2)^{\frac{\alpha}{2}+1}} \leqslant \frac{1}{\alpha(n - 2 - \alpha)} \int \frac{df(x)}{(1 + |x|^2)^{\frac{\alpha}{2}}}\,, \tag{3.34}$$

(3) *for all* $t > 0$,

$$\left(\int (|u(x)| - t)^+\,dx\right)^{n-2} \leqslant \frac{C_n}{t^2}\left(\int df(x)\right)^n \tag{3.35}$$

with

$$C_n = \left(\frac{n}{n-2}\right)^n \frac{1}{4^{n-1}\sigma_n^2}\,.$$

Proof. Since u is the limit of Newtonian potentials of $\rho_\varepsilon f$, using the Fatou–Lebesgue lemma we see that we can always suppose that f has compact support.

[58] In each estimate, we have the convention that if the integral on the right is finite, then the integral on the left is finite and we have the inequality.

In regularizing eventually, we can also suppose that f is regular, with the result that u is a classical solution of Poisson's equation

$$\Delta u = f \quad \text{on} \quad \mathbb{R}^n .$$

For $\alpha > 0$, we put $g_\alpha = (1 + |x|^2)^{-\alpha/2}$. We have

$$\begin{aligned} \operatorname{grad} g_\alpha &= -\alpha x g_{\alpha+2} \\ \Delta g_\alpha &= -\alpha g_{\alpha+4}(n + (n - 2 - \alpha)|x|^2) . \end{aligned}$$

Applying Green's formula on the ball $B(0, R)$, we have

$$\int_{B(0,R)} g_\alpha f \mathrm{d}x = \int_{B(0,R)} u \Delta g_\alpha \mathrm{d}x + I_1 - I_2$$

with

$$\begin{aligned} I_1 &= \int_{\partial B(0,R)} g_\alpha \frac{\partial u}{\partial n} \mathrm{d}\gamma \\ I_2 &= \int_{\partial B(0,R)} u \frac{\partial g_\alpha}{\partial n} \mathrm{d}\gamma . \end{aligned}$$

Using the values of g_α, grad g_α and the estimates of u and grad u at infinity, namely

$$(u(x) = O(1/|x|^{n-2}), |\operatorname{grad} u(x)| = O(1/|x|^{n-1})$$

(see Proposition 2) we have

$$|I_1| \leqslant CR^{-\alpha}, \qquad |I_2| \leqslant CR^{-\alpha}$$

where C depends on n, f, α but not on R.
Therefore I_1 and I_2 tend to 0 when $R \to \infty$.
Now $u = E_n * f$ is negative, so if $0 < \alpha \leqslant n - 2$

$$u \Delta g_\alpha = \alpha |u| g_{\alpha+4}(n + (n - 2 - \alpha)|x|^2) \geqslant 0 .$$

In the limit, we deduce

$$\int_{\mathbb{R}^n} g_\alpha f \mathrm{d}x = \alpha \int_{\mathbb{R}^n} |u| g_{\alpha+4}(n + (n - 2 - \alpha)|x|^2) \mathrm{d}x .$$

For $\alpha = n - 2$, we obtain (3.33) (with equality even).
For $0 < \alpha < n - 2$, $(n + (n - 2 - \alpha)|x|^2) \geqslant (n - 2 - \alpha)(1 + |x|^2)$, from which we have (3.34).
To prove (3.35), we use quite another method. First of all, we show that for every compact set K of $\mathbb{R}^n$,

$$(3.36) \qquad \int_K |E_n(x)| \mathrm{d}x \leqslant c_n \left(\frac{|K|}{\sigma_n} \right)^{2/n} \quad \text{with} \quad c_n = \frac{n}{2} \times (n - 2)^{2\left(\frac{1}{n} - 1\right)}$$

where $|K|$ denotes the Lebesgue measure of the compact set K.

In effect, for $R > 0$,

$$\int_K |E_n(x)|\,\mathrm{d}x \leqslant \int_{B(0,R)} |E_n(x)|\,\mathrm{d}x + \int_{\{x\in K;\,|x|\geqslant R\}} |E_n(x)|\,\mathrm{d}x$$

$$\leqslant \frac{\sigma_n}{|k_n|}\frac{R^2}{2} + \frac{|K|}{|k_n|R^{n-2}} \quad (k_n = -(n-2)\sigma_n)\,.$$

This being true for all $R > 0$,

$$\int_K |E_n(x)|\,\mathrm{d}x \leqslant \min_{R>0} \frac{1}{n-2}\left(\frac{R^2}{2} + \frac{|K|}{\sigma_n R^{n-2}}\right).$$

Calculating this minimum, we obtain (3.36)
Now we have $u = E_n * f$, so

$$\int_K |u(x)|\,\mathrm{d}x = \int_K \mathrm{d}x \int |E_n(x-y)|\,f(y)\,\mathrm{d}y = \int f(y)\,\mathrm{d}y \int_{K+y} |E_n(x)|\,\mathrm{d}x$$

$$\leqslant \left(\int f(y)\,\mathrm{d}y\right) c_n \left(\frac{|K|}{\sigma_n}\right)^{2/n}$$

since $|K + y| = |K|$.
In particular, applying this inequality with $K(t) = \{x;\,|u(x)| \geqslant t\}$, we have

$$t|K(t)| \leqslant \int_{K(t)} |u|\,\mathrm{d}x \leqslant \left(\int f(y)\,\mathrm{d}y\right) c_n \left(\frac{|K(t)|}{\sigma_n}\right)^{2/n},$$

from which we derive

$$|K(t)| \leqslant \left(\frac{c_n}{t}\int f(y)\,\mathrm{d}y\right)^{n/n-2} \times \frac{1}{\sigma_n^{2/n-2}}\,.$$

Finally, using the fact that

$$\frac{\mathrm{d}}{\mathrm{d}t}\int (|u| - t)^+\,\mathrm{d}x = -\,|K(t)|$$

we obtain

$$\int (|u| - t)^+\,\mathrm{d}x = \int_t^{+\infty} |K(s)|\,\mathrm{d}s \leqslant \left(c_n \int f(y)\,\mathrm{d}y\right)^{n/(n-2)} \times \frac{n-2}{2}\,\frac{1}{(\sigma_n t)^{2/(n-2)}}\,.$$

□

4b. Other Examples of Newtonian Potentials of Distributions Without Compact Support

In the preceding §4.a., we have characterised the positive measures, regular at infinity. This gives a criterion for an arbitrary measure (see Corollary 2), but *there exist measures (or functions), regular at infinity, without their absolute value being so*: this corresponds to the case of integrals or series which are semi-convergent (*i.e.* convergent, but not absolutely convergent).

We illustrate this by considering the case of a function f integrable on $\mathbb{R}$. We define its primitive:

$$F(x) = \int_{-\infty}^{x} f(y)\,dy\,.$$

Given $\rho \in \mathscr{D}(\mathbb{R})$ with $\rho = 1$ in the neighbourhood of 0, the Newtonian potential of $\rho_\varepsilon f$ is given by

$$u_\varepsilon(x) = \frac{1}{2}\int |x - y|\rho(\varepsilon y) f(y)\,dy\,.$$

Integrating by parts we may this as

$$u_\varepsilon(x) = \frac{1}{2}\int_{-\infty}^{x} \rho(\varepsilon y)F(y)\,dy - \frac{1}{2}\int_{x}^{\infty} \rho(\varepsilon y)F(y)\,dy - \frac{\varepsilon}{2}\int |x - y|\rho'(\varepsilon y)F(y)\,dy\,.$$

The convergence of u_ε *as* $\varepsilon \to 0$ *is assured by the integrability of* F *on* $\mathbb{R}$. In effect

$$|\varepsilon|x - y|\rho'(\varepsilon y)| \leqslant \varepsilon|x|\,\|\rho'\|_{L^\infty} + \|y\rho'(y)\|_{L^\infty}$$

with the result that then

$$\varepsilon \int |x - y|\rho'(\varepsilon y)F(y)\,dy \to 0 \quad \text{uniformly for } x \text{ bounded}$$

This condition of integrability of a primitive of f is different from the condition $\int |x||f(x)|\,dx < \infty$ which characterizes the regularity at infinity of the absolute value of f: for example

$$f(x) = \frac{\sin x^2}{x^2}$$

is integrable on $\mathbb{R}$; its absolute value is not regular at infinity[59], but f can be written in the form

$$f(x) = f_0(x) + \frac{d}{dx}F(x)$$

where f_0 is of absolute value regular at infinity and F is integrable[60]; it is therefore regular at infinity.
In a general manner, we prove in this direction the

---

[59] $\displaystyle\int\left|\frac{\sin x^2}{x}\right| dx = \int_0^\infty \frac{|\sin x|}{x}\,dx = \sum_0^\infty \int_0^\pi \frac{\sin x}{x + k\pi}\,dx = +\infty\,.$

[60] Choosing $\rho \in \mathscr{D}(\mathbb{R})$ with $\rho = 1$ in the neighbourhood of 0 (to eliminate the singularity at 0) take

$$F(x) = \frac{1}{2}(\rho(x) - 1)\frac{\cos x^2}{x^3}\,,$$

$$f_0(x) = -\rho(x)\frac{\sin x^2}{x^2} + \frac{3}{2}(1 - \rho(x))\frac{\cos x^2}{x^4} - \frac{1}{2}\rho'(x)\frac{\cos x^2}{x^3}\,.$$

Proposition 18. *Let f be a distribution on $\mathbb{R}^n$ and let us suppose that it can be written*

$$f = \sum_{|\alpha| \leqslant m} \frac{\partial^\alpha f_\alpha}{\partial x^\alpha}$$

where (f_α) is a family of measures on $\mathbb{R}^n$ satisfying:

$$(3.37)\quad \begin{cases} \text{for } n \geqslant 2 \text{ and } n + |\alpha| \geqslant 3, \text{ the measure } f_\alpha/(1 + |x|)^{n + |\alpha| - 2} \text{ is} \\ \text{bounded on } \mathbb{R}^n \\ \text{for } n = 2 \text{ and } \alpha = 0, \text{ the measure } \operatorname{Log}(1 + |x|) f_0 \\ \text{is bounded on } \mathbb{R}^n \\ \text{for } n = 1, \text{ the measures } x f_0 \text{ and } f_\alpha \ (\alpha \geqslant 1) \text{ are bounded on } \mathbb{R}^n \end{cases}$$

Then f is regular at infinity.

Proof. We fix $\rho, \zeta \in \mathscr{D}(\mathbb{R}^n)$ and consider

$$\begin{aligned} I_\varepsilon = \langle f, \rho_\varepsilon(E_n * \zeta)\rangle &= \sum_{|\alpha| \leqslant m} (-1)^{|\alpha|} \left\langle f_\alpha, \frac{\partial^\alpha}{\partial x^\alpha}(\rho_\varepsilon(E_n * \zeta)) \right\rangle \\ &= \sum_{|\alpha| \leqslant m} (-1)^{|\alpha|} \sum_{\beta \leqslant \alpha} C_\beta^\alpha \left\langle f_\alpha, \frac{\partial^\beta \rho_\varepsilon}{\partial x^\beta} \frac{\partial^{\alpha - \beta}}{\partial x^{\alpha - \beta}}(E_n * \zeta) \right\rangle^{61} \\ &= \sum_{|\alpha| \leqslant m} (-1)^{|\alpha|} \sum_{\beta \leqslant \alpha} C_\beta^\alpha \left\langle g_\alpha, \frac{\partial^\beta \rho_\varepsilon}{\partial x^\beta} v_{\alpha,\beta} \right\rangle, \end{aligned}$$

with

$$g_\alpha = \begin{cases} \dfrac{f_\alpha}{(1 + |x|)^{n + |\alpha| - 2}} & \text{for} \quad n \geqslant 2, \quad n + |\alpha| \geqslant 3 \\ (1 + \operatorname{Log}(1 + |x|)) f_0 & \text{for} \quad n = 2, \quad \alpha = 0 \end{cases}$$

$$v_{\alpha,\beta} = \begin{cases} (1 + |x|)^{n + |\alpha| - 2} \dfrac{\partial^{\alpha - \beta}}{\partial x^{\alpha - \beta}}(E_n * \zeta) & \text{for} \quad n \geqslant 2, \quad n + |\alpha| \geqslant 3 \\ \dfrac{E_2 * \zeta}{1 + \operatorname{Log}(1 + |x|)} & \text{for} \quad n = 2, \quad \alpha = 0 \end{cases}$$

(we shall consider the case $n = 1$ separately).
Using Proposition 2, for all $\alpha \geqslant \beta \geqslant 0$

$$v_{\alpha,\beta}(x) = O(|x|^{|\beta|}) \quad \text{when} \quad |x| \to \infty\,.$$

Now

$$\frac{\partial^\beta \rho_\varepsilon}{\partial x^\beta}(x) = \varepsilon^{|\beta|} \frac{\partial^\beta \rho}{\partial x^\beta}(\varepsilon x)\,.$$

---

[61] We use Leibnitz's formula: for $\alpha = (\alpha_1, \ldots, \alpha_n)$, $\beta = (\beta_1, \ldots, \beta_n)$, $\beta \leqslant \alpha$ means $\beta_i \leqslant \alpha_i$ for $i = 1, \ldots, n$; $\alpha - \beta = (\alpha_1 - \beta_1, \ldots, \alpha_n - \beta_n)$; $C_\beta^\alpha = \prod_i \dfrac{\alpha_i!}{\beta_i!(\alpha_i - \beta_i)!}$.

Hence $\dfrac{\partial^\beta \rho_\varepsilon}{\partial x^\beta} v_{\alpha,\beta}$ is bounded on $\mathbb{R}^n$, uniformly for $\varepsilon > 0$ and when $\varepsilon \to 0$.

$$\frac{\partial^\beta \rho_\varepsilon}{\partial x^\beta} v_{\alpha,\beta} \to \begin{cases} \rho(0) v_{\alpha,0} & \text{if} \quad \beta = 0 \\ 0 & \text{if} \quad |\beta| > 0 \,. \end{cases}$$

Since, by hypothesis, the measures g_α are bounded we deduce that

$$I_\varepsilon \to \rho(0) \sum_{|\alpha| \leqslant m} (-1)^{|\alpha|} \langle g_\alpha, v_{\alpha,0} \rangle \quad \text{when} \quad \varepsilon \to 0 \,.$$

This shows that f is regular at infinity; its Newtonian potential is defined by

$$u = \sum \frac{\partial^\alpha}{\partial x^\alpha} u_\alpha$$

where u_α is the locally integrable function on $\mathbb{R}^n$ defined by

$$u_\alpha(x) = \int \frac{\partial^\alpha}{\partial x^\alpha} E_n(x - y) \, \mathrm{d}f_\alpha(x) \,.$$

The case $n = 1$ is treated in the same way, noting that in this case, when $|x| \to \infty$

$$E_1 * \zeta(x) = O(|x|), \quad \frac{\mathrm{d}}{\mathrm{d}x}(E_1 * \zeta) = O(1) \,,$$

$$\frac{\mathrm{d}^k}{\mathrm{d}x^k}(E_1 * \zeta) = 0 \quad \text{for} \quad k \geqslant 2 \,. \qquad \square$$

This proposition can be applied to the study of the regularity at infinity of functions as we have done above in the case $n = 1$. It can also be used for distributions. We now give *the example of the distribution of order* 1 *defining a double layer potential*:

Example 4. Let Ω be a regular open set of $\mathbb{R}^n$, but with unbounded boundary Γ and $\varphi \in \mathscr{C}^0(\Gamma)$. For the distribution of order $1, f = -\operatorname{div}(\varphi n \mathrm{d}\gamma)$[62], the condition (3.37) can be written

$$\int_\Gamma \frac{|\varphi(x)| \, \mathrm{d}\gamma(x)}{(1 + |x|)^{n-1}} < \infty \,. \tag{3.38}$$

The distribution is then regular at infinity; its Newtonian potential is the double layer potential u_2, harmonic on $\mathbb{R}^n \backslash \Gamma$ and defined by

$$u_2(x) = \int_\Gamma \frac{\partial E_n}{\partial n}(t - x) \varphi(t) \, \mathrm{d}\gamma(t) \quad \text{for all} \quad x \in \mathbb{R}^n \backslash \Gamma \,.$$

If we suppose that Ω is uniformly regular (see Example 3), the condition (3.37) is

---

[62] Which defines the double layer potential of φ (see Definition 2).

satisfied, in particular, if

$$|\varphi(x)| \leqslant \frac{C}{|x|^{\varepsilon}} \quad \text{for} \quad |x| > r_0, x \in \Gamma \quad \text{with} \quad \varepsilon > 0 . \qquad \square$$

As we remarked at the beginning of this sect. 4 of §3, we can consider "natural" potentials for distributions, non-regular at infinity.

We indicate here the *method of subtraction of an "infinite constant"*. To present this method we consider *the example of a measure* $f = 1_{\mathbb{R}^{n-2}} \otimes f_2$ $(n \geqslant 3)$, where f_2 is a measure with compact support in $\mathbb{R}^2$. As we have seen (Example 2) if f_2 is positive, this measure is not regular at infinity.

We now prove the

Proposition 19. *Let f_2 be a measure with compact support on $\mathbb{R}^2$, u_2 the Newtonian potential of f_2, ρ a Borel function with compact support in $\mathbb{R}^{n-2} (n \geqslant 3)$ with $\rho = 1$ in the neighbourhood of 0. For all $\varepsilon > 0$, consider u^{ε} the Newtonian potential of the measure $\rho_{\varepsilon} \otimes f_2$ with compact support in $\mathbb{R}^n$, $\rho_{\varepsilon}(x) = \rho(\varepsilon x)$. Then setting*

$$C_{\varepsilon} = \frac{1}{k_n}\left(\int_{\mathbb{R}^2} \mathrm{d}f_2(y'')\right)\left(\int_{\mathbb{R}^{n-2}} \frac{\rho(\varepsilon y')}{(1 + |y'|^2)^{\frac{n}{2}-1}} \mathrm{d}y'\right),$$

we have for all $x' \in \mathbb{R}^{n-2}$ and almost all $x'' \in \mathbb{R}^2$

$$u^{\varepsilon}(x', x'') - C_{\varepsilon} \to u_2(x'') \quad \textit{when} \quad \varepsilon \to 0 .$$

More precisely, for all $R > 0$,

$$\lim_{\varepsilon \to 0} \int_{\{|x''| \leqslant R\}} \left(\sup_{\{|x'| \leqslant R\}} |u^{\varepsilon}(x', x'') - C_{\varepsilon} - u_2(x'')| \right) \mathrm{d}x'' = 0 .$$

Proof. We have

$$u^{\varepsilon}(x', x'') = \int_{\mathbb{R}^{n-2}} \rho(\varepsilon y')\,\mathrm{d}y' \int_{\mathbb{R}^2} E_n(x' - y', x'' - y'')\,\mathrm{d}f_2(y'') .$$

Making use of

$$E_n(x', x'') = -\frac{1}{\sigma_n} \int_{|x''|}^{\infty} \frac{t\,\mathrm{d}t}{(t^2 + |x'|^2)^{n/2}}$$

with the result that

$$u^{\varepsilon}(x', x'') = -\frac{1}{\sigma_n} \int_{\mathbb{R}^2} \mathrm{d}f_2(y'') \int_{|x'' - y''|}^{\infty} t\,\mathrm{d}t \int_{\mathbb{R}^{n-2}} \frac{\rho(\varepsilon y')}{(t^2 + |x' - y'|^2)^{n/2}} \mathrm{d}y' ,$$

and letting $\varepsilon \to 0$, we obtain

$$\frac{1}{\sigma_n} \int_{\mathbb{R}^{n-2}} \frac{\rho(\varepsilon y')}{(t^2 + |x' - y'|^2)^{n/2}} \mathrm{d}y' \to \frac{1}{\sigma_n} \int_{\mathbb{R}^{n-2}} \frac{\mathrm{d}y'}{(t^2 + |x' - y'|^2)^{n/2}}$$

$$= \frac{\sigma_{n-2}}{\sigma_n t^2} \int_0^{\infty} \frac{r^{n-3}\,\mathrm{d}r}{(1 + r^2)^{n/2}} = \frac{1}{2\pi t^2} ,$$

the convergence being uniform with respect to x' bounded in $\mathbb{R}^{n-2}$ and dominated with respect to t. Hence, since f_2 has compact support, we find when $\varepsilon \to 0$

$$\frac{1}{\sigma_n}\int_{\mathbb{R}^2} \mathrm{d}f_2(y'')\int_1^{|x''-y''|} t\,\mathrm{d}t \int_{\mathbb{R}^{n-2}} \frac{\rho(\varepsilon y')}{(t^2+|x'-y'|^2)^{n/2}}\mathrm{d}y'$$

$$\to \frac{1}{2\pi}\int_{\mathbb{R}^2} \mathrm{d}f_2(y'')\int_1^{|x''-y''|}\frac{\mathrm{d}t}{t} = \int_{\mathbb{R}^2} E_2(x''-y'')\mathrm{d}f_2(y'') = u_2(x''),$$

the convergence being uniform with respect to x' bounded in $\mathbb{R}^{n-2}$ and dominated with respect to $x'' \in \mathbb{R}^2$.
It is sufficient therefore to prove that

$$I_\varepsilon = \frac{1}{\sigma_n}\int_{\mathbb{R}^2} \mathrm{d}f_2(y'')\int_1^{\infty} t\,\mathrm{d}t \int_{\mathbb{R}^{n-2}} \frac{\rho(\varepsilon y')}{(t^2+|x'-y'|^2)^{n/2}}\mathrm{d}y' + C_\varepsilon \to 0$$

when $\varepsilon \to 0$, uniformly for x' bounded in $\mathbb{R}^{n-2}$.
By integration over t in I_ε and a change of variable, we obtain

$$I_\varepsilon = \left(-\frac{1}{k_n}\int_{\mathbb{R}^2} \mathrm{d}f_2(y'')\right)\left(\int_{\mathbb{R}^{n-2}} \frac{\rho(y'+\varepsilon x')-\rho(y')}{(\varepsilon^2+|y'|^2)^{(n/2)-1}}\mathrm{d}y'\right).$$

Using the hypothesis $\rho = 1$ in the neighbourhood of 0 and the continuity of the translation in L^1, we obtain the limit $I_\varepsilon \to 0$. □

Proposition 19 shows on the one hand that for the measure f_2, with compact support in $\mathbb{R}^2$ *the distribution* $f = 1_{\mathbb{R}^{n-2}} \otimes f_2$ *in* $\mathbb{R}^n$ $(n \geqslant 3)$ *is regular at infinity iff* $\int_{\mathbb{R}^2} \mathrm{d}f_2(x'') = 0$. On the other hand, when $\int_{\mathbb{R}^2} \mathrm{d}f_2(x'') \neq 0$, with the condition that when the constant C_ε (tending to infinity) is subtracted, the Newtonian potentials of $\rho_\varepsilon \otimes f_2$ converge when $\varepsilon \to 0$.
We shall show in the same way that for f_1 a measure with compact support in $\mathbb{R}$ and a bounded Borel function ρ with compact support in $\mathbb{R}^{n-1}$ $(n \geqslant 3)$ with $\rho = 1$ in the neighbourhood of 0, there exists C_ε such that

$$E_n * (\rho_\varepsilon \otimes f_1) + C_\varepsilon \to 1_{\mathbb{R}^{n-1}} \otimes (E_1 * f_1) \quad \text{when} \quad \varepsilon \to 0.$$

A calculation identical with that above will show that we can take

$$C_\varepsilon = \left(\int_{\mathbb{R}} \mathrm{d}f_1(x'')\right)\left(\int_{\mathbb{R}^{n-1}} \rho(\varepsilon x')E_n(x')\mathrm{d}x'\right),$$

and hence that $1_{\mathbb{R}^{n-1}} \otimes f_1$ is regular at infinity in $\mathbb{R}^n$ iff $\int_{\mathbb{R}} \mathrm{d}f_1(x'') = 0$.

We can show that, in the two preceding examples, the distributions f, not regular at infinity in $\mathbb{R}^n$ satisfy:
for all $\rho \in \mathscr{D}(\mathbb{R}^n)$, *there exists* C_ε, *tending to infinity when* $\varepsilon \to 0$, *such that* $E_n * (\rho_\varepsilon f) + C_\varepsilon$ *converges.*
The limit has to be independent of ρ (when the "infinite" constants C_ε depend on ρ); we call this limit the Newtonian potential of f. We emphasise however that *in this method the Newtonian potential is defined to within an additive constant* (we can

always add a "finite" constant to C_ε which will always be "infinite"). From a mathematical point of view, that has an important consequence: if the sum of two distributions is regular at infinity, the sum of their Newtonian potentials is not necessarily the Newtonian potential of their sum; it is only it to within an additive constant.

The method of subtraction of an infinite constant does not permit the definition of a Newtonian potential for any distribution whatsoever: for example, the function $f \equiv 1$ (which is not regular at infinity), does not admit a Newtonian potential by the method of subtraction of an infinite constant. That can be seen easily by remarking that for $\rho \in \mathscr{D}(\mathbb{R}^n)$,

$$\operatorname{grad}(\rho_\varepsilon * E_n)(x) = \rho_\varepsilon * \operatorname{grad} E_n(x) = -\frac{1}{\varepsilon}\int \operatorname{grad}(y + \varepsilon x)E_n(y)\,dy$$

diverges when $\varepsilon \to 0$. Now, if $f \in \mathscr{D}'(\mathbb{R}^n)$ admits a Newtonian potential to within an additive constant, then

$$\operatorname{grad}[(\rho_\varepsilon f) * E_n] \quad \text{converges when} \quad \varepsilon \to 0\,.$$

This convergence is *a priori* in $\mathscr{D}'(\mathbb{R}^n)$; we can show, in fact, in the examples, that it is stronger. For example under the hypotheses of Proposition 19, we have

$$\lim_{\varepsilon \to 0}\int_{\{|x''| < R\}}\left(\sup_{\{|x'| \leqslant R\}} |\operatorname{grad}_{x'} u^\varepsilon(x', x'')| + |\operatorname{grad}_{x''} u^\varepsilon(x', x'') - \operatorname{grad} u_2(x'')|\right)dx'' = 0.$$

5. Some Physical Interpretations (in Mechanics and Electrostatics)

Newtonian Potentials. In mechanics, let ρ be the density of matter distributed in a bounded domain of $\mathbb{R}^3$.

Let us assume that the gravitational force $\mathbf{f}$ which acts then on a point mass m placed at point x is the sum of the forces obeying Newton's law due to each "element of matter" $\rho(y)\,dy$. With the aid of Newton's law (see §2.6) we thus obtain:

$$\mathbf{f}(\mathbf{x}) = -\int_{\mathbb{R}^3} mk_g \operatorname{grad} E_3(\mathbf{x} - \mathbf{y})\rho(\mathbf{y})\,d\mathbf{y}\,.$$

This force is derived from a potential v_g given by:

$$v_g(\mathbf{x}) = k_g \int_{\mathbb{R}^3} E_3(\mathbf{x} - \mathbf{y})\rho(\mathbf{y})\,d\mathbf{y}$$

which is thus (to within the coefficient k_g) the Newtonian potential of the density of matter ρ, and which is therefore a solution of Poisson's equation

$$\Delta v_g = k_g \rho\,.$$

We stress the fact that we have thus chosen a particular solution of Poisson's equation, namely the one which tends to zero at infinity[63].

[63] The possibility of this choice means to a physicist the absence of mass at infinity.

We can act similarly in electricity, for a charge density ρ: $\mathbb{R}^3 \to \mathbb{R}$ with the Coulomb force. We again obtain, in a perfect homogeneous medium (or in the vacuum), a force $\mathbf{f}$ deriving from a potential

$$v_c(\mathbf{x}) = -k_c \int_{\mathbb{R}^3} E_3(\mathbf{x} - \mathbf{y})\rho(\mathbf{y})\,\mathrm{d}\mathbf{y} .$$

This potential is called *the Coulomb potential.*
Examples of the Coulomb potential are given by the simple and double layer potentials.
(i) A density of electric charge β distributed over a regular bounded surface Γ (for example the surface of a conductor) creates a simple layer (Coulomb) potential in the vacuum, this potential being given at a point $\mathbf{x} \in \mathbb{R}^3$ by:

$$v_c(x) = \frac{k_c}{4\pi}\int_\Gamma \frac{\beta(t)}{|t - x|}\mathrm{d}\gamma(t) = -k_c \int E_3(x - t)\beta(t)\,\mathrm{d}\gamma(t)^{64} .$$

The electric field $\mathbf{E}$ being related to the Coulomb potential v_c by: $\mathbf{E} = -\operatorname{grad} v_c$ the jump in the normal derivative of v_c across Γ (see Proposition 1) means a discontinuity in the normal component of the electric field across Γ, proportional to the charge density β:

$$[\mathbf{E}.\mathbf{n}]_\Gamma = \mathbf{E}^e.\mathbf{n} - \mathbf{E}^i.\mathbf{n} = -\left[\frac{\partial v_c}{\partial n}\right] = k_c\beta . \qquad (\text{for } \beta \in \mathscr{C}^0(\Gamma)) .$$

On the contrary, there is no jump in the components, tangential to Γ of the electric field (see Remark 9).
(ii) A density of electric moment α (see Jackson [1] p. 36, Jouget [1], p. 169) on a surface Γ, creates in the vacuum, at a point $\mathbf{x} \in \mathbb{R}^3$, a double layer (Coulomb) potential $v_c(\mathbf{x})$ given by:

$$\begin{aligned} v_c(\mathbf{x}) &= -\frac{k_c}{4\pi}\int_\Gamma \alpha(t)\frac{\cos(\mathbf{t} - \mathbf{x}, \mathbf{n}(\mathbf{t}))}{|\mathbf{t} - \mathbf{x}|^2}\mathrm{d}\gamma(\mathbf{t}) \\ &= -k_c \int_\Gamma \frac{\partial}{\partial n}E_3(\mathbf{t} - \mathbf{x})\boldsymbol{\alpha}(\mathbf{t})\,\mathrm{d}\gamma(\mathbf{t}) \end{aligned}$$

By Proposition 11, the potential is discontinuous across Γ:

$$[v_c]_\Gamma = (v_c^e - v_c^i)|_\Gamma = +k_c\alpha \qquad (\text{for } \alpha \in \mathscr{C}^0(\Gamma)) ,$$

implying a discontinuity of the tangential components of the electric field $\mathbf{E}$ (for $\alpha \in \mathscr{C}^{2+\varepsilon}(\Gamma)$), and then that the normal component of $\mathbf{E}$ is continuous across Γ.
Examples of simple and double layer potentials can also be found in magnetostatics, and in fluid mechanics (see Brard [1]).

Problems with Cylindrical (or Plane) Geometry. Especially in electricity, we are led to study distributions of charge which are cylindrical or plane (for example an

[64] Do not confuse E_3 elementary solution in $\mathbb{R}^3$ of the Laplacian with the third component of the electric field.

electric wire or the plates of a plane condenser); it is important to recognise the new difficulties due to the existence of charges at infinity when we take, in first approximation, the cylinder or the plates considered to be infinite. This is given by Proposition 19 (with $n = 3$).
The direct study of the case of a distribution of electric charges ρ constant on a rectilinear wire AB, infinitesimally thin, of length L is interesting. The Coulomb ("Newtonian") potential created at a point $\mathbf{x} \in \mathbb{R}^n$ is given by:

$$v_c(\mathbf{x}) = \frac{k_c}{4\pi} \rho \operatorname{Log} \frac{r_A + r_B + l}{r_A + r_B - l}$$

with

$$r_A = |\mathbf{x} - \mathbf{x}_A|, \qquad r_B = |\mathbf{x} - \mathbf{x}_B| .$$

Denoting by r the distance of the point $\mathbf{x}$ from the x_3-axis which carries the wire AB. For $l \to \infty$, the (Coulomb) potential $v_c(\mathbf{x})$ has an asymptotic form v_{as}^c given by

$$v_{as}^c(x) = \frac{k_c}{2\pi} \rho \log \frac{l}{r} = \frac{k_c}{2\pi} \rho \log l - \frac{k_c}{2\pi} \rho \log r$$[65]

which thus appears (to within the coefficient $-k_c \rho$) as the elementary solution $E_2(x)$ in $\mathbb{R}^2$.
Similarly a uniform distribution of charge in a plane leads to the occurrence of the elementary solution $E_1(x)$ in $\mathbb{R}$, as a consequence of Proposition 19.

§4. Classical Theory of Dirichlet's Problem

1. Generalities on Dirichlet's Problem $P(\Omega, \varphi)$ in the case Ω Bounded: Classical Solution, Examples, Outline of Perron's Method, Generalized Solutions, Regular Point of the Boundary, Barrier Function

We consider an open set Ω in $\mathbb{R}^n$ with boundary Γ and φ defined on Γ. We pose the problem of finding u satisfying

$$P(\Omega, \varphi) \begin{cases} u \text{ is harmonic on } \Omega \\ u = \varphi \text{ on } \Gamma \end{cases}$$

This is *Dirichlet's problem for Laplace's equation.* Within the classical framework of this §4, *we shall suppose that* $\varphi \in \mathscr{C}^0(\Gamma)$; the case of a less regular given function φ will be considered in §6. We pose the

[65] In the third member of this equation, the two numbers l and r are here considered to be dimensionless.

Definition 1. Given an open set Ω of $\mathbb{R}^n$ with boundary Γ and $\varphi \in \mathscr{C}^0(\Gamma)$; we call a function u a *classical solution of* $P(\Omega, \varphi)$ if u is harmonic on Ω and is such that

$$\lim_{\substack{x \to z \\ x \in \Omega}} u(x) = \varphi(z) \quad \text{for all} \quad z \in \Gamma .$$

In other words u is the *classical solution of* $P(\Omega, \varphi)$ if $u \in \mathscr{H}(\Omega) \cap \mathscr{C}^0(\bar{\Omega})$ and φ *is the trace of* u *on* Γ.

Now let us suppose that Ω *is bounded.* We know then (see Corollary 3 of §2) that for all $u \subset \mathscr{H}(\Omega) \cap \mathscr{C}^0(\bar{\Omega})$,

$$\sup_{\Omega} |u| = \max_{\Gamma} |u| ;$$

in other words, the (linear) map which takes u to its trace φ on Γ is an isometry of $\mathscr{H}(\Omega) \cap \mathscr{C}^0(\bar{\Omega})$ with the uniform convergence norm on Ω, that is to say

$$\|u\| = \sup_{\Omega} |u|$$

into $\mathscr{C}^0(\Gamma)$ given the uniform convergence norm on Γ:

$$\|\varphi\| = \max_{\Gamma} |\varphi| .$$

We have likewise seen (Proposition 10 of §2) that every uniform limit of harmonic functions on Ω was harmonic on Ω. It follows that the set of traces φ on Γ of the $u \in \mathscr{H}(\Omega) \cap \mathscr{C}^0(\bar{\Omega})$ is a closed vector sub-space of $\mathscr{C}^0(\Gamma)$. From the definition of a classical solution of $P(\Omega, \varphi)$ we have the

Proposition 1. *Let* Ω *be a bounded open set of* $\mathbb{R}^n$ *with boundary* Γ.
(1) *The set* $E(\Omega)$ *of the* $\varphi \in \mathscr{C}^0(\Gamma)$ *such that the Dirichlet problem* $P(\Omega, \varphi)$ *admits a classical solution is a closed vector sub-space of* $\mathscr{C}^0(\Gamma)$;
(2) *for all* $\varphi \in E(\Omega)$, *there exists one and only one classical solution* $u(\varphi)$ *of* $P(\Omega, \varphi)$; *the map* $\varphi \to u(\varphi)$ *is a (linear) isometry of* $E(\varphi)$, *with norm the uniform convergence norm on* Γ, *onto* $\mathscr{H}(\Omega) \cap \mathscr{C}^0(\bar{\Omega})$, *with norm the uniform convergence norm on* Ω.

From the principle of the maximum, we have for $\varphi \in E(\Omega)$

$$\min_{\Gamma} \varphi \leqslant u(\varphi) \leqslant \max_{\Gamma} \varphi \quad \text{on} \quad \Omega \tag{4.1}$$

and even more precisely if Ω_0 is a connected component of Ω with boundary Γ_0

$$\begin{cases} \varphi \text{ constant on } \Gamma_0 \Rightarrow u(\varphi) \text{ constant on } \Omega_0 \\ \varphi \text{ not constant on } \Gamma_0 \Rightarrow \min_{\Gamma_0} \varphi < u(\varphi) < \max_{\Gamma_0} \varphi \text{ sur } \Omega_0 . \end{cases} \tag{4.2}$$

Let us now give some examples.

Example 1. $\Omega =]a, b[$ *a bounded interval of* $\mathbb{R}$.
The boundary of Ω is $\Gamma = \{a, b\}$. The functions of $\mathscr{H}(\Omega)$ are the affine functions $u(x) = \lambda x + \mu$. It is immediate that the solution of $P(\Omega, \varphi)$ is

$$u(x) = \frac{\varphi(a)(b - x) + \varphi(b)(x - a)}{b - a} . \tag{4.3}$$

□

Example 2. $\Omega = B(x_0, r_0)$ *a ball in* $\mathbb{R}^n$.
Let $B = B(x_0, r_0)$ and $\varphi \in \mathscr{C}^0(\partial B)$. Poisson's integral formula[66] gives us an indication that the classical solution of $P(B, \varphi)$, if it exists, is given by

$$u(x) = \frac{1}{r_0 \sigma_n} \int_{\partial B} \frac{r_0^2 - |\mathbf{x} - \mathbf{x}_0|^2}{|\mathbf{t} - \mathbf{x}|^n} \varphi(\mathbf{t}) \, \mathrm{d}\gamma(\mathbf{t}) \,. \tag{4.4}$$

Let us now prove the

Proposition 2. *Being given* $B = B(x_0, r_0)$ *and* $\varphi \in \mathscr{C}^0(\partial B)$, *the function* u *defined by* (4.4) *is the classical solution of* $P(B, \varphi)$.

Proof. For all $t \in \partial B$, the function

$$x \to \frac{1}{\sigma_n} \frac{r_0^2 - |x - x_0|^2}{|t - x|^n}$$

is harmonic on B: this in effect is the function

$$\operatorname{grad} E_n(t - x) . (t + x - 2x_0)$$

a product each term of which is harmonic on B, hence of the Laplacian

$$2 \sum_i \left(\frac{\partial}{\partial x_i} \operatorname{grad} E_n(t - x) \right) . \frac{\partial}{\partial x_i} (t + x - 2x_0) = 2 \Delta E_n(t - x) = 0 \,.$$

Now from the proof of Poisson's formula[66], we have, for $x \in B$,

$$\frac{1}{\sigma_n} \frac{r_0^2 - |x - x_0|^2}{|t - x|^n} = \frac{\partial}{\partial n} \left(E_n(t - x) - \left(\frac{r_0}{|x - x_0|} \right)^{n-2} E_n(t - y) \right)$$

where

$$y = x_0 + \frac{r_0^2}{|x - x_0|^2} (x - x_0) \,.$$

We therefore see that

$$u(x) = u_2(x) - \left(\frac{r_0}{|x - x_0|} \right)^{n-2} u_2(y) \,,$$

where u_2 is the double layer potential of φ on ∂B (see Definition 2 and above all Proposition 11 of §3). Given $z \in \partial B$, we know that

$$\lim_{\substack{x \to z \\ x \in B}} u_2(x) = \int_{\partial B} \frac{\partial}{\partial n} E_n(t - z) \varphi(t) \, \mathrm{d}\gamma(t) + \frac{\varphi(z)}{2}$$

$$\lim_{\substack{y \to z \\ y \in \mathbb{R}^n \setminus B}} u_2(y) = \int_{\partial B} \frac{\partial}{\partial n} E_n(t - z) \varphi(t) \, \mathrm{d}\gamma(t) - \frac{\varphi(z)}{2} \,.$$

---

[66] See Proposition 9 of §2.

Hence

$$\lim_{\substack{x \to z \\ x \in B}} u(x) = \left(1 - \left(\frac{r_0}{|z - x_0|}\right)^{n-2}\right) \int_{\partial B} \frac{\partial}{\partial n} E_n(t - z)\varphi(t)\,d\gamma(t)$$
$$+ \left(1 + \left(\frac{r_0}{|z - x_0|}\right)^{n-2}\right) \frac{\varphi(z)}{2} = \varphi(z) . \qquad \square$$

Example 3. $\Omega = B(x_0, r_0)\backslash\{x_0\}$ *punctured ball*[67] *of* $\mathbb{R}^n$, $(n \geqslant 2)$.
Let $B = B(x_0, r_0)$ a ball of $\mathbb{R}^n$ and $\Omega = B\backslash\{x_0\}$ whose boundary is $\Gamma = \partial B \cup \{x_0\}$. Being given $\varphi \in E(\Omega)$, the solution $u(\varphi) \in \mathscr{H}(\Omega)$ and

$$\lim_{x \to x_0} u(\varphi, x) = \varphi(x_0)^{68}$$

We know (see Proposition 16 of §2) that the function u defined on B by

$$u(x) = \begin{cases} u(\varphi, x) & \text{for} \quad x \in \Omega \\ \varphi(x_0) & \text{for} \quad x = x_0 \end{cases}$$

is harmonic on B and hence the solution of $P(B, \varphi_0)$ where φ_0 is the restriction of φ to ∂B. From the formula of the mean:

$$\varphi(x_0) = \frac{1}{\sigma_n r_0^{n-1}} \int_{\partial B} \varphi(t)\,d\gamma(t) \tag{4.5}$$

and from Poisson's integral formula

$$u(\varphi, x) = \frac{1}{r_0 \sigma_n} \int_{\partial B} \frac{r_0^2 - |x - x_0|^2}{|t - x|^n} \varphi(t)\,d\gamma(t) \quad \text{for} \quad x \in \Omega . \tag{4.6}$$

The converse can be deduced from Proposition 2, *in the case of a punctured ball* $\Omega = B(x_0, r_0)\backslash\{x_0\}$ *of* $\mathbb{R}^n$ *with* $n \geqslant 2$, $E(\Omega) = \{\varphi \in \mathscr{C}^0(\partial\Omega)$ *satisfying* (4.5)$\}$ *and for all* $\varphi \in E(\varphi)$, $u(\varphi)$ *is given by* (4.6). $\square$

Example 4. *A Dirichlet problem in an annulus*

$$\Omega = \{x; r_1 < |x| < r_2\} .$$

Let $\Omega = \{x; r_1 < |x| < r_2\}$ with $0 < r_1 < r_2 < \infty$ a ring in $\mathbb{R}^n$ whose boundary is $\Gamma = \partial B(0, r_1) \cup \partial B(0, r_2)$.
Also let φ be *a radial function on* Γ, that is to say

$$\varphi \equiv \varphi(r_i) \quad \text{on} \quad \partial B(0, r_i) \quad (i = 1, 2) .$$

We look for a radial solution of $P(\Omega, \varphi)$: if it exists, then from uniqueness this will be the solution $u(\varphi)$. A radial function u, which is harmonic on Ω, is of the form (see Proposition 4 of §2)

$$u(x) = c_0 + c_1 E_n(x) \quad \text{on} \quad \Omega .$$

---

[67] Punctured = deprived of a point (the point x_0)

[68] We use here (and in the sequel) the notation:

$$u(\varphi, x) \overset{\text{def}}{=} u(\varphi)(x) .$$

The boundary condition can be written

$$\varphi(r_i) = c_0 + c_1 E_n(r_i) \quad (i = 1, 2)\,,$$

which determines the constants and gives

$$u(x) = \frac{\varphi(r_2)(E_n(x) - E_n(r_1)) + \varphi(r_1)(E_n(r_2) - E_n(x))}{E_n(r_2) - E_n(r_1)}\,. \tag{4.7}$$

Note that the problem for a non-radial function will be very much more complicated. □

Example 5. *A Dirichlet problem for a rectangle in* $\mathbb{R}^2$.
We consider a rectangle $\Omega =]a_1, a_2[\times]b_1, b_2[$ of $\mathbb{R}^2 = \{(x, y)\}$ with boundary $\Gamma = \Gamma_1 \cup \Gamma_2 \cup \Gamma_3 \cup \Gamma_4$ where

$$\Gamma_1 = [a_1, a_2] \times \{b_1\}, \qquad \Gamma_2 = [a_1, a_2] \times \{b_2\}\,,$$
$$\Gamma_3 = \{a_1\} \times [b_1, b_2] \quad \text{and} \quad \Gamma_4 = \{a_2\} \times [b_1, b_2]\,.$$

Let $\varphi \in \mathscr{C}^0(\Gamma)$ be *zero at the vertices* (a_i, b_j) $(i, j = 1, 2)$ *of* Ω with the result that $\varphi_i = \varphi\chi_{\Gamma_i}$. If, for $i = 1, 2, 3, 4$, u_i is the classical solution of $P(\Omega, \varphi_i)$ then $u = u_1 + u_2 + u_3 + u_4$ is the classical solution of $P(\Omega, \varphi)$. In other words, we can, within a change of notation reduce the problem to one in which $\varphi = 0$ on $\Gamma \backslash \{a_1\} \times]b_1, b_2[$. To within a translation, we can suppose that $\Omega =]0, a[\times]a, b[$.
We are thus led to the following problem: given $\varphi \in \mathscr{C}([0, b])$ with $\varphi(0) = \varphi(b) = 0$, to find u harmonic on $\Omega =]0, a[\times]0, b[$, continuous on $\bar{\Omega}$ and satisfying

$$\begin{cases} u(0, y) = \varphi(y) \quad \text{for} \quad y \in]0, b[\\ u(a, y) = u(x, 0) = u(x, b) = 0 \quad \text{for} \quad x \in [0, a] \quad \text{and} \quad y \in [0, b]\,. \end{cases}$$

We shall solve this problem by developing u in a Fourier series.
We seek $u(x, y)$ in the form

$$u(x, y) = \sum_{n \geq 1} u_n(x) \sin n\omega y \quad \text{with} \quad \omega = \frac{\pi}{b}\,. \tag{4.8}$$

The boundary conditions can be written

$$\varphi(y) = \sum u_n(0) \sin n\omega y \quad \text{that is to say} \quad u_n(0) = \frac{2}{b}\int_0^b \sin(n\omega y)\varphi(y)\,\mathrm{d}y$$
$$0 = \sum u_n(a) \sin n\omega y \quad \text{that is to say} \quad u_n(a) = 0\,.$$

The condition that u is harmonic can be written

$$\sum \left(\frac{\mathrm{d}^2 u_n}{\mathrm{d}x^2}(x) - (n\omega)^2 u_n(x)\right) \sin n\omega y = 0\,,$$

that is to say

$$\frac{\mathrm{d}^2 u_n}{\mathrm{d}x^2} = (n\omega)^2 u_n(x)\,;$$

from which by using the boundary conditions we derive the formula

$$u_n(x) = \frac{2}{b}\frac{\operatorname{sh} n\omega(a - x)}{\operatorname{sh} n\omega a}\int_0^b \sin(n\omega y)\varphi(y)\,dy\,. \tag{4.9}$$

We have carried out a *formal* calculation: however the reader familiar with the elements of Fourier series will see that by *supposing φ to be a little more than continuous (e.g. satisfying a Hölder condition), the series (4.8) where (u_n) is given by (4.9), converges uniformly on $\bar{\Omega}$ to the classical solution of the Dirichlet problem* $P(\Omega, \varphi\chi_{\{0\}\times[0,b]})$. □

Example 3 shows that there does not always exist a classical solution of $P(\Omega, \varphi)$: furthermore, it shows also that we can, at least in this case, consider a "natural" solution of $P(\Omega, \varphi)$ without it being a classical solution.
The use of the *subharmonic functions* developed in *Perron's method* will permit us to define a generalized solution of $P(\Omega, \varphi)$, for every bounded open Ω of $\mathbb{R}^n$ with boundary Γ and every $\varphi \in \mathscr{C}^0(\Gamma)$.
We introduce these notions here without proof; the development of this method will be treated in §4.6.
First, we give

Definition 2. Let Ω be an open set of $\mathbb{R}^n$. We say that $v \in \mathscr{C}^0(\Omega)$ is *sub-harmonic on* Ω if

$$\Delta v \geqslant 0 \quad \text{in} \quad \mathscr{D}'(\Omega)$$

that is to say

$$\langle \Delta v, \zeta\rangle = \int v\,\Delta\zeta\,dx \geqslant 0 \quad \text{for all} \quad \zeta \in \mathscr{D}(\Omega) \quad \text{with} \quad \zeta \geqslant 0\,.$$

We say that $w \in \mathscr{C}^0(\Omega)$ is *superharmonic on* Ω if $v = -w$ is sub-harmonic on Ω.
Note that we have defined here sub-harmonicity and super-harmonicity for a continuous function: we shall come back to this notion in a more general setting[69].
From Definition 2, it is immediate that a function $u \in \mathscr{C}^0(\Omega)$ is harmonic iff it is at the same time subharmonic and superharmonic; also *the set of the $v \in \mathscr{C}^0(\Omega)$ which are subharmonic on Ω is a closed cone of* $\mathscr{C}^0(\Omega)$, that is to say that

$$\text{for} \quad v_1, v_2 \in \mathscr{C}^0(\Omega),\ \lambda_1, \lambda_2 \geqslant 0\,,$$

$$v_1, v_2 \text{ subharmonic} \Rightarrow \lambda_1 v_1 + \lambda_2 v_2 \text{ sub-harmonic}$$

for a sequence (v_n) of $\mathscr{C}^0(\Omega)$ converging to $v \in \mathscr{C}^0(\Omega)$,

$$v_n \text{ subharmonic } \forall n \Rightarrow v \text{ sub-harmonic}\,.$$

---

[69] An example of a sub-harmonic function is a quadratic form

$$V(x) = \sum\sum a_{ij}x_i x_j \quad \text{with} \quad x = (x_1, \dots, x_i, \dots, x_j, \dots, x_n)\,,$$

with trace $\sum_{i=1}^{n} a_{ii}$ positive.

We shall principally make use of *the principle of the maximum for subharmonic continuous functions*, which we state here in the form:

Property 1. *Let Ω be a bounded open set of $\mathbb{R}^n$ with boundary Γ, $u \in \mathscr{H}(\Omega) \cap \mathscr{C}^0(\bar{\Omega})$ and $v \in \mathscr{C}^0(\Omega)$ subharmonic. Suppose that*

$$\limsup_{\substack{x \to z \\ x \in \Omega}} v(x) \leqslant u(z) \quad \textit{for all} \quad z \in \Gamma \,.$$

Then

$$v \leqslant u \leqslant \max_{\Gamma} u \quad \text{on} \quad \Omega \,.$$

This property will be proved in a more general setting in §4.6 (see Propositions 20 and 21): its localization is characteristic and will serve to give a definition to the notion of a general subharmonic function (see Definition 15). This property can be reformulated in terms of Dirichlet's problem: let Ω be a bounded open set in $\mathbb{R}^n$ with boundary Γ and $\varphi \in \mathscr{C}^0(\Gamma)$; we use the notation

$$(4.10) \qquad \begin{cases} \mathscr{I}_-(\varphi) \overset{\text{def}}{=} \{v \in \mathscr{C}^0(\Omega) \text{ sub-harmonic such that} \\ \limsup\limits_{x \to z} v(x) \leqslant \varphi(z), \quad \forall z \in \Gamma\} \\ \mathscr{I}_+(\varphi) \overset{\text{def}}{=} \{w \in \mathscr{C}^0(\Omega) \text{ super-harmonic such that} \\ \liminf\limits_{x \to z} w(x) \geqslant \varphi(z), \quad \forall z \in \Gamma\} \,. \end{cases}$$

The principle of the maximum (Property 1) shows immediately:

$$(4.11) \qquad v \leqslant w \qquad \forall\, v \in \mathscr{I}_-(\varphi) \quad \text{and} \quad w \in \mathscr{I}_+(\varphi)$$ [70]

and *u is the classical solution of $P(\Omega, \varphi)$ iff $u \in \mathscr{I}_-(\varphi) \cap \mathscr{I}_+(\varphi)$*. Note that $\mathscr{I}_-(\varphi)$ and $\mathscr{I}_+(\varphi)$ are not empty since they contain $v \equiv \min_\Gamma \varphi$ and $w \equiv \max_\Gamma \varphi$ respectively. The property (4.11) therefore permits us to consider for all $\varphi \in \mathscr{C}^0(\Gamma)$ and all $x \in \Omega$

$$(4.12) \qquad u_-(\varphi, x) \overset{\text{def}}{=} \sup_{v \in \mathscr{I}_-(\varphi)} v(x) \quad \text{and} \quad u_+(\varphi, x) \overset{\text{def}}{=} \inf_{w \in \mathscr{I}_+(\varphi)} w(x) \,.$$

The essential result of Perron's method is *Wiener's theorem* which we now state:

Property 2. *Let Ω be a bounded open set in $\mathbb{R}^n$ with boundary Γ and let $\varphi \in \mathscr{C}^0(\Gamma)$.*
(1) *for all $x \in \Omega$, (with the notation of* (4.12)), *we have*:

$$u_-(\varphi, x) = u_+(\varphi, x) \,;$$

[70] Apply the Property 1 to the sub-harmonic function $v - w$ and $u \equiv 0$.

(2) *the function* $u(\varphi)$ *defined on* Ω *by*

$$u(\varphi)(x) = u(\varphi, x) = u_-(\varphi, x) = u_+(\varphi, x), \quad x \in \Omega \tag{4.13}$$

is harmonic on Ω.

We shall prove this theorem in §4.6 (Proposition 22). Admitting it, we pose the

Definition 3. Given Ω a bounded open set in $\mathbb{R}^n$ with boundary Γ and $\varphi \in \mathscr{C}^0(\Gamma)$, we call the function $u(\varphi)$, harmonic on Ω, defined by (4.13), *the generalized solution of* $P(\Omega, \varphi)$.

We observe that the generalized solution of $P(\Omega, \varphi)$ is defined for all $\varphi \in \mathscr{C}^0(\Gamma)$. From the preceding remarks, it is clear that, if it exists, the classical solution of $P(\Omega, \varphi)$ is the generalized solution: in other words the map which with $\varphi \in \mathscr{C}^0(\Omega)$ associates the generalized solution $u(\varphi)$, extends the map which with every $\varphi \in E(\Omega)$ associates the classical solution $u(\varphi)$.
We show that this extension possesses interesting properties:

Proposition 3. *Given a bounded open set* Ω *in* $\mathbb{R}^n$ *with boundary* Γ,
(1) *for all* $\varphi \in \mathscr{C}^0(\Gamma)$

$$\min_\Gamma \varphi \leqslant u(\varphi) \leqslant \max_\Gamma \varphi\,; \tag{4.14}$$

(2) *The map* $\varphi \to u(\varphi)$ *is linear and continuous from* $\mathscr{C}^0(\Gamma)$ *into* $\mathscr{H}_b(\Omega)$ *the space of bounded harmonic functions on* Ω *with norm the uniform convergence norm.*

Proof. The point (1) follows from the definition since the functions

$$v = \min_\Gamma \varphi \quad \text{and} \quad w = \max_\Gamma \varphi$$

belong to $\mathscr{I}_-(\varphi)$ and $\mathscr{I}_+(\varphi)$ respectively.
We now note that from (4.10), (4.12), we have

$$\begin{cases} u_+(-\varphi) = -u_-(\varphi), \quad u_-(\lambda\varphi) = \lambda u_-(\varphi) \qquad \lambda \geqslant 0 \\ u_-(\varphi_1) + u_-(\varphi_2) \leqslant u_-(\varphi_1 + \varphi_2)\,. \end{cases} \tag{4.15}$$

Since $u(\varphi) = u_-(\varphi) = u_+(\varphi)$ it follows that the map $\varphi \to u(\varphi)$ is linear. The continuity is a consequence of (4.14)

$$\|u(\varphi)\| = \sup_\Omega |u(\varphi)| \leqslant \max_\Gamma |\varphi| = \|\varphi\|\,. \qquad \square$$

To return to the classical Dirichlet problem, we introduce the

Definition 4. Given an open set Ω in $\mathbb{R}^n$ with boundary Γ and $z \in \Gamma$ we say that z is a *regular boundary point of* Ω if there exists $r > 0$ and v satisfying

$$\begin{cases} v \text{ is continuous and sub-harmonic on } \Omega \cap B(z, r) \\ \text{for all } \ 0 < \delta < r, \quad \sup\limits_{\Omega \cap B(z,r)\setminus B(z,\delta)} v < 0 \\ \lim\limits_{x \to z} v(x) = 0\,. \end{cases} \tag{4.16}$$

A function v satisfying (4.16) will be called *a barrier function of z on $\Omega \cap B(z, r)$*. We shall specify it in §4.6 (see Proposition 23).

Property 3. *Let Ω be a bounded open set in $\mathbb{R}^n$ with boundary Γ and z a regular point of the boundary of Ω. Then, for all $\varphi \in \mathscr{C}^0(\Gamma)$, the generalized solution $u(\varphi)$ of $P(\Omega, \varphi)$ satisfies*

$$\lim_{x \to z} u(\varphi, x) = \varphi(z)$$

We deduce the

Theorem 1. *Let Ω be a bounded open set in $\mathbb{R}^n$ with boundary Γ. The following assertions are equivalent:*
(i) *for all $\varphi \in \mathscr{C}^0(\Gamma)$, $P(\Omega, \varphi)$ admits a classical solution:*
(ii) *all the points of the boundary of Ω are regular.*

Proof. First, we prove that (i) $\Rightarrow$ (ii). Given $z \in \Gamma$, using (i) we consider the classical solution u of $P(\Omega, \varphi)$ with $\varphi(x) = -|x - z|$. Then u is a barrier function of z on Ω: in effect $u \in \mathscr{H}(\Omega)$, $\lim_{x \to z} u(x) = 0$, and from the principle of the maximum we have, for all $\delta > 0$

$$\sup_{\Omega \setminus B(z, \delta)} u < 0$$

since $u \in \mathscr{C}^0(\bar{\Omega})$ and $u < 0$ on $\Gamma \setminus \{z\}$.
The implication (ii) $\Rightarrow$ (i) follows from Property 3: for $\varphi \in \mathscr{C}^0(\Gamma)$ the generalized solution of $P(\Omega, \varphi)$ satisfies, with the use of (ii),

$$\lim_{x \to z} u(\varphi, x) = \varphi(z) \quad \text{for all} \quad z \in \Gamma\,,$$

and hence is the classical solution of $P(\Omega, \varphi)$. □

We emphasize that the useful part of theorem, (ii) $\Rightarrow$ (i), is also the difficult part which springs from the Properties 1, 2 and 3 which we shall prove in §4.6. The interest of this result is to reduce the existence of a classical solution of Dirichlet's problem to the study of the regularity of boundary points of Ω, that is to say to *the search for barrier functions.*
We give here some elementary examples.

Example 6. *The case $n = 1$.* Given Ω an open set of $\mathbb{R}$ with boundary Γ and $z \in \Gamma$, the function $v(x) = -|x - z|$ is a barrier function of z on Ω. We deduce from Theorem 1, that *for every bounded open set Ω of $\mathbb{R}$ with boundary Γ and every $\varphi \in \mathscr{C}^0(\Gamma)$, there exists a (unique) classical solution of $P(\Omega, \varphi)$.* This can be shown directly: an open set Ω of $\mathbb{R}$ is the union of intervals $]a_i, b_i[$ two by two disjoint, their connected components; supposing that the intervals $]a_i, b_i[$ are bounded, we have $\{a_i, b_i\} \subset \Gamma$. Given $\varphi \in \mathscr{C}^0(\Gamma)$, we can, using the formula (4.3) of Example 1, consider the function $u \in \mathscr{H}(\Omega)$ defined by

$$u(x) = \frac{\varphi(a_i)(b_i - x) + \varphi(b_i)(x - a_i)}{b_i - a_i} \quad \text{for} \quad x \in]a_i, b_i[\,.$$

Let us show that *this function u is the classical solution of* $P(\Omega, \varphi)$. For that we consider $z \in \Gamma$ and a sequence (x_k) of Ω such that $x_k \to z$; if z is an isolated point of Γ, then it belongs to the boundary in at least one and at the most two intervals $]a_i, b_i[$ it is clear that $u(x_k) \to \varphi(z)$; if z is a limit point of Γ, considering the intervals $]a_{i_k}, b_{i_k}[$ containing x_k we shall have $a_{i_k} \to z$, $b_{i_k} \to z$; since φ is continuous, $\varphi(a_{i_k}) \to \varphi(z)$, $\varphi(b_{i_k}) \to \varphi(z)$; from which $u(x_k) \to \varphi(z)$ since

$$|u(x_k) - \varphi(z)| \leqslant \max(|\varphi(a_{i_k}) - \varphi(z)|, |\varphi(b_{i_k}) - \varphi(z)|) . \qquad \square$$

Example 7. *Exterior ball condition.* We say that *an open set Ω of $\mathbb{R}^n$ satisfies the exterior ball condition at a point z of its boundary* if there exists $x_0 \in \mathbb{R}^n \backslash \{z\}$ such that

$$B(x_0, |x_0 - z|) \cap \Omega = \varnothing .$$

Replacing x_0 by $\lambda x_0 + (1 - \lambda)z$ with $0 < \lambda < 1$, it comes to the same thing to suppose that

$$\bar{B}(x_0, |x_0 - z|) \cap \bar{\Omega} = \{z\} .$$

The function $v(x) = E_n(z - x_0) - E_n(x - x_0)$ is then a barrier function of z on Ω.

We note that, in particular, a convex open set satisfies the exterior ball condition at every point of its boundary: in effect from the Hahn–Banach theorem (in a finite dimensional space), an open set Ω is convex iff for every point z of its boundary there exists $x_0 \in \mathbb{R}^n \backslash \{z\}$ such that

$$\Omega \cap \{x \in \mathbb{R}^n; (x - z).(x_0 - z) \geqslant 0\} = \varnothing .$$

Hence from Theorem 1, we have

Corollary 1. *Given a* **convex** *bounded open set Ω with boundary Γ, for all $\varphi \in \mathscr{C}^0(\Gamma)$, there exists a (unique) classical solution of $P(\Omega, \varphi)$.*

This criterion of the exterior ball is very rough: we now give a finer criterion, distinguishing the cases $n = 2$ and $n \geqslant 3$.

Example 8. *Case $n = 2$: exterior segment condition.*

We say that *an open set Ω in $\mathbb{R}^n$ satisfies the external segment condition at a point z of its boundary* if there exists $x_0 \in \mathbb{R}^n \backslash \{z\}$ such that

$$\lambda x_0 + (1 - \lambda)z \notin \Omega \quad \text{for all} \quad \lambda \in [0, 1] .$$

In the case $n=2$, this condition is sufficient to show that z is a regular point of the boundary. In effect we specify a point x of $\mathbb{R}^2$ by its polar coordinates $r = |x - z|$, $\theta =$ angle $(z - x_0, x - z)$. The condition can be written

$$\Omega \cap B(z, r_0) \subset \{x; 0 < r < r_0, -\pi < \theta < \pi\} \quad \text{where} \quad r_0 = |x_0 - z| .$$

We can always suppose that $0 < r_0 < 1$ and we then verify without difficulty that

$$v(x) = \frac{\operatorname{Log} r}{(\operatorname{Log} r)^2 + \theta^2}$$

is a barrier function of z on $\Omega \cap B(z, r_0)$.

Using conformal transformations (see §2.5) we can obtain other criteria[71] for the regularity of a point on the boundary of an open set of $\mathbb{R}^2$.
In the case $n \geqslant 3$, the exterior segment condition is no longer sufficient for the regularity of a boundary point.

Example 9. *Exterior cone condition.* We say that *an open set in $\mathbb{R}^n$ satisfies the exterior cone condition at a point z of its boundary* if there exists an open convex cone with vertex O and $r_0 > 0$ such that

$$\Omega \cap (z + C) \cap B(z, r_0) = \varnothing ,$$

in other words if there exists $x_0 \in \mathbb{R}^n \backslash \{z\}$, $0 < \theta_0 < \pi$, $r_0 > 0$ such that

$$\Omega \cap B(z, r_0) \subset \{x; \text{angle}(z - x_0, x - z) < \theta_0\}^{72} . \tag{4.17}$$

We shall see that *this condition is sufficient to show that z is a regular point of the boundary of* Ω. For that, we must suppose that $n \geqslant 3$, and we shall seek a barrier function of the form

$$v(x) = - r^\lambda f(\theta) ,$$

where $r = |x - z|$, $\theta = \text{angle } (z - x_0, x - z) \in [0, \theta_0[$, with $\lambda > 0$ and f continuous and strictly positive on $[0, \theta_0]$, still to be determined, but such that $\Delta v \geqslant 0$ in $\mathscr{D}'(\Omega)$.
Denoting by x' the projection of x on the straight line $D = z + \mathbb{R}(x_0 - z)$,

$$\rho = r \sin \theta = \|x' - x\|, \qquad \xi = r \cos \theta = x' - z ,$$

and using the formula for the Laplacian in cylindrical coordinates (see §1.4) we have

$$\Delta v = \frac{\partial^2 v}{\partial \xi^2} + \frac{\partial^2 v}{\partial \rho^2} + \frac{n - 2}{\rho} \frac{\partial v}{\partial \rho} \quad \text{in} \quad \mathscr{D}'(\Omega \backslash D) ,$$

from which, passing to polar coordinates (r, θ) in the $\xi\rho$-plane, and taking account of $\dfrac{\partial v}{\partial \rho} = \dfrac{\partial v}{\partial r} \cdot \dfrac{\rho}{r} + \dfrac{\partial v}{\partial \theta} \cdot \dfrac{\cos \theta}{r}$, we have

[71] The open set Ω (cf. Fig. 2) does not satisfy the exterior segment condition (at z) but has a barrier function if the curve is analytic.
[72] The open set (cf. Fig. 3) does not satisfy the external ball condition at z. It satisfies the external cone condition at z iff $\theta > 0$.

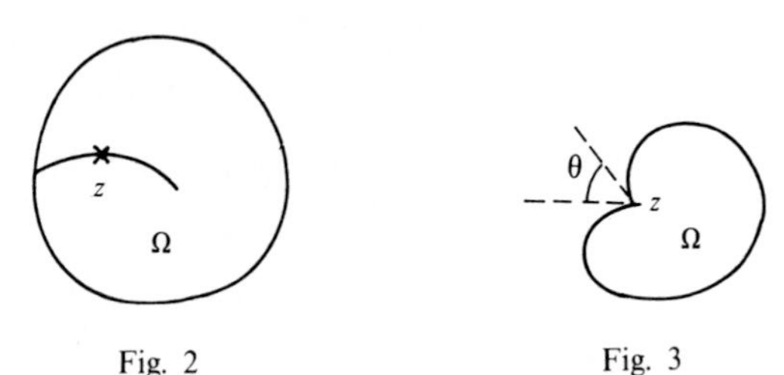

Fig. 2 Fig. 3

$$\begin{aligned}\Delta v &= \frac{\partial^2 v}{\partial r^2} + \frac{n-1}{r}\frac{\partial v}{\partial r} + \frac{1}{r^2}\left(\frac{\partial^2 v}{\partial \theta^2} + (n-2)\operatorname{cotg}\theta\frac{\partial v}{\partial \theta}\right)\\ &= \frac{1}{r^{n-1}}\frac{\partial}{\partial r}\left(r^{n-1}\frac{\partial v}{\partial r}\right) + \frac{1}{r^2\sin\theta^{n-2}}\frac{\partial}{\partial\theta}\left(\sin\theta^{n-2}\frac{\partial v}{\partial\theta}\right)\\ &= -r^{\lambda-2}\left(\lambda(\lambda+n-2)f(\theta) + \frac{1}{\sin\theta^{n-2}}\frac{\mathrm{d}}{\mathrm{d}\theta}\left(\sin\theta^{n-2}\frac{\mathrm{d}f}{\mathrm{d}\theta}(\theta)\right)\right).\end{aligned}$$

We fix $\theta_0 < \theta_0' < \pi$ and choose

$$f(\theta) = \int_\theta^{\theta_0'}\left(\frac{t}{\sin t}\right)^{n-2}\mathrm{d}t \tag{4.18}$$

with the result that

$$f \in \mathscr{C}^1([0,\theta_0]) \cap \mathscr{C}^2(]0,\theta_0]), \qquad f \geqslant f(\theta_0) > 0 \quad \text{on} \quad [0,\theta_0]\,,$$

and

$$\Delta v = r^{\lambda-2}\left(\frac{(n-2)\theta^{n-3}}{\sin\theta^{n-2}} - \lambda(\lambda+n-2)f(\theta)\right) \quad \text{on} \quad \Omega\backslash D\,.$$

Since $f(\theta) \leqslant f(0)$ and $\dfrac{(n-2)\theta^{n-3}}{\sin\theta^{n-2}} \geqslant \dfrac{n-2}{\theta_0}$, choosing

$$\lambda = \frac{n-2}{2}\left(\left(1 + \frac{4}{\theta_0^2 f(0)^2}\right)^{1/2} - 1\right), \tag{4.19}$$

we have $\lambda > 0$ and $\Delta v \geqslant 0$ on $\Omega\backslash D$. To complete the proof we must show that $\Delta v \geqslant 0$ in $\mathscr{D}'(\Omega)$. For that it is enough to show that

$$\lim_{\varepsilon\to 0}\int_{\{x;\,\varepsilon<\theta<\theta_0\}} v\Delta\zeta\,\mathrm{d}x \geqslant 0 \quad \text{for} \quad \zeta \in \mathscr{D}(\Omega)\,, \quad \zeta \geqslant 0$$

with ζ depending only on r and θ. Now

$$\mathrm{d}x = \sigma_{n-1}\rho^{n-2}\,\mathrm{d}\rho\,\mathrm{d}\xi = \sigma_{n-1}r^{n-1}\sin\theta^{n-2}\,\mathrm{d}r\,\mathrm{d}\theta\,.$$

Hence, using the formula of the Laplacian in the coordinates (r,θ) and integrating by parts, remembering that $\operatorname{supp}\zeta \subset\,]0,\infty[\times]\,0,\theta_0[$,

$$\begin{aligned}\frac{1}{\sigma_{n-1}}\int_{\{\varepsilon<\theta<\theta_0\}} v\,\Delta\zeta\,\mathrm{d}x &= \int_\varepsilon^{\theta_0}\sin\theta^{n-2}\,\mathrm{d}\theta\int_0^\infty v\frac{\partial}{\partial r}\left(r^{n-1}\frac{\partial\zeta}{\partial r}\right)\mathrm{d}r\\ &\quad + \int_0^\infty r^{n-3}\,\mathrm{d}r\int_\varepsilon^{\theta_0} v\frac{\partial}{\partial\theta}\left(\sin\theta^{n-2}\frac{\partial\zeta}{\partial\theta}\right)\mathrm{d}\theta\\ &= \frac{1}{\sigma_{n-1}}\int_{\{\varepsilon<\theta<\theta_0\}}\zeta\,\Delta v\,\mathrm{d}x\\ &\quad + \sin\varepsilon^{n-2}f(\varepsilon)\int_0^\infty\frac{\partial\zeta}{\partial\theta}(r,\varepsilon)r^{n+\lambda-3}\,\mathrm{d}r\\ &\quad + \varepsilon^{n-2}\int_0^\infty\zeta(r,\varepsilon)r^{n+\lambda-3}\,\mathrm{d}r\end{aligned}$$

from which

$$\lim_{\varepsilon \to 0} \int_{\{\varepsilon < \theta < \theta_0\}} v \, \Delta\zeta \, dx = \lim_{\varepsilon \to 0} \int_{\{\varepsilon < \theta < \theta_0\}} \zeta \, \Delta v \, dx \geqslant 0 .$$

We have therefore proved that, *under the hypothesis* (4.17) *the function* $v(x) = -r^{\lambda} f(\theta)$ *where* f *is given by* (4.18) *and* λ *by* (4.19) *is a barrier function of* z *on* $\Omega \cap B(0, r_0)$.
We sum up in a Proposition the results of Examples 6, 8 and 9:

Proposition 4. *Let* Ω *be an open set in* $\mathbb{R}^n$ *with boundary* Γ *and* $z \in \Gamma$.
(i) *Case* $n = 1$, *z is a regular point of* Γ;
(ii) *case* $n = 2$, *if* Ω *satisfies the exterior segment condition at* z *(see Example* 8) *then* z *is a regular point of* Γ;
(iii) *case* $n \geqslant 3$, *if* Ω *satisfies the exterior cone condition at* z *(see Example* 9) *then* z *is a regular point of* Γ.

When Ω is a regular open set (of class $\mathscr{C}^1$) (see §1.3.a.) it is clear that it satisfies the exterior cone condition $\forall z \in \Gamma$: we have also

$$\begin{cases} \text{for all } R, \text{ for all } \theta_0 > \dfrac{\pi}{2}, \text{ there exists } r_0 \text{ such that} \\ \Omega \cap B(z, r_0) \subset \{x; \text{angle}\,(-n(z), x - z) < \theta_0\} \\ \text{for all } z \in \Gamma \cap B(0, R) . \end{cases}$$

From Proposition 4 and Theorem 1, we therefore have the

Corollary 2. *Let* Ω *be a bounded regular open set (of class* $\mathscr{C}^1$) *of* $\mathbb{R}^n$ *with boundary* Γ. *Then for all* $\varphi \in \mathscr{C}^0(\Gamma)$, *there exists a unique classical solution of* $P(\Omega, \varphi)$.

We shall study later the regularity of the (classical) solution of $P(\Omega, \varphi)$ as a function of that of Ω and φ (see §6).
We complete this section by the

Remark 1. In the case $n \geqslant 2$, given an open set Ω in $\mathbb{R}^n$, and $x_0 \in \Omega$, x_0 is not a regular point of the boundary of $\Omega_0 = \Omega \backslash \{x_0\}$. More precisely, assuming Ω to be bounded (which it can always be made to be), for all $\varphi_0 \in \mathscr{C}^0(\partial\Omega_0)$ *the generalized solution of* $P(\Omega_0, \varphi_0)$ *is the restriction to* Ω_0 *of the generalized solution of* $P(\Omega, \varphi)$ *with* φ *the restriction of* φ_0 *to* $\partial\Omega$. In effect for $v \in \mathscr{I}_-(\varphi)$ and $\varepsilon > 0$ the function

$$v_0(x) = v(x) + \varepsilon\Big(E_n(x - x_0) - E_n\Big(\max_{\partial\Omega} |z - x_0|\Big)\Big)$$

is in $\mathscr{I}_1(\varphi_0)$; therefore

$$u(\varphi_0) \geqslant u(\varphi) + \varepsilon\Big(E_n(x - x_0) - E_n\Big(\max_{\partial\Omega} |z - x_0|\Big)\Big);$$

in the limit as $\varepsilon \to 0$, $u(\varphi_0) \geqslant u(\varphi)$; applying to $-\varphi$ (or repeating the argument with $w \in \mathscr{I}_+(\varphi)$), $u(\varphi_0) = u(\varphi)$ on Ω_0. This property will be extended in §5 to $\Omega \backslash K$ where K is a compact set of "zero capacity". □

2. Generalities on the Dirichlet Problem $P(\Omega, \varphi, f)$ and the Green's Function of Ω, a Bounded Open Set

Let us suppose that Ω is an open set in $\mathbb{R}^n$ with boundary Γ, φ defined on Γ, f defined on Ω. We now consider *Dirichlet's problem for Poisson's equation*

$$P(\Omega, \varphi, f) \quad \begin{cases} \Delta u = f & \text{on} \quad \Omega \\ u = \varphi & \text{on} \quad \Gamma . \end{cases}$$

Dirichlet's problem for Laplace's equation $P(\Omega, \varphi)$ is thus the particular case $P(\Omega, \varphi, 0)$. We shall consider also the particular case $P(\Omega, 0, f)$:

$$PH(\Omega, f) \quad \begin{cases} \Delta u = f & \text{on} \quad \Omega \\ u = 0 & \text{on} \quad \Gamma , \end{cases}$$

which we shall call *the homogeneous Dirichlet problem* (for Poisson's equation). Generalizing the Definition 1, we pose the

Definition 5. Let Ω be an open set in $\mathbb{R}^n$ with boundary Γ, $\varphi \in \mathscr{C}^0(\Gamma)$ and $f \in \mathscr{C}^0(\Omega)$. We call by the name *classical solution of* $P(\Omega, \varphi, f)$ every function $u \in \mathscr{C}^2(\Omega) \cap \mathscr{C}^0(\bar{\Omega})$ such that

$$\begin{cases} \Delta u(x) = f(x) & \text{for all} \quad x \in \Omega \\ u(z) = \varphi(z) & \text{for all} \quad z \in \Gamma . \end{cases}$$

As in the case of Poisson's equation (*see* §1.1) this (classical) notion of solution is not sufficient: it is necessary to introduce a concept of solution in the sense of distributions; however, we shall use in this classical setting only distributions on Ω. The introduction of distributions on the boundary Γ will be postponed until later (see §6). In a natural fashion in the classical setting, given $u \in \mathscr{D}'(\Omega)$ and $z \in \Gamma$, we shall say that

$$\lim_{x \to z} u(x) = l$$

if for all $\varepsilon > 0$, there exists $r > 0$ such that

$$l - \varepsilon \leqslant u \leqslant l + \varepsilon \quad \text{in} \quad \mathscr{D}'(\Omega \cap B(z, r)) ,$$

that is to say

$$|\langle u, \zeta \rangle - l| \leqslant \varepsilon \quad \text{for all} \quad \zeta \in \mathscr{D}(\Omega \cap B(z, r)) \quad \zeta \geqslant 0 , \quad \int \zeta \, dx = 1 .$$

If $u \in \mathscr{C}^0(\Omega)$ this idea of a limit clearly coincides with the classical topological idea; more generally if u is continuous in the neighbourhood of z in Ω, that is to say, if there exists $r_0 > 0$ and $\bar{u} \in \mathscr{C}^0(\Omega \cap B(z, r_0))$ such that $u = \bar{u}$ in $\mathscr{D}'(\Omega \cap B(z, r_0))$, then $\lim_{x\to z} u(x) = l$ in the sense specified above iff $\lim_{x\to z} \bar{u}(x) = l$ in the classical sense. We then pose the

Definition 6. Given Ω an open set in $\mathbb{R}^n$ with boundary Γ, $\varphi \in \mathscr{C}^0(\Gamma)$ and $f \in \mathscr{D}'(\Omega)$ a *quasi-classical solution of* $P(\Omega, \varphi, f)$ is the name given to a distribution u satisfying

$$\Delta u = f \quad \text{in} \quad \mathscr{D}'(\Omega)^{73}$$

$$\lim_{x\to z} u(x) = \varphi(z) \quad \text{for all} \quad z \in \Gamma^{74}$$

Remark 2. It is clear that u is a classical of $P(\Omega, \varphi, f)$ iff $u \in \mathscr{C}^2(\Omega)$ and is a quasi-classical solution of $P(\Omega, \varphi, f)$. In the case $n=1$, for $f \in \mathscr{C}^0(\Omega)$, the notions of classical and quasi-classical solutions coincide[75]; in the case $n \geqslant 2$, f continuous is no longer sufficient, but from the theorems on local regularity of the solutions of Poisson's equation (see Propositions 1 and 9 of §3) *if f is locally a Hölder function on Ω (and hence a fortiori if $f \equiv 0$ on Ω) the notions of classical and quasi-classical solutions of $P(\Omega, \varphi, f)$ coincide*[75]. □

Remark 3. Dirichlet's problem $P(\Omega, \varphi, f)$ is linear with respect to the data (φ, f): if u_i is a classical (resp. quasi-classical) solution of $P(\Omega, \varphi_i, f_i)$ and $\lambda_i \in \mathbb{R}$, then $u = \Sigma\lambda_i u_i$ is a classical (resp. quasi-classical) solution of $P(\Omega, \Sigma\lambda_i\varphi_i, \Sigma\lambda_i f_i)$. In particular

$$P(\Omega, \varphi, f) \simeq P(\Omega, \varphi) + PH(\Omega, f)$$

in the sense in which u_φ is a classical solution of $P(\Omega, \varphi)$ and u_f is a classical (resp. quasi-classical) solution of $PH(\Omega, f)$, $u = u_\varphi + u_f$ is a classical (resp. quasi-classical) solution of $P(\Omega, \varphi, f)$. In fact *Dirichlet's problem for Poisson's equation is the "sum" of Dirichlet's problem for Laplace's and for Poisson's equation: if $u_0 \in \mathscr{C}^2(\Omega)$ (resp. $u_0 \in \mathscr{D}'(\Omega)$), satisfying*

$$\lim_{x\to z} u_0(x) = \varphi_0(z) \quad \textit{exists for all} \quad z \in \Gamma\,,$$

is a solution of Poisson's equation $\Delta u_0 = f$ on Ω, then u is a classical (resp. quasi-classical) solution of $P(\Omega, \varphi, f)$ iff $u - u_0$ is a classical solution of $P(\Omega, \varphi - \varphi_0)$. □

We suppose now that Ω is bounded. First we note that there is then *uniqueness of a quasi-classical solution* of $P(\Omega, \varphi, f)$: in effect if u_1 and u_2 are two quasi-classical solutions of $P(\Omega, \varphi, f)$, by linearity $u = u_1 - u_2$ is a quasi-classical solution of $P(\Omega, 0, 0)$ and hence (see Remark 2) the classical solution $u \equiv 0$ of $P(\Omega, 0)$. Given $f \in \mathscr{D}'(\Omega)$, we shall denote

[73] See Definition 2 of §1.

[74] In the sense specified above.

[75] With the condition of identifying $u \in \mathscr{C}^2(\Omega)$ with the distribution which it defines.

$E(\Omega, f) = \{\varphi \in \mathscr{C}^0(\Gamma)$; there exists a quasi-classical solution of $P(\Omega, \varphi, f)\}$ and for $\varphi \in E(\Omega, f)$, $u(\varphi, f)$ denotes the quasi-classical solution of $P(\Omega, \varphi, f)$
From Remark 3, if $\varphi_0 \in E(\Omega, f)$

$$E(\Omega, f) = \varphi_0 + E(\Omega),$$

and for $\qquad \varphi \in E(\Omega, f), \qquad u(\varphi, f) = u(\varphi_0, f) + u(\varphi - \varphi_0)$

where $E(\Omega) = E(\Omega, 0)$ is a closed subspace of $\mathscr{C}^0(\Gamma)$ (see Proposition 1) and $u(\varphi - \varphi_0)$ is the classical solution of $P(\Omega, \varphi - \varphi_0)$. Now for all $\varphi \in \mathscr{C}^0(\Gamma)$ we can consider the generalized solution $u(\varphi - \varphi_0)$; the function $u(\varphi_0, f) + u(\varphi - \varphi_0)$ is independent of the choice of φ_0: indeed, if $\hat{\varphi}_0$ is another element of $E(\Omega, f)$

$$\begin{aligned} u(\varphi_0, f) + u(\varphi - \varphi_0) &= [u(\hat{\varphi}_0, f) + u(\varphi_0 - \hat{\varphi}_0)] + u(\varphi - \varphi_0) \\ &= u(\hat{\varphi}_0, f) + [u(\varphi_0 - \hat{\varphi}_0) + u(\varphi - \varphi_0)] \\ &= u(\hat{\varphi}_0, f) + u(\varphi - \hat{\varphi}_0) \end{aligned}$$

where we have used the linearity of the quasi-classical solution and the linearity of the generalized solution.
This permits us to define a generalized solution of $P(\Omega, \varphi, f)$ when $E(\Omega, f) \neq \varnothing$, that is to say when there exists $u_0 \in \mathscr{D}'(\Omega)$ a solution of Poisson's equation $\Delta u_0 = f$, and satisfying $\lim\limits_{x \to z} u_0(x)$ exists for all $z \in \Omega$.
We pose the

Definition 7. Let Ω be a bounded open set in $\mathbb{R}^n$ and let $f \in \mathscr{D}'(\Omega)$. We suppose that there exists $u_0 \in \mathscr{D}'(\Omega)$ such that

$$(4.20) \qquad \begin{cases} \Delta u_0 = f \quad \text{in} \quad \mathscr{D}'(\Omega) \\ \lim\limits_{x \to z} u_0(x) = \varphi_0(z) \quad \text{exists for all} \quad z \in \Gamma . \end{cases}$$

For all $\varphi_0 \in \mathscr{C}^0(\Gamma)$ we then define *the generalized solution of* $P(\Omega, \varphi, f)$ to be the function $u_0 + u(\varphi - \varphi_0)$ where $u(\varphi - \varphi_0)$ is the generalized solution of $P(\Omega, \varphi - \varphi_0)$ (see Definition 3).
As we have shown above this function depends only on Ω, φ and f; it is independent of the choice of u_0 satisfying (4.20).
We introduce now *the Green's function of* Ω. Given $y \in \Omega$ the distribution $u_0(x) = E_n(x - y)$ satisfies (4.20) with $f = \delta_y$ and $\varphi_0 = \varphi_y$ defined by

$$(4.21) \qquad \varphi_y(z) = E_n(z - y) \quad \text{for all} \quad z \in \Gamma .$$

The distribution $u_0 - u(\varphi_y)$ is the generalized solution of $PH(\Omega, \delta_y)$.
We pose the

Definition 8. Given Ω a bounded open set in $\mathbb{R}^n$, we call *the Green's function of* Ω, the function G defined on $\Omega \times \Omega \backslash D$, where $D = \{(x, y) \in \Omega \times \Omega;\, x = y\}$ is the diagonal of $\Omega \times \Omega$, by

$$G(x, y) = E_n(x - y) - u(\varphi_y, x) \tag{4.22}$$

where for all $y \in \Omega$, $u(\varphi_y)$ is the generalized solution of $P(\Omega, \varphi_y)$, φ_y being given by (4.21).

In other words, as we have seen, G is the function defined on $\Omega \times \Omega \backslash D$ by the property: *for all $y \in \Omega$, $G(., y)$ is the generalized solution of $PH(\Omega, \delta_y)$.*

Let us begin by giving two examples of Green's functions:

Example 10. *Green's function of a bounded interval* $]a, b[$

For $y \in]a, b[$, we have

$$\varphi_y(a) = \frac{1}{2}(y - a), \quad \varphi_y(b) = \frac{1}{2}(b - y);$$

from which, applying (4.3) we have

$$u(\varphi_y, x) = \frac{1}{2}\frac{(y - a)(b - x) + (b - y)(x - a)}{b - a},$$

and hence

$$G(x, y) = \frac{1}{2}\left\{|x - y| - \frac{(y - a)(b - x) + (b - y)(x - a)}{b - a}\right\};$$

which can be,written

$$G(x, y) = \begin{cases} \dfrac{(x - b)(y - a)}{b - a} & \text{if } x > y \\[2ex] \dfrac{(x - a)(y - b)}{b - a} & \text{if } x < y \end{cases} \tag{4.23}$$

or, alternatively

$$G(x, y) = \frac{(\max(x, y) - b)(\min(x, y) - a)}{b - a}. \qquad \square$$

Example 11. *Green's function of a Ball in* $\mathbb{R}^n$ $(n \geqslant 2)$.

Let $B = B(x_0, r_0)$. Using Poisson's integral formula, its Green's function is given by

$$G(x, y) = E_n(x - y) - \frac{1}{r_0 \sigma_n}\int_{\partial B} \frac{r_0^2 - |x - x_0|^2}{|t - x|^n} E_n(t - y)\,d\gamma(t).$$

We shall obtain a more explicit formula by using the inversion operator $I(x_0, r_0)$. Given $y \in B \backslash \{x_0\}$, the function

$$w(x) = E_n(x - I(x_0, r_0)y), \quad \text{where} \quad I(x_0, r_0)y = x_0 + \frac{r_0^2}{|y - x_0|^2}(y - x_0),$$

is harmonic in the neighbourhood of $\bar{B}$ since $I(x_0, r_0)y \notin \bar{B}$. For $z \in \partial B$ we have from the relation (2.12) of §2,

$$\begin{cases} \text{case} \quad n \geqslant 3, \quad w(z) = \left(\dfrac{|y - x_0|}{r_0}\right)^{n-2} E_n(z - y) \\ \text{case} \quad n = 2, \quad w(z) = E_2(z - y) - \dfrac{1}{2\pi} \operatorname{Log} \dfrac{|y - x_0|}{r_0}. \end{cases}$$

From the uniqueness of the Dirichlet problem $P(B, \varphi)$, we therefore have

$$\begin{cases} u(\varphi_y, x) = \left(\dfrac{r_0}{|y - x_0|}\right)^{n-2} w(x) & \text{in the case} \quad n \geqslant 3 \\ u(\varphi_y, x) = w(x) + \dfrac{1}{2\pi} \operatorname{Log} \dfrac{|y - x_0|}{r_0} & \text{in the case} \quad n = 2\,, \end{cases}$$

from which we deduce

$$\text{(4.24)} \quad \begin{cases} \text{case } n \geqslant 3, \quad G(x, y) = E_n(x - y) - E_n\left(\dfrac{|y - x_0|}{r_0}(x - x_0) - \dfrac{r_0}{|y - x_0|}(y - x_0)\right) \\ \text{case } n = 2, \quad G(x, y) = \dfrac{1}{2\pi} \operatorname{Log} \dfrac{|x - y|}{\left|\dfrac{|y - x_0|}{r_0}(x - x_0) - \dfrac{r_0}{|y - x_0|}(y - x_0)\right|}. \end{cases}$$

These formulae are valid for $y \in B \backslash \{x_0\}$ and $x \in B \backslash \{y\}$. But we note that G is symmetric (as in the case $n = 1$ where this is immediate); in effect, replacing $x - x_0$, $y - y_0$ by x, y respectively, that comes down to showing that

$$\left|\frac{|y|}{r_0} x - \frac{r_0}{|y|} y\right| = \left|\frac{|x|}{r_0} y - \frac{r_0}{|x|} x\right|$$

that we verify immediately by squaring both sides. Hence for $x \in B \backslash \{x_0\}$,

$$\text{(4.25)} \qquad G(x, x_0) = G(x_0, x) = \begin{cases} E_n(x - x_0) - E_n(r_0) & \text{if} \quad n \geqslant 3 \\ \dfrac{1}{2\pi} \operatorname{Log} \dfrac{|x - x_0|}{r_0} & \text{if} \quad n = 2\,. \end{cases} \qquad \square$$

We now give the general properties of the Green's functions.

Proposition 5. *Let Ω be a bounded open set in $\mathbb{R}^n$ and G its Green's function*
(1) *G is symmetric*:

$$\text{(4.26)} \qquad G(x, y) = G(y, x) \quad \textit{for all} \quad (x, y) \in \Omega \times \Omega \backslash D\,;$$

(2) *given $x, y \in \Omega$ not connected in Ω,*

$$G(x, y) = 0\,;$$

(3) *given* $x, y \in \Omega$, $x \neq y$, belonging to the same connected component Ω_0 *of* Ω

$$E_n(x - y) - E_n(\delta(x, y)) \leqslant G(x, y) < 0 \tag{4.27}$$

where
$$\delta(x, y) = \min\left(\max_{\partial\Omega_0}|x - z|, \max_{\partial\Omega_0}|y - z|\right);$$

(4) *G is analytic on* $\Omega \times \Omega \backslash D$ *and for every open set* Ω_0 *with* $\bar{\Omega}_0 \subset \Omega$, *the map* $y \to G(., y)$ *is analytic from* Ω_0 *into* $\mathscr{H}_b(\Omega \backslash \bar{\Omega}_0)$;

(5) *suppose that in addition*

$$\lim_{x \to z} G(x, y) = 0 \quad \text{for all} \quad (z, y) \in \Gamma \times \Omega\,; \tag{4.28}$$

then for all $\alpha \in \mathbb{N}^n$, $\dfrac{\partial^\alpha G}{\partial y^\alpha} \in \mathscr{C}^0(\bar{\Omega} \times \Omega \backslash D)$

and
$$\lim_{x \to z} \frac{\partial^\alpha G}{\partial y^\alpha}(x, y) = 0 \quad \text{for all} \quad (z, y) \in \Gamma \times \Omega\,.$$

Proof of the points (2) *to* (5). We shall in fact demonstrate the properties on the function $(x, y) \in \Omega \times \Omega \to u(\varphi_y, x)$, where φ_y is defined by (4.21) and $u(\varphi_y)$ is the generalized solution of $P(\Omega, \varphi_y)$.

It is clear that the function $y \to \varphi_y$ is analytic from Ω into $\mathscr{C}^0(\Gamma)$. Since the map $\varphi \to u(\varphi)$ is linear and continuous from $\mathscr{C}^0(\Gamma)$ into $\mathscr{H}_b(\Omega)$ (with norm the uniform convergence norm) (see Proposition 3), the map $y \to u(\varphi_y)$ is analytic from Ω into $\mathscr{H}_b(\Omega)$ and hence $(x, y) \to u(\varphi_y, x)$ is analytic on $\Omega \times \Omega$. This proves (4). We have also

$$\frac{\partial^\alpha}{\partial y^\alpha} u(\varphi_y, x) = u\left(\frac{\partial^\alpha \varphi_y}{\partial y^\alpha}, x\right).$$

The supplementary hypothesis comes down to saying that for all $y \in \Omega$, $u(\varphi_y)$ is the classical solution of $P(\Omega, \varphi_y)$, that is to say, $\varphi_y \in E(\Omega)$. Since $E(\Omega)$ is a closed subspace of $\mathscr{C}^0(\Gamma)$ and $\varphi \to u(\varphi)$ is an isometry of $E(\Omega)$ into $\mathscr{C}^0(\bar{\Omega})$, $\partial^\alpha \varphi_y/\partial y^\alpha \in E(\Omega)$ for all $\alpha \in \mathbb{N}^n$ and $y \in \Omega$ and the map $y \to u(\varphi_y)$ from Ω into $\mathscr{C}^0(\bar{\Omega})$ is analytic. This proves the point (5).

We consider now Ω_0 a connected component of Ω and $y \in \Omega$. The restriction u_0 of $u(\varphi_y)$ to Ω_0 is clearly the generalized solution of $P(\Omega_0, \varphi_0)$ where $\varphi_0(z) = E_n(z - y)$ is the restriction of φ_y to $\partial\Omega_0$. If $y \notin \Omega_0$, we have $y \notin \bar{\Omega}_0$ and $u_0(x) = E_n(x - y)$, proving the point (2). Now let us suppose $y \in \Omega_0$. From (4.14),

$$u_0 \leqslant \max_{\partial\Omega_0} \varphi_0 = E_n\left(\max_{\partial\Omega_0}|y - z|\right),$$

This proves the first inequality of (4.27), under the claim of the symmetry of G which we shall prove later.

Finally, we prove $u_0(x) > E_n(x - y)$ for all $x \in \Omega_0 \backslash \{y\}$. In the case $n = 1$, Ω_0 is an interval and this is satisfied immediately by the explicit formula (4.23). Let us suppose then that $n \geqslant 2$; then, firstly, $\Omega_0 \backslash \{y\}$ is connected: since $u_0(x) - E_n(x - y)$ is harmonic on $\Omega_0 \backslash \{y\}$, we see, from the principle of the maximum, that it is enough to show that $u_0(x) \geqslant E_n(x - y)$ for all $x \in \Omega_0 \backslash \{y\}$. By the definition of a generalized solution, it is enough to prove that

$$(4.29) \quad \begin{cases} \text{for a continuous function } w, \text{ superharmonic on } \Omega_0 \text{ satisfying} \\ \liminf\limits_{x \to z} w(x) \geqslant E_n(z - y) \qquad \text{for all } z \in \partial\Omega_0 \text{ we have} \\ \quad w(x) \geqslant E_n(x - y) \qquad \text{for all } x \in \Omega_0 \backslash \{y\}\,. \end{cases}$$

Since $\lim\limits_{x \to y} E_n(x - y) = -\infty$, there exists $0 < r < \operatorname{dist}(y, \partial\Omega_0)$ such that

$$w(x) \geqslant E_n(x - y) \quad \text{for} \quad x \in \bar{B}(y, r) \backslash \{y\}\,.$$

Now the function $v(x) = E_n(x - y) - w(x)$ is continuous and sub-harmonic on $\Omega_1 = \Omega_0 \backslash \bar{B}(y, r)$ and satisfies

$$\limsup_{x \to z} v(x) \leqslant 0 \quad \text{for all} \quad z \in \partial\Omega_1 = \partial\Omega_0 \cup \partial B(y, r)\,.$$

Hence from the principle of the maximum for sub-harmonic functions (see Property 1), $v(x) \leqslant 0$ on Ω_1. □

To prove the point (1) of Proposition 5, we shall make use of a result on the continuous dependence of the Green's function as a function of Ω:

Proposition 6. *Let (Ω_k) be an increasing sequence of sets in $\mathbb{R}^n$ with bounded union Ω. Denote by $\tilde{G}_k$ the function defined on $\Omega \times \Omega \backslash D$ by*

$$\tilde{G}_k(x, y) = \begin{cases} G_k(x; y) & \text{if} \quad (x, y) \in \Omega_k \times \Omega_k \\ 0 & \text{if} \quad (x, y) \notin \Omega_k \times \Omega_k \end{cases}$$

where G_k is the Green's function of Ω_k. Then for all $(x, y) \in \Omega \times \Omega \backslash D$ the sequence $(\tilde{G}_k(x, y))$ converges by decreasing to $G(x, y)$ the Green's function of Ω.

Let us admit this proposition, for the moment, to complete the proof of Proposition 5. For that we shall make use of two other results:

Lemma 1. *Let Ω be an open set in $\mathbb{R}^n$. There exists an increasing sequence of regular bounded open sets with boundaries of class $\mathscr{C}^\infty$ whose union is Ω.*

We leave to the reader the details of the proof of this lemma: using a covering of Ω by a sequence of balls B_k with $\bar{B}_k \subset \Omega$ we are led to prove that, given $\bar{B}_1, \ldots, \bar{B}_N$, closed balls contained in Ω, there exists Ω_0, a regular, bounded open set of class $\mathscr{C}^\infty$ containing $\bar{B}_1 \cup \ldots \cup \bar{B}_N$ with $\bar{\Omega}_0 \subset \Omega$; this is seen by induction over N.

Property 4. *Let Ω be a regular bounded open set in $\mathbb{R}^n$ with boundary of class $\mathscr{C}^{1+\varepsilon}$[76]. Then for all $y \in \Omega$, the function $x \to G(x, y)$ is in $\mathscr{C}^1_n(\bar{\Omega} \backslash \{y\})$.*

---

[76] See Definition 3 of §3.

We shall prove this result later using the integral method (see Corollary 8). Now we give the

Proof of Point (1) of Proposition 5. Making use of Lemma 1 and Proposition 6, we see that it is sufficient to prove the result for an open set Ω as regular as we wish. From Property 4, stated above, *we can therefore suppose* Ω *to be regular and for all* $y \in \Omega$, $G(., y) \in \mathscr{C}_n^1(\bar{\Omega}\backslash\{y\})$.
Let us then fix $y_1, y_2 \in \Omega$, $y_1 \neq y_2$ and

$$0 < r < \min(\operatorname{dist}(y_1, \Gamma), \operatorname{dist}(y_2, \Gamma), \frac{1}{2}|y_1 - y_2|)$$

with the result that $\Omega_0 = \Omega\backslash(\bar{B}(y_1, r) \cup \bar{B}(y_2, r))$ is regular with boundary

$$\Gamma \cup \partial B(y_1, r) \cup \partial B(y_2, r)\,.$$

The functions $v_1(x) = G(x, y_1)$ and $v_2(x) = G(x, y_2)$ are in $\mathscr{H}(\Omega_0) \cap \mathscr{C}_n^1(\bar{\Omega}_0)$. Since $v_1 = v_2 = 0$, on Γ, Green's formula applied to v_1 and v_2 on Ω_0 yields the result

$$\int_{\partial B(y_1, r)} \left(v_1 \frac{\partial v_2}{\partial n} - v_2 \frac{\partial v_1}{\partial n}\right) d\gamma = \int_{\partial B(y_2, r)} \left(v_2 \frac{\partial v_1}{\partial n} - v_1 \frac{\partial v_2}{\partial n}\right) d\gamma \tag{4.30}$$

in which the normal derivatives can all be taken to point inwards to Ω_0 with the result that they point outwards from $B(y_1, r)$ and $B(y_2, r)$.
Putting $u_1 = u(\varphi_{y_1})$ and applying Green's formula to u_1 and v_2 harmonic on $B(y_1, r)$ we have

$$\int_{\partial B(y_1, r)} \left(u_1 \frac{\partial v_2}{\partial n} - v_2 \frac{\partial u_1}{\partial n}\right) d\gamma = 0\,.$$

Since $v_1 = E_n(r) - u_1$ and $\dfrac{\partial v_1}{\partial n} = \dfrac{1}{\sigma_n r^{n-1}} - \dfrac{\partial u_1}{\partial n}$ on $\partial B(y_1, r)$, the left hand side I_1 of (4.30) is

$$I_1 = E_n(r) \int_{\partial B(y_1, r)} \frac{\partial v_2}{\partial n} d\gamma + \frac{1}{\sigma_n r^{n-1}} \int_{\partial B(y_1, r)} v_2\, \partial\gamma\,,$$

namely $I_1 = v_2(y_1)$, by applying Gauss' theorem and the formula of the mean to the function $v_2 \in \mathscr{H}(B(y_1, r)) \cap \mathscr{C}^1(\bar{B}(y_1, r))$. Similarly, the right hand side of (4.30) is $v_1(y_2)$ from which $G(y_1, y_2) = G(y_2, y_1)$. □

Proof of Proposition 6. From the points (2) and (3) of Proposition 5

$$G_k \leqslant 0 \quad \text{on} \quad \Omega_k \times \Omega_k\backslash D\,. \tag{4.31}$$

It is sufficient therefore to show that

$$G_k \leqslant G_l \quad \text{on} \quad \Omega_l \times \Omega_l\backslash D \quad \text{for} \quad l \leqslant k$$

and that for $(x, y) \in \Omega_l \times \Omega_l\backslash D$, $\lim\limits_{\substack{k \to \infty \\ k > l}} G_k(x, y) = G(x, y)$.

We can thus fix $y \in \Omega$ and without loss of generality, suppose that $y = 0 \in \Omega_0$. We denote by φ_k the trace of E_n on $\partial\Omega_k$ and by u_k the generalized solution of

$P(\Omega_k, \varphi_k)$. From (4.31) we have

(4.32) $$u_k \geqslant E_n \quad \text{on} \quad \Omega_k \,.$$

For $l \leqslant k$, u_k is therefore a harmonic function on Ω_l satisfying the condition

$$\liminf_{x \to z} u_k(x) \geqslant \varphi_l(z) \quad \text{for all} \quad z \in \partial\Omega_l \,.$$

By the definition of a generalized solution

$$u_k \geqslant u_l \quad \text{on} \quad \Omega_l \,.$$

Now $u_k \leqslant \max_{\partial\Omega_k} \varphi_k \leqslant E_n(\delta)$ where $\delta = \max_{\Omega} |x|$. Applying Harnack's theorem (see Corollary 9 of §2), the function $u(x) = \lim_{k \to \infty} u_k(x)$ is harmonic on Ω. From (4.32), $u \geqslant E_n$ on Γ. Now given w, continuous and superharmonic on Ω satisfying $\liminf_{x \to z} w(x) \geqslant E_n(z)$ for all $z \in \partial\Omega$, we have $w \geqslant E_n$ on Ω; this has been proved in the proof of Proposition 5(3) (see (4.28)). The case $n = 1$ is immediate directly, a continuous super-harmonic function on Ω, being concave on each component interval of Ω; hence $w \geqslant u_k$ on Ω_k for all k and in the limit $w \geqslant u$ on Ω. By definition u is therefore the generalized solution of $P(\Omega, \varphi)$. □

The principal interest of the Green's function resides in the *formula of the integral representation of the solutions of* $P(\Omega, \varphi, f)$. First, let us suppose

(4.33) Ω regular (of class $\mathscr{C}^1$) and for all $x \in \Omega$, $G(x, .) \in \mathscr{C}^1_n(\bar{\Omega} \setminus \{x\})$

or what comes to the same thing since G is symmetric,

$$\text{for all} \quad x \in \Omega, \quad u(\varphi_x) \in \mathscr{C}^1_n(\bar{\Omega}) \,.$$

From Proposition 5 of §3, for $u \in \mathscr{C}^2(\Omega) \cap \mathscr{C}^1_n(\bar{\Omega})$ with $\Delta u \in L^1(\Omega)$ we have for $x \in \Omega$,

$$u(x) = \int_\Omega E_n(t - x)\Delta u(t)\,\mathrm{d}t + \int_\Gamma \frac{\partial E_n}{\partial n}(t - x)u(t)\,\mathrm{d}\gamma(t) - \int_\Gamma E_n(t - x)\frac{\partial u}{\partial n}(t)\,\mathrm{d}\gamma(t) \,.$$

Also from Green's formula applied to $u(\varphi_x)$ and u, we have

$$0 = \int_\Omega u(\varphi_x, t)\Delta u(t)\,\mathrm{d}t + \int_\Gamma \frac{\partial u(\varphi_x)}{\partial n}(t)u(t)\,\mathrm{d}\gamma(t) - \int_\Gamma u(\varphi_x, t)\frac{\partial u}{\partial n}(t)\,\mathrm{d}\gamma(t) \,;$$

from which by subtracting from (4.22), since Ω being regular, $u(\varphi_x)$ is a classical solution of $P(\Omega, \varphi_x)$ (see Corollary 2), that is to say $u(\varphi_x, t) = E_n(t - x)$, $t \in \Gamma$, we obtain

(4.34) $$u(x) = \int_\Omega G(t, x)\,\Delta u(t)\,\mathrm{d}t + \int_\Gamma \frac{\partial G}{\partial n}(t, x)u(t)\,\mathrm{d}\gamma(t)$$

which is, for the moment, valid for every function

$$u \in \mathscr{C}^2(\Omega) \cap \mathscr{C}_n^1(\bar{\Omega}) \quad \text{with} \quad \Delta u \in L^1(\Omega)$$

and under the hypothesis (4.33).
Let us first extend this formula to the generalized solutions of $PH(\Omega, f)$ for an arbitrary bounded open set Ω.
From Proposition 5

$$E_n(x-y) - E_n(\delta) \leqslant G(x, y) \leqslant 0 \quad \text{on} \quad \Omega \times \Omega \backslash D \tag{4.35}$$

where δ is the diameter of Ω ($\delta = \max_{x, y \in \Omega} |x - y|$).

From Lemma 3 of §3 and Lebesgue's theorem, we deduce that, *being given f a bounded measure on Ω*

(*a*) for a.e. $x \in \Omega$, the function $G(x, .)$ is integrable with respect to f;
(*b*) *the function Gf defined (a.e. on Ω) by*

$$Gf(x) = \int_\Omega G(x, y) \, \mathrm{d}f(y) \tag{4.36}$$

is integrable on Ω. It is even in $L^q(\Omega)$ for all $q < n/(n-2)$; in the case $n = 1$, it is continuous on $\bar{\Omega}$. In the case $n \geqslant 2$, if $f \in L^p(\Omega)$ with $p > \frac{1}{2}n$, Gf is continuous and bounded on Ω; if in addition G satisfies (4.28), then

$$\lim_{x \to z} Gf(x) = 0 \quad \text{for all} \quad z \in \Gamma .$$

All of this follows from Lemma 3 of §3 and from (4.35) by the use of Lebesgue's theorem.
The function Gf defined in this way is a solution (in the sense of distributions) of Poisson's equation

$$\Delta u = f \quad \text{on} \quad \Omega . \tag{4.37}$$

Indeed by Fubini's theorem

$$\langle \Delta Gf, \zeta \rangle = \int Gf(x) \Delta\zeta(x) \, \mathrm{d}x = \int \mathrm{d}f(y) \int G(x, y) \Delta\zeta(x) \, \mathrm{d}x = \int \zeta(y) \, \mathrm{d}f(y) .$$

In particular, if $\zeta \in \mathscr{D}(\Omega)$, $G\zeta \in \mathscr{C}^\infty$; this allows us to *define for every distribution f with compact support in Ω, the distribution Gf on Ω by*

$$\langle Gf, \zeta \rangle = \langle f, G\zeta \rangle . \tag{4.38}$$

This definition clearly extends that given by (4.36), since by the symmetry of G, for a bounded measure f, the function Gf satisfies

$$\int Gf(x) \zeta(x) \, \mathrm{d}x = \int \mathrm{d}f(y) \int G(x, y) \zeta(x) \, \mathrm{d}x = \int G\zeta(y) \, \mathrm{d}f(y)$$

for all $\zeta \in \mathscr{D}(\Omega)$ (and even for $\zeta \in L^p(\Omega)$ with $p > \frac{1}{2}n$).

Hence more generally, Gf is defined by (4.38) for every distribution $f = f_0 + f_1$ where f_0 is a distribution with compact support in Ω and f_1 a bounded measure on Ω. We can consider this distribution Gf as a solution in a weak sense of $PH(\Omega, f)$; this point of view will be developed in §6.
We shall prove here the

Proposition 7. *Let Ω be a bounded open set in $\mathbb{R}^n$ with Green's function G and let $f = f_0 + f_1$ where f_0 is a distribution with compact support in Ω and $f_1 \in L^p(\Omega)$ with $p > \frac{1}{2}n$ (f_1 a bounded measure on Ω in the case $n = 1$). Then the distribution Gf is the generalized solution of $PH(\Omega, f)$. If, in addition, G satisfies* (4.28), *then Gf is a quasi-classical solution.*

Proof. For $f \in \mathscr{E}'(\Omega)$, the space of distributions with compact support in Ω, the Newtonian potential $E_n * f$ is harmonic on $\mathbb{R}^n \backslash \operatorname{supp} f$; we denote by φ_f its trace on Γ. The map $f \to \varphi_f$ is linear and continuous from $\mathscr{E}'(\Omega)$ into $\mathscr{C}^0(\Gamma)$; hence the map $f \to u(\varphi_f)$ the generalized solution of $P(\Omega, \varphi_f)$ is linear and continuous from $\mathscr{E}'(\Omega)$ into $\mathscr{H}_b(\Omega)$. The map $f \to Gf$ is, by definition, even linear and continuous from $\mathscr{E}'(\Omega)$ into $\mathscr{D}'(\Omega)$. For $f = \delta_y$ with $y \in \Omega$, $Gf = E_n * f - u(\varphi_f)$; since $\{\delta_y;\, y \in \Omega\}$ is total in $\mathscr{E}'(\Omega)$, this equality is true for all $f \in \mathscr{E}'(\Omega)$; hence Gf is a generalized solution of $PH(\Omega, f)$.
In the same way for $f \in L^p(\Omega)$ with $p > \frac{1}{2}n$, the Newtonian potential $E_n * \tilde{f}$ where

$$\tilde{f}(x) = \begin{cases} f(x) & \text{if} \quad x \in \Omega \\ 0 & \text{if} \quad x \in \mathbb{R}^n \backslash \Omega\,, \end{cases}$$

is continuous on $\bar{\Omega}$ (see Proposition 6 of §3). We can define its trace φ_f on Γ and the map $f \to u(\varphi_f)$ is continuous from $L^p(\Omega)$ into $\mathscr{H}_b(\Omega)$. Then from (4.35) and Lebesgue's theorem, $f \to Gf$ is continuous from $L^p(\Omega)$ into $\mathscr{C}_b^0(\Omega)$. Since

$$Gf = E_n * \tilde{f} - u(\varphi_f) \quad \text{for} \quad f \in L^p(\Omega) \cap \mathscr{E}'(\Omega)\,,$$

this is still true for all $f \in L^p(\Omega)$.
If G satisfies (4.28), $\varphi_y \in E(\Omega)$ for all $y \in \Omega$; $E(\Omega)$ being closed in $\mathscr{C}^0(\Gamma)$, by continuity of $f \to \varphi_f$ and density

$$\varphi_f \in E(\Omega) \quad \text{for all} \quad f \in \mathscr{E}'(\Omega) + L^p(\Omega)\,. \qquad \square$$

We now extend the representation formula (4.34) to every classical solution of $P(\Omega, \varphi)$:

Proposition 8. *Let Ω be a regular bounded open set in $\mathbb{R}^n$ with Green's function G satisfying $G(x, .) \in \mathscr{C}_n^1(\bar{\Omega} \backslash \{x\})$ for all $x \in \Omega$ and $\varphi \in \mathscr{C}^0(\Gamma)$. Then the (classical) solution $u(\varphi)$ is given by*

$$u(\varphi, x) = \int_\Gamma \frac{\partial G}{\partial n}(x, t)\varphi(t)\,\mathrm{d}\gamma(t) \quad \textit{for all} \quad x \in \Omega.$$

Proof. First of all, let us suppose that there exists $u_0 \in \mathscr{C}^2(\Omega) \cap \mathscr{C}_n^1(\bar{\Omega})$ with $\Delta u_0 \in L^p(\Omega)$, $p > \frac{1}{2}n$ such that $u_0 = \varphi$ on Γ.

From (4.34) we have

$$u_0(x) = \int_\Omega G(t, x) \Delta u_0(t) \, \mathrm{d}t + \int_\Gamma \frac{\partial G}{\partial n}(t, x) \varphi(t) \, \mathrm{d}\gamma(t) .$$

Now $u_0 - u(\varphi)$ is a quasi-classical solution of $PH(\Omega, \Delta u_0)$. From Proposition 7,

$$u_0(x) - u(\varphi, x) = \int_\Omega G(t, x) \Delta u_0(t) \, \mathrm{d}t$$

and hence

$$u(\varphi, x) = \int_\Gamma \frac{\partial G}{\partial n}(t, x) \varphi(t) \, \mathrm{d}\gamma(t) .$$

In particular this formula is true for the traces φ on Γ of the fu nctions $u_0 \in \mathscr{C}^2(\bar{\Omega})$. From the Stone–Weierstrass theorem, these traces form a dense part of $\mathscr{C}^0(\Gamma)$. Being given an arbitrary $\varphi \in \mathscr{C}^0(\Gamma)$, then therefore exists, $\varphi_k \in \mathscr{C}^0(\Gamma)$ $k \in \mathbb{N}$ such that

$$\varphi_k \to \varphi \quad \text{uniformly on } \Gamma \text{ for} \quad k \to \infty$$

$$u(\varphi_k, x) = \int_\Gamma \frac{\partial G}{\partial n}(t, x) \varphi_k(t) \, \mathrm{d}\gamma(t) .$$

From Proposition 1, for $k \to \infty$

$$u(\varphi_k, x) \to u(\varphi, x)$$

and since $\dfrac{\partial G}{\partial n}(., x) \in \mathscr{C}^0(\Gamma)$,

$$\int_\Gamma \frac{\partial G}{\partial n}(t, x) \varphi_k(t) \, \mathrm{d}\gamma(t) \to \int_\Gamma \frac{\partial G}{\partial n}(t, x) \varphi(t) \, \mathrm{d}\gamma(t) .$$

Hence in the limit

$$u(\varphi, x) = \int_\Gamma \frac{\partial G}{\partial n}(t, x) \varphi(t) \, \mathrm{d}\gamma(t) . \qquad \square$$

3. Generalities on Dirichlet's Problem in an Unbounded Open Set

We now take Ω to be an *unbounded* open set in $\mathbb{R}^n$ with boundary Γ and take $\varphi \in \mathscr{C}^0(\Gamma)$, $f \in \mathscr{D}'(\Omega)$. We consider Dirichlet's problem $P(\Omega, \varphi, f)$. We note firstly that if all the connected components of Ω are bounded, there is a *unique* quasi-classical solution of $P(\Omega, \varphi, f)$: in effect if u is such a solution, its restriction to each connected component of Ω will be determined and hence also u. It will not be the same if Ω admits an unbounded connected component: we have to introduce *a condition at infinity*.

First of all, let us give some examples:

Example 12. *Case of an interval* $]a, \infty[$ *of* $\mathbb{R}$.
Let $\Omega =]a, \infty[$ with $a > -\infty$ with the result that $\Gamma = \{a\}$. We take $\varphi \in \mathbb{R}$ and, for simplicity $f \in \mathscr{C}([a, \infty[)$. The (classical) solutions of $P(\Omega, \varphi, f)$ are given by

$$(4.39) \qquad u(x) = \varphi + \int_a^x (x - t) f(t)\,dt + c(x - a) \qquad x \in [a, \infty[$$

where c is an arbitrary constant.
This constant c is given as a function of u by

$$(4.40) \qquad c = \lim_{x \to \infty} \frac{u(x)}{x} - \int_a^x \left(1 - \frac{t}{x}\right) f(t)\,dt\,.$$

In particular **if** $f(t) \in L^1(]a, \infty[)$, then

$$c = \lim_{x \to \infty} \frac{u(x)}{x} - \int_a^\infty f(t)\,dt\,;$$

and, changing the notation, *for all* $c \in \mathbb{R}$, *there exists one and only one (classical) solution of* $P(\Omega, \varphi, f)$ *satisfying the condition at infinity*

$$(4.41) \qquad \lim_{x \to \infty} \frac{u(x)}{x} = c\,.$$

This solution is given by

$$u(x) = \varphi + \int_a^x (x - t) f(t)\,dt + \left(c - \int_a^\infty f(t)\,dt\right)(x - a)\,,$$

namely

$$(4.42) \qquad u(x) = \varphi - \int_a^\infty G(x, t) f(t)\,dt + c(x - a)$$

with

$$(4.43) \qquad G(x, y) = a - \min(x, y)\,.$$

This function will be *the Green's function of the unbounded interval* $]a, \infty[$. Note that *it is the limit when* $b \to \infty$ *of the Green's function of the bounded interval* $]a, b[$ (given by (4.23)). *It is characterised by*

$$(4.44) \qquad \begin{cases} \text{for all } y \in]a, \infty[,\ u(x) = G(x, y) \text{ is the quasi-classical solution} \\ \text{of } PH(]a, \infty[, \delta_y) \text{ satisfying } \lim\limits_{x \to \infty} \dfrac{u(x)}{x} = 0\,. \end{cases}$$

□

Example 13. *Case of the exterior of a ball* $\mathbb{R}^n$, $n \geqslant 2$.
Suppose we take $\Omega = \mathbb{R}^n \backslash \bar{B}$ the exterior of the ball $B = B(x_0, r_0)$ whose boundary is $\Gamma = \partial B$. Propositions 20 and 21 of §2, tell us that, given u a bounded classical solution of $P(\Omega, \varphi)$, then

$$c = \lim_{|x| \to \infty} u(x)$$

exists and

$$(4.45) \qquad u(x) = \left(1 - \left(\frac{r_0}{|x - x_0|}\right)^{n-2}\right) c + \frac{1}{r_0 \sigma_n} \int_{\partial B} \frac{|x - x_0|^2 - r_0^2}{|t - x|^n} \varphi(t)\,d\gamma(t)\,.$$

Conversely, we show as in Example 2 (Proposition 2), that the function u given by (4.45) is a classical solution of $P(\Omega, \varphi)$ (we can make use of a Kelvin transformation as in Proposition 21 of §2 to reduce the problem to Proposition 2). But for the behaviour of the solution at infinity we have to distinguish between two cases

(*i*) case $n \geqslant 3$, it is immediate that $u(x) \to c$ when $|x| \to \infty$
(*ii*) case $n = 2$, we see easily that

$$u(x) \to \frac{1}{2\pi r_0} \int_{\partial B} \varphi(t)\, d\gamma(t) \quad \text{when} \quad |x| \to \infty .$$

Hence *in the case* $n \geqslant 3$, *for all* $\varphi \in \mathscr{C}^0(\partial B)$ *and all* $c \in \mathbb{R}$, *there exists a unique classical solution of* $P(\Omega, \varphi)$ *satisfying*

$$\lim_{|x| \to \infty} u(x) = c$$

given by (4.45).
While *in the case* $n = 2$, *for all* $\varphi \in \mathscr{C}^0(\partial B)$, *there exists a unique classical solution of* $P(\Omega, \varphi)$ *satisfying the condition at infinity*:

u is bounded on Ω.

This solution will again be given by (4.45) (where c is of little importance), it will have a *limit at infinity* but which *is imposed by the giving of* φ contrarily to the case $n \geqslant 3$:

$$\lim_{|x| \to \infty} u(x) = \frac{1}{2\pi r_0} \int_{\partial B} \varphi(t)\, d\gamma(t) .$$

This difference between the cases $n = 2$ and $n \geqslant 3$ on the conditions at infinity appears clearly also in considering the radial solutions of $P(\Omega, 0)$: such a solution is necessarily of the form

case $n \geqslant 3$, $$u(x) = c\left(\frac{1}{r_0^{n-2}} - \frac{1}{|x - x_0|^{n-2}}\right)$$

and the constant c is determined by

$$c = r_0^{n-2} \lim_{|x| \to \infty} u(x) ;$$

case $n = 2$, $u(x) = c(\operatorname{Log}|x - x_0| - \operatorname{Log} r_0) = c \operatorname{Log} \dfrac{|x - x_0|}{r_0}$ and then

$$c = \lim_{|x| \to \infty} \frac{u(x)}{\operatorname{Log}|x|} . \tag{4.46}$$

We shall prove later in the case of almost any unbounded open set in $\mathbb{R}^2$ that this condition at infinity (4.46) determines the solution of $P(\Omega, \varphi)$ (see Corollary 3). □

Example 14. *Case of a half-space of* $\mathbb{R}^n$, $n \geqslant 2$.
We consider $\Omega = \mathbb{R}^{n-1} \times \mathbb{R}^+$ of which the boundary $\Gamma = \mathbb{R}^{n-1} \times \{0\}$ is identified with $\mathbb{R}^{n-1}$ and take φ to be a continuous function, bounded on $\mathbb{R}^{n-1}$. Poisson's integral formula for a half-space (see Proposition 22 of §2) shows that a *bounded* classical solution of $P(\Omega, \varphi)$ is of the form

$$u(x', x_n) = \frac{2x_n}{\sigma_n} \int_{\mathbb{R}^{n-1}} \frac{\varphi(t)\,dt}{(|t - x'|^2 + x_n^2)^{n/2}} = \frac{2}{\sigma_n} \int_{\mathbb{R}^{n-1}} \frac{\varphi(x' + x_n t)\,dt}{(1 + |t|^2)^{n/2}}. \tag{4.47}$$

Conversely the function u given by (4.47) is harmonic on Ω since

$$\frac{2x_n}{\sigma_n}(|t - x'|^2 + x_n^2)^{-n/2} = \frac{\partial}{\partial x_n} E_n(x' - t, x_n).$$

Using $\displaystyle\int_{\mathbb{R}^{n-1}} \frac{dt}{(1 + |t|^2)^{n/2}} = \frac{\sigma_n}{2}$, and then Lebesgue's theorem we see that u is a bounded classical solution of $P(\Omega, \varphi)$.
Hence for every continuous φ, *bounded on* $\mathbb{R}^{n-1}$, *there exists a unique classical solution of* $P(\mathbb{R}^{n-1} \times \mathbb{R}^+, \varphi)$ *satisfying the condition at infinity:*

$$u \text{ is bounded on } \mathbb{R}^{n-1} \times \mathbb{R}^+$$

given by (4.47). We note that this solution does not, in general, have a limit (in the topological sense) when $|x| \to \infty$; however

$$\lim_{|x| \to \infty} u(x) = \lim_{|x'| \to \infty} \varphi(x') \tag{4.48}$$

if this latter limit exists. □

We have seen appearing in these examples, various conditions at infinity assuring the existence and uniqueness of a solution of $P(\Omega, \varphi)$. Firstly, we prove a sufficiently general result on the uniqueness of a classical or quasi-classical solution of $P(\Omega, \varphi, f)$, grouping together the various conditions which have appeared:

Proposition 9. *Let* Ω *be an unbounded open set in* $\mathbb{R}^n$ *with boundary* Γ *and let* $\varphi \in \mathscr{C}^0(\Gamma)$, $f \in \mathscr{D}'(\Omega)$, $u_\infty \in \mathscr{D}'(\Omega)$ *and* $h \in \mathscr{H}(\Omega) \cap \mathscr{C}^0(\bar{\Omega})$ *with* $h(x) > 0$ *for* $|x|$ *large. Then there exists at the most one quasi-classical solution of* $P(\Omega, \varphi, f)$ *satisfying the condition at infinity:*

$$\lim_{|x| \to \infty} \frac{u(x)}{h(x)} - u_\infty(x) = 0^{77}. \tag{4.49}$$

---

[77] When $u/h - u_\infty$ is a distribution, the limit is taken in the sense (see just before Definition 6): for all $\varepsilon > 0$, there exists $R > 0$, such that

$$-\varepsilon \leqslant \frac{u}{h} - u_\infty \leqslant \varepsilon \quad \text{in} \quad \mathscr{D}'(\Omega \backslash \bar{B}(0, R))$$

Proof. Given two solutions, the difference u is a classical solution (see Remark 2) of $P(\Omega, 0)$ and satisfies

$$\lim_{|x| \to \infty} \frac{u(x)}{h(x)} = 0 .$$

From the hypothesis, we fix R_0 such that $h(x) > 0$ on $\Omega \backslash \bar{B}(0, R_0)$. Given $\varepsilon > 0$, there exists $R_\varepsilon \geqslant R_0$ such that

$$\left|\frac{u}{h}\right| \leqslant \varepsilon \quad \text{on} \quad \Omega \backslash \bar{B}(0, R_\varepsilon) .$$

The function $v = u - \varepsilon h$ is harmonic on Ω (u and h are!)

$$v \leqslant 0 \quad \text{on} \quad \Omega \backslash B(0, R_\varepsilon) \text{ (by hypothesis on } R_\varepsilon) ,$$

$$v \leqslant 0 \quad \text{on} \quad \Gamma \backslash B(0, R_0) \text{ (since } u = 0 \text{ and } h \geqslant 0) .$$

Hence from the maximum principle:

$$v \leqslant \max_{\Gamma \cap B(0,R_0)} v^+ = \varepsilon \max_{\Gamma \cap B(0,R_0)} h^- (h^- = (-h)^+ = \max(0, -h))$$

and

$$u \leqslant \varepsilon \left(h + \max_{\Gamma \cap B(0,R_0)} h^- \right).$$

We can show similarly that

$$u \geqslant -\varepsilon \left(h + \max_{\Gamma \cap B(0,R_0)} h^- \right)$$

and hence when $\varepsilon \to 0$, $u = 0$. □

Obviously this proposition contains the condition at infinity

$$\lim_{|x| \to \infty} u(x) = c .$$

But it also contains, in the case $n = 2$, the condition (4.46) with the reservation that $\bar{\Omega} \neq \mathbb{R}^2$: in effect then, (4.46) is equivalent

$$\lim_{|x| \to \infty} \frac{u(x)}{\operatorname{Log}|x - x_0|} - c = 0$$

where $x_0 \in \mathbb{R}^2 \backslash \bar{\Omega}$, $h(x) = \operatorname{Log}|x - x_0|$ satisfies the hypotheses. We therefore have the

Corollary 3. *Let Ω be an open set in $\mathbb{R}^2$ with $\bar{\Omega} \neq \mathbb{R}^2$ and let $\varphi_0 \in \mathscr{C}^0(\Gamma)$, $f \in \mathscr{D}'(\Omega)$ and $u_\infty \in \mathscr{D}'(\Omega)$. Then there exists at most one quasi-classical solution of $P(\Omega, \varphi, f)$, satisfying the condition at infinity*

$$\lim_{|x| \to \infty} \frac{u(x)}{\operatorname{Log}|x|} - u_\infty(x) = 0 . \tag{4.50}$$

In particular, there will exist at most one solution satisfying

$$u - u_\infty \text{ is bounded at infinity.} \tag{4.51}$$

This situation will occur for every open set Ω satisfying

(4.52) *there exists* $h \in \mathscr{H}(\Omega) \cap \mathscr{C}^0(\bar{\Omega})$ *such that* $\lim_{|x| \to \infty} h(x) = +\infty$.

We state the

Corollary 4. *Let Ω be an unbounded open set in $\mathbb{R}^n$ satisfying* (4.52), $\varphi \in \mathscr{C}^0(\Gamma)$, $f \in \mathscr{D}'(\Omega)$, $u_\infty \in \mathscr{D}'(\Omega)$. *Then there exists at most one quasi-classical solution of $P(\Omega, \varphi, f)$ satisfying the condition* (4.51).

Example 15. *Open sets satisfying the condition* (4.52).
(*a*) $n = 1$, every open set $\Omega \neq \mathbb{R}$: in effect if $x_0 \in \mathbb{R} \backslash \Omega$,

$$h(x) = |x - x_0| \quad \text{is suitable.}$$

(*b*) $n = 2$, every open set Ω with $\bar{\Omega} \neq \mathbb{R}^2$: in effect $x_0 \in \mathbb{R}^2 \backslash \bar{\Omega}$,

$$h(x) = \operatorname{Log}|x - x_0| \text{ is suitable.}$$

(*c*) $n \geqslant 3$, every open set Ω contained in a half-space.
In effect we can always suppose $\bar{\Omega} \subset \mathbb{R}^{n-1} \times \mathbb{R}^+ = \{(x', x_n);\ x_n > 0\}$. For all $0 < \alpha < 1$, the function

$$h_\alpha(x) = \int_{\mathbb{R}^{n-1}} \frac{|x' + x_n t|^\alpha}{(1 + |t|^2)^{n/2}} \, dt \tag{4.53}$$

is suitable (see Example 14, formula (4.47)). □

Finally we note that in the case $u_\infty \equiv 0$, the condition at infinity (4.49) is independent of h.

Corollary 5. *Let Ω be an unbounded open set in $\mathbb{R}^n$, $\varphi \in \mathscr{C}^0(\Gamma)$, $f \in \mathscr{D}'(\Omega)$. Then there exists at most one quasi-classical solution of $P(\Omega, \varphi, f)$ satisfying the condition at infinity*

$$\text{(4.54)} \quad \begin{cases} \textit{there exists } h \in \mathscr{H}(\Omega) \cap \mathscr{C}^0(\bar{\Omega}) \textit{ with } h(x) > 0 \textit{ for } |x| \textit{ large so that} \\ \lim_{|x| \to \infty} \dfrac{u(x)}{h(x)} = 0 . \end{cases}$$

Proof. Let u_1, u_2 be two quasi-classical solutions of $P(\Omega, \varphi, f)$ and $h_1, h_2 \in \mathscr{H}(\Omega) \cap \mathscr{C}^0(\bar{\Omega})$ such that

$$h_i(x) > 0 \quad \text{for } |x| \text{ large, we have } \lim_{|x| \to \infty} \frac{u_i(x)}{h_i(x)} = 0 \quad i = 1, 2 .$$

We have $h = h_1 + h_2 \in \mathscr{H}(\Omega) \cap \mathscr{C}^0(\bar{\Omega})$, $h(x) > 0$ for $|x|$ large, and since

$$0 < \frac{h_i(x)}{h(x)} < 1 \quad \text{for } |x| \text{ large,} \quad \lim_{|x| \to \infty} \frac{u_i(x)}{h(x)} = 0 \quad i = 1, 2$$

from which Corollary 5 follows. □

This leads us to frame the

Definition 9. Let Ω be an unbounded open set in $\mathbb{R}^n$. We say that $u \in \mathscr{D}'(\Omega)$ satisfies the *null condition at infinity* if it satisfies (4.54).

We note that a solution u of $P(\Omega, \varphi, f)$ satisfying the null condition at infinity does not necessarily satisfy

$$\lim_{|x| \to \infty} u(x) = 0 . \tag{4.55}$$

For example for $c \in \mathbb{R}$, *the function* $u \equiv c$ is a classical solution of $P(\Omega, c)$ and it *satisfies the null condition at infinity iff* $c = 0$, *or* Ω *satisfies* (4.52).
We note also that although this is the case for the exterior of a ball in $\mathbb{R}^2$ (see Example 13), in general a solution of $P(\Omega, \varphi, f)$ satisfying the null condition at infinity is not necessarily bounded at infinity. For example, for $\bar{\Omega} \subset \mathbb{R}^{n-1} \times \mathbb{R}^+$ and $0 < \alpha < 1$, the unbounded function $u = h_\alpha$ given by (4.53) solution of $P(\Omega, \varphi)$, where φ is the trace of h_α on Γ, satisfies the null condition at infinity: indeed $\lim_{|x| \to \infty} u(x)/h(x) = 0$ for $h = h_\beta$ with $\alpha < \beta < 1$.
Finally, we note that *the null condition at infinity is equivalent to the condition* (4.55) *iff* Ω *satisfies*

(4.55)′ *every minorized function of* $\mathscr{H}(\Omega) \cap \mathscr{C}^0(\bar{\Omega})$ *is bounded.*

Example 16. *The exterior* Ω *of a compact set of* $\mathbb{R}^n$ *with* $n \geqslant 3$ *satisfying* (4.55)′. In effect let $h \in \mathscr{H}(\Omega) \cap \mathscr{C}^0(\bar{\Omega})$ minorized. Let us take $\zeta \in \mathscr{D}(\mathbb{R}^n)$ such that $\zeta = 1$ in the neighbourhood of $\mathbb{R}^n \backslash \Omega$ and put

$$h_1 = h(1 - \zeta) \in \mathscr{C}^\infty(\mathbb{R}^n), \quad f = \Delta h_1 \in \mathscr{D}(\mathbb{R}^n), \quad \tilde{h} = h_1 - E_n * f$$

which is harmonic on $\mathbb{R}^n$. We know (see Proposition 2 of §3), since $n \geqslant 3$ that $E_n * f$ is bounded; since $h_1(x) = h(x)$ for $|x|$ large, $\tilde{h}$ is minorized and hence constant (see Proposition 7 of §2) we deduce that h_1 and hence h are bounded. □

Let us study now, *the existence of* $P(\Omega, \varphi, f)$.
If there exists a quasi-classical solution of $P(\Omega, \varphi, f)$ satisfying (4.49), *a fortiori* there exists $u_0 \in \mathscr{D}'(\Omega)$ satisfying

$$\begin{cases} \Delta u_0 = f \quad \text{in} \quad \mathscr{D}'(\Omega) \\ \lim_{|x| \to \infty} \dfrac{u_0(x)}{h(x)} - u_\infty(x) = 0 \\ \lim_{x \to z} u_0(x) = \varphi_0(z) \quad \text{exists for all } z \in \Gamma . \end{cases} \tag{4.56}$$

Conversely, *given* $v_0 \in \mathscr{D}'(\Omega)$ *satisfying* (4.56), u *is a solution of* $P(\Omega, \varphi, f)$ *satisfying* (4.49) *iff* $u - u_0$ *is a classical solution of* $P(\Omega, \varphi - \varphi_0)$ *satisfying the null condition at infinity.*
As in the bounded case, we shall obtain $u_0 \in \mathscr{D}'(\Omega)$ satisfying (4.56) with the help of Newtonian potentials. We note carefully that the condition at infinity (u_∞, h)

cannot be taken arbitrarily: the possible weights h are determined by the open set and u_∞ depends to within an arbitrary constant on f, φ and h.

We are thus led to *study the existence of a solution of* $P(\Omega, \varphi)$ *satisfying the null condition at infinity.*

We can, as in the bounded case develop *Perron's method* to define a *generalized solution.*

Given $\varphi \in \mathscr{C}^0(\Gamma)$, we define

$$(4.57)\qquad \begin{cases} \mathscr{I}_-(\varphi) = \Big\{ v \in \mathscr{C}^0(\Omega) \text{ sub-harmonic satisfying} \\ \limsup\limits_{x \to z} v(x) \leqslant \varphi(z) \text{ for all } z \in \Gamma \text{ and there exists} \\ h \in \mathscr{H}(\Omega) \cap \mathscr{C}^0(\bar{\Omega}) \text{ with } h(x) > 0 \text{ for } |x| \text{ large such} \\ \text{that } \limsup\limits_{|x| \to \infty} \dfrac{v(x)}{h(x)} \leqslant 0 \Big\} \\ \mathscr{I}_+(\varphi) = -\mathscr{I}_-(-\varphi)\,. \end{cases}$$

Using Property 1 and adapting the proofs of Proposition 9 and of Corollary 5, we show without difficulty that

$$(4.58)\qquad v \leqslant w \quad \text{for all} \quad v \in \mathscr{I}_-(\varphi) \quad \text{and} \quad w \in \mathscr{I}_+(\varphi)\,.$$

Unlike the bounded case $\mathscr{I}_-(\varphi)$ and $\mathscr{I}_+(\varphi)$ can be empty, and it is not true in general, when they are non-empty, that $\sup \mathscr{I}_-(\varphi) = \inf \mathscr{I}_+(\varphi)$.

We pose the

Definition 10. Let Ω be an unbounded open set in $\mathbb{R}^n$ with boundary Γ and $\varphi \in \mathscr{C}^0(\Gamma)$. We say that Dirichlet's problem $P(\Omega, \varphi)$ with null condition at infinity admits a generalized solution if the sets $\mathscr{I}_-(\varphi)$ and $\mathscr{I}_+(\varphi)$ defined by (4.57) are non-empty and if $\sup \mathscr{I}_-(\varphi) = \inf \mathscr{I}_+(\varphi)$. We call the function $u(\varphi) = \sup \mathscr{I}_-(\varphi) = \inf \mathscr{I}_+(\varphi)$ *the generalized solution of the problem* $P(\Omega, \varphi)$ *with the null condition at infinity.*

It is clear that a classical solution of $P(\Omega, \varphi)$ satisfying the null condition at infinity is a generalized solution of $P(\Omega, \varphi)$ with the null condition at infinity. We note that in general $u(\varphi)$, if it exists, does not satisfy the null condition at infinity, and it no longer satisfies

$$\lim_{x \to z} u(\varphi, x) = \varphi(z) \quad \text{for} \quad z \in \Gamma\,.$$

We see easily as in the bounded case that *the map* $\varphi \to u(\varphi)$ *is linear.* This follows from

$$\mathscr{I}_-(\lambda\varphi) = \lambda\,\mathscr{I}_-(\varphi) \quad \text{for} \quad \lambda > 0, \quad \mathscr{I}_-(\varphi_1) + \mathscr{I}_-(\varphi_2) \subset \mathscr{I}_-(\varphi_1 + \varphi_2)\,.$$

We have also immediately

$$\varphi_1 \leqslant \varphi_2 \quad \text{on} \quad \Gamma \Rightarrow u(\varphi_1) \leqslant u(\varphi_2) \quad \text{on} \quad \Omega\,.$$

We deduce that *when Ω satisfies* (4.52)

$$\inf_{\Gamma} \varphi \leqslant u(\varphi) \leqslant \sup_{\Gamma} \varphi \quad \text{on} \quad \Omega^{78} \tag{4.59}$$

since in this case $u(c) \equiv c$ for all $c \in \mathbb{R}$.
In the general case, we have merely

$$\min\left(0, \inf_{\Gamma} \varphi\right) \leqslant u(\varphi) \leqslant \max\left(0, \sup_{\Gamma} \varphi\right) \quad \text{on} \quad \Omega^{78} \tag{4.60}$$

which follows from $v \equiv \min\left(0, \min_{\Gamma} \varphi\right) \in \mathscr{I}_{-}(\varphi)$.
We shall not develop further Perron's method in the unbounded case. We shall rather take an interest in *Kelvin's inversion method* for which we shall need *to suppose* $\bar{\Omega} \neq \mathbb{R}^n$ *with* $n \geqslant 2$. We fix $x_0 \in \mathbb{R}^n \backslash \Omega$, $r_0 > 0$ and denote by Ω' the image $I(x_0, r_0)(\Omega)$ of Ω under the inversion $I(x_0, r_0)$; it is a bounded open set whose boundary $\partial\Omega'$ consists of $\Gamma' = I(x_0, r_0)(\Gamma)$ and $\{x_0\}$[79]. Given u, v, etc., defined on Ω we denote by u', v', etc., their Kelvin transforms defined on Ω'. We know already that $u \in \mathscr{H}(\Omega)$ iff $u' \in \mathscr{H}(\Omega')$. Given $v \in \mathscr{C}^0(\Omega)$, we have $v' \in \mathscr{C}^0(\Omega')$ and v is sub-harmonic on Ω iff v' is sub-harmonic on Ω': that follows immediately from Definition 2 and from Proposition 19 of §2. We fix $\varphi \in \mathscr{C}^0(\Gamma)$; its Kelvin transform is a function $\varphi' \in \mathscr{C}^0(\Gamma')$. For all $z \in \Gamma$, we have:

$$\limsup_{x \to z} v(x) \leqslant \varphi(z) \quad \text{iff} \quad \limsup_{x' \to z'} v'(x') \leqslant \varphi'(z')$$

where $z' = I(x_0, r_0)z \in \Gamma'$.
We now distinguish two cases.

Case 1. Γ is bounded (exterior problem). In this case $\tilde{\Omega}' = \Omega' \cup \{x_0\}$ is a bounded open set with boundary Γ'. We can consider $\tilde{u}'$ the generalized solution of $P(\tilde{\Omega}', \varphi')$ and u the Kelvin transform of the restriction u' of $\tilde{u}'$ to Ω'. We have

$$\lim_{|x| \to \infty} |x - x_0|^{n-2} u(x) = r_0^{n-2} \tilde{u}'(x_0).$$

Now by the definition of $\tilde{u}'$,

$u = \sup\{v \in \mathscr{C}^0(\Omega)$, sub-harmonic, $\limsup_{x \to z} v(x) \leqslant \varphi(z)$ for all $z \in \Gamma$, and v' extends to a continuous function, sub-harmonic on $\tilde{\Omega}'\}$;
hence

$$u \leqslant \sup \mathscr{I}_{-}(\varphi).$$

Similarly $u \geqslant \mathscr{I}_{+}(\varphi)$ and hence $u = u(\varphi)$, generalized solution of $P(\Omega, \varphi)$ with the null condition at infinity.
We state already the result

---

[78] *A priori* φ is not bounded on Γ; but the inequality is trivial when there is an ∞.
[79] We refer to §2.5 for all the ideas concerning Kelvin's transformation.

Proposition 10. *Let Ω be an unbounded set non-dense in $\mathbb{R}^n$ with $n \geqslant 2$ with bounded boundary Γ and $\varphi \in \mathscr{C}^0(\Gamma)$. Then the Dirichlet problem $P(\Omega, \varphi)$ with null condition at infinity admits a generalized solution $u(\varphi)$ satisfying the condition that*

$$\lim_{|x| \to \infty} |x|^{n-2} u(\varphi, x) \quad \textit{exists}\,. \tag{4.61}$$

Also for $x_0 \in \mathbb{R}^n \backslash \bar{\Omega}$ and $r_0 > 0$, $u(\varphi)$ is the Kelvin transform $H(x_0, r_0)u'$ of the restriction to $\Omega' = I(x_0, r_0)\Omega$ of the generalized solution of $P(\Omega' \cup \{x_0\}, \varphi')$.

Case 2. Γ is unbounded. In this case Γ' is dense in $\partial\Omega'$. Firstly, *let us suppose* that φ' is extended by continuity to $\tilde{\varphi}' \in \mathscr{C}^0(\partial\Omega')$, that is to say *that* $\lim\limits_{|z| \to \infty} |z - x_0|^{n-2}\varphi(z) = r_0^{n-2}\tilde{\varphi}'(x_0)$ *exists.* We can then consider the generalized solution u' of $P(\Omega', \tilde{\varphi}')$ and its Kelvin transform u.
By definition of u'

$$u = \sup\left\{ v \in \mathscr{C}^0(\Omega) \text{ sub-harmonic; } \limsup_{x \to z} v(x) \leqslant \varphi(z) \text{ for all } z \in \Gamma \text{ and } \limsup_{|x| \to \infty} |x - x_0|^{n-2} v(x) \leqslant r_0^{n-2}\tilde{\varphi}'(x_0)\} \leqslant \sup \mathscr{I}_-(\varphi) \right\}.$$

As above $u = u(\varphi)$. Finally, we note that, since u' is bounded, $|x - x_0|^{n-2}u(x)$ is bounded.
We consider now $\varphi \in \mathscr{C}^0(\Gamma)$ satisfying

$$\begin{cases} \text{there exists } h \in \mathscr{H}(\Omega) \cap \mathscr{C}^0(\bar{\Omega}) \quad \text{with} \quad \lim\limits_{|x| \to \infty} |x - x_0|^{n-2} h(x) = +\infty \\[2ex] \text{such that} \quad \lim\limits_{|z| \to \infty} \dfrac{\varphi(z)}{h(z)} = 0\,. \end{cases} \tag{4.62}$$

Even if it entails replacing h by $h + \dfrac{M}{|x - x_0|^{n-2}}$ with $M > \sup\limits_{\Omega} |x - x_0|^{n-2} h(x)^-$, we can always suppose that $h > 0$ on $\bar{\Omega}$. We take a sequence (φ_k) of continuous functions with compact support in Γ such that

$$\varepsilon_k = \|\varphi_k - \varphi\|_h = \max_{\Gamma} \frac{|\varphi_k - \varphi|}{h} \to 0 \quad \text{when} \quad k \to \infty\,.$$

From the preceding case $u_k = u(\varphi_k)$ exists and we have

$$u_k = \sup \mathscr{I}_-^h(\varphi_k) = \inf \mathscr{I}_+^h(\varphi_k)$$

where

$$\mathscr{I}_-^h(\varphi) = \left\{ v \in \mathscr{C}^0(\Omega) \text{ sub-harmonic: } \limsup_{x \to z} v(x) \leqslant \varphi(z)\, \forall z \in \Gamma \quad \text{and} \quad \limsup_{|x| \to \infty} \frac{v(x)}{h(x)} \leqslant 0 \right\}$$

and

$$\mathscr{I}_+^h(\varphi) = -\mathscr{I}_-^h(-\varphi).$$

We shall show that, given $\varepsilon > 0$

(4.63) $$u_k - u_l \leqslant \max_\Gamma (\varphi_k - \varphi_l - \varepsilon h)^+ + \varepsilon h \quad \text{on} \quad \Gamma$$

It follows immediately that

(4.64) $$u_k - u_l \leqslant (\varepsilon_k + \varepsilon_l)h \quad \text{on} \quad \Omega,$$

and hence

$$\|u_k - u_l\|_h = \sup_\Omega \frac{|u_k - u_l|}{h} \leqslant \varepsilon_k + \varepsilon_l,$$

that is to say u_k/h converges uniformly on Ω and in particular u_k converges uniformly on every bounded set of Ω. Denoting its limit by u we show that

$$u = \sup \mathscr{I}_-^h(\varphi) = \inf \mathscr{I}_+^h(\varphi).$$

Given $x \in \Omega$, for all l there exists $v_l \in \mathscr{I}_-^h(\varphi_l)$ such that $u_l(x) \leqslant v_l(x) + \varepsilon_l$; we have $v = v_l - \varepsilon_l h \in \mathscr{I}_-^h(\varphi)$ since $\varphi_l - \varepsilon_l h \leqslant \varphi$ by the definition of ε_l. Applying (4.64) in the limit when $k \to \infty$ we have

$$u(x) \leqslant v(x) + \varepsilon_l(2h(x) + 1).$$

This shows clearly that $u \leqslant \sup \mathscr{I}_-^h(\varphi)$; similarly, we can show that $u \geqslant \inf \mathscr{I}_+^h(\varphi)$. We shall therefore have proved

Proposition 11. *Let Ω be an open set, non dense in $\mathbb{R}^n$ with $n \geqslant 2$ and unbounded boundary Γ, $\varphi \in \mathscr{C}^0(\Gamma)$, and $h \in \mathscr{H}(\Omega) \cap \mathscr{C}^0(\bar{\Omega})$ such that*

(4.65) $$\lim_{|x| \to \infty} |x|^{n-2} h(x) = +\infty, \qquad \lim_{|z| \to \infty} \frac{\varphi(z)}{h(z)} = 0.$$

Then the Dirichlet problem $P(\Omega, \varphi)$ with null condition at infinity admits a generalized solution satisfying

(4.66) $$u(\varphi) \leqslant \max_\Gamma (\varphi - \varepsilon h)^+ + \varepsilon h \quad \text{on} \quad \Omega \quad \text{for all} \quad \varepsilon > 0.$$

If, in addition, $\lim_{|z| \to \infty} |z|^{n-2} \varphi(z)$ exists, then for all $x_0 \in \mathbb{R}^n \backslash \bar{\Omega}$ and $r_0 > 0$, $u(\varphi)$ is the Kelvin transform $H(x_0, r_0)u'$ of the generalized solution of $P(\Omega', \tilde{\varphi}')$ where $\Omega' = I(x_0, r_0)\Omega$ and

$$\tilde{\varphi}'(z') = \begin{cases} H(x_0, r_0)\varphi(z') & \text{for} \quad z' \in I(x_0, r_0)\Gamma \\ \lim\limits_{|z| \to \infty} \left(\dfrac{|z - x_0|}{r_0}\right)^{n-2} \varphi(z) & \text{for} \quad z' = x_0. \end{cases}$$

Remark 4. With the above notation, being given $\varphi' \in \mathscr{C}^0(\Gamma')$ satisfying

$$\begin{cases} \text{there exists} \quad h' \in \mathscr{H}(\Omega') \cap \mathscr{C}^0(\bar{\Omega}'\backslash\{x_0\}), \quad \text{with} \quad \lim\limits_{x' \to x_0} h'(x') = +\infty \\ \text{such that} \quad \lim\limits_{z' \to x_0} \dfrac{\varphi'(z')}{h'(z')} = 0\,, \end{cases}$$

the Kelvin transform $\varphi = H(x_0, r_0)\varphi'$ satisfies (4.62) with $h = H(x_0, r_0)h'$. We can therefore consider the Kelvin transform u' of the generalized solution $u(\varphi)$. This function is quite naturally a "generalized solution" of the problem $P(\Omega', \varphi')$ where φ' is not continuous on $\partial\Omega'$. We come back in §6, to the definition of generalized solutions of Dirichlet problems with data which are not continuous. □

Conclusion of the Proof of Proposition 11. Proposition 11 has been demonstrated above, with the exception of (4.66), under the condition (4.63). In fact (4.63) is a particular case of (4.66) ($u_k - u_l = u(\varphi_k - \varphi_l)$) if we can demonstrate this last result for all φ such that

$$\lim_{|x| \to \infty} \frac{\varphi(z)}{h(z)} = 0 \quad \text{and} \quad u(\varphi) = \sup \mathscr{I}^h_-(\varphi)\,.$$

For that it is sufficient to prove that, given $\varepsilon > 0$ and $v \in \mathscr{I}^h_-(\varphi)$

$$v \leqslant \max_{\Gamma}(\varphi - \varepsilon h)^+ + \varepsilon h \quad \text{on} \quad \Omega\,.$$

From the hypothesis, there exists R such that $v \leqslant \varepsilon h$ on $\Omega \backslash B(0, R)$. The function $v - \varepsilon h$ is sub-harmonic and continuous on $\Omega\backslash B(0, R)$ and

$$\limsup_{x \to z} v(x) - \varepsilon h(x) \leqslant (\varphi(z) - \varepsilon h(z))^+ \quad \text{on} \quad \partial(\Omega \cap B(0, R))\,.$$

The proof of the Property 1 is then complete. □

The method of Kelvin's transformation likewise provides a result on the existence of a classical solution: in effect with the above notation for $z \in \Gamma$ and $v \in \mathscr{C}^0(\Omega \cap B(z, r))$ is a barrier function[80] of z iff v' is a barrier function of $z' = I(x_0, r_0)z$, and hence:

> *z is a regular point of the boundary of Ω iff z' is a regular point of the boundary of Ω'.*

Applying Theorem 1, we deduce

Theorem 2. *Let Ω be an unbounded set, not dense in $\mathbb{R}^n$ ($n \geqslant 2$) with boundary Γ.*
(1) *The following assertions are equivalent:*
(i) *for every continuous φ with compact support in Γ, there exists a classical solution of $P(\Omega, \varphi)$ satisfying the null condition at infinity;*
(ii) *all the points of the boundary of Ω are regular*[80].

[80] See Definition 4.

(2) *When these assertions are satisfied, then:*
(a) *Case of Γ bounded (exterior problem): for all $\varphi \in \mathscr{C}^0(\Gamma)$, there exists a unique classical solution u of $P(\Omega, \varphi)$ satisfying the null condition at infinity.*
In addition

$$\lim_{|x| \to \infty} |x|^{n-2} u(x)$$

exists.
(b) *Case of Γ unbounded: for all $\varphi \in \mathscr{C}^0(\Gamma)$ and $h \in \mathscr{H}(\Omega) \cap \mathscr{C}^0(\bar{\Omega})$* satisfying (4.65), *there exists a unique classical solution u of $P(\Omega, \varphi)$ satisfying the null condition at infinity; in addition*

$$\lim_{|x| \to \infty} \frac{u(x)}{h(x)} = 0 .$$

Let us complete this section by the introduction of *Green's function of an unbounded open set*. Given $y \in \Omega$ we seek to solve the Dirichlet problem $PH\ (\Omega, \delta_y)$ with the null condition at infinity. Applying the general method, to take us to a Dirichlet problem $P(\Omega, \varphi)$ with null condition at infinity, we have to find $u_0 \in \mathscr{D}'(\Omega)$ solution of

$$\begin{cases} \Delta u_0 = \delta_y \text{ in } \mathscr{D}'(\Omega) , \\ u_0 \text{ satisfies the null condition at infinity} , \\ \lim\limits_{x \to z} u_0(x) \text{ exists for all } z \in \Gamma . \end{cases}$$

In the case $n \geqslant 3$, there is no problem it is enough to consider $u_0(x) = E_n(x - y)$. In the case $n = 1$, if $\Omega \neq \mathbb{R}$, we can take $u_0(x) = \frac{1}{2}|x - y| - \frac{1}{2}|x - x_0|$, where x_0 is a fixed point of $\mathbb{R}\backslash\Omega$.
In the case $n = 2$, to carry out the same operation, we must suppose that $\bar{\Omega} \neq \mathbb{R}^2$; we can take $u_0(x) = \dfrac{1}{2\pi} \operatorname{Log} \dfrac{|x - y|}{|x - x_0|}$ where x_0 is a fixed point of $\mathbb{R}^2\backslash\bar{\Omega}$.
Once u_0 is chosen, we have to solve $P(\Omega, \varphi_0)$ with null condition at infinity. We know how to solve this problem, by Propositions 10 and 11 when $\bar{\Omega} \neq \mathbb{R}^n$; in addition, when all the points of the boundary of Ω are regular, we shall obtain by Theorem 2, a classical solution.
We can equally well use *a method of passage to the limit*: we consider an increasing sequence (Ω_k) of bounded open sets with union Ω (for example, $\Omega_k = \Omega \cap B(0, R_k)$, with (R_k) an increasing sequence of positive real numbers $R_k \to \infty$ as $k \to \infty$). We can study the limit of the sequence (G_k) where G_k is the Green's function of the open set Ω_k; in the case in which Ω is bounded, this limit gives the Green's function of Ω (see Proposition 6); also in the Example 12, we have seen that the Green's functions of the intervals $]a, b[$ converge when $b \to \infty$ to a function which we have called the Green's function of $]a, \infty[$.
In a general way, we prove

Proposition 12. *Let (Ω_k) be an increasing sequence of bounded open sets in $\mathbb{R}^n$ with union Ω bounded or not. We define, as in Proposition 6, the extensions $\tilde{G}_k$ by O of the Green's functions G_k of the bounded open sets Ω_k.*

(1) *The following assertions are equivalent*:
(i) *for all* $(x, y) \in \Omega \times \Omega \backslash D$, *the sequence* $(\tilde{G}_k(x, y))$ *converges*;
(ii) *for every connected component* Ω_0 *of* Ω, *there exists* $(x, y) \in \Omega_0 \times \Omega_0 \backslash D$ *such that* $\tilde{G}_k(x, y)$ *is minorized*;
(iii) *there exists* $u \in \mathscr{H}(\Omega)$ *minorized such that*

$$\liminf_{|x| \to \infty} u(x) - E_n(x) > -\infty$$

(2) *When the assertions of* (1) *are satisfied, the function*

$$G(x, y) = \lim_{k \to \infty} \tilde{G}_k(x, y)$$

satisfies

$$(4.67) \qquad \begin{cases} \text{for all } y \in \Omega, \text{ the function } x \to G(x, y) \text{ is the largest solution} \\ v \in \mathscr{C}^0(\Omega \backslash \{y\}) \cap L^1_{\text{loc}}(\Omega) \text{ satisfying} \\ v \leqslant 0 \text{ on } \Omega \backslash \{y\}, \quad \Delta v = \delta_y \quad \text{in} \quad \mathscr{D}'(\Omega)\,, \end{cases}$$

or, which is equivalent:

$$(4.68) \qquad \begin{cases} \text{for all } y \in \Omega, \text{ the function } x \to E_n(x - y) - G(x, y) \text{ is the smallest} \\ \text{function } u, \text{ harmonic on } \Omega, \text{ satisfying} \\ u(x) \geqslant E_n(x - y) \quad \text{for all} \quad x \in \Omega\,. \end{cases}$$

The existence of a function G defined on $\Omega \times \Omega \backslash D$ satisfying (4.67) or (4.68), which is equivalent to it, implies the assertion (iii) of the point (1): in effect, for $y \in \Omega$ and $0 < r < \text{dist}(y, \partial\Omega)$, the function $x \to E_n(x - y)$ is minorized on $\Omega \backslash \bar{B}(y, r)$ by $E_n(r)$ and

$$\text{case} \quad n \geqslant 2, \quad \lim_{|x| \to \infty} E_n(x - y) - E_n(x) = 0$$

$$\text{case} \quad n = 1, \quad \liminf_{|x| \to \infty} E_n(x - y) - E_n(x) \geqslant -\frac{1}{2}|y|\,.$$

We pose the

Definition 11. Let Ω be an open set of $\mathbb{R}^n$. The function G defined on $\Omega \times \Omega \backslash D$, satisfying (4.67) or (4.68) *if it exists* is called *the Green's function of* Ω.

From Proposition 12, the Green's function of Ω exists iff the sequence of Green's functions of $\Omega \cap B(0, R_k)$ (extended by 0) converge when $R_k \to \infty$, the limit being then the Green's function of Ω. Also the Green's function of Ω exists iff the assertion (iii) of the point (1) of the proposition is satisfied.
In the case $n \geqslant 3$, *every open set of* $\mathbb{R}^n$ *admits a Green's function*: indeed, the function $u \equiv 0$ satisfies the assertion (iii).

In the case $n = 1$, *an open set* Ω *of* $\mathbb{R}$ *admits a Green's function iff* $\Omega \neq \mathbb{R}$; in effect if $x_0 \in \mathbb{R}\backslash\Omega$, $u(x) = \frac{1}{2}|x - x_0|$ satisfies assertion (iii); conversely a function u, harmonic on $\mathbb{R}$, is linear; it cannot therefore satisfy

$$\liminf_{|x| \to \infty} u(x) - \frac{1}{2}|x| > -\infty .$$

In the case $n = 2$, $\Omega = \mathbb{R}^2$, *does* **not** *admit a Green's function*: by Liouville's theorem (see Proposition 7, §2), a harmonic function u minorized on $\mathbb{R}^2$ is constant and therefore cannot satisfy $\lim_{|x| \to \infty} u(x) - \operatorname{Log}|x| > -\infty$. We have the same conclusion for $\Omega = \mathbb{R}^2\backslash\{x_0\}$; to prove it we make use of the following argument: the Green's function of $B(x_0, r)\backslash\{x_0\}$ is the restriction G_r of that of $B(x_0, r)$ (this follows from Remark 1). Now the explicit formula (4.24) shows that $\tilde{G}_r(x, y) \to -\infty$ when $r \to +\infty$. Finally, we note that *a non dense open set* Ω *of* $\mathbb{R}^2$ *admits a Green's function*: indeed if $x_0 \in \mathbb{R}^2\backslash\bar{\Omega}$, $u(x) = E_2(x - x_0)$ is harmonic and minorized on Ω and $\lim_{|x| \to \infty} u(x) - E_2(x) = 0$.

Proof of Proposition 12. Let us first fix y and suppose in the first instance that:

$$\text{(4.69)} \quad \begin{cases} \text{for every connected component } \Omega_0 \text{ of } \Omega, \text{ there exists } x \in \Omega_0 \\ \text{such that } \tilde{G}_k(x, y) \text{ is minorized} . \end{cases}$$

Putting $u_k(x) = E_n(x - y) - \tilde{G}_k(x, y)$, we recall from the proof of Proposition 6 that the sequence (u_k) is increasing. From the hypothesis (4.49) and Harnack's theorem (see Corollary 9 of §2, $u_k \in \mathscr{H}(\Omega_l)$, for all $k \geqslant l$), u_k converges uniformly on every compact subset of Ω; its limit u is harmonic on Ω; since $\tilde{G}_k \leqslant 0$, $u(x) \geqslant E_n(x - y)$ for all $x \in \Omega$; finally considering $\bar{u} \in \mathscr{H}(\Omega)$ and satisfying $\bar{u}(x) \geqslant E_n(x - y)$ for all $x \in \Omega$, we shall have $\bar{u} \geqslant u_k$ on Ω_k and hence $\bar{u} \geqslant u$. In summary, given $y \in \Omega$ satisfying (4.69), for all $x \in \Omega\backslash\{y\}$, $\tilde{G}_k(x, y)$ converges in decreasing and the function

$$u(x) = E_n(x - y) - \lim \tilde{G}_k(x, y)$$

is the smallest harmonic function on Ω verifying $u(x) \geqslant E_n(x - y)$. Also as we have seen above u is suitable for (*iii*). To conclude it is sufficient therefore to show that (*iii*) implies:

$$\tilde{G}_k(x, y) \text{ is minorized for all } (x, y) \in \Omega \times \Omega\backslash D ;$$

now if u satisfies (*iii*), $m(y) = \inf_{x \in \Omega} u(x) - E_n(x - y) > -\infty$, for $y \in \Omega$ fixed, the function $\bar{u}(x) = u(x) - m(y)$ is harmonic on Ω and $\bar{u}(x) \geqslant E_n(x - y)$ for all $x \in \Omega$; hence $\bar{u}(x) \geqslant E_n(x - y) - G_k(x - y)$ for *all* $x \in \Omega_k$; in short

$$\tilde{G}_k(x, y) \geqslant \min(0, E_n(x - y) + m(y) - u(x)) . \qquad \square$$

Finally let us give some examples.

Example 17. *Green's function of the exterior of a ball in* $\mathbb{R}^n$, $n \geqslant 2$.
Let $\Omega = \mathbb{R}^n \backslash \bar{B}(x_0, r_0)$. We look back to Example 11; the function G given by the formulae (4.24) is defined for all $(x, y) \in \mathbb{R}^n \times \mathbb{R}^n$ with $y \neq x_0$, $x \neq y$ and $x \neq I(x_0, r_0)y$ and satisfies

$$G = 0 \quad \text{on} \quad \partial B(x_0, r_0) \times (\mathbb{R}^n \backslash \{x_0\}) \backslash D$$

and for all $y \in \mathbb{R}^n \backslash \{x_0\}$, $\Delta_x G(x, y) = \delta_y - \dfrac{r_0}{|y - x_0|} \delta_{I(x_0, r_0)y}$. Finally

$$\text{case} \quad n \geqslant 3, \quad \lim_{|x| \to \infty} G(x, y) = 0$$

$$\text{case} \quad n = 2, \quad \lim_{|x| \to \infty} G(x, y) = -\frac{1}{2\pi} \operatorname{Log} \frac{|y - x_0|}{r_0} .$$

Hence the Green's function of $\Omega = \mathbb{R}^n \backslash \bar{B}(x_0, r_0)$ is given by the same formulae for all $(x, y) \in \Omega \times \Omega \backslash D$. □

Example 18. *Green's function of a half-space* $\mathbb{R}^{n-1} \times \mathbb{R}^+$.
The symmetry $T: (y', y_n) \to (y', -y_n)$ maps $\mathbb{R}^{n-1} \times \mathbb{R}^+$ onto $\mathbb{R}^{n-1} \times \mathbb{R}^-$. It is clear, identifying $\mathbb{R}^{n-1}$ with $\mathbb{R}^{n-1} \times \{0\}$ that

$$E_n(x' - y) = E_n(x' - Ty) \quad \text{for all} \quad x' \in \mathbb{R}^{n-1} .$$

We deduce immediately that the Green's function on $\mathbb{R}^{n-1} \times \mathbb{R}^+$

$$G(x, y) = E_n(x - y) - E_n(x - Ty) ,$$

that is to say

$$G((x', x_n), (y', y_n)) = E_n(x' - y', x_n - y_n) - E_n(x' - y', x_n + y_n) .$$

We note that in the case $n = 1$, we recover (4.43) (with $a = 0$). □

4. The Neumann Problem; Mixed Problem; Hopf's Maximum Principle; Examples

We take Ω to be an open set of $\mathbb{R}^n$ with boundary Γ and consider the problem

$$PN(\Omega, \psi) \begin{cases} u \text{ harmonic} & \text{on} \quad \Omega \\ \dfrac{\partial u}{\partial n} = \psi & \text{on} \quad \Gamma . \end{cases}$$

where ψ is given and defined on Γ.
This is *the Neumann problem for Laplace's equation*. In the classical setting, to define $\partial u / \partial n$, we must *suppose* Ω *to be regular*. We pose in a natural way the

Definition 12. Let Ω be a regular open set of $\mathbb{R}^n$ with boundary Γ and let $\psi \in \mathscr{C}^0(\Gamma)$. By a *classical solution of* $PN(\Omega, \psi)$ we mean any solution

$u \in \mathscr{H}(\Omega) \cap \mathscr{C}_n^1(\bar{\Omega})$ such that

$$\frac{\partial u}{\partial n}(x) = \psi(z) \quad \text{for all} \quad z \in \Gamma .$$

It is clear that there is *no unique solution of the Neumann problem*: if u is a classical solution of $PN(\Omega, \psi)$, for every connected component Ω_0 of Ω and all $c \in \mathbb{R}$, $u + c\chi_{\Omega_0}$ is also a classical solution of $PN(\Omega, \psi)$ since

$$\operatorname{grad}(u + c\chi_{\Omega_0}) = \operatorname{grad} u \quad \text{on} \quad \Omega .$$

On the other hand, *a solution does* **not** *exist for all* $\psi \in \mathscr{C}^0(\Gamma)$: from Gauss' theorem, if $PN(\Omega, \psi)$ admits a classical solution, then for every bounded connected component Ω_0 of Ω,

$$\int_{\partial\Omega_0} \psi \, d\gamma = 0 .$$

Making use of *Fredholm's integral method*, we shall prove the

Theorem 3. *Let Ω be a* **bounded** *regular open set of $\mathbb{R}^n$ with boundary Γ of class $\mathscr{C}^{1+\varepsilon}$ and let $\psi \in \mathscr{C}^0(\Gamma)$.*
(1) *The Neumann problem $PN(\Omega, \psi)$ admits a solution iff ψ satisfies Gauss' conditions*

$$\text{(4.70)} \qquad \textit{for every connected component } \Omega_0 \textit{ of } \Omega, \int_{\partial\Omega_0} \psi \, d\gamma = 0 .$$

This solution is then unique to within an additive constant on each component of Ω .
(2) *The exterior Neumann problem $PN(\mathbb{R}^n \backslash \bar{\Omega}, \psi)$ admits a solution iff ψ satisfies Gauss' exterior conditions:*

$$\text{(4.71)} \qquad \textit{for every bounded connected component } \Omega_0' \textit{ of } \mathbb{R}^n\bar{\Omega}, \int_{\partial\Omega_0'} \psi \, d\gamma = 0 .$$

Then
(*a*) *in the case $n \geqslant 3$, there exists a solution u of $PN(\mathbb{R}^n \backslash \bar{\Omega}, \psi)$ satisfying*

$$\lim_{|x| \to \infty} u(x) = 0 ;$$

this solution is unique to within an additive constant on each bounded connected component of $\mathbb{R}^n \backslash \bar{\Omega}$.
(*b*) *in the case $n \geqslant 2$, there exists a solution u of $PN(\mathbb{R}^n \backslash \bar{\Omega}, \psi)$ satisfying*

$$\lim_{|x| \to \infty} \operatorname{grad} u(x) = 0 .$$

This solution is unique to within an additive constant on each connected component (bounded or not) of $\mathbb{R}^n \backslash \bar{\Omega}$.

In the case $n = 1$, we shall determine the solutions explicitly (see Example 19). We shall prove this theorem in the course of §4.5 (see Theorems 4 and 5 and

Proposition 17). Fredholm's integral method of solution allows us to deal with, more generally, *mixed boundary value problems*. Given Ω an open set in $\mathbb{R}^n$ of boundary Γ we consider the problem

$$PM(\Omega, \Gamma_0, \varphi, \psi, \lambda) \begin{cases} u \text{ harmonic} \quad \text{on} \quad \Omega \\ u = \varphi \quad \text{on} \quad \Gamma_0 \\ \dfrac{\partial u}{\partial n} + \lambda u = \psi \quad \text{on} \quad \Gamma \backslash \Gamma_0 , \end{cases}$$

where Γ_0 is a part of the boundary Γ, φ is defined on Γ_0 and ψ and λ are defined on $\Gamma \backslash \Gamma_0$. We note that the Dirichlet problem $P(\Omega, \varphi)$ corresponds to $\Gamma_0 = \Gamma$ and the Neumann problem $PN(\Omega, \psi)$ to $\Gamma_0 = \varnothing$ and $\lambda \equiv 0$. In the *classical theory*, we can always assume that Γ_0 *is closed*: in effect if $u \in \mathscr{C}^0(\Omega)$ satisfies $u = \varphi$ on Γ_0, φ can be extended to a function $\bar{\varphi} \in \mathscr{C}^0(\bar{\Gamma}_0)$ and $u = \bar{\varphi}$ on $\bar{\Gamma}_0$. Also to give a meaning to $\partial u / \partial n$ on $\Gamma \backslash \Gamma_0$ we must suppose

$$\text{(4.72)} \qquad \begin{cases} \Gamma \backslash \Gamma_0 \text{ is a hypersurface (of class } \mathscr{C}^1\text{) with } \Omega \text{ locally} \\ \text{on the one side of } \Gamma \backslash \Gamma_0 \end{cases}$$

We shall then always be able for $u \in \mathscr{C}^1(\Omega)$ and $z \in \Gamma \backslash \Gamma_0$ to define $\partial u / \partial n(z)$ (see §1.3). We have the

Definition 13. Let Ω be an open set of $\mathbb{R}^n$, Γ_0 a closed part of its boundary Γ with $\Gamma \backslash \Gamma_0$ satisfying (4.72) and let $\varphi \in \mathscr{C}^0(\Gamma_0)$, $\psi, \lambda \in \mathscr{C}^0(\Gamma \backslash \Gamma_0)$.
We give the name *classical solution of* $PM(\Omega, \Gamma_0, \varphi, \psi, \lambda)$ to any function $u \in \mathscr{H}(\Omega) \cap \mathscr{C}^0(\bar{\Omega})$ satisfying

$$\begin{cases} u(z) = \varphi(z) \quad \text{for all} \quad z \in \Gamma_0 \\ \dfrac{\partial u}{\partial n}(z) \quad \text{exists and} \quad \dfrac{\partial u}{\partial n}(z) + \lambda(z) u(z) = \psi(z) \quad \text{for all} \quad z \in \Gamma \backslash \Gamma_0 . \end{cases}$$

This definition clearly generalizes, at the same time, the definition of a classical solution of a Dirichlet problem $P(\Omega, \varphi)$ and that of a classical solution of a Neumann problem $PN(\Omega, \psi)$. We prove a "uniqueness' result for the mixed boundary value problem; more generally, we prove the following *maximum principle*:

Proposition 13. *Let Ω be a connected open set of $\mathbb{R}^n$, Γ_0 a closed part of its boundary Γ satisfying* (4.72), $\varphi \in \mathscr{C}^0(\Gamma_0)$, ψ, $\lambda \in \mathscr{C}^0(\Gamma \backslash \Gamma_0)$ *and u a classical solution of* $PM(\Omega, \Gamma_0, \varphi, \psi, \lambda)$.
Suppose that $\varphi \geqslant 0$ on Γ_0, $\psi \geqslant 0$ and $\lambda \geqslant 0$ on $\Gamma \backslash \Gamma_0$ and in the case when Ω is unbounded

$$\liminf_{|x| \to \infty} u(x) \geqslant 0 .$$

Then u is constant on Ω or $u > 0$ on Ω.

We then have the

Corollary 6. *Let Ω be an open set in $\mathbb{R}^n$, Γ_0 a closed part of its boundary Γ with $\Gamma\backslash\Gamma_0$ satisfying* (4.72), $\varphi \in \mathscr{C}^0(\Gamma_0)$, ψ, $\lambda \in \mathscr{C}^0(\Gamma\backslash\Gamma_0)$ *and in the case when Ω is unbounded $u_\infty \in \mathscr{C}^0(\Omega)$. Let $\lambda \geqslant 0$ on $\Gamma\backslash\Gamma_0$. Then a solution u of $PM(\Omega, \Gamma_0, \varphi, \psi, \lambda)$ satisfying, in the case when Ω is unbounded the condition at infinity*

$$\lim_{|x|\to\infty} u(x) - u_\infty(x) = 0$$

is, if it exists, unique up to within an additive constant on each bounded connected component Ω_0 of Ω such that

$$\partial\Omega_0 \cap \Gamma_0 = \varnothing \quad \textit{and} \quad \lambda \equiv 0 \quad \textit{on} \quad \partial\Omega_0 \cap \Gamma\backslash\Gamma_0 . \tag{4.73}$$

Proof of the Corollary. By linearity, we need only consider u a solution of $PM(\Omega, \Gamma_0, 0, 0, \lambda)$ satisfying in the case of Ω unbounded $\lim_{|x|\to\infty} u(x) = 0$. From Proposition 13 u is constant on each connected component Ω_0 of Ω. We have

$$u = 0 \quad \text{on} \quad \partial\Omega_0 \cap \Gamma_0 \quad \text{and} \quad \{z \in \partial\Omega_0 \,;\, \lambda(z) > 0\}$$

and in the case of unbounded Ω_0, $\lim_{|x|\to\infty} u(x) = 0$; hence this constant is zero unless Ω_0 is bounded satisfying (4.73). □

Proof of the Proposition 13. From the maximum principle (see Proposition 8 of §2) u is constant on Ω on $u(x) > \inf_\Omega u$ for all $x \in \Omega$; we can therefore suppose that $\inf_\Omega < 0$. We take $0 < \varepsilon < -\inf_\Omega u$ and $\rho \in \mathscr{C}^1(\mathbb{R})$ with

$$\rho = 0 \text{ on } [-\varepsilon, \infty[\quad \text{and} \quad 0 < \rho' \leqslant 1, \quad -1 \leqslant \rho < 0 \quad \text{on} \quad]-\infty, -\varepsilon[.$$

We put $v = \rho(u)$. From the hypothesis

$$v(x) = 0 \quad \text{for} \quad x \in \Omega \text{ sufficiently near } \Gamma_0$$

and also in the case when Ω is unbounded

$$v(x) = 0 \quad \text{for} \quad |x| \text{ sufficiently large.}$$

Hence there exists a bounded open set $\Omega_0 \subset \Omega$ such that

$$\partial\Omega_0\backslash\Omega \subset \Gamma\backslash\Gamma_0 \quad \text{and} \quad v \equiv 0 \quad \text{on} \quad \Omega\backslash\Omega_0 .$$

Using the hypothesis (4.72), we can suppose that Ω_0 is regular. We then have

$$u \in \mathscr{H}(\Omega_0) \cap \mathscr{C}^1_n(\bar{\Omega}_0), \quad v \in \mathscr{C}^1_n(\bar{\Omega}_0), \quad \text{(see Proposition 5 of §1)}$$

$$\operatorname{grad} v . \operatorname{grad} u = \rho'(u)|\operatorname{grad} u|^2 \in L^1(\Omega_0)$$

$$v\frac{\partial u}{\partial n} \equiv 0 \quad \text{on} \quad \partial\Omega_0 \cap \Omega ,$$

$$v\frac{\partial u}{\partial n} = v\psi - \lambda v u \leqslant 0 \quad \text{on} \quad \partial\Omega_0\backslash\Omega .$$

Applying Green's formula for integration by parts (see Proposition 4 of §1),

$$\int_{\Omega_0} \rho'(u)|\operatorname{grad} u^2|\,dx = \int_{\partial\Omega_0\setminus\Omega} (v\psi - \lambda v u)\,d\gamma \leqslant 0 .$$

Hence $\rho'(u)|\operatorname{grad} u|^2 = 0$ on Ω_0. We deduce that

$$\operatorname{grad} v = \rho'(u)\operatorname{grad} u = 0 \quad \text{on} \quad \Omega_0 .$$

We can always suppose that Ω_0 is connected; hence v is constant on Ω_0. By hypothesis $\inf_\Omega v = \rho\left(\inf_\Omega u\right) < 0$; hence $v = \rho(u) = c < 0$ and $u \equiv \rho^{-1}(c)$. □

We note another means of approaching the principle of the maximum for boundary value problems by the use of *Hopf's maximum principle*; first, we state

Lemma 2. *Let Ω be a connected open set of $\mathbb{R}^n$, $z \in \partial\Omega$ and n a unit vector of $\mathbb{R}^n$ satisfying*

$$(4.74)\quad \begin{cases} \text{there exists an open set } \Omega_0 \subset \Omega \text{ and } v \in \mathscr{C}^0(\bar{\Omega}_0) \text{ sub-harmonic on } \Omega_0 \\ \text{such that } \{x \in \partial\Omega_0;\, v(x) > 0\} \text{ is relatively compact in } \Omega , \\ z - \lambda n \in \Omega_0 \text{ for } \lambda > 0 \text{ small and } \liminf\limits_{\lambda\downarrow 0} \dfrac{v(z-\lambda n)}{\lambda} > 0 . \end{cases}$$

Then, for all $u \in \mathscr{H}(\Omega) \cap \mathscr{C}^0(\bar{\Omega})$ non constant

$$u(z) = \min_{\bar{\Omega}} u \Rightarrow \limsup_{\lambda\downarrow 0} \frac{u(z) - u(z-\lambda n)}{\lambda} < 0 .$$

Proof. We can always suppose that $u(z) = \min_{\bar{\Omega}} u = 0$; from the maximum principle $u > 0$ on Ω and hence in particular

$$u \geqslant m > 0 \quad \text{on} \quad \{x \in \partial\Omega_0;\, v(x) > 0\} .$$

We put $\varepsilon = \dfrac{m}{\max_{\bar{\Omega}_0} v}$ if $\max_{\bar{\Omega}_0} v > 0$ and $\varepsilon = 1$ otherwise.

We have $\qquad \varepsilon v \leqslant m \leqslant u \quad \text{on} \quad \{x \in \partial\Omega_0;\, v(x) > 0\}$

and $\qquad \varepsilon v \leqslant 0 \leqslant u \quad \text{on} \quad \{x \in \partial\Omega_0;\, v(x) \leqslant 0\}$.

Therefore we deduce from the maximum principle for sub-harmonic functions (Property 1) that

$$\varepsilon v \leqslant u \quad \text{on} \quad \Omega_0$$

from which we have

$$\limsup_{\lambda\downarrow 0} \frac{u(z) - u(z-\lambda n)}{\lambda} = -\liminf_{\lambda\downarrow 0} \frac{u(z-\lambda n)}{\lambda}$$

$$\leqslant -\varepsilon \liminf_{\lambda\downarrow 0} \frac{v(z-\lambda n)}{\lambda} < 0 . \qquad □$$

The most common example in which the hypotheses of the lemma are satisfied is the case of an *interior ball condition*: an open set Ω of $\mathbb{R}^n$ satisfies an interior ball condition at $z \in \partial\Omega$ if there exists $x_0 \in \Omega$ such that $B(x_0, |x_0 - z|) \subset \Omega$. Putting

$$n = \frac{z - x_0}{|z - x_0|}, \qquad r = |x_0 - z|, \qquad \Omega_0 = B(x_0, r)\backslash \bar{B}\left(x_0, \frac{r}{2}\right);$$

for $\alpha \geqslant 2n/r^2$, the function $v(x) = \mathrm{e}^{-\alpha r^2} - \mathrm{e}^{-\alpha|x - x_0|^2}$ is sub-harmonic on Ω_0; in effect

$$\Delta v(x) = 2\alpha(2\alpha|x - x_0|^2 - n)\mathrm{e}^{-\alpha|x - x_0|^2}.$$

We have then $\{x \in \partial\Omega_0; v(x) > 0\} = \partial B\left(x_0, \frac{r}{2}\right)$ compact in Ω,

$$\lim_{\lambda \downarrow 0} \frac{v(z - \lambda n)}{\lambda} = 2\alpha r \mathrm{e}^{-\alpha r^2} > 0 .$$

We deduce from the lemma, the so-called *Hopf's maximum principle*:

Proposition 14. *Let Ω be a connected open set in $\mathbb{R}^n$ with boundary Γ, $u \in \mathscr{H}(\Omega) \cap \mathscr{C}^0(\bar{\Omega})$ and $c \in \mathbb{R}$. Let us suppose that*
(1) *for all $z \in \Gamma$ with $u(z) \leqslant c$, there exists a unit vector n satisfying* (4.74) *such that*
$$\limsup_{\lambda \downarrow 0} \frac{u(z) - u(z - \lambda n)}{\lambda} \geqslant 0 ;$$
(2) *there exists $x \in \Omega$ such that $u(x) > c$;*
(3) *When Ω is unbounded,* $\liminf_{|x| \to \infty} u(x) \geqslant c$;

then
$$u > c \quad on \quad \Omega .$$

Proof. From the maximum principle for harmonic functions, if the conclusion is false, u attains a minimum which is less than or equal to c at a point $z \in \Gamma$, which contradicts the lemma. □

This gives the

Corollary 7. *Let Ω be an open set of $\mathbb{R}^n$, Γ a closed part of its boundary Γ, $\varphi \in \mathscr{C}^0(\Gamma_0)$, ψ and λ defined on $\Gamma\backslash\Gamma_0$ with $\lambda \geqslant 0$ and in the case of Ω unbounded $u_\infty \in \mathscr{C}^0(\Omega)$. For all $z \in \Gamma\backslash\Gamma_0$, we choose a unit vector $n(z)$ satisfying* (4.74) *(we suppose that it exists). We then have the conclusion of Corollary* 6, *for a solution of $PM(\Omega, \Gamma_0, \varphi, \psi, \lambda)$ in the sense*

$$(4.75) \begin{cases} u \in \mathscr{H}(\Omega) \cap \mathscr{C}^0(\bar{\Omega}), \quad u(z) = \varphi(z) & \textit{for all} \quad z \in \Gamma_0, \\ \lim\limits_{\lambda \downarrow 0} \dfrac{u(z) - u(z - \lambda n(z))}{\lambda} + \lambda(z)u(z) = \psi(z) & \textit{for all} \quad z \in \Gamma\backslash\Gamma_0 . \end{cases}$$

The concept of a solution of $PM(\Omega, \Gamma_0, \varphi, \psi, \lambda)$ defined by (4.75) is weaker than that of the classical solution when, $\Gamma\backslash\Gamma_0$ satisfying (4.72), for every point $z \in \Gamma/\Gamma_0$

the exterior normal vector $n(z)$ satisfies (4.74); this is in particular the case if we have the interior ball condition at each point $z \in \Gamma \backslash \Gamma_0$.
Note that mixed problems appear naturally when we apply a *Kelvin transformation* to a Neumann problem. Let us take an open set Ω and $x_0 \in \mathbb{R}^n$, $r_0 > 0$; we denote $\Omega' = I(x_0, r_0)(\Omega \backslash \{x_0\})$ whose boundary $\partial\Omega'$ is

$$\Gamma' = I(x_0, r_0)(\partial\Omega \backslash \{x_0\}) \quad \text{or} \quad \Gamma' \cup \{x_0\}$$

according as Ω is bounded or not.
Given a closed part Γ_0 of $\Gamma = \partial\Omega$ with $\Gamma \backslash \Gamma_0$ satisfying (4.72), we find that

$$\Gamma'_0 = \partial\Omega' \backslash I(x_0, r_0)(\Gamma \backslash (\Gamma_0 \cup \{x_0\}))$$

is a closed part of $\partial\Omega'$ and $\partial\Omega' \backslash \Gamma'_0$ satisfies (4.72). Given $u \in \mathscr{C}^1(\Omega)$ its Kelvin transform[81] $u' = H(x_0, r_0)u \in \mathscr{C}^1(\Omega')$; if $\dfrac{\partial u}{\partial n}(z)$ is defined for $z \in \Gamma \backslash (\Gamma_0 \backslash \{x_0\})$ then $\dfrac{\partial u'}{\partial n'}(z')$ is defined for $z' = I(x_0, r_0)z$ and making use of the formula (2.25) of §2, we obtain

$$\frac{\partial u'}{\partial u'}(z') = H(x_0, r_0)\frac{\partial u}{\partial n}(z') + 2\frac{(z' - x_0).n'(z')}{r_0^2}u'(z').$$

Hence *if u is a classical solution of $PM(\Omega, \Gamma_0, \varphi, \psi, \lambda)$ satisfying, in the case when Ω is unbounded, a condition at infinity*

$$\lim_{|x| \to \infty} |x|^{n-2}u(x) = c,$$

its Kelvin transform $u' = H(x_0, r_0)u$ is a classical solution of the mixed problem $PM(\Omega', \Gamma'_0, \varphi', \psi', \lambda')$ where $\Omega' = I(x_0, r_0)(\Omega \backslash \{x_0\})$, with boundary

$$\Gamma'_0 = \partial\Omega' \backslash I(x_0, r_0)(\partial\Omega \backslash (\Gamma_0 \cup \{x_0\})), \qquad \psi' = H(x_0, r_0)\psi,$$

and the functions φ' and λ' defined by

$$\varphi' = \begin{cases} H(x_0, r_0)\varphi & \text{on } I(x_0, r_0)\Gamma_0, \\ c & \text{on } x_0 \text{ if } \Omega \text{ is bounded}. \end{cases}$$

and $$\lambda'(z') = \lambda(I(x_0, r_0)z') - 2\frac{(z' - x_0).n'(z')}{r_0^2}.$$

If $x_0 \in \Omega$, then $\lim_{|x'| \to \infty} |x'|^{n-2}u'(x') = c'$ with $c' = u(x_0)$; it is the same in the case $x_0 \in \Gamma_0$ with $c' = \varphi(x_0)$. Finally, in the case $x_0 \in \partial\Omega \backslash \Gamma_0$ we shall likewise have a mixed condition at infinity.
Let us conclude this section with some examples:

---

[81] See §2.5.

Example 19. *Case $n = 1$.*
Let us consider $\Omega =]a, b[\subset \mathbb{R}$. A function u harmonic on Ω is of the form $u(x) = \alpha x + \beta$. The boundary Γ of Ω is $\{a, b\}$, $\{a\}$ or $\{b\}$ according as Ω bounded, $\Omega =]a, \infty[$ or $\Omega =]-\infty, b[$. A part Γ_0 of Γ could be ϕ, $\{a\}$, $\{b\}$ or the whole of Γ. We give in Table 1 the solutions of $PM(]a, b[, \Gamma_0, \varphi, \psi, \lambda)$ with φ defined on Γ_0, ψ and λ defined on $\Gamma \setminus \Gamma_0$ in the different possible cases.

Example 20. *The radial case.*
Let us consider a radial connected open set in $\mathbb{R}^n$ with $n \geqslant 2$. Excluding the cases $\Omega = \mathbb{R}^n$ and $\Omega = \mathbb{R}^n \setminus \{0\}$, it is one of the following four types:
$\Omega = B(0, r)$ an open ball with boundary $\Gamma = \partial B(0, r)$,
$\Omega = B(0, r) \setminus \{0\}$ a punctured open ball with boundary $\Gamma = \partial B(0, r) \cup \{0\}$,
$\Omega = \{x; r_1 < |x| < r_2$ with $0 < r_1 < r_2 < \infty$, a ring with boundary

$$\Gamma = \partial B(0, r_1) \cup \partial B(0, r_2)\,.$$

$\Omega = \mathbb{R}^n \setminus \bar{B}(0, r)$, the exterior of a ball with boundary $\Gamma = \partial B(0, r)$.
A radial part Γ_0 of the boundary can be ϕ, $\{0\}$, $\partial B(0, r)$ or the whole of Γ according to type. We take φ to be a radial function defined on Γ_0, ψ and λ radial functions defined on $\Gamma \setminus \Gamma_0$. From uniqueness (possibly to within an additive constant), a solution of the mixed problem is necessarily radial and hence of the form:

$$u(x) = c_0 + c_1 E_n(|x|)\,.$$

We shall give in Table 2 of the following page, the solutions in the different possible cases. □

Example 21. *Neumann problem in the half-space.*
We take $\Omega = \mathbb{R}^{n-1} \times \mathbb{R}^+$ with boundary Γ identified with $\mathbb{R}^{n-1}$ and $\psi \in \mathscr{C}^0(\Gamma)$. By definition u is a solution of $PN(\Omega, \psi)$ iff $u \in \mathscr{H}(\Omega)$ and $-\partial u/\partial x_n$ is a solution of the Dirichlet problem $P(\Omega, \psi)$. From Example 14, supposing ψ to be bounded, u is a solution of $PN(\Omega, \psi)$ with $\partial u/\partial x_n$ bounded iff

$$-\frac{\partial u}{\partial x_n}(x) = \frac{2}{\sigma_n}\int_{\mathbb{R}^{n-1}} \frac{\psi(x' + t x_n)}{|t - x|^n}\,dt = 2\int_{\mathbb{R}^{n-1}} \frac{\partial E_n}{\partial x_n}(x - t)\psi(t)\,dt\,,$$

or, on integration

$$u(x', x_n) = u(x', \bar{x}_n) - 2\int_{\mathbb{R}^{n-1}} (E_n(x' - t, x_n) - E_n(x' - t, \bar{x}_n))\psi(t)\,dt\,. \tag{4.76}$$

We now distinguish between the cases $n \geqslant 3$ and $n = 2$.
Case $n \geqslant 3$. Let us suppose

$$\begin{cases} \text{there exists } \varepsilon\colon \mathbb{R}^+ \to \mathbb{R}^+ \text{ decreasing with } \displaystyle\int_0^\infty \varepsilon(r)\,dr < \infty \\ \text{and such that } |\psi(t)| \leqslant \varepsilon(|t|)\,. \end{cases} \tag{4.77}$$

Table 1

			Solutions of $PM(\Omega, \Gamma_0, \varphi, \psi, \lambda)$	Remarks
$\Omega =]a, b[$ bounded		$\Gamma_0 = \Gamma = \{a, b\}$	$\dfrac{\varphi(a)(b - x) + \varphi(b)(x - a)}{b - a}$	Dirichlet's problem
	$\Gamma = \{a\}$	$\lambda(b) \neq -\dfrac{1}{b - a}$	$\dfrac{\varphi(a)(1 + \lambda(b)(b - x)) + \psi(b)(x - a)}{1 + \lambda(b)(b - a)}$	existence and uniqueness
		$\lambda(b) = -\dfrac{1}{b - a}$	$\varphi(a) + c(x - a)$	c arbitrary constant existence iff $\varphi(a) + (b - a)\psi(b) = 0$
	$\Gamma_0 = \varnothing$	$\lambda(a)\lambda(b) \neq \dfrac{\lambda(b) - \lambda(a)}{b - a}$	$\dfrac{\psi(a) - \psi(b) + \psi(a)\lambda(b)(b - x) + \psi(b)\lambda(a)(x - a)}{\lambda(a)\lambda(b)(b - a) - (\lambda(b) - \lambda(a))}$	existence and uniqueness
		$-\lambda(a)\lambda(b) = \dfrac{\lambda(b) + \lambda(a)}{b - a} \neq 0$	$\dfrac{\psi(a)}{\lambda(a)} + c\left(x - a + \dfrac{1}{\lambda(a)}\right)$	c arbitrary constant existence iff $\psi(b)/\lambda(b) = \psi(a)/\lambda(a)$
		$\lambda(a) = \lambda(b) = 0$	$\psi(a)(a - x) + c$	c arbitrary constant existence iff $\psi(a) + \psi(b) = 0$
$\Omega =]a, \infty[$		$\Gamma_0 = \Gamma = \{a\}$	$\varphi(a) + c(a - x)$	c arbitrary constant no condition for existence
		$\Gamma_0 = \{a\}$	$\psi(a)(a - x) + c[1 + \lambda(a)(x - a)]$	c arbitrary constant no condition for existence

Table 2. The radical case (Example 20)

Ω	Γ_0	Condition	Solution(s) of $PM(\Omega, \Gamma_0, \varphi, \psi, \lambda)$	Remarks
$\Omega = B(0, r)$	$\Gamma_0 = \Gamma = \partial B(0, r)$		$\varphi(r)$	Dirichlet's problem
	$\Gamma_0 = \varnothing$	$\lambda(r) \neq 0$	$-\dfrac{\psi(r)}{\lambda(r)}$	existence and uniqueness
		$\lambda(r) = 0$	c	c an arbitrary constant existence iff $\psi(r) = 0$
$\Omega = B(0, r) \setminus \{0\}$	$\Gamma_0 = \Gamma = \partial B(0, r) \cup \{0\}$		$\varphi(r)$	generalized solution of $P(\Omega, \varphi)$ is classical iff $\varphi(0) = \varphi(r)$
	$\Gamma_0 = \{0\}$		$\varphi(0)$	uniqueness and existence iff $\lambda(r)\varphi(0)+\psi(r)=0$
$\Omega = B(0, r_2) \setminus B(0, r_1)$	$\Gamma_0 = \Gamma = \partial B(0, r_1) \cup \partial B(0, r_2)$		$\dfrac{\varphi(r_1)(E_n(r_2)-E_n(x))+\varphi(r_2)(E_n(x)-E_n(r_1))}{E_n(r_2)-E_n(r_1)}$	Dirichlet's problem
	$\Gamma_0 = \partial B(0, r_1)$	$\lambda(r_2)E_n(r_1) \neq E_n(r_2)+\dfrac{1}{\sigma_n r_2^{n-1}}$	$\varphi(r_1) + \dfrac{(\psi(r_2) - \lambda(r_2)\varphi(r_1))(E_n(x) - E_n(r_1))}{E_n(r_2) + \dfrac{1}{\sigma_n r_2^{n-1}} - \lambda(r_2)E_n(r_1)}$	existence and uniqueness
		$\lambda(r_2)E_n(r_1) = E_n(r_2)+\dfrac{1}{\sigma_n r_2^{n-1}}$	$\varphi(r_1) + c(E_n(x) - E_n(r_1))$	c arbitrary constant existence iff $\psi(r_2) = \lambda(r_2)\varphi(r_1)$
	$\Gamma_0 = \varnothing$	$\sigma_n\lambda(r_1)\lambda(r_2)(E_n(r_2) - E_n(r_1)) + \dfrac{\lambda(r_1)}{r_2^{n-1}} + \dfrac{\lambda(r_2)}{r_1^{n-1}} \neq 0$	$u(x)^{(1)}$	existence and uniqueness
		$\sigma_n\lambda(r_1)\lambda(r_2)(E_n(r_1) - E_n(r_2)) = \dfrac{\lambda(r_1)}{r_2^{n-1}} + \dfrac{\lambda(r_2)}{r_1^{n-1}} \neq 0$	$\dfrac{\psi(r_1)}{\lambda(r_1)} + c\left(E_n(x) - E_n(r_1) - \dfrac{1}{\sigma_n\lambda(r_1)r_1^{n-1}}\right)$	c an arbitrary constant; existence iff $\dfrac{\psi(r_1)}{\lambda(r_1)} = \dfrac{\psi(r_2)}{\lambda(r_2)}$
		$\lambda(r_1) = \lambda(r_2) = 0$	$\sigma_n r_1^{n-1}\psi(r_1)(E_n(r_1) - E_n(x)) + c$	c an arbitrary constant; existence iff $r_1^{n-1}\psi(r_1) + r_2^{n-1}\psi(r_2)=0$
$\Omega = \mathbb{R}^n \setminus B(0, r)$	$\Gamma_0 = \Gamma = \partial B(0, r)$		$\varphi(r) + c(E_n(x) - E_n(r))$	c arbitrary constant no condition for existence
	$\Gamma_0 = \varnothing$		$\sigma_n r^{n-1}(E_n(x) - E_n(r))(c\lambda(r) - \psi(r)) + c$	c arbitrary constant no condition for existence

(1) with
$$u(x) = \left(\sigma_n\lambda(r_1)\lambda(r_2)(E_n(r_2) - E_n(r_1)) + \frac{\lambda(r_1)}{r_2^{n-1}} + \frac{\lambda(r_2)}{r_1^{n-1}}\right)^{-1} \times$$
$$\times \left(\sigma_n\lambda(r_1)\psi(r_2)(E_n(x) - E_n(r_1)) + \sigma_n\lambda(r_2)\psi(r_1)(E_n(r_2) - E_n(x)) + \frac{\psi(r_1)}{r_2^{n-1}} - \frac{\psi(r_2)}{r_1^{n-1}}\right).$$

Then for all $x \in \mathbb{R}^n$, the function $t \to E_n(x - t)\psi(t)$ is integrable on $\mathbb{R}^{n-1}$ and

$$\lim_{|x| \to \infty} \int_{\mathbb{R}^{n-1}} E_n(x - t)\psi(t)\,dt = 0 .$$

In effect

$$\int_{\mathbb{R}^{n-1}} |E_n(x - t)\psi(t)|\,dt \leqslant \frac{1}{(n-2)\sigma_n} \int_{\mathbb{R}^{n-1}} \frac{\varepsilon(|t|)\,dt}{|x' - t|^{n-2}}$$

$$= \frac{1}{(n-2)\sigma_n} \int_{\{|x'-t| \geqslant |t|\}} \left(\frac{\varepsilon(|t|)}{|x'-t|^{n-2}} + \frac{\varepsilon(|x'-t|)}{|t|^{n-2}} \right) dt$$

$$\leqslant \frac{2}{(n-2)\sigma_n} \int_{\mathbb{R}^{n-1}} \frac{\varepsilon(|t|)}{|t|^{n-2}}\,dt = \frac{2\sigma_{n-1}}{(n-2)\sigma_n} \int_0^\infty \varepsilon(r)\,dr .$$

Making $\bar{x}_n \to \infty$ in (4.76), taking account of Corollary 6, *in the case* $n \geqslant 3$, *for* $\psi \in \mathscr{C}^0(\mathbb{R}^{n-1})$ *satisfying* (4.77), we see that *there exists one and only one solution u of the Neumann problem* $PN(\mathbb{R}^{n-1} \times \mathbb{R}^+, \psi)$ *tending to zero at infinity, given by*

$$u(x', x_n) = \frac{2x_n}{(n-2)\sigma_n} \int_{\mathbb{R}^{n-1}} \frac{\psi(x' + x_n t)}{(1 + |t|^2)^{n/2 - 1}}\,dt . \tag{4.78}$$

Case $n = 2$. Let us suppose that

$$\int_{\mathbb{R}} \operatorname{Log}(1 + t^2)|\psi(t)|\,dt < \infty \quad \text{and that} \quad \int_{\mathbb{R}} \psi(t)\,dt = 0 . \tag{4.79}$$

Then for all $(x, y) \in \mathbb{R}^2$ the function $t \to E_2(x - t, y)\psi(t)$ is integrable, since $t \to E_2(x - t, y)$ is locally integrable and

$$E_2(x - t, y) = O(\operatorname{Log}(1 + t^2)) \quad \text{when} \quad |t| \to \infty .$$

Now for $x \in \mathbb{R}$,

$$\lim_{y \to \infty} \int E_2(x - t, y)\psi(t)\,dt = 0 ,$$

since

$$\int \psi(t)\,dt = 0 ,$$

$$\int E_2(x - t, y)\psi(t)\,dt = \frac{1}{\pi} \int \operatorname{Log}\left(1 + \frac{(x-t)^2}{y^2}\right)\psi(t)\,dt$$

and

$$\operatorname{Log}\left(1 + \frac{(x - t)^2}{y^2}\right) \searrow 0 \quad \text{when} \quad y \nearrow \infty .$$

Passing to the limit in (4.76), we see then that *in the case* $n = 2$, *for* ψ *continuous and bounded on* $\mathbb{R}$ *satisfying* (4.79), *there exists one and only one solution u of* $PN(\mathbb{R}^{n-1} \times \mathbb{R}^+, \psi)$ *satisfying*

$$\frac{\partial u}{\partial y} \text{ is bounded on } \mathbb{R} \times \mathbb{R}^+ , \quad \lim_{y \to \infty} u(x, y) = 0 \quad \text{for all } x \in \mathbb{R} , \tag{4.80}$$

given by

$$u(x, y) = -\frac{y}{2\pi}\int_{\mathbb{R}} \psi(x + yt)\,\mathrm{Log}(1 + t^2)\,\mathrm{d}t\,. \tag{4.81}$$ □

Example 22. *Kelvin transform of the Neumann problem in a half space.*
We consider a ball $B = B(x_0, r_0)$ and $z_0 \in \partial B$. The image of B by the inversion $I(z_0, 2r_0)$ is the half-space

$$\Omega' = \{x' \in \mathbb{R}^n\,; (x' + z_0 - 2x_0)(x_0 - z_0) > 0\}\,.$$

Let us take $\psi \in \mathscr{C}^0(\partial B\backslash\{z_0\}$ and consider its Kelvin transform $\psi' = H(z_0, 2r_0)\psi$ defined on $\partial\Omega'$. As we have explained in the general case above, if we are given u' a solution of the Neumann problem $PN(\Omega', \psi')$ its Kelvin transform

$$u = H(z_0, 2r_0)u' \in \mathscr{H}(B) \cap \mathscr{C}^1_n(\bar{B}\backslash\{z_0\})$$

and satisfies

$$\frac{\partial u}{\partial n}(z) + \frac{(z_0 - z)(z - x_0)}{2r_0^3}\,u(z) = \psi(z) \quad \text{for all} \quad z \in \partial B\backslash\{z_0\} \tag{4.82}$$

Also if u' tends to zero at infinity, then

$$\lim_{x \to z_0} |x - z_0|^{n-2}\,u(x) = 0 \tag{4.83}$$

We restrict ourselves to the case $n \geqslant 3$. The condition (4.77) on $\psi' \in \mathscr{C}^0\,(\partial\Omega')$, can be expressed as:

$$\begin{cases} \textit{there exists } C\colon \mathbb{R}^+ \to \mathbb{R}^+ \textit{ increasing with } \displaystyle\int_0^\infty C(r)\frac{\mathrm{d}r}{r^2} < +\infty \\ \textit{such that } \quad |\psi(z)| \leqslant \dfrac{C(|z - z_0|)}{|z - z_0|^{n-2}}\,. \end{cases} \tag{4.84}$$

By Kelvin's transformation, *in the case $n \geqslant 3$, for all $\psi \in \mathscr{C}^0(\partial B\backslash\{z_0\})$ satisfying* (4.84), *there exists a unique solution u of the mixed boundary value problem specified by* (4.82) *and* (4.83).

We notice that the function $\lambda(z) = \dfrac{(z_0 - z).(z - x_0)}{2r_0^3} < 0$ on $\partial B\backslash\{z_0\}$. □

Example 23. *A mixed problem in a rectangle.*
We take a rectangle $\Omega =]0, a[\times]0, b[$ and consider the mixed problem:

$$\begin{cases} u \text{ harmonic in } \Omega \\ \dfrac{\partial u}{\partial x}(0, y) + \lambda u(0, y) = \psi(y) \quad \text{for} \quad y \in]0, b[\\ u(a, y) = u(x, 0) = u(x, b) = 0 \quad \text{for} \quad x \in [0, a],\, y \in]0, b[\end{cases} \tag{4.85}$$

where ψ is continuous and bounded on $]0, b[$ and $\lambda \in \mathbb{R}$.

We apply the same method as in Example 5, looking for a solution in the form

$$u(x, y) = \sum_{n \geq 1} u_n(x) \sin n\omega y \quad \textit{with} \quad \omega = \frac{\pi}{b} . \tag{4.86}$$

We must have

$$\begin{cases} \dfrac{\mathrm{d}^2 u_n}{\mathrm{d}x^2} = (n\omega)^2 u_n \\ u(a) = 0 \\ \dfrac{\mathrm{d}u_n}{\mathrm{d}x}(0) + \lambda u_n(0) = \dfrac{2}{b} \displaystyle\int_0^b \sin n\omega y \, \psi(y) \mathrm{d}y . \end{cases} \tag{4.87}$$

If $\lambda \neq -n\omega \coth(n\omega a)$, (4.87) admits one and only one solution

$$u_n(x) = \frac{2}{b} \frac{\operatorname{sh} n\omega(x - a)}{n\omega \operatorname{ch} n\omega a + \lambda \operatorname{sh} n\omega a} \int_0^b \sin n\omega y \, \psi(y) \mathrm{d}y . \tag{4.88}$$

If $\lambda = -n\omega \coth(n\omega a)$, (4.87) admits a solution iff

$$\int_0^b \sin n\omega y \, \psi(y) \mathrm{d}y = 0 ,$$

and then the solutions are

$$u_n(x) = c \operatorname{sh} n\omega(x - a) \quad \text{or} \quad c, \text{ an arbitrary constant}$$

To sum up: *for all* $\lambda \in \mathbb{R} \backslash \{-k\omega \coth k\omega a; k = 1, 2, \ldots\}$, *the problem* (4.85) *admits one and only one solution given by* (4.86) *where, for all* n, $u_n(x)$ *is given by* (4.88); *if* $\lambda = -k\omega \coth k\omega a$, *for an integer* $k \geq 1$, *then* (4.85) *admits a solution iff*

$$\int_0^b \sin k\omega y \, \psi(y) \mathrm{d}y = 0 ,$$

and when this condition is satisfied, the solutions of (4.85) *are given by* (4.86) *where for all* $n \neq k$, $u_n(x)$ *is given by* (4.88) *and*

$$u_k(x) = c \operatorname{sh} k\omega(x - a) \quad \textit{with } c \textit{ an arbitrary constant.}$$

As we have remarked in Example 5, the formal calculation can be justified: we verify in particular that for ψ being locally a Hölder function on $]0, a[$, the solution found is clearly a classical solution of the problem. □

5. Solution by Simple and Double Layer Potentials: Fredholm's Integral Method

In this section we consider Ω *a regular bounded open set in* $\mathbb{R}^n$ *with* $n \geq 2$. We shall suppose that *the boundary* Γ *of* Ω *is of class*[82] $\mathscr{C}^{1+\varepsilon}$ so that we can apply the results

[82] See Definition 3 of §3.

of §3.3 on simple and double layer potentials. We seek to find solutions of the Dirichlet problem, the Neumann problem, or more generally, the mixed boundary value problem, in the form of simple layer or double layer potentials. First we recall the results of §3.3 and introduce further notation.

Given $\alpha \in \mathscr{C}^0(\Gamma)$, from Propositions 10 and 12 of §3, *the interior simple layer potential* defined on Ω by

$$u(x) = \int_\Gamma E_n(x - t)\alpha(t)\,\mathrm{d}\gamma(t) \quad x \in \Omega \tag{4.89}$$

belongs to $\mathscr{H}(\Omega) \cap \mathscr{C}^1_n(\bar{\Omega})$; for all $z \in \Gamma$, the functions

$$t \to E_n(z - t)\alpha(t) \quad \text{and} \quad t \to \operatorname{grad} E_n(z - t).n(z)\alpha(t)$$

are integrable on Γ and

$$u(z) = \int_\Gamma E_n(z - t)\alpha(t)\,\mathrm{d}\gamma(t)$$

$$\frac{\partial u}{\partial n}(z) = \int_\Gamma \operatorname{grad} E_n(z - t).n(z)\alpha(t)\,\mathrm{d}\gamma(t) - \frac{\alpha(z)}{2}.$$

For all $\alpha \in \mathscr{C}^0(\Gamma)$, we denote by $L\alpha$ the function defined on Γ by

$$L\alpha(z) = \int_\Gamma E_n(z - t)\alpha(t)\,\mathrm{d}\gamma(t) \tag{4.90}$$

and *we denote* by $J\alpha$ the function defined on Γ by

$$\begin{cases} J\alpha(z) & = 2\displaystyle\int_\Gamma \operatorname{grad} E_n(z - t).n(z)\alpha(t)\,\mathrm{d}\gamma(t) \\ & = \dfrac{2}{\sigma_n}\displaystyle\int_\Gamma \dfrac{(z - t).n(z)}{|z - t|^n}\alpha(t)\,\mathrm{d}\gamma(t)\,. \end{cases} \tag{4.91}$$

From what is recalled above $L\alpha$ and $J\alpha$ are continuous on Γ; also the maps $\alpha \to L\alpha$ and $\alpha \to J\alpha$: *L and J are linear operators on $\mathscr{C}^0(\Gamma)$.*

Also what was recalled above can be expressed in the following manner:

(1) *solution of the Dirichlet problem $P(\Omega, \varphi)$ by a simple layer potential*: being given $\varphi \in \mathscr{C}^0(\Gamma)$ and $\alpha \in \mathscr{C}^0(\Omega)$, the interior simple layer potential defined by α (given by equation (4.89)) is the solution of $P(\Omega, \varphi)$ iff

$$\varphi(z) = L\alpha(z) = \int_\Gamma E_n(z - t)\alpha(t)\,\mathrm{d}\gamma(t) \quad \text{for all} \quad z \in \Gamma\,; \tag{4.92}$$

(2) *solution of the Neumann problem $PN(\Omega, \psi)$ by a simple layer potential*: given $\psi \in \mathscr{C}^0(\Gamma)$ and $\alpha \in \mathscr{C}^0(\Gamma)$, the interior simple layer potential defined by α is the solution of $PN(\Omega, \psi)$ iff

(4.93)

$$\psi(z) = \frac{J\alpha(z) - \alpha(z)}{2} = \int_\Gamma \frac{(z - t).n(z)}{\sigma_n |z - t|^n}\alpha(t)\,\mathrm{d}\gamma(t) - \frac{\alpha(z)}{2} \quad \text{for all} \quad z \in \Gamma;$$

(3) more generally given a closed part Γ_0 of Γ,

$$\varphi \in \mathscr{C}^0(\Gamma_0), \quad \psi, \lambda \in \mathscr{C}^0(\Gamma \backslash \Gamma_0) \quad \text{and} \quad \alpha \in \mathscr{C}^0(\Gamma)$$

the interior simple layer potential defined by α is the solution of the mixed problem $PM(\Omega, \Gamma_0, \varphi, \psi, \lambda)$ iff

$$(4.94) \qquad \begin{cases} L\alpha(z) = \varphi(z) \quad \text{for all} \quad z \in \Gamma_0 \\ \dfrac{J\alpha(z) + 2\lambda(z) L\alpha(z) - \alpha(z)}{2} = \psi(z) \quad \text{for all} \quad z \in \Gamma \backslash \Gamma_0 . \end{cases}$$

The Propositions 10 and 12 of §3, indicates to us that also the *exterior simple layer potential* defined on $\mathbb{R}^n \backslash \bar{\Omega}$ by

$$(4.95) \qquad u_e(x) = \int_\Gamma E_n(x - t)\alpha(t) \mathrm{d}\gamma(t) \quad x \in \mathbb{R}^n \backslash \bar{\Omega}$$

belongs to $\mathscr{C}^1_n(\mathbb{R}^n \backslash \Omega)$ and with the notation

$$(4.96) \qquad \begin{cases} u_e(z) = L\alpha(z) & \text{for all } z \in \Gamma \\ \dfrac{\partial u_e}{\partial n}(z) = -\dfrac{J\alpha(z) + \alpha(z)}{2} & \text{for all } z \in \Gamma . \end{cases}$$

Also from Proposition 2 of §3

$$(4.97) \qquad \begin{cases} u_e(x) = \left(\displaystyle\int_\Gamma \alpha \, \mathrm{d}\gamma\right) E_n(x) + O\left(\dfrac{1}{|x|^{n-1}}\right) & \text{when} \quad |x| \to \infty \\ \operatorname{grad} u_e(x) = \left(\displaystyle\int_\Gamma \alpha \, \mathrm{d}\gamma\right) \dfrac{x}{\sigma_n |x|^n} + O\left(\dfrac{1}{|x|^n}\right) & \text{when} \quad |x| \to \infty . \end{cases}$$

Also, given $\beta \in \mathscr{C}^0(\Gamma)$, from Proposition 11 of §3, the interior double layer potential

(4.98)

$$v(x) = \int_\Gamma \frac{\partial E_n}{\partial n}(t - x)\beta(t) \mathrm{d}\gamma(t) = \int_\Gamma \frac{(t - x).n(t)}{\sigma_n |t - x|^n} \beta(t) \mathrm{d}\gamma(t) , \qquad x \in \Omega ,$$

belongs to $\mathscr{H}(\Omega) \cap \mathscr{C}^0(\bar{\Omega})$; for all $z \in \Gamma$, the function $t \to \dfrac{\partial E_n}{\partial n}(t - z)\beta(t)$ is integrable on Γ and

$$v(z) = \int_\Gamma \frac{\partial E_n}{\partial n}(t - z)\beta(t) \mathrm{d}\gamma(t) + \frac{\beta(z)}{2} .$$

For all $\beta \in \mathscr{C}^0(\Gamma)$, *we denote by* $K\beta$ the function defined on Γ by

$$(4.99)\quad K\beta(z) = 2\int_\Gamma \frac{\partial E_n}{\partial n}(t - z)\beta(t)\mathrm{d}\gamma(t) = \frac{2}{\sigma_n}\int_\Gamma \frac{(t - z).n(t)}{|t - z|^n}\beta(t)\mathrm{d}\gamma(t)\,.$$

The function $K\beta$ is continuous on Γ and the map $\beta \to K\beta$ is linear: K *is a linear operator on* $\mathscr{C}^0(\Gamma)$. We have the *solution of the Dirichlet problem* $P(\Omega, \varphi)$ *by a double layer potential*: given $\varphi \in \mathscr{C}^0(\Gamma)$ and $\beta \in \mathscr{C}^0(\Gamma)$, the interior double layer potential defined by β (given by (4.98)) is the solution of $P(\Omega, \varphi)$ iff

$$(4.100)\quad \begin{cases} \varphi(z) = \dfrac{K\beta(z) + \beta(z)}{2} = \displaystyle\int_\Gamma \frac{(t - z).n(t)}{\sigma_n|t - z|^n}\beta(t)\mathrm{d}\gamma(t) + \frac{\beta(z)}{2} \\ \text{for all } z \in \Gamma\,. \end{cases}$$

Proposition 11 of §3, indicates to us also that the exterior double layer potential defined on $\mathbb{R}^n\backslash\bar{\Omega}$ by

$$(4.101)\quad v_e(x) = \int_\Gamma \frac{\partial E_n}{\partial n}(t - x)\beta(t)\mathrm{d}\gamma(t)\,,\quad x \in \mathbb{R}^n\backslash\bar{\Omega}$$

belongs to $\mathscr{H}(\mathbb{R}^n\backslash\bar{\Omega}) \cap \mathscr{C}^0(\mathbb{R}^n\backslash\Omega)$ and

$$(4.102)\quad v_e(z) = \frac{K\beta(z) - \beta(z)}{2}\quad \text{for all}\quad z \in \Gamma\,.$$

In addition, from Proposition 2 of §3,

(4.103)

$$v_e(x) = O\left(\frac{1}{|x|^{n-1}}\right),\quad \operatorname{grad} v_e(x) = O\left(\frac{1}{|x|^n}\right)\quad \text{when}\quad |x| \to \infty$$

since v_e is the Newtonian potential of the distribution f with compact support

$$\langle f, \zeta\rangle = \int_\Gamma \frac{\partial\zeta}{\partial n}(t)\beta(t)\mathrm{d}\gamma(t)$$

which is such that $\langle f, 1\rangle = 0$.

These references back, with the notation introduced, show that *the solution of Dirichlet, Neumann, or mixed problems, interior or exterior, by simple or double layer potentials leads to the solution of the following integral equations*:

$$\varphi = L\alpha\,,\quad \varphi = \frac{K\beta \pm \beta}{2}\,,\qquad -\psi = \frac{J\alpha \pm \alpha}{2}\,,\ \text{etc}\ldots$$

We shall therefore study these equations, that is to say the operators L, K and J First we note some properties of these operators[83]. We introduce the following

[83] Other properties will be developed in §5.

notation:

$$(4.104)\quad \begin{cases} \Omega_1, \ldots, \Omega_N \text{ the connected components of } \Omega \\ \Omega'_1, \ldots, \Omega'_{N'} \text{ the bounded connected components of } \mathbb{R}^n \backslash \bar{\Omega} \\ \Omega'_0 \text{ the unbounded connected component of } \mathbb{R}^n \backslash \bar{\Omega} \\ \Gamma_1, \ldots, \Gamma_N, \Gamma'_0, \Gamma'_1, \ldots, \Gamma_{N'} \text{ the boundaries of } \Omega_1, \ldots, \Omega_N, \\ \Omega'_0, \Omega'_1, \ldots \Omega'_{N'} \text{ respectively .} \end{cases}$$

Note that $\{\Gamma_1, \ldots, \Gamma_N\}$ and $\{\Gamma'_0, \Gamma'_1, \ldots, \Gamma'_{N'}\}$ are *two partitions of the boundary* Γ *of* Ω (which is also the boundary of $\mathbb{R}^n \backslash \bar{\Omega}$).

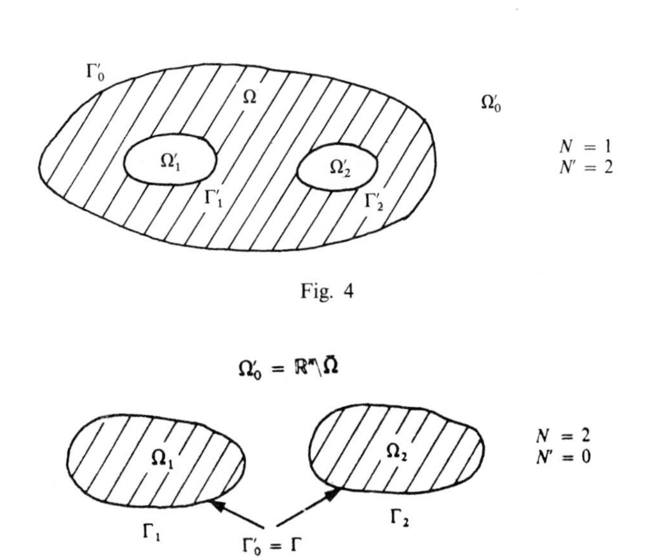

Fig. 4

Fig. 5

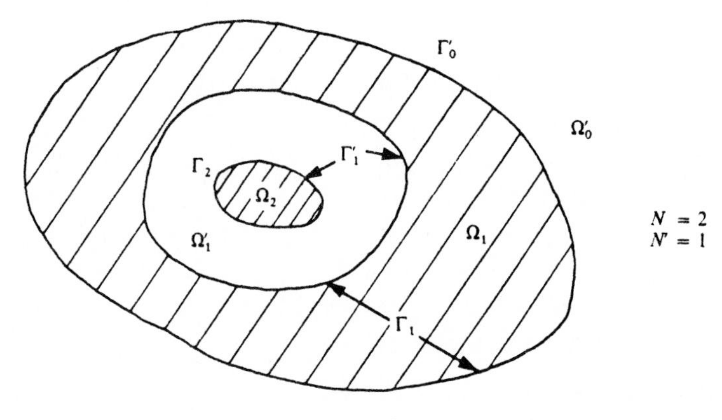

Fig. 6

It must be emphasised that the connectedness of Ω and $\mathbb{R}^n\backslash\bar{\Omega}$ are independent. With the notation of (4.104): *Ω connected iff $N = 1$ and $\mathbb{R}^n\backslash\bar{\Omega}$ connected iff $N' = 0$.* Figs. 4 to 6 give examples of different situations.

By construction, Ω being regular the set $\Gamma_i \cap \Gamma_j$, with $i = 1, \ldots, N$, and $j = 0, \ldots, N'$ *is open and closed in* Γ with the result that *the characteristic functions $\chi_{\Gamma_i \cap \Gamma_j}$ are continuous on* Γ[84]. (We shall show later that the number of connected components of Γ is exactly $N + N'$). It is obviously the same for χ_{Γ_i} and χ_{Γ_j}.

Proposition 15. *The operators L, J, K on $\mathscr{C}^0(\Gamma)$ being defined by* (4.90), (4.91) *and* (4.99)

(1) *L is symmetric, that is*

$$\int_\Gamma \alpha L\beta \,\mathrm{d}\gamma = \int_\Gamma \beta L\alpha \,\mathrm{d}\gamma \quad \textit{for all} \quad \alpha, \beta \in \mathscr{C}^0(\Gamma)\,;$$

(2) *J and K are adjoints one with the other, that is*

$$\int_\Gamma \alpha K\beta \,\mathrm{d}\gamma = \int_\Gamma \beta J\alpha \,\mathrm{d}\gamma \quad \textit{for all} \quad \alpha, \beta \in \mathscr{C}^0(\Gamma)\,;$$

(3) *With the notation* (4.104), *we have*

$$\text{(4.105)} \quad K\chi_{\Gamma_i} = \chi_{\Gamma_i}, \int_{\Gamma_i'} J\alpha \,\mathrm{d}\gamma = \int_{\Gamma_i'} \alpha \,\mathrm{d}\gamma \quad \textit{for} \quad i = 1, \ldots, N, \alpha \in \mathscr{C}^0(\Gamma)$$

$$\text{(4.106)} \quad K\chi_{\Gamma_j'} = -\chi_{\Gamma_j'}, \int_{\Gamma_j'} J\alpha \,\mathrm{d}\gamma = -\int_{\Gamma_j'} \alpha \,\mathrm{d}\gamma \quad \textit{for} \quad j = 1, \ldots, N', \alpha \in \mathscr{C}^0(\Gamma)\,;$$

(4) *L is a bijection from $\mathscr{C}^0(\Gamma)$ onto* $\operatorname{Im} L$[85], subject in the case $n=2$ to the *following hypothesis*:

$$\text{(4.107)} \quad \textit{there exists} \quad x_0 \in \mathbb{R}^n\backslash\bar{\Omega}_0' \quad \textit{such that} \quad \Gamma_0' \cap \partial B(x_0, 1) = \varnothing\,.$$

Proof. The points (1) and (2) follow from the definitions of L, J, K and Fubini's theorem:

$$\begin{aligned}\int_\Gamma \alpha L\beta \,\mathrm{d}\gamma &= \int_\Gamma \alpha(z)\,\mathrm{d}\gamma(z) \int_\Gamma E_n(z - t)\beta(t)\,\mathrm{d}\gamma(t) \\ &= \int_\Gamma \beta(t)\,\mathrm{d}\gamma(t) \int_\Gamma E_n(t - z)\alpha(z)\,\mathrm{d}\gamma(z) = \int_\Gamma \beta L\alpha \,\mathrm{d}\gamma\end{aligned}$$

---

[84] Given a part A, the characteristic function $\chi_A(x) = \begin{cases} 1 & \text{if} \quad x \in A\,, \\ 0 & \text{if} \quad x \notin A\,. \end{cases}$

[85] Given a linear map $T\colon X \to Y$ where X, Y are vector spaces,

$$\operatorname{Im} T = \{T\alpha; \alpha \in X\}\,, \qquad \ker T = \{\alpha \in X;\ T\alpha = 0\}\,.$$

$$\int_\Gamma \alpha K \beta \, d\gamma = \int_\Gamma \alpha(z) \, d\gamma(z) \frac{2}{\sigma_n} \int_\Gamma \frac{(t-z).n(t)}{|t-z|^n} \beta(t) \, d\gamma(t)$$

$$= \int_\Gamma \beta(t) \, d\gamma(t) \frac{2}{\sigma_n} \int_\Gamma \frac{(t-z).n(t)}{|t-z|^n} \alpha(z) \, d\gamma(z) = \int_\Gamma \beta J \alpha \, d\gamma .$$

Now given $\alpha \in \mathscr{C}^0(\Gamma)$, the interior simple layer potential u (resp. exterior potential u_e) defined by α is a solution of

$$PN\left(\Omega, \frac{J\alpha - \alpha}{2}\right)\left(\text{resp. } PN\left(\mathbb{R}^n \backslash \bar{\Omega}, -\frac{J\alpha + \alpha}{2}\right)\right).$$

Hence for $i = 1, \ldots, N$ (resp. $j = 1, \ldots, N'$), the restriction of u to Ω_i (resp. u_e to Ω'_j) is a solution of $PN(\Omega_i, \psi)$ (resp. $PN(\Omega'_j, \psi)$) where ψ is the restriction of $\frac{1}{2}(J\alpha - \alpha)$ (resp. $-\frac{1}{2}(J\alpha + \alpha)$) to Γ_i (resp. Γ'_j). From Gauss' theorem

$$\int_{\Gamma_i} \frac{J\alpha - \alpha}{2} d\gamma = 0 \quad \left(\text{resp. } \int_{\Gamma'_j} \frac{J\alpha + \alpha}{2} d\gamma = 0\right),$$

that is

$$\int_{\Gamma_i} J\alpha \, d\gamma = \int_{\Gamma_i} \alpha \, d\gamma \quad \left(\text{resp. } \int_{\Gamma'_j} J\alpha \, d\gamma = -\int_{\Gamma'_j} \alpha \, d\gamma\right).$$

To complete the proof of point (3), we use (2): we have

$$\int_\Gamma \alpha K \chi_{\Gamma_i} d\gamma = \int_\Gamma \chi_{\Gamma_i} J\alpha \, d\gamma = \int_{\Gamma_i} J\alpha \, d\gamma = \int_\Gamma \alpha \chi_{\Gamma_i} d\gamma$$

$$\left(\text{resp. } \int_\Gamma \alpha K \chi_{\Gamma'_j} d\gamma = \int_\Gamma \chi_{\Gamma'_j} J\alpha \, d\gamma = \int_\Gamma \alpha(-\chi_{\Gamma'_j}) \, d\gamma\right)$$

for all $\alpha \in \mathscr{C}^0(\Gamma)$. The stated result follows immediately.
Finally we prove the point (4): Let us consider $\alpha \in \mathscr{C}^0(\Gamma)$ such that $L\alpha \equiv 0$ on Γ. The interior simple layer potential u defined by α being a solution of the Dirichlet problem $P(\Omega, L\alpha)$ is identically zero: we have already deduced

$$\frac{\partial u}{\partial n} = \frac{J\alpha - \alpha}{2} \equiv 0 \quad \text{on} \quad \Gamma,$$

that is

$$J\alpha \equiv \alpha \quad \text{on} \quad \Gamma . \tag{4.108}$$

The exterior simple layer potential u_e is a solution of the Dirichlet problem $P(\mathbb{R}^n \backslash \bar{\Omega}, L\alpha)$: in particular for $j = 1, \ldots, N'$, $u_e \equiv 0$ on Ω'_j and therefore

$$\frac{\partial u_e}{\partial n_e} = -\frac{J\alpha + \alpha}{2} = 0 \quad \text{on} \quad \Gamma'_j .$$

Making use of (4.108), we have

$$\alpha \equiv 0 \quad \text{on} \quad \Gamma \backslash \Gamma'_0 .$$

In the case $n \geqslant 3$, from (4.97), $\lim_{|x| \to \infty} u_e(x) = 0$; we have thus also $u_e \equiv 0$ on Ω'_0, from which as above, $\alpha \equiv 0$ on Γ'_0. In the case $n = 2$ if $\int_\Gamma \alpha \, d\gamma = 0$, then still from (4.97), we shall have again $\lim_{|x| \to \infty} u_e(x) = 0$ and hence $\alpha \equiv 0$ on Γ'_0. To complete the proof in the case $n = 2$, let us suppose, for example, that $\int \alpha \, d\gamma > 0$.

Given $x_0 \in \mathbb{R}^n \backslash \bar{\Omega}'_0$, we have with (4.97)

$$\lim_{|x| \to \infty} \frac{u_e(x)}{E_2(x - x_0)} = \int \alpha \, d\gamma > 0 ,$$

and hence adapting the proof of Proposition 9,

$$u_e \geqslant 0 \quad \text{on} \quad \Omega'_0 ;$$

from which we deduce

$$\frac{\partial u_e}{\partial n_e} \leqslant 0 \quad \text{on} \quad \Gamma'_0 ,$$

and hence

$$\alpha \geqslant 0 \quad \text{on} \quad \Gamma'_0 .$$

Now, we have, since $x_0 \in \mathbb{R}^n \backslash \bar{\Omega}'_0 \subset \bar{\Omega} \cup \bar{\Omega}'_1 \cup \ldots \cup \bar{\Omega}'_N$,

$$\int_{\Gamma'_0} \alpha(t) E_2(x_0 - t) \, d\gamma(t) = 0 .$$

Choosing x_0 such that $\Gamma'_0 \cap \partial B(x_0, 1) = \emptyset$, on making use of (4.107), $E_2(x_0 - t)$ is not identically zero for $t \in \Gamma'_0$ and hence keeps the same sign: we deduce $\alpha \equiv 0$ for Γ'_0. □

Remark 5. From the proof, we see that we always have

$$\ker L \cap \left\{ \alpha \in \mathscr{C}^0(\Gamma) ; \int_\Gamma \alpha \, d\gamma = 0 \right\} = \{0\} . \tag{4.109}$$

It is false in general in the case $n = 2$ that L is injective: for example, considering $\Omega = B(0, 1)$, we have

$$L1 \equiv 0 \quad \text{on} \quad \Sigma = \partial\Omega .$$

This particular feature of the case $n = 2$ arises from the normalization of the elementary solution E_2 of the Laplacian in $\mathbb{R}^2$: we can in an equivalent manner suppress the hypothesis (4.107) by replacing, in the definition of the simple layer potential, the elementary solution $E_2(x)$ by $E_2(x/\delta) = E_2(x) - E_2(\delta)$ where $\delta > 0$ is chosen as follows:

$$\delta > \min_{\bar{\Omega}} \max_{\Gamma} |x - z| \quad \text{or} \quad \delta < \max_{\bar{\Omega}} \min_{\Gamma} |x - z| .$$

In other words, *we can suppress the hypothesis* (4.107) *by changing the unit of length in the plane.*

Remark 6. *Alexander's Relation.* An interesting application of the formulae (4.105) and (4.106) is the equality (called Alexander's relation: $N + N' =$ *number of connected components of* Γ. Let us take, in effect, Γ_0, a connected component of Γ; there exists a bounded open set Ω_0 such that $\Gamma_0 = \partial\Omega_0$; from Lemma 7 of §3, the interior (resp. exterior) double layer potential of χ_{Γ_0} is constant on Ω_0 and zero on $\mathbb{R}^n \backslash \bar{\Omega}_0$ and hence *a fortiori* zero on Ω'_0 and constant on $\Omega_1, \ldots, \Omega_N, \Omega'_1, \ldots, \Omega'_{N'}$; we deduce from (4.100) and (4.102) that

$$(I + K)\chi_{\Gamma_0} \text{ is constant on } \Gamma_1, \ldots, \Gamma_N$$

$$(I - K)\chi_{\Gamma_0} \text{ is zero on } \Gamma'_0 \text{ and constant on } \Gamma'_1, \ldots, \Gamma'_{N'} .$$

Hence, denoting by F the space of the functions which are constant on each connected component of Γ, $V = \sum_{i=1}^{N} \chi_{\Gamma_i} \mathbb{R}$ and $V' = \sum_{j=1}^{N'} \chi_{\Gamma'_j} \mathbb{R}$, we have thus shown that

$$(I + K)F \subset V \quad \text{and} \quad (I - K)F \subset V' ;$$

from which

$$F \subset V + V' .$$

On the other hand from (4.105) and (4.106), we have

$$K\varphi = \varphi \quad \text{for} \quad \varphi \in V \quad \text{and} \quad K\varphi = -\varphi \quad \text{for} \quad \varphi \in V' .$$

Hence $V \cap V' = \{0\}$. Since obviously V and V' are contained in F, $F = V \oplus V'$, from which we have $N + N' = \dim F =$ number of connected components of F. □

We shall now make use of the *Riesz–Fredholm theory*. Let us recall the principal results (see, for example, Lang [1]).

Definition 14. Let X be a Banach space and T a map of X into itself. We say that T is a *compact operator* if for every bounded sequence (u_n) in X, there exists a subsequence (u_{n_k}) such that the sequence Tu_{n_k}converges in X when $k \to \infty$. Given a compact linear operator T, the operator $I - T$: $u \to u - Tu$ is a *Fredholm operator*: we shall now state the theorem of Riesz and Schauder.

Lemma 3 (The Riesz–Schauder Theorem). *Let X be a Banach space and T a compact linear operator of X into itself; then*
(1) $\operatorname{Im}(I - T)$ *is a closed sub-space of X*;
(2) $\ker(I - T)$ *is a sub-space of X of finite dimension*;
(3) $\dim\ker(I - T) = \operatorname{codim}\operatorname{Im}(I - T)$[86].

We shall obviously apply this theory with $X = \mathscr{C}^0(\Gamma)$ provided with the uniform

---

[86] The codimension of a sub-space F of a vector space X is the dimension of the quotient space X/F; we have codim $F = N$ iff there exists a sub-space X_0 of dimension N such that $X = F \oplus X_0$.

convergence norm

$$\|\alpha\| = \sup_{\Gamma} |\alpha| .$$

The essential point is

Proposition 16. *The operators L, J, K defined by* (4.90), (4.91) *and* (4.99) *are compact operators on* $\mathscr{C}^0(\Gamma)$.

To prove this proposition, we shall make use of Ascoli's theorem (see Dieudonné [1], Vol. 1, Schwartz [2]) which we state in the form:

Lemma 4 (Ascoli's Theorem). *A map T of $\mathscr{C}^0(\Gamma)$ into itself is compact iff for every bounded sequence (α_k) of functions $\alpha_k \in \mathscr{C}^0(\Gamma)$ and every $z_0 \in \Gamma$,*

(i) $\sup_k |T\alpha_k(z_0)| < \infty$

(ii) $\sup_k |T\alpha_k(z) - T\alpha_k(z_0)| \to 0 \quad \text{when} \quad z \to z_0 .$

Proof of Proposition 16. Let us take a bounded sequence of functions $(\alpha_k) \in \mathscr{C}^0(\Gamma)$ and $z_0 \in \Gamma$. To prove the compactness of L we go back to the proof of Proposition 10 of §3.

Let us consider a normal parametric representation $(R, U, \mathcal{O}, \alpha)$ in the neighbourhood of z_0. We can always suppose that $B(0, r_0) \subset \mathcal{O}$, $z_0 = (0, 0)$. For $0 < r \leqslant r_0$, $x' \in B(0, r)$, $z = (x', \alpha(x'))$ we put

$$L\alpha_k(z) = \varphi_k^r(z) + I_k^r(x') ,$$

with

$$\varphi_k^r(z) = \int_{\Gamma \setminus \Gamma_r} E_n(t - z)\alpha_k(t)\,\mathrm{d}\gamma(t) ,$$

where

$$\Gamma_r = \{(t', \alpha(t')) ; \qquad t' \in B(0, r)\}$$

and[87]

$$I_k^r(x') = \int_{B(0,r)} E_n(t' - x', \alpha(t') - \alpha(x'))\alpha_k(t', \alpha(t'))(1 + |\operatorname{grad} \alpha(t')|^2)^{1/2}\,\mathrm{d}t' .$$

We have

$$|I_k^r(x')| \leqslant C_k \int_{B(0,r)} g(t' - x')\,\mathrm{d}t' \leqslant C_k \int_{B(0,2r)} g(t')\,\mathrm{d}t'$$

with

$$C_k = \|\alpha_k\| \sup_{B(0,r_0)} (1 + |\operatorname{grad} \alpha(t')|^2)^{1/2}$$

$$g(t') = \begin{cases} \dfrac{1}{(n-2)\sigma_n |t'|^{n-2}} & \text{if} \quad n \geqslant 3 \\[2ex] \dfrac{1}{2\pi} \operatorname{Log} \dfrac{1}{|t'|} & \text{if} \quad n = 2 . \end{cases}$$

---

[87] For the details, see the proof of Proposition 10 of §3.

Since g is locally integrable on $\mathbb{R}^{n-1}$

$$\sup_{k \geqslant 1} \sup_{x' \in B(0,r)} |I_k^r(x')| \to 0 \quad \text{when} \quad r \to 0 .$$

On the other hand,

$$|\varphi_k^r(z_0)| \leqslant \|\alpha_k\| \int_{\Gamma \backslash \Gamma_r} |E_n(t - z_0)| \mathrm{d}\gamma(t)$$

$$|\varphi_k^r(z) - \varphi_k^r(z_0)| \leqslant \|\alpha_k\| \int_{\Gamma \backslash \Gamma_k} |E_n(t - z) - E_n(t - z_0)| \mathrm{d}\gamma(t)$$

with the result that for $0 < r \leqslant r_0$ fixed

$$\sup_k |\varphi_k^r(z_0)| < \infty$$

$$\sup_k |\varphi_k^r(z) - \varphi_k^r(z_0)| \to 0 \quad \text{when} \quad z \to z_0 .$$

We verify thus the points (i) and (ii) of Lemma 4, from which we deduce the compactness of L. The compactness of K and J will be proved in the same way as in Propositions 11 and 12. □

We are now in a position to prove the

Theorem 4. *Let Ω be a regular bounded open set in $\mathbb{R}^n$ with $n \geqslant 2$, with boundary Γ of class $\mathscr{C}^{1+\varepsilon}$ and, in the case $n = 2$, also satisfying the hypothesis* (4.107). *On the other hand, let $\lambda, \psi \in \mathscr{C}^0(\Gamma)$ with $\lambda \geqslant 0$ on Γ. Then the following properties are equivalent*:
(i) *there exists a classical solution of the mixed problem*

$$(4.110) \qquad \begin{cases} u \text{ harmonic on } \Omega \\ \dfrac{\partial u}{\partial n} + \lambda u = \psi \text{ on } \Gamma ; \end{cases}$$

(ii) *there exists $\alpha \in \mathscr{C}^0(\Gamma)$ solution of the integral equation*

$$(4.111) \qquad \frac{J\alpha - \alpha}{2} + \lambda L\alpha = \psi \quad \text{on} \quad \Gamma ,$$

where the operators L and J are defined by (4.90), (4.91);
(iii) *For $i = 1, \ldots, N$*

$$\lambda \equiv 0 \quad \text{on} \quad \Gamma_i \Rightarrow \int_{\Gamma_i} \psi \, \mathrm{d}\gamma = 0 ,$$

where $\Gamma_1, \ldots, \Gamma_N$ are defined by (4.104).
In addition, the solution of (4.110) *is unique to within an additive constant on each Ω_i, $i = 1, \ldots, N$ such that $\lambda \equiv 0$ on Γ_i; for every solution α of* (4.111) *the simple layer potential u defined by α (by* (4.89)*) is a solution of* (4.110) *and the map $\alpha \to u$ is a bijection from the set of solutions of* (4.111) *onto the set of solutions of* (4.110).

Proof. Given $\lambda \in \mathscr{C}^0(\Gamma)$ fixed, we denote by T the operator on $\mathscr{C}^0(\Gamma)$ defined by

$$T\alpha(z) = J\alpha(z) + 2\lambda(z)L\alpha(z)\,, \quad \alpha \in \mathscr{C}^0(\Gamma)\,, \quad z \in \Gamma\,,$$

and B the map of $\mathscr{H}(\Omega) \cap \mathscr{C}^1_n(\bar{\Omega})$ into $\mathscr{C}^0(\Gamma)$ defined by

$$Bu(z) = \frac{\partial u}{\partial n}(z) + \lambda(z)u(z)\,, \quad u \in \mathscr{H}(\Omega) \cap \mathscr{C}^1_n(\bar{\Omega})\,, \quad z \in \Gamma\,.$$

With the notation (4.104), we can always suppose: $\lambda \equiv 0$ on $\Gamma_1, \ldots, \Gamma_r$ and $\lambda \equiv 0$ on $\Gamma_{r+1}, \ldots, \Gamma_N$ with $0 \leqslant r \leqslant N$. We denote

$$F = \left\{\psi \in \mathscr{C}^0(\Gamma)\,;\, \int_{\Gamma_i} \psi\, \mathrm{d}\gamma = 0 \quad \text{for} \quad i = 1, \ldots, r\right\}.$$

With this notation for $\psi \in \mathscr{C}^0(\Gamma)$, (4.110) is written $Bu = \psi$ and (4.111) is written $\frac{1}{2}(T\alpha - \alpha) = \psi$; in other words the equivalence of (i), (ii) and (iii) is expressed by

$$\text{(4.112)} \qquad \operatorname{Im}\frac{T - I}{2} = \operatorname{Im} B = F\,.$$

From Gauss' theorem, $\operatorname{Im} B \subset F$. From what has been recalled above (see (4.94)) for $\alpha \in \mathscr{C}^0(\Gamma)$, the interior simple layer potential u defined by α is a solution of (4.110) with $\psi = (T\alpha - \alpha)/2$; we deduce

$$\text{(4.113)} \qquad \operatorname{Im}\frac{T - I}{2} \subset \operatorname{Im} B \subset F\,.$$

From Proposition 15(4), the map $\alpha \to u$, the interior simple layer potential of α, is L: for a bounded sequence (α_n) of $\mathscr{C}^0(\Gamma)$, we can extract a sub-sequence α_{n_k} such that the sequences $J\alpha_{n_k}$ and $L\alpha_{n_k}$ converge and hence, immediately, so does $T\alpha_{n_k}$. Hence $\operatorname{Im}(I - T) = \operatorname{Im}(T - I)/2$ is a closed sub-space of finite codimension equal to $\dim\ker(I - T)$ (see Lemma 3). On the other hand, using for example, the projection theorem in the Hilbert space $L^2(\Gamma)$ on which is defined the scalar product

$$(\varphi, \psi) = \int_\Gamma \varphi\psi\, \mathrm{d}\gamma\,,$$

since

$$F = \{\psi \in \mathscr{C}^0(\Gamma)\,; (\chi_{\Gamma_i}, \psi) = 0 \quad i = 1, \ldots, r\}\,,$$

we have

$$\mathscr{C}^0(\Gamma) = F \oplus \chi_{\Gamma_1}\mathbb{R} \oplus \cdots \oplus \chi_{\Gamma_r}\mathbb{R}$$

and hence codim $F = r$. From (4.113), the equality (4.112) is therefore equivalent to

$$\text{(4.114)} \qquad \dim\ker(I - T) \leqslant r$$

From Proposition 15(4), the map $\alpha \to u$, the interior simple layer potential of α, is injective; in effect $L\alpha$ is the trace of u on Γ with the result that $u \equiv 0$ on Ω implies that $L\alpha \equiv 0$ on Γ. Now, if $\alpha \in \ker(I - T)$, $\psi = (T\alpha - \alpha)/2 \equiv 0$ on Γ with the result that, from Corollary 6, when $\lambda \geqslant 0$ on Γ, u is constant on $\Omega_1, \ldots, \Omega_7$ and

$u \equiv 0$ on $\Omega_{r+1}, \ldots, \Omega_N$. In other words, when $\lambda \geqslant 0$ on Γ, the map $\alpha \to u$ is an injection from $\ker(I - T)$ into the space $\chi_{\Omega_1}\mathbb{R} \oplus \ldots \oplus \chi_{\Omega_r}\mathbb{R}$.
This proves (4.114). □

Similarly, we have for *the exterior problem in the case* $n \geqslant 3$:

Theorem 5. *With the same hypotheses as in Theorem 4 with* $n \geqslant 3$, *the following properties are equivalent*:
(i) *there exists a classical solution of the exterior mixed problem*

$$(4.115)\qquad \begin{cases} u_e \text{ harmonic on } \mathbb{R}^n\backslash\bar{\Omega} \\ \dfrac{\partial u_e}{\partial n_e} + \lambda u_e = \psi \text{ on } \Gamma \\ \lim\limits_{|x| \to \infty} u_e(x) = 0\,; \end{cases}$$

(ii) *there exists* $\alpha \in \mathscr{C}^0(\Gamma)$ *solution of the integral equation*

$$(4.116)\qquad \lambda L\alpha - \frac{J\alpha + \alpha}{2} = \psi \text{ on } \Gamma\,;$$

(iii) *for* $j = 1, \ldots, N'$

$$\lambda \equiv 0 \text{ on } \Gamma'_j \Rightarrow \int_{\Gamma'_j} \psi \, d\gamma = 0\,,$$

where $\Gamma'_1, \ldots, \Gamma'_{N'}$ *are defined by* (4.104).
In addition the solution of (4.115) *is unique to within an additive constant on each* Ω'_j, $j = 1, \ldots, N'$ *such that* $\lambda \equiv 0$ *on* Γ'_j; *for every solution* α *of* (4.116) *the exterior simple layer potential* u_e *defined by* α (*by* (4.95)) *is solution of* (4.115) *and the map* $\alpha \to u_e$ *is a bijection of the set of solutions of* (4.116) *onto the set of solutions of* (4.115).

The proof is absolutely identical with that of Theorem 4, making use of what was recalled above concerning exterior simple layer potentials (see (4.96), (4.97)).
Theorems 4 and 5 admits as a particular case, corresponding to $\lambda \equiv 0$ on Γ, the points (1) and (2a) of Theorem 3. The conclusion of Theorem 3 follows immediately from

Proposition 17. *Let* Ω *be a regular bounded open set in* $\mathbb{R}^n$ *with* $n \geqslant 2$, *with boundary* Γ *of class* $\mathscr{C}^{1+\varepsilon}$ *and* $\psi \in \mathscr{C}^0(\Gamma)$.
Then the following assertions are equivalent:
(i) *there exists a classical solution* u_e *of the exterior Neumann problem* $PN(\mathbb{R}^n\backslash\bar{\Omega}, \psi)$ *satisfying*

$$(4.117)\qquad \lim_{|x| \to \infty} \operatorname{grad} u_e(x) = 0\,;$$

(ii) *there exists a solution* $\alpha \in \mathscr{C}^0(\Gamma)$ *of the integral equation*

$$(4.118)\qquad -\frac{J\alpha + \alpha}{2} = \psi \text{ on } \Gamma\,;$$

(iii) *For* $j = 1, \ldots, N'$, $\int_{\Gamma'_j} \psi \, \mathrm{d}\gamma = 0$; *where* $\Gamma'_1, \ldots, \Gamma'_{N'}$ *are defined by* (4.104).
In addition, for every classical solution u_e *of* $PN(\mathbb{R}^n \backslash \bar{\Omega}, \psi)$ *satisfying*

$$\text{(4.119)} \quad \operatorname{grad} u_e(x) = \left(-\int_\Gamma \psi \, \mathrm{d}\gamma\right) \frac{x}{\sigma_n |x|^n} + O\left(\frac{1}{|x|^n}\right) \quad \textit{when} \quad |x| \to \infty$$

$$\text{(4.120)} \qquad c = \lim_{|x| \to \infty} \left(u_e(x) + \left(\int_\Gamma \psi \, \mathrm{d}\gamma\right) E_n(x)\right) \quad \textit{exists}\,,$$

and there exists a unique solution α *of* (4.118) *such that*

$$\text{(4.121)} \qquad u_e(x) = c + \int_\Gamma E_n(x - t)\alpha(t) \, \mathrm{d}\gamma(t)\,.$$

Proof of Proposition 17. Fist we take u_e to be a classical solution of $PN(\mathbb{R}^n \backslash \bar{\Omega}, \psi)$ satisfying (4.117). Given $\zeta \in \mathscr{C}^\infty(\mathbb{R}^n)$ with $\zeta \equiv 0$ in the neighbourhood of $\bar{\Omega}$, $\zeta \equiv 1$ for $|x| \geqslant R_0$,

$$v = \zeta u_e \in \mathscr{C}^\infty(\mathbb{R}^n) \quad \text{and} \quad f = \Delta v \in \mathscr{D}(\mathbb{R}^n)\,.$$

Since $v = u_e$ for $|x| \geqslant R_0$ it satisfies also (4.117). From Proposition 4 of §2, there exists c such that

$$v = E_n * f + c\,.$$

We have therefore from Proposition 2 of §3 that when $|x| \to \infty$

$$u_e(x) = c + \left(\int f \mathrm{d}x\right) E_n(x) + O\left(\frac{1}{|x|^{n-1}}\right)$$

$$\operatorname{grad} u_e(x) = \left(\int f \mathrm{d}x\right) \frac{x}{\sigma_n |x|^n} + O\left(\frac{1}{|x|^n}\right).$$

But

$$\int f \mathrm{d}x = \int_{B(0, R_0)} \Delta v \, \mathrm{d}x = \int_{\partial B(0, R_0)} \frac{\partial v}{\partial n} \mathrm{d}\gamma = \int_{\partial B(0, R_0)} \frac{\partial u_e}{\partial n} \mathrm{d}\gamma = -\int_\Gamma \psi \, \mathrm{d}\gamma\,,$$

the last equality coming from Gauss' theorem applied to u_e harmonic on $B(0, R_0) \backslash \bar{\Omega}$.
In short, every classical solution u_e of $PN(\mathbb{R}^n \backslash \bar{\Omega}, \psi)$ satisfying (4.117) satisfies (4.119) and (4.121).
On the other hand, given α solution of (4.118) and $c \in \mathbb{R}$, the function u_e given by (4.121) is a classical solution of $PN(\mathbb{R}^n \backslash \bar{\Omega}, \psi)$ satisfying (4.117). In addition identifying (4.97) with (4.119) and (4.120), we have

$$\text{(4.122)} \qquad \int_\Gamma \psi \, \mathrm{d}\gamma = -\int_\Gamma \alpha \, \mathrm{d}\gamma\,.$$

We are hence led to showing that the assertions (ii) and (iii) are equivalent to:
(i)' there exists a classical solution of $PN(\mathbb{R}^n \backslash \bar{\Omega}, \psi)$ satisfying

$$\text{(4.123)} \qquad \lim_{|x| \to \infty} u_e(x) - \left(\int_\Gamma \psi \, \mathrm{d}\gamma\right) E_n(x) = 0\,.$$

In the case $n \geqslant 3$, this is a corollary of Theorem 5. In the case $n = 2$, it is necessary to take up again the reasoning of the proof of Theorems 4 and 5. We have

$$\operatorname{Im}(I+J) \subset \left\{\frac{\partial u_e}{\partial n_e};\, u_e \in \mathscr{H}(\mathbb{R}^n\backslash\bar{\Omega}) \cap \mathscr{C}_n^1(\mathbb{R}^n\backslash\Omega) \text{ satisfying } (4.117)\right\}$$

$$\subset \left\{\psi \in \mathscr{C}^0(\Gamma);\, \int_{\Gamma'_j} \psi\, \mathrm{d}\gamma = 0 \text{ for } j = 1, \ldots, N'\right\},$$

with the result that by the Riesz–Schauder theorem, we are led to show that $\dim\ker(I + J) \leqslant N'$. But the map $\alpha \to u_e$, the exterior simple layer potential, is an injection of $\ker(I + J)$ into $\chi_{\Omega'_1}\mathbb{R} \oplus \ldots \oplus \chi_{\Omega'_{N'}}\mathbb{R}$; indeed if $\alpha \in \ker(I + J)$, then u_e is a solution of $PN(\mathbb{R}^n\backslash\bar{\Omega}, 0)$ and satisfies $\lim\limits_{|x| \to \infty} u_e(x) = 0$ hence $u_e \equiv 0$ on $\mathbb{R}^n\backslash\bar{\Omega}$. □

We have thus proved Theorem 3 which had been stated without proof in sect. 4.4 and more precisely that *Neumann's problem*, when it admits a solution *is solved by a simple layer potential.* It is not always the same for the Dirichlet problem: more precisely let us prove the

Proposition 18. *Let Ω be a regular bounded open set Ω in $\mathbb{R}^n$ ($n \geqslant 2$) with boundary Γ of class $\mathscr{C}^{1+\varepsilon}$ and also in the case $n = 2$ the hypothesis* (4.107). *On the other hand, let $\varphi \in \mathscr{C}^0(\Gamma)$.*
Then the following assertions are equivalent:
(i) $\varphi \in \operatorname{Im} L$;
(ii) *the solution of $P(\Omega, \varphi)$ is in $\mathscr{C}_n^1(\bar{\Omega})$;*
(iii) *there exists a solution of $P(\mathbb{R}^n\backslash\bar{\Omega}, \varphi)$ in $\mathscr{C}_n^1(\mathbb{R}^n\backslash\Omega)$;*
(iv) *every solution of $P(\mathbb{R}^n\backslash\bar{\Omega}, \varphi)$ is in $\mathscr{C}_n^1(\mathbb{R}^n\backslash\Omega)$.*

We notice that $\varphi \in \operatorname{Im} L$ iff the solution $u(\varphi)$ of $P(\Omega, \varphi)$ can be written as the single layer potential of a function $\alpha \in \mathscr{C}^0(\Gamma)$, which, from point (4) of Proposition 15 is unique; for the exterior problem we have also $\varphi \in \operatorname{Im} L$ iff $P(\mathbb{R}^n\backslash\bar{\Omega}, \varphi)$ admits a solution which is written as an exterior simple layer potential: in the case $n \geqslant 3$, it is the solution tending to zero at infinity which is no longer true in general in the case $n = 2$.

Proof of Proposition 18. Supposing $\varphi \in \operatorname{Im} L$, the solution $u(\varphi)$ of $P(\Omega, \varphi)$ is in $\mathscr{C}_n^1(\bar{\Omega})$ since it can be written as a simple layer potential; conversely if the solution $u(\varphi)$ of $P(\Omega, \varphi)$ is in $\mathscr{C}_n^1(\bar{\Omega})$, the function $\psi = \partial u(\varphi)/\partial n \in \mathscr{C}^0(\Gamma)$ and satisfies the Gauss conditions; from Theorem 4, $u(\varphi)$ solution of $PN(\Omega, \psi)$ can hence be written as a simple layer potential, that is to say that $\varphi \in \operatorname{Im} L$. Hence (i) $\Leftrightarrow$ (ii). We also have (i) $\Rightarrow$ (iii) and (iv) $\Rightarrow$ (i); for this latter point, it is enough to apply Proposition 17, noting that the constants are in $\operatorname{Im} L$ which follows from the equivalence (i) $\Leftrightarrow$ (ii).
To complete the proof, i.e. to prove that (iii) $\Rightarrow$ (iv), we are led, by linearity, to prove that

(4.124) every solution of $P(\mathbb{R}^n\backslash\bar{\Omega}, 0)$ is in $\mathscr{C}_n^1(\mathbb{R}^n\backslash\Omega)$.

Let us consider, therefore, a solution u of $P(\mathbb{R}^n\backslash\bar{\Omega}, 0)$; we fix $\rho \in \mathscr{D}(\mathbb{R}^n)$ such that $\rho = 1$ in the neighbourhood of $\bar{\Omega}$: we have $\rho u \in \mathscr{C}^\infty(\mathbb{R}^n\backslash\bar{\Omega})$ and $\Delta(\rho u) \in \mathscr{D}(\mathbb{R}^n\backslash\bar{\Omega})$.
Let us denote by f the extension of $\Delta(\rho u)$ by 0 to the whole of $\mathbb{R}^n$. We have $f \in \mathscr{D}(\mathbb{R}^n)$ and the Newtonian potential of f is harmonic in the neighbourhood of $\bar{\Omega}$; hence the trace φ of $E_n * f$ is in $\operatorname{Im} L$. The exterior simple layer potential u_e of $\alpha = L^{-1}\varphi$ is in $\mathscr{C}_n^1(\mathbb{R}^n\backslash\bar{\Omega})$: the function $v = \rho u - E_n * f + u_e$ is solution of $P(\mathbb{R}^n\backslash\bar{\Omega}, 0)$.

In the case $n \geqslant 3$, $\lim_{|x|\to\infty} v(x) = 0$ and so $v(x) \equiv 0$ on $\mathbb{R}^n\backslash\bar{\Omega}$; we deduce that $u = E_n * f - u_e$ in the neighbourhood of Γ in $\mathbb{R}^n\backslash\bar{\Omega}$ and so $u \in \mathscr{C}_n^1(\mathbb{R}^n\backslash\Omega)$.
In the case $n = 2$, this is again true; obviously, we always have $v \equiv 0$ on $\Omega'_1, \ldots, \Omega'_N$; also from Proposition 2 of §2 (see also (4.97)) when $|x| \to \infty$

$$v(x) = CE_2(x) + O\left(\frac{1}{|x|}\right) \quad \text{with} \quad C = \int_\Gamma \alpha\, d\gamma - \int f dx .$$

If $\Omega'_0 = \mathbb{R}^2\backslash\bar{B}(x_0, r_0)$, then, from the uniqueness of the Dirichlet problem (see Proposition 9), v is radial with respect to x_0 on Ω'_0; since $r_0 \neq 1$ from the hypothesis (4.107), necessarily $C = 0$ and $v \equiv 0$.
In the general case, we take $R > 1$ such that $B(0, R) \supset \bar{\Omega}$ and put $\Omega_1 = (\mathbb{R}^2\backslash\bar{\Omega}) \cap B(0, R)$ and v_1 defined on $\mathbb{R}^2\backslash\bar{\Omega}_1 = \Omega \cup (\mathbb{R}^2\backslash\bar{B}(0, R))$ by

$$v_1(x) = \begin{cases} 0 & \text{on} \quad \Omega \\ v(x) & \text{on} \quad \mathbb{R}^2\backslash\bar{B}(0, R) . \end{cases}$$

It is clear that v_1 is harmonic on $\mathbb{R}^2\backslash\bar{\Omega}_1$ and belongs to $\mathscr{C}_n^1(\mathbb{R}^2\backslash\Omega)$; since the result is established when the unbounded component is the exterior of a ball, the restriction of v to Ω_1, solution of $P(\Omega_1, \varphi_1)$ where φ_1 is the trace of v_1 on $\partial\Omega_1$, is in $\mathscr{C}_n^1(\bar{\Omega}_1)$ and hence again $u \in \mathscr{C}_n^1(\mathbb{R}^2\backslash\Omega)$. □

We notice the

Corollary 8. *Let Ω be a regular bounded open set in $\mathbb{R}^n$ with boundary Γ of class $\mathscr{C}^{1+\varepsilon}$. Then the Green's function*[88] *G of Ω* (resp. $\mathbb{R}^n\backslash\bar{\Omega}$) *satisfies*

$$\begin{cases} \textit{for all } y \in \Omega \text{ (resp. } \mathbb{R}^n\backslash\bar{\Omega}), \textit{ the function } x \to G(x, y) \\ \textit{is in } \ \mathscr{C}_n^1(\bar{\Omega}\backslash\{y\}) \textit{ (resp. } \mathscr{C}_n^1(\mathbb{R}^n\backslash(\Omega \cup \{y\})). \end{cases}$$

Proof. We have[88] $G(x, y) = E_n(x - y) - u(\varphi_y, x)$ (resp. $u_e(\varphi_y, x)$) where $\varphi_y(z) = E_n(z - y)$.
Since the function $x \to E_n(x - y)$ is a solution of $P(\mathbb{R}^n\backslash\bar{\Omega}, \varphi_y)$ (resp. $P(\Omega, \varphi_y)$), and is $\mathscr{C}^\infty$ in the neighbourhood of Γ, from the proposition, $u(\varphi_y)$ (resp. $u_e(\varphi_y)$) is in $\mathscr{C}_n^1(\bar{\Omega})$ (resp. $\mathscr{C}_n^1(\mathbb{R}^n\backslash\Omega)$)[89]. From which we have the result. □

---

[88] See §§4.2 and 4.3.
[89] In the case $n = 2$, we can always make the hypothesis (4.107).

We now give the *solution of Dirichlet's problem by a double layer potential.* From Proposition 18, the functions

$$\chi_{\Gamma_1}, \ldots, \chi_{\Gamma_N}, \chi_{\Gamma'_1}, \ldots, \chi_{\Gamma'_{N'}} \in \operatorname{Im} L \quad \text{where} \quad \Gamma_i, \Gamma'_j \quad \text{are defined by (4.104)}$$

Taking account of point (4) of Proposition 15 we can therefore define

$$\left\{\begin{array}{ll} \alpha_1 = L^{-1}\chi_{\Gamma_i} & i = 1, \ldots, N \\ \alpha'_j = L^{-1}\chi_{\Gamma'_j} & j = 1, \ldots, N' . \end{array}\right. \tag{4.125}$$

The interior (resp. exterior) simple layer potential defined by α_i (resp. α'_j) is a solution of the Dirichlet problem $P(\Omega, \chi_{\Gamma_i})$ (resp. $P(\mathbb{R}^n\backslash\bar{\Omega}, \chi'_{\Gamma'_j})$) hence χ_{Ω_i} (resp. $\chi_{\Omega'_j}$); from which we deduce

$$\left\{\begin{array}{lll} J\alpha_i = \alpha_i & \text{for} & i = 1, \ldots, N \\ J\alpha'_j = -\alpha'_j & \text{for} & j = 1, \ldots, N' . \end{array}\right. \tag{4.126}$$

In other words, taking into account Theorems 4, 5 and Proposition 17 in which we have shown that

$$\dim \ker(I - J) = N \quad \text{and} \quad \dim \ker(I + J) = N',$$

we see that $\{\alpha_1, \ldots, \alpha_N\}$ (resp. $\{\alpha'_1, \ldots, \alpha'_{N'}\}$) *form a base of* $\ker(I - J)$ (resp. $\ker(I + J)$)
Using point (2) of Proposition 15 we are able to prove:

Proposition 19. *Let Ω be a bounded open set of $\mathbb{R}^n$ with $n \geqslant 2$ with boundary Γ of class $\mathscr{C}^{1+\varepsilon}$ satisfying the hypothesis* (4.107) *when $n = 2$. Also let $\varphi \in \mathscr{C}^0(\Gamma)$.*
(1) *The following assertions are equivalent*:
(i) *there exists $\beta \in \mathscr{C}^0(\Gamma)$ such that the solution of $P(\Omega, \varphi)$ be the interior double layer potential of β given by* (4.98);
(ii) *φ satisfies*

$$\int_\Gamma \varphi(t)\alpha'_j(t)\,\mathrm{d}\gamma(t) = 0 \quad \textit{for} \quad j = 1, \ldots, N'$$

where α'_j is defined by (4.125).
In addition, then β is defined to within an additive constant on each Γ'_j, $j = 1, \ldots, N'$.
(2) *The following assertions are equivalent*:
(i) *there exists $\beta \in \mathscr{C}^0(\Gamma)$ such that the exterior double layer potential of β given by* (4.101) *is a solution of $P(\mathbb{R}^n\backslash\bar{\Omega}, \varphi)$*;
(ii) *φ satisfies*

$$\int_\Gamma \varphi(t)\alpha_i(t)\,\mathrm{d}\gamma(t) = 0 \quad \textit{for} \quad i = 1, \ldots, N$$

where α_i is defined by (4.125).
In addition, β is defined then to within an additive constant on each Γ_i, $i = 1, \ldots, N$.

Proof. Given a part M of $\mathscr{C}^0(\Gamma)$, let us denote

$$M^\perp = \{\varphi \in \mathscr{C}^0(\Gamma); \int_\Gamma \alpha\varphi \,\mathrm{d}\gamma = 0 \quad \text{for all} \quad \alpha \in M\}.$$

It is clear, by the projection theorem in $L^2(\Gamma)$ that for a sub-space of finite dimension M of $\mathscr{C}^0(\Gamma)$,

$$\mathscr{C}^0(\Gamma) = M \oplus M^\perp$$

in other words

$$\operatorname{codim} M^\perp = \dim M .$$

The operators K and J being compact (see Proposition 16), $\ker(I + K)$ and $\ker(I + J)$ are spaces of finite dimension (see Lemma 3) and

$$\operatorname{codim} \operatorname{Im}(I \pm K) = \dim \ker (I \pm K)$$

$$\operatorname{codim} \operatorname{Im}(I \pm J) = \dim \ker (I \pm J) .$$

Now from Proposition 15(2), K and J are adjoints the one of the other, from which we derive immediately that

$$\operatorname{Im}(I \pm K) \subset \ker(I \pm J)^\perp$$

$$\operatorname{Im}(I \pm J) \subset \ker(I \pm K)^\perp .$$

We deduce that

$$\dim\ker(I \pm J) = \operatorname{codim}\ker(I \pm J)^\perp \leqslant \operatorname{codim}\operatorname{Im}(I \pm K) =$$

$$= \dim\ker(I \pm K) = \operatorname{codim}\ker(I \pm K)^\perp \leqslant \operatorname{codim}\operatorname{Im}(I \pm J);$$

from which we have the equalities

$$\operatorname{codim}\ker(I \pm J)^\perp = \operatorname{codim}\operatorname{Im}(I \pm K),$$

and

$$\operatorname{Im}(I \pm K) = \ker(I \pm J)^\perp .$$

That is to say, account being taken of what was recalled above

$$\operatorname{Im}(I + K) = \{\alpha'_1, \ldots, \alpha'_{N'}\}^\perp$$

$$\operatorname{Im}(I - K) = \{\alpha_1, \ldots, \alpha_N\}^\perp .$$

From Proposition 15 (3)

$$\begin{cases} \ker(I + K) \supset \chi_{\Gamma'_1}\mathbb{R} \oplus \ldots \oplus \chi_{\Gamma'_{N'}}\mathbb{R} \\ \ker(I - K) \supset \chi_{\Gamma_1}\mathbb{R} \oplus \ldots \oplus \chi_{\Gamma_N}\mathbb{R} . \end{cases} \tag{4.127}$$

But from what was proved above

$$\dim\ker(I + K) = \operatorname{codim}\operatorname{Im}(I + K) = N'$$

$$\dim\ker(I - K) = \operatorname{codim}\operatorname{Im}(I - K) = N ;$$

from which we have equality in the inclusions (4.127).
This, together with (4.100) and (4.102) proves the proposition. □

Remark 7. *If, in addition to the hypotheses of the proposition, $\mathbb{R}^n \backslash \bar{\Omega}$ is connected,* i.e. $N' = 0$, *for all $\varphi \in \mathscr{C}^0(\Gamma)$, the solution of $P(\Omega, \varphi)$ can be written in a unique way as a double layer potential.* We shall see in §7, that in the case Ω convex (which implies $\mathbb{R}^n \backslash \bar{\Omega}$ connected, the solution of $P(\Omega, \varphi)$ can be obtained by an iterative procedure with the aid of the integral operator K: this is *Neumann's method.*
In the general case, for all $\varphi \in \mathscr{C}^0(\Gamma)$, there exist $\lambda_1, \ldots, \lambda_N \in \mathbb{R}$ unique and such that the solution u of $P(\Omega, \varphi)$ is of the form

$$u = \lambda_1 u_1 + \ldots + \lambda_{N'} u_{N'} + v$$

where u_j is the interior simple layer potential of α_j' and v is an interior double layer potential. □

6. Sub-Harmonic Functions. Perron's Method

In this section, we develop the solution of Dirichlet's problem by the method of Perron of which we have given the broad outlines in §4.1. As we have indicated, this method is based on the properties of sub-harmonic functions. First, let us prove

Proposition 20. *Let Ω be an open set in $\mathbb{R}^n$ and v a distribution on Ω. The following assertions are equivalent:*
(i) *$\Delta v \geqslant 0$ in $\mathscr{D}'(\Omega)$, i.e.*

$$\langle \Delta v, \zeta \rangle = \langle v, \Delta \zeta \rangle \geqslant 0 \quad \textit{for all} \quad \zeta \in \mathscr{D}(\Omega), \zeta \geqslant 0\,;$$

(ii) *v is a (locally integrable) function satisfying*

$$v(x) \leqslant \frac{n}{\sigma_n r^n} \int_{B(x,r)} v(t)\,\mathrm{d}t \quad \text{a.e.} \quad x \in \Omega \quad \textit{and for all} \quad 0 < r < \operatorname{dist}(x, \partial\Omega)\,;$$

(iii) *there exists $\bar{v}: \Omega \to [-\infty, +\infty[$ u.s.c.*[90] *and locally integrable satisfying*

$$\bar{v}(x) \leqslant \frac{n}{\sigma_n r^n} \int_{B(x,r)} \bar{v}(t)\,\mathrm{d}t \quad \textit{for all} \quad x \in \Omega \quad \textit{and} \quad 0 < r < \operatorname{dist}(x, \partial\Omega)\,, \tag{4.128}$$

such that $v = \bar{v}$ in $\mathscr{D}'(\Omega)$, that is

$$\langle v, \zeta \rangle = \int \bar{v}(t)\zeta(t)\,\mathrm{d}t \quad \textit{for all} \quad \zeta \in \mathscr{D}(\Omega)\,.$$

In addition, then, $\bar{v}$ is unique.

Proposition 21. *Let Ω be an open set in $\mathbb{R}^n$ and a map $v: \Omega \to [-\infty, \infty[$ u.s.c. The following assertions are equivalent:*
(i) *for every open set $\omega \subset \Omega$ and $u \in \mathscr{H}(\omega) \cap \mathscr{C}^0(\bar{\omega})$ satisfying*

$$\limsup_{x \to z} v(x) \leqslant u(z) \quad \text{for all} \quad z \in \partial\omega\,,$$

---

[90] u.s.c. is the abbreviation for upper semi-continuous: $\{x \in \Omega; \bar{v}(x) < \lambda\}$ is open for all $\lambda \in \mathbb{R}$, or again in an equivalent way, $\limsup_{x \to x_0} \bar{v}(x) = \bar{v}(x_0)$, for all $x_0 \in \Omega$.

and in the case when ω *is unbounded*

$$\limsup_{\substack{|x| \to \infty \\ x \in \omega}} v(x) - u(x) \leqslant 0 ,$$

we have

$$v(x) \leqslant u(x) \quad \textit{for all} \quad x \in \omega ;$$

(ii) *for all* $x_0 \in \Omega$, $0 < r < \operatorname{dist}(x_0, \partial\Omega)$

$$(4.129) \qquad v(x) \leqslant \frac{1}{r\sigma_n} \int_{\partial B(x_0, r)} \frac{r^2 - |x - x_0|^2}{|t - x|^n} v(t) \, d\gamma(t)^{91} \quad \textit{for all} \quad x \in B(x_0, r) ;$$

(iii) *for all* $x \in \Omega$ *and* $0 < \varepsilon \leqslant \operatorname{dist}(x, \partial\Omega)$, *there exists* $0 < r < \varepsilon$ *such that*

$$v(x) \leqslant \frac{n}{\sigma_n r^n} \int_{B(x, r)} v(t) \, dt^{91} .$$

These characterisations lead us to frame the definitions:

Definition 15. Let Ω be an open set in $\mathbb{R}^n$:
(1) a *distribution* v on Ω is said to be *subharmonic* if $\Delta v \geqslant 0$ in $\mathscr{D}'(\Omega)$ (see Proposition 20(i)).
(2) a *map* $v: \Omega \to [-\infty, \infty]$ is said to be *subharmonic* if it satisfies the condition (i) of Proposition 21.
A distribution (resp. map) w is said to be *superharmonic* if $v = -w$ subharmonic.

Propositions 20 and 21 show that given a distribution v, subharmonic on Ω, there exists a unique map $\bar{v}: \Omega \to [-\infty, \infty[$ u.s.c., locally integrable and subharmonic, such that $v = \bar{v}$ in $\mathscr{D}'(\Omega)$: $\bar{v}$ is the *sub-harmonic u.s.c. representative of* v.
For a function $v \in \mathscr{C}^0(\Omega)$, more generally for $v: \Omega \to [-\infty, \infty[$ u.s.c., locally integrable, the definitions of sub-harmonic in so far as the map or distribution coincide; in particular the Definition 15 generalizes the Definition 2 and contains the property 1 given in §4.1.
In the case $n = 1$ and $\Omega =]a, b[$, a map $v:]a, b[\to [-\infty, \infty]$ is sub-harmonic iff the epigraph of v, i.e. the set

$$\{(x, y) \in]a, b[\times \mathbb{R}; v(x) \leqslant y\} ,$$

is convex.
In other words, v is sub-harmonic iff it satisfies:

(i) $I = \{x \in]a, b[; v(x) < \infty\}$ is an interval
(ii) $v \equiv -\infty$ on I or v is finite and convex on I,

---

[91] The function v being u.s.c. for every compact set K of Ω, v is majorized on K and even attains its maximum on K (see, for example, Bourbaki [1], Chap. IV). It follows that for every positive Borel measure μ on K, we can define $\int_K v(t) \, d\mu(t)$, putting $\int_K v(t) \, d\mu(t) = -\infty$ when v is not integrable on K.

This follows immediately from the fact that the harmonic functions on an interval are the affine functions on this interval. *In the case* $n = 1$, *sub-harmonic coincides with convex*; this is obviously no longer true in the case $n \geqslant 2$. Also, in the case $n = 1$, a function $v:]a, b[\to [-\infty, \infty[$ (not taking the value $+\infty$) sub-harmonic and not identically $-\infty$ is finite and continuous. This is no longer true either in the case $n \geqslant 2$.

Example 24. For $n \geqslant 2$, the map $v: \mathbb{R}^n \to [-\infty, \infty[$ defined by

$$v(x) = \begin{cases} E_n(x) & \text{if } x \neq 0 \\ -\infty & \text{if } x = 0 \end{cases}$$

is sub-harmonic. This follows in fact from the characterisation of Proposition 20 (v is u.s.c., locally integrable, and $\Delta v = \delta \geqslant 0$) as well as of the characterisation (iii) of Proposition 21).

Proof of Proposition 20. Let us note that the assertion (i) implies that Δv is a (positive) Radon measure on Ω; from Proposition 6 of §3, if v satisfies (i), it is then a (locally integrable)[92] function on Ω. We can therefore, without loss of generality, suppose v to be locally integrable on Ω. Let us take $\rho \in \mathscr{D}(\mathbb{R}^n)$, $\rho \geqslant 0$, $\int \rho(x)\,dx = 1$ and $\operatorname{supp}\rho \subset \bar{B}(0, 1)$ and put for all $k \geqslant 1$,

$$\Omega_k = \{x \in \Omega; \operatorname{dist}(x, \partial\Omega) < k^{-1}\},$$

$$v_k(x) = \int \rho(t) v(x + k^{-1}t)\,dt \quad \text{for} \quad x \in \Omega_k .$$

We have

$$v_k \in \mathscr{C}^\infty(\Omega_k), \qquad \Delta v_k(x) = \langle \Delta v(x + k^{-1}t), \rho(t) \rangle$$

$$v_k \to v \quad \text{in} \quad L^1_{\text{Loc}}(\Omega_l) \quad \text{when} \quad k \to \infty \quad \text{for all} \quad l \geqslant 1 .$$

We deduce immediately that v satisfies (i) iff for all $k \geqslant 1$

$$\Delta v_k(x) \geqslant 0 \quad \text{for all} \quad x \in \Omega_k .$$

Also

$$\int_{B(x,r)} v_k(s)\,ds = \int \rho(t)\,dt \int_{B(x + k^{-1}t, r)} v(s)\,ds ;$$

so v satisfies (ii) iff for all $k \geqslant 1$

$$v_k(x) \leqslant \frac{n}{\sigma_n r^n} \int_{B(x,r)} v_k(s)\,ds \quad \text{for} \quad x \in \Omega_k , \quad 0 < r < \operatorname{dist}(x, \partial\Omega_k) .$$

In other words it is enough to prove the equivalence (i) $\Leftrightarrow$ (ii) when $v \in \mathscr{C}^\infty(\Omega)$. Hence, let us suppose that $v \in \mathscr{C}^2(\Omega)$ satisfying $\Delta v \geqslant 0$ on Ω. We put for $x \in \Omega$ and

---

[92] We have even $v \in L^p_{\text{loc}}(\Omega)$ for $p < \dfrac{n}{n-2}$ and $\dfrac{\partial v}{\partial x_i} \in L^p_{\text{loc}}(\Omega)$ for $p < \dfrac{n}{n-1}$.

$0 < r < r(x) = \operatorname{dist}(x, \partial\Omega)$

$$M(x, r) = \frac{n}{\sigma_n r^n}\int_{B(x,r)} v(t)\,\mathrm{d}t = \frac{n}{\sigma_n}\int_{B(0,1)} v(x + rt)\,\mathrm{d}t\,.$$

We have (see (2.21) in §2)

$$\Delta_x M(x, r) = \frac{n}{\sigma_n}\int_{B(0,1)} \Delta v(x + rt)\,\mathrm{d}t = \frac{1}{r^{n-1}}\frac{\partial}{\partial r}\left(r^{n+1}\frac{\partial M}{\partial r}(x, r)\right)$$

and therefore
$$\frac{\partial}{\partial r}\left(r^{n+1}\frac{\partial M}{\partial r}(x, r)\right) \geqslant 0\,.$$

In particular:

$$r^{n+1}\frac{\partial M}{\partial r}(x, r) \geqslant \lim_{r\to 0} r^{n+1}\frac{\partial M}{\partial r}(x, r)$$
$$= \lim_{r\to 0}\frac{nr^{n+1}}{\sigma_n}\int_{B(0,1)} \operatorname{grad} v(x + rt).t\,\mathrm{d}t = 0\,,$$

from which we deduce $\frac{\partial M}{\partial r} \geqslant 0$ and hence

$$M(x, r) \geqslant \lim_{r\to 0} M(x, r) = v(x)\,.$$

This proves (i) ⇒ (ii). Conversely, let us consider now $v \in \mathscr{C}^2(\Omega)$ satisfying (ii), and let us put $\omega = \{x \in \Omega;\ \Delta v(x) < 0\}$. Since v satisfies (ii) and $-v$ satisfies (i) and hence (ii) on ω, we have

$$v(x) = M(x, r) \quad \text{for all} \quad x \in \omega \quad \text{and} \quad 0 < r < \operatorname{dist}(x, \omega)\,.$$

From Proposition 15 of §2, v is therefore harmonic on ω, i.e. $\Delta v = 0$ on ω. This is contradictory unless $\omega = \varnothing$, i.e. $\Delta v \geqslant 0$ on Ω.

The implication (iii) ⇒ (ii) being trivial, it remains to prove (ii) ⇒ (iii) and the uniqueness of $\bar{v}$. Let us suppose that v satisfies (ii). We prove first the uniqueness of $\bar{v}$. Let, first of all, $\bar{v}$ satisfy (iii): we shall then have for $x \in \Omega$

$$\bar{v}(x) \leqslant \frac{n}{\sigma_n r^n}\int_{B(x,r)} \bar{v}(t)\,\mathrm{d}t = \frac{n}{\sigma_n r^n}\int_{B(x,r)} v(t)\,\mathrm{d}t \quad \text{for all} \quad 0 < r < r(x)$$

since $v = \bar{v}$ a.e. on Ω; hence

$$\bar{v}(x) \leqslant \inf_{0<r<r(x)}\frac{n}{\sigma_n r^n}\int_{B(x,r)} v(t)\,\mathrm{d}t\,.$$

Now since $\bar{v}$ is u.s.c. for $\lambda > \bar{v}(x)$, there exists $0 < r < r(x)$ such that $\bar{v} \leqslant \lambda$ on $B(x, r)$ and so

$$\frac{n}{\sigma_n r^n}\int_{B(x,r)} v(t)\,\mathrm{d}t = \frac{n}{\sigma_n r^n}\int_{B(x,r)} \bar{v}(t)\,\mathrm{d}t \leqslant \lambda\,.$$

We deduce that necessarily

$$\bar{v}(x) = \inf_{0<r<r(x)} \frac{n}{\sigma_n r^n} \int_{B(x,r)} v(t)\,dt\,. \tag{4.130}$$

Conversely, the function $\bar{v}$ defined by (4.130) satisfies (iii). In effect the function $x \to r(x)$ is continuous. For $x_0 \in \Omega$, fixed and $0 < r < r(x_0)$ fixed, the function

$$x \to \frac{n}{\sigma_n r^n} \int_{B(x,r)} v(t)\,dt = \frac{n}{\sigma_n} \int_{B(0,1)} v(x + rt)\,dt$$

is defined and continuous in the neighbourhood of x_0 from the continuity of translations in L^1_{loc}. Hence $\bar{v}$ the lower envelope of continuous functions is u.s.c. On the other hand, from the hypothesis (ii), for almost all $x \in \Omega$

$$v(x) \leqslant \frac{n}{\sigma_n r^n} \int_{B(x,r)} v(t)\,dt \quad \text{for all} \quad 0 < r < r(x)\,.$$

Also, from Lebesgue's differentiation theorem, for almost all $x \in \Omega$,

$$v(x) = \lim_{r \to 0} \frac{n}{\sigma_n r^n} \int_{B(x,r)} v(t)\,dt\,.$$

Hence $\bar{v}(x) = v(x)$ for almost all $x \in \Omega$.
Finally by the definition of $\bar{v}$, for all $x \in \Omega$ and $0 < r < r(x)$,

$$\bar{v}(x) \leqslant \frac{n}{\sigma_n r^n} \int_{B(x,r)} v(t)\,dt = \frac{n}{\sigma_n r^n} \int_{B(x,r)} \bar{v}(t)\,dt\,;$$

which concludes the proof of the proposition. □

Proof of Proposition 21. To prove (i) ⇒ (ii), we note that given $x_0 \in \Omega$, $0 < r < r(x_0)$ and $\varphi \in \mathscr{C}^0(\partial B(x_0, r))$, the function u defined on $\omega = B(x_0, r)$ by

$$u(x) = \frac{1}{r\sigma_n} \int_{\partial B(x_0,r)} \frac{r^2 - |x - x_0|^2}{|t - x|^n} \varphi(t)\,d\gamma(t)$$

is a solution of $P(\omega, \varphi)$ (see Example 2); hence if v satisfies (i) and $v \leqslant \varphi$ on $\partial B(x_0, r)$, we have $v \leqslant u$ on $B(x_0, r)$. Now, since v is u.s.c.[93], there exists $\varphi_k \in \mathscr{C}^0(\partial B(x_0, r))$ such that

$$\varphi_k(z) \downarrow v(z) \quad \text{when} \quad k \uparrow \infty \quad \text{for all} \quad z \in \partial B(x_0, r)\,.$$

We deduce that (i) ⇒ (ii).
The proof of (ii) ⇒ (iii) is identical to that of Corollary 2 of §2 since (4.129) implies

$$v(x_0) \leqslant \frac{1}{\sigma_n r^{n-1}} \int_{\partial B(x_0,r)} v(t)\,d\gamma(t) \quad \text{for all} \quad 0 < r < r(x_0)\,.$$

Finally let us prove (iii) ⇒ (i). Let us take $\omega \subset \Omega$ and $u \in \mathscr{H}(\omega) \cap \mathscr{C}^0(\bar{\omega})$

---

[93] See, for example, Dieudonné [1] vol. 1, Bourbaki [1], Chap. IV.

satisfying

$$(4.131)\qquad \begin{cases} \limsup\limits_{x \to z} v(x) \leqslant u(z) & \text{for all} \quad z \in \partial\omega \\ \limsup\limits_{|x| \to \infty} v(x) - u(x) \leqslant 0 & \text{when} \quad \omega \text{ is unbounded} . \end{cases}$$

Supposing that (iii) is satisfied, it is the same for the restriction of v to ω and, by use of the formula of the mean for $u \in \mathscr{H}(\omega)$, for $v - u$ on ω. In other words, we can suppose $u \equiv 0$ on ω; we have to show $v \leqslant 0$ on ω. By contradiction, suppose $\sup\limits_{\omega} v > 0$; from (4.131) and the u.s.c. hypothesis

$$F = \left\{ x \in \omega; v(x) = \sup_{\omega} v \right\}$$

is a compact non-empty subset of ω; considering $x_0 \in \partial F$ and $\varepsilon = \operatorname{dist}(x_0, \partial\omega)$, from hypothesis (iii), there exists $0 < r < \varepsilon$ such that

$$\sup_{\omega} v = v(x_0) \leqslant \frac{n}{\sigma_n r^n} \int_{B(x_0, r)} v(t) \, dt \leqslant \sup_{B(x_0, r)} v .$$

Hence, $v \equiv \sup\limits_{\omega} v$ on $B(x_0, r)$, a contradiction with the choice of $x_0 \in \partial F$. □

Remark 8

(1) *The set of sub-harmonic distributions on* Ω *is a cone* of $\mathscr{D}'(\Omega)$ the space of distributions on Ω; similarly the set of maps (resp. u.s.c.), sub-harmonic on Ω into $[-\infty, \infty]$ (resp. $[-\infty, \infty[$) is a cone.

The map which to a sub-harmonic distribution v makes correspond its sub-harmonic u.s.c. representative $\bar{v}$ satisfies

$$\overline{\lambda_1 v_1 + \lambda_2 v_2} = \lambda_1 \overline{v_1} + \lambda_2 \overline{v_2} \quad \text{for all} \quad \lambda_1 \text{ and } \lambda_2 \geqslant 0 .$$

(2) *The cone of sub-harmonic distributions is closed* for convergence in the sense of distributions: if (v_k) is a sequence of sub-harmonic distributions such that

$$\lim_{k \to z} \langle v_k, \zeta \rangle \quad \text{exists for all} \quad \zeta \in \mathscr{D}(\Omega) ,$$

then $\langle v, \zeta \rangle = \lim\limits_{k \to \infty} \langle v_k, \zeta \rangle$ is a sub-harmonic distribution. In general, however, the sub-harmonic u.s.c. representative of v is not the (point) limit of the sub-harmonic u.c.s. representatives of the v_k.

(3) *The cone of sub-harmonic maps of* Ω *into* $[-\infty, \infty]$ *is stable under the upper envelope*: if (v_i) is a family of sub-harmonic maps of Ω into $[-\infty, \infty]$, then the map

$$v : x \in \Omega \to \sup_i v_i(x)$$

is sub-harmonic.

In general, however, the upper envelope of sub-harmonic u.s.c. maps of Ω into $[-\infty, \infty[$ is not u.s.c.

(4) *The increasing convex functions act in the cone of sub-harmonic u.s.c. maps of* Ω *into* $[-\infty, \infty[$: if $\Phi: \mathbb{R} \to \mathbb{R}$ is increasing and convex and $v: \Omega \to [-\infty, \infty]$ is sub-harmonic u.s.c., then $\Phi(v)$ is subharmonic u.s.c., with the convention $\Phi(-\infty) = \inf \Phi(\mathbb{R})$. This follows from the characterisation (iii) and Jensen's inequality

$$\Phi\left(\frac{1}{\mu(K)} \int_K v(t)\,\mathrm{d}\mu(t)\right) \leqslant \frac{1}{\mu(K)} \int_K \Phi(v(t))\,\mathrm{d}\mu(t)\,.$$

This same inequality shows that *given* $\Phi: \mathbb{R} \to \mathbb{R}$ *convex* (arbitrary) *and* $u \in \mathscr{H}(\Omega)$, $\Phi(u)$ *is sub-harmonic (continuous) on* Ω.

Let us now develop Perron's method for the solution of Dirichlet's problem. We consider a bounded open set Ω in $\mathbb{R}^n$ with boundaty Γ and $\varphi \in \mathscr{C}^0(\Gamma)$. We take up again the notation (4.10):

$$\mathscr{I}_-(\varphi) = \{v \in \mathscr{C}^0(\Omega);\ v \text{ subharmonic on } \Omega \text{ and}$$

$$\limsup_{x \to z} v(x) \leqslant \varphi(z) \quad \text{for all} \quad z \in \Gamma\}$$

$$\mathscr{I}_+(\varphi) = \{w \in \mathscr{C}^0(\Omega);\ w \text{ subharmonic on } \Omega \text{ and}$$

$$\liminf_{x \to z} w(x) \geqslant \varphi(z) \quad \text{for all} \quad z \in \Gamma\}\,;$$

we have

$$v \leqslant w \quad \text{on} \quad \Omega \quad \text{for all} \quad v \in \mathscr{I}_-(\varphi) \quad \text{and} \quad w \in \mathscr{I}_+(\varphi)$$

and $\mathscr{I}_-(\varphi)$ and $\mathscr{I}_+(\varphi)$ are non-empty since they contain

$$v \equiv \min_\Gamma \varphi \quad \text{and} \quad w \equiv \max_\Gamma \varphi \text{ respectively}\,.$$

We can thus define, with the notation (4.12), for all $x \in \Omega$

$$u_-(\varphi, x) = \sup_{v \in \mathscr{I}_-(\varphi)} v(x), \quad u_+(\varphi, x) = \inf_{w \in \mathscr{I}_+(\varphi)} w(x)\,. \tag{4.132}$$

We then prove Wiener's theorem (see Property 2 of §4.1):

Proposition 22. *Let* Ω *be a bounded open set in* $\mathbb{R}^n$ *with boundary* Γ *and* $\varphi \in \mathscr{C}^0(\Gamma)$. *The functions* $u_-(\varphi)$ *and* $u_+(\varphi)$ *defined by* (4.132) *are harmonic and identical on* Ω.

First, we state some lemmas:

Lemma 5. *Let* Ω *be an open set in* $\mathbb{R}^n$, $v \in \mathscr{C}^0(\Omega)$ *sub-harmonic on* Ω *and a ball* $B = B(x_0, r_0)$ *with* $\bar{B} \subset \Omega$. *Let us construct the function.*

$$T_B v(x) = \begin{cases} v(x) & \text{if} \quad x \in \Omega \backslash B \\ \dfrac{1}{r_0 \sigma_n} \displaystyle\int_{\partial B} \frac{r_0^2 - |x - x_0|^2}{|t - x|^n} v(t)\,\mathrm{d}\gamma(t) & \text{if} \quad x \in B\,. \end{cases}$$

Then
(1) $v \leqslant T_B v$ *on* Ω;
(2) $T_B v \in \mathscr{C}^0(\Omega)$ *is subharmonic on* Ω;
(3) $T_B v$ *is harmonic on* B.

Proof of Lemma 5. The point (1) is a result of the characterisation (ii) of Proposition 21. The final point and the continuity of $T_B v$ follows from the fact that, from Example 12, $T_B v$ is a solution of $P(B, v)$. Hence, it remains to prove that $T_B v$ is sub-harmonic on Ω: it is clearly so in the neighbourhood of each point of $\Omega \backslash \partial B$. Now for $x \in \partial B$ and $0 < r < \operatorname{dist}(x, \partial\Omega)$, we have

$$T_B v(x) = v(x) \leqslant \frac{n}{\sigma_n r^n} \int_{B(x, r)} v(t)\, dt \leqslant \frac{n}{\sigma_n r^n} \int_{B(x, r)} T_B v(t)\, dt$$

from which the result follows. □

Lemma 6. *Let* Ω *be an open set in* $\mathbb{R}^n$, $B = B(x_0, r_0)$ *a ball with* $\bar{B} \subset \Omega$ *and* $\mathscr{F}$ *a family of subharmonic functions* $v \in \mathscr{C}^0(\Omega)$. *We suppose*

$$\begin{cases} \text{for all } v \in \mathscr{F},\ T_B v \in \mathscr{F}\ ; \\ \text{for all } v_1, v_2 \in \mathscr{F},\ \max(v_1, v_2) \in \mathscr{F}\ ; \\ \{v(x_0); v \in \mathscr{F}\} \text{ is majorized in } \mathbb{R}\ . \end{cases}$$

Then, for all $x \in B$, $\{v(x); v \in \mathscr{F}\}$ *is majorized in* $\mathbb{R}$, *and the function*

$$x \in B \to \sup_{v \in \mathscr{F}} v(x)$$

is harmonic on B.

Proof of Lemma 6. We put, for all $x \in B$

$$u(x) = \sup_{v \in \mathscr{F}} v(x) \in]-\infty, \infty]\ .$$

Let us consider a sequence of points (x_k) of B, dense in B. For all k, there exists $v_k \in \mathscr{F}$ such that

$$\begin{cases} v_k(x_k) + k^{-1} \geqslant u(x_k) & \text{if } u(x_k) \leqslant \infty \\ v_k(x_k) \geqslant k & \text{if } u(x_k) = \infty\ . \end{cases}$$

We then construct the sequence $(\tilde{v}_k)$ according to the recurrence relation

$$\tilde{v}_1 = T_B v_1\ , \quad \tilde{v}_{k+1} = T_B(\max(\tilde{v}_k, v_{k+1}))\ .$$

From the hypotheses and Lemma 5, we have for all k

$$\tilde{v}_k \in \mathscr{F}\ , \qquad \tilde{v}_k \text{ harmonic on } B\ , \qquad \tilde{v}_k \geqslant v_k$$

and the sequence $(\tilde{v}_k)$ is increasing.
Using Corollary 9 of §2, since $\tilde{v}_k(x_0)$ is bounded, $(\tilde{v}_k)$ converges uniformly on every compact set of B. We denote its limit by $\tilde{u}$ and show that $u = \tilde{u}$. We have $\tilde{u} \leqslant u$ by the definition of u.

Now, for all k, $\tilde{u} \geqslant v_k$ and therefore

$$(4.133)\qquad \begin{cases} \tilde{u}(x_k) + k^{-1} \geqslant u(x_k) & \text{if} \quad u(x_k) < \infty \\ \tilde{u}(x_k) \geqslant k & \text{if} \quad u(x_k) = \infty . \end{cases}$$

Let us suppose that at point $x \in B$, $\tilde{u}(x) < u(x)$; since u is an upper envelope of continuous functions, it is u.s.c.; there exists, therefore, $r > 0$ and $\delta > 0$ such that $u \geqslant \tilde{u} + \delta$ on $\bar{B}(x, r) \subset B$. The sequence (x_k) being dense, there exists $k_l \to \infty$ such that $x_{k_l} \in \bar{B}(x, r)$ for all $l \in \mathbb{N}$. Since $\tilde{u}$ is bounded on $\bar{B}(x, r)$, from (4.133) for l sufficiently large, $u(x_{k_l}) < \infty$ and

$$u(x_{k_l}) \leqslant \tilde{u}(x_{k_l}) + k_l^{-1} < \tilde{u}(x_{k_l}) + \delta$$

from which we have a contradiction. ◻

Proof of Proposition 22. We should notice that these two lemmas prove the first part of the proposition: the functions $u_-(\varphi)$ and $u_+(\varphi)$ are harmonic on Ω; it is sufficient to apply Lemma 6 to

$$\mathscr{F} = \mathscr{I}_-(\varphi) \quad \text{and} \quad \mathscr{F} = \{ -w; w \in \mathscr{I}_+(\varphi)\}$$

for every ball B such that $\bar{B} \subset \Omega$.
Now we have immediately by definition of $u_-(\varphi)$ and $u_+(\varphi)$

$$(4.134)\qquad \begin{cases} u_-(\varphi) \leqslant u_+(\varphi) = -u_-(-\varphi), u_-(\lambda\varphi) = \lambda u_-(\varphi) \quad \text{for all} \quad \lambda \geqslant 0 \\ u_-(\varphi_1) + u_-(\varphi_2) \leqslant u_-(\varphi_1 + \varphi_2), \\ u_-(\varphi + c) = u_-(\varphi) + c \quad \text{for every constant } c \\ \varphi_1 \leqslant \varphi_2 \Rightarrow u_-(\varphi_1) \leqslant u_-(\varphi_2) . \end{cases}$$

We derive

$$u_+(\varphi) + u_-(\psi) \leqslant u_+\left(\psi + \sup_\Gamma(\varphi - \psi)\right) + u_-\left(\varphi + \sup_\Gamma(\psi - \varphi)\right)$$

$$= u_+(\psi) + u_-(\varphi) + \sup_\Gamma(\varphi - \psi) + \sup_\Gamma(\psi - \varphi),$$

from which we have

$$(4.135)\qquad u_+(\varphi) - u_-(\varphi) \leqslant u_+(\psi) - u_-(\psi) + 2\|\varphi - \psi\| .$$

We deduce that the set

$$E = \{\varphi \in \mathscr{C}^0(\Gamma); u_-(\varphi) = u_+(\varphi)\}$$

is a closed vector sub-space of $\mathscr{C}^0(\Gamma)$. To prove the proposition it is sufficient therefore to show that the set E contains a dense part of $\mathscr{C}^0(\Gamma)$. We observe that E contains the traces on Γ of the functions $v \in \mathscr{C}^0(\bar{\Omega})$ which are sub-harmonic on Ω; in effect such a function v with trace φ is trivially in $\mathscr{I}_-(\varphi)$ with the result that

$$v \leqslant u_-(\varphi) .$$

It follows that for all $z \in \Gamma$

$$\liminf_{x \to z} u_-(\varphi, x) \geqslant \lim_{x \to z} v(x) = \varphi(z)\,,$$

and hence, since $u_-(\varphi)$ is harmonic (from which *a priori* subharmonic) on Ω, $u_-(\varphi) \in \mathscr{I}_+(\varphi)$ and $u_-(\varphi) = u_+(\varphi)$.
Given $u \in \mathscr{C}^2(\bar{\Omega})$, we have $u = v_1 - v_2$ with

$$v_1(x) = \lambda |x|^2\,, \quad v_2(x) = \lambda |x|^2 - u(x)\,;$$

let us choose $\lambda = \dfrac{1}{2n} \sup_{\Omega} (\Delta u(x))^+$ with the result that

$$\Delta v_1 = 2\lambda n \geqslant 0 \quad \text{and} \quad \Delta v_2 = 2\lambda n - \Delta u \geqslant 0\,;$$

the functions $v_1, v_2 \in \mathscr{C}^0(\bar{\Omega})$ sub-harmonic on Ω have their traces in E; hence E contains the traces of the functions $u \in \mathscr{C}^2(\bar{\Omega})$. From the Stone–Weierstrass theorem, the result is proved. □

Wiener's theorem led us to define the function $u(\varphi) = u_-(\varphi) = u_+(\varphi)$ as the generalized solution of $P(\Omega, \varphi)$ (see Definition 3). We have next introduced the notion of a regular point of the boundary of an open set (see Definition 4); to conclude the proof of Theorem 1, it remains to us to prove Property 3 of §4.1.

Proposition 23. *Let Ω be a bounded open set of $\mathbb{R}^n$ and z a regular point of the boundary Γ of Ω. Then for all $\varphi \in \mathscr{C}^0(\Gamma)$, the generalized solution $u(\varphi)$ of $P(\Omega, \varphi)$ satisfies*

$$\lim_{x \to z} u(\varphi, x) = \varphi(z)\,.$$

Proof. Let us take a barrier function $v \in \mathscr{C}^0(\Omega \cap B(z, r))$ (see Definition 4). For $\varepsilon > 0$, let us choose $0 < r_\varepsilon < r$ such that

$$|\varphi - \varphi(z)| \leqslant \varepsilon \quad \text{on} \quad \Gamma \cap \bar{B}(z, r_\varepsilon)$$

and put

$$m_\varepsilon = \sup_{\Omega \cap B(z,r) \backslash B(z,r_\varepsilon)} v\,.$$

By hypothesis $m_\varepsilon < 0$; we then define

$$k_\varepsilon = \frac{1}{|m_\varepsilon|} \sup_{\Gamma} |\varphi - \varphi(z)|$$

$$v_\varepsilon(x) = \begin{cases} \varphi(z) - \varepsilon + k_\varepsilon \max(v(x), m_\varepsilon) & \text{on} \quad \Omega \cap B(z, r_\varepsilon) \\ \varphi(z) - \varepsilon + k_\varepsilon m_\varepsilon & \text{on} \quad \Omega \backslash B(z, r_\varepsilon) \end{cases}$$

$$w_\varepsilon(z) = \begin{cases} \varphi(z) + \varepsilon - k_\varepsilon \max(v(x), m_\varepsilon) & \text{on} \quad \Omega \cap B(z, r_\varepsilon) \\ \varphi(z) + \varepsilon - k_\varepsilon m_\varepsilon & \text{on} \quad \Omega \backslash B(z, r_\varepsilon)\,. \end{cases}$$

By construction $v_\varepsilon \in \mathscr{I}_-(\varphi)$ and $w_\varepsilon \in \mathscr{I}_+(\varphi)$; hence

$$v_\varepsilon \leqslant u(\varphi) \leqslant w_\varepsilon\,,$$

and in particular

$$|u(\varphi) - \varphi(z)| \leqslant \varepsilon - k_\varepsilon v \quad \text{on} \quad \Omega \cap B(z, r_\varepsilon)\,,$$

which proves

$$\limsup_{x \to z} |u(\varphi, x) - \varphi(z)| \leqslant \varepsilon \quad \text{for all} \quad \varepsilon > 0\,. \qquad \square$$

§5. Capacities*

1. Interior and Exterior Capacity Operators

We take *a regular bounded open set* Ω *of* $\mathbb{R}^n$ with boundary Γ. Given $\varphi \in \mathscr{C}^0(\Gamma)$, we can consider the solution $u(\varphi)$ of the Dirichlet problem $P(\Omega, \varphi)$ (see §4); supposing $u(\varphi) \in \mathscr{C}^1_n(\bar{\Omega})$, we can consider $\psi = \partial u(\varphi)/\partial n \in \mathscr{C}^0(\Gamma)$; the map $\varphi \to \psi$ defines an operator C of $D(C)$ into $\mathscr{C}^0(\Gamma)$ where

$$D(C) = \{\varphi \in \mathscr{C}^0(\Gamma); u(\varphi) \in \mathscr{C}^1_n(\bar{\Omega})\}\,. \tag{5.1}$$

We shall call this operator *the interior capacity operator of* $\mathscr{C}^0(\Gamma)$. This terminology will be jusfied by the physical interpretation which we shall make in Sect. 2. We observe that C *is an unbounded linear operator* (not defined everywhere) *of* $\mathscr{C}^0(\Gamma)$. By definition also, given $\varphi, \psi \in \mathscr{C}^0(\Gamma)$, $\varphi \in D(C)$ and $\psi = C\varphi$ iff *the solution of the Dirichlet problem* $P(\Omega, \varphi)$ *is a solution of the Neumann Problem* $PN(\Omega, \psi)$.
Similarly, given $\varphi \in \mathscr{C}^0(\Gamma)$, we can consider the solution $u_e(\varphi)$ of the exterior Dirichlet problem $P(\mathbb{R}^n \backslash \bar{\Omega}, \varphi)$ satisfying the null condition at infinity (see §4.3); supposing that $u_e(\varphi) \in \mathscr{C}^1_n(\mathbb{R}^n \backslash \Omega)$ we can consider $\psi = \partial u_e(\varphi)/\partial n$: the map $\varphi \to \psi$ defines an operator C_e of $D(C_e)$ into $\mathscr{C}^0(\Gamma)$, where

$$D(C_e) = \{\varphi \in \mathscr{C}^0(\Gamma); u_e(\varphi) \in \mathscr{C}^1_n(\mathbb{R}^n \backslash \Omega)\}\,. \tag{5.2}$$

We call this operator, *the exterior capacity operator of* $\mathscr{C}^0(\Gamma)$. By definition, *given* $\varphi, \psi \in \mathscr{C}^0(\Gamma)$, $\varphi \in D(C_e)$ *and* $\psi = C_e \varphi$ *iff the solution of the exterior Dirichlet problem* $P(\mathbb{R}^n \backslash \bar{\Omega}, \varphi)$ *satisfying the null condition at infinity is a solution of* $PN(\mathbb{R}^n \backslash \bar{\Omega}, \psi)$.
We have defined the interior and exterior capacities in $\mathscr{C}^0(\Gamma)$ for Γ the boundary of a regular bounded open set in $\mathbb{R}^n$; this choice of $\mathscr{C}^0(\Gamma)$ corresponds to the classical theory. We can equally well consider these operators in other contexts (see §6 and Chap. XI): the capacity operators appear in fact as operators on the boundary Γ. We note however that they are neither differential operators nor integral operators: they are in fact *pseudo-differential operators*. We shall not enlarge on this remark, contenting ourselves to explaining it by an example on the unit circle.

Example 1. *Capacity operators on the unit circle in* $\mathbb{R}^2$. We consider in $\mathbb{R}^2$, referred to polar coordinates, the unit disk Ω with boundary the unit circle. We can identify functions on Σ with periodic functions of θ of period 2π. Given $\varphi = \Sigma\, c_n e^{in\theta}$, the solution of $P(\Omega, \varphi)$ is $u(r, \theta) = \Sigma c_n r^{|n|} e^{in\theta}$ whose normal derivative is

$$\psi(\theta) = \frac{\partial u}{\partial r}(1, \theta) = \Sigma |n| c_n e^{in\theta} ;$$

we are assured that $\varphi \in D(C)$ if $\Sigma |nc_n| < \infty$, but more generally for all $\varphi \in D(C)$, $\psi = C\varphi$ is defined by

$$\psi(\theta) = \Sigma |n| c_n e^{in\theta} \quad \text{where} \quad c_n = \frac{1}{2\pi} \int_0^{2\pi} \varphi(\theta) e^{-in\theta} \, d\theta .$$

In other words, *C is the restriction to* $\mathscr{C}^0(\Sigma)$ *of the pseudo-differential operator* $(-d^2/d\theta^2)^{1/2}$.

Similarly for $\varphi(\theta) = \Sigma c_n e^{in\theta}$, the solution of $P(\mathbb{R}^n \backslash \bar{\Omega})$ satisfying the null condition at infinity is $u(r, \theta) = \Sigma c_n r^{-|n|} e^{-in\theta}$ of which the normal derivative (exterior to $\mathbb{R}^n \backslash \bar{\Omega}$) is

$$\psi(\theta) = -\frac{\partial u}{\partial r}(1, \theta) = \Sigma |n| c_n e^{in\theta} .$$

In other words, *in the case of the unit circle in* $\mathbb{R}^2$, $C = C_e$.

It should be emphasized that this equality $C = C_e$ is particular to the unit circle in $\mathbb{R}^2$ (see Propositions 2 and 3). □

As usual, we consider directly the case $n = 1$.

Example 2. *Case* $n = 1$. We take a regular bounded open set of $\mathbb{R}$, that is to say a finite union of bounded open intervals, two by two disjoint

$$\Omega =]a_1, a_2[\cup \ldots \cup]a_{N-1}, a_N[\quad \text{with} \quad a_1 < a_2 < a_3 < \ldots < a_N (N \text{ even}) ;$$

we have $\Gamma = \{a_1, a_2, \ldots, a_N\}$ and $\mathscr{C}^0(\Gamma)$ is identified with $\mathbb{R}^N$. Given $\varphi \in \mathbb{R}^N$ the derivative of the solution u of $P(\Omega, \varphi)$ is

$$u'(x) = \frac{\varphi_{k+1} - \varphi_k}{a_{k+1} - a_k} \quad \text{on} \quad]a_k, a_{k+1}[$$

since u is affine on $]a_k, a_{k+1}[$ (k odd). Hence

$$(5.3) \qquad (C\varphi)_k = \begin{cases} -\dfrac{\varphi_{k+1} - \varphi_k}{a_{k+1} - a_k} & \text{if } k \text{ is odd} \\[2ex] \dfrac{\varphi_k - \varphi_{k-1}}{a_k - a_{k-1}} & \text{if } k \text{ is even} \end{cases}$$

Similarly, the derivative of the solution u_e of $P(\mathbb{R} \backslash \bar{\Omega}, \varphi)$ is

$$\begin{cases} u'(x) = 0 & \text{on }]-\infty, a_1[\cup]a_N, \infty[\\[1ex] u'(x) = \dfrac{\varphi_{k+1} - \varphi_k}{a_{k+1} - a_k} & \text{on }]a_k, a_{k+1}[\ (k \text{ even}) \end{cases}$$

Hence

$$(5.4) \qquad (C_e\varphi)_k = \begin{cases} 0 & \text{if } k = 1 \text{ and } k = N \\ -\dfrac{\varphi_{k+1} - \varphi_k}{a_{k+1} - a_k} & \text{if } k \text{ is even}, k < N \\ \dfrac{\varphi_k - \varphi_{k-1}}{a_k - a_{k-1}} & \text{if } k \text{ is odd}, k > 1 . \end{cases} \qquad \square$$

We now state the

Proposition 1. *Let Ω be a regular bounded open set in $\mathbb{R}^n$ of boundary Γ.*
(1) For all $\varphi \in D(C)$, the solution $u(\varphi)$ of $P(\Omega, \varphi)$ satisfies $|\operatorname{grad} u(\varphi)|^2 \in L^1(\Gamma)$ and

$$(5.5) \qquad \int_\Omega |\operatorname{grad} u(\varphi)|^2 \, dx = \int_\Gamma \varphi C\varphi \, d\gamma .$$

More generally for $\varphi_1, \varphi_2 \in D(C)$

$$(5.6) \qquad \int_\Gamma \varphi_1 C\varphi_2 \, d\gamma = \int_\Omega \operatorname{grad} u(\varphi_1) . \operatorname{grad} u(\varphi_2) \, dx ,$$

and in particular C is positive and symmetric.
(2) For all $\varphi \in D(C_e)$, the solution $u_e(\varphi)$ of the exterior problem $P(\mathbb{R}^n \backslash \bar{\Omega}, \varphi)$ satisfying the null condition at infinity, satisfies $|\operatorname{grad} u_e(\varphi)|^2 \in L^1(\mathbb{R}^n \backslash \bar{\Omega})$ and

$$(5.7) \qquad \int_{\mathbb{R}^n \backslash \Omega} |\operatorname{grad} u_e(\varphi)|^2 \, dx = \int_\Gamma \varphi C_e \varphi \, d\gamma .$$

More generally for $\varphi_1, \varphi_2 \in D(C_e)$

$$(5.8) \qquad \int_\Gamma \varphi_1 C_e \varphi_2 \, d\gamma = \int_\Omega \operatorname{grad} u_e(\varphi_1) . \operatorname{grad} u_e(\varphi_2) \, dx$$

and in particular C_e is symmetric and positive.

Proof. The point (1) is an immediate corollary of Propositions 4 and 5 of §1 and of the definition of the interior capacity operator. For the point (2), from the same propositions, for R sufficiently large such that $\bar{\Omega} \subset B(0, R)$, we have

$$|\operatorname{grad} u_e(\varphi)|^2 \in L^1(B(0, R) \backslash \bar{\Omega})$$

and

$$\int_{B(0,R) \backslash \bar{\Omega}} |\operatorname{grad} u_e(\varphi)|^2 \, dx = \int_\Gamma \varphi C_e \varphi \, d\gamma + \int_{\partial B(0,R)} \frac{\partial u_e(\varphi)}{\partial n} u_e(\varphi) \, d\gamma .$$

In the case $n \geqslant 3$, we have, (see the proof of Proposition 17 of §4)

$$u_e(\varphi, x) = O\left(\frac{1}{|x|^{n-2}}\right), \quad |\operatorname{grad} u_e(\varphi, x)| = O\left(\frac{1}{|x|^{n-1}}\right)$$

when $|x| \to \infty$ with the result that

$$\int_{\partial B(0,R)} \frac{\partial u_e(\varphi)}{\partial n} u_e(\varphi)\,\mathrm{d}\gamma = O\left(R^{n-1} \times \frac{1}{R^{n-1}} \times \frac{1}{R^{n-2}}\right) = O\left(\frac{1}{R^{n-2}}\right)$$

when $R \to \infty$ and

$$\lim_{R \to \infty} \int_{B(0,R)\backslash\bar{\Omega}} |\operatorname{grad} u_e(\varphi)|^2\,\mathrm{d}x = \int_\Gamma \varphi C_e \varphi\,\mathrm{d}\gamma\,.$$

which proves (5.7). We prove (5.8) in the same way.

In the case $n = 1$, we can verify it directly by the formulae (5.4) of Example 2; we have moreover in this case $\frac{\mathrm{d}}{\mathrm{d}x} u_e(\varphi, x) = 0$ for $|x|$ large.

In the case $n = 2$, we know (see also Proposition 17 of §4) that

$$c = \lim u_e(\varphi, x) \text{ exists and } |\operatorname{grad} u_e(\varphi, x)| = O(1/|x|)$$

with the result that

$$\lim_{R \to \infty} \left[\int_{\partial B(0,R)} \frac{\partial u_e(\varphi)}{\partial n} u_e(\varphi)\,\mathrm{d}\gamma - c\int_{\partial B(0,R)} \frac{\partial u_e}{\partial n}(\varphi)\,\mathrm{d}\gamma\right] = 0\,.$$

Now, applying Gauss' theorem to $u_e(\varphi)$ on $B(0, R)\backslash\bar{\Omega}$

$$\int_{\partial B(0,R)} \frac{\partial u_e}{\partial n}(\varphi)\,\mathrm{d}\gamma = -\int_\Gamma C_e\varphi\,\mathrm{d}\gamma\,.$$

The point (2) is then deduced from the following lemma. □

Lemma 1. *Let Ω be a regular bounded open set in $\mathbb{R}^2$. For all $\varphi \in D(C_e)$*

$$\int_\Gamma C_e\varphi\,\mathrm{d}\gamma = 0\,.$$

Proof. We have (see Proposition 17 of §4)

$$\lim_{|x| \to \infty} \left(u_e(\varphi, x) + E_2(x)\int_\Gamma (C_e\varphi)\,\mathrm{d}\gamma\right) \text{ exists}$$

Since by hypothesis $\lim\limits_{|x| \to \infty} u_e(\varphi, x)$ exists then necessarily $\int_\Gamma C_e\varphi\,\mathrm{d}\gamma = 0.$ □

From Example 2, the result of Lemma 1 is true for every open set in $\mathbb{R}$; on the contrary when $n \geqslant 3$, the result of Lemma 1 is false for every open set in $\mathbb{R}^n$; this appears clearly in the following Proposition. We recall the notation of §4.5 (see (4.104) of §4):

$$(5.9)\quad \begin{cases} \Omega_1, \dots, \Omega_N \text{ denote the connected components of } \Omega\,, \\ \Omega'_1, \dots, \Omega'_{N'} \text{ the bounded connected components of } \mathbb{R}^n\backslash\bar{\Omega}\,, \\ \Omega'_0 \text{ the unbounded connected component of } \mathbb{R}^n\backslash\bar{\Omega}\,(n \geqslant 2)\,, \\ \Gamma_1, \dots, \Gamma_N, \Gamma'_0, \Gamma'_1, \dots, \Gamma'_{N'} \text{ the boundaries of } \Omega_1, \dots, \Omega_N, \Omega'_0, \dots, \Omega'_{N'}\,. \end{cases}$$

Proposition 2. *Let Ω be a regular bounded open set of $\mathbb{R}^n$. Then*

(1) $$\ker C = \chi_{\Gamma_1}\mathbb{R} \oplus \ldots \oplus \chi_{\Gamma_N}\mathbb{R}\,;$$

if $n \geqslant 3$, $$\ker C_e = \chi_{\Gamma'_1}\mathbb{R} \oplus \ldots \oplus \chi_{\Gamma'_{N'}}\mathbb{R}\,;$$

if $n = 2$, $$\ker C_e = \chi_{\Gamma'_0}\mathbb{R} \oplus \ldots \oplus \chi_{\Gamma'_{N'}}\mathbb{R}\,.$$

(2) *Let us suppose Γ is of class $\mathscr{C}^{1+\varepsilon}$*[94]

$$\operatorname{Im} C = \left\{\psi \in \mathscr{C}^0(\Gamma);\ \int_{\Gamma_i} \psi \, \mathrm{d}\gamma = 0 \quad \textit{for} \quad i = 1, \ldots, N\right\};$$

if $n \geqslant 3$, $$\operatorname{Im} C_e = \left\{\psi \in \mathscr{C}^0(\Gamma);\ \int_{\Gamma'_j} \psi \, \mathrm{d}\gamma = 0 \quad \textit{for} \quad j = 1, \ldots, N'\right\},$$

if $n = 2$, $$\operatorname{Im} C_e = \left\{\psi \in \mathscr{C}^0(\Gamma);\ \int_{\Gamma'_j} \psi \, \mathrm{d}\gamma = 0 \quad \textit{for} \quad j = 0, \ldots, N'\right\}.$$

Proof. The solution of $P(\Omega, \chi_{\Gamma_i})$ is χ_{Ω_i} which belongs to $\mathscr{C}^1_n(\bar{\Omega})$ with normal derivative zero on Γ. Hence $\chi_{\Gamma_i} \in D(C)$ and $C\chi_{\Gamma_i} \equiv 0$. Conversely if $\varphi \in \ker C$, from Proposition 1, the solution $u(\varphi)$ of $P(\Omega, \varphi)$ satisfies

$$|\operatorname{grad} u(\varphi)|^2 \in L^1(\Omega)$$

and

$$\int_\Omega |\operatorname{grad} u(\varphi)|^2 \, \mathrm{d}x = \int_\Gamma \varphi C \varphi \, \mathrm{d}\gamma = 0\,.$$

Hence $\operatorname{grad} u(\varphi) \equiv 0$ on Ω and $u(\varphi)$ is constant on each connected component of Ω; therefore φ is constant on each Γ_i. The kernel of the exterior capacity operator is determined in the same manner by noting:
(a) if $n \geqslant 3$, the solution $u_e(\varphi)$ of $P(\mathbb{R}^n \backslash \bar{\Omega}, \varphi)$ which satisfies the null condition at infinity tends to zero at infinity; if therefore $u_e(\varphi)$ is constant on each connected component of $\mathbb{R}^n \backslash \bar{\Omega}$, it is null on Ω'_0 and hence $\varphi \equiv 0$ on Γ'_0;
(b) if $n = 2$, the solution of $P(\mathbb{R}^2 \backslash \bar{\Omega}, \chi_{\Gamma'_0})$ satisfying the null condition at infinity is $\chi_{\Omega'_0}$; hence $\chi_{\Gamma'_0} \in \ker C_e$.
It is clear that $\operatorname{Im} C$ is the set of the $\psi \in \mathscr{C}^0(\Gamma)$ such that Neumann's problem $PN(\Omega, \psi)$ admits a classical solution, Similarly, $\operatorname{Im} C_e$ is the set of the $\psi \in \mathscr{C}^0(\Gamma)$ such that $PN(\mathbb{R}^n \backslash \bar{\Omega}, \psi)$ admits a classical solution satisfying the null condition at infinity. Hence, the point (2) follows from Theorem 3 of §4. □

Theorem 3 of §4 has been proved by the integral method with the help of the *integral operators L, J, K* (see §4.5). We recall (see Proposition 15 of §4) that L is a bijection of $\mathscr{C}^0(\Gamma)$ onto $\operatorname{Im} L$[95]; *we shall denote by L^{-1}, the inverse bijection from* $D(L^{-1}) = \operatorname{Im} L$ *onto* $\mathscr{C}^0(\Gamma)$
We then state:

[94] See Definition 3 of §3.
[95] With the reservation that in the case $n = 2$ we also have the hypothesis (4.107) of §4.

Proposition 3. *Let Ω be a regular bounded open set of $\mathbb{R}^n$ with boundary $\partial\Omega$ of class $\mathscr{C}^{1+e}$ satisfying in addition in the case $n = 2$ the hypothesis* (4.107) *of* §4. *Then* (1) $D(C) = D(C_e) = \operatorname{Im} L$ *and, for all* $\varphi \in \operatorname{Im} L$

$$C\varphi = \frac{1}{2}(J - I)L^{-1}\varphi\,; \tag{5.10}$$

$$\text{if } n \geqslant 3, \quad C_e\varphi = -\frac{1}{2}(J + I)L^{-1}\varphi\,, \tag{5.11}$$

$$\text{if } n = 2, \quad C_e\varphi = -\frac{1}{2}(J + I)L^{-1}\varphi + \left(\int_\Gamma (L^{-1}\varphi)\,\mathrm{d}\gamma\right)\frac{\alpha_0}{\displaystyle\int_\Gamma \alpha_0\,\mathrm{d}\gamma}\,, \tag{5.12}$$

$$\text{with} \quad \alpha_0 = L^{-1}1\,.$$

(2) *For all* $\varphi \in \operatorname{Im} L$, $K\varphi \in \operatorname{Im} L$ *and*

$$C\varphi = \frac{1}{2}L^{-1}(K - I)\varphi\,; \tag{5.13}$$

$$\text{if } n \geqslant 3, \quad C_e\varphi = -\frac{1}{2}L^{-1}(K + I)\varphi \tag{5.14}$$

$$\text{if } n = 2, \quad C_e\varphi = -\frac{1}{2}L^{-1}(K + I)\varphi + \left(\int_\Gamma (L^{-1}\varphi)\,\mathrm{d}\gamma\right)\frac{\alpha_0}{\displaystyle\int_\Gamma \alpha_0\,\mathrm{d}\gamma}\,. \tag{5.15}$$

Proof. The first part of the point (1) follows from Proposition 18 of §4, account being taken of the definitions of $D(C)$ and $D(C_e)$. Now, for $\varphi \in \operatorname{Im} L$, the interior simple layer potential of $\alpha = L^{-1}\varphi$ is a solution of $P(\Omega, \varphi)$ (see (4.92) of §4) and of $PN(\Omega, \frac{1}{2}(J\alpha - \alpha))$ (see (4.93) of §4); hence, by the definition of the interior capacity operator

$$C\varphi = \frac{1}{2}(J - I)\alpha = \frac{1}{2}(J - I)L^{-1}\varphi\,.$$

In the case $n \geqslant 3$, the exterior simple layer potential of $\alpha = L^{-1}\varphi$ is a solution of $P(\mathbb{R}^n\backslash\bar{\Omega}, \varphi)$ and $PN(\mathbb{R}^n\backslash\bar{\Omega}, -\frac{1}{2}(\alpha + J\alpha))$ and tends to zero at infinity (see (4.96), (4.97) of §4); hence

$$C_e\varphi = -\frac{1}{2}(J + I)\alpha = -\frac{1}{2}(J + I)L^{-1}\varphi\,.$$

In the case $n = 2$, the exterior simple layer potential u_e of $\alpha = L^{-1}\varphi$ is always a solution of $P(\mathbb{R}^2\backslash\bar{\Omega}, \varphi)$ and $PN(\mathbb{R}^2\backslash\bar{\Omega}, -\frac{1}{2}(J\alpha + \alpha))$; but then

$$u_e(x) = \left(\int_\Gamma \alpha\,\mathrm{d}\gamma\right)E_2(x) + O\left(\frac{1}{|x|}\right) \quad \text{when} \quad |x| \to \infty\,.$$

Let us consider $x_0 \in \Omega$ and $\varphi_0(z) = E_2(z - x_0)$. The function:

$$u_e(x) - \left(\int_\Gamma \alpha\,\mathrm{d}\gamma\right)E_2(x - x_0) \quad \text{is a solution of} \quad P\left(\mathbb{R}^2\backslash\bar{\Omega}, \varphi - \left(\int_\Gamma \alpha\,\mathrm{d}\gamma\right)\varphi_0\right)$$

and tends to zero at infinity; hence

$$C_e\left(\varphi - \left(\int_\Gamma \alpha \, d\gamma\right)\varphi_0\right) = \frac{\partial u_e}{\partial n} - \left(\int_\Gamma \alpha \, d\gamma\right)\frac{\partial E_2}{\partial n_e}(z - x_0)\,;$$

is, by the linearity of C_e,

$$C_e\varphi = -\frac{J\alpha + \alpha}{2} + \left(\int_\Gamma \alpha \, d\gamma\right)\psi_0 \tag{5.16}$$

with $$\psi_0 = C_e\varphi_0 - \frac{\partial}{\partial n_e} E_2(z - x_0)\,.$$

But the formula (5.16), where $\alpha = L^{-1}\varphi$ is true for all $\varphi \in \operatorname{Im} L = D(C_e)$; taking $\varphi \equiv 1$, we obtain

$$C_e 1 = 0 = -\frac{1}{2}(J + I)\alpha_0 + \left(\int_\Gamma \alpha_0 \, d\gamma\right)\psi_0\,.$$

This gives (5.12), since by (5.10), $J\alpha_0 = \alpha_0$.
For the point (2), we prove first that for $\varphi \in D(C)$ we have

$$LC\varphi = \frac{1}{2}(K\varphi - \varphi)\,. \tag{5.17}$$

We now consider the solution $u(\varphi)$ of $P(\Omega, \varphi)$. Since $u(\varphi) \in \mathscr{C}^1_n(\bar{\Omega})$, from Proposition 5 of §3,

$$u(\varphi) = u_1 + u_2 \quad \text{on} \quad \Omega\,, \tag{5.18}$$

where u_1 is the interior simple layer potential of $-\dfrac{\partial u(\varphi)}{\partial n} = -C\varphi$ and u_2 the interior double layer potential of φ.
Now u_1 is a solution of $P(\Omega, L(-C\varphi))$ and u_2 is a solution of $P(\Omega, \frac{1}{2}(K\varphi + \varphi))$ (see (4.100) of §4), so that passing to the limit in (5.18), we have

$$\varphi = -LC\varphi + \frac{K\varphi + \varphi}{2} \quad \text{on} \quad \Gamma\,,$$

from which we have (5.17).
We deduce that for $\varphi \in \operatorname{Im} L = D(C)$, we have

$$K\varphi = \varphi + 2LC\varphi \in \operatorname{Im} L \quad \text{and} \quad (5.13)\,.$$

Identifying (5.10) and (5.13), we deduce that

$$JL^{-1}\varphi = L^{-1}K\varphi \quad \text{for all} \quad \varphi \in \operatorname{Im} L\,. \tag{5.19}$$

Taking this value of $JL^{-1}\varphi$ into (5.11) and (5.12), we obtain (5.14) and (5.15).
From the equality (5.19) we derive immediately

$$LJ\alpha = KL\alpha \quad \text{for all} \quad \alpha \in \mathscr{C}^0(\Gamma)\,. \tag{5.20}$$

This shows in particular that *the operator L exchanges the kernels and the images of*

$I \pm J$ and $I \pm K$; more precisely, we have

(5.21)

$$\begin{cases} L(\ker(I - J)) & = \ker(I - K) = \chi_{\Gamma_1}\mathbb{R} \oplus \ldots \oplus \chi_{\Gamma_N}\mathbb{R} \\ L(\ker(I + J)) & = \ker(I + K) = \chi_{\Gamma'_1}\mathbb{R} \oplus \ldots \oplus \chi_{\Gamma'_{N'}}\mathbb{R} \\ L^{-1}(\operatorname{Im}(I - K)) & = \operatorname{Im}(I - J) = \ker(I - K)^{\perp} \\ & = \left\{\psi \in \mathscr{C}^0(\Gamma)\,;\, \int_{\Gamma_i} \psi \, \mathrm{d}\gamma = 0 \quad \text{for} \quad i = 1, \ldots, N\right\} \\ L^{-1}(\operatorname{Im}(I + K)) & = \operatorname{Im}(I + J)) = \ker(I + K)^{\perp} \\ & = \left\{\psi \in \mathscr{C}^0(\Gamma);\, \int_{\Gamma'_j} \psi \, \mathrm{d}\gamma = 0 \quad \text{for} \quad j = 1, \ldots, N'\right\}. \end{cases}$$

These equalities result from §4.5 (see the remarks preceding Proposition 19). Taking account of Proposition 2, we have also

$$(5.22) \qquad \begin{cases} \ker(I - K) = \ker C, \quad \operatorname{Im}(I - K) = \operatorname{Im} C\,; \\ \text{if } n \geqslant 3, \quad \ker(I + K) = \ker C_e, \quad \operatorname{Im}(I + K) = \operatorname{Im} C \\ \text{if } n = 2, \quad \ker C_e = \ker(I + K) \oplus \chi_{\Gamma'_0}\mathbb{R} \\ \text{and } \operatorname{Im} C_e = \operatorname{Im}(I + K) \cap \left\{\psi \in \mathscr{C}^0(\Gamma);\, \int_\Gamma \psi \, \mathrm{d}\gamma = 0\right\}. \end{cases}$$

We note also the immediate consequences of (5.10) and (5.15):

$$(5.23) \qquad \begin{cases} \text{for all } \varphi \in D(C) = D(C_e) = D(L^{-1})\,; \\ \text{if } n \geqslant 3, C\varphi + C_e\varphi = -L^{-1}\varphi \\ \text{if } n = 2, C\varphi + C_e\varphi = -L^{-1}\varphi + \left(\int_\Gamma (L^{-1}\varphi)\,\mathrm{d}\gamma\right) \dfrac{\alpha_0}{\int_\Gamma \alpha_0 \, \mathrm{d}\gamma}, \end{cases}$$

also, because of (5.19)

$$(5.24) \qquad \begin{cases} \text{for all } \varphi \in D(C) = D(C_e), K\varphi \in D(C) = D(C_e) \\ CK\varphi = JC\varphi, C_eK\varphi = JC_e\varphi\,. \end{cases}$$

The equality for C_e in (5.24) is immediate from (5.14) in the case $n \geqslant 3$; in the case $n = 2$, it must be noted that in addition $J\alpha_0 = \alpha_0$ (which follows from (5.10)) and

$$\int_\Gamma (L^{-1}\varphi)\,\mathrm{d}\gamma = \int_\Gamma (L^{-1}K\varphi)\,\mathrm{d}\gamma\,,$$

which follows from (5.19) and

$$\int_\Gamma J\alpha \, \mathrm{d}\gamma = \int_\Gamma \alpha \, \mathrm{d}\gamma$$

(see Proposition 15(3) of §4).

In §4.5 we have considered the problems:

$$(5.25)\qquad \begin{cases} \Delta u = 0 \quad \text{on} \quad \Omega \\ \dfrac{\partial u}{\partial n} + \lambda u = \psi \quad \text{on} \quad \Gamma\,; \end{cases}$$

$$(5.26)\qquad \begin{cases} \Delta u_e = 0 \quad \text{on} \quad \mathbb{R}^n\backslash\bar{\Omega} \\ \dfrac{\partial u_e}{\partial n_e} + \lambda u_e = \psi \quad \text{on} \quad \Gamma \\ u_e \text{ satisfies the null condition at infinity} \end{cases}$$

where $\lambda \in \mathscr{C}^0(\Gamma)$ was supposed positive or zero. We have solved them (see Theorems 4 and 5) using the capacity operators J and L. We note that the problems (5.25) (resp. (5.26)) can be written in terms of the capacity operators in the following manner: u (resp. u_e) is solution of (5.25) (resp. (5.26)) iff u (resp. u_e) is the solution of $P(\Omega, \varphi)$ (resp. $P(\mathbb{R}^n\backslash\bar{\Omega}, \varphi)$ satisfying the null condition at infinity) where $\varphi \in D(C)$ (resp $D(C_e)$) is a solution of

$$(5.27)\qquad C\varphi + \lambda\varphi = \psi \quad (\text{resp. } C_e\varphi + \lambda\varphi = \psi) \quad \text{on} \quad \Gamma\,.$$

We have thus reduced the boundary value differential problems on Ω to pseudo-differential problems on the boundary $\Gamma = \partial\Omega$.

The study of the equations (5.27) will be greatly facilitated by the numerous properties of the capacity operators: we have already seen (Proposition 1) that C and C_e are symmetric and positive, that is to say that for all $\varphi \in D(C)$

$$\int_\Gamma \varphi C\varphi \, d\gamma \geqslant 0\,, \quad \int_\Gamma \varphi C_e\varphi \, d\gamma \geqslant 0\,.$$

This property will be exploited in all its force in the use of variational methods. We note here two other properties directly applicable in the classical theory: *the compactness and positivity of the resolvants $(\lambda I + C)^{-1}$ with $\lambda > 0$.* □

First of all we give the following version of Hopf's maximum principle (see Proposition 14 of §4):

Proposition 4. *Suppose that Ω is a regular bounded open set in $\mathbb{R}^n$ with boundary Γ satisfying the condition of the interior (resp. exterior) ball*[96]*, $\varphi \in D(C)$ (resp. $D(C_e)$) and $z \in \Gamma_i$ (resp. Γ'_i) such that*

$$\varphi(z) = \min_{\Gamma_i} \varphi \left(\textit{resp. } \varphi(z) = \min_{\Gamma'_j} \varphi \qquad \textit{with} \quad \varphi(z) \leqslant 0 \quad \textit{when} \quad j = 0 \right).$$

Then

either $\quad \varphi \equiv \varphi(z)$ *and* $C\varphi \equiv 0$ *(resp.* $C_e\varphi \equiv 0$*) on* Γ_i *(resp.* Γ'_j*)*

---

[96] For all $z \in \Gamma$, there exists $x_0 \in \Omega$ (resp. $\mathbb{R}^n \setminus \bar{\Omega}$) such that $B(x_0, |x_0 - z|) \subset \Omega$, (resp. $\mathbb{R}^n \setminus \bar{\Omega}$) (see §4, Example 7 and commentary on Lemma 5).

or $$C\varphi(z) < 0 \quad (resp.\ C_e\varphi(z) < 0)$$ [97]

Proof. It is sufficient to apply Lemma 2 of §4 with the connected open set Ω_i (resp. Ω'_j) and u the restriction to Ω_i (resp. Ω'_j) of the solution of $P(\Omega, \varphi)$ (resp. $P(\mathbb{R}^n \backslash \bar{\Omega}, \varphi)$ satisfying the null condition at infinity). From the classical maximum principle (see Corollary 3 of §2 and §4.3), $\varphi(z) = \min_{\Omega_i} u$ (resp. $\varphi(z) = \min_{\Omega'_j} u$). From the condition of the interior (exterior) ball, the hypothesis (4.74) of §4 is satisfied, so the conclusion of Lemma 2 of §4 indicates to us that

either $$u \equiv \varphi(z) \quad \text{on} \quad \Omega_i \quad (\text{resp. } \Omega'_j),$$

or $$\frac{\partial u}{\partial n}(z) < 0,$$

from which follows the conclusion of the Proposition. □

We obtain then the

Proposition 5. *Let Ω be a regular bounded open set in $\mathbb{R}^n$ with boundary Γ satisfying the hypotheses of Propositions 3 and 4 and $\lambda \in \mathscr{C}^0(\Gamma)$ satisfying*

$$\begin{cases} \lambda \geqslant 0 \text{ on } \Gamma \text{ and } \lambda \text{ not identically zero on } \Gamma_i \text{ (resp. } \Gamma'_j) \\ \text{for } i = 1, \ldots, N \text{ (resp. } j = 0, 1, \ldots, N'). \end{cases}$$

(1) *The operator $\lambda I + C$ (resp. $\lambda I + C_e$) is a bijection of $D(C)$ (resp. $D(C_e)$) onto $\mathscr{C}^0(\Gamma)$.*

(2) *The inverse operator $(\lambda I + C)^{-1}$ (resp. $(\lambda I + C_e)^{-1}$) is a compact operator on $\mathscr{C}^0(\Gamma)$.*

(3) *For all $\psi \in \mathscr{C}^0(\Gamma)$ with $\psi \geqslant 0$ on Γ,*

$$(\lambda I + C)^{-1}\psi \geqslant 0 \quad (resp.\ (\lambda I + C_e)^{-1}\psi \geqslant 0) \quad on \quad \Gamma.$$

If, in addition, ψ is not identically zero on Γ_i (resp. Γ'_j), then

$$\min_{\Gamma_i} (\lambda I + C)^{-1}\psi > 0 \quad \left(\text{resp. } \min_{\Gamma'_j} (\lambda I + C_e)^{-1}\psi > 0\right).$$

Proof. From Theorem 4 (resp. 5) of §4, we know that for all $\psi \in \mathscr{C}^0(\Gamma)$ there exists a unique solution u (resp. u_e) of the problem (5.25) (resp. (5.26)). From the equivalence of these problems with the equation (5.27), the point (1) of the Proposition appears as a corollary of this theorem.

Let us prove the point (3). Given $\psi \in \mathscr{C}^0(\Gamma)$ with $\psi \geqslant 0$ on Γ, we put $\varphi = (\lambda I + C)^{-1}\psi$ (resp. $(\lambda I + C_e)^{-1}\psi$) and consider for $i \in \{1, \ldots, N\}$ (resp. $j \in \{0, \ldots, N'\}$) fixed, $z \in \Gamma_i$ (resp. Γ'_j) such that $\varphi(z) = \min_{\Gamma_i} \varphi$ (resp. $\min_{\Gamma'_j} \varphi$).

From Proposition 4

either $$\varphi \equiv \varphi(z) \quad \text{and} \quad C\varphi \equiv 0 \quad (\text{resp. } C_e\varphi \equiv 0) \quad \text{on} \quad \Gamma_i \text{ (resp. } \Gamma'_j)$$

or $$C\varphi(z) < 0 \quad (\text{resp. } C_e\varphi(z) < 0).$$

---

[97] Hence (e.g. in the interior case) if $z \in \Gamma_i$ is such that $\varphi(z) = \max_{\Gamma_i} \varphi$, then either $\varphi \equiv \varphi(z)$ and $C\varphi = 0$, or $C\varphi(z) > 0$.

Hence, since $\lambda\varphi + C\varphi = \psi$ (resp. $\lambda\varphi + C_e\varphi = \psi$) on Γ,

either $\varphi \equiv \varphi(z)$ and $\lambda\varphi \equiv \psi$ on Γ_i (resp. Γ'_j),

or
$$\lambda(z)\varphi(z) > \psi(z) .$$

From the hypotheses on λ, since $\psi \geqslant 0$ on Γ_i (resp. Γ'_j), necessarily in one or the other case $\varphi(z) \geqslant 0$: more precisely:
if $\lambda(z)\varphi(z) > \psi(z)$, then $\varphi(z) > 0$;
if $\varphi \equiv \varphi(z)$ and $\lambda\varphi \equiv \psi$ on Γ_i (resp. Γ'_j) then $\varphi(z) > 0$ or $\psi \equiv 0$ on Γ_i (resp. Γ'_j).
This proves the point (3).
From the point (3) we deduce the continuity of the operator

$$(\lambda I + C)^{-1} \quad (\text{resp.}\ (\lambda I + C_e)^{-1}) .$$

In effect, putting

$\varphi_0 = (\lambda I + C)^{-1}1$ (resp. $(\lambda I + C_e)^{-1}1$), for all $\psi \in \mathscr{C}^0(\Gamma)$ the function φ defined by $\varphi = (\lambda I + C)^{-1}\psi$ (resp. $(\lambda I + C_e)^{-1}\psi$) satisfies

$$\min\left(0, \min_{\Gamma} \psi\right)\varphi_0 \leqslant \varphi \leqslant \max\left(0, \max_{\Gamma} \psi\right)\varphi_0 \quad \text{on} \quad \Gamma \tag{5.28}$$

and hence, in particular

$$\|\varphi\| = \max_{\Gamma} |\varphi| \leqslant \left(\max_{\Gamma} \varphi_0\right)\left(\max_{\Gamma} |\psi|\right) = \left(\max_{\Gamma} \varphi_0\right) \|\psi\| .$$

Finally, we prove the compactness of the operator $(\lambda I + C)^{-1}$ (resp. $(\lambda I + C_e)^{-1}$). From the continuity proved above that reduces to proving:

$$\left\{\begin{array}{l}\text{for every sequence } (\varphi_n) \text{ of elements of } D(C) \text{ (resp. } D(C_e)) \text{ bounded in } \mathscr{C}^0_n(\Gamma),\\ \text{with } (C\varphi_n) \text{ (resp. } C_e\varphi_n), \text{ bounded in } \mathscr{C}^0(\Gamma), \text{ we can extract a sub-}\\ \text{sequence } (n_k) \text{ such that } (\varphi_{n_k}) \text{ converges in } \mathscr{C}^0(\Gamma).\end{array}\right.$$

Given such a sequence (φ_n), using the compactness of the operators K and L (see Proposition 16 of §4) we can extract a subsequence (φ_{n_k}) such that

$$(K\varphi_{n_k}) \quad \text{and} \quad (LC\varphi_{n_k})\,(\text{resp.}\ (LC_e\varphi_{n_k})) \quad \text{converge in} \quad (\mathscr{C}^0(\Gamma)) .$$

From Proposition 3 (2), we deduce that the sequence

$$\varphi_{n_k} = K\varphi_{n_k} - 2LC\varphi_{n_k} \quad \text{converges in} \quad \mathscr{C}^0(\Gamma)$$

(resp. if $n \geqslant 3$, $\varphi_{n_k} = -(K\varphi_{n_k} + 2LC_e\varphi_{n_k})$ converges in $\mathscr{C}^0(\Gamma)$; if $n = 2$,

$$\varphi_{n_k} - \left(\int_{\Gamma} \alpha_0\, d\gamma\right)^{-1}\left(\int_{\Gamma} (L^{-1}\varphi_{n_k})\, d\gamma\right)$$

converges in $\mathscr{C}^0(\Gamma)$ and hence again after possibly extracting a subsequence φ_{n_k} converges in $\mathscr{C}^0(\Gamma)$.

□

2. Electrical Equilibrium. Coefficients of Capacitance

2a. Electrical Equilibrium: Electrostatic Problem: Capacitors

We say that a *homogeneous ohmic conductor*[98] K (see Chap. I, §4.2.6) is in *electric equilibrium* if the electric field $\mathbf{E}$ in the interior of K[99] is null. Hence, *if K is in electric equilibrium, it can have non-zero charges only on the boundary of K and (supposing K to be connected) the electrostatic potential is constant on K.*

Let us consider now a system S of homogeneous ohmic conductors placed in a domain D of the space $\mathbb{R}^3$. We say that the system S is in electrical equilibrium if each of the components of S is in electrical equilibrium. We suppose that the domain $D \setminus S$ is occupied by a perfect dielectric; therefore if the system S is in electrical equilibrium

(i) the electrostatic potential v_c in D is constant on each homogeneous component (assumed to be connected) of S. Denoting the components of S by $K_1, \ldots, K_m$ we have

$$v_c = v_i \quad \text{on} \quad K_i \quad \text{for} \quad i = 1, \ldots, m \tag{5.29}$$

where $v_1, \ldots, v_m$ are constants.

(ii) the electric charges in D are distributed on the boundaries of the conductors K_i. We denote by ρ_i *the (surface) charge density on the boundary ∂K_i of K_i*. The total electric charge of the conductor K_i is

$$q_i = \int_{\partial K_i} \rho_i(z)\, \mathrm{d}\gamma_i(z)$$

where $\mathrm{d}\gamma_i$ is the surface measure of the boundary ∂K_i, assumed to be regular.

Let us consider a problem where *a priori the densities ρ_i and the potentials v_i are again unknown.* They are related through the equation (see §1.2)

$$\Delta v_c = -\frac{1}{\varepsilon} \sum_i \rho_i \, \mathrm{d}\gamma_i \quad \text{in} \quad D \tag{5.30}$$

where ε is the permittivity of the medium occupying $D \setminus S$. We can rewrite (5.29) and (5.30) in the form

$$\begin{cases} \Delta v_c = 0 \quad \text{in the open set } D \setminus (K_1 \cup \ldots \cup K_n) \\ \left. \begin{aligned} & v_c = v_i \\ & \frac{\partial v_c}{\partial n_i} = -\frac{1}{\varepsilon} \rho_i \end{aligned} \right\} \quad \text{on} \quad \partial K_i \quad \text{for} \quad i = 1, \ldots, m \end{cases} \tag{5.31}$$

where $\partial/\partial n_i$ is the normal derivative exterior to K_i.

---

[98] The word homogeneous employed here indicates that the material used in every element K_i is the same at every point of K_i, and therefore that the electrical properties are the same at every point of K_i. This qualification will be implied in the following pages.

[99] This holds at every instant t, being equivalent to the absence of current in K. The presence of a non-zero current in an ohmic conductor is linked, through Ohm's law, to a dissipation of energy, from which we deduce the impossibility of a stationary regime without an external source of energy (see Fournet [1]).

We see appearing in (5.31) the notion of a capacity operator introduced in Sect. 1. In order to make this more precise, we distinguish between two situations:

Case 1: *The exterior problem.* The domain D *is supposed to be the whole space* $\mathbb{R}^3$. *We then impose on the electrostatic potential* v_c *the condition that it is zero at infinity. We denote by* Ω_i *the* (connected) interior of the conductor K_i, $\Gamma_i = \partial\Omega_i = \partial K_i$, Ω the union of the Ω_i whose boundary Γ is the union of the Γ_i; $\Omega_1, \ldots, \Omega_m$ are the connected components of Ω.

From (5.31), v_c is the solution of the exterior problem $P(\mathbb{R}^3\backslash\bar{\Omega}, \Sigma v_i\chi_{\Gamma_i})$ tending to zero at infinity and $\partial v_c/\partial n_e = (1/\varepsilon)\rho_i$ on Γ_i, where $\partial/\partial n_e$ is the exterior normal derivative to $\mathbb{R}^n\backslash\bar{\Omega}$. In other words

$$\rho_i = \varepsilon C_e(\Sigma\, v_j\chi_{\Gamma_j}) \quad \text{on} \quad \Gamma_i\,,$$

where C_e is the exterior capacity operator of $\mathscr{C}^0(\Gamma)$. Using the linearity

$$\rho_i = \varepsilon \sum v_j C_e \chi_{\Gamma_j} \qquad \text{on} \quad \Gamma_i = \partial K_i \tag{5.32}$$

and the total electric charge q_i of the conductor K_i is

$$q_i = \int_{\Gamma_i} \rho_i(z)\,\mathrm{d}\gamma(z) = \varepsilon \sum_j v_j \int_{\Gamma_i} C_e\chi_{\Gamma_j}\,\mathrm{d}\gamma\,,$$

namely

$$q_i = \sum_j C^e_{i,j} v_j \tag{5.33}$$

where

$$C^e_{i,j} = \varepsilon \int_{\Gamma_i} C_e \chi_{\Gamma_j}\,\mathrm{d}\gamma\,. \tag{5.34}$$

We call $C^e_{i,j}$ the *exterior coefficient of capacitance* of the conductor K_i on the conductor K_j in the system S placed in the dielectric of permittivity ε (occupying all of the space $\mathbb{R}^3\backslash S$).

Case 2: *The interior problem.* The domain D *is supposed bounded containing all of* the cavity *of a uniform ohmic conductor* K_0 *in equilibrium* (see Fig. 7).

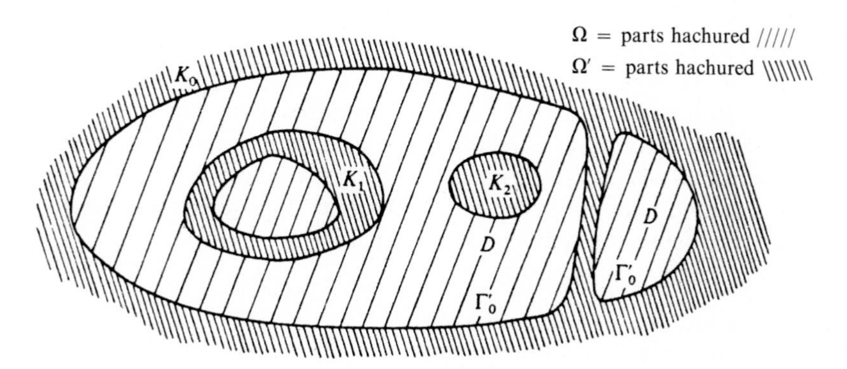

Fig. 7

We denote by Ω'_0 the exterior of D ($\Omega'_0 = \mathbb{R}^3\backslash\bar{D}$) whose boundary is $\Gamma'_0 = \partial K_0 \cap \bar{D}$; for $j = 1, \ldots, m$, Ω'_j the (connected) interior of the conductor K_j,

$$\Gamma'_j = \partial\Omega'_j = \partial K_j; \quad \Omega' = \Omega'_0 \cup \Omega'_1 \cup \ldots \cup \Omega'_m$$

and $\Omega = \mathbb{R}^3\backslash\bar{\Omega}'$ whose common boundary is $\Gamma = \Gamma'_0 \cup \Gamma'_1 \cup \ldots \cup \Gamma'_m$; the connected components of Ω' are denoted by $\Omega'_0, \Omega'_1, \ldots, \Omega'_m$.

The conductor K_0 being supposed in equilibrium, the electrostatic potential is constant in K_0 and hence

$$v_c = v_0 \quad \text{on} \quad \partial K_0 \,.$$

In other words, from (5.31), v_c is the solution of the interior problem $P\left(\Omega, \sum_{j=0}^{m} v_j \chi_{\Gamma'_j}\right)$. Hence

$$\rho_i = \varepsilon C\left(\sum_{j=0}^{m} v_j \chi_{\Gamma'_j}\right) \quad \text{on} \quad \Gamma'_i$$

where C is the interior capacity operator of $\mathscr{C}^0(\Gamma)$.

From the linearity of C

$$\rho_i = \varepsilon \sum_{j=0}^{m} v_j C \chi_{\Gamma'_j} \quad \text{on} \quad \Gamma'_i = \partial K_i \tag{5.35}$$

and the total charge of the conductor K_i is

$$q_i = \sum_{j=0}^{m} C_{i,j} v_j \tag{5.36}$$

where

$$C_{i,j} = \varepsilon \int_{\Gamma'_i} C \chi_{\Gamma'_j} \, d\gamma \,. \tag{5.37}$$

The equalities (5.35), (5.36) are also true for $i = 0$, ρ_0 denoting the density of the electric charges on Γ'_0 and q_0 the total electric charge on Γ'_0[100].

Let us consider the system S consisting of the set of m ohmic conductors, $K_1, \ldots, K_m$ placed in the dielectric D filling the cavity of an ohmic conductor K_0. We shall call $C_{i,j}$, defined by (5.37), the *interior coefficient of capacitance*[101] of the conductor K_i on the conductor K_j (for $i, j = 0, \ldots, m$).

We observe that in both the exterior case and the interior case, *the coefficient of capacitance of K_i on K_j is the total charge of the conductor K_i*[102], *when the conductor K_j is at unit potential and the other conductors are at zero potential.* The coefficients of influence therefore have dimension QV^{-1}, where Q is the dimension of electric charge and V the dimension of electrostatic potential.

---

[100] Which is **not** *a priori* the total electric charge of K_0.

[101] We also say, quite simply, "coefficient of capacitance".

[102] In the case of an interior problem for $i = 0$, this is the total charge of Γ_0, the interior boundary of K_0.

In the S.I. system of units (see §1.2), the unit of coefficients of capacitance is the *Farad* (F):

$$1F = 1\,\text{coulomb/volt} = 1\,\text{m}^{-2}\,\text{kg}^{-1}\,\text{s}^4\,\text{A}^2\,.$$

In *electricity*, we call a set of several conductors, which strongly influence one another, a *capacitor*.
We shall call a system of an arbitrary number of conductors a capacitor (in the wide sense).

2b. Properties of the Matrix of the Coefficients of Capacitances

We have defined the exterior (and interior) coefficients of capacitance by the formulae (5.34) and (5.37), where in the two cases Ω denotes a bounded open set and Γ_i, Γ'_j are defined as in §5.1 (see (5.9))[103]. To within the multiplicative constant ε, the matrix of the interior (resp. exterior) coefficients of capacitance is the matrix

$$\left(\int_{\Gamma'_i} C\chi_{\Gamma'_j}\,\mathrm{d}\gamma\right)\left(\text{resp.}\left(\int_{\Gamma_i} C_e\chi_{\Gamma_j}\,\mathrm{d}\gamma\right)\right)$$

that we can consider for a regular bounded open set Ω in $\mathbb{R}^n$ with $n \geqslant 2$. The case $n = 2$ is particularly interesting for applications in electrostatics since it corresponds to cylindrical capacitors (see §5.2).
In the general case, we state the

Proposition 6. *Let Ω be a regular bounded open set in $\mathbb{R}^n$. With the notation* (5.9) *we put*[104]

$$c_{i,j} = \int_{\Gamma'_i} C\chi_{\Gamma'_j}\,\mathrm{d}\gamma \quad \textit{for} \quad i,j = 0,\ldots,N'$$

$$c^e_{i,j} = \int_{\Gamma_i} C_e\chi_{\Gamma_j}\,\mathrm{d}\gamma \quad \textit{for} \quad i,j = 1,\ldots,N\,;$$

(1) *the matrices $(c_{i,j})_{i,j=0,\ldots,N'}$ and $(c^e_{i,j})_{i,j=1,\ldots,N}$ are symmetric and positive;*
(2) *given $i \neq j = 0,\ldots,N'$* (resp. $1,\ldots,N$), *if Γ'_i and Γ'_j (resp. Γ_i and Γ_j) are connected in Ω (resp. $\mathbb{R}^n\backslash\bar{\Omega}$) then $c_{i,j} < 0$ (resp. $c^e_{i,j} < 0$), otherwise $c_{ij} = 0$ (resp. $c^e_{ij} = 0$);*

(3) $c_{i,i} = -\sum\limits_{\substack{j=0,\ldots,N'\\ j\neq i}} c_{i,j}$ *for* $i = 0,\ldots,N'$;

$$\textit{if}\quad n \geqslant 3 \quad \textit{and} \quad \Gamma_i \cap \Gamma'_0 \neq \varnothing,\; c^e_{i,i} > -\sum_{\substack{j=1,\ldots,N\\ j\neq i}} c^e_{i,j}$$

$$\textit{if}\quad n = 2 \quad \textit{and} \quad \Gamma_i \cap \Gamma'_0 = \varnothing,\; c^e_{i,i} = -\sum_{\substack{j=1,\ldots,N\\ j\neq i}} c^e_{i,j}\,;$$

---

[103] We observe that the interpretation of Ω is different in the exterior case from that in the interior case.
[104] In the case $n = 3$, $c_{i,j}$ and $c^e_{i,j}$ are, to within the constant ε, the coefficients of influence.

(4) *the matrix* $(c_{i,j})_{i,j=0,\dots,N'}$ *is of rank* N';

if $n \geqslant 3$, *the matrix* $(c^e_{i,j})_{i,j=1,\dots,N}$ *is invertible* ,

if $n = 2$, *the matrix* $(c^e_{i,j})_{i,j=1,\dots,N}$ *is of rank* $N-1$.

Proof. Given $(\lambda_i), (\mu_i) \in \mathbb{R}^{N'+1}$ (resp. $\mathbb{R}^N$), we have from Proposition 1,

$$\sum\sum c_{i,j}\lambda_i\mu_j = \sum\sum \lambda_i\mu_j \int_\Gamma \chi_{\Gamma'_i} C\chi_{\Gamma'_j}\,d\gamma$$

$$= \int_\Gamma \left(\sum \lambda_i\chi_{\Gamma'_i}\right) C\left(\sum u_j\chi_{\Gamma'_j}\right) d\gamma$$

$$= \int_\Omega \operatorname{grad} u(\varphi)\,.\,\operatorname{grad} u(\psi)\,dx$$

$$\left(\text{resp. } \sum\sum c^e_{i,j}\lambda_i\mu_j = \int_{\mathbb{R}^n\setminus\bar\Omega} \operatorname{grad} u_e(\varphi)\,.\,\operatorname{grad} u_e(\psi)\,dx\right),$$

where $\varphi = \sum\lambda_i\chi_{\Gamma'_i}$, $\psi = \sum\mu_i\chi_{\Gamma'_i}$ (resp. $\varphi = \sum\lambda_i\chi_{\Gamma'_i}$, $\psi = \sum\mu_i\chi_{\Gamma'_i}$), $u(\varphi)$ (resp. $u_e(\varphi)$) is the solution of $P(\Omega, \varphi)$ (resp. $P(\mathbb{R}^n\setminus\Omega, \varphi)$ satisfying the null condition at infinity). This proves the point (1).
Let us fix $i \neq j = 0, \dots, N'$ (resp. $1, \dots, N$); given Ω_k (resp. Ω'_k) a connected component of Ω (resp. $\mathbb{R}^n\setminus\bar\Omega$) whose boundary meets Γ'_i (resp. Γ_i); we distinguish two cases:

Case 1: $\Gamma_k \cap \Gamma'_j \neq \varnothing$ (resp. $\Gamma'_k \cap \Gamma_j \neq \varnothing$) From the maximum principle

$$0 < u(\chi_{\Gamma'_j}) < 1 \text{ on } \Omega_k \text{ (resp. } 0 < u_e(\chi_{\Gamma_j}) < 1 \quad \text{on} \quad \Omega'_k) \tag{5.38}$$

and

$$u(\chi_{\Gamma'_j}) = \min_{\bar\Omega_k} u(\chi_{\Gamma'_j}) \quad \text{on} \quad \Gamma'_i \cap \Gamma_k$$

$$\left(\text{resp. } u_e(\chi_{\Gamma_j}) = \min_{\bar\Omega'_k} u_e(\chi_{\Gamma_j}) \quad \text{on} \quad \Gamma_i \cap \Gamma'_k\right).$$

From Hopf's principle of the maximum (see Proposition 4)[105],

$$\begin{cases} C\chi_{\Gamma'_j} = \dfrac{\partial}{\partial n} u(\chi_{\Gamma'_j}) < 0 & \text{on} \quad \Gamma'_i \cap \Gamma_k \\ \left(\text{resp. } C_e\chi_{\Gamma_j} = \dfrac{\partial}{\partial n} u_e(\chi_{\Gamma_j}) < 0 \right. & \left.\text{on} \quad \Gamma_i \cap \Gamma'_k\right). \end{cases} \tag{5.39}$$

---

[105] The use of Proposition 4 demands (in the interior case) the hypothesis of the interior ball. We can in fact, construct a direct argument without making use of Hopf's maximum principle. We have in effect (without any hypothesis of regularity), $C\chi_{\Gamma'_j} \leqslant 0$ on $\Gamma'_i \cap \Gamma_k$ and $C\chi_{\Gamma'_j}$ not identically zero on $\Gamma'_i \cap \Gamma_k$ as otherwise (see Corollary 6 of §4) $u(\chi_{\Gamma'_j})$ would be zero on Ω_k; hence $\int_{\Gamma'_i\cap\Gamma_k} C_{\chi\Gamma'_j}\,d\gamma < 0$ which suffices to prove (2).

Case 2: $\Gamma_k \cap \Gamma'_j = \varnothing$ (resp. $\Gamma'_i \cap \Gamma_j = \varnothing$), then $u(\chi_{\Gamma'_j}) = 0$ on Γ_k (resp. $u_e(\chi_{\Gamma_j}) = 0$ on Γ'_k); hence $u(\chi_{\Gamma'_j}) = 0$ on Ω_k (resp. $u_e(\chi_{\Gamma_j}) = 0$ on Ω_k) and

$$(5.40) \qquad C\chi_{\Gamma'_j} \equiv 0 \quad \text{on} \quad \Gamma_k \quad (\text{resp. } C\chi_{\Gamma_j} \equiv 0 \quad \text{on} \quad \Gamma'_k)\,.$$

Since

$$c_{i,j} = \sum_{k=1}^{N} \int_{\Gamma'_i \cap \Gamma_k} C\chi_{\Gamma'_j}\,\mathrm{d}\gamma \left(\text{resp. } c^e_{i,j} = \sum_{k=0}^{N'} \int_{\Gamma_i \cap \Gamma'_k} C_e\chi_{\Gamma_j}\,\mathrm{d}\gamma \right)$$

the point (2) follows from (5.39) of (5.40).
From Gauss's theorem, for $i = 0, \ldots, N'$

$$\sum_{j=0}^{N'} c_{i,j} = \sum \int_{\Gamma'_j} C\chi_{\Gamma'_i}\,\mathrm{d}\gamma = \int_{\Gamma} \frac{\partial}{\partial n} u(\chi_{\Gamma'_i})\,\mathrm{d}\gamma = 0\,;$$

hence

$$c_{i,i} = - \sum_{\substack{j=0,\ldots,N' \\ j \neq i}} c_{i,j}\,.$$

We have also for $i = 1, \ldots, N$

$$\sum_{j=1}^{N} c^e_{i,j} = \int_{\Gamma} C_e\chi_{\Gamma_i}\,\mathrm{d}\gamma = \int_{\Gamma} \frac{\partial}{\partial n} u_e(\chi_{\Gamma_i})\,\mathrm{d}\gamma\,.$$

In the case $n = 2$, from Lemma 1, the integral is zero.
In the case $n \geqslant 3$, from Gauss' theorem

$$\int_{\Gamma'_k} \frac{\partial}{\partial n} u_e(\chi_{\Gamma_i})\,\mathrm{d}\gamma = 0 \quad \text{for} \quad k = 1, \ldots, N'\,,$$

with the result that

$$(5.41) \qquad \sum_{j=1}^{N} c^e_{i,j} = \int_{\Gamma'_0} \frac{\partial}{\partial n} u_e(\chi_{\Gamma_i})\,\mathrm{d}\gamma\,;$$

case $\Gamma'_0 \cap \Gamma_i = \varnothing$: $u_e(\chi_{\Gamma_i}) \equiv 0$ on Ω'_0 and $\sum_{j=1}^{N} c^e_{i,j} = 0$;
case $\Gamma'_0 \cap \Gamma_i \neq \varnothing$: since $u_e(\chi_{\Gamma_i})$ tends to zero at infinity, from the maximum principle,

$$(5.42) \qquad 0 < u_e(\chi_{\Gamma_i}) < 1 \quad \text{on} \quad \Omega'_0\,.$$

Let us fix R sufficiently large to ensure that $B(0, R) \supset \mathbb{R}^n \backslash \Omega'_0$; we have from Gauss' theorem, for all $r > R$

$$(5.43) \qquad \int_{\Gamma'_0} \frac{\partial}{\partial n} u_e(\chi_{\Gamma_i})\,\mathrm{d}\gamma = - \int_{\partial B(0,r)} \frac{\partial}{\partial n} u_e(\chi_{\Gamma_i})\,\mathrm{d}\gamma = - r^{n-1}\tilde{u}'(r)$$

where $\tilde{u}$ is the radialized of $u(\chi_{\Gamma_i})$ (see §1.4)

$$\tilde{u}(r) = \int_{\Sigma} u(\chi_{\Gamma_i})(r\sigma)\,\mathrm{d}\sigma\,.$$

Being the radialized of a harmonic function, $\tilde{u}$ can be written (see Proposition 6 of §1 and Proposition 4 of §2)

$$\tilde{u}(r) = \frac{a}{r^{n-2}} + b \quad \text{with} \quad a, b \in \mathbb{R} .$$

Since $u_e(\chi_{\Gamma_i}) > 0$ (see (5.42)) and tends to zero at infinity, $b = 0$ and $a > 0$. Hence $\tilde{u}'(r) = -\dfrac{(n-2)a}{r^{n-1}} < 0$, which, from (5.41) and (5.43), shows that

$$c^e_{i,i} > - \sum_{\substack{j=1,\ldots,N \\ j \neq 1}} c^e_{i,j} .$$

This concludes the proof of the point (3).

We know already from the point (3) that the matrix $(c_{i,j})$, $i, j = 0, \ldots, N'$, is of rank $\leqslant N'$ and also in the case $n = 2$ that the matrix $(c^e_{i,j})$, $i, j = 1, \ldots, N$ of rank $< N$. Let us consider $(\lambda_i) \in \mathbb{R}^{N'}$ (resp. $\mathbb{R}^N$); since the matrix $(c_{i,j})$, $i, j = 1, \ldots, N'$ (resp. $c^e_{i,j}$ $i, j = 1, \ldots, N$) is symmetric and positive,

$$\sum_{j=1}^{N'} \lambda_j c_{i,j} = 0 \quad \text{for} \quad i = 1, \ldots, N' \left(\text{resp.} \sum_{j=1}^{N} \lambda_j c^e_{i,j} = 0 \quad \text{for} \quad i = 1, \ldots, N \right)$$

iff $\sum\sum \lambda_i \lambda_j c_{i,j} = 0$ (resp. $\sum\sum \lambda_i \lambda_j c^e_{i,j} = 0$); namely using

$$\varphi = \sum_{i=1,\ldots,N'} \lambda_i \chi_{\Gamma'_i} \left(\text{resp.}\ \varphi = \sum_{i=1,\ldots,N} \lambda_i \chi_{\Gamma_i} \right),$$

from the proof of the point (1) above, iff

$$\int_\Omega |\text{grad}\, u(\varphi)|^2 \, \mathrm{d}x = 0 \quad \left(\text{resp.} \int_{\mathbb{R}^n \setminus \bar{\Omega}} |\text{grad}\, u_e(\varphi)|^2 \, \mathrm{d}x \right),$$

that is to say iff $u(\varphi)$ is constant on $\Omega_1, \ldots, \Omega_N$. (resp. $u_e(\varphi)$ is constant on $\Omega'_0, \ldots, \Omega'_{N'}$), i.e. iff

$$\varphi \in \sum_{i=1}^{N} \chi_{\Gamma_i} \mathbb{R} \quad \left(\text{resp., in the case} \quad n = 2, \varphi \in \sum_{i=0}^{N'} \chi_{\Gamma'_i} \mathbb{R} , \right.$$

and in the case $n \geqslant 3$, $\varphi \in \sum_{i=1}^{N'} \chi_{\Gamma'_i} \mathbb{R}$, since $u_e(\varphi)$ tends to zero at infinity).

Using the fact that

$$\sum_{i=1}^{N} \chi_{\Gamma_i} \mathbb{R} \cap \sum_{i=1}^{N'} \chi_{\Gamma'_i} \mathbb{R} = \{0\} ,$$

we deduce that the matrix $(c_{i,j})$, $i, j = 1, \ldots, N'$, is invertible and hence that the matrix $(c_{i,j})$, $i, j = 0, \ldots, N'$, is of rank N'; we deduce also in the case $n \geqslant 3$, that the matrix $(c^e_{i,j})$, $i j = 1, \ldots, N$ is invertible and in the case $n = 2$ that the matrix (c^e_{ij}) is of rank $N - 1$. □

Remark 1. Returning to the *problem of electrical equilibrium* presented in §5.2*a*, the results of Proposition 6 are interpreted in the following manner.

(1) *Two conductors of a system influence each other if and only if they are connectable* (in the dielectric without crossing other conductors). In other words, given a system in equilibrium, if two conductors are connectable, every change in the potential of the one implies a change in the total charge of the other; conversely if two conductors are not connectable, we can change the potential of the one without destroying the equilibrium of the other. This follows from point (2) of the Proposition.

(2) *The influence of two conductors of a system is symmetrical.* This follows from the symmetry of the matrices of the coefficients of capacitance (point (1) of the Proposition).

(3) *Case of the exterior problem*: given a system S of conductors $K_1, \ldots, K_m$ placed in a dielectric occupying all of the space $\mathbb{R}^3 \setminus S$, for all $q_1, \ldots, q_m$, *there exists one and only one state of equilibrium such that the total charge of each conductor* K_i *is* q_i. That is to say that there exists a unique set $v_1, \ldots, v_m$ ($v_i \in \mathbb{R}$) such that the system is in equilibrium when K_i is at the potential v_i with the total charge density q_i. This follows from the invertibility of the matrix of the exterior coefficients of capacitance (point (4) of Proposition 6).

(4) *Case of the interior problem.* Given a system formed by conductors $K_1, \ldots, K_m$ placed in a dielectric in the cavity of a conductor K_0, *for all* $q_1, \ldots, q_m$ *and* v_0, *there exists one and only one state of equilibrium such that the total charge on each conductor* K_i *is* q_i *and* K_0 *is at the potential* v_0. We note that then, the total charge on the surface Γ'_0 of the interior boundary of K_0 is $q_0 = -\sum_{z=1}^{m} q_i$. This follows from points (3) and (4) of the Proposition. We observe that we should be able also to fix the potential of one of the conductors K_{i_0} and the total charge of each of the conductors K_i, $i = 1, \ldots, m$, $i \neq i_0$ and of the interior boundary of K_0.

(5) As we have just seen, the equilibrium of a system of conductors is determined by the prescription of the total charge of each of the conductors in the exterior case and of that of the internal conductors and of the potential of external conductor in the interior case. The system being in equilibrium, the distribution of electrical charges ρ_i on the boundary ∂K_i of the conductor K_i is then perfectly determined: it is given by the formulae (5.32) and (5.35) according to the exterior or interior case. The argument of the proof of the point (2) of the Proposition indicates that given a system of conductors in equilibrium, on the one hand the charge is zero on every part of the boundary of a conductor which is not connectable (in the dielectric) to any other conductor, while, on the other hand, that the charge is strictly negative (resp. positive) on every part of the boundary of a conductor K_i whose potential is minimum (resp. maximum) with conditions that this part is connectable to a conductor of strictly higher (resp. lower) potential. This is true in the interior case and also in the exterior case by considering at infinity a fictitious conductor at zero potential. □

We conclude this sub-section by the explanation of the form of the matrix of the coefficients of capacitance and of the distributions of charge in some simple cases:

Example 3. *System of two conductors.*
We have several possible situations:

(1) *An exterior problem*

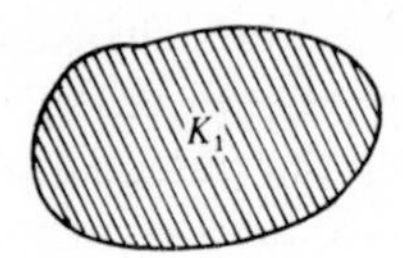

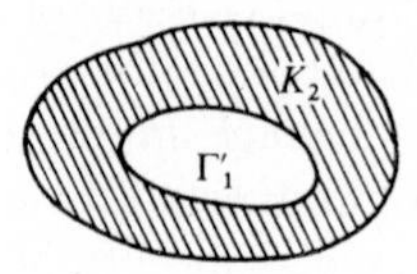

Fig. 8

The matrix of the coefficients $\begin{pmatrix} C_{1,1} & C_{1,2} \\ C_{2,1} & C_{2,2} \end{pmatrix}$ has to satisfy

$$\min(C_{1,1}, C_{2,2}) > -C_{1,2} = -C_{2,1} > 0 .$$

In a state of equilibrium, K_1 and K_2 being at the potentials v_1, v_2 respectively:
(i) the interior boundary Γ'_1 of the hollow conductor K_2 will not be charged: this hollow conductor will behave like the solid conductor $\tilde{K}_2$ obtained by filling in the interior;
(ii) if $v_1 \leqslant 0 \leqslant v_2$, $v_1 \neq v_2$, the charge will be strictly negative (resp. positive) on every part of the boundary of K_1 (resp. of the exterior boundary of K_2);
(iii) if $0 < v_1 < v_2$, the charge will be strictly positive on every part of the exterior boundary of K_2 and there will be positive charges and negative charges on the boundary of K_1;
(iv) if $0 < v_1 = v_2$, the charge will be strictly positive on the boundary of K_1 and the exterior boundary of K_2.
(2) *Two examples of interior problems.*

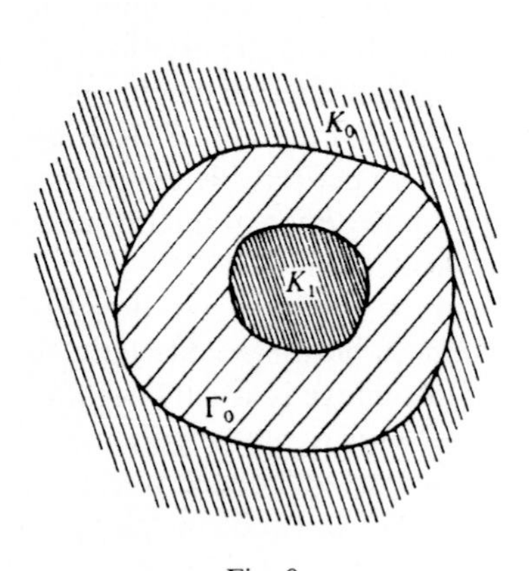

Fig. 9

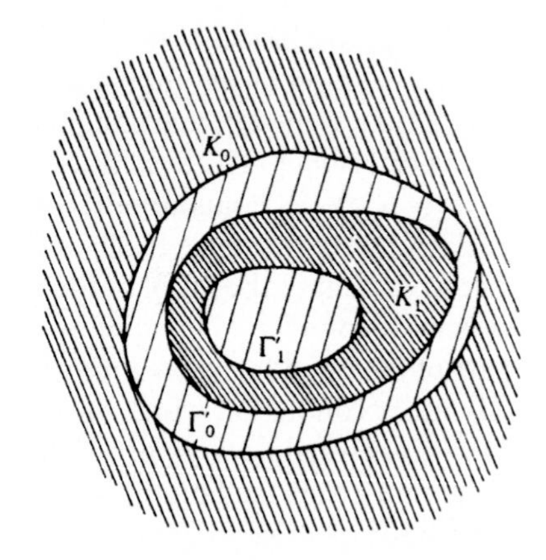

Fig. 10

In the two cases, the matrix of the coefficients of capacitance $\begin{pmatrix} C_{0,0} & C_{0,1} \\ C_{1,0} & C_{1,1} \end{pmatrix}$ must satisfy

$$C_{0,0} = -C_{0,1} = -C_{1,0} = C_{1,1} > 0 .$$

It only depends therefore on a single constant C, called *the capacity of the capacitor*, the matrix of the coefficients of capacitance being $\begin{pmatrix} C & -C \\ -C & C \end{pmatrix}$.

In a state of equilibrium, in the case of Fig. 10, the interior boundary Γ'_1 of K_1 will not be charged: the capacity of the capacitor will depend only on the exterior boundary of the interior conductor (and also only on the interior boundary of the exterior conductor).

(3) *Another exterior problem.*

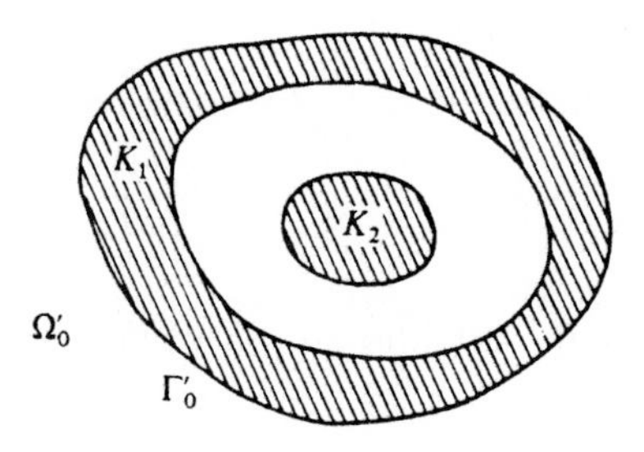

Fig. 11

In this example the "capacitor" (K_1, K_2) is immersed in the dielectric occupying all of $\mathbb{R}^3 \backslash (K_1 \cup K_2)$. If we denote by C the capacity of the capacitor (K_1, K_2), it is clear that the exterior coefficients of influence of K_2 on itself, and of K_2 on K_1 will be C and $-C$. As for the exterior coefficient of influence of K_1 on itself, it will be $C^e_{1,1} = C + C^e$, where

$$C^e = \int_{\Gamma'_0} \frac{\partial u_e}{\partial n} \mathrm{d}\gamma$$ [106]

where Γ'_0 is the exterior boundary of K_1, boundary of the exterior Ω'_0 of the condenser and u_e the solution of $P(\Omega'_0, 1)$ tending to zero at infinity. We have (see proof of the point (3) of Proposition 6)

$$C^e > 0 .$$

In brief, the matrix of the interior coefficients of capacitance of this system is

$$\begin{pmatrix} C + C^e & -C \\ -C & C \end{pmatrix}. \qquad \square$$

2c. Energy Balance. Examples of Capacities

We consider a system of conductors immersed in a dielectric, The electrostatic field $\mathbf{E}$ is given by (see Chap. IA, §4)

$$\mathbf{E} = -\operatorname{grad} v_c \quad \text{in} \quad D$$

[106] Here the permittivity ε of the exterior electrical medium is supposed to be equal to 1.

where v_c is the electrostatic potential. The electrical energy[107] of the system is given by

$$(5.44) \qquad \mathscr{E} = \frac{\varepsilon}{2}\int_D |\mathbf{E}|^2\, dx = \frac{\varepsilon}{2}\int_D |\operatorname{grad} v_c|^2\, dx\,.$$

We distinguish two cases.

Case of the exterior problem. With the notation of §5.2*a*, v_c is the solution of $P(\mathbb{R}^n \backslash \bar{\Omega}, \Sigma\, v_i \chi_{\Gamma_i})$ tending to zero at infinity with the result that (see Proposition 1),

$$\int_{\mathbb{R}^n \backslash \bar{\Omega}} |\operatorname{grad} v_c|^2\, dx = \int_\Gamma \left(\sum v_i \chi_{\Gamma_i}\right) C_e \left(\sum v_j \chi_{\Gamma_j}\right) d\gamma$$

$$= \sum\sum v_i v_j \int_{\Gamma_i} C_e \chi_{\Gamma_j}\, d\gamma\,.$$

Hence, taking account of the definition of the exterior coefficients of capacitance (see (5.34)), we have

$$(5.45) \qquad \mathscr{E} = \frac{1}{2}\sum\sum C^e_{i,j} v_i v_j\,.$$

Case of the interior problem. With the notation of §5.2.*a*, v_c is the solution of $P\left(\Omega, \sum_{i=0}^{m} v_i \chi_{\Gamma'_i}\right)$ with the result that as above

$$(5.46) \qquad \mathscr{E} = \frac{1}{2}\sum\sum C_{i,j} v_i v_j$$

where $C_{i,j}$ are the coefficients of influence of the capacitor (see (5.37)).

In these formulae the (*numerical values of the*) *coefficients of capacitance* $C^e_{i,j}$ *and* $C_{i,i}$ *appear as twice* (*the numerical values*) *of the energy of the system in equilibrium when the conductor* K_i *is at unit potential, the others being at zero potential.* For that reason[108], we call $C^e_{i,i}$ the *exterior capacity of the conductor* K_i *in the* "exterior" *system and* $C_{i,i}$ the *interior capacity of the conductor* K_i *in the* "interior" system. This coincides totally with the definition of Example 3.2. More generally, let S_0 be a family of conductors of the system; we call the capacity (exterior or interior) of S_0 in the system, twice the energy of the system when we put the conductors of the family S_0 at unit potential and the remaining conductors at zero potential. In particular, the *exterior capacity of an entire system* is double the energy of the system when all the conductors are at unit potential; it is therefore

$$C^e = \sum\sum C^e_{i,j}\,.$$

Also the *interior capacity of the* "*interior*" *system* also called the *capacity of the*

---

[107] In fact, the *free* energy for perfect media; see Chap. IA, §4 where we used the notation W for $\mathscr{E}$.

[108] They measure the capacity to store energy for a given potential. The capacities (of conductors in a system) are coefficients of capacitance: the *unit of capacity in the S.I. is then the Farad.*

capacitor is twice the energy of the system when all the *interior* conductors $K_1, \ldots, K_m$ are at unit potential, namely

$$C = \sum_{i,j=1,\ldots,m} C_{i,j} \,.$$

We note that $C = C_{0,0}$, capacity of the conductor K_0 in the "interior" system: this follows from the relations $\sum_{j=1,\ldots,m} C_{i,j} = - C_{i,0}$ for $i = 0$ to m.
We observe that *the capacity* (exterior or interior) *of a family* of conductors in a system *is always less than the sum of the capacities of its components.* This follows from that the coefficients of influence $C^e_{i,j}$ or $C_{i,j}$ (for $i \neq j$) are negative. This capacity can even be less than the maximum of the capacities of its components: this is clear for the system of Example 3(3) (Fig. 11), where the exterior capacity of the system (K_1, K_2) is

$$C + C^e + (-C) + (-C) + C = C^e$$

which is strictly less than the exterior capacity of K_1 in the system. The notion of capacity has very many applications going far beyond the electrostatic problem: we shall study it in Sect. 3. We shall content ourselves here with giving some examples.

Example 4. *Spherical capacities.* We consider a radial system of conductors, i.e. a system in which all the conductors of the system are in the form of concentric spherical layers (see Fig. 12).
(1) *Exterior capacity of the system.* The conductors being all at unit potential, the electrostatic potential v_c is equal to 1 on $\bar{B}(0, r)$ where *r is the radius of the exterior conductor*: the exterior capacity of the system is the same as that of a solid

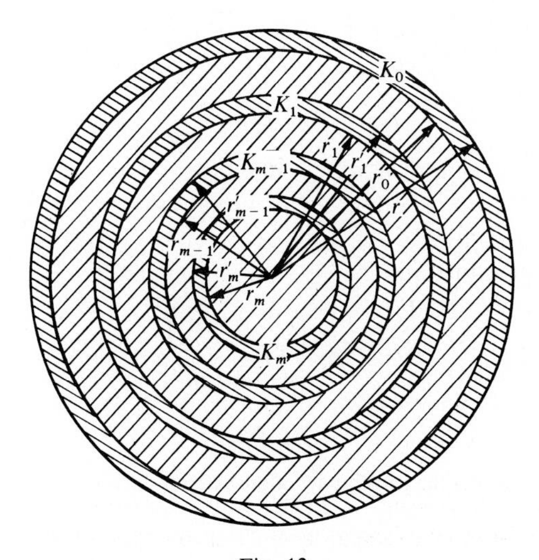

Fig. 12

conductor occupying all of the ball $B(0, r)$; by definition it is

$$C^e = \varepsilon \int_{\partial B(0, r)} \frac{\partial u_e}{\partial n_e} \, d\gamma$$

where u_e is the solution of $P(\mathbb{R}^3 \backslash \bar{B}(0, r), 1)$ tending to zero at infinity. We have $u_e(x) = r/|x|$ on $\mathbb{R}^3$, so

$$\frac{\partial u_e}{\partial n_e}(z) = \frac{r}{|z|^2} = \frac{1}{r} \quad \text{on} \quad \partial B(0, r) ,$$

and

$$C^e = 4\pi r \varepsilon^{109} . \tag{5.47}$$

(2) *Capacity of the capacitor.* This is the capacity of the exterior conductor in the "interior" system. The interior conductors all being at zero potential, the electrostatic potential is zero on $\bar{B}(0, r'_1)$ where r'_1 *is the radius of the exterior boundary of the largest interior conductor*: the capacity of the capacitor is the same as that of a capacitor with two conductors where the interior conductor occupies the whole ball $\bar{B}(0, r_1)$. Denoting by r_0 the radius of the interior boundary of the exterior conductor, we have

$$C = \varepsilon \int_{\partial B(0, r_0)} \frac{\partial u}{\partial n} \, d\gamma$$

where u is the solution of $P(\{r'_1 < |x| < r_0\}, \chi_{\partial B(0, r_0)})$. From Table 2 of Example 20, §4, we have

$$u(x) = \left(\frac{1}{r'_1} - \frac{1}{|x|}\right)\left(\frac{1}{r'_1} - \frac{1}{r_0}\right)^{-1} \quad \text{on} \quad \{r'_1 < |x| < r_0\}$$

from which

$$\frac{\partial u}{\partial n}(z) = \frac{1}{r_0^2}\left(\frac{1}{r'_1} - \frac{1}{r_0}\right)^{-1} \quad \text{on} \quad \partial B(0, r_0)$$

and

$$C = \frac{4\pi r_0 r'_1 \varepsilon}{r_0 - r'_1} \tag{5.48}$$

or again

$$\frac{1}{C} = \frac{1}{4\pi\varepsilon}\left(\frac{1}{r'_1} - \frac{1}{r_0}\right)^{110} \tag{5.49}$$

[109] In this expression, ε denotes the permittivity of the exterior dielectric medium $\{|x| > r\}$, and n_e is directed towards the centre of the ball.

[110] In this expression, ε denotes the permittivity of the interior dielectric medium which in general is different from that of the exterior dielectric medium.

(3) *Capacities and coefficients of capacitance of the components of the system.* Let us consider first the exterior conductor K_0: its interior capacity (in the capacitor) is given by (5.48); from Example (3.3) the exterior capacity (of K_0 in the system) is $C^e_{0,0} = C + C^e$ where C is given by (5.48) and C^e by (5.47). Finally K_0 influences only the largest interior conductor K_1 and (see Example 3(2) and (3)) $C^e_{0,1} = C_{0,1} = -C$.

The exterior and interior coefficients of capacitance of the interior conductors are equal. The smallest conductor K_m is influenced only by the smallest conductor containing it $- K_{m-1}$; its capacity and its coefficient of capacitance on K_{m-1} are the same as if it were the interior conductor of the capacitor (K_{m-1}, K_m). Hence, from (5.48),

$$C_{m,m} = -C_{m,m-1} = \frac{4\pi r_{m-1} r'_m \varepsilon}{r_{m-1} - r'_m}$$

where for a capacitor K_i, r_i (resp. r'_i) denotes the radius of the interior (resp. exterior) boundary of K_i.

Finally for $0 < i < m$, the conductor K_i is influenced by K_{i-1} and K_{i+1}: its capacity in the system is the sum of the capacities of the capacitors (K_{i-1}, K_i) and (K_i, K_{i+1}); its coefficient of capacitance on K_{i-1} (resp. K_{i+1}) is the same as it would be if it were the interior (resp. exterior) conductor of the capacitor (K_{i-1}, K_i) (resp. (K_i, K_{i+1})), i.e. the opposite of the capacity of that capacitor. Therefore

$$C_{i,i} = 4\pi\varepsilon\left(\frac{r_{i-1} r'_i}{r_{i-1} - r'_i} + \frac{r_i r'_{i+1}}{r_i - r'_{i+1}}\right)$$

$$C_{i,i-1} = -\frac{4\pi\varepsilon r_{i-1} r'_i}{r_{i-1} - r'_i}, \quad C_{i,i+1} = -\frac{4\pi\varepsilon r_i r'_{i+1}}{r_i - r'_{i+1}}. \quad \square$$

Remark 2. The above considerations on the coefficients of capacitance of the conductors K_i relatively to the capacities of the capacitors (K_{i-1}, K_i) are obviously valid for any system of conductors whatsoever in which they are contained the one within the other. $\square$

Example 5. *Plane capacitor.* Let us consider a system of two identical plane conductors K_1 and K_2 placed parallel to each other at a distance d apart (see Fig. 13).

The system being in electrical equilibrium, each conductor K_i is at the constant potential v_i. If the plates were infinite, the field between each plate would be

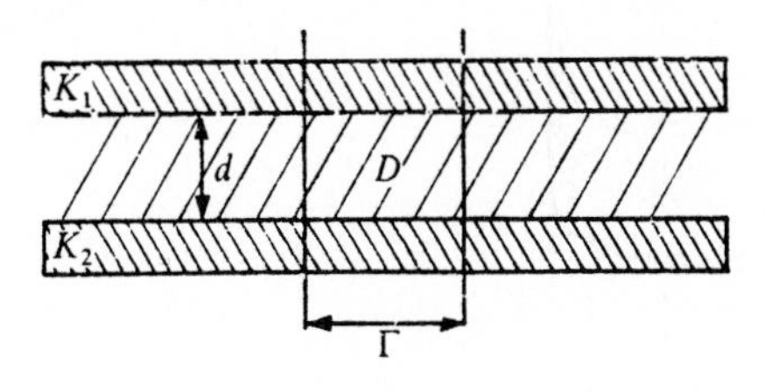

Fig. 13

uniform and given by

$$\mathbf{E} = \frac{v_1 - v_2}{d}\mathbf{n}$$

where $\mathbf{n}$ is the unit vector, normal to the direction of the plates, directed from K_1 to K_2.

The electrical energy in the domain D of the dielectric determined by the surface Γ is therefore

$$\mathscr{E} = \frac{\varepsilon}{2} \times |E|^2 \times \text{Volume of } D = \frac{1}{2}(v_1 - v_2)^2 \frac{\varepsilon}{d} \times \text{surface of } \Gamma .$$

The quantity ε/d *is the capacity per unit area of surface of the plane capacitor* (K_1, K_2). The quantity $\varepsilon S/d$ where S is the surface area of each of the plane conductors K_1, K_2 *is the capacity of the idealized plane capacitor*. This theoretical capacity differs from the real capacity – twice the electrical energy in the dielectric contained between the two plates when the potential difference between the plates is unity: indeed the field is not really uniform between the plates because of the "edge effects"; however, the difference is negligible in many applications (if $\sqrt{S} \gg d$). □

Example 6. *Cylindrical capacitors.* We consider the system of two coaxial circular cylindrical conductors K_0 and K_1 of the same length l (see Fig. 14). Supposing nevertheless that the cylinders are infinite, the system being in equilibrium, the electric field will be orthogonal to Oz and independent of z. Hence

$$\mathbf{E} = -\operatorname{grad} v_c \quad \text{where} \quad v_c(x, y, z) = u(x, y)$$

with u the solution of the Dirichlet problem $P(\Omega, v_0\chi_{\Gamma_0} + v_1\chi_{\Gamma_1'})$ where Ω is the

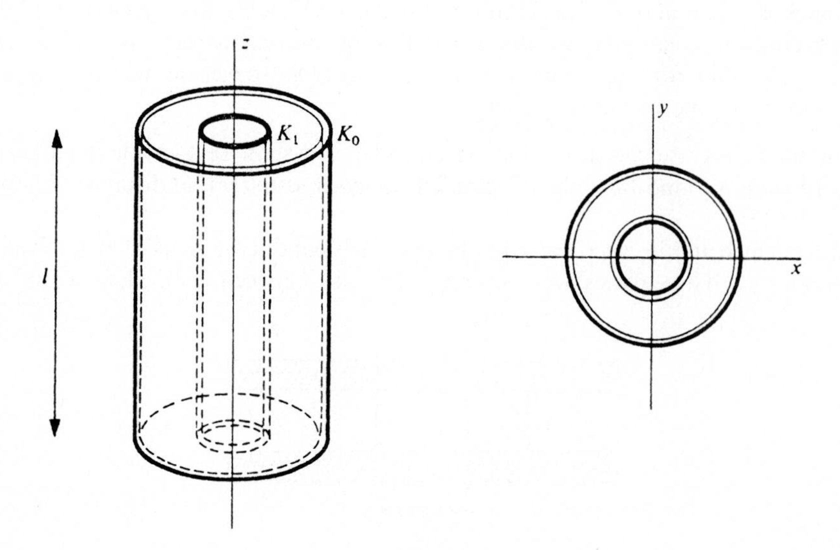

Fig. 14

annulus of $\mathbb{R}^2$ contained between the circles Γ_0 and Γ'_1 respectively the cross-sections of the outer and the inner conductors. In the present case, where we have circular symmetry $\Omega = \{r'_1 < r < r_0\}$ and u is given by (see Example 20 of §4)

$$u(x, y) = \left(v_1 \operatorname{Log}\frac{r_0}{r} + v_0 \operatorname{Log}\frac{r}{r'_1}\right)\left(\operatorname{Log}\frac{r_0}{r'_1}\right)^{-1}$$

where $r = (x^2 + y^2)^{\frac{1}{2}}$.
Hence the electrical energy in the domain of the dielectric of length l is

$$\mathscr{E} = \frac{\varepsilon}{2} \times \int_\Omega |\operatorname{grad} u|^2 \, dx\, dy \times l = \frac{\varepsilon}{2} \times 2\pi l \times \left(\operatorname{Log}\frac{r_0}{r'_1}\right)^{-1} (v_0 - v_1)^2 .$$

The quantity $2\pi\varepsilon/\operatorname{Log}(r_0/r'_1)$ is the *capacity per unit length of the cylindrical capacitor* (K_0, K_1). The quantity $2\pi\varepsilon l/\operatorname{Log}(r_0/r'_1)$ *is the theoretical capacity of the cylindrical capacitor of length l*. We can show, in this case also, that it differs little from the real capacity in many applications. □

Condensers, generally consisting of two conductors separated by an insulating or dielectric medium, are used in electrical circuits (continuous, variable (alternating, ...)). In electronic circuits, their great utility is linked to the fact that they can absorb or part with large quantities of electric charge, without much changing their potential.
The S.I. unit of capacity is, as we have seen, the Farad (1 Farad = 1 Coulomb/volt). In practice, charges are much weaker than the Coulomb and measurements are currently made in terms ranging from the picofarad (10^{-12} F) in high frequency circuits to some microfarads (10^{-6} F) in the filters of electrical power supply. □

3. Capacity of a Part of an Open Set in $\mathbb{R}^n$

3a. Capacity of a Compact Set: Definition, Properties, Examples

In this part we introduce the mathematical idea of the capacity of a part of $\mathbb{R}^n$. We shall call by a *regular compact set* of $\mathbb{R}^n$ every $K = \bar{\omega}$ where ω is a regular bounded open set of $\mathbb{R}^n$. Given K a regular compact set of $\mathbb{R}^n$ and Ω a regular bounded open set containing K, $\Omega \backslash K$ is a regular bounded open set. We denote by $u_{K,\Omega}$ the function defined on $\mathbb{R}^n$ by

$$(5.50) \qquad \begin{cases} u_{K,\Omega} = 1 \quad \text{on} \quad K \\ u_{K,\Omega} = 0 \quad \text{on} \quad \mathbb{R}^n \backslash \bar{\Omega} \\ u_{K,\Omega} \text{ is continuous on } \mathbb{R}^n \text{ and harmonic on } \Omega \backslash K . \end{cases}$$

In other words, $u_{K,\Omega}$ is defined on $\Omega \backslash K$ as the solution of the Dirichlet problem

$P(\Omega\backslash K, \chi_K)$. We have

$$u_{K,\Omega} \in \overline{\mathscr{C}^1_n(\Omega\backslash K)} \quad \text{(see Proposition 18 of §4)}$$

$$|\text{grad}\, u_{K,\Omega}|^2 \in L^1(\Omega) \quad \text{and (see Proposition 5 of §1)}$$

$$\int_{\Omega\backslash K} |\text{grad}\, u_{K,\Omega}|^2 \, \mathrm{d}x = \int_{K \cap \partial(\Omega\backslash K)} \frac{\partial}{\partial n} u_{K,\Omega} \, \mathrm{d}\gamma \, .$$

From Gauss' theorem and account being taken of

$$(5.51) \quad \begin{cases} \dfrac{\partial}{\partial n} u_{K,\Omega} \geqslant 0 \quad \text{on} \quad K \cap \partial(\Omega\backslash K) \quad \text{and} \quad \dfrac{\partial}{\partial n} u_{K,\Omega} \leqslant 0 \quad \text{on} \quad \partial\Omega \, , \\ \displaystyle\int_{K \cap \partial(\Omega\backslash K)} \frac{\partial}{\partial n} u_{K,\Omega} \, \mathrm{d}\gamma = \int_{\partial\Omega} - \frac{\partial}{\partial n} u_{K,\Omega} \, \mathrm{d}\gamma = \frac{1}{2} \int_{\partial(\Omega\backslash K)} \left| \frac{\partial}{\partial n} u_{K,\Omega} \right| \mathrm{d}\gamma \, . \end{cases}$$

This can be interpreted in the sense of distributions:
The derivatives $\partial u_{K,\Omega}/\partial x_i$ (in the sense of distributions) are functions of $L^2(\mathbb{R}^n)$.
$\Delta u_{K,\Omega}$ (in the sense of distributions) is a Radon measure on $\mathbb{R}^n$, supported by $\partial(\Omega\backslash K)$, negative on Ω, positive on $\mathbb{R}^n\backslash K$ and

$$(5.52) \quad \begin{cases} \displaystyle\int_{\mathbb{R}^n} |\text{grad}\, u_{K,\Omega}|^2 \, \mathrm{d}x & = \text{total mass of} - \Delta u_{K,\Omega} \text{ in } \Omega \\ & = \text{total mass of } \Delta u_{K,\Omega} \text{ in } \mathbb{R}^n\backslash K \\ & = \tfrac{1}{2} \times \text{ the total variation of } \Delta u_{K,\Omega} \text{ in } \mathbb{R}^n \, . \end{cases}$$

By definition, this quantity is *the capacity of K on Ω*, which we denote by $\text{cap}_\Omega K$[111]:

$$\text{cap}_\Omega K = \int_{\mathbb{R}^n} |\text{grad}\, u_{K,\Omega}|^2 \, \mathrm{d}x \, .$$

Example 7. *Case $n = 1$.* We note that we have trivially in the case n arbitrary

$$(5.53) \qquad \text{cap}_\Omega K = \sum \text{cap}_{\Omega_i} K_i$$

where $\Omega_1, \ldots, \Omega_m$ are the connected components of Ω and $K_i = K \cap \Omega_i$. *We can therefore always suppose Ω is connected*, that is to say in the case $n = 1$, $\Omega = \,]a, b[$. A regular compact set of $\mathbb{R}$ is of the form

$$K = [a_1, b_1] \cup \ldots \cup [a_m, b_m] \quad \text{with} \quad a_1 < b_1 < a_2 < \ldots < b_m \, .$$

[111] If K is the system of ohmic conductors placed in the dielectric D occupying the domain Ω surrounded by an ohmic conductor K_0, $\text{cap}_\Omega K$ is to within the multiplicative constant ε, the capacity of the condenser thus constructed (see §5.20*a* and *c*).

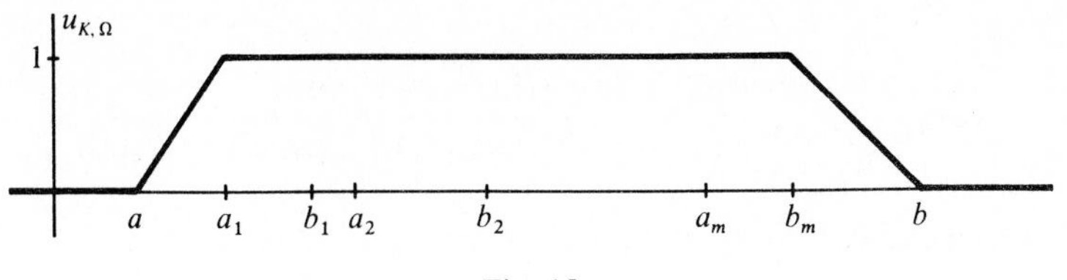

Fig. 15

It is clear that the graph of $u_{K,\Omega}$ is given by Fig. 15 and hence

$$\text{cap}_\Omega K = \frac{1}{a_1 - a} + \frac{1}{b - b_m} \tag{5.54}$$

where, we recall, $\Omega =]a, b[$, $a_1 = \min K$, $b_m = \max K$. □

Example 8. *Radial case.* From (5.53), we can suppose that Ω is connected. A radial bounded connected open set is either a ball or a ring. A radial regular compact set is of the form

$$K = \bigcup_{i=1}^{m} \{r_i \leqslant |x| \leqslant r'_i\} \quad \text{with} \quad 0 \leqslant r_1 < r'_1 < r_2 < \ldots < r'_m .$$

Making use of the results of Example 20 of §4,
Case (a). $\Omega = B(0, r_0)$: $u_{K,\Omega}$ is given (on Ω) by

$$u_{K,\Omega}(x) = \begin{cases} 1 & \text{on} \quad \bar{B}(0, r'_m) \\ \dfrac{E_n(r_0) - E_n(x)}{E_n(r_0) - E_n(r'_m)} & \text{on} \quad \Omega \backslash \bar{B}(0, r'_m) \end{cases}$$

and

$$\frac{\partial}{\partial n} u_{K,\Omega} = \frac{1}{\sigma_n r_0^{n-1}} \times \frac{1}{E_n(r'_m) - E_n(r_0)} \quad \text{on} \quad \partial\Omega$$

and

$$\text{cap}_{B(0, r_0)} K = \frac{1}{E_n(r_0) - E_n(r)} \tag{5.55}$$

where K being radial, $r = r'_m = \max\limits_K |x|$;

case (b). $\Omega = \{r_0 < |x| < r'_0\}$

$$u_{K,\Omega} = \begin{cases} \dfrac{E_n(r_0) - E_n(x)}{E_n(r_0) - E_n(r'_1)} & \text{on} \quad \{r_0 < |x| < r'_1\} \\ 1 & \text{on} \quad \{r'_1 \leqslant |x| \leqslant r'_m\} \\ \dfrac{E_n(r'_0) - E_n(x)}{E_n(r'_0) - E_n(r'_m)} & \text{on} \quad \{r'_m < |x| < r'_0\} \end{cases}$$

and

$$\text{(5.56)} \qquad \operatorname{cap}_{\{r_0 < |x| < r'_0\}} K = \frac{1}{E_n(r) - E_n(r_0)} + \frac{1}{E_n(r'_0) - E_n(r')}$$

where K being radial $r = r'_1 = \min_K |x|$, $r' = r'_m = \max_K |x|$. □

We shall now prove the fundamental result:

Proposition 7. *Let K be a regular compact set of $\mathbb{R}^n$ and Ω a regular bounded open set of $\mathbb{R}^n$ containing K. Then*

$$\text{(5.57)} \qquad \operatorname{cap}_\Omega K = \min_{\mathscr{S}(K,\Omega)} \int_{\mathbb{R}^n} |\operatorname{grad} u|^2 \, dx$$

where $\mathscr{S}(K, \Omega)$ is the set of the $u \in \mathscr{C}^0(\mathbb{R}^n)$ such that the derivatives $\partial u/\partial x_i$ (in the sense of distributions) are in $L^2(\mathbb{R}^n)$ and $u \geqslant 1$ on K, $u \leqslant 0$ on $\mathbb{R}^n \backslash \Omega$.

To simplify the writing we introduce $\mathscr{I}(u) = \int_{\mathbb{R}^n} |\operatorname{grad} u|^2 \, dx$ defined on the space

$$B^1(\mathbb{R}^n) = \{u \in L^2_{\text{loc}}(\mathbb{R}^n); \textit{the derivatives} \frac{\partial u}{\partial x_i} \in L^2(\mathbb{R}^n) \textit{ (in the sense of distributions)}\}.$$

The proposition expresses that $u_{K,\Omega}$ minimises the energy functional $\mathscr{I}(u)$ on the set $\mathscr{S}(K, \Omega) = \{u \in \mathscr{C}^0(\mathbb{R}^n) \cap B^1(\mathbb{R}^n); u \geqslant 1 \text{ on } K, u \leqslant 0 \text{ on } \mathbb{R}^n \backslash \Omega\}$.

Proof. Since $u_{K,\Omega} \in \mathscr{S}(K, \Omega)$, it is sufficient to prove that, for $u \in \mathscr{S}(K, \Omega)$,

$$\operatorname{cap}_\Omega K = \mathscr{I}(u_{K,\Omega}) \leqslant \mathscr{I}(u).$$

Let us put $v = u - u_{K,\Omega}$; we have

$$\text{(5.58)} \qquad v \in \mathscr{C}^0(\mathbb{R}^n) \cap B^1(\mathbb{R}^n), \quad v \geqslant 0 \text{ on } K, \quad v \leqslant 0 \text{ on } \mathbb{R}^n \backslash \Omega.$$

On the other hand

$$\begin{aligned} \mathscr{I}(u) - \mathscr{I}(u_{K,\Omega}) &= \mathscr{I}(v) + 2 \int_{\mathbb{R}^n} \operatorname{grad} u_{K,\Omega} \cdot \operatorname{grad} v \, dx \\ &\geqslant 2 \int_{\Omega \backslash K} \operatorname{grad} u_{K,\Omega} \cdot \operatorname{grad} v \, dx \\ &= \int_{\partial(\Omega \backslash K)} v \frac{\partial}{\partial n} u_{K,\Omega} \, d\gamma . \end{aligned}$$

From (5.51) and (5.58) we therefore have $\mathscr{I}(u) \geqslant \mathscr{I}(u_{K,\Omega})$. □

This proposition leads us to frame the

Definition 1. *Let Ω be an (arbitrary) open set in $\mathbb{R}^n$ and K a compact set in Ω. We define the capacity of K in Ω to be*

$$\operatorname{cap}_\Omega K = \inf_{\mathscr{S}(K,\Omega)} \mathscr{I}(u),$$

where $\mathscr{S}(K, \Omega)$ *is the set of the* $u \in \mathscr{C}^0(\mathbb{R}^n) \cap B^1(\mathbb{R}^n)$ *with* $u \geqslant 1$ *on* K, *and* $u \leqslant 0$ *on* $\mathbb{R}^n \setminus \Omega$, $\limsup_{|x| \to \infty} u(x) \leqslant 0$.

This definition clearly generalizes that given in the case where Ω is a regular bounded set and K is regular. In general the energy function $\mathscr{I}(u)$ does not attain its minimum on $\mathscr{S}(K, \Omega)$; this is due to the fact that the Dirichlet problem $P(\Omega \setminus K, \chi_K)$ does not always admit of a classical solution (i.e. continuous on $\overline{\Omega \setminus K}$); however in the case where Ω is bounded, the Dirichlet problem $P(\Omega \setminus K, \chi_K)$ always has a *generalized solution* (see Definition 3 of §4); in the case of Ω unbounded, we can also consider, if it exists, the generalized solution of $P(\Omega/K, \chi_K)$ satisfying the null condition at infinity (see Definition 10 of §4). We define *the function* $u_{K,\Omega}$ *by*

$$
(5.59) \quad \begin{cases} u_{K,\Omega} = 1 \text{ on } K \\ u_{K,\Omega} = 0 \text{ on } \mathbb{R}^n \setminus \Omega \\ u_{K,\Omega} \text{ is equal on } \Omega \setminus K \text{ to the generalized solution of } P(\Omega/K, \chi_K) \\ \text{satisfying, in the case of unbounded } \Omega, \text{ the null condition at infinity.} \end{cases}
$$

We note that *this function* $u_{K,\Omega}$ *is well defined on* $\mathbb{R}^n$ in the case of Ω bounded, and if Ω is unbounded, if the Dirichlet problem $P(\Omega \setminus K, \chi_K)$ admits a generalized solution satisfying the null condition at infinity. By definition[112] $u_{K,\Omega}$ is defined on $\Omega \setminus K$ as the lower envelope of $\mathscr{I}_+(\Omega \setminus K, \chi_K)$ the set of the continuous functions w_0 superharmonic on $\Omega \setminus K$ satisfying

$$
\liminf_{x \to z} w_0(x) \geqslant \chi_K(z) \quad \text{for all} \quad z \in \partial(\Omega \setminus K),
$$

and in the case where Ω is unbounded, $\min(w_0, 0)$ satisfies the null condition at infinity (relative to Ω/K). Such a function w_0 is positive or zero on $\Omega \setminus K$ and

$$
\lim_{x \to z} \min(w_0(x), 1) = 1 \quad \text{for all} \quad z \in \partial K .
$$

Hence, the function defined on Ω by

$$
w = 1 \quad \text{on} \quad K, \quad w = \min(w_0, 1) \quad \text{on} \quad \Omega \setminus K,
$$

is positive or zero on Ω; it is also superharmonic on Ω (see Remark 8(3) of §4). It follows that $u_{K,\Omega}$ *is defined on* Ω *as the lower envelope of the set* $\mathscr{S}^+(K, \Omega)$ *of positive, continuous functions, superharmonic on* Ω, *satisfying* $w \geqslant 1$ *on* K.
This leads us to formulate

Definition 2. Let Ω be an (arbitrary) open set of $\mathbb{R}^n$ and K an (arbitrary) compact set in Ω. We denote by $u_{K,\Omega}$ the function defined on $\mathbb{R}^n$ by

$$
\begin{cases} u_{K,\Omega} = 0 \quad \text{on} \quad \mathbb{R}^n \setminus \Omega \\ u_{K,\Omega}(x) = \inf\limits_{\mathscr{S}^+(K,\Omega)} w(x) \quad \text{for all} \quad x \in \Omega \end{cases}
$$

[112] See §4.1 in the case in which Ω is bounded and (4.3) in the case in which Ω is unbounded.

where $\mathscr{S}^+(K, \Omega)$ is the set of functions w, continuous, superharmonic and positive on Ω with $w \geqslant 1$ on K[113].

In the case where the solution of (5.59) is well defined, as we have just seen, this solution coincides with the function $u_{K,\Omega}$ of Definition 2. These definitions 1 and 2 are justified by the

Proposition 8. *Let Ω be an (arbitrary) open set in $\mathbb{R}^n$ and K an (arbitrary) compact set in Ω. With the definitions*

(1) $\operatorname{cap}_\Omega K = \inf_{\mathscr{D}(K,\Omega)} \mathscr{I}(u)$, *where*

$$\mathscr{D}(K, \Omega) = \{u \in \mathscr{D}(\Omega);\ 0 \leqslant u \leqslant 1,\ u = 1 \text{ in the neighbourhood of } K\}\ ;$$

(2) $u_{K,\Omega} \in B^1(\mathbb{R}^n)$[114] *and* $\operatorname{cap}_\Omega K = \mathscr{I}(u_{K,\Omega})$;
(3) *The distribution $u_{K,\Omega}$ is subharmonic on Ω[115], harmonic on $\Omega \backslash \partial K$ and*

$$\operatorname{cap}_\Omega K = \int_{\partial K} \mathrm{d}(-\Delta u_{K,\Omega})^{116}.$$

In the proof of this proposition we shall make use of the following essential lemma which will be proved at the end of this sub-section 3.a.

Lemma 2. *Let Ω be an open set of $\mathbb{R}^n$ and $u \in L^1_{\text{loc}}(\Omega)$ with $\dfrac{\partial u}{\partial x_i} \in L^1_{\text{loc}}(\Omega)$ (in the sense of distributions). Then*

$$\frac{\partial}{\partial x_i} u^+ \in L^1_{\text{loc}}(\Omega) \quad \textit{and} \quad \frac{\partial}{\partial x_i} u^+ = \chi_{\{u>0\}} \frac{\partial u}{\partial x_i} \quad \text{a.e. } \textit{on } \Omega^{117} \tag{5.60}$$

$$\frac{\partial u}{\partial x_i} = 0 \quad \text{a.e. } \textit{on} \quad \{x \in \Omega;\ u(x) = 0\}. \tag{5.61}$$

Proof of Proposition 8.(1) We have $\mathscr{D}(K, \Omega) \subset \mathscr{S}(K, \Omega)$ with the result that

$$\operatorname{cap}_\Omega K \leqslant \inf_{\mathscr{D}(K,\Omega)} \mathscr{I}(u)\,.$$

We now consider $u_0 \in \mathscr{S}(K, \Omega)$. For $\varepsilon > 0$, we put

$$u_\varepsilon = ((1 + 2\varepsilon)u_0 - \varepsilon)^+ \wedge 1^{118}$$

---

[113] Note that $\mathscr{S}^+(K, \Omega)$ is not the cone of the positive functions of $\mathscr{S}(K, \Omega)$.
[114] The function $u_{K,\Omega}$ is a bounded Borel function on $\mathbb{R}^n$ and hence in the class $L^2_{\text{loc}}(\mathbb{R}^n)$.
[115] That is $-\Delta u_{K,\Omega} \geqslant 0$ in $\mathscr{D}'(\Omega)$ (see Definition of §4).
[116] $-\Delta u_{K,\Omega}$ is a positive Radon measure on Ω supported by ∂K.
[117] $u^+ = \max(u, 0)$, $\chi_{\{u>0\}}$ is the characteristic function of the set $\{x \in \Omega;\ u(x) > 0\}$.
[118] For two functions f, g: $f \wedge g = \min(f, g) = f - (f - g)^+$.

We have $u_\varepsilon \in \mathscr{C}^0(\mathbb{R}^n)$; from Lemma 2 the derivatives $\dfrac{\partial u_\varepsilon}{\partial x_i} \in L^1_{\text{loc}}(\mathbb{R}^n)$ and

$$\frac{\partial u_\varepsilon}{\partial x_i} = \chi_{\{1 > (1+2\varepsilon)u_0 - \varepsilon > 0\}}(1 + 2\varepsilon)\frac{\partial u_0}{\partial x_i} \quad \text{a.e. on} \quad \mathbb{R}^n ,$$

with the result that $u_\varepsilon \in B^1(\mathbb{R}^n)$ and

$$\mathscr{I}(u_\varepsilon) \leqslant (1 + 2\varepsilon)^2 \mathscr{I}(u_0) .$$

Now $0 \leqslant u_\varepsilon \leqslant 1$ and

$$u_\varepsilon = 1 \quad \text{on} \quad U_\varepsilon = \left\{u_0 \geqslant \frac{1+\varepsilon}{1+2\varepsilon}\right\}$$

$$u_\varepsilon = 0 \text{ on } \quad V_\varepsilon = \left\{u_0 \leqslant \frac{\varepsilon}{1+2\varepsilon}\right\}.$$

From the hypothesis on u_0, there are $r > 0$, $R > 0$ such that

$$u_\varepsilon = 1 \quad \text{on} \quad K + \bar{B}(0, r)$$

$$u_\varepsilon = 0 \quad \text{on} \quad (\mathbb{R}^n \backslash (\Omega \cap B(0, R))) + \bar{B}(0, r) .$$

Let us take $\rho \in \mathscr{D}(B(0, r))$ with $\rho \geqslant 0$, $\displaystyle\int_{\mathbb{R}^n} \rho \, dx = 1$.

We have: $\rho * u_\varepsilon \in \mathscr{D}(K, \Omega)$ and

$$\mathscr{I}(\rho * u_\varepsilon) = \|\rho * \operatorname{grad} u_\varepsilon\|^2_{L^2} \leqslant \mathscr{I}(u_\varepsilon) .$$

Therefore

$$\inf_{\mathscr{D}(K, \Omega)} \mathscr{I}(u) \leqslant (1 + 2\varepsilon)^2 \mathscr{I}(u_0) .$$

This being true for all $\varepsilon > 0$, and for all $u_0 \in \mathscr{S}(K, \Omega)$, we obtain point (1). □

To continue the proof of Proposition 8, we shall use the property of the continuity of $\text{cap}_\Omega K$ and $u_{K,\Omega}$ with respect to K and Ω. We group together different properties of $\text{cap}_\Omega K$ and $u_{K,\Omega}$ in the

Proposition 9. *We have the following properties of* $\text{cap}_\Omega K$ and $u_{K,\Omega}$:
(1) $\text{cap}_\Omega K$ *is increasing with respect to* K *and decreasing with respect to* Ω :

$$K_1 \subset K_2 \subset \Omega_2 \subset \Omega_1 \Rightarrow \text{cap}_\Omega K_1 \leqslant \text{cap}_{\Omega_2} K_2 ;$$

(2) $u_{K,\Omega}$ *is increasing with respect to* K *and* Ω:

$$K_1 \subset \Omega_1 \subset \Omega_2, K_1 \subset K_2 \subset \Omega_2 \Rightarrow u_{K_1,\Omega_1} \leqslant u_{K_2,\Omega_2} .$$

(3) *For every increasing sequence of open sets* Ω_k *with union* Ω *and every decreasing sequence of compact sets* $K_k \subset \Omega_k$ *with intersection* K

$$\text{cap}_\Omega K = \lim \text{cap}_{\Omega_k} K_k$$

and $\lim u_{K_k,\Omega_k}(x) = u_{K,\Omega}(x)$, *for all* $x \in \mathbb{R}^n$ *with the reservation that if* $\Omega_k \neq \Omega$ *that* $u_{K,\Omega}$ *be the solution of* (5.59).

(4) *The capacity is strongly sub-additive with respect to K and Ω*

$$\mathrm{cap}_{\Omega_1 \cup \Omega_2} K_1 \cup K_2 + \mathrm{cap}_{\Omega_1 \cap \Omega_2} K_1 \cap K_2 \leqslant \mathrm{cap}_{\Omega_1} K_1 + \mathrm{cap}_{\Omega_2} K_2 .$$

(5) *For every family of open sets Ω_i, two by two disjoint, with union Ω and every compact set K in Ω.*

$$\mathrm{cap}_{\Omega} K = \sum \mathrm{cap}_{\Omega_i}(K \cap \Omega_i)$$

$$u_{K,\Omega} = u_{K \cap \Omega_i, \Omega_i} \quad on \quad \Omega_i \text{ for all } i.$$

(6) *Given a diffeomorphism h of class $\mathscr{C}^1$ of an open set U of $\mathbb{R}^n$ into $\mathbb{R}^n$ and Ω a relatively compact open set of U, there exists c such that*

$$\mathrm{cap}_{\Omega} K \leqslant c\ \mathrm{cap}_{h(\Omega)} h(K) .$$

for every compact set K in Ω.

Proof of Proposition 9. The point (1) follows from Definition 1, since

$$K_1 \subset K_2 \subset \Omega_2 \subset \Omega_1 \Rightarrow \mathscr{S}(K_2, \Omega_2) \subset \mathscr{S}(K_1, \Omega_1)$$

The point (2) follows from Definition 2, since under the hypothesis

$$K_1 \subset \Omega_1 \subset \Omega_2 \quad \text{and} \quad K_1 \subset K_2 \subset \Omega_1 ,$$

for all $w \in \mathscr{S}^+(K_2, \Omega_2)$ the restriction of w to Ω_1 is in $\mathscr{S}^+(K_1, \Omega_1)$.
For the point (3), with its notation, we easily see that $\mathrm{cap}_{\Omega} K = \lim \mathrm{cap}_{\Omega_k} K_k$ by using point (1) of Proposition 8 (proved above) and the fact that the sequence of sets $\mathscr{D}(K_k, \Omega_k)$ is increasing with union $\mathscr{D}(K, \Omega)$; besides we note that $\mathrm{cap}_{\Omega k} K_k$ converges by decreasing to $\mathrm{cap}_{\Omega} K$. The situation is not so simple for the sequence of functions $u_k = u_{K_k, \Omega_k}$.
We note first that $\limsup u_k(x) \leqslant u_{K,\Omega}(x)$ for all x. It is trivial for $x \in \mathbb{R}^n \backslash \Omega$; for $x \in \Omega$ and $\varepsilon > 0$ from the definition there exists $w \in \mathscr{S}^+(K, \Omega)$ such that $w(x) \leqslant u_{K,\Omega}(x) + \varepsilon$; since w is continuous on Ω and $\geqslant 1$ on K, $\{w \geqslant (1 + \varepsilon)^{-1}\}$ is a neighbourhood of K and hence contains K_k for k sufficiently large; in other words $(1 + \varepsilon)w \in \mathscr{S}^+(K_k, \Omega)$ and

$$u_k(x) \leqslant u_{K_k,\Omega}(x) \leqslant (1 + \varepsilon)w(x) \leqslant (1 + \varepsilon)(u_{K,\Omega}(x) + \varepsilon)$$

for k sufficiently large, which proves the statement.
When $\Omega_k = \Omega$, $u_k \geqslant u_{K,\Omega}$ and hence the sequence $u_k(x)$ converges to $u_{K,\Omega}(x)$ for all x (decreasing moreover). Now if $u_{K,\Omega}$ is the solution of (5.59), it is also defined on $\Omega \backslash K$ as the upper envelope of the set $\mathscr{I}_-(\Omega \backslash K, \chi_K)$ of continuous functions v, subharmonic on $\Omega \backslash K$ satisfying the inequality

$$\limsup_{x \to z} v(x) \leqslant \chi_K(z) \quad \text{for all} \quad z \in \partial(\Omega \backslash K) ,$$

and in the case of Ω unbounded, there exists $h \in \mathscr{H}(\Omega \backslash K) \cap \mathscr{C}^0(\overline{\Omega \backslash K})$ with $h(x) > 0$ for $|x|$ large so that $\displaystyle\limsup_{|x| \to \infty} \frac{v(x)}{h(x)} \leqslant 0.$

For such a function v and $\varepsilon > 0$, we put $v_\varepsilon = \max(0, v - \varepsilon h_0)$ where $h_0 = 1$ in the case of Ω bounded; in the case of Ω unbounded $h_0 = h + 1 - \min h$, h being the function defining the null condition at infinity. We verify without difficulty that v_ε is continuous, subharmonic on $\Omega \backslash K$ with compact support contained in Ω and $\limsup_{x \to z} v_\varepsilon(x) \leqslant 1$ for all $z \in \partial K$, with the result that for k sufficiently large $v_\varepsilon(x) \leqslant w$ on $\Omega_k \backslash K_k$ for all $w \in \mathscr{S}^+(K_k, \Omega_k)$ and hence $v_\varepsilon \leqslant u_k$ on $\Omega_k \backslash K_k$. We deduce that $v_\varepsilon(x) \leqslant \liminf u_k(x)$ for all $x \in \Omega \backslash K$. Letting $\varepsilon \to 0$ and taking the upper envelope of the v, we have

$$u_{K,\Omega}(x) \leqslant \liminf u_k(x) \quad \text{for all} \quad x \in \Omega \backslash K\,,$$

which completes the proof of the point (3).
For the point (4) we consider $u_i \in \mathscr{S}(K_i, \Omega_i)$, $i = 1, 2$ and put

$$v = \max(u_1, u_2) = u_1 + (u_2 - u_1)^+, w = \min(u_1, u_2) = u_2 - (u_2 - u_1)^+ \,.$$

We have

$$v, w \in \mathscr{C}^0(\mathbb{R}^n), \quad \lim_{|x| \to \infty} v(x) = \lim_{|x| \to \infty} w(x) = 0\,,$$

$$v \geqslant 1 \quad \text{on} \quad K_1 \cup K_2\,, \qquad v \leqslant 0 \quad \text{on} \quad (\mathbb{R}^n \backslash \Omega_1) \cap (\mathbb{R}^n \backslash \Omega_2) = \mathbb{R}^n \backslash (\Omega_1 \cup \Omega_2)$$

$$w \geqslant 1 \quad \text{on} \quad K_1 \cap K_2 \quad \text{and} \quad w \leqslant 0 \quad \text{on} \quad \mathbb{R}^n \backslash (\Omega_1 \cap \Omega_2)\,.$$

On the other hand, from Lemma 2, the derivatives $\dfrac{\partial v}{\partial x_i}, \dfrac{\partial w}{\partial x_i} \in L^1_{\text{loc}}(\mathbb{R}^n)$ and

$$\frac{\partial v}{\partial x_i} = \frac{\partial u_1}{\partial x_i}(1 - \chi_{\{u_2 > u_1\}}) + \frac{\partial u_2}{\partial x_i} \chi_{\{u_1 > u_2\}} \quad \text{a.e on } \mathbb{R}^n$$

$$\frac{\partial w}{\partial x_i} = \frac{\partial u_1}{\partial x_i} \chi_{\{u_2 > u_1\}} + \frac{\partial u_2}{\partial x_i}(1 - \chi_{\{u_2 > u_1\}})\,;$$

from which we have

$v \in \mathscr{S}(K_1 \cup K_2, \Omega_1 \cup \Omega_2)$, $\quad w \in \mathscr{S}(K_1 \cap K_2, \Omega_1 \cap \Omega_2)$ and noting that

$$1 - \chi_{\{u_2 > u_1\}} = \chi_{\{u_2 \leqslant u_1\}}\,,$$

we have

$$|\text{grad}\, v|^2 + |\text{grad}\, w|^2 = |\text{grad}\, u_1|^2 + |\text{grad}\, u_2|^2\,. \tag{5.62}$$

Therefore

$$\text{cap}_{\Omega_1 \cup \Omega_2} K_1 \cup K_1 + \text{cap}_{\Omega_1 \cap \Omega_1} K_1 \cap K_2 \leqslant \mathscr{I}(v) + \mathscr{I}(w) = \mathscr{I}(u_1) + \mathscr{I}(u_2)\,.$$

Minimizing on $u_i \in \mathscr{S}(K_i, \Omega_i)$, we establish the point (4).
The point (5) is immediate and for the point (6) it is sufficient to note that for $u \in \mathscr{D}(h(K), h(\Omega))$ the function v defined on $\mathbb{R}^n$ by

$$v(x) = u(h(x)) \quad \text{for} \quad x \in \Omega\,, \qquad v(x) = 0 \quad \text{for} \quad x \in \mathbb{R}^n \backslash \Omega$$

is of class $\mathscr{C}^1$, $v \geqslant 1$ on K and

$$|\text{grad}\, v(x)| \leqslant |\text{grad}\, u(h(x))|\,|h'(x)|$$

with the result that

$$\operatorname{cap}_\Omega K \leqslant \mathscr{I}(v) \leqslant \int |\operatorname{grad} u(h(x))|^2 |h'(x)|^2 \, dx \leqslant c\mathscr{I}(u)$$

with
$$c = \max_{\bar{\Omega}} \frac{|h'(x)|^2}{|\det h'(x)|} . \qquad \square$$

We can now give the

Proof of Proposition 8.(2) and (3). Suppose, in the first instance, that $u_{K,\Omega}$ is the solution of (5.59) and take an increasing sequence of regular bounded open sets Ω_k containing K, with union Ω, and a decreasing sequence of compact sets $K_k \subset \Omega_k$ with intersection $\overline{K}$[119]. From the hypothesis of regularity, the functions $u_k = u_{K_k, \Omega_k} \in \mathscr{C}^1_n(\Omega_k \backslash K_k)$ and from Proposition 7

$$\operatorname{cap}_{\Omega_k} K_k = \mathscr{I}(u_k) = -\int_{\partial K_k} \frac{\partial u_k}{\partial n} \, d\gamma .$$

From the Proposition 9.(3)

$$\operatorname{cap}_{\Omega_k} K_k \to \operatorname{cap}_\Omega K \quad \text{and} \quad u_k(x) \to u_{K,\Omega}(x) \quad \text{for all} \quad x \in \mathbb{R}^n .$$

We deduce[120] that $u_{K,\Omega} \in B^1(\mathbb{R}^n)$ and

$$\mathscr{I}(u_{K,\Omega}) \leqslant \liminf \mathscr{I}(u_k) = \operatorname{cap}_\Omega K . \tag{5.63}$$

Also for $\zeta \in \mathscr{D}(\Omega)$, $\zeta \geqslant 0$, since $\partial u_k / \partial n \leqslant 0$ on ∂K_k

$$\langle - \Delta u_{K,\Omega}, \zeta \rangle = \lim \langle - \Delta u_k, \zeta \rangle = \lim - \int_{\partial K_k} \frac{\partial u_k}{\partial n} \zeta \, d\gamma \geqslant 0$$

i.e. the distribution $u_{K,\Omega}$ is superharmonic on Ω. For $\zeta \in \mathscr{D}(K, \Omega)$, $\zeta = 1$ on ∂K_k for k sufficiently large, and hence

$$\int_{\partial K} d(-\Delta u_{K,\Omega}) = \langle - \Delta u_{K,\Omega}, \zeta \rangle = \lim - \int_{\partial K_k} \frac{\partial u_k}{\partial n} \, d\gamma = \operatorname{cap}_\Omega K .$$

Also, always for $\zeta \in \mathscr{D}(K, \Omega)$,

$$\langle - \Delta u_{K,\Omega}, \zeta \rangle = \int \operatorname{grad} u_{K,\Omega} \operatorname{grad} \zeta \, dx \leqslant \| \operatorname{grad} u_{K,\Omega} \|_{L^2} \| \operatorname{grad} \zeta \|_{L^2}$$

---

[119] Such sequences exist (see Lemma 1 of §4).

[120] For $\zeta \in \mathscr{D}(\mathbb{R}^n)^n$,

$$\langle \operatorname{grad} u_{K,\Omega}, \zeta \rangle = \int u_{K,\Omega} \operatorname{div} \zeta \, dx = \lim \int u_k \operatorname{div} \zeta \, dx = \lim \int \operatorname{grad} u_k . \zeta \, dx$$
$$\leqslant \liminf \| \operatorname{grad} u_k \|_{L^2} \| \zeta \|_{L^2} = (\operatorname{cap}_\Omega K)^{1/2} \| \zeta \|_{L^2} .$$

Hence $\operatorname{grad} u_{K,\Omega} \in L^2(\mathbb{R}^n)^n$ and $\| \operatorname{grad} u_{K,\Omega} \|_{L^2} \leqslant (\operatorname{cap}_\Omega K)^{1/2}$.

and therefore

$$\operatorname{cap}_\Omega K = \langle -\Delta u_{K,\Omega}, \zeta\rangle \leqslant \mathscr{I}(u_{K,\Omega})^{1/2}\mathscr{I}(\zeta)^{1/2}.$$

Minimizing on $\zeta \in \mathscr{D}(K, \Omega)$, we have, from the point (1)

$$\operatorname{cap}_\Omega K \leqslant \mathscr{I}(u_{K,\Omega})^{1/2}(\operatorname{cap}_\Omega K)^{1/2}$$

from which we deduce that $\operatorname{cap}_\Omega K = \mathscr{I}(u_{K,\Omega})$ with (5.63).
In the case where $u_{K,\Omega}$ is not a solution of (5.59), Ω is necessarily unbounded. We approach K by a decreasing sequence of compact sets K_k with non-empty interior; from Proposition 11 of §4, there exists a generalized solution of $P(\Omega\backslash K_k, \chi_{K_k})$ satisfying the null condition at infinity. We can therefore apply the results of the proposition to $u_{K_k,\Omega}$. Since, from Proposition 9(3), $u_{K_k,\Omega}(x) \to u_{K,\Omega}(x)$ we can pass to the limit as in the preceding case. □

Remark 3. As the limit of subharmonic functions (or distributions) in the neighbourhood of each point of $\mathbb{R}^n\backslash K$, the distribution $u_{K,\Omega}$ is sub-harmonic on $\mathbb{R}^n\backslash K$, $\Delta u_{K,\Omega}$ is a positive Radon measure on $\mathbb{R}^n\backslash K$ carried by $\partial\Omega$. *In the case of Ω bounded*, choosing $\zeta \in \mathscr{D}(\mathbb{R}^n)$ with $\zeta = 1$ in the neighbourhood of $\bar{\Omega}$, we have

$$\int_{\partial\Omega} \mathrm{d}(\Delta u_{K,\Omega}) + \int_{\partial K} \mathrm{d}(\Delta u_{K,\Omega}) = \langle \Delta u_{K,\Omega}, \zeta\rangle = \int u_{K,\Omega}\Delta\zeta\,\mathrm{d}x = 0\,.$$

Hence, we have the equality which makes (5.52) precise

$$\operatorname{cap}_\Omega K = \int_{\partial\Omega} \mathrm{d}(\Delta u_{K,\Omega})\,. \tag{5.64}$$

In the case of Ω unbounded, this equality is again *true when $n = 1$ or 2*. In effect, considering $\rho \in \mathscr{D}(\bar{B}(0,1), B(0,2))$, the function $\rho_k(x) = \rho(x/k) \to 1$ and therefore by Lebesgue's theorem

$$\int_{\partial\Omega} \mathrm{d}(\Delta u_{K,\Omega}) - \operatorname{cap}_\Omega K = \lim \langle \Delta u_{K,\Omega}, \rho_k\rangle = -\lim \int_{\mathbb{R}^n} \operatorname{grad} u_{K,\Omega}, \operatorname{grad} \rho_k \mathrm{d}x\,.$$

By Schwarz's inequality,

$$\left|\int_{\mathbb{R}^n} \operatorname{grad} u_{K,\Omega}.\operatorname{grad}\rho_k \mathrm{d}x\right| \leqslant \left(\int_{\{k<|x|<2k\}} |\operatorname{grad} u_{K,\Omega}|^2 \mathrm{d}x\right)^{1/2} \mathscr{I}(\rho_k)^{1/2}\,.$$

We have $\mathscr{I}(\rho_k) = k^{n-2}\mathscr{I}(\rho)$ and by Lebesgue

$$\int_{\{k<|x|<2k\}} |\operatorname{grad} u_{K,\Omega}|^2 \mathrm{d}x \to 0\,.$$

We deduce (5.64) in the case $n=1$ or 2 where $\mathscr{I}(\rho_k)$ is bounded; in the case $n \geqslant 3$, $\mathscr{I}(\rho_k) \to \infty$ and the argument falls. In fact *the equality* (5.64) *is not in general true in the case $n \geqslant 3$*. Indeed, considering $\Omega = \mathbb{R}^n$, $\partial\Omega = \varnothing$ and so (5.64) can be written $\operatorname{cap}_{\mathbb{R}^n} K = 0$; now, as we shall see in the example below, when $n \geqslant 3$, the capacity of a compact set in $\mathbb{R}^n$ is often non-zero. □

Remark 4. As the lower envelope of continuous superharmonic functions, *the function $u_{K,\Omega}$ is an u.s.c. map, superharmonic on Ω* (see Remark 8 of §4). Now the distribution $u_{K,\Omega}$ being superharmonic on Ω, defines a unique *l.s.c. map $\tilde{u}_{K,\Omega}$, superharmonic on Ω*, given by

$$\tilde{u}_{K,\Omega}(x) = \lim_{r\to 0} \frac{n}{\sigma_n r^n} \int_{B(x,r)} u_{K,\Omega}(t)\,dt\,. \tag{5.65}$$

We have obviously, $\tilde{u}_{K,\Omega} = u_{K,\Omega}$ on $\Omega \backslash \partial K$; but in general, the set

$$\tilde{K} = \{x \in K;\, \tilde{u}_{K,\Omega}(x) = 1\} \tag{5.66}$$

is strictly contained in K.
We note that the set $\tilde{K}$ is the set of the $x \in K$ points of continuity of $u_{K,\Omega}$. In particular, in addition to the interior of K, $\tilde{K}$ contains all the $z \in \partial K$, regular points of the boundary of $\Omega \backslash K$ (in the sense of Definition 4 of §4).
We note also that $\tilde{K}$ is not in general is compact set; in the following sub-section, we shall define the capacity of an arbitrary part in an open set in $\mathbb{R}^n$, and, with the help of Proposition 12, we can see that $\operatorname{cap}_\Omega K = \operatorname{cap}_\Omega \tilde{K}$: *the capacity of K depends only on the part $\tilde{K}$ of K.* □

Remark 5. Given K a compact set in an open set Ω in $\mathbb{R}^n$

$$\operatorname{cap}_\Omega K = \operatorname{cap}_\Omega \partial K \qquad u_{K,\Omega} = u_{\partial K,\Omega}\,. \tag{5.67}$$

In effect if w is continuous, superharmonic on $\Omega \supset K$,

$$w \geqslant \min_{\partial K} w \quad \text{on} \quad K \text{ therefore } \mathscr{S}^+(K,\Omega) = \mathscr{S}^+(\partial K,\Omega)\,,$$

$u_{K,\Omega} = u_{\partial K,\Omega}$ and from Proposition 8, $\operatorname{cap}_\Omega K = \operatorname{cap}_\Omega \partial K$. With the notation of Remark 4, we obviously have also $\tilde{u}_{K,\Omega} = \tilde{u}_{\partial K,\Omega}$ and therefore

$$\widetilde{\partial K} = \tilde{K} \cap \partial K\,. \tag{5.68}$$

□

Finally let us give some examples.

Example 9. *Capacity of a compact set in an open set of $\mathbb{R}$.* From Proposition 9 (5), we can restrict ourselves to calculating the capacity of a compact set K in an open interval $\Omega =]a, b[$ of $\mathbb{R}$.
Putting $a_1 = \min K$, $b_1 = \max K$ we have

$$\operatorname{cap}_\Omega K \leqslant \operatorname{cap}_\Omega [a_1, b_1] = \operatorname{cap}_\Omega \{a_1, b_1\} \leqslant \operatorname{cap}_\Omega K\,,$$

the first and last inequalities resulting from the increase of the capacity with respect to K (Proposition 9.(1)) and the equality in the middle from Remark 5.
From Example 7, we thus have for every compact set $K \subset]a, b[$

$$\operatorname{cap}_{]a,b[} K = \operatorname{cap}_{]a,b[} [\min K, \max K] = \frac{1}{\min K - a} + \frac{1}{b - \max K}\,. \tag{5.69}$$

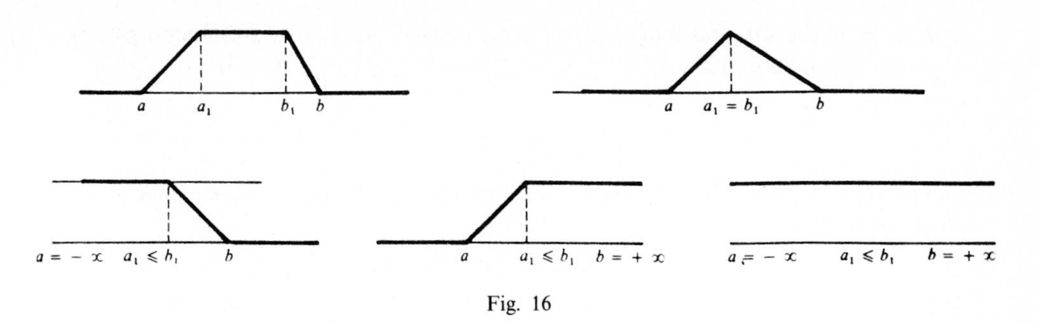

Fig. 16

In fact, this equality has been proved for

$$-\infty < a < \min K < \max K < b < \infty\,,$$

but from Proposition 9(3) it is still true in the general case.
Similarly $u_{K,\Omega} = u_{[a_1,b_1],\Omega}$ whose graph is given in the different cases in Fig. 16. Finally, we note (see Remark 4) that $\tilde{u}_{K,\Omega} = u_{K,\Omega}$ and $\tilde{K} = K$. □

Example 10. Some compact sets of zero or non-zero capacity.
(*a*) *In the case $n = 1$, from Example 9 above*

$$\operatorname{cap}_\Omega K = 0 \quad \textit{iff} \quad \Omega = \mathbb{R} \quad \textit{or} \quad K = \varnothing\,.$$

(*b*) *In the case $n = 2$, it follows from Remark 3 above that*

$$\operatorname{cap}_{\mathbb{R}^2} K = 0 \quad \textit{for every compact set } K \textit{ of } \mathbb{R}^2\,.$$ [121]

This follows also because of Example 8 and of Proposition 9, on putting $r = \max_K |x|$, with

$$\operatorname{cap}_{\mathbb{R}^2} K \leqslant \operatorname{cap}_{\mathbb{R}^2} \bar{B}(0, r) = \lim_{r_0 \to \infty} \operatorname{cap}_{B(0,r_0)} \bar{B}(0, r) = \lim_{r_0 \to \infty} \frac{2\pi}{\operatorname{Log}\dfrac{r_0}{r}} = 0\,.$$

(*c*) *In the case $n \geqslant 3$*, for every point x_0 of an open set Ω of $\mathbb{R}^n$, $\operatorname{cap}_\Omega\{x_0\} = 0$. In effect, considering $r_0 > 0$ with $B(x_0, r_0) \subset \Omega$,

$$\operatorname{cap}_\Omega\{x_0\} \leqslant \operatorname{cap}_{B(x_0,r_0)}\{x_0\} = \lim_{r \to 0} \operatorname{cap}_{B(x_0,r_0)} \bar{B}(x_0, r)$$

$$= \lim_{r \to 0} \frac{1}{E_n(r_0) - E_n(r)} = 0\,.$$

More generally, *given V a regular manifold of dimension $n - 2$ in an open set Ω of $\mathbb{R}^n$ and $K \subset V$, then* $\operatorname{cap}_\Omega K = 0$.

---

[121] In $\mathbb{R}$ and $\mathbb{R}^2$, we must not confuse the sets of zero capacity with the polar sets for the H^1 norm (for definition see later): because a compact set has zero capacity in $\mathbb{R}$ and $\mathbb{R}^2$, even though it is **not** obviously polar in general.

Indeed, from the sub-additivity, we can suppose that there exists a diffeomorphism h of an open neighbourhood U of K in Ω on $\mathbb{R}^n = \{(x', x''); x' \in \mathbb{R}^{n-2}, x'' \in \mathbb{R}^2\}$ such that

$$h(U \cap V) = \{(x', 0); x' \in \mathbb{R}^{n-2}\}.$$

Considering a bounded open set Ω_0 containing K with $\bar{\Omega}_0 \subset U$, we have from Proposition 9(6)

$$\operatorname{cap}_{\Omega} K \leqslant \operatorname{cap}_{\Omega_0} K \leqslant c \operatorname{cap}_{h(\Omega_0)} h(K).$$

We are thus led to proving the result in the case $\Omega = \Omega' \times \Omega''$, $K = K' \times \{0\}$, where K' is a compact set of an open set Ω' of $\mathbb{R}^{n-2}$ and Ω'' is an open neighbourhood of $\{0\}$ in $\mathbb{R}^2$. Let us fix $u' \in \mathscr{D}(K', \Omega')$ and for $0 < \varepsilon < \operatorname{dist}(0, \partial\Omega'')$, take

$$u'' \in \mathscr{D}(\{0\}, B(0, \varepsilon)) \quad \text{such that} \quad \int_{\mathbb{R}^2} |\operatorname{grad} u''(x'')|^2 \mathrm{d}x'' < \varepsilon ;$$

such a u'' exists since $\operatorname{cap}_{B(0,\varepsilon)}\{0\}$. Putting $u(x', x'') = u'(x')u''(x'')$, we have $u \in \mathscr{D}(K, \Omega)$ and

$$\begin{aligned} \mathscr{I}(u) &= \left(\int_{\mathbb{R}^{n-2}} |\operatorname{grad} u'(x')|^2 \mathrm{d}x'\right)\left(\int_{\mathbb{R}^2} u''(x'')^2 \mathrm{d}x''\right) \\ &\quad + \left(\left(\int_{\mathbb{R}^{n-2}} u'(x')^2 \mathrm{d}x'\right)\left(\int_{\mathbb{R}^2} |\operatorname{grad} u''(x'')|^2 \mathrm{d}x''\right)\right) \\ &\leqslant \pi\varepsilon^2 \int_{\mathbb{R}^{n-2}} |\operatorname{grad} u'(x')|^2 \mathrm{d}x' + \varepsilon \int_{\mathbb{R}^{n-2}} u'(x')^2 \mathrm{d}x' \end{aligned}$$

Letting $\varepsilon \to 0$, we have $\operatorname{cap}_{\Omega} K = 0$. □

(*d*) *On the contrary, let V be a regular hypersurface (manifold of dimension $n - 1$) K a compact set containing V and Ω a bounded open set containing K. Then* $\operatorname{cap}_{\Omega} K > 0$. Transforming a piece of V by a diffeomorphism, we can always suppose that K contains the hemisphere

$$\Sigma_+ = \{x = (x', x_n); x_n = (1 - |x'|^2)^{\frac{1}{2}}, |x'| < 1\}$$

and hence that $K \cup (-K)$ contains the unit sphere Σ. Considering $r_0 = \sup_{\bar{\Omega}} |x|$, we have

$$\begin{aligned} 0 < \operatorname{cap}_{B(0,r_0)} \bar{B}(0, 1) &= \operatorname{cap}_{B(0,r_0)} \Sigma \leqslant \operatorname{cap}_{\Omega \cup (-\Omega)} K \cup (-K) \\ &\leqslant \operatorname{cap}_{\Omega} K + \operatorname{cap}_{(-\Omega)}(-K) = 2 \operatorname{cap}_{\Omega} K, \end{aligned}$$

where, in order, we have used the relation (5.55), Remark 5, Proposition 9, (1) then (4) and the obvious fact that

$$\operatorname{cap}_{(-\Omega)}(-K) = \operatorname{cap}_{\Omega} K. \qquad \square$$

(*e*) *In the case $n \geqslant 3$, we have the result of (d) for an unbounded open set: for every compact set K of $\mathbb{R}^n$ containing a regular hypersurface,* $\operatorname{cap}_{\mathbb{R}^n} K > 0$. In effect, with the same reasoning, we are led to $\mathrm{cap}_{\mathbb{R}^n}\bar{B}(0, 1) > 0$; now from (5.55)

$$\mathrm{cap}_{\mathbb{R}^n}\bar{B}(x_0, r) = \lim_{r_0 \to \infty} \mathrm{cap}_{B(x_0, r_0)}\bar{B}(x_0, r)$$

$$= \lim_{r_0 \to \infty} \frac{1}{E_n(r_0) - E_n(r)} = (n - 2)\sigma_n r^{n-2} . \quad \square$$

3b. Capacity of an Arbitrary Part. Applications

Given an open set Ω of $\mathbb{R}^n$, we have defined $\mathrm{cap}_\Omega K$ for every compact set K of Ω; this set function $K \to \mathrm{cap}_\Omega K$ presents some *some similarities with a measure*: it is increasing, continuous for decreasing sequences of compact sets, sub-additive (see Proposition 9). But note clearly that *this is* **not** *a measure*: *it is not additive* (that appears clearly in Remark 5 and in Examples 9 and 10).
We propose, given a part A of an open set Ω of $\mathbb{R}^n$, to define $\mathrm{cap}_\Omega A$ as the infimum of the energy functional $\mathscr{I}(u)$ on a family $\mathscr{F}(A, \Omega)$ of elements $u \in B^1(\mathbb{R}^n)$ satisfying "in a certain sense"

$$u \leqslant 0 \quad \text{on} \quad \mathbb{R}^n\backslash\Omega , \qquad u \geqslant 1 \quad \text{on} \quad A .$$

We obviously want this definition to generalize that given when A is compact; but we also want the set function $A \in \mathscr{P}(\Omega) \to \mathrm{cap}_\Omega A$ to be an *exterior measure* (in the sense of measure theory), that is, satisfies the following condition:

$$\text{(5.70)} \quad \begin{cases} \text{for every increasing sequence of parts } A_k \text{ of } \Omega \text{ with union } A, \\ \text{the sequence } \mathrm{cap}_\Omega A_k \text{ is increasing and converges to } \mathrm{cap}_\Omega A. \end{cases}$$

We note that *this condition imposes the value* $\mathrm{cap}_\Omega U$ *for every open set* U *of* Ω or more generally for each part U, the denumerable union of compact sets[122] of Ω: we must necessarily have

$$\text{(5.71)} \qquad \mathrm{cap}_\Omega U = \sup_{K \subset U} \mathrm{cap}_\Omega K \quad (K \text{ compact}) .$$

We note also that in the case $n \geqslant 2$, this condition imposes that *we cannot, in general, define* $\mathrm{cap}_\Omega A$ as $\inf_{\mathscr{S}(A, \Omega)} \mathscr{I}(u)$, where

$$\mathscr{S}(A, \Omega) = \{u \in \mathscr{C}^0(\mathbb{R}^n) \cap B^1(\mathbb{R}^n); u \leqslant 0 \quad \text{on} \quad \mathbb{R}^n\backslash\Omega ,$$

$$\limsup_{|x| \to \infty} u(x) \leqslant 0 \text{ and } u \geqslant 1 \text{ on } A\} .$$

It is clear, in effect, that $\mathscr{S}(A, \Omega) = \mathscr{S}(\bar{A}, \Omega)$; given a compact set K of Ω and a sequence (x_k) dense in K, the sequence of compact sets $K_k = \{x_1, \ldots, x_k\}$ is increasing with union A dense in K; the sequence $\mathrm{cap}_\Omega K_k = 0$ does not converge, in general, to

$$\inf_{\mathscr{S}(A, \Omega)} \mathscr{I}(u) = \inf_{\mathscr{S}(K, \Omega)} \mathscr{I}(u) = \mathrm{cap}_\Omega K .$$

---

[122] We say then that U is a K_σ.

We must therefore to define $\text{cap}_\Omega K$, at least in the case $n \geqslant 1$, go out of the space of continuous functions. The problem is then posed with the meaning:
$u \leqslant 0$ on $\mathbb{R}^n \backslash \Omega$ and $u \geqslant 1$ on A. In effect, *a priori*, $u \in B^1(\mathbb{R}^n)$ is defined almost everywhere on $\mathbb{R}^n$; on the other hand the *inequality* $u \leqslant 0$ *a.e. on* $\mathbb{R}^n \backslash \Omega$ *or* $u \geqslant 1$ *a.e. on* A *is not, in general, sufficient.* We see this easily in the case $n = 1$:
(1) $u \equiv 0$ satisfies $u \geqslant 1$ a.e. on $\{0\}$ since the point 0 is negligible and $\mathscr{I}(u) = 0 < \text{cap}_\Omega\{0\}$ for every open set $\Omega \neq \mathbb{R}$, $0 \in \Omega$.
(2) Taking $\Omega =]-2, 2[\backslash\{0\}$ and $K = \{-1, +1\}$, let us consider u whose graph is shown in Fig. 17. We have $u \leqslant 0$ a.e. on $\mathbb{R}\backslash\Omega$ since 0 is negligible and

$$\mathscr{I}(u) = 2 < \text{cap}_\Omega K = \text{cap}_{]-2,0[}\{-1\} + \text{cap}_{]0,2[}\{1\} = 4\,.$$

In order to specify the family $\mathscr{F}(A, \Omega)$ on which we shall minimize the energy functional we must introduce a definition of a limit in the space $B^1(\mathbb{R}^n)$. We shall say that a sequence (u_k) converges to u in $B^1(\mathbb{R}^n)$ if

$$(5.72)\qquad \begin{cases} \lim \displaystyle\int_{B(0,R)} |u_k - u|^2 \mathrm{d}x = 0 \quad \text{for all} \quad R > 0 \quad \text{and} \\ \lim \displaystyle\int_{\mathbb{R}^n} |\text{grad}\, u_k - \text{grad}\, u|^2 \mathrm{d}x = 0\,. \end{cases}$$

It is clear that

$$u_k \to u \quad \text{in} \quad B^1(\mathbb{R}^n) \Rightarrow \mathscr{I}(u_k) \to \mathscr{I}(u)\text{[123]}.$$

Therefore for every family $\mathscr{F} \subset B^1(\mathbb{R}^n)$,

$$(5.73)\qquad \lim_{\mathscr{F}} \mathscr{I}(u) = \lim_{\bar{\mathscr{F}}} \mathscr{I}(u)$$

where $\bar{\mathscr{F}}$ is the closure of $\mathscr{F}$ in $B^1(\mathbb{R}^n)$, that is, the set of limit points in $B^1(\mathbb{R}^n)$ of sequences of elements of $\mathscr{F}$.
On the other hand, let us recall Lemma 2, which we shall complete by the

Lemma 3. *For all* $u \in B^1(\mathbb{R}^n)$, $u^+ = \max(u, 0) \in B^1(\mathbb{R}^n)$ *and the map* $u \to u^+$ *of* $B^1(\mathbb{R}^n)$ *into itself*[124] *is continuous.*

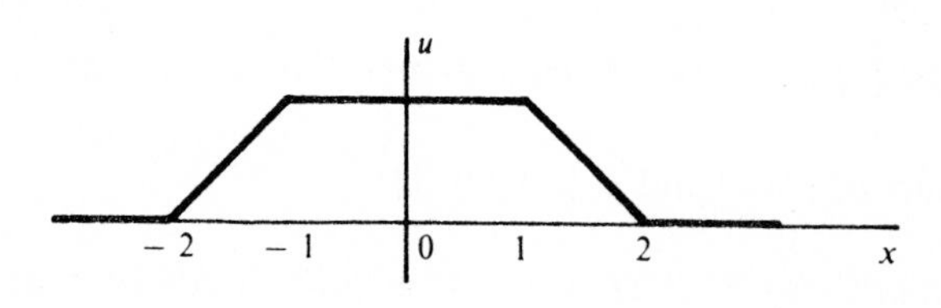

Fig. 17

[123] The converse is true with the reservation that there be quite a small measure of convergence of u_k to u. To the precise: we can show that if there exists a non-empty open set ω such that $u_k \to u$ in $\mathscr{D}'(\omega)$ then $\mathscr{I}(u_k) \to \mathscr{I}(u) \Rightarrow u_k \to u$ in $B^1(\mathbb{R}^n)$.
[124] We note that this is not a linear map.

We shall prove this lemma with Lemma 2 at the end of this section. In order to express the condition $u \leqslant 0$ on $\mathbb{R}^n \backslash \Omega$, we now introduce the *Beppo–Levi space* $B_0^1(\Omega)$, as the closure of $\mathscr{D}(\Omega)$ in $B^1(\mathbb{R}^n)$. *In the case when Ω is bounded* from Poincaré's inequality[125], we have

$$\|u\|_{L^2} \leqslant C(\Omega)\|\operatorname{grad} u\|_{L^2} \quad \text{for all} \quad u \in \mathscr{D}(\Omega)$$

the space $B_0^1(\Omega)$ coincides with the Sobolev space $H_0^1(\Omega)$, the closure of $\mathscr{D}(\Omega)$ in $H^1(\mathbb{R}^n)$[126]; but the situation is not the same in the case in which Ω is unbounded[127].

We now give the

Definition 3. Let Ω be an open set of $\mathbb{R}^n$, and A an arbitrary part of Ω.
(1) We denote by $\mathscr{F}(A, \Omega)$ the set of $u \in B^1(\mathbb{R}^n)$ satisfying:

$$u^+ \in B_0^1(\Omega) \quad \text{and} \quad u \geqslant 1 \text{ a.e. in the neighbourhood of } A.$$

(2) We define the *capacity of A in Ω*

$$\operatorname{cap}_\Omega A = \inf_{\mathscr{F}(A,\Omega)} \mathscr{I}(u)$$

with the convention

$$\operatorname{cap}_\Omega A = +\infty \quad \text{if} \quad \mathscr{F}(A, \Omega) = \varnothing\,.$$

In the case of a compact part K, this definition coincides with that of Definition 1: this is what we prove in the

Proposition 10
(1) *For a compact part K of an open set Ω of $\mathbb{R}^n$, $\operatorname{cap}_\Omega K$ in the sense of Definition* 1 *coincides with that in the sense of Definition* 3.
(2) *For every increasing sequence of parts (A_k) of an open set Ω, with union A, the sequence $(\operatorname{cap}_\Omega A_k)$ is increasing and converges to $\operatorname{cap}_\Omega A$.*

Proof
Proof of the point (1). We have trivially $\mathscr{D}(K, \Omega) \subset \mathscr{F}(K, \Omega)$. We take therefore $u_0 \in \mathscr{F}(K, \Omega)$; let U be an open neighbourhood of K in Ω such that $u_0 \geqslant 1$ a.e. on

---

[125] See Chap. IV, §7.

[126] $H^1(\mathbb{R}^n) = \{u \in L^2(\mathbb{R}^n);\ \dfrac{\partial u}{\partial x_i} \in L^2(\mathbb{R}^n) \text{ for } i=1,\dots,n\} = B^1(\mathbb{R}^n) \cap L^2(\mathbb{R}^n)$ is a Hilbert space with norm $\|u\|_1$ defined by

$$\|u\|_{H^1}^2 = \int_{\mathbb{R}^n} (|u|^2 + |\operatorname{grad} u|^2)\,dx \quad \text{(see Chap. IV).}$$

[127] In the case $n \geqslant 3$, $B_0^1(\mathbb{R}^n) = B^1(\mathbb{R}^n) \cap L^{2n/(n-2)}(\mathbb{R}^n)$; that follows from Sobolev's inequality (see J. Lions [1], p. 33)

$$\|u\|_{L^{2n/(n-2)}} \leqslant C\|\operatorname{grad} u\|_{L^2}, \quad \forall u \in \mathscr{D}(\mathbb{R}^n)\,.$$

The situation is more delicate in the case $n = 1$ and $n = 2$.

U and $u_k \in \mathscr{D}(\Omega)$ such that $u_k \to u_0^+$ in $B^1(\mathbb{R}^n)$; we fix $\rho \in \mathscr{D}(K, U)$ and put

$$v_k = u_k \vee \rho = u_k + (\rho - u_k)^+ .$$

We have $v_k \in \mathscr{S}(K, \Omega)$ and from Lemma 3,

$$v_k \to u_0^+ + (\rho - u_0)^+ = u_0^+ \quad \text{in} \quad B^1(\mathbb{R}^n) .$$

Therefore

$$\inf_{\mathscr{S}(K, \Omega)} \mathscr{I}(u) \leqslant \liminf \mathscr{I}(v_k) = \mathscr{I}(u_0^+) \leqslant \mathscr{I}(u_0) . \qquad \square$$

Proof of the point (2). It is immediate that $\text{cap}_\Omega A$ is increasing with respect to A. In the sequence $\text{cap}_\Omega A_k$ is increasing with limit $l \leqslant \text{cap}_\Omega A$. Let us suppose $l < \text{cap}_\Omega A$ and fix $\varepsilon > 0$; since $c_k = \text{cap}_\Omega A_k \leqslant l < \infty$, there exists $u_k \in \mathscr{F}(A_k, \Omega)$ such that $\mathscr{I}(u_k) \leqslant c_k + \varepsilon 2^{-k}$; we can suppose $0 \leqslant u_k \leqslant 1$ on replacing u_k by $\min(u_k^+, 1)$. We then define by recurrence

$$v_1 = u_1 , \quad v_{k+1} = \max(v_k, u_{k+1}) , \quad w_k = \min(v_k, u_{k+1}) .$$

We have $v_1 \leqslant v_2 \leqslant \ldots \leqslant v_k \leqslant \ldots \leqslant 1$, $v_k \in \mathscr{F}(A_k, \Omega)$, $w_k \in \mathscr{F}(A_k, \Omega)$.
Now from (5.62),

$$\mathscr{I}(v_{k+1}) + \mathscr{I}(w_k) = \mathscr{I}(v_k) + \mathscr{I}(u_{k+1}) .$$

so

$$\mathscr{I}(v_{k+1}) \leqslant \mathscr{I}(v_k) + c_{k+1} + \varepsilon 2^{-(k+1)} - c_k$$

and by recurrence

$$\mathscr{I}(v_k) \leqslant c_k + \varepsilon \sum_{i=1}^{k} 2^{-i} .$$

We deduce $v = \lim v_k \in B_0^1(\Omega)$ and $\mathscr{I}(v) \leqslant l + \varepsilon$. On the other hand $v \geqslant v_k \geqslant 1$ a.e. in the neighbourhood of A_k, for all k. Hence $v \geqslant 1$ a.e. in the neighbourhood of A. $\square$

Remark 6. By definition $\mathscr{F}(A, \Omega)$ is the union of the $\mathscr{F}(U, \Omega)$ for all the open sets U of Ω containing A, hence by definition

$$\text{(5.74)} \qquad \text{cap}_\Omega A = \inf_{A \subset U \subset \Omega} \text{cap}_\Omega U \quad (U \text{ open}) .$$

the capacity of an open set U of Ω being given by (5.71)[128]

$$\text{cap}_\Omega U = \sup_{K \subset U} \text{cap}_\Omega K \quad (K \text{ compact}) .$$

We say that a part A *is capacitable in* Ω *if*

$$\text{cap}_\Omega A = \sup_{K \subset A} \text{cap}_\Omega K \quad (K \text{ compact}) ,$$

---

[128] Since from the point (2) of Proposition 10, cap_Ω is an exterior measure.

that is, more explicitly, if

$$(5.75)\quad \begin{cases} \text{for all } \varepsilon > 0, \text{ there exists a compact set } K_\varepsilon \text{ and an open set } U_\varepsilon \text{ of } \Omega \\ \text{such that } K_\varepsilon \subset A \subset U_\varepsilon \text{ and } \operatorname{cap}_\Omega U_\varepsilon - \operatorname{cap}_\Omega K_\varepsilon \leqslant \varepsilon. \end{cases}$$

Every open set, or more generally, every K_σ[129] of Ω is capacitable.

The capacitability theorem of Choquet[130] shows that *every part of Ω which is a Borel set is capacitable in Ω.* □

Remark 7. We can generalize the properties (1), (4), (5) and (6) of Proposition 9. In particular, *the capacity is strongly sub-additive*: for every part A_i of an open set Ω_i, $i = 1, 2$,

$$(5.76)\quad \operatorname{cap}_{\Omega_1 \cup \Omega_2} A_1 \cup A_2 + \operatorname{cap}_{\Omega_1 \cap \Omega_2} A_1 \cap A_2 \leqslant \operatorname{cap}_{\Omega_1} A_1 + \operatorname{cap}_{\Omega_2} A_2 .$$

The proof is identical with that of Proposition 9 by noting that, from Lemma 3, for $u_i \in B^1(\mathbb{R}^n)$, $i = 1, 2$

$$u_i^+ \in B_0^1(\Omega_i) \Rightarrow \max(u_1, u_2)^+ \in B_0^1(\Omega_1 \cup \Omega_2) \quad \text{and}$$
$$\min(u_1, u_2)^+ \in B_0^1(\Omega_1 \cap \Omega_2)$$

On the other hand, *Property* (3) *of Proposition* 9 *does* **not** *generalize to a decreasing sequence of arbitrary parts of Ω*: for example let us consider in the case $n = 1$

$$\Omega =]-2, +2[\quad \text{and} \quad U_k = \left]-1, -1 + \frac{1}{k}\right[\cup \left]1 - \frac{1}{k}, 1\right[;$$

we have (using Example 7 or 9 and (5.71)), $\operatorname{cap}_\Omega U_k = 2$, even though the sequence (U_k) is decreasing with empty intersection. □

Given A a part of an open set Ω of $\mathbb{R}^n$, by definition $\operatorname{cap}_\Omega A = 0$ iff

$$(5.77)\quad \begin{cases} \text{for all } \varepsilon > 0, \text{ there exists } u_\varepsilon \in B_0^1(\Omega) \text{ such that} \\ u_\varepsilon \geqslant 1 \text{ a.e. in the neighbourhood of } A \text{ and } \mathscr{I}(u_\varepsilon) \leqslant \varepsilon. \end{cases}$$

We note that *this property is local*:

$$(5.78)\quad \begin{cases} \operatorname{cap}_\Omega A = 0 \text{ iff for all } x \in \Omega, \text{ there exists an open neighbourhood } U \\ \text{of } x \text{ in } \Omega \text{ such that } \operatorname{cap}_\Omega(U \cap A) = 0, \end{cases}$$

since, Ω being the denumerable union of compact sets,

$$\operatorname{cap}_\Omega A = 0 \quad \text{iff} \quad \operatorname{cap}_\Omega(A \cap K) = 0 \text{ for every compact set } K \text{ in } \Omega$$

and from the sub-additivity (see (5.76)), if $U_1, \ldots, U_k$ cover K

$$\operatorname{cap}_\Omega A \cap K \leqslant \sum_{i=1}^{k} \operatorname{cap}_\Omega U_i .$$

---

[129] That is to say (see the beginning of this section 3b) a denumerable union of compact sets of Ω.

[130] See Choquet [1].

When A is capacitable in Ω and, in particular, thus from Choquet's theorem (see Remark 6), *when A is a Borel set*,

$$(5.79)\quad \begin{cases} \text{cap}_\Omega A = 0 \text{ iff for all } x \in A, \text{ there exists an open neighbourhood } U \\ \text{of } x \text{ in } \Omega \text{ such that } \text{cap}_\Omega(U \cap A) = 0\,. \end{cases}$$

In effect, then by the definition of capacitability

$$\text{cap}_\Omega A = 0 \quad \text{iff} \quad \text{cap}_\Omega K = 0 \quad \text{for all compact } K \subset A.$$

This shows that *the property* $\text{cap}_\Omega A = 0$ *is independent of the bounded open set* Ω *containing* A (for A bounded).
More precisely, we have

Proposition 11. *Let A be a bounded part of $\mathbb{R}^n$. The following assertions are equivalent*:
(*i*) *there exists a bounded open set Ω containing A such that* $\text{cap}_\Omega A = 0$;
(*ii*) *for every open set Ω of $\mathbb{R}^n$*, $\text{cap}_\Omega(A \cap \Omega) = 0$;
(*iii*) *there exists a Borel part $\tilde{A}$ containing A such that for all $\varepsilon > 0$, there exists $u \in H^1(\mathbb{R}^n)$ satisfying*

$$u \geqslant 1 \text{ a.e. in the neighbourhood of } \tilde{A} \text{ and } \|u\|_{H^1} \leqslant \varepsilon.$$

Proof. The implication (*ii*) $\Rightarrow$ (*i*) is trivial. For the implication (*i*) $\Rightarrow$ (*iii*) we note first, from (5.74), that if $\text{cap}_\Omega A = 0$, there exists a sequence of open sets $U_k \supset A$ such that $\text{cap}_\Omega U_k \to 0$; $\tilde{A} = \cap U_k$ is a Borel set containing A with $\text{cap}_\Omega \tilde{A} = 0$. Applying Poincaré's inequality[131], we have (*iii*).
Finally we prove (*iii*) $\Rightarrow$ (*ii*). Replacing A by $\tilde{A}$ we can suppose A is a Borel set; from (5.79) (and (5.77)), it is sufficient therefore to prove that for every open set Ω of $\mathbb{R}^n$, all $x_0 \in \Omega \cap A$ and all $\varepsilon > 0$, there exists $u \in B_0^1(\Omega)$ and $r > 0$ such that $u \geqslant 1$ a.e. on $B(x_0, r)$ and $\mathscr{I}(u) \leqslant \varepsilon$. Let us fix Ω, $x_0 \in \Omega \cap A$, $0 < r_1 < r_0 = \text{dist}(x_0, \partial\Omega)$ and $\zeta \in \mathscr{D}(\bar{B}(x_0, r_1), B(x_0, r_0))$; given $\varepsilon > 0$ (using (*iii*)), there exists $u_\varepsilon \in H^1(\mathbb{R}^n)$ and $r_\varepsilon > 0$ such that $u_\varepsilon \geqslant 1$ a.e. on $B(x_0, r_\varepsilon)$ and $\|u_\varepsilon\|_{H^1} \leqslant \dfrac{\varepsilon^{1/2}}{\|\text{grad}\,\zeta\|_{L^\infty}}$. It is then clear that $u = \zeta u_\varepsilon$ and $r = \min(r_\varepsilon, r_1)$ are suitable choices, since

$$\mathscr{I}(u) = \int |u_\varepsilon \text{grad}\,\zeta + \zeta\,\text{grad}\,u_\varepsilon|^2\,dx \leqslant \|\text{grad}\,\zeta\|_{L^\infty}^2\,\|u_\varepsilon\|_{H^1}^2 \leqslant \varepsilon\,. \qquad \square$$

This leads us to frame the

Definition 4. (1) We say that a part A of $\mathbb{R}^n$ is *polar*[132] if $\text{cap}_\Omega(A \cap \Omega) = 0$, for every open set Ω of $\mathbb{R}^n$.
(2) We say that a *map u: $D \subset \mathbb{R}^n \to \bar{\mathbb{R}}$ is zero quasi-everywhere* (we denote this by $u=0$ q.e. on D) if there exists a polar part A such that $u = 0$ on D/A.
Similarly, we define $u_1 = u_2$ q.e. on D, $u_1 \leqslant u_2$ q.e. on D, etc. . . .

---

[131] See Chap. IV, §7.
[132] It is polar with respect to $H^1(\mathbb{R}^n)$ (see J. L. Lions [1]).

Remark 8. The idea of a polar set and of equality quasi-everywhere is closely related to that of a negligible set and of equality almost everywhere. In effect, *every denumerable union of polar sets is polar* and also *every polar set is contained in a polar Borel set. A polar set is obviously negligible and the converse is obviously false.* In the case $n = 1$, a polar part is empty and equality quasi-everywhere is equality everywhere. In the case $n \geqslant 2$, a polar part A contained in a regular hypersurface V is negligible for the surface measure of this hypersurface (this follows from the trace theorems[133]), but the converse is false[134]; also a part contained in a regular manifold of dimension $n - 2$ is polar (that results from Example 10c). □

We now give the application of these ideas to the *obstacle problem.*

Proposition 12. *Let Ω be an open set of $\mathbb{R}^n$ and A a part of Ω with finite capacity;*
(1) $\operatorname{cap}_\Omega A = \min_{\mathscr{F}(A,\Omega)} \mathscr{I}(u)$ *where $\mathscr{F}^+(A,\Omega)$ is the set of the $u \in B_0^1(\Omega)$ superharmonic on Ω and whose l.s.c. superharmonic representative $\tilde{u}$*[135] satisfies $\tilde{u} \geqslant 1$ q.e. on A;
(2) *if* $\operatorname{cap}_\Omega A > 0$, *there exists a unique* $u_{A,\Omega}$ *in* $\mathscr{F}^+(A,\Omega)$ *such that* $\operatorname{cap}_\Omega A = \mathscr{I}(u_{A,\Omega})$.
In addition $u_{A,\Omega}$ is characterized by

$$u_{A,\Omega} \in \mathscr{F}^+(A,\Omega)\,, \quad \int_{\Omega\setminus A} \mathrm{d}(-\Delta u_{A,\Omega}) = 0\,,$$

and satisfies $0 \leqslant u_{A,\Omega} \leqslant 1$ *a.e.*
Also $\tilde{u}_{A,\Omega}$ is characterized as the smallest l.s.c. superharmonic map. $\tilde{u}\colon \Omega \to [0,1]$ *satisfying* $\tilde{u} = 1$ *q.e. on A.*

We shall not prove this proposition, referring the reader, for example, to Kinderlehrer–Stampacchia [1][136] for this and for other developments of this problem.
Let us give now, still without proof,[136] *Wiener's criterion* of regularity of the points on the boundary of an open set in $\mathbb{R}^n$.
Given a bounded open set Ω in $\mathbb{R}^n$ with $n \geqslant 3$, and z a point of its boundary Γ, the following assertions are equivalent:
(i) for all $\varphi \in \mathscr{C}^0(\partial\Omega)$, the generalized solution $u(\varphi)$ of $P(\Omega,\varphi)$ satisfies

$$\lim_{x\to z} u(\varphi, x) = \varphi(z)$$

(ii) $\displaystyle\sum_{k=1}^{\infty} 2^{k(n-2)} \operatorname{cap}_{\mathbb{R}^n}(\mathbb{R}^n\setminus\bar{\Omega} \cap A(z, 2^{-k})) = +\infty$

where:
$$A(z,r) = \bar{B}(z,r)\setminus B(z,r/2)\,.$$

---

[133] For all $u \in \mathscr{D}(\mathbb{R}^n)$, $\|u(.,0)\|_{L^2(\mathbb{R}^{n-1})} \leqslant \|u(.,0)\|_{H^{1/2}(\mathbb{R}^{n-1})} \leqslant \|u\|_{H^1(\mathbb{R}^n)}$ (see Chap. IV, §4).
[134] This follows from the theorems on lifting of trace; in particular, we show that the Cantor set K, negligibly compact on $\mathbb{R}$, is, identified with $K \times \{0\}$, of strictly positive capacity in every bounded open set Ω containing it.
[135] See Definition 15 of §4. Note that this implies that $\tilde{u} \geqslant 0$.
[136] See Wiener [1].

Finally we conclude this section with the

Proof of Lemma 2. Given $u \in L^1_{loc}(\Omega)$ with $\dfrac{\partial u}{\partial x_i} \in L^1_{loc}(\Omega)$, we have to show that for all $\zeta \in \mathscr{D}(\Omega)$,

$$\int \frac{\partial \zeta}{\partial x_i} u^+ \,dx = - \int_{\{u > 0\}} \zeta \frac{\partial u}{\partial x_i} \,dx . \tag{5.80}$$

Applying to $-u$ and subtracting, we obtain

$$\int \frac{\partial \zeta}{\partial x_i} u \,dx = - \int_{\{u \neq 0\}} \zeta \frac{\partial u}{\partial x_i} \,dx ;$$

thus

$$\int_{\{u = 0\}} \zeta \frac{\partial u}{\partial x_i} \,dx = 0 ,$$

which will prove (5.61).

First, let us prove that for each bounded continuous function $p: \mathbb{R} \to \mathbb{R}$

$$\int \frac{\partial \zeta}{\partial x_i} j(u) \,dx = - \int p(u) \frac{\partial u}{\partial x_i} \zeta \,dx \tag{5.81}$$

where $j(r) = \displaystyle\int_0^r p(s)\,ds$ is thus a Lipschitzian function of class $\mathscr{C}^1$. The relation (5.81) is obviously true if $u \in \mathscr{C}^1(\Omega)$ for then

$$j(u) \in \mathscr{C}^1(\Omega) \quad \text{and} \quad \frac{\partial}{\partial x_i} j(u) = p(u) \frac{\partial u}{\partial x_i} .$$

To obtain with the hypotheses $u \in L^1_{loc}(\Omega)$, $\partial u/\partial x_i \in L^1_{loc}(\Omega)$, it is sufficient to regularize u; there exists a sequence of functions u_k such that $u_k \to u$, $\partial u_k/\partial x_i \to \partial u/\partial x_i$ in $L^1_{loc}(\Omega)$ and a.e. on Ω; since j is Lipschitzian, $j(u_k) \to j(u)$ in $L^1_{loc}(\Omega)$ and

$$\int \frac{\partial \zeta}{\partial x_i} j(u_k) \,dx \to \int \frac{\partial \zeta}{\partial x_i} j(u) \,dx ;$$

p being continuous, $p(u_k) \to p(u)$ a.e. on Ω; p also being bounded, we have by Lebesgue's theorem $p(u_k) \dfrac{\partial u_k}{\partial x_i} \to p(u) \dfrac{\partial u}{\partial x_i}$ in $L^1_{loc}(\Omega)$ and

$$\int p(u_k) \frac{\partial u_k}{\partial x_i} \zeta \,dx \to \int p(u) \frac{\partial u}{\partial x_i} \zeta \,dx .$$

Passing to the limit in the equation (5.81) for u_k, we obtain it for u. Equation (5.81) is still true for every simple bounded function $p: \mathbb{R} \to \mathbb{R}$, i.e. $p(r) = \lim p_k(r)$ for all $r \in \mathbb{R}$ where $p_k \in \mathscr{C}^0(\mathbb{R})$, which we can always suppose is uniformly bounded. In effect then, by Lebesgue's theorem,

$$p_k(u)\frac{\partial u}{\partial x_i} \to p(u)\frac{\partial u}{\partial x_i} \quad \text{in} \quad L^1_{\text{loc}}(\Omega)$$

and

$$\int p_k(u)\frac{\partial u}{\partial x_i}\zeta\,\mathrm{d}x \to \int p(u)\frac{\partial u}{\partial x_i}\zeta\,\mathrm{d}x\,;$$

also $j(r) = \int_0^r p(s)\,\mathrm{d}s = \lim j_k(r)$ are uniformly Lipschitzian, with the result that $j_k(u) \to j(u)$ in $L^1_{\text{loc}}(\Omega)$ and

$$\int \frac{\partial \zeta}{\partial x_i} j_k(u)\,\mathrm{d}x \to \int \frac{\partial \zeta}{\partial x_i} j(u)\,\mathrm{d}x\,.$$

Passing to the limit in the equalities (5.81) for p_k we obtain them for p.

Applying (5.81) to

$$p(r) = \begin{cases} 1 & \text{if} \quad r > 0 \\ 0 & \text{if} \quad r \leqslant 0\,, \end{cases}$$

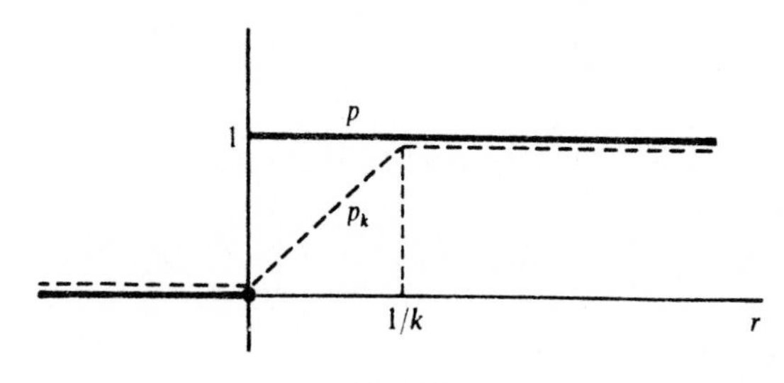

Fig. 18

we obtain (5.80). Note that p is clearly a simple function as is shown by its graph (see Fig. 18) and that

$$j(r) = \int_0^r p(s)\,\mathrm{d}s = r^+\,. \qquad \square$$

Remark 9

(1) Applying (5.81) with $p = \chi_{\{c\}}$ which is a simple function as is shown by its graph in Fig. 19.

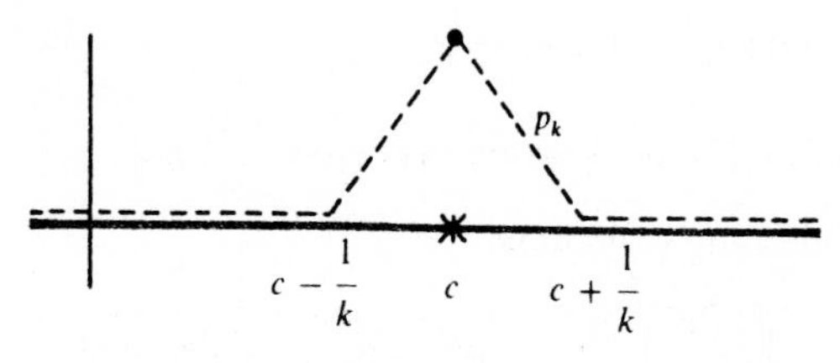

Fig. 19

and hence $j(r) = \int_0^r p(s)\,ds \equiv 0$, we obtain again $\int_{\{u = c\}} \frac{\partial u}{\partial x_i} \zeta\, dx = 0$.

In fact (5.81) is true for every Borel function and even for every measurable function. Applying to $p = \chi_K$ where K *is a negligible set of* $\mathbb{R}$, we obtain $\partial u/\partial x_i = 0$ *a.e. on* $\{x: u(x) \in K\}$

(2) The same argument as in the proof of Lemma 2, leads to *Kato's inequality*: given $u \in L^1_{\text{loc}}(\Omega)$ with $\Delta u \in L^1_{\text{loc}}(\Omega)$ (in the sense of distributions), $\Delta(u^+)$ is a Radon measure on Ω, minorized by the function $\chi_{\{u > 0\}} \Delta u$; in other words

$$\Delta(u^+) \geqslant \chi_{\{u > 0\}} \Delta u \quad \text{in } \mathscr{D}'(\Omega) . \tag{5.82}$$

Using the same function p as in Lemma 2, Kato's inequality (5.82) is a deduction from the more general inequality

$$\Delta j(u) \geqslant p(u) \Delta u \quad \text{in} \quad \mathscr{D}'(\Omega) , \tag{5.83}$$

true for every *increasing* bounded function $p: \mathbb{R} \to \mathbb{R}$.
In the case $u \in \mathscr{C}^2(\Omega)$

$$\Delta j(u) = p(u)\Delta u + p'(u)|\text{grad}\, u|^2 \geqslant p(u)\Delta u \quad \text{on} \quad \Omega .$$

By double approximation as in the proof of Lemma 2, we obtain (5.83) in the general case. □

Finally, we give the

Proof of Lemma 3. Let $u_k \to u$ in $B^1(\mathbb{R}^n)$. We have

$$|u_k^+ - u^+| \leqslant |u_k - u| \quad \text{and hence} \quad \lim \int_{B(0,\,R)} |u_k^+ - u^+|^2\, dx = 0 .$$

Now from Lemma 2

$$\frac{\partial}{\partial x_i} u_k^+ - \frac{\partial}{\partial x_i} u^+ = \left(\frac{\partial u_k}{\partial x_i} - \frac{\partial u}{\partial x_i}\right) \chi_{\{u_k > 0\}} + \frac{\partial u}{\partial x_i} (\chi_{\{u_k > 0\}} - \chi_{\{u > 0\}})$$

from which it follows that

$$\|\text{grad}\, u_k^+ - \text{grad}\, u^+\|_{L^2} \leqslant \|\text{grad}\, u_k - \text{grad}\, u\|_{L^2} + \left(\int \left|\frac{\partial u}{\partial x_i}\right|^2 |\chi_{\{u_k > 0\}} - \chi_{\{u > 0\}}|^2\, dx\right)^{\frac{1}{2}}.$$

But we can suppose that $u_k \to u$ a.e. with the result that

$$\limsup |\chi_{\{u_k > 0\}} - \chi_{\{u > 0\}}| \leqslant \chi_{\{u = 0\}} \quad \text{a.e.}$$

Making use of point (2) of Lemma we have $\frac{\partial u}{\partial x_i} \chi_{\{u = 0\}} = 0$ a.e. and, as a consequence, by Lebesgue's theorem,

$$\int \left(\frac{\partial u}{\partial x_i}\right)^2 |\chi_{\{u_k > 0\}} - \chi_{\{u > 0\}}|^2\, dx \to 0 .$$ □

§6. Regularity*

1. Regularity of the Solutions of Dirichlet and Neumann Problems

In §3.2, we have studied the local regularity of solutions of Poisson's equation

$$\Delta u = f \quad \text{on} \quad \Omega$$

as a function of that of f. In particular, we have shown (see Proposition 6 of §3)

$$(6.1) \qquad f \in L^p_{\text{loc}}(\Omega) \text{ with } p > \frac{n}{2} \text{ (resp. } p > n) \Rightarrow u \in \mathscr{C}^0(\Omega) \text{ (resp. } \mathscr{C}^1(\Omega))$$

and also (see Proposition 9 of §3) that for $m \in \mathbb{N}$ and $0 < \theta < 1$,

$$(6.2) \qquad f \in \mathscr{C}^{m,\theta}(\Omega) \Rightarrow u \in \mathscr{C}^{m+2,\theta}(\Omega)^{137} .$$

These results establish the *interior regularity* of the solutions of the Dirichlet and Neumann problems. We propose now to study the *regularity on the boundary* of the problems $P(\Omega, \varphi, f)$ and $PN(\Omega, \psi, f)$ as a function of the data Ω, φ, ψ, f. An important aspect is that, just as with interior regularity, *regularity at the boundary is a local property*. To clarify this assertion, we state the

Theorem 1. *Let Ω be an open set of $\mathbb{R}^n$, Γ_0 a regular open part of class $\mathscr{C}^\infty$ of its boundary $\varphi \in \mathscr{C}^\infty(\Gamma_0)$ (resp. $\psi \in \mathscr{C}^\infty(\Gamma_0)$), $f \in \mathscr{C}^\infty(\Omega \cup \Gamma_0)^{138}$ and*

$$u \in \mathscr{C}^0(\Omega \cup \Gamma_0) \quad (\text{resp. } \mathscr{C}^1_n(\Omega \cup \Gamma_0)^{139}$$

solution

$$(6.3) \qquad \begin{cases} \Delta u = f & \text{in} \quad \mathscr{D}'(\Omega) \\ u = \varphi \left(\text{resp. } \dfrac{\partial u}{\partial n} = \psi \right) & \text{on} \quad \Gamma_0 . \end{cases}$$

Then $u \in \mathscr{C}^\infty(\Omega \cup \Gamma_0)$.

We shall prove this theorem and other regularity results which we shall state precisely as they arise in the course of this section. We, first of all, change the statement of this theorem. Under the above hypotheses, we know already that $u \in \mathscr{C}^\infty(\Omega)$ (see Proposition 1 of §3); it is sufficient, therefore, to show that the derivatives of u extend by continuity to each point $z \in \Gamma_0$. This result being local, we can always suppose that

$$(6.4) \qquad \begin{cases} \Omega = \{(x', \alpha(x') - t; x' \in \mathcal{O}, \quad 0 < t < \delta\} \\ \Gamma_0 = \{(x', \alpha(x')); x' \in \mathcal{O}\} \end{cases}$$

---

[137] See Proposition 9 of §3 for the first use in this chapter of functions of class $\mathscr{C}^{m,\theta}$.

[138] That is, $f \in \mathscr{C}^\infty(\Omega)$ and all its derivatives extend by continuity to Γ_0.

[139] That is, $u \in \mathscr{C}^1(\Omega)$ and $\partial u / \partial n(z)$ exist for all $z \in \Gamma_0$ (see §1.3b).

where $\mathcal{O}$ is an open set of $\mathbb{R}^{n-1}$, $\delta > 0$, $\alpha \in \mathscr{C}^0(\mathcal{O})$ and (x', x_n) denotes the coordinates in an orthonormal reference frame in $\mathbb{R}^n$. We can always take z to be the origin in this frame and take $n(z)$ to be the last of the base vectors in the reference system, with the result that

$$0 \in \mathcal{O} . \quad \alpha(0) = 0 , \quad \operatorname{grad}\alpha(0) = 0 . \tag{6.5}$$

We put
$$\bar{u}(x', t) = u(x', \alpha(x') - t)$$

We have
$$\bar{u} \in \mathscr{C}^\infty(\mathcal{O} \times]0, \delta[) \text{ and}$$

$$\sum_{i=1}^{n-1} \frac{\partial}{\partial x_i}\left(\frac{\partial \bar{u}}{\partial x_i} + \frac{\partial \alpha}{\partial x_i}\frac{\partial \bar{u}}{\partial t}\right) - \frac{\partial}{\partial t}\left(\sum_{i=1}^{n-1} \frac{\partial \alpha}{\partial x_i}\frac{\partial \bar{u}}{\partial x_i}\right) + (1 + |\operatorname{grad}\alpha|^2)\frac{\partial^2 \bar{u}}{\partial t^2} = \bar{f} \tag{6.6}$$

with
$$\bar{f}(x', t) = f(x', \alpha(x') - t) .$$

In the case of the Dirichlet problem, we have $\bar{u} \in \mathscr{C}^0(\mathcal{O} \times]0, \delta[)$ and

$$\bar{u}(x', 0) = \bar{\varphi}(x') = \varphi(x', \alpha(x')) \quad \text{for all} \quad x' \in \mathcal{O} .$$

In the case of the Neumann problem, by the definition of $\partial/\partial n$, denoting

$$B\bar{u} = \sum_{i=1}^{n-1} \frac{\partial \alpha}{\partial x_i}\frac{\partial \bar{u}}{\partial x_i} + (1 + |\operatorname{grad}\alpha|^2)\frac{\partial \bar{u}}{\partial t} ,$$

we have

$$B\bar{u} \in \mathscr{C}^0(\mathcal{O} \times [0, \delta[) \quad \text{and} \quad B\bar{u}(x', 0) = \bar{\psi}(x') \quad \text{for all} \quad x' \in \mathcal{O}$$

where
$$\bar{\psi}(x') = -(1 + |\operatorname{grad}\alpha(x')|^2)^{1/2}\bar{\psi}(x', \alpha(x')) .$$

Using the hypothesis (6.5) and changing the notation, we see that the theorem therefore reduces to the following lemma:

Lemma 1. *Let $\mathcal{O}$ be an open neighborhood of O in $\mathbb{R}^{n-1}$, $\delta > 0$, $a_{ij} = a_{ji}$ and f all belonging to $\mathscr{C}^\infty(\mathcal{O} \times [0, \delta[)$, $\varphi \in \mathscr{C}^\infty(\mathcal{O})$ (resp. $\psi \in \mathscr{C}^\infty(\mathcal{O})$) and $u \in \mathscr{C}^\infty(\mathcal{O} \times]0, \delta[$ satisfying*

$$\sum\sum \frac{\partial}{\partial x_i}\left(a_{ij}\frac{\partial u}{\partial x_j}\right) = f \quad on \quad \mathcal{O} \times]0, \delta[$$

$$u \in \mathscr{C}^0(\mathcal{O} \times [0, \delta[) \quad and \quad u(x', 0) = \varphi(x') \quad for\ all \quad x' \in \mathcal{O}$$

$$(resp. \quad Bu = \sum a_{in}\frac{\partial u}{\partial x_i} \in \mathscr{C}^0(\mathcal{O} \times [0, \delta[) \quad and \quad Bu(x', 0) = \psi(x')) .$$

We suppose that $a_{ij}(0) = \delta_{ij}$ (the Kronecker delta). Then u is of class $\mathscr{C}^\infty$ in the neighbourhood of O.

For the theorem, from (6.6), we shall apply the lemma with

$$\begin{cases} a_{ij} = \delta_{ij} & for \quad i, j = 1, \ldots, n-1 \\ a_{in} = a_{ni} = \dfrac{\partial \alpha}{\partial x_i} & for \quad i = 1, \ldots, n-1 \\ a_{nn} = 1 + |\operatorname{grad}\alpha|^2 . \end{cases} \tag{6.7}$$

There are many approaches to the proof of Lemma 1. We shall commence, for Dirichlet's problem, by the use of the Agmon–Douglis–Niremberg method, based on the Calderon–Zygmund estimate, which we recall (see Proposition 8 of §3): for $1 < p < \infty$.

$$\left\| \frac{\partial^2 u}{\partial x_i \partial x_j} \right\|_{L^p} \leqslant C \| \Delta u \|_{L^p} \quad \text{for all} \quad u \in \mathscr{D}(\mathbb{R}^n) \tag{6.8}$$

where C is a constant depending only on p and n. We shall now prove the

Lemma 2. *Let $n < p < \infty$ and $a_{ij} = a_{ji} \in L^\infty(\mathbb{R}^n_+)$*[140] *satisfying the condition that $\delta_{ij} - a_{ij}$ is of bounded support, and*

$$\sum\sum \| \delta_{ij} - a_{ij} \|_{L^1} < (2C)^{-1}$$

where C is the constant of (6.8). *Then for all $f_0, \ldots, f_n \in L^p(\mathbb{R}^n_+)$ with compact support, there exists one and only one solution u of the problem*

$$\begin{cases} u \in \mathscr{C}^0(\mathbb{R}^{n-1} \times \mathbb{R}^+) \ \text{ and its derivatives } \ \dfrac{\partial u}{\partial x_i} \in L^2_{\text{loc}}(\mathbb{R}^n_+) \\[2ex] \sum\sum \dfrac{\partial}{\partial x_i}\left(a_{ij} \dfrac{\partial u}{\partial x_j} \right) = f_0 + \sum \dfrac{\partial f_k}{\partial x_k} \quad \text{in } \ \mathscr{D}'(\mathbb{R}^n_+)^{141} \\[2ex] u(x', 0) = 0 \ \text{ for all } \ x' \in \mathbb{R}^{n-1} \ \text{ and } \ \lim\limits_{|x| \to \infty} u(x) = 0 \,. \end{cases} \tag{6.9}$$

In addition

(a) the derivatives $\dfrac{\partial u}{\partial x_i} \in L^p(\mathbb{R}^n_+)$,

(b) If, for $m \in \mathbb{N}$, $f_0 + \sum \dfrac{\partial f_k}{\partial x_k} \in W^{m,p}(\mathbb{R}^n_+)$ and

$$a_{ij} \in W^{m+1,\infty}(\mathbb{R}^n_+) \quad \text{for} \quad i, j = 1, \ldots, n \,,$$

then $u \in \mathscr{C}^{m+1}(\mathbb{R}^{n-1} \times \mathbb{R}^+)$ and $u \in W^{m+2,p}(\mathbb{R}^n_+)$[142].

Let us show, first of all, that this lemma implies Lemma 1 and hence Theorem 1 for the Dirichlet problem:

Proof of Lemma 1 (The Dirichlet problem). We shall note first that we can always suppose that $\varphi = 0$.

---

[140] $\mathbb{R}^n_+ = \{(x', x_n) : x' \in \mathbb{R}^{n-1}, x_n > 0\}$ and $\mathbb{R}^+ = [0, \infty[$.

[141] That is to say $\sum\sum \int a_{ij} \dfrac{\partial u}{\partial x_j} \dfrac{\partial \zeta}{\partial x_j} \mathrm{d}x + \int f_0 \zeta \, \mathrm{d}x = \sum \int f_k \dfrac{\partial \zeta}{\partial x_k} \mathrm{d}x, \quad \forall \zeta \in \mathscr{D}(\mathbb{R}^n_+)$

[142] $W^{m,p}(\Omega)$ is the space of the functions $f \in L^p(\Omega)$ such that the derivatives (in the sense of distributions) $\partial^{|\beta|} f / \partial x^\beta \in L^p(\Omega)$ for all $\beta = (\beta_1, \ldots, \beta_n) \in \mathbb{N}^n$ with $|\beta| = \beta_1 + \ldots + \beta_n \leqslant m$.

In effect, the function $\bar{u}(x', x_n) = u(x', x_n) - \varphi(x')$ satisfies

$$\sum\sum \frac{\partial}{\partial x_i}\left(a_{ij}\frac{\partial \bar{u}}{\partial x_j}\right) = f - \sum_{i,j=1}^{n-1}\sum \frac{\partial}{\partial x_i}\left(a_{ij}\frac{\partial \varphi}{\partial x_j}\right)$$

and $\bar{u}(x', 0) = 0$. Finally u is $\mathscr{C}^\infty$ in the neighbourhood of O iff this is true of $\bar{u}$. We can thus suppose that $\varphi \equiv 0$ and this we now do.
We fix $n < p < \infty$ and $0 < \varepsilon < (2C)^{-1}$ where C is the constant of (6.8). Using the continuity of a_{ij} and the hypothesis $a_{ij}(0) = \delta_{ij}$, $\forall i, j = 1, \ldots, n$ we see that there exists

$$0 < r_1 < \operatorname{dist}(0, \partial\mathcal{O}) \quad \text{and} \quad 0 < \delta_1 < \delta$$

such that $|a_{ij} - \delta_{ij}| \leqslant \varepsilon$ on $\bar{B}(0, r_1) \times [0, \delta_1]$. Let us take $0 < r_0 < r_1$, $0 < \delta_0 < \delta_1$ and $\rho_0 \in \mathscr{C}^\infty(\mathbb{R}^{n-1} \times \mathbb{R}^+)$ with $0 \leqslant \rho_0 \leqslant 1$,

$$\operatorname{supp}\rho_0 \subset \bar{B}(0, r_1) \times [0, \delta_1] \quad \text{and} \quad \rho_0 \equiv 1 \quad \text{on} \quad \bar{B}(0, r_0) \times [0, \delta_0]\,.$$

We put $\bar{a}_{ij} = \delta_{ij} + \rho_0(a_{ij} - \delta_{ij})$ with the result that $\bar{a}_{ij} \in \mathscr{C}^\infty(\mathbb{R}^{n-1} \times \mathbb{R}^+)$,

$$\operatorname{supp}(\bar{a}_{ij} - \delta_{ij}) \subset \operatorname{supp}\rho_0,\ \text{bounded},$$

$$\|\bar{a}_{ij} - \delta_{ij}\|_{L^\infty} \leqslant \|a_{ij} - \delta_{ij}\|_{L^\infty(\operatorname{supp}\rho_0)} \leqslant \varepsilon\,.$$

We shall prove that $u \in \mathscr{C}^\infty(B(0, r_0) \times [0, \delta_0[)$ or, what is its equivalent, that

$$\text{(6.10)} \quad \begin{cases} \text{for all } m \in \mathbb{N} \quad \text{and all} \quad \rho \in \mathscr{C}^\infty(\mathbb{R}^{n-1} \times \mathbb{R}^+) \text{ with} \\ \operatorname{supp}\rho \subset B(0, r_0) \times [0, \delta_0[, \quad \rho u \in \mathscr{C}^m(\mathbb{R}^{n-1} \times \mathbb{R}^+)\,. \end{cases}$$

Given $\rho \in \mathscr{C}^\infty(\mathbb{R}^{n-1} \times \mathbb{R}^+)$ with $\operatorname{supp}\rho \subset B(0, r_0) \times [0, \delta_0[$, the function $\bar{u} = \rho u$ satisfies

$$\bar{u} \in \mathscr{C}^\infty(\mathbb{R}^n_+) \cap \mathscr{C}^0(\mathbb{R}^{n-1} \times \mathbb{R}^+), \quad \bar{u}(x', 0) = 0\,,$$

$\operatorname{supp}\bar{u} \subset B(0, r) \times [0, \delta_0[$, and hence, in particular, $\lim\limits_{|x| \to \infty} \bar{u}(x) = 0$

$$\sum\sum \frac{\partial}{\partial x_i}\left(\bar{a}_{ij}\frac{\partial \bar{u}}{\partial x_j}\right) = \bar{f}_0 + \sum \frac{\partial \bar{f}_k}{\partial x_k}$$

with

$$\text{(6.11)} \quad \begin{cases} \bar{f}_0 = \rho f - u\sum\sum \dfrac{\partial}{\partial x_i}\left(a_{ij}\dfrac{\partial \rho}{\partial x_j}\right) \\[2ex] \bar{f}_k = 2u\sum a_{kj}\dfrac{\partial \rho}{\partial x_j} \end{cases}$$

(single ρ is involved since $\rho_0 \equiv 1$ in the neighbourhood of $\operatorname{supp}\rho$).
From Lemma 2, $\bar{u}$ is the unique solution of (6.9) corresponding to $\bar{a}_{ij}$ and $\bar{f}_k$. Hence the derivatives $\dfrac{\partial \bar{u}}{\partial x_i} \in L^p(\mathbb{R}^n_+)$. This being true for all ρ, we have

$$\frac{\partial u}{\partial x_i} \in L^p_{\text{loc}}(B(0, r_0) \times [0, \delta_0[)\,.$$

Using a recurrence argument and supposing that for a certain $m \geqslant 1$, we would have $u \in W^{m,p}_{\text{loc}}(B(0, r_0) \times [0, \delta_0[)$. Then for arbitrary fixed ρ,

$$\bar{f}_0 + \sum \frac{\partial \bar{f}_k}{\partial x_k} = \rho f + u \sum\sum \frac{\partial}{\partial x_j}\left(a_{ij}\frac{\partial \rho}{\partial x_i}\right) \tag{6.12}$$

$$+ 2\sum\sum a_{ij}\frac{\partial u}{\partial x_i}\frac{\partial \rho}{\partial x_j} \in W^{m-1,p}(\mathbb{R}^n_+)\,,$$

and hence from Lemma 2(b), $\bar{u}$ is of class $\mathscr{C}^m$ on $\mathbb{R}^{n-1} \times \mathbb{R}^+$ and $\bar{u} \in W^{m+1,p}(\mathbb{R}^n_+)$. This being true for all ρ,

$$u \in \mathscr{C}^\infty(B(0, r_0) \times [0, \delta_0[) \cap W^{m+1,p}_{\text{loc}}(B(0, r_0) \times [0, \delta_0[)\,.$$

By induction on m, we therefore have the stated result. □

Going back to the reduction of Theorem 1 to Lemma 1 and the above proof, we have in fact proved the

Proposition 1. *Let Ω be an open set of $\mathbb{R}^n$, Γ_0 a regular open part of its boundary, $n < p < \infty$ and $u \in \mathscr{C}^0(\Omega \cup \Gamma_0)$ satisfying $u = 0$ on Γ_0.*
(1) Let us suppose that Γ_0 is of class $W^{2,p}$ and that

$$\Delta u = f_0 + \sum \frac{\partial f_k}{\partial x_k} \quad \textit{in} \quad \mathscr{D}'(\Omega)$$

with $f_0, \ldots, f_n \in L^p_{\text{loc}}(\Omega \cup \Gamma_0)$[143]. Then $u \in W^{1,p}_{\text{loc}}(\Omega \cup \Gamma_0)$.
(2) For $m \in \mathbb{N}$, let us suppose that Γ_0 is of class $W^{m+2,\infty}$ and

$$\Delta u = f \quad \textit{in} \quad \mathscr{D}'(\Omega)$$

with $f \in W^{m,p}_{\text{loc}}(\Omega \cup \Gamma_0)$. Then $u \in \mathscr{C}^{m+1}(\Omega \cup \Gamma_0) \cap W^{m+2,p}_{\text{loc}}(\Omega \cup \Gamma_0)$.

Remark 1. Proposition 1 expresses *the regularity $W^{m,p}$ up to the boundary for the homogeneous Dirichlet problem* (the interior regularity has been seen in Proposition 8 of § 3. We note that the restriction $n < p < \infty$ is not essential: it can be generalized to the case $1 < p \leqslant n$; note however that we have the inclusion[144] $W^{m,p}_{\text{loc}}(\Omega \cup \Gamma_0) \subset \mathscr{C}^0(\Omega \cup \Gamma_0)$ iff $mp > n$; to generalize Proposition 1 (and Lemma 2) for $1 < p \leqslant n$, it will be necessary to interpret the condition $u = 0$ on Γ_0 in a weak sense.
The case of the inhomogeneous Dirichlet problem, with a condition $u = \varphi$ on Γ_0 can be deduced immediately, as we have seen in the proof of Lemma 1, if there exists a "lifting" u_0 having the desired regularity $W^{m,p}_{\text{loc}}(\Omega \cup \Gamma_0)$ such that $u_0 = \varphi$ on Γ_0. The characterization of the functions φ defined on Γ_0 admitting a lifting in

[143] Given a part E of $\mathbb{R}^n$, $L^p_{\text{loc}}(E)$ is the space of functions f (defined a.e.) measurable on E such that $|f|^p$ is integrable (resp. f bounded if $p = +\infty$) in the neighbourhood of *each point of E*. The space $L^p_{\text{loc}}(\Omega \cup \Gamma_0)$ must not be confused with $L^p_{\text{loc}}(\Omega)$: the first is the set of $f \in L^p_{\text{loc}}(\Omega)$ such that $|f|^p$ is integrable (resp. f bounded) in the neighbourhood of each point of Γ_0. The space $W^{m,p}_{\text{loc}}(\Omega \cup \Gamma_0)$ is the space of the $f \in L^p_{\text{loc}}(\Omega \cup \Gamma_0)$ all of whose derivatives (in the sense of distributions in Ω) up to order m are in $L^p_{\text{loc}}(\Omega \cup \Gamma_0)$. A regular manifold Γ_0 is of class $W^{m,p}$ if in the local parametric representations $(U, R, \mathcal{O}, \alpha)$ we have $\alpha \in W^{m,p}_{\text{loc}}(\mathcal{O})$.
[144] See Chap. IV.

$W^{m,p}_{\text{loc}}(\Omega \cup \Gamma_0)$, involves fractional Sobolev spaces: these concern trace theorems which we shall not take up here (see Chap. IV).
Proposition 1 has its equivalent for Hölder regularity up to the boundary; we note that for this regularity, the trace theorems are simple. We have the following result which we shall prove for the Neumann problem (see Lemma 4):

Given $m \in \mathbb{N}$ *and* $0 < \theta < 1$ *(supposing* Γ_0 *of class* $\mathscr{C}^{m+2,\theta}$*), a function* $u \in \mathscr{C}^0(\Omega \cup \Gamma_0)$*, solution of*

$$\Delta u = f \quad \text{in} \quad \mathscr{D}'(\Omega)\,, \qquad u = \varphi \quad \text{on} \quad \Gamma_0$$

is in $\mathscr{C}^{2m+2,\theta}(\Omega \cup \Gamma_0)$ *iff* $f \in \mathscr{C}^{m,\theta}(\Omega \cup \Gamma_0)$ *and* $\varphi \in \mathscr{C}^{m+2,\theta}(\Gamma_0)$. □

Finally, we note that when Ω is bounded and Γ_0 is the whole boundary of Ω, these results give a global regularity on $\bar{\Omega}$: we express this in:

Corollary 1. *Let* $n < p < \infty$; Ω *a regular bounded open set in* $\mathbb{R}^n$ *with boundary* Γ *of class* $W^{2,p}$ *and* $f_0, f_1, \ldots, f_n \in L^p(\Omega)$. *There exists a (unique) quasi-classical*[145] *solution u of the homogeneous Dirichlet problem* $PH(\Omega, f_0 + \Sigma\, \partial f_k/\partial x_k)$. *In addition* $u \in W^{1,p}(\Omega)$ *and*

$$\|\text{grad}\, u\|_{L^p} \leqslant C \sum \|f_k\|_{L^p}\,.$$

If for $m \in \mathbb{N}$, Γ *is of class* $W^{m+2,\infty}$ *and* $f = f_0 + \sum \dfrac{\partial f_k}{\partial x_k} \in W^{m,p}(\Omega)$, *then* $u \in W^{m+2,p}(\Omega)$ *and* $\|u\|_{W^{m,p+2}} \leqslant C \|f\|_{W^{m,p}}$.

Proof of Corollary 1. The problem $PH(\Omega, f_0 + \Sigma\, \partial f_k/\partial x_k)$ has a (unique) quasi-classical solution: considering $u_0, \ldots, u_n$ the Newtonian potentials of $f_0 \chi_\Omega, \ldots, f_n \chi_\Omega$, we know that $u_k \in \mathscr{C}^1(\mathbb{R}^n)$ with the result that $v = u_0 + \Sigma\, \partial u_k/\partial x_k \in \mathscr{C}^0(\mathbb{R}^n)$. Considering w the solution of $P(\Omega, \varphi)$ where φ is the trace of v on Γ, $u = v - w$ is the quasi-classical solution of $PH(\Omega, f_0 + \Sigma\, \partial f_k/\partial x_k)$

From Proposition 1, $u \in W^{1,p}(\Omega) = W^{1,p}_{\text{loc}}(\Omega)$ and if Γ is of class $W^{m+2,\infty}$ and $f = f_0 + \Sigma\, \partial f_k/\partial x_k \subset W^{m,p}(\Omega)$, then $u \in W^{m+2,p}(\Omega)$. The existence of a constant C follows from the closed graph theorem; similarly, we could make direct use of Lemma 2 with a partition of unity. □

Let us now give the

Proof of Lemma 2 (1st stage). Let us first demonstrate the *uniqueness* of a solution of (6.9); taking account of the linearity, we consider a function $u \in \mathscr{C}^0(\mathbb{R}^{n-1} \times \mathbb{R}^+)$ with derivatives $\partial u/\partial x_i \in L^2_{\text{loc}}(\mathbb{R}^n_+)$ satisfying $u(x', 0) = 0$ for all $x' \in \mathbb{R}^{n-1}$, $\lim_{|x| \to \infty} u(x) = 0$

$$\sum\sum \iint a_{ij} \frac{\partial u}{\partial x_j} \frac{\partial \zeta}{\partial x_i}\, dx = 0 \quad \text{for all} \quad \zeta \in \mathscr{D}(\mathbb{R}^n_+) \tag{6.13}$$

and show that $u \equiv 0$. For that let us take $F \in \mathscr{C}^1(\mathbb{R})$ with $F' \geqslant 0$ and $F \equiv 0$ in the neighbourhood of O. The function $w = F(u)$ is continuous on $\mathbb{R}^{n-1} \times \mathbb{R}^+$ with

[145] See Definition 6 of §4.

compact support in $\mathbb{R}^n_+$; we have $\partial w/\partial x_i = F'(u)\partial u/\partial x_i \in L^2(\mathbb{R}^n_+)$. Since $a_{ij}\partial u/\partial x_i \in L^2_{\text{loc}}(\mathbb{R}^n_+)$, on regularizing w and passing to the limit, we see that we can apply (6.13) with $\zeta = w$; that is to say

$$\int F'(u)\left(\sum\sum a_{ij}\frac{\partial u}{\partial x_j}\frac{\partial u}{\partial x_i}\right)\mathrm{d}x = 0\,. \tag{6.14}$$

Now

$$\begin{aligned}\sum\sum a_{ij}\frac{\partial u}{\partial x_j}\frac{\partial u}{\partial x_i} &= |\operatorname{grad} u|^2 + \sum\sum(a_{ij} - \delta_{ij})\frac{\partial u}{\partial x_i}\frac{\partial u}{\partial x_j}\\ &\geqslant |\operatorname{grad} u|^2\left(1 - \sum\sum\|a_{ij} - \delta_{ij}\|_{L^\infty}\right) \geqslant \frac{1}{2}|\operatorname{grad} u|^2\end{aligned}$$

on using the hypothesis and noting that, necessarily, the constant C of (6.8) satisfies $C \geqslant 1$.
Substituting in (6.14), we obtain

$$\int F'(u)|\operatorname{grad} u|^2\,\mathrm{d}x = 0 \quad \text{from which} \quad F'(u)|\operatorname{grad} u|^2 = 0\,,$$

then $\operatorname{grad} w = 0$ and hence $F(u) = w = 0$ on $\mathbb{R}^{n-1} \times \mathbb{R}^+$. This being true for every function $F \in \mathscr{C}^1(\mathbb{R})$ with $F' > 0$, $F \equiv 0$ in the neighbourhood of O, we obtain $u \equiv 0$ on $\mathbb{R}^{n-1} \times \mathbb{R}$.

Proof of Lemma 2 (2nd stage). We prove now the *existence* of a solution u of (6.9) *satisfying point* (a), $\partial u/\partial x_i \in L^p(\mathbb{R}^n_+)$. Using the hypothesis, we consider a compact

---

[146] This estimate (6.8) has been stated for $w \in \mathscr{D}(\mathbb{R}^n)$; by regularisation, it is still true for all $w \in W^{2,p}(\mathbb{R}^n)$ with compact support; to extend it to the Newtonian potential of $g \in L^p(\mathbb{R}^n)$ with compact support let us consider $\zeta \in \mathscr{D}(\mathbb{R}^n)$: we have

$$\zeta\frac{\partial^2 w}{\partial x_i \partial x_j} = \frac{\partial^2}{\partial x_i \partial x_j}(\zeta w) - \left(\frac{\partial \zeta}{\partial x_i}\frac{\partial w}{\partial x_j} + \frac{\partial \zeta}{\partial x_j}\frac{\partial w}{\partial x_i} + w\frac{\partial^2 \zeta}{\partial x_i \partial x_j}\right)$$

Applying (6.8) to ζw, we have therefore

$$\left\|\zeta\frac{\partial^2 w}{\partial x_i \partial x_j}\right\|_{L^p} \leqslant C\|\Delta(\zeta w)\|_{L^p} + \left\|\frac{\partial \zeta}{\partial x_i}\frac{\partial w}{\partial x_j}\right\|_{L^p} + \left\|\frac{\partial \zeta}{\partial x_j}\frac{\partial w}{\partial x_i}\right\|_{L^p} + \left\|w\frac{\partial^2 \zeta}{\partial x_i \partial x_j}\right\|_{L^p};$$

we also have with $\zeta_0 \equiv 1$ on $B(0, 1)$ and $\operatorname{supp}\zeta_0 \subset B(0, 2)$ and applying the inequalities with $\zeta_k(x) = \zeta_0(x/k)$. Using

$$\|\Delta(\zeta w)\|_{L^p} \leqslant \|\zeta\Delta w\|_{L^p} + \sum\left(2\left\|\frac{\partial \zeta}{\partial x_i}\frac{\partial w}{\partial x_i}\right\|_{L^p} + \left\|w\frac{\partial^2 \zeta}{\partial x_i^2}\right\|_{L^p}\right).$$

Let us fix $\zeta_0 \in \mathscr{D}(\mathbb{R}^n)$

$$w(x) = O\left(\frac{1}{|x|^{n-2}}\right), \quad \frac{\partial w}{\partial x_i}(x) = O\left(\frac{1}{|x|^{n-1}}\right),$$

$$\operatorname{supp}(\operatorname{grad}\zeta_k) \subset B(0, 2)\setminus B(0, 1)\,, \quad \operatorname{grad}\zeta_k(x) = \frac{1}{k}\operatorname{grad}\zeta_0\left(\frac{x}{k}\right)$$

we find that $\left\|\frac{\partial \zeta_k}{\partial x_i}\frac{\partial w}{\partial x_j}\right\|_{L^p}$ and $\left\|w\frac{\partial^2 \zeta_k}{\partial x_i \partial x_j}\right\|_{L^p}$ are $O(1/k^{\frac{p-1}{p}n})$ and therefore tend to zero as $k \to \infty$. Passing to the limit we obtain the stated estimate.

set K of $\mathbb{R}^{n-1} \times \mathbb{R}^+$ containing a neighbourhood of the supports of $a_{ij} - \delta_{ij}$ and f_k. We denote by $L^p(K)$ the space of functions $g \in L^p(\mathbb{R}^n)$ with support contained in K. Given $g \in L^p(K)$, we consider the Newtonian potential w of g. From Propositions 1, 2, 6 and 8 of §3: $w \in \mathscr{C}^1(\mathbb{R}^n)$, w is harmonic on $\mathbb{R}^n \backslash K$, when $|x| \to \infty$

$$w(x) = \left(\int g \, \mathrm{d}x\right) E_n(x) + O\left(\frac{1}{|x|^{n-1}}\right)$$

$$\frac{\partial w}{\partial x_i}(x) = \left(\int g \, \mathrm{d}x\right) \frac{\partial E_n}{\partial x_i}(x) + O\left(\frac{1}{|x|^n}\right) = O\left(\frac{1}{|x|^{n-1}}\right)$$

and the derivatives $\partial^2 w/\partial x_i \partial x_j$ with the estimate[146]

$$\left\| \frac{\partial^2 w}{\partial x_i \partial x_j} \right\|_{L^p} \leqslant C \|g\|_{L^p} .$$

The functions $w_0(x', x_n) = w(x', x_n) - w(x', -x_n)$ and

$$w_k(x', x_n) = \frac{\partial w}{\partial x_k}(x', x_n) - \frac{\partial w}{\partial x_k}(x', -x_n) \quad \textit{for } k = 1, \ldots, n$$

satisfy $w_0 \in \mathscr{C}^1(\mathbb{R}^{n-1} \times \mathbb{R}^+)$, $w_1, \ldots, w_n \in \mathscr{C}^0(\mathbb{R}^{n-1} \times \mathbb{R}^+)$,

$$w_k(x', 0) = 0 \quad \text{for all} \quad x' \in \mathbb{R}^{n-1} ,$$

$$\lim_{|x| \to \infty} w_k(x) = 0 \quad \text{for} \quad k = 0, \ldots, n .$$

Also $\partial^2 w_0/\partial x_i \partial x_j \in L^p(\mathbb{R}^n_+)$ with $\|\partial^2 w_0/\partial x_i \partial x_j\|_{L^p} \leqslant 2C \|g\|_{L^p}$

$$\frac{\partial w_k}{\partial x_i} \in L^p(\mathbb{R}^n_+) \quad \text{with} \quad \left\| \frac{\partial w_k}{\partial x_i} \right\|_{L^p} \leqslant 2C \|g\|_{L^p} .$$

Finally since w is harmonic on $\mathbb{R}^{n-1} \times]-\infty, 0[$,

$$\Delta w_0 = g \quad \text{and} \quad \Delta w_k = \frac{\partial g}{\partial x_k} \quad \text{in} \quad \mathscr{D}'(\mathbb{R}^n_+) .$$

We now denote by $w_k(g)$ the function defined in this way starting from $g \in L^p(K)$. Now we put $\bar{f}_k = f_k + \sum_j (\delta_{kj} - a_{kj}) \dfrac{\partial w_0(f_0)}{\partial x_j}$ for $k = 1, \ldots, n$. A function u is a solution of (6.9), iff $\bar{u} = u - w_0(f)$ is a solution corresponding to $\bar{f}_0 = 0$, $\bar{f}_1, \ldots, \bar{f}_n$ which belong to $L^p(K)$. For $g = (g_1, \ldots, g_n) \in L^p(K)^n$, we consider

$$v(g) = \sum_{k=1}^n w_k(g_k) \quad \text{and} \quad T_i(g) = \sum_j (\delta_{ij} - a_{ij}) \frac{\partial v(g)}{\partial x_j} .$$

We have $v(g) \in \mathscr{C}^0(\mathbb{R}^{n-1} \times \mathbb{R}^+)$, $v(g)(x', 0) = 0$ for all $x' \in \mathbb{R}^{n-1}$,

$$\lim_{|x| \to \infty} v(g)(x) = 0 , \quad \frac{\partial v(g)}{\partial x_j} \in L^p(\mathbb{R}^n_+)$$

and

$$\sum\sum \frac{\partial}{\partial x_i}\left(a_{ij}\frac{\partial v(g)}{\partial x_j}\right) = \sum \frac{\partial}{\partial x_i}(g_i - T_i(g)) \quad \text{in} \quad \mathscr{D}'(\mathbb{R}^n_+)\,.$$

We are therefore assured that $\bar{u} = v(g)$ is a solution of (6.9) corresponding to $O_1, \bar{f}_1, \ldots, \bar{f}_n$ if

$$g_i - T_i(g) = \bar{f}_i \quad \text{for} \quad i = 1, \ldots, n\,.$$

Now for all $g \in L^p(K)^n$, $T_i(g) \in L^p(K)$ and

$$\| T_i(g)\|_{L^p} \leqslant \sum_j \|\delta_{ij} - a_{ij}\|_{L^\infty}\left(\sum_k \left\|\frac{\partial w_k(g_k)}{\partial x_j}\right\|_{L^p}\right),$$

from which

$$\sum_i \| T_i(g)\|_{L^p} \leqslant c\sum_k \|g_k\|_{L^p}$$

with

$$c = 2C\sum\sum \|\delta_{ij} - a_{ij}\|_{L^\infty}\,.$$

From the hypothesis, $c > 1$; hence the map

$$T(g) = (T_1(g), \ldots, T_n(g))$$

is a (strict) linear contraction of $L^p(K)^n$ into itself. Hence $I - T$ is an isomorphism of $L^p(K)^n$ onto itself and there exists a unique $g \in L^p(K)^n$ such that $g - Tg = (\bar{f}_1, \ldots, \bar{f}_n)$. For this g, the function $u = w_0(f_0) + v(g)$ is the solution of (6.9). Finally note that $\partial u/\partial x_j \in L^p(\mathbb{R}^n_+)$; we know that $\dfrac{\partial v(g)}{\partial x_j} \in L^p(\mathbb{R}^n_+)$ and $w_0(f_0) \in \mathscr{C}^1(\mathbb{R}^{n-1} \times \mathbb{R}^+)$; but $\dfrac{\partial w_0(g_0)}{\partial x_i} = O\left(\dfrac{1}{|x|^{n-1}}\right)$ when $|x| \to \infty$, so for $p > n \geqslant 2$

$$\frac{\partial w_0(g_0)}{\partial x_i} \in L^p(\mathbb{R}^n)^{147}\,.$$

Proof of Lemma 2 (3rd stage). We shall prove *the point* (b) of the lemma *for* $m = 0$.

By hypothesis $f = f_0 + \sum \partial f_k/\partial x_k \in L^p(K)$ and $a_{ij} \in W^{1,\infty}(\mathbb{R}^n_+)$. The solution u of (6.9) can be written $u = \bar{u} + w_0(f)$ where $\bar{u}$ is the solution of (6.9) corresponding to $\bar{f}_0 = 0, \bar{f}_k = \sum(\delta_{kj} - a_{kj})\partial w_0(f)/\partial x_j$. Since

$$w_0(f) \in \mathscr{C}^1(\mathbb{R}^{n-1} \times \mathbb{R}^+) \quad \text{and} \quad \frac{\partial^2 w_0(f)}{\partial x_i \partial x_j} \in L^p(\mathbb{R}^n_+)\,,$$

we have $\bar{f}_k \in W^{1,p}(\mathbb{R}^n_+)$ and the conclusion

$$u \in \mathscr{C}^1(\mathbb{R}^{n-1} \times \mathbb{R}^+) \quad \text{and} \quad \frac{\partial^2 u}{\partial x_i \partial x_j} \in L^p(\mathbb{R}^n_+)$$

---

[147] We obviously suppose that $n \geqslant 2$, the case $n = 1$ being trivial throughout this section.

holds iff it does for $\bar{u}$. In other words, *we can always suppose* $f_0 = 0$ and $f_k \in W^{1,p}(\mathbb{R}^n_+)$, which we are doing. With the notation of the proof of the 2nd stage, the solution u is equal to $v((I - T)^{-1}(f_1, \ldots, f_n))$ with the result that we have the estimate

$$\left\| \frac{\partial u}{\partial x_j} \right\|_{L^p} \leqslant C' \sum \| f_k \|_{L^p} \quad \text{for} \quad j = 1, \ldots, n \tag{6.15}$$

with $C' = 2C(1 - c)^{-1}$, with $c = 2C \sum \sum \| a_{ij} - \delta_{ij} \|_{L^\infty}$[148].

We now prove that for $i = 1, \ldots, n - 1$ and all $j = 1, \ldots, n$ we have $\partial^2 u / \partial x_i \partial x_j \in L^p(\mathbb{R}^n_+)$. For that we shall use the *method of translations*: given $h = (h', 0) \in \mathbb{R}^{n-1} \times \{0\}$, let us consider $\bar{u}(x) = u(x + h) - u(x)$. The function $\bar{u}$ is a solution of (6.9) corresponding to $\bar{f}_0 = 0$ (we have supposed that $f_0 = 0$) and

$$\bar{f}_k(x) = f_k(x + h) - f_k(x) + \sum_l (a_{kl}(x + h) - a_{kl}(x)) \frac{\partial u}{\partial x_l}(x + h) .$$

Applying (6.15) to $\bar{u}$, we have for $j = 1, \ldots, n$

$$\begin{aligned} \left\| \frac{\partial u}{\partial x_j}(. + h) - \frac{\partial u}{\partial x_j} \right\|_{L^p} = \left\| \frac{\partial \bar{u}}{\partial x_j} \right\|_{L^p} &\leqslant C' \Big\{ \sum \| f_k(. + h) - f_k \|_{L^p} \\ &+ \sum \sum \| a_{kl}(. + h) - a_{kl} \|_{L^\infty} \left\| \frac{\partial u}{\partial x_l} \right\|_{L^p} \Big\} \end{aligned}$$

and hence, h' being arbitrary in $\mathbb{R}^{n-1}$, for $i = 1, \ldots, n - 1$[149]

$$\frac{\partial^2 u}{\partial x_i \partial x_j} \in L^p(\mathbb{R}^n_+)$$

and

$$\left\| \frac{\partial^2 u}{\partial x_i \partial x_j} \right\|_{L^p} \leqslant C' \Big\{ \sum \left\| \frac{\partial f_k}{\partial x_i} \right\|_{L^p} + \sum \sum \left\| \frac{\partial a_{kl}}{\partial x_i} \right\|_{L^\infty} \left\| \frac{\partial u}{\partial x_l} \right\|_{L^p} \Big\} .$$

We have also $\partial^2 u / \partial x_n^2 \in L^p(\mathbb{R}^n_+)$. Indeed

$$\left\{ \begin{aligned} a_{nn} \frac{\partial^2 u}{\partial x_n^2} &= \sum \frac{\partial f_k}{\partial x_k} - \sum_{i=1}^{n-1} \sum_{j=1}^{n-1} \left(a_{ij} \frac{\partial^2 u}{\partial x_i \partial x_j} + \frac{\partial a_{ij}}{\partial x_i} \frac{\partial u}{\partial x_i^j} \right) \\ &- \frac{\partial a_{nn}}{\partial x_n} \frac{\partial u}{\partial x_n} - 2 \sum_{i=1}^{n-1} \left(a_{in} \frac{\partial^2 u}{\partial x_i \partial x_n} + \frac{\partial a_{in}}{\partial x_n} \frac{\partial u}{\partial x_i} \right) . \end{aligned} \right. \tag{6.16}$$

---

[148] If T is a strict linear contraction of a Banach space X,

$$(I - T)^{-1} = \sum_{m \geqslant 0} T^m \quad \text{and} \quad \|(I - T)^{-1}\| \leqslant \sum \| T \|^m = (1 - \| T \|)^{-1} .$$

[149] Given $1 < p \leqslant \infty$ and $f \in L^p(\Omega)$ we have $\partial f / \partial x_i \in L^p(\Omega)$ if and only if $\| h \|^{-1} \| f(. + h) - f \|_{L^p(\Omega \cap (\Omega - h))}$ is bounded for h a vector in the i-th direction.

From what we have proved $a_{nn}\partial^2 u/\partial x_n^2 \in L^p(\mathbb{R}^n_+)$. But

$$a_{nn} \geqslant 1 - |a_{nn} - 1| \geqslant 1 - (2C)^{-1} \geqslant \frac{1}{2}, \quad \text{therefore} \quad \frac{\partial^2 u}{\partial x_n^2} \in L^p(\mathbb{R}^n_+) .$$

It remains to prove that $u \in \mathscr{C}^1(\mathbb{R}^{n-1} \times \mathbb{R}^+)$. The function u is the solution of the homogeneous Dirichlet problem $PH(\mathbb{R}^n_+, f)$ tending to zero at infinity with $f = \sum \frac{\partial f_k}{\partial x_k} + \sum\sum \frac{\partial}{\partial x_i}(\delta_{ij} - a_{ij})\frac{\partial u}{\partial x_j} \in L^p(K)$.

Therefore
$$u = w_0(f) \in \mathscr{C}^1(\mathbb{R}^{n-1} \times \mathbb{R}^+) .$$

Proof of Lemma 2 (Final stage). We shall prove the *point (b) for arbitrary* $m \in \mathbb{N}$ by induction.
By hypothesis

$$f = f_0 + \sum \frac{\partial f_k}{\partial x_k} \in W^{m,p}(\mathbb{R}^n_+) \quad \text{and} \quad a_{ij} \in W^{m+1,\infty}(\mathbb{R}^n_+)$$

For $l = 1, \ldots, n-1$, the function $\bar{u} = \partial u/\partial x_l$ is a solution of (6.9) (we know that $u \in \mathscr{C}^1(\mathbb{R}^{n-1} \times \mathbb{R}^+)$ and $\partial^2 u/\partial x_i \partial x_j \in L^p(\mathbb{R}^n_+)$) with

$$\bar{f}_0 = \frac{\partial f}{\partial x_l}, \qquad \bar{f}_k = -\sum \frac{\partial a_{kj}}{\partial x_l}\frac{\partial u}{\partial x_j} .$$

Using the induction hypothesis we have

$$u \in W^{m+1,p}(\mathbb{R}^n_+) \quad \text{and hence} \quad \bar{f}_0 + \sum \frac{\partial \bar{f}_k}{\partial x_k} \in W^{m-1,p}(\mathbb{R}^n_+) .$$

Applying the same hypothesis to $\bar{u}$, we have $\partial u/\partial x_l \in W^{m+1,p}(\mathbb{R}^n_+)$.
Thus we have shown that all the derivatives of u of order $\leqslant m+2$ were in $L^p(\mathbb{R}^n_+)$, with the exception of $\partial^{m+2}u/\partial x_n^{m+2}$. But from (6.16), $\partial^2 u/\partial x_n^2 \in W^{m,p}(\mathbb{R}^n_+)$, from which the result follows. □

Lemma 2 and hence Proposition 1 can be extended to the case of the *Neumann problem.* We refer to Agmon–Douglis–Nirenberg [1] for the general study of the regularity $W^{m,p}$. Here, we shall prove Lemma 1 and Theorem 1 for the Neumann problem using *Schauder's method* based on the *Hölder estimates.* Given $0 < \theta < 1$, for a function u defined on a part Ω of $\mathbb{R}^n$, we denote

$$[u]_{\Omega,\theta} = \sup_{\substack{x,y \in \Omega \\ x \neq y}} \frac{|u(x) - u(y)|}{|x - y|^\theta} . \tag{6.17}$$

We have $[u]_{\Omega,\theta} < \infty$ iff u satisfies a Hölder condition of order θ uniformly on Ω[150]. We notice that $[.]_{\Omega,\theta}$ is not a norm since $[u]_{\Omega,\theta} = 0$ iff u is constant on Ω. However, if Ω is bounded, $[.]_{\Omega,\theta}$ is a norm on the space $\{u \in \mathscr{C}^\theta(u); u = 0 \text{ on } \partial\Omega\}$;

[150] We then denote: $u \in \mathscr{C}^\theta(\bar{\Omega})$. We also use the notation $u \in \mathscr{C}^{0,\theta}(\bar{\Omega})$, (see §3, Prop. 9).

this follows from the more general inequality

(6.17)′
$$\max_{\bar{\Omega}}|u| \leqslant \max_{\partial\Omega}|u| + R^{\theta}[u]_{\Omega,\theta}$$

where
$$R = \max_{x \in \Omega} \operatorname{dist}(x, \partial\Omega)\,.$$

Proposition 9 of §3 shows that for every function $f \in \mathscr{C}^{\theta}(\mathbb{R}^n)$ with compact support, the Newtonian potential u of f satisfies

(6.18)
$$\left[\frac{\partial^2 u}{\partial x_i \partial x_j}\right]_{\mathbb{R}^n,\theta} \leqslant C[f]_{\mathbb{R}^n,\theta}$$

where because of the homogeneity (see also the proof of Proposition 9 of §3), C depends only on n and θ. We notice also from Proposition 2 of §3 that $\lim_{|x|\to\infty} \partial^2 u/\partial x_i \partial x_j(x) = 0$; the function $\partial^2 u/\partial x_i \partial x_j$, being harmonic on $\mathbb{R}^n \setminus \operatorname{supp} f$ attains its maximum on $\operatorname{supp} f$. An explicit calculation or the use of the closed graph theorem and of a homogeneity argument, shows that

(6.18)′
$$\max_{\mathbb{R}^n}\left|\frac{\partial^2 u}{\partial x_i \partial x_j}\right| \leqslant CR^{\theta}[f]_{\mathbb{R}^n,\theta}$$

where C depends only on n and θ, and R is the radius of $\operatorname{supp} f$

$$R = \frac{1}{2}\max_{x,y \in \operatorname{supp} f}|x - y| = \min_{x_0 \in \operatorname{supp} f}\ \max_{x \in \operatorname{supp} f}|x_0 - x|\,.$$

Let us state

Lemma 3. *Let $0 < \theta < 1$, $R > 0$, $f_1, \ldots, f_n \in \mathscr{C}^{\theta}(\mathbb{R}^{n-1} \times \mathbb{R}^+)$ with supports contained in $\bar{B}(O, R)$ and let $\psi \in \mathscr{C}^{\theta}(\mathbb{R}^{n-1})$ with $\operatorname{supp}\psi \subset \bar{B}(0, R)$. Then there exists one and only one quasi-classical solution u of the Neumann problem $PN(\mathbb{R}^n_+, \psi, \Sigma\, \partial f_k/\partial x_k)$ satisfying*

$$\lim_{|x|\to\infty} u(x) + 2\left(\int \psi\, dx'\right)E_n(x) = 0\,.$$

This solution $u \in \mathscr{C}^{1,\theta}(\mathbb{R}^{n-1} \times \mathbb{R}^+)$ and we have an estimate (in which C depends only on n and θ):–

(6.19)
$$\left[\frac{\partial u}{\partial x_i}\right]_{\mathbb{R}^n_+,\theta} + R^{-\theta}\max_{\mathbb{R}^{n-1}\times\mathbb{R}^+}\left|\frac{\partial u}{\partial x_i}\right| \leqslant C\left\{\sum [f_k]_{\mathbb{R}^n_+,\theta} + [\psi]_{\mathbb{R}^{n-1},\theta}\right\}$$

Proof of Lemma 3. The uniqueness follows from Corollary 6 of §4 (see also the lemma below). For the existence and the estimate (6.19), we consider separately the problems

$$PN\left(\mathbb{R}^n_+, 0, \frac{\partial f_k}{\partial x_k}\right) \quad \text{and} \quad PN(\mathbb{R}^n_+, \psi, 0)\,.$$

For $k = 1, \ldots, n-1$, we consider $\tilde{f}_k(x', x_n) = f_k(x', |x_n|)$. We have $\tilde{f}_k \in \mathscr{C}^{\theta}(\mathbb{R}^n)$

with $\operatorname{supp}\tilde{f}_k \subset \bar{B}(0, R)$. The Newtonian potential $\tilde{u}_k$ of $\tilde{f}_k$ is even with respect to x_n; hence the restriction u_k of $\partial\tilde{u}/\partial x_k$ is solution of $PN(\mathbb{R}^n_+, 0, \partial f_k/\partial x_k)$. Applying (6.18) and (6.18)′, we obtain (6.19) for u_k.

For $k = n$, let us consider $\tilde{f}_n(x', x_n) = \dfrac{x_n}{|x_n|}(f_n(x', x_n) - f_n(x', 0))$.

Again we have $\tilde{f}_n \in \mathscr{C}^\theta(\mathbb{R}^n)$ and $\partial\tilde{f}/\partial x_n = \partial f/\partial x_n$ in $\mathscr{D}'(\mathbb{R}^n_+)$. The Newtonian potential $\tilde{u}_n$ of $\tilde{f}_n$ being odd, the restriction u_n of $\partial\tilde{u}/\partial x_n$ is the solution of $PN(\mathbb{R}^n_+, 0, \partial f_n/\partial x_n)$.

Finally, following Example 21 of §4, we find the function $u(x', x_n)$ defined by

$$u(x', x_n) = -2\int_{\mathbb{R}^{n-1}} E_n(x' - t, x_n)\psi(t)\,dt$$

is the solution of $PN(\mathbb{R}^n_+, \psi, 0)$ satisfying

$$\lim_{|x|\to\infty} u(x) + 2\left(\int \psi\,dx'\right)E_n(x) = 0\,.$$

We have
$$\frac{\partial u}{\partial x_i}(x', x_n) = -2\int_{\mathbb{R}^{n-1}} \frac{\partial E_n}{\partial x_i}(x' - t, x_n)\psi(t)\,dt\,,$$

$$\frac{\partial u}{\partial x_n}(x', x_n) = -\frac{2}{\sigma_n}\int_{\mathbb{R}^{n-1}} \frac{\psi(x' + tx_n)\,dt}{(1 + |t|^2)^{n/2}}$$

$$\frac{\partial u}{\partial x_i}(x', x_n) = -\frac{2}{\sigma_n}\int_{\mathbb{R}^{n-1}} \frac{\psi(x' + tx_n)t_i\,dt}{(1 + |t|^2)^{n/2}}\,.$$

It is clear that $\partial u/\partial x_n$ satisfies a Hölder condition:

$$\begin{aligned}\left|\frac{\partial u}{\partial x_n}(x', x_n) - \frac{\partial u}{\partial x_n}(y', y_n)\right| &\leqslant \frac{2}{\sigma_n}[\psi]_{\mathbb{R}^{n-1},\theta}\int \frac{(|x' - y'|^2 + |t|^2(x_n - y_n)^2)^{\theta/2}\,dt}{(1 + |t|^2)^{n/2}} \\ &\leqslant \frac{2\sigma_{n-1}}{\sigma_n}\left(\int_0^1 \frac{r^{n-2}\,dr}{(1 + r^2)^{n/2}} + \int_1^\infty \frac{r^{n-2+\theta}}{(1 + r^2)^{n/2}}\,dr\right) \\ &\quad \times [\psi]_{\mathbb{R}^{n-1},\theta}|x - y|^\theta\,.\end{aligned}$$

We also have immediately

$$\left|\frac{\partial u}{\partial x_n}(x', x_n)\right| \leqslant \max_{\mathbb{R}^{n-1}}|\psi| \leqslant R^{-\theta}[\psi]_{\mathbb{R}^{n-1},\theta}\,.$$

To prove the same results for $\partial u/\partial x_i$, $i = 1, \ldots, n - 1$, we make use of $\displaystyle\int_{B(0,R)} \frac{t_i}{(1 + |t|^2)^{n/2}}\,dt = 0$. We shall not give the details of the proof which run along identical (if not simpler) lines to those of Proposition 9 of §3 (see Gilbarg–Trudinger [1] and also Appendix “Singular integrals” Vol. 4). □

Let us take $a_{ij} = a_{ji} \in \mathscr{C}^\theta(\mathbb{R}^{n-1} \times \mathbb{R}^+)$ with $\operatorname{supp}(a_{ij} - \delta_{ij}) \subset B(0, R)$. Given $f_1, \ldots, f_n$, ψ as in Lemma 3, let us consider u the solution of

$PN(\mathbb{R}^n_+, \psi, \sum \partial f_k/\partial x_k)$. We have:

$$\begin{cases} \sum\sum \dfrac{\partial}{\partial x_i}\left(a_{ij}\dfrac{\partial u}{\partial x_j}\right) = \sum \dfrac{\partial}{\partial x_k}(f_k + \bar{f}_k) \quad \text{in} \quad \mathscr{D}'(\mathbb{R}^n_+) \\ \sum a_{in}(x',0)\dfrac{\partial u}{\partial x_i}(x',0) = -(\psi(x') + \bar{\psi}(x')) \quad \text{for all} \quad x' \in \mathbb{R}^{n-1} \end{cases}$$

with

$$\bar{f}_k = \sum (a_{kj} - \delta_{kj})\frac{\partial u}{\partial x_j}, \qquad \bar{\psi}(x') = \sum (\delta_{in} - a_{in}(x',0))\frac{\partial u}{\partial x_i}(x',0).$$

We have $(a_{kj} - \delta_{kj})\dfrac{\partial u}{\partial x_j} \in \mathscr{C}^\theta(\mathbb{R}^{n-1} \times \mathbb{R}^+)$ with support contained in $\bar{B}(0,R)$, and taking account of (6.17)′ we find that

$$\left[(a_{kj} - \delta_{kj})\frac{\partial u}{\partial x_j}\right]_{\mathbb{R}^n_+,\theta} \leqslant R^\theta [a_{kj}]_{\mathbb{R}^n_+,\theta}\left(\left[\frac{\partial u}{\partial x_j}\right]_{\mathbb{R}^n_+,\theta} + R^{-\theta}\max\left|\frac{\partial u}{\partial x_j}\right|\right).$$

Reasoning similarly with $(\delta_{in} - a_{in}(.,0))\,\partial u/\partial x_i(.,0)$, we see that $T: (f_1, \ldots, f_n, \psi) \mapsto (\bar{f}_1, \ldots, \bar{f}_n, \bar{\psi})$ is a linear map of the space $X = \{(f_1, \ldots, f_n, \psi)\}$ into itself; in addition, norming X by $\Sigma[f_k]_{\mathbb{R}^n_+,\theta} + [\psi]_{\mathbb{R}^{n-1},\theta}$ which makes it a Banach space, T is continuous with norm less than or equal to

$$CR^\theta\left(\sum\sum [a_{ij}]_{\mathbb{R}^n_+,\theta} + \sum [a_{in}(.,0)]_{\mathbb{R}^{n-1},\theta}\right).$$

Following then the proof of Lemma 2, step by step, we shall prove the

Lemma 4. *Let* $0 < \theta < 1$, $R > 0$, $a_{ij} = a_{ji} \in \mathscr{C}^\theta(\mathbb{R}^{n-1} \times \mathbb{R}^+)$ *satisfying* $\operatorname{supp}(a_{ij} - \delta_{ij}) \subset \bar{B}(0,R)$ *and*

$$c = CR^\theta\left(\sum\sum [a_{ij}]_{\mathbb{R}^n_+,\theta} + \sum [a_{in}(.,0)]_{\mathbb{R}^{n-1},\theta}\right) < 1 \tag{6.20}$$

where C is the constant of (6.19). Then for all $f_0, f_1, \ldots, f_n \in \mathscr{C}^\theta(\mathbb{R}^{n-1} \times \mathbb{R}^+)$ with supports continuous in $\bar{B}(0,R)$, and $\psi \in \mathscr{C}^\theta(\mathbb{R}^{n-1})$, there exists one and only one solution of the problem:

$$\begin{cases} u \in \mathscr{C}^1(\mathbb{R}^n_+), \qquad Bu = \sum a_{in}\dfrac{\partial u}{\partial x_i} \in \mathscr{C}^0(\mathbb{R}^{n-1} \times \mathbb{R}^+) \\ \sum\sum \dfrac{\partial}{\partial x_i}\left(a_{ij}\dfrac{\partial u}{\partial x_j}\right) = f_0 + \sum \dfrac{\partial f_k}{\partial x_k} \quad \textit{in} \quad \mathscr{D}'(\mathbb{R}^n_+) \\ Bu(x',0) = \psi(x') \quad \textit{for all} \quad x' \in \mathbb{R}^{n-1} \\ \lim\limits_{|x|\to\infty} u(x) + 2\left(\displaystyle\int_{\mathbb{R}^{n-1}} \psi\,dx' - \int_{\mathbb{R}^n_+} f_0\,dx\right)E_n(x) = 0\,. \end{cases} \tag{6.21}$$

In addition
(*a*) $u \in \mathscr{C}^{1,\theta}(\mathbb{R}^{n-1} \times \mathbb{R}^+)$;
(*b*) *if for* $m \in \mathbb{N}$, $a_{ij} \in \mathscr{C}^{m+1,\theta}(\mathbb{R}^{n-1} \times \mathbb{R}^+)$ *then*

$$u \in \mathscr{C}^{m+2,\theta}(\mathbb{R}^{n-1} \times \mathbb{R}^+) \quad \textit{iff} \quad f_0 + \sum \frac{\partial f_k}{\partial x_k} \in \mathscr{C}^{m,\theta}(\mathbb{R}^{n-1} \times \mathbb{R}^+)$$

and
$$\psi \in \mathscr{C}^{m+1,\theta}(\mathbb{R}^{n-1}) .$$

Proof of Lemma 1 (Neumann problem). Reasoning as for the Dirichlet problem, we have only to prove that there exists $R > 0$ and $\rho \in \mathscr{C}^\infty(\mathbb{R}^{n-1} \times \mathbb{R})$ with compact support contained in $\mathscr{O} \times [0, \delta[\cap \bar{B}(0, R)$ and $\rho \equiv 1$ in the neighbourhood of O such that $\bar{a}_{ij} = \delta_{ij} + \rho(a_{ij} - \delta_{ij})$ satisfies (6.20). We fix $\rho_0 \in \mathscr{C}^\infty(\mathbb{R}^{n-1} \times \mathbb{R}^+)$ with $\operatorname{supp} \rho_0 \subset B(0, 2)$ and $\rho_0 = 1$ on $\mathbb{R}^{n-1} \times \mathbb{R}^+ \cap B(0, 1)$ and look for ρ of the form $\rho(x) = \rho_0(x/R)$. We have $[\rho]_{\mathbb{R}^n_+,\theta} = R^{-\theta}[\rho_0]_{\mathbb{R}^n_+,\theta}$ and $\|\rho\|_{L^\infty} = \|\rho_0\|_{L^\infty}$, so

$$[\bar{a}_{ij}]_{\mathbb{R}^n_+,\theta} \leqslant R^{-\theta}[\rho_0]_{\mathbb{R}^n_+,\theta} \|a_{ij} - \delta_{ij}\|_{L^\infty(B(0,R))} + \|\rho_0\|_{L^\infty}[a_{ij}]_{\mathbb{R}^n_+ \cap B(0,R),\theta} .$$

From the hypothesis $a_{ij}(0) = \delta_{ij}$, for $R > 0$ sufficiently small $R^\theta[\bar{a}_{ij}]_{\mathbb{R}^n_+,\theta}$ can be made as small as we please and hence satisfy (6.20). □

As we have said, the proof of Lemma 4 follows, step by step, that of Lemma 2; we leave the details to the reader (see also Gilbarg–Trudinger [1]). We shall give exactly only the proof of uniqueness:

Proof of Lemma 4 (Uniqueness). By linearity we are led to show that a solution u of (6.21) corresponding to $f_0 = f_1 = \cdots = f_n = 0$, $\psi = 0$ is identically zero. We take $F \in \mathscr{C}^1(\mathbb{R}^n_+)$ with $F'(u) > 0$ and $F = 0$ in the neighbourhood of O. Then $w = F(u) \in \mathscr{C}^1(\mathbb{R}^n_+)$ and is of bounded support. Given $\zeta_n \in \mathscr{C}^1(\mathbb{R}^+)$ with $\operatorname{supp} \mathrm{f}_n \subset]0, \infty[$ the function $\zeta(x', x_n) = w(x', x_n)\zeta_n(x_n)$ is of class $\mathscr{C}^1$ on $\mathbb{R}^n_+$ with compact support in $\mathbb{R}^n_+$, therefore

$$\sum\sum \int a_{ij} \frac{\partial u}{\partial x_j} \frac{\partial \zeta}{\partial x_i} \mathrm{d}x = 0 ,$$

which can also be written

$$\int F'(u)\zeta_n(x_n) \sum\sum a_{ij} \frac{\partial u}{\partial x_j} \frac{\partial u}{\partial x_i} \mathrm{d}x + \int \zeta_n'(x_n) w Bu \, \mathrm{d}x = 0 .$$

Use of the hypothesis (6.20), gives $\sum\sum a_{ij} \dfrac{\partial u}{\partial x_i} \dfrac{\partial u}{\partial x_j} \geqslant (1 - c)|\operatorname{grad} u|^2$. Choosing

$\zeta_n(x_n) = \zeta(x_n/\varepsilon)$ where $\zeta \in \mathscr{C}^1(\mathbb{R}^+)$ satisfies $\zeta' \geqslant 0$, $\zeta = 0$ on $[0, 1]$, $\zeta = 1$ on $[2, \infty[$, we have

$$\int_{\mathbb{R}^{n-1} \times]2\varepsilon, +\infty[} F'(u)|\operatorname{grad} u|^2 \, \mathrm{d}x \leqslant -\frac{1}{\varepsilon} \int_{\mathbb{R}^{n-1} \times]\varepsilon, 2\varepsilon[} \zeta'\left(\frac{x_n}{\varepsilon}\right) w Bu \, \mathrm{d}x .$$

Since $Bu \in \mathscr{C}^0(\mathbb{R}^{n-1} \times \mathbb{R}^+)$ and $Bu(x', 0) = 0$, passing to the limit as $\varepsilon \to 0$, we have $\int_{\mathbb{R}^n_+} F'(u)|\operatorname{grad} u|^2 \, dx \leqslant 0$. We conclude as in the proof of Lemma 2 (1st stage). ☐

2. Analytic Regularity and Trace on the Boundary of a Harmonic Function

Let us take an open set of Ω of $\mathbb{R}^n$ and Γ_0 a regular open part of its boundary which we shall suppose to be *analytic*[151]. We shall use the *Cauchy–Kovalevsky theorem* (see Chapter V, §1.5): there exists an open neighbourhood U of Γ_0 such that for all functions f, φ, ψ analytic on $\Omega \cup \Gamma_0$ [152], there exists one and only one solution u, analytic on $(U \cap \Omega) \cup \Gamma_0$, of the *Cauchy problem*

$$
\text{(6.22)} \qquad \begin{cases} \Delta u = f & \text{on} \quad \Omega \cap U \\ u = \varphi & \text{on} \quad \Gamma_0 \\ \dfrac{\partial u}{\partial n} = \psi & \text{on} \quad \Gamma_0 \,. \end{cases}
$$

Notice the difference between the Cauchy problem, in which u **and** $\partial u / \partial n$ are given on a part of the boundary of Ω, and the Dirichlet (resp. Neumann) problem in which u (resp. $\partial u/\partial n$) is given over the whole boundary of Ω. We notice first the *analytic regularity* of the solutions of the Cauchy and Neumann problems.

Proposition 2. *Let Ω be an open set in $\mathbb{R}^n$, Γ_0 an analytic regular open part of its boundary $\partial\Omega$, and let φ (resp. ψ) be an analytic function on Γ_0, f an analytic function on $\Omega \cup \Gamma_0$ and $u \in \mathscr{C}^0(\Omega \cup \Gamma_0)$ (resp. $\mathscr{C}^1_n(\Omega \cup \Gamma_0)$) solution of*

$$
\begin{cases} \Delta u = f \quad \text{in} \quad \mathscr{D}'(\Omega) \\ u(z) = \varphi(z) \quad \left(\text{resp.} \ \dfrac{\partial u}{\partial n}(z) = \psi(z) \right) \quad \text{for all} \quad z \in \Gamma_0 \,. \end{cases}
$$

Then u is analytic on $\Omega \cup \Gamma_0$.

Proof. We know already (see Theorem 1) that u is of class $\mathscr{C}^\infty$ on $\Omega \cup \Gamma_0$. Using the Cauchy–Kovalevsky theorem we see that there exists an open neighbourhood U of Γ_0 and u_0 analytic on $(\Omega \cap U) \cup \Gamma_0$ and solution of the Cauchy problem

$$
\begin{cases} \Delta u_0 = f \quad \text{on} \quad \Omega \cap U \\ u_0 = \varphi \quad (\text{resp. } u_0 = 0) \quad \text{on} \quad \Gamma_0 \\ \dfrac{\partial u_0}{\partial n} = 0, \quad \left(\text{resp.} \ \dfrac{\partial u}{\partial n} = \psi \right) \quad \text{on} \quad \Gamma_0 \,. \end{cases}
$$

---

[151] That is to say that for every local parametric representation $(R, U, \mathcal{O}, \alpha)$ of Γ_0, the function α is analytic on $\mathcal{O}$.

[152] I.e. can be developed as a power series in the neighbourhood of each point of $\Omega \cup \Gamma_0$.

The function $\bar{u} = u - u_0$ is therefore harmonic on $\Omega \cap U$, of class $\mathscr{C}^\infty$ on $(\Omega \cup U) \cup \Gamma_0$ and satisfies

$$\bar{u} = 0 \quad \left(\text{resp. } \frac{\partial \bar{u}}{\partial n} = 0\right) \quad \text{on} \quad \Gamma_0 .$$

We are therefore led, at least for analytic regularity in the neighbourhood of the boundary, to the case $f \equiv 0$ and $\varphi \equiv 0$ (resp. $\psi \equiv 0$), which we now assume.
In the case where $\Omega = \mathcal{O} \times]0, \delta[$ and $\Gamma_0 = \mathcal{O} \times \{0\}$, $\mathcal{O}$ being an open set of $\mathbb{R}^{n-1}$, the analyticity on $\mathcal{O} \times]0, \delta[$ of a solution $u \in \mathscr{C}^1(\Omega \cup \Gamma_0)$ of

$$\begin{cases} \Delta u = 0 \quad \text{on} \quad \Omega \\ u = 0 \quad \left(\text{resp. } \dfrac{\partial u}{\partial x_n} = 0\right) \quad \text{on} \quad \Gamma_0 \end{cases}$$

is immediate. Indeed, the function $\tilde{u}(x', x_n) = \dfrac{x_n}{|x_n|} u(x', |x_n|)$ (resp. $\tilde{u}(x', x_n) = u(x', |x_n|)$) is clearly of class $\mathscr{C}^1$ on $\mathcal{O} \times]-\delta, \delta[$; being harmonic on $\mathcal{O} \times]-\delta, \delta[\setminus \Gamma_0$ it is so on all of $\mathcal{O} \times]-\delta, \delta[$ (see Proposition 17 of §2); hence u extends to an analytic function in the neighbourhood of Γ_0. Using a Kelvin transformation (see §2.5) we shall have the same result if Γ_0 is contained in a sphere of $\mathbb{R}^n$.
We now come to the general case. Reasoning as in Theorem 1, by the use of a local parametric representation, we are led to prove that, given $\mathcal{O}$ an open neighbourhood of O in $\mathbb{R}^{n-1}$, $\delta > 0$, and $a_{ij} = a_{ji}$ analytic on $\mathcal{O} \times [-\delta, \delta[$ with $a_{ij}(0) = \delta_{ij}$, every solution $u \in \mathscr{C}^\infty(\mathcal{O} \times [0, \delta[)$ of

$$\begin{cases} \sum \dfrac{\partial}{\partial x_i}\left(a_{ij} \dfrac{\partial u}{\partial x_j}\right) = 0 \quad \text{on} \quad \mathcal{O} \times [0, \delta[\\ u(x', 0) = 0 \quad \left(\text{resp. } \sum a_{in}(x') \dfrac{\partial u}{\partial x_i}(x', 0) = 0\right) \quad \text{for} \quad x' \in \mathcal{O} \end{cases}$$

is analytic in the neighbourhood of O in $\mathcal{O} \times [0, \delta[$. In fact, from the Cauchy–Kovalevsky theorem, it is sufficient to prove that $\partial u / \partial x_n(x', 0)$ (resp. $u(x', 0)$) is analytic in the neighbourhood of O in $\mathcal{O}$.
Let us consider the Dirichlet problem; we fix $r_0 > 0$ which we shall specify below, and $k = 1, \ldots, n-1$; for all $\beta \in \mathbb{N}$ we write $u_\beta = \dfrac{r_0^\beta}{\beta!} \dfrac{\partial^\beta u}{\partial x_k^\beta}$. Differentiating the equation and using Leibniz's formula u_β satisfies

$$\begin{cases} \sum\sum \dfrac{\partial}{\partial x_i}\left(a_{ij} \dfrac{\partial u_\beta}{\partial x_j}\right) = \sum \dfrac{\partial}{\partial x_i} f_{i,\beta} \quad \text{on} \quad \mathcal{O} \times [0, \delta[\\ u_\beta(x', 0) = 0 \quad \text{for} \quad x' \in \mathcal{O} \end{cases}$$

with
$$f_{i,\beta} = -\sum_j \sum_{0<\gamma\leqslant\beta} \frac{r_0^\gamma}{\gamma!} \frac{\partial^\gamma a_{ij}}{\partial x_k^\gamma} u_{\beta-\gamma}\,.$$

Using the hypothesis of analyticity of the coefficients, for r_0 chosen sufficiently small

$$\frac{r_0^\gamma}{\gamma!}\left|\sum_j \frac{\partial^\gamma a_{ij}}{\partial x_k^\gamma}\right| \leqslant 1 \quad \text{on} \quad B(0,r_0)\,, \tag{6.23}$$

and hence supposing $r_0 \leqslant \delta$,

$$|f_{i,\beta}| \leqslant \sum_{\gamma<\beta} |u_\gamma| \quad \text{on} \quad B(0,r_0)\times[0,r_0[\,. \tag{6.24}$$

Given $\rho \in \mathscr{C}^\infty(\mathbb{R}^{n-1}\times\mathbb{R}^+)$ with $\operatorname{supp}\rho \subset B(0,r_0)\times[0,r_0[$ we multiply the equation by $\rho^2 u_\beta$ and integrate by parts; we obtain

$$\int \sum_i \left(\sum_j a_{ij}\frac{\partial u_\beta}{\partial x_j} - f_{i,\beta}\right)\left(\rho^2\frac{\partial u_\beta}{\partial x_i} + 2u_\beta\rho\frac{\partial\rho}{\partial x_i}\right)\mathrm{d}x = 0\,.$$

Using the hypothesis $a_{ij}(0) = \delta_{ij}$, we can always suppose that

$$\sum\sum a_{ij}\xi_i\xi_j \geqslant \frac{1}{2}|\xi|^2 \quad \text{on} \quad B(0,r_0) \quad \text{for} \quad \xi\in\mathbb{R}^n\,.$$

We deduce, using Schwarz's inequality, that

$$\frac{1}{2}\|\rho\operatorname{grad}u_\beta\|_{L^2}^2 \leqslant \|\rho\operatorname{grad}u_\beta\|_{L^2}\left(\int\sum_i\left(\rho f_{i,\beta} - 2u_\beta\sum_j a_{ij}\frac{\partial\rho}{\partial x_j}\right)^2\mathrm{d}x\right)^{\frac{1}{2}}$$
$$+ \int\left(u_\beta^2|\operatorname{grad}\rho|^2 + \rho^2\sum f_{i,\beta}^2\right)\mathrm{d}x$$

from which, we have

$$\|\rho\operatorname{grad}u_\beta\|_{L^2} \leqslant c\left(\sum\|\rho f_{i,\beta}\|_{L^2} + \|u_\beta\operatorname{grad}\rho\|_{L^2}\right) \tag{6.25}$$

with c depending only on the $\|a_{ij}\|_{L^\infty(B(0,r_0))}$.
For $r>0$, we denote $Q_r = B(0,r)\times[0,r[$. Given $0<r<r+\varepsilon<r_0$ there exists $\rho\in\mathscr{C}^\infty(\mathbb{R}^{n-1}\times\mathbb{R}^+)$ with $0\leqslant\rho\leqslant 1$, $\operatorname{supp}\rho\subset Q_{r+\varepsilon}$, $\rho\equiv 1$ on Q_r and $|\operatorname{grad}\rho|\leqslant 2\varepsilon^{-1}$. With (6.25) with this function ρ, taking account of (6.24), we have for all $0<r<r+\varepsilon<r_0$,

$$\|\operatorname{grad}u_\beta\|_{L^2(Q_r)} \leqslant C\varepsilon^{-1}\sum_{\gamma\leqslant\beta}\|u_\gamma\|_{L^2(Q_{r+\varepsilon})} \tag{6.26}$$

where C does **not** depend on β, r and ε.
In particular we have

$$\|u_{\beta+1}\|_{L^2(Q_r)} = \frac{r_0}{\beta+1}\left\|\frac{\partial u_\beta}{\partial x_k}\right\|_{L^2(Q_r)} \leqslant \frac{Cr_0}{(\beta+1)\varepsilon}\sum_{\gamma\leqslant\beta}\|u_\gamma\|_{L^2(Q_{r+\varepsilon})}\,,$$

from which we deduce by induction[153]

$$\sum_{\gamma \leqslant \beta} \| u_\gamma \|_{L^2(Q_r)} \leqslant \frac{1}{\beta!} \left(\frac{2Cr_0\beta}{\varepsilon} \right)^\beta \| u \|_{L^2(Q_{r+\varepsilon})} . \tag{6.27}$$

Returning to (6.26) and to the definition of u_β, we have

$$\begin{aligned} \left\| \frac{\partial^\beta}{\partial x_k^\beta} \operatorname{grad} u \right\|_{L^2(Q_{r_0/3})} &= \frac{\beta!}{r_0^\beta} \| \operatorname{grad} u_\beta \|_{L^2(Q_{r_0/3})} \\ &\leqslant \frac{\beta!}{r_0^\beta} C \left(\frac{r_0}{3} \right)^{-1} \sum_{\gamma \leqslant \beta} \| u_\gamma \|_{L^2(Q_{2r_0/3})} \leqslant \frac{3C}{r_0} \left(\frac{6C\beta}{r_0} \right)^\beta \| u \|_{L^2(Q_{r_0})} \end{aligned}$$

The direction x_k in $\mathbb{R}^{n-1}$ being arbitrary, we have in fact proved by changing the notation, that for r_0 sufficiently small, for every multi-index $\beta = (\beta_1, \ldots, \beta_{n-1})$

$$\left\| \frac{\partial^\beta}{\partial x'^\beta} \operatorname{grad} u \right\|_{L^2(Q_{r_0})} \leqslant (C|\beta|)^{|\beta|} \text{ [154]} . \tag{6.28}$$

Returning to the equation, we can write

$$\frac{\partial^2 u}{\partial x_n^2} = \sum_{(i,j) \neq (n,n)} \sum c_{ij} \frac{\partial^2 u}{\partial x_i \partial x_j} + \sum_i c_i \frac{\partial u}{\partial x_i}$$

where the coefficients c_{ij}, c_i are analytic on $B(0, r_0)$. Differentiating with respect to x' using Leibniz's formula, using (6.28) and the analyticity of the coefficients, we deduce that for $\beta \in \mathbb{N}^{n-1}$

$$\left\| \frac{\partial^2}{\partial x_n^2} \frac{\partial^\beta u}{\partial x'^\beta} \right\|_{L^2(Q_{r_0})} \leqslant (C|\beta|)^{|\beta|} .$$

Using a trace theorem[155]

$$\left\| \frac{\partial^\beta}{\partial x'^\beta} \left(\frac{\partial u}{\partial x_n}(x', 0) \right) \right\|_{L^2(B(0, r_0))} \leqslant (C|\beta|)^{|\beta|}$$

---

[153] For $0 < \varepsilon_1 < \varepsilon$,

$$\sum_{\gamma \leqslant \beta + 1} \| u_\gamma \|_{L^2(Q_r)} \leqslant \sum_{\gamma \leqslant \beta} \| u_\gamma \|_{L^2(Q_r)} + \frac{Cr_0}{(\beta + 1)\varepsilon_1} \sum_{\gamma \leqslant \beta} \| u_\gamma \|_{L^2(Q_{r+\varepsilon_1})} .$$

Applying with $\varepsilon_1 = \beta\varepsilon/(\beta + 1)$ and using the induction hypothesis, we obtain (6.27).

[154] In this formula and those which follow C denotes a constant with respect to β.

[155] For $v \in \mathscr{C}^2(\bar{Q}_r)$, $\left| \frac{\partial v}{\partial x_n}(x', 0) \right| \leqslant \left| \frac{\partial v}{\partial x_n}(x', x_n) \right| + \int_0^r \left| \frac{\partial^2 v}{\partial x_n^2}(x', x_n) \right| dx_n$ from which we derive

$$\left\| \frac{\partial v}{\partial x_n}(x', 0) \right\|_{L^2(B(0,r))} \leqslant \frac{1}{\sqrt{r}} \left\| \frac{\partial v}{\partial x_n} \right\|_{L^2(Qr)} + \left\| \frac{\partial^2 v}{\partial x_n^2} \right\|_{L^2(Qr)}$$

from which we have the analyticity of $\partial u/\partial x_n(x', 0)$ in the neighbourhood of 0 in $\mathbb{R}^{n-1}$, by using Sobolev's inclusions[156] (see Vol. 2, Chapter 4, § 7).
The Neumann problem can be treated in the same way: a supplementary difficulty comes from the necessity of also deriving the condition on the boundary. We refer the reader to Lions–Magenes [1] for a proof and for many more general results. $\square$

We propose now to use this result to define the *trace on Γ_0 of an arbitrary harmonic function u on Ω*. For simplicity, we shall now suppose that Ω is bounded and that $\Gamma_0 = \partial\Omega$.
Hence we take Ω to be a bounded open set with boundary Γ which we suppose to be analytic. Given $u \in \mathscr{H}(\Omega) \cap \mathscr{C}^1_n(\bar{\Omega})$ with trace φ on Γ (in the classical sense) from the classical Green's formula (see Proposition 4 of § 1), we have

$$\int_\Gamma \varphi \frac{\partial \zeta}{\partial n} \mathrm{d}\gamma = \int_\Omega u \Delta \zeta \, \mathrm{d}x \tag{6.29}$$

for every function $\zeta \in \mathscr{C}^1_n(\bar{\Omega}) \cap \mathscr{C}^2(\Omega)$ with $\Delta\zeta \in L^1(\Omega)$ satisfying $\zeta = 0$ on Γ.
Given arbitrary $u \in \mathscr{H}(\Omega)$, the right side of (6.29) is defined for all $\zeta \in \mathscr{C}^2(\Omega)$ with $\Delta\zeta$ of compact support in Ω, *i.e.* harmonic in the neighbourhood of Γ in Ω. More generally, the right side of (6.29) can be defined for every distribution ζ on Ω harmonic in the neighbourhood of Γ in Ω.
From Proposition 2, a distribution ζ on Ω harmonic in the neighbourhood of Γ in Ω and satisfying $\zeta = 0$ on Γ (in the classical sense: $\lim_{x \to z} \zeta(x) = 0$ for all $z \in \Gamma$) is necessarily analytic up to the boundary Γ and in particular $\partial\zeta/\partial n$ exists and is an analytic function on Γ. In other words, denoting by $\mathcal{A}(\Gamma)$, the space of functions analytic on Γ, for every distribution ζ on Ω, harmonic in the neighbourhood of Γ in Ω and satisfying $\zeta = 0$ on Γ, we have $\partial\zeta/\partial n \in \mathcal{A}(\Gamma)$.
We prove the

Proposition 3. *Let Ω be a bounded open set with analytic boundary Γ, and u a harmonic function on Ω. Denote by $\mathcal{A}(\Gamma)$ the space of functions analytic on Γ.*
(1) *There exists a unique linear form φ on $\mathcal{A}(\Gamma)$ such that*

$$\int_\Omega u \Delta \zeta \, \mathrm{d}x = \left\langle \varphi, \frac{\partial \zeta}{\partial n} \right\rangle \tag{6.30}$$

for every function $\zeta \in \mathscr{C}^\infty(\bar{\Omega})$ with $\Delta\zeta$ of compact support in Ω and satisfying $\zeta = 0$ on Γ.
(2) *For every distribution ζ harmonic in the neighbourhood of Γ and satisfying $\zeta = 0$ on Γ,*

$$\langle \Delta\zeta, u \rangle = \left\langle \varphi, \frac{\partial \zeta}{\partial n} \right\rangle . \tag{6.31}$$

---

[156] If Ω is an open set of $\mathbb{R}^n$, for $m > \frac{1}{2}n$ we have $H^m(\Omega) \in \mathscr{C}^0(\Omega)$. More precisely

$$\|u\|_{L^\infty(B(x_0, r_0/2))} \leqslant C \sum_{|\beta| \leqslant m} \left\| \frac{\partial^\beta u}{\partial x^\beta} \right\|_{L^2(B(x_0, r_0))}$$

In particular

$$(6.32)\qquad u(x) = \left\langle \varphi, \frac{\partial G}{\partial n}(x, .) \right\rangle \quad \textit{for all} \quad x \in \Omega$$

where G is the Green's function of the open set Ω.
(3) If $u \in \mathscr{C}^0(\bar{\Omega})$, then φ is defined by

$$(6.33)\qquad \langle \varphi, \psi \rangle = \int_\Gamma u(t)\psi(t)\,\mathrm{d}\gamma(t) \qquad \text{for all} \quad \psi \in \mathcal{A}(\Gamma) .$$

This leads us naturally to set up the

Definition 1. Let Ω be a bounded open set with analytic boundary Γ, and u an arbitrary harmonic function on Ω. We call *the trace of u on* Γ, the linear form φ on the space $\mathcal{A}(\Gamma)$ of analytic functions on Γ defined by (6.30).

From the point (3) of the Proposition, *this concept of trace generalizes clearly the classical notion when* $u \in \mathscr{C}^0(\bar{\Omega})$, with the condition of identifying the continuous function $\varphi \in \mathscr{C}^0(\Gamma)$ with the linear form on $\mathcal{A}(\Gamma)$

$$\psi \in \mathcal{A}(\Gamma) \to \int_\Gamma \varphi(t)\psi(t)\,\mathrm{d}\gamma(t) .$$

From Weierstrass' theorem we can carry out this identification: in effect $\mathcal{A}(\Gamma)$ being dense in $\mathscr{C}^0(\Gamma)$, given $\varphi_1, \varphi_2 \in \mathscr{C}^0(\Gamma)$, $\varphi_1 = \varphi_2$ on Γ iff

$$\int_\Gamma \varphi_1(t)\psi(t)\,\mathrm{d}\gamma(t) = \int_\Gamma \varphi_2(t)\psi(t)\,\mathrm{d}\gamma(t) \quad \text{for all} \quad \psi \in \mathcal{A}(\Gamma) .$$

Proof of Proposition 3. Given ζ a distribution on Ω, harmonic in the neighbourhood of Γ and satisfying $\zeta = 0$ on Γ, we show first that $\langle \Delta\zeta, u \rangle$ depends only on the normal derivative $\psi = \partial\zeta/\partial n$ on Γ. In other words, by linearity, we must prove

$$\frac{\partial \zeta}{\partial n} = 0 \quad \text{on} \quad \Gamma \Rightarrow \langle \Delta\zeta, u \rangle = 0 .$$

Now from the uniqueness in the Cauchy–Kovalevsky theorem, if ζ is harmonic in the neighbourhood of Γ and satisfies $\zeta = 0$, $\partial\zeta/\partial n = 0$ on Γ, necessarily it is identically zero on a neighbourhood U of Γ. Let us take $\rho \in \mathscr{D}(\Omega)$ such that $\rho \equiv 1$ on a neighbourhood of $\Omega \setminus U$. We have $\rho \equiv 1$ in the neighbourhood of $\operatorname{supp} \Delta\zeta$, from which we have

$$\langle \Delta\zeta, u \rangle = \langle \Delta\zeta, \rho u \rangle = \langle \zeta, \Delta(\rho u) \rangle ;$$

but

$$\Delta(\rho u) = \rho \Delta u + 2 \operatorname{grad} \rho . \operatorname{grad} u + u \Delta\rho = 0$$

in the neighbourhood of $\operatorname{supp} \zeta \subset \Omega \setminus \bar{U}$. Hence $\langle \Delta\zeta, u \rangle = 0$.
Now, given $\psi \in \mathcal{A}(\Gamma)$, from the Cauchy–Kovalevsky theorem, there exists ζ_0 harmonic on a neighbourhood U of Γ satisfying $\zeta_0 = 0$, $\partial\zeta_0/\partial n = 0$ on Γ. Let us

take $\rho_0 \in \mathscr{D}(\Omega)$ with $\rho_0 = 1$ in the neighbourhood of $\Omega \backslash U$. The function $\zeta = (1 - \rho_0)\zeta_0 \in \mathscr{C}^\infty(\bar{\Omega})$ and $\zeta = \zeta_0$ in the neighbourhood of Γ. Hence $\Delta\zeta$ has compact support in Ω, $\zeta = 0$ and $\partial\zeta/\partial n = \psi$ on Γ. The value $\int_\Omega u\Delta\zeta \, dx$, from the preceding point depends only on ψ so we can denote it by $\langle \varphi, \psi \rangle$. We have thus proved the point (1) and (6.31).

The formula (6.32) can be deduced immediately from (6.31) since by definition, for $x \in \Omega$, $\zeta = G(x, .)$ is a solution of $\Delta\zeta = \delta_x$ on Ω. The formula (6.33) reduces in fact to (6.29) and follows immediately from Green's formula if $u \in \mathscr{C}^1_n(\bar{\Omega})$; to prove it in the case where u is only continuous on $\bar{\Omega}$, it is sufficient to consider $u_k \in \mathscr{C}^1_n(\bar{\Omega}) \cap \mathscr{H}(\Omega)$ such that $u_k \to u$ in $\mathscr{C}^0(\bar{\Omega})$ and to pass to the limit in (6.29). □

It is immediate that the map $u \in \mathscr{H}(\Omega) \to$ trace of u on Γ is linear. From (6.32), this map is a bijection of $\mathscr{H}(\Omega)$ on the set of linear forms φ on $\mathscr{A}(\Gamma)$ satisfying

$$x \in \Omega \to \left\langle \varphi, \frac{\partial G}{\partial n}(x, .) \right\rangle \quad \text{is harmonic on} \quad \Omega$$

We can *provide $\mathscr{A}(\Gamma)$ with a topology* such that these linear forms are exactly the continuous linear forms on $\mathscr{A}(\Gamma)$. We refer the reader to Lions–Magenes [1] for a precise definition of this topology; here we note only that it can be characterized by the following notion of limit:

(6.34) a sequence[157] (ψ_k) of elements of $\mathscr{A}(\Gamma)$ converges to zero if and only if there exists an open neighbourhood U of Γ such that for all k we can solve, in a unique manner in U the Cauchy problem

$$\Delta\zeta_k = 0 \quad \text{in } U\,, \qquad \zeta_k = 0 \quad \text{and} \quad \frac{\partial\zeta}{\partial n} = \psi_k \text{ on } \Gamma$$

and the solution ζ_k converges to zero in $\mathscr{H}(U)$.

From now on we shall suppose that $\mathscr{A}(\Gamma)$ is given this topology; we shall denote by $\mathscr{A}'(\Gamma)$ the *dual of* $\mathscr{A}(\Gamma)$ and we shall call the elements of $\mathscr{A}'(\Gamma)$ *analytic functional.*

Proposition 4. *Let Ω be a bounded open set with analytic boundary Γ. For all $u \in \mathscr{H}(\Omega)$, the trace of u on Γ is an analytic functional; the map $u \to \varphi$ is an isomorphism from $\mathscr{H}(\Omega)$ onto $\mathscr{A}'(\Gamma)$*[158].

Proof. Let $\psi_k \to 0$ in $\mathscr{A}(\Gamma)$ and U, ζ_k as in (6.34); considering $\rho_0 \in \mathscr{D}(\Omega)$ with $\rho_0 \equiv 1$ in the neighbourhood of $\Omega \backslash U$, we have

$$\langle \varphi, \psi_k \rangle = \int_\Omega u\Delta[(1 - \rho_0)\zeta_k] \, dx \quad \text{(see proof of Proposition 3).}$$

---

[157] Possibly generalized.

[158] $\mathscr{H}(\Omega)$ is given the uniform convergence topology on every compact set and $\mathscr{A}'(\Gamma)$ the strong topology.

Since $\zeta_k \to 0$ in $\mathscr{H}(U)$,

$$\Delta[(1 - \rho_0)\zeta_k] = -2 \operatorname{grad} \rho_0 . \operatorname{grad} \zeta_k - \zeta_k \Delta\rho_0 \to 0$$

uniformly on Ω with $\operatorname{supp} \Delta[(1 - \rho_0)\zeta_k] \subset \operatorname{supp} \rho_0 \cap \operatorname{supp}(1 - \rho_0)$ a compact set in Ω. Hence $\int_\Omega u\Delta((1 - \rho_0)\zeta_k)\,dx \to 0$. This shows that $\varphi \in \mathfrak{a}(\Gamma)$.
Using (6.34) and the definition of the Green's function, it is clear that the map $x \to \frac{\partial G}{\partial n}(x, .)$ of Ω into $\mathfrak{a}(\Gamma)$ is continuous; more precisely this is a harmonic map since for all $t \in \Gamma$, $x \to \frac{\partial G}{\partial n}(x, t)$ is harmonic. Hence, for all $\varphi \in \mathfrak{a}'(\Gamma)$, the function u defined by (6.32) is harmonic on Ω. This shows that the map $u \to \varphi$ of $\mathscr{H}(\Omega)$ onto $\mathfrak{a}'(\Gamma)$ is a bijection.
The continuity of the maps $u \to \varphi$ and $\varphi \to u$ can be proved without difficulty by the use of the topologies of $\mathscr{H}(\Omega)$ and $\mathfrak{a}'(\Gamma)$: we leave the details to the reader (see Lions–Magenes [1]). □

Proposition 4 can also be stated: *for every analytic functional φ on Γ, there exists one and only one solution of the Dirichlet problem $P(\Omega, \varphi)$*, where, obviously, $u = \varphi$ on Γ is taken in the sense of Definition 1; *also this solution u is given by* (6.33) *and the map $\varphi \to u$ of $\mathfrak{a}'(\Gamma)$ into $\mathscr{H}(\Omega)$ is continuous*
It is clear that *every distribution on Γ is an analytic functional on Γ*; more precisely we can identify $\mathscr{D}'(\Gamma)$ with a sub-space of $\mathfrak{a}'(\Gamma)$ since from Weierstrass' theorem, $\mathfrak{a}(\Gamma)$ is dense in $\mathscr{C}^\infty(\Gamma)$[159]. In particular for every distribution φ on Γ there exists one and only one solution of the Dirichlet problem $P(\Omega, \varphi)$. We note, however, that for an arbitrary harmonic function u on Ω, its trace on Γ is not, in general, a distribution on Γ.
Let us give some examples.

Example 1. For $z \in \mathbb{R}^n \backslash \Omega$ let us consider $u_z(x) = E_n(x - z)$. This is a harmonic function on Ω, continuous on $\bar{\Omega} \backslash \{z\}$; but if $z \in \Gamma$, it is not continuous at z. The map $z \in \mathbb{R}^n \backslash \Omega \to \varphi_z \in \mathscr{H}(\Omega)$ is continuous (and even harmonic); for $z \notin \Gamma$, the trace of u_z in the classical sense or in that of Definition 1 coincide in identifying the function $\varphi_z(t) = E_n(t - z)$ with the analytic functional

$$\langle \varphi_z, \psi \rangle = \int_\Gamma E_n(t - z)\psi(t)\,d\gamma(t) .$$

For all $z \in \mathbb{R}^n \backslash \Omega$, the function $\varphi_z(t) = E_n(t - z)$ is integrable on Γ and the map $z \in \mathbb{R}^n \backslash \Omega \to \varphi_z \in L^1(\Gamma)$ is continuous. Hence *for all $z \in \Gamma$ the trace of u_z on Γ is the analytic functional defined by the function $\varphi_z(t) = E_n(t - z)$*. We note that

$$\langle \varphi_z, \psi \rangle = \int_\Gamma E_n(t - z)\psi(t)\,d\gamma(t)$$

is the value at z of the simple layer potential defined by ψ. □

---

[159] $\mathscr{C}^\infty$ is provided with the topology of uniform convergence of all the derivatives.

Example 2. Now let us consider $u_z^i(x) = \dfrac{\partial E_n}{\partial x_i}(x - z)$. As above for $z \in \mathbb{R}^n \backslash \Omega$, $u_z^i \in \mathscr{H}(\Omega) \cap \mathscr{C}^0(\bar{\Omega} \backslash \{z_0\})$ and the map $z \to u_z^i$ of $\mathbb{R}^n \backslash \Omega$ into $\mathscr{H}(\Omega)$ is continuous. But for $z \in \Gamma$, the function $\dfrac{\partial E_n}{\partial x_i}(t - z)$ is not, in general integrable on Γ in the neighbourhood of $t = z$. We denote by φ_z^i the trace of u_z^i on Γ; for $\psi \in \mathcal{A}(\Gamma)$ the function $z \to \langle \varphi_z^i, \psi \rangle$ is continuous on $\mathbb{R}^n \backslash \Omega$. When $z \in \mathbb{R}^n \backslash \bar{\Omega}$, $u_z^i \in \mathscr{C}^0(\bar{\Omega})$ and

$$\langle \varphi_z^i, \psi \rangle = \int_\Gamma \frac{\partial E_n}{\partial x_i}(t - z)\psi(t)\,\mathrm{d}\gamma(t) = -\frac{\partial}{\partial x_i} u_\psi(z)$$

where u_ψ is the simple layer potential of ψ (see §3),

$$u_\psi(z) = \int_\Gamma E_n(t - z)\psi(t)\,\mathrm{d}\gamma(t)\,.$$

Suppose for simplicity that $n \geqslant 3$ and that $\mathbb{R}^n \backslash \bar{\Omega}$ is connected; we consider $\alpha \in \mathscr{C}^0(\Gamma)$ such that $\displaystyle\int_\Gamma E_n(t - z)\alpha(t)\,\mathrm{d}\gamma(t) = 1$ for all $z \in \Gamma$.

Such a function exists (see Proposition 18 of §4); its interior (resp. exterior) simple layer potential is the solution of the Dirichlet problem $P(\Omega, 1)$ (resp. $P(\mathbb{R}^n \backslash \bar{\Omega}, 1)$ tending to zero at infinity); in other words the function

$$u_\alpha(z) = \int_\Gamma E_n(t - z)\alpha(t)\,\mathrm{d}\gamma(t)$$

continuous on $\mathbb{R}^n$ satisfies $u_\alpha \equiv 1$ on $\bar{\Omega}$ and from the maximum principle $0 < u_\alpha < 1$. Now from Proposition 12 of §3, $\alpha(z)$ is the jump in the normal derivative of u_α at $z \in \Gamma$; hence

$$\alpha(z) = \lim_{t \downarrow 0} \frac{u_\alpha(z + tn(z)) - u_\alpha(z)}{t}\,;$$

also by Hopf's maximum principle (see Lemma 2 of §4), $\alpha(z) < 0$ for all $z \in \Gamma$. From Proposition 2, u_α is analytic up to the boundary of $\mathbb{R}^n \backslash \Omega$ and hence $\alpha \in \mathfrak{a}(\Gamma)$. Finally since $u_\alpha = 1$ on Γ, the tangential derivatives of u_α are zero and for all $z \in \Gamma$, $(\partial u_\alpha / \partial x_i)(z) = n_i(z)\alpha(z)$ where $n_i(z)$ is the i-th component of the normal $n(z)$. This being specified, we write for $z \in \Gamma$

$$\langle \varphi_z^i, \psi \rangle = \left\langle \varphi_z^i, \psi - \frac{\psi(z)}{\alpha(z)}\alpha \right\rangle + \frac{\psi(z)}{\alpha(z)} \langle \varphi_z^i, \alpha \rangle\,.$$

From what we have just seen, we deduce that

$$\langle \varphi_z^i, \alpha \rangle = -\frac{\partial u_\alpha}{\partial x_i}(z) = -n_i(z)\alpha(z)\,.$$

Also since $\left(\psi - \dfrac{\psi(z)}{\alpha(z)}\alpha\right)(z) = 0$, we have

$$\left\langle \varphi_z^i, \psi - \frac{\psi(z)}{\alpha(z)}\alpha \right\rangle = \int_\Gamma \frac{\partial E_n}{\partial x_i}(t - z)\left[\psi(t) - \frac{\psi(z)}{\alpha(z)}\alpha(t)\right]\mathrm{d}\gamma(t)\,,$$

the integral being defined since when $\psi \in \mathscr{C}^1(\Gamma)$ and $\psi(z) = 0$

$$\left|\frac{\partial E_n}{\partial x_i}(t - z)\psi(t)\right| \leqslant \frac{C}{|t - z|^{n-1}} .$$

In short, the trace of $u_z^i(x) = \dfrac{\partial E_n}{\partial x_i}(x - z)$ is defined by

$$\text{(6.35)} \qquad \langle \varphi_z^i, \psi \rangle = \int_\Gamma u_z^i(t)\left[\psi(t) - \frac{\psi(z)}{\alpha(z)}\alpha(t)\right]\mathrm{d}\gamma(t) - \psi(z)n_i(z) ,$$

where α is the normal derivative of the solution of the exterior Dirichlet problem $P(\mathbb{R}^n \backslash \bar{\Omega}, 1)$.
We note that the trace φ_z^i is a distribution of order at most 1. □

Example 3. Let us take $\Omega = \{(x, y) \in \mathbb{R}^2; x^2 + y^2 \leqslant 1\}$ which we identify with the unit disk $D = \{z \in \mathbb{C}; |z| < 1\}$ of the complex plane $\mathbb{C}$. A harmonic function $u(x, y)$ on Ω is the real part of a holomorphic function $f(z)$ on D (see Proposition 1 of §2). The function can be developed in a power series

$$f(z) = \sum_{n=0}^{\infty} a_n z^n, \quad z \in D .$$

A function on $\Gamma = \partial\Omega$ is identified with a periodic function $g(\theta)$.
If $\sum |a_n| < \infty$, the function f is continuous on $\bar{D}$ and its trace on Γ is the continuous function $g(\theta) = \sum a_n \mathrm{e}^{in\theta}$. We show that in the genera 1 case, *the trigonometrical series $\sum a_n \mathrm{e}^{in\theta}$ converges in $\mathscr{A}'(\Gamma)$ and the trace of f is the sum of this series.* Let ψ be analytic on Γ; we can extend it to a holomorphic function $\varphi(z)$ on an annulus $\{r_1 < |z| < r_1^{-1}\}$; this function is developable in a Laurent series (see Cartan [1]):

$$\varphi(z) = \sum_{n=-\infty}^{+\infty} b_n z^n, \quad r_1 < |z| < r_1^{-1} .$$

Considering $r_1 < r < 1$, since the series $\sum b_n r^n$ converges, there exists M such that $|b_n| \leqslant M r^{-n}$. We then have

$$\sum_{n \geqslant 0} |a_n b_{-n}| \leqslant M \sum_{n \geqslant 0} |a_n| r^n < \infty ;$$

and we deduce

$$\sum_{n\geqslant 0} |\langle a_n e^{in\theta}, \psi \rangle| = \sum_{n \geqslant 0} |a_n| \left|\sum_m b_m \int_0^{2\pi} e^{i(n+m)\theta}\,\mathrm{d}\theta\right| = \sum_{n\geqslant 0} 2\pi |a_n||b_{-n}| < \infty$$

and

$$\sum_{n=0}^{\infty} \langle a_n e^{in\theta}, \psi \rangle = 2\pi \sum a_n b_{-n} .$$

□

Example 4. Let us again consider $\Omega = \{(x, y) \in \mathbb{R}^2; x^2 + y^2 < 1\}$ and

$$u(x, y) = \left[\exp\left(\frac{x}{x^2 + (y - 1)^2}\right)\right]\left[\cos\left(\frac{y - 1}{x^2 + (y - 1)^2}\right)\right] .$$

This function is harmonic on Ω as it is the real part of the holomorphic function defined on $\mathbb{C}\backslash\{i\}$ by

$$f(z) = \exp\frac{1}{z - i}.$$

This function is continuous on $\bar{\Omega}\backslash\{(0, 1)\}$ and its trace $\varphi(\theta)$ is defined for $\theta \neq \frac{1}{2}\pi$ by

$$\varphi(\theta) = \cos\left(\frac{1}{2}\right)\exp\frac{1}{2}\left(\frac{\cos\theta}{1 - \sin\theta}\right).$$

The function φ, admitting an essential singularity for $\theta = \frac{1}{2}\pi$ cannot be extended to a distribution on Γ. This gives an *example in which the trace of a harmonic function is not a distribution*, but only an analytic functional. □

We conclude this section by noting that Definition 1 permits us likewise to give a meaning to the *normal derivative* (and also to the tangential derivatives) of a harmonic function: if $u \in \mathscr{H}(\Omega)$, the derivatives $\partial u/\partial x_i$ are also harmonic on Ω and we can consider their traces φ_i on Γ. On the other hand the components $n_i(t)$ of the unit interior normal vector $n(t)$ are analytic functions of $t \in \Gamma$; we can therefore consider the analytic functional

$$\langle \sum n_i\varphi_i, \psi\rangle = \sum \langle \varphi_i, n_i\psi\rangle .$$

We prove the

Proposition 5. *Let Ω be a bounded open set with analytic boundary Γ and let $u \in \mathscr{H}(\Omega)$. Denote by $\varphi, \varphi_1, \ldots, \varphi_n$ the traces of $u, \partial u/\partial x_1, \ldots, \partial u/\partial x_n$ respectively.* (1) *We then have the Green's formula*

$$\langle \Delta\zeta, u\rangle = \left\langle \varphi, \frac{\partial\zeta}{\partial n}\right\rangle - \sum \langle \varphi_i, n_i\zeta\rangle \tag{6.36}$$

for every distribution ζ on Ω, harmonic in the neighbourhood of Γ in $\mathbb{R}^n$; (2) *$u \in \mathscr{C}_n^1(\bar{\Omega})$ iff the analytic functional $\sum n_i\varphi_i$ is continuous on Γ.*

Proof. To prove (6.36) we first note that we can suppose that $\zeta \in \mathscr{H}(\Omega)$: it is sufficient to replace ζ by $\bar{\zeta}$ the solution of the Dirichlet problem $P(\Omega, \zeta|_\Gamma)$ by applying the formula (6.31) to $\zeta - \bar{\zeta}$. We shall then use a density argument: (6.36) is true for $u \in \mathscr{C}^1(\bar{\Omega})$ since it reduces then to the elementary Green's formula; to prove it for arbitrary $u \in \mathscr{H}(\Omega)$, it is enough therefore, from the continuity of the trace, to show that $\mathscr{C}^1(\bar{\Omega}) \cap \mathscr{H}(\Omega)$ is dense in $\mathscr{H}(\Omega)$. This last point is easily seen for particular open sets; in the general case, from the isomorphism between $\mathscr{H}(\Omega)$ and $\mathscr{A}'(\Gamma)$, it suffices to prove that $\mathscr{C}^\infty(\Gamma)$ is dense in $\mathscr{A}'(\Gamma)$; using the fact that $\mathscr{A}'(\Gamma)$ is reflexive (see Lions–Magenes [1]), that leads to: $\mathscr{A}(\Gamma)$ is dense in $\mathscr{D}'(\Gamma)$, which follows from the Weierstrass theorem.
To prove the point (2), we note first that, given Ω_0, a connected component of Ω, then applying (6.36) with $\zeta = \chi_{\Omega_0}$, we obtain Gauss' theorem

$$\sum_i \langle \varphi_i, n_i\chi_{\partial\Omega_0}\rangle = 0 .$$

Supposing $\psi = \sum n_i \varphi_i$ to be continuous on Γ we find, from Theorem 3 of §4, that there exists $\bar{u} \in \mathscr{C}_n^1(\bar{\Omega}) \cap \mathscr{H}(\Omega)$ a solution of $PN(\Omega, \psi)$. Replacing u by $u - \bar{u}$, we are therefore led to proving the result by assuming $\sum n_i \varphi_i = 0$, and hence, using (6.36) again, $\langle \varphi, \partial\zeta/\partial n \rangle = 0$ for every function ζ, harmonic in the neighbourhood of $\bar{\Omega}$. We can always suppose that Ω is connected; given $\psi \in \mathscr{A}(\Gamma)$ with $\int_\Gamma \psi \, \mathrm{d}\gamma = 0$, there exists a solution ζ of the Neumann problem $PN(\Omega, \psi)$, which, from Proposition 2 is analytic up to the boundary Γ and hence can be extended to a function harmonic in the neighbourhood of $\bar{\Omega}$.
Applying (6.36) to this function ζ, we obtain

$$\langle \varphi, \psi \rangle = 0 \quad \text{for all} \quad \psi \in \mathscr{A}(\Gamma) \quad \text{with} \quad \int_\Gamma \psi \, \mathrm{d}\gamma = 0 \,,$$

hence

$$\langle \varphi, \psi \rangle = \langle \varphi, 1 \rangle \frac{1}{|\Gamma|} \int_\Gamma \psi \, \mathrm{d}\gamma \qquad \text{for all} \quad \psi \in \mathscr{A}(\Gamma)$$

where $|\Gamma|$ is the total surface of Γ. In other words, φ is the constant function $\langle \varphi, 1 \rangle / |\Gamma|$ and u the solution of the Dirichlet problem $P(\Omega, \varphi)$ is constant on Ω. □

3. Dirichlet Problem with Given Measures or Discontinuous Functions. Herglotz's Theorem

We consider a positive concave function u on an interval $]a, b[$ of $\mathbb{R}$. It is elementary to see the limits of u at the points a and b exist and that we have the representation formula[160]

$$u(x) = \frac{x - a}{b - a}\left(u(b) + \int_x^b (b - t)\,\mathrm{d}\mu(t)\right) + \frac{b - x}{b - a}\left(u(a) + \int_a^x (t - a)\,\mathrm{d}\mu(t)\right)$$

where $\mu = -\mathrm{d}^2u/\mathrm{d}x^2$ in the sense of distributions on $]a, b[$, is a positive Radon measure. In particular, we have

$$\int_x^b (b - t)\,\mathrm{d}\mu(t) < \infty \quad \text{and} \quad \int_a^x (t - a)\,\mathrm{d}\mu(t) < \infty \quad \text{for all} \quad x \in \,]a, b[$$

which we can summarize in

$$\int_a^b \min(t - a, b - t)\,\mathrm{d}\mu(t) < \infty \,.$$

We propose to generalize this result to positive super-harmonic functions on an open set Ω of $\mathbb{R}^n$. We take an open set $\Omega \subset \mathbb{R}^n$ which, for simplicity we suppose

[160] In the case $]a, b[$ bounded; see Example 12 of §4 for the unbounded case.

satisfies

(6.37)

Ω is a connected regular bounded open set with boundary Γ of class $W^{2,\infty}$ [161].

It follows from Proposition 1, that G, the Green's function of Ω is of class $\mathscr{C}^1$ on $\{(x, y) \in \bar{\Omega} \times \bar{\Omega};\, x \neq y\}$. Also we show easily that Ω satisfies the condition of the interior (and also exterior) ball at every point of the boundary Γ (see example 7 of §4) and hence from the maximum principle of Hopf (see Lemma 2 of §4) we deduce that:

$$\frac{\partial G}{\partial n}(x, z) > 0 \quad \text{at every point} \quad x \in \Omega \quad \text{and} \quad z \in \Gamma . \tag{6.38}$$

We recall that for a function u sufficiently regular, we have Green's representation formula,

$$u(x) = \int_\Omega G(x, t)\,\Delta u(t)\,\mathrm{d}t + \int_\Gamma \frac{\partial G}{\partial n}(x, t) u(t)\,\mathrm{d}\gamma(t), \quad x \in \Omega . \tag{6.39}$$

We have proved this formula (see Propositions 7 and 8 of §4) for $u \in \mathscr{C}^0(\bar{\Omega})$ with $\Delta u \in L^p(\Omega)$ where $p > \frac{1}{2}n$.

We note first that for all $x \in \Omega$, the function $t \to \dfrac{\partial G}{\partial n}(x, t)$ being continuous on Γ, for every Radon measure v on Γ we can define

$$u(v, x) = \int_\Gamma \frac{\partial G}{\partial n}(x, t)\,\mathrm{d}v(t), \quad x \in \Omega . \tag{6.40}$$

The function $u(v)$: $x \to u(v, x)$ is harmonic on Ω since for all $t \in \Gamma$ the function $x \to \dfrac{\partial G}{\partial n}(x, t)$ is harmonic on Ω. If v is the measure of density φ with respect to the surface measure of Γ, where $\varphi \in \mathscr{C}^0(\Gamma)$ that is to say $\mathrm{d}v = \varphi \mathrm{d}\gamma$, then the function $u(v)$ given by (6.40) is the classical solution of the Dirichlet problem $P(\Omega, \varphi)$. For this reason we introduce the

Definition 2. Given Ω an open set of $\mathbb{R}^n$ satisfying (6.37) and v a Radon measure on Γ, we call *the solution of the Dirichlet problem* $P(\Omega, v)$ the function $u(v)$ defined by (6.40).

Remark 2. When Γ is analytic, we have defined in the preceding section the trace of an arbitrary harmonic function on Ω; the Definition 2 is clearly compatible with this notion: for a Radon measure v on Γ, which is *a fortiori* an analytic functional, the solution $u = u(v)$ of $P(\Omega, v)$ in the sense of Definition 2 is the harmonic function u on Ω admitting v as trace on Γ in the sense of Definition 1. □

---

[161] That is to say, a little weaker than the class $\mathscr{C}^2$.

It is immediate that the map $\nu \to u(\nu)$ is an increasing linear map of the space $\mathfrak{M}(\Gamma)$ of Radon measures on Γ into the space $\mathscr{H}(\Omega)$ of functions harmonic on Ω; this map is increasing since $\dfrac{\partial G}{\partial n}(x, z) \geqslant 0$ for all $x \in \Omega$ and $z \in \Gamma$.

From (6.38), we have also for $\nu_1, \nu_2 \in \mathfrak{M}(\Gamma)$

$$\nu_1 \geqslant \nu_2, \quad \nu_1 \neq \nu_2 \Rightarrow u(\nu_1) > u(\nu_2) \quad \text{on} \quad \Omega.$$

We state the first form of Herglotz's theorem:

Proposition 6. *Given Ω an open set in $\mathbb{R}^n$ satisfying* (6.37), *the map $\nu \to u(\nu)$ is a bijection of the cone $\mathfrak{M}^+(\Gamma)$ of positive Radon measures on Γ onto the cone $\mathscr{H}^+(\Omega)$ of positive harmonic functions on Ω.*

Given that every Radon measure is the difference of two positive measures (see Dieudonné [2], we have the

Corollary 2. *Given Ω an open set of $\mathbb{R}^n$ satisfying* (6.37), *the map $\nu \to u(\nu)$ is a linear bijection of the space $\mathfrak{M}(\Gamma)$ of Radon measures on Γ onto the space $\mathscr{H}^+(\Omega) - \mathscr{H}^+(\Omega)$ of the differences of positive harmonic functions on Ω; also this bijection is bi-increasing, i.e. for $\nu_1, \nu_2 \in \mathfrak{M}(\Gamma)$*

$$\nu_1 \geqslant \nu_2 \quad \text{on} \quad \Gamma \Leftrightarrow u(\nu_1) \geqslant u(\nu_2) \quad \text{on} \quad \Omega\,.$$

Given $\varphi \in L^1(\Gamma)$, we can consider the measure ν of density φ with respect to the surface measure $\mathrm{d}\nu = \varphi \mathrm{d}\gamma$; identifying ν and φ we denote by $u(\varphi)$ the harmonic function on Ω defined by

$$u(\varphi, x) = \int_\Gamma \frac{\partial G}{\partial n}(x, t)\varphi(t)\,\mathrm{d}\gamma(t), \quad x \in \Omega\,. \tag{6.41}$$

For a constant function $\varphi \equiv c$, we have $u(\varphi) \equiv c$. Hence for $\nu \in \mathfrak{M}(\Gamma)$

$$\nu \geqslant c \quad \text{on} \quad \Gamma \Leftrightarrow u(\nu) \geqslant c \quad \text{on} \quad \Omega\,.$$

We deduce:

Corollary 3. *Let Ω be an open set in $\mathbb{R}^n$ satisfying* (6.37) *and $u \in \mathscr{H}(\Omega)$. The function u is bounded on Ω if and only if there exists $\varphi \in L^\infty(\Gamma)$ such that $u = u(\varphi)$ given by* (6.41). *Also, φ is unique (a.e. on Γ) and*

$$\sup_\Omega u = \sup_\Gamma \operatorname{ess} \varphi, \quad \inf_\Omega u = \inf_\Gamma \operatorname{ess} \varphi\,.$$

Before proving Herglotz's theorem we note

Lemma 5. *Let Ω be an open set of $\mathbb{R}^n$ satisfying* (6.37), *$u \in \mathscr{H}(\Omega)$ and $\nu \in \mathfrak{M}(\Gamma)$. Then $u = u(\nu)$ iff $u \in L^1(\Omega)$ and*

$$\int_\Omega u\,\Delta v\,\mathrm{d}x = \int_\Gamma \frac{\partial v}{\partial n}\mathrm{d}\nu \tag{6.42}$$

for every function $v \in W^{2,\infty}(\Omega)$ with $v \doteq 0$ on Γ.

Proof of Lemma 5. Let $n < p < \infty$. From Corollary 1, given $f \in L^p(\Omega)$ we have that the (classical) solution $v(f)$ of the homogeneous Dirichlet problem $PH(\Omega, f)$ is in $\mathscr{C}^1(\bar{\Omega})$ and therefore, in particular, $\dfrac{\partial v(f)}{\partial n} \in \mathscr{C}^0(\Gamma)$, and

$$\left\| \frac{\partial v(f)}{\partial n} \right\|_{\mathscr{C}^0(\Gamma)} \leqslant C \| f \|_{L^p(\Omega)} \tag{6.43}$$

where C depends only on p and Ω.

Now using the same argument as in the proof of Proposition 4, $u = u(v)$ iff we have (6.42) for every function $v = v(f)$ with $f \in \mathscr{D}(\Omega)$. Suppose, therefore, that $u = u(v)$; for all $f \in \mathscr{D}(\Omega)$, making use of (6.42) with $v = v(f)$ and (6.43), we have

$$\left| \int_\Omega u f \, \mathrm{d}x \right| \leqslant C \| f \|_{L^p(\Omega)} \| v \|_{\mathfrak{M}(\Gamma)} \,.$$

Hence $u^q(\Omega)$ where q is the index conjugate to p. Finally for $v \in W^{2,p}(\Omega)$ with $v = 0$ on Γ, approximating Δv in $L^p(\Omega)$ by the functions $f_k \in \mathscr{D}(\Omega)$, we have from (6.43), $\partial v(f_k)/\partial n \to \partial v/\partial n$ in $\mathscr{C}^0(\Gamma)$ and hence (6.42) in the limit for v. □

Proof of Proposition 6. From Lemma 5, the fact that the map $v \to u(v)$ is injective comes from the density in $\mathscr{C}^0(\Gamma)$ of

$$\left\{ \frac{\partial v}{\partial n}; \quad v \in W^{2,\infty}(\Omega); \quad v = 0 \quad \text{on} \quad \Gamma \right\}.$$

Making use of a theorem on the lifting of traces (see Chap. IV), given $\psi \in W^{1,\infty}(\Gamma)$, there exists $v \in W^{2,\infty}(\Omega)$ such that $v = 0$ and $\partial v/\partial n = \psi$ on Γ. From Weierstrass' theorem, $W^{1,\infty}(\Gamma)$ is dense in $\mathscr{C}^0(\Gamma)$; hence the result. It remains to show that the map is surjective. We shall, in the first instance, suppose that the open set Ω is star-shaped, that is to say satisfies

$$\begin{cases} \text{there exists } x_0 \in \Omega \text{ such that } \lambda x + (1 - \lambda) x_0 \in \Omega \\ \text{for all } x \in \Omega \text{ and all } \lambda \in [0, 1] \,. \end{cases}$$

By translation we can always suppose that $x_0 = 0$. Given $u \in \mathscr{H}^+(\Omega)$, for all $\lambda \in]0, 1[$, the function $u_\lambda(x) = u(\lambda x)$ is harmonic in the neighbourhood of $\bar{\Omega}$; we therefore have

$$u(\lambda x) = \int_\Gamma \frac{\partial G}{\partial n}(x, t) u_\lambda(t) \, \mathrm{d}\gamma(t), \quad x \in \Omega \,. \tag{6.44}$$

As a result of (6.38),

$$c = \min_{z \in \Gamma} \frac{\partial G}{\partial n}(0, z) > 0$$

and hence

$$\int_\Gamma u_\lambda(t) \, \mathrm{d}\gamma(t) \leqslant c^{-1} \int_\Gamma \frac{\partial G}{\partial n}(0, t) u_\lambda(t) \, \mathrm{d}\gamma(t) = c^{-1} u(0) \,.$$

We deduce that the positive Radon measures ν_λ on Γ defined by $\mathrm{d}\nu_\lambda = u_\lambda \mathrm{d}\gamma$, are of total bounded masses; we can therefore find $\lambda_k \to 1$ such that (in a vaguely defined way) the measures ν_{λ_k} converge to a positive measure ν on Γ[162]. Passing to the limit in (6.44), we see that $u = u(\nu)$; this proves the theorem in the case of an open star-shaped set.

We now consider the general case of an open set satisfying (6.37). We can find a finite covering of star-shaped open sets $\Omega_1, \ldots, \Omega_N$ contained in Ω such that $\Gamma = \Gamma_1 \cup \ldots \Gamma_N$ where $\Gamma_i = \partial\Omega_i \cap \Gamma$. Given $u \in \mathscr{H}^+(\Omega)$, the restrictions of u to Ω_i are obviously in $\mathscr{H}^+(\Omega_i)$; hence $u \in L^1(\Omega_i)$ and there exists $\nu_i \in \mathfrak{M}^+(\partial\Omega_i)$ such that

$$\int_{\Omega_i} u \, \Delta v \, \mathrm{d}x = \int_{\partial\Omega_i} \frac{\partial v}{\partial n} \mathrm{d}\nu_i \,,$$

for all $v \in W^{2,\infty}(\Omega_i)$ with $v = 0$ on $\partial\Omega_i$, and hence, denoting by $\bar{\nu}_i$ the restrictions of ν_i to Γ_i,

$$\int_{\Omega_i} u \, \Delta v \, \mathrm{d}x = \int_{\Gamma_i} \frac{\partial v}{\partial n} \mathrm{d}\bar{\nu}_i$$

for all $v \in W^{2,\infty}(\Omega_i)$ with $v = 0$ on Γ_i and $\operatorname{supp} v \subset \Omega_i \cup \Gamma_i$.

Using partitions of unity we can "glue" together the measures $\bar{\nu}_i$ to define a positive measure ν on Γ such that $u = u(\nu)$. □

Given $\varphi \in \mathscr{C}^0(\Gamma)$, the solution $u = u(\varphi)$ is continuous on $\bar{\Omega}$ and

$$\varphi(z) = \lim_{\substack{x \to z \\ x \in \Omega}} u(x) \quad \text{for all} \quad z \in \Gamma \,.$$

More generally, given $\nu \in \mathfrak{M}(\Gamma)$ and $z \in \Gamma$, $l = \lim_{\substack{x \to z \\ x \in \Omega}} u(\nu, x)$ exists iff "ν is continuous at z, with value l" in the sense in which there exists U a neighbourhood of z and φ: $U \cap \Gamma \to \mathbb{R}$ bounded and measurable with $\lim_{\substack{t \to z \\ t \in U \cap \Gamma}} \varphi(t) = l$ such that $\mathrm{d}\nu = \varphi \mathrm{d}\gamma$ on $U \cap \Gamma$.

This follows easily from the local character of the trace ν of $u(\nu)$ obvious from Lemma 5 and from Corollary 3.

A measure $\nu \in \mathfrak{M}(\Gamma)$ can very well fail to be continuous at any point $z \in \Gamma$ and hence for a function $u \in \mathscr{H}^+(\Omega)$, $\lim_{\substack{x \to z \\ x \in \Omega}} u(x)$ can very well not exist for any $z \in \Gamma$.

However, we can show that the non-tangential limit exists for almost all $z \in \Gamma$. More precisely we have:

---

[162] See Dieudonné [12]. A sequence (ν_n) of Radon measures converges vaguely to the Radon measure ν if $\langle \nu_n, f \rangle \to \langle \nu, f \rangle$ for all continuous f with compact support.

Proposition 7. *Let Ω be an open set of $\mathbb{R}^n$ satisfying (6.37) and $v \in \mathfrak{M}(\Gamma)$. Then for all $0 < \theta \leqslant 1$*

$$\varphi(z) = \lim_{\substack{x \to z \\ x \in \Omega \\ \cos(z - x, n(z)) \geqslant \theta}} u(v, x)$$

exists for almost all $z \in \Gamma$.
Also $\varphi \in L^1(\Gamma)$ and is the density of the regular part of v[163].

We shall not prove this result which appeals to delicate arguments of the theory of integral operators (see Halmos [1], Chap. VI and Helms [1], Chap. III). We only note its significance in some examples.

Example 5. Given an open set Ω of $\mathbb{R}^n$ satisfying (6.37), let us consider $z \in \Gamma$ and $v = \delta_z$, Dirac mass at z. We have $u(\delta_z, x) = \dfrac{\partial G}{\partial n}(x, z)$. The function $u(\delta_z)$ is continuous (and even of class $\mathscr{C}^1$) on $\bar{\Omega} \backslash \{z\}$ with $u(\delta_z) = 0$ on $\Gamma \backslash \{z\}$.
We shall now prove that for all $\theta \in]0, 1]$,

$$\lim_{\substack{x \to z \\ x \in \Omega \\ \cos(x - z, n(z)) \geqslant \theta}} u(\delta_z, x) = +\infty . \tag{6.45}$$

Notice, first, that we can always suppose that Ω is a ball: in effect, because of the regularity (6.37) there exists $r > 0$ such that $B = B(z - rn(z), r) \subset \Omega$, the Green's function G_B of the ball B satisfies from the maximum principle, $G_B \leqslant G$ on $B \times B$; hence for $x \in B$

$$\frac{\partial G_B}{\partial n}(x, z) = \lim_{t \downarrow 0} \frac{G_B(x, z - tn(z))}{t} \leqslant \frac{\partial G}{\partial n}(x, z) .$$

By translation and homothety, we can therefore suppose that $\Omega = B(0, 1)$. From Poisson's integral formula (see Proposition 9 of §2) we then have

$$u(\delta_z, x) = \frac{\partial G}{\partial n}(x, z) = \frac{1}{\sigma_n} \frac{1 - |x|^2}{|z - x|^n} = \frac{1}{\sigma_n} \frac{2\cos(z - x, z) - |x - z|}{|x - z|^{n-1}} ;$$

In this form, we easily verify (6.45). □

Example 6. Let us consider in the plane $\mathbb{R}^2 = \{(x, y)\}$ the unit disk

$$\Omega = \{(x, y);\; x^2 + y^2 < 1\}$$

and for φ the characteristic function of the semi-circle

$$\Sigma^+ = \{(x, y);\; x^2 + y^2 = 1 ,\; y > 0\} .$$

Using the semi-polar coordinates (r, θ) where $-1 < r < +1$, $0 \leqslant \theta < \pi$, the

[163] From the Lebesgue decomposition, v can be written in a unique manner as the sum of a measure v_Γ with density with respect to the surface measure and of a singular measure v_s (see Dieudonné [1]).

function $u = u(\varphi)$ is given by

$$u(r, \theta) = \frac{1 - r^2}{2\pi} \int_{-\theta}^{\pi - \theta} \frac{d\tau}{1 - 2r\cos\tau + r^2} = \frac{1}{\pi} \operatorname{Arctg}\left(\frac{1 + r}{1 - r} \operatorname{tg} \frac{\tau}{2}\right)\Bigg|_{-\theta}^{\pi - \theta}.$$

The function u is continuous on

$$\bar{\Omega}\backslash\{(\pm 1, 0)\} = \{(r, \theta); \quad -1 \leqslant r \leqslant 1\,, \quad 0 < \theta < \pi\}$$

with $u(1, \theta) = 1$, $u(-1, \theta) = 0$, i.e. $u = \varphi$ on $\Sigma\backslash\{(\pm 1, 0)\}$.
Now let us consider the limit of u at the point $z = (1, 0)$ of the boundary. We have $u(r, 0) = \frac{1}{2}$: the normal limit at the boundary is $\frac{1}{2}$ which is moreover half the sum of the tangential limits. Given $\zeta \in \Omega$, we denote by α the angle $(z - \zeta, z)$ (see Fig. 20) and show that for fixed α

$$\lim_{\substack{\zeta \to z \\ (z - \zeta, z) = \alpha}} u(\zeta) = \frac{1}{2} + \frac{\alpha}{\pi}.$$

Because of the symmetry we have $u(x, y) + u(x, -y) = 1$, with the result that we can restrict α to $]0, \frac{1}{2}\pi[$.

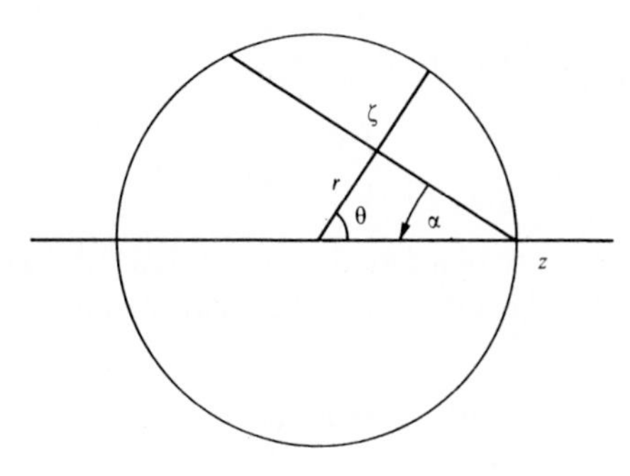

Fig. 20

Making use of polar coordinates, we then have

$$\operatorname{tg}\alpha = \frac{r\sin\theta}{1 - r\cos\theta}$$

from which we derive

$$\frac{1 + r}{1 - r} = \frac{\sin(\theta + \alpha) + \sin\alpha}{\sin(\theta + \alpha) - \sin\alpha}.$$

When $\zeta \to z$, we have $\theta \to 0$ and $\dfrac{1 + r}{1 - r} \to +\infty$ with $\dfrac{1 + r}{1 - r} \sim \dfrac{2\operatorname{tg}\alpha}{\theta}$.

Hence,

$$u(r, \theta) = \frac{1}{\pi} \operatorname{Arctg} \frac{1+r}{1-r} \times \frac{1}{\operatorname{tg}\frac{\theta}{2}} + \frac{1}{\pi} \operatorname{Arctg} \frac{1+r}{1-r} \times \operatorname{tg}\frac{\theta}{2}$$

$$\to \frac{1}{\pi} \operatorname{Arctg}(+\infty) + \frac{1}{\pi} \operatorname{Arctg}(\operatorname{tg}\alpha) = \frac{1}{2} + \frac{\alpha}{\pi}. \qquad \square$$

Example 7. We can obviously consider the Dirichlet problem $P(\Omega, \nu)$ for an unbounded open set Ω and a Radon measure ν on the boundary Γ of Ω, with a condition of growth at infinity. Let us consider the case of the half-space $\Omega = \mathbb{R}^{n-1} \times \mathbb{R}^+ = \{(x', x_n)\}$ with boundary $\Gamma = \mathbb{R}^{n-1} = \{x'\}$. Poisson's integral formula (see Proposition 22 of §2) tells us that for φ continuous and bounded on $\mathbb{R}^{n-1}$, the bounded solution of $P(\Omega, \varphi)$ is given by

$$u(x', x_n) = \frac{2x_n}{\sigma_n} \int_{\mathbb{R}^{n-1}} \frac{\varphi(t)\,dt}{(|t - x'|^2 + x_n^2)^{n/2}}. \tag{6.46}$$

This formula again has a meaning for $\varphi \in L^\infty(\mathbb{R}^{n-1})$. More generally we can define the solution of $P(\Omega, \nu)$ by

$$u(x', x_n) = \frac{2x_n}{\sigma_n} \int_{\mathbb{R}^{n-1}} \frac{d\nu(t)}{(|t - x'|^2 + x_n^2)^{n/2}}$$

for every Radon measure ν on $\mathbb{R}^{n-1}$ satisfying

$$\int_{\mathbb{R}^{n-1}} \frac{d|\nu|(t)}{(1 + |t|^2)^{n/2}} < \infty .$$

Let us now return to the case of a function $\varphi \in L^\infty(\mathbb{R}^{n-1})$ and show that for every Lebesgue point of φ[164] and every $c > 0$, the solution u given by the formula (6.46) satisfies

$$\lim_{\substack{x_n \to 0 \\ |x' - z| \leqslant c x_n}} u(x', x_n) = \varphi(z) .$$

Since almost every point $z \in \mathbb{R}^{n-1}$ is a Lebesgue point (see Sadosky [1]) this will prove Proposition 7 in this particular case.

We can always suppose that $z = 0$ and $\varphi(z) = 0$. By a change of variables, we have

$$u(x', x_n) = \frac{2}{\sigma_n} \int_{\mathbb{R}^{n-1}} \frac{\varphi(x' + x_n t)\,dt}{(1 + |t|^2)^{n/2}},$$

and hence for fixed $R > 0$

$$|u(x', x_n)| \leqslant \frac{2}{\sigma_n} \int_{B(0,R)} |\varphi(x' + x_n t)|\,dt + \frac{2}{\sigma_n} \|\varphi\|_{L^\infty} \int_{|t| \geqslant R} \frac{dt}{(1 + |t|^2)^{n/2}} .$$

---

[164] That is to say $\lim_{r \to 0} \frac{1}{r^{n-1}} \int_{B(z,r)} |\varphi(t) - \varphi(z)|\,dt = 0$.

Since, if we suppose $|x'| \leqslant c x_n$,

$$\int_{B(0,R)} |\varphi(x' + x_n t)| \,\mathrm{d}t = x_n^{-n+1} \int_{B(x', R x_n)} |\varphi(t)| \,\mathrm{d}t \leqslant x_n^{-n+1} \int_{B(0,(R+c)x_n)} |\varphi(t)| \,\mathrm{d}t ,$$

from the hypothesis: 0 is a Lebesgue-point of φ and $\varphi(0) = 0$, so

$$\limsup_{\substack{x_n \to 0 \\ |x'| \leqslant c x_n}} |u(x', x_n)| \leqslant \frac{2}{\sigma_n} \|\varphi\|_{L^\infty} \int_{|t| \geqslant R} \frac{\mathrm{d}t}{(1 + |t|^2)^{n/2}} .$$

Letting $R \to \infty$, we obtain the required result. □

Let us consider now the homogeneous Dirichlet problem $PH(\Omega, \mu)$ for a Radon measure μ on Ω. For all $x \in \Omega$, the function $t \to G(x, t)$ is u.s.c. negative on Ω (taking the value $-\infty$ for $t = x$), hence, for every positive Radon measure μ on Ω, we can define the function

$$v(\mu, x) = \int_\Omega G(x, t) \,\mathrm{d}\mu(t) , \qquad x \in \Omega \tag{6.47}$$

on putting $v(\mu, x) = -\infty$ if the function $-G(x, .)$ is not integrable with respect to μ; the function $v(\mu)$ is u.s.c. negative on Ω. Let us prove

Lemma 6. *Let Ω be an open set of $\mathbb{R}^n$ satisfying (6.37), μ a positive Radon measure on Ω. The function $v(\mu)$ defined by (6.47) is locally integrable on Ω (with respect to the Lebesgue measure) iff μ satisfies*

$$\int_\Omega \operatorname{dist}(t, \partial\Omega) \,\mathrm{d}\mu(t) < \infty .$$

Then $v(\mu)$ is sub-harmonic (u.s.c. negative) on Ω, and $\Delta v(\mu) = \mu$ in $\mathscr{D}'(\Omega)$.

Proof. Given $\varepsilon > 0$, we put $\Omega_\varepsilon = \{x \in \Omega;\ \operatorname{dist}(x, \partial\Omega) > \varepsilon\}$ and define the function

$$v_\varepsilon(x) = \int_{\Omega_\varepsilon} G(x, t) \,\mathrm{d}\mu(t) , \qquad x \in \Omega .$$

We have, for all $x \in \Omega$,

$$v_\varepsilon(x) \downarrow v(\mu, x) \quad \text{when} \quad \varepsilon \downarrow 0 .$$

Using the inequality (see (4.27))

$$E_n(x - t) - E_n(\delta) \leqslant G(x, t)$$

where δ is the diameter of Ω, we see that

$$v_\varepsilon(x) \geqslant w_\varepsilon(x) - E_n(\delta) \int_{\Omega_\varepsilon} \mathrm{d}\mu(t)$$

where w_ε is the Newtonian potential of the measure with compact support $\mu_\varepsilon = \mu \chi_{\Omega_\varepsilon}$ on $\mathbb{R}^n$. From Lemma 3 of §3, v_ε is integrable on Ω: it is also in $L^p(\Omega)$ for all $p < n/(n - 2)$. Also, making use of Fubini's theorem, we find that for all

$\zeta \in \mathscr{D}(\Omega)$

$$\int v_\varepsilon \, \Delta\zeta \, dx = \int_{\Omega_\varepsilon} d\mu(t) \int G(x, t) \, \Delta\zeta(x) \, dx = \int_{\Omega_\varepsilon} \zeta(t) \, d\mu(t) \, .$$

Hence $\Delta v_\varepsilon = \mu_\varepsilon$ in $\mathscr{D}'(\Omega)$ and in particular v_ε is sub-harmonic on Ω (see Definition 15 of §4).

Given a compact set K in Ω, for $\varepsilon > 0$ sufficiently small, $K \subset \Omega_\varepsilon$ and hence G is of class $\mathscr{C}^1$ on $K \times \bar{\Omega}\backslash\Omega_\varepsilon$. Using (6.38), we see there exists a constant $c > 0$ such that, for $\varepsilon > 0$ sufficiently small

$$C \operatorname{dist}(t, \partial\Omega) \leqslant |G(x, t)| \leqslant C^{-1} \operatorname{dist}(t, \partial\Omega) \, , \qquad x \in K, \qquad t \in \Omega\backslash\Omega_\varepsilon \, .$$

We deduce that for $\varepsilon > 0$ sufficiently small

$$C \int_{\Omega\backslash\Omega_\varepsilon} \operatorname{dist}(t, \partial\Omega) \, d\mu(t) \leqslant v_\varepsilon(x) - v(\mu, x) \leqslant C^{-1} \int_{\Omega\backslash\Omega_\varepsilon} \operatorname{dist}(t, \partial\Omega) \, d\mu(t) \, , \quad x \in K$$

from which the result follows. □

Remark 3. From the above proof, for a positive measure μ, either $v(\mu) \in L^p(\Omega) + L^\infty_{\text{loc}}(\Omega)$ for all $p < n/(n-2)$, *or* $v(\mu, x) = -\infty$, a.e. $x \in \Omega$. □

Let us now consider a Radon measure μ on Ω (not necessarily positive) satisfying

$$\int_\Omega \operatorname{dist}(t, \partial\Omega) \, d|\mu|(t) < \infty \tag{6.48}$$

where $|\mu|$ is the total variation of the measure μ. From the lemma and Fubini's theorem, for almost all $x \in \Omega$, the function $t \to G(x, t)$ is μ-integrable, the function

$$v(\mu, x) = \int_\Omega G(x, t) \, d\mu(t) \quad \text{a.e.} \quad x \in \Omega$$

is locally integrable on Ω and

$$\Delta v(\mu) = \mu \quad \text{in} \quad \mathscr{D}'(\Omega) \, .$$

We can now state the complete form of Herglotz's theorem.

Theorem 2. *Let Ω be an open set in $\mathbb{R}^n$ satisfying* (6.37) *and* u: $\Omega \to [-\infty, +\infty]$.
(1) *The following assertions are equivalent*:
(*i*) *there exist w_1, w_2: $\Omega \to [0, \infty]$ l.s.c. locally integrable superharmonic such that $u = w_1 - w_2$ a.e. on Ω*;
(*ii*) *there exists a Radon measure μ on Ω satisfying* (6.48) *and ν a Radon measure on Γ such that*

$$u(x) = \int_\Omega G(x, t) \, d\mu(t) + \int_\Gamma \frac{\partial G}{\partial n}(x, t) \, d\nu(t) \quad \text{a.e.} \quad x \in \Omega \, ;$$

(2) *when the properties* (*i*) *and* (*ii*) *are satisfied*,
(*a*) *the measures μ and ν of* (*ii*) *are unique and $\Delta u = \mu$ in $\mathscr{D}'(\Omega)$*;

(*b*) *the functions* w_+ *and* $w_-: \Omega \to [0, \infty]$ *defined by*[165]

$$w_\pm(x) = -\int_\Omega G(x, t)\,d\mu^\mp(t) + \int_\Gamma \frac{\partial G}{\partial n}(x, t)\,d\nu^\pm(t), \quad x \in \Omega$$

are l.s.c., locally integrable and superharmonic and $u = w_+ - w_-$ *a.e. on* Ω;
(*c*) *given* w_1, w_2 *satisfying* (i)

$$w_+(x) \leqslant w_1(x) \quad and \quad w_-(x) \leqslant w_2(x) \quad for\ all \quad x \in \Omega$$

$$\Delta w_+ \geqslant \Delta w_1 \quad and \quad \Delta w_- \geqslant \Delta w_2 \quad in \quad \mathscr{D}'(\Omega)\,.$$

Proof. Let us suppose that (ii) is satisfied. From Lemma 6, the functions $v(\mu^+)$ and $v(\mu^-)$ are u.s.c., negative, locally integrable and subharmonic; the functions $u(\nu^\pm)$, solutions in the sense of Definition 2 of the Dirichlet problems $P(\Omega, \nu^\pm)$ are positive and harmonic. Hence the functions $w_\pm = -v(\mu^\mp) + u(\nu^\pm)$ are l.s.c., positive, locally integrable and superharmonic; by definition $u = w_+ - w_-$ a.e. on Ω. We have thus proved the implication (ii) $\Rightarrow$ (i) of point (1) as well as (*b*) of point (2). Since $\Delta v(\mu^\pm) = \mu^\pm$ in $\mathscr{D}'(\Omega)$ and $u(\nu^\pm) \in \mathscr{H}(\Omega)$, we have $\Delta u = \mu^+ - \mu^- = \mu$ in $\mathscr{D}'(\Omega)$; hence μ is completely determined by u. From Proposition 6 (see Corollary 2), ν is therefore also completely determined; this proves (a) of the point (2).
We consider now $w: \Omega \to [0, \infty]$ u.s.c., locally integrable and superharmonic. From Propositions 20 and 21 of §4, $\mu = -\Delta w$ (in the sense of distributions) is a positive Radon measure on Ω. For $\varepsilon > 0$, the function

$$w_\varepsilon(x) = -\int_{\Omega_\varepsilon} G(x, t)\,d\mu(t)$$

is a quasi-classical solution of $P(\Omega, -\mu\chi_{\Omega_\varepsilon})$. Since

$$\Delta w_\varepsilon = -\mu\chi_{\Omega_\varepsilon} \geqslant -\mu = \Delta w \quad \text{in} \quad \mathscr{D}'(\Omega)$$

and

$$w_\varepsilon(z) = 0 \leqslant \liminf_{x \to z} w(x) \quad \text{for all} \quad z \in \Gamma\,,$$

we have $w_\varepsilon \leqslant w$ a.e. on Ω. Passing to the limit

$$-v(\mu) = \lim_{\varepsilon \to 0} w_\varepsilon \leqslant w \quad \text{a.e. on} \quad \Omega\,.$$

In particular $v(\mu)$ is locally integrable and so from Lemma 6, μ satisfies (6.48). Also, we have $\Delta v(\mu) = \mu = -\Delta w$ in $\mathscr{D}'(\Omega)$, so $w + v(\mu)$ is a positive harmonic distribution on Ω; from Proposition 6, there exists ν a positive Radon measure on Γ such that $w + v(\mu) = u(\nu)$ a.e. on Ω. But the functions w and $-v(\mu) + u(\nu)$ being l.s.c., super-harmonic and locally integrable, the equality almost everywhere implies that they are identically equal (see Proposition 20 of §4).

[165] μ^+, μ^- (resp ν^+, ν^-) denote the positive and negative variations of μ on Ω (resp. ν on Γ) (that is, for the measure μ, $\mu^+ = \frac{1}{2}(|\mu| + \mu)$, $\mu^- = \frac{1}{2}(|\mu| - \mu)$, where $|\mu|$ is the total variation of μ).

Applying the result to w_1 and w_2 we have thus proved the implication (*i*) ⇒ (*ii*) of the point (1).
Finally let, for $i = 1, 2$,

$$(\mu_i, \nu_i) \in \mathfrak{M}^+(\Omega) \times \mathfrak{M}^+(\Gamma) \qquad w_i = -v(\mu_i) + u(\nu_i)\,.$$

From uniqueness, we have $\mu = \Delta u = \mu_2 - \mu_1$ and $\nu = \nu_1 - \nu_2$; from which, by definition, $\mu^+ \leqslant \mu_2, \mu^- \leqslant \mu_1$ on Ω and $\nu^+ \leqslant \nu_1, \nu^- \leqslant \nu_2$ on Γ. We deduce (*c*) of the point (2). □

For $0 \leqslant \theta \leqslant 1$, we denote by $\mathfrak{M}_\theta(\Omega)$ the space of Radon measures μ on Ω satisfying

$$\int_\Omega \operatorname{dist}(t, \partial\Omega)^\theta \, \mathrm{d}|\mu|(t) < \infty\,. \tag{6.49}$$

Theorem 2 shows that the map $(\mu, \nu) \to -v(\mu) + u(\nu)$ is a (linear) bijection of $\mathfrak{M}_1(\Omega) \times \mathfrak{M}(\Gamma)$ onto the space of the differences of positive locally integrable functions (defined a.e. on Ω) and sub-harmonic; this map is *bi-increasing*: for $(\mu_1, \nu_1), (\mu_2, \nu_2) \in \mathfrak{M}_1(\Omega) \times \mathfrak{M}(\Gamma)$. $\mu_1 \geqslant \mu_2$ on Ω and $\nu_1 \geqslant \nu_2$ on Γ iff

$$-v(\mu_1) + u(\nu_1) \geqslant -v(\mu_2) + u(\nu_2) \quad \text{on} \quad \Omega$$

and similarly

$$(\mu_1, \nu_1) \neq (\mu_2, \nu_2), \quad \mu_1 \geqslant \mu_2 \quad \text{on} \quad \Omega \quad \text{and} \quad \nu_1 \geqslant \nu_2 \quad \text{on} \quad \Gamma$$
$$\Leftrightarrow -v(\mu_1) + u(\nu_1) > -v(\mu_2) + u(\nu_2) \quad \text{on} \quad \Omega\,.$$

At the beginning of this section we studied the harmonic function $u(\nu)$ for $\nu \in \mathfrak{M}(\Gamma)$ which we defined as the solution of the Dirichlet problem $P(\Omega, \nu)$. Now we study the function $v(\mu)$, $\mu \in \mathfrak{M}_1(\Omega)$, considered as the solution of the homogeneous Dirichlet problem $PH(\Omega, \mu)$.
We know that if $\mathrm{d}\mu = f\mathrm{d}t$ with $f \in L^p(\Omega)$, where $p > \frac{1}{2}n$, then $v(\mu) \in \mathscr{C}^0(\bar{\Omega})$ and is a quasi-classical solution of $PH(\Omega, f)$; for $\mu \in \mathfrak{M}_1(\Omega)$ this is no longer obviously true: even if $\mathrm{d}\mu = f\mathrm{d}t$ with $f \in \mathscr{C}^\infty(\Omega)$ (in which case $v(\mu) \in \mathscr{C}^\infty(\Omega)$), the hypothesis $\mu \in \mathfrak{M}_1(\Omega)$ (which then be written $\int_\Omega \operatorname{dist}(t, \partial\Omega)|f(t)|\,\mathrm{d}t < \infty$) is not sufficient to ensure that $v(\mu) = 0$ on Γ, in the classical sense. To specify in what sense $v(\mu) = 0$ on Γ, we shall use the Sobolev spaces $W_0^{1,p}(\Omega)$. We recall[166] that for $1 \leqslant p \leqslant \infty$, $W^{1,p}(\Omega) = \{u \in L^p(\Omega)$; the derivatives $\partial u/\partial x_i$ (in the sense of distributions) are functions of $L^p(\Omega)\}$; it is a Banach space, which for $1 \leqslant p < \infty$ has the norm

$$\|u\|_{W^{1,p}} = \left(\int_\Omega \left(|u|^p + \sum \left|\frac{\partial u}{\partial x_i}\right|^p\right) \mathrm{d}x\right)^{1/p}.$$

By definition $W_0^{1,p}(\Omega)$ is the closure of $\mathscr{D}(\Omega)$ in $W^{1,p}(\Omega)$.

[166] See Chap. IV.

In fact, in the case considered where Ω is of class $W^{2,\infty}$, $W_0^{1,p}(\Omega)$ can also be identified with the space of the $u \in W^{1,p}(\mathbb{R}^n)$ such that $u = 0$ a.e. on $\mathbb{R}^n \setminus \Omega$ or again $u = 0$ in $\mathscr{D}'(\mathbb{R}^n \setminus \bar{\Omega})$ (see Chap. IV).

Let us now prove

Proposition 8. *Let Ω be an open set in $\mathbb{R}^n$ satisfying (6.37) and μ a Radon measure on Ω.*

(1) *Given $0 < \theta < 1$, let us suppose that $\mu \in \mathfrak{M}_\theta(\Omega)$ (i.e. satisfies (6.49)). Then the solution u of $PH(\Omega, \mu)$ satisfies $u = v(\mu) \in W_0^{1,p}(\Omega)$ with $p = (n - \theta)/(n - 1)$.*

(2) *Suppose $\mu \in \mathfrak{M}_\theta(\Omega)$ for $0 \leqslant \theta < 1$. Then the function $v = v(\mu)$ is characterised by*

$$v \in W_0^{1,1}(\Omega) \quad \textit{and} \quad \Delta v = \mu \quad \textit{in} \quad \mathscr{D}'(\Omega)\,.$$

Proof. Given $\zeta_1, \ldots, \zeta_n \in \mathscr{D}(\Omega)$ let us consider the classical solution ζ of $PH\left(\Omega, \sum \dfrac{\partial \zeta_i}{\partial x_i}\right)$. We know from Corollary 1 and the hypothesis (6.37) that $\zeta \in \mathscr{C}^1(\bar{\Omega})$ and for $n < q < \infty$

$$\|\operatorname{grad} \zeta\|_{L^q} \leqslant C \sum \|\zeta_i\|_{L^q}$$

where C depends only on q, n and Ω. Using the Sobolev inclusion[167] $W^{1,q} \subset \mathscr{C}^\theta$ with $\theta = \dfrac{q - n}{q - 1}$, since $\zeta = 0$ on $\partial\Omega$ we have, for $0 < \theta < 1$

$$|\zeta(x)| \leqslant C \operatorname{dist}(x, \partial\Omega)^\theta \sum \|\zeta_i\|_{L^{\frac{n-\theta}{1-\theta}}}$$

with a constant C (possibly different from the preceding one), depending only on θ, n and Ω.

Let us suppose, first of all, that μ has compact support in Ω. Then $v = v(\mu)$ is the quasi-classical solution of $P(\Omega, \mu)$; from the theorem of local regularity (see Proposition 6 of §3), we know that $v \in W_{\text{loc}}^{1,p}(\Omega)$ for all $p < n/(n - 1)$; from Proposition 1 (applied to $\Omega \setminus \operatorname{supp} \mu$ and $\Gamma_0 = \partial\Omega$) we know that $v \in \mathscr{C}^1(\bar{\Omega}/K)$; also $v = 0$ on $\partial\Omega$ with the result that by Green's formula, we have

$$\int_\Omega \sum \frac{\partial v}{\partial x_i} \zeta_i \, dx = -\int_\Omega u \sum \frac{\partial \zeta_i}{\partial x_i} dx = -\int_\Omega u \, \Delta\zeta \, dx$$

$$= -\int_\Omega \zeta \, d\mu \leqslant C \sum \|\zeta_i\|_{L^{\frac{n-\theta}{1-\theta}}} \int_\Omega \operatorname{dist}(x, \partial\Omega)^\theta \, d|\mu|(x)\,.$$

This being true for the functions $\zeta_1, \ldots, \zeta_n \in \mathscr{D}(\Omega)$ arbitrary

$$\|\operatorname{grad} v(\mu)\|_{L^p} \leqslant C \int_\Omega \operatorname{dist}(x, \partial\Omega)^\theta \, d|\mu|(x) \tag{6.50}$$

where p is the index conjugate to $q = \dfrac{n - \theta}{1 - \theta}$, that is $p = \dfrac{n - \theta}{n - 1}$.

---

[167] See Adams [1].

Now let us take $\mu \in \mathfrak{M}_\theta(\Omega)$ with $0 < \theta < 1$. We apply (6.50) to $\mu_\varepsilon = \mu\chi_{\Omega_\varepsilon}$

$$\|\operatorname{grad} v(\mu_\varepsilon)\|_{L^p} \leqslant C\int_\Omega \operatorname{dist}(x, \partial\Omega)^\theta \, \mathrm{d}|\mu_\varepsilon|(x) \leqslant C\int_\Omega \operatorname{dist}(x, \partial\Omega)^\theta \, \mathrm{d}|\mu|(x) .$$

Now from the growth of the map $\mu \to -v(\mu)$,

$$|v(\mu_\varepsilon)| \leqslant |v(\mu)| \quad \text{a.e. on} \quad \Omega .$$

From the definition of $v(\mu)$, therefore

$$v(\mu_\varepsilon) \to v(\mu) \quad \text{in} \quad L^p_{\mathrm{loc}}(\Omega) \quad \text{and a.e. on} \quad \Omega .$$

In the limit we thus have again (6.50) for $v(\mu)$ and, using the reflexivity of the Banach space $W^{1,p}(\Omega)$, $v(\mu_\varepsilon)$ converges weakly to $v(\mu)$ in $W^{1,p}(\Omega)$. Since $v(\mu_\varepsilon) \in W^{1,p}_0(\Omega)$, the same is true for $v(\mu)$.
It remains to show that $v = v(\mu)$ is characterized by the property

$$v \in W^{1,1}_0(\Omega) \quad \text{and} \quad \Delta v = \mu \quad \text{in} \quad \mathscr{D}'(\Omega) .$$

From point (1) and the linearity, we are led to prove the nullity of a harmonic function $v \in W^{1,1}_0(\Omega)$. Let, therefore, $v \in W^{1,1}_0(\Omega)$ be harmonic on Ω; given $f \in \mathscr{D}(\Omega)$, from Corollary 1, the function $v(f)$ is in $\mathscr{C}^1(\bar{\Omega})$. Since $v \in W^{1,1}_0(\Omega)$ on approximating by the functions $v_k \in \mathscr{D}(\Omega)$ converging to v in $W^{1,1}(\Omega)$, we have

$$\int_\Omega vf \, \mathrm{d}x = \int_\Omega v \, \Delta v(f) \, \mathrm{d}x = -\int_\Omega \operatorname{grad} v \,.\, \operatorname{grad} v(f) \, \mathrm{d}x .$$

Now let us approximate $v(f)$ by the functions $u_k \in \mathscr{D}(\Omega)$ satisfying: $\operatorname{grad} u_k$ is bounded in $L^\infty(\Omega)$ and $\operatorname{grad} u_k \to \operatorname{grad} v(f)$ a.e. in Ω.
Then we shall have by dominated convergence

$$\int_\Omega \operatorname{grad} v \,.\, \operatorname{grad} v(f) \, \mathrm{d}x = \lim \int_\Omega \operatorname{grad} v \,.\, \operatorname{grad} u_k \, \mathrm{d}x .$$

But since v is harmonic and $u_k \in \mathscr{D}(\Omega)$

$$\int_\Omega \operatorname{grad} v \,.\, \operatorname{grad} u_k \, \mathrm{d}x = -\int_\Omega u_k \, \Delta v \, \mathrm{d}x = 0 .$$

Therefore

$$\int_\Omega vf \mathrm{d}x = -\int_\Omega \operatorname{grad} v \,.\, \operatorname{grad} v(f) \, \mathrm{d}x = 0$$

This being true for all $f \in \mathscr{D}(\Omega)$, v is identically zero. □

Remark 4. As appears in the proof, for $\mu \in \mathfrak{M}_\theta(\Omega)$ with $0 < \theta < 1$ and $v \in L^1_{\mathrm{loc}}(\Omega)$, the following assertions are equivalent:
(i) $v = v(\mu)$ a.e. on Ω;
(ii) $v \in W^{1,1}_0(\Omega)$ and $\Delta v = \mu$ in $\mathscr{D}'(\Omega)$;

(*iii*) $v \in W_0^{1,p}(\Omega)$ with $p = \dfrac{n-\theta}{n-1}$ and $\displaystyle\int_\Omega \operatorname{grad} v \,.\, \operatorname{grad} \zeta \, dx + \int_\Omega \zeta \, d\mu = 0$, for every function $\zeta \in W_0^{1,q}(\Omega)$ with $q = \dfrac{n-\theta}{1-\theta}$[168]. □

4. Neumann Problem with Given Measures

We recall the classical Green's formula: given Ω, a regular bounded open set of $\mathbb{R}^n$, u and $v \in \mathscr{C}_n^1(\bar{\Omega}) \cap \mathscr{C}^2(\Omega)$ with $u\Delta v - v\Delta u \in L^1(\Omega)$,

$$\int_\Omega (u\,\Delta v - v\,\Delta u)\,dx = \int_\Gamma \left(u\frac{\partial v}{\partial n} - v\frac{\partial u}{\partial n}\right)d\gamma\,.$$

We also have the formula of integration by parts: for $u \in \mathscr{C}_n^1(\bar{\Omega}) \cap \mathscr{C}^2(\Omega)$ and $v \in \mathscr{C}^0(\bar{\Omega}) \cap \mathscr{C}^1(\Omega)$ with

$$v\,\Delta u + \operatorname{grad} v \,.\, \operatorname{grad} u \in L^1(\Omega)\,,$$

$$\int_\Omega (v\,\Delta u + \operatorname{grad} v \,.\, \operatorname{grad} u)\,dx = \int_\Gamma v\frac{\partial u}{\partial n}\,d\gamma\,.$$

In Sects. 2 and 3, we have used these formulae to define the trace on the boundary of a function, harmonic in Ω: from the Definition 2 and Lemma 5, for example, when Ω satisfies (6.37), given a function $u \in \mathscr{H}(\Omega)$, a Radon measure ν on Γ is the trace of u on Γ iff

$$\int_\Omega u\,\Delta v\,dx = \int_\Gamma \frac{\partial v}{\partial n}\,d\nu$$

for every function $v \in \mathscr{C}_n^1(\bar{\Omega})$ with $\Delta v \in \mathscr{D}(\Omega)$ and $v = 0$ on Γ. In other words, the trace of $u \in \mathscr{H}(\Omega)$ on Γ, is defined as the measure such that Green's formula is true for all *v varying in a space of "test functions"*. Such a *definition* will be called *variational.*

The variational method will be developed in Chap. VII in an abstract context. Here we propose, for Neumann problems, to relate a variational notion of solution of these problems to more classical ideas; notice also that the functions considered here will vary in classical spaces of measures or continuous functions which, not being reflexive, would take badly to an abstract theory. First of all we note:–

Proposition 9. *Let Ω be an open set of $\mathbb{R}^n$ satisfying (6.37), μ a bounded Radon measure on Ω and u the solution of the homogeneous Dirichlet problem*[169]

$$u \in W_0^{1,1}(\Omega)\,, \quad \Delta u = \mu \quad in \quad \mathscr{D}'(\Omega)\,.$$

[168] We recall that $W_0^{1,q}(\Omega) \subset \mathscr{C}^0(\bar{\Omega})$ for $q = (n-\theta)/(1-\theta)$ which shows that $\int_\Omega \zeta\,d\mu$ is effectively defined for $\zeta \in W_0^{1,q}(\Omega)$ and $\mu \in \mathfrak{M}_\theta(\Omega)$.

[169] See Proposition 8.

There exists a unique $\psi \in L^1(\Gamma)$ *such that*

$$\int_\Gamma v\psi \,d\gamma = \int_\Omega v \,d\mu + \int_\Omega \operatorname{grad} v \,.\, \operatorname{grad} u \,dx \tag{6.51}$$

for all $v \in \mathscr{C}'(\bar{\Omega})$.
Also

(6.52) (i) $$\|\psi\|_{L^1(\Gamma)} \leqslant \int_\Omega d|\mu|$$

(6.52) (ii) $$\mu \not\equiv 0, \quad \mu \geqslant 0 \quad \text{on} \quad \Omega \Rightarrow \inf_\Gamma \operatorname{ess} \psi > 0 \,.$$

Proof. We notice first that the uniqueness of $\psi \in L^1(\Gamma)$ satisfying (6.51) is immediate since, taking account of the regularity of Ω,

$$\int_\Omega v\psi \,d\gamma = 0 \quad \text{for all} \quad v \in \mathscr{C}^1(\bar{\Omega}) \Rightarrow \psi = 0 \quad \text{a.e. on} \quad \Gamma \,.$$

Let us suppose first that the measure μ *has compact support* $K \subset \Omega$. From Proposition 8, u is the quasi-classical solution of $PH(\Omega, \mu)$ and hence, from Proposition 1, $u \in \mathscr{C}^1(\bar{\Omega} \backslash K)$. We prove that $\psi = \partial u/\partial n$ satisfies the Proposition. The relation (6.51) is true for every function $v \in \mathscr{C}^1(\bar{\Omega})$ with compact support contained in Ω: it can then in effect be written

$$\int v \,d\mu = -\int \operatorname{grad} v \,.\, \operatorname{grad} u \,dx$$

which is the definition of $\Delta u = \mu$ in $\mathscr{D}'(\Omega)$. The relation (6.51) is also true for every function $v \in \mathscr{C}^1(\bar{\Omega})$ with support disjoint from K: the field $p = v \operatorname{grad} u$ is continuous on $\bar{\Omega}$, of class $\mathscr{C}^1$ on Ω and

$$\operatorname{div} p = \operatorname{grad} v \,.\, \operatorname{grad} u \in L^1(\Omega)$$

with the result that (6.51) amounts to the classical Ostrogradski formula (see Proposition 4 of §1). The relation (6.51) is thus true for every function $v \in \mathscr{C}^1(\bar{\Omega})$ which can always be written as the sum of a function with compact support in Ω and of a function with support disjoint from K.
When $\mu \geqslant 0$ on Ω we have $u \leqslant 0$ on Ω and hence $\psi = \partial u/\partial n \geqslant 0$ on Γ; more precisely, if $\mu \not\equiv 0$ on Ω, from Hopf's maximum principle (see Lemma 2 of §4) $\psi > 0$ on Γ and since ψ is continuous, $\min_\Gamma \psi > 0$. Considering μ^+ and μ^-, the positive and negative parts of the measure μ, we denote by u_+ and u_-, the (quasi-classical) solutions of $PH(\Omega, \mu^+)$ and $PH(\Omega, \mu^-)$. We have $u = u_+ - u_-$ and hence for $\psi = \partial u/\partial n$

$$\int_\Gamma |\psi| \,d\gamma \leqslant \int_\Gamma \left(\left|\frac{\partial u_+}{\partial n}\right| + \left|\frac{\partial u_-}{\partial n}\right| \right) d\gamma = \int_\Gamma \frac{\partial u_+}{\partial n} d\gamma + \int_\Gamma \frac{\partial u_-}{\partial n} d\gamma$$

$$= \int_\Omega d\mu^+ + \int_\Omega d\mu^- = \int_\Omega d|\mu| \,.$$

The second-last equality is a consequence of Gauss' theorem

$$\int_\Gamma \frac{\partial u}{\partial n} \mathrm{d}\gamma = \int_\Omega \mathrm{d}\mu$$

which is merely (6.51) with $v \equiv 1$ on Ω.
Let us now consider the *general case* of an arbitrary bounded Radon measure μ on Ω. We take an increasing sequence of compact sets K_k of union Ω and denote $\mu_k = \mu\chi_{K_k}$ and u_k the (quasi-classical) solution of $PH(\Omega, \mu_k)$. Using Lebesgue's theorem, we have

$$\int_\Omega \mathrm{d}|\mu_k - \mu| = \int_{\Omega \backslash K_k} \mathrm{d}|\mu| \to 0 \quad \text{when} \quad k \to \infty .$$

From Proposition 8 we deduce that $u_k \to u$ in $W^{1,1}(\Omega)$ – and even $W^{1,p}(\Omega)$ for all $1 \leqslant p < n/(n-1)$. Now, applying (6.52) (i) to the measure $\mu_k - \mu$ (with compact support in Ω), we have for $k \geqslant 1$

$$\int_\Omega \left| \frac{\partial u_k}{\partial n} - \frac{\partial u_l}{\partial n} \right| \mathrm{d}\gamma \leqslant \int_\Omega \mathrm{d}|\mu_k - \mu_l| = \int_{K_k \backslash K_l} \mathrm{d}|\mu| .$$

We deduce that $(\partial u_k / \partial n)$ is a Cauchy sequence in $L^1(\Gamma)$; we denote its limit by ψ. Passing to the limit in (6.51) and (6.52) (*i*) applied to $\psi_k = \partial u_k / \partial n$, we obtain (6.51) and (6.52) (i) for ψ. Finally, in the case $\mu \geqslant 0$ on Ω, the sequence (u_k) is increasing with the result that so is the sequence $(\partial u_k / \partial n)$; hence

$$\inf_\Gamma \operatorname{ess} \psi \geqslant \min_\Gamma \frac{\partial u_k}{\partial n} \quad \text{for all} \quad k ;$$

if, in addition, $\mu \not\equiv 0$, then there exists k such that $\mu_k \not\equiv 0$ and hence $\min_\Gamma \dfrac{\partial u_k}{\partial n} > 0$.

□

Remark 5. Proposition 9 states[170] that the normal derivative of a function $u \in W_0^{1,1}(\Omega)$ whose Laplacian is a bounded measure on Ω is an integrable function on Γ. We notice that *in general, this normal derivative is not more than integrable on* Γ. In effect, let us suppose that there exists $p > 1$ such that for every bounded measure μ on Ω, the normal derivative ψ of the solution u of $PH(\Omega, \mu)$ is in $L^p(\Gamma)$; from the closed graph theorem, there will exist C such that

$$\| \psi \|_{L^p(\Gamma)} \leqslant C \int_\Omega \mathrm{d}|\mu|$$

and hence, in particular, applying with $\mu = \delta_x$, Dirac mass at $x \in \Omega$,

$$\left\| \frac{\partial G}{\partial n}(x, .) \right\|_{L^p} \leqslant C \quad \text{for all} \quad x \in \Omega , \tag{6.53}$$

[170] See Definition 3 below.

where G is the Green's function of Ω. We can easily verify in the case of a ball that the relation (6.53) is impossible for $p > 1$, by using the explicit value of $\frac{\partial G}{\partial n}(x, z)$ given by Poisson's formula. Another argument showing the impossibility of (6.53) consists of noting that if (6.53) is true, then for all $\varphi \in \mathscr{C}^0(\Gamma)$ the solution $u(\varphi)$ of $P(\Omega, \varphi)$ will satisfy

$$|u(\varphi, x)| = \left|\int_\Gamma \frac{\partial G}{\partial n}(x, z)\varphi(z)\,\mathrm{d}\gamma(z)\right| \leqslant C\|\varphi\|_{L^q} \quad x \in \Omega\,,$$

where q is the index conjuge to p, and therefore

$$\max_\Gamma |\varphi| = \sup_\Omega |u(\varphi)| \leqslant C\|\varphi\|_{L^q}$$

which is a contradiction with $q < \infty$, that is, with $p > 1$. □

Proposition 9, and Proposition 10, lead us to formulate the

Definition 3. Given Ω a bounded open set in $\mathbb{R}^n$ and $u \in W^{1,1}(\Omega)$ whose Laplacian is a bounded measure μ on Ω, we call *normal derivative (in the sense of distributions) of u on Γ* the distribution v_1 defined on $\mathbb{R}^n$ by

$$\langle v_1, \zeta\rangle = \int_\Omega \zeta\,\mathrm{d}\mu + \int_\Omega \operatorname{grad}\zeta\,.\operatorname{grad} u\,\mathrm{d}x\,, \quad \zeta \in \mathscr{D}(\mathbb{R}^n)\,. \tag{6.54}$$

Note that the distribution v_1 defined by (6.54) is of compact support in the boundary Γ of Ω: if $\operatorname{supp}\zeta \cap \Gamma = \varnothing$, then

$$\zeta\chi_\Omega \in \mathscr{D}(\Omega) \quad \text{and} \quad \operatorname{grad}(\zeta\chi_\Omega) = (\operatorname{grad}\zeta)\chi_\Omega$$

with the result that by the definition of the Laplacian in the sense of distributions

$$\int_\Omega \zeta\,\mathrm{d}\mu = \langle \Delta u, \zeta\chi_\Omega\rangle = -\int \operatorname{grad}(\zeta\chi_\Omega\,.\operatorname{grad} u\,\mathrm{d}x = -\int_\Omega \operatorname{grad}\zeta\,.\operatorname{grad} u\,\mathrm{d}x\,.$$

Proposition 9 can be stated with this definition: for every bounded Radon measure μ on Ω, the solution u of $PH(\Omega, \mu)$, which is in $W_0^{1,p}(\Omega)$ for all $p \in [1, n/(n-1)[$ (see Proposition 8) has for normal derivative on Γ a measure with density with respect to the surface measure of Γ: $\mathrm{d}v_1 = \psi\mathrm{d}\gamma$ where $\psi \in L^1(\Gamma)$ is defined by (6.51). We are interested now in the case where v_1 is a Radon measure (carried by Γ); let us prove:

Proposition 10. *Let Ω be an open set of $\mathbb{R}^n$ satisfying (6.37), μ a bounded Radon measure on Ω and v_1 a Radon measure on Γ. Let us now consider the Neumann problem $PN(\Omega, \mu, v_1)$*

$$\begin{cases} u \in W^{1,1}(\Omega),\ \Delta u = \mu \quad \textit{in} \quad \mathscr{D}'(\Omega) \\ v_1 \textit{ is the normal derivative on } \Gamma \textit{ of } u \end{cases} \tag{6.55}$$

(1) *There exists a solution of $PN(\Omega, \mu, v_1)$ iff*

$$\int_\Gamma \mathrm{d}v_1 = \int_\Omega \mathrm{d}\mu\,. \tag{6.56}$$

(2) *When* (6.56) *is satisfied, the solution of* (6.55) *is defined to within an additive constant*[171] *and* $u \in W^{1,p}(\Omega)$ *for all* $1 \leqslant p < n/(n-1)$.

Proof. The Gauss condition (6.56) is obviously necessary: more generally, for $u \in W^{1,1}(\Omega)$ with Laplacian μ a bounded measure on Ω, the normal derivative ν_1 (with compact support) satisfies

$$\langle \nu_1, 1 \rangle = \int_\Omega d\mu$$

(apply (6.54) with $\zeta \equiv 1$ in the neighbourhood of $\bar{\Omega}$).
We prove now the solution of $PN(\Omega, \mu, \nu_1)$, if it exists, is unique to within an additive constant; by linearity, this comes down to showing that a solution of

$$(6.57) \qquad u \in W^{1,1}(\Omega), \int_\Omega \operatorname{grad} \zeta . \operatorname{grad} u \, dx = 0 \quad \text{for all} \quad \zeta \in \mathscr{D}(\mathbb{R}^n)$$

is constant on Ω. Given $f \in \mathscr{D}(\Omega)$ with $\int_\Omega f dx = 0$, we have the existence of a solution of the classical homogeneous Neumann problem

$$v \in \mathscr{C}^1(\bar{\Omega}) \cap \mathscr{C}^2(\Omega), \quad \Delta v = f \quad \text{on} \quad \Omega, \quad \frac{\partial v}{\partial n} = 0 \quad \text{on} \quad \Gamma$$

(this follows from Theorem 2 of §4 and from the regularity results of §6.1). If u is the solution of (6.57), we have

$$(6.58) \qquad \int_\Omega f u \, dx = - \int_\Omega \operatorname{grad} v . \operatorname{grad} u \, dx = 0 .$$

The first equality in (6.58) is obtained by approximating u by the functions $u_k \in \mathscr{C}^1(\bar{\Omega})$ such that $u_k \to u$ in $W^{1,1}(\Omega)$; the second equality is obtained from (6.57) by now approximating v by functions $\zeta_k \in \mathscr{D}(\mathbb{R}^n)$ such that $\zeta_k \to v$ in $\mathscr{C}^1(\bar{\Omega})$. These approximations are possible due to the regularity hypothesis on Ω. By the density of $\mathscr{D}(\Omega)$ in $L^1(\Omega)$, we therefore have from (6.58)

$$\int_\Omega f u \, dx = 0 \quad \text{for all} \quad f \in L^\infty(\Omega) \quad \text{with} \quad \int_\Omega f dx = 0$$

from which we deduce that u is constant on Ω.
Finally we prove the existence of a solution u of $PN(\Omega, \mu, \nu_1)$ in $W^{1,p}(\Omega)$ for all $1 \leqslant p < n/(n-1)$. From Propositions 8 and 9, we can reduce to the case $\mu \equiv 0$ on Ω: in effect considering $\bar{u}$ the solution of $PH(\Omega, \mu)$, u is the solution of $PN(\Omega, \mu, \nu_1)$ iff $u - \bar{u}$ is solution of $PN(\Omega, 0, \nu_1 - \bar{\nu}_1)$ where $\bar{\nu}_1$ is the normal derivative of $\bar{u}$, a Radon measure with density, from Proposition 9. Suppose, first that ν_1 is a measure with density $\psi \in \mathscr{C}^0(\Gamma)$ satisfying the Gauss condition

[171] For simplicity, we supposed in (6.37) that Ω was connected; in the case where Ω has several connected components, we have the same result with the Gauss condition (6.56) and to within an additive constant on each component.

$\int_\Gamma \psi \, d\gamma = 0$. From Theorem 2 of §4, there exists a classical solution u of the Neumann problem

$$u \in \mathscr{C}_n^1(\bar{\Omega}) \cap \mathscr{H}(\Omega), \quad \frac{\partial u}{\partial n} = \psi \quad \text{on} \quad \Gamma .$$

We know (see Proposition 5 of §1) that $|\operatorname{grad} u|^2 \in L^1(\Omega)$ and hence *a fortiori* $u \in W^{1,1}(\Omega)$. Given $\zeta_1, \ldots, \zeta_n \in \mathscr{D}(\Omega)$, then as we have just seen above there exists a solution v of

$$v \in \mathscr{C}^1(\bar{\Omega}) \cap \mathscr{C}^2(\Omega), \quad \Delta v = \sum \frac{\partial \zeta_i}{\partial x_i} \quad \text{on} \quad \Omega, \quad \frac{\partial n}{\partial n} = 0 \quad \text{on} \quad \Gamma,$$

and, by integration by parts, we have

$$\int_\Omega \sum \zeta_i \left| \frac{\partial u}{\partial x_i} \, dx = - \int_\Omega u \, \Delta v \, dx = \int_\Omega \operatorname{grad} u \,.\, \operatorname{grad} v \, dx = \int_\Gamma \frac{\partial u}{\partial n} v \, d\gamma \, .\right.$$

From the analogue for the homogeneous Neumann problem of Proposition 1, for all $n < q < \infty$

$$\| v \|_{L^\infty} \leqslant C \sum \| \zeta_i \|_{L^q}$$

where C depends only on Ω and q; hence

$$\left| \int_\Omega \sum \zeta_i \frac{\partial u}{\partial x_i} \, dx \right| \leqslant C \left\| \frac{\partial u}{\partial n} \right\|_{L^1(\Gamma)} \sum \| \zeta_i \|_{L^q} ,$$

from which we deduce

$$\| \operatorname{grad} u \|_{L^p} \leqslant C \left\| \frac{\partial u}{\partial n} \right\|_{L^1(\Gamma)}$$

where p is the index conjugate to q.
The function u being defined to within an additive constant, we can always impose on it $\int_\Omega u \, dx = 0$; then we have from Poincaré's inequality[172]

$$\| u \|_{W^{1,p}} \leqslant C_0 \| \operatorname{grad} u \|_{L^p} \leqslant C' \left\| \frac{\partial u}{\partial n} \right\|_{L^1(\Gamma)} .$$

We conclude by approximating an arbitrary measure ν_1 on Γ by measures with continuous density and passing to the limit. □

5. Dependence of Solutions of Dirichlet Problems as a Function of the Open Set: Hadamard's Formula

In §4 we proved that, given an increasing sequence of open sets (Ω_k) with union Ω, the Green's function G of Ω is the decreasing limit of the Green's functions G_k of Ω_k

---

[172] See Chap. IV. (Vol. 2, 126, 379).

(see Propositions 6 and 12 of §4). Here, we propose to study more completely the dependence with respect to a parameter λ of the Green's function G_λ of an open set Ω_λ depending regularly on λ and, more generally, to study solutions of Dirichlet problems $P(\Omega_\lambda, \varphi_\lambda, f_\lambda)$.

We take a parameter λ varying in a domain Λ which, to fix ideas, we take to be real. We take a family Ω_λ of open sets in $\mathbb{R}^n$ with boundary $\Gamma_\lambda = \partial\Omega_\lambda$ and we shall suppose that *the open sets satisfy* (6.37) *uniformly with respect to* $\lambda \in \Lambda$ which we shall express by the existence of U, a bounded open set in $\mathbb{R}^n$ containing all the boundaries Γ_λ and Φ: $(\lambda, x) \in \Lambda \times U \to \mathbb{R}$ of class $W^{2,\infty}$ with respect to $x \in U$ uniformly for $\lambda \in \Lambda$ such that

$$\begin{cases} \operatorname{grad}_x \Phi(\lambda, x) \neq 0 & \text{for all} \quad (\lambda, x) \in \Lambda \times U \\ \Omega_\lambda \cap U = \{x \in U;\ \ \Phi(\lambda, x) < 0\} & \text{for all} \quad \lambda \in \Lambda \end{cases}$$

with the result that for all $\lambda \in \Lambda$

$$\Gamma_\lambda = \{x \in U;\ \ \Phi(\lambda, x) = 0\}\,.$$

We denote by G_λ, the Green's function of the open set Ω_λ extended to zero outside of $\bar{\Omega}_\lambda \times \bar{\Omega}_\lambda$. From Proposition 8, for $x \in \Omega_\lambda$, the function $v_{\lambda,x}(y) = G_\lambda(x, y)$ is characterized by

$$v_{\lambda,x} \in W^{1,1}(\mathbb{R}^n)\,, \quad v_{\lambda,x} = 0 \quad \text{in} \quad \mathscr{D}'(\mathbb{R}^n \backslash \bar{\Omega}_\lambda)\,,$$

$$\Delta v_{\lambda,x} = \delta_x \quad \text{in} \quad \mathscr{D}'(\Omega_\lambda)\,.$$

In addition, for all $p \in [1, n/(n-1)[$, $v_{\lambda,x} \in W^{1,p}(\mathbb{R}^n)$ and, account being taken of the uniformity with respect to λ of the regularity of Ω_λ,

$$\|v_{\lambda,x}\|_{W^{1,p}} \leqslant C\,, \quad \text{independent of } \lambda \text{ and } x\,.$$

We deduce that the *map* $(\lambda, x) \to v_{\lambda,x} = G_\lambda(x, .)$ *is continuous in* $\{(\lambda, x);\ \lambda \in \Lambda, x \in \Omega_\lambda\}$ *in* $W^{1,p}(\mathbb{R}^n)$ *with the weak topology, subject to the continuity of the function* Φ *with respect to* λ which ensures that for all $\lambda_0 \in \Lambda$ and $\zeta \in \mathscr{D}(\Omega_{\lambda_0})$ (resp. $\mathscr{D}(\mathbb{R}^n \backslash \bar{\Omega}_{\lambda_0})$, we have $\zeta \in \mathscr{D}(\Omega_\lambda)$ (resp. $\mathscr{D}(\mathbb{R}^n \backslash \Omega_\lambda)$) for λ sufficiently close to λ_0. This allows us to prove

Proposition 11. *With the above data in supposing Φ continuous with respect to λ, let us consider $\varphi: \Lambda \times U \to \mathbb{R}$ continuous and denote for all $\lambda \in \Lambda$, $\varphi_\lambda: z \in \Gamma_\lambda \to \varphi(\lambda, z)$ and u_λ the (classical) solution of $P(\Omega_\lambda, \varphi_\lambda)$. Then the function $(\lambda, x) \to u_\lambda(x) = u(\lambda, x)$ is continuous on the set $\{(\lambda, x);\ \lambda \in \Lambda, x \in \bar{\Omega}_\lambda\}$.*

Proof. Given $\lambda_k \to \lambda_0$ and $x_k \to x_0$ with $x_k \in \Omega_{\lambda_k}$, we wish to prove that $u_{\lambda_k}(x_k) \to u_{\lambda_0}(x_0)$. From the principle of the maximum, denoting $\varphi_k: z \in \Gamma_{\lambda_k} \to \varphi(\lambda_0, z)$ and u_k the solution of $P(\Omega_{\lambda_k}, \varphi_k)$, we have

$$|u_{\lambda_k}(x_k) - u_k(x_k)| \leqslant \max_{z \in \Gamma_{\lambda_k}} |\varphi(\lambda_k, z) - \varphi(\lambda_0, z)|\,.$$

From the continuity of φ on $\Lambda \times U$ it is sufficient therefore to prove that $u_k(x_k) \to u_{\lambda_0}(x_0)$; in other words, we can suppose $\varphi(\lambda, x) = \varphi(x)$ independent of λ.

Using the same maximum principle and the density of $\mathscr{C}^\infty$ in $\mathscr{C}^0$, we can suppose that $\varphi \in \mathscr{C}^\infty$. Finally since only the values of φ on a compact set of U are involved, we can suppose that $\varphi(\lambda, x) = \varphi(x)$ on $\Lambda \times U$ with $\varphi \in \mathscr{C}^\infty(\mathbb{R}^n)$. We then have

$$u_\lambda(x) = \varphi(x) - \int G_\lambda(x, y)\, \Delta\varphi(y)\, dy\,, \qquad x \in \bar{\Omega}_\lambda\,;$$

the result is then deduced from the continuous dependence of $G_\lambda(x, .)$ proved above. □

Returning to the Green's function G_λ, we have

$$G_\lambda(x, y) = E_n(x - y) - u_\lambda(\varphi_x) \quad \text{where} \quad \varphi_x(y) = E_n(x - y) \quad \text{on} \quad \Gamma_\lambda\,;$$

we then deduce the

Corollary 4. *With the above data and supposing Φ continuous with respect to λ the function $(\lambda, x, y) \to G_\lambda(x, y)$ is continuous on $\{(\lambda, x, y);\ \lambda \in \Lambda,\ x \neq y\}$.*

In the same way we shall have results on continuous dependence of the solution of $P(\Omega_\lambda, \varphi_\lambda, f_\lambda)$; we leave it to the reader to supply the details. We shall now study differentiability with respect to λ. We state *Hadamard's formula*:

Proposition 12. *With the above data, let us suppose Φ is of class $\mathscr{C}^1$ with respect to λ. Then the function $(\lambda, x, y) \to G_\lambda(x, y)$ is of class $\mathscr{C}^1$ on*

$$\{(\lambda, x, y);\quad \lambda \in \Lambda, x, y \in \Omega_\lambda\,,\quad x \neq y\}$$

and

$$\frac{\partial}{\partial\lambda} G_\lambda(x, y) = \int_{\Gamma_\lambda} \frac{\partial G_\lambda}{\partial n}(x, z)\, \frac{\partial G_\lambda}{\partial n}(y, z)\, \rho(\lambda, z)\, d\gamma_\lambda(z) \tag{6.59}$$

where

$$\rho(\lambda, x) = \frac{\dfrac{\partial\Phi}{\partial\lambda}(\lambda, x)}{|\mathrm{grad}_x\, \Phi(\lambda, x)|}\,.$$

We note that from Corollary 4 and Proposition 1, account being taken of the uniformity with respect to λ of the regularity of the open sets Ω_λ, the function $G_\lambda(x, y)$ is of class $\mathscr{C}^1$ with respect to (x, y) on

$$\{(\lambda, x, y);\quad \lambda \in \Lambda, x, y \in \bar{\Omega}_\lambda, x \neq y\}\,.$$

This shows in particular that the term on the right in Hadamard's formula (6.59) is continuous in (λ, x, y). It is sufficient therefore for $\lambda \in \Lambda$, $(x, y) \in \Omega_\lambda$ with $x \neq y$ to show that

$$\lim_{h \to 0} \frac{1}{h}\left(G_{\lambda + h}(x, y) - G_\lambda(x, y)\right)$$

exists and is given by (6.59); similarly we content ourselves with proving it for

$\lim\limits_{\substack{h \to 0 \\ h > 0}}$ or $\lim\limits_{\substack{h \to 0 \\ h < 0}}$ since a function differentiable on the right and with continuous right-derivative is continuously differentiable. We shall be content here to prove Proposition 12 under the hypothesis that the map $\lambda \to \Omega_\lambda$ is monotonic (increasing or decreasing); we refer the reader to Garabedian–Schiffer [1] for a proof in the general case.

Proof under the hypothesis of monotonicity of $\lambda \to \Omega_\lambda$. We can suppose that $\lambda \to \Omega_\lambda$ is monotonic increasing. For $\lambda \in \Lambda$, $x, y \in \Omega_\lambda$, $x \neq y$ let us prove that

$$\lim_{\substack{h \to 0 \\ h > 0}} \frac{1}{h}(G_{\lambda + h}(x, y) - G_\lambda(x, y))$$

exists and is given by (6.59). We fix $\lambda \in \Lambda$, and $y \in \Omega_\lambda$; for simplicity of writing we put:

$$\Omega = \Omega_\lambda\,, \quad \Gamma = \Gamma_\lambda\,, \quad G = G_\lambda\,, \quad \rho(z) = \rho(\lambda, z)\,, \quad u(x) = G(x, y)\,.$$

For all $h > 0$, small, we put $u_h(x) = G_{\lambda + h}(x, y)$. Since $\Omega_{\lambda + h} \supset \Omega_\lambda = \Omega$, the function $v_h = h^{-1}(u_h - u)$ is harmonic on Ω and continuous on $\bar{\Omega}$ with $v_h(z) = h^{-1}u_h(z)$ for all $z \in \Gamma$.

We fix $z \in \Gamma$. The exterior normal $n(z)$ is given by $n(z) = \dfrac{\operatorname{grad}_x \Phi(\lambda, z)}{|\operatorname{grad}_x \Phi(\lambda, z)|}$. From the implicit function theorem, for h small there exists a unique $t(h)$ in a neighbourhood of O, such that $z + t(h)n(z) \in \Gamma_{\lambda + h}$, that is to say

$$\Phi(\lambda + h, z + t(h)n(z)) = 0\,; \tag{6.60}$$

The function $t(h)$ is of class $\mathscr{C}^1$ and by the development of (6.60)

$$\frac{dt}{dh}(0) = -\frac{\dfrac{\partial \Phi}{\partial \lambda}(\lambda, z)}{\operatorname{grad}_x \Phi(\lambda, z).n(z)} = -\rho(z)\,.$$

Hence by the mean value theorem

$$0 = u_h(z + t(h)n(z)) = u_h(z) + t(h)\operatorname{grad} u_h(z + \theta(h)n(z)).n(z)$$

with

$$0 < \frac{\theta(h)}{t(h)} < 1\,.$$

From which on regrouping, we have

$$u_h(z) = h\rho(z)\frac{\partial u}{\partial n}(z) + R(h, z)$$

with

$$R(h, z) = t(h)(\operatorname{grad} u(z) - \operatorname{grad} u_h(z + \theta(h)n(z))).n(z) - \frac{\partial u}{\partial n}(z)(t(h) - ht'(0))\,.$$

From what we have seen before the beginning of this proof, the function $(h, x) \to \operatorname{grad} u_h(x)$ is continuous with the result that $R(h, z) = o(h)$ uniformly with respect to $z \in \Gamma$. Using the principle of the maximum we see that $v_h \to v$ in $\mathscr{C}^0(\bar{\Omega})$ where v is the solution of $P\left(\Omega, \rho \dfrac{\partial u}{\partial n}\right)$. □

The Hadamard formula (6.59) allows us to obtain the derivative with respect to λ of the solution u_λ of

$$P(\Omega_\lambda, \varphi_\lambda, f_\lambda) \quad \text{where} \quad \varphi_\lambda(z) = \varphi(\lambda, z) \quad \text{and} \quad f_\lambda(x) = f(\lambda, x)$$

with φ and f functions sufficiently regular on $\Lambda \times \mathbb{R}^n$. Considering, for $\lambda \in \Lambda$ fixed, v_h the solution of $P(\Omega_{\lambda+h}, \varphi_h, f_h)$ with

$$\varphi_h(z) = \varphi(\lambda, z)\,, \quad f_h(x) = f(\lambda, x)\,,$$

$\dfrac{u_{\lambda+h} - v_h}{h}$ is solution of $P\left(\Omega_{\lambda+h}, \dfrac{\varphi(\lambda+h, .) - \varphi(\lambda, .)}{h}, \dfrac{f(\lambda+h, .) - f(\lambda, .)}{h}\right)$;

from Proposition 11, $\dfrac{u_{\lambda+h} - v_h}{h} \to w$ solution of

$$P\left(\Omega_\lambda, \frac{\partial \varphi}{\partial \lambda}(\lambda, .), \frac{\partial f}{\partial \lambda}(\lambda, .)\right).$$

We can therefore reduce the problem to the case in which φ and f are independent of λ. Let us state

Corollary 5. *Under the hypotheses of Proposition* 12 *let*

$$\varphi \in W^{2,p}_{\text{loc}}(\mathbb{R}^n) \quad \textit{and} \quad f \in L^p_{\text{loc}}(\mathbb{R}^n) \quad \textit{for} \quad p > n\,.$$

For all $\lambda \in \Lambda$, denote by u_λ the solution of $P(\Omega_k, \varphi\chi_{\Gamma_\lambda}, f\chi_{\Omega_\lambda})$.
The function $(\lambda, x) \to u_\lambda(x)$ is of class $\mathscr{C}^1$ on $\{(\lambda, x); \lambda \in \Lambda, x \in \Omega_\lambda\}$ and

$$\frac{\partial}{\partial \lambda} u_\lambda(x) = \int_{\Gamma_\lambda} \frac{\partial G_\lambda}{\partial n}(x, z) \frac{\partial v_\lambda}{\partial n}(z) \rho(\lambda, z)\, \mathrm{d}\gamma_\lambda(z)$$

where v_λ is the solution of $PH(\Omega_\lambda, (f - \Delta\varphi)\chi_{\Omega_\lambda})$.

Proof. We have $u_\lambda = \varphi + v_\lambda$; replacing f by $f - \Delta\varphi$ we can therefore suppose that $\varphi \equiv 0$. We then have

$$u_\lambda(x) = \int_{\Omega_\lambda} G_\lambda(x, y) f(y)\, \mathrm{d}y, \quad x \in \Omega_\lambda\,;$$

from which for $\lambda \in \Lambda$ and $x \in \Omega_\lambda$ fixed, we deduce

$$\frac{u_{\lambda+h}(x) - u_\lambda(x)}{h} = \frac{1}{h} \int_{\Omega_{\lambda+h} \setminus \Omega_\lambda} G_{\lambda+h}(x, y) f(y)\, \mathrm{d}y + \int_{\Omega_\lambda} \frac{G_{\lambda+h}(x, y) - G_\lambda(x, y)}{h} f(y)\, \mathrm{d}y\,.$$

Using Corollary 4 and $G_{\lambda+h}(x, y) = 0$ for $y \in \Gamma_{\lambda+h}$ the first integral tends to zero.

From Proposition 12, the second integral tends to

$$\int_{\Omega_\lambda} f(y)\,dy \int_{\Gamma_\lambda} \frac{\partial G_\lambda}{\partial n}(x, z) \frac{\partial G_\lambda}{\partial n}(y, z)\rho(\lambda, z)\,d\gamma_\lambda(z)$$

$$= \int_{\Gamma_\lambda} \frac{\partial G_\lambda}{\partial n}(x, z)\rho(\lambda, z)\,d\gamma_\lambda(z) \int_{\Omega_\lambda} \frac{\partial G_\lambda}{\partial n}(y, z) f(y)\,dy$$

$$= \int_{\Gamma_\lambda} \frac{\partial G_\lambda}{\partial n}(x, z) \frac{\partial u_\lambda}{\partial n}(z)\rho(\lambda, z)\,d\gamma_\lambda(z)\,. \qquad \square$$

§7. Other Methods of Solution of the Dirichlet Problem*

1. Case of a Convex Open Set: Neumann's Integral Method

We take up again the solution of the problems of Dirichlet and Neumann by the integral method when the open set Ω is convex: we shall obtain the solution by an iterative method due to Neumann.[173] First we recall the notation of the integral method (see §4.15).

Let Ω be an open set in $\mathbb{R}^n$ which, in the first instance, we suppose to be regular bounded with boundary of class $\mathscr{C}^{1+\varepsilon}$; for $\alpha \in \mathscr{C}^0(\Gamma)$ we define (see (4.99))

$$K\alpha(z) = \frac{2}{\sigma_n} \int_\Gamma \frac{(t - z).n(t)}{|t - z|^n} \alpha(t)\,d\gamma(t), \quad z \in \Gamma\,. \tag{7.1}$$

The function $K\alpha \in \mathscr{C}^0(\Gamma)$, and the interior (resp. exterior) double layer potential

$$u(x) = \frac{1}{\sigma_n} \int_\Gamma \frac{(t - x).n(t)}{|t - x|^n} \alpha(t)\,d\gamma(t), \quad x \in \Omega\,(\text{resp. } \mathbb{R}^n\backslash\bar{\Omega})\,, \tag{7.2}$$

is the solution of the Dirichlet problem $P(\Omega, \frac{1}{2}(K\alpha + \alpha))$ (respectively $P(\mathbb{R}^n\backslash\bar{\Omega}, \frac{1}{2}(K\alpha - \alpha))$ with the null condition at infinity)[173].

In other words, given $\varphi \in \mathscr{C}^0(\Gamma)$, to solve the Dirichlet problem $P(\Omega, \varphi)$ (resp. $P(\mathbb{R}^n\backslash\bar{\Omega}, \varphi)$ with the null condition at infinity) by a double layer potential we are led to solve the integral equation

$$\frac{K\alpha + \alpha}{2} = \varphi \quad \left(\text{resp. } \frac{K\alpha - \alpha}{2} = \varphi\right). \tag{7.3}$$

Now let us *suppose that the open set Ω is convex*, which can be expressed analytically as

$$(t - x).n(t) \geqslant 0 \quad \text{for all} \quad t \in \Gamma \quad \text{and} \quad x \in \Omega. \tag{7.4}$$

[173] See §4.5 and Proposition 11 of §3.

The operator K is then *increasing*:

$$\alpha_1 \geqslant \alpha_2 \quad \text{on} \quad \Gamma \Rightarrow K\alpha_1 \geqslant K\alpha_2 \quad \text{on} \quad \Gamma \,. \tag{7.5}$$

Using the equality (see Proposition 15 of § 4)

$$K\chi_\Gamma \equiv 1 \quad \text{on} \quad \Gamma \,, \tag{7.6}$$

we deduce that

$$\min_\Gamma \alpha \leqslant K\alpha \leqslant \max_\Gamma \alpha \quad \text{on} \quad \Gamma \tag{7.7}$$

and hence, in particular, that

$$\max_\Gamma |K\alpha| \leqslant \max_\Gamma |\alpha| \,. \tag{7.8}$$

We note, furthermore, that for an open set Ω, the properties (7.5), (7.7) and (7.8) are all equivalent to the convexity of Ω.
Since Ω is convex, the open sets Ω and $\mathbb{R}^n \backslash \bar{\Omega}$ are both connected; we know by Fredholm's method (see Proposition 19 of § 4) that for all $\varphi \in \mathscr{C}^0(\Gamma)$ (resp. satisfying $\int_\Gamma \alpha\varphi \, \mathrm{d}\gamma = 0$[174]) the integral equation (7.3) has a unique solution (resp. to within an additive constant). We shall develop an iterative method to determine these solutions: besides, this method will prove the existence of these without any appeal being made to Fredholm theory.
Given $\alpha \in \mathscr{C}^0(\Gamma)$ we put

$$c_\alpha = \frac{\max \alpha + \min \alpha}{2} \,, \quad \tilde{\alpha} = \alpha - c_\alpha \,. \tag{7.9}$$

We have

$$\widetilde{\alpha + c} = \tilde{\alpha} \quad \text{for all} \quad c \in \mathbb{R}$$

and

$$\max|\tilde{\alpha}| = \max \tilde{\alpha} = -\min \tilde{\alpha} = \frac{\max \alpha - \min \alpha}{2} = \min_c \max_\Gamma |\alpha + c| \,. \tag{7.10}$$

We prove:

Proposition 1. *Let Ω be a* **convex** *regular bounded open set of $\mathbb{R}^n$ with boundary Γ of class $\mathscr{C}^{1+\varepsilon}$ and $\varphi \in \mathscr{C}^0(\Gamma)$. We define recursively the sequence of functions $\alpha_k \in \mathscr{C}^0(\Gamma)$ by*

$$\alpha_0 = \tilde{\varphi} \,, \quad \alpha_k = \widetilde{K\alpha_{k-1}} \tag{7.11}$$

where the map $\alpha \to \tilde{\alpha}$ is defined by (7.9) and the operator K by (7.1).

---

[174] $\alpha = L^{-1}\chi_\Gamma$ is the normal derivative of the solution of the exterior Dirichlet problem $P(\mathbb{R}^n \backslash \bar{\Omega}, \chi_\Gamma)$ satisfying the null condition at infinity.

(1) $\sum_k \max_\Gamma |\alpha_k| < \infty$ *and hence, in particular, we can define*

$$\alpha^i = 2\sum_k (-1)^k \alpha_k \qquad \alpha^e = -2\sum_k \alpha_k \tag{7.12}$$

the series being uniformly convergent on Γ.
(2) *There exists a unique* $c^i \in \mathbb{R}$ *such that the interior double layer potential defined by* (7.2) *with* $\alpha = \alpha^i + c^i$ *is the solution of the Dirichlet problem* $P(\Omega, \varphi)$.
(3) *There exists a unique* $c^e \in \mathbb{R}$ *such that the exterior double layer potential defined by* (7.2) *with* $\alpha = \alpha^e$ *is the solution of the Dirichlet problem* $P(\mathbb{R}^n \backslash \bar{\Omega}, \varphi + c^e)$ *tending to zero*

The proof rests on

Lemma 1. *Let* Ω *be a regular bounded convex open set in* $\mathbb{R}^n$ *with boundary* Γ *of class* $\mathscr{C}^{1+\varepsilon}$. *There exists* $\delta \in]0, 1[$ *such that*

$$\max_\Gamma |K\alpha| \leqslant \delta \max_\Gamma |\tilde{\alpha}| \quad \textit{for all} \quad \alpha \in \mathscr{C}^0(\Gamma)\,.$$

Proof of Proposition 1. Using Lemma 1, we have

$$\max|\alpha_k| = \max|K\alpha_{k-1}| \leqslant \delta \max|\tilde{\alpha}_{k-1}| = \delta \max|\alpha_{k-1}|$$

and hence

$$\sum_k \max|\alpha_k| \leqslant \max|\alpha_0| \sum \delta^k = \frac{\max|\alpha_0|}{1-\delta}$$

which proves the point (1).
Since the operator K maps $\mathscr{C}^0(\Gamma)$ continuously into itself, we deduce from (7.8) that

$$\frac{K\alpha^i + \alpha^i}{2} = \sum (-1)^k (K\alpha_k + \alpha_k)\,.$$

But

$$K\alpha_k = \alpha_{k+1} - c_k \quad \text{with} \quad c_k = \frac{\max K\alpha_k + \min K\alpha_k}{2}\,;$$

also

$$\alpha_0 = \varphi + c_\varphi \quad \text{with} \quad c_\varphi = \frac{\max \varphi + \min \varphi}{2}\,.$$

Hence

$$\sum_{l=0}^{k} (-1)^l (K\alpha_l + \alpha_l) = \varphi + c_\varphi - \sum_{l=0}^{k} (-1)^l c_l + (-1)^{k+1} \alpha_{k+1}\,.$$

We deduce that $\sum_k (-1)^k c_k$ is convergent and

$$\frac{K\alpha^i + \alpha^i}{2} = \varphi - c^i \quad \text{with} \quad c^i = \sum (-1)^k c_k - c_\varphi\,.$$

Similarly the series Σc_k is convergent and

$$\frac{K\alpha^e - \alpha^e}{2} = \varphi + c^e \quad \text{with} \quad c^e = \sum c_k + c_\varphi . \qquad \square$$

Proof of Lemma 1. Given $\alpha \in \mathscr{C}^0(\Gamma)$, we have from (7.6) that $K\alpha = K\tilde{\alpha}$ and from (7.10) that

$$\max|\widetilde{K\alpha}| = \frac{1}{2}(\max K\tilde{\alpha} - \min K\tilde{\alpha}) .$$

We put

$$\Gamma^+(\tilde{\alpha}) = \{t \in \Gamma;\ \tilde{\alpha}(t) > 0\}, \quad \Gamma^-(\tilde{\alpha}) = \{t \in \Gamma;\ \tilde{\alpha}(t) < 0\} ;$$

and obtain

$$K\tilde{\alpha}(z) \leqslant \frac{2}{\sigma_n}\int_{\Gamma^+(\tilde{\alpha})} \frac{(t - z).n(t)}{|t - z|^n}\tilde{\alpha}(t)\,\mathrm{d}\gamma(t) \leqslant \delta(\Gamma^+(\tilde{\alpha}))\max\tilde{\alpha}$$

where for every measurable part Γ_0 of Γ

$$\delta(\Gamma_0) = \max_{\Gamma}\frac{2}{\sigma_n}\int_{\Gamma_0}\frac{(t - z).n(t)}{|t - z|^n}\,\mathrm{d}\gamma(t) .$$

Similarly, we have

$$K\tilde{\alpha}(z) \geqslant \delta(\Gamma^-(\tilde{\alpha}))\min\tilde{\alpha} ,$$

with the result that, using (7.10), we obtain

$$\max|K\tilde{\alpha}| \leqslant \frac{\delta(\Gamma^+(\tilde{\alpha})) + \delta(\Gamma^-(\tilde{\alpha}))}{2}\max|\tilde{\alpha}| . \tag{7.13}$$

We note now that all of the preceding can be extended to a function $\alpha \in L^\infty(\Gamma)$: we can define

$$K\alpha(z) = \frac{2}{\sigma_n}\int_{\Gamma}\frac{(t - z).n(t)}{|t - z|^n}\alpha(t)\,\mathrm{d}\gamma(t) \quad \text{for all} \quad z \in \Gamma .$$

From the compactness of the operator K of $\mathscr{C}^0(\Gamma)$ into itself (see Proposition 16 of §4), $K\alpha \in \mathscr{C}^0(\Gamma)$ and the map $\alpha \to K\alpha$ is continuous on the bounded sets of $L^\infty(\Gamma)$ provided with the weak-star topology, into $\mathscr{C}^0(\Gamma)$ with the uniform convergence topology. We can define $\tilde{\alpha}$ replacing $\max_{\Gamma}$ and $\min_{\Gamma}$ by $\sup_{\Gamma}\text{ess}$ and $\inf_{\Gamma}\text{ess}$ respectively. From the compactness of the unit ball of $L^\infty(\Gamma)$ for the weak-star topology, there exists $\alpha_0 \in L^\infty(\Gamma)$ with $\|\tilde{\alpha}_0\|_{L^\infty} \leqslant 1$ such that

$$\max|\widetilde{K\alpha_0}| = \sup_{\substack{\alpha\in\mathscr{C}^0(\Gamma)\\ \alpha\neq 0}}\frac{\max|\widetilde{K\alpha}|}{\max|\tilde{\alpha}|} .$$

From (7.13) it is sufficient, therefore, to prove the lemma by showing that given $\alpha \in L^\infty(\Gamma)$, $\alpha \not\equiv 0$, we must have

$$\frac{\delta(\Gamma^+(\alpha)) + \delta(\Gamma^-(\alpha))}{2} < 1 ,$$

where $\delta(\Gamma^{\pm}(\alpha))$ is well defined, since $\Gamma^{\pm}(\alpha)$ is defined to within a negligible set; moreover $\delta(\Gamma^{\pm}(\alpha)) = \max_{\Gamma} K\chi_{\Gamma^{\pm}(\alpha)}$.

Given a measurable part Γ_0 of Γ, $K\chi_{\Gamma_0} = 1 - K\chi_{\Gamma\backslash\Gamma_0}$, with the result that $\delta(\Gamma_0) \leqslant 1$ and there is equality iff there exists $z_0 \in \Gamma$ such that $K\chi_{\Gamma\backslash\Gamma_0}(z_0) = 0$, that is to say account being taken of the convexity of Ω,

$$\Gamma\backslash\Gamma_0 \subset H_0 = \{t \in \mathbb{R}^n; (t - z_0).n(z_0) = 0\} \quad \text{(to within a negligible set).}$$

Given two measurable parts Γ_+ and Γ_- of Γ, we therefore have

$$\frac{\delta(\Gamma_+) + \delta(\Gamma_-)}{2} \leqslant 1$$

and there is equality iff there exist hyperplanes H_+ and H_- such that

$$\Gamma\backslash\Gamma_{\pm} \subset H_{\pm} \quad \text{(to within a negligible set)} \tag{7.14}$$

In the case $\Gamma_{\pm} = \Gamma^{\pm}(\alpha)$ with $\alpha \not\equiv 0$, we have $\Gamma_+ \cap \Gamma_- = \varnothing$ and Γ_+, Γ_- not both negligible, with the result that (7.14) is impossible. ☐

Remark 1. In the proof of Lemma 1, we have used the extension of the integral operator K to $L^\infty(\Gamma)$. Given $\varphi \in L^\infty(\Gamma)$, we can define as in Proposition 1 the sequence (α_k) by (7.11), the sums α^i, α^e by (7.12), as well as the constants c^i and c^e. By a continuity argument, we show that the interior (resp. exterior) double layer potential of $\alpha = \alpha^i + c^i$ (resp. $\alpha = \alpha^e$) is the solution of $P(\Omega, \varphi)$ (resp. $P(\mathbb{R}^n\backslash\bar{\Omega}, \varphi + c^e)$ tending to zero at infinity) in the sense of Definition 2 of §6. ☐

In §4.5, we have likewise considered the operator J defined by

$$J\alpha(z) = \frac{2}{\sigma_n}\int_\Gamma \frac{(z - t).n(z)}{|z - t|^n}\alpha(t)\,\mathrm{d}\gamma(t)\,, \quad z \in \Gamma\,.$$

For $\alpha \in \mathscr{C}^0(\Gamma)$, we have $J\alpha \in \mathscr{C}^0(\Gamma)$ and the interior (resp. exterior) single layer potential

$$u(x) = \int_\Gamma E_n(x - t)\alpha(t)\,\mathrm{d}\gamma(t)\,, \quad x \in \Omega\,(\text{resp. } \mathbb{R}^n\backslash\bar{\Omega})$$

is the solution of the Neumann problem $PN(\Omega, \frac{1}{2}(J\alpha - \alpha))$ (respectively $PN(\mathbb{R}^n\backslash\bar{\Omega}, -\frac{1}{2}(J\alpha + \alpha))$ with $u(x) = \left(\int_\Gamma \alpha\,\mathrm{d}\gamma\right)E_n(x) + O\left(\frac{1}{|x|^{n-1}}\right)$ when $|x| \to \infty$).

When Ω is convex, the operator J is increasing:

$$\alpha_1 \geqslant \alpha_2 \quad \text{on} \quad \Gamma \Rightarrow J\alpha_1 \geqslant J\alpha_2 \quad \text{on} \quad \Gamma\,.$$

Let us now prove

Lemma 2. *Let Ω be a regular bounded convex open set in $\mathbb{R}^n$ with boundary Γ of class $\mathscr{C}^{1+\varepsilon}$ and μ a Radon measure on Γ.*

(1) *For $\mathrm{d}\gamma$-almost all $z \in \Gamma$ the function $t \to \dfrac{(z - t).n(z)}{|z - t|^n}$ is $|\mu|$-integrable on Γ; and*

the function

$$J\mu(z) = \frac{2}{\sigma_n}\int_\Gamma \frac{(z-t).n(z)}{|z-t|^n}\,\mathrm{d}\mu(t) \quad \mathrm{d}\gamma \quad \text{a.e.} \quad z \in \Gamma$$

is $\mathrm{d}\gamma$*-integrable.*
(2) *The interior (resp. exterior) simple layer potential*

$$u(x) = \int_\Gamma E_n(x-t)\,\mathrm{d}\mu(t) \quad x \in \Omega \qquad (\textit{resp. } \mathbb{R}^n\backslash\bar{\Omega})$$

is solution of the Neumann problem $PN(\Omega, \frac{1}{2}(J\mu - \mu))$ (resp. $PN(\mathbb{R}^n\backslash\bar{\Omega}, \frac{1}{2}(-J\mu + \mu))$ *with* $u(t) = \left(\int_\Gamma \mathrm{d}\mu\right)E_n(x) + O\left(\frac{1}{|x|^{n-1}}\right)$ *when* $|x| \to \infty$) *in the sense of Definition* 3 *of* § 6.
(3) *When* $\int_\Gamma \mathrm{d}\mu = 0$, *we have* $\int_\Gamma J\mu(z)\,\mathrm{d}\gamma(z) = 0$ *and*

$$\|J\mu\|_{L^1(\Gamma)} \leqslant \delta \int_\Gamma \mathrm{d}|\mu|$$

where δ *is the constant of Lemma* 1.

Proof. Since Ω is convex, the function $t \to \frac{(z-t).n(z)}{|z-t|^n}$ is positive. From Fubinis theorem

$$\int_\Gamma \mathrm{d}\gamma(z)\int_\Gamma \frac{2(z-t).n(z)}{\sigma_n|z-t|^n}\,\mathrm{d}|\mu|(t) = \int_\Gamma \mathrm{d}|\mu|(t)\int_\Gamma \frac{2(z-t).n(z)}{\sigma_n|z-t|^n}\,\mathrm{d}\gamma(z)$$

$$= \int_\Gamma K\chi_\Gamma(t)\,\mathrm{d}|\mu|(t) = \int_\Gamma \mathrm{d}|\mu|(t)$$

giving point (1). More generally, we have for all $\alpha \in \mathscr{C}^0(\Gamma)$,

$$\int_\Gamma \alpha(z)\,J\mu(z)\,\mathrm{d}\gamma(z) = \int_\Gamma K\alpha(t)\,\mathrm{d}\mu(t)\,. \tag{7.15}$$

In particular $\int_\Gamma J\mu(z)\,\mathrm{d}\gamma(z) = \int_\Gamma K\chi_\Gamma(t)\,\mathrm{d}\mu(t) = \int_\Gamma \mathrm{d}\mu(t)\,.$

When $\int_\Gamma \mathrm{d}\mu = 0$, we have $\int_\Gamma J\mu\,\mathrm{d}\gamma = 0$ and for all $\alpha \in \mathscr{C}^0(\Gamma)$,

$$\int_\Gamma \alpha\,J\mu\,\mathrm{d}\gamma = \int_\Gamma K\alpha\,\mathrm{d}\mu = \int_\Gamma \widetilde{K\alpha}\,\mathrm{d}\mu \leqslant \delta\max_\Gamma|\tilde{\alpha}|\int_\Gamma \mathrm{d}|\mu| \leqslant \delta\max_\Gamma|\alpha|\int_\Gamma \mathrm{d}|\mu|$$

from which point (3) follows.
Finally for point (2), we consider the interior (resp. exterior) simple layer potential

$$u(x) = \int_\Gamma E_n(x-t)\,\mathrm{d}\mu(t) \quad x \in \Omega \quad (\textit{resp. } \mathbb{R}^n\backslash\bar{\Omega})$$

When μ is the measure of density $\alpha \in \mathscr{C}^0(\Gamma)$, then u is the classical solution of $PN(\Omega, \frac{1}{2}(J\alpha - \alpha))$ (resp. $PN(\mathbb{R}^n \backslash \bar{\Omega}, -\frac{1}{2}(J\alpha + \alpha))$); then, for all $\zeta \in \mathscr{D}(\mathbb{R}^n)$

$$\int_\Gamma \frac{J\alpha - \alpha}{2} \zeta \, d\gamma = \int_\Omega \operatorname{grad} \zeta . \operatorname{grad} u \, dx \tag{7.16}$$

$$\left(\text{resp.} \int_\Gamma \frac{J\alpha + \alpha}{2} \zeta \, d\gamma + \int_{\mathbb{R}^n \backslash \bar{\Omega}} \operatorname{grad} \zeta . \operatorname{grad} u \, dx = 0 \right).$$

From Proposition 10 of §6, for $1 \leqslant p < \dfrac{n}{n-1}$, we have

$$\| u \|_{W^{1,p}(\Omega)} \leqslant C_p \left\| \frac{J\alpha - \alpha}{2} \right\|_{L^1(\Gamma)} \leqslant C_p \| \alpha \|_{L^1(\Gamma)} \tag{7.17}$$

(resp. given R such that $\bar{\Omega} \subset B(0, R)$

$$\| u \|_{W^{1,p}(B(0,R) \backslash \bar{\Omega})} \leqslant C_{p,R} \| \alpha \|_{L^1(\Gamma)}) .$$

Approximating an arbitrary measure μ by measures of densities $\alpha_k \in \mathscr{C}^0(\Gamma)$ such that

$$\int_\Gamma \zeta \alpha_k \, d\gamma \to \int_\Gamma \zeta \, d\mu \quad \text{for all} \quad \zeta \in \mathscr{C}^0(\Gamma)$$

(and hence $\| \alpha_k \|_{L^1(\Gamma)}$ bounded), the corresponding simple layer potentials converge in $\mathscr{H}(\mathbb{R}^n \backslash \Gamma)$ to u; from (7.16), $\| u_k \|_{W^{1,p}(\Omega)}$ (resp. $\| u_k \|_{W^{1,p}(B(0,R) \backslash \bar{\Omega})}$) is bounded. Finally, taking account of (7.15) and (7.16), for $\zeta \in \mathscr{D}(\mathbb{R}^n)$, we have

$$\begin{aligned}
\int_\Gamma \zeta \frac{J\mu \, d\gamma - d\mu}{2} &= \int_\Gamma \frac{K\zeta - \zeta}{2} d\mu \\
&= \lim \int_\Gamma \frac{K\zeta - \zeta}{2} \alpha_k \, d\gamma = \lim \int_\Gamma \zeta \frac{J\alpha_k - \alpha_k}{2} d\gamma \\
&= \lim \int_\Omega \operatorname{grad} \zeta . \operatorname{grad} u_k \, dx = \int_\Omega \operatorname{grad} \zeta . \operatorname{grad} u \, dx
\end{aligned}$$

$$\left(\text{resp. similarly} \int_\Gamma \zeta \frac{J\mu \, d\gamma + d\mu}{2} + \int_{\mathbb{R}^n \backslash \bar{\Omega}} \operatorname{grad} \zeta . \operatorname{grad} u \, dx = 0 \right). \qquad \square$$

Remark 2. Points (1) and (2) of Lemma 2 remain valid in the non-convex case: indeed, the operator K being a compact map of $\mathscr{C}^0(\Gamma)$ into itself, its transpose ${}^tK$ is a compact map of $\mathfrak{M}(\Gamma)$ into itself (see Dieudonné [1]). $\square$

Now let us prove

Proposition 2. *Let Ω be a regular bounded convex open set of $\mathbb{R}^n$ with boundary Γ of class $\mathscr{C}^{1+\varepsilon}$ and v_1 a Radon measure on Γ with zero integral. We define recursively the sequence of functions $\alpha_k \in L^1(\Gamma)$*

$$\alpha_1 = J v_1 , \quad \alpha_{k+1} = J\alpha_k$$

where J is the extension considered in Lemma 2.1 of the classical operator

(1) $\sum_{k \geqslant 1} \|\alpha_k\|_{L^1(\Gamma)} < \infty$ *and hence in particular we can define*

$$\alpha^i = -2\sum_{k=1}^{\infty} \alpha_k\,, \qquad \alpha^e = 2\sum_{k=1}^{\infty} (-1)^{k+1}\alpha_k\,,$$

the series converging in mean on Γ.
(2) *The interior (resp. exterior) simple layer potential defined by the measure* $\mu^i = \alpha^i - 2v_1$ *(resp.* $\mu^e = \alpha^e - 2v_1$*) is the solution of the Neumann problem* $PN(\Omega, v_1)$ *(resp.* $PN(\mathbb{R}^n\backslash\bar{\Omega}, v_1)$ *tending to zero at infinity).*

Proof. From the point (3) of Lemma 2, we have for all $k \geqslant 1$,

$$\int_\Gamma \alpha_k\,\mathrm{d}\gamma = \int_\Gamma \mathrm{d}v_1 = 0 \quad \text{and} \quad \|\alpha_k\|_{L^1(\Gamma)} \leqslant \delta^k \int_\Gamma \mathrm{d}|v_1|\,.$$

Point (1) of the Proposition follows.
By the continuity of J in $L^1(\Gamma)$, we have

$$\begin{cases} J\alpha^i = -2\sum_{k \geqslant 1} \alpha_{k+1} = 2\alpha_1 + \alpha^i \\ J\alpha^e = 2\sum_{k \geqslant 1} (-1)^{k+1}\,\alpha_{k+1} = 2\alpha_1 - \alpha^e\,. \end{cases}$$

Therefore

$$\frac{J\mu^i - \mu^i}{2} = v_1 \quad \text{and} \quad -\frac{J\mu^e + \mu^e}{2} = v_1\,.$$

Account being taken of Lemma 2(2), this proves the Proposition. □

Remark 3. We can generalize the preceding results to the case of an arbitrary convex open set Ω of $\mathbb{R}^n$. We note first that the boundary Γ of Ω satisfies locally a Lipschitz condition (and Ω is situated on one side of its boundary); this allows us to define the surface measure γ of its boundary and the exterior normal to $\mathrm{d}\gamma$ a.e. $z \in \Gamma$; besides, at every point $z \in \Gamma$, we can define

$$N(z) = \{n \in \mathbb{R}^n\,;\, |n| = 1\,, (z - x).n \geqslant 0 \quad \text{for all} \quad x \in \Omega\}$$

the set of exterior unit "normals" to Ω at z.
The "multi-map" $z \to N(z)$ is "continuous" in the sense

$$\begin{cases} \forall z \in \Gamma\,, \quad \forall \varepsilon > 0\,, \quad \exists \delta > 0 \quad \text{such that} \\ t \in \Gamma \cap B(z, \delta)\,, n \in N(t) \Rightarrow \mathrm{dist}(n, N(z)) < \varepsilon \end{cases}$$

and $\mathrm{d}\gamma$. a.e. $z \in \Gamma$, $N(z)$ is reduced to a point $n(z)$.
Using this (very weak) continuity and the (very strong) property $(z - x).n(z) \geqslant 0$, we can take up again the theory of simple and double layer potentials (see § 3.3) and generalize Propositions 1 and 2. □

We note, however, that the operators K and J are no longer compact when the open set Ω admits a corner (irregular point); these operators are then singular

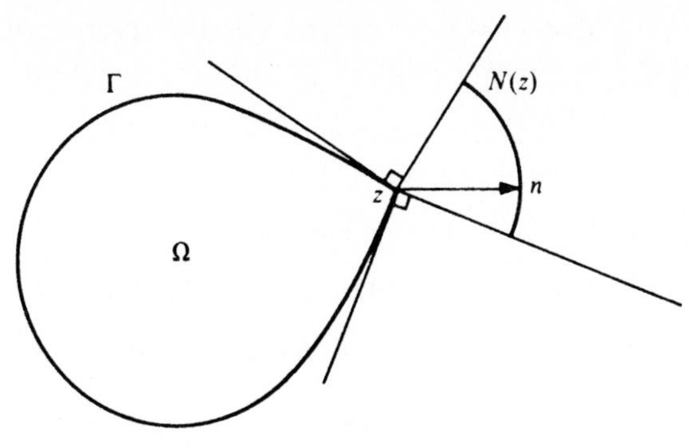

Fig. 21

integrals of the type:

$$K\alpha(z) = \lim_{\varepsilon \to 0} \int_{\Gamma \setminus B(z,\varepsilon)} \frac{(t - z) . n(t)}{|t - z|^n} \alpha(t) \, d\gamma(t)$$

(see Y. Meyer [1] for a study of these operators in the $L^2(\Gamma)$ spaces).

2. Alternating Procedure of Schwarz

We take two open sets Ω_1, Ω_2 of $\mathbb{R}^n$ with boundaries Γ_1, Γ_2 respectively and consider $\Omega = \Omega_1 \cup \Omega_2$ whose boundary is denoted by Γ. The *alternating procedure of Schwarz* gives an *approximation method* of solving the Dirichlet problem $P(\Omega, \varphi)$ from the solutions of the Dirichlet problems $P(\Omega_1, \varphi_1)$ and $P(\Omega_2, \varphi_2)$. We denote

$$\Gamma'_1 = \Gamma_1 \cap \Gamma \,, \quad \Gamma'_2 = \Gamma_2 \cap \Gamma$$

with the result that

$$\Gamma = \Gamma'_1 \cup \Gamma'_2 \,.$$

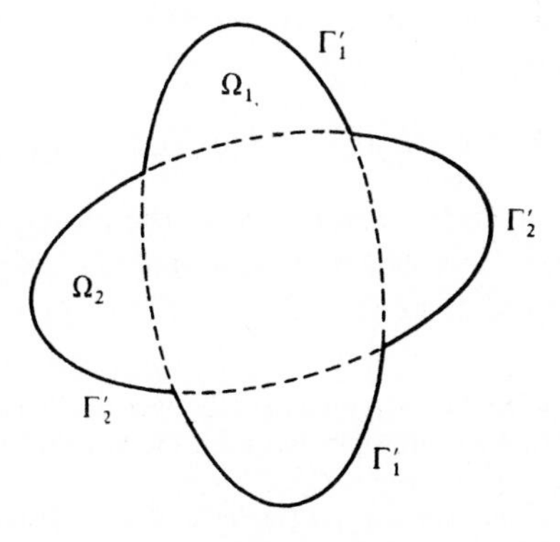

Fig. 22

Given $\varphi \in \mathscr{C}^0(\Gamma)$, it is clear that there exists a classical solution of $P(\Omega, \varphi)$ iff there exist $\varphi_i \in \mathscr{C}^0(\Gamma_i)$ and u_i classical solution of $P(\Omega_i, \varphi_i)$ such that

$$\begin{cases} \varphi_i = \varphi & \text{on} \quad \Gamma'_i \quad \text{for} \quad i = 1, 2 \\ u_1 = u_2 & \text{on} \quad \Omega_1 \cap \Omega_2 \,. \end{cases}$$

Such a solution u of $P(\Omega, \varphi)$ is then defined by

$$u = u_i \quad \text{on} \quad \Omega_i \quad \text{for} \quad i = 1, 2 \,.$$

We restrict ourselves to the *case of Ω bounded* and we *fix* $\varphi \in \mathscr{C}^0(\Gamma)$. For $i = 1, 2$ we denote

$$E_i(\varphi) = \{\varphi_i \in \mathscr{C}^0(\Gamma_i)\,; \quad \varphi_i = \varphi \quad \text{on} \quad \Gamma'_i\}$$

and *we suppose that for all $\varphi_i \in E_i(\varphi)$, the Dirichlet problem $P(\Omega_i, \varphi_i)$ admits a classical solution denoted by $u_i(\varphi_i)$*[175].

Given $\varphi_1 \in E_1(\varphi)$, we can associate with it the function $T_2\varphi_1$ defined on Γ_2 by

$$T_2\varphi_1(z) = \begin{cases} \varphi(z) & \text{on} \quad \Gamma'_2 \\ u_1(\varphi_1, z) & \text{on} \quad \Gamma_2 \cap \Omega_1 \,. \end{cases} \tag{7.18}$$

It is clear that $T_2\varphi_1 \in E_2(\varphi)$ since

$$\Gamma_2 = \Gamma'_2 \cup (\Gamma_2 \cap \Omega_1) \quad \text{and} \quad \Gamma'_2 \cap \overline{\Gamma_2 \cap \Omega_1} = \Gamma_1 \cap \Gamma_2 \,.$$

Similarly with $\varphi_2 \in E_2(\varphi)$ we associate the function $T_1\varphi_2 \in E_1(\varphi)$ defined by

$$T_1\varphi_2(z) = \begin{cases} \varphi(z) & \text{on} \quad \Gamma'_1 \\ u_2(\varphi_2, z) & \text{on} \quad \Gamma_1 \cap \Omega_2 \,. \end{cases} \tag{7.19}$$

Now for $\varphi_i \in E_i(\varphi)$,

$$u_1(\varphi_1) = u_2(\varphi_2) \quad \text{on} \quad \Omega_1 \cap \Omega_2$$

iff[176]

$$u_1(\varphi_1) = u_2(\varphi_2) \quad \text{on} \quad (\Gamma_1 \cap \Omega_2) \cup (\Gamma_2 \cap \Omega_1)$$

that is, with the above notation, iff

$$T_2\varphi_1 = \varphi_2 \quad \text{and} \quad T_1\varphi_2 = \varphi_1 \,.$$

In other words, *with the above notation, the Dirichlet problem $P(\Omega, \varphi)$ admits a classical solution iff the map T*

$$(\varphi_1, \varphi_2) \in E_1(\varphi) \times E_2(\varphi) \to T(\varphi_1, \varphi_2) = (T_1\varphi_2, T_2\varphi_1)$$

admits a fixed point[177]. From the uniqueness of the Dirichlet problem $P(\Omega, \varphi)$, if it exists, *this fixed point (φ_1, φ_2) will be unique and the solution of $P(\Omega, \varphi)$ will be defined by*: $u = u_i(\varphi_i)$ on Ω_i for $i = 1, 2$.

---

[175] In particular this is true if all the boundary points of the open sets Ω_i are regular. (see Theorem 1, §4).

[176] This follows from the maximum principle and from the fact that the boundary of $\Omega_1 \cap \Omega_2$ is contained in $(\Gamma_1 \cap \Omega_2) \cup (\Gamma_2 \cap \Omega_1) \cup (\Gamma_1 \cap \Gamma_2)$.

[177] Given a map T of a set E into itself, a fixed point of T is an element of E mapped onto itself (invariant under T).

We note that (φ_1, φ_2) is a fixed point of T iff φ_1 (resp. φ_2) is a fixed point of the map $S_1 = T_1 T_2$ (resp. $S_2 = T_2 T_1$) and $\varphi_2 = T_2 \varphi_1$ (resp. $\varphi_1 = T_1 \varphi_2$). There exist many ways of obtaining fixed points; the simplest method is certainly the *method of successive approximations*: starting from $\varphi_1^0 \in E_1(\varphi)$ (resp. $\varphi_2^0 \in E_2(\varphi)$), we construct by recurrence

$$\varphi_1^k = T_1 T_2 \varphi_1^{k-1} \quad (\text{resp. } \varphi_2^k = T_2 T_1 \varphi_2^{k-1})$$

that is to say

$$\varphi_1^0 \in E_1(\varphi) \text{ given}, \quad \varphi_2^0 = T_2 \varphi_1^0, \quad \varphi_1^1 = T_1 \varphi_2^0, \quad \varphi_2^1 = T_2 \varphi_1^1, \quad \text{etc} \ldots$$

$$(\text{resp. } \varphi_2^0 \in E_2(\varphi) \text{ given}, \quad \varphi_1^0 = T_1 \varphi_2^0, \varphi_2^1 = T_2 \varphi_1^0, \varphi_1^1 = T_1 \varphi_2^1, \text{etc} \ldots)$$

hence the name "*alternating procedure*": starting from a given value on the boundary of one of the open sets having the value φ on the part common with Γ, we construct, by solving a Dirichlet problem in this open set, a function on the boundary of the other open set with the value φ on the part in common with Γ and we recommence working alternately with each open set.

Given $\varphi_1, \bar{\varphi}_1 \in E_1(\varphi)$, the function $v_1 = u_1(\varphi_1) - u_1(\bar{\varphi}_1)$, harmonic on Ω_1, continuous on $\bar{\Omega}_1$ and satisfies

$$v_1 = 0 \quad \text{on} \quad \Gamma_1', \quad v_1 = \varphi_1 - \bar{\varphi}_1 \quad \text{on} \quad \Gamma_1 \cap \Omega_2 .$$

From the maximum principle, supposing $\varphi_1 \not\equiv \bar{\varphi}_1$, we have

$$|v_1| < \max_{\Gamma_1} |\varphi_1 - \bar{\varphi}_1| \quad \text{on} \quad \Omega_1$$

and therefore since $T_2 \varphi_1 = T_2 \bar{\varphi}_1 = \varphi$ on $\Gamma_2' \supset \Gamma_2 \backslash \Omega_1$

$$|T_2 \varphi_1 - T_2 \bar{\varphi}_1| < \max_{\Gamma_1} |\varphi_1 - \bar{\varphi}_1| \quad \text{on} \quad \Gamma_2 .$$

In short, *using the uniform norm in* $\mathscr{C}^0(\Gamma_i)$[178],

$$\| T_2 \varphi_1 - T_2 \bar{\varphi}_1 \| < \| \varphi_1 - \bar{\varphi}_1 \| \quad \text{for all} \quad \varphi_1, \bar{\varphi}_1 \in E_1(\varphi), \quad \varphi_1 \not\equiv \bar{\varphi}_1$$

and similarly

(7.20)
$$\| T_1 \varphi_2 - T_1 \bar{\varphi}_2 \| < \| \varphi_2 - \bar{\varphi}_2 \| \quad \text{for all} \quad \varphi_2, \bar{\varphi}_2 \in E_2(\varphi), \quad \varphi_2 \not\equiv \bar{\varphi}_2 .$$

Although strict, these inequalities are not, in general, sufficient to ensure the convergence of the successive approximations (φ_i^k).

For that reason we shall introduce the quantity

(7.21)
$$\delta(\Omega_1, \Omega_2) = \sup_{x \in \Omega_1 \cap \Gamma_2} v_1^0(x)$$

where v_1^0 is the solution of the Dirichlet problem $P(\Omega_1, \chi_{\Gamma_1 \cap \Omega_2})$.

[178] $\| \varphi_i \| = \max_{\Gamma_i} |\varphi_i|$.

The function $\chi_{\Gamma_1 \cap \Gamma_2}$ not being continuous on Γ_1, v_1^0 must be considered in a weak sense: when Ω_1 is sufficiently regular, we can put (see Definition 2 of §6)

$$(7.22) \qquad v_1^0(x) = \int_{\Gamma_1 \cap \Omega_2} \frac{\partial G_1}{\partial n}(x, z)\, d\gamma(z), \quad x \in \Omega$$

where G_1 is the Green's function of Ω_1; in the general case, it is sufficient to take

$$(7.23) \quad v_1^0(x) = \sup\{v_1(x); v_1 \in \mathscr{H}(\Omega_1) \cap \mathscr{C}^0(\bar{\Omega}_1), v_1 = 0 \quad \text{on} \quad \Gamma'_1, 0 \leqslant v_1 \leqslant 1\}\,.$$

This clearly generalizes that given by (7.22) in the case in which Ω_1 is regular, since then, for all $v_1 \in \mathscr{H}(\Omega_1) \cap \mathscr{C}^0(\bar{\Omega}_1)$

$$v_1(x) = \int_{\Gamma} \frac{\partial G_1}{\partial n}(x, z) v_1(z)\, d\gamma(z)\,.$$

By linearity, we have for all $v_1 \in \mathscr{H}(\Omega_1) \cap \mathscr{C}^0(\bar{\Omega}_1)$ with $v_1 \equiv 0$ on Γ'_1,

$$|v_1| \leqslant v_1^0 \max_{\Gamma_1} |v_1| \quad \text{on} \quad \Omega_1$$

and therefore

$$|v_1| \leqslant \delta(\Omega_1, \Omega_2) \max_{\Gamma_1} (v_1) \quad \text{on} \quad \Omega_1 \cap \Gamma_2$$

from which we deduce

$$\| T_2\varphi_1 - T_2\bar{\varphi}_1 \| \leqslant \delta(\Omega_1, \Omega_2) \| \varphi_1 - \bar{\varphi}_1 \| \quad \text{for all} \quad \varphi_1, \bar{\varphi}_1 \in E_1(\varphi)\,.$$

Similarly, we have

$$\| T_1\varphi_2 - T_1\bar{\varphi}_2 \| \leqslant \delta(\Omega_2, \Omega_1) \| \varphi_2 - \bar{\varphi}_2 \| \quad \text{for all} \quad \varphi_2, \bar{\varphi}_2 \in E_2(\varphi)$$

and hence

$$\| S_i\varphi_i - S_i\bar{\varphi}_i \| \leqslant \delta(\Omega_1, \Omega_2)\delta(\Omega_2, \Omega_1) \| \varphi_i - \bar{\varphi}_i \|$$
$$\text{for all} \quad \varphi_i, \bar{\varphi}_i \in E_i(\varphi)\,,$$

where, we recall, $S_1 = T_1T_2$, $S_2 = T_2T_1$.
Supposing that $\delta = \delta(\Omega_1, \Omega_2)\delta(\Omega_2, \Omega_1) < 1$, the successive approximations φ_i^k converge uniformly on Γ_i; their limit φ_i is the fixed point of S_i and we have the estimate

$$\| \varphi_i^k - \varphi_i \| \leqslant \frac{\delta^k}{1 - \delta} \| \varphi_i^1 - \varphi_i^0 \|\,.$$

The Dirichlet problem $P(\Omega, \varphi)$ admits a classical solution u which, by the maximum principle, satisfies

$$(7.24) \qquad |u - u_i(\varphi_i^k)| \leqslant \frac{\delta^k}{1 - \delta} \| \varphi_i^1 - \varphi_i^0 \| \quad \text{on} \quad \Omega_i\,.$$

We have thus proved:

Proposition 3. *Let Ω_1, Ω_2 be two bounded open sets in $\mathbb{R}^n$ with boundaries Γ_1, Γ_2, $\Omega = \Omega_1 \cup \Omega_2$ with boundary Γ and $\varphi \in \mathscr{C}^0(\Gamma)$. We put*

$$\Gamma'_i = \Gamma_i \cap \Gamma \quad \textit{and} \quad E_i(\varphi) = \{\varphi_i \in \mathscr{C}^0(\Gamma_i); \varphi_i = \varphi \quad \textit{on} \quad \Gamma'_i\}\,.$$

We suppose that for all $\varphi_i \in E_i(\varphi)$, the Dirichlet problem $P(\Omega_i, \varphi_i)$ admits a classical solution denoted by $u_i(\varphi_i)$. Finally, we suppose that $\delta = \delta(\Omega_1, \Omega_2)\delta(\Omega_2, \Omega_1) < 1$, where $\delta(\Omega_1, \Omega_2)$ is defined by (7.21). *Then the Dirichlet problem $P(\Omega, \varphi)$ admits a classical solution u which can be approximated by $u_i(\varphi_i^k)$ uniformly on Ω_i with the estimate* (7.24), *where the sequences (φ_i^k) are defined recursively by*

$$\varphi_i^0 \in E_i(\varphi) \quad \textit{given}, \quad \varphi_2^k = T_2\varphi_1^k, \varphi_1^{k+1} = T_1\varphi_2^k \quad \textit{for all} \quad k \geqslant 0$$

with T_1, T_2 given by (7.18) *and* (7.19).

We have always $\delta(\Omega_1, \Omega_2) \leqslant 1$, with the result that $\delta(\Omega_1, \Omega_2)\delta(\Omega_2, \Omega_1) < 1$ iff $\delta(\Omega_1, \Omega_2) < 1$ or $\delta(\Omega_2, \Omega_1) < 1$. We shall now give *geometrical hypotheses sufficient for the condition $\delta(\Omega_1, \Omega_2) < 1$ to be satisfied.* We shall establish the results in the case $n = 2$:

Proposition 4. *Let Ω_1 and Ω_2 be two bounded open sets in $\mathbb{R}^2$ with boundaries Γ_1 and Γ_2 respectively. We suppose that for all $z \in \Gamma_1 \cap \overline{(\Gamma_2 \cap \Omega_1)}$,*
(1) Ω_1 and Ω_2 *satisfies the condition of the exterior segment at z, that is to say that there exists $x_0 \in \mathbb{R}^2$ such that*

$$\lambda x_0 + (1 - \lambda)z \notin \Omega_1 \cup \Omega_2 \quad \textit{for all} \quad \lambda \in [0, 1]\,,$$

(2) $\Omega_1 \cap \Omega_2$ *satisfies the condition of the interior cone at z, that is to say that there exist y_1 and y_2 in $\mathbb{R}^2$, not aligned with z, such that*

$\lambda_1 y_1 + \lambda_2 y_2 + (1 - \lambda_1 - \lambda_2)z \in \Omega_1 \cap \Omega_2$ *for all* $\lambda_1, \lambda_2 > 0$ *with* $\lambda_1 + \lambda_2 < 1$.
Then $\delta_1(\Omega_1, \Omega_2) < 1$.

Proof. Considering v_1^0 the solution of $P(\Omega_1, \chi_{\Gamma_1 \cap \Omega_2})$ we have to show that $\delta_1 = \sup\limits_{\Omega_1 \cap \Gamma_2} v_1^0 < 1$. By the maximum principle, $v_1^0 < 1$ on $\Omega_1 \cap \Gamma_2$ with the result that it is sufficient to show that:

for all $\qquad z \in \overline{\Omega_1 \cap \Gamma_2} \backslash (\Omega_1 \cap \Gamma_2) = \Gamma_1 \cap \overline{(\Omega_1 \cap \Gamma_2)}\,,$

$$\limsup_{\substack{x \to z \\ x \in \Omega_1 \cap \Gamma_2}} v_1^0(x) < 1\,.$$

Hence we fix $z \in \Gamma_1 \cap \overline{(\Omega_1 \cap \Gamma_2)}$; the hypothesis is represented by Fig. 23. For $r > 0$ sufficiently small, we denote by Ω_1' the disk sector

$$B(z, r) \backslash \{z + \lambda_0(x_0 - z) + \lambda_1(y_1 - z); \lambda_0, \lambda_1 \geqslant 0\}$$

and by v_1' the solution of $P(\Omega_1', 1 - \chi_S)$ where S is the segment

$$\{\lambda x_0 + (1 - \lambda)z; \lambda \in [0, 1]\}\,.$$

The boundary of $\Omega_1 \cap \Omega_1'$ is made up of

$\Gamma_1 \cap \bar{B}(z, r)$ which for r sufficiently small is disjoint from Ω_2 with the result that $v_1^0 = 0 \leqslant v_1'$ on this part of the boundary;
$\partial\Omega_1' \cap \Omega_1$ which is disjoint from S with the result that $v_1^0 \leqslant 1 = v_1'$ on this part of the boundary.

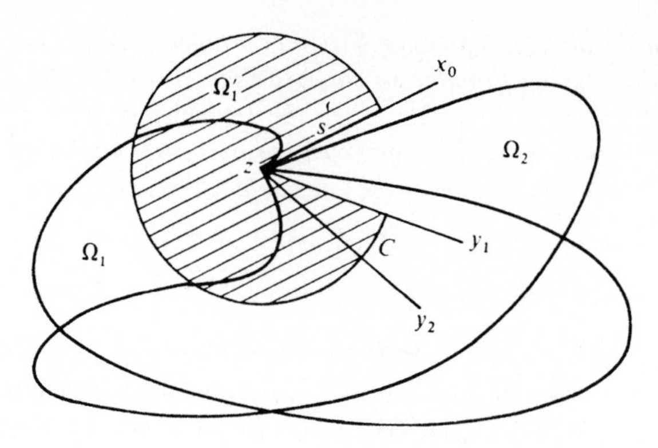

Fig. 23

By the maximum principle, $v_1^0 \leqslant v_1'$ on $\Omega_1 \cap \Omega_1'$. Now, for r sufficiently small, $\Gamma_2 \cap \Omega_1 \cap B(z, r)$ is disjoint from the cone

$$C = \{z + \lambda_1(y_1 - z) + \lambda_2(y_2 - z); \lambda_1, \lambda_2 > 0\}$$

and therefore

$$\limsup_{\substack{x \to z \\ x \in \Gamma_2 \cap \Omega_1}} v_1^0(x) \leqslant \limsup_{\substack{x \to z \\ x \in \Omega_1' \setminus C}} v_1'(x) .$$

In other words, we are led to prove that

$$\limsup_{\substack{x \to z \\ x \in \Omega_1' \setminus C}} v_1'(x) < 1 .$$

To within a similitude, this is a result of the following lemma. □

Lemma 3. *For $0 < \alpha < 2\pi$, we denote by Ω_α the sector of the unit disk in $\mathbb{R}^2$ formed by the points of polar angle $\theta \in]0, \alpha[$; let v_α be the solution of $P(\Omega_\alpha, 1 - \chi_S)$ where S is the segment $[0, 1]$ of the axis of the abscissae. Then, for all $0 < \beta < \alpha$,*

$$\limsup_{\substack{x \to 0 \\ x \in \Omega_\beta}} v_\alpha(x) \leqslant \frac{\beta}{\alpha} .$$

Proof. We shall identify $\mathbb{R}^2$ with the complex plane $\mathbb{C}$. Given $0 < \alpha, \beta < 2\pi$, the conformal transformation $z \to z^{\beta/\alpha}$ transforms Ω_α to Ω_β; using the invariance of harmonic functions under a conformal transformation (see Proposition 23 of § 2), we have $v_\alpha(z) = v_\beta(z^{\beta/\alpha})$ for all $z \in \Omega_\alpha$.
In particular

$$v_\alpha(z) = v_\pi(z^{\pi/\alpha}) \quad \text{for all} \quad z \in \Omega_\alpha .$$

An explicit calculation of v_π gives (see Sect. 5, (7.99))

$$v_\pi(\rho e^{i\varphi}) = \frac{1}{\pi}\left[\pi - \operatorname{Arctg}\left(\frac{1 - \rho\cos\varphi}{\rho\sin\varphi}\right) + \operatorname{Arctg}\left(\frac{\rho - \cos\varphi}{\sin\varphi}\right)\right] \tag{7.25}$$

a formula by means of which we verify that for $0 < \varphi_0 < \pi$

$$\limsup_{\substack{\rho \to 0 \\ 0 < \varphi < \varphi_0}} v_\pi(\rho e^{i\varphi}) \leqslant \lim_{\rho \to 0} v_\pi(\rho e^{i\varphi_0}) = \frac{1}{\pi}\left[\frac{\pi}{2} - \operatorname{Arctg}\operatorname{cotg}\varphi_0\right] = \frac{\varphi_0}{\pi}.$$

Returning to v_α, we find for $0 < \beta < \alpha$

$$\limsup_{\substack{z \to 0 \\ z \in \Omega_\beta}} v_\alpha(z) = \limsup_{\substack{\rho = |z|^{\pi/\alpha} \to 0 \\ \varphi = \frac{\pi}{\alpha}\arg z < \frac{\pi\beta}{\alpha}}} v_\pi(\rho e^{i\varphi}) \leqslant \frac{\pi\beta}{\alpha\pi} = \frac{\beta}{\alpha}. \qquad \square$$

In the case $n \geqslant 3$, the situation is more delicate. We can show that $\delta(\Omega_1, \Omega_2) < 1$ if for all $z \in \Gamma_1 \cap \Gamma_2$, Γ_1 and Γ_2 are regular in the neighbourhood of z with exterior unit normal vectors $n_1(z)$ and $n_2(z)$ satisfying

$$|n_1(z) . n_2(z)| < 1 .$$

More generally we can moreover suppose that Ω_1, Ω_2 are, in the neighbourhood of z, the union of regular open sets $\Omega_1^1, \ldots, \Omega_1^r, \Omega_2^1, \ldots, \Omega_2^S$ with boundaries passing through z and with unit exterior normal vectors

$$n_1^1(z), \ldots, n_1^r(z) , \qquad n_2^1(z), \ldots, n_2^s(z)$$

linearly independent (which implies that $r + s \leqslant n$).
In the case $n = 3$, admitting as intuitive the notions of vertex and edge, these conditions express that
(1) Γ_1 (resp. Γ_2) contains no vertex of Γ_2 (resp. Γ_1),
(2) an edge of Γ_1 (resp. Γ_2) which meets Γ_2 (resp. Γ_1) cuts it in a regular point.
We shall not prove these results (see Courant–Hilbert [1]).
The alternating procedure of Schwarz can be applied to problems other than that of Dirichlet. For example, let us consider the mixed problem:

$$(7.26) \qquad \begin{cases} \Delta u = 0 & \text{on} \quad \Omega = \Omega_1 \cup \Omega_2 \\ u = \varphi & \text{on} \quad \Gamma_1' = \Gamma_1 \cap \Gamma \\ \dfrac{\partial u}{\partial n} = \psi & \text{on} \quad \Gamma \backslash \Gamma_1' . \end{cases}$$

For $\varphi_1 \in E_1(\varphi)$, we shall consider the solution of the mixed problem on Ω_2:

$$(7.27) \qquad \begin{cases} \Delta u_2 = 0 & \text{on} \quad \Omega_2 \\ u_2 = u_1(\varphi_1) & \text{on} \quad \Gamma_2 \cap \bar{\Omega}_1 \\ \dfrac{\partial u_2}{\partial n} = \psi & \text{on} \quad \Gamma \backslash \Gamma_1' = \Gamma_2 \backslash \bar{\Omega}_1 \end{cases}$$

where $u_1(\varphi_1)$ is the solution of $P(\Omega_1, \varphi_1)$. We shall suppose that (7.27) admits a classical solution for all $\varphi_1 \in E_1(\varphi)$; we recall that it is then unique (see Corollary 6 of § 4). We shall then define $S\varphi_1$ by

$$(7.28) \qquad S\varphi_1 = \begin{cases} \varphi & \text{on } \Gamma'_1 \\ u_2 & \text{on } \Gamma_1 \backslash \Gamma'_1 = \Gamma_1 \cap \Omega_2 . \end{cases}$$

Given $\varphi_1, \bar{\varphi}_1 \in E_1(\varphi)$, u_2, $\bar{u}_2$ the corresponding solutions of (7.27), the function $v_2 = u_2 - \bar{u}_2$ is the solution of the mixed problem

$$\begin{cases} \Delta v_2 = 0 & \text{on } \Omega_2 \\ v_2 = v_1 & \text{on } \Gamma_2 \cap \bar{\Omega}_1 \\ \dfrac{\partial v_2}{\partial n} = 0 & \text{on } \Gamma_2 \backslash \bar{\Omega}_1 \end{cases}$$

where $v_1 = u_1(\varphi_1) - u_1(\bar{\varphi}_1)$.
From the principle of the maximum for mixed problems (see Proposition 13 of § 4) we have

$$|v_2| \leqslant \max_{\Gamma_2 \cap \bar{\Omega}_1} |v_1| \leqslant \delta(\Omega_1, \Omega_2) \| \varphi_1 - \bar{\varphi}_1 \| .$$

Therefore

$$\| S\varphi_1 - S\bar{\varphi}_1 \| \leqslant \delta(\Omega_1, \Omega_2) \| \varphi_1 - \bar{\varphi}_1 \| .$$

In other words we have proved:

Proposition 5. *Let Ω_1, Ω_2 be two bounded open sets of $\mathbb{R}^n$ with boundaries Γ_1, Γ_2, $\Omega = \Omega_1 \cup \Omega_2$ with boundary Γ, $\Gamma'_1 = \Gamma \cap \Gamma_1$, $\varphi \in \mathscr{C}^0(\Gamma'_1)$, $\psi \in \mathscr{C}^0(\Gamma \backslash \Gamma'_1)$. Suppose that for every $\varphi_1 \in \mathscr{C}^0(\Gamma_1)$ with $\varphi_1 = \varphi$ on Γ'_1, the Dirichlet problem $P(\Omega_1, \varphi_1)$ admits a classical solution $u_1(\varphi_1)$, the mixed problem* (7.27) *admits a classical solution $u_2(\varphi_1)$ and $\delta = \delta(\Omega_1, \Omega_2) < 1$. Then the mixed problem* (7.26) *admits a classical solution u which can be approximated uniformly on Ω_1 and Ω_2 by the solutions $u_1(\varphi_1^k)$ and $u_2(\varphi_1^k)$ with the estimates*

$$|u - u_i(\varphi_1^k)| \leqslant \frac{\delta^k}{1 - \delta} \| \varphi_1^1 - \varphi_1^0 \| \quad \text{on} \quad \Omega_i$$

where (φ_1^k) is the sequence constructed recursively by: $\varphi_1^0 \in E_1(\varphi)$ given, $\varphi_1^{k+1} = S\varphi_1^k$ where S is given by (7.28).

Remark 4. The alternating procedure of Schwarz gives a method of approximate calculation for fairly general open sets arising from the knowledge of the solutions for elementary open sets (ball, cube, etc. . .), since every open set is the union of an increasing sequence of finite unions of elementary open sets. For complementary results and a probabilistic interpretation, see P.L. Lions [1]. This method can be exploited at the level of numerical calculation within the context of the so-called methods of "sub-domains" (see Frailong–Paklesa [1], Glowinski–Periaux–Dinh [1] and Miellou [1][179]).

---

[179] In this last article, it is a question of work directed towards parallel calculation relative to the use of multiprocessors.

3. Method of Separation of Variables. Harmonic Polynomials. Spherical Harmonic Function

Let us first of all consider an open set Ω of $\mathbb{R}^n$ of the form $\Omega = \Omega_1 \times \Omega_2$ where Ω_1, Ω_2 are open sets of $\mathbb{R}^{n_1}$, $\mathbb{R}^{n_2}$ respectively with $1 \leqslant n_2 \leqslant n_1 < n_1 + n_2 = n$. Denoting by x_1, x_2 the variables of $\mathbb{R}^{n_1}$, $\mathbb{R}^{n_2}$ respectively, *a function $u(x_1, x_2)$ on Ω is said to have separable variables* if it can be written in the form

$$u(x_1, x_2) = u_1(x_1) u_2(x_2), \qquad x_1 \in \Omega_1, x_2 \in \Omega_2 .$$

Such a function with separable variables is harmonic on Ω iff

$$u_2(x_2) \Delta u_1(x_1) + u_1(x_1) \Delta u_2(x_2) = 0 \quad \textit{for all} \quad x_1 \in \Omega_1 , x_2 \in \Omega_2 ,$$

that is to say iff *there exists a constant λ such that*

$$\begin{cases} \Delta u_1(x_1) = \lambda u_1(x_1) & \textit{for all} \quad x_1 \in \Omega_1 \\ \Delta u_2(x_2) + \lambda u_2(x_2) = 0 & \textit{for all} \quad x_2 \in \Omega_2 . \end{cases}$$

In other words, *the set of functions with separable variables, harmonic on $\Omega = \Omega_1 \times \Omega_2$ is obtained by solving, for every constant λ, the equations*

$$\Delta u_1 = \lambda u_1 \quad \text{on} \quad \Omega_1 , \qquad \Delta u_2 + \lambda u_2 = 0 \quad \text{on} \quad \Omega_2 . \tag{7.29}$$

The equations (7.29) are a little more complicated than Laplace's equation (corresponding to $\lambda = 0$); by contrast the open sets Ω_i are of dimension n_i strictly less than n, which *a priori* makes the search for harmonic functions on Ω with variables separable much easier. For example, *in the case $n = 2$, we have $n_1 = n_2 = 1$* and *we are led to consider the pair of ordinary differential equations*

$$\frac{d^2 u_1}{dx_1^2} = \lambda u_1 , \qquad \frac{d^2 u_2}{dx_2^2} + \lambda u_2 = 0 .$$

In the case $n = 3$, we have $n_1 = 2$, $n_2 = 1$ and we are led to consider

(i) an ordinary differential equation $\dfrac{d^2 u_2}{dx_2^2} + \lambda u_2 = 0$,

(ii) a partial differential equation $\Delta u_1 = \lambda u$ on an open set Ω_1 of $\mathbb{R}^2$, for which we can use the special techniques of theory of functions of a complex variable.

The interest in the search for harmonic functions with separable variables on Ω rests in *Weierstrass' theorem*: every function continuous on $\Omega = \Omega_1 \times \Omega_2$ is the limit (uniformly on every compact set of Ω) of a sequence of polynomials and hence of a sequence of the sum of functions with variables separable. Indeed, we shall see, at least in a certain number of examples, that *every harmonic function u on Ω can be expressed with the aid of harmonic functions, with variables separable on Ω and even often can be written*

(i) as the sum of a series $\sum_k u_k$,

or

(ii) as an integral $\int u_\lambda \, \mathrm{d}\lambda$,

where

$$u_k(x_1, x_2) = u_1^k(x_1)u_2^k(x) , \qquad u_\lambda(x_1, x_2) = u_1^\lambda(x_1)u_2^\lambda(x_2)$$

are harmonic functions with separable variables.
To illustrate this statement, we shall consider examples in the case $n_2 = 1$. We put $\Omega = \Omega' \times I$ *where* I *is an open interval of* $\mathbb{R}$ *and* Ω' *an open set in* $\mathbb{R}^{n-1}$. A function with variables separable $u(x', t) = u'(x')v(t)$ is harmonic on Ω iff there exists a constant λ such that

$$\Delta u' = \lambda u' \quad \text{on} \quad \Omega' , \qquad \frac{\mathrm{d}^2 v}{\mathrm{d}t^2} + \lambda v = 0 \quad \text{on} \quad I .$$

The second (ordinary differential) equation can be solved classically:

$$v(t) = \alpha \mathrm{e}^{i\omega t} + \beta \mathrm{e}^{-i\omega t} , \qquad t \in I ,$$

where ω is a square root (in $\mathbb{C}$) of λ and α, β are arbitrary constants. In other words, *a harmonic function with variables separable on* $\Omega = \Omega' \times I$ *can be written in the form*

$$u(x', t) = u'(x')(\alpha \mathrm{e}^{i\omega t} + \beta \mathrm{e}^{-i\omega t}) \tag{7.30}$$

where $\omega, \alpha, \beta \in \mathbb{C}$ *and* u' *is a solution of*

$$\Delta u' = \omega^2 u' \quad \text{on} \quad \Omega' . \tag{7.31}$$

We now show in examples how a harmonic function on Ω can be developed with the aid of harmonic functions with variables separable.

Example 1. *Harmonic function on* $\Omega = \Omega' \times]0, l[$ *(with* $0 < l < \infty$*) continuous on* $\Omega' \times [0, l]$ *and zero on* $\Omega' \times \{0, l\}$. Let us consider first a function u with variables separable given by (7.30); it is continuous on $\Omega' \times [0, l]$. u is not identically zero on $\Omega' \times]0, l[$ and zero on $\Omega' \times \{0, l\}$ iff

$$u' \text{ is not identically zero on } \Omega', \omega \neq 0 ,$$

$$|\alpha| + |\beta| \neq 0, \alpha + \beta = 0 \quad \text{and} \quad \alpha \mathrm{e}^{i\omega l} + \beta \mathrm{e}^{-i\omega l} = 0 ,$$

namely iff

$$u' \text{ is not identically zero on } \Omega', \beta = -\alpha \neq 0$$

$$\omega \neq 0 \quad \text{and} \quad \sin \omega l = 0 .$$

Now $\sin \omega l = 0$ iff $\omega = k\pi/l$ with $k \in \mathbb{Z}$.
Hence, *the harmonic functions with separable variables on* $\Omega' \times]0, l[$ *and zero on* $\Omega' \times \{0, l\}$

$$u(x', t) = u_k'(x') \sin \frac{k\pi t}{l} \tag{7.32}$$

with $k = 1, 2, \ldots$ *and* u'_k *solution of*

$$\Delta u'_k = \left(\frac{k\pi}{l}\right)^2 u'_k \quad on \quad \Omega' . \tag{7.33}$$

Now, let us consider an arbitrary function u harmonic on $\Omega' \times]0, l[$ continuous on $\Omega' \times [0, l]$ and zero on $\Omega' \times \{0, l\}$. We can extend it (in a unique manner) to a function $\tilde{u}$ on $\Omega' \times \mathbb{R}$ odd and periodic with period $2l$ with respect to t. The function $\tilde{u}$ is harmonic on $\Omega' \times \mathbb{R}$ as it is so on $\Omega' \times (\mathbb{R} \backslash \mathbb{Z}\pi/l)$ and at each point $(x', k\pi l)$:

$$\tilde{u}\left(x', \frac{k\pi}{l} + 0\right) = \tilde{u}\left(x', \frac{k\pi}{l} - 0\right) = 0 , \quad \frac{\partial \tilde{u}}{\partial t}\left(x', \frac{k\pi}{l} + 0\right)$$
$$= \frac{\partial \tilde{u}}{\partial t}\left(x', \frac{k\pi}{l} - 0\right).$$

For all $x' \in \Omega'$, we develop the periodic function of period $2l$, $\tilde{u}_{x'}(t) = \tilde{u}(x', t)$ as a Fourier series; since it is odd we have

$$\tilde{u}_{x'}(t) = \sum_{k=1}^{\infty} u'_k(x') \sin \frac{k\pi t}{l}$$

with

$$u'_k(x') = \frac{1}{l}\int_{-l}^{+l} \tilde{u}(x', t) \sin \frac{k\pi t}{l} \mathrm{d}t = \frac{2}{l}\int_0^l u(x', t) \sin \frac{k\pi t}{l} \mathrm{d}t ;$$

we have

$$\Delta u'_k(x') = \frac{1}{2}\int_{-l}^{+l} \Delta_{x'} \tilde{u}(x', t) \sin \frac{k\pi t}{l} \mathrm{d}t = -\frac{1}{2}\int_{-l}^{+l} \frac{\partial^2 \tilde{u}}{\partial t^2}(x', t) \sin \frac{k\pi t}{l} \mathrm{d}t$$
$$= \left(\frac{k\pi}{l}\right)^2 \frac{1}{2}\int_{-l}^{+l} \tilde{u}(x', t) \sin \frac{k\pi t}{l} \mathrm{d}t = \left(\frac{k\pi}{l}\right)^2 u'_k(x') .$$

Hence *every harmonic function on* $\Omega' \times]0, l[$ *continuous on* $\Omega' \times [0, l]$ *and zero on* $\Omega' \times \{0, l\}$ *can be written*

$$u(x', t) = \sum_{k=1}^{\infty} u'_k(x') \sin \frac{k\pi t}{l} , \tag{7.34}$$

with u'_k a solution of (7.33).

We note that the series $\sum_{k=1}^{\infty} u'_k(x') \sin \frac{k\pi t}{l}$ converges uniformly on $K' \times [0, l]$ for every compact set $K' \subset \Omega'$ or, more precisely, from the classical theory of Fourier series on $D' \times [0, l]$ for each part D' of Ω' such that the modulus of uniform continuity

$$\varepsilon(r) = \sup_{x' \in D'} \sup_{\substack{t, s \in [0, l] \\ |t - s| \leqslant r}} |u(x', t) - u(x', s)|$$

satisfies $\int_0^l \varepsilon(r) \frac{\mathrm{d}r}{r} < \infty$. □

Example 2. *Function harmonic on $\Omega = \Omega' \times \mathbb{R}$.* We consider a function with variables separable given by (7.30),

$$u(x', t) = u'(x')(\alpha e^{i\omega t} + \beta e^{-i\omega t}) \quad \text{with} \quad \alpha, \beta, \omega \in \mathbb{C}\,.$$

Two cases arise:

(*i*) if $\omega \in \mathbb{R}$, then u is bounded on $D' \times \mathbb{R}$ for every part $D' \subset \Omega'$ on which u' is bounded;

(*ii*) if $\omega \notin \mathbb{R}$, then for all $x' \in \Omega'$ the function $u_{x'}(t) = u(x', t)$ is neither identically zero nor unbounded on $\mathbb{R}$ and even in this case there will exist $c > 0$ such that $e^{-c|t|}u_{x'}(t)$ is unbounded and hence for all $m \in \mathbb{N}$, $u_{x'}(t)/(1 + t^2)^m$ will not be bounded.

In other words:

A function with separable variables, harmonic on $\Omega' \times \mathbb{R}$ satisfying (7.35) *for all $x' \in \Omega'$, there exists $m \in \mathbb{N}$ such that $\sup\limits_{t\in\mathbb{R}} \dfrac{|u(x', t)|}{(1 + t^2)^m} < \infty$, is of the form*

$$u(x', t) = u'(x')(\alpha e^{i\omega t} + \beta e^{-i\omega t})$$

with $\omega \in \mathbb{R}$ and u' a solution of $\Delta u' = \omega^2 u'$ on Ω'.

Let us now consider an arbitrary function (not with variables separable), harmonic on $\Omega' \times \mathbb{R}$. We suppose, in the first place that $u \in \mathscr{C}^0(\Omega'; L^1(\mathbb{R}))$ with the result that for all $x' \in \Omega'$, we can define the classical Fourier transform

$$\hat{u}(x', \omega) = \int e^{-i\omega t} u(x', t)\, dt\,, \qquad \omega \in \mathbb{R}\,. \tag{7.36}$$

Supposing that we also have $\hat{u} \in \mathscr{C}^2(\Omega'; L^1(\mathbb{R}))$, we can apply the classical Fourier inversion formula

$$u(x', t) = \frac{1}{2\pi} \int \hat{u}(x', \omega) e^{i\omega t}\, d\omega\,.$$

Now, assuming $u \in \mathscr{C}^0(\Omega', L^1(\mathbb{R}))$ and using (7.36), we have

$$\Delta_{x'}\hat{u}(x', \omega) = \int e^{-i\omega t} \Delta_{x'} u(x', t)\, dt = -\int e^{-i\omega t} \frac{\partial^2 u}{\partial t^2}(x', t)\, dt$$

$$= \omega^2 \int e^{-i\omega t} u(x', t)\, dt = \omega^2 \hat{u}(x', \omega)\,.$$

In short, if we denote $E_1(\mathbb{R}) = \{v \in L^1(\mathbb{R}); \ \hat{v} \in L^1(\mathbb{R})\}$ *every function $u \in \mathscr{C}^0(\Omega'; E_1(\mathbb{R})) \cap \mathscr{C}^2(\Omega'; L^1(\mathbb{R}))$ harmonic on $\Omega' \times \mathbb{R}$ can be written*

$$u(x', t) = \int u'_\omega(x') e^{i\omega t}\, d\omega$$

with for all $\omega \in \mathbb{R}$, $\Delta u'_\omega = \omega^2 u'_\omega$.

The important restriction on $u(x', t)$ for $|t| \to \infty$ can be lifted by use of the generalized Fourier transform. Let us consider u harmonic on $\Omega' \times \mathbb{R}$ satisfying only (7.35). Then for all $x' \in \Omega'$ we can define the generalized Fourier transform of $u_{x'}(t) = u(x', t)$:

$\hat{u}_{x'}$ is the distribution defined by

$$(7.37)\qquad \langle \hat{u}_{x'}, \zeta \rangle = \int u(x', s)\,\mathrm{d}s \int \mathrm{e}^{-its}\zeta(t)\,\mathrm{d}t\,, \qquad \zeta \in \mathscr{D}(\mathbb{R})\,.$$

The right hand side clearly has a meaning: in effect, on the one hand, since

$$\int \frac{|u(x', t)|}{(1 + t^2)^{m+1}}\,\mathrm{d}t \leqslant \pi \sup_{t \in \mathbb{R}} \frac{|u(x', t)|}{(1 + t^2)^m}$$

from (7.35),

$$(7.38)\qquad \text{for all } x' \in \Omega', \text{ there exists } m \in \mathbb{N} \text{ such that } \int \frac{|u(x', t)|}{(1 + t^2)^m}\,\mathrm{d}t < \infty\,.$$

On the other hand

$$\left|\int \mathrm{e}^{-its}\zeta(t)\,\mathrm{d}t\right| = \left|(1 + s^2)^{-m} \int \zeta(t)\left(I - \frac{\mathrm{d}^2}{\mathrm{d}t^2}\right)^m \mathrm{e}^{-its}\,\mathrm{d}t\right|$$

$$= (1 + s^2)^{-m}\left|\int \mathrm{e}^{-its}\left(I - \frac{\mathrm{d}^2}{\mathrm{d}t^2}\right)^m \zeta(t)\,\mathrm{d}t\right| \leqslant (1 + s^2)^{-m}\left\|\left(I - \frac{\mathrm{d}^2}{\mathrm{d}t^2}\right)^m \zeta\right\|_{L^1(\mathbb{R})}$$

Now we have the Fourier inversion formula: $u_{x'} = \frac{1}{2\pi}\hat{\hat{u}}_{x'}$, namely

$$(7.39)\qquad \int u(x', s)\,\xi(s)\,\mathrm{d}s = \langle \hat{u}_{x'}(\omega), \frac{1}{2\pi}\int \mathrm{e}^{i\omega s}\,\xi(s)\,\mathrm{d}s \rangle\,, \qquad \xi \in \mathscr{D}(\mathbb{R})\,.$$

This formula clearly has a meaning, for, as we have seen in justifying (7.37) $\langle \hat{u}_{x'}, \zeta \rangle$ is defined for every function $\zeta \in \mathscr{C}^{2m}(\mathbb{R})$ with $\left(I - \frac{\mathrm{d}^2}{\mathrm{d}t^2}\right)^m \zeta \in L^1(\mathbb{R})$ where m is an integer corresponding to x' in (7.38). Now for $\xi \in \mathscr{D}(\mathbb{R})$, the function $\zeta(t) = \frac{1}{2\pi}\int \mathrm{e}^{its}\,\xi(s)\,\mathrm{d}s$ is $\mathscr{C}^\infty$ and for all $m \in \mathbb{N}$,

$$\left\|\left(I - \frac{\mathrm{d}^2}{\mathrm{d}t^2}\right)^m \zeta\right\|_{L^1} = \frac{1}{2\pi}\int \mathrm{d}t \left|\int (1 + s^2)^m \mathrm{e}^{its}\,\xi(s)\,\mathrm{d}s\right|$$

$$= \frac{1}{2\pi}\int \frac{\mathrm{d}t}{1 + t^2}\left|\int \mathrm{e}^{its}\left(I - \frac{\mathrm{d}^2}{\mathrm{d}s^2}\right)[(1 + s^2)^n\,\xi(s)]\,\mathrm{d}s\right|$$

$$\leqslant \frac{1}{2}\left\|\left(I - \frac{\mathrm{d}^2}{\mathrm{d}s^2}\right)(1 + s^2)^m\,\xi\right\|_{L^1} < \infty\,.$$

This calculation shows that the two members of (7.39) are perfectly defined and equal for every function $\xi \in \mathscr{C}^2(\mathbb{R})$ with $\left(I - \frac{\mathrm{d}^2}{\mathrm{d}s^2}\right)(1 + s^2)^m\xi \in L^1(\mathbb{R})$ where m is an integer corresponding to x' in (7.38).

Now, let us fix $\rho \in \mathscr{D}(\mathbb{R})$ with $\rho(0) = 1$. The Fourier transform $\hat{\rho}$ then satisfies

$$\left(I - \frac{\mathrm{d}^2}{\mathrm{d}s^2}\right)^r (1 + s^2)^m \hat{\rho} \in L^1(\mathbb{R}) \quad \text{for all} \quad r, m \in \mathbb{N}$$

and $\displaystyle\int \hat{\rho}(s)\,\mathrm{d}s = 2\pi$. We deduce

$$u(x', t) = \lim_{k \to \infty} u_k(x', t) \quad \text{for all} \quad (x', t) \in \Omega' \times \mathbb{R} \tag{7.40}$$

with

$$u_k(x', t) = \frac{1}{2\pi} \int u\left(x', t + \frac{s}{k}\right) \hat{\rho}(s)\,\mathrm{d}s\,.$$

In effect

$$u\left(x', t + \frac{s}{k}\right) \hat{\rho}(s) \to u(x', t)\hat{\rho}(s) \quad \text{for all} \quad s \in \mathbb{R}\,,$$

and we can apply Lebesgue's theorem since

$$\left|u\left(x', t + \frac{s}{k}\right)\hat{\rho}(s)\right| = \frac{\left|u\left(x', t + \frac{s}{k}\right)\right|}{\left(1 + \left(t + \frac{s}{k}\right)^2\right)^m} \left(\frac{1 + \left(t + \frac{s}{k}\right)^2}{1 + s^2}\right)^m (1 + s^2)^m |\hat{\rho}(s)|$$

$$\leqslant (1 + 2t^2)^m \sup_{\tau \in \mathbb{R}} \frac{|u(x', \tau)|}{(1 + \tau^2)^m} (1 + s^2)^m |\hat{\rho}(s)|\,.$$

But, by a change of variables

$$u_k(x', t) = \frac{1}{2\pi} \int u(x', s)\, k\hat{\rho}(k(s - t))\,\mathrm{d}s$$

and hence applying (7.39) with $\xi(s) = k\hat{\rho}(k(s - t))$, noting that

$$\frac{1}{2\pi} \int \mathrm{e}^{i\omega s}\, k\hat{\rho}(k(s - t))\,\mathrm{d}s = \frac{\mathrm{e}^{i\omega t}}{2\pi} \int \mathrm{e}^{i\frac{\omega}{k}s}\, \hat{\rho}(s)\,\mathrm{d}s = \mathrm{e}^{i\omega t}\rho\left(\frac{\omega}{k}\right)$$

we obtain

$$u_k(x', t) = \left\langle \frac{1}{2\pi} \hat{u}_{x'}(\omega), \mathrm{e}^{i\omega t}\rho\left(\frac{\omega}{k}\right)\right\rangle.$$

It is clear that the map $u' : x' \to \dfrac{1}{2\pi} \hat{u}_{x'}$ is continuous (and even analytic) from Ω into $\mathscr{D}'(\mathbb{R})$ and we have

$$\Delta u'(x') = \omega^2 u'(x') \quad \text{in} \quad \mathscr{D}'(\mathbb{R}) \quad \text{for all} \quad x' \in \Omega'\,. \tag{7.41}$$

In brief: *every function u, harmonic on* $\Omega' \times \mathbb{R}$ *satisfying* (7.35) *can be written in the form*

$$u(x', t) = \langle u'(x')(\omega)\,,\ \mathrm{e}^{i\omega t}\rangle \tag{7.42}$$

with the (analytic) map u' of Ω' into $\mathscr{D}'(\mathbb{R})$ satisfying (7.41), *the equality* (7.42) *being taken in the sense*

$$u(x', t) = \lim_{k \to \infty} \left\langle u'(x')(\omega), e^{i\omega t} \rho\left(\frac{\omega}{k}\right) \right\rangle$$

for each function $\rho \in \mathscr{D}(\mathbb{R})$ with $\rho(0) = 1$.
The examples given above can be generalized for the study of harmonic functions on $\Omega = \Omega_1 \times \Omega_2$ satisfying *linear conditions independent of $x_1 \in \Omega_1$ on the boundary $\Omega_1 \times \partial\Omega_2$*. In effect the method of separation of variables leads to the study of the constants λ and the functions u_2 on Ω_2 satisfying

$$\begin{cases} \Delta u_2 + \lambda u_2 = 0 \quad \text{on} \quad \Omega_2 \\ \text{boundary conditions on } \partial\Omega_2 \,. \end{cases} \tag{7.43}$$

In other words, considering the (linear) space X_2 of the functions u_2 on Ω_2, satisfying the boundary conditions on $\partial\Omega_2$, we are led to seeking the *eigenvalues λ and the eigenfunctions u_2 of the operator $-\Delta$ on X_2*. The cases considered in Examples 1, 2 will be found in the general context which will be studied in § 8. By means of theorems of *spectral resolution*, every harmonic function on $\Omega_1 \times \Omega_2$ satisfying the conditions on the boundary $\Omega_1 \times \partial\Omega_2$ can be written as a sum (possibly in a generalized sense, as in Example 2) of harmonic functions with separable variables $u_1^\lambda(x_1)u_2^\lambda(x_2)$, where u_2^λ is a solution of (7.43) and u_1^λ is a solution of

$$\Delta u_1^\lambda = \lambda u_1^\lambda \quad \text{on} \quad \Omega_1 \,.$$

We note clearly that we have not imposed any condition on the boundary $\partial\Omega_1$ and that u_1^λ is not, strictly speaking, an eigenfunction of the Laplacian.
We shall now *generalize the method of separation of variables* which will lead us to study harmonic polynomials.
Let us consider an open set Ω of $\mathbb{R}^n$ of the form $\Omega = h(M_1 \times M_2)$ where *M_1, M_2 are differentiable manifolds of dimension n_1, n_2 respectively with $1 \leqslant n_2 \leqslant n_1 < n_1 + n_2 = n$ and h a diffeomorphism of $M_1 \times M_2$ on Ω*: $(y_1, y_2) \in M_1 \times M_2$ is a system of *curvilinear coordinates* of $x = h(y_1, y_2)$. The function u defined on Ω is *with separable variables in this system of curvilinear coordinates* if

$$u(h(y_1, y_2)) = u_1(y_1)u_1(y_2) \quad \text{for all} \quad (y_1, y_2) \in M_1 \times M_2 \tag{7.44}$$

where u_1, u_2 are functions defined on M_1, M_2 respectively.
In the general case, the change of variables will give a very complicated form to $(\Delta u)(h(y_1, y_2))$; *let us suppose that for every function $u \in \mathscr{C}^2(\Omega)$*

$$(\Delta u)(h(y_1, y_2)) = c(y_1, y_2)\left\{ P_1\left(y_1, \frac{\partial}{\partial y_1}\right) + P_2\left(y_2, \frac{\partial}{\partial y_2}\right)\right\}(u \circ h)(y_1, y_2) \tag{7.45}$$

where P_1, P_2 are differential operators (of order 2) on M_1 and M^{180}, respectively

[180] See Chap. V, §1.3 for the definition of a differential operator on a manifold.

and c is a non-zero function defined on $M_1 \times M_2$. Then a function with variables separable u satisfying (7.44) is harmonic iff there exists a constant λ such that

$$\begin{cases} P_1\left(y_1, \dfrac{\partial}{\partial y_1}\right) u_1(y_1) = \lambda u_1(y_1) & \text{on } M_1 \\ P_2\left(y_2, \dfrac{\partial}{\partial y_2}\right) u_2(y_2) + \lambda u_2(y_2) = 0 & \text{on } M_2 . \end{cases}$$

The hypothesis (7.45) depends on the parametrization h of Ω. It generalizes the case in which $\Omega = \Omega_1 \times \Omega_2$: when $c \equiv 1$, $P_1 = \Delta_{y_1}$, $P_2 = \Delta_{y_2}$. An important case where (7.45) is satisfied is the parametrization in *polar coordinates*

$$(\sigma, r) \in M \times I \to r\sigma \in \Omega ,$$

where I is an open interval of $]0, \infty[$, M an open set of the unit sphere Σ of $\mathbb{R}^n$, $\Omega = \{r\sigma;\, r \in I, \sigma \in M\}$ is a *sector of an annulus in* $\mathbb{R}^n$. We know then that for all $u \in \mathscr{C}^2(\Omega)$ (see (1.27) of § 1)

$$(\Delta u)(r\sigma) = \left(\frac{\partial^2}{\partial r^2} + \frac{(n-1)}{r}\frac{\partial}{\partial r} + \frac{1}{r^2}\Delta_\sigma\right) u(r\sigma)$$

where Δ_σ is the Laplace–Beltrami operator on the unit sphere Σ.
A function with variables separable in the parametrization in polar coordinates is of the form

$$u(x) = c(r)v(\sigma) , \qquad x \in \Omega , \qquad r = |x| , \qquad \sigma = \frac{x}{r} .$$

Such a function will be harmonic iff there exists λ such that

$$\text{(7.46)} \qquad \begin{cases} \dfrac{d^2 c}{dr^2} + \dfrac{n-1}{r}\dfrac{dc}{dr} + \dfrac{\lambda}{r^2} c = 0 & \text{on } I \\ \Delta_\sigma v = -\lambda v & \text{on } M . \end{cases}$$

The first equation of (7.46) is solved classically: we put

$$\text{(7.47)} \qquad \tau_0 = 1 - \frac{n}{2} .$$

(1) If $\lambda \neq \tau_0^2$, then $c(r) = r^{\tau_0}(\alpha r^\tau + \beta r^{-\tau})$ where τ is a square root of $\tau_0^2 - \lambda$.
(2) If $\lambda = \tau_0^2$, then $c(r) = r^{\tau_0}(\alpha + \beta \operatorname{Log} r)$.
Hence *the functions with separable variables* (in the parametrization in polar coordinates) *harmonic on* $\Omega = \{x;\, r = |x| \in I, \sigma = x/r \in M\}$ *are of the form*

$$\text{(7.48)} \qquad r^{\tau_0}(\alpha + \beta \operatorname{Log} r) v(\sigma) \quad \textit{with} \quad \Delta_\sigma v = \tau_0^2 v \quad \textit{on} \quad M$$

or for $\tau \in \mathbb{C}$ *with* $\tau \neq 0$,

$$\text{(7.49)} \qquad r^{\tau_0}(\alpha r^\tau + \beta r^{-\tau}) v(\sigma) \quad \textit{with} \quad \Delta_\sigma v = (\tau_0^2 - \tau^2) v \quad \textit{on} \quad M$$

where τ_0 *is given by* (7.47) *and* $\alpha, \beta \in \mathbb{C}$ *are arbitrary constants.*

We notice the *invariance under Kelvin's transformation*: if $u(x) = c(r)v(\sigma)$ is a function with variables separable (in polar coordinates) on $M \times I$, its Kelvin transform[181] $H(0, 1)u$ is $u'(x') = |x'|^{2-n} u(I(0, 1)x')$ where $I(0, 1)$ is the inversion leaving the unit sphere Σ invariant; that is to say

$$u'(x') = c'(r')v(\sigma) \quad \text{with} \quad c'(r') = r'^{2-n}c(r'^{-1}) .$$

In particular, if $c(r) = r^{\tau_0+\tau}$(resp. $r^{\tau_0-\tau}$, r^{τ_0} Log r), then

$$c'(r') = r'^{\tau_0-\tau} \text{ (resp. } r'^{\tau_0+\tau}, \ -r'^{\tau_0} \operatorname{Log} r') .$$

We note also that the radial functions $u(x) = c(r)$ are obviously with separable variables: we recover *the characterisation of the harmonic radial functions* (see Proposition 4 of §2) by taking $v(\sigma) \equiv 1$ in (7.48) and (7.49);
(a) in (7.48), $v(\sigma) \equiv 1$ is the solution of $\Delta_\sigma v = \tau_0^2 v$ iff $\tau_0^2 = 0$, that is to say $n = 2$ and the harmonic radial functions are of the form

$$\alpha + \beta \operatorname{Log} r$$

(b) in (7.49), $v(\sigma) \equiv 1$ is the solution of $\Delta_\sigma v = (\tau_0^2 - \tau^2)v$ iff $\tau^2 = \tau_0^2$, namely, since $\tau \neq 0$, in the case $n \neq 2$ and the harmonic radial functions are of the form

$$\alpha r^{2\tau_0} + \beta = \frac{\alpha}{r^{n-2}} + \beta .$$

We note finally that the *harmonic functions on the ball* $B(0, r_0)$ are identified with the bounded harmonic functions on the punctured ball

$$B(0, r_0)\backslash\{0\} = \{r\sigma; 0 < r < r_0, \sigma \in \Sigma\}$$

(see Proposition 16 of §2). Hence from (7.48) and (7.49) the functions with separable variables, harmonic on $B(0, r_0)$ are of the form $r^{\tau_0+\tau}v(\sigma)$ with $\tau \in \mathbb{C}$, Re $\tau \geqslant \frac{1}{2}n - 1$ and v a solution of $\Delta_\sigma v = (\tau_0^2 - \tau^2)v$ on Σ. In fact we shall see later (see Proposition 6) that *the harmonic functions with separable variables on* $B(0, r_0)$ *are the harmonic homogeneous polynomials.*
We can develop the same argument as in Examples 1 and 2 by imposing a condition on the boundary $M \times \partial I$; we leave the details of such a study to the reader, preferring to develop the discussion of the second equation of the pair (7.46).
To introduce the fundamental ideas, let us commence with the case $n = 2$.

Example 3. *Harmonic function on a sector of an annulus of* $\mathbb{R}^2$.
In the case $n = 2$, the unit circle is parametrized by the polar angle θ;

$$\Sigma = \{(\cos\theta, \sin\theta); \theta \in \mathbb{R}\}$$

and

$$(\Delta_\sigma v)(\cos\theta, \sin\theta) = \frac{d^2}{d\theta^2} v(\cos\theta, \sin\theta) .$$

In other words, a function v defined on M is a solution of the second equation of the

[181] See §2.5.

pair (7.46) iff $w(\theta) = v(\cos\theta, \sin\theta)$ is a solution of

$$\frac{d^2w}{d\theta^2} = \lambda w \,. \tag{7.50}$$

We ought, however, to distinguish two cases:
case of a complete annulus: $\Omega = \{(r\cos\theta, r\sin\theta); r_1 < r < r_2, \theta \in \mathbb{R}\}$,
case of a strict sector of an annulus:

$$\Omega = \{(r\cos\theta, r\sin\theta); r_1 < r < r_2, \theta_1 < \theta < \theta_2\} \quad \text{with} \quad \theta_2 - \theta_1 \leqslant 2\pi \,.$$

In the second case the parametric representation

$$(r, \theta) \in \,]r_1, r_2[\, \times \,]\theta_1, \theta_2[\, \to (r\cos\theta, r\sin\theta) \in \Omega$$

is bijective, whereas in the first case there exists no parametric representation bijective from the unit circle onto an open interval of $\mathbb{R}$.
We identify the plane $\mathbb{R}^2 = \{(x, y)\}$ with the complex plane $\mathbb{C} = \{z = x + iy\}$, the unit circle Σ with $\{e^{i\theta}; \theta \in \mathbb{R}\}$.
(*1*) *Case of a strict sector of an annulus*. Since $\tau_0 = 1 - \frac{1}{2}n = 0$, the solutions of $\Delta_\sigma = \tau_0^2 v$ (resp. $\Delta_\sigma v = (\tau_0^2 - \tau^2)v$ with $\tau \neq 0$) on $M = \{e^{i\theta}; \theta_1 < \theta < \theta_2\}$ where $\theta_2 - \theta_1 \leqslant 2\pi$, are $v(e^{i\theta}) = \alpha' + \beta'\theta$ (resp. $v(e^{i\theta}) = \alpha' e^{i\tau\theta} + \beta' e^{-i\tau\theta}$). Hence from (7.48) and (7.49), *the functions with variables separable, harmonic on the sector* $\{re^{i\theta}; r_1 < r < r_2, \theta_1 < \theta < \theta_2\}$ *with* $\theta_2 - \theta_1 \leqslant 2\pi$ *are of the form*

$$\begin{cases} (\alpha + \beta \operatorname{Log} r)(\alpha' + \beta'\theta) \\ (\alpha r^\tau + \beta r^{-\tau})(\alpha' e^{i\tau\theta} + \beta' e^{-i\tau\theta}) \quad \text{for} \quad \tau \in \mathbb{C}\,, \quad \tau \neq 0\,. \end{cases} \tag{7.51}$$

We can impose a condition on the boundary $\Gamma_1 \cup \Gamma_2$ where

$$\Gamma_j = \{re^{i\theta_j}; r_1 < r < r_2\}\,, \qquad j = 1, 2\,.$$

To change from Example 1, let us suppose that $\theta_2 - \theta_1 < 2\pi$ and consider the mixed boundary condition:

$$u(re^{i\theta_1}) = 0, \frac{\partial}{\partial\theta} u(re^{i\theta_2}) = 0\,, \qquad r \in \,]r_1, r_2[\,. \tag{7.52}$$

A function with separable variables $c(r)v(\theta)$, not identically zero, satisfies this condition iff

$$v(e^{i\theta_1}) = 0\,, \qquad \frac{\partial}{\partial\theta} v(e^{i\theta_2}) = 0\,. \tag{7.53}$$

A function $v(e^{i\theta}) = \alpha' + \beta'\theta$ cannot satisfy (7.53) without being identically zero; a function $v(e^{i\theta}) = \alpha' e^{i\tau\theta} + \beta' e^{-i\tau\theta}$ satisfies (7.53) iff

$$\begin{cases} \alpha' e^{i\tau\theta_1} + \beta' e^{-i\tau\theta_1} = 0 \\ \alpha' e^{i\tau\theta_2} - \beta' e^{-i\tau\theta_2} = 0\,, \end{cases} \tag{7.54}$$

a system which admits a non-zero solution (α', β') iff

$$\cos\tau(\theta_2 - \theta_1) = 0$$

namely iff $\tau = \left(k + \frac{1}{2}\right)\frac{\pi}{\theta_2 - \theta_1}$ with $k \in \mathbb{Z}$. Then the solutions of (7.54) are proportional to

$$\alpha' = \frac{1}{2}e^{-i\tau\theta_2}, \qquad \beta' = \frac{1}{2}e^{i\tau\theta_2}$$

to which corresponds

$$v(e^{i\theta}) = \cos\tau(\theta_2 - \theta)\,.$$

In summary, therefore, account taken of (7.51), *the functions with variables separable, harmonic on the sector* $\Omega: \{re^{i\theta}; r_1 < r < r_2, \theta_1 < \theta < \theta_2\}$ *with* $\theta_2 - \theta_1 < 2\pi$ *and satisfying the boundary conditions* (7.52) *are of the form*

$$(\alpha r^{\tau_k} + \beta r^{-\tau_k})\cos\tau_k(\theta_2 - \theta)$$

with $\tau_k = \left(k + \frac{1}{2}\right)\frac{\pi}{\theta_2 - \theta_1}$ *for* $k = 0, 1, \ldots$.
Using a development in Fourier series, we see, as in Example 1, that every harmonic function u on Ω satisfying (7.52) can be written

$$\sum_{k=0}^{\infty} (\alpha_k \tau^k + \beta_k \tau^{-k})\cos\tau_k(\theta_2' - \theta)\,.$$

(2) *Case of a complete annulus.* The functions v defined on the complete unit circle are identified with periodic functions of period 2π on $\mathbb{R}$. The differential equation (7.50) admits a periodic solution of period 2π, not identically zero iff $\lambda = -k^2$ with $k \in \mathbb{Z}$; the solutions are then of the form $w(\theta) = \alpha' e^{ik\theta} + \beta' e^{-ik\theta}$. Hence from (7.48) and (7.49), *the functions with separable variables, harmonic on the annulus* $\{re^{i\theta}, r_1 < r < r_2, \theta \in \mathbb{R}\}$ *are of the form*

$$(7.55)\qquad \begin{cases} \alpha + \beta \operatorname{Log} r \quad \text{(radial function)} \\ \text{or } (\alpha r^k + \beta r^{-k})(\alpha' e^{ik\theta} + \beta' e^{-ik\theta}), \quad k = 1, 2, \ldots. \end{cases}$$

In other words, putting aside the radial functions $\alpha + \beta \operatorname{Log} r$ the functions with variables separable, which are harmonic on the annulus $\{z \in \mathbb{C}; r_1 < |z| < r_2\}$ are of the form

$$\alpha z^k + \alpha' \bar{z}^k + \beta z^{-k} + \beta' \bar{z}^{-k}, \qquad k = 1, 2, \ldots.$$

There it is a question of functions with complex values ($\alpha, \alpha', \beta, \beta' \in \mathbb{C}$); the functions with real values are of the form

$$r^k(\alpha\cos k\theta + \beta\sin k\theta) + r^{-k}(\alpha'\cos k\theta + \beta'\sin k\theta)$$

which can also be written in the classical manner

$$\alpha r^k \cos k(\theta - \varphi) + \alpha' r^{-k}\cos k(\theta - \varphi')$$

with $\alpha, \alpha', \varphi, \varphi' \in \mathbb{R}$.
Every function harmonic on the annulus $\{z \in \mathbb{C}; r_1 < |z| < r_2\}$ can be written in the form

$$\beta_0 \operatorname{Log}|z| + \sum_{k=-\infty}^{+\infty} (\alpha_k z^k + \alpha_k' \bar{z}^k)$$

or again in the case of real-valued functions

$$\alpha_0 + \beta_0 \operatorname{Log} r + \sum_{k=1}^{\infty} (\alpha_k r^k \cos k(\theta - \varphi_k) + \alpha'_k r^{-k} \cos k(\theta - \varphi'_k)) .$$

This can be seen either as a development in Fourier series, or as a Laurent series of a holomorphic function on an annulus in the complex plane. □

We propose now to extend to the case of the *arbitrary dimension n* the results we have just obtained for harmonic functions on an annulus in $\mathbb{R}^2$. We first prove:

Lemma 4. *Every polynomial homogeneous of degree k in n variables* $x_1, \ldots, x_n$ *can be written in the form*

$$(7.56) \qquad \sum_{0 \leq p \leq k/2} (x_1^2 + \cdots + x_n^2)^p h_{k-2p}(x_1, \ldots, x_n)$$

where h_{k-2p} *is a homogeneous harmonic polynomial of degree* $k - 2p$ (p an integer).

Proof. This comes down to showing the every homogeneous polynomial f of degree k can be written in the form

$$(7.57) \quad f(x_1, \ldots, x_n) = h(x_1, \ldots, x_n) + (x_1^2 + \ldots + x_n^2) g(x_1, \ldots, x_n)$$

where h is a homogeneous harmonic polynomial of degree k and g a homogeneous polynomial of degree $k-2$. Using Euler's identity

$$f(x_1, \ldots, x_n) = \frac{1}{k} \sum_{i=1}^{n} x_i \frac{\partial f}{\partial x_i} (x_1, \ldots, x_n) ,$$

and reasoning by recurrence on k, we are led to proving (7.57) for a polynomial $f(x_1, \ldots, x_n) = x_i h_i(x_1, \ldots, x_n)$ where h_i is a homogeneous harmonic polynomial of degree $k - 1$.

$$h(x_1, \ldots, x_n) = x_i h_i(x_1, \ldots, x_n) - \lambda(x_1^2 + \cdots + x_n^2) \frac{\partial h_i}{\partial x_i} (x_1, \ldots, x_n)$$

we have, since h_i and $\dfrac{\partial h_i}{\partial x_i}$ are harmonic,

$$\Delta h = 2 \frac{\partial h_i}{\partial x_i} - \lambda \left[2n \frac{\partial h_i}{\partial x_i} + 4 \sum_{j=1}^{n} x_j \frac{\partial^2 h_i}{\partial x_i \partial x_j} \right] .$$

But since $\partial h / \partial x_i$ is homogeneous of degree $k - 2$, we have, from Euler's identity

$$\sum_{j=1}^{n} x_j \frac{\partial^2 h_i}{\partial x_i \partial x_j} = (k-2) \frac{\partial h_i}{\partial x_i} ,$$

therefore

$$\Delta h = (2 - \lambda(2n + 4(k-2))) \frac{\partial h_i}{\partial x_i} ;$$

choosing $\lambda = 1/(n + 2(k-2))$, we see that h is harmonic. □

A homogeneous function is a function with separable variables (in the parametrization in polar coordinates); from (7.48) and (7.49), *a homogeneous harmonic polynomial h of degree k* can therefore be written

$$h(x) = r^k Y(\sigma)$$

with
$$\Delta_\sigma Y = (\tau_0^2 - (k - \tau_0)^2) Y \quad \text{on} \quad \Sigma$$

namely

(7.58)
$$\Delta_\sigma Y = - k(k + n - 2) Y \quad \text{on} \quad \Sigma$$

Let us now prove

Proposition 6

(1) *For all k, $(k = 0, 1, \ldots)$, the set $\mathscr{Y}_k^n$ of the functions Y solutions of (7.58) where Σ is the unit sphere in $\mathbb{R}^n$ is a finite-dimensional vector space; we have*

$$\dim \mathscr{Y}_0^n = 1 , \qquad \dim \mathscr{Y}_1^n = n$$

and

in the case $n = 2, \dim \mathscr{Y}_k^2 = 2$ *for all* $k \geqslant 1$

in the case $n = 3, \dim \mathscr{Y}_k^3 = 2k + 1$ *for all* $k \geqslant 0$

in the case $n = 4, \dim \mathscr{Y}_k^4 = (k + 1)^2$ *for all* $k \geqslant 0$.

In general

(7.59)
$$\dim \mathscr{Y}_k^n = \sum_{l=0}^{k} \dim \mathscr{Y}_l^{n-1}$$

(2) *The spaces $\mathscr{Y}_k = \mathscr{Y}_k^n$ are two-by-two orthogonal in $L^2(\Sigma)$, and $L^2(\Sigma)$ is the Hilbert sum of the subspaces $\mathscr{Y}_k$*[181a].

(3) *Every harmonic function on a ball $B(0, r_0)$ (resp. an annulus $B(0, r_2)\backslash\bar{B}(0, r_1)$) can be written in a unique manner in the form*

$$\sum_{k=0}^{\infty} r^k Y_k \quad (\sigma) \quad \left(\text{resp.} \sum_{-\infty}^{+\infty} r^k Y_k(\sigma) + \delta_{n,2} \beta_0 \operatorname{Log} r\right)$$

with $Y_k \in \mathscr{Y}_k$ (resp. $Y_k \in \mathscr{Y}_k$ if $k \geqslant 0$, $Y_k = 0$ if $2 - n < k < 0$ and $Y_k \in \mathscr{Y}_{2-n-k}$ if $k \leqslant 2 - n$), the series being uniformly convergent on every compact set of $B(0, r_0)$ (resp. $B(0, r_2)\backslash B(0, r_1)$).

Proof. Identifying $Y \in \mathscr{Y}_k^n$ with the function $h(x) = r^k Y(\sigma)$, we see from (7.49), that $\mathscr{Y}_k^n$ is identified with the space of all homogeneous functions (of degree k) harmonic on $\mathbb{R}^n$; we know that every tempered distribution, harmonic on $\mathbb{R}^n$, is a harmonic polynomial (see Proposition 3 of § 2); hence $\mathscr{Y}_k$ *is identified with the space of homogeneous harmonic polynomials of degree k.*

Now let $Y_k \in \mathscr{Y}_k$, $Y_l \in \mathscr{Y}_l$ with $k \neq l$.

[181a] Every element v of $L^2(\Sigma)$ can be written $v = \sum_{k=0}^{\infty} \operatorname{Proj}_{\mathscr{Y}_k} v$ where $\operatorname{Proj}_{\mathscr{Y}_k}$ is the orthogonal projection on the sub-space $\mathscr{Y}_k$.

By homogeneity (Y_k is identified with the harmonic polynomial $r^k Y_k(\sigma)$) we have, from Euler's identity

$$k\,Y_k(\sigma) = \sum_{i=1}^{n} \sigma_i \frac{\partial Y_k}{\partial x_i}(\sigma) = \frac{\partial Y_k}{\partial n}(\sigma)$$

where $\partial/\partial n$ is differentiation external to $B(0, 1)$. Since Y_k and Y_l are harmonic, we have, from Green's formula

$$\begin{aligned} k\int_\Sigma Y_k(\sigma)\,Y_l(\sigma)\mathrm{d}\sigma &= \int_\Sigma \frac{\partial Y_k}{\partial n}(\sigma)\,Y_l(\sigma)\mathrm{d}\sigma \\ &= \int_\Sigma Y_k(\sigma)\frac{\partial Y_l}{\partial n}(\sigma)\mathrm{d}\sigma = l\int_\Sigma Y_k(\sigma)\,Y_l(\sigma)\mathrm{d}\sigma\,. \end{aligned}$$

Therefore $\int_\Sigma Y_k(\sigma)\,Y_l(\sigma)\mathrm{d}\sigma = 0$.

From Lemma 4, for every homogeneous polynomial $r^k v(\sigma)$ we have: $v \in \sum_{0 \leqslant p \leqslant k/2} \mathscr{Y}_{k-2p}$. From the Stone–Weierstrass theorem $\sum_k \mathscr{Y}_k$ is dense in $\mathscr{C}^0(\Sigma)$ and hence also in $L^2(\Sigma)$; hence $L^2(\Sigma)$ is clearly the Hilbert sum of the sub-spaces $\mathscr{Y}_k$. Also the space $\mathscr{P}_k^n$ of homogeneous polynomials of degree k can be written as the direct sum of the $\mathscr{Y}_{k-2p}^n$ for $0 \leqslant p \leqslant \frac{1}{2}k$ with the result that

$$\dim \mathscr{P}_k^n = \sum_{0 \leqslant p \leqslant k/2} \dim \mathscr{Y}_{k-2p}^n\,.$$

Since

$$\dim \mathscr{P}_k^n = \sum_{l=0}^{k} \dim \mathscr{P}_l^{n-1}\,,$$

we deduce (7.59) and the values $\dim \mathscr{Y}_k^n$ for $n = 2, 3, 4$ beginning with $\dim \mathscr{Y}_k^1 = 0$ for $k \geqslant 2$.

Given a function $u \in L^2_{\mathrm{loc}}(\{r_1 < |x| < r_2\})$, it can be decomposed in the unique manner

$$u(x) = \sum_{k=0}^{\infty} u_k(x) \tag{7.60}$$

where for all $k \in \mathbb{N}$ and $r_1 < r < r_2$, the function $u_k(r\sigma)$ is the orthogonal projection of $u(r\sigma)$ on $\mathscr{Y}_k$; if $(Y_k^i)_{i=1,\dots,I_k}$ is an orthonormal base of $\mathscr{Y}_k$, $I_k = \dim \mathscr{Y}_k$,

$$u_k(x) = \sum_{i=1}^{I_k} c_k^i(r)\,Y_k^i(\sigma)\,, \qquad r = |x|\,, \qquad \sigma = \frac{x}{r} \tag{7.61}$$

with

$$c_k^i(r) = \int_\Sigma u(r\sigma)\,Y_k^i(\sigma)\mathrm{d}\sigma\,. \tag{7.62}$$

Supposing u to be harmonic, we shall have

$$\left(\frac{d^2}{dr^2} + \frac{n-1}{r}\frac{d}{dr}\right)c_k^i(r) = \int_\Sigma -\frac{1}{r^2}\Delta_\sigma u(r\sigma)\,Y_k^i(\sigma)\,d\sigma$$

$$= -\frac{1}{r^2}\int_\Sigma u(r\sigma)\Delta_\sigma Y_k^i(\sigma)\,d\sigma = \frac{k(k+n-2)}{r^2}\int_\Sigma u(r\sigma)\,Y_k^i(\sigma)\,d\sigma$$

$$= \frac{(k-\tau_0)^2 - \tau_0^2}{r^2}\,c_k^i(r)$$[182]

where $\tau_0 = 1 - (n/2)$; going back to the proof of (7.48) and (7.49), we deduce

$$c_k^i(r) = \alpha_k^i r^k + \beta_k^i r^{-(k+n-2)}$$

unless $n = 2$, $k = 0$ ($\dim \mathscr{Y}_0 = 1$) in which case $c_0 = \alpha_0 + \beta_0 \log r$. Hence

$$u_k(x) = r^k \sum_{i=1}^{I_k} \alpha_k^i\, Y_k^i(\sigma) + r^{-(n-2+k)} \sum_{i=1}^{I_k} \beta_k^i\, Y_k^i(\sigma)$$

unless $n = 2$, $k = 0$ when $u_0(x) = \alpha_0 + \beta_0 \operatorname{Log} r$.
If u is harmonic on $B(0, r_0)$, then the functions $c_k^i(r)$ given by (7.62) on $]0, r_0[$, are bounded as $r \to 0$ and hence

$$c_k^i(r) = \alpha_k^i r^k \quad \text{and} \quad u_k(x) \equiv r^k \sum \alpha_k^i\, Y_k^i(\sigma)\,.$$

Finally the uniform convergence follows from Harnack's theorems (see §2.3) since the functions u_k are harmonic. □

In the case $n = 2$, the sub-spaces $\mathscr{Y}_k^2$ are of dimension 2 for all $k \geqslant 1$. As we have seen in Example 3, in the complex plane

$$\mathscr{Y}_k^2 = \{\alpha e^{ik\theta} + \beta e^{-ik\theta};\, \alpha, \beta \in \mathbb{C}\}$$

while in real terms

$$\mathscr{Y}_k^2 = \{\alpha \cos k\theta + \beta \sin k\theta;\, \alpha, \beta \in \mathbb{R}\}$$

which we can rewrite as:

$$\mathscr{Y}_k^2 = \{\alpha \cos k(\theta - \varphi)\,;\quad \alpha \in \mathbb{R}\,,\quad \varphi \in \mathbb{R}\ (\text{modulo } 2\pi)\}\,.$$

In other words the functions of $\mathscr{Y}_k^2$ are to within a multiplicative constant the *circular functions* $Y(\theta) = \cos k(\theta - \varphi)$.
In a general way we pose the

---

[182] We have used the formula of the Laplacian in polar coordinates (see (1.27))

$$(\Delta u)(r\sigma) = \left(\frac{\partial^2}{\partial r^2} + \frac{n-1}{r}\frac{\partial}{\partial r} + \frac{1}{r^2}\Delta_\sigma\right)u(r\sigma)$$

and the fact deduced from it that Δ_σ is a symmetric operator

$$\int_\Sigma v_1(\sigma)\Delta_\sigma v_2(\sigma)\,d\sigma = \int_\Sigma v_2(\sigma)\Delta_\sigma v_1(\sigma)\,d\sigma\,.$$

Definition 1. The elements of the space $\mathscr{Y}_k^n$ are called *the spherical harmonic functions of order k in* $\mathbb{R}^n$.

Let us determine the spherical harmonics in the space $\mathbb{R}^3 = \{x, y, z\}$. The unit sphere Σ of $\mathbb{R}^3$ is represented by

$$\left\{\begin{aligned} x &= \sin\theta\cos\varphi \\ y &= \sin\theta\sin\varphi \\ z &= \cos\theta \end{aligned}\right. \tag{7.63}$$

where $\theta \in [0, \pi]$ is the angle of $\sigma(x, y, z)$ with the polar axis Oz and φ the polar angle of (x, y) in the plane $\mathbb{R}^2$ defined modulo 2π (and even undefined at the poles $\theta = 0$ and $\theta = \pi$. We prove

Proposition 7. *The space $\mathscr{Y}_k^3$ of spherical harmonic functions of order k in the space $\mathbb{R}^3$ admits for orthonormal base the functions $(Y_k^m)_{-k \leqslant m \leqslant k}$ defined in the parametric representation (7.63) by*

$$Y_k^m(\sigma) = P_k^{|m|}(\cos\theta)e^{im\varphi} \tag{7.64}$$

where for $0 \leqslant m \leqslant k$, $P_k^m(z)$ is the solution of

$$(1 - z^2)\frac{d^2P}{dz^2} - 2z\frac{dP}{dz} + \left(k(k+1) - \frac{m^2}{1 - z^2}\right)P = 0\,, \tag{7.65}$$

satisfying the normalization

$$\int_{-1}^{+1} |P_k^m(z)|^2\,dz = \frac{1}{2\pi}\,. \tag{7.66}$$

Such a function is given by

$$P_k^m(z) = \alpha_k^m \sum_{0 \leqslant l \leqslant \frac{k-m}{2}} \frac{(-1)^l(1 - z^2)^{l + \frac{m}{2}} z^{k-m-2l}}{4^l\, l!\,(l + m)!\,(k - m - 2l)!} \tag{7.67}$$

where the constant α_k^m is chosen so that (7.66) is satisfied.

Proof. We note first of all that the functions Y_k^m defined by (7.64) where P_k^m satisfies (7.66) form an orthonormal system on $L^2(\Sigma)$. We have in effect

$$\begin{aligned}\int_\Sigma Y_k^{m_1}(\sigma)\overline{Y_k^{m_2}(\sigma)}\,d\sigma &= \int_0^{2\pi} e^{i(m_1 - m_2)\varphi}\,d\varphi \int_0^\pi P_k^{|m_1|}(\cos\theta)P_k^{|m_2|}(\cos\theta)\sin\theta\,d\theta \\ &= 2\pi\,\delta_{m_1, m_2}\int_{-1}^{+1} P_k^{|m_1|}(z)P_k^{|m_2|}(z)\,dz\,.\end{aligned}$$

Now we know that, expressed in the parametric representation[183] (7.63) the

[183] See (1.28) of §1.

operator Δ_σ becomes

$$\Delta_\sigma = \frac{1}{\sin\theta}\frac{\partial}{\partial\theta}\left(\sin\theta\frac{\partial}{\partial\theta}\right) + \frac{1}{\sin^2\theta}\frac{\partial^2}{\partial\varphi^2}.$$

Hence a function. $Y(\sigma) = P(\cos\theta)e^{im\varphi}$ is in $\mathscr{Y}_k^3$ iff

$$\Delta_\sigma Y = \left(P''(\cos\theta)\sin^2\theta - 2P'(\cos\theta)\cos\theta - \frac{m^2}{\sin^2\theta}P(\cos\theta)\right)e^{im\varphi}$$

$$= -k(k+1)P(\cos\theta)e^{im\varphi}$$

namely iff P is a solution of (7.65).

To determine the explicit formula (7.67), we use a method other than verification. We know that the spherical harmonic functions of order k are the trace on Σ of the homogeneous harmonic polynomials of order k. A homogeneous polynomial of degree k can always be written

$$h(x, y, z) = \sum_{\substack{p,q\in\mathbb{N}\\ p+q\leqslant k}} \alpha_{p,q}(x+iy)^p(x-iy)^q z^{k-p-q}. \tag{7.68}$$

We have

$$\Delta h(x,y,z) = \sum_{p+q\leqslant k}\alpha_{p,q}[4pq(x+iy)^{p-1}(x-iy)^{q-1}z^{k-p-q}$$
$$+ (k-p-q)(k-p-q-1)(x+iy)^p(x-iy)^q z^{k-p-q-2}].$$

Hence h given by (7.68) is harmonic iff

$$4pq\alpha_{p,q} + (k+2-p-q)(k+1-p-q)\alpha_{p-1,q-1} = 0. \tag{7.69}$$

Putting, for $-k\leqslant m\leqslant k$ and $0\leqslant l\leqslant \dfrac{k-|m|}{2}$,

$$\alpha_l^m = \begin{cases}\alpha_{l+m,l} & m\geqslant 0\\ \alpha_{l,l-m} & m\leqslant 0\end{cases}$$

we have $\alpha_{p,q} = \alpha^{p-q}_{\min(p,q)}$ and (7.69) can be written

$$4l(l+|m|)\alpha_l^m + (k+2-|m|-2l)(k+1-|m|-2l)\alpha_{l-1}^m = 0$$

that is

$$\alpha_l^m = (-1)^l\alpha_0^m\prod_{j=1}^{l}\frac{(k+2-|m|-2j)(k+1-|m|-2j)}{4j(j+|m|)}$$
$$= (-1)^l\alpha_0^m(k-|m|)!\,|m|!\,c_l^m$$

with

$$\frac{1}{c_l^m} = 4^l l!(l+|m|)!(k-|m|-2l)! \tag{7.70}$$

To sum up, passing to the trace on Σ specified by (7.63) we find that the functions of

$\mathscr{Y}_k^3$ are of the form

$$\sum_{-k \leqslant m \leqslant k} \alpha_m e^{im\varphi} \sum_{0 \leqslant l \leqslant \frac{k-|m|}{2}} (-1)^l c_l^m (\sin\theta)^{2l+|m|} (\cos\theta)^{k-|m|-2l}$$

from which (7.67) follows. □

We note that $P_k^k(z) = \alpha(1 - z^2)^{k/2}$ with

$$\frac{1}{2\pi\alpha^2} = \int_{-1}^{+1} (1 - z^2)^k \mathrm{d}z = \frac{2^{2k+1}(k!)^2}{(2k+1)!}\,[184] .$$

Therefore

$$(7.71) \qquad Y_k^k(\sigma) = \frac{\sqrt{(2k+1)!\,\pi^{-1}}}{2^{k+1}k!} (\sin\theta)^k e^{ik\varphi} .$$

Thus $P_k^{k-1}(z) = \alpha(1 - z^2)^{\frac{k-1}{2}} z$ with

$$\frac{1}{2\pi\alpha^2} = \int_{-1}^{+1} (1 - z^2)^{k-1} z^2 \mathrm{d}z = \frac{1}{2k}\int_{-1}^{+1} (1 - z^2)^k \mathrm{d}z = \frac{2^{2k}(k!)^2}{k(2k+1)!} .$$

Hence

$$(7.72) \qquad Y_k^{k-1}(\sigma) = \frac{\sqrt{(2k+1)!\,2k\pi^{-1}}}{2^{k+1}k!} (\sin\theta)^{\frac{k-1}{2}} \cos\theta\, e^{i(k-1)\varphi} .$$

Propositions 6 and 7 show that *every function $u(x, y, z)$ harmonic on $B(0, r_0)$ can be written in a unique manner*.

$$(7.73) \qquad u(x, y, z) = \sum_{k=0}^{\infty} \sum_{-k \leqslant m \leqslant k} c_k^m r^k P_k^m(\cos\theta) e^{im\varphi}$$

with

(7.74)

$$c_k^m = r^{-k} \int_0^{2\pi} e^{-im\varphi} \mathrm{d}\varphi \int_0^{\pi} P_k^m(\cos\theta)\, u(r\sin\theta\cos\varphi, r\sin\theta\sin\varphi, r\cos\theta) \sin\theta\, \mathrm{d}\theta .$$

We note that although r appears in the expression giving c_k^m, this is in fact independent of r; in the case where $k = m = 0$, this independence is the formula of the mean

$$u(0, 0, 0) = \frac{c_0^0}{2\sqrt{\pi}} = \frac{1}{4\pi}\int_0^{2\pi} \mathrm{d}\varphi \int_0^{\pi} u(r\sin\theta\cos\varphi, r\sin\theta\sin\varphi, r\cos\theta) \sin\theta\, \mathrm{d}\theta .$$

In this case, *we can consider (7.74) as generalizing the formula of the mean.*

---

[184] $I_k = \int_0^1 (1 - z^2)^k \mathrm{d}z = \frac{1}{2}\int_{-1}^1 (1 - z^2)^k \mathrm{d}z$ satisfies the recurrence relation $\left(1 + \frac{1}{2k}\right) I_k = I_{k-1}$; namely, with $I_0 = 1$, $I_k = \prod_{l=1}^{k} \frac{2l}{2l+1}$.

We have considered an orthonormal base of $\mathscr{Y}_k^3$ in the complex domain; *in the real domain, an orthonormal base of $\mathscr{Y}_k^3$ is formed of the $(2k + 1)$ functions*

$$P_k^0(\cos\theta)\,, \quad \sqrt{2}P_k^m(\cos\theta)\cos m\varphi, \quad \sqrt{2}P_k^m(\cos\theta)\sin m\varphi, \quad m = 1, \ldots, k\,.$$

We note that *every real harmonic function on $B(0, r_0)$ can be written*

$$(7.75) \qquad u(x, y, z) = \sum_{k=0}^{\infty} \sum_{m=0}^{k} a_k^m r^k P_k^m(\cos\theta)\cos m(\varphi - \varphi_k^m)\,,$$

where $a_k^0 = c_k^0$ (real) and for $m = 1, \ldots, k$

$$a_k^m = 2|c_k^m|\,, \qquad \varphi_k^m = -\frac{\arg c_k^m}{m}\,,$$

c_k^m being given by (7.74). It is enough to use (7.73) noting that $c_k^{-m} = \overline{c_k^m}$. A particular case concerns *the harmonic solutions u on $B(0, r_0)$ with cylindrical symmetry*, that is to say invariant under rotations about the polar axis Oz. We then will have $c_k^m = 0$ for all $m \neq 0$. Hence *every harmonic function u with cylindrical symmetry can be written*

$$u(x, y, z) = \sum_{k=0}^{\infty} c_k^0 (x^2 + y^2 + z^2)^{k/2} P_k^0(z)$$

where $P_k^0(z)$ is given from (7.67) by

$$P_k^0(z) = \alpha_k \sum_{0 \leqslant l \leqslant k/2} \frac{(-1)^l (1 - z^2)^l z^{k-2l}}{4^l (l!)^2 (k - 2l)!}\,.$$

To within a multiplicative constant, P_k^0 is the Legendre polynomial of degree k, it is also the *Gegenbauer polynomial $C_k^{\frac{1}{2}}$*. More generally the Gegenbauer polynomials

$$(7.76) \qquad C_k^p(t) = \sum_{0 \leqslant l \leqslant k/2} \frac{(-1)^l (2t)^{k-2l} \Gamma(p + k - l)}{k!(k - 2l)!\Gamma(p)}$$

solutions of the differential equation

$$(1 - t^2)\frac{\mathrm{d}^2 P}{\mathrm{d}t^2} - (2p + 1)t\frac{\mathrm{d}P}{\mathrm{d}t} + k(k + 2p)P = 0$$

allow us to express the spherical harmonic functions of $\mathbb{R}^n$ with cylindrical symmetry: we can show (see Vilenkin [1]) that *the spherical harmonic functions of $\mathbb{R}^n = \{(x', x_n); x' \in \mathbb{R}^{n-1}, x_n \in \mathbb{R}\}$ of order k with cylindrical symmetry* (that is to say invariant under rotations about the axis Ox_n) *are proportional to $C_k^{\frac{1}{2}n-1}(x_n)$*. Finally we note that to within a multiplicative constant $P_k^m(z)$ is the polynomial $(1 - z^2)^{m/2}\dfrac{\mathrm{d}^m}{\mathrm{d}z^m} P_k^0(z)$. This is easily verified by change of the unknown function $P = (1 - z^2)^{\frac{1}{2}m} u$ in (7.65) and use of Leibniz's formula in (7.65) taken for $m = 0$. The functions $P_k^m(z)$ are also called the associated Legendre functions of the first kind (see Robin [1]).

We shall not develop further the theory of spherical harmonic functions: we note that it is *intimately linked with the theory of the group of orthogonal transformations of* $\mathbb{R}^n$; in effect it is clear that $\mathscr{Y}_k^n$ *is invariant under the group of orthogonal transformations*, for if T is a linear transformation of $\mathbb{R}^n$ and h a homogeneous polynomial of $\mathbb{R}^n$, the image $Th = h \circ T$ is a homogeneous polynomial of the same degree; now if h is harmonic and T orthogonal, then Th is harmonic (see Proposition 2 of § 1), from which the above statement follows. We refer the reader to Vilenkin [1] for a systematic study of symmetric harmonic functions based on the theory of groups.

4. Dirichlet's Method

Let us first consider Dirichlet's problem $P(\Omega, \varphi)$ where Ω is a set of $\mathbb{R}^n$ with boundary Γ and $\varphi \in \mathscr{C}^0(\Gamma)$. We suppose first of all that Ω is regular and bounded and that the solution u of $P(\Omega, \varphi)$[185] is in $\mathscr{C}_n^1(\bar{\Omega})$. Then we know[186] that $|\operatorname{grad} u|^2 \in L^1(\Omega)$ and that for every function $v \in \mathscr{C}^1(\Omega) \cap \mathscr{C}^0(\bar{\Omega})$ with $|\operatorname{grad} v|^2 \in L^1(\Omega)$, we have

$$\int_\Omega \operatorname{grad} u \,.\, \operatorname{grad} v \, \mathrm{d}x = \int_\Gamma v \frac{\partial u}{\partial n} \, \mathrm{d}\gamma \,.$$

In particular

$$(7.77) \quad \begin{cases} \displaystyle\int_\Omega \operatorname{grad} u \,.\, \operatorname{grad} v \, \mathrm{d}x = 0 \quad \text{for every function } v \in \mathscr{C}^1(\Omega) \cap \mathscr{C}^0(\bar{\Omega}) \\ \text{satisfying } |\operatorname{grad} v|^2 \in L^1(\Omega) \quad \text{and} \quad v = 0 \quad \text{on} \quad \Gamma \,. \end{cases}$$

This property (7.77) can be expressed in an equivalent manner by

$$(7.78) \quad \begin{cases} \displaystyle\int_\Omega |\operatorname{grad} u|^2 \mathrm{d}x \leqslant \int_\Omega |\operatorname{grad} w|^2 \mathrm{d}x \quad \text{for every function } w \in \mathscr{C}^1(\Omega) \cap \mathscr{C}^0(\bar{\Omega}) \\ \text{satisfying} \quad |\operatorname{grad} w|^2 \in L^1(\Omega) \quad \text{and} \quad w = \varphi \quad \text{on} \quad \Gamma \,. \end{cases}$$

In effect applying (7.77) with $v = w - u$ and using

$$|\operatorname{grad} u|^2 = |\operatorname{grad} w|^2 - 2 \operatorname{grad} u \,.\, \operatorname{grad} v - |\operatorname{grad} v|^2 \,,$$

we obtain (7.78). Conversely applying (7.78) with $w = u + \lambda v$, ($\lambda \in \mathbb{R}$), we obtain

$$\int_\Omega |\operatorname{grad} u|^2 \mathrm{d}x \leqslant \int_\Omega (|\operatorname{grad} u|^2 + 2\lambda \operatorname{grad} u \,.\, \operatorname{grad} v + \lambda^2 |\operatorname{grad} v|^2) \mathrm{d}x$$

from which we deduce that if $\lambda > 0$ (resp. $\lambda < 0$)

$$\int_\Omega \operatorname{grad} u \,.\, \operatorname{grad} v \, \mathrm{d}x \geqslant (\text{resp.} \leqslant) - \frac{\lambda}{2} \int_\Omega |\operatorname{grad} v|^2 \mathrm{d}x \,.$$

Letting $\lambda \to 0$, we obtain (7.77).

---

[185] Which exists from Theorem 1 of §4.

[186] See Proposition 5 of § 1.

In other words, *if Ω is a regular bounded open set of $\mathbb{R}^n$ and if the solution of $P(\Omega, \varphi)$ is in $\mathscr{C}^1_n(\bar{\Omega})$, then that solution minimises the energy functional*

$$\mathscr{I}(u) = \int_\Omega |\operatorname{grad} u|^2 \,\mathrm{d}x$$

on $\{u \in \mathscr{C}^1(\Omega) \cap \mathscr{C}^0(\bar{\Omega}); |\operatorname{grad} u|^2 \in L^1(\Omega)$ *and* $u = \varphi$ *on* $\Gamma\}$.
We note that we have already applied this result in the study of the capacity of a regular compact set K contained in a regular bounded open set (see Proposition 7 of §5). Dirichlet's method consists of using this property of the minimization of the energy functional to define a solution of $P(\Omega, \varphi)$. Before developing the method in a general framework, let us recall that given Ω, *an arbitrary open set of* $\mathbb{R}^n$, *we denote*

$$B^1(\Omega) = \left\{ u \in L^2_{\mathrm{loc}}(\Omega);\ \textit{the derivatives}\ \frac{\partial u}{\partial x_i} \in L^2(\Omega)\ \text{(in the sense of distributions)} \right\},$$

and by $B^1_0(\Omega)$ the closure of $\mathscr{D}(\Omega)$ *in* $B^1(\Omega)$ *which is given naturally the topology defined by the convergence*

$$u_k \to u \quad \text{in} \quad B^1(\Omega) \quad \text{iff} \quad \begin{cases} u_k \to u & \text{in} \quad L^2_{\mathrm{loc}}(\Omega) \\ \operatorname{grad} u_k \to \operatorname{grad} u & \text{in} \quad L^2(\Omega)\,. \end{cases}$$

In general, $B^1(\Omega)$ contains strictly the Sobolev space[187]

$$H^1(\Omega) = \left\{ u \in L^2(\Omega);\ \text{the derivatives}\ \frac{\partial u}{\partial x_i} \in L^2(\Omega) \right\} = B^1(\Omega) \cap L^2(\Omega)\,.$$

If Ω is bounded (or, more generally, of finite measure), then $B^1_0(\Omega)$ coincides with $H^1_0(\Omega)$, the closure of $\mathscr{D}(\Omega)$ in $H^1(\Omega)$[188]. *The energy functional* $\mathscr{I}(u) = \int_\Omega |\operatorname{grad} u|^2 \,\mathrm{d}x$ *is defined for all* $u \in B^1(\Omega)$.
We now prove

Proposition 8. *Let Ω be an open set of $\mathbb{R}^n$ and $\varphi \in B^1(\Omega)$. We suppose only $\Omega \neq \mathbb{R}$ if $n = 1$ and $\bar{\Omega} \neq \mathbb{R}^2$ if $n = 2$.*
(1) *There exists a unique function u minimizing the energy functional on $\{\varphi + v;\ v \in B^1_0(\Omega)\}$;*
(2) *u is the unique solution of the problem:*

$$(7.79) \qquad \begin{cases} u \text{ is harmonic on } \Omega\,, \\ u - \varphi \in B^1_0(\Omega)\,. \end{cases}$$

[187] Even in the case Ω bounded: for example $\Omega = \bigcup_{k=1}^{\infty} \left]\frac{1}{k+1}, \frac{1}{k}\right[$ an open set of $\mathbb{R}$

$$u = \sum k \chi_{]\frac{1}{k+1}, \frac{1}{k}[} \in B^1(\Omega) \backslash H^1(\Omega)\,.$$

[188] With the norm $\|u\|_{H^1(\Omega)} = \left(\int_\Omega \left(|u|^2 + |\operatorname{grad} u|^2 \right) \mathrm{d}x \right)^{\frac{1}{2}}$.

Proof. Consider first u minimizing $\mathscr{I}$ on $\varphi + B_0^1(\Omega)$.
Since $u + B_0^1(\Omega) = \varphi + B_0^1(\Omega)$, we have

$$\mathscr{I}(u) \leqslant \mathscr{I}(u + \lambda v) \quad \text{for all} \quad v \in B_0^1(\Omega) \quad \text{and} \quad \lambda \in \mathbb{R}$$

from which we deduce, as above, that

$$\int_\Omega \operatorname{grad} u . \operatorname{grad} v \, dx = 0 \quad \text{for all} \quad v \in B_0^1(\Omega) . \tag{7.80}$$

Conversely if $u \in \varphi + B_0^1(\Omega)$ satisfies (7.80), then it minimizes $\mathscr{I}$ on $\varphi + B_0^1(\Omega)$.
We note also that by definition of $B_0^1(\Omega)$, for $u \in B^1(\Omega)$, (7.80) is equivalent to

$$\langle \Delta u, \zeta \rangle = - \int \operatorname{grad} u . \operatorname{grad} \zeta \, dx = 0 \quad \text{for all} \quad \zeta \in \mathscr{D}(\Omega) ,$$

that is to say u is harmonic on Ω.
To prove existence, we consider a sequence (v_k) of $B_0^1(\Omega)$ such that

$$m = \inf_{\varphi + B_0^1(\Omega)} \mathscr{I} = \lim_{k \to \infty} \mathscr{I}(\varphi + v_k) ;$$

we have

$$\begin{aligned} \frac{1}{2}\mathscr{I}(v_k - v_l) &= \mathscr{I}(\varphi + v_k) + \mathscr{I}(\varphi + v_l) - 2\mathscr{I}\left(\varphi + \frac{v_k + v_l}{2}\right) \\ &\leqslant \mathscr{I}(\varphi + v_k) + \mathscr{I}(\varphi + v_l) - 2m , \end{aligned}$$

so that $\lim_{k,l \to \infty} \mathscr{I}(v_k - v_l) = 0$. To conclude the proof of existence, it must be shown that (v_k) (or a sub-sequence extracted from it) converges in $L^2_{\text{loc}}(\Omega)$; then (v_k) will converge in $B^1(\Omega)$ and, considering its limit v, the function $\varphi + v$ will minimize $\mathscr{I}$ on $\varphi + B_0^1(\Omega)$.
In the case of Ω bounded (or of finite measure) we can make use of Poincaré's inequality (see Chap. IV, §7 – Vol. 2, p. 126; see also Adams [1], Gilbarg–Trudinger [1]) to obtain

$$\| v_k - v_l \|_{L^2}^2 \leqslant C(\Omega) \mathscr{I}(u_k - v_l)$$

which proves that (v_k) converges in $L^2(\Omega)$.

In the case $n \geqslant 3$, Ω being arbitrary, we can use Sobolev's inequality (see Gilbarg–Trudinger [1], Adams [1])

$$\| v_k - v_l \|_{L^{\frac{2n}{n-2}}}^2 \leqslant C \mathscr{I}(v_k - v_l)$$

which proves that (v_k) converges in $L^{2n/(n-2)}(\Omega)$.
In the case $n = 1$, if $\Omega \neq \mathbb{R}$ we have

$$|(v_k - v_l)(x)|^2 \leqslant \operatorname{dist}(x, \partial\Omega) \mathscr{I}(v_k - v_l)$$

and hence (v_k) converges uniformly on every bounded set of Ω.

In the case $n \neq 2$, if $\bar{\Omega} = \mathbb{R}^2$ we can always suppose $O \notin \bar{\Omega}$. For $v \in B_0^1(\Omega)$ let us consider

$$w(y) = v\left(\frac{y}{|y|^2}\right) \qquad \text{(image of } v \text{ under the inversion } I(0,1)) ;$$

using the fact that $I(0,1)'(y) = |y|^{-2} S R(0, 2\theta)$, where $S = \begin{pmatrix} 1 & 0 \\ 0 & -1 \end{pmatrix}$ is the symmetry with respect to the Oy_1-axis, $R(0, 2\theta)$ is the rotation about O, with angle 2θ (see §2.25), we have

$$|\operatorname{grad} w(y)| = \frac{1}{|y|^2}\left|\operatorname{grad} v\left(\frac{y}{|y|^2}\right)\right|$$

and

$$\int_\Omega |\operatorname{grad} v(x)|^2 \, dx = \int_{\Omega'} |\operatorname{grad} w(y)|^2 \, dy \quad \text{with} \quad \Omega' = I(0,1)\Omega .$$

Since Ω' is bounded, using Poincare's inequality we obtain

$$\int_{\Omega'} |w(y)|^2 \, dy = \int_\Omega \frac{|v(x)|^2}{|x|^4} \, dx \leqslant C(\Omega') \mathscr{I}(v) ;$$

we deduce that the sequence $(v_k |x|^{-2})$ converges in $L^2(\Omega)$ and hence (v_k) converges in $L^2_{\text{loc}}(\Omega)$.
From the different cases considered, we see that, generally, for $v \in B_0^1(\Omega)$,

$$\mathscr{I}(v) = 0 \Rightarrow v = 0 ;$$

this proves the uniqueness of a function minimizing $\mathscr{I}$ on $\varphi + B_0^1(\Omega)$. □

Remark 5. *In the case $\Omega = \mathbb{R}$ or $\Omega = \mathbb{R}^2$, $B_0^1(\Omega) = B^1(\Omega)$.*
We show first that $B_0^1(\Omega)$ contains the constants: given $\zeta \in \mathscr{D}(\mathbb{R})$ (resp. $\mathscr{D}(\mathbb{R}^2)$), the sequence of functions $\zeta_k(x) = \zeta(x/k)$ converges in effect in $B^1(\mathbb{R})$ (resp. weakly in $B^1(\mathbb{R}^2)$) to the constant function $v(x) \equiv \zeta(0)$; because clearly

$$\zeta_k \to v \quad \text{in} \quad L^2_{\text{loc}}(\mathbb{R}) \qquad (\text{resp. } L^2_{\text{loc}}(\mathbb{R}^2))$$

and

$$\mathscr{I}(\zeta_k) = \frac{1}{k^2}\int \zeta'\left(\frac{x}{k}\right)^2 dx = \frac{1}{k}\int \zeta'(x)^2 \, dx \to 0$$

$$\left(\text{resp. } \mathscr{I}(\zeta_k) = \frac{1}{k^2}\iint |\operatorname{grad} \zeta(x/k)|^2 \, dx = \iint |\operatorname{grad} \zeta(x)|^2 \, dx \text{ is bounded}\right).$$

We consider now $\varphi \in B^1(\Omega)$ and consider a sequence (v_k) of $B_0^1(\Omega)$ such that $\mathscr{I}(\varphi + v_k) \to \inf_{\varphi + B_0^1(\Omega)} \mathscr{I}$; we know that $\lim_{k,l \to \infty} \mathscr{I}(v_k - v_l) = 0$ (see the proof of Proposition 8).
In the case $\Omega = \mathbb{R}$, we have $\varphi \in \mathscr{C}^0(\mathbb{R})$ and $v_k \in \mathscr{C}^0(\mathbb{R})$; replacing v_k by $v_k - v_k(0) - \varphi(0) \in B_0^1(\mathbb{R})$, we can suppose that $\varphi(0) + v_k(0) = 0$; we then have by the Schwarz inequality

$$|v_k(x) - v_l(x)|^2 \leqslant |x| \mathscr{I}(v_k - v_l)$$

and hence (v_k) converges uniformly on every compact set of $\mathbb{R}$; the limit $v \in B_0^1(\mathbb{R})$ and $u = v + \varphi$ minimizes $\mathscr{I}$ on $\varphi + B_0^1(\mathbb{R})$ and satisfies $u(0) = 0$; as we have seen u satisfies (7.79), and in particular, is harmonic on $\mathbb{R}$; necessarily $u \equiv 0$ and therefore $\varphi \in B_0^1(\mathbb{R})$.

In the case $\Omega = \mathbb{R}^2$, we can suppose $\int_{B(0,1)} v_k(x)\,dx = -\int_{B(0,1)} \varphi(x)\,dx$; from Poincaré's inequality[189], since

$$\int_{B(0,1)} (v_k(x) - v_l(x))\,dx = 0\,,$$

we have for all $R > 1$

$$\int_{B(0,R)} |v_k(x) - v_l(x)|^2\,dx \leqslant C(R)\mathscr{I}(v_k - v_l)$$

and hence (v_k) converges in $L^2_{\text{loc}}(\mathbb{R}^2)$; the limit $v \in B_0^1(\mathbb{R}^2)$ and $u = v + \varphi$ is harmonic on $\mathbb{R}^2$; since $\int_{B(0,1)} u(x)\,dx = 0$, necessarily $u \equiv 0$[190]. □

Proposition 8 and this Remark 5 lead us to frame:

Definition 2. Given an open set Ω in $\mathbb{R}^n$ and $\varphi \in B^1(\Omega)$ we call a *generalized solution* (in the sense of Dirichlet) of $P(\Omega, \varphi)$ every solution u of (7.79), or what is equivalent every function u minimizing the energy functional

$$\mathscr{I}(u) = \int_\Omega |\operatorname{grad} u|^2\,dx \quad \text{on} \quad \varphi + B_0^1(\Omega)\,.$$

Unlike the classical Dirichlet problem we assume that we are given φ on Ω and not on the boundary Γ of Ω. We note however that given $\varphi_1, \varphi_2 \in B^1(\Omega)$ with $\varphi_1 - \varphi_2 \in B_0^1(\Omega)$, *$u$ is a generalized solution of $P(\Omega, \varphi_1)$ iff it is a solution of $P(\Omega, \varphi_2)$; in other words the notion of a generalized solution (in the Dirichlet sense)*

[189] Poincare's inequality can be stated: given Ω a connected regular bounded open set and $w \in L^2(\Omega)$, $w \neq 0$, there exists $C(\Omega, w)$ such that, for $u \in H^1(\Omega)\ (= B^1(\Omega))$

$$\|u - \bar{u}\|_{L^2} \leqslant C(\Omega, w)\,\|\operatorname{grad} u\|_{L^2}$$

where

$$\bar{u} = \int uw\,dx \Big/ \int w\,dx$$

(see Lemma 5 below).

[190] $B^1(\mathbb{R}^n) \cap \mathscr{H}(\mathbb{R}^n)$ is reduced to constant function; in effect the derivatives

$$v_i = \frac{\partial u}{\partial x_i} \in L^2(\mathbb{R}^n) \cap \mathscr{H}(\mathbb{R}^n)$$

are since their Fourier transforms $\hat{v}_i \in L^2(\mathbb{R}^n)$ and satisfy

$$|\xi|^2 \hat{v}_i(\xi) = 0 \quad \text{a.e.} \quad \xi \in \mathbb{R}^n\,.$$

of $P(\Omega, \varphi)$ depends only on "trace" of φ on Γ and in the case of Ω unbounded, at infinity. Here, the notion of trace is *the equivalence class in $B'^1(\Omega)$ modulo $B_0^1(\Omega)$*; we note that Proposition 8 can be stated: the map $u \mapsto$ *class of u in $B^1(\Omega)$ modulo $B_0^1(\Omega)$ is a bijection of $\mathscr{H}(\Omega) \cap B^1(\Omega) = \{u$ harmonic on Ω with energy $\mathscr{I}(u) = \int_\Omega |\operatorname{grad} u|^2 \, dx$ finite$\}$ onto the quotient space $B'(\Omega)/B_0^1(\Omega)$.*

When Ω is a regular, bounded open set,[191] $B^1(\Omega) = H^1(\Omega)$ and the trace theorem (see Chap. IV) states that we can extend the map:

$$u \in H^1(\Omega) \cap \mathscr{C}^0(\bar{\Omega}) \to \varphi = \text{trace of } u \text{ on } \Gamma \text{ (in the classical sense)}$$

to a continuous linear map, called the trace map, of $H^1(\Omega)$ onto $H^{\frac{1}{2}}(\Gamma)$[192]; on the other hand we prove that

$$u \in H_0^1(\Omega) = B_0^1(\Omega) \Leftrightarrow u \in H^1(\Omega) \text{ and the trace of } u \text{ on } \Gamma = 0 \text{ (in } H^{\frac{1}{2}}(\Gamma)) .$$

It follows that, *in the case Ω regular, bounded, $B^1(\Omega)/B_0^1 \simeq H^{\frac{1}{2}}(\Gamma)$ and more precisely that the map $u \to$ trace of u on Γ is a bijection of $\mathscr{H}(\Omega) \cap B^1(\Omega)$ onto $H^{\frac{1}{2}}(\Gamma)$.* We shall not develop this aspect further, but content ourselves with making the link with the classical theory presented in §4.
First, we prove:

Proposition 9. *Let Ω be an open set of $\mathbb{R}^n$, with only $\Omega \neq \mathbb{R}$ if $n = 1$, and $\bar{\Omega} \neq \mathbb{R}^2$ if $n = 2$, $\varphi \in B^1(\Omega)$ and u the generalized solution (in the Dirichlet sense) of $P(\Omega, \varphi)$.*
(1) *If $\varphi \geqslant 0$ a.e. on Ω, then $u \geqslant 0$ on Ω; more generally*

$$\inf_\Omega \operatorname{ess} \varphi \leqslant u(x) \leqslant \sup_\Omega \operatorname{ess} \varphi \quad \text{for all} \quad x \in \Omega .$$

(2) *Let (Ω_k) be an increasing sequence of open sets with union Ω, and u_k the generalized solution of $P(\Omega_k, \varphi|_{\Omega_k})$ for all $k = 1, \ldots$ Then $u_k \to u$ in $B^1(\Omega)$.*

Proof. Considering $m \geqslant \sup_\Omega \operatorname{ess}$, the function $u \wedge m = \min(u, m)$, as a result of Lemmas 2 and 3 of §5, satisfies

$$(7.81) \quad u \wedge m \in B^1(\Omega), \mathscr{I}(u \wedge m) \leqslant \mathscr{I}(u) \quad \text{and} \quad u \wedge m - \varphi \in B_0^1(\Omega) ,$$

from which $u \wedge m = u$ implies $u \leqslant m$. For the last point of (7.81), note that if a sequence (v_k) of $\mathscr{D}(\Omega)$ is such that

$$v_k \to u - \varphi \quad \text{in} \quad B^1(\Omega) ,$$

then $[(v_k + \varphi) \wedge m] - \varphi$ is a function of $B^1(\Omega)$ with support contained in $\operatorname{supp} v_k$,

---

191 It is sufficient to suppose Ω to have a boundary satisfying a Lipschitz condition.
192 We can characterize $H^{\frac{1}{2}}(\Gamma)$ in many ways; for example

$$H^{\frac{1}{2}}(\Gamma) = \left\{ \varphi \in L^2(\Gamma); \iint_{\Gamma \times \Gamma} \frac{|\varphi(x) - \varphi(y)|^2}{|x - y|} d\gamma(x) d\gamma(y) < \infty \right\} .$$

hence a function of $B_0^1(\Omega)$ and

$$[(v_k + \varphi) \wedge m] - \varphi \to u \wedge m - \varphi \quad \text{in} \quad B^1(\Omega)\,.$$

This proves point (1).
For point (2), by definition of u_k

$$\int_{\Omega_k} |\operatorname{grad} u_k|^2 \,\mathrm{d}x \leqslant \int_{\Omega_k} |\operatorname{grad} \varphi|^2 \,\mathrm{d}x \leqslant \int_{\Omega} |\operatorname{grad} \varphi|^2 \,\mathrm{d}x$$

and there exists $v_k \in \mathscr{D}(\Omega_k)$ such that

$$\|\operatorname{grad}(u_k - \varphi - v_k)\|_{L^2(\Omega_k)} \leqslant \frac{1}{k}$$

and for every compact set K of Ω_k there exists C_K such that

$$\|u_k - \varphi - v_k\|_{L^2(K)} \leqslant \frac{C_K}{k}\,.$$

To prove that $u_k \to u$ in $B^1(\Omega)$, it is sufficient to prove that $v_k \to u - \varphi$ in $B^1(\Omega)$. We have

$$\limsup \|\operatorname{grad}(v_k + \varphi)\|_{L^2(\Omega)} \leqslant \|\operatorname{grad} u\|_{L^2(\Omega)}\,. \tag{7.82}$$

In effect since $u - \varphi \in B_0^1(\Omega)$, there exists a sequence $w_k \in \mathscr{D}(\Omega)$ such that

$$w_k \to u - \varphi \quad \text{in} \quad B^1(\Omega)\,;$$

we can always suppose that $w_k \in \mathscr{D}(\Omega_k)$ and hence that

$$\|\operatorname{grad} u_k\|_{L^2(\Omega_k)} \leqslant \|\operatorname{grad}(\varphi + w_k)\|_{L^2(\Omega_k)} \leqslant \|\operatorname{grad}(\varphi + w_k)\|_{L^2(\Omega)}\,.$$

We then have

$$\begin{aligned}
\|\operatorname{grad}(v_k + \varphi)\|_{L^2(\Omega)} &\leqslant \|\operatorname{grad} \varphi\|_{L^2(\Omega \backslash \Omega_k)} + \|\operatorname{grad}(v_k + \varphi)\|_{L^2(\Omega_k)} \\
&\leqslant \|\operatorname{grad} \varphi\|_{L^2(\Omega \backslash \Omega_k)} + \|\operatorname{grad} u_k\|_{L^2(\Omega_k)} + \frac{1}{k} \\
&\leqslant \|\operatorname{grad} \varphi\|_{L^2(\Omega \backslash \Omega_k)} + \|\operatorname{grad}(\varphi + w_k))\|_{L^2(\Omega)} + \frac{1}{k}\,,
\end{aligned}$$

from which we obtain (7.82) in the limit.
It follows from (7.82) that grad v_k is bounded in $L^2(\Omega)$; from the proof of Proposition 8 we see that v_k is bounded in $L^2_{\text{loc}}(\Omega)$. Given a sub-sequence (k_l) such that

$$v_{k_l} \rightharpoonup v \quad \text{in} \quad B^1(\Omega)\,,$$

we have that the function $\varphi + v$ is the generalized solution of $P(\Omega, \varphi)$: in effect $v \in B_0^1(\Omega)$ and for $\zeta \in \mathscr{D}(\Omega)$,

$$\int_{\Omega} \operatorname{grad}(\varphi + v) . \operatorname{grad} \zeta \,\mathrm{d}x = \lim_{l \to \infty} \int_{\Omega} \operatorname{grad}(\varphi + v_{k_l}) . \operatorname{grad} \zeta \,\mathrm{d}x = 0$$

since for l sufficiently large, $\operatorname{supp}\zeta \subset \Omega_{k_l}$ and hence

$$\int_{\Omega_{k_l}} \operatorname{grad} u_{k_l} \cdot \operatorname{grad}\zeta \, dx = 0 .$$

In other words, $v_k \rightharpoonup u - \varphi$ in $B^1(\Omega)$; using (7.82) again, we have

$$\operatorname{grad}(v_k + \varphi) \to \operatorname{grad} u \quad \text{in} \quad L^2(\Omega) ,$$

namely

$$\operatorname{grad} v_k \to \operatorname{grad}(u - \varphi) \quad \text{in} \quad L^2(\Omega) ;$$

we deduce that

$$v_k \to u - \varphi \quad \text{in} \quad B^1(\Omega)$$

which concludes the proof of point (2). □

We can then prove

Proposition 10. *Let Ω be a bounded open set of $\mathbb{R}^n$ and $\varphi \in B^1(\Omega) \cap \mathscr{C}^0(\bar{\Omega})$. Then the generalized solution (in the Dirichlet sense) of $P(\Omega, \varphi)$ coincides with the generalized solution of $P(\Omega, \varphi|_\Gamma)$ (in the sense of Definition 3 of §4). In particular the generalized solution (in the Dirichlet sense) u of $P(\Omega, \varphi)$ satisfies*

$$\lim_{x \to z} u(x) = \varphi(z)$$

for every regular point z of the boundary (in the sense of Definition 4 of §4).

We observe that Dirichlet's method allows us to obtain a solution of $P(\Omega, \varphi)$ for all $\varphi \in B^1(\Omega)$, whereas the classical method would give a solution for all $\varphi \in \mathscr{C}^0(\Gamma)$. Proposition 10 shows that the two notions coincide for $\varphi \in B^1(\Omega) \cap \mathscr{C}^0(\bar{\Omega})$. *When Ω is regular*, we can identify $B^1(\Omega)/B^1_0(\Omega)$ with $H^{\frac{1}{2}}(\Gamma)$, and *the Dirichlet method and the classical method give the same solution for $\varphi \in \mathscr{C}^0(\Gamma) \cap H^{\frac{1}{2}}(\Gamma)$.*

Proof. Let us consider u the generalized solution (in the Dirichlet sense) of $P(\Omega, \varphi)$. From the definition of the generalized solution of $P(\Omega, \varphi|_\Gamma)$, we have to prove that given $w \in \mathscr{C}^0(\Omega)$ superharmonic satisfying

$$\liminf_{x \to z} w(x) \geqslant \varphi(z) \quad \text{for all} \quad z \in \Gamma$$

we have

$$w(x) \geqslant u(x) \quad \text{for all} \quad x \in \Omega .$$

Let us consider for $k = 1, \ldots$

$$U_k = \left\{ x \in \Omega ; \mathrm{d}(x, \Gamma) < \frac{1}{k}, w(x) > \varphi(x) - \frac{1}{k} \right\} .$$

The set $\Omega \backslash U_k$ is a compact set of Ω; we can always find a regular open set $\Omega_k \supset \Omega \backslash U_k$ with boundary Γ_k of class $\mathscr{C}^\infty$ contained in U_k.

We consider now $\rho \in \mathscr{D}(\mathbb{R}^n)$ with $\rho \geqslant 0$, $\operatorname{supp}\rho \subset \{|x| < 1\}$ and $\displaystyle\int \rho(x)\,dx = 1$.

Given $\varepsilon > 0$, we put

$$\varphi_\varepsilon(x) = \int \varphi(x + \varepsilon t)\rho(t)\,\mathrm{d}t\,.$$

The function φ_ε is $\mathscr{C}^\infty$ on $\{x \in \Omega;\ \mathrm{dist}(x, \Gamma) > \varepsilon\}$.
For $0 < \varepsilon < \mathrm{dist}(\Omega_k, \Gamma)$, let us consider the classical solution $u_{k,\varepsilon}$ of $P(\Omega_k, \varphi_\varepsilon|_{\Gamma_k})$; we know that $u_{k,\varepsilon}$ is $\mathscr{C}^\infty(\bar{\Omega})$ (see Theorem 1 of §6).
The function $w(x) + \dfrac{1}{k} + \max\limits_{\Gamma_k} |\varphi - \varphi_\varepsilon|$ is superharmonic on Ω_k, continuous on $\bar{\Omega}_k$ and greater than or equal to $\varphi_\varepsilon = u_{k,\varepsilon}$ on Γ_k. Hence

$$w(x) + \frac{1}{k} + \max_{\Gamma_k} |\varphi - \varphi_\varepsilon| \geqslant u_{k,\varepsilon}(x) \quad \text{for all} \quad x \in \Omega_k\,.$$

Since Ω_k and $u_{k,\varepsilon}$ are regular, $u_{k,\varepsilon}$ is the generalized solution (in the Dirichlet sense) of $P(\Omega_k, \varphi_\varepsilon|_{\Omega_k})$; denoting by u_k the generalized solution (in the Dirichlet sense) of $P(\Omega_k, \varphi|_{\Omega_k})$, from point (1) of Proposition 9,

$$\sup_{\Omega_k} |u_k - u_{k,\varepsilon}| \leqslant \sup_{\Omega_k} |\varphi - \varphi_\varepsilon|\,.$$

Since $\varphi_\varepsilon \to \varphi$ uniformly on every compact set of Ω,

$$w(x) + \frac{1}{k} \geqslant u_k(x) \quad \text{for all} \quad x \in \Omega_k\,;$$

using point (2) of Proposition 9, in the limit as $k \to \infty$,

$$w(x) \geqslant u(x) \quad \text{for all} \quad x \in \Omega\,. \qquad \square$$

Proposition 10 extends to the case of an unbounded open set Ω; with reservation $\Omega \neq \mathbb{R}$ if $n = 1$ and $\bar{\Omega} \neq \mathbb{R}^2$ if $n = 2$, for every $\varphi \in B^1(\Omega) \cap \mathscr{C}^0(\bar{\Omega})$ satisfying the null condition at infinity (see Definition 9 of §4), the generalized solution (in the sense of Dirichlet) of $P(\Omega, \varphi)$ coincides with the generalized solution satisfying the null condition at infinity of $P(\Omega, \varphi|_\Gamma)$ (see Definition 10 of §4). We leave to the reader the proof of this assertion. Dirichlet's method is linked with the notion of capacity and of the capacity operator. First, let us take *two open sets* Ω_1, Ω_2 *of* $\mathbb{R}^n$ *such that the set* $(\Omega_1 \backslash \Omega_2) \cup (\Omega_2 \backslash \Omega_1)$ *is a polar part* (for the norm $H^1(\mathbb{R}^n)$, see Definition 4 of §5). *Then* $B^1(\Omega_1) = B^1(\Omega_2)$; this equation clearly has a sense, since every polar part being negligible (see Remark 8 of §5), a function defined a.e. on Ω_1 is defined a.e. on Ω_2; we prove that the equation $B^1(\Omega_1) = B^1(\Omega_2)$ is true. For that, it suffices, given $\varphi \in B^1(\Omega_1 \cap \Omega_2)$ and $\varphi_i = \partial\varphi/\partial x_i$ in $\mathscr{D}'(\Omega_1 \cap \Omega_2)$, to show that $\varphi \in L^2_{\mathrm{loc}}(\Omega_1)$ (resp. $L^2_{\mathrm{loc}}(\Omega_2)$) and

$$\int \varphi_i \zeta\,\mathrm{d}x = -\int \varphi \frac{\partial \zeta}{\partial x_i}\mathrm{d}x \quad \text{for all} \quad \zeta \in \mathscr{D}(\Omega_1) \quad (\text{resp. } \mathscr{D}(\Omega_2))\,; \tag{7.83}$$

we note, in effect, that $\varphi_i \in L^2(\Omega_1) = L^2(\Omega_2) = L^2(\Omega_1 \cap \Omega_2)$. We suppose first that $\varphi \in L^2_{\mathrm{loc}}(\Omega_1)$; given $\zeta \in \mathscr{D}(\Omega_1)$, $K = \mathrm{supp}\,\zeta \backslash \Omega_2$ is a polar compact set and hence of zero capacity in every open set containing it (see Definition 4 of §5). From the characterization of the capacity of a compact set in an open set (see Proposition 8 of §5), given $\varepsilon > 0$, there exists $\rho_\varepsilon \in \mathscr{D}(\Omega_1)$ such that $0 \leqslant \rho_\varepsilon \leqslant 1$, $\rho_\varepsilon = 1$ in the neighbourhood of K,

$$\operatorname{supp}\rho_\varepsilon \subset K_\varepsilon = \{x; d(x, K) \leqslant \varepsilon\} \quad \text{and} \quad \|\operatorname{grad}\rho_\varepsilon\|_{L^2} \leqslant \varepsilon\,;$$

the function $\zeta_\varepsilon = (1 - \rho_\varepsilon)\zeta$ belongs to $\mathscr{D}(\Omega_1 \cap \Omega_2)$ and

$$\begin{aligned}\|\zeta - \zeta_\varepsilon\|_{H^1} &\leqslant \|\rho_\varepsilon\zeta\|_{L^2} + \|\rho_\varepsilon \operatorname{grad}\zeta\|_{L^2} + \|\zeta\operatorname{grad}\rho_\varepsilon\|_{L^2}\\ &\leqslant [(\|\zeta\|_{L^\infty} + \|\operatorname{grad}\zeta\|_{L^\infty})|K|^{\frac{1}{2}+\frac{1}{n}} + \|\zeta\|_{L^\infty}]\varepsilon^{193}.\end{aligned}$$

In other words for $\zeta \in \mathscr{D}(\Omega_1)$ and $\delta > 0$, there exists $\zeta_\delta \in \mathscr{D}(\Omega_1 \cap \Omega_2)$ such that $\|\zeta - \zeta_\delta\|_{H_1} \leqslant \delta$; we hence prove easily (7.83), since by definition

$$\int \varphi_i \zeta_\delta \, dx = -\int \varphi \frac{\partial \zeta_\delta}{\partial x_i}\, dx\,.$$

It remains to prove that $\varphi \in L^2_{\text{loc}}(\Omega_1)$; for that, let us consider for $k > 0$, the truncation $\varphi_k = \max(\min(\varphi, -k), -k)$. We have $\varphi_k \in L^2_{\text{loc}}(\Omega_1)$ and $\varphi_k \in B^1(\Omega_1 \cap \Omega_2)$ with $\dfrac{\partial\varphi_k}{\partial x_i} = \dfrac{\partial\varphi}{\partial x_i}\chi_{\{|\varphi| < k\}}$ (see Lemma 2 of §5). From the preceding $\varphi_k \in B^1(\Omega_1)$ and $\dfrac{\partial\varphi_k}{\partial x_i} = \dfrac{\partial\varphi}{\partial x_i}\chi_{\{|\varphi| < k\}}$ in $\mathscr{D}'(\Omega_1)$, from which in particular we have $\|\operatorname{grad}\varphi_k\|_{L^2(\Omega_1)} \leqslant \|\operatorname{grad}\varphi\|_{L^2(\Omega_1 \cap \Omega_2)}$; it follows that (φ_k) is bounded in $L^2_{\text{loc}}(\Omega_1)$[194] thus that $\varphi \in L^2_{\text{loc}}(\Omega_1)$.
The above argument proves similarly, which besides is simpler, that $B^1_0(\Omega_1) = B^1_0(\Omega_2)$[195]. By definition of the generalized solution in the Dirichlet sense, we can state:

Proposition 11. *Let Ω_1 and Ω_2 be two open sets of $\mathbb{R}^n$ such that $(\Omega_1\backslash\Omega_2) \cup (\Omega_2\backslash\Omega_1)$ is a polar part. For all $\varphi \in B^1(\Omega_1) = B^1(\Omega_2)$, the generalized solution (in the Dirichlet sense) of $P(\Omega_1, \varphi)$ and $P(\Omega_2, \varphi)$ coincide on $\Omega_1 \cap \Omega_2$ and are thus the extensions by continuity to Ω_1 and Ω_2 respectively of the generalized solution (in the Dirichlet sense) of $P(\Omega_1 \cap \Omega_2, \varphi)$.*

Corollary 1. *Let Ω be an open set of $\mathbb{R}^n$, F a closed polar part of Ω and $u \in \mathscr{H}(\Omega\backslash F)$. Then u extends to a harmonic function on Ω iff $\int_{K\backslash F} |\operatorname{grad} u|^2\, dx < \infty$ for every compact set K of Ω.*

Corollary 2. *Let Ω be a bounded open set of $\mathbb{R}^n$ with boundary Γ, F a closed polar part of Ω and $\varphi \in \mathscr{C}^0(\Gamma \cup F)$. The generalized solution (in the sense of Definition 4*

---

[193] We have used $\|\rho_\varepsilon\|_{L^1} \leqslant |K_\varepsilon|^{\frac{1}{2}+\frac{1}{n}} \|\operatorname{grad}\rho_\varepsilon\|_{L^2}$ which is deduced from Sobolev's inequality.
[194] Otherwise, there would exist $k_l \to \infty$ and a ball $\bar{B}(x_0, r_0) \subset \Omega_1$ such that $t_l = \|\varphi_{k_l}\|_{L^2(B(x_0, r_0))} \to \infty$ and we should thus have for a sequence of functions $v_l = \varphi_{k_l}/t_l$ the contradiction $\operatorname{grad} v_l \to 0$ in $L^2(B(x_0, r_0))$, $v_l \to 0$ a.e. on $B(x_0, r_0)$ and $\|v_l\|_{L^2(B(x_0, r_0))} = 1$ (using the compact ness of $H^1(B(x_0, r_0))$ in $L^2(B(x_0, r_0))$.
[195] Also that $H^1(\Omega_1) = H^1(\Omega_2)$ and $H^1_0(\Omega_1) = H^1_0(\Omega_2)$.

of §4) of $P(\Omega \backslash F, \varphi)$ extends by continuity to Ω the generalized solution of $P(\Omega, \varphi|_\Gamma)$. In particular, the generalized solution of $P(\Omega \backslash F, \varphi)$ does not depend on the values of φ on Γ.

Proof of Corollary 1. The condition is obviously necessary; let us show that it is sufficient. The problem being local, we can always suppose that

$$\int_{\Omega \backslash F} |\operatorname{grad} u|^2 \, dx < \infty$$

and therefore $u \in B^1(\Omega \backslash F) = B^1(\Omega)$; it is therefore the generalized solution of $P(\Omega \backslash F, u)$ which from Proposition 11 extends by continuity the generalized solution of $P(\Omega, u)$ harmonic on Ω. ◻

Proof of Corollary 2. From Proposition 10, Corollary 2 reduces to Proposition 11 if there exists $\tilde{\varphi} \in B^1(\Omega) \cap \mathscr{C}^0(\bar{\Omega})$ such that $\tilde{\varphi} \equiv \varphi$ on $\Gamma \cup F$. Now from Weierstrass' theorem, there exists a sequence (φ_n) of polynomials converging to φ on $\Gamma \cup F$; the corollary is then deduced from the continuity of the generalized solution with respect to the data on the boundary (see Proposition 3 of §4). ◻

Let us now consider a Neumann problem, or more generally a mixed problem:

$$(7.84) \qquad \begin{cases} u \text{ harmonic on } \Omega \\ u = \varphi \quad \text{on} \quad \Gamma_0 \\ \dfrac{\partial u}{\partial n} = \psi \quad \text{on} \quad \Gamma_1 \end{cases}$$

where *Γ_0 is a closed part of the boundary Γ of Ω and $\Gamma_1 = \Gamma \backslash \Gamma_0$.*
If Ω is bounded and regular and $u \in \mathscr{C}^1_n(\bar{\Omega})$ is the solution of (7.84), we have from Green's formula

$$(7.85) \qquad \begin{cases} \displaystyle\int_\Omega \operatorname{grad} u \, . \operatorname{grad} v \, dx = \int_{\Gamma_1} \psi v \, d\gamma \\ \text{for every function } v \in B^1(\Omega) \cap \mathscr{C}^0(\bar{\Omega}) \quad \text{with} \quad v = 0 \quad \text{on} \quad \Gamma_0 \,. \end{cases}$$

Using the same reasoning as in the proof of Proposition 8, we see that (7.85) is equivalent to

$$(7.86) \qquad \begin{cases} u \text{ minimises the functional} \\ \displaystyle\frac{1}{2}\int_\Omega |\operatorname{grad} u|^2 \, dx - \int_{\Gamma_1} \psi u \, d\gamma \\ \text{on } \{u \in B^1(\Omega) \cap \mathscr{C}^0(\bar{\Omega});\, u = \varphi \quad \text{on} \quad \Gamma_0\} \,. \end{cases}$$

We give a formulation of this problem in the case of an arbitrary open set Ω. We denote by $V(\Omega, \Gamma_0)$ the closure in $B^1(\Omega)$ of $\{v \in B^1(\Omega);\, v \equiv 0$ *in the neighbourhood of Γ_0 in Ω*$\}$.

Heuristically, $V(\Omega, \Gamma_0)$ is the space of the $v \in B^1(\Omega)$ such that $v = 0$ on Γ_0; in the case where Ω is regular, $V(\Omega, \Gamma_0)$ is exactly the set $\{v \in B^1(\Omega)$; trace of v on Γ is zero on $\Gamma_0\}$; in the case with Ω bounded and $\Gamma_0 = \Gamma$, $V(\Omega, \Gamma) = B_0^1(\Omega)$. In the case Ω unbounded, we do not make a hypothesis at infinity on $v \in V(\Omega, \Gamma_0)$; we can say that, *in the case of Ω unbounded*, the space $V(\Omega, \Gamma_0)$ corresponds to a *Neumann condition at infinity*. When Ω is bounded and regular; for $\psi \in \mathscr{C}^0(\Gamma)$, the map

$$(7.87) \qquad u \in B^1(\Omega) \cap \mathscr{C}^0(\bar{\Omega}) \to \int_{\Gamma_1} \psi u \, d\gamma$$

extends to a continuous linear form on $B^1(\Omega)$: in effect as we have recalled above (see just after Definition 2), the map $u \in B^1(\Omega) \cap \mathscr{C}^0(\bar{\Omega}) \to u|_\Gamma$ extends to a continuous linear map of $B^1(\Omega)$ onto $H^{\frac{1}{2}}(\Gamma) \subset L^2(\Gamma)$[196].
In the general case we can hence replace the given ψ on Γ by a continuous linear form on $B^1(\Omega)$ and consider the following *generalization of the mixed problem* (7.84): given $\varphi \in B^1(\Omega)$ a ψ a continuous linear form on $B^1(\Omega)$,

$$(7.88) \qquad \begin{cases} u \text{ minimises the functional} \\ \dfrac{1}{2}\displaystyle\int_\Omega |\operatorname{grad} u|^2 \, dx - \langle \psi, u \rangle \\ \text{on } \{\varphi + v; v \in V(\Omega, \Gamma_0)\} . \end{cases}$$

We shall interpret this general problem later; first of all, we prove

Proposition 12. *Let Ω be an arbitrary open set of $\mathbb{R}^n$, Γ_0 a closed part of Γ the boundary of Ω, $\varphi \in B^1(\Omega)$ and ψ a continuous linear form on $B^1(\Omega)$. The problem* (7.88) *admits a solution iff*

$$(7.89) \qquad \begin{cases} \text{for every connected component } \Omega_0 \text{ of } \Omega , \\ \chi_{\Omega_0} \in V(\Omega, \Gamma_0) \Rightarrow \langle \psi, \chi_{\Omega_0} \rangle = 0 . \end{cases}$$

Then the problem (7.88) *admits a solution depending on as many arbitrary constants as there are connected components Ω_0 of Ω such that $\chi_{\Omega_0} \in V(\Omega, \Gamma_0)$.*

The condition (7.89) is the *generalized form of Gauss' theorem* (see §2.2). We shall make the property $\chi_{\Omega_0} \in V(\Omega_0, \Gamma_0)$ clear later.
For the proof of Proposition 12, we shall make use of

Lemma 5. *Let Ω_0 be a connected open set, ψ_0 a continuous linear form on $B^1(\Omega_0)$ with $\langle \psi_0, \chi_{\Omega_0} \rangle \neq 0$. Then, for every compact set K_0, there exists a constant C_{K_0} such that*

$$\left\| u - \frac{\langle \psi_0, u \rangle}{\langle \psi_0, \chi_{\Omega_0} \rangle} \right\|_{L^2(K_0)} \leqslant C_{K_0} \| \operatorname{grad} u \|_{L^2(\Omega_0)} .$$

---

[196] From the Sobolev inclusions, $H^{\frac{1}{2}}(\Gamma) \subset L^q(\Gamma)$ for all $q < \infty$ if $n = 2$ and all $q \leqslant \dfrac{2n-2}{n-2}$ if $n \geqslant 3$ (in particular $q = 4$ if $n = 3$); hence if $\psi \in L^p(\Gamma_i)$ with $p > 1$ if $n = 2$ and $p \geqslant 2\left(\dfrac{n-1}{n}\right)$ if $n \geqslant 3$, the map defined by (7.87) extends to a continuous linear form on $B^1(\Omega)$.

We shall make use in the proof of this lemma, which is a generalization of Poincaré's inequality, of the structure of the continuous linear forms on $B^1(\Omega)$; the following lemma will help us similarly to interpet the problem (7.88):

Lemma 6. *Let Ω be an arbitrary open set of $\mathbb{R}^n$. The continuous linear forms on $B^1(\Omega)$ are of the form*

$$\langle \psi, u\rangle = \int_\Omega \left(f_0 u + f_1 \frac{\partial u}{\partial x_1} + \cdots + f_n \frac{\partial u}{\partial x_n}\right) \mathrm{d}x$$

where $f_0, f_1, \ldots, f_n \in L^2(\Omega)$ and f_0 is with compact support.

Proof of Lemma 6. The map $u \in B^1(\Omega) \to \left(u, \dfrac{\partial u}{\partial x_1}, \ldots \dfrac{\partial u}{\partial x_n}\right)$ is an isomorphism of $B^1(\Omega)$ onto a sub-space of $L^2_{\mathrm{loc}}(\Omega) \times L^2(\Omega)^n$; from the Hahn–Banach theorem, a continuous bilinear form on $B^1(\Omega)$ is therefore of the form

$$\langle \psi, u\rangle = \langle L_0, u\rangle + \left\langle L_1, \frac{\partial u}{\partial x_1}\right\rangle + \cdots + \left\langle L_1, \frac{\partial u}{\partial x_n}\right\rangle$$

where L_0 is a continuous linear form on $L^2_{\mathrm{loc}}(\Omega)$ and $L_1, \ldots, L_n$ continuous linear forms on $L^2(\Omega)$. Hence the lemma, account being taken of the representation of continuous linear forms on $L^2(\Omega)$. □

Proof of Lemma 5. By contradiction, given K_0 a compact set of Ω_0, let us suppose that for all $k \in \mathbb{N}$, there exists $u_k \in B^1(\Omega_0)$ such that

$$\left\| u_k - \frac{\langle \psi_0, u_k\rangle}{\langle \psi_0, \chi_{\Omega_0}\rangle}\right\|_{L^2(K_0)} > k \|\operatorname{grad} u_k\|_{L^2(\Omega_0)}\,.$$

From Lemma 6, there exist $f_0, \ldots, f_n \in L^2(\Omega_0)$ with f_0 having compact support such that

$$\langle \psi_0, u\rangle = \int_{\Omega_0} \left(f_0 u + f_1 \frac{\partial u}{\partial x_1} + \cdots + f_n \frac{\partial u}{\partial x_n}\right) \mathrm{d}x\,.$$

Let us consider an open set Ω_1, connected, regular, bounded and such that

$$K_0 \cup \operatorname{supp} f_0 \subset \Omega_1 \subset \bar{\Omega}_1 \subset \Omega_0$$

and put

$$t_k = \left\| u_k - \frac{\langle \psi_0, u_k\rangle}{\langle \psi_0, \chi_{\Omega_0}\rangle}\right\|_{L^2(\Omega_1)} \qquad v_k = \frac{1}{t_k}\left(u_k - \frac{\langle \psi_0, u_k\rangle}{\langle \psi_0, \chi_{\Omega_0}\rangle}\right),$$

we have $\langle \psi_0, v_k\rangle = 0$, $\|v_k\|_{L^2(\Omega_1)} = 1$ and $\|\operatorname{grad} v_k\|_{L^2(\Omega_0)} < 1/k$.
From the compact injection of $H^1(\Omega_1)$ into $L^2(\Omega_1)$[197], there exists a sub-sequence (k_l) such that

$$v_{k_l} \to v_0 \quad \text{in} \quad L^2(\Omega_1)\,.$$

Since $\operatorname{grad} v_{k_l} \to 0$ in $L^2(\Omega_1)$, $\operatorname{grad} v_0 = 0$ and hence, Ω_1, being connected, $v_0 = c$,

[197] See Chap. IV; we recall that Ω_1 is a regular bounded set.

on Ω_1. Since

$$\| v_{k_l} \|_{L^2(\Omega_1)} = 1 , \quad \text{then} \quad |c|(\operatorname{mes} \Omega_1)^{\frac{1}{2}} = \| v_0 \|_{L^2(\Omega_1)} = 1$$

and, in particular, $c \neq 0$.
On the other hand, since $\operatorname{supp} f_0 \subset \Omega_1$,

$$\int_\Omega f_0 v_{k_l} \, dx \to c \int_\Omega f_0 \, dx ,$$

and since $\| \operatorname{grad} v_k \|_{L^2(\Omega)} \to 0$,

$$\int_\Omega \left(f_1 \frac{\partial v_k}{\partial x_1} + \cdots + f_n \frac{\partial v_k}{\partial x_n} \right) dx \to 0 .$$

Hence $\langle \psi_0, v_{k_l} \rangle = 0 \to c \int_\Omega f_0 dx = c \langle \psi_0, \chi_{\Omega_0} \rangle$, a contradiction since $c \neq 0$, $\langle \psi_0, \chi_{\Omega_0} \rangle \neq 0$. □

Proof of Proposition 12. We denote

$$\mathscr{I}_\psi(u) = \frac{1}{2} \int_\Omega |\operatorname{grad} u|^2 \, dx - \langle \psi, u \rangle .$$

Suppose first that (7.88) admits a solution and hence, *a fortiori*,

$$m = \inf_{\varphi + V(\Omega, \Gamma_0)} \mathscr{I}_\psi(u) > -\infty .$$

Given Ω_0 a connected component of Ω with $\chi_{\Omega_0} \in V(\Omega, \Gamma_0)$, we have, for all $\lambda \in \mathbb{R}$

$$m \leqslant \mathscr{I}_\psi(\varphi + \lambda \chi_{\Omega_0}) = \mathscr{I}_\psi(\varphi) - \lambda \langle \psi, \chi_{\Omega_0} \rangle ,$$

namely

$$\lambda \langle \psi, \chi_{\Omega_0} \rangle \leqslant \mathscr{I}_\psi(\varphi) - m , \qquad \forall \lambda \in \mathbb{R}$$

and hence $\langle \psi, \chi_{\Omega_0} \rangle = 0$.
Conversely, let us suppose that the condition (7.89) is satisfied. Let us denote by $(\Omega_i)_{i \in I}$ the family of connected components of Ω[198] and

$$I_0 = \{ i \in I ; \chi_{\Omega_i} \in V(\Omega, \Gamma_0) \} ;$$

from (7.89), for every family $(\lambda_i)_{i \in I_0}$ of real numbers

$$u \in \varphi + V(\Omega, \Gamma_0) \Leftrightarrow u \sum_{I_0} \lambda_i \chi_{\Omega_i} \in \varphi + V(\Omega, \Gamma_0)$$

and

$$\mathscr{I}_\psi\left(u + \sum_{I_0} \lambda_i \chi_{\Omega_i} \right) = \mathscr{I}_\psi(u)$$

and hence, in particular,

$$u \text{ is the solution of (7.88)} \Leftrightarrow u + \sum_{I_0} \lambda_i \chi_{\Omega_i} \text{ is the solution of (7.88)} .$$

---

[198] If Ω is regular, bounded, the family $(\Omega_i)_{i \in I}$ is finite; in the general case it is at most denumerable.

For all $i \in I_0$, we fix B_i a compact set of Ω_i of measure $|B_i| > 0$; applying Lemma 5 to the linear form

$$u \to \frac{1}{|B_i|}\int_{B_i} u\,dx\,, \quad \text{for every compact} \quad K_i \quad \text{of} \quad \Omega_i\,,$$

there exists C_{K_i} such that for all $u \in B^1(\Omega)$

$$\left\| u - \frac{1}{|B_i|}\int_{B_i} u\,dx \right\|_{L^2(K_i)} \leqslant C_{K_i} \|\operatorname{grad} u\|_{L^2(\Omega_i)}\,.$$

For $i \in I\backslash I_0$, since $\chi_{\Omega_i} \notin V(\Omega, \Gamma_0)$, we have, from the Hahn–Banach theorem, that there exists a continuous linear form ψ_i on $B^1(\Omega_i)$ such that

$$\langle \psi_i, v\chi_{\Omega_i} \rangle = 0 \quad \text{for all} \quad v \in V(\Omega, \Gamma_0)$$
$$\langle \psi_i, \chi_{\Omega_i} \rangle = 1\,.$$

Applying Lemma 5, we see that for every compact set K_i of Ω_i, there exists C_{K_i} such that for all $v \in V(\Omega, \Gamma_0)$

$$\|v\|_{L^2(K_i)} \leqslant C_{K_i} \|\operatorname{grad} v\|_{L^2(\Omega_i)}\,.$$

Regrouping the two results, we find that for every compact K of Ω, there exists C_K such that

$$\|v\|_{L^2(K)} \leqslant C_K \|\operatorname{grad} v\|_{L^2(\Omega)} \tag{7.90}$$

for all $v \in V(\Omega, \Gamma_0)$ with $\int_{B_i} v\,dx = 0$ for all $i \in I_0$. It is sufficient to take $C_K = \max\{C_{K_i}; i \in I, K_i = K \cap \Omega_i \neq \varnothing\}$ in noting that there is only a finite number of $i \in I$ such that $K \cap \Omega_i \neq \varnothing$.

After these preliminaries, we consider a sequence (v_k) in $V(\Omega, \Gamma_0)$ such that

$$\mathscr{I}_\psi(\varphi + v_k) \to m = \inf_{\varphi + V(\Omega, \Gamma_0)} \mathscr{I}_\psi(u) \quad \text{when} \quad k \to \infty\,.$$

We can always suppose $\int_{B_i} v_k\,dx = 0$ for all $i \in I_0$: it is sufficient to replace v_k by w_k where

$$w_k = v_k - \sum_{i \in I_0} \chi_{\Omega_i} |B_i|^{-1} \int_{B_i} v_k\,dx\,,$$

since

$$\mathscr{I}_\psi(\varphi + v_k) = \mathscr{I}_\psi(\varphi + w_k)\,.$$

We note first that $m > -\infty$. Using Lemma 6, with its notation

$$\begin{aligned}\mathscr{I}_\psi(u) &= \frac{1}{2}\int_\Omega |\operatorname{grad} u|^2\,dx - \int_\Omega \left(f_0 u + f_1 \frac{\partial u}{\partial x_1} + \cdots + f_n \frac{\partial u}{\partial x_n}\right) dx \\ &\geqslant \frac{1}{4}\int_\Omega |\operatorname{grad} u|^2\,dx - \int_\Omega (f_0 u + f_1^2 + \cdots + f_n^2)\,dx\,.\end{aligned}$$

On the other hand, let us suppose $\mathscr{I}_\psi(\varphi + v_k) \to -\infty$ when $k \to +\infty$; we can always suppose

$$\mathscr{I}_\psi(\varphi + v_k) + \int_\Omega (f_0 \varphi + f_1^2 + \cdots + f_n^2)\,\mathrm{d}x \leqslant 0\,,$$

with the result that

$$\frac{1}{4}\int_\Omega |\operatorname{grad}(\varphi + v_k)|^2\,\mathrm{d}x \leqslant \int_\Omega f_0 v_k\,\mathrm{d}x \leqslant \|f_0\|_{L^2(\Omega)} C_K \|\operatorname{grad} v_k\|_{L^2(\Omega)}$$

where we have applied (7.90) with $K = \operatorname{supp} f_0$. We deduce

$$\int_\Omega f_0 v_k\,\mathrm{d}x \leqslant C_K \|f_0\|_{L^2(\Omega)}\left(\|\operatorname{grad}\varphi\|_{L^2(\Omega)} + 2\left(\int_\Omega f_0 v_k\,\mathrm{d}x\right)^{\frac{1}{2}}\right),$$

namely

$$\int_\Omega f_0 v_k\,\mathrm{d}x \leqslant 2C_K \|f_0\|_{L^2(\Omega)}(\|\operatorname{grad}\varphi\|_{L^2(\Omega)} + 2C_K \|f_0\|_{L^2(\Omega)})\,,$$

and hence $\mathscr{I}_\psi(\varphi + v_k)$ is minorized – a contradiction.

We now show that (v_k) converges in $B^1(\Omega)$. We have

$$\frac{1}{4}\int_\Omega |\operatorname{grad}(v_k - v_l)|^2\,\mathrm{d}x \leqslant \mathscr{I}_\psi(\varphi + v_k) + \mathscr{I}_\psi(\varphi + v_l) - 2\mathscr{I}_\psi\left(\varphi + \frac{v_k + v_l}{2}\right)$$
$$\leqslant \mathscr{I}_\psi(\varphi + v_k) + \mathscr{I}_\psi(\varphi + v_l) - 2m\,.$$

Therefore $\|\operatorname{grad}(v_k - v_l)\|_{L^2(\Omega)} \to 0$ as $k, l \to \infty$; taking account of (7.90), the sequence (v_k) converges in $B^1(\Omega)$. Its limit v belongs to $V(\Omega, \Gamma_0)$ and by continuity of $\mathscr{I}_\psi$, $u = \varphi + v$ minimises $\mathscr{I}_\psi$ on $\varphi + V(\Omega, \Gamma_0)$.

To conclude the proof we have to prove uniqueness, to within arbitrary constants on the Ω_i for $i \in I_0$. Given

$$v_1, v_2 \in V(\Omega, \Gamma_0) \quad \text{such that} \quad \mathscr{I}_\psi(\varphi + v_1) = \mathscr{I}_\psi(\varphi + v_2) = m\,,$$

we have

$$\frac{1}{4}\int_\Omega |\operatorname{grad}(v_1 - v_2)|^2\,\mathrm{d}x \leqslant \mathscr{I}_\psi(\varphi + v_1) + \mathscr{I}_\psi(\varphi + v_2)$$
$$- 2\mathscr{I}_\psi\left(\varphi + \frac{v_1 + v_2}{2}\right) \leqslant 0\,,$$

so $v_1 - v_2$ is constant on each Ω_i, $i \in I$. But

$$(v_1 - v_2)\chi_{\Omega_i} \in V(\Omega, \Gamma_0) \quad \text{for all} \quad i \in I$$

and so $v_1 - v_2 = 0$ on Ω_i for all $i \in I \backslash I_0$; in other words

$$v_1 = v_2 + \sum_{i \in I_0} \lambda_i \chi_{\Omega_i}\,. \qquad \square$$

Let us conclude this section by explaining *the significance of the problem* (7.88). From Lemma 6,

$$\langle \psi, u \rangle = \int_{\Omega_i} \left(f_0 u + f_1 \frac{\partial u}{\partial x_1} + \cdots + f_n \frac{\partial u}{\partial x_n} \right) \mathrm{d}x$$

where $f_0, f_1, \ldots, f_n \in L^2(\Omega)$, f_0 with compact support in Ω. Note carefully that $f_0, f_1, \ldots, f_n$ are not unique.
As we have seen in Proposition 8 for the Dirichlet problem, u is a solution of (7.88) iff $u \in \varphi + V(\Omega, \Gamma_0)$ and for all $v \in V(\Omega, \Gamma_0)$

$$\int_\Omega \operatorname{grad} u . \operatorname{grad} v \, \mathrm{d}x = \langle \psi, v \rangle , \tag{7.91}$$

that is to say

$$\int_\Omega \operatorname{grad} u . \operatorname{grad} v \, \mathrm{d}x = \int_\Omega \left(f_0 v + f_1 \frac{\partial v}{\partial x_1} + \cdots + f_n \frac{\partial v}{\partial x_n} \right) \mathrm{d}x .$$

In particular, applying with $v \in \mathscr{D}(\Omega)$, we see that

$$\Delta u = \frac{\partial f_1}{\partial x_1} + \cdots + \frac{\partial f_n}{\partial x_n} - f_0 \quad \text{in} \quad \mathscr{D}'(\Omega) .$$

Going back, first of all to the case $\varphi = 0$ and

$$f_0 = \frac{\partial f_1}{\partial x_1} + \cdots + \frac{\partial f_n}{\partial x_n} \quad \text{in} \quad \mathscr{D}'(\Omega) .$$

It is clear that $V(\Omega, \Gamma)$ is contained in $V(\Omega, \Gamma_0)$; in particular if $V(\Omega, \Gamma_0)$ satisfies (7.89), it is the same for $V(\Omega, \Gamma)$ and hence there exists u_0 a solution of

$$u_0 \text{ minimises } \mathscr{I}_\psi \quad \text{on} \quad \varphi + V(\Omega, \Gamma) . \tag{7.92}$$

Putting $\bar{u} = u - u_0$, it is clear for example by using the characterization (7.91) that u is the solution of (7.88) iff

$$\bar{u} \text{ minimises } \mathscr{I}_{\bar{\psi}} \quad \text{on} \quad V(\Omega, \Gamma_0) \tag{7.93}$$

where $\bar{\psi}$ is defined by

$$\langle \bar{\psi}, v \rangle = \langle \psi, v \rangle - \int_\Omega \operatorname{grad} u_0 . \operatorname{grad} v \, \mathrm{d}x . \tag{7.94}$$

In other words, u the solution of (7.88) can be written as the sum of u_0 solution of (7.92) and $\bar{u}$ solution of (7.93) where $\bar{\psi}$ is given by (7.94).
In the case in which Ω is bounded, $V(\Omega, \Gamma)$ coincides with $B_0^1(\Omega)$: the solution u_0 of (7.92) is characterized by $u_0 \in \varphi + B_0^1(\Omega)$ and

$$\int_\Omega \operatorname{grad} u_0 . \operatorname{grad} v \, \mathrm{d}x = \int_\Omega \left(f_0 v + f_1 \frac{\partial v}{\partial x_1} + \cdots + f_n \frac{\partial v}{\partial x_n} \right) \mathrm{d}x$$

for all $v \in \mathscr{D}(\Omega)$. In other words, *in the case Ω bounded, u_0 is the solution of* (7.92) *iff*

$u_0 - \varphi \in B_0^1(\Omega)$ and

$$\Delta u = \frac{\partial f_1}{\partial x_1} + \cdots + \frac{\partial f_n}{\partial x_n} - f_0 \quad \textit{in} \quad \mathscr{D}'(\Omega)\,;$$

the solution of (7.92) *is the generalized solution* (*in the Dirichlet sense*) *of the Dirichlet problem* $P\left(\Omega, \varphi, \frac{\partial f_1}{\partial x_1} + \cdots + \frac{\partial f_n}{\partial x_n} - f_0\right)$. We note that there is *uniqueness of this solution*: if u_1, u_2 are two solutions, $u_1 - u_2 \in B_0^1(\Omega) \cap \mathscr{H}(\Omega)$ and so from Proposition 8, $u_1 = u_2$.

In the case of Ω unbounded, $V(\Omega, \Gamma)$ no longer coincides, in general, with $B_0^1(\Omega)$: a solution u_0 of (7.92) is then a (generalized) solution of the Dirichlet problem $P\left(\Omega, \varphi, \frac{\partial f_1}{\partial x_1} + \cdots + \frac{\partial f_n}{\partial x_n} - f_0\right)$ satisfying a *Neumann condition at infinity*. There is not necessarily uniqueness: for example, if $\Omega = \mathbb{R}^n$, $V(\Omega, \Gamma) = B^1(\mathbb{R}^n)$ and the solutions of (7.92) are defined to within an additive constant; however, there is uniqueness if the unbounded connected components of Ω have a boundary which is not a polar part of $\mathbb{R}^n$.

Let us now consider the problem (7.93). First of all, we note that

$$\langle \bar{\psi}, v \rangle = \int_\Omega \left(f_0 v + \bar{f}_1 \frac{\partial v}{\partial x_1} + \cdots + \bar{f}_n \frac{\partial v}{\partial x_n} \right) \mathrm{d}x$$

with $\bar{f}_i = f_i - \dfrac{\partial u_0}{\partial x_i}$ and from the construction of u_0

$$\langle \bar{\psi}, v \rangle = 0 \quad \text{for all} \quad v \in V(\Omega, \Gamma)$$

that is to say that $\bar{\psi}$ *is a continuous linear form on the "trace space $B^\circ(\Omega)/V(\Omega, \Gamma)$"*.

In the case in which Ω is a regular bounded open set and if the functions $\bar{f}_1, \ldots, \bar{f}_n$ are regular, we have from Ostrogradski's formula (see §1.3), and denoting by $\bar{f}$ the field $(\bar{f}_1, \ldots, \bar{f}_n)$

$$\int_\Omega \left(\bar{f}_1 \frac{\partial v}{\partial x_1} + \cdots + \bar{f}_n \frac{\partial v}{\partial x_n} \right) \mathrm{d}x = \int_\Gamma v \bar{f}.n \,\mathrm{d}\gamma - \int_\Omega v \left(\frac{\partial \bar{f}_1}{\partial x_1} + \cdots + \frac{\partial \bar{f}_n}{\partial x_n} \right) \mathrm{d}x\,;$$

thus, since

$$\frac{\partial \bar{f}_1}{\partial x_1} + \cdots + \frac{\partial \bar{f}_n}{\partial x_n} = f_0\,,$$

$$\langle \bar{\psi}, v \rangle = \int_\Gamma v \bar{f}.n \,\mathrm{d}\gamma\,.$$

Therefore $\bar{u}$ is a solution of (7.93) iff $\bar{u} \in V(\Omega_0, \Gamma_0)$ and

$$\int_\Omega \operatorname{grad} \bar{u} . \operatorname{grad} v \,\mathrm{d}x = \int_\Gamma v \bar{f}.n \,\mathrm{d}\gamma \quad \text{for all} \quad v \in V(\Omega, \Gamma_0)\,.$$

If $\bar{u} \in \mathscr{C}^0(\bar{\Omega}) \cap \mathscr{C}_n^1(\bar{\Omega} \backslash \Gamma_0)$, this expresses that $\bar{u}$ is the classical solution of the mixed problem

$$(7.95)\qquad \begin{cases} \Delta\bar{u} = 0 & \text{on} \quad \Omega \\ \bar{u} = 0 & \text{on} \quad \Gamma_0 \\ \dfrac{\partial\bar{u}}{\partial n} = \bar{f}.n & \text{on} \quad \Gamma\backslash\Gamma_0 . \end{cases}$$

In the general case, we shall say that $\bar{u}$ is a *generalized solution* (*in the Dirichlet sense*) *of the mixed problem* (7.95).

We notice that the passage from the notion of generalized solution to that of classical solution requires a regularity of the data Ω and $\bar{f}$ and of the solution $\bar{u}$. The Dirichlet method will be generalized in the rest of this book: it is in fact a particular case of the variational method; we shall not develop further this method in the present chapter.

5. Symmetry Methods and Method of Images

We recall that the Laplace operator is invariant under displacements of the space: given U an open set of $\mathbb{R}^n$ and T a displacement leaving U invariant – i.e. $T[U] = U$ – *for all* $g \in \mathscr{D}'(U)$, v is a solution of Poisson's equation

$$\Delta v = g \quad \text{in} \quad \mathscr{D}'(U)$$

iff $w = Tv$ is a solution of

$$\Delta w = Tg \quad \text{in} \quad \mathscr{D}'(U)$$

where for a function v, the function Tv is $v \circ T^{-1}$ and for a distribution g, the distribution Tg is defined by

$$\langle Tg, \zeta \rangle = \langle g, \zeta \circ T \rangle$$

(where T is a displacement).

We now consider U, an open set of $\mathbb{R}^n$, symmetric with respect to a hyperplane H and put $\Omega = U \cap H^+$ where H^+ is one of the open half-spaces defined by H (cf.

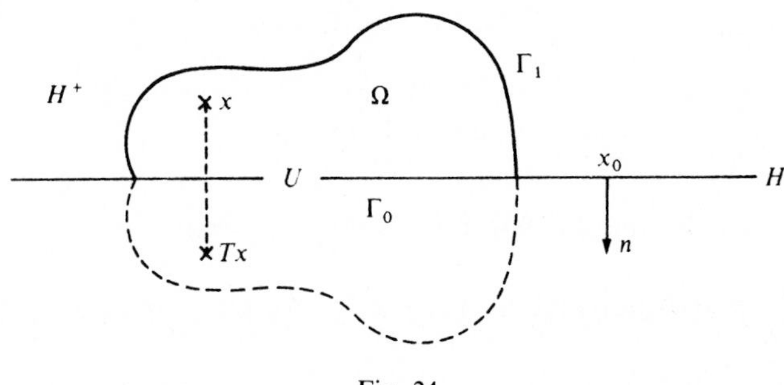

Fig. 24

Fig. 24). The boundary Γ of Ω is made up of the plane part $\Gamma_0 = U \cap H$ and of

$$\Gamma_1 = \Gamma \backslash \Gamma_0 = \Gamma \cap \partial U$$

where ∂U is the boundary of U.
Let us show to *reduce a problem*

$$\text{(7.96)} \quad \begin{cases} \Delta u = f \quad \text{on} \quad \Omega \\ u = 0 \quad \left(\text{resp.} \dfrac{\partial u}{\partial n} = 0\right) \quad \text{on} \quad \Gamma_0 \\ a_0 u + a_1 \dfrac{\partial u}{\partial n} = \varphi \quad \text{on} \quad \Gamma_1 \end{cases}$$

to a problem

$$\text{(7.97)} \quad \begin{cases} \Delta v = g \quad \text{on} \quad U \\ b_0 v + b_1 \dfrac{\partial v}{\partial n} = \psi \quad \text{on} \quad \partial U \end{cases}$$

by using the symmetry T with respect to the hyperplane H.
We shall work within the *classical framework* although this method of symmetry can also be developed in a variational framework. Let us state:

Proposition 13. *Let $\Omega = U \cap H^+$, where U is an open set of $\mathbb{R}^n$ symmetric with respect to a hyperplane H and H^+ one of the open half-spaces determined by H; the boundary Γ of Ω is made up of $\Gamma_0 = U \cap H$ and $\Gamma_1 = \partial U \cap \Gamma$. Let f be a distribution on Ω written in the form $f = f_0 + f_1$ where $f_0 \in L^p(\Omega)$ with $p > \frac{1}{2}n$ (resp. $p > n$) and f_1 is a distribution with compact support in Ω; we denote by g the distribution on U extending f by 0 on $U \backslash \Omega$[199]. Finally let $a_0, a_1, \varphi \in \mathscr{C}^0(\Gamma_1)$ with*

$$a_0 \geqslant 0, \quad a_1 \geqslant 0, \quad a_0 + a_1 > 0 \quad \text{on} \quad \Gamma_1, \varphi = 0 \quad \text{on} \quad \Gamma_1 \cap H;$$

we denote by b_0, b_1, ψ the continuous extensions of a_0, a_1, φ to ∂U by

$$b_0(z) = a_0(Tz), \quad b_1(z) = a_1(Tz), \quad \psi(z) = 0 \quad \textit{for} \quad z \in \partial U \backslash \Gamma_1{}^{200}.$$

We suppose that the open part of the boundary ∂U, $\{z \in \partial U; b_1(z) > 0\}$ is regular (of class $\mathscr{C}^1$) and that the problem (7.97) admits a quasi-classical solution v. Then the restriction u to Ω of $v - Tv$ (resp. $v + Tv$) is a quasi-classical solution of the problem (7.96)[201].

[199] This is possible because of the form of f:

$$\langle g, \zeta \rangle = \int_\Omega f_0(x)\zeta(x)\,dx + \langle f_1, \zeta\chi_\Omega \rangle, \quad \text{for all} \quad \zeta \in \mathscr{D}(U).$$

[200] We note that $b_0, b_1, \psi \in \mathscr{C}^0(\partial U)$, $b_0 \geqslant 0$, $b_1 \geqslant 0$, $b_0 + b_1 > 0$ on ∂U.

[201] This is true again if Ω is unbounded; we could then introduce a condition at infinity: for example if v is a solution of (7.97) satisfying the null condition at infinity (see §4.3), it is the same for Tv and hence the restriction of $v - Tv$ (resp. $v + Tv$) to Ω satisfies the null condition at infinity.

We note that account taken of the hypothesis that $\{z \in \partial U; b_1(z) > 0\}$ is a regular part of the boundary ∂U, the notion of a quasi-classical solution of (7.97) is clear (see Definition 13 of §4): the (mixed) condition on ∂U can again be written

$$\begin{cases} \dfrac{\partial v}{\partial n} + \dfrac{b_0}{b_1} v = \dfrac{\psi}{b_1} & \text{on } \{b_1 > 0\} \\ v = \dfrac{\psi}{b_0} & \text{on } \{b_1 = 0\} \ (\text{contained in } \{b_0 > 0\}) . \end{cases}$$

For the problem (7.96), the condition on the boundary is well defined in a classical fashion on Γ_0 (regular open part of Γ), on $\{z \in \Gamma_1; a_1(z) = 0\}$ (the condition is of Dirichlet type, written $u = \varphi/a_0$) and on $\{z \in \Gamma_1 \cap H^+; a_1(z) > 0\}$ (from the hypothesis, this is a regular open part). Now at a point $z \in \Gamma_1 \cap H$ with $a_1(z) > 0$, it is necessary to specify in a classical sense the condition on u, since from all the evidence Γ is not of class $\mathscr{C}^1$ in the neighbourhood of the points of $\Gamma_1 \cap H$: we shall take up this discussion again below; first let us give the

Proof of Proposition 13. The function $w = v - Tv$ (resp. $v + Tv$) is the solution of Poisson's equation

$$\Delta w = g - Tg \quad (\text{resp.}\, g + Tg) \quad \text{in} \quad \mathscr{D}'(U) .$$

But the restriction of $g \pm Tg$ to $U_0 = U \backslash (\operatorname{supp} f_1 \cup T(\operatorname{supp} f_1))$ is in $L^p(U_0)$ with $p > \frac{1}{2}n$ (resp. $p > n$); hence (see Proposition 6 of §3), w is continuous (resp. of class $\mathscr{C}^1$) on U_0. The restriction u of w to Ω is therefore continuous (resp. of class $\mathscr{C}^1$) on $\bar{\Omega} \cap U_0 = (\Omega \backslash \operatorname{supp} f_1) \cup \Gamma_0$ and, since $Tw = -w$ (resp. $Tw = w$) we obtain

$$u = 0 \quad \left(\text{resp.}\, \frac{\partial u}{\partial n} = 0\right) \quad \text{on} \quad \Gamma_0 .$$

Now the restriction of $g \pm Tg$ to Ω is f, so

$$\Delta u = f \quad \text{in} \quad \mathscr{D}'(\Omega) .$$

Finally w satisfies classically the boundary condition

$$b_0 w + b_1 \frac{\partial w}{\partial n} = \psi - T\psi \quad (\text{resp.}\, \psi + T\psi) \quad \text{on} \quad \partial U .$$

Since the restriction of $\psi \pm T\psi$ to $\Gamma_1 = \partial U \cap \Gamma$ is φ, u satisfies classically the boundary condition

$$a_0 u + a_1 \frac{\partial u}{\partial n} = \varphi \quad \text{on} \quad \Gamma_1 . \qquad \square$$

We note:

Corollary 3. *Let $\Omega = U \cap H^+$ where U is a bounded open set of $\mathbb{R}^n$ with all its boundary points regular*[202]*, symmetric with respect to a hyperplane H and H^+ one of*

[202] See Definition 4 of §4; this hypothesis is introduced to ensure the existence of a classical Green's function G_U; the formula (7.98) is in fact true for an arbitrary symmetric open set U.

the half-spaces determined by H. Then the Green's function G_Ω of Ω is given by

(7.98) $$G_\Omega(x, y) = G_U(x, y) - G_U(Tx, y)\,, \quad x, y \in \Omega\,, \quad x \neq y$$

where G_U is the Green's function of U and T the symmetry with respect to H.

Proof. For $y \in \Omega$, it is sufficient to apply Proposition 13 with $f = \delta_y$, $a_0 = 1$, $a_1 = \varphi = 0$; in effect then $g = \delta_y, b_0 = 1, b_1 = \psi = 0$ and $v(x) = G_U(x, y)$ is a classical solution of (7.97). □

We show by some examples the use of the method of symmetry.

Example 4. *Case of a half-ball.* We consider a half-ball $\Omega = B(x_0, r_0) \cap H^+$ of $\mathbb{R}^n$, where H is a hyperplane passing through x_0. From (7.98),

$$G_\Omega(x, y) = G_{B(x_0, r_0)}(x, y) - G_{B(x_0, r_0)}(Tx, y)\;;$$

thus, by use of the value of the Green's function of a ball (see Example 11 of §4),

$$G_\Omega(x, y) = E_n(x - y) - E_n(Tx - y) - \left(\frac{r_0}{(|x - x_0|)}\right)^{n-2}$$
$$[E_n(I(x_0, r_0)x - y) - E_n(I(x_0, r_0)Tx - y)]$$

where $I(x_0, r_0)$ is the inversion with pole x_0 and power r_0. We can observe, for example by applying (7.98) with $U = \mathbb{R}^n$ (see also Example 18 of §4), that $E_n(x - y) - E_n(Tx - y)$ is the Green's function $G_{H^+}(x, y)$ of the half-space H_+. Since the symmetry T commutes with the inversion $I(x_0, r_0)$,

$$G_{H^+ \cap B(x_0, r_0)}(x, y) = G_{H^+}(x, y) - \left(\frac{r_0}{|x - x_0|}\right)^{n-2} G_{H^+}(I(x_0, r_0)x, y)\,.$$

We can derive this formula by the method of images (see below). In particular in the case $n = 2$, identifying $\mathbb{R}^2$ with the complex plane, the Green's function of the unit semi-disk

$$\Omega = \{z; |z| < 1, \operatorname{Im} z > 0\} = \{re^{i\theta}; 0 < r < 1, 0 < \theta < \pi\}\,,$$

is

$$G_\Omega(z_1, z_2) = \frac{1}{2\pi}\operatorname{Log}(Q(z_1, z_2)Q(I(z_1), \bar{z}_2))$$

where

$$Q(z_1, z_2) = \frac{|z_1 - z_2|}{|z_1 - \bar{z}_2|}\,, \qquad I(z) = \frac{z}{|z|^2}$$

since

$$E_2(z) = \frac{1}{2\pi}\operatorname{Log}|z| \quad \text{and} \quad Tz = \bar{z}\,.$$

We use this value of the Green's function to solve the Dirichlet problem $P(\Omega, \chi_S)$ where S is the segment $[0, 1]$ of the real axis. The solution is given by

$$u(z) = -\int_0^1 \frac{\partial G_\Omega}{\partial y_2}(z, x_2)\,dx_2$$

where (x_2, y_2) are the coordinates of z_2 in $\mathbb{R}^2$.

Now

$$\frac{\partial G_\Omega}{\partial y_2}(re^{i\theta}, x + iy_2)|_{y_2=0} = \frac{r\sin\theta}{\pi}\left[\frac{1}{|e^{i\theta} - rx|^2} - \frac{1}{|re^{i\theta} - x|^2}\right]$$

so

$$u(re^{i\theta}) = \frac{1}{\pi}\left[\int_0^{1/r\sin\theta} \frac{dt}{1 + (t - \operatorname{cotg}\theta)^2} - \int_0^{r/\sin\theta} \frac{dt}{1 + (t - \operatorname{cotg}\theta)^2}\right]$$

$$(7.99)\qquad = \frac{1}{\pi}\left(\operatorname{Arctg}\left(\frac{1}{r\sin\theta} - \operatorname{cotg}\theta\right) - \operatorname{Arctg}\left(\frac{r}{\sin\theta} - \operatorname{cotg}\theta\right)\right)$$

which is therefore the solution of $P(\Omega, \chi_s)$.

The symmetry method (Proposition 13) allows us to treat other problems. For example, let us consider the mixed problem

$$(7.100)\qquad \begin{cases} \Delta u = 0 & \text{on} \quad B(x_0, r_0) \cap H^+ \\ \dfrac{\partial u}{\partial n} = \psi & \text{on} \quad B(x_0, r_0) \cap H \\ u = \varphi & \text{on} \quad \partial B(x_0, r_0) \cap \overline{H^+} \end{cases}$$

where $\psi \in \mathscr{C}^0(H)$ and $\varphi \in \mathscr{C}^0(\partial B(x_0, r_0)$ are given.

To simplify the notation, we can always suppose that

$$x_0 = 0, \qquad H^+ = \mathbb{R}^{n-1} \times \mathbb{R}^+ .$$

We begin by reducing to the case $\psi = 0$ by considering the solution v of $PN(H^+, \psi)$; we know (see Example 21 of §4) that v is given by

$$(7.101)\qquad \begin{cases} v(x', x_n) = \dfrac{2x_n}{(n-2)\sigma_n}\displaystyle\int_{\mathbb{R}^{n-1}} \frac{\psi(x' + x_n t)}{(1 + |t|^2)^{\frac{n}{2}-1}}\,dt & \text{if } n \geqslant 3 \\ v(x, y) = -\dfrac{y}{2\pi}\displaystyle\int_{\mathbb{R}} \psi(x + yt)\operatorname{Log}(1 + t^2)\,dt & \text{if } n = 2 \end{cases}$$

these formulae being justified if ψ satisfies (4.77) if $n \geqslant 3$ and (4.79) if $n = 2$; in fact only the values of ψ on $B(x_0, r_0)$ interest us with the result that we can always suppose that ψ is of compact support satisfying, in addition, $\int_{\mathbb{R}^{n-1}} \psi(t)\,dt = 0$ and hence that (4.77) and (4.79) are satisfied.

Putting $w = u - v$ we see that u satisfies (7.100) iff w is a solution of

$$(7.102)\qquad \begin{cases} \Delta w = 0 & \text{on} \quad B(x_0, r_0) \cap H^+ \\ \dfrac{\partial w}{\partial n} = 0 & \text{on} \quad B(x_0, r_0) \cap H \\ w = \varphi - v & \text{on} \quad \partial B(x_0, r_0) \cap \overline{H^+} . \end{cases}$$

To solve (7.102), we use the symmetry method: let us consider g the function defined on $\partial B(x_0, r_0)$ by

$$g(x', x_n) = g(x', -x_n) \quad \text{for all} \quad (x', x_n) \in \partial B(x_0, r_0)$$

$$g = \varphi - v \quad \text{on} \quad \partial B(x_0, r_0) \cap \overline{H^+} ;$$

it is continuous and the solution of the Dirichlet problem $P(B(x_0, r_0), g)$ is even with respect to x_n with the result that its restriction to $B(x_0, r_0) \cap H^+$ is the classical solution of (7.102); from Poisson's integral formula (see §2.3) we have

$$w(x) = \frac{1}{r_0 \sigma_n} \int_{\partial B(x_0, r_0)} \frac{r_0^2 - |x - x_0|^2}{|z - x|^n} g(z) \mathrm{d}\gamma(z)$$

which, from the definition of g, can be written

$$(7.103) \qquad w(x) = \frac{r_0^2 - |x - x_0|^2}{r_0 \sigma_n} \int_{\partial B(x_0, r_0) \cap H^+} (|z - x|^{-n} + |z - Tx|^{-n})(\varphi(z) - v(z)) \mathrm{d}\gamma(z)$$

where T is the symmetry with respect to H.
In brief, *the mixed system* (7.100) *admits a classical solution*[203] $u = v + w$ *where* v *is given by* (7.101) *and* w *by* (7.103).
Besides the possibility of obtaining explicit solutions as in the above example, one of the interests of the symmetry method is to yield regularity results for Dirichlet, Neumann or mixed problems in non-regular open sets. We illustrate this in:

Example 5. *Case of a cube.* Let us take a cube of $\mathbb{R}^n$ or more generally $\Omega = \prod_{i=1}^{n}]a_i, b_i[$ where $]a_i, b_i[$ are bounded intervals of $\mathbb{R}$. We take $f \in \mathscr{C}^\infty(\bar{\Omega})$ and consider the problem

$$(7.104) \qquad \begin{cases} \Delta u = f & \text{on} \quad \Omega \\ u = 0 & \text{on} \quad \Gamma_0 \\ \dfrac{\partial u}{\partial n} = 0 & \text{on} \quad \Gamma_1 \end{cases}$$

where Γ_0 is a union of (open) faces of Ω and Γ_1 is the union of the other faces[204] with the result that $\Gamma_0 \cap \Gamma_1 = \varnothing$ and $\Gamma \backslash (\Gamma_0 \cup \Gamma_1)$ are the edges of Ω. Using, for example, the Dirichlet method (see §7.4) we shall show that if $\Gamma_0 \neq \varnothing$ (resp. $\Gamma_0 = \varnothing$ and $\int_\Omega f \mathrm{d}x = 0$), there exists a unique (resp. to within an additive constant) generalized solution (in the Dirichlet sense) of (7.104). In fact using the results of local regularity (see §3.2) as well as the regularity at the regular boundary (see §6.1), we verify that this solution is classical and is in $\mathscr{C}^\infty(\Omega \cap \Gamma_0 \cup \Gamma_1)$. The

[203] Unique (see Corollary 6, §4).

[204] Ω admits $2n$ faces such that $\prod_{i=1}^{n}]a_i, b_i[\times \{a_n\}$.

real problem is *the regularity in the neighbourhood of the points of the boundary of the faces*. Using the symmetry method, we prove that *the solution* $u \in \mathscr{C}^1(\bar{\Omega})$ and even $u \in W^{2,p}(\Omega)$ for all $1 < p < \infty$[205]; in general however $u \notin \mathscr{C}^2(\bar{\Omega})$, although this is the case if f is zero on $\Gamma \backslash (\Gamma_0 \cup \Gamma_1)$; more generally *if* f *is with support contained in* $\Omega \cup \Gamma_0 \cup \Gamma_1$, *then* $u \in \mathscr{C}^\infty(\bar{\Omega})$.
So as to throw light on these properties, we shall restrict ourselves to the case $n = 2$. We can suppose that $\Omega =]0, 1[\times]0, 1[$ and study the *regularity of u in the neighbourhood of* $(0, 0)$. We can always suppose

$$\Gamma_0 \supset \{0\} \times]0, 1[\cup]0, 1[\times \{0\} .$$

even if we replace u by $\dfrac{\partial u}{\partial x}, \dfrac{\partial u}{\partial y}$ or $\dfrac{\partial^2 u}{\partial x \partial y}$ according as Γ_1 contains $\{0\} \times]0, 1[$, $]0, 1[\times \{0\}$ or both.
We consider then $\tilde{\Omega} =]0, 1[\times]-1, 1[$, $\tilde{f}$ defined on $\tilde{\Omega}$ by

$$\tilde{f} = f \quad \text{on} \quad \Omega \quad \text{and} \quad \tilde{f}(x, -y) = -f(x, y) \quad \text{for all} \quad (x, y) \in \tilde{\Omega}$$

and $\tilde{\Gamma}_0 = \{0\} \times]-1, 1[$ joined possibly to $\{1\} \times]-1, 1[$ and $]0, 1[\times \{-1, 1\}$ according as Γ_0 contains $\{1\} \times]0, 1[$ and $]0, 1[\times \{1\}$. Given $1 < p < \infty$, under the hypothesis $f \in L^p(\Omega)$, the function $\tilde{f} \in L^p(\tilde{\Omega})$ with the result that the solution $\tilde{u}$ of the corresponding problem (7.104) is in $W^{2,p}_{\text{loc}}(\tilde{\Omega} \cup \tilde{\Gamma}_0)$ (see §3.2).
Now

$$\tilde{u}(x, -y) = -\tilde{u}(x, y)$$

with the result that u is the restriction of $\tilde{u}$ to Ω and hence u is $W^{2,p}$ in the neighbourhood of $(0, 0)$.
Even within the hypothesis $f \in \mathscr{C}^\infty(\bar{\Omega})$, in general $\tilde{f}$ is not even continuous on $\tilde{\Omega}$: we have this property, however, if

$$f(x, 0) = 0 \qquad \text{for all} \quad x \in]0, 1[\;;$$

then $\tilde{f}$ satisfies a Lipschitz condition on $\tilde{\Omega}$ and hence $\tilde{u} \in W^{3,p}_{\text{loc}}(\tilde{\Omega} \cup \tilde{\Gamma}_0)$ and u is $W^{3,p}$ in the neighbourhood of $(0, 0)$ for all $1 < p < \infty$.
In fact *the solution u is of class* $\mathscr{C}^2$ *in the neighbourhood of* $(0, 0)$ *iff* $f(0, 0) = 0$. Let us prove this property. First of all, since $u(x, 0) = u(0, y) = 0$, then $\dfrac{\partial^2 u}{\partial x^2}(x, 0) = \dfrac{\partial^2 u}{\partial y^2}(0, y) = 0$ for all $(x, y) \in \Omega$ and hence u of class $\mathscr{C}^2$ in the neighbourhood of $(0, 0)$ implies $f(0, 0) = 0$. To prove the converse, we consider $\tilde{v}$ defined on $\tilde{\Omega}$ by

$$\tilde{v} = \frac{\partial u}{\partial x} \quad \text{on} \quad \Omega\,, \qquad \tilde{v}(x, -y) = -\tilde{v}(x, y) \quad \text{for all} \quad (x, y) \in \tilde{\Omega}\,.$$

Since

$$u \in \mathscr{C}^1(\Omega \cup [0, 1[\times \{0\} \cup \{0\} \times [0, 1[) \quad \text{and} \quad \frac{\partial u}{\partial x} = 0 \quad \text{on} \quad [0, 1[\times \{0\},$$

we have $\tilde{v} \in \mathscr{C}^0(\tilde{\Omega} \cup \{0\} \times]-1, +1[)$.

---

[205] Under the sole hypothesis $f \in L^p(\Omega)$ (see Proof below).

We have

$$\Delta \tilde{v} = \widetilde{\frac{\partial f}{\partial x}} \quad \text{on} \quad \tilde{\Omega}$$

and also

$$\frac{\partial \tilde{v}}{\partial x} = \widetilde{\frac{\partial^2 u}{\partial x^2}} = \widetilde{f - \frac{\partial^2 u}{\partial y^2}} = \tilde{f} \quad \text{on} \quad \{0\} \times]-1, +1[)$$

since $u = 0$ on $\{0\} \times [0, 1[$. Given $0 < \alpha < 1$, under the hypothesis $f \in \mathscr{C}^\alpha(\bar{\Omega})$, since $f(0, 0) = 0$, the function $\tilde{f}(0, .) \in \mathscr{C}^\alpha([-1, +1])$; from the regularity theorems (see §6.1) we have therefore

$$\tilde{v} \in \mathscr{C}^{1+\alpha}_{\text{loc}}(\tilde{\Omega} \cup \{0\} \times]-1, +1[)$$

and hence $\dfrac{\partial u}{\partial x}$ of class $\mathscr{C}^{1+\alpha}$ in the neighbourbood of (0, 0); using the equation $\dfrac{\partial^2 u}{\partial y^2} = f - \dfrac{\partial^2 u}{\partial x^2}$, u is therefore of class $\mathscr{C}^{2+\alpha}$ in the neighbourhood of (0, 0). To sum up, we have proved the property and even that *if* $f \in \mathscr{C}^\alpha(\bar{\Omega})$ *with* $f(0, 0) = 0$, *then* u *is* $\mathscr{C}^{2+\alpha}$ *in the neighbourhood of* (0, 0).

This example, not only puts in evidence the phenomena of the problems of regularity in the neighbourhood of a non-regular point of the boundary, but is fundamental for the study of the problems of regularity in the corners. We shall not deal here with the subject referring the reader to Grisvard [1], Kondratiev [1]. We conclude this section by showing that the symmetry method is, in a certain sense a particular case of what is called *the method of images*.

Let us consider an open set Ω in $\mathbb{R}^n$ with boundary Γ. Given $x \in \Omega$, we give the name (electrostatic) *image of the point* x *with respect to* Γ to a distribution $T(x) \in \mathscr{E}'(\mathbb{R}^n \backslash \bar{\Omega})$ such that the Newtonian potential of $T(x)$ coincides on Γ with that of δ_x, namely

$$E_n * T(x)(z) = E_n(x - z) \quad \text{for all} \quad z \in \Gamma .$$

In the case of a half-space H^+, it is clear that for all $x \in H^+$, $\delta_{T(x)}$ is the image of x with respect to H where T is the symmetry with respect to H.

Knowledge of the image of the points of Ω permit the determination of the Green's function of Ω: in effect if $T(x)$ is the image of x with respect to Γ,

$$G_\Omega(x, y) = E_n(x - y) - (E_n * T(x))(y)$$

since the function $E_n * T(x)$ is harmonic on Ω and by definition takes the value $E_n(x - .)$ on Γ.

We have effectively used the method of images to determine the Green's function of a half-space H^+:

$$G_{H^+}(x, y) = E_n(x - y) - E_n(T(x) - y)$$

where T is the symmetry with respect to H.

Similarly we have used the method of images to determine the Green's function of a ball $B(x_0, r_0)$: considering $I(x_0, r_0)$ inversion with pole x_0 and power r_0, we have seen (see Example 11 of §4) that

Case $n \geqslant 3$. For all $x \in B(x_0, r_0)\backslash\{x_0\}$, we have

$$E_n(I(x_0, r_0)x - z) = \left(\frac{|x - x_0|}{r_0}\right)^{n-2} E_n(x - z) \quad \text{for all} \quad z \in \partial B(x_0, r_0)$$

hence the distribution $\left(\dfrac{r_0}{|x - x_0|}\right)^{n-2} \delta_{I(x_0, r_0)x}$ is the image of x with respect to $\partial B(x_0, r_0)$ and as we have seen

$$G_{B(x_0, r_0)}(x, y) = E_n(x - y) - \left(\frac{r_0}{|x - x_0|}\right)^{n-2} E_n(I(x_0, r_0)x - y).$$

We note that the Newtonian potential of a uniform distribution μ on a sphere $\partial B(x_0, r)$ is invariant under rotation about x_0 and its value on the sphere $\partial B(x_0, r_0)$ is $\mu\sigma_n r^{n-1} E_n(r_0)$. Hence, for $r > r_0$, the uniform distribution of density $1/\sigma_n r^{n-1}$ on $\partial B(x_0, r)$ is an image of the point x_0 with respect to $\partial B(x_0, r_0)$.
Case: $n = 2$. For all $x \in B(x_0, r_0)\backslash\{x_0\}$

$$\begin{aligned} E_2(I(x_0, r_0)x - z) &= \frac{1}{2\pi} \operatorname{Log}|I(x_0, r_0)x - z| \\ &= E_2(x - z) + \frac{1}{2\pi} \operatorname{Log}\frac{r_0}{|x - x_0|} \quad \text{for all} \quad z \in \partial B(x_0, r_0). \end{aligned}$$

As we have seen above the constant potential c on the circle $\partial B(x_0, r_0)$ is created by a uniform distribution of density $\mu = \dfrac{c}{r \operatorname{Log} r_0}$ on the circle $\partial B(x_0, r)$. Hence an image of x with respect to $\partial B(x_0, r_0)$ is

$$\delta_{I(x_0, r_0)x} - \frac{1}{2\pi r}\left(1 - \frac{\operatorname{Log}|x - x_0|}{\operatorname{Log} r_0}\right)\delta_{\partial B(x_0, r)} \quad \text{with} \quad r > r_0 \text{ arbitrary.}$$

§8. Elliptic Equations of the Second Order*

In the preceding sections of this chapter, we have studied Laplace's equation and Poisson's equation and the Dirichlet, Neumann or mixed problems for these equations. In Chap. V (vol. 2), we shall study, and more particularly, we shall classify general linear partial differential equations. In this section we propose to show whether or not the properties of Laplace's equation generalize to more general elliptic equations of the second order.
Given an open set Ω in $\mathbb{R}^n$ we consider a partial differential equation of the second order

$$(8.1) \qquad -\sum_{i,j=1}^{n} a_{ij}\frac{\partial^2 u}{\partial x_i \partial x_j} + \sum_{i=1}^{n} b_i \frac{\partial u}{\partial x_i} + cu = f \quad \text{on} \quad \Omega$$

in which the coefficients a_{ij}, b_i, c and f are real functions defined on Ω. A function

$u \in \mathscr{C}^2(\Omega)$ satisfying (8.1) for all $x \in \Omega$ is called *a classical solution* of equation (8.1). Given that for $u \in \mathscr{C}^2(\Omega)$, $\dfrac{\partial^2 u}{\partial x_i \partial x_j} = \dfrac{\partial^2 u}{\partial x_j \partial x_i}$, even if we replace a_{ij} by $\frac{1}{2}(a_{ij} + a_{ji})$, *we can always suppose that* $a_{ij} = a_{ji}$. We say that the equation (8.1) *is elliptic on* Ω[206] *if for all* $x \in \Omega$

$$\lambda(x) = \min_{\substack{\xi \in \mathbb{R}^n \\ |\xi| = 1}} \sum_{i,j=1}^{n} a_{ij}(x)\xi_i\xi_j > 0\,; \tag{8.2}$$

we shall say that it is *strictly elliptic on* Ω if $\inf_{x \in \Omega} \lambda(x) > 0$. We observe that under the assumption $a_{ij} = a_{ji}$, $\lambda(x)$ is the smallest (real) eigenvalue of the symmetric matrix $(a_{ij}(x))_{i,j}$, while

$$\Lambda(x) \stackrel{\text{def}}{=} \max_{\substack{\xi \in \mathbb{R}^n \\ |\xi| = 1}} \sum_{i,j=1}^{n} a_{ij}(x)\xi_i\xi_j$$

is the largest eigenvalue. We say that (8.1) is *uniformly elliptic on* Ω if it is elliptic and $\inf_{x \in \Omega} \dfrac{\lambda(x)}{\Lambda(x)} > 0$.

When the *coefficients* a_{ij}, b_i, c are *constants* we can always reduce the equation[207] by change of the coordinate axes and of the function f to the equation

$$-\Delta u + cu = f \tag{8.3}$$

where $c \in \mathbb{R}$. The case $c = 0$ is Poisson's equation, the case $c < 0$ is *Helmholtz's equation* which we rather write

$$\Delta u + k^2 u = f \tag{8.4}$$

where k is real constant that we can always choose to be positive. We shall study this equation in greater depth in Sect. 7 on account of its importance as much mathematical as for modelling numerous physical problems.

The case $c > 0$, which we shall rather write

$$-\Delta u + k^2 u = f \tag{8.5}$$

with k a real positive constant, will present practically all the properties of Poisson's equation as we shall see below: we shall sometimes give it the name of *the equation of neutronics*, where it occurs as a model (see Chap. IA, §5); the model closest to reality is rather the equation:

$$-\operatorname{div}(a \operatorname{grad} u) + k^2 u = f \quad \text{on} \quad \Omega \tag{8.6}$$

where a and k are positive functions defined on Ω.

---

[206] See Chap. V for a more general definition of ellipticity.

[207] In the elliptic case (see Chap. V).

1. The Divergence Form, Green's Formula

We can put every equation (8.1) in the *divergence form*

$$(8.7)\qquad -\sum_{i=1}^{n}\frac{\partial}{\partial x_i}\left(\sum_{j=1}^{n}a_{ij}\frac{\partial u}{\partial x_j}+c_i u\right)+\sum_{i=1}^{n}d_i\frac{\partial u}{\partial x_i}+du=f \quad\text{on}\quad \Omega$$

the relation between the coefficients (b_i, c) of (8.1) and (c_i, d_i, d) of (8.7) being given by

$$(8.8)\qquad b_i = d_i - c_i - \sum_{j=1}^{n}\frac{\partial a_{ij}}{\partial x_j}, \qquad c = d - \sum_{i=1}^{n}\frac{\partial c_i}{\partial x_i}.$$

We notice that *the divergence form* (8.7) *is not unique*. Setting aside the fact that physical problems are habitually modelled in divergence form, this provides many advantages from the mathematical point of view. First of all it allows us to generalize simply *Green's formula*; for that we introduce some new concepts.

Definition 1. Let $L = -\sum_{i=1}^{n}\frac{\partial}{\partial x_i}\left(\sum_{j=1}^{n}a_{ij}\frac{\partial}{\partial x_j}+c_i\right)+\sum_{i=1}^{n}d_i\frac{\partial}{\partial x_i}+d$ be a differential operator of the second order in divergence form on an open set Ω of $\mathbb{R}^n$. We denote by ∇_L the differential field associated with L, whose components are

$$(8.9)\qquad \nabla_L^i = \sum_{j=1}^{n}a_{ij}\frac{\partial}{\partial x_j}+c_i \qquad i=1,\ldots,n.$$

By the *co-normal derivative associated with* L we mean the trace operator on the boundary Γ of Ω

$$(8.10)\qquad \frac{\partial}{\partial n_L} = n.\nabla_L$$

where n is the unit vector normal to Γ exterior to Ω.
We give the name (formal) *adjoint operator of* L[208], to the operator

$$(8.11)\qquad L^* = -\sum_{i=1}^{n}\frac{\partial}{\partial x_i}\left(\sum_{j=1}^{n}a_{ji}\frac{\partial}{\partial x_j}+d_i\right)+\sum_{i=1}^{n}c_i\frac{\partial}{\partial x_i}+d.$$

Within the *classical framework*, the coefficients a_{ij}, c_i, d_i, d will be supposed continuous on Ω; we shall denote by $\mathscr{C}_L^2(\Omega)$ *the set of functions* $u \in \mathscr{C}^1(\Omega)$ *such* $p = \nabla_L u \in \mathscr{C}^1(\Omega)^n$ *and* $\mathscr{C}^1_{n_L}(\bar{\Omega})$ *the set of functions* $u \in \mathscr{C}^1(\Omega) \cap \mathscr{C}^0(\bar{\Omega})$ *such that* $\frac{\partial u}{\partial n_L}(z) = p.n(z)$ *exists for all* $z \in \Gamma$ *in the classical sense* considered in §1.3[209].

---

[208] Or *transposed*.

[209] It is clear that if $a_{ij}, c_i \in \mathscr{C}^1(\Omega)$ then $\mathscr{C}^2(\Omega) \subset \mathscr{C}_L^2(\Omega)$; we have even $\mathscr{C}_L^2(\Omega) = \mathscr{C}^2(\Omega)$ if L is elliptic or, more generally, if the matrix $(a_{ij}(x))_{i,j}$ is invertible for all $x \in \Omega$. Also if a_{ij}, $c_i \in \mathscr{C}^0(\bar{\Omega})$, then $\mathscr{C}^1(\bar{\Omega}) \subset \mathscr{C}^1_{n_L}(\bar{\Omega})$; we obviously do not have equality except in dimension $n = 1$.

Given $u \in \mathscr{C}^2_L(\Omega)$ and $v \in \mathscr{C}^1(\Omega)$, we have

$$\operatorname{div}(v\,\nabla_L u) = \nabla v \,.\, \nabla_L u + v \operatorname{div} \nabla_L u = -\,v\,LU + \sum_{i,j=1}^{n} a_{ij}\frac{\partial u}{\partial x_j}\frac{\partial v}{\partial x_i} + \sum_{i=1}^{n}\left(c_i u\frac{\partial u}{\partial x_i} + d_i v\frac{\partial u}{\partial x_i}\right) + duv\,.$$

We introduce *the bilinear differential operator associated with L*

$$(8.12)\qquad L(u,v) = \sum_{i,j=1}^{n} a_{ij}\frac{\partial u}{\partial x_j}\frac{\partial v}{\partial x_i} + \sum_{i=1}^{n}\left(c_i u\frac{\partial v}{\partial x_i} + d_i\frac{\partial u}{\partial x_i}v\right) + duv\,,$$

with the result that we have

$$\operatorname{div}(v\,\nabla_L u) = L(u,v) - vLu\,,$$

from which by application of Ostrogradski's formula (see Proposition 4, §1) with the reservation that $u \in \mathscr{C}^1_{n_L}(\bar{\Omega})$, $v \in \mathscr{C}^0(\bar{\Omega})$ and $L(u,v) - vLu \in L^1(\Omega)$

$$(8.13)\qquad \int_\Omega (L(u,v) - vLu)\,\mathrm{d}x = \int_\Gamma v\frac{\partial u}{\partial n_L}\mathrm{d}\gamma$$

which we shall rather write in the form

$$(8.14)\qquad \int_\Omega vLu\,\mathrm{d}x + \int_\Omega v\frac{\partial u}{\partial n_L}\,\mathrm{d}\gamma = \int_\Omega L(u,v)\,\mathrm{d}x$$

under the condition that $vLu \in L^1(\Omega)$.
We notice now that by the definition of the adjoint operator

$$L(u,v) = L^*(v,u)\,.$$

Hence under adequate hypothesis

$$\int_\Omega L(u,v)\,\mathrm{d}x = \int_\Omega L^*(v,u)\,\mathrm{d}x = \int_\Omega uL^*v\,\mathrm{d}x + \int_\Gamma u\frac{\partial u}{\partial n_{L^*}}\mathrm{d}\gamma\,,$$

from which with (8.14), we deduce Green's formula

$$(8.15)\qquad \int_\Omega (uL^*v - vLu)\,\mathrm{d}x = \int_\Gamma\left(v\frac{\partial u}{\partial n_L} - u\frac{\partial v}{\partial n_{L^*}}\right)\mathrm{d}\gamma\,.$$

To sum up, we have

Proposition 1. *Given* $L = -\sum_{i,j=1}^{n}\frac{\partial}{\partial x_i}\left(a_{ij}\frac{\partial}{\partial x_j} + c_i\right) + \sum_{i=1}^{n} d_i\frac{\partial}{\partial x_i} + d$ *a differential operator of second order in divergence form on a regular bounded open set* Ω *of* $\mathbb{R}^n$ *with boundary* Γ. *With the notation of Definition* 1 *above, for all*

$$u \in \mathscr{C}^2_L(\Omega) \cap \mathscr{C}^1_{n_L}(\bar{\Omega}) \quad \textit{and} \quad v \in \mathscr{C}^1(\Omega) \cap \mathscr{C}^0(\bar{\Omega})$$

(resp. $v \in \mathscr{C}^2_{L^*}(\Omega) \cap \mathscr{C}^1_{n_{L^*}}(\Omega)$*) with* $L(u,v)$, $vLu \in L^1(\Omega)$ *(resp. and* $uL^*v \in L^1(\Omega)$*), we have Green's formula* (8.14) *for integration by parts (resp. Green's formula* (8.15)*).*

We notice as a corollary of Proposition 1, *Gauss' theorem*, corresponding to the particular case $v \equiv 1$.

Corollary 1. *With the data of Proposition* 1, *for all* $u \in \mathscr{C}_L^2(\Omega) \cap \mathscr{C}_{n_L}^1(\bar{\Omega})$ *with* $L_0 u = -\operatorname{div} \nabla_L u \in L^1(\Omega)$, *we have*

$$\int_\Gamma \frac{\partial u}{\partial n_L} \mathrm{d}\gamma + \int_\Omega L_0 u \,\mathrm{d}x = 0 .$$

Following Proposition 5 of §1, for

$$v \in \mathscr{C}^1(\Omega) \cap \mathscr{C}^0(\bar{\Omega}) \quad \text{and} \quad u \in \mathscr{C}_L^2(\Omega) \cap \mathscr{C}_{n_L}^1(\bar{\Omega}) ,$$

the integrability of $L(u, v) - vLu$ is assured since this function is minorized by an integrable function. Let us suppose that the operator L is elliptic and denote by $\lambda(x)$ the smallest eigenvalue of the symmetric matrix $(a_{ij}(x))$. For all $u \in \mathscr{C}^1(\Omega)$ we have

$$L(u, u) \geqslant \lambda |\operatorname{grad} u|^2 - (\textstyle\sum (c_i + d_i)^2)^{\frac{1}{2}} |u| |\operatorname{grad} u| + du^2 ,$$

and for all $t : \Omega \to [0, 1]$

$$L(u, u) \geqslant (1 - t)\lambda |\operatorname{grad} u|^2 + \left(d - \frac{1}{4t\lambda} \sum (c_i + d_i)^2\right) u^2 . \tag{8.16}$$

We deduce

Corollary 2. *With the data of Proposition* 1, *given a measurable function* $t: \Omega \to]0, 1]$, *we suppose* L *is elliptic and*

$$\left(\frac{1}{4\lambda t} \sum (c_i + d_i)^2 - d\right)^+ \in L^1(\Omega) \tag{8.17}$$

where $\lambda(x)$ *is the smallest eigenvalue of the symmetric matrix* $(a_{ij}(x))$. *Then for every function* $u \in \mathscr{C}_L^2(\Omega) \cap \mathscr{C}_{n_L}^1(\bar{\Omega})$ *with* $Lu \in L^1(\Omega)$ *we have*

$$(1 - t)\lambda |\operatorname{grad} u|^2 \in L^1(\Omega) , \qquad \left(\frac{1}{4\lambda t} \sum (c_i + d_i)^2 - d\right) u^2 \in L^1(\Omega)$$

and

$$\begin{cases} \displaystyle\int_\Omega (1 - t)\lambda |\operatorname{grad} u|^2 \,\mathrm{d}x \leqslant \int_\Gamma u \frac{\partial u}{\partial n_L} \mathrm{d}\gamma \\ \displaystyle\quad + \int_\Omega \left(uLu + \left(\frac{1}{4\lambda t} \sum (c_i + d_i)^2 - d\right) u^2\right) \mathrm{d}x . \end{cases} \tag{8.18}$$

Supposing always that L is elliptic, we denote by $\Lambda(x)$ the largest eigenvalue of the symmetric matrix $(a_{ij}(x))$; it is classical that

$$\left|\sum_{i,j} a_{ij}(x) \xi_i \eta_j\right| \leqslant \Lambda(x) |\xi| |\eta| \quad \text{for all} \quad \xi, \eta \in \mathbb{R}^n .$$

Hence for $u, v \in \mathscr{C}^1(\Omega)$, we have

$$\begin{aligned}|L(u,v)| &\leqslant \Lambda|\operatorname{grad} u||\operatorname{grad} v| + (\textstyle\sum c_i^2)^{\frac{1}{2}}|u||\operatorname{grad} v| \\ &\quad + (\textstyle\sum d_i^2)^{\frac{1}{2}}|v||\operatorname{grad} u| + |d||u||v| \\ &\leqslant \left(\frac{\Lambda}{\lambda} + 1\right)\left(\frac{\lambda}{2}|\operatorname{grad} u|^2 + \frac{\lambda}{2}|\operatorname{grad} v|^2\right) \\ &\quad + \left(\frac{\sum c_i^2}{\lambda} + |d|\right)\frac{u^2}{2} + \left(\frac{\sum d_i^2}{\lambda} + |d|\right)\frac{v^2}{2}\,.\end{aligned}$$

Applying Corollary 2 with $t = \frac{1}{2}$, we deduce:

Corollary 3. *With the data of Proposition* 1, *let L be elliptic and*

$$\frac{\Lambda}{\lambda} \in L^\infty(\Omega)\,, \qquad \frac{c_i^2}{\lambda}, \frac{d_i^2}{\lambda}, d \in L^1(\Omega)\,,$$

where $\lambda(x)$, $\Lambda(x)$ are respectively the smallest and the largest eigenvalue of the symmetric matrix $(a_{ij}(x))$[210]. *Then for every $u \in \mathscr{C}_L^2(\Omega) \cap \mathscr{C}_{n_L}^1(\bar{\Omega})$ and $v \in \mathscr{C}_{L^*}^2(\Omega) \cap \mathscr{C}_{n_{L^*}}^1(\bar{\Omega})$ with Lu, $L^*v \in L^1(\Omega)$, we have $L(u,v) \in L^1(\Omega)$* and *the Green's formulae*

$$\int_\Omega L(u,v)\,\mathrm{d}x = \int_\Gamma v\frac{\partial u}{\partial n_L}\,\mathrm{d}\gamma + \int_\Omega vLu\,\mathrm{d}x = \int_\Gamma u\frac{\partial v}{\partial n_{L^*}}\,\mathrm{d}\gamma + \int_\Omega uL^*v\,\mathrm{d}x\,.$$

We recall that the divergence form $Lu = f$ on Ω of equation (8.1) is not unique with the result that the conormal derivatives $\frac{\partial}{\partial n_L}$, $\frac{\partial}{\partial n_{L^*}}$ are not associated with the differential operator

$$P = -\sum_{i,j=1}^n a_{ij}\frac{\partial^2}{\partial x_i \partial x_j} + \sum b_i\frac{\partial}{\partial x_i} + c$$

defining (8.1), but with its particular divergence form L. We notice however that the principal parts of $\frac{\partial}{\partial n_L}$ and $\frac{\partial}{\partial n_{L^*}}$ are the same, this is $\sum_{i,j=1}^n a_{ij}n_i\frac{\partial}{\partial x_j}$, and so they are associated with P.
Now let us give another statement of Green's formula

Proposition 2. *Let Ω be a regular bounded open set of $\mathbb{R}^n$ of boundary Γ of class $\mathscr{C}^{m+2}$ where $m \geqslant 0$,*

$$P = -\sum_{i,j=1}^n a_{ij}\frac{\partial^2}{\partial x_i \partial x_j} + \sum_{i=1}^n b_i\frac{\partial}{\partial x_i} + c$$

a differential operator of order 2 *on Ω with*

$$a_{ij} = a_{ji} \in \mathscr{C}^{m+1}(\bar{\Omega})\,, \qquad b_i \in \mathscr{C}^m(\bar{\Omega}) \cap \mathscr{C}^1(\bar{\Omega})\,, \qquad c \in \mathscr{C}^0(\bar{\Omega})\,,$$

[210] That is to say that we suppose that L is uniformly elliptic on Ω.

and $B = \beta_0 + \sum_{j=1}^{n} \beta_j \frac{\partial}{\partial x_j}$ *a differential operator of order* 1 *on* Γ *with* $\beta_0 \in \mathscr{C}^m(\Gamma)$, $\beta_1, \ldots, \beta_n \in \mathscr{C}^{m+1}(\Gamma)$ *satisfying*

$$\sum_{j=1}^{n} \beta_j(z) n_j(z) \neq 0 \quad \textit{for all} \quad z \in \Gamma . \tag{8.19}$$

Then for all $u, v \in \mathscr{C}^2(\Omega) \cap \mathscr{C}^1(\bar{\Omega})$ *with* $Pu, P^*v \in L^1(\Omega)$

$$\int_\Omega (vPu - uP^*v)\,\mathrm{d}x = \int_\Gamma (uB^*v - \alpha v Bu)\,\mathrm{d}\gamma , \tag{8.20}$$

where

$$P^* = -\sum_{i,j} a_{ij} \frac{\partial^2}{\partial x_i \partial x_j} - \sum b_j^* \frac{\partial}{\partial x_j} + c^* , \tag{8.21}$$

with

$$b_i^* = b_i + 2\sum_j \frac{\partial a_{ij}}{\partial x_j}, \qquad c^* = c - \sum_i \frac{\partial b_i}{\partial x_i} - \sum_{i,j} \frac{\partial^2 a_{ij}}{\partial x_i \partial x_j}, \tag{8.21)'$$

and

$$\alpha = \Big(\sum_j \beta_j n_j\Big)^{-1} \Big(\sum_{i,j} a_{ij} n_i n_j\Big), \tag{8.22}$$

$$B^* = \alpha B + T \quad \textit{with} \quad T = \tau_0 + \sum_{j=1}^{n} \tau_j \frac{\partial}{\partial x_j} \tag{8.23}$$

a differential operator of order 1 *on* Γ *with coefficients* $\tau_0 \in \mathscr{C}^m(\Gamma)$ *and* $\tau_1, \tau_2, \ldots, \tau_n \in \mathscr{C}^{m+1}(\Gamma)$, *satisfying*

$$\sum_{j=1}^{n} \tau_j(z) . n_j(z) = 0 \quad \textit{for all} \quad z \in \Gamma . \tag{8.24}$$

In other words, for P *a differential operator of order* 2 *on* Ω *and* B *a differential operator* 1 *on* Γ, *non-tangential* (*that is satisfying* (8.19)), *there exists a function* α *on* Γ *and* T *a differential operator of order* 1 *on* Γ, *tangential* (that is satisfying (8.24)) *such that we have Green's formula* (8.20) *with* P^* *the "adjoint differential operator" of* P *on* Ω (given by 8.21) *and* $B^* = \alpha B + T$. We notice that if P^* depends only on P, α and T depends on P, B and Γ; we can further specify $T = \tau_0 + \sum \tau_j \frac{\partial}{\partial x_j}$ with

$$\tau_0 = \sum_i \Big(b_i + \sum_j \frac{\partial a_{ij}}{\partial x_j} \Big) n_i - \frac{1}{2} \operatorname{div}_\Gamma (\tau_1, \ldots, \tau_n) \tag{8.25}$$

$$\tau_j = 2\Big(\sum_i a_{ij} n_i - \alpha \beta_j \Big), \qquad j = 1, \ldots, n , \tag{8.26}$$

where $\operatorname{div}_\Gamma(\tau_1, \ldots, \tau_n)$ is defined by the formula for integration by parts on Γ.

Lemma 1. *Let* Ω *be a regular bounded open set with boundary* Γ *of class* $\mathscr{C}^{m+2}$ *and* $\tau_1, \ldots, \tau_n \in \mathscr{C}^{m+1}(\Gamma)$ *satisfying* (8.24). *There exists a function*

$\operatorname{div}_\Gamma(\tau_1, \ldots, \tau_n) \in \mathscr{C}^m(\Gamma)$ *such that*

$$\int_\Gamma \left(\sum_j \frac{\partial u}{\partial x_j} \tau_j\right) \mathrm{d}\gamma = \int_\Gamma u \operatorname{div}_\Gamma (\tau_1, \ldots, \tau_n) \mathrm{d}\gamma \tag{8.27}$$

for all $u \in \mathscr{C}^1(\bar{\Omega})$.

We leave to the reader the verification of this lemma by the use of local maps or the application of general formulae of differential geometry (see Choquet–Bruhat [1]). Let us now give the

Proof of Proposition 2. Denote $L = -\sum_{i,j} \frac{\partial}{\partial x_i}\left(a_{ij} \frac{\partial}{\partial x_j}\right) + \sum d_i \frac{\partial}{\partial x_i} + c$ with $d_i = b_i + \sum_j \frac{\partial a_{ij}}{\partial x_j}$.

For $u \in \mathscr{C}^2(\Omega) \cap \mathscr{C}^1(\bar{\Omega})$ we have $u \in \mathscr{C}^2_L(\Omega) \cap \mathscr{C}^1_{n_L}(\bar{\Omega})$, $Lu = Pu$ on Ω and $\frac{\partial u}{\partial n_L} = \gamma . \operatorname{grad} u$ on Γ where for all $z \in \Gamma$, $\gamma(z)$ is the vector with components

$$\gamma_j(z) = \sum_{i=1}^n a_{ij}(z) n_i(z) .$$

With α given by (8.22) and $\tau_1, \ldots, \tau_n$ given by (8.26) we have

$$\frac{\partial}{\partial n_L} = \alpha(B - \beta_0) + \frac{1}{2} \sum_j \tau_j \frac{\partial}{\partial x_j} .$$

On the other hand $L^* = -\sum_{i,j} \frac{\partial}{\partial x_i}\left(a_{ij} \frac{\partial}{\partial x_j} + d_i\right) + c$,

$$\frac{\partial}{\partial n_{L^*}} = \frac{\partial}{\partial n_L} + \sum_i d_i n_i = \alpha(B - \beta_0) + \sum_i d_i n_i + \frac{1}{2} \sum \tau_j \frac{\partial}{\partial x_j} .$$

For $v \in \mathscr{C}^2(\Omega) \cap \mathscr{C}^1(\bar{\Omega})$, we have $v \in \mathscr{C}^2_{n_{L^*}}(\Omega) \cap \mathscr{C}^1_{n_{L^*}}(\bar{\Omega})$, $L^* v = P^* v$ and

$$u \frac{\partial v}{\partial n_{L^*}} - v \frac{\partial u}{\partial n_{L^*}} = \alpha(uBv - vBu) + \left(\sum_i d_i n_i\right) uv + \frac{1}{2} \sum \tau_j \left(u \frac{\partial v}{\partial x_j} - v \frac{\partial u}{\partial x_j}\right) .$$

Noting that

$$\frac{1}{2} \sum_i \tau_j \left(u \frac{\partial v}{\partial x_j} - v \frac{\partial u}{\partial x_j}\right) = u \left(\sum_j \tau_j \frac{\partial v}{\partial x_j}\right) - \frac{1}{2} \sum_j \tau_j \frac{\partial}{\partial x_j} (uv) ,$$

we therefore have from (8.27)

$$\int_\Gamma \left(u \frac{\partial v}{\partial n_{L^*}} - v \frac{\partial u}{\partial n_L}\right) \mathrm{d}\gamma = \int_\Gamma \left\{\alpha(uBv - vBu) + u \sum_j \tau_j \frac{\partial v}{\partial x_j} + \left(\sum_i d_i n_i - \frac{1}{2} \operatorname{div}_\Gamma (\tau_1, \ldots, \tau_n)\right) uv\right\} \mathrm{d}\gamma .$$

Green's formula (8.20) is deduced from Proposition 1. □

2. Different Concepts of Solutions, Boundary Value Problems, Transmission Conditions

We have defined what we mean naturally by a classical solution of the equation (8.1). As we have seen in the case of Poisson's equation, this notion is not sufficient; in the case of a general equation of the form (8.1) or of the divergence form (8.7), we shall be led to consider several concepts of solution weakening that of a classical solution.

First of all, we take a differential operator

$$P = -\sum_{i,j} a_{ij} \frac{\partial^2}{\partial x_i \partial x_j} + \sum_i b_i \frac{\partial}{\partial x_i} + c$$

with coefficients a_{ij}, b_i, $c \in \mathscr{C}^\infty(\Omega)$. We have defined by the formula (8.21), the adjoint operator

$$P^* = -\sum_{i,j} a_{ij} \frac{\partial^2}{\partial x_i \partial x_j} - \sum_i b_i^* \frac{\partial}{\partial x_i} + c^*$$

with

$$b_i^* = b_i + 2\sum_j \frac{\partial a_{ij}}{\partial x_j}, \qquad c^* = c - \sum_i \frac{\partial b_i}{\partial x_i} - \sum_{i,j} \frac{\partial^2 a_{ij}}{\partial x_i \partial x_j}$$

Given $u \in \mathscr{C}^2(\Omega)$, we have from the construction of P^*

$$\int_\Omega \xi P u \, dx = \int_\Omega u P^* \xi \, dx \qquad \text{for all} \quad \xi \in \mathscr{D}(\Omega)\,.$$

This, quite naturally, leads us to pose the

Definition 2. Given $P = -\sum_{i,j} a_{ij} \frac{\partial^2}{\partial x_i \partial x_j} + \sum_i b_i \frac{\partial}{\partial x_i} + c$ a differential operator of order 2 on an open set Ω of $\mathbb{R}^n$ with coefficients a_{ij}, b_i, $c \in \mathscr{C}^\infty(\Omega)$ and $f \in \mathscr{D}'(\Omega)$; by *a solution in the sense of distributions* of the equation

$$Pu = f \quad \text{on} \quad \Omega$$

we mean every u on Ω satisfying

$$\langle u, P^* \xi \rangle = \langle f, \xi \rangle \quad \text{for all} \quad \xi \in \mathscr{D}(\Omega)$$

where P^* is the adjoint operator defined by (8.21).

We notice that this definition has a meaning since $P^* \xi \in \mathscr{D}(\Omega)$ for all $\xi \in \mathscr{D}(\Omega)$. It generalizes clearly that given of a solution in the sense of distributions of Poisson's equation $\Delta u = f$ on Ω since Δ is formally self-adjoint. The hypothesis of $\mathscr{C}^\infty$-coefficients can be weakened by imposing a minimum regularity on u: in particular, if we restrict ourselves to considering solutions $u \in L^p_{\text{loc}}(\Omega)$ with $1 \leqslant p \leqslant \infty$, it is sufficient to suppose that the coefficients of P^* are in $L^q_{\text{loc}}(\Omega)$ with $q = p/(p-1)$ the conjugate index of p, with the result that $uP^*\xi \in L^1(\Omega)$ for all $\xi \in \mathscr{D}(\Omega)$. Although this aspect is interesting in certain applications, we shall rather develop the notion of a weak solution, for which we restrict ourselves to consider functions $u \in W^{1,1}_{\text{loc}}(\Omega)$[211] in using the divergence form

$$L = -\sum_i \frac{\partial}{\partial x_i}\left(\sum_j a_{ij}\frac{\partial}{\partial x_j} + c_i\right) + \sum_i d_i \frac{\partial}{\partial x_i} + d\,.$$

If the coefficients of P and L are regular on Ω and linked by the relations (8.8), we have, for all $u \in \mathscr{C}^2(\Omega)$,

(8.28) $$\int_\Omega \xi P u \,\mathrm{d}x = \int_\Omega L(u, \xi)\,\mathrm{d}x \quad \text{for all} \quad \xi \in \mathscr{D}(\Omega)$$

where $L(\cdot, \cdot)$ is the bilinear differential operator defined by (8.12). The regularity of the coefficients is not necessary for considering $L(u, \xi)$ for $\xi \in \mathscr{D}(\Omega)$ and $u \in W^{1,p_0}_{\text{loc}}(\Omega) \cap L^{p_1}_{\text{loc}}(\Omega)$; as we see below, it is interesting to consider operators with discontinuous coefficients. Although we can consider coefficients which are only measurable (see, for example, Trudinger [1], we suppose here, for simplicity that the coefficients are in $L^\infty_{\text{loc}}(\Omega)$.
We therefore pose the

Definition 3. Let $L = -\sum_i \frac{\partial}{\partial x_i}\left(\sum_j a_{ij}\frac{\partial}{\partial x_j} + c_i\right) + \sum_i d_i \frac{\partial}{\partial x_i} + d$ be a differential operator of order 2 in divergence form on an open set Ω of $\mathbb{R}^n$, whose coefficients are in $L^\infty_{\text{loc}}(\Omega)$ and $f \in \mathscr{D}'(\Omega)$. By a *weak solution* of the equation

$$Lu = f \quad \text{on} \quad \Omega$$

we mean any function $u \in W^{1,1}_{\text{loc}}(\Omega)$ satisfying

(8.29) $$\int_\Omega L(u, \xi)\,\mathrm{d}x = \langle f, \xi\rangle \quad \text{for all} \quad \xi \in \mathscr{D}(\Omega)\,.$$

If the coefficients of L are of class $\mathscr{C}^\infty$, it is clear that every weak solution of $Lu = f$ on Ω is a solution in the sense of distributions of $Pu = f$ on Ω, where P is the differential operator of the second order whose coefficients are given by (8.8). We note also that under the hypotheses of Definition 3, if there exists a weak solution of $Lu = f$ on Ω, the distribution f is of the form $f_0 + \Sigma\, \partial f_i/\partial x_i$ with $f_0, f_1, \ldots, f_n \in L^1_{\text{loc}}(\Omega)$; *the notion of a weak solution does not permit the solution of equations with arbitrary data and distribution* f[212]. Finally, from the relation (8.28) *if the coefficients are regular on Ω, then for $f \in \mathscr{C}^0(\Omega)$ u is the classical solution of $Pu = f$ on Ω iff $u \in \mathscr{C}^2(\Omega)$ and is a weak solution of $Lu = f$ on Ω*. In fact, the

---

[211] That is to say $u \in L^1_{\text{loc}}(\Omega)$ and its derivatives $\dfrac{\partial u}{\partial x_i}$ (in the sense of distributions) are in $L^1_{\text{loc}}(\Omega)$.

[212] We are limited by $f \in W^{-1,1}_{\text{loc}}(\Omega)$, which, all the same allows Radon measures on Ω; when the coefficients are continuous on Ω, we can consider $L(u, \xi)$ for every function $u \in BV_{\text{loc}}(\Omega)$ (that is to say $u \in L^1_{\text{loc}}(\Omega)$ and its derivatives $\partial u/\partial x_i$ are Radon measures on Ω) and define for an arbitrary distribution of order 1, a weak solution of $Lu = f$.

equation (8.28) can be verified under much more general hypotheses: without considering the most general case, let us suppose that the coefficients of P are in $L^\infty_{\text{loc}}(\Omega)$, with the result that $Pu \in L^1_{\text{loc}}(\bar{\Omega})$ for all $u \in W^{2,1}_{\text{loc}}(\Omega)$. Now to suppose (8.8) to be true in the sense of distributions comes down to supposing that (8.28) is true for $u \in \mathscr{C}^\infty(\Omega)$. Using the density of $\mathscr{C}^\infty(\Omega)$ in $W^{2,1}_{\text{loc}}(\Omega)$ and, under the hypothesis that the coefficients of P and L are in $L^\infty_{\text{loc}}(\Omega)$ the continuity of the maps $u \to L(u, \zeta)$, $u \to \zeta Pu$ of $W^{2,1}_{\text{loc}}(\Omega)$, into $L^1_{\text{loc}}(\Omega)$ we deduce that (8.28) is true for all $u \in W^{2,1}_{\text{loc}}(\Omega)$.
This leads us to pose the

Definition 4. Let $P = -\sum_{i,j} a_{ij} \dfrac{\partial^2}{\partial x_i \partial x_j} + \sum_i b_i \dfrac{\partial}{\partial x_i} + c$ be a differentiable operator of order 2 on an open set Ω of $\mathbb{R}^n$, whose coefficients are in $L^\infty_{\text{loc}}(\Omega)$, and $f \in L^1_{\text{loc}}(\Omega)$; by a *strong solution* of the equation $Pu = f$ on Ω, we call every function $u \in W^{2,1}_{\text{loc}}(\Omega)$ such that

$$Pu(x) = f(x) \quad \text{a.e.} \quad x \in \Omega\,. \tag{8.30}$$

We have shown above that, the coefficients of P and L being in $L^\infty_{\text{loc}}(\Omega)$ and (8.8) being satisfied in the sense of distributions, *u is a strong solution of $Pu = f$ on Ω iff $u \in W^{2,1}_{\text{loc}}(\Omega)$ and is a weak solution of $Lu = f$ on Ω.*
The different concepts of solutions that we have introduced for equation (8.1) or its divergence form (8.7) will have their correspondence for the boundary value problem

$$\begin{cases} Pu = f \quad (\text{resp. } Lu = f) \quad \text{on} \quad \Omega \\ Bu = g \quad \text{on} \quad \Gamma \end{cases} \tag{8.31}$$

where P (resp. L) is a differential operator of order 2 (resp. with divergence form) on Ω and B a differential operator of order less than or equal to 1.

Definition 5. Given Ω an open set in $\mathbb{R}^n$ with boundary Γ,

$$P = -\sum_{i,j} a_{ij} \frac{\partial^2}{\partial x_i \partial x_j} + \sum_i b_i \frac{\partial}{\partial x_i} + c$$

a differential operator of order 2 with coefficients a_{ij}, b_i, $c \in \mathscr{C}^0(\Omega)$ (resp. $\mathscr{C}^\infty(\Omega)$), $B = \beta_0 + \sum_{j=1}^n \beta_j \dfrac{\partial}{\partial x_j}$ a differential operator of order less than or equal to 1 with coefficients $\beta_j \in \mathscr{C}^0(\Omega)$ (resp. $\mathscr{C}^\infty(\Omega)$)[213], $f \in \mathscr{C}^0(\Omega)$ (resp. $\mathscr{D}'(\Omega)$) and $g \in \mathscr{C}^0(\Gamma)$, we call a classical (resp. *quasi-classical*) solution of the boundary value problem (8.31) every function $u \in \mathscr{C}^2(\Omega)$ (resp. distribution $u \in \mathscr{D}'(\Omega)$) such that
(1) u is a classical (resp. in the sense of distributions) of the equation $Pu = f$ on Ω;

---

[213] Taking account of the definition below, we find that it is sufficient to define $\beta_{0,j}$ in the neighbourhood of Γ in Ω.

(2) $\lim_{x \to z} Bu(x) = g(z)$ for all $z \in \Gamma$ in the classical sense (resp. in the sense specified in Definition 6 of §4)[214].

This definition clearly generalizes those given in §4 for the concepts of classical and quasi-classical solutions of a Dirichlet, Neumann, or even mixed problem for Poisson's equation. We note that a Dirichlet (resp. Neumann) condition

$$u = g \left(\text{resp. } \frac{\partial u}{\partial n} = g \right) \quad \text{on} \quad \Gamma$$

corresponds to $Bu = u$ (resp. $B = \sum_{j=1}^{n} n_i \frac{\partial}{\partial x_i}$, where $n = (n_1, \ldots, n_n)$ is a continuous extension to $\bar{\Omega}$ of the field of the unit normal vectors to Γ, exterior to Ω)[215]. We take as given the boundary condition in the form of a differential operator $B = \beta_0 + \sum_{j=1}^{n} \beta_i \frac{\partial}{\partial x_j}$ with coefficients in $\mathscr{C}^0(\Omega)$: it is obvious from the definition that the condition depends only on the values of the β_j in the neighbourhood of the boundary Γ, or more precisely of the "trace of the operator B on Γ" defined as an equivalence class modulo

$$\left\{ \beta_0 + \sum \beta_j \frac{\partial}{\partial x_j}; \quad \beta_j \in \mathscr{C}^0(\Omega), \quad \beta_j = 0 \quad \text{in the neighbourhood of } \Gamma \right\}^{216}.$$

There exist many ways of defining a notion of a weak solution of the boundary value problem (8.31), the weakening being able to bear moreover as much on the equation $Pu = f$ (resp. $Lu = f$) on Ω as on the condition on the boundary. We shall be content here to discuss the idea of a weak solution for a mixed boundary value in the case of an equation in divergence form:

$$\text{(8.32)} \qquad \begin{cases} Lu = f & \text{on} \quad \Omega \\ u = \varphi & \text{on} \quad \Gamma_0 \\ \dfrac{\partial u}{\partial n_L} = \psi & \text{on} \quad \Gamma \backslash \Gamma_0 \end{cases}$$

---

[214] That is to say, for all $\varepsilon > 0$, there exists $r > 0$ such that

$$|\langle Bu, \zeta \rangle - g(z)| \leqslant \varepsilon \quad \text{for all} \quad \zeta \in \mathscr{D}(\Omega \cap B(z, r)), \zeta \geqslant 0, \int \zeta \, dx = 1,$$

where

$$\langle Bu, \zeta \rangle = \langle u, B^* \zeta \rangle = \left\langle u, \beta_0 \zeta - \sum_{j=1}^{n} \frac{\partial}{\partial x_j} (\beta_j \zeta) \right\rangle.$$

[215] Such an extension always exists if we suppose Ω to be regular (with boundary of class $\mathscr{C}^1$) and we have seen (see §1.3) that for $u \in \mathscr{C}_n^1(\bar{\Omega})$, $\lim_{x \to z} \operatorname{grad} u(x) . n(x)$ was independent of the choice of this extension.

[216] See §7.4 for another example of trace as an equivalence class.

where L is a differential operator of the second order in divergence form on the open set Ω, Γ_0 is a closed part of the boundary Γ of Ω.
Supposing Ω to be bounded, the (open) part of the boundary $\Gamma\backslash\Gamma_0$ regular, the coefficients of L as well as that of f continuous on Ω and $\varphi \in \mathscr{C}^0(\Gamma_0)$, $\psi \in \mathscr{C}^0(\Gamma\backslash\Gamma_0)$ we can define, in a natural way, what we call a *classical solution of* (*8.32*): with the notation of Definition 1, this is a function

$$u \in \mathscr{C}_L^2(\Omega) \cap \mathscr{C}^0(\Omega \cup \Gamma_0) \cap \mathscr{C}_{n_L}^1(\bar{\Omega}\backslash\Gamma_0)$$

satisfying the equalities of (8.32) at every point of Ω, Γ_0, $\Gamma\backslash\Gamma_0$ respectively. Now, from Proposition 1, for such a classical solution u, we have

$$\int_\Omega L(u, v)\,\mathrm{d}x = \int_\Omega fv\,\mathrm{d}x + \int_{\Gamma\backslash\Gamma_0} \psi v\,\mathrm{d}\gamma$$

for every function $v \in \mathscr{C}^1(\Omega) \cap \mathscr{C}^0(\bar{\Omega})$ with $\operatorname{supp} v \cap \Gamma_0 = \varnothing$, fv and $L(u, v) \in L^1(\Omega)$[217]. More precisely, given

$$u \in \mathscr{C}_L^2(\Omega) \cap \mathscr{C}^0(\Omega \cup \Gamma_0) \cap \mathscr{C}_{n_L}^1(\bar{\Omega}\backslash\Gamma_0) \quad \text{with} \quad u = \varphi \quad \text{on} \quad \Gamma_0\,,$$

the above equalities characterize the property "u is a classical solution of (8.32)" subject to the density in $\mathscr{C}^0(\Gamma\backslash\Gamma_0)$ of the set of traces on $\Gamma\backslash\Gamma_0$ of the functions $v \in \mathscr{C}^1(\Omega) \cap \mathscr{C}^0(\bar{\Omega})$ with $\operatorname{supp} v \cap \Gamma_0 = \varnothing$, fv and $L(u, v) \in L^1(\Omega)$.
To place ourselves in the generality of the Definition 3, given Ω an arbitrary open set in $\mathbb{R}^n$, Γ_0 a closed part of its boundary, $1 \leqslant p < \infty$, *we denote by* $V^p(\Omega, \Gamma_0)$ *the closure in* $W_{\mathrm{loc}}^{1,p}(\bar{\Omega})$[218] *of the set of the restrictions to* Ω *of the* $v \in \mathscr{D}(\mathbb{R}^n\backslash\Gamma_0)$.

Definition 6. Let Ω be an arbitrary open set in $\mathbb{R}^n$, Γ_0 a closed part of its boundary and

$$L = -\sum_i \frac{\partial}{\partial x_i}\left(\sum_j a_{ij}\frac{\partial}{\partial x_j} + c_i\right) + \sum d_i \frac{\partial}{\partial x_i} + d$$

a differential operator of order 2 in divergence form with coefficients in $L_{\mathrm{loc}}^\infty(\bar{\Omega}\backslash\Gamma_0)$; finally, let $1 \leqslant p < \infty$, $\varphi \in W_{\mathrm{loc}}^{1,1}(\bar{\Omega}\backslash\Gamma_0)$, f a distribution on Ω of the form $f_0 + f_1$ with f_0 a Radon measure on Ω, bounded on $\Omega \cap K$ for all compact K of $\bar{\Omega}\backslash\Gamma_0$ and $f_1 \in \mathscr{E}'(\Omega \cup \Gamma_0)$, $\psi \in \mathscr{D}'(\mathbb{R}^n\backslash\Gamma_0)$ with support in $\Gamma\backslash\Gamma_0$. By a *weak solution relative to* $W_{\mathrm{loc}}^{1,p}(\bar{\Omega})$ *of the mixed problem* (8.32) we mean every function $u \in \varphi + V^p(\Omega, \Gamma_0)$

---

[217] We apply Proposition 1 to a regular open set Ω_1 contained in Ω such that $\bar{\Omega}_1 \cap \Gamma_0 = \varnothing$ and $\operatorname{supp} v \subset \Omega_1$.

[218] We recall that by an abuse of notation, for an open set Ω and Γ_0 a closed part of its boundary:

$$L_{\mathrm{loc}}^p(\bar{\Omega}\backslash\Gamma_0) = \{u \in L_{\mathrm{loc}}^1(\Omega)\,;\, u\chi_K \in L^p(\Omega) \text{ for every compact set } K \text{ of } \bar{\Omega}\backslash\Gamma_0\}\,,$$

$$W_{\mathrm{loc}}^{1,p}(\bar{\Omega}\backslash\Gamma_0) = \{u \in L_{\mathrm{loc}}^p(\bar{\Omega}\backslash\Gamma_0), \text{ the derivatives (in the sense of distributions on } \Omega) \ \frac{\partial u}{\partial x_i} \text{ are in } L_{\mathrm{loc}}^p(\bar{\Omega}\backslash\Gamma_0)\}.$$

Obviously $W_{\mathrm{loc}}^{1,p}(\bar{\Omega})$ corresponds to $\Gamma_0 = \varnothing$; if Ω is bounded, $L_{\mathrm{loc}}^p(\bar{\Omega}) = L^p(\Omega)$, $W_{\mathrm{loc}}^{1,p}(\bar{\Omega}) = W^{1,p}(\Omega)$ and $V^p(\Omega, \Gamma) = W_0^{1,p}(\Omega)$.

satisfying

$$(8.33) \quad \int_\Omega L(u, \zeta)\,dx = \int_\Omega f_0 \zeta\,dx + \langle f_1, \zeta \rangle + \langle \psi, \zeta \rangle, \quad \text{for all} \quad \zeta \in \mathscr{D}(\mathbb{R}^n \backslash \Gamma_0)$$

where $L(.,.)$ is the bilinear differential operator defined by (8.12).
We note that from the hypotheses, (8.33) has a meaning for all $\zeta \in \mathscr{D}(\mathbb{R}^n \backslash \Gamma_0)$. As we have done it in the preceding definitions, the Dirichlet condition on Γ_0 is given by a function $\varphi \in W^{1,1}_{\text{loc}}(\bar{\Omega} \backslash \Gamma_0)$; it is clear that the condition $u \in \varphi + V^p(\Omega, \Gamma_0)$ depends only on the trace of φ on Γ_0, defined here as the equivalence class modulo $V^p(\Omega, \Gamma_0)$. We notice also that in the case when Ω is unbounded, this definition of a weak solution and, moreover, that of a classical or a quasi-classical solution (Definition 5) do not involve a condition at infinity: that condition can be introduced in different ways as we have seen for Laplace's equation or Poisson's equation (see §4.3).
The notion of a weak solution like that we have defined above makes evident the duality between the problem (8.32) and the dual mixed boundary value problem

$$(8.34) \quad \begin{cases} L^* u^* = f^* & on \quad \Omega \\ u^* = \varphi^* & on \quad \Gamma_0 \\ \dfrac{\partial u^*}{\partial n_{L^*}} = \psi^* & on \quad \Gamma \backslash \Gamma_0 . \end{cases}$$

Restricting our attention to the case in which Ω is bounded but not necessarily regular, we can state in a precise fashion:

Proposition 3. *Let Ω be a bounded set of $\mathbb{R}^n$, Γ_0 a closed part of its boundary, $1 < p, p^* < \infty$ with $p^{-1} + p^{*-1} = 1$*

$$L = -\sum \frac{\partial}{\partial x_i}\left(a_{ij}\frac{\partial}{\partial x_j} + c_i\right) + \sum d_i \frac{\partial}{\partial x_i} + d$$

a linear differential operator of order 2 in divergence form with coefficients in $L^\infty(\Omega)$

Given $\quad \varphi \in W^{1,p}(\Omega) \quad (\text{resp. } \varphi^* \in W^{1,p^*}(\Omega))$

and

$$f_0, \ldots, f_n \in L^p(\Omega) \quad (\text{resp. } f_0^*, \ldots, f_n^* \in L^{p^*}(\Omega))$$

with $$\sum_{j=1}^n \frac{\partial f_i}{\partial x_i}\left(\text{resp. } \sum_{i=1}^n \frac{\partial f_i^*}{\partial x_i^*}\right) \in \mathscr{E}'(\Omega \cup \Gamma_0),$$

$$g_0, \ldots, g_n \in L^p(\Omega) \quad (\text{resp. } g_0^*, \ldots, g_n^* \in L^{p^*}(\Omega))$$

with $$g_0 = \sum_{i=1}^n \frac{\partial g_i}{\partial x_i} \quad \left(\text{resp. } g_0^* = \sum_{i=1}^n \frac{\partial g_i^*}{\partial x_i}\right) \quad \text{in} \quad \mathscr{D}'(\Omega),$$

u (resp. u^*) a weak solution relative to $W^{1,p}(\Omega)$ (resp. $W^{1,p^*}(\Omega)$) of (8.32)

(resp. (8.34)) with $f = f_0 - \sum_{i=1}^n \frac{\partial f_i}{\partial x_i}\left(\text{resp. } f^* = f_0^* - \sum_{i=1}^n \frac{\partial f_i^*}{\partial x_i}\right)$ *in* $\mathscr{D}'(\Omega)$ and

$$\psi = g_0 - \sum_i \frac{\partial g_i}{\partial x_i} \left(\text{resp. } \psi^* = g_0^* - \sum \frac{\partial g_i^*}{\partial x_i} \right) \quad \textit{in} \quad \mathscr{D}'(\mathbb{R}^n \backslash \Gamma_0), \quad \textit{we have}$$

(8.35)

$$\begin{cases} \displaystyle\int_\Omega (L(\varphi, u^*) - L(u, \varphi^*))\,\mathrm{d}x = \int_\Omega \{(f_0 + g_0)(u^* - \varphi^*) - (f_0^* + g_0^*)(u - \varphi)\}\,\mathrm{d}x \\ \displaystyle + \sum_{j=1}^n \int_\Omega (f_i + g_i)\left(\frac{\partial u^*}{\partial x_i} - \frac{\partial \varphi^*}{\partial x_i}\right) - (f_i^* + g_i^*)\left(\frac{\partial u}{\partial x_i} - \frac{\partial \varphi}{\partial x_i}\right)\Bigg\}\,\mathrm{d}x\,. \end{cases}$$

Proof. By definition of a weak solution of (8.32)

$$\text{(8.36)} \qquad \int_\Omega L(u, \zeta)\,\mathrm{d}x = \int_\Omega (f_0 + g_0)\zeta\,\mathrm{d}x + \sum_{i=1}^n \int_\Omega (f_i + g_i)\frac{\partial \zeta}{\partial x_i}\mathrm{d}x$$

for all $\zeta \in \mathscr{D}(\mathbb{R}^n \backslash \Gamma_0)$. Since $u^* \in \varphi^* + V^{p^*}(\Omega, \Gamma_0)$ the relation (8.36) will again be true for $\zeta = u^* - \varphi^*$: in effect using the definition of $V^{p^*}(\Omega, \Gamma_0)$ there exists a sequence (ζ_n), $\zeta_n \in \mathscr{D}(\mathbb{R}^n \backslash \Gamma_0)$

$$\zeta_n \to u^* - \varphi^* \quad \text{in} \quad W^{1,p^*}(\Omega)\,.$$

From the hypotheses on the data, we can therefore pass to the limit in (8.36). Similarly, we shall have

$$\text{(8.37)} \qquad \int_\Omega L^*(u^*, \zeta^*)\,\mathrm{d}x = \int_\Omega (f_0^* + g_0^*)\zeta^*\,\mathrm{d}x + \sum_{i=1}^n \int_\Omega (f_i^* + g_i^*)\frac{\partial \zeta^*}{\partial x_i}\mathrm{d}x$$

for $\zeta^* = u - \varphi$. Using $L^*(u^*, \zeta^*) = L(\zeta^*, u^*)$, and subtracting the equalities (8.36) and (8.37), we obtain (8.35). □

We note that (8.35) is only the expression of Green's formula (8.15) for weak solutions of the dual mixed problems (8.32), (8.34).

Finally, in this sub-section, we show how *operators with discontinuous coefficients* can be involved, and how to interpret the notions of *weak and strong solutions.* We take a domain Ω separated into two parts by a regular hypersurface $S \subset \Omega$ (see Fig. 25); we denote by Ω_1, Ω_2 the two domains thus determined. Given $u_1 \in L^1_{\text{loc}}(\Omega_1 \cup S)$, $u_2 \in L^1_{\text{loc}}(\Omega_2 \cup S)$ we can define $u = \chi_{\Omega_1} u_1 + \chi_{\Omega_2} u_2 \in L^1_{\text{loc}}(\Omega)$ that

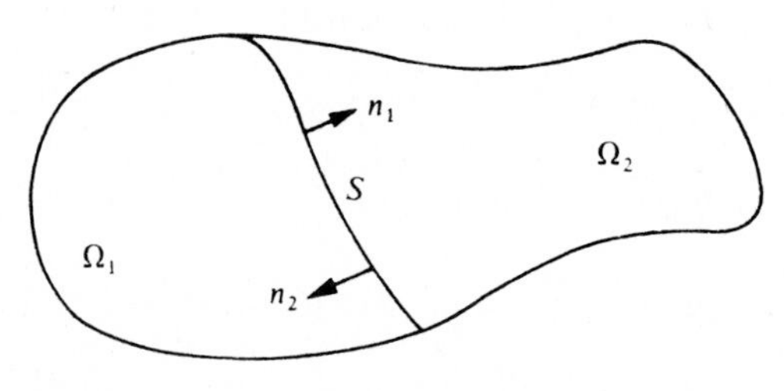

Fig. 25

is to say u is defined a.e. on Ω by

$$u = u_1 \quad \text{on} \quad \Omega_1, \quad u = u_2 \quad \text{on} \quad \Omega_2 .$$

Even if u_1, u_2 are very regular on Ω_1, Ω_2 right up to the interface S, the regularity of u on Ω depends on the coincidence of u_1, u_2 and of their derivatives on S. In a precise fashion, supposing within the classical framework thật $u_i \in \mathscr{C}^1(\Omega_i) \cap \mathscr{C}^0(\Omega_i \cup S)$ with $\operatorname{grad} u_i \in L^1_{\text{loc}}(\Omega_i \cup S)$ $(i = 1, 2)$ we have for $\zeta \in \mathscr{D}(\Omega)$

$$\begin{aligned}\int_\Omega u \operatorname{grad} \zeta \, dx &= \int_{\Omega_1} u_1 \operatorname{grad} \zeta \, dx + \int_{\Omega_2} u_2 \operatorname{grad} \zeta \, dx \\ &= -\int_{\Omega_1} \zeta \operatorname{grad} u_1 \, dx - \int_{\Omega_2} \zeta \operatorname{grad} u_2 \, dx + \int_S \zeta (u_1 n_1 + u_2 n_2) \, d\gamma\end{aligned}$$

where n_1, n_2 are the unit normal vectors to S exterior to Ω_1, Ω_2 respectively. In other words, the gradient of u in the sense of distributions on Ω, is the sum of

$$\chi_{\Omega_1} \operatorname{grad} u_1 + \chi_{\Omega_2} \operatorname{grad} u_2 \in L^1_{\text{loc}}(\Omega)$$

and of the measure carried by S, $-(u_1 n_1 + u_2 n_2) d\gamma$; the hypothesis $u_i \in \mathscr{C}^1(\Omega_i) \cap \mathscr{C}^0(\Omega_i \cup S)$ is in fact superfluous and it is sufficient to suppose $u_i \in W^{1,1}_{\text{loc}}(\Omega_i \cup S)$, which allows us to define the trace of u_i on S; we have thus shown that given $u_i \in W^{1,1}_{\text{loc}}(\Omega_i \cup S)$ $(i = 1, 2)$, *the function* $u = \chi_{\Omega_1} u_1 + \chi_{\Omega_2} u_2$ *belongs to* $W^{1,1}_{\text{loc}}(\Omega)$ *iff* $u_1 = u_2$ *on* S, *and then* $\operatorname{grad} u = \chi_{\Omega_1} \operatorname{grad} u_1 + \chi_{\Omega_2} \operatorname{grad} u_2$.
Let us consider now, on each open set, Ω_1, Ω_2 the equations

$$\left\{\begin{array}{l} P_1 u_1 \text{ (resp. } L_1 u_1) = f_1 \quad \text{on} \quad \Omega_1 \\ P_2 u_2 \text{ (resp. } L_2 u_2) = f_2 \quad \text{on} \quad \Omega_2 \end{array}\right. \tag{8.38}$$

where P_1, P_2 (resp. L_1, L_2) are the differential operators of order 2 (resp. in divergence form) on Ω_1, Ω_2.
Let us denote by P (resp. L) the differential operator on Ω,

$$\chi_{\Omega_1} P_1 + \chi_{\Omega_2} P_2$$
$$(\text{resp. } -\operatorname{div}(\chi_{\Omega_1} V_{L_1} + \chi_{\Omega_2} V_{L_2}) + \chi_{\Omega_1}(L_1 + \operatorname{div} V_{L_1}) + \chi_{\Omega_2}(L_2 + \operatorname{div} V_{L_2}))$$

whose coefficients a_{ij}, b_i, c (resp. a_{ij}, c_i, d_i, d) are defined a.e. on Ω by the values of the corresponding coefficients of P_1, P_2 (resp. L_1, L_2) on Ω_1, Ω_2 respectively. *Even if the coefficients of* P_1, P_2 *(resp.* L_1, L_2*) are very regular* (or even constants) *on* Ω_1, Ω_2, *the coefficients of* P *(resp.* L*) are, in general discontinuous on* Ω and we can consider only the strong (resp. weak) solutions of the equation

$$Pu \text{ (resp. } Lu) = f \quad \text{on} \quad \Omega , \tag{8.39}$$

where $f = \chi_{\Omega_1} f_1 + \chi_{\Omega_2} f_2$.

It is obvious that if u is a strong (resp. weak) solution of (8.39), its restrictions u_1, u_2 to Ω_1, Ω_2 are strong (resp. weak) solutions of (8.38) and $u = \chi_{\Omega_1} u_1 + \chi_{\Omega_2} u_2$.

In the case of a strong solution, $u \in W^{2,1}_{\text{loc}}(\Omega)$; hence $u_i \in W^{2,1}_{\text{loc}}(\Omega_i \cup S)$ and from what we have seen above

$$u_1 = u_2, \quad \frac{\partial u_1}{\partial x_k} = \frac{\partial u_2}{\partial x_k} \quad (k = 1, \ldots, n) \quad \text{on} \quad S \tag{8.40}$$

In fact if $u_1 = u_2$ on S, the tangential derivatives of u_1, u_2 coincide on S, with the result that (8.40) reduces to

$$u_1 = u_2\,, \quad \frac{\partial u_1}{\partial n_1} + \frac{\partial u_2}{\partial n_2} = 0 \quad \text{on} \quad S \tag{8.41}$$

Conversely, if $u_i \in W^{2,1}_{\text{loc}}(\Omega_i \cup S)$, $i = 1, 2$ are strong solutions of (8.38) and satisfy the conditions (8.41), it is clear that $u = \chi_{\Omega_1} u_1 + \chi_{\Omega_2} u_2$ is a strong solution of (8.39).
In the case of a weak solution, $u \in W^{1,1}_{\text{loc}}(\Omega)$; hence $u_i \in W^{1,1}_{\text{loc}}(\Omega_i \cup S)$ and $u_1 = u_2$ on S. We must have

$$\int_\Omega L(u, \zeta)\,\mathrm{d}x = \int_{\Omega_1} f_1 \zeta\,\mathrm{d}x + \int_{\Omega_2} f_2 \zeta\,\mathrm{d}x \quad \text{for all} \quad \zeta \in \mathscr{D}(\Omega)\,.$$

Now

$$\int_\Omega L(u, \zeta)\,\mathrm{d}x = \int_{\Omega_1} L_1(u_1, \zeta)\,\mathrm{d}x + \int_{\Omega_2} L_2(u_2, \zeta)\,\mathrm{d}x$$

and

$$\int_{\Omega_i} L_i(u_i, \zeta)\,\mathrm{d}x = \int_{\Omega_i} f_i \zeta\,\mathrm{d}x + \left\langle \frac{\partial u_i}{\partial n_{L_i}}, \zeta \right\rangle$$

where $\dfrac{\partial u_i}{\partial n_{L_i}}$ is the conormal derivative taken in the weak sense of Definition 6.
Hence u is a weak solution of (8.39) iff its restrictions u_i to Ω_i are weak solutions of (8.38) in $W^{1,1}_{\text{loc}}(\Omega \cup S)$ and satisfy the conditions

$$u_1 = u_2, \quad \frac{\partial u_1}{\partial n_{L_1}} + \frac{\partial u_2}{\partial n_{L_2}} = 0 \quad \text{on} \quad S\,. \tag{8.42}$$

We sum up this discussion in

Proposition 4. *With the notation and hypothesis of Definition* 4 (*resp. Definition* 3), *we consider S a regular hypersurface separating Ω into two open sets Ω_1, Ω_2, and denote by P_1, P_2 (resp. L_1, L_2) linear differential operators of order* 2 (*resp. in divergence form*) *on Ω_1, Ω_2 whose coefficients are the restrictions to Ω_1, Ω_2 of the coefficients of P (resp. L). A function u is a strong (resp. weak) solution of* (8.39) *iff its restrictions u_1, u_2 to Ω_1, Ω_2 are strong (resp. weak) solutions of* (8.38) *satisfying the conditions* (8.41) (*resp.* (8.42)).

Conditions (8.41), (8.42) are called *transmission conditions*: the introduction of an operator with discontinuous coefficients enables us therefore to interpret a problem made up of a problem of a partial differential equation on each open set Ω_1, Ω_2 and of transmission across their common boundary S as a differential problem on Ω. We note *the difference between the transmission conditions* (8.41) *corresponding to the strong solutions on* Ω *and the transmission conditions* (8.42) *corresponding to the weak solutions.*

3. General Results on the Regularity of Elliptic Problems of the Second Order

First of all we give some results on local regularity and regularity up to the boundary in the case of operators with regular coefficients.

Proposition 5. *Let* Ω *be an open set of* $\mathbb{R}^n$, $P = -\sum_{i,j} a_{ij} \frac{\partial^2}{\partial x_i \partial x_j} + \sum_i b_i \frac{\partial}{\partial x_i} + c$, *a differential operator of order* 2 *with* $\mathscr{C}^\infty$- *(resp. analytic) coefficients on* Ω *and* $f \in \mathscr{C}^\infty(\Omega)$ *(resp. analytic on* Ω*).*
(1) *Let us suppose* P *is elliptic on* Ω, *that is to say*

$$\sum_{i,j} a_{ij}(x)\xi_i\xi_j > 0 \quad \text{for all} \quad x \in \Omega,\ \xi \in \mathbb{R}^n \quad \text{with} \quad \xi \neq 0\,.$$

Then every solution in the sense of distributions of the equation

$$Pu = f \quad \text{on} \quad \Omega$$

is $\mathscr{C}^\infty$ *(resp. analytic on* Ω*)*[219].
(2) *Let* Γ_0 *be a regular open part of class* $\mathscr{C}^\infty$ *(resp. analytic) of the boundary of* Ω; *we suppose that the coefficients of* P *and* f *are* $\mathscr{C}^\infty$ (resp. *analytic on* Γ_0) on $\Omega \cup \Gamma_0$ *and* P *strictly elliptic on* Ω. *On the other hand, let* $g \in \mathscr{C}^\infty(\Gamma_0)$ (resp. *analytic on* Γ_0) *and* $B = \beta_0 + \sum_{j=1}^n \beta_j \frac{\partial}{\partial x_j}$ *a differential operator of order less than or equal to* 1 *with coefficients belonging to* $\mathscr{C}^\infty(\Omega \cup \Gamma_0)$ (resp. *analytic on* $\Omega \cup \Gamma_0$).
We suppose

$$\beta_0 \neq 0 \qquad \beta_1 = \cdots = \beta_n \equiv 0 \quad \text{on} \quad \Gamma_0 \quad (\textit{Dirichlet problem})$$

or

$$\sum_{j=1}^n \beta_j n_j \neq 0 \quad \text{on} \quad \Gamma_0 \quad (\textit{Neumann problem})^{220}.$$

Then every classical solution of the problem

$$\begin{cases} Pu = f & \text{on} \quad \Omega \\ Bu = g & \text{on} \quad \Gamma_0 \end{cases}$$

is $\mathscr{C}^\infty$ (resp. analytic) on $\Omega \cup \Gamma_0$.

---

[219] And hence a classical solution of $Pu = f$ on Ω.
[220] n_j is the j-th component of the unit normal to Γ_0 exterior to Ω. This condition corresponds to a Neumann problem with possibly an oblique derivative (see Proposition 2).

This proposition has been proved to within the details stated in §6 except what concerns the Neumann problem where by the given proof we are confined to the case of an operator $B = \sum_j \gamma_j \dfrac{\partial}{\partial x_j}$ where $\gamma_j = \sum_j a_{ij} n_i$[221].
We note that further we have shown results of regularity $W^{m+2,p}$ or $\mathscr{C}^{m+2,\alpha}$ when the data have the corresponding regularity. Finally, we note that the point (1) is proved for an elliptic operator in Chap. V. We refer to Agmon–Douglis–Nirenberg [1] for a complete proof of the Proposition in the $W^{m,p}$ framework and to Gilbarg–Trudinger [1] for the $\mathscr{C}^{m,\alpha}$ context. In all of these results, the continuity of the coefficients a_{ij} is essential since we use principally that in the neighbourhood of x_0, the operator $\sum a_{ij}(x)\partial^2/\partial x_i \partial x_j$ "is not very different" from the operator with constant coefficients $\sum a_{ij}(x_0)\partial^2/\partial x_i \partial x_j$ which is, to within a change of coordinates, the Laplace operator.
In the case of operators with discontinuous coefficients, we cannot hope to have the regularity $\mathscr{C}^2$ (resp. $\mathscr{C}^1$) of strong (resp. weak) solution; to convince ourselves of this, it is enough to consider in 1 dimension, the problem:

$$a \frac{\mathrm{d}^2 u}{\mathrm{d}x^2}\left(\text{resp. } \frac{\mathrm{d}}{\mathrm{d}x}\left(a \frac{\mathrm{d}u}{\mathrm{d}x}\right)\right) = f \quad \text{on} \quad]-1, 1[\tag{8.43}$$

where

$$a(x) = \begin{cases} a_+ & \text{on} \quad]0, 1[\\ a_- & \text{on} \quad]-1, 0[\,, \end{cases}$$

a_+, a_- being distinct negative constants; if f is continuous on $]-1, +1[$ (resp. $f \in L^1_{\text{loc}}(]-1, +1[)$) a strong (resp. weak) solution of (8.43) cannot be of class $\mathscr{C}^2$ (resp. $\mathscr{C}^1$) since $\dfrac{\mathrm{d}^2 u}{\mathrm{d}x^2} = \dfrac{f}{a}\left(\text{resp. } \dfrac{\mathrm{d}u}{\mathrm{d}x} = \dfrac{1}{a}\displaystyle\int f(x)\,\mathrm{d}x\right)$ is discontinuous at O.
The De Georgi–Nash theorem shows that we have, however, even in the case of operators with discontinuous coefficients, a Hölder continuity of the weak solutions; in precise form, we state[222]:

Proposition 6. *Let Ω be an open set in $\mathbb{R}^n$, $L = -\sum_i \dfrac{\partial}{\partial x_i}\left(\sum_j a_{ij}\dfrac{\partial}{\partial x_j} + c_i\right)$ $+ \sum_i d_i \dfrac{\partial}{\partial x_i} + d$ a linear differential operator of order 2 with coefficients in $L^\infty(\Omega)$ and strictly elliptic, that is to say satisfying, with a constant $c_0 > 0$*

$$\sum_{i,j} a_{ij}(x)\xi_i\xi_j \geqslant c_0|\xi|^2 \quad \text{a.e.} \quad x \in \Omega, \quad \textit{for all} \quad \xi \in \mathbb{R}^n\,;$$

furthermore, let $p > n$ and $f_0, \ldots, f_n \in L^p(\Omega)$.

---

[221] Or, more generally an operator B whose tangential terms τ_j given by (8.26) are "small".
[222] See Trudinger [1] for the results under more general hypotheses.

(1) *Every weak solution of*

$$Lu = f_0 - \sum \frac{\partial f_i}{\partial x_i} \quad \text{on} \quad \Omega \tag{8.44}$$

satisfies a Hölder condition locally on Ω.
(2) *Suppose* Ω *is bounded with Lipschitz boundary and let* $\varphi \in W^{1,p}(\Omega)$ (*resp.* $g_0, \ldots, g_n \in L^p(\Omega)$ *with* $g_0 = \sum \frac{\partial g_i}{\partial x_i}$ *on* $\mathscr{D}'(\Omega)$ *and suppose also that* $\sum \frac{\partial f_i}{\partial x_i} \in \mathscr{E}'(\Omega)$).
Then every weak solution relative to $W^{1,2}(\Omega)$ *of the Dirichlet* (*resp. Neumann*) *problem*

$$\begin{cases} Lu = f_0 - \sum \dfrac{\partial f_i}{\partial x_i} & \text{on} \quad \Omega \\ u = \varphi \quad \text{on} \quad \Gamma \\ \left(\text{resp.} \dfrac{\partial u}{\partial n_L} = g_0 - \sum \dfrac{\partial g_i}{\partial x_i} \quad \text{on} \quad \Gamma\right) \end{cases} \tag{8.45}$$

satisfies a Hölder condition on $\bar{\Omega}$.

In the case of the Poisson equation, $-\Delta u = f_0 - \sum \partial f_i/\partial x_i$ on Ω, we know that for $f_0, \ldots, f_n \in L^p_{\text{loc}}(\Omega)$ with $p > n$, the solution satisfies locally a Hölder condition of order $(1/n) - (1/p)$ on Ω[223]; this is still true for an elliptic operator with regular coefficients. In the case of De Georgi–Nash theorem, we affirm only that a solution of $Lu = f_0 - \sum \partial f_i/\partial x_i$ on Ω satisfies a Hölder condition locally on Ω: its order on a compact set K of Ω depends on K, norms in $L^\infty(\Omega)$, of the coefficients of $(1/c_0)L$ and on $p > n$. We refer the reader to Trudinger [1], Stampacchia [1] or Ladyzenskaya–Ural'ceva [1] for a complete proof of the De Giorgi–Nash theorem as well as different developments on elliptic problems of the second order with discontinuous coefficients. We shall give here the essential elements of a proof of this theorem so as to show the constituents used which are very different from those used in the case of continuous coefficients: an essential technique is the *truncation method* introduced by De Giorgi. In this section we shall use it to obtain global estimates of a weak solution relative to $W^{1,2}(\Omega)$ of the mixed problem (8.45). We shall show in Sect. 5 below how to localize this method to obtain a Harnack inequality (see Propositions 15 and 16) from which we can deduce a proof of Proposition 6.
First of all we prove:

Lemma 2. *Let* Ω *be an open set in* $\mathbb{R}^n$, L *a differential operator of order 2 in divergence form with coefficients in* $L^\infty(\Omega)$ *and strictly elliptic,* Γ_0 *a part of the boundary of* Ω $p \to 2$, $f_0, \ldots, f_n \in L^p(\Omega)$ *with* $\sum \frac{\partial f_i}{\partial x_i} \in \mathscr{E}'(\Omega \cup \Gamma_0)$,

---

[223] In effect $u \in W^{1,p}_{\text{loc}}(\Omega)$ (see §3) and $W^{1,p}_{\text{loc}}(\Omega) \subset \mathscr{C}^{\frac{1}{n}-\frac{1}{p}}_{\text{loc}}(\Omega)$ (see Brézis [1]).

$g_0, \ldots, g_n \in L^p(\Omega)$ with $g_0 = \sum \frac{\partial g_i}{\partial x_i}$ in $\mathscr{D}'(\Omega)$, $\varphi \in W^{1,p}(\Omega)$ and u a weak solution relative to $W^{1,2}_{\mathrm{loc}}(\bar{\Omega})$ of the mixed problem (8.32) with $f = f_0 - \sum \frac{\partial f_i}{\partial x_i}$ in $\mathscr{D}'(\Omega)$ and $\psi = g_0 - \sum \frac{\partial g_i}{\partial x_i}$ in $\mathscr{D}'(\mathbb{R}^n \backslash \Gamma_0)$. For every function θ, increasing on $\mathbb{R}$ with $r\theta(r) \geqslant 0$ such that, putting $w = u - \varphi$, we have

$$w_1 = w|\theta(w)|^{1/2} \in L^2(\Omega) \quad \text{and} \quad w_2 = |\theta(w)|^{1/2} \in L^{2p/p-2}(\Omega)\,,$$

then the function $v = \int_0^w |\theta(r)|^{1/2}\,\mathrm{d}r$ belongs to $V^2(\Omega, \Gamma_0)$ and

$$\text{(8.46)} \quad \|v\|_{W^{1,2}} \leqslant C\{1 + \|\varphi\|_{W^{1,p}} + \frac{1}{c_0}\sum_{i=0}^{n} \|f_i + g_i\|_{L^p}\}(\|w_1\|_{L^2} + \|w_2\|_{L^{2p/p-2}})$$

where C depends only on the L^∞-norms of the coefficients of $(1/c_0)\,L$ where c_0 is the constant of strict ellipticity of L.

Proof. By definition $w \in V^2(\Omega, \Gamma_0)$ and

$$\text{(8.47)} \quad \int_\Omega L(w, \zeta)\,\mathrm{d}x = \int_\Omega \left(\bar{f}_0 \zeta + \sum \bar{f}_i \frac{\partial \zeta}{\partial x_i}\right)\mathrm{d}x \quad \text{for all} \quad \zeta \in \mathscr{D}(\mathbb{R}^n \backslash \Gamma_0)$$

where

$$\bar{f}_0 = f_0 + g_0 - \left(\sum d_i \frac{\partial \varphi}{\partial x_i} + d\varphi\right)$$

$$\bar{f}_i = f_i + g_i - \left(\sum a_{ij} \frac{\partial \varphi}{\partial x_j} + c_i\varphi\right) \quad i = 1, \ldots, n\,.$$

In the limit, (8.47) is again true for all $\zeta \in V^2(\Omega, \Gamma_0)$. Supposing first θ to be continuous and bounded on $\mathbb{R}$, we can therefore apply (8.47) with $\zeta = \int_0^w |\theta(r)|\,\mathrm{d}r$; using the growth of θ, we have $|\zeta| \leqslant w\theta(w)$ and the application of (8.47) gives

$$\int_\Omega |\theta(w)| \sum_{i,j} a_{ij} \frac{\partial w}{\partial x_i}\frac{\partial w}{\partial x_j}\,\mathrm{d}x \leqslant \int_\Omega (|\bar{f}_0| - d|w|)w\theta(w)\,\mathrm{d}x$$
$$+ \int_\Omega \sum_i \left|\frac{\partial w}{\partial x_i}\right| |\theta(w)|(|\bar{f}_i| + (|c_i| + |d_i|)|w|)\,\mathrm{d}x\,.$$

Using the ellipticity and majorizing the second member by Young's inequality we obtain

$$\frac{c_0}{2}\int_\Omega |\theta(w)| \sum \left(\frac{\partial w}{\partial x_i}\right)^2 \mathrm{d}x \leqslant \int_\Omega \frac{|\theta(w)|}{c_0} \sum_{i=0}^{n} \bar{f}_i^2\,\mathrm{d}x$$
$$+ \int_\Omega w^2|\theta(w)|\left(\left(\frac{c_0}{4} - d\right)^+ + \frac{1}{c_0}\sum_{i=1}^{n}(|c_i| + |d_i|)^2\right)\mathrm{d}x,$$

from which (8.46) follows noting that $|\theta(w)|\left(\frac{\partial w}{\partial x_i}\right)^2 = \left(\frac{\partial v}{\partial x_i}\right)^2$ and $v^2 \leqslant w^2|\theta(w)| = w_1^2$.
The explanation of the constant C offers nothing of interest, but we see clearly that it depends only on the L^∞ norms of the coefficients of $(1/c_0)L$.
To prove the lemma in the case of a general function θ, it is sufficient to note that there exists a sequence (θ_k) of increasing bounded continuous functions such that

$$|\theta_k(r)| \leqslant |\theta(r)| \quad \text{for all} \quad k \quad \text{and all} \quad r \in \mathbb{R}$$

$$\theta_k(r) \to \theta(r) \quad \text{when} \quad k \to \infty\,, \quad \text{a.e.} \quad r \in \mathbb{R}\,.$$

Applying (8.46) for θ_k, we obtain a majorization of

$$v_k = \int_0^w |\theta_k(r)|^{1/2}\,\mathrm{d}r \quad \text{in} \quad W^{1,2}(\Omega)$$

by the term on the right of (8.46) (corresponding to θ); since $v_k \to v$ a.e. on Ω, we deduce that $v \in V^2(\Omega, \Gamma_0)$ with the estimate (8.46). □

When Ω is bounded with a Lipschitz boundary in $\mathbb{R}^n$ with $n \geqslant 3$ (resp. $n = 2$), we have the Sobolev inclusion $W^{1,2}(\Omega) \subset L^{2m/(m-2)}$ with $m \geqslant n$ (resp. $m > 2$) which allows us to write (8.46):

(8.48)

$$\|v\|_{L^{2m/m-2}} \leqslant C\left\{1 + \|\varphi\|_{W^{1,p}} + \frac{1}{c_0}\sum \|f_i + g_i\|_{L^p}\right\}(\|w_1\|_{L^2} + \|w_2\|_{L^{2p/(p-2)}})$$

where we have included the constant of the Sobolev inclusion in C which thus now depends on m and Ω[224].
Such an inequality (8.48) allows us to obtain estimates on u.

Lemma 3. *Let $m, p \geqslant 2$ and $w \in L^2(\Omega)$[225]. We suppose that there exists a constant C_0 such that for every increasing function θ on $\mathbb{R}$ with $r\theta(r) \geqslant 0$*

$$w_1 = w|\theta(w)|^{1/2} \in L^2(\Omega) \quad \textit{and} \quad w_2 = |\theta(w)|^{1/2} \in L^{2p/(p-2)}(\Omega)\,,$$

the function $v = \int_0^w |\theta(r)|^{1/2}\,\mathrm{d}r \in L^{2m/(m-2)}(\Omega)$ and satisfies

(8.49)
$$\|v\|_{L^{2m/m-2}} \leqslant C_0(\|w_1\|_{L^2} + \|w_2\|_{L^{2p/(p-2)}})\,.$$

If $p < m$ (resp. $p > m$), then $w \in L^{pm/(m-p)}(\Omega)$ (resp. $w \in L^\infty(\Omega)$).

We deduce:

---

[224] We have $v \in V^2(\Omega, \Gamma_0)$; in the case in which $\Gamma_0 = \Gamma$, $V^2(\Omega, \Gamma_0) = W_0^{1,2}(\Omega)$ and the constant C depends only on m and $|\Omega|^{\frac{1}{n}-\frac{1}{m}}$ where $|\Omega|$ is the measure of Ω. In particular in this case, the hypothesis that the boundary satisfies a Lipschitz condition is not necessary, also if $m = n \geqslant 3$, the assumption that Ω is bounded is superfluous.

[225] This is a lemma from the theory of integration: Ω can denote an arbitrary measure space.

Proposition 7. *With the data of Lemma 2, let us suppose Ω is bounded with Lipschitz boundary. If $p < n$ (resp. $p > n$), then $u \in L^{pn/(n-p)}(\Omega)$ (resp. $u \in L^{\infty}(\Omega)$) and*

$$(8.50) \qquad \|u\|_{L^{np/(n-p)}} \quad (\text{resp. } \|u\|_{L^\infty}) \leqslant C\left\{\|\varphi\|_{W^{1,p}} + \frac{1}{c_0}\sum_{i=0}^{n}\|f_i + g_i\|_{L^p}\right\}$$

where C depends only on the norms in L^∞ of the coefficients of $(1/c_0)L$, on p and on the open set Ω.

This Proposition is clearly a consequence of Lemmas 2 and 3, since, Ω being bounded with a boundary which satisfies a Lipschitz condition, if $p < n$ (resp. $p > n$) we have $W^{1,p}(\Omega) \subset L^{pn/(n-p)}(\Omega)$ (resp. $W^{1,p}(\Omega) \subset L^\infty(\Omega)$) with the result that $u \in L^{pn/(n-p)}(\Omega)$ (resp. $L^\infty(\Omega)$) iff it is thus of the form $w = u - \varphi$. The estimate (8.50) is obtained from (8.48) and from Lemma 3 by a homogeneity argument.

Proof of Lemma 3. Given $k \geqslant 0$, we denote by

$$\Omega_k = \{x \in \Omega; |w(x)| > k\} \quad \text{and} \quad \varphi(k) = \|(|w| - k)^+\|_{L^1}.$$

The function φ is convex, decreasing on $\mathbb{R}^+$ with derivative $\varphi'(k) = -|\Omega_k|$, the measure of the set Ω_k. Given $\alpha \geqslant 0$, we apply (8.49) with the function[226]

$$\theta(r) = (\alpha + 1)^2 \quad (\operatorname{sign} r)(|r| - k)^{+2\alpha}$$

We have then

$$w_1 = (\alpha + 1)w(|w| - k)^{+\alpha}, \qquad w_2 = (\alpha + 1)(|w| - k)^{+\alpha}$$

and

$$v = (\operatorname{sign} w)(|w| - k)^{+\alpha+1};$$

so if $w \in L^{p_\alpha}(\Omega)$ where $p_\alpha = \max\left(2(\alpha + 1), \dfrac{2\alpha p}{p - 2}\right)$, then $w \in L^{q_\alpha}(\Omega)$ where $q_\alpha = 2(\alpha + 1)\dfrac{m}{m - 2}$ and for all $k \geqslant 0$,

$$\|(|w| - k)^+\|_{L^{q_\alpha}}^{\alpha+1} \leqslant C_0(\alpha + 1)(\|w(|w| - k)^{+\alpha}\|_{L^2} + \|(|w| - k)^{+\alpha}\|_{L^{2p/(p-2)}}).$$

By Hölder's inequality

$$\begin{aligned}\|w(|w| - k)^{+\alpha}\|_{L^2} &\leqslant \|(|w| - k)^{+\alpha+1}\|_{L^2} + k\|(|w| - k)^{+\alpha}\|_{L^2} \\ &\leqslant \|(|w| - k)^+\|_{L^{q_\alpha}}^{\alpha+1}|\Omega_k|^{1/m} + k\|(|w| - k)^+\|_{L^{q_\alpha}}^{\alpha}|\Omega_k|^{(\frac{1}{2} + \frac{\alpha}{m})\alpha+1}\end{aligned}$$

and subject to $p_\alpha \leqslant q_\alpha$,

$$\|(|w| - k)^{+\alpha}\|_{L^{2p/(p-2)}} \leqslant \|(|w| - k)^+\|_{L^{q_\alpha}}^{\alpha}|\Omega_k|^{(\frac{1}{2} + \frac{\alpha}{m} - \frac{\alpha+1}{p})/\alpha+1}$$

Hence on condition that $w \in L^{p_\alpha}(\Omega)$, $p_\alpha \leqslant q_\alpha$ and $C_0(\alpha + 1)|\Omega_k|^{1/m} \leqslant \frac{1}{2}$, we shall have

$$(8.51) \qquad \|(|w| - k)^+\|_{L^{q_\alpha}} \leqslant 2C_0(\alpha + 1)|\Omega_k|^{(\frac{1}{2} + \frac{\alpha}{m} - \frac{\alpha+1}{p})/\alpha+1}(1 + k|\Omega_k|^{1/p}).$$

---

[226] We denote here $(|r| - k)^{+2\alpha}$ for $((|r| - k)^+)^{2\alpha}$.

We note that $k|\Omega_k|^{1/2} \leqslant \|w\|_{L^2}$ with the result that the condition $C_0(\alpha + 1)|\Omega_k|^{1/m} \leqslant \frac{1}{2}$ is in particular satisfied if

$$k \geqslant k_\alpha = (2C_0(\alpha + 1))^{m/2} \|w\|_{L^2} .$$

When $p < m$, $\{\alpha \geqslant 0,\ p_\alpha \leqslant q_\alpha\} = \left[0, \dfrac{m(p-2)}{2(m-p)}\right]$ and for $\alpha = \dfrac{m(p-2)}{2(m-p)}$, $p_\alpha = q_\alpha = \dfrac{mp}{m-p}$; therefore, by continuity of the norm $w \in L^{mp/(m-p)}(\Omega)$ and (8.51) gives an estimate of $\|w\|_{L^{mp/(m-p)}}$ noting that for $q \geqslant 2$

$$\|w\|_{L^q} \leqslant 2k^{1-\frac{2}{q}} \|w\|_{L^2}^{2/q} + \|(|w| - k)^+\|_{L^q} ;$$

The reader will verify that if we use $k = k_n$ we have

$$\|w\|_{L^q} \leqslant 3\left(C_0 p \frac{m-2}{m-p}\right)^{1 + m(\frac{1}{2} - \frac{1}{p})} \|w\|_{L^2} + C_0 p \frac{m-2}{m-p} .$$

When $p > m$, to prove that $w \in L^\infty(\Omega)$ we shall use (8.51) with $\alpha = 0$. We then have for all $k \geqslant k_0 = (2C_0)^{m/2} \|w\|_{L^2}$

$$\|(|w| - k)^+\|_{L^{2m/m-2}} \leqslant C_1 k |\Omega_k|^{\frac{1}{2} - \frac{1}{p}}$$

with
$$C_1 = 2C_0\left(\frac{1}{k_0} + |\Omega_{k_0}|^{1/p}\right).$$

Then by Hölder's inequality we have

$$\varphi(k) = \|(|w| - k)^+\|_{L^1} \leqslant C_1 k |\Omega_k|^{1 + \frac{1}{m} - \frac{1}{p}} ;$$

which by putting $\beta = 1 - \left(1 + \dfrac{1}{m} - \dfrac{1}{p}\right)^{-1} > 0$, and recalling that $|\Omega_k| = -\varphi'(k)$ we can rewrite as

$$\frac{\mathrm{d}}{\mathrm{d}k}(C_1^{1-\beta}\varphi(k)^\beta + k^\beta) \leqslant 0 \quad \text{for} \quad k \geqslant k_0 ,$$

from which we deduce that $\varphi(k) = 0$ for

$$k^\beta \geqslant C_1^{1-\beta}\varphi(k_0)^\beta + k_0^\beta$$

namely that again $w \in L^\infty(\Omega)$ and

$$\|w\|_{L^\infty} \leqslant (k_0^\beta + C_1^{1-\beta}\varphi(k_0)^\beta)^{1/\beta} . \qquad \square$$

We have detailed the proof of the estimates of Proposition 7 to convince the reader of the power of the truncation method: using this method we shall obtain the De Georgi–Nash theorem; as we have already stated we shall take up this local technique in Sect. 3 to prove Harnack's inequality. The above proof also makes obvious the estimates of Proposition 7; we shall state without proof the corresponding local result:

Proposition 8. *Let Ω be an open set of $\mathbb{R}^n$, L a linear differential operator of order 2 in divergence form with coefficients in $L^\infty(\Omega)$ and strictly elliptic, $p \geqslant 2$, $f_0, \ldots, f_n \in L^p(\Omega)$ and u a weak solution in $W^{1,2}_{\text{loc}}(\Omega)$ of $Lu = f$ on Ω. If $p < n$ (resp. $p > n$), then $u \in L^{pn/(n-p)}_{\text{loc}}$ (resp. $u \in L^\infty_{\text{loc}}(\Omega)$) and for every compact set K in Ω*

$$\|u\|_{L^{pn/n-p}(K)} (\text{resp. } \|u\|_{L^\infty(K)}) \leqslant C \sum_{i=0}^{n} \|f_i\|_{L^p(\Omega)}$$

where C depends only on the L^∞-norms of the coefficients of L, on the constant of uniform ellipticity and on the distance from K to the boundary of Ω.

As we have shown in the preceding section, operators with discontinuous coefficients allow us, in particular, to express problems on two different media Ω_1, Ω_2 with transmission conditions on the common interface S to Ω_1 and Ω_2. As we have seen, the global regularity of a solution in the passage to S is limited; however, supposing the operators on Ω_1 and Ω_2 to have coefficients regular right up to the interface S, we can show that the solution will likewise be regular on Ω_1 and Ω_2 up to the interface S. We state the result in the case of an operator in divergence form:

Proposition 9. *Let Ω be an open set of $\mathbb{R}^n$ separated into two open sets Ω_1 and Ω_2 by a hypersurface S of class $\mathscr{C}^\infty$, L a linear differential operator of order 2 in divergence form, strictly elliptic on Ω and f a function on Ω. We suppose the restrictions of f and of the coefficients of L to Ω_1 and Ω_2 are $\mathscr{C}^\infty(\Omega_1 \cup S)$ and $\mathscr{C}^\infty(\Omega_2 \cup S)$ respectively. Then every weak solution in $W^{1,2}_{\text{loc}}(\Omega)$ of*

$$Lu = f \quad on \quad \Omega$$

has its restrictions u_1, u_2 to Ω_1, Ω_2 in $\mathscr{C}^\infty(\Omega_1 \cup S)$, $\mathscr{C}^\infty(\Omega_2 \cup S)$ respectively and therefore in particular classical solutions of the transmission problem

$$\begin{cases} L_1 u_1 = f & on \quad \Omega_1 \\ L_2 u_2 = f & on \quad \Omega_2 \\ u_1 = u_2 & on \quad S \\ \dfrac{\partial u_1}{\partial n_{L_1}} + \dfrac{\partial u_2}{\partial n_{L_2}} = 0 & on \quad S \end{cases}$$

where L_1, L_2 are the restrictions of L to Ω_1, Ω_2 respectively.

Proof. The problem being local, we can always suppose that, after a change of variables, $S = \{(x', 0); x' \in \mathbb{R}^{n-1}\} \cap \Omega$. From Proposition 6 we know already that u is continuous on Ω with the result that u_1 (resp. u_2) is a quasi-classical solution of

$$\begin{cases} Pv = f & \text{on} \quad \Omega_1 (\text{resp. } \Omega_2) \\ v = u & \text{on} \quad S \end{cases}$$

where P is the differential operator of order 2 on $\Omega \backslash S$ strictly elliptic with $\mathscr{C}^\infty$-coefficients defined by (8.8) in terms of those of L. From the proof of Theorem 1 of

§6, if the derivatives with respect to x', $\dfrac{\partial^\alpha u}{\partial x'^\alpha}$ belong to $L^2_{\text{loc}}(\Omega)$ for all $|\alpha| \leqslant m$, then the restrictions u_1 (resp. u_2) of u to Ω_1 (resp. Ω_2) are in $H^m_{\text{loc}}(\Omega_1 \cup S)$ (resp. $H^m_{\text{loc}}(\Omega_2 \cup S)$). Since $\bigcap_{M \geqslant 0} H^m_{\text{loc}}(\Omega_i \cup S) = \mathscr{C}^\infty(\Omega_i \cup S)$ (see Chap. IV), it is enough to prove that $\dfrac{\partial^\alpha u}{\partial x'^\alpha} \in L^2_{\text{loc}}(\Omega)$ for all α.

We shall make use of the method of translations. Given $h \in \mathbb{R}$, $h \neq 0$ $k = 1, \ldots, n-1$, for a function v defined on Ω we shall use the notation

$$\frac{\partial^h v}{\partial x_k}(x_1, \ldots, x_n) = \frac{1}{h}(v(x_1, \ldots, x_k + h, \ldots, x_n) - v(x_1, \ldots, x_n))$$

defined on $\Omega^h_k = \{(x_1, \ldots, x_n) \in \Omega;\, (x_1, \ldots, x_k + h, \ldots, x_n) \in \Omega\}$. It is clear that if $v \in L^1(\mathbb{R}^n)$, then $\displaystyle\int \frac{\partial^h v}{\partial x_k}(x)\,\mathrm{d}x = 0$; if $v \in W^{1,2}(\Omega)$, then $\dfrac{\partial^h v}{\partial x_k} \in W^{1,2}(\Omega^h_k)$; finally for $1 < p \leqslant \infty$, $\dfrac{\partial v}{\partial x_k} \in L^p(\Omega)$ iff $\dfrac{\partial^h v}{\partial x_k}$ is bounded in $L^p(\Omega^h_k)$ when $h \to 0$, and that then

$$\lim_{h \to 0} \left\| \frac{\partial^h v}{\partial x_k} - \frac{\partial v}{\partial x_k} \right\|_{L^p(\Omega^h_k)} = 0\,.$$

Given $\zeta \in \mathscr{D}(\Omega^h_k)$, $L(u, \zeta) \in L^1(\mathbb{R}^n)$ and

$$\frac{\partial^h}{\partial x_k} L(u, \zeta) = L\left(\frac{\partial^h u}{\partial x_k}, \zeta\right) + L\left(u, \frac{\partial^h \zeta}{\partial x_k}\right) + \frac{\partial^h L}{\partial x_k}(u, \zeta)$$

where $\dfrac{\partial^h L}{\partial x_k}$ is the operator whose coefficients are the finite differences $\dfrac{\partial^h}{\partial x_k}$ of the coefficients of L. Since u is a weak solution of $Lu = f$ on Ω, we have

$$\int_\Omega L\left(u, \frac{\partial^h \zeta}{\partial x_k}\right)\mathrm{d}x = \int_\Omega f \frac{\partial^h \zeta}{\partial x_k}\mathrm{d}x = -\int_\Omega \frac{\partial^{-h} f}{\partial x_k}\zeta\,\mathrm{d}x$$

and hence
$$\int_\Omega L\left(\frac{\partial^h u}{\partial x_k}, \zeta\right)\mathrm{d}x = \int_\Omega \left\{\frac{\partial^{-h} f}{\partial x_k} - \frac{\partial^h L}{\partial x_k}(u, \zeta)\right\}\mathrm{d}x\,.$$

In other words $v_h = \dfrac{\partial^h u}{\partial x_k}$ is a weak solution of

$$L v_h = g^h_0 + \sum_{i=1}^n \frac{\partial g^h_i}{\partial x_i} \quad \text{on} \quad \Omega^h_k$$

with
$$g^h_0 = \frac{\partial^{-h} f}{\partial x_k} - \left(\sum_j \frac{\partial^h d_j}{\partial x_k}\frac{\partial u}{\partial x_j} + \frac{\partial^h d}{\partial x_k} u\right)$$

and
$$g^h_i = \sum_j \frac{\partial^h a_{ij}}{\partial x_k}\frac{\partial u}{\partial x_j} + \frac{\partial^h c_i}{\partial x_k} u\,.$$

We have
$$|g^h_i| \leqslant C\left(1 + |u| + \sum_j \left|\frac{\partial u}{\partial x_j}\right|\right) \quad \text{for} \quad i = 0, \ldots, n$$

where C depends only on the norms in $L^\infty(\Omega\setminus S)$ of the derivatives $\partial f/\partial x_k$ of f and of the coefficients of L[227]. Given Ω_0 relatively compact in Ω, we have, from the strict ellipticity

$$\|\operatorname{grad} v_h\|_{L^2(\Omega_0)} \leqslant C(1 + \|u\|_{W^{1,2}(\Omega)})\,.$$

In the limit, when $h \to 0$, we shall have $\operatorname{grad}\dfrac{\partial u}{\partial x_k} \in L^2(\Omega_0)$. Also $v = \dfrac{\partial u}{\partial x_k}$ will be a weak solution of

$$Lv = g_0 + \sum \frac{\partial g_i}{\partial x_i} \quad \text{on} \quad \Omega_0$$

where g_i are the limits of the g_i^h. By induction, we obtain the result. □

4. Results on Existence and Uniqueness of Solutions of Strictly Elliptic Boundary Value Problems of the Second Order on a Bounded Open Set

The quickest method of stating general results on existence and uniqueness is certainly the variational method joined with the Fredholm theory: this method will be developed within an abstract framework in Chap. VII. We content ourselves here with applying it to the study of a mixed boundary value problem for a partial differential equation in divergence form.

Proposition 10. *Let Ω be a bounded open set with a Lipschitz boundary of $\mathbb{R}^n$, L a differential operator of order 2 in divergence form with coefficients in $L^\infty(\Omega)$ and strictly elliptic, Γ_0 a closed part of the boundary Γ of Ω.*
(1) The set $N(L, \Gamma_0)$ of the u, weak solutions relative to $W^{1,2}(\Omega)$ of the mixed boundary value problem

$$\begin{cases} Lu = 0 & \text{on} \quad \Omega \\ u = 0 & \text{on} \quad \Omega \\ \dfrac{\partial u}{\partial n_L} = 0 & \text{on} \quad \Omega \end{cases}$$

is a sub-space of finite dimension of $V^2(\Omega, \Gamma_0)$. In addition

$$\dim N(L, \Gamma_0) = \dim N(L^*, \Gamma_0)$$

where L^* is the operator in divergence form, adjoint of L.
(2) Given $f_0, \ldots, f_n \in L^2(\Omega)$ with $\sum \dfrac{\partial f_i}{\partial x_i} \in \mathscr{E}'(\Omega \cup \Gamma_0)$, $g_0, \ldots, g_n \in L^2(\Omega)$ with $g_0 = \sum \dfrac{\partial g_i}{\partial x_i}$ in $\mathscr{D}'(\Omega)$ and $\varphi \in W^{1,2}(\Omega)$, the boundary value problem

[227] We can always suppose that the derivatives on $\Omega\setminus S$ of f and the coefficients of L are bounded.

$$(8.52) \qquad \begin{cases} Lu = f_0 - \sum \dfrac{\partial f_i}{\partial x_i} & on \quad \Omega \\ u = \varphi \quad on \quad \Gamma_0 \\ \dfrac{\partial u}{\partial n_L} = g_0 - \sum \dfrac{\partial g_i}{\partial x_i} & on \quad \Gamma \backslash \Gamma_0 \end{cases}$$

admits a weak solution relative to $W^{1,2}(\Omega)$ *iff*

$$(8.53) \qquad \int_\Omega L(\varphi, u^*)\,\mathrm{d}x = \int_\Omega \left\{ (f_0 + g_0)u^* + \sum_i (f_i + g_i) \frac{\partial u^*}{\partial x_i} \right\} \mathrm{d}x$$

for all $u^* \in N(L^*, \Gamma_0)$.

Proof. The map $(u, u^*) \in V^2(\Omega, \Gamma_0) \times V^2(\Omega, \Gamma_0) \to \int_\Omega L(u, u^*)\,\mathrm{d}x$ is a continuous bilinear form on the Banach space $V^2(\Omega, \Gamma_0)$. We have

$$\int_\Omega L(u, u)\,\mathrm{d}x = \int_\Omega \left[\sum_{i,j} a_{ij} \frac{\partial u}{\partial x_i} \frac{\partial u}{\partial x_j} + \sum_i (c_i + d_i) u \frac{\partial u}{\partial x_i} + du^2 \right] \mathrm{d}x$$

$$\geqslant \frac{c_0}{2} \int_\Omega \left(\sum_i \left(\frac{\partial u}{\partial x_i} \right)^2 + u^2 \right) \mathrm{d}x + \int_\Omega \left[d - \frac{1}{2c_0} \left(c_0^2 + \sum_i (c_i + d_i)^2 \right) \right] u^2\,\mathrm{d}x\,;$$

hence considering $\lambda_0 = \left\| \left[\frac{1}{2c_0} \left(c_0^2 + \sum_i (c_i + d_i)^2 \right) - d \right]^+ \right\|_{L^\infty}$, we see that the bilinear form $(u, u^*) \to \int_\Omega (L(u, u^*) + \lambda_0 u u^*)\,\mathrm{d}x$ is coercive on $V^2(\Omega, \Gamma_0)^2$.
From the Lax–Milgram theorem (see Chaps. VI and VII) for every continuous linear form l on $V^2(\Omega, \Gamma_0)$, there exists a unique $u \in V^2(\Omega, \Gamma_0)$

$$\int_\Omega (L(u, \zeta) + \lambda_0 u \zeta)\,\mathrm{d}x = l(\zeta) \quad \text{for all} \quad \zeta \in V^2(\Omega, \Gamma_0)\,.$$

In particular, given $f \in L^2(\Omega)$, there exists a unique $u \in V^2(\Omega, \Gamma_0)$ such that

$$\int_\Omega (L(u, \zeta) + \lambda_0 u \zeta)\,\mathrm{d}x = \int_\Omega f \zeta\,\mathrm{d}x \quad \text{for all} \quad \zeta \in V^2(\Omega, \Gamma_0)\,.$$

We denote by G_0 the map $f \to u$; it is linear, continuous from $L^2(\Omega)$ into $V^2(\Omega, \Gamma_0)$ and therefore a compact operator on $L^2(\Omega)$[228].
By definition we have $u \in N(L, \Gamma_0)$ iff $u = \lambda_0 G_0 u$. In other words $N(L, \Gamma_0)$ is the kernel of the operator $I - \lambda_0 G_0$ on $L^2(\Omega)$. From the Riesz–Schauder theorem (see §4), we know that $N(L, \Gamma_0)$ is of finite dimension, that $\mathrm{Im}(I - \lambda_0 G_0)$ is closed with finite codimension and

$$\dim N(L, \Gamma_0) = \operatorname{codim} \mathrm{Im}(I - \lambda_0 G_0)\,.$$

---

[228] Taking account of the hypothesis that Ω is bounded with a Lipschitz boundary, we see that the canonical injection of $W^{1,2}(\Omega)$ into $L^2(\Omega)$ is compact (see Chap. IV).

Now $\operatorname{Im}(I - \lambda_0 G_0)$ is the orthogonal in $L^2(\Omega)$ of the kernel $I - \lambda_0 G_0^*$ where G_0^* is the adjoint operator of G_0 in $L^2(\Omega)$: as we see immediately G_0^* is the operator associated with L^* as G_0 is that associated with L. Hence $\operatorname{Im}(I - \lambda_0 G_0)$ is the orthogonal of $N(L^*, \Gamma_0)$ in $L^2(\Omega)$.
We deduce, in particular, that

$$\dim N(L^*, \Gamma_0) = \operatorname{codim} \operatorname{Im}(I - \lambda_0 G_0)$$

which completes the proof of point (1).
For the point (2), we put

$$\bar{f_0} = f_0 + g_0 - \left(\sum d_i \frac{\partial \varphi}{\partial x_i} + d\varphi\right)$$

$$\bar{f_i} = f_i + g_i - \left(\sum a_{ij} \frac{\partial \varphi}{\partial x_j} + c_i \varphi\right).$$

By definition u is a weak solution relative to $W^{1,2}(\Omega)$ of (8.52) iff $u = \varphi + w$ with $w \in V^2(\Omega, \Gamma_0)$ and

$$\int_\Omega L(w, \zeta)\,dx = \int_\Omega \left(\bar{f_0}\zeta + \sum_i \bar{f_i} \frac{\partial \zeta}{\partial x_i}\right) dx \quad \text{for all} \quad \zeta \in \mathscr{D}(\mathbb{R}^n \backslash \Gamma_0).$$

There exists a (*unique*) $v \in V^2(\Omega, \Gamma_0)$ such that

$$\int_\Omega (L(v, \zeta) + \lambda_0 v\zeta)\,dx = \int_\Omega \left(\bar{f_0}\zeta + \sum_i \bar{f_i} \frac{\partial \zeta}{\partial x_i}\right) dx \quad \text{for all} \quad \zeta \in V^2(\Omega, \Gamma_0).$$

Hence u is a weak solution relative to $W^{1,2}(\Omega)$ of (8.52) iff $u = \varphi + w$ with $w \in V^2(\Omega, \Gamma_0)$ and $w = v + \lambda_0 G_0 w$. There exists therefore a weak solution relative to $W^{1,2}(\Omega)$ of (8.52) iff $v \in \operatorname{Im}(I - \lambda_0 G_0)$. Since $\operatorname{Im}(I - \lambda_0 G_0)$ is the orthogonal of $N(L^*, \Gamma_0)$, we have $v \in \operatorname{Im}(I - \lambda_0 G_0)$ iff

$$\int_\Omega v u^*\,dx = 0 \quad \text{for all} \quad u^* \in N(L^*, \Gamma_0).$$

But, for $u^* \in N(L^*, \Gamma_0)$, $\int_\Omega L(v, u^*)\,dx = 0$; if $\lambda_0 \neq 0$, $v \in \operatorname{Im}(I - \lambda_0 G_0)$ if and only if, for all $u^* \in N(L^*, \Gamma_0)$

$$\int_\Omega \left(\bar{f_0} u^* + \sum \bar{f_i} \frac{\partial u^*}{\partial x_i}\right) dx = \int_\Omega (L(v, u^*) + \lambda_0 v u^*)\,dx = 0$$

which is clearly the condition (8.53); if $\lambda_0 = 0$, the result is trivial. □

Combined with the results on regularity of the preceding section, this result has a vast field of applications although it is limited as we shall see below. It shows that in a very general context the solution of the boundary value problem (8.52) leads to the study of the dual homogeneous problem. Let us give some examples.

Example 1. *Case of dimension 1.* Let us give a complete discussion of an elliptic boundary value problem on a bounded interval of $\mathbb{R}$. We can always reduce the interval to $]0, l[$[229]. Let $a, b, c, d, f, \varphi, \psi$ be functions defined on $I =]0, l[$, Γ_0 a part of $\Gamma = \{0, l\}$; we consider the boundary value problem:

$$(8.54) \quad \begin{cases} -\dfrac{d}{dx}\left(a\dfrac{du}{dx} + bu\right) + c\dfrac{du}{dx} + du = f \quad on \quad I \\ u = \varphi \quad on \quad \Gamma_0 \\ a\dfrac{du}{dx} + bu = \psi \quad on \quad \Gamma\backslash\Gamma_0\,; \end{cases}$$

The homogeneous adjoint problem can be written:

$$(8.55) \quad \begin{cases} -\dfrac{d}{dx}\left(a\dfrac{dv}{dx} + cv\right) + b\dfrac{dv}{dx} + dv = 0 \quad \text{on} \quad I \\ v = 0 \quad \text{on} \quad \Gamma_0 \\ a\dfrac{dv}{dx} + cv = 0 \quad \text{on} \quad \Gamma\backslash\Gamma_0\,. \end{cases}$$

Supposing the coefficients to be sufficiently regular and $a > 0$ on $\bar{I}$, we change the dependent variable from v to w, where

$$v(x) = \frac{e^{\int_0^x \tau(s)\,ds}}{\sqrt{a(x)}}\,w(x) \quad \text{with} \quad \tau = \frac{b - c}{2a}\,.$$

The problem (8.55) can then be written

$$(8.56) \quad \begin{cases} \dfrac{d^2w}{dx^2} = kw \quad \text{on} \quad I \\ w = 0 \quad \text{on} \quad \Gamma_0 \\ w_x = \lambda w \quad \text{on} \quad \Gamma\backslash\Gamma_0 \end{cases}$$

with

$$(8.57) \quad k = \lambda' + \lambda^2 + \frac{ad - bc}{a^2}\,, \qquad \lambda = \frac{a' - (b + c)}{2a}\,.$$

The set $N(\Gamma_0)$ of the solutions of (8.56) is either reduced to the null function, or to a space of dimension 1. In effect considering w_1, w_2 solutions of $\dfrac{d^2w}{dx^2} = kw$ satisfying the initial conditions

$$w_1(0) = 0\,, \qquad w_1'(0) = 1\,, \qquad w_2(0) = 1\,, \qquad w_2'(0) = 0\,,$$

we have the cases shown in Table 3.

---

[229] We could suppose that $l = 1$, but here we prefer to retain the length of the interval as a parameter.

Table 3

Γ_0	Condition for $N(\Gamma_0) \neq \{0\}$	Particular solution of $N(\Gamma_0)$
Γ	$w_1(l) = 0$	w_1
$\{0\}$	$w_1'(l) = \lambda(l) w_1(l)$	w_1
$\{l\}$	$\lambda(0) w_1(l) + w_2(l) = 0$	$\lambda(0) w_1 + w_2$
$\varnothing$	$\lambda(0) w_1'(l) + w_2'(l) = \lambda(l)(\lambda(0) w_1(l) + w_2(l))$	$\lambda(0) w_1 + w_2$

The application of Proposition 10 then gives:
case $N(\Gamma_0) = \{0\}$; for all functions f, φ, ψ there exists a unique solution of (8.54);
case $N(\Gamma_0) \neq \{0\}$. Then $N(\Gamma_0) = \mathbb{R} w_0$, w_0 being a particular (non null) solution of (8.56). Given f, φ, ψ there exists a solution of (8.5) iff according to the cases quoted $\int_0^l v_0(x) f(x)\,\mathrm{d}x$ is equal to

$$\begin{array}{ll} \Gamma_0 = \Gamma & a(0) v_0'(0) \varphi(0) - a(l) v_0'(l) \varphi(l) \\ \Gamma_0 = \{0\} & a(0) v_0'(0) \varphi(0) + v_0(l) \psi(l) \\ \Gamma_0 = \{l\} & - v_0(0) \psi(0) - a(l) v_0'(l) \varphi(l) \\ \Gamma_0 = \varnothing & - v_0(0) \psi(0) + v_0(l) \psi(l) \end{array}$$

where $$v_0(x) = \sqrt{a(x)}\, \mathrm{e}^{\int_0^x \tau(s)\,\mathrm{d}s} w_0(x) \quad \text{with} \quad \tau = \frac{b - c}{2a} .$$

Moreover, then, the solution of (8.45) is unique to within the addition of a constant times the function

$$u_0(x) = \sqrt{a(x)}\, \mathrm{e}^{-\int_0^x \tau(s)\,\mathrm{d}s} w_0(x) .$$

The last point comes from the fact that k and λ, defined by (8.57), are symmetric in b and c with the result that we shall reduce the homogeneous problem of (8.54) to (8.56). The explanation of the condition of existence and uniqueness, that is to say $N(\Gamma_0) = \{0\}$, or of a particular solution w_0 in the case $N(\Gamma_0) \neq \{0\}$, obviously depends on the knowledge of the particular solutions w_1, w_2 of $\dfrac{\mathrm{d}^2 w}{\mathrm{d}x^2} = kw$ which we cannot integrate by quadratures in the general case.
If k is constant, that is simple:

$$k = \omega^2 > 0, \qquad w_1 = \frac{\operatorname{sh} \omega x}{\omega}, \qquad w_2 = \operatorname{ch} \omega w$$

Table 4. Condition for $N(\Gamma_0) \neq \varnothing$

Γ_0	$k = \omega^2 > 0$	$k = 0$	$k = -\omega^2 < 0$
Γ	$N(\Gamma) = \{0\}$	$N(\Gamma) = \{0\}$	$\omega = \frac{m\pi}{l}$, $w_0 = \sin \omega x$
$\{0\}$	$\lambda(l) \operatorname{th} \omega l = \omega$ $w_0 = \operatorname{sh} \omega x$	$\lambda(l) l = 1$ $w_0 = x$	$\lambda(l) \operatorname{tg} \omega l = \omega$ $w_0 = \sin \omega x$
$\{l\}$	$\lambda(0) \operatorname{th} \omega l + \omega = 0$ $w_0 = \operatorname{sh} \omega(l - x)$	$1 + l\lambda(0) = 0$ $w_0 = l - x$	$\lambda(0) \operatorname{tg} \omega l + \omega = 0$ $w_0 = \sin \omega(l - x)$
$\varnothing$	$\omega[\lambda(0) + \omega \operatorname{th} \omega l]$ $= \lambda(l)[\omega + \lambda(0) \operatorname{th} \omega l]$ $w_0 = \operatorname{sh} \omega(x + \varphi)$ $\varphi = \frac{1}{\omega} \operatorname{Arg\,th} \frac{\omega}{\lambda(0)}$	$\lambda(l) - \lambda(0) = \lambda(0)\lambda(l)l$ $w_0 = 1 + \lambda(0)x$	$\omega[\lambda(0) - \omega \operatorname{tg} \omega l]$ $= \lambda(l)[\omega + \lambda(0) \operatorname{tg} \omega l]$ $w_0 = \sin \omega(x + \varphi)$ $\varphi = \frac{1}{\omega} \operatorname{Arc\,tg} \frac{\omega}{\lambda(0)}$

$$k = 0\,, \qquad w_1 = x\,, \qquad w_2 = 1$$

$$k = -\omega^2 < 0\,,\ w_1 = \frac{\sin \omega x}{\omega}\,, \qquad w_2 = \cos \omega x\,.$$

Table 3 expresses itself then in the following Table 4 giving the condition for $N(\Gamma_0) \neq \{0\}$ and in this case a particular solution w_0. Read upside down, this table exhibits the eigenvalues of the operator d^2/dx^2 having as domain

$$D(\Gamma_0, \lambda) = \left\{ w \in W^{2,2}(]0, l[);\ \ w = 0 \ \text{ on } \ \Gamma_0, \frac{dw}{dx} = \lambda w \ \text{ on } \ \Gamma \backslash \Gamma_0 \right\}.$$

In particular we can make clear *the largest eigenvalue* $k\,(\Gamma_0, \lambda)$ according to the values of Γ_0 and λ; this value $k(\Gamma_0, \lambda)$ is specially interesting as we know[230] that, the operator d^2/dx^2 being symmetric on $D(\Gamma_0, \lambda)$, the operator $k(\Gamma_0, \lambda) - d^2/dx^2$ is positive on $D(\Gamma_0, \lambda)$: we deduce that for every function k such that

$$k \geqslant k(\Gamma_0, \lambda), \quad k \not\equiv k(\Gamma_0, \lambda)\,, \tag{8.58}$$

the homogeneous problem (8.56) admits only the null solution; in effect if w is a solution of (8.56) we shall have

$$\int_0^l (k - k(\Gamma_0, \lambda))w^2 \, dx = \int_0^l \left(\frac{d^2 w}{dx^2} - k(\Gamma_0, \lambda) w \right) w \, dx \leqslant 0$$

and hence from (8.58), $w = 0$.

---

[230] See Chap. VI.

Returning to the problem (8.54), we see therefore that if *the coefficients a, b, c, d satisfy*

(8.59) $$\frac{\mathrm{d}\lambda}{\mathrm{d}x} + \lambda^2 + \frac{ad - bc}{a^2} \geqslant k(\Gamma_0, \lambda) \quad \textit{where} \quad \lambda = \frac{\mathrm{d}}{\mathrm{d}x} \operatorname{Log}\sqrt{a} - \frac{b + c}{2a}$$

and $k(\Gamma_0, \lambda)$ *is the largest eigenvalue of the operator* $\mathrm{d}^2/\mathrm{d}x^2$ *having as domain*

$$D(\Gamma_0, \lambda) = \left\{ w \in W^{2,2}(]0, l[);\ w = 0 \quad on \quad \Gamma_0, \frac{\mathrm{d}w}{\mathrm{d}x} = \lambda w \quad on \quad \Gamma \backslash \Gamma_0 \right\},$$

then for all functions f, φ, ψ, *the boundary value problem* (8.54) *admits one and only one solution.*

We conclude this example by explaining $k(\Gamma_0, \lambda)$ in Table 4:

case $\Gamma = \Gamma_0$ (pure Dirichlet problem): $k(\Gamma_0) = -(\pi/l)^2$;

case $\Gamma = \{0\}$ (resp. $\Gamma = \{l\}$), putting $\mu = l\lambda(l)$ (resp. $-l\lambda(0)$),

(a) if $\mu > 1, k(\Gamma_0, \lambda) = (\theta/l)^2$ where θ is the unique positive solution of $\mu \operatorname{th} \theta = \theta$;
(b) if $\mu = 1$, $k(\Gamma_0, \lambda) = 0$;
(c) if $\mu < 1, k(\Gamma_0, \lambda) = (\theta/l)^2$ where θ is the unique solution in $]0, \pi[$ of $\mu \operatorname{tg} \theta = \theta$;

case $\Gamma = \varnothing$, putting $\mu_0 = l\lambda(l)$, $\mu_1 = = - l\lambda(0)$

(a) if $\mu_0 + \mu_1 > \mu_0\mu_1, k(\Gamma_0, \lambda) = (\theta/l)^2$ where θ is the unique positive solution of $\operatorname{th} \theta = \dfrac{(\mu_0 + \mu_1)\theta}{\theta^2 + \mu_0\mu_1}$;
(b) if $\mu_0 + \mu_1 = \mu_0\mu_1$, $k(\Gamma_0, \lambda) = 0$;
(c) if $\mu_0 + \mu_1 < \mu_0\mu_1$, $k(\Gamma_0, \lambda) = (\theta/l)^2$ where θ is the unique solution in $]0, \pi[$ of $\operatorname{tg} \theta = (\mu_0 + \mu_1)\theta/(\theta^2 - \mu_0\mu_1)$.

Example 2. *A quasi-classical Dirichlet problem.* We consider the Dirichlet problem

$$\begin{cases} -\sum \dfrac{\partial}{\partial x_i}\left(\sum a_{ij} \dfrac{\partial u}{\partial x_j}\right) + ku = f \quad \text{on} \quad \Omega \\ u = g \quad \text{on} \quad \Gamma \end{cases}$$

which we propose to solve in the "quasi-classical" sense as follows:

(8.60) $$\begin{cases} u \in W^{1,2}(\Omega) \cap \mathscr{C}^0(\bar{\Omega}), \\ -\sum \dfrac{\partial}{\partial x_i}\left(\sum a_{ij} \dfrac{\partial u}{\partial x_j}\right) + ku = f \quad \text{in} \quad \mathscr{D}'(\Omega) \\ u = g \quad \text{on} \quad \Gamma \text{ (in the classical sense)} \end{cases}$$

We state

Proposition 11. *Let* Ω *be a bounded open set in* $\mathbb{R}^n$ *with* $n \geqslant 2$, *with a Lipschitz boundary*, $a_{ij} = a_{ji} \in L^\infty(\Omega)$, *with* $\sum a_{ij}\xi_i\xi_j \geqslant c_0|\xi|^2$ *for all* $\xi \in \mathbb{R}^n$ *where* $c_0 > 0$, $k \in L^{\frac{n}{2}+\varepsilon}(\Omega)$.

(1) *The set* $N(a_{ij}, k)$ *of solutions of the homogeneous problem* (8.60) (*corresponding to* $f = 0$ *on* Ω, $g = 0$ *on* Γ) *is a finite-dimensional space.*

(2) *For* $f \in L^{\frac{n}{2}+\varepsilon}(\Omega)$, *and* $g = \varphi$ *on* Γ *where* $\varphi \in W^{1,n+\varepsilon}(\Omega)$[231] ($\varepsilon > 0$) *there exists a solution of* (8.60) *iff for all* $\zeta \in N(a_{ij}, k)$

$$\int_\Omega \left(\sum_{i,j} a_{ij} \frac{\partial \varphi}{\partial x_i} \frac{\partial \zeta}{\partial x_j} + k\varphi\zeta \right) \mathrm{d}x = \int_\Omega f\zeta \,\mathrm{d}x .$$

(3) *With the notation* $c(\Omega) = \inf\limits_{\zeta \in \mathscr{D}(\Omega)} \dfrac{\|\operatorname{grad} \zeta\|_{L^2}^2}{\|\zeta\|_{L^2}^2}$, $N(\delta_{ij}, -c(\Omega))$ *is a space of dimension 1, namely* $\mathbb{R}u(\Omega)$. *Now supposing that*

$$k \geqslant -c_0 c(\Omega)$$

we have $N(a_{ij}, k) = \{0\}$ *except if*

$$k \equiv -c_0 c(\Omega) \text{ and } \sum_j a_{ij} \frac{\partial u(\Omega)}{\partial x_j} \equiv c_0 \frac{\partial u(\Omega)}{\partial x_i}, \; i = 1, \ldots, n ,$$

in which case $N(a_{ij}, k) = N(\delta_{ij}, -c(\Omega)) = \mathbb{R}u(\Omega)$.

Proof. We note first that if $w \in W^{1,2}(\Omega) \cap \mathscr{C}^0(\bar{\Omega})$, then $w \in H_0^1(\Omega)$ iff $w = 0$ on Γ[232]; hence, given $g = \varphi$ on Γ where $\varphi \in W^{1,2}(\Omega)$, we have u a solution of (8.60) iff $u \in \mathscr{C}^0(\bar{\Omega})$ and is a weak solution relative to $W^{1,2}(\Omega)$ *of*

$$\text{(8.61)} \qquad Lu = f - ku \quad \text{on} \quad \Omega , \quad u = g \quad \text{on} \quad \Gamma$$

with

$$L = -\sum \frac{\partial}{\partial x_i} \left(\sum a_{ij} \frac{\partial}{\partial x_j} \right) .$$

We note now the hypothesis $f \in L^{\frac{n}{2}+\varepsilon}(\Omega)$ that ensures the existence of $f_1, \ldots, f_n$ in $L^{n+\varepsilon}(\Omega)$ such that $f = \sum \partial f_i/\partial x_i$ in $\mathscr{D}'(\Omega)$[233]. Using the De Georgi–Nash theorem (Proposition 6), when $k \in L^\infty(\Omega), f \in L^{\frac{n}{2}+\varepsilon}(\Omega)$ and $\varphi \in W^{1,n+\varepsilon}(\Omega)$: a weak solution relative to $W^{1,2}(\Omega)$ of (8.61) is continuous on $\bar{\Omega}$; hence, thanks to this theorem, u is a solution of (8.60) iff u is a weak solution relative to $W^{1,2}(\Omega)$ of (8.61).
In other words, when $k \in L^\infty$ the points (1) and (2) of Proposition 11 are a particular case of Proposition 10.

To treat the general hypothesis $k \in L^{\frac{n}{2}+\varepsilon}(\Omega)$ we use the following argument. First we consider $N(a_{ij}, 0) = \{0\}$: this is an obvious particular case of point (3) which we

---

[231] We recall that $W^{1,n+\varepsilon}(\Omega) \subset \mathscr{C}^0(\bar{\Omega})$. The condition $g = \varphi$ on Γ with $\varphi \in W^{1,n+\varepsilon}(\Omega)$ can be expressed in an equivalent manner by: $g \in \mathscr{C}^0(\Gamma)$ and the (classical) solution of the Dirichlet problem $\Delta\varphi = 0$ on Ω, $\varphi = g$ on Γ is in $W^{1,n+\varepsilon}(\Omega)$; by using the trace theorems we can also say that $g \in W^{\frac{n-1}{n}+\varepsilon; n+\varepsilon}(\Gamma)$ (even if to change $\varepsilon > 0$).

[232] See Chap. IV.

[233] It is sufficient to take $f_i = \dfrac{\partial v}{\partial x_i}$ on Ω where v is the Newtonian potential of $f\chi_\Omega$.

shall prove below. We fix $g = \varphi$ on Γ with $\varphi \in W^{1,n+\varepsilon}(\Omega)$ and for all $f \in L^{\frac{n}{2}+\varepsilon}(\Omega)$ we denote by $G_0 f$ the solution of the problem (8.60) corresponding to the case $k \equiv 0$. It is clear that u is a solution of (8.60) iff

$$u = G_0(f - ku) .$$

Now, from the De Georgi–Nash theorem, we know that $G_0 f$ is not only continuous on $\bar{\Omega}$ but even satisfies a Hölder condition. As we have noted above, the modulus of continuity of $G_0 f$ depends only on Ω, $\| a_{ij} \|_{L^\infty}$, c_0, $\| \varphi \|_{W^{1,n+\varepsilon}}$ and $\| f \|_{L^{\frac{1}{2}n+\varepsilon}}$ and therefore G_0 will be a compact operator on $L^{\frac{n}{2}+\varepsilon}(\Omega)$ into $\mathscr{C}^0(\bar{\Omega})$. Since $k \in L^{\frac{n}{2}+\varepsilon}(\Omega)$, the map $u \to f - ku$ is continuous on $\mathscr{C}^0(\bar{\Omega})$ into $L^{\frac{n}{2}+\varepsilon}(\Omega)$ and therefore $u \to G_0(f - ku)$ is a compact map of $\mathscr{C}^0(\bar{\Omega})$ into itself. It suffices then to apply the Riesz–Schauder theorem to conclude this part of the proof.
Finally, let us prove Point (3). We refer to the Appendix to Chap. VIII (vol. 3) for the proof that $N(\delta_{ij}, -c(\Omega))$ is of dimension 1, although one of the essential arguments is used in what follows. Hence, we take $u \in N(a_{ij}, k)$; we have by definition of $c(\Omega)$

$$0 = \int_\Omega \left(\sum_{i,j} a_{ij} \frac{\partial u}{\partial x_i} \frac{\partial u}{\partial x_j} + ku^2 \right) \mathrm{d}x \geqslant \int_\Omega \left(\sum_{i,j} a_{ij} \frac{\partial u}{\partial x_i} \frac{\partial u}{\partial x_j} - c_0 \sum \left(\frac{\partial u}{\partial x_i} \right)^2 \right) \mathrm{d}x + \int_\Omega (k + c_0 c(\Omega)) u^2 \, \mathrm{d}x .$$

Hence if $k \geqslant -c_0 c(\Omega)$, we have

$$k \equiv -c_0 c(\Omega) \quad \text{a.e. on} \quad \{x \in \Omega; u(x) \neq 0\}$$

and

$$\sum_{i,j} a_{ij} \frac{\partial u}{\partial x_i} \frac{\partial u}{\partial x_j} = c_0 \sum \left(\frac{\partial u}{\partial x_i} \right)^2 \quad \text{a.e. on} \quad \Omega$$

But c_0 minorizes the smallest eigenvalue of the symmetric matrix $(a_{ij}(x))$; hence the second relation implies

$$\sum_j a_{ij} \frac{\partial u}{\partial x_j} = c_0 \frac{\partial u}{\partial x_i} \quad \text{a.e. on} \quad \Omega \quad \text{for} \quad i = 1, \ldots, n .$$

Supposing $u \not\equiv 0$, we always suppose that

$$\Omega_+ = \{x \in \Omega; u(x) > 0\}$$

is non-empty; we then have $u \in W^{1,2}(\Omega_+) \cap \mathscr{C}^0(\bar{\Omega}_+)$, $u = 0$ on $\partial\Omega_+$ and

$$\Delta u = \frac{1}{c_0} \sum \frac{\partial}{\partial x_i} \left(\sum a_{ij} \frac{\partial u}{\partial x_j} \right) = -c(\Omega) u \quad \text{in} \quad \mathscr{D}'(\Omega_+) .$$

In particular, we have

$$\| \operatorname{grad} u \|^2_{L^2(\Omega_+)} = c(\Omega) \| u \|^2_{L^2(\Omega_+)}$$

and hence $c(\Omega_+) = c(\Omega)$ (we always have $c(\Omega) \leqslant c(\Omega_+)$. We deduce that $\Omega \backslash \Omega_+$

has zero capacity and $u \in W_0^{1,2}(\Omega)$ and $\Delta u = -c(\Omega)u$ in $\mathscr{D}(\Omega)$. In other words, to within a multiplicative constant, $u = u(\Omega)$. □

The proof shows that, for the points (1) and (2), we can replace $L = -\sum_i \frac{\partial}{\partial x_i}\left(\sum_j a_{ij}\frac{\partial}{\partial x_j}\right)$ by a more general operator of the form

$$L = -\sum_i \frac{\partial}{\partial x_i}\left(\sum_j a_{ij}\frac{\partial}{\partial x_j} + c_i\right) + \sum d_i \frac{\partial}{\partial x_i} \quad \text{with} \quad c_i, d_i \in L^\infty(\Omega)\,;$$

we could even refine the result by assuming only that c_i, $d_i \in L^{n+\varepsilon}(\Omega)$ (see Ladyzenskaya–Ural'ceva [1]); in the existence condition of point (2) we obviously ought to make use of $N(L^*, k)$.

When L contains the "transport terms" $-\sum \frac{\partial}{\partial x_i} c_i + \sum d_i \frac{\partial}{\partial x_i}$, the uniqueness condition $k + c_0 c(\Omega) \geqslant 0$ of point (3) is no longer valid as we see in the case of dimension 1 (see Table 4). We must replace it by the condition

$$k + c_0 c(\Omega) \geqslant \sum \frac{\partial}{\partial x_i}\left(\frac{c_i + d_i}{2}\right) \quad \text{in} \quad \mathscr{D}'(\Omega)\,,$$

noting that if $u \in N(L, k)$ we have formally

$$\begin{aligned} 0 &= \int_\Omega \left(\sum a_{ij}\frac{\partial u}{\partial x_i}\frac{\partial u}{\partial x_j} + \sum (c_i + d_i) u \frac{\partial u}{\partial x_i} + k u^2\right) \mathrm{d}x \\ &= \int_\Omega \left(\sum a_{ij}\frac{\partial u}{\partial x_i}\frac{\partial u}{\partial x_j} + \left(k - \sum \frac{\partial}{\partial x_i}\left(\frac{c_i + d_i}{2}\right)\right) u^2\right) \mathrm{d}x\,. \end{aligned}$$

Other conditions can be developed as is shown by the following example:

Example 3. *A classical Neumann problem.* We consider the Neumann problem

$$\text{(8.62)} \qquad \begin{cases} -\sum_{i,j} a_{ij}\dfrac{\partial^2 u}{\partial x_i \partial x_j} + \sum_i b_i \dfrac{\partial u}{\partial x_i} + cu = f & \text{on} \quad \Omega \\ \sum_{i,j} a_{ij}\dfrac{\partial u}{\partial x_i} n_j + \lambda u = g & \text{on} \quad \Gamma \end{cases}$$

which we shall solve in the classical sense supposing that

$$a_{ij} = a_{ji},\, b_i,\, c, f \in \mathscr{C}^\infty(\bar\Omega),\ \sum_{i,j} a_{ij}\xi_i\xi_j \geqslant c_0|\xi|^2 \quad \text{on} \quad \bar\Omega\,,$$

Ω being a regular bounded open set with boundary Γ of class $\mathscr{C}^\infty$ with λ, $g \in \mathscr{C}^\infty(\Gamma)$. We state

Proposition 12. *Under the above hypotheses*
(1) *The set $N(a_{ij}, b_i, c, \lambda)$ of the solutions $u \in \mathscr{C}^\infty(\bar\Omega)$ of the homogeneous problem (8.62) (with $f = 0$ on Ω, $g = 0$ on Γ) is a finite dimensional space.*

(2) *The problem* (8.62) *admits a solution in* $\mathscr{C}^\infty(\Omega)$ *iff for all* $\zeta \in N(a_{ij}, b_i^*, c^*, \lambda^*)$,

$$\int_\Omega f\zeta \,\mathrm{d}x + \int_\Gamma g\zeta \,\mathrm{d}\gamma = 0$$

where b_i^*, c^* *are given by* (8.21)′ *and* $\lambda^* = \lambda + \sum_i \left(\sum_j \frac{\partial a_{ij}}{\partial x_j} + b_i \right) n_i$.

(3) *When* $\lambda \geqslant 0$ *on* Γ *and* $c > 0$ *on* $\bar{\Omega}$ *we have* $N(a_{ij}, b_i, c, \lambda) = N(a_{ij}, b_i^*, c^*, \lambda^*) = \{0\}$.

Proof. We can write (8.62) in divergence form

$$(8.63) \quad \begin{cases} Lu \equiv -\sum \frac{\partial}{\partial x_i}\left(\sum a_{ij} \frac{\partial u}{\partial x_j} + c_i u \right) + \sum d_i \frac{\partial u}{\partial x_i} + du = f \quad \text{on} \quad \Omega \\ \frac{\partial u}{\partial n_L} = g \quad \text{on} \quad \Gamma \end{cases}$$

by choosing c_i, d_i, d satisfying (8.8) and

$$\sum c_i n_i = \lambda \quad \text{on} \quad \Gamma .$$

The best way is to choose $\mu \in \mathscr{C}^\infty(\bar{\Omega})$ such that

$$\int_\Omega \mu \,\mathrm{d}x = \int_\Gamma \lambda \,\mathrm{d}\gamma$$

and to solve the Neumann problem

$$\Delta w = \mu \quad \text{on} \quad \Omega, \quad \frac{\partial w}{\partial n} = \lambda \quad \text{on} \quad \Gamma .$$

It admits a solution $w \in \mathscr{C}^\infty(\bar{\Omega})$[234]; we put $c_i = \partial w/\partial x_i$ with the result that $\sum c_i n_i = \lambda$ on Γ and $\sum \partial c_i/\partial x_i = \mu$ on Ω; putting then $d_i = b_i + c_i + \sum_j \partial a_{ij}/\partial x_j$, $d = c + \mu$, we obtain equivalence between the problems (8.62) and (8.63). From the theorems on regularity, a weak solution, relative to $W^{1,2}(\Omega)$, of (8.63) will be in $\mathscr{C}^\infty(\bar{\Omega})$. Points (1) and (2) result from Proposition 10.

To prove point (3) we make use of a truncation method. Given $u \in N(a_{ij}, b_i, c, \lambda)$, we have for $\theta \in \mathscr{C}^1(\mathbb{R})$ with $\theta' \geqslant 0$

$$\begin{aligned} 0 = \int_\Omega \theta(u)\left[-\sum a_{ij} \frac{\partial^2 u}{\partial x_i \partial x_j} + \sum b_i \frac{\partial u}{\partial x_i} + cu \right] \mathrm{d}x = -\int_\Gamma \theta(u) \sum a_{ij} \frac{\partial u}{\partial x_i} n_j \,\mathrm{d}\gamma \\ + \int_\Omega \left[\theta'(u) \sum a_{ij} \frac{\partial u}{\partial x_i}\frac{\partial u}{\partial x_j} + \theta(u) \sum_i \frac{\partial u}{\partial x_i} \left(\sum_j \frac{\partial a_{ij}}{\partial x_j} + b_i \right) + cu\theta(u) \right] \mathrm{d}x \\ \geqslant \int_\Omega \left[c - \frac{w}{2c_0^2} \sum_i \left(\sum_j \frac{\partial a_{ij}}{\partial x_j} + b_i \right)^2 \right] u\theta(u) \,\mathrm{d}x , \end{aligned}$$

where w is a function such that $\theta(u)^2 \leqslant wu\theta(v)\theta'(u)$. Choosing $\theta(r) = r|r|^\alpha$ we can

---

[234] We can always suppose Ω to be connected.

take $w = 1/\alpha$ and hence for α sufficiently large, in fact $\alpha > \left\| \sum_i \frac{1}{2c_0^2 c} \left(\sum_j \frac{\partial a_{ij}}{\partial x_j} + b_i \right)^2 \right\|_{L^\infty}$, we have $c > \frac{w}{2c_0^2} \sum_i \left(\sum_j \frac{\partial a_{ij}}{\partial x_j} + b_i \right)^2$ and hence

$$u\theta(u) = |u|^{\alpha+2} = 0 . \qquad \square$$

Proposition 10 gives the existence of weak solutions of (8.52) for the given $f_0, \ldots, f_n, g_0, \ldots, g_n \in L^2(\Omega)$, $\varphi \in W^{1,2}(\Omega)$: these conditions are optimal since we seek a weak solution in $W^{1,2}(\Omega)$. In particular, this Proposition does not permit the solution of the equations $Lu = \delta_y$ where δ_y is the Dirac mass at y: in effect, such a distribution cannot be written in the form $f_0 - \sum \partial f_i/\partial x_i$ with $f_0, \ldots, f_n \in L^2(\Omega)$ (except in dimension 1). The following result provides an answer to this problem: for simplicity, we shall restrict our attention to the case of a homogeneous problem.

Proposition 13. *Let Ω be a regular bounded open set of $\mathbb{R}^n$ with $W^{2,\infty}$-boundary L a differential operator of order 2 in divergence form with coefficients $a_{ij} = a_{ji} \in W^{1,\infty}(\Omega)$, c_i, d_i, $d \in L^\infty(\Omega)$ and strictly elliptic.*
(1) For $f \in L^{n+\varepsilon}(\Omega)$, every weak solution relative to $W^{1,1}(\Omega)$ of the homogeneous Dirichlet (resp. Neumann) problem

$$(8.64) \qquad Lu = f \quad \text{on} \quad \Omega\,, \qquad u = 0 \quad \left(\textit{resp.}\ \frac{\partial u}{\partial n_L} = 0 \right) \quad \textit{on} \quad \Gamma$$

is of class $\mathscr{C}^1$ on $\bar{\Omega}$. In particular, with the notation of Proposition 10, $N(L, \Gamma)$ (resp. $N(L, \varnothing)$) is contained in $\mathscr{C}^1(\bar{\Omega})$.
(2) Given f, a bounded Radon measure on Ω, the Dirichlet (resp. Neumann) problem (8.64) admits a weak solution relatively to $W^{1,1}(\Omega)$ iff

$$\int_\Omega \zeta f \mathrm{d}x = 0 \quad \textit{for all} \quad \zeta \in N(L^*, \Gamma) \quad (\textit{resp.}\ N(L^*, \varnothing))^{235} .$$

In addition, when it exists, this is unique, to within the addition of an element of $N(L, \Gamma)$ (resp. $N(L, \varnothing)$) and is in $W^{1,p}(\Omega)$ for all $p < n/(n-1)$.

Proof. For the point (1), we note first that a weak solution relative to $W^{1,2}(\Omega)$ is of class $\mathscr{C}^1$ on $\bar{\Omega}$. From Proposition 6, we know already that it is continuous; the regularity of class $\mathscr{C}^1$ will be proved by the method of translations (see proof of Proposition 9), the hypotheses $\Gamma \in W^{2,\infty}$, $a_{ij} \in W^{1,\infty}(\Omega)$ securing a recurrence method. We leave the details to the reader[236]. By Fredholm's method, used as in Propositions 11 or 12, we can always reduce the problem to the case where L is coercive. Given $g_0, \ldots, g_n \in \mathscr{D}(\Omega)$ let us consider v the solution of

$$L^* v = g_0 - \sum \frac{\partial g_i}{\partial x_i} \quad \text{on} \quad \Omega,\ v = 0 \quad \left(\text{resp.}\ \frac{\partial v}{\partial n} = 0 \right) \quad \text{on} \quad \Gamma$$

---

[235] We recall that we use $\int_\Omega \zeta f \mathrm{d}x$ to denote the integral of ζ with respect to the measure f.

[236] See also the proof of Theorem 1 of §6. We can, if we wish, suppose Γ, f and the coefficients of L to be of class $\mathscr{C}^\infty$: the important point of this Proposition is the possibility of using weak solutions relative to $W^{1,1}(\Omega)$ (rather than $W^{1,2}(\Omega)$).

We know that $v \in \mathscr{C}^1(\bar{\Omega})$ and also from Proposition 7, for $q > n$

$$\|v\|_{L^\infty} \leqslant C \sum_{i=0}^{n} \|g_i\|_{L^q}$$

where C depends only on the coefficients, Ω and q.
Let us take f to be a bounded measure on Ω. If u is a weak solution relative to $W^{1,1}(\Omega)$ of (8.64) we have

$$\int_\Omega fv\,\mathrm{d}x = \int_\Omega L(u, v)\,\mathrm{d}x = \int_\Omega u\left(g_0 - \sum \frac{\partial g_i}{\partial x_i}\right)\mathrm{d}x = \int_\Omega \left(ug_0 + \sum g_i \frac{\partial u}{\partial x_i}\right)\mathrm{d}x\,.$$

Therefore

$$\left|\int_\Omega \left(ug_0 + \sum g_i \frac{\partial u}{\partial x_i}\right)\mathrm{d}x\right| \leqslant \|v\|_{L^\infty} \int |f|\,\mathrm{d}x \leqslant C\left(\int |f|\,\mathrm{d}x\right) \sum_{i=0}^{n} \|g_i\|_{L^q}\,.$$

This being true for all $g_0, \ldots, g_n \in \mathscr{D}(\Omega)$, $u \in W^{1,p}(\Omega)$ and

$$\|u\|_{W^{1,p}} \leqslant C \int |f|\,\mathrm{d}x$$

where p is the index, conjugate to q.
This relation shows, first of all the uniqueness of a weak solution relative to $W^{1,1}(\Omega)$ (take $f = 0$). It also proves existence: approximate f by regular functions f_k such that

$$\|f_k\|_{L^1} \leqslant \int |f|\,\mathrm{d}x \quad \text{and} \quad \int_\Omega \zeta f_k\,\mathrm{d}x \to \int_\Omega \zeta f\,\mathrm{d}x$$

for all $\zeta \in \mathscr{D}(\Omega)$ (resp. $\mathscr{D}(\mathbb{R}^n)$). Considering the corresponding solutions u_k, they are bounded in $W^{1,p}(\Omega)$; we can therefore suppose, possibly after the extraction of a sub-sequence, that u_k converges weakly to u in $W^{1,p}(\Omega)$; it is clear that in the limit that u is a weak solution, relative to $W^{1,p}(\Omega)$, of (8.64). □

We note the

Proposition 14. *Under the hypotheses of Proposition* 13, *let us take* $(e_1, \ldots, e_r)$, $(e_1^*, \ldots, e_r^*)$ *as orthonormal bases in* $N(L, \Gamma)$, $N(L^*, \Gamma)$ (*resp.* $N(L, \varnothing)$, $N(L^*, \varnothing)$) *respectively. There exists a unique function* $G\colon \bar{\Omega} \times \bar{\Omega} \to \bar{\mathbb{R}}$ *of class* $\mathscr{C}^1$ *on* $\bar{\Omega} \times \bar{\Omega} \backslash \{(x, y); x = y\}$ *and such that for all* $y \in \Omega$, *the function* $u(x) = G(x, y)$ *satisfies*

(1) *u is a weak solution relative to* $W^{1,1}(\Omega)$ *of the Dirichlet* (*resp. Neumann*) *problem* (8.64) *with* $f = \delta_y - \sum_{k=1}^{n} e_k^*(y) e_k^*$

(2) $\int_\Omega u(x) e_k(x)\,\mathrm{d}x = 0$ *for* $k = 1, \ldots, r$.

This function G is the Green's function of the Dirichlet (resp. Neumann) problem for the operator L on Ω; it is independent of the choice of the orthonormal bases

$(e_1, \ldots, e_r)$, $(e_1^*, \ldots, e_r^*)$. We shall come back to the idea of a Green's function in Sect. 6. □

5. Harnack's Inequality and the Maximum Principle

There exist several approaches to the generalization of the principle of the maximum for subharmonic functions which we have proved in §3, based on the formula of the mean. We shall, in the first instance, approach that based on a generalization of Harnack's inequality. We introduce the

Definition 7. Given Ω an arbitrary open set, L a differential operator of order 2 in divergence form with coefficients in $L^\infty(\Omega)$ and $f \in \mathscr{D}'(\Omega)$. A *subsolution* (resp. *supersolution*) of the equation $Lu = f$ on Ω is any function $u \in W^{1,1}_{\text{loc}}(\Omega)$ such that

$$\int_\Omega L(u, \zeta)\,dx \leqslant (\text{resp. } \geqslant) \langle f, \zeta \rangle \quad \text{for all} \quad \zeta \in \mathscr{D}(\Omega), \zeta \geqslant 0\,.$$

When $L = -\Delta$, a subsolution (resp. supersolution) of $-\Delta u = 0$ on Ω is a subharmonic (resp. superharmonic) distribution on Ω (see Definition 15 of §4). First of all, let us state, *the generalized Harnack inequality for subsolutions.*

Proposition 15. *Let L be a differential operator of order 2 in divergence form with coefficients in $L^\infty(\Omega)$ and strictly elliptic, a subsolution u of $Lu = 0$ on Ω which we suppose to be in $W^{1,2}_{\text{loc}}(\Omega)$. Then $u^+ \in L^\infty_{\text{loc}}(\Omega)$ and for every compact set K of Ω, $0 < r < \operatorname{dist}(K, \partial\Omega)$ and $p > 1$,*

$$u \leqslant \left[C\left(1 + \frac{2}{r}\right)\frac{p^2}{p-1}\right]^{n/p} \|u^+\|_{L^p(K + B(0,r))} \quad \text{a.e. on} \quad K \tag{8.65}$$

where C depends only on n and the L^∞-norms of the coefficients of $(1/c_0)L$.

The proof rests on the

Lemma 4. *Under the hypotheses of Proposition 15, for all $\alpha > 1$*

$$\|u^+\|_{L^{n\alpha/n-1}(K)} \leqslant C(\alpha, r)\,\|u^+\|_{L^\alpha(K + B(0,r))} \tag{8.66}$$

with $C(\alpha, r) = \left[C\left(1 + \frac{1}{r}\right)\frac{\alpha^2}{\alpha - 1}\right]^{1/\alpha}$ *where C depends only on the L^∞-norms of $(1/c_0)L$.*

Proof of Lemma 4. We put $\Omega_r = K + B(0, r)$ and $\eta(x) = \dfrac{\operatorname{dist}(x, \Omega\backslash\Omega_r)}{\operatorname{dist}(x, \Omega\backslash\Omega_r) + \operatorname{dist}(x, K)}$. The function η satisfies a Lipschitz condition on Ω with index $1/r$, with compact support $\bar{\Omega}_r$ and satisfying $0 \leqslant \eta \leqslant 1, \eta \equiv 1$ on K. On the other hand, we take a bounded Borel set θ on $\mathbb{R}$ satisfying $\theta \equiv 0$ on $]-\infty, 0[$ and $\theta > 0$ on $]0, \infty[$ and we put $\zeta = \eta^2 \int_{-\infty}^u \theta(r)\,dr$; we have $\zeta \in W^{1,2}(\Omega)$, $\operatorname{supp}\zeta \subset \bar{\Omega}_r$ with the result that in the limit in the definition of a subsolution $\int_\Omega L(u, \zeta)\,dx \leqslant 0$. Developing $L(u, \zeta)$, using the strict ellipticity and Young's inequality, we obtain

$$\frac{c_0}{2}\int_\Omega \eta^2\theta(u)|\operatorname{grad} u|^2\,dx \leqslant \int_\Omega \frac{1}{2c_0}\sum_i\left[w_1\left(\eta d_i + 2\frac{\partial\eta}{\partial x_i}\sum_i a_{ij}\right) + w_2 c_i\eta\right]^2 dx - \int_\Omega w_3\left(2\eta\sum_i c_i\frac{\partial\eta}{\partial x_i} + d\eta^2\right)dx$$

where we have put

$$w_1 = \theta(u)^{-1/2}\int_{-\infty}^{u}\theta(r)\,dr\,, \qquad w_2 = u\theta(u)^{1/2}\,, \qquad w_3 = u^+\int_{-\infty}^{u}\theta(r)\,dr\,.$$

We put $v = \int_{-\infty}^{u}\theta(r)^{1/2}\,dr$, $w = w_1 + w_2$; noting that $v^2 + w_3 \leqslant w^2$ and recalling that $\eta \leqslant 1$, $|\operatorname{grad}\eta| \leqslant 1/r$, we obtain

$$\|\operatorname{grad}(\eta v)\|_{L^2} \leqslant C\left(1 + \frac{1}{r}\right)\|w\|_{L^2(\Omega_r)}\,.$$

Making use now of the Sobolev inclusion in the Gagliardo–Nirenberg form[237] and noting that $\|v\|_{L^2(K)} \leqslant \|\eta v\|_{L^2} \leqslant \|w\|_{L^2(\Omega_1)}$ we have

$$\|v\|_{L^{2n/(n-1)}(K)} \leqslant \left[C\left(1 + \frac{1}{r}\right)\right]^{1/2}\|w\|_{L^2(\Omega_r)} \tag{8.67}$$

where C depends always on the L^∞-norms of $(1/c_0)L$.
In the limit we can apply this relation with

$$\theta(r) = \chi_{\{r>0\}}\frac{\alpha^2}{4}r^{\alpha-2}\,;$$

we then have $v = u^{+^{\alpha/2}}$ and $w = \dfrac{\alpha^2}{2(\alpha-1)}u^{+^{\alpha/2}}$ from which (8.66) follows[238]. □

Proof of Proposition 15. We shall use *Moser's iterative method.* We fix K a compact sub-set of Ω, $0 < r < \operatorname{dist}(K, \partial\Omega)$, $p > 1$; we put

$$\delta_k = r2^{-k}\,, \qquad r_k = \delta_1 + \cdots + \delta_k = r(1 - 2^{-k})\,, \qquad \alpha_k = p\left(\frac{n}{n-1}\right)^k.$$

For k fixed, we define recursively the compact sets K_i by $K_k = K$ and $K_{i-1} = K_i + \bar{B}(0, \delta_i)$ with the result that $K_0 = K + \bar{B}(0, r_k)$. Applying (8.66) to the compact set K_i with $\alpha = \alpha_{i-1}$ and $r = \delta_i$, we have

$$\|u^+\|_{L^{\alpha_i}(K_i)} \leqslant C_i\|u^+\|_{L^{\alpha_{i-1}}(K_{i-1})}$$

---

[237] We have $\|\zeta\|_{L^{n/(n-1)}} \leqslant \|\operatorname{grad}\zeta\|_{L^1}$ and hence

$$\|\zeta\|^2_{L^{2n/(n-1)}} = \|\zeta^2\|_{L^{n/(n-1)}} \leqslant 2\|\zeta\operatorname{grad}\zeta\|_{L^1} \leqslant 2\|\zeta\|_{L^2}\|\operatorname{grad}\zeta\|_{L^2}$$

[238] We recall that $u^{+^{\alpha/2}}$ denotes $(u^+)^{\alpha/2}$.

with

$$C_i = \left[C\left(1 + \frac{1}{\delta_i}\right)\frac{\alpha_{i-1}^2}{\alpha_{i-1} - 1}\right]^{1/\alpha_{i-1}}$$

and hence

$$\|u^+\|_{L^{\alpha_k}(K)} \leqslant \left(\prod_{i=1}^{k} C_i\right)\|u^+\|_{L^{\alpha_0}(K + B(0, r_k))} . \tag{8.68}$$

Using $C_i \leqslant C\left(p, \frac{r}{2}\right)^{\left(\frac{\pi - 1}{n}\right)^{i-1}} \times \left(\frac{2n}{n-1}\right)^{\frac{i-1}{p}\left(\frac{n-1}{n}\right)^{i-1}}$, we obtain

$$\prod_{i=1}^{k} C_i \leqslant C\left(p, \frac{r}{2}\right)^{n\left(1 - \left(\frac{n-1}{n}\right)^k\right)} \times \left(\frac{2n}{n-1}\right)^{\frac{n(n-1)}{p}\left(1 + (k-1)\left(\frac{n-1}{n}\right)^k - k\left(\frac{n-1}{n}\right)^{k-1}\right)}$$

Passing to the limit as $k \to \infty$ in (8.68), we therefore have

$$\|u^+\|_{L^\infty(K)} \leqslant C\left(p, \frac{r}{2}\right)^n \left(\frac{2n}{n-1}\right)^{\frac{n(n-1)}{p}} \|u^+\|_{L^p(K + B(0, r))}$$

that is to say (8.65). □

We now prove the *generalized Harnack inequality for positive supersolutions.*

Proposition 16. *Let L be a differential operator of order 2 in divergence form with coefficients in* $L^\infty(\Omega)$ *and strictly elliptic, a supersolution u of* $Lu = 0$ *on* Ω *which we suppose to be in* $W^{1,2}_{\text{loc}}(\Omega)$. *We suppose* Ω *to be connected,* $u \geqslant 0$ *on* Ω *and* $u \not\equiv 0$ *on* Ω.
Then $1/u \in L^\infty_{\text{loc}}(\Omega)$ *and for all* $x_0 \in \Omega, 0 < r \leqslant r_0 < \text{dist}(x_0, \partial\Omega)$ *and* $p < n/(n-1)$,

$$\|u\|_{L^p(B(x_0, r))} \leqslant Cr^{n/p}u \quad \text{a.e. on} \quad B\left(x_0, \frac{r}{2}\right) \tag{8.69}$$

where C depends only on the L^∞*-norms of the coefficients of* $(1/c_0)L$, *on n, p,* r_0 *and* $\text{dist}(x_0, \partial\Omega) - r_0$.

Proof. Given a compact set K of Ω and $0 < r < \text{dist}(K, \partial\Omega)$ we choose η as in the proof of Lemma 4 and fix $\varepsilon > 0$. Applying the definition of a supersolution with the test function $\zeta = \dfrac{\eta^2}{(\alpha + 1)(u + \varepsilon)^{\alpha + 1}}$ with $\alpha > -1$; expanding $L(u, \zeta)$ we obtain

$$\frac{c_0}{2}\int_\Omega \frac{\eta^2 |\text{grad}\, u|^2}{(u + \varepsilon)^{\alpha + 2}}\text{d}x \leqslant \int_\Omega \frac{1}{2c_0}\sum_i\left[\frac{1}{(\alpha + 1)(u + \varepsilon)^{\alpha/2}}\left(2\sum_j a_{ij}\frac{\partial\eta}{\partial x_j} + d_i\eta\right)\right.$$
$$\left. - c_i\eta\frac{u}{(u + \varepsilon)^{\frac{\alpha}{2} + 1}}\right]^2\text{d}x + \int_\Omega \frac{\eta u}{(\alpha + 1)(u + \varepsilon)^{\alpha + 1}}\left(d\eta + 2\sum_i c_i\frac{\partial\eta}{\partial x_i}\right)\text{d}x .$$

In the case $\alpha = 0$, we obtain

$$\left\|\frac{\eta\, \text{grad}\, u}{u + \varepsilon}\right\|_{L^2} \leqslant C(\|\eta\|_{L^2} + \|\text{grad}\,\eta\|_{L^2}) . \tag{8.70}$$

In the case $\alpha > 0$, we obtain by the same reasoning as in the proof of Lemma 4,

$$(8.71) \qquad \|(u + \varepsilon)^{-1}\|_{L^{\alpha n/n-1}(K)} \leqslant \left(C\left(1 + \frac{1}{r}\right)\alpha\right)^{1/\alpha} \|(u + \varepsilon)^{-1}\|_{L^{\alpha}(K + B(0, r))} .$$

Finally, in the case $-1 < \alpha < 0$, we shall have by putting $\beta = -\alpha$

$$(8.72) \qquad \|u + \varepsilon\|_{L^{\beta n/n-1}(K)} \leqslant \left[C\left(1 + \frac{1}{r}\right)\frac{\beta}{1 - \beta}\right]^{1/\beta} \|u + \varepsilon\|_{L^{\beta}(K + B(0, r))} ,$$

where in all of these inequalities, C depends only on the $L^\infty(\Omega)$ norms of the coefficients of $(1/c_0)L$.

Applying Moser's iteration method to (8.71) as in the proof of Proposition 15, we deduce for all $\alpha > 0$,

$$(8.73) \qquad \|(u + \varepsilon)^{-1}\|_{L^{\infty}(K)} \leqslant \left(C\left(1 + \frac{1}{r}\right)\alpha\right)^{1/\alpha} \|(u + \varepsilon)^{-1}\|_{L^{\alpha}(K + B(0, r))} .$$

Using (8.70), we see that $\operatorname{Log} u \in W^{1,2}_{\mathrm{loc}}(\Omega)$. In effect, fixing a compact set K_0 of non-zero measure, we put

$$k_\varepsilon = \frac{1}{|K_0|}\int_{K_0} \operatorname{Log}(u + \varepsilon)\,\mathrm{d}x \quad \text{and} \quad v_\varepsilon = \operatorname{Log}(u + \varepsilon) - k_\varepsilon ;$$

we have $v_\varepsilon \in W^{1,2}_{\mathrm{loc}}(\Omega)$, $\int_{K_0} v_\varepsilon\,\mathrm{d}x = 0$ and from (8.70), $\operatorname{grad} v_\varepsilon$ is bounded in $L^2(K)$ for every compact set in Ω; from Poincaré's inequality we deduce that v_ε is relatively compact in $L^2_{\mathrm{loc}}(\Omega)$; now $k_\varepsilon \geqslant \operatorname{Log} u - \overline{\lim_{\varepsilon \to 0}}\, v_\varepsilon$ a.e. on Ω and hence k_ε is bounded since $u \not\equiv 0$ on Ω. We deduce that

$$\operatorname{Log} u = \lim v_\varepsilon + \lim k_\varepsilon \in W^{1,2}_{\mathrm{loc}}(\Omega) .$$

Given a ball $\bar{B}(x_0, r_0) \subset \Omega$, we put $v = \operatorname{Log} u - k$ where

$$k = \frac{1}{|B(x_0, r_0)|}\int_{B(x_0,r_0)} \operatorname{Log} u\,\mathrm{d}x .$$

We have $v \in W^{1,2}(B(x_0, r_0))$, $\int_{B(x_0,r_0)} v\,\mathrm{d}x = 0$ and from (8.70), for every ball B of $\mathbb{R}^n$

$$\|\operatorname{grad} v\|_{L^2(B \cap B(x_0, r_0))} \leqslant C(r_0, \operatorname{dist}(x_0, \partial\Omega))|B|^{\frac{1}{2} - \frac{1}{n}} .$$

Using Lemma 5 below

$$\int_{B(x_0,r_0)} e^{\alpha_0|v|}\,\mathrm{d}x \leqslant C_0|B(x_0, r_0)|$$

where $p_0 = \alpha_0 C(r_0, \operatorname{dist}(x_0, \partial\Omega))$ and C_0 depending only on n. Taking account of

$$\|u\|_{L^{\alpha_0}(B(x_0, r_0))} \leqslant \frac{1}{\|u^{-1}\|_{L^{\alpha_0}(B(x_0, r_0))}}\left(\int_{B(x_0,r_0)} e^{\alpha_0|v|}\,\mathrm{d}x\right)^{2/\alpha_0}$$

and using (8.73) and (8.72) which we iterate a sufficient number of times, we deduce that for all $p < n/(n-1)$ and $0 < r_1 < r_0$

$$\|u^{-1}\|_{L^\infty(B(x_0, r_1))} \|u\|_{L^p(B(x_0, r_0))} \leqslant C$$

where C depends only on the $L^\infty(\Omega)$-norms of the coefficients of $(1/c_0)L$, of n, r_0, $\operatorname{dist}(x_0, \partial\Omega) - r_0$ and $r_0 - r_1$. By a dilatation argument, we deduce the Proposition. □

We state Lemma 5, which we have used above, referring the reader to John and Nirenberg [1]:

Lemma 5. *Let $B(x_0, r_0)$ be a ball of $\mathbb{R}^n$ and $v \in W^{1,1}(B(x_0, r_0))$ satisfying $\int_{B(x_0, r_0)} v\,dx = 0$ and $\|\operatorname{grad} v\|_{L^1(B \cap B(x_0, r_0))} \leqslant C|B|^{1-\frac{1}{n}}$ for every ball B. Then*

$$\int_{B(x_0, r_0)} e^{p_0 C^{-1}|v|}\,dx \leqslant C_0 |B(x_0, r_0)| \tag{8.74}$$

where p_0, C_0 depend only on n.

We note now, as an immediate corollary of Propositions 15 and 16, *Harnack's inequality* for positive solutions.

Corollary 4. *Let L be a differential operator of order 2 in divergence form with coefficients in $L^\infty(\Omega)$ and strictly elliptic, and a positive weak solution of $Lu = 0$ on Ω which we suppose is in $W^{1,2}_{\text{loc}}(\Omega)$.*

(1) *For all $x_0 \in \Omega$, $0 < r \leqslant r_0 < \operatorname{dist}(x_0, \partial\Omega)$*

$$\sup_{B(x_0, r)} u \leqslant C \inf_{B(x_0, r)} u \tag{8.75}$$

where C depends only on the $L^\infty(\Omega)$-norms of the coefficients $(1/c_0)L$, and on n, r_0, $\operatorname{dist}(x_0, \partial\Omega) - r_0$.

(2) *Suppose Ω is connected; for every compact set K in Ω*

$$\max_K u \leqslant C \min_K u \tag{8.76}$$

where C depends on $(1/c_0)L$, K and $\operatorname{dist}(K, \partial\Omega)$.

The inequality (8.75) is immediate from (8.65) and (8.69); as for (8.76), we deduce it from (8.75) as in the case of Harnack's inequality for harmonic functions. Furthermore, we note that by the same reasoning we have the analogue for a compact set of the inequality (8.69) for the positive supersolutions.
In the inequalities (8.75) and (8.76), we have used sup, max, etc., in place of sup ess, etc.; in effect from the De Giorgi–Nash, we know that the weak solution u is continuous on Ω. In the case

$$L = -\sum \frac{\partial}{\partial x_i}\left(a_{ij}\frac{\partial}{\partial x_j}\right) + \sum d_i \frac{\partial}{\partial x_i},$$

this continuity results from Harnack's inequality as we have seen in the proof of Proposition 13 of §2[239]. In the general case, we refer the reader to Ladyzenskaya–Ural'ceva [1] as well as for complementary results such as, for example, the semi-continuity of subsolutions and supersolutions.

Another immediate corollary of Harnack's inequalities is the *principle of the strong maximum.*

Proposition 17. *Let* $L = -\sum \frac{\partial}{\partial x_i}\left(\sum a_{ij}\frac{\partial}{\partial x_j} + c_i\right) + \sum d_i \frac{\partial}{\partial x_i} + d$ *be a differential operator of order 2 with coefficients in* $L^\infty(\Omega)$ *and strictly elliptic,* u *a subsolution of* $Lu = 0$ *on* Ω *which we shall suppose to be in* $W^{1,2}_{\mathrm{loc}}(\Omega)$. *We suppose* Ω *to be connected and that we are in one of the following two cases:*

case 1: $d = \sum \frac{\partial c_i}{\partial x_i}$ *in* $\mathscr{D}'(\Omega)$

case 2: $d \geqslant \sum \frac{\partial c_i}{\partial x_i}$ *in* $\mathscr{D}'(\Omega)$ $\sup\operatorname{ess}_{\Omega} u \geqslant 0$.

---

[239] To apply the method of the proof of Proposition 13 of §2, we must sharpen the dependence of the constant C in (8.75): in particular, if $0 < 4r_0 < \operatorname{dist}(x_0, \partial\Omega)$ C depends only on the L^∞-norms of the coefficients of L, n and r_0 (see Theorem 8.20 in Gilbarg–Trudinger [1]).
Then, for $0 < r \leqslant r_0 < \operatorname{dist}(x_0, \partial\Omega)$ with the notation

$$m(r) = \inf_{B(x_0,r)}\operatorname{ess} u, \qquad M(r) = \sup_{B(x_0,r)}\operatorname{ess} u, \qquad \omega(r) = M(r) - m(r)\,;$$

applying this inequality to the functions $u - m(r)$ and $M(r) - u$, we obtain

$$M(r/4) - m_F \leqslant C(m(r/4) - m(r)), \quad M(r) - m(r/4) \leqslant C(M(r) - M(r/4))$$

where C does not depend on r. Thus

$$\omega(r/4) + \omega(r) \leqslant C(\omega(r) - \omega(r/4))$$

namely

$$\omega(r/4) \leqslant \frac{C-1}{C+1}\omega(r)$$

and by induction

$$\omega\left(\frac{r}{4^k}\right) \leqslant \left(\frac{C-1}{C+1}\right)^k \omega(r)\,.$$

Hence for $\frac{r_0}{4^{k+1}} \leqslant r < \frac{r_0}{4^k}$, we have

$$\omega(r) \leqslant \omega\left(\frac{r_0}{4^k}\right) \leqslant \left(\frac{C+1}{C-1}\right)^{-k} \omega(r_0) \leqslant M_0 r^{\alpha_0}$$

with

$$M_0 = \frac{C+1}{C-1} r_0^{-\alpha_0}, \qquad \alpha_0 = \frac{\operatorname{Log}\frac{C+1}{C-1}}{\operatorname{Log} 4}\,.$$

Therefore, this proves that u satisfies a Hölder condition.

Then either u is constant on Ω, or for every compact set K of Ω

$$\sup_K \operatorname{ess} u < \sup_\Omega \operatorname{ess} u .$$

Proof. Let us consider $m = \sup\operatorname{ess}_\Omega u$. If $m = +\infty$, the result is trivial; let us suppose $m < \infty$ and put $v = m - u$. It is clear that $v \in W^{1,2}_{\text{loc}}(\Omega)$, $v \geqslant 0$ on Ω and for all $\zeta \in \mathscr{D}(\Omega)$ with $\xi \geqslant 0$

$$\int_\Omega L(v, \zeta)\,\mathrm{d}x \geqslant \int_\Omega L(m, \zeta)\,\mathrm{d}x = m \int_\Omega \left(d\zeta + \sum c_i \frac{\partial \zeta}{\partial x_i}\right)\mathrm{d}x .$$

As much as in the case 1 as in the case 2, v is thus a supersolution of $Lv = 0$ on Ω. From Proposition 16, if v is not identically zero on Ω, for every compact set K of Ω, $\inf_K \operatorname{ess} v > 0$. □

The principle of the strong maximum gives immediately as in the case of the Laplacian, the uniqueness of a quasi-classical solution of the Dirichlet problem

$$\begin{cases} Lu = f & \text{on} \quad \Omega \\ u = g & \text{on} \quad \Gamma . \end{cases} \tag{8.77}$$

More precisely, as a corollary of Proposition 17, given $L = -\sum_i \frac{\partial}{\partial x_i}\left(\sum_j a_{ij}\frac{\partial}{\partial x_j} + c_i\right) + \sum_i d_i \frac{\partial}{\partial x_i} + d$ a differential operator of order 2 in divergence form on an arbitrary bounded[240] open set Ω with coefficients in $L^\infty_{\text{loc}}(\Omega)$ and strictly elliptic on every compact set of Ω, if $d \geqslant \sum \partial c_i/\partial x_i$ in $\mathscr{D}'(\Omega)$, then L satisfies the *principle of the weak positive maximum on* Ω namely:

$$\left\{\begin{array}{l} \textit{every subsolution } u \in W^{1,2}_{\text{loc}}(\Omega) \textit{ of } Lu = 0 \textit{ on } \Omega \textit{ satisfying} \\ \overline{\lim_{x \to z}}\, u(x) \leqslant 0 \textit{ for all } z \in \Gamma \textit{ satisfies } u \leqslant 0 \textit{ on } \Omega. \end{array}\right. \tag{8.78}$$

It is clear that the principle of the weak positive maximum on Ω implies the uniqueness of a quasi-classical solution of (8.77): more precisely this solution, if it exists, will be obtained by *Perron's method* as the outer (resp. inner) envelope of the set of subsolutions v (resp. supersolutions w) of $Lu = f$ on Ω satisfying

$$\overline{\lim_{x \to z}}\, v(x) \leqslant g(z) \quad (\text{resp. } \underline{\lim_{x \to z}}\, v(x) \leqslant g(z)) \quad \text{for all} \quad z \in \Gamma .$$

Let us content ourselves with proving

[240] We restrict our attention to the case in which, for simplicity, Ω is bounded; the case of Ω unbounded can be considered in the same way by taking a condition at infinity.

Proposition 18. *Let Ω be a bounded open set with Lipschitz boundary and L a differential operator of order 2 in divergence form with coefficients in $L^\infty(\Omega)$ and strictly elliptic. We suppose that L satisfies the principle of the weak positive maximum* (8.78); *then for all $g \in \mathscr{C}^0(\Gamma)$, there exists a unique solution $u \in W^{1,2}_{\text{loc}}(\Omega) \cap \mathscr{C}^0(\bar{\Omega})$ of the problem*

$$(8.79) \qquad \begin{cases} Lu = 0 & \text{on} \quad \Omega \quad \textit{(in the weak sense)} \\ u = g & \text{on} \quad \Gamma \quad \textit{(in the classical sense)} \end{cases}$$

In addition, we have

$$(8.80) \qquad e(x)\min_\Gamma(g \wedge 0) \leqslant u(x) \leqslant e(x)\max_\Gamma(g \vee 0) \quad \textit{for all} \quad x \in \Omega$$

where e is the solution of (8.79) *corresponding to $g \equiv 1$ on Γ.*

Before proving this Proposition, we note that $e > 0$ on $\bar{\Omega}$; also if $\sum \frac{\partial c_i}{\partial x_i} \geqslant d$ $\left(\text{resp. } \sum \frac{\partial c_i}{\partial x_i} \leqslant d\right)$ in $\mathscr{D}'(\Omega)$, then $e \leqslant 1$ (resp. $e \geqslant 1$) on Ω; moreover $e \equiv 1$ on Ω iff $\sum \frac{\partial c_i}{\partial x_i} = d$ and then we have for the solution u of (8.79)

$$(8.81) \qquad \min_\Gamma g \leqslant u(x) \leqslant \max_\Gamma g \quad \text{for all} \quad x \in \Omega$$

as in the case $L = -\Delta$. We note finally that if Ω is connected, and if g is not identically zero on Γ (resp. if g is not constant on Γ) we then have strict inequalities in (8.80) (resp. (8.81)). All of these remarks arise immediately from the principle of the weak positive maximum (8.78) and from Harnack's inequality (Corollary 4). We can besides sum up these remarks by noting that from Harnack's inequality, *if L is a differential operator of order 2 in divergence form on a connected bounded open set Ω with coefficients in $L^\infty_{\text{loc}}(\Omega)$, strictly elliptic on every compact set in Ω and satisfying the principle of the weak positive maximum* (8.78) *then*

$$(8.82) \qquad \begin{cases} \textit{every subsolution } u \in W^{1,2}_{\text{loc}}(\Omega) \textit{ of } Lu = 0 \textit{ on } \Omega \textit{ satisfying} \\ \overline{\lim_{x \to z}}\, u(x) \leqslant 0 \textit{ for all } z \in \Gamma \textit{ satisfies } u \equiv 0 \textit{ on } \Omega \textit{ or } \sup\operatorname{ess}_K u < 0 \\ \textit{for every compact set } K \textit{ of } \Omega\,. \end{cases}$$

Proof of Proposition 18. The uniqueness follows immediately from (8.78). In particular, with the notation of Proposition 10, $N(L, \Gamma) = \{0\}$; from this Proposition and the De Giorgi–Nash theorem (Proposition 6), there exists, therefore, a solution of (8.79) when $g = \varphi$ on Γ where $\varphi \in W^{1,n+\varepsilon}(\Omega)$. Thus we obtain the existence of e corresponding to $g \equiv 1$ on Γ. Applying (8.78) to $u - (\max_\Gamma g \vee 0)e$ and $(\min_\Gamma g \wedge 0)e - u$, we find the relation (8.80) is true for every solution of (8.79).

Let us now consider $g \in \mathscr{C}^0(\Gamma)$ arbitrary and approximated uniformly on Γ by a sequence of functions $g_k = \varphi_k$ on Γ where $\varphi_k \in W^{1,n+\varepsilon}(\Omega)$. There exists a unique solution u_k corresponding to g_k and we have from (8.80),

$$\max_{\bar{\Omega}} |u_k - u_l| \leqslant (\max_{\Gamma} |g_k - g_l|)(\max_{\bar{\Omega}} e) .$$

Also from the local estimates for every compact set K of Ω we have

$$\| \operatorname{grad}(u_k - u_l) \|_{L^2(K)} \leqslant C_K \| u_k - u_l \|_{L^2(\Omega)}$$

(see the proof of Proposition 15 for example). We deduce that the sequence (u_k) converges uniformly on $\bar{\Omega}$ and in $W^{1,2}_{\text{loc}}(\Omega)$. The limit is the solution of (8.79). □

An important tool for the study of subsolutions and of the principle of the weak positive maximum is *Kato's inequality*.

Proposition 19. *Let* $L = -\sum_i \frac{\partial}{\partial x_i}\left(\sum_j a_{ij} \frac{\partial}{\partial x_j} + c_i\right) + \sum d_i \frac{\partial}{\partial x_i} + d$ *be a differential operator of order 2 in divergence form with coefficients in* $L^\infty_{\text{loc}}(\Omega)$ *and weakly elliptic, that is to say satisfying* $\sum_{i,j} a_{ij}\xi_i\xi_j \geqslant 0$ a.e. *on* Ω *for all* $\xi \in \mathbb{R}^n$. *Let* $f \in L^1_{\text{loc}}(\Omega)$ *and let* $u \in W^{1,2}_{\text{loc}}(\Omega)$ *be a subsolution of* $Lu = f$.
Then the function $v = u^+ \in W^{1,2}_{\text{loc}}(\Omega)$ *is a subsolution of*

$$Lv = (\chi_{\{u>0\}} + \mu\chi_{\{u=0\}}) f$$

on Ω *for all* $\mu \in [0, 1]$.

Proof. Let us fix $\mu \in [0, 1]$ and take $\theta \in \mathscr{C}^1(\mathbb{R})$ with $\theta' \geqslant 0$, $\theta(0) = \mu$, $\theta \equiv 0$ on $]-\infty, -1[$, $\theta \equiv 1$ on $[1, +\infty[$. We take $\eta \in \mathscr{D}(\Omega)$ with $\eta \geqslant 0$ and for $\varepsilon > 0$ let us use the test function $\zeta_\varepsilon = \theta(u/\varepsilon)\eta$. We have

$$L(u, \zeta_\varepsilon) = \sum_{i,j} a_{ij} \frac{\partial u}{\partial x_j} \theta\left(\frac{u}{\varepsilon}\right) \frac{\partial \eta}{\partial x_i} + \sum_i c_i u \theta\left(\frac{u}{\varepsilon}\right) \frac{\partial \eta}{\partial x_i} + \sum_i d_i \frac{\partial u}{\partial x_i} \theta\left(\frac{u}{\varepsilon}\right) \eta$$
$$+ du\theta\left(\frac{u}{\varepsilon}\right)\eta + \frac{\eta}{\varepsilon}\theta'\left(\frac{u}{\varepsilon}\right) \sum_{i,j} a_{ij} \frac{\partial u}{\partial x_i}\frac{\partial u}{\partial x_j} + \eta \frac{u}{\varepsilon}\theta'\left(\frac{u}{\varepsilon}\right) \sum_i c_i \frac{\partial u}{\partial x_i} .$$

The second last term is positive or zero, from the hypothesis of (weak) ellipticity. When $\varepsilon \to 0$

$$\theta\left(\frac{u}{\varepsilon}\right) \to \chi_{\{u>0\}} + \mu\chi_{\{u=0\}} \quad \text{a.e. on} \quad \Omega$$

and

$$\frac{u}{\varepsilon}\theta'\left(\frac{u}{\varepsilon}\right) \to 0 \quad \text{a.e. on} \quad \Omega$$

with

$$0 \leqslant \theta\left(\frac{u}{\varepsilon}\right) \leqslant 1 \quad \text{and} \quad \left|\frac{u}{\varepsilon}\theta'\left(\frac{u}{\varepsilon}\right)\right| \leqslant \|\theta'\|_{L^\infty} .$$

Now

$$\frac{\partial u}{\partial x_i}\chi_{\{u>0\}} = \frac{\partial u^+}{\partial x_i} \quad \text{a.e. on } \Omega \quad \frac{\partial u}{\partial x_i} = 0 \quad \text{a.e. on} \quad \{u = 0\} .$$

Hence in the limit

$$\int_\Omega (\chi_{\{u>0\}} + \mu\chi_{\{u=0\}}) f\eta \, dx \geqslant \int_\Omega L(u^+, \eta)\, dx ,$$

which can be considered as the weak form of Kato's inequality (we compare it with (5.82) in the case in which $L = -\Delta$). □

As a particular case of Proposition 19, we have that, if u is a subsolution of $Lu = 0$ on Ω, then it is the same for the function u^+. Because of that *the principle of the weak positive maximum* (8.78) *can be expressed in an equivalent way for an operator L satisfying the hypotheses of Proposition* 19:

$$\text{(8.83)} \qquad \begin{cases} \textit{every subsolution } u \in W^{1,2}_{\text{loc}}(\Omega) \textit{ of } Lu = 0 \textit{ on } \Omega \textit{ satisfying} \\ u \geqslant 0 \textit{ a.e. on } \Omega \textit{ and } \lim\limits_{x \to z} u(x) = 0 \textit{ for all } z \in \Gamma , \end{cases}$$

is identically zero on Ω.

In particular if L *satisfies the principle of the weak positive maximum the same is true of* $L + \lambda$ *for all* $\lambda \geqslant 0$.

This permits us to obtain the following characterization:

Proposition 20. *Let Ω be a bounded open set with Lipschitz boundary Γ and L a differential operator of order 2 in divergence form with L^∞-coefficients and strictly elliptic. Then L satisfies the principle of the weak positive maximum if and only if $N(L + \lambda, \Gamma) = \{0\}$ for all $\lambda \geqslant 0$*[241]*. If also L is formally self-adjoint, then these properties are equivalent to the coercivity of L in $W^{1,2}_0(\Omega)$, that is to say*

$$\text{(8.84)} \qquad \textit{there is } c > 0 \textit{ such that } \int_\Omega L(u, u)\, dx \geqslant c\|u\|^2_{W^{1,2}}$$

$$\textit{for all} \quad u \in W^{1,2}_0(\Omega) .$$

Proof. If L satisfies (8.78), then for all $\lambda \geqslant 0$, as we shall see, $L + \lambda$ satisfies (8.78) and hence $N(L + \lambda, \Gamma) = \{0\}$. For the converse we shall show that:

(*a*) the coercivity (8.84) implies (8.78) always;
(*b*) if $L + \lambda$ satisfies (8.78) for all $\lambda > 0$ and $N(L, \Gamma) = \{0\}$, then L satisfies (8.78);
(*c*) if L satisfies (8.78), then there exists $\mu < 0$ such that $L + \lambda$ satisfies (8.78) for all $\lambda > \mu$.

[241] We recall that $N(L, \Gamma) = \{u \in W^{1,2}_0(\Omega); Lu = 0 \text{ on } \Omega\}$ (see Proposition 10).

Considering then $\Lambda = \{\lambda \in \mathbb{R}; L + \lambda$ satisfies (8.78)$\}$; from (*a*), $\Lambda \neq \emptyset$ (for λ large, $L + \lambda$ is coercive; from (*c*), $\Lambda =]\lambda_0, \infty[$ and from (*b*) $N(L + \lambda_0, \Gamma) \neq \{0\}$ which clearly proves that $\lambda_0 < 0$ if $N(L + \lambda, \Gamma) = \{0\}$ for all $\lambda \geq 0$.

Let us prove (*a*) *first of all.* Let $u \in W^{1,2}_{\text{loc}}(\Omega)$ subsolution of $Lu = 0$ on Ω and satisfying $\overline{\lim_{x \to z}} u(x) \leq 0$ for all $z \in \Gamma$. For $\varepsilon > 0$, let us use the test function $\zeta = (u - \varepsilon)^+$ which is in $W^{1,2}(\Omega)$ with compact support in Ω.

Expanding $L(u, (u - \varepsilon)^+)$ we obtain

$$L(u, \zeta) = L((u - \varepsilon)^+, (u - \varepsilon)^+) + \varepsilon\left(\sum_i c_i \frac{\partial (u - \varepsilon)^+}{\partial x_i} + d(u - \varepsilon)^+\right),$$

from which, using (8.84) and $\int_\Omega L(u, \zeta)\,dx \leq 0$, we have

$$c\,\|(u - \varepsilon)^+\|_{W^{1,2}} \leq \varepsilon\left(\int\left(\sum c_i^2 + d^2\right)dx\right)^{1/2}.$$

In the limit when $\varepsilon \to 0$, $u^+ = 0$.

Let us now consider (b). First, let $u \in W^{1,2}(\Omega)$ be a subsolution of $Lu = 0$ on Ω with compact support in Ω. Given $\lambda > 0$, we can consider the solution $u_\lambda \in W^{1,2}_0(\Omega)$ of $Lu_\lambda + \lambda u_\lambda = Lu$ on Ω; since $\operatorname{supp} Lu \subset \operatorname{supp} u$, from the De Giorgi–Nash theorem $u_\lambda \in \mathscr{C}(\bar{\Omega} \backslash \operatorname{supp} u)$ and from the hypothesis, since $Lu \leq 0$ on Ω, $u_\lambda \leq 0$ on Ω. Now $u - u_\lambda \in W^{1,2}_0(\Omega)$ and $L(u - u_\lambda) = \lambda u_\lambda$ on Ω; since $N(L, \Gamma) = \{0\}$, L is an isomorphism of $W^{1,2}_0(\Omega)$ onto $W^{-1,2}(\Omega)$ and hence *a fortiori*

$$\lambda\,\|u_\lambda\|_{W^{1,2}} \geq c\,\|u - u_\lambda\|_{W^{1,2}} \quad \text{with} \quad c > 0\,;$$

we deduce that $u_\lambda \to u$ in $W^{1,2}(\Omega)$, from which $u \leq 0$.

In the general case $u \in W^{1,2}_{\text{loc}}(\Omega)$ subsolution of $Lu = 0$ on Ω, satisfying $\overline{\lim_{x \to z}} u(x) \leq 0$ for all $z \in \Gamma$, we consider $u_\varepsilon = (u - \varepsilon e)^+$ where e is the solution of $Le = 0$ on Ω, $e \equiv 1$ on Γ. From Kato's inequality, u_ε is also a subsolution of $Lu = 0$ on Ω and $u_\varepsilon \in W^{1,2}(\Omega)$ with compact support. Applying the above result, $u_\varepsilon = 0$ on Ω and hence $u \leq \varepsilon e$; in the limit when $\varepsilon \to 0$, $u \leq 0$ on Ω.

To prove (*c*), supposing that L satisfies (8.78), let us consider the solution u_0 of $Lu = 1$ on Ω, $u = 0$ on Γ. We have $u_0 \in \mathscr{C}^0(\bar{\Omega})$ and $u_0 > 0$ on Ω; let us put $\mu = -1/\|u_0\|_{L^\infty}$. We consider $0 > \lambda > \mu$ and $u \in W^{1,2}_{\text{loc}}(\Omega)$ a subsolution of $Lu + \lambda u = 0$ on Ω satisfying $\overline{\lim_{x \to z}} u(x) \leq 0$ for all $z \in \Gamma$. From Proposition 15, $u^+ \in L^\infty(\Omega)$. The function $v = u + \lambda\|u^+\|_{L^\infty} u_0$ is then in $W^{1,2}_{\text{loc}}(\Omega)$, subsolution of $Lv = \lambda(\|u^+\|_{L^\infty} - u) \leq 0$ and satisfies $\overline{\lim_{x \to z}} v(x) \leq 0$ for all $z \in \Gamma$. Therefore $u \leq -\lambda\|u^+\|_{L^\infty}$ on Ω from which we have $\|u^+\|_{L^\infty} = 0$ from the choice of λ.

To complete the proof of the Proposition, we must in the symmetric case, show that $N(L + \lambda, \Gamma) = \{0\}$ for all $\lambda \geq 0$ implies the coercivity (8.84). We note that the coercivity in $W_0^{1,2}(\Omega)$ is equivalent to the coercivity in $L^2(\Omega)$:

(8.85) there exists $c > 0$ such that $\displaystyle\int_\Omega L(u, u)\,\mathrm{d}x \geqslant c\,\|u\|_{L^2}^2$

for all $u \in W_0^{1,2}(\Omega)$;

we have in effect for $\varepsilon > 0$,

$$(8.86)\quad L(u, u) \geqslant (1 - \varepsilon)L(u, u) + \varepsilon\left[\frac{c_0}{2}|\operatorname{grad} u|^2 + \left(d - \frac{1}{2c_0}\sum_i (c_i + d_i)^2\right)u^2\right];$$

therefore, assuming (8.85) and choosing

$$\varepsilon = c\left(c + \frac{c_0}{2} + \left\|\frac{1}{2c_0}\sum_i (c_i + d_i)^2 - d\right\|_{L^\infty}\right)^{-1}$$

we shall have $\displaystyle\int L(u, u)\,\mathrm{d}x \geqslant \frac{\varepsilon c_0}{2}\,\|u\|_{W^{1,2}}^2$.

Let us consider now $\displaystyle\mu = \inf_{\substack{u \in W_0^{1,2}(\Omega)\\ \|u\|_{L^2} = 1}} \int_\Omega L(u, u)\,\mathrm{d}x$.

Using (8.86) with $\varepsilon = 1$, it is clear that $\mu > -\infty$; also using the compact injection of $W_0^{1,2}(\Omega)$ into $L^2(\Omega)$ we see that there exists $u \in W_0^{1,2}$ with $\|u\|_{L^2} = 1$ such that $\displaystyle\int_\Omega L(u, u)\,\mathrm{d}x = \mu$. Using

$$\int_\Omega L(u + \zeta, u + \zeta)\,\mathrm{d}x \geqslant \int_\Omega L(u, u)\,\mathrm{d}x \quad \text{for all} \quad \zeta \in W_0^{1,2}(\Omega),$$

we see that u is a weak solution of $Lu - \mu u = 0$ on Ω. Hence if $N(L + \lambda, \Gamma) = \{0\}$ for all $\lambda \geqslant 0$, necessarily, we have $\mu > 0$. □

In the case where L is not formally self-adjoint, the principle of the weak positive maximum does not necessarily imply the coercivity of L, as is shown by the following example in dimension $n = 1$: taking

$$L = -\frac{\mathrm{d}^2}{\mathrm{d}x^2} + b\frac{\mathrm{d}}{\mathrm{d}x} + d \quad \text{where} \quad d \in L^\infty(]0, 1[),\ b \in W^{1,1}(]0, 1[)\,;$$

L is coercive iff $d + \pi^2 \geqslant \frac{b'}{2}$ and $d + \pi^2 \not\equiv \frac{b'}{2}$; L will satisfy the principle of the weak positive maximum iff $\frac{1}{4}b^2 + d + \pi^2 \geqslant \frac{1}{2}b'$ and $\frac{1}{4}b^2 + d + \pi^2 \equiv \frac{1}{2}b'$ (see Example 1).

From Proposition 20, *if L satisfies the principle of the weak positive maximum, it is the same for its* adjoint L^*. In particular, with the usual notation, if $d \geqslant \sum \partial d_i/\partial x_i$ in $\mathscr{D}'(\Omega)$, then L satisfies the principle of the weak positive maximum since it is so for its adjoint from Proposition 17. In fact in this case L satisfies a principle of the maximum in L^1, dual of the classical maximum principle.

Let us state this precisely

Proposition 21. *Let Ω be an arbitrary bounded open set of $\mathbb{R}^n$*

$$L = -\sum \frac{\partial}{\partial x_i}\left(\sum a_{ij}\frac{\partial}{\partial x_j} + c_i\right) + \sum d_i \frac{\partial}{\partial x_i} + d$$

a differential operator of order 2 in divergence form with coefficients in L^∞ and strictly elliptic, $f \in L^1(\Omega)$ and u a subsolution of $Lu = f$ on Ω which we shall suppose to be in $W^{1,2}_{\text{loc}}(\Omega)$. We suppose $d \geqslant \sum \frac{\partial d_i}{\partial x_i}$ in $\mathscr{D}'(\Omega)$ $\overline{\lim_{x \to z}}\, u(x) \leqslant 0$ for all $z \in \Gamma$. Then

$$\int_{\{u>0\}} f\,\mathrm{d}x \geqslant 0\,.$$

Formally, denoting $\Omega_+ = \{u > 0\}$, we have

$$\int_{\Omega_+} f\,\mathrm{d}x \geqslant \int_{\Omega_+} Lu\,\mathrm{d}x = -\int_{\partial\Omega_+} \frac{\partial u}{\partial n_L}\,\mathrm{d}\gamma + \int_{\Omega_+}\left(\sum d_i \frac{\partial u}{\partial x_i} + du\right)\mathrm{d}x\,.$$

Now if $u \leqslant 0$ on Γ, we have $u = 0$ on $\partial\Omega_+$; hence $\frac{\partial u}{\partial n_L} \leqslant 0$ on $\partial\Omega_+$ and since $d \geqslant \sum \frac{\partial d_i}{\partial x_i}$ in $\mathscr{D}'(\Omega)$, $\int_{\Omega+}\left(\sum d_i \frac{\partial u}{\partial x_i} + du\right)\mathrm{d}x \geqslant 0$. To be justified, this reasoning would demand that the coefficients of L and u be sufficiently regular, which we might possibly suppose, but also that Ω_+ be regular which is not, in general true. Let us give a general proof based on the same argument as that used in establishing Kato's inequality.

Proof. Let us take $\theta \in \mathscr{C}^1(\mathbb{R})$ with $\theta' \geqslant 0$, $\theta \equiv 0$ on $]-\infty, 1]$ and $\theta \equiv 1$ on $[2, \infty[$ and use the test function $\zeta_\varepsilon = \theta\left(\frac{u}{\varepsilon}\right)$ which from the hypothesis is in $W^{1,2}(\Omega)$ with compact support in Ω. Using the strict ellipticity, we have

$$(8.87)\quad \left\{\begin{aligned} &\int_\Omega f\zeta_\varepsilon\,\mathrm{d}x \geqslant \int_\Omega L(u,\zeta_\varepsilon)\,\mathrm{d}x \geqslant \frac{c_0}{2\varepsilon}\int_\Omega \theta'\left(\frac{u}{\varepsilon}\right)|\operatorname{grad} u|^2\,\mathrm{d}x \\ &\quad + \int_\Omega\left(\sum_i d_i \frac{\partial w_\varepsilon}{\partial x_i} + dw_\varepsilon\right)\mathrm{d}x - \frac{1}{2c_0}\int_\Omega \frac{u}{\varepsilon}\theta'\left(\frac{u}{\varepsilon}\right)\sum_i c_i^2\,\mathrm{d}x \\ &\quad + \int_\Omega d\left(u\theta\left(\frac{u}{\varepsilon}\right) - w_\varepsilon\right)\mathrm{d}x\,, \end{aligned}\right.$$

where $w_\varepsilon = \int_0^u \theta\left(\frac{r}{\varepsilon}\right)\mathrm{d}r \in W^{1,2}(\Omega)$ with compact support. In the term on the right of (8.87), the first integral is obviously positive or zero; the second term also thanks to the hypothesis $d \geqslant \sum \frac{\partial d_i}{\partial x_\varepsilon}$ in $\mathscr{D}'(\Omega)$. Now when $\varepsilon \to 0$

$$\zeta_\varepsilon \to \chi_{\{u>0\}}\,, \quad \frac{u}{\varepsilon}\theta'\left(\frac{u}{\varepsilon}\right) \to 0\,, \qquad w_\varepsilon \to u^+ \quad \text{a.e. on} \quad \Omega$$

with

$$0 \leqslant \zeta_\varepsilon \leqslant 1\,, \qquad 0 \leqslant \frac{u}{\varepsilon}\theta'\left(\frac{u}{\varepsilon}\right) \leqslant \|\theta'\|_{L^\infty}\,, \qquad 0 \leqslant u\zeta_\varepsilon - w_\varepsilon \leqslant u^+\,.$$

Therefore in the limit in (8.87), $\int_\Omega f\chi_{\{u>0\}}\,\mathrm{d}x \geqslant 0$. □

Remark 1. When the coefficients $a_{ij} \in W^{1,\infty}(\Omega)$, all the results of Propositions 15 to 21 remain valid for the subsolutions (or supersolutions) $u \in W^{1,1}_{\text{loc}}(\Omega)$ without the restriction $u \in W^{1,2}_{\text{loc}}(\Omega)$ which we have imposed. In effect, given $u \in W^{1,1}_{\text{loc}}(\Omega)$ subsolution of $Lu = 0$ on Ω, for every bounded open set Ω_0 with $\bar{\Omega}_0 \subset \Omega$, there exists a sequence of subsolutions (u_k) in $\mathscr{C}^1(\bar{\Omega}_0)$ of $Lu = 0$ on Ω_0 such that $u_k \rightharpoonup u$ in $W^{1,1}(\Omega_0)$; that follows from Proposition 14. If u is a subsolution of $Lu = 0$ on Ω, the distribution Lu is a negative Radon measure on Ω with the result that the trace f_0 of Lu on Ω_0 is a (negative) bounded measure. Supposing Ω_0 with Lipschitz boundary, $u \in W^{1,1}(\Omega_0)$ admits a trace $g_0 \in L^1(\partial\Omega_0)$ on the boundary of Ω_0 (see Nečas [1]). When $\partial\Omega_0$ is a $W^{2,\infty}$ boundary, we know (see Proposition 14) that $N(L^*, \partial\Omega_0) \subset \mathscr{C}^1(\bar{\Omega}_0)$ and we shall have

$$\int_{\Omega_0} \zeta f_0\,\mathrm{d}x = \int_{\partial\Omega_0} g_0 \frac{\partial\zeta}{\partial n_{L^*}}\,\mathrm{d}\gamma \quad \text{for all} \quad \zeta \in N(L^*, \partial\Omega_0)\,;$$

we can always approximate (f_0, g_0) by a sequence of regular functions (f_k, g_k) such that $f_k \leqslant 0$ on Ω_0 and

$$\int_{\Omega_0} \zeta f_k\,\mathrm{d}x = \int_{\partial\Omega_0} g_k \frac{\partial\zeta}{\partial n_{L^*}}\,\mathrm{d}\gamma \quad \text{for all} \quad \zeta \in N(L^*, \partial\Omega_0)$$

with the result that we can solve the problem $Lu_k = f_k$ on Ω_0, $u_k = g_k$ on $\partial\Omega_0$ by imposing on the solution the condition of verifying

$$\int_{\Omega_0} u_k\zeta\,\mathrm{d}x = \int_{\Omega_0} u\zeta\,\mathrm{d}x \quad \text{for all} \quad \zeta \in N(L, \partial\Omega_0)\,;$$

from Proposition 14, $u_k \in \mathscr{C}^1(\bar{\Omega}_0)$ and $u_k \rightharpoonup u$ in $W^{1,1}(\Omega_0)$. □

Let us consider now a differential operator of order 2,

$$P = -\sum a_{ij}\frac{\partial^2}{\partial x_i\,\partial x_j} + \sum b_i \frac{\partial}{\partial x_i} + c\,.$$

Supposing the coefficients $a_{ij} = a_{ji} \in W^{1,\infty}(\Omega)$, b_i, $c \in L^\infty(\Omega)$, we can put it in divergence form

$$L = -\sum_i \frac{\partial}{\partial x_i}\left(\sum_j a_{ij}\frac{\partial}{\partial x_j}\right) + \sum_i\left(b_i + \sum_j \frac{\partial a_{ij}}{\partial x_j}\right)\frac{\partial}{\partial x_i} + c \tag{8.88}$$

with coefficients in $L^\infty(\Omega)$. We can then apply all the preceding results; in particular *if P is uniformly elliptic and $c = 0$ (resp. $c \geqslant 0$) a.e. on Ω supposed connected, every function $u \in W^{2,1}_{\text{loc}}(\Omega)$ satisfying $Pu \leqslant 0$ a.e. on Ω (resp. and* $\sup\operatorname{ess}_\Omega u > 0$) *satisfies* *the principle of the maximum*:

$$u \equiv \sup_{\Omega} \operatorname{ess} u \quad \text{or} \quad \sup_{K} \operatorname{ess} u < \sup_{\Omega} \operatorname{ess} u \quad \text{for every compact set } K \text{ of } \Omega.$$

In effect if $u \in W^{2,1}_{\text{loc}}(\Omega)$ satisfies $Pu \leqslant 0$ *a.e.* on Ω, then u is a subsolution (in the sense of Definition 7) of $Lu = 0$ on Ω where L is given by (8.88); from which, taking account of Remark 1, we have the result from Proposition 17.
We can obtain the same property without the hypothesis $a_{ij} \in W^{1,\infty}(\Omega)$. We state the following *maximum property* due to *Bony*:

Proposition 22. *Let* $P = -\sum_{ij} a_{ij} \dfrac{\partial^2}{\partial x_i \partial x_j} + \sum b_i \dfrac{\partial}{\partial x_i} + c$ *be a differential operator of order 2 with coefficients in* $L^\infty(\Omega)$ *and strictly elliptic and* $u \in W^{2,p}_{\text{loc}}(\Omega)$ *with* $p > n$ *such that* $Pu \leqslant 0$ *a.e. on* Ω; *we suppose that* Ω *is connected and* $c = 0$ *a.e. on* Ω (*resp.* $c \geqslant 0$ *a.e. on* Ω *and* $\sup_{\Omega} u \geqslant 0$). *Then if* u *is not constant on* Ω,

$$u(x) < \sup_{\Omega} u \quad \textit{for all} \quad x \in \Omega\,.$$

The proof will make use, in an essential way, the following result:

Lemma 6. *Let* $F \in W^{1,p}_{\text{loc}}(\Omega; \mathbb{R}^n)$ *with* $p > n$. *Then the image under* F *of a negligible set* (*for the Lebesgue measure*) *of* Ω *is negligible in* $\mathbb{R}^n$.

We refer the reader to J. M. Bony [1] for the proof of this lemma based on the Sobolev inclusion $W^{1,p}_{\text{loc}}(\Omega) \subset \mathscr{C}^{1-\frac{n}{p}}(\Omega)$.

Proof of Proposition 22. We shall first give a proof *supposing* $c \geqslant \delta > 0$ *a.e. on* Ω *that* P *satisfies the principle of the weak positive maximum*:

$$\text{(8.89)} \qquad \begin{cases} \textit{for every bounded open set } \Omega_0 \quad \textit{with} \quad \overline{\Omega_0} \subset \Omega\,, \\ u \leqslant 0 \quad \textit{on} \quad \partial\Omega_0 \Rightarrow u \leqslant 0 \quad \textit{on} \quad \Omega_0\,. \end{cases}$$

Let us suppose $u \leqslant 0$ on $\partial\Omega_0$ and $\sup_{\Omega_0} u > 0$; then there exists $x_0 \in \Omega_0$ such that $u(x_0) = \sup_{\Omega_0} u > 0$; since $c \geqslant \delta > 0$, $b \in L^\infty$ and $\operatorname{grad} u(x_0) = 0$, there exists $r_0 > 0$ and $\delta_0 > 0$ such that

$$\sum b_i \frac{\partial u}{\partial x_i} + cu \geqslant \delta_0 \quad \text{a.e. on} \quad B(x_0, r_0)\,,$$

namely, since $Pu \leqslant 0$,

$$\sum a_{ij} \frac{\partial^2 u}{\partial x_i \partial x_j} \geqslant \delta_0 \quad \text{a.e. on} \quad B(x_0, r_0)\,.$$

Let us put $v(x) = u(x) - \frac{\alpha}{2}|x - x_0|^2$ where $\alpha > 0$ satisfies $\alpha \sum a_{ii} < \delta_0$ with the result that

$$\sum a_{ij} \frac{\partial^2 v}{\partial x_i \partial x_j} = \sum a_{ij} \frac{\partial^2 u}{\partial x_i \partial x_j} - \alpha \sum a_{ii} > 0 \quad \text{a.e. on} \quad B(x_0, r_0) .$$

To obtain a contradiction, we show that the set M of the points $x \in B(x_0, r_0)$ such that

$$v(y) \leqslant v(x) + \operatorname{grad} v(x).(y - x) \quad \text{for all} \quad y \in B(x_0, r_0)$$

is non-negligible. This will conclude the proof since at a point x of M where v is twice differentiable, the symmetric matrix $\left(\frac{\partial^2 v}{\partial x_i \partial x_j}(x)\right)$ is negative and therefore $\sum a_{ij}(x) \frac{\partial^2 v}{\partial x_i \partial x_j}(x) \leqslant 0.$

To prove that M is non-negligible, we note from Lemma 6 that it is enough to show that $\{\operatorname{grad} v(x);\ x \in M\}$ is non-negligible. Now, given $\xi \in \mathbb{R}^n$, let us consider $\lambda(\xi) = \max_{\bar{B}(x_0, r_0)} v(x) - \xi.(x - x_0)$ and $x(\xi) \in \bar{B}(x_0, r_0)$ such that

$$\lambda(\xi) = v(x(\xi)) - \xi.(x(\xi) - x_0) ;$$

we have

$$v(y) \leqslant v(x(\xi)) + \xi.(y - x(\xi)) \quad \text{for all} \quad y \in \bar{B}(x_0, r_0)$$

with the result that

$$x(\xi) \in B(x_0, r_0) \Rightarrow \operatorname{grad} v(x(\xi)) = \xi \quad \text{and} \quad x(\xi) \in M .$$

Now $v(x) \leqslant v(x_0) - \frac{\alpha}{2}|x - x_0|^2$ and $\lambda(\xi) \geqslant v(x_0)$; therefore

$$\xi.(x(\xi) - x_0) + \lambda(\xi) \leqslant \lambda(\xi) - \frac{\alpha}{2}|x(\xi) - x_0|^2$$

and

$$|x(\xi) - x_0| \leqslant \frac{2}{\alpha}|\xi| .$$

To sum up: $\{\operatorname{grad} v(x); x \in M\}$ contains $\left\{\xi \in \mathbb{R}^n, |\xi| < \frac{\alpha r_0}{2}\right\}$ which completes this part of the proof.

Now (8.89) is true with only $c \geqslant 0$ a.e. on Ω, but for the open sets Ω_0 of sufficiently small diameter. Let us suppose that $\Omega_0 \subset B(0, r)$ and put for $\alpha > 0$, $v = e^{\alpha |x|^2} u$.

We have

$$Pu = e^{-\alpha |x|^2} Qv, \quad \text{where} \quad Q = - \sum_{i,j} a_{ij} \frac{\partial^2}{\partial x_i \partial x_j} + \sum_i \bar{b}_i \frac{\partial}{\partial x_i} + \bar{c} ,$$

where $\bar{c} = e^{\alpha|x|^2} P(e^{-\alpha|x|^2}) = - 4\alpha^2 \sum a_{ij} x_i x_j + 2\alpha \sum a_{ii} - 2\alpha \sum_i b_i x_i + c ;$

from which, using $a_{ii} \geqslant c_0$ the constant of strict ellipticity, we deduce

$$\bar{c} \geqslant 2\alpha[nc_\infty - r(\|(\sum b_i^2)^{1/2}\|_{L^\infty} + 2\alpha r \sum \|a_{ij}\|_{L^\infty})] \quad \text{on} \quad \Omega_0 .$$

If $r < nc_0 \|(\sum b_i^2)^{1/2}\|_{L^\infty}^{-1}$, for $\alpha > 0$ sufficiently small we therefore have $\inf\operatorname{ess}_{\Omega_0} \bar{c} > 0$. Since $Qv = e^{\alpha|x|^2} Pu \leqslant 0$ on Ω_0 and $v \leqslant 0$ on $\partial\Omega_0$, from (8.89) applied to Q and v, we have $v \leqslant 0$ on Ω_0 and so $u \leqslant 0$ on Ω_0.
Now let us show under the hypothesis $c \geqslant 0$ a.e. on Ω,

$$(8.90)\quad \begin{cases} \text{for every ball } B(x_0, r) \text{ with } \bar{B}(x_0, r) \subset \Omega \text{ and } r \text{ sufficiently small,} \\ \text{and } z \in \partial B(x_0, r) \\ u(z) > u(x) \quad \text{for all} \quad x \in B(x_0, r) \text{ and } u(z) \geqslant 0 \\ \Rightarrow \lim\limits_{t \downarrow 0} \dfrac{u(z) - u(tx_0 + (1-t)z)}{t} > 0\,; \end{cases}$$

We can always suppose that $x_0 = 0$; applying (8.89) to the function $v = u - u(z) + \varepsilon(e^{-\alpha|x|^2} - e^{-\alpha r^2})$ on the open set $\Omega_0 = B(0, r)\backslash\bar{B}(0, \frac{1}{2}r)$; we have from the above calculation, using $c \geqslant 0$ and the strict ellipticity

$$Pv \leqslant \varepsilon P(e^{-\alpha|x|^2}) \leqslant \varepsilon e^{-\alpha|x|^2}[-\alpha^2 c_0 r^2$$
$$+ 2\alpha(\|\sum a_{ii}\|_{L^\infty} + r\sum \|b_i\|_{L^\infty}) + \|c\|_{L^\infty}] \quad \text{on} \quad \Omega_0 .$$

We can therefore choose α sufficiently large, independently of $\varepsilon > 0$, such that $Pv \leqslant 0$ on Ω_0. We now put $m = \max\limits_{\partial B(0,r/2)} u$; from the hypothesis $u(x) < u(z)$ for all $x \in B(0, r)$, we have $m < u(z)$; let us now choose $\varepsilon = (u(z) - m)(e^{-\alpha r^2/4} - e^{-\alpha r^2})^{-1} > 0$, with the result that $v \leqslant 0$ on $\partial B(0, \frac{1}{2}r)$; we have $v = 0$ on $\partial B(0, r)$ and hence by application of (8.89), $v \leqslant 0$ on Ω_0.
In particular, for $t > 0$ sufficiently small,

$$\frac{u(z) - u(tx_0 + (1-t)z)}{t} \geqslant \varepsilon \frac{e^{-\alpha r^2(1-t)^2} - e^{-\alpha r^2}}{t}$$

and in the limit

$$\lim_{t \downarrow 0} \frac{u(z) - u(tx_0 + (1-t)z)}{t} \geqslant 2\varepsilon\alpha r e^{-\alpha r^2} > 0 .$$

We finally complete the proof of the proposition. We suppose that

$$F = \left\{x \in \Omega; u(x) = \sup_{\Omega} u\right\}$$

is non-empty and distinct from Ω. Since Ω is connected, there exists $x_1 \in \partial F \cap \Omega$; let us choose $x_0 \in \Omega\backslash F$ such that $r = \operatorname{dist}(x_0, F)$ is strictly less than $\operatorname{dist}(x_0, \partial\Omega)$ with the result that $\bar{B}(x_0, r) \subset \Omega$, and on the other hand, sufficiently small for us to apply (8.90); there exists $z \in \partial B(x_0, r) \cap F$ and $u(x) < u(z)$ for all $x \in B(x_0, r) \subset \Omega\backslash F$. Since $u(z) = \max u$, $\operatorname{grad} u(z) = 0$; but from (8.90), $\operatorname{grad} u(z).(x_0 - z) > 0$ which is a contradiction. □

We notice that in the above proof we have passed to the end through the medium of *Hopf's maximum principle* which we now state in the general case.

Proposition 23. *Let* $P = -\sum a_{ij} \dfrac{\partial^2}{\partial x_i \partial x_{ij}} + \sum b_i \dfrac{\partial}{\partial x_i} + c$ *be a differential operator of order 2 with coefficients in* L^∞ *and strictly elliptic,* $u \in W^{2,p}_{\mathrm{loc}}(\Omega)$ *with* $p > n$ *such that* $Pu \leqslant 0$ *a.e. on* Ω *and* $z \in \partial\Omega$. *We suppose that there exists a ball* $B(x_0, r_0) \subset \Omega$ *such that* $z \in \partial B(x_0, r_0)$, *that* $u(z) = \lim\limits_{x \to z} u(x)$ *exists and is equal to* $\sup\limits_\Omega u$, *and* $c = 0$ *a.e. on* Ω *(resp.* $c \geqslant 0$ *a.e. on* Ω *and* $\sup\limits_\Omega u \geqslant 0$*).*
Then u is constant on the connected component containing x_0, *or for all* $\theta \in [0, \pi/2[$

$$\varliminf_{\substack{x \to z \\ x \in C_\theta}} \frac{u(z) - u(x)}{|z - x|} > 0$$

where $C_\theta = \{x \in \Omega;\ (x - z)(x_0 - z) \geqslant |x - z||x_0 - z|\cos\theta\}$.

Proof. We can suppose Ω to be connected; if u is not constant, from Proposition 22, $u(x) < u(z) = \sup u_0$ for all $x \in \Omega$. It is sufficient then to take up again the proof of (8.90), noting that

$$\varliminf_{\substack{x \to z \\ x \in C_\theta}} \frac{\mathrm{e}^{-\alpha|x - x_0|^2} - \mathrm{e}^{-\alpha r^2}}{|x - z|} \geqslant 2\alpha r \mathrm{e}^{-\alpha r^2} \cos\theta\,. \qquad \square$$

As in the case of the Laplacian (see §4.4), Hopf's maximum principle provides the proof of the uniqueness in a Neumann problem:

$$\begin{cases} Pu = f & \text{on } \Omega \\ Bu = g & \text{on } \Gamma \end{cases}$$

where $B = \sum \beta_i \dfrac{\partial}{\partial x_i}$ is a homogeneous operator of order 1 non-tangential on Γ.

6. Green's Functions

First of all, let us take Ω a bounded open set of $\mathbb{R}^n$ with Lipschitz boundary Γ $L = -\sum \dfrac{\partial}{\partial x_i}\left(\sum a_{ij} \dfrac{\partial}{\partial x_j} + c_i\right) + \sum d_i \dfrac{\partial}{\partial x_i} + d$ a differential operator of order 2 in divergence form with coefficients in $L^\infty(\Omega)$ and strictly elliptic on Ω, Γ_0 a closed part of Γ. Let us consider the mixed problem

$$\text{(8.91)} \qquad \begin{cases} Lu = f & \text{on } \Omega \\ u = \varphi & \text{on } \Gamma_0 \\ \dfrac{\partial u}{\partial n_L} = \psi & \text{on } \Gamma \backslash \Gamma_0\,. \end{cases}$$

Following Proposition 10, we denote by $N(L, \Gamma_0)$ the set of weak solutions (relative to $W^{1,2}(\Omega)$) of (8.91) corresponding to $f = 0$, $\varphi = 0$, $\psi = 0$; we know that this is a finite dimensional space of the same dimension as $N(L^*, \Gamma_0)$ where L^* is the formal adjoint of L.
Supposing that $\varphi = 0$, $\psi = 0$ we know from Proposition 10 that, given $f \in L^2(\Omega)$ (8.91) admits a solution iff

$$\int_\Omega fv\,\mathrm{d}x = 0 \quad \text{for all} \quad v \in N(L^*, \Gamma_0)$$

that is to say iff f belongs to $N(L^*, \Gamma_0)^\perp$, the orthogonal subspace in $L^2(\Omega)$ of $N(L^*, \Gamma_0)$; finally, when this is the case a solution of (8.91) is defined to within the addition of an element of $N(L, \Gamma_0)$: we can thus consider that, if it exists, the unique solution of (8.91) is in $N(L, \Gamma_0)^\perp$. Finally, we recall (see Proposition 6), that if $f \in L^p(\Omega)$ with $p > \frac{1}{2}n$, a solution of (8.91), with $\varphi = 0$, $\psi = 0$ is continuous and bounded in Ω. In particular the elements of $N(L, \Gamma_0)$ and $N(L^*, \Gamma_0)$ are continuous and bounded on Ω.
We now prove

Proposition 24. *Under the hypotheses and notation introduced above, there exists a unique function G defined to be continuous on $\{(x, y) \in \Omega \times \Omega;\ x \neq y\}$ with $G \in L^1(\Omega \times \Omega)$* satisfying

(i) *for all $f \in N(L^*, \Gamma_0)^\perp \cap L^\infty(\Omega)$, the unique solution u in $N(L, \Gamma_0)^\perp$* of (8.91) *corresponding to $\varphi = 0$, $\psi = 0$ is given by*

$$u(x) = \int_\Omega G(x, y) f(y)\,\mathrm{d}y \quad a.e. \quad x \in \Omega \tag{8.92}$$

(ii) *for all $v \in N(L^*, \Gamma_0)$,*

$$\int_\Omega G(x, y) v(y)\,\mathrm{d}y = 0 \quad a.e. \quad x \in \Omega$$

In addition, for all $y \in \Omega$, the function $G(., y)$ is a weak solution relative to $W^{1,r}(\Omega)$ for all $r < \dfrac{n}{n-1}$ of

$$\begin{cases} Lu = f_y & on \quad \Omega \\ u = 0 & on \quad \Gamma_0 \\ \dfrac{\partial u}{\partial n_L} = 0 & on \quad \Gamma \backslash \Gamma_0 \end{cases} \tag{8.93}$$

where f_y is the unique bounded measure on Ω such that

$$\begin{cases} f_y - \delta_y \in N(L^*, \Gamma_0) \\ \displaystyle\int_\Omega v(x)\,\mathrm{d}f_y(x) = v(y) \quad for\ all \quad v \in N(L^*, \Gamma_0). \end{cases} \tag{8.94}$$

Definition 8. Under the hypotheses and with the notation introduced above, the function G considered in Proposition 24 is called *the Green's function of the mixed problem relative to* (Ω, Γ_0, L).

Proof of Propostion 24. There is uniqueness of the function G for if G_1 and G_2 are two solutions of the problem, the difference $G = G_1 - G_2$ defined to be continuous on $\{(x, y) \in \Omega \times \Omega; x \neq y\}$ and integrable on $\Omega \times \Omega$ will satisfy

$$\int_\Omega G(x, y) f(y) \mathrm{d}y = 0 \quad \text{for all} \quad f \in L^\infty(\Omega), \quad \text{a.e.} \quad x \in \Omega$$

Therefore $G(x, y) = 0$ for a.e. $y \in \Omega$ and a.e. $x \in \Omega$ and by Fubini's theorem and continuity $G(x, y) = 0$ for all $(x, y) \in \Omega \times \Omega$ with $x \neq y$.
To prove existence we first of all denote by G the map which with $f \in L^2(\Omega)$ associates the unique solution $u \in N(L, \Gamma_0)^\perp$ of (8.91) corresponding to $\varphi = 0$, $\psi = 0$ and the projection (orthogonal in $L^2(\Omega)$) of f on $N(L^*, \Gamma_0)^\perp$. We say that G *is the Green's operator in* $L^2(\Omega)$ *of the mixed problem relative to* (Ω, Γ_0, L).
As we have seen in the proof of Proposition 10, G is a compact operator in $L^2(\Omega)$ and its adjoint G^* is the Green's operator in $L^2(\Omega)$ of the mixed problem relative to (Ω, Γ_0, L^*). Also, as we have recalled above, for $p > \frac{1}{2}n$, G and G^* map $L^p(\Omega)$ into the space $\mathscr{C}_b^0(\Omega)$ of bounded continuous functions on Ω.
Given $y \in \Omega$, the map $g \in L^p(\Omega) \to (G^*g)(y)$ is a continuous linear form on $L^p(\Omega)$ for $p > \frac{1}{2}n$; thus, there exists $G_y \in L^q(\Omega)$ for $q < n/(n-2)$ such that

$$\int_\Omega G_y(x) g(x) \mathrm{d}x = (G^*g)(y) \quad \text{for all} \quad g \in L^p(\Omega) .$$

We shall show that the function G_y is continuous on $\Omega \backslash \{y\}$ or more precisely that, given K a compact set in Ω, Ω_0 a neighbourhood of K in Ω, the function G_y is continuous on $\Omega \backslash \Omega_0$ uniformly for $y \in K$.
We can therefore consider the function G defined on $\{(x, y) \in \Omega \times \Omega; x \neq y\}$ by $G(x, y) = G_y(x)$; since from the definition of G_y, the map $y \to G_y$ is weakly continuous from Ω into $L^q(\Omega)$, the function G is continuous. Also G is bounded in $L^q(\Omega)$ uniformly for $y \in \Omega$ and therefore *a fortiori* G is integrable on $\Omega \times \Omega$. Given $f \in L^\infty(\Omega)$, we have for all $g \in L^\infty(\Omega)$

$$\begin{aligned} \int_\Omega g(x) \mathrm{d}x \int_\Omega G(x, y) f(y) \mathrm{d}y &= \int_\Omega f(y) \mathrm{d}y \int_\Omega G_y(x) g(x) \mathrm{d}x \\ &= \int_\Omega f(y)(G^*g)(y) \mathrm{d}y = \int_\Omega (Gf)(x) g(x) \mathrm{d}x \end{aligned}$$

and hence

$$\int_\Omega G(x, y) f(y) \mathrm{d}y = Gf(x) \quad \text{a.e.} \quad x \in \Omega .$$

By the definition of Green's operator G, we see therefore that the function G satisfies the properties (i) and (ii).

To complete the proof let us take $\rho \in \mathscr{D}(\mathbb{R}^n)$ with $\rho \geqslant 0$, $\int \rho(x)dx = 1$, $\operatorname{supp} \rho \subset B(0, 1)$ and consider for $y \in \Omega$, the sequence of functions $u_k = Gf_k$ where $f_k(z) = k^n \rho(k(z - y))$. We have $u_k \rightharpoonup G_y$ in $L^q(\Omega)$ when $k \to \infty$, since for all $g \in L^p(\Omega)$,

$$\int_\Omega u_k(x) g(x) dx = \int_\Omega f_k(z)(G^* g)(z) dz \to (G^* g)(y) .$$

Let us consider the projection $\tilde{f}_k$ of f_k on $N(L^*, \Gamma_0)^\perp$; the function u_k is the solution in $N(L, \Gamma_0)^\perp$ of

$$\begin{cases} Lu_k = \tilde{f}_k & \text{on} \quad \Omega \\ u_k = 0 & \text{on} \quad \Gamma_0 \\ \dfrac{\partial u_k}{\partial n_L} = 0 & \text{on} \quad \Gamma \backslash \Gamma_0 . \end{cases}$$

The sequence of the functions $\tilde{f}_k$ is bounded in $L^1(\Omega)$ and converges vaguely to the measure f_y defined by (8.94). By duality, as in the proof of Proposition 13, we show that the sequence (u_k) is bounded in $W^{1,r}(\Omega)$ for all $r < n/(n-1)$ and so its limit G_y is a weak solution relative to $W^{1,r}(\Omega)$ of (8.93). Finally, considering K a compact set in Ω and Ω_0 a neighbourhood of K in Ω, making use of the fact that $N(L^*, \Gamma_0)$ is of finite dimension, $\tilde{f}_k$ is bounded on $\Omega \backslash \Omega_0$ uniformly for $k \in \mathbb{N}$, $y \in K$ and with the limit G_y bounded on $\Omega \backslash \Omega_0$ uniformly. From the De Giorgi–Nash theorem, since f_y is bounded and continuous on $\Omega \backslash \Omega_0$ uniformly for $y \in K$, G_y satisfies a Hölder condition on $\Omega \backslash \Omega_0$ uniformly for $y \in K$. □

Remark 2. As we have seen in Proposition 14 when the coefficients $a_{ij} \in W^{1,\infty}(\Omega)$ and in the pure Dirichlet case ($\Gamma_0 = \Gamma$) or in the pure Neumann case ($\Gamma_0 = \emptyset$), there exists a unique weak solution relative to $W^{1,r}(\Omega)$ of (8.93); this problem then defines completely the Green's function. In particular, Definition 8 generalizes that given in §4: the Green's function of a bounded open set Ω (with boundary satisfying a Lipschitz condition) is the Green's function relative to the Dirichlet problem ($\Gamma_0 = \Gamma$) in Ω for the operator $L = -\Delta$.
Let us now give some properties of the Green's function generalizing those proved in §4 in the case of the Dirichlet problem for the Laplacian.

Proposition 25. *Under the preceding hypotheses and notation, let us consider G the Green's function of the mixed problem relative to* (Ω, Γ_0, L).
(1) *The Green's function* G^* *of the mixed problem relative to* (Ω, Γ_0, L^*) *is given by*

$$G^*(x, y) = G(y, x) \quad \textit{for all} \quad (x, y) \in \Omega \times \Omega, x \neq y .$$

In particular, if L is formally self-adjoint ($L^* = L$), *then G is symmetric*:

$$G(x, y) = G(y, x) \quad \textit{for all} \quad (x, y) \in \Omega \times \Omega, x \neq y.$$

(2) *Let us suppose that L is coercive in* $V^2(\Omega, \Gamma_0)$ *(that is to say that there exists* $c > 0$ *such that* $\int_\Omega L(u, u) dx \geqslant c \|u\|^2_{W^{1,2}}$ *for all* $u \in V^2(\Omega, \Gamma_0)$ *then* $G(x, y) > 0$

for all $(x, y) \in \Omega \times \Omega$, *x and y two distinct points of the same connected component of* Ω.

Proof of Proposition 25. Point (1) follows immediately from the construction of G given in the proof of Proposition 24. For point (2), we notice first of all the hypothesis of coercivity implies that $N(L, \Gamma_0) = N(L^*, \Gamma_0) = \{0\}$ since, given $u \in N(L, \Gamma_0)$, we have $\int_\Omega L(u, u)\mathrm{d}x = 0$ by definition of $N(L, \Gamma_0)$. From Proposition 20 and (8.82), it is sufficient to prove that $G(x, y) \geqslant 0$; if account is taken of the proof of Proposition 24, it is sufficient to prove that the Green's operator G of the mixed problem relative to (Ω, Γ_0, L) is positive in the sense

$$f \geqslant 0 \quad \text{a.e. on } \Omega \Rightarrow Gf \geqslant 0 \quad \text{a.e. on } \Omega$$

Let us consider $f \geqslant 0$ a.e. on Ω and let $u = Gf$; given $\varepsilon > 0$, we consider $v = (-u)^+$ which is $V^2(\Omega, \Gamma_0)$: we have $L(u, v) = -L(v, v)$ a.e. on Ω, so

$$c\,\|v\|_{W^{1,2}} \leqslant \int_\Omega L(v, v)\,\mathrm{d}x = -\int_\Omega L(u, v)\,\mathrm{d}x = -\int_\Omega fv\,\mathrm{d}x \leqslant 0$$

and $v = 0$, that is to say $u \geqslant 0$ a.e. on Ω. □

As we shall see in Example 4 below, in general the Green's function is not positive; in the case of a Dirichlet problem, the positivity of the Green's function is equivalent to the principle of the weak positive maximum (8.78).
The Green's function has been introduced as the kernel in the solution of the homogeneous mixed problem ($\varphi = 0$, $\psi = 0$). In the case of a general mixed problem, we have *Green's representation formula of the solution in* $N(L, \Gamma_0)^\perp$ *of* (8.91) (if it exists):

$$(8.95) \qquad \begin{cases} u(x) = \displaystyle\int_\Omega G(x, y)f(y)\mathrm{d}y - \int_{\Gamma_0} \frac{\partial G}{\partial n_{L^*}}(x, z)\varphi(z)\mathrm{d}\gamma(z) \\ \qquad + \displaystyle\int_{\Gamma\backslash\Gamma_0} G(x, z)\psi(z)\mathrm{d}\gamma(z)\,. \end{cases}$$

Formally this is deduced from Green's formula (8.15); considering $x \in \Omega$, we put $v_x(y) = G(x, y)$ so we have from (8.93) and point (1) of Proposition 25

$$L^* v_x = g_x$$

where g_x is defined by

$$g_x - \delta_x \in N(L, \Gamma_0)$$

$$\int_\Omega w(y)\mathrm{d}g_x(y) = 0 \quad \text{for all} \quad w \in N(L, \Gamma_0)$$

and

$$v_x = 0 \quad \text{on} \quad \Gamma_0\,, \qquad \frac{\partial v_x}{\partial n_{L^*}} = 0 \quad \text{on} \quad \Gamma\backslash\Gamma_0\,.$$

Applying (8.15) formally, we obtain for the solution u in $N(L, \Gamma_0)^\perp$ of (8.91):

$$\begin{aligned} u(x) &= \int_\Omega u(y) d\delta_x(y) = \int_\Omega u(y) dg_x(y) = \int_\Omega u L^* v_x \, dx \\ &= \int_\Omega v_x Lu \, dx + \int_\Omega \left(v_x \frac{\partial u}{\partial n_L} - u \frac{\partial v_x}{\partial n_{L^*}} \right) d\gamma \\ &= \int_\Omega G(x, y) f(y) dy + \int_{\Gamma \backslash \Gamma_0} G(x, z) \psi(z) d\gamma(z) - \int_{\Gamma_0} \frac{\partial G}{\partial n_{L^*}}(x, z) \varphi(z) d\gamma(z) \end{aligned}$$

To justify this calculation, it is sufficient, following the notation of the proof of Proposition 24, to approximate v_x by

$$v_k = G^* g_k \quad \text{where} \quad g_k(y) = k^n \rho(k(x - y)) .$$

The passages to the limit

$$\int_\Omega u L^* v_k \, dx = \int_\Omega u g_k \, dx \to u(x)$$

$$\int_\Omega v_k Lu \, dy \to \int_\Omega G(x, y) f(y) \, dy$$

do not pose problems; for the passages to the limit

$$\int_{\Gamma \backslash \Gamma_0} v_k \psi \, d\gamma \to \int_{\Gamma \backslash \Gamma_0} G(x, z) \psi(z) \, d\gamma(z)$$

$$\int_{\Gamma_0} \frac{\partial v_k}{\partial n_{L^*}} \varphi \, d\gamma \to \int_{\Gamma_0} \frac{\partial G}{\partial n_{L^*}} (x, z) \varphi(z) \, d\gamma(z)$$

we shall need some justification of regularity up to the boundary. Although this is not the most general such result, we state the following proposition:

Proposition 26. *With the preceding hypotheses and notation, let us suppose $a_{ij} \in W^{1,\infty}(\Omega)$, Ω with $W^{2,\infty}$ and $\overline{\Gamma \backslash \Gamma_0} = \Gamma \backslash \Gamma_0$*[242]*. The Green's function G of the problem relative to (Ω, Γ_0, L) is in $\mathscr{C}^1(\{(x, y) \in \Omega \times \Omega; x \neq y\})$ and the solution of (8.91) in $N(L, \Gamma_0)^\perp$ is, if it exists, given by (8.95).*

Proof. It suffices to make use of the $\mathscr{C}^1$-regularity of the solutions of (8.91) for $\varphi = 0$, $\psi = 0$ when $f \in L^p(\Omega)$ with $p > n$ (see Proposition 13). □

Let us now give examples of Green's functions.

Example 4. *The case of a symmetric operator with constant coefficients in dimension* 1. Let us consider $\Omega =]0, l[$ a bounded open interval of $\mathbb{R}$ and $L = -\dfrac{d^2}{dx^2} + k$. The study of Table 4 of Example 1[243] leads to the determination of $N(L, \Gamma_0) = N(L^*, \Gamma_0)$ according to the values of k and

[242] In particular, $\Gamma_0 = \Gamma$ (Dirichlet problem) and $\Gamma_0 = \varnothing$ (Neumann problem).
[243] With $\lambda \equiv 0$.

$\Gamma_0 \subset \Gamma = \{0, l\}$. We recall that $N(L, \Gamma_0)$ is either red uced to $\{0\}$, or is of dimension 1 generated by an eigenfunction w_0. When $N(L, \Gamma_0) = \{0\}$, given $y \in]0, l[$, the function $u(x) = G(x, y)$ is a solution of

$$-u'' + ku = \delta_y \quad \text{in} \quad \mathscr{D}'(]0, l[)$$

which we can rewrite as

$$\text{(8.96)} \qquad \begin{cases} u'' = ku & \text{on} \quad]0, y[\cup]y, l[\\ u(y-) = u(y+), & u'(y-) = u'(y+) + 1 \,. \end{cases}$$

When $k = \omega^2 > 0$ (resp. $k = 0, k = -\omega^2 < 0$), the solutions of (8.96) are given by

$$u(x) = ae^{\omega x} + be^{-\omega x} - \frac{\operatorname{sh} \omega(x - y)^+}{\omega}$$

$$\left(\text{resp. } ax + b - (x - y)^+, ae^{i\omega x} + be^{-i\omega x} - \frac{\sin \omega(x - y)^+}{\omega}\right).$$

Incorporating the boundary conditions we obtain the Green's functions (see Table 5).

Table 5. Green's function associated with $\left(]0, l[, \Gamma_0, -\dfrac{d^2}{dx^2} + k\right)$ when $N\left(-\dfrac{d^2}{dx^2} + k, \Gamma_0\right) = \{0\}$

	$(\Gamma_0 = \Gamma)$
$k = \omega^2$	$G(x, y) = \dfrac{1}{\omega \operatorname{sh} \omega l}[\operatorname{sh} \omega x \operatorname{sh} \omega(l - y) - \operatorname{sh} \omega l \operatorname{sh} \omega(x - y)^+]$
$k = 0$	$G(x, y) = \dfrac{1}{l}[(l - y)x - l(x - y)^+]$
$k = -\omega^2$	$\omega \neq \dfrac{m\pi}{l}$,
	$G(x, y) = \dfrac{1}{\omega \sin \omega l}[\sin \omega x \sin \omega(l - y) - \sin \omega l \sin \omega(x - y)^+]$
	$(\Gamma_0 = \{0\})$
$k = \omega^2$	$G(x, y) = \dfrac{1}{\omega \operatorname{ch} \omega l}[\operatorname{sh} \omega x \operatorname{ch} \omega(l - y) - \operatorname{ch} \omega l \operatorname{sh} \omega(x - y)^+]$
$k = 0$	$G(x, y) = x - (x - y)^+$
$k = -\omega^2$	$\omega \neq \dfrac{\pi}{l}\left(m + \dfrac{1}{2}\right)$,
	$G(x, y) = \dfrac{1}{\omega \cos \omega l}[\sin \omega x \cos \omega(l - y) - \cos \omega l \sin \omega(x - y)^+]$
	$(\Gamma_0 = \varnothing)$
$k = \omega^2$	$G(x, y) = \dfrac{1}{\omega \operatorname{sh} \omega l}[\operatorname{ch} \omega x \operatorname{ch} \omega(l - y) - \operatorname{sh} \omega l \operatorname{sh} \omega(x - y)^+]$
$k = -\omega^2$	$\omega \neq \dfrac{m\pi}{l}$,
	$G(x, y) = \dfrac{1}{\omega \sin \omega l}[\cos \omega x \cos \omega(l - y) - \sin \omega l \sin \omega(x - y)^+]$

We now consider the case in which $N(L, \Gamma_0)$ is of dimension 1 generated by the eigenfunction w_0 which we can always assume to be real and normalized $\left(\int_0^l |w_0|^2 \mathrm{d}x = 1\right)$. Given $y \in]0, l[$, the measure f_y defined by (8.94) is $\delta_y - w_0(y) w_0$ with the result that $u = G(x, y)$ is the solution of

$$\begin{cases} - u'' + ku = \delta_y - w_0(y) w_0 \quad \text{in} \quad \mathscr{D}'(]0, l[) \\ u = 0 \quad \text{on} \quad \Gamma_0, \quad u' = 0 \quad \text{on} \quad \Gamma \backslash \Gamma_0 \\ \displaystyle\int_0^l u w_0 \, \mathrm{d}x = 0 \,. \end{cases}$$

Given a particular solution u_0 of the equation

$$- u_0'' + k u_0 = w_0 \quad \text{on} \quad]0, l[\tag{8.97}$$

the function $\bar{u} = u + w_0(y) u_0$ is the solution of

$$\begin{cases} - \bar{u}'' + k\bar{u} = \delta_y \quad \text{in} \quad \mathscr{D}'(]0, l[) \\ \bar{u} = w_0(y) u_0 \quad \text{on} \quad \Gamma_0, \bar{u}' = w_0(y) u_0' \quad \text{on} \quad \Gamma \backslash \Gamma_0 \\ \displaystyle\int_0^l \bar{u} w_0 \, \mathrm{d}x = w_0(y) \int_0^l u_0 w_0 \, \mathrm{d}x \,. \end{cases} \tag{8.98}$$

The case $N(L, \Gamma_0) \neq \{0\}$ occurs if

$$k = 0 \quad \text{and} \quad \Gamma_0 = \varnothing \,.$$

or:

$$k = - \left(\frac{m\pi}{l}\right)^2 \quad \text{and} \quad \operatorname{card} \Gamma_0 \neq 1 \,,$$

or:

$$k = - \left(\left(m + \frac{1}{2}\right)\frac{\pi}{l}\right)^2 \quad \text{and} \quad \operatorname{card} \Gamma_0 = 1 \,.$$

When $k = 0$ and $\Gamma_0 = \varnothing$, we can take $w_0 = 1/\sqrt{l}$ and $u_0 = - \dfrac{x^2}{2\sqrt{l}}$; the solution $\bar{u}$ of (8.98) is then

$$\bar{u} = c - (x - y)^+ \,,$$

where the constant c is determined by the last condition

$$cl = \frac{(l - y)^2}{2} - \frac{l^2}{6} = \frac{l^2}{3} - ly + \frac{y^2}{2} \,.$$

To sum up: $u = \dfrac{l}{3} - y + \dfrac{y^2 - x^2}{2l} - (x - y)^+$.

The other calculations carried out in the same way give the results shown in Table 6.

Table 6. Green's function associated with $\left(]0, l[, \Gamma_0, -\frac{d^2}{dx^2} + k\right)$ when $N\left(-\frac{d^2}{dx^2} + k, \Gamma_0\right) \neq 0$

Problem	
$k = 0$ and $\Gamma_0 = \varnothing$ (Neumann)	$G(x, y) = \frac{l}{3} - y + \frac{x^2 + y^2}{2l} - (x - y)^+$
$k = -\omega^2$ with $\Gamma_0 = \Gamma$ (Dirichlet) and $\omega = \frac{m\pi}{l}$, $m = 1, \ldots$ or $\Gamma_0 = \{0\}$ (mixed) and $\omega = \left(m + \frac{1}{2}\right)\frac{\pi}{l}$, $m = 0, \ldots$	$G(x, y) = \frac{\sin \omega x}{\omega}\left[\left(1 - \frac{y}{l}\right)\cos \omega y + \frac{\sin \omega y}{\omega l}\right] - \frac{x}{\omega l}\sin \omega y \cos \omega x - \frac{1}{\omega}\sin \omega(x - y)^+$
$k = -\omega^2$ $\omega = \frac{m\pi}{l}$, $m = 1, \ldots$ and $\Gamma_0 = \varnothing$ (Neumann)	$G(x, y) = \frac{1}{\omega}\left[\frac{x}{l}\cos \omega y \sin \omega x - \left(1 - \frac{y}{l}\right)\sin \omega y \sin \omega x - \sin \omega(x - y)^+\right]$

Example 5. *Green's function of the Neumann problem for the Laplacian.* Given Ω, a bounded set, we consider the Neumann problem ($\Gamma_0 = \varnothing$) for the Laplacian ($L = -\Delta$). Let us suppose that Ω is connected with the result that $N(L, \Gamma_0) = N(L^*, \Gamma_0) = \{\text{constant functions on } \Omega\}$ (see §4.4).

Given $y \in \Omega$, the measure f_y defined by (8.94) is $\delta_y - \frac{1}{|\Omega|}$ where $|\Omega|$ is the Lebesgue measure of the open set Ω. Therefore, the function $u(x) = G(x, y)$ is the solution of the problem:

$$\text{(8.99)} \qquad \begin{cases} \Delta u = \dfrac{1}{|\Omega|} - \delta_y \quad \text{on} \quad \Omega \\ \dfrac{\partial u}{\partial n} = 0 \quad \text{on} \quad \Gamma \\ \displaystyle\int_\Omega u \, dx = 0 \,. \end{cases}$$

Let us put u in the form

$$u(x) = \frac{|x - x_0|^2}{2n|\Omega|} - E_n(x - y) + c + v(x)$$

where E_n is the elementary solution of the Laplacian (see §2), x_0 is an arbitrary point and $c = \int_\Omega \left(E_n(\xi - y) - \frac{|\xi - x_0|^2}{2n|\Omega|}\right) d\xi$. The function v is then a solution of the problem

$$\text{(8.100)} \quad \begin{cases} \Delta v = 0 \quad \text{on} \quad \Omega \\ \frac{\partial v}{\partial n}(z) = \left(\frac{z - y}{\sigma_n |z - y|^n} - \frac{z - x_0}{n|\Omega|}\right) . n(z) \quad \text{for} \quad z \in \Gamma \\ \int_\Omega v(x) dx = 0 . \end{cases}$$

We shall obtain the function v and hence the Green's function G in the *case of a ball B of $\mathbb{R}^n$*. We can always suppose that $B = B(0, 1)$ and put $y = r\sigma$ with $0 \leqslant r < 1$ and $\sigma \in \Sigma$; we denote by $v(r, x)$ the solution of (8.100). We can always take $x_0 = 0$ with the result that $v(r, x)$ satisfies

$$\text{(8.101)} \quad \begin{cases} \Delta v(r, x) = 0 \quad \text{on} \quad x \in \Omega \\ \frac{\partial v}{\partial n}(r, z) = \frac{1}{\sigma_n}\left[\frac{1 - r\sigma . z}{|r\sigma - z|^n} - 1\right] \quad \text{for} \quad z \in \Sigma \\ \int_B v(r, x) dx = 0 . \end{cases}$$

We put $w(r, x) = E_n(rx - \sigma)$, so that we have

$$\text{(8.102)} \quad \begin{cases} \Delta w(r, x) = 0 \quad \text{for} \quad x \in \Omega \\ \frac{\partial w}{\partial n}(r, z) = \frac{r}{\sigma_n} \times \frac{r - \sigma . z}{|rz - \sigma|^n} \quad \text{for} \quad z \in \Sigma . \end{cases}$$

Noting that $z, \sigma \in \Sigma$ we have

$$|rz - \sigma| = |z - r\sigma| \quad \text{and} \quad r(r - \sigma . z) = |z - r\sigma|^2 - (1 - r\sigma . z) ,$$

we deduce that the function $h(r, x) = v(r, x) + w(r, x)$ is a solution of

$$\begin{cases} \Delta h(r, x) = 0 \quad \text{for } x \in \Omega \\ \frac{\partial h}{\partial n}(r, z) = \frac{1}{\sigma_n}\left[\frac{1}{|r\sigma - z|^{n-2}} - 1\right] \quad \text{for} \quad z \in \Sigma . \end{cases}$$

Differentiating with respect to $r \in \left]0, 1\right[$ we see that $k(r, x) = \frac{\partial h}{\partial r}(r, x)$ is a solution of

$$\begin{cases} \Delta k(r, x) = 0 \quad \text{for} \quad x \in \Omega \\ \frac{\partial k}{\partial n}(r, z) = \frac{n - 2}{\sigma_n} \times \frac{\sigma . z - r}{|r\sigma - z|^n} \quad \text{for} \quad z \in \Sigma . \end{cases}$$

In other words from (8.102),

$$\text{(8.103)} \quad \frac{\partial h}{\partial r}(r, x) = k(r, x) = -\frac{(n - 2) w(r, x)}{r} + c(r) .$$

Now for $r = 0$, we have trivially that $v(0, x) \equiv 0$; but by the construction of v

$$0 \equiv v(0, x) = G(x, 0) - \frac{|x|^2}{2\sigma_n} + E_n(x) + \int_B \left(\frac{|\xi|^2}{2\sigma_n} - E_n(\xi)\right) d\xi$$

and for $r \in]0, 1[$

$$v(r, 0) = G(0, r\sigma) + E_n(r) + \int_B \left(\frac{|\xi|^2}{2\sigma_n} - E_n(\xi - r\sigma)\right) d\xi .$$

Using the symmetry of G (see Proposition 25) we have

$$G(0, r\sigma) = G(r\sigma, 0) = \frac{r^2}{2\sigma_n} - E_n(r) - \int_B \left(\frac{|\xi|^2}{2\sigma_n} - E_n(\xi)\right) d\xi$$

and hence that

$$v(r, 0) = \frac{r^2}{2\sigma_n} + \int_B (E_n(\xi) - E_n(\xi - r\sigma))\, d\xi . \tag{8.104}$$

Since $w(r, 0) = E_n(1)$,

$$c(r) = \frac{(n-2)w(r, 0)}{r} + \frac{\partial h}{\partial r}(r, 0) = \frac{(n-2)E_n(1)}{r} + \frac{\partial v}{\partial r}(r, 0)$$

from which from (8.103), integrating $k(r, x)$ with respect to r, we find that

$$\begin{cases} v(r, x) = v(r, 0) + E_n(1) - E_n(rx - \sigma) \\ \qquad + (n-2)\displaystyle\int_0^r (E_n(1) - E_n(\rho x - \sigma)) \frac{d\rho}{\rho} ; \end{cases} \tag{8.105}$$

the formulae (8.104) and (8.105) therefore give us an explicit expression for $v(r, x)$ so that, regrouping, we obtain

$$\begin{cases} G(x, y) = \dfrac{|x|^2 + |y|^2}{2\sigma_n} + E_n(x - y) - E_n\left(|y|x - \dfrac{y}{|y|}\right) \\ \qquad + (n-2)\displaystyle\int_0^r \left(E_n(1) - E_n\left(\rho x - \frac{y}{|y|}\right)\right) \frac{d\rho}{\rho} \\ \qquad + \displaystyle\int_B \left(E_n(\xi) - \frac{|\xi|^2}{2\sigma_n}\right) d\xi . \end{cases} \tag{8.106}$$

We note that in the case $n = 2$ the first integral vanishes and that for $n \geqslant 3$, this integral converges for all $(x, y) \in B \times B$ with $y \neq 0$, since

$$E_n(1) - E_n\left(\rho x - \frac{y}{|y|}\right) = \frac{1}{(n-2)\sigma_n}\left[\frac{1}{\left|\rho x - \dfrac{y}{|y|}\right|^{n-2}} - 1\right] = O(\rho) \quad \text{as} \quad \rho \to 0 .$$

□

This example has shown us the part which we can extract from the method of images for more general problems than the Dirichlet problem for the Laplacian. Let us know how to use this method for the operator $-\Delta + k^2$:

Example 6. *Green's function for the operator* $-\Delta + k^2$. The operator $-\Delta + k^2$, where $k > 0$ is *coercive on* $W^{1,2}(\Omega)$ *for every open set* Ω *of* $\mathbb{R}^n$ for any dimension n whatsoever. This allows to solve in a unique fashion every problem of the form

$$\begin{cases} -\Delta u + k^2 u = f & \text{on} \quad \Omega \\ u = \varphi & \text{on} \quad \Gamma_0 \\ \dfrac{\partial u}{\partial n} = \psi & \text{on} \quad \Gamma \backslash \Gamma_0 \end{cases}$$

for any open set Ω of $\mathbb{R}^n$ with the data f, φ, ψ not having too great a rate of growth at infinity. This allows us also to define a Green's function of the mixed problem relative to $(\Omega, \Gamma_0, -\Delta + k^2)$ for every open set Ω.

Without developing a complete theory for the operator $-\Delta + k^2$, we shall be content with some particular results.

Proposition 27. *Let* $k > 0$.

(1) *For every tempered distribution*[244] f *on* $\mathbb{R}^n$, *there exists a unique tempered distribution* u *on* $\mathbb{R}^n$ *such that*

$$-\Delta u + k^2 u = f \quad \text{in} \quad \mathscr{D}'(\mathbb{R}^n). \tag{8.107}$$

In particular, there exists a unique tempered distribution E *on* $\mathbb{R}^n$ *satisfying*

$$-\Delta E + k^2 E = \delta \quad \text{in} \quad \mathscr{D}'(\mathbb{R}^n)$$

where δ *is the distribution at the origin. This distribution* E *is analytic on* $\mathbb{R}^n \backslash \{0\}$ *and radial; it is the elementary solution of* $-\Delta + k^2$ *given by*

$$E(x) = \frac{k^{n-2}}{\sigma_n (n-2)!} \int_1^\infty e^{-k|x|t} (t^2 - 1)^{\frac{n-3}{2}} \, dt \qquad (n \geqslant 2) \tag{8.108}$$

or again

$$E(x) = \frac{1}{\pi} \left(\frac{k}{2\pi |x|} \right)^{\frac{n}{2} - 1} K_{\frac{n}{2} - 1}(k|x|) \qquad (\textit{if } n \geqslant 3) \tag{8.109}$$

where K *denotes a Bessel function of the second kind (in standard notation); in the particular case* $n = 3$ *we have the particularly simple form*

$$E(x) = \frac{1}{4\pi |x|} e^{-k|x|}. \tag{8.110}$$

(2) *Given* $p \in [1, \infty]$, *if* $f \in L^p(\mathbb{R}^n)$, *the solution* u *of* (8.107) *is in* $L^p(\mathbb{R}^n)$ *and*

$$\| u \|_{L^p} \leqslant \frac{1}{k^2} \| f \|_{L^p}.$$

If $1 < p < \infty$, *then* $u \in W^{2,p}(\mathbb{R}^n)$.

(3) *If* f *is a tempered distribution, positive or zero (and hence a Radon measure), then* $u \in W^{1,p}_{\text{loc}}(\mathbb{R}^n)$ *for all* $p < n/(n-1)$ *and* $u > 0$ *on* $\mathbb{R}^n$ *(unless* $f \equiv 0$ *on* $\mathbb{R}^n$*) in the*

---

[244] See Appendix "Distributions" Vol. 2, p. 506.

sense that

$$\inf_{B(0,R)} \operatorname{ess} u > 0 \quad \text{for all} \quad R > 0.$$

(4) *If f is a distribution with compact support, then u has exponential decay of order k at infinity, that is to say*

$$u(x) = o(e^{-k|x|}) \quad \text{when} \quad |x| \to \infty .$$

Proof Considering u and f to be tempered distributions we have (8.107) iff their Fourier transforms $\hat{u}$ and $\hat{f}$ satisfy

$$\xi^2 \hat{u} + k^2 \hat{u} = \hat{f}$$

that is iff

$$\hat{u} = \frac{\hat{f}}{\xi^2 + k^2} .$$

Given that if f is a tempered distribution, the distribution $\hat{f}/(\xi^2 + k^2)$ is also a tempered distribution, there exists a unique tempered distribution u, solution of (8.107), which is the inverse Fourier transform of $\hat{f}/(\xi^2 + k^2)$. In particular if $f = \delta$, the solution E is the inverse Fourier transform of $(\xi^2 + k^2)^{-1}$. The formulae (8.108) and (8.109) are then deduced classically (see Schwartz [1]).
Supposing that $f \in \mathscr{D}(\mathbb{R}^n)$ we find that the solution $u \in \mathscr{C}^\infty(\mathbb{R}^n)$, and that u and all of its derivatives are in the space $L^1(\mathbb{R}^n) \cap L^\infty(\mathbb{R}^n)$. If we multiply equation (8.107) by $\theta(u)$, where $\theta \in \mathscr{C}^1(\mathbb{R})$, and integrate by parts, we obtain

$$\int \theta'(u)|\operatorname{grad} u|^2 \,\mathrm{d}x + k^2 \int u\theta(u)\,\mathrm{d}x = \int f\theta(u)\,\mathrm{d}x$$

from which we deduce that if $\theta' \geqslant 0$,

$$k^2 \int \theta(u) u \,\mathrm{d}x \leqslant \int f\theta(u)\,\mathrm{d}x . \tag{8.111}$$

Applying this result with $\theta(r) = |r|^{p-1} r$ (approximating with a sequence of increasing functions of class $\mathscr{C}^1$, if $1 \leqslant p < 2$), we have

$$k^2 \int |u|^p \mathrm{d}x \leqslant \int f|u|^{p-1} u \,\mathrm{d}x \leqslant \|f\|_{L^p} \|u\|_{L^p}^{p-1}$$

from which we have the estimate of point (2). The result is then deduced by a density argument. Now applying (8.111) with $\theta(r) = \min(r, 0) = -u^-$, we obtain

$$k^2 \int (u^-)^2 \mathrm{d}x \leqslant - \int f u^- \mathrm{d}x ,$$

from which if $f \geqslant 0$, $\int (u^-)^2 \mathrm{d}x = 0$ and $u \geqslant 0$. Point (3) is deduced by density and use of Harnack's formula for subsolutions (see Proposition 16)[245].

[245] In the present we can prove Harnack's formula directly as in the case of the Laplace operator.

For point (4), we notice, first of all, that $u \in \mathscr{C}^\infty(\mathbb{R}^n \backslash \operatorname{supp} f)$ and is given by

$$u(x) = \langle f(y), E(x - y) \rangle \quad \text{for} \quad x \in \mathbb{R}^n \backslash \operatorname{supp} f.$$

We can develop the same method as that used in the proof of Proposition 2 of §3 to obtain the estimate at infinity. Another method consists of reducing to the radial case: considering R_0 such that

$$\operatorname{supp} f \subset B(0, R_0), \qquad M = \max_{|x| = R_0} |u(x)|,$$

we have $|u(x)| \leqslant \bar{u}(|x|)$ for $|x| > R_0$ where $\bar{u}$ is the solution of

$$\begin{cases} -\dfrac{1}{r^{n-1}} \dfrac{\mathrm{d}}{\mathrm{d}r}\left(r^{n-1} \dfrac{\mathrm{d}\bar{u}}{\mathrm{d}r}\right) + k^2 \bar{u} = 0 \quad \text{on} \quad]R_0, \infty[\\ \bar{u}(R_0) = M, \bar{u}(+\infty) = 0 . \end{cases}$$

Multiplying the equation by $2r^{2(n-1)} \dfrac{\mathrm{d}\bar{u}}{\mathrm{d}r}$, we obtain

$$\frac{\mathrm{d}}{\mathrm{d}r}\left[-\left(r^{n-1} \frac{\mathrm{d}\bar{u}}{\mathrm{d}r}\right)^2 + (kr^{n-1}\bar{u})^2 \right] = 2(n-1)k^2 r^{2n-3} (\bar{u})^2 .$$

From this, having noted that $\bar{u} \geqslant 0$ and $\mathrm{d}\bar{u}/\mathrm{d}r \leqslant 0$, we deduce that

$$\frac{\mathrm{d}\bar{u}}{\mathrm{d}r} + k\bar{u} \leqslant 0 \quad \text{and} \quad \bar{u}(r) \leqslant M e^{k(R_0 - r)} . \qquad \square$$

Remark 3. The functions $u \in \mathscr{C}^2(\Omega)$ satisfying $-\Delta u + k^2 u = 0$ are called *meta-harmonic functions*; we can develop for these functions a theory similar to that for harmonic functions; for example, we can prove a *formula of the mean* (see Garnir [1]):

$$c(n, r, k) u(x_0) = \frac{1}{\sigma_n r^{n-1}} \int_{\partial B(x_0, r)} u \, \mathrm{d}\gamma \tag{8.112}$$

where

$$c(n, r, k) = \frac{2}{\sqrt{\pi}} \frac{\Gamma\left(\dfrac{n}{2}\right)}{\Gamma\left(\dfrac{n-1}{2}\right)} \int_0^1 \operatorname{ch}(krt)(1 - t^2)^{\frac{n-3}{2}} \mathrm{d}t . \tag{8.113}$$

In particular in the case $n = 3$

$$u(x_0) = \frac{k}{4\pi r \operatorname{sh}(kr)} \int_{\partial B(x_0, r)} u \, \mathrm{d}\gamma \tag{8.114}$$

which also gives by integration

$$u(x_0) = \frac{k^3}{4\pi[kr \operatorname{ch}(kr) - \operatorname{sh}(kr)]} \int_{B(x_0, r)} u \, \mathrm{d}x . \tag{8.115}$$

Finally, let us give the *Green's function of a half-space* $\mathbb{R}^{n-1} \times \mathbb{R}^+$ *for the operator* $-\Delta + k^2$. We note that given f a tempered distribution on $\mathbb{R}^n$ with support in $\mathbb{R}^{n-1} \times \mathbb{R}^+$, there exists a unique quasi-classical tempered solution u of the Dirichlet (resp. Neumann) problem

$$\begin{cases} -\Delta u + k^2 u = f \quad \text{on} \quad \mathbb{R}^{n-1} \times \mathbb{R}_+ \\ u(.,0) = 0 \quad \left(\text{resp.} \dfrac{\partial u}{\partial x_n}(.,0) = 0\right) \quad \text{on} \quad \mathbb{R}^{n-1}. \end{cases}$$

In effect u is a solution iff its odd (resp. even) in x_n extension $\tilde{u}$ to $\mathbb{R}^n$ by symmetry is a solution of

$$-\Delta\tilde{u} + k^2\tilde{u} = \tilde{f} \quad \text{on} \quad \mathbb{R}^n$$

where $\tilde{f}$ is the extension of f.
It suffices to apply point (1) of the Proposition 27. In particular the Green's function of the Dirichlet (resp. Neumann) problem in the half-space $\mathbb{R}^{n-1} \times \mathbb{R}^+$ for the operator $-\Delta + k^2$ is

$$G(x, y) = E(x - y) - E(x - \bar{y}) \quad (\text{resp. } E(x - y) + E(x - \bar{y}))$$

where for $y = (y', y_n)$, $\bar{y} = (y', -y_n)$. □

Example 7. *Green's function of a Robin problem for the Laplacian.* We consider the problem

$$(8.116) \qquad \begin{cases} \Delta u = f \quad \text{on} \quad \Omega \\ \dfrac{\partial u}{\partial n} + \alpha u = g \quad \text{on} \quad \Gamma. \end{cases}$$

This problem can be put into the form of a Neumann problem

$$\begin{cases} Lu = -f \quad \text{on} \quad \Omega \\ \dfrac{\partial u}{\partial n_L} = g \quad \text{on} \quad \Gamma \end{cases}$$

by the introduction of the operator

$$L = -\sum \frac{\partial}{\partial x_i}\left(\frac{\partial}{\partial x_i} + \alpha n_i\right) + \sum \alpha n_i \frac{\partial}{\partial x_i} + d \quad \text{with} \quad d = \sum \frac{\partial(\alpha n_i)}{\partial x_i},$$

where n_i are the components of a vector field, defined on Ω by the unit exterior normal on Γ. If G denotes the Green's function of this problem, the solution of (8.116) will be given by[246]

$$u(x) = \int_\Gamma G(x, z) g(z) \mathrm{d}\gamma(z) - \int_\Omega G(x, y) f(y) \mathrm{d}y\,.$$

---

[246] If it exists, this will be the unique solution orthogonal to the set of solutions of $\Delta u = 0$ on Ω, $\dfrac{\partial u}{\partial n} + \alpha u = 0$ on Γ.

We shall specify the Green's function G when $\Omega = \mathbb{R}^{n-1} \times \mathbb{R}_+$ and α constant. For $\alpha = 0$, the Green's function G_0 is given by the method of images (see §7.5):

$$G_0(x, y) = E_n(x - y) + E_n(x - \bar{y})$$

where $\bar{y} = (y', -y_n)$ if $y = (y', y_n)$.
We denote by $G(\alpha, x, y)$ the solution of

$$\text{(8.117)} \qquad \begin{cases} -\Delta G(\alpha, x, y) = \delta(x - y)\,, & x \in \Omega \\ \dfrac{\partial G}{\partial n}(\alpha, z, y) + \alpha G(\alpha, z, y) = 0\,, & z \in \Gamma\,. \end{cases}$$

The open set Ω being unbounded, we must specify a condition at infinity. For $\alpha \geqslant 0$ the null condition at infinity allows us to ensure the existence and uniqueness of the solution of this problem; this solution depends regularly on α, as can be shown *a priori* but which will appear *a posteriori* in the explicit expression for $G(\alpha, x, y)$. Let us suppose that we can expand

$$G(\alpha, x, y) = \sum_{m \geqslant 0} \alpha^m G_m(x, y)\,.$$

Then substituting this form in (8.117) and equating powers of α, we have, for $m \geqslant 1$

$$\begin{cases} -\Delta G_m(x, y) = 0\,, & x \in \Omega \\ \dfrac{\partial G_m}{\partial n}(z, y) + G_{m-1}(z, y) = 0, & z \in \Gamma\,. \end{cases}$$

We note that for $z \in \Gamma$

$$G_0(z, y) = 2E_n(z - y) = 2E_n(z - \bar{y})$$

so that

$$\frac{\partial G_1}{\partial x_n}(z, y) = -\frac{\partial G_1}{\partial n}(z, y) = 2E_n(z - \bar{y})\,.$$

Since $\Delta \dfrac{\partial G_1}{\partial x_n}(x, y) = 0 = \Delta(2E_n(x - \bar{y}))$ for $x \in \Omega$, we have

$$\frac{\partial G_1}{\partial x_n}(x, y) = 2E_n(x - \bar{y})\,.$$

For $m \geqslant 2$, we have

$$\Delta \frac{\partial G_m}{\partial x_n}(x, y) = 0 = \Delta G_{m-1}(x, y) \quad \text{for} \quad x \in \Omega$$

and

$$\frac{\partial G_m}{\partial x_n}(z, y) = -\frac{\partial G_m}{\partial n}(z, y) = G_{m-1}(z, y) \quad \text{for} \quad z \in \Gamma\,;$$

hence $\dfrac{\partial G_m}{\partial x_n} = G_{m-1}\,.$

Regrouping we have

$$\frac{\partial G}{\partial x_n}(\alpha, x, y) = \sum_{m \geq 0} \alpha^m \frac{\partial G_m}{\partial x_n}(x, y) = \frac{\partial G_0}{\partial x_n}(x, y) + 2\alpha E_n(x - \bar{y})$$

$$+ \sum_{m \geq 2} \alpha^m G_{m-1}(x, y) = \alpha G(\alpha, x, y) + 2\alpha E_n(x - \bar{y})$$

$$+ \frac{\partial G_0}{\partial x_n}(x, y) - \alpha G_0(x, y),$$

namely, after an integration in x_n

$$G(\alpha, x, y) = G_0(x, y) - 2\int_{x_n}^{\infty} e^{\alpha(x_n - t)} E_n(x' - y', t + y_n)\mathrm{d}t$$

which we can rewrite

$$(8.118) \quad \begin{cases} G(\alpha, x, y) = E_n(x' - y', x_n - y_n) + E_n(x' - y', x_n + y_n) \\ \qquad - 2\displaystyle\int_{x_n + y_n}^{+\infty} e^{\alpha(x_n + y_n - t)} E_n(x' - y', t)\,\mathrm{d}t\,; \end{cases}$$

the reader should verify that this formula defines for $\alpha > 0$ the solution of (8.117).

7. Helmholtz's Equation

Helmholtz's equation is the name given to the equation

$$(8.119) \qquad \Delta u + k^2 u = 0 \quad \text{on} \quad \Omega$$

or, more generally, to the equation

$$(8.120) \qquad \Delta u + k^2 u = f \quad \text{on} \quad \Omega$$

where k is a positive real constant ($k > 0$).

This equation occurs as a model of numerous physical problems, notably in vector form (both u, f are vector fields) in the theory of the *radiation from electromagnetic sources* (see Chap. IA, §4 and Roubine [1]).

The link with *the wave equation*

$$(8.121) \qquad \frac{\partial^2 w}{\partial t^2} - \Delta w + g = 0 \quad \text{on} \quad \Omega \times \mathbb{R}$$

is clear: let us suppose that $g(x, t) = \mathrm{e}^{ikt} f(x)$ on $\Omega \times \mathbb{R}$, then $u(x)$ *is a solution of Helmholtz's equation* (8.120) *iff* $w(x, t) = \mathrm{e}^{ikt} u(x)$ *is a solution of the wave equation* (8.121): such a solution is a *stationary wave*. In the case of the equation of free waves

$$(8.122) \qquad \frac{\partial^2 w}{\partial t^2} - \Delta w = 0 \quad \text{on} \quad \Omega \times \mathbb{R},$$

the stationary waves are the solutions with variables separable $w(x, t) = \rho(t)u(x)$

of (8.122) satisfying

$$\inf_{x \in \Omega} \sup_{t \in \mathbb{R}} |w(x, t)| < \infty \tag{8.123}$$ [247].

We notice that Helmholtz's equation is similarly involved in the *method of the separation of variables in the study of Laplace's equation* (see §7.3).
In the case $n = 1$, Helmholtz' equation

$$u'' + k^2 u = 0$$

is the *equation of the vibrating string* whose general solution is

$$u(x) = \lambda e^{ikx} + \mu e^{-ikx} .$$

As we have already seen above, the solutions that we shall consider will be *complex-valued.* We shall use, as for the case of Laplace's or Poisson's equation (see Definitions 1 and 2 of §1) the terminology of classical solutions and of solutions in the sense of distributions as well as in the case of boundary value problems, those of quasi-classical and weak solutions (see Definitions 5 and 6)[248]. The Helmholtz operator is elliptic (with constant coefficients) with the result that every solution of (8.119) or of (8.120) with f analytic (resp. $\mathscr{C}^\infty$) is analytic (resp. $\mathscr{C}^\infty$) (see Proposition 5 as well as Chap. V). Similarly, we shall have regularity results up to the boundary of solutions of boundary value problems with regular data. On the other hand, the example of dimension 1 shows immediately that *the real solutions of Helmholtz's equation* (8.119) *do not, in general, satisfy the maximum principle.*

7a. Radial Solutions. Elementary Solutions

Just as with the Laplace operator, we find that the Helmholtz operator is invariant under a Euclidean transformation. For that reason, it is especially interesting to study the radial solutions and the spherical solutions.
Given v defined on an interval I of $\mathbb{R}^+$, the radial function $u(x) = v(|x|)$ is a solution of (8.119) on the annulus $\Omega = \{x \in \mathbb{R}^n;\ |x| \in I\}$ iff

$$\frac{d^2 v}{dr^2} + \frac{n-1}{r}\frac{dv}{dr} + k^2 v = 0 \quad \text{on} \quad I. \tag{8.124}$$

Putting $w = r^{\frac{1}{2}n - 1} v$, we transform equation (8.124) to *Bessel's equation*

$$\frac{d^2 w}{dr^2} + \frac{1}{r}\frac{dw}{dr} + \left(k^2 - \left(\frac{n-2}{2r}\right)^2\right) w = 0 . \tag{8.125}$$

We shall use as the independent particular solutions of (8.125) the Hankel function $H^{(1)}_{\frac{n}{2}-1}(kr)$ and its conjugate $H^{(2)}_{\frac{n}{2}-1}(kr)$.

---

[247] In effect $w(x, t) = \rho(t)u(x)$ is a solution of (8.122) iff $\rho''(t)u(x) = \rho(t)\Delta u(x)$ namely iff $\rho'' = c\rho$ on $\mathbb{R}$ and $\Delta u = cu$ on Ω for a certain constant c; the condition (8.123) expresses that ρ is bounded on $\mathbb{R}$ and this, in turn, imposes the condition $c < 0$.

[248] Equation (8.120) can be written in divergence form

$$Lu = -f \quad \text{with} \quad L = -\sum \frac{\partial}{\partial x_i}\left(\delta_{ij}\frac{\partial}{\partial x_j}\right) - k^2 .$$

We recall that

$$H_\alpha^{(1)} = \mathscr{I}_\alpha + iN_\alpha\,, \qquad H_\alpha^{(2)} = \mathscr{I}_\alpha - iN_\alpha\,, \tag{8.126}$$

where $\mathscr{I}_\alpha$ is the Bessel function of the first kind

$$\mathscr{I}_\alpha(r) = \frac{r^\alpha}{2^\alpha \Gamma\left(\frac{1}{2}\right)\Gamma\left(\alpha + \frac{1}{2}\right)} \int_0^\pi \mathrm{e}^{ir\cos\theta} \sin^{2\alpha}\theta\, \mathrm{d}\theta\,, \tag{8.127}$$

and N_α is the Bessel function of the second kind or Weber's function:

$$N_\alpha(r) = \frac{\mathscr{I}_\alpha(r)\cos\alpha\pi - \mathscr{I}_{-\alpha}(r)}{\sin\alpha\pi} \tag{8.128}$$

extended by continuity for all integers α[249].
With this notation, the radial solutions of Helmholtz's equation (8.119) are therefore the functions

$$u(x) = \left(\frac{k}{|x|}\right)^{\frac{n}{2}-1} \left(\lambda H^{(1)}_{\frac{n}{2}-1}(k|x|) + \mu H^{(2)}_{\frac{n}{2}-1}(k|x|)\right) \tag{8.129}$$

where λ, μ are scalar constants.
Let us now seek the elementary radial solutions of Helmholtz's equation. An elementary solution E is by definition a solution (in the sense of distributions) of the equation

$$\Delta E + k^2 E = \delta \quad \text{on} \quad \mathbb{R}^n\,,$$

or again, account being taken of Green's formula and the regularity of the solutions of (8.119), a function $E \in \mathscr{C}^\infty(\mathbb{R}^n \backslash \{0\}) \cap L^1_{\mathrm{loc}}(\mathbb{R}^n)$ satisfying

$$\Delta E + k^2 E = 0 \quad \text{on} \quad \mathbb{R}^n \backslash \{0\} \tag{8.130}$$

and for a regular bounded open set Ω containing O

$$\int_{\partial\Omega} \frac{\partial E}{\partial n} \mathrm{d}\gamma + k^2 \int_\Omega E\, \mathrm{d}x = 1\,. \tag{8.131}$$

This relation will then be true for every regular bounded open set Ω containing O. In the case of a radial function $E(|x|)$, the relation (8.131) can be written for $\Omega = B(0, r)$

$$r^{n-1}E'(r) + k^2 \int_0^r \rho^{n-1} E(\rho)\mathrm{d}\rho = \frac{1}{\sigma_n}\,,$$

and since this relation is true for all $r > 0$,

$$\lim_{r \to 0} r^{n-1} E'(r) = \frac{1}{\sigma_n}\,. \tag{8.132}$$

---

[249] See G. Petiau [1] whose notation we have adopted; N_α is sometimes denoted Y_α.

Using the relation $\mathrm{d}/\mathrm{d}r(r^{-\alpha}H_\alpha^{(1)}(r)) = -r^{-\alpha}H_{\alpha+1}^{(1)}(r)$[250], we see that

(8.133) $$r^{n-1}\frac{\mathrm{d}}{\mathrm{d}r}\left[\left(\frac{k}{r}\right)^{\frac{n}{2}-1} H_{\frac{n}{2}-1}^{(1)}(kr)\right] = -(kr)^{\frac{n}{2}} H_{\frac{n}{2}-1}^{(1)}(kr)$$

therefore

$$\lim_{r\to 0} r^{n-1}\frac{\mathrm{d}}{\mathrm{d}r}\left[\left(\frac{k}{r}\right)^{\frac{n}{2}-1} H_{\frac{n}{2}-1}^{(1)}(kr)\right] = -\lim_{r\to 0} r^{\frac{n}{2}} H_{\frac{n}{2}-1}^{(1)}(r)$$

$$= \frac{i(n+1)!}{\sqrt{\pi}\,\Gamma\left(\frac{n+1}{2}\right)2^{\frac{n}{2}-1}}\ {}^{251}$$

Using (8.129) and (8.132), we see therefore that *the radial elementary solutions of Helmholtz's equation are the functions*

(8.134) $$E(x) = \lambda L_k^{(n)}(|x|) + (1-\lambda)\overline{L_k^{(n)}}(|x|)$$

where λ is an arbitrary constant and

(8.134)′ $$L_k^{(n)}(r) = \frac{\sqrt{\pi}\,\Gamma\left(\frac{n+1}{2}\right)}{i\sigma_n(n-1)!}\left(\frac{2k}{r}\right)^{\frac{n}{2}-1} H_{\frac{n}{2}-1}^{(1)}(kr)\,.$$

Let us explain

(8.135) $$\textit{case } n = 2\,,\qquad L_k^{(2)}(r) = \frac{H_0^{(1)}(kr)}{4i}$$

(8.136) $$\textit{case } n = 3\,,\qquad L_k^{(3)}(r) = -\frac{\mathrm{e}^{ikr}}{4\pi r}$$

where we have used $H_{\frac{n}{2}}^{(2)}(r) = -i\sqrt{\frac{2}{\pi r}}\,\mathrm{e}^{ir}$.

We can further find the case $n = 3$ directly by the *method of descent*: in effect if v is a solution of (8.124), the function $w = \frac{1}{r}\frac{\mathrm{d}v}{\mathrm{d}r}$ is a solution of

$$\frac{\mathrm{d}^2 w}{\mathrm{d}r^2} + \frac{n+1}{r}\frac{\mathrm{d}w}{\mathrm{d}r} + k^2 w = 0\,.$$

---

[250] See G. Petiau [1], page 81.

[251] Using, for example

$$H_\alpha^{(1)}(r) = \sqrt{\frac{2}{\pi r}}\,\frac{\mathrm{e}^{-i\left(\alpha+\frac{1}{2}\right)\frac{\pi}{2}}}{\Gamma\left(\alpha+\frac{1}{2}\right)}\int_0^\infty \mathrm{e}^{ir-t}\left(1+\frac{it}{2r}\right)^{\alpha-\frac{1}{2}} t^{\alpha-\frac{1}{2}}\,\mathrm{d}t\,.$$

Since we know the solutions $\lambda e^{ikr} + \mu e^{-ikr}$ of (8.124) for $n = 1$, the solutions for $n = 3$ are $\frac{1}{r}(\lambda e^{ikr} + \mu e^{-ikr})$; since $\lim_{r \to 0} r^2 \frac{d}{dr}\left(\frac{e^{ikr}}{r}\right) = -1$, the radial elementary solutions in the case $n = 3$ are clearly the functions $-\frac{1}{4\pi r}(\lambda e^{ikr} + (1 - \lambda)e^{ikr})$.

We note also that in the case $n = 1$ the radial elementary solutions are always given by (8.134) with

$$L_k^{(1)}(r) = \frac{e^{ikr}}{2ik}. \tag{8.137}$$

7b. Solutions with Spherical Symmetry

We shall now look for solutions of Helmholtz's equation (8.119) with spherical symmetry, that is to say, solutions of the form $u(x) = h(|x|)y\left(\frac{x}{|x|}\right)$.

Making use of the expression for the Laplacian in spherical polar coordinates, (see §1.4), we see that u is a solution of (8.119) iff we have

$$\left(\frac{d^2h}{dr^2}(r) + \frac{n-1}{r}\frac{dh}{dr}(r)\right)y(\sigma) + \frac{h(r)}{r^2}\Delta_\sigma y(\sigma) + k^2 h(r)y(\sigma) = 0$$

where Δ_σ is the Laplace–Beltrami operator of the unit sphere Σ. This equation splits into

$$\begin{cases} \Delta_\sigma y + cy = 0 \\ \dfrac{d^2h}{dr^2} + \dfrac{n-1}{r}\dfrac{dh}{dr} + \left(k^2 - \dfrac{c}{r^2}\right)h = 0 \end{cases} \tag{8.138}$$

where c is a constant.

We know the first equation of the pair (8.138) admits non-constant solutions only if $c = l(l + n - 2)$ with $l \in \mathbb{N}$ and then $\mathscr{Y}_l^n$ the space of solutions is of finite dimension (see Proposition 6 of §7).

Transforming the second equation of the pair (8.138) by changing the dependent variable from $h(r)$ to $\tilde{h}(r) = r^{\frac{1}{2}n-1}h(r)$, we therefore have for $c = l(l + n - 2)$ that $\tilde{h}$ is a solution of Bessel's equation

$$\frac{d^2\tilde{h}}{dr^2} + \frac{1}{r}\frac{d\tilde{h}}{dr} + \left(k^2 - \left(\frac{\alpha}{r}\right)^2\right)\tilde{h} = 0$$

with

$$\alpha^2 = \left(\frac{n}{2} - 1\right)^2 + l(l + n - 2) = \left(l + \frac{1}{2}n - 1\right)^2. \tag{8.139}$$

To sum up: *the solutions with spherical symmetry of Helmholtz's equation* (8.119) *are of the form*

$$\left(\frac{k}{|x|}\right)^{\frac{n}{2}-1}(\lambda H_\alpha^{(1)}(k|x|) + \mu H_\alpha^{(2)}(k|x|))\,Y_l\left(\frac{x}{|x|}\right) \tag{8.140}$$

where for $l \in \mathbb{N}^*$, Y_l is a spherical harmonic function of order l in $\mathbb{R}^n$ and H^1_α, H^2_α are the Hankel functions of order α given by (8.139).

Let us explain the cases $n = 2$ and $n = 3$.

Case $n = 2$. Then $\mathscr{Y}^2_l$ is of dimension 2 admitting for $l \neq 0$ the base $\{e^{il\theta}, e^{-il\theta}\}$ where θ is the polar angle θ. In (8.139), $\alpha = l$. Hence the solutions with circular symmetry of Helmholtz's equation (8.139) are for $l \in \mathbb{N}^*$ the functions

$$(8.141)\qquad u(r\cos\theta, r\sin\theta) = (\lambda H^{(1)}_l(kr) + \mu H^{(2)}_l(kr))(\alpha e^{il\theta} + \beta e^{-il\theta})$$

with λ, μ, α, β arbitrary constants.

Case $n = 3$. Then $\mathscr{Y}^3_l$ is of dimension $2l + 1$ admitting the base $\{Y^m_l;\ -l \leqslant m \leqslant l\}$ defined by (7.64). In (8.139), $\alpha = l + \frac{1}{2}$. Hence the spherically symmetric solutions of Helmholtz's equation (8.119) are the functions

$$(8.142)\qquad |x|^{-\frac{1}{2}}(\lambda H^{(1)}_{l+\frac{1}{2}}(k|x|) + \mu H^{(2)}_{l+\frac{1}{2}}(k|x|))\left(\sum_{m=-l}^{l} \alpha_m Y^m_l\left(\frac{x}{|x|}\right)\right).$$

We observe that for $\alpha > 0$, $N_\alpha(r)$ has a singularity $r^{-\alpha}$ and $r^{-\alpha}\mathscr{J}_\alpha(r)$ is bounded as $r \to 0$. It follows that

$$\frac{1}{|x|^{\frac{n}{2}-1}}(\lambda H^{(1)}_\alpha(|x|) + \mu H^{(2)}_\alpha(|x|)) = (\lambda + \mu)\frac{\mathscr{J}_\alpha(|x|)}{|x|^{\frac{n}{2}-1}} + i(\lambda - \mu)\frac{N_\alpha(|x|)}{|x|^{\frac{n}{2}-1}}$$

is in $L^1_{\text{loc}}(\mathbb{R}^n)$ iff $\lambda = \mu$ or $\alpha < \frac{1}{2}n + 1$; in particular, for α given by (8.139), this is not true except for $l = 1$.

7c. Outgoing and Incoming Flux of Energy

Unlike the case of the Laplace operator where when $n \geqslant 3$ there existed one and only one elementary solution, null at infinity, in the case of the Helmholtz equation we have an infinity of such elementary solutions; it will be necessary to fid criteria to distinguish among them. We introduce now a first criterion enabling us to classify them into two categories by using the stationary wave which defines them. Let us consider in a general manner a wave, that is to say a solution w of the wave equation

$$\frac{\partial^2 w}{\partial t^2} - \Delta w + g = 0 \quad \text{on} \quad \Omega \times \mathbb{R}\,.$$

We suppose that Ω is a regular bounded set and w sufficiently regular to justify the following calculations. We multiply the equation by $\overline{w'} = \dfrac{\partial \bar{w}}{\partial t}$ and integrate over Ω; using Green's formula and the identities

$$\frac{\partial}{\partial t}\frac{1}{2}\left|\frac{\partial w}{\partial t}\right|^2 = \operatorname{Re}\frac{\partial^2 w}{\partial t^2}\frac{\partial \bar{w}}{\partial t}, \frac{\partial}{\partial t}\frac{1}{2}|\operatorname{grad} w|^2 = \operatorname{Re}\operatorname{grad} w \,.\, \operatorname{grad}\frac{\partial \bar{w}}{\partial t}$$

we obtain

$$\frac{d}{dt}\frac{1}{2}\int_\Omega\left(\left|\frac{\partial w}{\partial t}\right|^2 + |\operatorname{grad} w|^2\right)dx - \operatorname{Re}\int_{\partial\Omega}\frac{\partial w}{\partial n}\frac{\partial \bar{w}}{\partial t}d\gamma + \operatorname{Re}\int_\Omega g\frac{\partial \bar{w}}{\partial t}dw = 0\,.$$

By definition the quantity

$$E_\Omega(w, t) = \frac{1}{2}\int_\Omega\left(\left|\frac{\partial w}{\partial t}(x, t)\right|^2 + |\mathrm{grad}\, w(x, t)|^2\right)\mathrm{d}x$$

is the energy of the wave in the domain Ω at the instant t,

$$I_{\partial\Omega}(w, t) = -\ \mathrm{Re}\int_{\partial\Omega}\frac{\partial w}{\partial n}(z, t)\frac{\partial \bar{w}}{\partial t}(z, t)\mathrm{d}\gamma(z)$$

is *the flux of energy of the outgoing wave across $\partial\Omega$ at the instant t*, and finally the term $-\ \mathrm{Re}\int_\Omega g(x, t)\frac{\partial \bar{w}}{\partial t}(x, t)\mathrm{d}x$ is the energy applied to the domain Ω at the instant t. The *energy balance* can be written

$$\frac{\mathrm{d}}{\mathrm{d}t}E_\Omega(w, t) + I_{\partial\Omega}(w, t) = -\ \mathrm{Re}\int_\Omega g(x, t)\frac{\partial \bar{w}}{\partial t}(x, t)\mathrm{d}x$$

that is to say: *the variation of the energy plus the outgoing flux of energy across $\partial\Omega$ is equal to the energy applied at each instant.*
We consider now a stationary wave $w(x, t) = \mathrm{e}^{ikt}u(x)$ with $g(x, t) = \mathrm{e}^{ikt}f(x)$ and u is a solution of the Helmholtz equation (8.120). The energies are then constant (from which we have the name stationary wave):

$$E_\Omega(w, t) = \frac{1}{2}\int_\Omega(|\mathrm{grad}\, u|^2 + k^2|u|^2)\mathrm{d}x$$

$$I_{\partial\Omega}(w, t) = \mathrm{Re}\, ik\int_{\partial\Omega}\frac{\partial u}{\partial n}\bar{u}\,\mathrm{d}\gamma$$

and the energy balance can be written

$$\mathrm{Re}\, ik\int_{\partial\Omega}\frac{\partial u}{\partial n}\bar{u}\,\mathrm{d}\gamma = -\ \mathrm{Re}\, ik\int_\Omega f\bar{u}\,\mathrm{d}x\,.$$

We can state:

Proposition 28. *Let Ω be a regular bounded open set and $u \in \mathscr{C}^2(\Omega)\cap\mathscr{C}^1_n(\bar{\Omega})$ a classical solution of Helmholtz's equation (8.120) with $f \in L^1(\Omega)$. The energy flux going out across $\partial\Omega$*

$$I_{\partial\Omega}(u) = \mathrm{Re}\, ik\int_{\partial\Omega}\frac{\partial u}{\partial n}\bar{u}\,\mathrm{d}\gamma \tag{8.143}$$

of the stationary wave $\mathrm{e}^{ikt}u(x)$[252] *is equal to* $-\ \mathrm{Re}\, ik\int_\Omega f\bar{u}\mathrm{d}x$. *In particular, if $f \equiv 0$ on Ω, then $I_{\partial\Omega} = 0$.* □

[252] $I_{\partial\Omega}(u)$ is the energy flux of the wave $\mathrm{e}^{ikt}u(x)$; some authors associate with u the wave $\mathrm{e}^{-ikt}u(x)$; in this case the energy flux will have the opposite sign and the terms outgoing/incoming will be reversed.

The precision given by the regularity in the classical theory[253] justifies the calculations involved in the proof of the proposition. We have the corollary:

Corollary 5. *Let Ω be an arbitrary open set in $\mathbb{R}^n$, K a compact set in Ω and u a solution of Helmholtz's equation (8.119) on $\Omega' = \Omega \backslash K$. Then for every regular bounded open set Ω_0 with $K \subset \Omega_0 \subset \bar{\Omega}_0 \subset \Omega$, the energy flux $I_{\partial\Omega_0}$(defined by (8.143)) is independent of Ω_0.*

Proof. We notice first of all that $u \in \mathscr{C}^\infty(\Omega')$ and $\partial\Omega_0 \subset \Omega'$ with the result that $I_{\partial\Omega_0}(u)$ is defined. Let Ω_1, Ω_2 be two regular bounded open sets satisfying $K \subset \Omega_i \subset \overline{\Omega_i} \subset \Omega$; there exists a regular bounded open set Ω_0 such that $\overline{\Omega_1 \cup \Omega_2} \subset \Omega_0 \subset \overline{\Omega_0} \subset \Omega$. Now applying the proposition with the open set $\Omega_0 \backslash \bar{\Omega}_i$, which is relatively compact in Ω'; we have $I_{\partial(\Omega_0 \backslash \bar{\Omega}_i)}(u) = 0$; but

$$\partial(\Omega_0 \backslash \overline{\Omega_i}) = \partial\Omega_0 \cup \partial\Omega_i \quad \text{and} \quad I_{\partial(\Omega_0 \backslash \bar{\Omega}_i)}(u) = I_{\partial\Omega_0}(u) - I_{\partial\Omega_i}(u)\,,$$

from which we deduce that

$$I_{\partial\Omega_i}(u) = I_{\partial\Omega_0}(u) \quad \text{for} \quad i = 1, 2\,. \qquad \square$$

This leads us to pose

Definition 9. Let K be a compact set in $\mathbb{R}^n$, Ω an open set containing K and u a solution of Helmholtz's equation (8.119) on $\Omega' = \Omega \backslash K$. We shall call[254] by the name *energy flux of u* to the quantity

$$I(u) = \operatorname{Re} ik \int_{\partial\Omega_0} \frac{\partial u}{\partial n} \bar{u} \, \mathrm{d}\gamma$$

for a regular bounded open set Ω_0 with $K \subset \Omega_0 \subset \overline{\Omega_0} \subset \Omega$ ($I(u) = I_{\partial\Omega_0}(u)$ is independent of Ω_0). We shall say that the flux energy of u is *outgoing* (resp. *incoming*) if $I(u) \geqslant 0$ (resp. $I(u) \leqslant 0$); we shall say also that u defines an *outgoing* (resp. *incoming*) *wave*[255].

In an obvious way $I(\bar{u}) = -I(u)$: the energy flux of u is outgoing iff that of $\bar{u}$ is incoming. In particular, if u is real, $I(u) = 0$.

In the terminology and notation of Definition 9, we have not involved the given sets K and Ω; in effect it is clear that $I(u)$ depends only on the values of u in the neighbourhood of K in $\mathbb{R}^n \backslash K$; also it does not depend on K in the measure where, given another compact set, $K_1 \subset \Omega$, if we can extend u to a solution u_1 of the homogeneous Helmholtz equation on $\Omega'_1 = \Omega \backslash (K_1 \cap K)$, $I(u_1) = I(u)$. The reason for this independence is the analyticity of the solutions of the homogeneous Helmholtz equation: on every connected component of Ω'_1 meeting Ω', the extension u_1 of u is unique; also for a connected component Ω'_0 of Ω'_1 not meeting Ω', we have $\Omega'_0 \cap K_1 = \varnothing$ and hence if Ω_1 is a regular bounded open set with

---

[253] We can give a formulation within a framework of a weak solution in $W^{1,2}(\Omega)$.

[254] By abuse of language.

[255] As we have said, this terminology is linked to the choice of the stationary wave associated with u, here $e^{ikt}u(x)$; it will be reversed for the choice $e^{-ikt}u(x)$.

$K_1 \subset \Omega_1 \subset \overline{\Omega_1} \subset \Omega$, u_1 is a solution of the homogeneous Helmholtz equation on Ω_1 and $I_{\partial\Omega_0}(u_1) = 0$.

To clarify these notions we consider the *radial case*. Let $u(x) = v(|x|)$ be a solution of the homogeneous Helmholtz equation on an annulus

$$\Omega' = B(0, r_1) \backslash \bar{B}(0, r_0) \quad (0 \leqslant r_0 < r_1 \leqslant +\infty)\,.$$

The energy flux is given by

$$I(u) = \operatorname{Re}(ik\sigma_n r^{n-1} v'(r)\bar{v}(r))\,.$$

We know (see (8.129)) that

$$v(r) = \left(\frac{k}{r}\right)^{\frac{n}{2}-1} (\lambda H^{(1)}_{\frac{n}{2}-1}(kr) + \mu H^{(2)}_{\frac{n}{2}-1}(kr))$$

from which, due to (8.133), we have

$$r^{n-1} v'(r) = -(kr)^{\frac{n}{2}} (\lambda H^{(1)}_{\frac{n}{2}}(kr) + \mu H^{(2)}_{\frac{n}{2}}(kr))$$

and

$$I(u) = (|\lambda|^2 - |\mu|^2) k^n \sigma_n r \operatorname{Im}(H^{(1)}_{\frac{n}{2}}(kr) H^{(2)}_{\frac{n}{2}-1}(kr))\,.$$

Now

$$\begin{aligned}\operatorname{Im}(H^{(1)}_{\alpha+1}(r) H^{(2)}_{\alpha}(r)) &= N_{\alpha+1}(r)\mathscr{I}_\alpha(r) - N_\alpha(r)\mathscr{I}_{\alpha+1}(r) \\ &= \mathscr{I}_\alpha(r)\frac{\mathrm{d}N_\alpha}{\mathrm{d}r}(r) - N_\alpha(r)\frac{\mathrm{d}\mathscr{I}_\alpha}{\mathrm{d}r}(r) = -\frac{2}{\pi r}\end{aligned}$$

so

$$I(u) = \frac{2k^n\sigma_n}{\pi}(|\mu|^2 - |\lambda|^2)\,. \tag{8.144}$$

To sum up: the energy flux of the radial solution given by (8.129) is given by (8.144); hence the energy flux is outgoing (resp. incoming) iff $|\mu| \geqslant |\lambda|$ (resp. $|\mu| \leqslant |\lambda|$). In particular with the notation (8.134), *the stationary wave* $e^{ikt} L_k^{(n)}(x)$ *(resp.* $e^{ikt} \overline{L_k^{(n)}}(x)$*) is incoming (resp. outgoing) and a superposition* $e^{ikt}(\lambda L_k^{(n)}(x) + \mu \overline{L_k^{(n)}}(x))$ *is outgoing iff the intensity* $|\mu|$ *of the outgoing wave* $\mu e^{ikt} \overline{L_k^{(n)}}(x)$ *exceeds the intensity* $|\lambda|$ *of the incoming wave* $\lambda e^{ikt} L_k^{(n)}(x)$.

In the case $n = 3$, $L_k^{(n)}(x) = -\dfrac{e^{ik|x|}}{4\pi|x|}$ and $e^{ikt} L_k^{(3)}(x) = -\dfrac{e^{ik(t+|x|)}}{4\pi|x|}$ is the incoming wave while $e^{ikt}\overline{L_k^{(3)}}(x) = -\dfrac{e^{ik(t-|x|)}}{4\pi|x|}$ is the outgoing wave. The former wave is also said to be *convergent* and the latter to be *divergent*. This terminology can be employed instead of incoming and outgoing respectively.

7d. Sommerfeld's Radiation Conditions

The notion of an outgoing or an incoming wave allowed us to classify the radial elementary solutions of the Helmholtz operator but we did not characterise the solutions $L_k^{(n)}$ and $\overline{L_k^{(n)}}$, although their importance was made apparent. This characterisation will be effected because of conditions of uniqueness, formulated by Sommerfeld, called radiation conditions.

In the case $n = 3$, the elementary solution $L_k^{(3)}(x) = -\dfrac{e^{ik|x|}}{4\pi|x|}$ immediately satisfies

$$\frac{\partial L_k^{(3)}}{\partial r}(x) - ikL_k^{(3)}(x) = \frac{e^{ik|x|}}{4\pi|x|^2}. \tag{8.145}$$

Use of the asymptotic expansion of Hankel functions[256]

$$H_\alpha^{(1)}(r) = \left(\frac{2}{\pi r}\right)^{\frac{1}{2}} \frac{e^{ir}}{i^{\alpha+\frac{1}{2}}}\left[1 + O\left(\frac{1}{r}\right)\right] \quad \textit{when} \quad r \to \infty \tag{8.146}$$

and the differentiation formula (8.133) show that for arbitrary n

$$\frac{d}{dr}\left(\frac{k}{r}\right)^\alpha H_\alpha^{(1)}(kr) - ik\left(\frac{k}{r}\right)^\alpha H_\alpha^{(1)}(kr) = -k\left(\frac{k}{r}\right)^\alpha [H_{\alpha+1}^{(1)}(kr)$$
$$+ iH_\alpha^{(1)}(kr)] = O\left(\frac{1}{r^{\alpha+\frac{3}{2}}}\right)$$

and hence using the definition of the $L_k^{(n)}$,

$$\frac{\partial L_k^{(n)}}{\partial r}(x) - ikL_k^{(n)}(x) = O\left(\frac{1}{|x|^{\frac{n+1}{2}}}\right) \quad \text{when} \quad |x| \to \infty. \tag{8.147}$$

Given f a distribution with compact support on $\mathbb{R}^n$, the distribution $u = L_k^n * f$ is a solution (in the sense of distributions) on $\mathbb{R}^n$ of the Helmholtz equation (8.120) and hence $u \in \mathscr{C}^\infty(\Omega')$ with $\Omega' = \mathbb{R}^n \backslash \operatorname{supp} f$ and solution on Ω' of the homogeneous Helmholtz equation (8.119). We point out the

Lemma 7. *Let f be a distribution with compact support on $\mathbb{R}^n$. Then $u = L_k^{(n)} * f$ has the asymptotic behaviour*

$$\frac{\partial u}{\partial r}(x) - iku(x) = O\left(\frac{1}{|x|^{\frac{n+1}{2}}}\right) \quad \textit{when} \quad |x| \to \infty. \tag{8.148}$$

We shall prove this lemma later, this condition permitting the characterisation of the solution $L_k^{(n)} * f$ of the Helmholtz equation (8.120). Moreover we shall consider two other weaker conditions,

$$\frac{\partial u}{\partial r}(x) - iku(x) = o\left(\frac{1}{|x|^{\frac{n-1}{2}}}\right) \quad \text{when} \quad |x| \to \infty \tag{8.149}$$

$$\lim_{r \to \infty} \int_{\partial B(0,r)} \left|\frac{\partial u}{\partial r} - iku\right|^2 d\gamma = 0. \tag{8.150}$$

---

[256] See G. Petiau [1], p. 133.

We say that (8.149) (resp. (8.148), (8.150)) is the *incoming classical* (*resp. strong, weak*) *radiation condition of Sommerfeld*. It is clear that the classical condition (8.149) implies the weak condition, since denoting

$$\varepsilon(r) = \sup_{|x| \geqslant r} \left| \frac{\partial u}{\partial r}(x) - iku(x) \right| |x|^{\frac{n-1}{2}},$$

we have

$$\int_{\partial B(0,r)} \left| \frac{\partial u}{\partial r} - iku \right|^2 \mathrm{d}\gamma \leqslant \varepsilon(r) .$$

We can similarly define the *outgoing* radiation conditions of Sommerfeld – strong, classical and weak – by

$$\frac{\partial u}{\partial r}(x) + iku(x) = O\left(\frac{1}{|x|^{\frac{n+1}{2}}}\right) \quad \text{when} \quad |x| \to \infty , \tag{8.148b}$$

$$\frac{\partial u}{\partial r}(x) + iku(x) = o\left(\frac{1}{|x|^{\frac{n-1}{2}}}\right) \quad \text{when} \quad |x| \to \infty , \tag{8.149b}$$

$$\lim_{r \to \infty} \int_{\partial B(0,r)} \left| \frac{\partial u}{\partial r} + iku \right|^2 \mathrm{d}\gamma = 0 . \tag{8.150b}$$

The terminology incoming and outgoing corresponds closely to the notions defined previously thanks to the

Proposition 29. *Let K be a compact set and u a solution of the homogeneous Helmholtz equation on $\Omega' = \mathbb{R}^n \backslash K$. Let us suppose that u satisfies the incoming (resp. outgoing) weak (8.150) (resp. 8.150b) radiation condition of Sommerfeld. Then the energy flux of u is incoming (resp. outgoing); more precisely*

$$-I(u) \quad (\textit{resp. } I(u)) = k^2 \lim_{r \to \infty} \int_{\partial B(0,r)} |u|^2 \mathrm{d}\gamma. \tag{8.151}$$

Proof. We have for all r sufficiently large

$$I(u) = \operatorname{Re} ik \int_{\partial B(0,r)} \frac{\partial u}{\partial r} \bar{u} \, \mathrm{d}\gamma = \operatorname{Re} ik \int_{\partial B(0,r)} \left(\frac{\partial u}{\partial r} - iku \right) \bar{u} \, \mathrm{d}\gamma - k^2 \int_{\partial B(0,r)} |u|^2 \mathrm{d}\gamma.$$

Now by Schwarz's inequality

$$\left| \operatorname{Re} ik \int_{\partial B(0,r)} \left(\frac{\partial u}{\partial r} - iku \right) \bar{u} \, \mathrm{d}\gamma \right| \leqslant k \left(\int_{\partial B(0,r)} \left| \frac{\partial u}{\partial r} - iku \right|^2 \mathrm{d}\gamma \right)^{\frac{1}{2}} \left(\int_{\partial B(0,r)} |u|^2 \mathrm{d}\gamma \right)^{\frac{1}{2}} ;$$

first of all, we deduce from these two relations

$$k \left(\int_{\partial B(0,r)} |u|^2 \mathrm{d}\gamma \right)^{\frac{1}{2}} \leqslant \left(\int_{\partial B(0,r)} \left| \frac{\partial u}{\partial r} - iku \right|^2 \mathrm{d}\gamma \right)^{\frac{1}{2}} + \sqrt{|I(u)|} \tag{8.152}$$

then (8.151). □

We shall now state the result of *the uniqueness of exterior Dirichlet or Neumann problems with a radiation condition at infinity.*

Proposition 30. *Let Ω' be a connected open set containing the exterior of a compact set of $\mathbb{R}^n$ and whose boundary Γ, if it is non-empty*[257], *is regular of class $\mathscr{C}^{1+\varepsilon}$. Then the function $u \equiv 0$ on Ω' is the unique classical solution of the problem*

$$\begin{cases} \Delta u + k^2 u = 0 \quad \text{on} \quad \Omega \\ u = 0 \quad \left(\text{resp.} \dfrac{\partial u}{\partial n} = 0\right) \quad \text{on} \quad \Gamma \end{cases} \tag{8.153}$$

satisfying an (incoming or outgoing) radiation condition at infinity.

Proof. We take u to be solution of the problem and we have to prove that $u \equiv 0$. We note that, because of the hypothesis of regularity, $u \in \mathscr{C}^1_n(\Omega')$ and the energy flux of u can be calculated on Γ; the hypothesis of a homogeneous Dirichlet or Neumann problem hence implies that $I(u) = 0$. From Proposition 29, we deduce

$$\lim_{r \to \infty} \int_{\partial B(0,r)} |u|^2 \, d\gamma = 0 . \tag{8.154}$$

Let us fix $r_0 > 0$ such that $B(0, r_0) \supset K = \mathbb{R}^n \backslash \Omega'$. For $r \geqslant r_0$; Green's formula (or Gauss' theorem) assures us that for every solution L of the homogeneous Helmholtz equation

$$C(L) = \int_{\partial B(0,r)} \left(u \frac{\partial L}{\partial r} - \frac{\partial u}{\partial r} L \right) d\gamma \text{ is constant for } r > r_0. \tag{8.155}$$

Let us suppose that u were to satisfy an incoming condition. We can write

$$\begin{aligned} |C(L)| &= \left| \int_{\partial B(0,r)} u \left(\frac{\partial L}{\partial r} - ikL \right) d\gamma - \int_{\partial B(0,r)} L \left(\frac{\partial u}{\partial r} - iku \right) d\gamma \right| \\ &\leqslant \max_{\partial B(0,r)} \left| \frac{\partial L}{\partial r} - ikL \right| \int_{\partial B(0,r)} |u| \, d\gamma + \max_{\partial B(0,r)} |L| \int_{\partial B(0,r)} \left| \frac{\partial u}{\partial r} - iku \right| d\gamma . \end{aligned}$$

From Schwarz's inequality we have that, for every function w

$$\int_{\partial B(0,r)} |w| \, d\gamma \leqslant r^{\frac{n-1}{2}} \sigma_n^{1/2} \left(\int_{\partial B(0,r)} |w|^2 \, d\gamma \right)^{\frac{1}{2}} . \tag{8.156}$$

From the radiation condition and (8.154), we can conclude that $C(L) = 0$, if we choose a function L satisfying

$$L(x) = O\left(\frac{1}{|x|^{\frac{n-1}{2}}} \right), \quad \frac{\partial L}{\partial r}(x) = O\left(\frac{1}{|x|^{\frac{n-1}{2}}} \right) \quad \text{when} \quad |x| \to \infty . \tag{8.157}$$

First of all let us choose $L = L_k^{(n)}$. It is clear that (8.157) is satisfied and hence $C(L) = 0$ which can be written

$$v_0(r) \frac{\partial L_k^{(n)}}{\partial r}(r) - v_0'(r) L_k^{(n)}(r) = 0 \quad \text{for} \quad r > r_0$$

[257] We include the case $\Omega' = \mathbb{R}^n$; then $\Gamma' = \varnothing$ as well as the boundary condition in (8.153).

where

$$v_0(r) = \frac{1}{r^{n-1}} \int_{\partial B(0,r)} u \, d\gamma = \int_{\Sigma} u(r\sigma) \, d\sigma \, .$$

We have therefore $v_0(r) = cL_k^{(n)}(r)$. But from (8.156) at (8.154), we have

$$\lim_{r \to \infty} r^{\frac{n-1}{2}} v_0(r) = 0 \, .$$

Taking account of (8.146) and (8.134)′ we find that $\lim_{r \to \infty} r^{\frac{n-1}{2}} L_k^{(n)}(r) \neq 0$ and hence that $c = 0$. Thus we have proved that

$$\int_{\Sigma} u(r\sigma) \, d\sigma = 0 \quad \text{for all} \quad r > r_0 \, .$$

We now carry out the same reasoning with

$$L(x) = \left(\frac{k}{|x|}\right)^{\frac{n}{2}-1} H_\alpha^{(1)}(k|x|) \, Y_l\left(\frac{x}{|x|}\right)$$

putting

$$v_l(r) = \int_{\Sigma} u(r\sigma) \, Y_l(\sigma) \, d\sigma$$

where Y_l is a spherical harmonic function of order l and α is given by (8.139). We know (see (8.140)) that L is a solution of the homogeneous Helmholtz equation; it is clear from (8.146) that $L(x) = O\left(\frac{1}{|x|^{\frac{n-1}{2}}}\right)$.

Also $\frac{\partial L}{\partial r}(x) = O\left(\frac{1}{|x|^{\frac{n-1}{2}}}\right)$ since

$$\frac{d}{dr}\left(\frac{1}{r^{\frac{n}{2}-1}} H_\alpha^{(1)}(r)\right) = \left(\frac{n}{2} - 1 - \alpha\right) \frac{1}{r^{\frac{n}{2}}} H_\alpha^{(1)}(r) - \frac{1}{r^{\frac{n}{2}-1}} H_{\alpha+1}^{(1)}(r) \, .$$

Therefore $v_l(r) = c\left(\frac{k}{r}\right)^{\frac{n}{2}-1} H_\alpha^{(1)}(kr)$ and we conclude that $c = 0$ as above.

The set of spherical harmonic functions being complete (see Proposition 6 of §7) we conclude that

$$u(r\sigma) \equiv 0 \quad \text{for all} \quad \sigma \in \Sigma \quad \text{and all} \quad r > r_0 \, .$$

The connexity of Ω' and the analyticity of u leads to the conclusion. □

Remark 4. The above proof shows in fact that *a solution u of the homogeneous Helmholtz equation on a connected set with complementary compact set, whose energy flux is zero and which satisfies a radiation condition is identically zero.* The regularity of the boundary of the open set is involved only to ensure that the null condition on the boundary implies the vanishing of the energy flux. Going back to

the proof of Proposition 28 we see this is true, without a regularity assumption, if u is a weak solution relative to $W^{1,2}_{\text{loc}}(\overline{\Omega'})$ of (8.153); we then have the same uniqueness result.
In the case of the Dirichlet problem with Γ a Lipschitz boundary, this is called *Meixner's uniqueness condition.* □

7e. Retarded Potentials

Given a distribution f with compact support on $\mathbb{R}^n$, we call the distribution $\overline{L_k^{(n)}} * f$ the *retarded potential of f*. As a corollary of Sommerfeld's uniqueness theorem we have

Proposition 31. *Let f be a distribution with compact support in $\mathbb{R}^n$. The retarded potential of f is the unique solution of the Helmholtz equation satisfying the outgoing weak radiation condition.* □

The uniqueness is Proposition 30 with $\Omega' = \mathbb{R}^n$. We have already stated in Lemma 7 that $L_k^{(n)} * f$ satisfies the incoming radiation condition and hence $\overline{L_k^{(n)}} * f = \overline{L_k^{(n)} * \bar{f}}$ satisfies the outgoing condition. Now we give the

Proof of Lemma 7. We can always write $f = \sum_{|\alpha| \leqslant m} \frac{\partial^\alpha f_\alpha}{\partial x^\alpha}$ for a family (f_α) of bounded continuous functions with compact support contained in $\bar{B}(0, r_0)$. For $|x| > r_0$ we have

$$u(x) = \langle f(y), \overline{L_k^{(n)}}(x - y) \rangle = \sum_{|\alpha| \leqslant m} \int_{\mathbb{R}^n} \frac{\partial^\alpha \overline{L_k^{(n)}}}{\partial x^\alpha}(x - y) f_\alpha(y)\, dy \,.$$

Therefore

$$\frac{\partial u}{\partial r}(x) + iku(x) = \sum_{|\alpha| \leqslant m} \left[\int_{\mathbb{R}^n} \frac{\partial^\alpha F}{\partial x^\alpha}(x - y) f_\alpha(y)\, dy + R_\alpha(x) \right]$$

where

$$F = \frac{\partial \overline{L_k^{(n)}}}{\partial r} + ik \overline{L_k^{(n)}}$$

and

$$- R_\alpha(x) = \sum_{\substack{\beta \leqslant \alpha \\ \beta \neq 0}} C_\alpha^\beta \sum_j \int_{\mathbb{R}^n} \frac{\partial^\beta}{\partial x^\beta} \omega_j(x - y) \frac{\partial^{\alpha - \beta}}{\partial x^{\alpha - \beta}} \frac{\partial \overline{L_k^{(n)}}}{\partial x_j}(x - y) f_\alpha(y)\, dy$$

with

$$\omega_j(x) = \frac{x_j}{|x|} \,.$$

Now when $|x| \to \infty$, we have $\frac{\partial^\beta}{\partial x^\beta} \omega_j(x) = O\left(\frac{1}{|x|^{|\beta|}}\right)$.

In the case $n = 3$, $\overline{L_k^{(3)}}(x) = \dfrac{-\mathrm{e}^{-ik|x|}}{4\pi|x|}$, $F(x) = \dfrac{\mathrm{e}^{-ik|x|}}{4\pi|x|^2}$ from which when $|x| \to \infty$

$$\frac{\partial^\alpha}{\partial x^\alpha} L_k^{(3)}(x) = O\left(\frac{1}{|x|}\right), \qquad \frac{\partial^\alpha}{\partial x^\alpha} F(x) = O\left(\frac{1}{|x|^2}\right)$$

In the case n arbitrary, we shall verify, with the help of the properties of the Hankel functions, that

$$\frac{\partial^\alpha}{\partial x^\alpha} L_k^{(n)}(x) = O\left(\frac{1}{|x|^{\frac{n-1}{2}}}\right), \qquad \frac{\partial^\alpha F}{\partial x^\alpha}(x) = O\left(\frac{1}{|x|^{\frac{n+1}{2}}}\right).$$

The lemma is then easily deduced as in Proposition 2 of §3. □

We take note of the

Corollary 6. *Let u be a solution of the homogeneous Helmholtz equation on $\Omega' = \mathbb{R}^n \backslash K$ with compact K. Then the outgoing (resp. incoming) strong* (8.148*b*) *(resp.* (8.148)), *classical* (8.149*b*) *(resp.* (8.149)) *and weak* (8.150*b*) *(resp.* (8.150)) *radiation conditions are equivalent.*

Proof. It is enough to prove that the weak condition implies the strong condition. Let $\rho \in \mathscr{D}(\mathbb{R}^n)$ with $\rho = 1$ in the neighbourhood of $\mathbb{R}^n \backslash \Omega'$. We put $v = (1 - \rho)u$, $f = \Delta v + k^2 v$. Since $v(x) = u(x)$ for $|x|$ large, v satisfies the weak radiation condition and hence is the retarded potential of f. From Lemma 7, v therefore satisfies the strong condition and it is the same for u. □

Corollary 7. *Let Ω be a regular bounded open set and u a solution of the homogeneous Helmholtz equation on $\Omega' = \mathbb{R}^n \backslash \bar{\Omega}$. Suppose $u \in \mathscr{C}_n^1(\overline{\Omega'})$, then there exists a unique decomposition $u = u_1 + u_2$ on Ω' where*
u_1 is a solution of the homogeneous Helmholtz equation on $\mathbb{R}^n$,
u_2 is a solution of the homogeneous Helmholtz equation on Ω' and satisfies the outgoing radiation condition.

Proof. Let us first prove the uniqueness; considering two decompositions, $u = u_1 + u_2 = v_1 + v_2$, we shall have $v_1 - u_1 = u_2 - v_2$. Since u_2, v_2 satisfy the radiation condition, it is the same for $v_1 - u_1$ which satisfying the homogeneous Helmholtz equation on the whole space $\mathbb{R}^n$ is therefore the null solution. Hence $v_1 = u_1$ and $v_2 = u_2$.
For existence, let us put

$$u_2(x) = -\int_\Gamma \left[u(t) \frac{\partial}{\partial n} \overline{L_k^{(n)}}(t - x) - \frac{\partial u}{\partial n}(t) \overline{L_k^{(n)}}(t - x) \right] \mathrm{d}\gamma(t)$$

that is to say that u_2 is the retarded potential of the distribution with support in Γ

$$f = \frac{\partial u}{\partial n} \mathrm{d}\gamma + \frac{\partial}{\partial n}(u \, \mathrm{d}\gamma) \quad \text{(see Prop. 5 of §3, or Def. 2, §3)}$$

Now from Green's formula

$$u_1(x) = u(x) - u_2(x) = \int_{\partial B(0,r)} \left[u(t) \frac{\partial}{\partial n} \overline{L_k^{(n)}}(t - x) - \frac{\partial u}{\partial n}(t) \overline{L_k^{(n)}}(t - x) \right] \mathrm{d}\gamma(t)$$

for all r such that $B(0, r) \supset \bar{\Omega}$ and $x \in B(0, r) \backslash \bar{\Omega}$. □

In the neighbourhood of 0, $L_k^{(n)}(x) \sim E_n(x)$ is an elementary solution of the Laplacian. It follows that *all the results on the local regularity of Newtonian potentials* (see §3.2) *are valid for the retarded potentials.* In particular, if f is a Radon measure with compact support, its retarded potential $u \in L_{\mathrm{loc}}^p(\mathbb{R}^n)$ for all $p < \dfrac{n}{(n-2)^+}$ and is given by

$$u(x) = \int_{\mathbb{R}^n} \overline{L_k^{(n)}}(x - y) \mathrm{d}f(y) \quad \text{a.e.} \quad x \in \mathbb{R}^n .$$

In the *case* $n = 3$, $u \in L_{\mathrm{loc}}^p(\mathbb{R}^n)$ for all $p < 3$ and

$$u(x) = -\int_{\mathbb{R}^3} \frac{\mathrm{e}^{-ik|x-y|}}{4\pi|x - y|} \mathrm{d}f(y) ;$$

in particular

$$|u(x)| \leqslant \int_{\mathbb{R}^3} \frac{\mathrm{d}|f|(y)}{4\pi|x - y|} \quad \text{a.e.} \quad x \in \mathbb{R}^3 ;$$

in other words, *in the case* $n = 3$, *the retarded potential of a Radon measure with compact support is dominated in modulus by the opposite of the Newtonian potential of the variation of* this measure (that is to say the modulus of the density when f is with density).

We can also define *simple and double layer retarded potentials* given in the case $n = 3$ by

$$u_1(x) = -\int_\Gamma \frac{\mathrm{e}^{-ik|z-x|}}{4\pi|z - x|} \varphi(z) \mathrm{d}\gamma(z)$$

$$u_2(x) = \int_\Gamma (ik|z - x| - 1) \mathrm{e}^{-ik|z-x|} \frac{(z - x).n(z)}{4\pi|z - x|^3} \varphi(z) \mathrm{d}\gamma(z)$$

where Γ is the boundary of a regular bounded open set Ω and $\varphi \in \mathscr{C}^0(\Gamma)$. We shall obtain the same regularity for these retarded potentials as we did in the case of the Newtonian potentials (see §3.3). In effect denoting by u_1^0, u_2^0 the simple and double layer Newtonian potentials

$$u_1(x) = u_1^0(x) + \int_\Gamma R_1(z - x) \varphi(z) \mathrm{d}\gamma(z)$$

$$u_2(x) = u_2^0(x) + \int_\Gamma R_2(z, x) \varphi(z) \mathrm{d}\gamma(z)$$

with

$$R_1(x) = \frac{1 - e^{-ik|x|}}{4\pi|x|}$$

$$R_2(z, x) = [(ik|z - x| - 1)e^{-ik|z-x|} - 1]\frac{(z - x).n(z)}{4\pi|z - x|^3}.$$

The kernels $R_1(x)$, $|x|$ grad $R_1(x)$, $|x - z|R_2(z, x)$ are bounded. From Lebesgue's theorem we deduce that $u_1 - u_1^0 \in \mathscr{C}^1(\mathbb{R}^n)$ and $u_2 - u_2^0 \in \mathscr{C}^0(\mathbb{R}^n)$. We can develop an integral method of solution of Dirichlet and Neumann problems (interior and exterior) for the Helmholtz equation as in the case of Laplace's equation (see Chap. XIB §3 and Kleinmann–Roach [1]). Observe, however, that unlike the case of Laplace's equation, *there is, in general, neither uniqueness nor existence of the solution of the interior Dirichlet problem*: this obviously complicates the discussion of Fredholm's method. □

Remark 5. In the usual way in physics, we call the retarded potential of a given charge density $\rho(x, t)$ in $\mathbb{R}_x^3 \times \mathbb{R}_t$

$$u(x, t) = \frac{1}{4\pi}\int_{\mathbb{R}^3} \frac{1}{|x - x'|}\rho\left(x', t - \frac{|x - x'|}{c}\right)dx' \tag{8.158}$$

(where c is the speed of light) which translates the fact that the (density of) charge at (x', t') influences the point (x, t) only if $|x - x'| = c(t - t')$. In the sequel we take $c = 1$.

We denote by $\hat{u}(x, k)$ (resp. $\hat{\rho}(x, k)$) the Fourier transform in t of $u(x, t)$ (resp. $\rho(x, t)$), supposing ρ is such this has a meaning. We have

$$\begin{cases} \hat{u}(x, k) = \displaystyle\int_{\mathbb{R}_t} u(x, t)e^{-ikt}\,dt = \int_{\mathbb{R}^3 \times \mathbb{R}} \frac{e^{-ik(\tau + |x - x'|)}}{4\pi|x - x'|}\rho(x', \tau)\,dx'\,d\tau \\ \qquad = \displaystyle\int_{\mathbb{R}^3 \times \mathbb{R}} \frac{e^{-ik|x - x'|}}{4\pi|x - x'|}\rho(x', \tau)e^{-ik\tau}\,dx'\,d\tau\,; \end{cases}$$

from which, with (8.136):

$$\hat{u}(x, k) = \overline{L_k^{(3)}} \underset{x}{*} (-\hat{\rho}(.,k)), \tag{8.159}$$

which *corresponds to the definition given at the beginning of this section* of the retarded potential for $f(x) = -\hat{\rho}(x, k)$.

We verify that since $\hat{u}$ is a solution of Helmholtz's equation

$$(\Delta + k^2)\hat{u}(x, k) = -\hat{\rho}(x, k),$$

u is a solution of the wave equation

$$\frac{\partial^2 u}{\partial t^2} - \Delta u = \rho.$$

Denoting by $\tilde{E}_3$ the elementary solution of the wave equation with support the future cone (see Chap. XIV, §3 formulae (3.45), (3.46)) which we write:

$$\tilde{E}_3(x, t) = \frac{1}{2\pi} Y(t)\delta(t^2 - x^2) = \frac{1}{4\pi|x|}\delta(t - |x|)\,;$$

the retarded potential (8.158) can be written

$$u = \tilde{E}_3 \underset{x,t}{*} \rho\,. \tag{8.160}$$

A particular case of interest is that of the retarded potential due to a charged point particle, of charge e whose motion is given by a function $x_0(t)$ with $x_0 \in \mathscr{C}^1(\mathbb{R}, \mathbb{R}^3)$ and $\left|\frac{dx_0}{dt}(t)\right| < c = 1$. The charge density $\rho(x, t)$ is then given by:

$$\rho(x, t) = e\delta(x - x_0(t))\,.$$

Let us calculate the retarded potential of such a charge density.
For all $\varphi \in \mathscr{D}(\mathbb{R}^4)$ we have

$$\left\{\begin{aligned} \langle \tilde{E}_3 * \rho, \varphi\rangle &= \langle \tilde{E}_3(x', \tau') \otimes \rho(x, \tau), \varphi(x' + x, \tau' + \tau)\rangle \\ &= \frac{e}{2\pi}\int_{\mathbb{R}^4} \frac{1}{2|x'|}\varphi(x' + x_0(\tau), |x'| + \tau)\,dx'\,d\tau\,. \end{aligned}\right. \tag{8.161}$$

We now make the change of variable θ: $(x', \tau) \to (x, t)$ defined by

$$\left\{\begin{aligned} x &= x' + x_0(\tau) \\ t &= |x'| + \tau\,. \end{aligned}\right.$$

The Jacobian $J(\theta)$ of this transformation is

$$J(\theta) = 1 - \frac{v_0(\tau).x'}{|x'|}, \quad \text{where} \quad v_0(\tau) \overset{\text{def}}{=} \frac{dx_0}{d\tau}(\tau)\,.$$

We verify that this transformation is invertible, with inverse:

$$\theta^{-1}\colon \left\{\begin{aligned} x' &= x - x_0(\tau_{x,t}) \\ \tau &= t - |x - x_0(\tau_{x,t})| \end{aligned}\right.$$

where $\tau_{x,t}$ is the solution of the equation $t - \tau_{x,t} = |x - x_0(\tau_{x,t})|$[258]. Hence

$$\begin{aligned} J(\theta)^{-1} &= \left(1 - \frac{v_0(\tau_{x,t}).(x - x_0(\tau_{x,t}))}{|x - x_0(\tau_{x,t})|}\right)^{-1} \\ &= \left(1 - \frac{v_0(\tau_{x,t}).(x - x_0(\tau_{x,t}))}{t - \tau_{x,t}}\right)^{-1} \end{aligned} \tag{8.162}$$

---

[258] Thus $(x_0(\tau_{x,t}), \tau_{x,t})$ are the coordinates of the trajectory of the particle considered with the past cone with apex (x, t), that is to say the set $\{(x', t') \in \mathbb{R}^4;\ t - t' = |x - x'|\}$.

and (8.161) may be written in terms of the new variable θ:

$$\langle \tilde{E}_3 * \rho, \varphi \rangle = \frac{e}{4\pi} \int_{\mathbb{R}^4} \frac{1}{|x - x_0(\tau_{x,t})|} \varphi(x, t) J(\theta)^{-1} \, dx \, dt ,$$

which shows that the retarded potential of the charge density $e\delta(x - x_0(t))$ is given, with (8.162), by

$$(8.163) \quad \begin{cases} u(x, t) = \tilde{E}_3 \underset{x,t}{*} \rho = \dfrac{e}{4\pi|x - x_0(\tau_{x,t})|} J(\theta)^{-1} \\ \quad = \dfrac{e}{4\pi} \times \dfrac{1}{|x - x_0(\tau_{x,t})| - v_0(\tau_{x,t}).(x - x_0(\tau_{x,t}))}{}^{259} . \end{cases} \quad \square$$

Review of Chapter II

As we indicated at the beginning, the primary objective of the chapter is to present the classical theory of *the harmonic operator*, of *harmonic functions* and of the *Newtonian potential*. This has been attained in §§1, 2 and 3.
We have applied that to the "classical" Dirichlet problem, i.e. the Dirichlet problem in a "good" open set in spaces of "regular enough" functions. This was carried out in §4, then followed, in this same section by the solution of "sharper" problems: unbounded or irregular open sets, other boundary conditions. . . . To proceed much further we have to introduce the notion of *capacity*; this is done in §5 (which may be omitted on a first reading). Various extensions of *regularity* can also be obtained; this is the object of §6 (which can similarly be omitted on a first reading). All of that is based (more or less directly) on the *Principle of the Maximum* (a systematic exposition of which will be found in Vol. 2, Chap. V, §5). Without using the methods of functional analysis (introduced in Vol. 2), other methods are indicated in §7 (certain of which are important for numerical applications). Finally, §8 gives various extensions of these ideas to more general elliptic operators of the second order. These two sections §§7 and 8 can be omitted initially by the reader. Methods based on the theory of holomorphic functions (specific to two dimensional problems) will be seen later in the context of numerical techniques.

[259] This retarded potential is called the Lienard–Wiechert potential.

Bibliography

Adams, R. A.

[1] *Sobolev Spaces.* Academic Press, New York 1975

Agmon, S., Douglis, A., Nirenberg, L.

[1] Estimates near the Boundary for Solutions of Elliptic Partial Differential Equations Satisfying General Boundary Conditions.
I. *Comm. Pure Appl. Math. 12* (1959) 623–727
II. *Comm. Pure Appl. Math. 17* (1964) 35–92

Amouyal, A., Benoist, P., Horowitz, J.

[1] Nouvelle methode de détermination du facteur d'utilisation thermique d'une cellule. *J. Nuclear Energy 6* (1957) 79–98

Arnold, V. I.

[1] *Mathematical methods in classical mechanics.* Springer-Verlag, Berlin 1978

Balescu, R.

[1] *Equilibrium and Nonequilibrium Statistical Mechanics.* Wiley-Interscience, New York 1975

Balian, R.

[1] *Du microscopique au macroscopique. Cours de physique statistique de l'École Polytechnique.* Tomes 1 & 2. École Polytechnique, Paris 1982

Bandle, C.

[1] *Isoperimetric Inequalities and Applications.* Pitman, London 1980

Bass, J.

[1] *Cours de mathématiques.* Tomes I & II. Masson, Paris 1978

Baur, A., Bourdet, L., Dejonghe, G., Gonnord, J., Monnier, A., Nimal, J. C., Vergnaud, T.

[1] *Programme de Monte-Carlo polycinétique à trois dimensions. Tripoli 02.* Tomes I–IV. Notes CEA (to appear)

Bell, G., Glasstone, S.

[1] *Nuclear Reactor theory,* Van Nostrand, New York 1970

Benoist, P.

[1] *Cours de Troisième Cycle. Théorie du transport.* Report CEA-R-4778, 1964

[2] Integral transport theory formalism for diffusion coefficient calculations in Wigner-Seitz cells. *Nucl. Sci. Engrg. 77* (1981) 1–12

Bernadou, M., Boisserie, J. M.

[1] *The finite element method in thin shell theory: Application to arch dam simulations.* Birkhauser, Boston 1982

Bogolubov, N. N., Logunov, A. A., Todorov, I. T.

[1] *Introduction to axiomatic quantum field theory.* Benjamin, Reading 1975

Bony, J. M.

[1] Principe du maximum, inégalité de Harnack et unicité des problèmes de Cauchy pour les opérateurs elliptiques dégénérés. *Ann. Inst. Fourier 19* (1969) 277–304

Bossavit, A.

[1] Application de quelques techniques de calcul opérationnel et de perturbations au problème de l'échauffement d'un câble en court-circuit. EDF *Bull. Dir. Etudes Recherches Ser.* C1 (1977), 59–68.

[2] Deux équations d'évolution non linéaires à constantes de temps très différentes: la méthode du changement de fréquence. EDF *Bull. Dir. Etudes Recherches Ser.* C2 (1979) 1–14

[3] Le problème des courants de Foucault. EDF *Bull. Dir. Etudes Recherches Ser.* C1 (1980) 5–14

[4] Définition et calcul d'une perméabilité équivalente pour l'acier saturé. EDF *Bull. Dir. Etudes Recherches. Ser.* C2 (1976) 45–58

Bouchard, J., Kavenoky, A., Reuss, P.

[1] *Le développement et la qualification du système Neptune de calcul des réacteurs à eau.* European Nuclear Conference, Hamburg, 6–11 May 1979

Boujot, J.

[1] *Analyse numérique des problèmes linéaires et non linéaires de la physique des plasmas.* Thesis, Orsay 1975

Bouligand, G., Giraud, G., Delens, F.

[1] *Le problème de la dérivée oblique dans la théorie du potentiel.* Hermann, Paris 1935

Bourbaki, N.

[1] *General topology.* 1974. Reprinted 1988, Springer-Verlag Berlin Heidelberg 1974

[2] *Intégration.* 1968. CCLS, Paris

Brard, R.

[1] *Cours de mathématiques appliquées de l'Ecole Polytechnique*

Brezis, H.

[1] Analyse fonctionnelle. Théorie et applications. Masson, Paris 1983 (English edition: Springer-Verlag, to appear).

Brillouin, L., Parodi, M.

[1] *Propagation des ondes dans les milieux périodiques.* Masson, Paris 1956

Brun, E. A., Martinot-Lagarde, A., Mathieu, J.

[1] *Mécanique des fluides.* Tomes 1 & 2, Dunod, Paris, 1968; Tome 3, Dunod, Paris 1970

Bussac, J., Reuss, P.

[1] *Traité de neutronique.* Hermann, Paris 1978

Butzer, P. L., Berens, H.

[1] *Semi-groups of operators and approximation theory.* Springer-Verlag, Berlin 1967

Cadilhac, M., Horowitz, J., Soule, J. L., Tretiakoff, O.

[1] Some mathematical and physical remarks on neutron thermalization in infinite homogeneous systems. *BNL 719*, 2 (1962) 439–463

Cadilhac, M., Soule, J. L., Tretiakoff, O.

[1] *Thermalisationn et spectres de neutrons.* Conference, Geneva 1964, P/73, 153–178

Cartan, H.

[1] *Elementary theory of functions of one or several complex variables.* Addison-Wesley, Reading 1963

Case, K. M., de Hoffmann, F., Placzek, G.

[1] *Introduction to the theory of neutron diffusion.* Los Alamos Scientific Laboratory 1953

Cessenat *et al.*

[1] *Méthodes probabilistes des équations de la physique.* Masson, Paris 1989

Chadwick, R. S., Cole, J. D.

[1] Modes and Waves in the Cochlea. *Mech. Res. Comm. 6* (1979) 177–184

Choquet, G.

[1] Theory of capacities. *Ann. Inst. Fourier 5* (1953–1954) 131–292

Choquet-Bruhat, Y.

[1] *Géométrie différentielle et systèmes extérieurs.* Dunod, Paris 1968

Choquet-Bruhat, Y., Dewitt-Morette, C., Dillard-Bleick, M.

[1] *Analysis, Manifolds and Physics.* North Holland, Amsterdam 1977

Ciarlet, P. G.

[1] *The Finite Element Method for Elliptic Problems.* North Holland, Amsterdam 1978

Cohen-Tannoudji, C., Diu, B., Laloe, F.

[1] *Mécanique quantique.* Tomes I & II. Hermann, Paris 1977

Colleter, P., Lederer, P.

[1] *Optimal Operation Feedbacks for the French Hydropower System.* CORS-TIMS-ORSA Joint National Meeting, Toronto, 3–6 May, 1981

Conley, C., Smoller, J.

[1] Remarks on the Steady-State Solutions of Reaction–Diffusion Equations. In: *Bifurcation Phenomena in Mathematical Physics and Related Topics.* Ed: Bardos C., Bessis D., pp. 47–56. Reidel, Dordrecht 1980

Conwell, E. M.
[1] *High Field Transport in Semiconductors*. Academic Press. New York 1967
Courant, R., Hilbert, D.
[1] *Methods of Mathematical Physics*.
Vol. I Wiley, New York 1953
Vol. II Wiley, New York 1962
Dautray, R.
[1] *Méthodes probabilistes pour les équations de la physique*. Collection CEA, Paris 1989
Davies, E. B., Lewis, J. T.
[1] An Operational Approach to Quantum Probability. *Comm. Math. Physics 17* (1970) 239–260
Davison, B., Sykes, J. B.
[1] *Neutron Transport Theory*. Oxford, University Press 1957
Delcroix, J. L.
[1] *Physique des Plasmas*. Tome I. Dunod, Paris, 1963. Tome II. Dunod, Paris 1966
Delsarte, J., Lions, J-L.
[1] Moyennes generalisées. *Comment. Math. Helv. 34* (1959) 59–69
Dieudonné, J.
[1] *Treatise on Analysis*.
Vol. I Academic Press, New York 1969
Vol. II Academic Press, New York 1970
Vol. III Academic Press, New York 1972
Vol. IV Academic Press, New York 1974
Vol. V Academic Press, New York 1977
Duderstadt, J. J., Martin, W. R.
[1] *Transport Theory*. Wiley-Interscience, New York 1979
Durand, E.
[1] *Magnétostatique*. Masson, Paris 1968
Duvaut, G., Lions, J-L.
[1] *Inequalities in Mechanics and Physics*. Springer, Berlin 1976
Felici, N.
[1] *Electrostatique. Etude du champ électrique et applications*. Gauthier-Villars, Paris 1962
Ferziger, J. H., Kaper, H. G.
[1] *Mathematical Theory of Transport Processes in Gases*. North Holland, Amsterdam 1972
Fichera, G.
[1] Sul problema della derivata obliqua e sul problema misto per l'équation di Laplace. *Boll. Un. Mat. Ital. 7* (1952) 367–377
Fife, P. C.
[1] *Mathematical Aspects of Reacting and Diffusion Systems*. Lecture Notes in Biomethematics, Vol. 28, Springer-Verlag, Berlin 1979
Foias, C., Temam, R.
[1] Structure of the Set of Stationary Solutions of the Navier–Stokes Equations. *Comm. Pure Appl. Math. 30* (1977) 149–164
Fournet, G.
[1] *Electromagnétisme à partir des équations locales*. Masson, Paris 1979
Frailong, J. M., Paklesa, J. G.
[1] Resolution of General Partial Differential Equations on a Fixed Size. SIMD/MIMD Large Cellular Processor. Proceedings of the IMAL'S International Congress, Sorrento, Italy (September 1979)
Freudenthal, A. M., Geiringer, J. G.
[1] The Mathematical Theories of the Inelastic Continuum. In: *Encyclopedia of Physics*, Ed. Flügge, S. Vol. VI, Elasticity and Plasticity. Springer-Verlag, Berlin 1958
Fung, Y. C.
[1] *Biomechanics. Mechanical Properties of Living Tissues*. Springer-Verlag, Berlin 1981
Garabedian, P. R., Schiffer, M.
[1] Convexity of Domain Functionals. *J. Anal. Math. 2* (1952–1953) 281–368

Garnir, H. G.
[1] *Les problèmes aux limites de la physique mathématique.* Birkhäuser, Basel 1958
Germain, P.
[1] *Mécanique des milieux continus.* Masson, Paris 1962
[2] *Cours de mécanique des milieux continus.* Masson, Paris 1973
[3] *Cours de mécanique de l'Ecole Polytechnique.* Edition polycopiée de l'Ecole 1979
[4] La méthode des puissances virtuelles en mécanique des milieux continus. *J. Mécanique 12* (1973) 235–274
Gilbarg, D., Trudinger, N. S.
[1] *Elliptic Partial Differential Equations of the Second Order.* Springer-Verlag, Berlin 1977
Glowinski, R., Periaux, J., Dinh, Y. V.
[1] *Domain Decomposition Methods for Nonlinear Problems in Fluid Mechanics.* Research Report INRIA, no. 187 (July, 1982)
Gottfried
[1] *Quantum Mechanics*, Vol. 1 *Fundamentals.* Benjamin, New York 1966
Grisvard, P.
[1] Behaviour of an Elliptic Boundary Value Problem in a Polygonal or Polyhedral Domain. In: *Numerical Solutions of Partial Differential Equations III.* Ed: Hubbard B., pp. 207–274. Academic Press, New York 1976
Gudder, S. P.
[1] A Survey of Axiomatic Quantum Mechanics. In: *The Logico-Algebraic Approach to Quantum Mechanics.* Vol. II. Contemporary Consolidation Ed: Hooker C. A., pp. 323–363. Reidel, Dordrecht 1979
[2] Four Approaches to Axiomatic Quantum Mechanics. In: *The Uncertainty Principle and Foundations of Quantum Mechanics. A Fifty Years' Survey*, Ed. Price W. G. and Chissick S. S., pp. 247–276. Wiley, New York 1977
Gunther, N.
[1] *Potential Theory and its Application to Basic Problems of Mathematical Physics.* Ungar, New York 1967
Gurtin, M. E., Sternberg, E.
[1] On the Linear Theory of Viscoelasticity. *Arch. Rational Mech. Anal. 11* (1962) 291–356
Haag, R., Kastler, D.
[1] An Algebraic Approach to Quantum Field Theory. *J. Math. Phys. 5* (1964) 848–861
Halmos, P. R.
[1] *Measure Theory.* Van Nostrand. New York 1950
Helms, L. L.
[1] *Introduction to Potential Theory.* Wiley, New York 1969
Hoffmann, A., Jeanpierre, F., Kavenoky, A., Livolant, M., Lorain, H.
[1] Appolo. Code multigroupé de résolution de l'équation du transport pour les neutrons thermiques et rapides. *Note CEA-N*-160, 1973
Holmes, M. H.
[1] Low Frequency Asymptotics for a Hydroelastic Model of the Cochlea. *SIAM J. Appl. Math. 38* (1980) 445–456
Horowitz, J., Tretiakoff, O.
[1] *Effective Cross-sections for Thermal Reactors.* EANDC (E) *14* (1960)
Itzykson, C., Zuber, J. B.
[1] *Quantum Field Theory.* McGraw-Hill, New York 1980
Jackson, J. D.
[1] *Classical Electrodynamics.* Wiley, New York 1975
Jauch, J. M.
[1] *Foundations of Quantum Mechanics.* Addison-Wesley, Reading 1968
Jeanpierre, J., Gibert, R. J., Hoffmann, A., Livolant, M.
[1] *Description of a General Method to Compute the Fluid-Structure Interaction.* 5th SMIRT. Berlin, August 1979, B4-1

John, F., Nirenberg, L.
[1] On Functions of Bounded Mean Oscillations. *Comm. Pure Appl. Math. 14* (1961) 415–426
Jones, D. S.
[1] *The Theory of Electromagnetism.* Pergamon, Oxford 1964
Joseph, D. D.
[1] *Stability of Fluid Motions.* Vol. I & II. Springer-Verlag, Berlin 1976
Jouguet, M.
[1] *Traité d'électricité théorique.* Tome 1. *Electrostatique.* Gauthier-Villars, Paris 1952
Kastler, D.
[1] Foundements de la mécanique statistique. In: *Algèbres d'opérateurs et leur applications en physique mathématique.* CNRS, No. 274, CNRS, Paris 1979
Kinderlehrer, D., Stampacchia, G.
[1] *Introduction to Variational Inequalities and their Applications.* Academic Press, New York 1980
Kato, T.
[1] *Perturbation Theory for Linear Operators.* Springer-Verlag, Berlin 1976
Kittel, C.
[1] *Introduction to Solid State Physics.* Wiley, New York 1976
Kleinman, R. E., Roach, G. F.
[1] Boundary Integral Equations for the Three-Dimensional Helmholtz Equation. *SIAM Rev. 16* (1974) 214–236
Kober, H.
[1] *Dictionary of Conformal Representation.* Dover, New York 1957
Kondratiev, V. A.
[1] Boundary Problems for Elliptic Equations in Domains with Conical or Angular Points. *Trudy Markov. Mat. Obšč. 16* (1967) 209–292
Krall, N. A., Trivelpiece, A. W.
[1] *Principles of Plasma Physics.* Mc-Graw Hill, New York 1973
Ladyzhenskaya, O. A., Uraltseva, N. N.
[1] Linear and Quasilinear Elliptic Equations. Academic Press, New York 1968
Landau, L., Lifschitz, E.
[1] *Mechanics*, Pergamon, Oxford 1960
[2] *Classical Theory of Fields*, Pergamon, Oxford 1959
[3] *Fluid Mechanics.* Pergamon, Oxford 1959
[4] *Theory of Elasticity*, Pergamon, Oxford 1959
[5] *Electrodynamics of Continuous Media.* Pergamon, Oxford 1960
[6] *Quantum Mechanics: Non-Relativistic Theory*, Pergamon, Oxford 1958
Lang, S.
[1] *Analysis II.* Addison-Wesley, Reading 1969
Lavrentiev, M., Shabat, B.
[1] Methods of the theory of functions of one complex variable
Lienard
[1] Le problème plan de la derivée oblique en théorie de potentiel. *J. Ecole Polytechnique*, 3rd series (5) pp. 35–158, pp. 177–226
Lions, J. L.
[1] *Problèmes aux limites dans les équations aux dérivées partielles.* Presses Univ. Montréal 1967
[2] Sur les problèmes aux limites du type dérivée oblique. *Ann. of Math. 64* (1956) 207–239
Lions, J. L., Magenes, E.
[1] *Problèmes aux limites non homogènes et applications.* Vol. I, Dunod, Paris 1968
Lions, P. L.
[1] Interprétation stochastique de la méthode alternée de Schwarz. *C.R. Acad. Sci. Paris Sér. A. 286* (1978) 235–328
Mackey, G. W.
[1] *The Mathematical Foundations of Quantum Mechanics*, Benjamin, Reading 1963

Marcuse, D.
[1] *Principles of Quantum Electronics*. Academic Press, New York 1980
Marle, C. M.
[1] *Mesures et probabilités*. Hermann, Paris 1974
Martin, P. A.
[1] *Modèles en mécanique statistique des processus irréversibles*. Lecture Notes in Physics. Vol. 103. Springer-Verlag, Berlin 1979
Messiah, A.
[1] *Quantum Mechanics* (2 vols.) North Holland Pub. Co., Amsterdam, 1970
Meyer, Y.
[1] *Théorie du potentiel dans les domaines lipchitziens d'après G. C. Verchota*. Séminaire Goulaouic-Meyer-Schwartz, 1982–1983. Equations aux dérivées partielles. Expose n^{0}V. Ecole Polytechnique, Paris
Miellou, J. C.
[1] *Variantes synchrones et asynchrones de la méthode alternée de Schwartz*. Rapport de recherche ERA 077 0654 (Besançon, 1982)
Miller, W.
[1] *Symmetry Groups and their Applications*. Academic Press, New York 1972
Mock, M. S.
[1] On Equations Describing Steady-State Carrier Distributions in a Semiconductor Device. *Comm. Pure Appl. Math.* *25* (1972) 781–792
[2] An Initial Value Problem for Semiconductor Device Theory. *SIAM J. Math. Anal.* *5* (1974) 597–612
Morrey, C. B.
[1] *Multiple Integrals in the Calculus of Variations*. Springer-Verlag, Berlin 1966
Mow, V. C., Lai, W. M.
[1] Recent Developments in Synovial Joint Biomechanics. *SIAM Rev.* *22* (1980) 275–317
Müller, C.
[1] *Foundations of the Mathematical Theory of Electromagnetic Waves*. Springer-Verlag, Berlin 1969
Muller, H. G.
[1] *An Introduction to Food Rheology*. Heinemann, London 1973
Nag, B. R.
[1] *Electron Transport in Compound Semiconductor*. Springer-Verlag, Berlin 1980
Nečas, J.
[1] *Les méthodes directes en théorie des équations elliptiques*. Masson, Paris 1967
Nehari, Z.
[1] *Conformal Mapping*. McGraw-Hill, New York 1952
Normand, J. M.
[1] *A Lie Group: Rotations in Quantum Mechanics*. North-Holland, Amsterdam 1980
Nussenzveig, H. M.
[1] *Causality and Dispersion Relations*. Academic Press, New York 1972
Petiau, G.
[1] *Applications des théories mathématiques. II. La théorie des fonctions de Bessel exposée en vue de ses applications à la physique mathématique*. Éditions du CNRS, 1955
Price, W. G., Chissick, S. S.
[1] *The Uncertainty Principle and Foundations of Quantum Mechanics. A Fifty Year's Survey*, Wiley, New York 1977
Prugovecki, E.
[1] *Quantum Mechanics in Hilbert Space*. Academic Press, New York 1971
Richtmyer, R. D.
[1] *Principles of Advanced Mathematical Physics*. Vol. I. Springer-Verlag, Berlin 1978
Robin, L.
[1] *Fonctions sphériques de Legendre et fonctions sphéroïdales*. Tome I. Gauthier-Villars, Paris 1957

Rocard, Y.
[1] *Électricité. Masson*, Paris 1956
Rodean, H. C.
[1] *Nuclear Explosion Seismology*. U.S. Atomic Energy Commission, 1971
Roubine, E.
[1] *Compléments d'électromagnétisme*. École supérieure d'électricité, 1980
Roubine, E., Bolomey, J. C.
[1] *Antenne*. Tome I. Introduction générale. Masson, Paris 1978
Roubine, E., Drabowitch, S., Ancona, C.
[1] *Antenne*. Tome II. Applications. Masson, Paris 1978
Rudin, W.
[1] *Real and Complex Analysis*. McGraw-Hill, New York 1966
Ruelle, D.
[1] *Statistical Mechanics. Rigorous Results*. Benjamin, New York 1969
Sadosky, C.
[1] *Interpolation of Operators and Singular Integral*. Dekker, New York 1979
Schiff, L. I.
[1] *Quantum Mechanics*. McGraw-Hill, New York 1968
Schulenberger, J. R.
[1] The Debye Potential: a Scalar Factorization for Maxwell's Equations, *J. Math. Anal. Appl. 63* (1978) 502–520
Schwartz, L.
[1] *Théorie des distributions à valeurs vectorielles.*
I. Ann. Inst. Fourier *7* (1957) 1–141
II. Ann. Inst. Fourier *8* (1958) 1–209
[2] *Théorie des distributions*. Hermann, Paris 1967
[3] *Cours d'analyse de l'école Polytechnique*. 2 volumes. Hermann, Paris 1971
[4] *Cours de l'école Polytechnique*. École Polytechnique, 1981
Segal, I. E.
[1] Postulates for General Quantum Mechanics. *Ann. of Math. 48* (1947) 930–948
Slater, J. C.
[1] *Quantum Theory of Atomic Structure*. Vol. I and II. McGraw-Hill, New York 1960
Souriau, J. M.
[1] *Structures des systèmes dynamiques*. Dunod, Paris 1970
Stampacchia, G.
[1] *Équations elliptiques du second ordre à coefficients discontinus*. Presses Univ. Montréal 1966
Strieder, W., Aris, R.
[1] *Variational Methods Applied to Problems of Diffusion and Reaction*. Springer-Verlag, Berlin 1973
Suquet, P.
[1] *Plasticité et homogénéisation*. Thèse de Doctorat d'état. Université Pierre et Marie Curie, 1982
Thirring, W.
[1] *A Course in Mathematical Physics*. Vol. 2, *Classical Field Theory*. Springer-Verlag, Berlin 1979
Treves, F.
[1] *Basic Linear Partial Differential Equations*. Academic Press, New York 1975
Trudinger, N. S.
[1] On Harnack Type Inequalities and their Application to Quasilinear Elliptic Equations. *Comm. Pure Appl. Math. 20* (1967) 721–747
Tye, R. P.
[1] *Thermal Conductivity*. Vol. I and II. Academic Press, New York 1969
Van der Waerden, B. L.
[1] *Group Theory and Quantum Mechanics*. Springer-Verlag, Berlin 1974
van Kampen, N. G., Felderhof, B. U.
[1] *Theoretical Methods in Plasma Physics*. North-Holland, Amsterdam 1967

van Roosbroeck, W.
[1] Theory of the Flow of Electrons and Holes in Germanium and other Semiconductors. *Bell System Techn. J. 29* (1950) 560–607
Vapaille, A.
[1] *Physique des dispositifs à semiconducteurs.* Masson, Paris 1970
Vilenkin, N. J.
[1] *Special Functions and the Theory of Group Representations.* A.M.S., Providence R. I. 1968
Weinberg, A. M., Wigner, E. P.
[1] *The Physical Theory of Neutron Chain Reactors.* Univ. Chicago Press 1958
Weinberger, H. F.
[1] *First Course in Partial Differential Equations with Complex Variables and Transform Methods.* Wiley, New York 1965
Wiener, N.
[1] The Dirichlet Problem. *J. Math. Phys. 3* (1924) 127–146
von Westenholz, C.
[1] *Differential Forms in Mathematical Physics.* North-Holland, Amsterdam 1978
Wilson, H. L., Scott, W. H., Pomraning, G. C.
[1] Neutron transport in Moving Media. *SIAM J. Appl. Math. 36* (1979) 230–262
Yariv, A.
[1] *Quantum Electronics.* Wiley, New York 1975
Yvon, J.
[1] *Oeuvre scientifique*
Zernike, F., Midwinter, J. E.
[1] *Applied Nonlinear Optics.* Wiley, New York 1973

Table of Notations

The Theorems, Propositions, Lemmas, Definitions, Remarks, Examples and Formula are numbered by Sections. Figures and footnotes are numbered by Chapters.

Generalities

$\overset{\text{def}}{=}$	equality by definition
$\rightarrow$ $\longmapsto$	arrows of mapping (see A below)
$\Rightarrow$	implication sign
δ_{ij}, δ^i_j	Kronecker delta (0 if $i \neq j$, 1 if $i = j$, i and $j \in \mathbb{N}$)
$\square$	denotes the end of a passage forming a logical sequence (remark, proof, . . .) if necessary for the clarity of the text
$\forall$	for all
$\exists$	there exist
iff	if and only if
i.e.	that is
$\varnothing$	the empty set

A. Notations Relating to Sets and Mappings

Let E, F, G be three sets and let A be subset of E. We denote by:

$\{x \in E; P\}$	the subset of E consisting of the elements possessing the property P
$E \times F$	the cartesian property of E and F
E^n	the n-th power of E (n a positive integer)
$E \setminus A$	the complement of A in E
$f\colon \left.\begin{matrix} E \rightarrow F \\ x \mapsto f(x) \end{matrix}\right\}$	the mapping of E into F which is such that to the element $x \in E$ there corresponds $f(x) \in F$[1]

or: $x \in E \rightarrow f(x) \in F$

[1] We often write fx instead of $f(x)$ when f is a linear operator.

I_E or I	the identity mapping in E
$f \circ g$	the composite mapping of $f: F \to G$ and $g: E \to F$[2] i.e $x \in E \to (f \circ g)(x) \equiv f(g(x)) \in G$
f^n	the composite mapping of order n of $f: E \to E$ (n a positive integer): $f^n = f \circ f \circ \ldots f$ (n factors)
$\mathrm{Im} f = f(E)$	the image of $f: E \to F$ in F[3]
f^{-1}	the inverse of f if f is injective[4]; $f^{-1}: f(E) \to E \Rightarrow f^{-1}(y) = x \in E$ iff $f(x) = y \in f(E)$; thus $f^{-1} \circ f = I_E$ and $f \circ f^{-1} = I_{f(E)}$[5]
$f\|_A$	the restriction of $f: E \to F$ to the subset A of E
$f(\cdot, t)$	the function defined by $x \mapsto f(x, t)$ for fixed t
$\{a_k\}$ or (a_k)	the sequence $a_1, a_2, \ldots, a_k, \ldots$
(a_{ij})	the matrix whose ij-th element is a_{ij}

B. Notations Relating to Topology

Ω	usually denotes an open set in a topological space
$B(x, \alpha)$	the open ball with centre x and radius $\alpha > 0$
$F \subsetneq E$	the set F is contained in E with continuous injection
$F \subset\subset E$	the set F is contained in E with compact injection

Let A be a proper subset of a topological space E. We denote by:

$\bar{A}$	the closure of A
$\mathring{A}$	the interior of A
∂A or Γ	the boundary of A
$\sup A$	the supremum or upper bound of A in $\mathbb{R}$
$\inf A$	the infimum or lower bound of A in $\mathbb{R}$
$\mathrm{diam}\ A$	the diameter of $A \subset \mathbb{R}^n$
$d(A, B)$	the distance between $A, B \subset \mathbb{R}^n$
$\sup\limits_{x \in B} f(x)$ (resp. $\inf\limits_{x \in B} f(x)$)	for $\sup A$ (resp. $\inf A$) where f is a mapping of set B into $\mathbb{R}$, and where $A = f(B) \subset \mathbb{R}$[6]
$\underline{\lim}\, u_n$ (or $\lim\inf u_n$)	the lower limit $\lim\inf u_n = \sup\limits_{n \in \mathbb{N}} (\inf\limits_{m \geqslant n} u_m)$

---

[2] We often write fg instead of $f \circ g$ when f and g are linear operators.

[3] The mapping f is said to be *surjective*, or a *surjection*, if $f(E) = F$ (i.e. $\forall y \in F$, $\exists x \in E$ such that $f(x) = y$.

[4] The mapping f is said to be *injective*, or *an injection* if $f(x) = f(x') \Rightarrow x = x'$.

[5] If the mapping f is both surjective and injective it is called *bijective* or a *bijection*.

[6] We also use the notation $\max\limits_{x \in B} f(x)$ to indicate that $\sup f(x)$ is attained at a point of B.

$\overline{\lim}\, u_n$ (or $\limsup u_n$)	the upper limit $\limsup u_n = \inf_{n \in \mathbb{N}} (\sup_{m \geqslant n} u_m)$
$\limsup_{\lvert x\rvert \to \infty} f(x)$	the upper limit of the real function f when $\lvert x\rvert \to \infty$ $\limsup_{\lvert x\rvert \to \infty} f(x) = \inf_{r \geqslant 0} \sup_{\lvert x\rvert > r} f(x)$
$\liminf_{\lvert x\rvert \to \infty} f(x)$	the lower limit of the real function f when $\lvert x\rvert \to \infty$ $\liminf_{\lvert x\rvert \to \infty} f(x) = \sup_{r \geqslant 0} \inf_{\lvert x\rvert > r} f(x)$
$\lim_{x \to +0} f(x)$	the limit of $f: \mathbb{R} \to E$ when x tends to zero through
$\lim_{x \to -0} f(x))$	positive (resp. negative) values
$f(x^+)$ (resp. $f(x^{-1})$)	the limit to the right (resp. left) of f at $x \in \mathbb{R}$
$O(x)$	a function satisfying $\lvert O(x)/x\rvert \leqslant k$, a positive constant.
$o(x)$	a function satisfying $\lvert o(x)/x\rvert \to 0$ as $x \to 0$
$f \sim g$ for $x \to 0$	functions f and g are real and such that $f(x) = g(x)\{1 + \varepsilon(x)\}$ where $\varepsilon(x) \to 0$ as $x \to 0$

C. Notations Relating to Numbers

C_1 Sub-sets of $\mathbb{R}^n$ and $\mathbb{C}^n$

$\mathbb{N}$, $\mathbb{Z}$, $\mathbb{R}$ and $\mathbb{C}$ denote the sets of natural numbers, (positive and negative) integers, real numbers and complex numbers.

A^*	$A^* \overset{\text{def}}{=} A \backslash \{0\}$ with A a subset of $\mathbb{R}^n$ or $\mathbb{C}^n$ containing 0
$[a, b] =$	$\{x \in \mathbb{R}; a \leqslant x \leqslant b\}$
$]a, b[=$	$\{x \in \mathbb{R}; a < x < b\}$
$]a, b] =$	$\{x \in \mathbb{R}; a < x \leqslant b\}$
$[a, b[=$	$\{x \in \mathbb{R}; a \leqslant x < b\}$
$\mathbb{R}_+, \mathbb{R}_-, \mathbb{R}^+$	$\mathbb{R}_+ \overset{\text{def}}{=}]0, +\infty[$, $\mathbb{R}_- \overset{\text{def}}{=}]-\infty, 0[$, $\mathbb{R}^+ \overset{\text{def}}{=} [0, +\infty[$
$\bar{\mathbb{R}}$	$\bar{\mathbb{R}} \overset{\text{def}}{=} \mathbb{R} \cup \{-\infty, +\infty\}$
$\mathbb{R}^n_x$	the n-th power of $\mathbb{R}$, when the generic element is denoted by x
$\mathbb{R}^n_+$ (resp $\bar{\mathbb{R}}^n_+$)	the half-space $\{x = (x_1, \ldots, x_n) \in \mathbb{R}^n; x_n > 0$ (resp. $x_n \geqslant 0)\}$
T or S^1	the one-dimensional torus or the unit circle in $\mathbb{R}^2$
S^{n-1}	the unit sphere in $\mathbb{R}^n$

C_2 Numbers

Let $n \in \mathbb{N}^*$, $\alpha = (\alpha_1, \ldots, \alpha_n) \in \mathbb{N}^n$, $x = (x_1, \ldots, x_n) \in \mathbb{R}^n$, $z \in \mathbb{C}$

$\lvert\alpha\rvert$	the modulus of α, $\lvert\alpha\rvert \overset{\text{def}}{=} \alpha_1 + \ldots + \alpha_n$
x^α	$x^\alpha = x_1^{\alpha_1} \ldots x_n^{\alpha_n}$
$\mathbf{x} \times \mathbf{y}$	vector product of $\mathbf{x}$ and $\mathbf{y}$ belonging to $\mathbb{R}^3$
$x . y$	scalar product of x, y with x and y belonging to $\mathbb{R}^n$
$\lvert x\rvert$	the euclidean norm of x: $\lvert x\rvert \overset{\text{def}}{=} \sqrt{(x.x)}$ (sometimes denoted by r in the text)
$\operatorname{Re}(z)$ or $\operatorname{Re} z$	the real part of z
$\operatorname{Im}(z)$ or $\operatorname{Im} z$	the imaginary part of z
$\arg(z)$ or $\arg z$	the argument of z
$\bar{z}$	the complex conjugate of z
$\lvert z\rvert$	the modulus of z
σ_n or Ω_n	the surface area of the unit sphere in $\mathbb{R}^n$
k_n	defined by $k_2 = 2\pi$, and $k_n = (2 - n)\sigma_n$ for $n > 2$
$[m]$	the integral part of $m \in \mathbb{R}$

D. Notations Relating to Functions and Distributions

D_1 Principal notations

Y (or $Y(x)$)	the Heaviside function on $\mathbb{R}_x$
δ (or $\delta(x)$)	the Dirac distribution, sometimes denoted by δ_0 or $\delta_0(x)$
δ_a (or $\delta_a(x)$)	the Dirac distribution concentrated at $a \in \mathbb{R}^n$

$$\delta_a(x) \overset{\text{def}}{=} \delta(x - a)$$

vp	Cauchy principal value (of an integral)
Pf	finite part (of an integral)

Let f and g be two (possibly vector) functions or distributions on $\Omega \subset \mathbb{R}^n$; we denote by:

$\mathscr{F} f$ or $\hat{f}$ — the Fourier transform of f; if $f\colon \mathbb{R}^n \to \mathbb{C}$ is integrable on $\mathbb{R}^n$, we have

$$\hat{f}(y) \overset{\text{def}}{=} \int_{\mathbb{R}^n} e^{-ixy} f(x)\, dx$$

$\bar{\mathscr{F}} f$ — the Fourier co-transform of f:

$$\bar{\mathscr{F}} f(x) \overset{\text{def}}{=} \int_{\mathbb{R}^n} e^{ixy} f(y)\, dy$$

if f is integrable on $\mathbb{R}^n$.

$\mathscr{F}^{-1}f$	the inverse Fourier transform of f: $$\mathscr{F}^{-1}f(x) \overset{\text{def}}{=} (2\pi)^{-n}\int_{\mathbb{R}^n} e^{ixy}f(y)\,dy$$ i.e. $\mathscr{F}^{-1} = (2\pi)^{-n}\bar{\bar{\mathscr{F}}}$.
$f * g$	the convolution product of f and g. If f and g are continuous with compact support in $\mathbb{R}^n$, we have $$(f * g)(x) \overset{\text{def}}{=} \int_{\mathbb{R}^n} f(x-y)g(y)\,dy$$
$f \otimes g$	the tensor product of f and g. If f and g are two real functions, we have $$(f \otimes g)(x, y) \overset{\text{def}}{=} f(x)g(y)$$
$\operatorname{supp} f$	the support of f
$\operatorname{sing\,supp} f$	the singular support of f
$\operatorname{sing\,supp}_a f$	the singular analytic support of f
$f \geqslant 0$	positive function or distribution
$f \gg 0$	function or distribution of positive type

D_2 Differential calculus and linear differential operators

Let f be a (possibly vector) function on $\mathbb{R}$ or $\Omega \subset \mathbb{R}^n$ with values in a Banach space X. We denote by:

$f'(a)$	the derivative of f: $\mathbb{R} \to X$ at the point $a \in \mathbb{R}$ [7]
$f^{(n)}(a)$	the derivative of order n of f at $a \in \mathbb{R}$. Notice that $$f^{(0)}(a) = f(a),\ f^{(1)}(a) = f'(a),\ f^{(2)}(a) = f''(a)$$
$Df(a)$	The Fréchet derivative of f: $\mathbb{R}^n \to X$ at the point $a \in \mathbb{R}^n$: $$f(a+h) = f(a) + Df(a).h + \lvert h\rvert o(h)$$
$D^k f(a)$	The Fréchet derivative of order k of f: $\mathbb{R}^n \to X$ at the point $a \in \mathbb{R}^n$

[7] In mechanics, we also use $\dot{f}$ for $\dfrac{\partial f}{\partial t}$ and $f_{,i}$ for $\dfrac{\partial f}{\partial x_i}$.

$D_k f$ or $\dfrac{\partial f}{\partial x_k}$ (or $\partial_k f$)	the partial derivative of $f: \mathbb{R}^n \to X$ with respect to the variable x_k [7, 8]
$D^\alpha f$ or $\dfrac{\partial^{\lvert\alpha\rvert} f}{\partial x^\alpha}$ (or $\partial_\alpha f$)	the partial derivative of order $\lvert\alpha\rvert$ of $f: \mathbb{R}^n \to X$ $$D^\alpha f \overset{\text{def}}{=} (D_1)^{\alpha_1}.(D_2)^{\alpha_2} \dots (D_n)^{\alpha_n} f,$$ where $\alpha = (\alpha_1, \alpha_2, \dots, \alpha_n)$
$\dfrac{\partial f}{\partial z}$ and $\dfrac{\partial f}{\partial \bar{z}}$	f being a function of x and y (both real) $$\frac{\partial f}{\partial z} \overset{\text{def}}{=} \frac{1}{2}\left(\frac{\partial f}{\partial x} - i\frac{\partial f}{\partial y}\right); \quad \frac{\partial f}{\partial \bar{z}} \overset{\text{def}}{=} \frac{1}{2}\left(\frac{\partial f}{\partial x} + i\frac{\partial f}{\partial y}\right)$$
grad f or ∇f (or Df)	the gradient of $f: \Omega \to \mathbb{R}$ or X: $\nabla f \overset{\text{def}}{=} \left(\dfrac{\partial f}{\partial x_1}, \dots, \dfrac{\partial f}{\partial x_n}\right)$
div f or $\nabla . f$	the divergence of $f: \Omega \to \mathbb{R}^n$ or $\mathbb{C}^n$, $$\nabla . f = \frac{\partial f}{\partial x_1} + \dots + \frac{\partial f}{\partial x_n}$$
curl $\mathbf{v}$ or $\nabla \times \mathbf{v}$ (or rot $\mathbf{v}$)	the curl of $\mathbf{v}: \mathbb{R}^3 \to \mathbb{R}^3$: $$\operatorname{curl} \mathbf{v} \overset{\text{def}}{=} \left(\frac{\partial v_3}{\partial x_2} - \frac{\partial v_2}{\partial x_3}, \frac{\partial v_1}{\partial x_3} - \frac{\partial v_3}{\partial x_1}, \frac{\partial v_2}{\partial x_1} - \frac{\partial v_1}{\partial x_2}\right)$$ More generally, if $f: \Omega \to \mathbb{R}^n$ $$\nabla \times f = \left(\frac{\partial f_i}{\partial x_j} - \frac{\partial f_j}{\partial x_i}\right)_{i,j=1,\dots,n}$$
Δf or $\Delta_n f$	the Laplacian of $f: \Omega \to \mathbb{R}^n$: $$\Delta f = \frac{\partial^2 f}{\partial x_1^2} + \dots + \frac{\partial^2 f}{\partial x_n^2}$$
$\Box f$ or $\Box_n f$	the d'Alembertian of $f: \Omega \to \mathbb{R}$: $\Box f = \dfrac{\partial^2 f}{\partial t^2} - \Delta f$, with here $\Omega \subset \mathbb{R}^n_x \times \mathbb{R}_t$
J_f	the Jacobian of $f: \mathbb{R}^n \to \mathbb{R}^n$: $J_f(x) \overset{\text{def}}{=} \lvert\det(\partial_j f_i(x))\rvert$

Let ν be the external normal to $\partial\Omega$ with j-th direction cosine $\cos(\nu, x_j)$[9]; we

[8] In the case in which f is a function of the space-variable $x \in \mathbb{R}^n$ and of the time-variable $t \in \mathbb{R}$, with values in X, we also denote the partial derivative of f with respect to t by $\dfrac{\partial f}{\partial t}$; we consider then f to be a function of time $t \mapsto f(\cdot, t)$ with values in a space of functions (or of distributions) of $x \in \mathbb{R}^n$ into X.

[9] We also often use the notation n to denote the normal to $\partial\Omega$.

denote by:

$f\|_{\partial\Omega}$ or $f\|_{\Gamma}$ or $\gamma_0 f$	the trace of order zero of: $\Omega \to \mathbb{R}$ (or $\mathbb{C}$ or X)
$\frac{\partial f}{\partial \nu}$ or $\partial_\nu f$ (or $\gamma_1 f$)	the trace of order 1 of $f:\Omega \to \mathbb{R}$ (i.e. the partial derivative with respect to ν)[10]
$\gamma_n f$	the trace of order n of $f:\Omega \to \mathbb{R}$, $\gamma_n f = \left.\frac{\partial^n f}{\partial \nu^n}\right\|_{\Omega}$
$\frac{\partial f}{\partial \nu_A}$	with $f:\mathbb{R}^n_x \times \mathbb{R}_t \to \mathbb{R}$ and A an operator: $A = \sum_{i,j=1}^{n} \partial_i(a_{ij}(x,t)\partial_j) + \sum a_i(x,t)\partial_i + a_0(x,t)$ we have $\frac{\partial f}{\partial \nu_A} \overset{\text{def}}{=} \sum_{i,j=1}^{n} a_{ij}(x,t)\partial_j f.\cos(\nu, x_i)$
D^α	the differential operator $D_j = \frac{\partial}{\partial x_j}$ and $D^\alpha = D_1^{\alpha_1}\dots D_n^{\alpha_n}$ with $\alpha = (\alpha_1, \dots, \alpha_n)$
$P(D)$	the differential operator $P(D) = \sum_{\alpha\in\mathbb{N}^n} a_\alpha D^\alpha$ where (a_α) is a locally finite family
$p(y)$	the characteristic polynomial associated with $P(D)$: $p(y) = \sum_{\alpha\in\mathbb{N}^n} a_\alpha (iy)^\alpha = P(iy)$[11]
$P^*(D)$	the formal adjoint of $P(D)$
$P^\bullet(D)$	the principal part of $P(D)$

E. Notations Relating to Spaces of Continuous Functions and of Distributions

E_1 Spaces of continuous functions

(*a*) *Functions with real or complex values.*
Suppose that Ω is an open set of $\mathbb{R}^n$, K a compact set of $\mathbb{R}^n$, $p_{\alpha,\mathscr{A}}$ the seminorm[12]

---

[10] We also use $\frac{\partial f}{\partial \nu^+}$ and $\frac{\partial f}{\partial \nu^-}$ where ν^+ and ν^- are respectively the external and internal normals to $\partial\Omega$.

[11] We also find in publications the alternative notations:

$D = \frac{1}{i}\frac{\partial}{\partial x}$; in this case the characteristic polynomial is $p(y) = P(y)$;

$D = i\frac{\partial}{\partial x}$; in this case the characteristic polynomial is $p(y) = P(-y)$.

[12] See §1 of Chap. VI.

defined by $f \mapsto \sup |D^\alpha f(x)|$, ($\alpha \in \mathbb{N}^n$ and $\mathscr{A} \subset \mathbb{R}^n$) and $k \in \mathbb{N}$ we denote generally by:

$\mathscr{C}_b(\mathbb{R}^n)$ or $\mathscr{B}(\mathbb{R}^n)$ — the space of bounded continuous functions on $\mathbb{R}^n$. It is a Banach space for the norm $p_{0,\mathbb{R}^n}$

$\mathscr{C}_0(\mathbb{R}^n)$ or $\mathscr{B}_0(\mathbb{R}^n)$ — the space of continuous functions on $\mathbb{R}^n$ tending to zero at infinity. It is a Banach space for the norm $p_{0,\mathbb{R}^n}$

$\mathscr{B}^k(\mathbb{R}^n)$ — the space of functions of class $\mathscr{C}^k$ on $\mathbb{R}^n$ which, together with all derivatives of orders $\leqslant k$, tend to zero at infinity. It is a Banach space for the norm

$$f \mapsto \sup_{|\alpha| \leqslant k} (p_{\alpha,\mathbb{R}^n}(f)).$$

$\mathscr{C}(K)$ or $\mathscr{C}^0(K)$ — the space of functions continuous on K. It is a Banach space for the norm $p_{0,K}$

$\mathscr{C}^k(\Omega)$ — the space of functions of class $\mathscr{C}^k$ on Ω, provided with the seminorms

$$\{p_{\alpha,K}, |\alpha| \leqslant k, K \subset \Omega\}$$

$\mathscr{C}_0^k(\Omega)$ — the space of functions of class $\mathscr{C}^k(\Omega)$ with compact support contained in Ω

$\mathscr{C}^{0,\alpha}(\bar{\Omega}), 0 < \alpha \leqslant 1$ — the space of Hölder functions of order α on $\bar{\Omega}$, i.e. the space of functions f continuous on Ω such that

$$\sup_{x,y\in\Omega} \frac{|f(x) - f(y)|}{|x - y|^\alpha} < +\infty.$$

$\mathscr{C}^{k,\alpha}(\bar{\Omega})$ — the space of functions $f \in \mathscr{C}^k(\Omega)$ such that

$$D^j f \in \mathscr{C}^{0,\alpha}(\bar{\Omega}), \quad \forall j, |j| \leqslant k.$$

$\mathscr{C}^\infty(\Omega)$ or $\mathscr{E}(\Omega)$ — the space of the functions $\mathscr{C}^\infty$ on Ω, provided with the semi-norms

$$\{p_{\alpha K}, \alpha \in \mathbb{N}^n, K \subset \Omega\}$$

$\mathscr{S}(\mathbb{R}^n)$ — the space of rapidly decreasing functions of $\mathscr{E}(\mathbb{R}^n)$ such that all the derivatives satisfy: $|x|^k |D^\alpha f(x)| \to 0$ as $|x| \to \infty$, $\forall k \in \mathbb{N}$, $\forall \alpha \in \mathbb{N}^n$; this space can be provided with the semi-norms

$$\{f \mapsto \sup_{x\in\mathbb{R}^n} (|x|^k D^\alpha f(x)), k \in \mathbb{N}, \alpha \in \mathbb{N}^n\}$$

$\mathscr{O}_M(\mathbb{R}^n)$ — the space of slowly increasing functions of $\mathscr{E}(\mathbb{R}^n)$, together with all their derivatives (space of multiplicators of $\mathscr{S}$)

$\mathscr{D}(\Omega)$ — the space of those functions of $\mathscr{E}(\Omega)$ which have compact support in Ω. The sequence (f_k) of $\mathscr{D}(\Omega)$

tends to zero in $\mathscr{D}(\Omega)$ if $\bigcup_{k \in \mathbb{N}} \operatorname{supp} f_k \subset K$ is compact in Ω and

$$p_{\alpha, K}(f_k) \to 0, \quad \forall \alpha \in \mathbb{N}^n$$

$\mathscr{D}([a, b])$ — the space of functions on $[a, b]$ which can be extended to functions of $\mathscr{D}(\mathbb{R})$. The sequence (f_k) of $\mathscr{D}([a, b])$ tends to zero in $\mathscr{D}[a, b]$ if $p_{\alpha, [a,b]}(f_k) \to 0$, $\forall \alpha \in \mathbb{N}^n$ when $k \to \infty$

$\mathscr{D}_K(\Omega)$ — the space of the functions of $\mathscr{D}(\Omega)$ with support in K, provided with the semi-norms $\{p_{\alpha, K}, \alpha \in \mathbb{N}\}$ [13]

If $\Omega = \mathbb{R}$ or $\mathbb{R}^n = \mathbb{R}$ we denote by $\mathscr{B}, \mathscr{B}_0, \ldots, \mathscr{D}, \mathscr{D}_K$ the sets $\mathscr{B}(\mathbb{R}), \mathscr{B}_0(\mathbb{R}), \ldots, \mathscr{D}(\mathbb{R}), \mathscr{D}_K(\mathbb{R})$

(*b*) *Functions with values in a Banach space X, with norm* $\|\,.\,\|$.

The preceding definitions can be generalized to functions with values in X. The semi-norm $p_{\alpha,\mathscr{A}}$ then becomes

$$f \mapsto \sup_{x \in \mathscr{A}} \| D^\alpha f(x) \|.$$

We denote the corresponding spaces by

$$\mathscr{B}(\mathbb{R}^n, X),\ \mathscr{B}_0(\mathbb{R}^n, X),\ \mathscr{B}^k(\mathbb{R}^n, X),\ \mathscr{C}(K, X) \text{ or } \mathscr{C}^0(K, X) \text{ etc.} \ldots$$

In the same way if $\Omega = \mathbb{R}$ or $\mathbb{R}^n = \mathbb{R}$, we denote

$$\mathscr{B}(\mathbb{R}, X),\ \mathscr{B}_0(\mathbb{R}, X) \text{ etc.} \ldots \text{ by } \mathscr{B}(X),\ \mathscr{B}_0(X), \text{ etc.} \ldots$$

E_2 Spaces of integrable functions

Let Ω be an open set in $\mathbb{R}^n$, and s a real number greater than or equal to 1.

(*a*) *Functions with real or complex values*

$L^s(\Omega)$ — the space of classes of measurable functions on Ω such that $x \mapsto |f(x)|^s$ is integrable on Ω. This is a Banach space for the norm

$$f \mapsto \left[\int_\Omega |f(x)|^s \mathrm{d}x \right]^{1/s}$$

$L^\infty(\Omega)$ — the space of classes of measurable functions on Ω such that $x \mapsto |f(x)|$ is essentially bounded. This is a Banach space for the norm

$$f \mapsto \operatorname*{sup\,ess}_{x \in \Omega} f(x)$$

$L^s_{\mathrm{loc}}(\Omega)$ — the space of classes of measurable functions on Ω such that $x \mapsto |f(x)|^s$ is locally integrable

[13] We also use the notation $\mathscr{D}(\bar{\Omega})$ to denote the set of the restrictions to Ω of the functions of $\mathscr{D}(\mathbb{R}^n)$.

$L^s_\rho(\Omega)$ — the space with weight $\rho:\Omega \to \mathbb{R}^+$, locally integrable, of the classes of measurable functions such that $x \mapsto \rho(x)|f(x)|^s$ is integrable on Ω. It is a Banach space for the norm

$$f \mapsto \left[\int_\Omega |f(x)|^s \rho(x)\,\mathrm{d}x\right]^{1/s}$$

$L^2_\mu(\mathbb{R})$ — the space of the classes of square integrable functions for the measure μ

If $\Omega =]\,a, b\,[$, these spaces are denoted by $L^s(a, b)$, $L^\infty(a, b), \ldots, L^2_\mu(a, b)$. We also use the notation $L^s, L^\infty, \ldots, L^2_\mu$ when there is no risk of confusion about the domain Ω.

(*b*) *Functions with vector values in a Banach space.*

The preceding definitions can easily be generalised to functions with values in X. We denote the corresponding spaces by:

$$L^s(\Omega, X), L^\infty(\Omega, X) \text{ etc.}$$

If X is a Hilbert space, so is the space $L^2(\Omega, X)$.

E_3 Spaces of distributions

(*a*) *Distributions with real or complex values.*

$\mathscr{D}'(\Omega)$ — the space of distributions on Ω, i.e. the set of continuous linear forms on $\mathscr{D}(\Omega)$:

$$\mathscr{D}'(\Omega) \overset{\text{def}}{=} \mathscr{L}(\mathscr{D}(\Omega), \mathbb{R} \text{ or } \mathbb{C})$$

$\mathscr{E}'(\mathbb{R}^n)$ — the space of distributions with compact support on $\mathbb{R}^n$, i.e. the set of continuous linear forms on $\mathscr{E}(\mathbb{R}^n)$:

$$\mathscr{E}'(\mathbb{R}^n) \overset{\text{def}}{=} \mathscr{L}(\mathscr{E}(\mathbb{R}^n), \mathbb{R} \text{ or } \mathbb{C})$$

$\mathscr{S}'(\mathbb{R}^n)$ — the space of tempered distributions on $\mathbb{R}^n$, i.e. the set of continuous linear forms on $\mathscr{S}(\mathbb{R}^n)$:

$$\mathscr{S}'(\mathbb{R}^n) = \mathscr{L}(\mathscr{S}(\mathbb{R}^n), \mathbb{R} \text{ or } \mathbb{C})$$

$\mathscr{O}'_c(\mathbb{R}^n)$ — the space of tempered distributions on $\mathbb{R}^n$ which, with all their derivatives, are rapidly decreasing (or the space of convolutors of $\mathscr{S}(\mathbb{R}^n)$).

If $\Omega = \mathbb{R}$ or $\mathbb{R}^n = \mathbb{R}$ we denote these spaces by $\mathscr{D}'$, $\mathscr{E}'$, $\mathscr{S}'$ and $\mathscr{O}'_c$.

(*b*) *Distributions with vector values in X, a Banach space*

The above definitions can be generalised and the corresponding spaces introduced:

$$\mathscr{D}'(\Omega, X), \mathscr{E}'(\mathbb{R}^n, X), \mathscr{S}'(\mathbb{R}^n, X) \text{ and } \mathscr{O}'_C(\mathbb{R}^n, X).$$

E_4 Sobolev spaces

Let Ω be an open set in $\mathbb{R}^n$ and that $m \in \mathbb{N}$, $1 \leqslant p < \infty$, $s \in \mathbb{R}$. We denote by:

$H^m(\Omega)$ — $H^m(\Omega) \stackrel{\text{def}}{=} \{f \in L^2(\Omega); D^\alpha f \in L^2(\Omega), \forall \alpha \in \mathbb{N}^n, |\alpha| \leqslant m\}$. This is a Hilbert space with scalar product

$$(f, g) \stackrel{\text{def}}{=} \sum_{|\alpha| \leqslant m} \int_\Omega D^\alpha f(x) . \overline{D^\alpha g(x)} \, \mathrm{d}x$$

$H_0^m(\Omega)$ — The closure of $\mathscr{D}(\Omega)$ in $H^m(\Omega)$

$H^{-m}(\Omega)$ — The dual space of $H_0^m(\Omega)$: $H^{-m}(\Omega) \stackrel{\text{def}}{=} \mathscr{L}(H_0^m(\Omega), \mathbb{R})$. This is a Hilbert space with the norm

$$F \mapsto \sup_{f \in H_0^m} \frac{|\langle F, f \rangle|}{\|f\|}.$$

$H^s(\mathbb{R}^n)$ — $H^s(\mathbb{R}^n) \stackrel{\text{def}}{=} \{f \in \mathscr{S}'(\mathbb{R}^n); (1 + |y|^2)^{s/2} \mathscr{F} f \in L^2(\mathbb{R}_y^n)\}$
If $s = m$ and $\Omega = \mathbb{R}^n$, $H^m(\Omega)$ and $H^s(\mathbb{R}^n)$ coincide.

$H_{\text{loc}}^s(\Omega)$ — $H_{\text{loc}}^s \stackrel{\text{def}}{=} \{f \in \mathscr{D}'(\Omega); \forall \varphi \in \mathscr{D}(\Omega), f . \varphi \in H^s(\Omega)\}$

$W^{m,p}(\Omega)$ — $W^{m,p}(\Omega) \stackrel{\text{def}}{=} \{f \in L^p(\Omega); \forall \alpha \in \mathbb{N}^n, |\alpha| \leqslant m \Rightarrow D^\alpha f \in L^p(\Omega)\}$. This a Banach space for the norm

$$f \mapsto \left[\sum_{|\alpha| \leqslant m} \| D^\alpha f \|^p \right]^{1/p}$$

$W_0^{m,p}(\Omega)$ — the closure of $\mathscr{D}(\Omega)$ in $W^{m,p}(\Omega)$

If $\Omega = \mathbb{R}$ or $\mathbb{R}^n = \mathbb{R}$ we denote these spaces also by H^m, H_0^m, etc. . . .

F. Notations Relating to Linear Operators

Generalities on the spaces relative to the operators considered

In a general way, we denote by

$A + B$	the sum of two parts A and B of a vector space
$F \oplus G$	the direct sum of two vector spaces F and G
$F \otimes G$	the tensor product of two vector spaces F and G
F/G	the quotient space of F by G if $G \subset F$
$H^\perp$	the orthogonal complement of H in a prehilbert space
$\bigoplus_{n=1}^{\infty} H_n$	the exterior Hilbert sum of a sequence (H_n) of Hilbert spaces

Let X be a Banach space with dual X' (see Sect. F_2) we denote by:

$\dim X$	the algebraic dimension of X
$\|x\|_X$ or $\|x\|$ or $\lvert x\rvert_X$ or $\lvert x\rvert$	the norm of $x \in X$
$\|x'\|'$	the norm of $x' \in X'$
$\langle x, x'\rangle$	the bracket of duality between $x \in X$ and $x' \in X'$[14]
I_x or id_x or I	the identity operator in X

F_1 Linear operators

Let X, Y be Banach spaces and $A: X \to Y$ a linear mapping; we adopt the following notations (when the entities exist)

$\ker A$ (or $N(A)$)	the kernel of A
$\mathrm{Im}\, A$ (or $R(A)$)	the image of A
$D(A)$	the domain of A
$G(A)$	the graph of A: $G(A) \stackrel{\text{def}}{=} \{(x, y) \in X \times Y; x \in D(A), y = Ax\}$
$G^s(A)$	the inverse graph of A: $G^s(A) \stackrel{\text{def}}{=} \{(y, x); (x, y) \in G(A)\}$
$\sigma(A)$ or $S(A)$	the spectrum of A
$\rho(A)$	the resolvent set of A: $\rho(A) = \mathbb{C} \backslash \sigma(A)$
A^{-1}	the inverse of A
A^*	the adjoint of A
${}^tA$	the transpose of A
$\bar{A}$	the closure of A
$R(\lambda, A)$ or $R(\lambda)$	the resolvent operator of A: $R(\lambda, A) \stackrel{\text{def}}{=} (A - \lambda I)^{-1}$ for $\lambda \in \rho(A)$[15]
U_A	the Cayley transform of $A: U_A = (A + iI)(A - iI)^{-1}$
$\alpha(A)$	the nullity index of $A \stackrel{\text{def}}{=} \dim N(A)$
$\beta(A)$	the deficiency index of $A \stackrel{\text{def}}{=} \operatorname{codim} R(A)$
$\mathrm{rang}\, A$ or $\mathrm{rg}(A)$	the rank of A
$\det A$	the determinant of A
$\mathrm{tr}\, A$ or $\mathrm{Tr}\, A$	the trace of A
$r(A)$	the spectral radius of A
$\|A\|$	the norm of A
$A_2 \supset A_1$	the operator A_2 is an extension of the operator A_1

---

[14] If H is a Hilbert space we denote the scalar product of x and y in H by $(x, y)_H$ or (x, y) or $((x, y))$.
[15] We sometimes adopt the definitions $R(\lambda, A) = (\lambda I - A)^{-1}$.

F_2 Spaces of linear operators

Let X and Y be two topological vector spaces, real or complex. We denote by:

$\mathscr{L}(X, Y)$	the set of continuous (or bounded) linear mappings of X with values in Y. If X and Y are both Banach spaces, then so is $\mathscr{L}(X, Y)$, provided with the norm: $$\|A\|_{\mathscr{L}(X,Y)} = \sup_{\|x\|_X = 1} \|Ax\|$$
$\mathscr{L}(X)$	$\mathscr{L}(X) \stackrel{\text{def}}{=} \mathscr{L}(X, X)$
X'	the dual space of X, i.e. $\mathscr{L}(X, \mathbb{R} \text{ or } \mathbb{C})$, the set of continuous linear forms on X[16]
X''	the bidual space of X, i.e. the dual of X' for the topology of the norm $$\|x\|'' = \sup_{\substack{x' \in X' \\ \|x'\| = 1}} |\langle x', x \rangle|.$$

Let H be a complete separable complex Hilbert space; we denote by $\mathscr{L}^1(H)$, the trace class, i.e. the set of nuclear operators in $\mathscr{L}(H)$. This is a Banach space with norm $S \mapsto \operatorname{tr} |S|$.[17]

---

[16] We also frequently denote by X' the antidual of X, i.e. the set of continuous antilinear forms on X.

[17] We shall see in Chaps. VI and VIII that $|S| \stackrel{\text{def}}{=} (S^*S)^{\frac{1}{2}}$, $S \in \mathscr{L}(H)$.

Index

Contents of Volumes 2–6

Volume 2 Functional and Variational Methods

Volume 3 Spectral Theory and Applications

Volume 4 Integral Equations and Numerical Methods

Volume 5 Evolution Problems I

Volume 6 Evolution Problems II: The Navier-Stokes and Transport Equations in Numerical Methods